DESIGN OF CONCRETE STRUCTURES

Eleventh Edition

Arthur H. Nilson

Professor of Structural Engineering
Cornell University

George Winter

Late Professor of Structural Engineering
Cornell University

A textbook originated by

L.C. Urquhart and C.E. O'Rourke

McGraw-Hill, Inc.

New York St. Louis San Francisco Auckland Bogotá
Caracas Lisbon London Madrid Mexico City Milan
Montreal New Delhi San Juan Singapore
Sydney Tokyo Toronto

This book was set in Times Roman by Publication Services.
The editors were B. J. Clark and John M. Morriss;
the production supervisor was Annette Mayeski.
The cover was designed by Joseph Gillians.
Project supervision was done by Publication Services.

Cover Photo: The East Huntington Bridge, a Cable-Stayed Concrete Girder Bridge in Huntington, West Virginia. (*Photo courtesy of David Goodyear and Arvid Grant Associates, Inc.*)

DESIGN OF CONCRETE STRUCTURES

4 5 6 7 8 9 10 11 12 13 14 BKMBKM 9 9 8 7 6 5 4

ISBN 0-07-046567-3

Library of Congress Cataloging-in-Publication Data

Nilson, Arthur H.
Design of concrete structures / Arthur H. Nilson, George Winter.—11th ed.
p. cm.
"A textbook originated by L.C. Urquhart and C.E. O'Rourke."
Winter's name appears first on the earlier edition.
Includes bibliographical references and index.
ISBN 0-07-046567-3
1. Reinforced concrete construction. 2. Prestressed concrete construction. I. Winter, George, (date). II. Urquhart, Leonard Church, (date). Design of concrete structures. III. Title.
TA683.2.W48 1991
624.1′8341—dc20 90-46528

ABOUT THE AUTHORS

Arthur H. Nilson has been engaged in research, teaching, and consulting relating to structural concrete for over 35 years. He has been a member of the faculty of the College of Engineering at Cornell University since 1956, where he is in charge of undergraduate and graduate courses in the design of reinforced concrete and prestressed concrete structures. He served as Chairman of the Department of Structural Engineering from 1978 to 1985. Dr. Nilson has served on many professional committees, including Building Code Subcommittee 318D of the American Concrete Institute (ACI). He was awarded the ACI Wason Medal for materials research in 1974 and the ACI Wason Medal for best technical paper in 1986 and 1987. Professor Nilson has been elected to the grade of fellow in ACI and in the American Society of Civil Engineers (ASCE) and has been honored by the civil engineering student body at Cornell for outstanding teaching. He has held research or lecturer appointments at the University of Manchester, Salford University, and Milan Technical University. He is a registered professional engineer in several states and, prior to entering teaching, was for six years engaged in full-time professional practice. He received a B.S. degree from Stanford University in 1948, an M.S. from Cornell in 1956, and a Ph.D. from the University of California at Berkeley in 1967.

The late George Winter was Class of 1912 Professor of Engineering at Cornell University. He received the Diploma of Engineering degree from the Technical University of Munich in 1930. He came to the United States after eight years of design and consulting practice in Europe and earned his Ph.D. at Cornell in 1940. He served as Chairman of the Department of Structural Engineering for 22 years, from 1948 to 1970. His many honors include election to the National Academy of Engineering, an honorary doctorate from the Technical University of Munich, and election to honorary memberships in ASCE and ACI. He received the ACI Wason Research Medal in 1965, Henry C. Turner Medal in 1972, and Joe W. Kelly Award in 1979. In 1982 he received the International Award of Merit in Structural Engineering from the International Association for Bridge and Structural Engineering. He was instrumental in the development of design codes and specifications for reinforced concrete, structural steel, and light gage cold-formed steel. Professor Winter served on ACI Building Code Committee 318 for many years and was chairman of Subcommittee 318D during the 1960s, pioneering the shift from allowable stress to ultimate strength design. He died in November of 1982.

CONTENTS

PREFACE

This edition represents a substantial expansion as well as an update of the previous work. It maintains the same basic approach: first to establish a firm understanding of the behavior of reinforced concrete, then to develop proficiency in the methods used in current design practice, with particular reference to the provisions of the 1989 American Concrete Institute (ACI) Building Code.

It is generally recognized that mere training in special design skills and codified procedures is inadequate for successful professional practice. These procedures are subject to frequent changes. To understand and keep abreast of these rapid developments, the engineer needs a thorough grounding in the basic performance of concrete and steel as structural materials and in the behavior of reinforced concrete members and structures. On the other hand, the main business of the structural engineer is to design structures safely, economically, and efficiently. Hence, with this basic understanding as a firm foundation, familiarity with current design procedures is essential. This edition, like the preceding ones, addresses both needs.

The text not only presents the basic mechanics of structural concrete and methods for the design of individual members for bending, shear, axial forces, and other actions but also provides much detail pertaining to applications, such as the various types of building systems, footings and foundations, retaining walls, and composite building systems. The five-chapter treatment of slab systems is particularly comprehensive. A chapter on bridge design is included, as in preceding editions.

Much present-day design is carried out using computer programs, either general-purpose, commercially available software or programs written by the individual for his or her particular needs. Step-by-step design procedures are given throughout the book to guide the student through the increasingly complex methodology of current design. These can easily be converted to flowcharts to aid in computer programming.

Several new chapters have been added. The crucial importance of joint design to structural safety has been recognized with a new chapter on joint detailing that incorporates the latest research results and ACI Code provisions. Slender columns, much more common than before because of increased material strengths and more sophisticated design concepts, are given a greatly expanded treatment in a separate chapter. A third new chapter offers a detailed presentation of the Hillerborg strip method, a powerful tool for design of slabs of irregular shape, boundary, or support conditions.

Much new material will be found in other chapters as well. The basic concepts of the strut-and-tie model are emphasized wherever appropriate, to aid in visualizing behavior and to provide a sound basis for design; this model is employed particularly in joint detailing, design for shear and torsion, and design of brackets and deep beams. The chapter on materials includes an expanded treatment of high-strength concrete. The extensive changes of the 1989 ACI Code relating to bond and anchorage have resulted in an extensive rewriting of Chapter 5. New ACI Code provisions for "structural integrity" have been incorporated and explained. Appropriate attention is also given to new methods for the design of flat plates floor systems for shear, particularly stud shear reinforcement. Finally, the chapter on prestressed concrete has been completely rewritten and expanded.

The teacher will find the text suitable for either a one- or two-semester course in the design of concrete structures. If the curriculum permits only a single course (probably taught in the fourth undergraduate year), the introduction and treatment of materials found in Chapters 1 and 2, respectively; the material on flexure, shear, and anchorage in Chapters 3, 4, and 5; Chapter 6 on serviceability; Chapter 8 on short columns; and the introduction to one- and two-way slabs in Chapter 12 will provide a good basis. Time may or may not permit classroom coverage of frame analysis and building systems, Chapters 11 and 20, but these could well be assigned as independent reading, concurrent with the earlier work of the course. In the writer's experience, such complementary outside reading tends to enhance student motivation.

The text is more than adequate for a second course, most likely taught in the first year of graduate study. The second course should include an introduction to the increasingly important topics of torsion, Chapter 7; slender columns, Chapter 9; and the design and detailing of joints, Chapter 10. It also offers the opportunity for a much-expanded study of slabs, including the ACI approach to column-supported slabs, Chapter 13, and the methods of analysis and design based on plastic theory, Chapters 14 and 15. Other topics appropriate to a second course include slabs on grade, Chapter 16; composite construction, Chapter 17; foundations and retaining walls, Chapters 18 and 19; and the material on bridges in Chapter 22. Prestressed concrete is sufficiently important to justify a separate course. If the curriculum does not permit this, the treatment of Chapter 21 provides an introduction, based on a separate text on prestressed concrete by the author, and can, in itself, be used as the text for a one-credit short course on that topic.

At the end of each chapter, the user will find greatly expanded reference lists to provide an entry into the literature for those wishing to increase their knowledge through individual study.

The present volume is the eleventh edition of a textbook originated in 1923 by Leonard C. Urquhart and Charles E. O'Rourke, both professors of structural engineering at Cornell University at that time. The second, third, and fourth editions firmly established the work as a leading text for elementary courses in the subject area. Professor George Winter, also of Cornell, collaborated with Urquhart in preparing the fifth and sixth editions, and Winter and I were responsible for the seventh, eighth, and ninth editions, which substantially expanded both the scope and the depth of the presentation. The tenth and present eleventh editions were prepared subsequent to Professor Winter's passing in 1982.

McGraw-Hill and I would like to thank the following reviewers for their many helpful comments and suggestions: Dan Branson, University of Iowa; Charles Dolan, University of Delaware; Kurt Gerstle, University of Colorado; Louis Geschwidner, Penn State University; Wayne Klaiber, Iowa State University; John Stanton, University of Washington; and James Wight, University of Michigan.

I gladly acknowledge my indebtedness to the original authors. Although it is safe to say that neither Urquhart nor O'Rourke would recognize very much of the detail, the approach to the subject and the educational philosophy that did so much to account for the success of earlier editions of this unique book would be familiar. I acknowledge with particular gratitude the influence of Professor Winter. My long professional and personal relationship with him had a profound effect in developing a point of view that has shaped all my work in the chapters that follow.

Arthur H. Nilson

DESIGN OF CONCRETE STRUCTURES

CHAPTER 1

INTRODUCTION

1.1 CONCRETE, REINFORCED CONCRETE, AND PRESTRESSED CONCRETE

Concrete is a stonelike material obtained by permitting a carefully proportioned mixture of cement, sand and gravel or other aggregate, and water to harden in forms of the shape and dimensions of the desired structure. The bulk of the material consists of fine and course aggregate. Cement and water interact chemically to bind the aggregate particles into a solid mass. Additional water, over and above that needed for this chemical reaction, is necessary to give the mixture the workability that enables it to fill the forms and surround the embedded reinforcing steel prior to hardening. Concretes in a wide range of strength properties can be obtained by appropriate adjustment of the proportions of the constituent materials. Special cements (such as high early strength cements), special aggregates (such as various lightweight or heavyweight aggregates), admixtures (such as plasticizers and air-entraining agents), and special curing methods (such as steam-curing) permit an even wider variety of properties to be obtained.

These properties depend to a very substantial degree on the proportions of the mix, on the thoroughness with which the various constituents are intermixed, and on the conditions of humidity and temperature in which the mix is maintained from the moment it is placed in the forms until it is fully hardened. The process of controlling these conditions is known as *curing*. To protect against the unintentional production of substandard concrete, a high degree of skillful control and supervision is necessary throughout the process, from the proportioning by weight of the individual components, through mixing and placing, until the completion of curing.

The factors that make concrete a universal building material are so pronounced that it has been used, in more primitive kinds and ways than at present, for thousands of years, probably beginning in Egyptian antiquity. The facility with which, while plastic, it can be deposited and made to fill forms or molds of almost any practical shape is one of these factors. Its high fire and weather resistance are evident advantages. Most of the constituent materials, with the possible exception of cement, are usually available at low cost locally or at small distances from the construction site. Its compressive strength, like that of natural stones, is high, which makes it suitable for members primarily subject to compression, such as columns and arches. On the other hand, again as in natural stones, it is a relatively brittle material whose tensile strength is small compared with its compressive strength. This prevents its economical use in structural members that are subject to tension either entirely (such as in tie rods) or over part of their cross sections (such as in beams or other flexural members).

To offset this limitation, it was found possible, in the second half of the nineteenth century, to use steel with its high tensile strength to reinforce concrete, chiefly in those places where its small tensile strength would limit the carrying capacity of the member. The reinforcement, usually round steel rods with appropriate surface deformations to provide interlocking, is placed in the forms in advance of the concrete. When completely surrounded by the hardened concrete mass, it forms an integral part of the member. The resulting combination of two materials, known as *reinforced concrete*, combines many of the advantages of each: the relatively low cost, good weather and fire resistance, good compressive strength, and excellent formability of concrete and the high tensile strength and much greater ductility and toughness of steel. It is this combination that allows the almost unlimited range of uses and possibilities of reinforced concrete in the construction of buildings, bridges, dams, tanks, reservoirs, and a host of other structures.

In more recent times, it has been found possible to produce steels, at relatively low cost, whose yield strength is of the order of 4 times and more that of ordinary reinforcing steels. Likewise, it is possible to produce concrete 3 to 4 times as strong in compression as the more ordinary concretes. These high-strength materials offer many advantages, including smaller member cross sections, reduced dead load, and longer spans. However, there are limits to the strengths of the constituent materials beyond which certain problems arise. To be sure, the strength of such a member would increase roughly in proportion to those of the materials. However, the high strains that result from the high stresses that would otherwise be permissible would lead to large deformations and deflections of such members under ordinary loading conditions. Equally important, the large strains in such high-strength reinforcing steel would induce large cracks in the surrounding low tensile strength concrete, cracks that would not only be unsightly but which would expose the steel reinforcement to corrosion by moisture and other chemical action. This limits the useful yield strength of high-strength reinforcing steel to about 80 ksi† compared with 40 to 60 ksi for conventional reinforcing steel.

† Abbreviation for kips per square inch, or thousands of pounds per square inch.

A special way has been found, however, to use steels and concretes of very high strength in combination. This type of construction is known as *prestressed concrete*. The steel, usually in the form of wires or strands but sometimes as bars, is embedded in the concrete under high tension that is held in equilibrium by compressive stresses in the concrete after hardening. Because of this precompression, the concrete in a flexural member will crack on the tension side at a much larger load than when not so precompressed. Prestressing greatly reduces both the deflections and the tensile cracks at ordinary loads in such structures, and thereby enables these high-strength materials to be used effectively. Prestressed concrete has extended, to a very significant extent, the range of spans of structural concrete and the types of structures for which it is suited.

1.2 STRUCTURAL FORMS

The figures that follow show some of the principal structural forms of reinforced concrete. Pertinent design methods for many of them are discussed later in this volume.

Floor-support systems for buildings include the monolithic slab-and-beam floor shown in Fig. 1.1, the one-way joist system of Fig. 1.2, and the flat plate

FIGURE 1.1
One-way reinforced concrete floor slab with monolithic supporting beams.

FIGURE 1.2
One-way joist floor system, with closely spaced ribs supported by monolithic concrete beams; transverse ribs provide for lateral distribution of localized loads. (*Concrete Reinforcing Steel Institute*.)

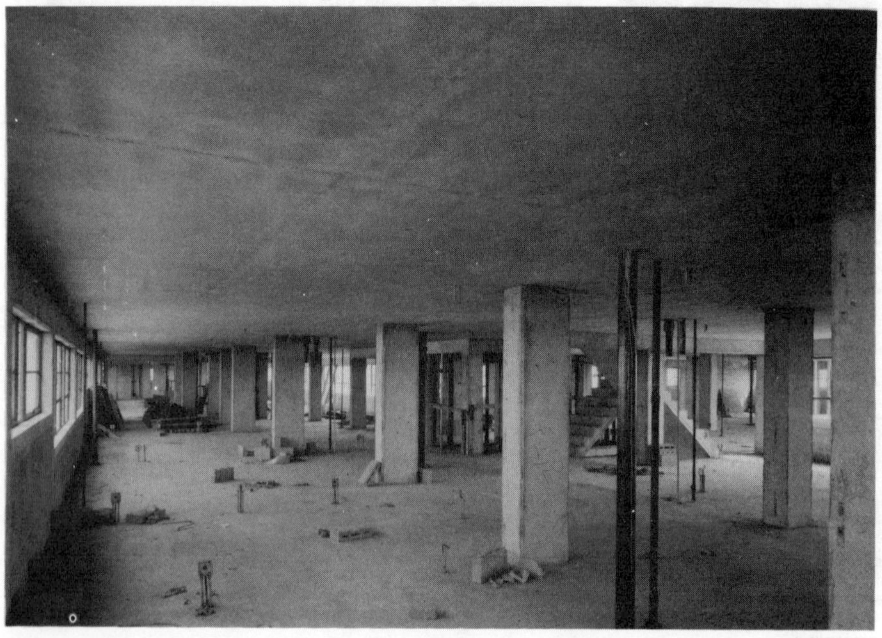

FIGURE 1.3
Flat plate floor slab, carried directly by columns without beams or girders. (*Portland Cement Association*.)

FIGURE 1.4
Flat slab floor system, without beams but with slab thickness increased at the columns and with flared column tops to provide for local concentration of forces. (*Portland Cement Association.*)

FIGURE 1.5
Folded plate roof of 125 ft span that, in addition to carrying ordinary roof loads, carries the second floor as well from a system of cable hangers; the ground floor is kept free of columns.

FIGURE 1.6
Precast concrete cylindrical shell roof supported by precast concrete rigid frames.

slab, without beams or girders, shown in Fig. 1.3. The flat slab floor of Fig. 1.4, frequently used for more heavily loaded buildings such as warehouses, is similar to the flat plate floor, but makes use of increased slab thickness in the vicinity of the columns, as well as flared column tops, to reduce stresses and increase strength in the support region. The choice among these and other systems for floors and roofs depends upon functional requirements, loads, spans, and permissible member depths, as well as on cost and esthetic factors.

Where long clear spans are required for roofs, concrete shells permit use of extremely thin surfaces, often thinner, relatively, than an eggshell. The folded plate roof of Fig. 1.5 is simple to form because it is composed of flat surfaces; such roofs have been employed for spans of 200 ft and more. The cylindrical shell of Fig. 1.6 is also relatively easy to form because it has only a single curvature; it is similar to the folded plate in its structural behavior and range of spans and loads.

FIGURE 1.7
St. Edmund's Episcopal Church, Elm Grove, Wisconsin; a hyperbolic paraboloid shell.

FIGURE 1.8
Spherical domes, supported only at three corner points; a part of the Olympic sports facility at Chamonix.

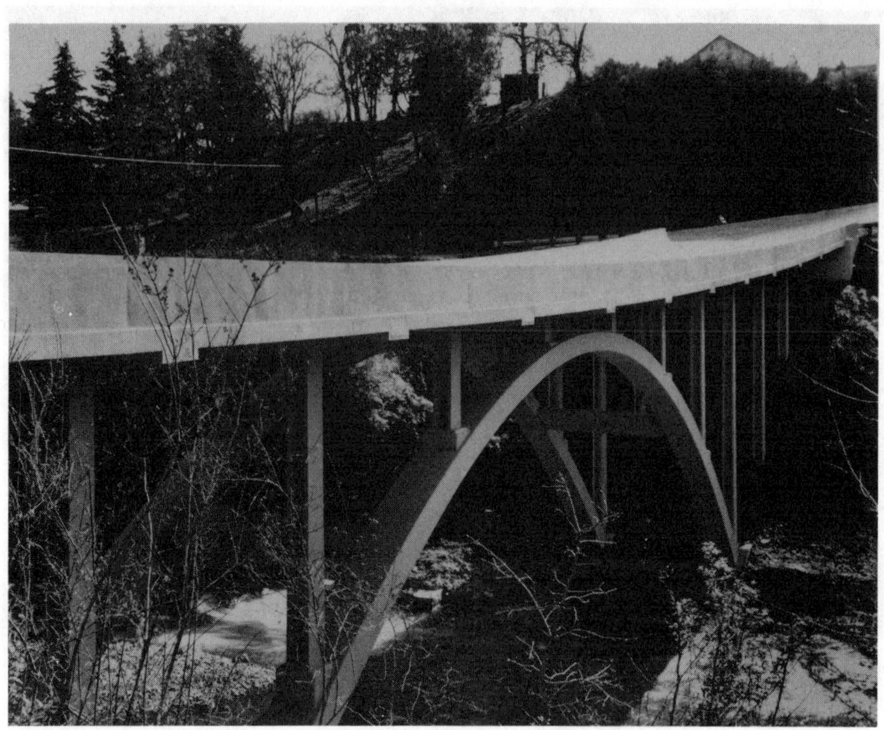

FIGURE 1.9
The Fairoaks Bridge, a short-span concrete arch structure. (*Prestressed Concrete Institute.*)

FIGURE 1.10
The East Huntington Bridge, a cable-stayed concrete girder bridge in Huntington, West Virginia. (*David Goodyear and Arvid Grant Associates.*)

FIGURE 1.11
Circular concrete tanks used as a part of the wastewater purification facility at Howden, on the River Tyne in England; design by Northumbrian Water Authority with Luder and Jones, Architects. (*Cement and Concrete Association*.)

Doubly curved roof shells may often be defined by mathematical curves. The hyperbolic paraboloid of Fig. 1.7 requires structural support at only two opposite corners. While it is a double-curved surface, it has the interesting property that it contains two systems of straight-line generators permitting use of straight forming lumber. Visually appealing spherical domes as in Fig. 1.8 provide long clear spans over assembly halls and athletic facilities.

Just two of the many bridge forms are shown in Figs. 1.9 and 1.10. In the first case, an arch has been used. This structure has a relatively short span, but concrete arch bridges have been constructed with spans in excess of 1200 ft. Cable-stayed girder bridges, such as shown in Fig. 1.10, are economical for medium to long spans.

Cylindrical concrete tanks are widely used for storage of water or as a part of waste purification plants. The award-winning project shown in Fig. 1.11 is proof that a sanitary engineering facility can be esthetically pleasing as well as functional. Cylindrical tanks are often prestressed circumferentially to maintain compression in the concrete and eliminate cracking that would otherwise result from internal pressure.

The forms of Figs. 1.1 through 1.11 do not constitute a complete range but are merely typical of shapes appropriate to the properties of reinforced or prestressed concrete. They illustrate the adaptability of the material to a great variety of one-dimensional (beams, girders, columns), two-dimensional (slabs, arches, rigid frames), and three-dimensional (shells, tanks) structures and structural components. This variability allows the shape of the structure to be adapted to its function in an economical manner, and furnishes the architect and design engineer with a wide variety of possibilities for esthetically satisfying structural solutions.

1.3 LOADS

Loads that act on structures can be divided into three broad categories: dead loads, live loads, and environmental loads.

Dead loads are those that are constant in magnitude and fixed in location throughout the lifetime of the structure. Usually the major part of the dead load is the weight of the structure itself. This can be calculated with good accuracy from the design configuration, dimensions of the structure, and density of the material. For buildings, floor fill, finish floors, and plastered ceilings are usually included as dead loads, and an allowance is made for suspended loads such as piping and lighting fixtures. For bridges, dead loads may include wearing surfaces, sidewalks, and curbing, and an allowance is made for piping and other suspended loads.

Live loads consist chiefly of occupancy loads in buildings and traffic loads on bridges. They may be either fully or partially in place or not present at all, and may also change in location. Their magnitude and distribution at any given time are uncertain, and even their maximum intensities throughout the lifetime of the structure are not known with precision. The minimum live loads for which the floors and roof of a building should be designed are usually specified in the building code that governs at the site of construction. Representative values of minimum live loads to be used in a wide variety of buildings are found in *Minimum Design Loads for Buildings and Other Structures* (Ref. 1.1), a portion of which is reprinted in Table 1.1. The table gives uniformly distributed live loads for various types of occupancies; these include impact provisions where necessary. These loads are expected maxima and considerably exceed average values.

In addition to these uniformly distributed loads, it is recommended that, as an alternative to the uniform load, floors be designed to support safely certain concentrated loads if these produce a greater stress. For example, according to Ref. 1.1, office floors are to be designed to carry a load of 2000 lb distributed over an area 2.5 ft square, to allow for the weight of a safe or other heavy equipment, and stair treads must safely support a 300 lb load applied on the center of the tread. Certain reductions are often permitted in live loads for members supporting large areas, on the premise that it is not likely that the entire area would be fully loaded at one time (Refs. 1.1 and 1.2).

Tabulated live loads cannot always be used. The type of occupancy should be considered and the probable loads computed as accurately as possible. Warehouses for heavy storage may be designed for loads as high as 500 psf or more; unusually heavy operations in manufacturing buildings may require a large increase in the 125 psf value specified in Table 1.1; special provisions must be made for all definitely located heavy concentrated loads.

Table 1.1 Minimum uniformly distributed live loads

Occupancy or use	Live load, psf[a]	Occupancy or use	Live load, psf[a]
Apartments (see residential)		Manufacturing	
Armories and drill rooms	150	Light	125
Assembly areas and theaters		Heavy	250
Fixed seats (fastened to floor)	60	Marquees and canopies	75
Lobbies	100	Office buildings	
Movable seats	100	File and computer rooms shall	
Platforms (assembly)	100	be designed for heavier loads	
Stage floors	150	based on anticipated occupancy	
Balconies (exterior)	100	Lobbies	100
On one- and two-family residences		Offices	50
only, and not exceeding 100 ft^2	60	Penal institutions	
Bowling alleys, poolrooms, and		Cell blocks	40
similar recreational areas	75	Corridors	100
Corridors		Residential	
First floor	100	Dwellings (one- and two- family)	
Other floors same as occupancy		Uninhabitable attics	
served except as indicated		without storage	10
Dance halls and ballrooms	100	Uninhabitable attics	
Decks (patio and roof)		with storage	20
Same as area served, or for the		Habitable attics, sleeping areas	30
type of occupancy accomodated		All other areas	40
Dining rooms and restaurants	100	Hotels and multifamily houses	
Dwellings (see residential)		Private rooms and corridors	
Fire escapes	100	serving them	40
On single family dwellings only	40	Public rooms and corridors	
Garages (passenger cars only)	50	serving them	100
For trucks and buses use AASHTO[b]		Schools	
lane loads (certain additional		Classrooms	40
requirements for concentrated		Corridors above first floor	80
loads may control)		Sidewalks, vehicular driveways,	
Grandstands (see stadium and arena		and yards, subject to trucking[d]	250
bleachers)		Stadium and arena bleachers[e]	100
Gymnasiums, main floors, and		Stairs and exitways	100
balconies	100	Storage warehouses	125
Hospitals		Light	125
Operating rooms, laboratories	60	Heavy	250
Private rooms	40	Stores	
Wards	40	Retail	
Corridors above first floor	80	First floor	100
Hotels (see residential)		Upper floors	75
Libraries		Wholesale, all floors	125
Reading rooms	60	Walkways and elevated platforms	
Stack rooms—not less than[c]	150	(other than exitways)	60
Corridors above first floor	80	Yards and terraces (pedestrians)	100

[a] Pounds per square foot.

[b] American Association of State Highway and Transportation Officials.

[c] The weight of books and shelving shall be computed using an assumed density of 65 lb/ft^3 (pounds per cubic foot, sometimes abbreviated pcf) and converted to a uniformly distributed load; this load shall be used if it exceeds 150 psf.

[d] AASHTO lane loads should also be considered where appropriate.

[e] For detailed recommendations, see American National Standard for Assembly Seating, Tents, and Air-Supported Structures, ANSI/NFPA 102-1978.

Source: From Ref 1.1. Used by permission of the American National Standards Institute.

Service live loads for highway bridges are specified by the American Association of State Highway and Transportation Officials (AASHTO) in its *Standard Specifications for Highway Bridges* (Ref. 1.3). For railway bridges, the American Railway Engineering Association (AREA) has published the *Manual of Railway Engineering* (Ref. 1.4), which specifies traffic loads.

Environmental loads consist mainly of snow loads, wind pressure and suction, earthquake loads (i.e., inertia forces caused by earthquake motions), soil pressures on subsurface portions of structures, loads from possible ponding of rainwater on flat surfaces, and forces caused by temperature differentials. Like live loads, environmental loads at any given time are uncertain both in magnitude and distribution. Reference 1.1 contains much information on environmental loads, which is often modified locally depending, for instance, on local climatic or seismic conditions.

Figure 1.12, from the 1972 edition of Ref. 1.1, gives snow loads for the continental United States, and is included here for illustration. The 1982 edition of Ref. 1.1 gives much more detailed information. In either case, specified values represent not average values, but expected upper limits. A minimum roof load of 20 psf is often specified to provide for construction and repair loads and to ensure reasonable stiffness.

Much progress has been made in recent years in developing rational methods for predicting horizontal forces on structures due to wind and seismic action.

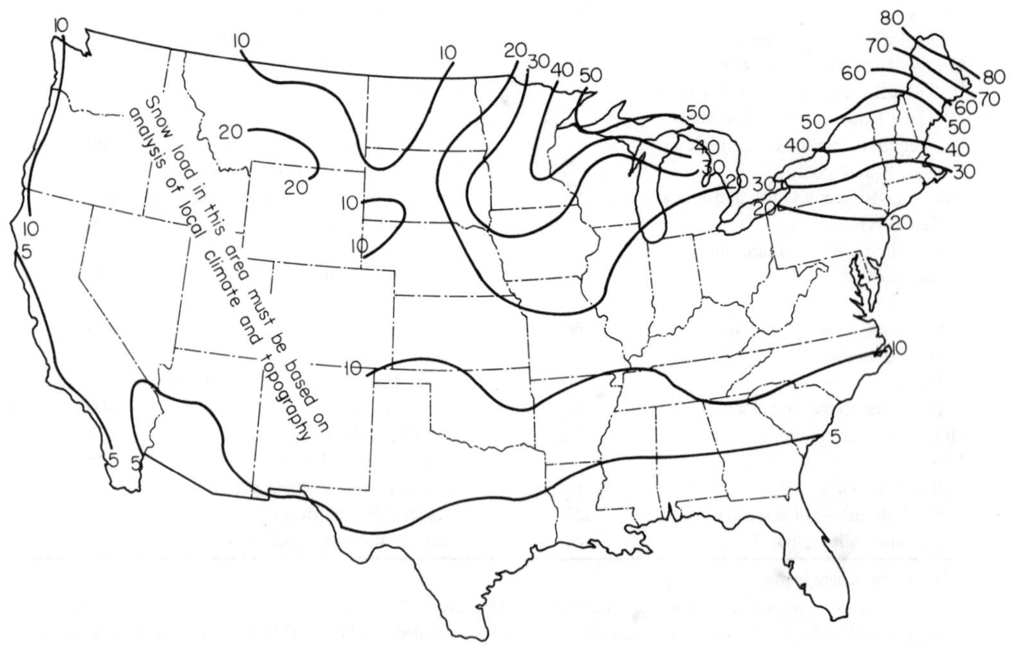

FIGURE 1.12
Snow loads in pounds per square foot (psf) on the ground, 50-year mean recurrence interval.

Reference 1.1 summarizes current thinking regarding wind forces, and has much information pertaining to earthquake loads as well. Reference 1.5 presents detailed recommendations for lateral forces from earthquakes.

Most building codes specify design wind pressures per square foot of vertical wall surface. Depending upon locality, these equivalent static forces vary from about 10 to 50 psf. Factors considered in more up-to-date standards include probable wind velocity, exposure (urban vs. open terrain, for example), height of the structure, the importance of the structure (i.e., consequences of failure), and gust-response factors to account for the fluctuating nature of the wind and its interaction with the structure.

Seismic forces may be found for a particular structure by elastic or inelastic dynamic analysis, considering expected ground accelerations and the mass, stiffness, and damping characteristics of the construction. However, often the design is based on equivalent static forces calculated from provisions such as those of Refs. 1.1 and 1.5. The base shear is found by considering such factors as location, type of structure and its occupancy, total dead load, and the particular soil condition. The total lateral force is distributed to floors over the entire height of the structure in such a way as to approximate the distribution of forces obtained from a dynamic analysis.

1.4 SERVICEABILITY, STRENGTH, AND STRUCTURAL SAFETY

To serve its purpose, a structure must be safe against collapse and serviceable in use. Serviceability requires that deflections be adequately small, that cracks, if any, be kept to tolerable limits, that vibrations be minimized etc. Safety requires that the strength of the structure be adequate for all loads that may foreseeably act on it. If the strength of a structure, built as designed, could be predicted accurately, and if the loads and their internal effects (moments, shears, axial forces) were known accurately, safety could be ensured by providing a carrying capacity just barely in excess of the known loads. However, there are a number of sources of uncertainty in the analysis, design, and construction of reinforced concrete structures. These sources of uncertainty, which require a definite margin of safety, may be listed as follows:

1. Actual loads may differ from those assumed.
2. Actual loads may be distributed in a manner different from that assumed.
3. The assumptions and simplifications inherent in any analysis may result in calculated load effects—moments, shears, etc.—different from those that, in fact, act in the structure.
4. The actual structural behavior may differ from that assumed, owing to imperfect knowledge.
5. Actual member dimensions may differ from those specified.
6. Reinforcement may not be in its proper position.
7. Actual material strength may be different from that specified.

In addition, in the establishment of a safety specification, consideration must be given to the consequences of failure. In some cases, a failure would merely be an inconvenience. In other cases, loss of life and significant loss of property may be involved. A further consideration should be the nature of the failure, should it occur. A gradual failure with ample warning permitting remedial measures is preferable to a sudden, unexpected collapse.

It is evident that the selection of an appropriate margin of safety is not a simple matter. However, progress has been made toward rational safety provisions in design codes (Refs. 1.6 and 1.9).

a. Variability of Loads

Since the maximum load that will occur during the life of a structure is uncertain, it can be considered a random variable. In spite of this uncertainty, the engineer must provide an adequate structure. A probability model for the maximum load can be devised by means of a probability density function for loads, as represented by the frequency curve of Fig. 1.13*a*. The exact form of this distribution curve, for any particular type of loading such as office loads, can be determined only on the basis of statistical data obtained from large-scale load surveys. A number of such surveys have been carried out or are in progress. For types of loads for which such data are scarce, fairly reliable information can be obtained from experience, observation, and judgment.

In such a frequency curve (Fig. 1.13*a*), the area under the curve between two abscissas, such as loads Q_1 and Q_2, represents the probability of occurrence of loads Q of magnitude $Q_1 < Q < Q_2$. A specified service load Q_d for design is selected conservatively in the upper region of Q in the distribution curve, as shown. The probability of occurrence of loads larger than Q_d is then given by the shaded area to the right of Q_d. It is seen that this specified service load is considerably larger than the mean load $\overline{Q}$ acting on the structure. This mean load is much more typical of average load conditions than the design load Q_d.

b. Strength

The strength of a structure depends on the strength of the materials from which it is made. For this purpose, minimum material strengths are specified in standardized ways. Actual material strengths cannot be known precisely and therefore also constitute random variables (see Sec. 2.6). Structural strength depends, furthermore, on the care with which a structure is built, which in turn reflects the quality of supervision and inspection. Members sizes may differ from specified dimensions, reinforcement may be out of position, poorly placed concrete may show voids, etc.

Strength of the entire structure or of a population of repetitive structures, e.g., highway overpasses, can also be considered a random variable with a probability density function of the type shown in Fig. 1.13*b*. As in the case of loads, the exact form of this function cannot be known but can be approximated from known data, such as statistics of actual, measured materials and member strengths and similar information. Considerable information of this type has been, or is being, developed and utilized.

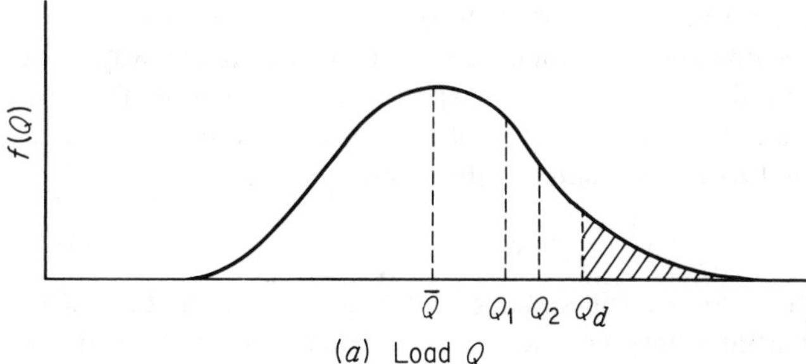

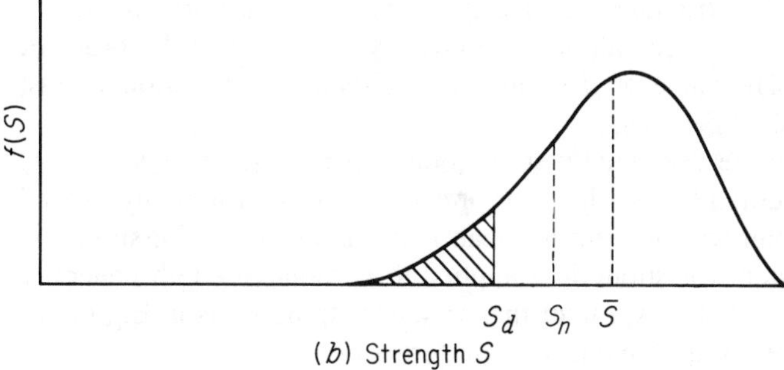

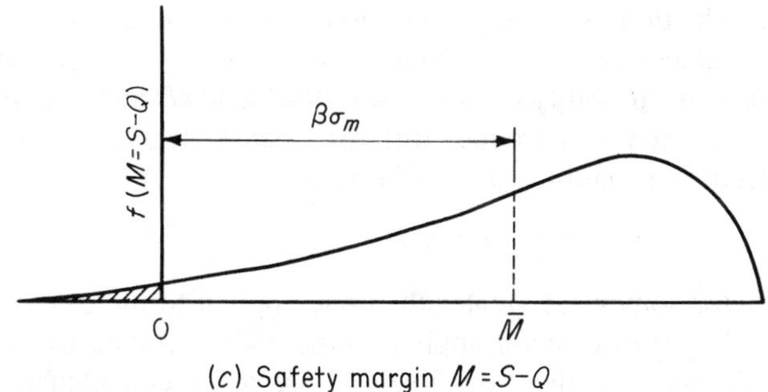

FIGURE 1.13
Frequency curves for (*a*) loads, Q; (*b*) strengths, S; and (*c*) safety margin, M.

c. Structural Safety

A given structure has a *safety margin M* if

$$M = S - Q > 0 \tag{1.1}$$

i.e., if the strength of the structure is larger than the load acting on it. Since S and Q are random variables, the safety margin $M = S - Q$ is also a random variable. A plot of the probability function of M may appear as in Fig. 1.13*c*. Failure occurs when M is less than zero. Thus, the probability of failure is represented by the shaded area in the figure.

Even though the precise form of the probability density functions for S and Q, and therefore for M, is not known, much can be achieved in the way of a rational approach to structural safety. One such approach is to require that the mean safety margin M be a specified number β of standard deviations σ_m above zero. It can be demonstrated that this results in the requirement that

$$\psi_s \overline{S} \geq \psi_L \overline{Q} \tag{1.2}$$

where ψ_s is a partial safety coefficient smaller than one applied to the mean strength $\overline{S}$, and ψ_L is a partial safety coefficient larger than one applied to the mean load $\overline{Q}$. The magnitude of each partial safety coefficient depends on the variance of the quantity to which it applies, S or Q, and on the chosen value of β, the safety index of the structure. As a general guide, a value of the safety index β between 3 and 4 corresponds to a probability of failure of the order of 1 : 100,000 (Ref. 1.8). The value of β is often established by calibration against well-proved and established designs.

In practice, it is more convenient to introduce partial safety coefficients with respect to code-specified loads which, as already noted, considerably exceed average values, rather than with respect to mean loads as in Eq. (1.2); similarly, the partial safety coefficient for strength is applied to nominal strength generally computed somewhat conservatively, rather than to mean strengths as in Eq. (1.2). A restatement of the safety requirement in these terms is

$$\phi S_n \geq \gamma Q_d \tag{1.3a}$$

in which ϕ is a strength reduction factor applied to nominal strength S_n and γ is a load factor applied to calculated or code-specified design loads Q_d. Furthermore, recognizing the differences in variability between, say, dead loads D and live loads L, it is both reasonable and easy to introduce different load factors for different types of loads. The preceding equation can thus be written

$$\phi S_n \geq \gamma_d D + \gamma_l L \tag{1.3b}$$

in which γ_d is a load factor somewhat greater than one applied to the calculated dead load D, and γ_l is a larger load factor applied to the code-specified live load L. When additional loads, such as the wind load W, are to be considered, the reduced probability that maximum dead, live, and wind or other loads will act simultaneously can be incorporated by including a factor α less than 1 such that

$$\phi S_n \geq \alpha(\gamma_d D + \gamma_l L + \gamma_w W + \cdots) \tag{1.3c}$$

Present U.S. design specifications follow the format of Eqs. (1.3*b*) and (1.3*c*).

1.5 DESIGN BASIS

The single most important characteristic of any structural member is its actual strength, which must be large enough to resist, with some margin to spare, all foreseeable loads that may act on it during the life of the structure, without failure or other distress. It is logical, therefore, to proportion members, i.e., to select concrete dimensions and reinforcement, so that member strengths are adequate to resist forces resulting from certain hypothetical overload stages, significantly above loads expected actually to occur in service. This design concept is known as *strength design*.

For reinforced concrete structures at loads close to and at failure, one or both of the materials, concrete and steel, are invariably in their nonlinear inelastic range. That is, concrete in a structural member reaches its maximum strength and subsequent fracture at stresses and strains far beyond the initial elastic range in which stresses and strains are fairly proportional. Similarly, steel close to and at failure of the member is usually stressed beyond its elastic domain into and even beyond the yield region. Consequently, the nominal strength of a member must be calculated on the basis of this inelastic behavior of the materials.

A member designed by the strength method must also perform in a satisfactory way under normal service loading. For example, beam deflections must be limited to acceptable values, and the number and width of flexural cracks at service loads must be controlled. Serviceability limit conditions are an important part of the total design, although attention is focused initially on strength.

As an alternative to the strength design method, members are sometimes proportioned so that stresses in the steel and concrete resulting from normal service loads are within specified limits. These limits, known as *allowable stresses*, are only fractions of the failure stresses of the materials. Concrete responds reasonably elastically up to compression stresses not exceeding about half its strength, while steel remains elastic practically up to the yield stress. Hence members may be designed on an elastic basis as long as the stresses under service loads remain below these limits.

If members are proportioned on such a service load basis, the needed margin of safety is provided by stipulating allowable stresses under service loads that are appropriately small fractions of the compressive concrete strength and the steel yield stress. This basis for design is known as the *service load design*. Allowable stresses, in practice, are set at about one-half the concrete compressive strength and one-half the yield stress of the steel.

In the older service load design method, all types of loads are treated the same no matter how different their individual variability and uncertainty. Also, stresses are calculated on an elastic basis, while in reality the strength of a member depends on the inelastic stress-strain behavior near and at failure. Hence, the service load design method does not permit an explicit evaluation of the margin of safety. In contrast, in the newer strength design method, individual load factors may be adjusted to represent different degrees of uncertainty for the various types of loads, and strength reduction factors likewise may be adjusted to the precision with which various types of strength (bending, shear, torsion, etc.) are calculable, and the strength itself in each case is calculated with explicit regard for inelastic action. By service load design methods, serviceability with respect to deflections and cracking is customarily considered only implicitly, through limits on service load stresses.

Because of these differences in realism and reliability, over the past 25 years the strength design method has rapidly displaced the older service load design method, both here and abroad. However, the older method is still in occasional use. Throughout this text, strength design is presented almost exclusively.

1.6 DESIGN CODES AND SPECIFICATIONS

The design of concrete structures such as those of Figs. 1.1 to 1.11 is generally done within the framework of codes giving specific requirements for materials, structural analysis, member proportioning, etc. In contrast with many other highly developed countries, the United States does not have an official, national code governing structural concrete. The responsibility for producing and maintaining design specifications rests with various professional groups, trade associations, and technical institutes that have produced the needed documents.

The American Concrete Institute (ACI) has long been a leader in such efforts. As one part of its activity, the American Concrete Institute has published the widely recognized *Building Code Requirements for Reinforced Concrete* (Ref. 1.10), which serves as a guide in the design and construction of reinforced concrete buildings. The ACI Code has no official status in itself. However, it is generally regarded as an authoritative statement of current good practice in the field of reinforced concrete. As a result, it has been incorporated by law into countless municipal and regional building codes that do have legal status. Its provisions thereby attain, in effect, legal standing. Most reinforced concrete buildings and related construction in the United States are designed in accordance with the current ACI Code. It has also served as a model document for many other countries. A second ACI publication, *Commentary on Building Code Requirements for Reinforced Concrete* (Ref. 1.11), provides background material and rationale for the Code provisions. The American Concrete Institute also publishes important journals and standards, as well as recommendations for the analysis and design of special types of concrete structures such as the tanks of Fig. 1.11.

Most highway bridges in the United States are designed according to the requirements of the AASHTO bridge specifications (Ref. 1.3) which not only contain the provisions relating to loads and load distributions mentioned earlier, but also include detailed provisions for the design and construction of concrete bridges. Many of the provisions follow ACI Code provisions closely, although differences will be found.

The design of railway bridges is done according to the specifications of the AREA *Manual of Railway Engineering* (Ref. 1.4). It, too, is patterned after the ACI Code in most respects, but it contains much additional material pertaining to railway structures of all types.

No code or design specification can be construed as a substitute for sound engineering judgment in the design of concrete structures. In structural practice, special circumstances are frequently encountered where code provisions can serve only as a guide, and the engineer must rely upon a firm understanding of the basic principles of structural mechanics applied to reinforced or prestressed concrete, and an intimate knowledge of the nature of the materials.

1.7 SAFETY PROVISIONS OF THE ACI CODE

The safety provisions of the ACI Code are given in the form of Eqs. (1.3*b*) and (1.3*c*) utilizing strength reduction factors and load factors. These factors are based to some extent on statistical information but to a larger degree on experience, engineering judgment, and compromise. In words, the design strength ϕS_n of a structure or member must be at least equal to the required strength U calculated from the factored loads, i.e.,

$$\text{Design strength} \geq \text{required strength}$$

or

$$\phi S_n \geq U \tag{1.4}$$

The nominal strength S_n is computed (usually somewhat conservatively) by accepted methods. The required strength U is calculated by applying appropriate load factors to the respective service loads: dead load D, live load L, wind load W, earthquake load E, earth pressure H, fluid pressure F, impact allowance I, and environmental effects T that may include settlement, creep, shrinkage, and temperature change. Loads are defined in a general sense, to include either loads or the related internal effects such as moments, shears, and thrusts. Thus, in specific terms for a member subjected, say, to moment, shear, and thrust:

$$\phi M_n \geq M_u \tag{1.5a}$$

$$\phi V_n \geq V_u \tag{1.5b}$$

$$\phi P_n \geq P_u \tag{1.5c}$$

where the subscripts n denote the nominal strengths in flexure, shear, and thrust, respectively, and the subscripts u denote the factored load moment, shear, and thrust. In computing the factored load effects on the right, load factors may be applied either to the service loads themselves or to the internal load effects calculated from the service loads.

The load factors specified in the ACI Code, to be applied to calculated dead loads and those live and environmental loads specified in the appropriate codes or standards, are summarized in Table 1.2. These are consistent with the

Table 1.2 Factored load combinations for determining required strength U in the ACI Code

Condition	Factored load or load effect U
Basic	$U = 1.4D + 1.7L$
Winds	$U = 0.75(1.4D + 1.7L + 1.7W)$ and include consideration of $L = 0$
	$U = 0.9D + 1.3W$
	$U = 1.4D + 1.7L$
Earthquake	$U = 0.75(1.4D + 1.7L + 1.87E)$ and include consideration of $L = 0$
	$U = 0.9D + 1.43E$
	$U = 1.4D + 1.7L$
Earth pressure	$U = 1.4D + 1.7L + 1.7H$
	$U = 0.9D + 1.7H$
	$U = 1.4D + 1.7L$
Fluids	Add $1.4F$ to all loads that include L
Impact	Substitute $(L + I)$ for L
Settlement, creep, shrinkage, or temperature change effects	$U = 0.75(1.4D + 1.4T + 1.7L)$
	$U = 1.4(D + T)$

Table 1.3 Strength reduction factors in the ACI Code

Kind of strength	Strength reduction factor ϕ
Flexure, without axial load	0.90
Axial load, and axial load with flexure	
Axial tension, and axial tension with flexure	0.90
Axial compression, and axial compression with flexure	
Members with spiral reinforcement	0.75
Other members	0.70
except that for low values of axial load ϕ may be increased in accordance with the following:[a]	
For members in which f_y does not exceed 60,000 psi, with symmetrical reinforcement, and with $(h - d' - d_s)/h$ not less than 0.70, ϕ may be increased linearly to 0.90 as ϕP_n decreases from $0.10 f'_c A_g$ to zero	
For other reinforced members, ϕ may be increased linearly to 0.90 as ϕP_n decreases from $0.10 f'_c A_g$ or ϕP_{nb}, whichever is smaller, to zero	
Shear and torsion	0.85
Bearing on concrete	0.70

[a] The details of, and reasons for, these permissible increases are discussed in Chap. 8.

concepts introduced in Sec. 1.4. For individual loads, lower factors are used for loads known with greater certainty, e.g., dead load, compared with loads of greater variability, e.g., live loads. Further, for load combinations such as dead plus live loads plus wind forces, a reduction coefficient is applied that reflects the improbability that an excessively large live load coincides with an unusually high windstorm. The factors also reflect, in a general way, uncertainties with which internal load effects are calculated from external loads in systems as complex as highly indeterminate, inelastic reinforced concrete structures which, in addition, consist of variable-section members (because of tension cracking, discontinuous reinforcement, etc.). Finally, the load factors also distinguish between two situations, particularly when horizontal forces are present in addition to gravity, i.e., the situation where the effects of all simultaneous loads are additive, as distinct from that in which various load effects counteract each other. For example, in a retaining wall the soil pressure produces an overturning moment, and the gravity forces produce a counteracting stabilizing moment.

In all cases in Table 1.2 the controlling equation is the one that gives the largest factored load effect U.

The strength reduction factors ϕ in the ACI Code are given different values depending on the state of knowledge, i.e., the accuracy with which various strengths can be calculated. Thus the value for bending is higher than that for shear or bearing. Also, ϕ values reflect the probable importance, for the survival of the structure, of the particular member and of the probable quality control achievable. For both these reasons, a lower value is used for columns than for beams. Table 1.3 gives the ϕ values specified in the ACI Code.

The joint application of strength reduction factors (Table 1.3) and load factors (Table 1.2) is aimed at producing approximate probabilities of understrength of the order of 1/100 and of overloads of 1/1000. This results in a probability of structural failure of the order of 1/100,000.

The main body of the ACI Code is formulated in terms of strength design, with the specific load and strength reduction factors just presented. A special Code Appendix A, "Alternate Design Method," permits the use of service load design for those who prefer this older method. This appendix specifies allowable stresses for flexure, shear, bearing, etc., that are to be used in connection with the internal effects (M, V, P, etc.) of the unfactored dead and specific service loads. For many situations, particularly with today's higher-strength steels and concretes, this alternative design method is less economical than strength design.

1.8 FUNDAMENTAL ASSUMPTIONS FOR REINFORCED CONCRETE BEHAVIOR

The chief task of the structural engineer is the design of structures. By *design* is meant the determination of the general shape and all specific dimensions of a particular structure so that it will perform the function for which it is created and will safely withstand the influences that will act on it throughout its useful life. These influences are primarily the loads and other forces to which it will be subjected, as well as other detrimental agents, such as temperature fluctuations, foundation settlements, and corrosive influences. *Structural mechanics* is one of the main tools in this process of design. As here understood, it is the body of scientific knowledge that permits one to predict with a good degree of certainty how a structure of given shape and dimensions will behave when acted upon by known forces or other mechanical influences. The chief items of behavior that are of practical interest are (1) the strength of the structure, i.e., that magnitude of loads of a given distribution which will cause the structure to fail, and (2) the deformations, such as deflections and extent of cracking, that the structure will undergo when loaded under service conditions.

The fundamental propositions on which the mechanics of reinforced concrete is based are as follows:

1. The internal forces, such as bending moments, shear forces, and normal and shear stresses, at any section of a member are in equilibrium with the effects of the external loads at that section. This proposition is not an assumption but a fact, because any body or any portion thereof can be at rest only if all forces acting on it are in equilibrium.
2. The strain in an embedded reinforcing bar (unit extension or compression) is the same as that of the surrounding concrete. Expressed differently, it is assumed that perfect bonding exists between concrete and steel at the interface, so that no slip can occur between the two materials. Hence, as the one deforms, so must the other. With modern deformed bars (see Sec. 2.12) a high degree of mechanical interlocking is provided in addition to the natural surface adhesion, so this assumption is very close to correct.

3. Cross sections which were plane prior to loading continue to be plane in the member under load. Accurate measurements have shown that when a reinforced concrete member is loaded close to failure, this assumption is not absolutely accurate. However, the deviations are usually minor, and the results of theory based on this assumption check well with extensive test information.
4. In view of the fact that the tensile strength of concrete is only a small fraction of its compressive strength (see Sec. 2.8), the concrete in that part of a member which is in tension is usually cracked. While these cracks, in well-designed members, are generally so narrow as to be hardly visible (they are known as *hairline* cracks), they evidently render the cracked concrete incapable of resisting tension stress. Correspondingly, it is assumed that concrete is not capable of resisting any tension stress whatever. This assumption is evidently a simplification of the actual situation because, in fact, concrete prior to cracking, as well as the concrete located between cracks, does resist tension stresses of small magnitude. Later in discussions of the resistance of reinforced concrete beams to shear, it will become apparent that under certain conditions this particular assumption is dispensed with and advantage is taken of the modest tension strength which concrete can develop.
5. The theory is based on the actual stress-strain relationships and strength properties of the two constituent materials (see Secs. 2.7 and 2.12) or some reasonable simplifications thereof. The fact that nonelastic behavior is reflected in modern theory, that concrete is assumed to be ineffective in tension, and that the joint action of the two materials is taken into consideration results in analytical methods which are considerably more complex, and also more challenging, than those which are adequate for members made of a single, substantially elastic material.

These five assumptions permit one to predict by calculation the performance of reinforced concrete members only for some simple situations. Actually, the joint action of two materials as dissimilar and complicated as concrete and steel is so complex that it has not yet lent itself to purely analytical treatment. For this reason, methods of design and analysis, while utilizing these assumptions, are very largely based on the results of extensive and continuing experimental research. They are modified and improved as additional test evidence becomes available.

1.9 BEHAVIOR OF MEMBERS SUBJECT TO AXIAL LOADS

Many of the fundamentals of the behavior of reinforced concrete, through the full range of loading from zero to ultimate, can be illustrated clearly in the context of members subject to simple axial compression or tension. The basic concepts illustrated here will be recognized in later chapters in the analysis and design of beams, slabs, eccentrically loaded columns, and other members subject to more complex loadings.

a. Axial Compression

In members that sustain chiefly or exclusively axial compression loads, such as building columns, it is economical to make the concrete carry most of the load. Still, some steel reinforcement is always provided for various reasons. For one, very few members are truly axially loaded; steel is essential for resisting any bending that may exist. For another, if part of the total load is carried by steel with its much greater strength, the cross-sectional dimensions of the member can be reduced—the more so, the larger the amount of reinforcement.

The two chief forms of reinforced concrete columns are shown in Fig. 1.14. In the square column, the four longitudinal bars serve as main reinforcement. They are held in place by transverse small-diameter steel ties which prevent displacement of the main bars during construction operations and counteract any tendency of the compression-loaded bars to buckle out of the concrete by bursting the thin outer cover. On the left is shown a round column with eight main reinforcing bars. These are surrounded by a closely spaced spiral which serves the same purpose as the more widely spaced ties but also acts to confine the concrete within it, thereby increasing its resistance to axial compression. The discussion which follows applies to tied columns.

When axial load is applied, the compression strain is the same over the entire cross section, and in view of the bonding between concrete and steel, is the same in the two materials (see propositions 2 and 3 in Sec. 1.8). To illustrate the action of such a member as load is applied, Fig. 1.15 shows two typical stress-strain curves, one for a concrete with compressive strength $f_c' = 4000$ psi and the other for a steel with yield stress $f_y = 60,000$ psi. The curves for the two materials are drawn on the same graph using different vertical stress scales. Curve b has the shape which would be obtained in a concrete cylinder test. The rate of loading in

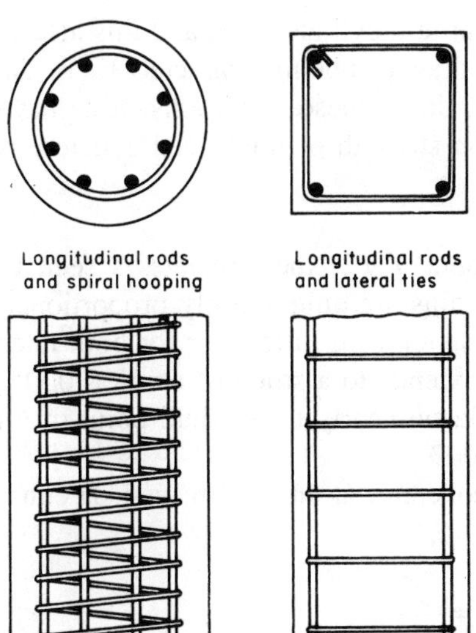

FIGURE 1.14
Reinforced concrete columns.

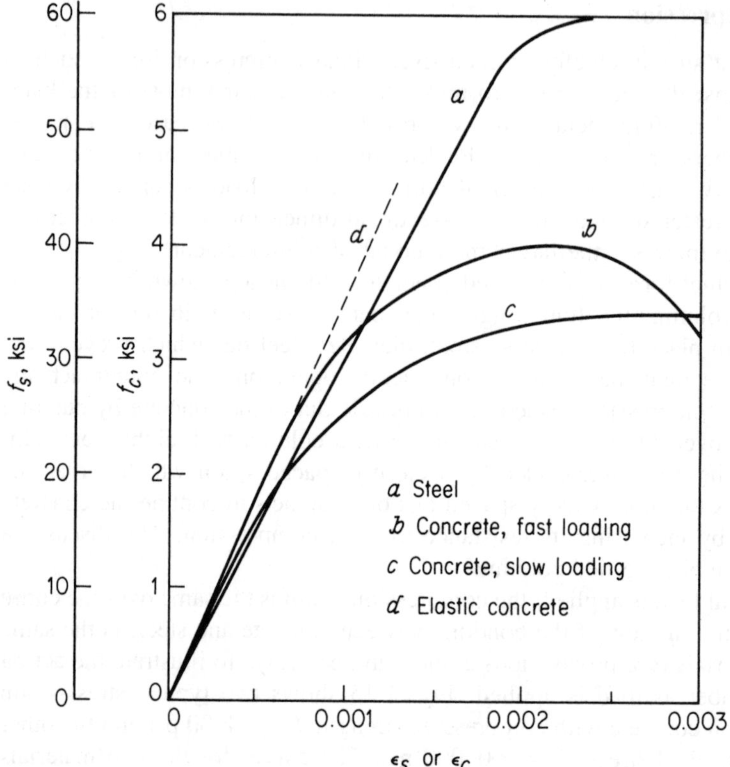

FIGURE 1.15
Concrete and steel stress-strain curves.

most structures is considerably slower than that in a cylinder test, and this affects the shape of the curve. Curve c, therefore, is drawn as being characteristic of the performance of concrete under slow loading. Under these conditions, tests have shown that the maximum reliable compression strength of reinforced concrete is about $0.85f'_c$, as shown.

Elastic Behavior. At low stresses, up to about $f'_c/2$, the concrete is seen to behave nearly elastically, i.e., stresses and strains are quite closely proportional; the straight line d represents this range of behavior with little error for both rates of loading. For the given concrete the range extends to a strain of about 0.0005. The steel, on the other hand, is seen to be elastic nearly to its yield point of 60 ksi, or to the much greater strain of about 0.002.

Because the compression strain in the concrete, at any given load, is equal to the compression strain in the steel,

$$\epsilon_c = \frac{f_c}{E_c} = \epsilon_s = \frac{f_s}{E_s}$$

from which the relation between the steel stress f_s, and the concrete stress f_c is obtained as

$$f_s = \frac{E_s}{E_c} f_c = n f_c \tag{1.6}$$

where $n = E_s/E_c$ is known as the *modular ratio*.

let A_c = net area of concrete, i.e., gross area minus area occupied by reinforcing bars
A_g = gross area
A_s = area of reinforcing bars
P = axial load

Then

$$P = f_c A_c + f_s A_s = f_c A_c + n f_c A_s$$

or

$$P = f_c(A_c + nA_s) \tag{1.7}$$

The term $A_c + nA_s$ can be interpreted as the area of a fictitious concrete cross section, the so-called *transformed area*, which when subjected to the particular concrete stress f_c results in the same axial load P as the actual section composed of both steel and concrete. This transformed concrete area is seen to consist of the actual concrete area plus n times the area of the reinforcement. It can be visualized as shown in Fig. 1.16. That is, in Fig. 1.16*b* the three bars along each of the two faces are thought of as being removed and replaced, at the same distance from the axis of the section, with added areas of fictitious concrete of total amount nA_s. Alternatively, as shown in Fig. 1.16*c*, one can think of the area of the steel bars as replaced with concrete, in which case one has to add to the gross concrete area A_g so obtained only $(n - 1)A_s$ in order to obtain the same total transformed area. Therefore, alternatively,

$$P = f_c[A_g + (n - 1)A_s] \tag{1.8}$$

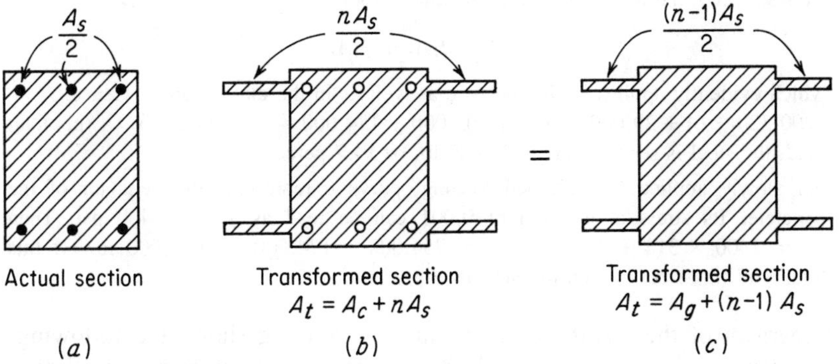

FIGURE 1.16
Transformed section in axial compression.

If load and cross-sectional dimensions are known, the concrete stress can be found by solving Eq. (1.7) or (1.8) for f_c, and the steel stress can be calculated from Eq. (1.6). These relations hold in the range in which the concrete behaves nearly elastically, i.e., up to about 50 to 60 percent of f'_c. For reasons of safety and serviceability, concrete stresses in structures under normal conditions are kept within this range. Therefore, these relations permit one to calculate *service load stresses*.

Example 1.1. A column of the materials defined in Fig. 1.15 has a cross section of 16 x 20 in. and is reinforced by six No. 9 bars disposed as shown in Fig. 1.16. (See Tables A.1 and A.2 of Appendix A for bar diameters and areas.) Determine the axial load that will stress the concrete to 1200 psi. The modular ratio n may be assumed equal to 8. (In view of the scatter inherent in E_c, it is customary and satisfactory to round off the value of n to the nearest integer.)

Solution. One finds $A_g = 16 \times 20 = 320 \text{ in}^2$, and from App. A, Table A.2, $A_s = 6.00$ in. The load on the column, from Eq. (1.8), is $P = 1200[320 + (8-1)6.00] = 434{,}000$ lb. Of this total load the concrete is seen to carry $P_c = f_c A_c = f_c(A_g - A_s) = 1200(320 - 6) = 377{,}000$ lb, and the steel $P_s = f_s A_s = (nf_c)A_s = 9600 \times 6 = 57{,}600$ lb, which is 13.3 percent of the total axial load.

Inelastic Range. Inspection of Fig. 1.15 shows that the elastic relationships which have been utilized so far cannot be used beyond a strain of about 0.0005 for the given concrete. To obtain information on the behavior of the member at larger strains and, correspondingly, at larger loads, it is therefore necessary to make direct use of the information of Fig. 1.15.

Example 1.2. One may want to calculate the magnitude of the axial load which will produce a strain or unit shortening $\epsilon_c = \epsilon_s = 0.0010$ in the column of the preceding example. At this strain the steel is seen to be still elastic, so that the steel stress $f_s = \epsilon_s E_s = 0.001 \times 29{,}000{,}000 = 29{,}000$ psi. The concrete is in the inelastic range, so that its stress cannot be directly calculated, but it can be read from the stress-strain curve for the given value of strain.

1. If the member has been loaded at a fast rate, curve b holds at the instant when the entire load is applied. The stress for $\epsilon = 0.001$ can be read as $f_c = 3200$ psi. Consequently, the total load can be obtained from

$$P = f_c A_c + f_s A_s \tag{1.9}$$

which evidently applies in the inelastic as well as in the elastic range. Hence, $P = 3200(320 - 6) + 29{,}000 \times 6 = 1{,}005{,}000 + 174{,}000 = 1{,}179{,}000$ lb. Of this total load, the steel is seen to carry 174,000 lb, or 14.7 percent.

2. For slowly applied or sustained loading, curve c represents the behavior of the concrete. Its stress at a strain of 0.001 can be read as $f_c = 2400$ psi. Then $P = 2400 \times 314 + 29{,}000 \times 6 = 754{,}000 + 174{,}000 = 928{,}000$ lb. Of this total load, the steel is seen to carry 18.8 percent.

Comparison of the results for fast and slow loading shows the following. Owing to creep of concrete, a given shortening of the column is produced by a smaller load when slowly applied or sustained over some length of time than when quickly applied. More important, the farther the stress is beyond the proportional limit of the concrete, and the more slowly the load is applied or the longer it is sustained, the smaller the share of the total load carried by the concrete, and the larger the share carried by the steel. In the sample column the steel was seen to carry 13.3 percent of the load in the elastic range, 14.7 percent for a strain of 0.001 under fast loading, and 18.8 percent at the same strain under slow or sustained loading.

Strength. The one quantity of chief interest to the structural designer is the *ultimate strength*, i.e., the maximum load which the structure or member will carry. Information on stresses, strains, and similar quantities serves chiefly as a tool for determining carrying capacity. The performance of the column discussed so far indicates two things: (1) in the range of large stresses and strains which precede attainment of ultimate strength and subsequent failure, elastic relationships cannot be utilized; (2) the member behaves differently under fast and under slow or sustained loading and shows less resistance to the latter than to the former. In usual construction, many types of loads, such as the weight of the structure and any permanent equipment housed therein, are sustained, and others are applied at slow rates. For this reason, to calculate a reliable magnitude of ultimate strength, curve *c* of Fig. 1.15 must be used as far as the concrete is concerned.

As for the steel, it reaches its ultimate strength (peak of the curve) at strains of the order of 0.08 (see Fig. 2.12). Concrete, on the other hand, fails by crushing at the much smaller strain of about 0.003 and, as seen from Fig. 1.15 (curve *c*), reaches its ultimate strength in the strain range of 0.002 to 0.003. Because the strains in steel and concrete are equal in axial compression, the load at which the steel begins to yield can be calculated from the information of Fig. 1.15.

If the small knee prior to yielding of the steel is disregarded., i.e., if the steel is assumed to be sharp-yielding, the strain at which it yields is

$$\epsilon_y = \frac{f_y}{E_s} \tag{1.10}$$

or

$$\epsilon_y = \frac{60,000}{29,000,000} = 0.00207$$

At this strain, curve *c* of Fig. 1.15 indicates a stress of 3200 psi in the concrete; therefore, by Eq. (1.9), the load in the member when the steel starts yielding is $P_y = 3200 \times 314 + 60,000 \times 6 = 1,365,000$ lb. At this load the concrete has not yet reached its ultimate strength, which, as mentioned before, can be assumed as $0.85 f_c' = 3400$ psi for slow or sustained loading, and therefore the load on the member can be further increased. During this stage of loading, the steel keeps yielding at constant stress. Finally, the ultimate load† of the member

† Throughout this book quantities which refer to the ultimate strength of members, calculated by accepted strength analysis methods, are furnished with the subscript *n*, which stands for "nominal." This notation is in agreement with the 1989 edition of the ACI Code. It is intended to convey that the actual ultimate strength of any member is bound to deviate to some extent from its calculated, nominal value because of inevitable variations of dimensions, materials properties, and other parameters. Design in all cases is based on this nominal strength, which represents the best available estimate of the actual member strength.

is reached when the concrete crushes while the steel yields, i.e.,

$$P_n = 0.85f_c'A_c + f_yA_s \tag{1.11}$$

Numerous careful tests have shown the reliability of Eq. (1.11) in predicting the ultimate strength of a concentrically loaded reinforced concrete column, provided its slenderness ratio is small so that buckling will not reduce it strength.

For the particular numerical example, $P_n = 3400 \times 314 + 60{,}000 \times 6 = 1{,}068{,}000 + 360{,}000 = 1{,}428{,}000$ lb. At this stage the steel carries as much as 25 percent of the load.

Summary. In the elastic range of low stresses the steel carries a relatively small portion of the total load of an axially compressed member. As ultimate strength is approached, there occurs a redistribution of the relative shares of the load resisted, respectively, by concrete and steel, the latter taking an increasing amount. The ultimate load at which the member is on the point of failure consists of the contribution of the steel when it is stressed to the yield point plus that of the concrete when its stress has attained the ultimate strength of $0.85f_c'$, as reflected in Eq. (1.11).

b. Axial Tension

The tension strength of concrete is only a small fraction of its compressive strength. It follows that reinforced concrete is not well suited for use in tension members because the concrete will contribute little, if anything, to their strength. Still, there are situations in which reinforced concrete is stressed in tension, chiefly in tie rods in structures such as arches. Such members consist of one or more bars embedded in concrete in a symmetrical arrangement similar to compression members (cf. Figs. 1.14 and 1.16).

When the tension force in the member is small enough for the stress in the concrete to be considerably below its tensile strength, both steel and concrete behave elastically. In this situation all the expressions derived for elastic behavior in compression in Sec. 1.9*a* are identically valid for tension. In particular, Eq. (1.7) becomes

$$P = f_{ct}(A_c + nA_s) \tag{1.12}$$

where f_{ct} is the tensile stress in the concrete.

However, when the load is further increased, the concrete reaches its tensile strength at a stress and strain of the order of one-tenth of what it could sustain in compression. At this stage the concrete cracks across the entire cross section. When this happens, it ceases to resist any part of the applied tension force, since, evidently, no force can be transmitted across the air gap in the crack. At any load larger than that which caused the concrete to crack, the steel is called upon to resist the entire tension force. Correspondingly, at this stage,

$$P = f_sA_s \tag{1.13}$$

With further increased load, the tensile stress f_s in the steel reaches the yield point f_y. When this occurs, the tension members cease to exhibit small, elastic deformations but instead stretch a sizable and permanent amount at substantially constant load. This does not impair the strength of the member. Its elongation, however, becomes so large (of the order of 1 percent or more of its length) as to render it useless. Therefore, the maximum useful strength P_{nt} of a tension member is that force which will just cause the steel stress to reach the yield point. That is,

$$P_{nt} = f_y A_s \tag{1.14}$$

To provide adequate safety, the force permitted in a tension member under normal service loads should be of the order of $\frac{1}{2}P_{nt}$. Because the concrete has cracked at loads considerably smaller than this, it does not contribute to the carrying capacity of the member in service. It does serve, however, as a fire and corrosion proofing and often improves the appearance of the structure.

There are situations, though, in which reinforced concrete is used in axial tension under conditions in which the occurrence of tension cracks must be prevented. A case in point is a circular tank (see Fig. 1.11). To provide watertightness, the hoop tension caused by the fluid pressure must be prevented from causing the concrete to crack. In this case, Eq. (1.12) can be used to determine a safe value for the axial tension force P by using, for the concrete tension stress f_{ct}, an appropriate fraction of the tensile strength of the concrete, i.e., of that stress which would cause the concrete to crack.

REFERENCES

1.1. *Minimum Design Loads for Buildings and Other Structures*, ANSI A58.1-1982, American National Standards Institute, New York, 1982.
1.2. *Uniform Building Code*, 1988 ed., International Conference of Building Officials, Whittier, CA, 1988.
1.3. *Standard Specifications for Highway Bridges*, 14th ed., American Association of State Highway and Transportation Officials (AASHTO), Washington, DC, 1989.
1.4. *Manual of Railway Engineering*, American Railway Engineering Association (AREA), Washington, DC, 1973.
1.5. *Recommended Lateral Force Requirements and Commentary*, Report by Seismology Committee, Structural Engineers Association of California (SEAOC), 1989.
1.6. J. G. MacGregor, S. A. Mirza, and B. Ellingwood, "Statistical Analysis of Resistance of Reinforced and Prestressed Concrete Members," *J. ACI*, vol. 80, no. 3, 1983, pp. 167–176.
1.7. J. G. MacGregor, "Load and Resistance Factors for Concrete Design," *J. ACI*, vol. 80, no. 4, pp. 279–287.
1.8. J. G. MacGregor, "Safety and Limit States Design for Reinforced Concrete," *Can. J. Civ. Eng.*, vol. 3, no. 4, 1976, pp. 484–513.
1.9. G. Winter, "Safety and Serviceability Provisions of the ACI Building Code," ACI-CEB-FIP-PCI Symposium, *ACI Special Publication SP-59*, 1979.
1.10. *Building Code Requirements for Reinforced Concrete*, ACI 318-89, American Concrete Institute, Detroit, 1989.
1.11. *Commentary on Building Code Requirements for Reinforced Concrete*, ACI 318R-89, American Concrete Institute, Detroit, 1989 (published as a part of ref. 1.10).
1.12. F. E. Richart and R. L. Brown, "An Investigation of Reinforced Concrete Columns," *Univ. Ill. Eng. Exp. Sta. Bull. 267*, 1934.

PROBLEMS

1.1 A 16 × 20 in. column is made of the same concrete and reinforced with the same six No. 9 bars as the column in Examples 1.1 and 1.2, except that a steel with yield strength $f_y = 40$ ksi is used. The stress-strain diagram of this reinforcing steel is shown in Fig. 2.12 for $f_y = 40$ ksi. For this column determine (*a*) the axial load that will stress the concrete to 1200 psi; (*b*) the load at which the steel starts yielding; (*c*) the ultimate strength; (*d*) the share of the total load carried by the reinforcement at these three stages of loading. Compare results with those calculated in the examples for $f_y = 60$ ksi, keeping in mind, in regard to relative economy, that the price per pound for reinforcing steels with 40 and 60 ksi yield points is about the same.

1.2 A square concrete column with dimensions 22 × 22 in. is reinforced with a total of eight No. 10 bars arranged uniformly around the column perimeter. Material strengths are $f_y = 60$ ksi and $f'_c = 4000$ psi, with stress-strain curves as given by curves *a* and *c* of Fig. 1.15. Calculate the percentages of total load carried by the concrete and by the steel as load is gradually increased from 0 to failure, which is assumed to occur when the concrete strain reaches a limit value of 0.0030. Determine the loads at strain increments of 0.0005 up to the failure strain and graph your results, plotting load percentages vs. strain. The modulur ratio may be assumed at $n = 8$ for these materials.

CHAPTER 2

MATERIALS

2.1 INTRODUCTION

The structures and component members treated in this text are composed of concrete reinforced with steel bars, and in some cases prestressed with steel wire, strand, or alloy bars. An understanding of the materials characteristics and behavior under load is fundamental to understanding the performance of structural concrete, and to safe, economical, and serviceable design of concrete structures. Although prior exposure to the fundamentals of material behavior is assumed, a brief review is presented in this chapter, as well as a description of the types of bar reinforcement and prestressing steels in common use. Numerous references are given as a guide for those seeking more information on any of the topics discussed.

2.2 CEMENT

A cementitious material is one that has the adhesive and cohesive properties necessary to bond inert aggregates into a solid mass of adequate strength and durability. This technologically important category of materials includes not only cements proper but also limes, asphalts, and tars as they are used in road building, and others. For making structural concrete, so-called *hydraulic cements* are used exclusively. Water is needed for the chemical process (hydration) in which the cement powder sets and hardens into one solid mass. Of the various hydraulic cements that have been developed, *Portland cement*, which was first patented in England in 1824, is by far the most common.

Portland cement is a finely powdered, grayish material that consists chiefly of calcium and aluminum silicates.† The common raw materials from which it is made are limestones, which provide CaO, and clays or shales, which furnish SiO_2 and Al_2O_3. These are ground, blended, fused to clinkers in a kiln, cooled, and ground to the required fineness. The material is shipped in bulk or in bags containing 94 lb of cement. Concretes made with Portland cement generally need about two weeks to reach sufficient strength so that forms of beams and slabs can be removed and reasonable loads applied; they reach their design strength after 28 days and continue to gain strength thereafter at a decreasing rate. To speed construction when needed, *high early strength cements* have been developed; they are more costly than ordinary Portland cement, but reach, within about 7 to 14 days, the strength a Portland cement would have after 28 days. They have the same basic composition as Portland cements, but are more carefully blended and more finely ground, both before and after clinkering.

When cement is mixed with water to form a soft paste, it gradually stiffens until it becomes a solid. This process is known as *setting* and *hardening*. The cement is said to have set when it has gained sufficient rigidity to support an arbitrarily defined pressure, after which it continues for a long time to harden, i.e., to gain further strength. The water in the paste dissolves material at the surfaces of the cement grains and forms a gel which gradually increases in volume and stiffness. This leads to a rapid stiffening of the paste 2 to 4 hours after water has been added to the cement. *Hydration* continues to proceed deeper into the cement grains, at decreasing speed, with continued stiffening and hardening of the mass. In ordinary concrete the cement is probably never completely hydrated. The gel structure of the hardened paste seems to be the chief reason for the volume changes that are caused in concrete by variations in moisture, such as the shrinkage of concrete as it dries.

For complete hydration of a given amount of cement, according to H. Rüsch, an amount of water equal to about 25 percent of that of cement, by weight—i.e., a *water-cement ratio* of 0.25—is needed chemically. An additional amount must be present, however, to provide mobility for the water in the cement paste during the hydration process so that it can reach the cement particles and to provide the necessary workability of the concrete mix. For normal concretes, the water-cement ratio is generally in the range of about 0.40 to 0.60, although for high-strength concretes, ratios as low as 0.25 have been used. In this case, the needed workability is obtained through the use of admixtures.

Any amount of water above the 25 percent consumed in the chemical reaction produces pores in the cement paste. The strength of the hardened paste decreases in inverse proportion to the fraction of the total volume occupied by pores. Put differently, since only the solids, and not the voids, resist stress, strength increases directly as the fraction of the total volume occupied by the solids. That is why the

† See ASTM C150, "Standard Specification for Portland Cement." This and other ASTM references are published and periodically updated by the American Society for Testing and Materials, Philadelphia.

strength of the cement paste depends primarily on, and decreases directly with, an increasing water-cement ratio.

The chemical process involved in the setting and hardening liberates heat, known as *heat of hydration*. In large concrete masses, such as dams, this heat is dissipated very slowly and results in a temperature rise and volume expansion of the concrete during hydration, with subsequent cooling and contraction. To avoid the serious cracking and weakening that may result from this process, special measures must be taken for its control.

2.3 AGGREGATES

In ordinary structural concretes the aggregates occupy about 70 to 75 percent of the volume of the hardened mass. The remainder consists of hardened cement paste, uncombined water (i.e., water not involved in the hydration of the cement), and air voids. The latter two evidently do not contribute to the strength of the concrete. In general, the more densely the aggregate can be packed, the better the strength, weather resistance, and economy of the concrete. For this reason the gradation of the particle sizes in the aggregate, to produce close packing, is of considerable importance. It is also important that the aggregate has good strength, durability, and weather resistance; that its surface is free from impurities such as loam, silt, and organic matter which may weaken the bond with cement paste; and that no unfavorable chemical reaction takes place between it and the cement.

Natural aggregates are generally classified as fine and coarse. *Fine aggregate*, or *sand*, is any material that will pass a No. 4 sieve, i.e., a sieve with four openings per linear inch. Material coarser than this is classified as *coarse aggregate*, or *gravel*. When favorable gradation is desired, aggregates are separated by sieving into two or three size groups of sand and several size groups of coarse aggregate. These can then be combined according to grading charts to result in a densely packed aggregate. The *maximum size of coarse aggregate* in reinforced concrete is governed by the requirement that it shall easily fit into the forms and between the reinforcing bars. For this purpose it should not be larger than one-fifth of the narrowest dimension of the forms or one-third of the depth of slabs, nor three-quarters of the minimum distance between reinforcing bars. Requirements for satisfactory aggregates are found in ASTM C33, "Standard Specification for Concrete Aggregates," and authoritative information on aggregate properties and their influence on concrete properties, as well as guidance in selection, preparation, and handling of aggregate, is found in Ref. 2.1.

The unit weight of *stone concrete*, i.e., concrete with natural stone aggregate, varies from about 140 to 152 pounds per cubic foot (pcf) and can generally be assumed to be 145 pcf. For special purposes, lightweight concretes, on the one hand, and heavy concretes, on the other, are being used with increasing frequency.

A variety of *lightweight* aggregates is available. Some unprocessed aggregates, such as pumice or cinders, are suitable for insulating concretes, but for structural lightweight concrete, *processed aggregates* are used because of better control. These consist of expanded shales, clays, slates, slags, or pelletized fly ash. They are light in weight because of the porous, cellular structure of the individual aggregate particle, which is achieved by gas or steam formation in processing the aggregates in rotary kilns at high temperatures (generally in excess of 2000°F). Requirements for satisfactory lightweight aggregates are found in ASTM C330, "Standard Specification for Lightweight Aggregates for Structural Concrete."

Three classes of lightweight concrete are distinguished in Ref. 2.2: low-density concretes, which are chiefly employed for insulation and whose unit weight rarely exceeds 50 pcf; moderate strength concretes, with unit weights from about 60 to 85 pcf and compressive strengths of 1000 to 2500 psi, which are chiefly used as fill, e.g., over light-gage steel floor panels; and structural concretes, with unit weights from 90 to 120 pcf and compressive strengths comparable to those of stone concretes. Similarities and differences in structural characteristics of lightweight and stone concretes are discussed in Secs. 2.7 and 2.8.

Heavyweight concrete is sometimes required for shielding against gamma and x-radiation in nuclear reactors and similar installations, for protective structures, and for special purposes, such as counterweights of lift bridges. Heavy aggregates are used for such concretes. These consist of heavy iron ores or barite (barium sulfate) rock crushed to suitable sizes. Steel in the form of scrap, punchings, or shot (as fines) is also used. Unit weights of heavyweight concretes with natural heavy rock aggregates range from about 200 to 230 pcf; if iron punchings are added to high density ores, weights as high as 270 pcf are achieved. The weight may be as high as 330 pcf if ores are used for the fines only and steel for the coarse aggregate.

2.4 PROPORTIONING AND MIXING OF CONCRETE

The various components of a mix are proportioned so that the resulting concrete has adequate strength, proper workability for placing, and low cost. The third calls for use of the minimum amount of cement (the most costly of the components) that will achieve adequate properties. The better the gradation of aggregates, i.e., the smaller the volume of voids, the less cement paste is needed to fill these voids. In addition to the water required for hydration, water is needed for wetting the surface of the aggregate. As water is added, the plasticity and fluidity of the mix increases (i.e., its workability improves), but the strength decreases because of the larger volume of voids created by the free water. To reduce the free water while retaining the workability, cement must be added. Therefore, as for the cement paste, the *water-cement ratio* is the chief factor that controls the strength of the concrete. For a given water-cement ratio, one selects the minimum amount of cement that will secure the desired workability.

Fig. 2.1 shows the decisive influence of the water-cement ratio on the compressive strength of concrete. Its influence on the tensile strength, as measured by the nominal flexural strength or modulus of rupture, is seen to be pronounced but much smaller than its effect on the compression strength. This seems to be so because, in addition to the void ratio, the tensile strength depends strongly on the strength of bond between coarse aggregate and cement mortar (i.e., cement paste plus fine aggregate). According to tests at Cornell University, this bond strength is relatively slightly affected by the water-cement ratio (Ref. 2.3).

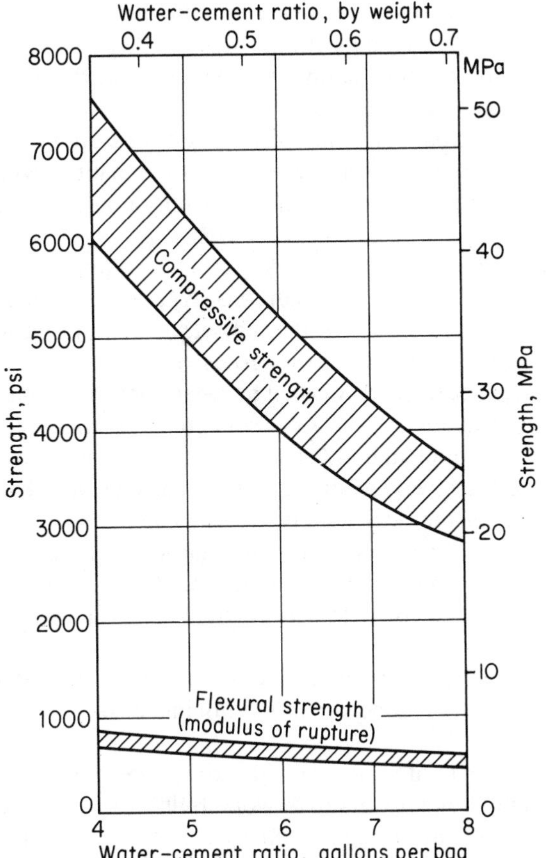

FIGURE 2.1
Effect of water-cement ratio on 28-day compressive and flexural tensile strength. *(Adapted from Ref. 2.4.)*

It has been customary to define the *proportions* of a concrete mix by the ratio, by volume or weight, of cement to sand to gravel, for example, 1:2:4. This method refers to the solid components only and, unless the water-cement ratio is specified separately, is insufficient to define the properties of the resulting concrete either in its fresh state or when set and hardened. For a complete definition of proportions it is now customary to specify, per 94 lb bag of cement, the weight of water, sand, and coarse aggregate. Thus, a mix may be defined as containing (per 94 lb bag of cement) 45 lb of water, 230 lb of sand, and 380 lb of coarse aggregate. Alternatively, batch quantities are often defined in terms of the total weight of each component needed to make up 1 yd^3 of wet concrete, such as 517 lb of cement, 300 lb of water, 1270 lb of dry sand, and 1940 lb of dry coarse aggregate.

Various methods of proportioning are used to obtain mixes of the desired properties from the cements and aggregates at hand. One is the so-called *trial-batch method*. Selecting a water-cement ratio from information such as that in Fig. 2.1, one produces several small trial batches with varying amounts of aggregate to obtain the required strength, consistency, and other properties with a minimum amount of paste. Concrete *consistency* is most frequently measured by the *slump test*. A metal mold in the shape of a truncated cone 12 in. high is filled with fresh concrete in a carefully specified manner. Immediately upon being filled, the mold is lifted off, and the slump of the concrete is measured as the difference in height between the mold and the pile of concrete. The slump is a good measure of the total water content in the mix and should be kept as low as is compatible with workability. Slumps for concretes in building construction generally range from 2 to 6 in.

The so-called ACI method of proportioning makes use of the slump test in connection with a set of tables that, for a variety of conditions (types of structures, dimensions of members, degree of exposure to weathering, etc.), permit one to estimate proportions that will result in the desired properties (Ref. 2.4). These preliminary selected proportions are checked and adjusted by means of trial batches to result in concrete of the desired quality. Inevitably, strength properties of a concrete of given proportions scatter from batch to batch. It is therefore necessary to select proportions that will furnish an average strength sufficiently greater than the specified design strength for even the accidentally weaker batches to be of adequate quality (for details, see Sec. 2.6). Discussion in detail of practices for proportioning concrete is beyond the scope of this volume; this topic is treated fully in Refs. 2.5 and 2.6, both for stone concrete and for lightweight aggregate concrete.

If the results of trial batches or field experience are not available, the ACI Code contains a conservative method of concrete proportioning based on the water-cement ratio.

In addition to the main components of concretes, *admixtures* are often used for special purposes. There are admixtures to improve workability, to accelerate or retard setting and hardening, to aid in curing, to improve durability, to add color, and to impart other properties. The beneficial effects of some admixtures are well established, but the claims for others should be viewed with caution. *Air-entraining agents* at present are the most important and most widely used admixtures. They cause the entrainment of air in the form of small dispersed bubbles in the concrete. These improve workability and durability, chiefly resistance to freezing and thawing, and reduce segregation during placing. They decrease density because of the increased void ratio and thereby decrease strength; however, this decrease can be partially offset by reduction of mixing water without loss of workability. The chief use of air-entrained concretes is in pavements, but they are also used for structures, particularly for exposed elements. *Plasticizers* and so-called *superplasticizers* are used increasingly, particularly for high-strength concretes (see Sec. 2.11), because they permit significant water reduction while maintaining high slumps needed for proper placement and compaction of the concrete. Useful design information pertaining to air-entraining agents, plasticizers, and other admixtures will be found in Refs. 2.7 and 2.8.

On all but the smallest jobs, *batching* is carried out in special batching plants. Separate hoppers contain cement and the various fractions of aggregate. Proportions are controlled, by weight, by means of manually operated or automatic dial scales connected to the hoppers. The mixing water is batched either by measuring tanks or by water meters.

The principal purpose of *mixing* is to produce an intimate mixture of cement, water, fine and coarse aggregate, and possible admixtures of uniform consistency throughout each batch. This is achieved in machine mixers of the revolving-drum type. Minimum mixing time is 1 min for mixers of not more than 1 yd^3 capacity, with an additional 15 sec for each additional $\frac{1}{2}$ yd^3. Mixing can be continued for a considerable time without adverse effect. This fact is particularly important in connection with ready-mixed concrete.

On large projects, particularly in the open country where ample space is available, movable mixing plants are installed and operated at the site. On the other hand, in construction under congested city conditions, on smaller jobs, and frequently in highway construction, *ready-mixed concrete* is used. Such concrete is batched in a stationary plant and then hauled to the site in trucks in one of three ways: (1) mixed completely at the stationary plant and hauled in a truck agitator, (2) transit-mixed, i.e., batched at the plant but mixed in a truck mixer, or (3) partially mixed at the plant with mixing completed in a truck mixer. Concrete should be discharged from the mixer or agitator within $1\frac{1}{2}$ hours after the water is added to the batch.

Much information on proportioning and other aspects of design and control of concrete mixtures will be found in Ref. 2.9.

2.5 CONVEYING, PLACING, COMPACTING, AND CURING

Conveying of most building concrete from the mixer or truck to the form is done in bottom-dump buckets or in wheelbarrows or buggies or by pumping through steel pipelines. The chief danger during conveying is that of *segregation*. The individual components of concrete tend to segregate because of their dissimilarity. In overly wet concrete standing in containers or forms, the heavier gravel components tend to settle, and the lighter materials, particularly water, tend to rise. Lateral movement, such as flow within the forms, tends to separate the coarse gravel from the finer components of the mix. The danger of segregation has caused the discarding of some previously common means of conveying, such as chutes and conveyor belts, in favor of methods that minimize this tendency.

Placing is the process of transferring the fresh concrete from the conveying device to its final place in the forms. Prior to placing, loose rust must be removed from reinforcement, forms must be cleaned, and hardened surfaces of previous concrete lifts must be cleaned and treated appropriately. Placing and compacting are critical in their effect on the final quality of the concrete. Proper placement must avoid segregation, displacement of forms or of reinforcement in the forms, and poor bond between successive layers of concrete. Immediately upon placing, the concrete should be *compacted* by means of hand tools or vibrators. Such compacting prevents honeycombing, ensures close contact with forms and reinforcement, and serves as a partial remedy to possible prior segregation. Compacting is achieved by hand tamping with a variety of special tools, but now more commonly and successfully with high frequency, power-driven *vibrators*. These are of the *internal* type, immersed in the concrete, or of the *external* type, attached to the forms. The former are preferable but must be supplemented by the latter where narrow forms or other obstacles make immersion impossible (Ref. 2.10).

Fresh concrete gains strength most rapidly during the first few days and weeks. Structural design is generally based on the *28-day strength*, about 70 percent of which is reached at the end of the first week after placing. The final concrete strength depends greatly on the conditions of moisture and temperature during this initial period. The maintenance of proper conditions during this time is known as *curing*. Thirty percent of the strength or more can be lost by premature drying out of the concrete; similar amounts may be lost by permitting the concrete temperature to drop to 40°F or lower during the first few days unless the concrete is kept continuously moist for a long time thereafter. Freezing of fresh concrete may reduce its strength by as much as 50 percent.

To prevent such damage, concrete should be protected from loss of moisture for at least 7 days and, in more sensitive work, up to 14 days. When high early strength cements are used, curing periods can be cut in half. Curing can be achieved by keeping exposed surfaces continually wet through sprinkling, ponding, or covering with plastic film or by the use of sealing compounds, which, when properly used, form evaporation-retarding membranes. In addition to improving strength, proper moist-curing provides better shrinkage control. To protect the concrete against low temperatures during cold weather, the mixing water and, occasionally, the aggregates are heated, temperature insulation is used where possible, and special admixtures, particularly calcium chloride, are employed. When air temperatures are very low, external heat may have to be supplied in addition to insulation (Refs. 2.9, 2.11, and 2.12.)

2.6 QUALITY CONTROL

The quality of mill-produced materials, such as structural or reinforcing steel, is guaranteed by the producer, who must exercise systematic quality controls, usually specified by pertinent ASTM standards. Concrete, in contrast, is produced at or close to the site, and its final qualities are affected by a number of factors, which have been discussed briefly. Thus, systematic quality control must be instituted at the construction site.

The main measure of the structural quality of concrete is its *compressive strength*. Tests for this property are made on cylindrical specimens of height equal to twice the diameter, usually 6 × 12 in. Impervious molds of this shape are filled with concrete during the operation of placement as specified by ASTM C172, "Standard Method of Sampling Freshly Mixed Concrete," and ASTM C31, "Standard Practice for Making and Curing Concrete Test Specimens in the Field." The cylinders are moist-cured at about 70°F, generally for 28 days, and then tested in the laboratory at a specified rate of loading. The compressive strength obtained from such tests is known as the *cylinder strength* f'_c and is the main property specified for design purposes.

To provide structural safety, continuous control is necessary to ensure that the strength of the concrete as furnished is in satisfactory agreement with the value called for by the designer. The ACI Code specifies that a pair of cylinders shall be tested for each 150 yd^3 of concrete or for each 5000 ft^2 of surface area actually placed, but not less than once a day. As mentioned in Sec. 2.4, the results of strength tests of different batches mixed to identical proportions show inevitable scatter. The scatter can be reduced by closer control, but occasional tests below the cylinder strength specified in the design cannot be avoided. To ensure adequate concrete strength in spite of such scatter, the ACI Code stipulates that concrete quality is satisfactory if (1) no individual strength test result (the average of a pair of cylinder tests) falls below the required f'_c by more than 500 psi and (2) the average of all sets of three consecutive strength tests is equal to or exceeds the required f'_c.

It is evident that, if concrete were proportioned so that its mean strength were just equal to the required strength f'_c, it would not pass these quality requirements, because about half of its strength test results would fall below the required f'_c. It is therefore necessary to proportion the concrete so that its mean strength f'_{cr}, used as the basis for selection of suitable proportions, exceeds the required design strength f'_c by an amount sufficient to ensure that the two quoted requirements are met. The minimum amount by which the required mean strength must exceed f'_c can be determined only by statistical methods because of the random nature of test scatter. Requirements have been derived, based on statistical analysis, to be used as a guide to proper proportioning of the concrete at the plant so that the probability of strength deficiency at the construction site is acceptably low.

The basis for these requirements is illustrated in Fig. 2.2, which shows three normal frequency curves giving the distribution of strength test results. The specified design strength is f'_c. The curves correspond to three different degrees of quality control, curve A representing the best control, i.e., the least scatter, and curve C the worst control, with the most scatter. The degree of control is measured statistically by the standard deviation σ (σ_a for curve A, σ_b for curve B, and σ_c for curve C), which is relatively small for producer A and relatively large for producer C. All three distributions have the same probability of strength less than the specified value f'_c, i.e., each has the same fractional part of the total area under the curve to the left of f'_c. For any normal distribution curve, that fractional part is defined by the index β_s, a multiplier applied to the standard deviation σ; β_s is the same for all three distributions of Fig. 2.2. It is seen that, in order to satisfy the requirement that, say, 1 test in 100 will fall below f'_c (with the value of β_s thus determined), for producer A with the best quality control the mean strength f'_{cr} can be much closer to the specified f'_c than for producer C with the most poorly controlled operation.

On the basis of such studies the ACI Code requires that concrete production facilities maintain records from which the standard deviation achieved in the particular facility can be determined. It then stipulates the minimum amount by which the average strength f'_{cr}, aimed at when selecting concrete proportions, shall exceed the specified design strength f'_c, depending on the standard deviation σ as follows:

$$f'_{cr} = f'_c + 1.34\sigma \tag{2.1}$$

$$f'_{cr} = f'_c + 2.33\sigma - 500 \tag{2.2}$$

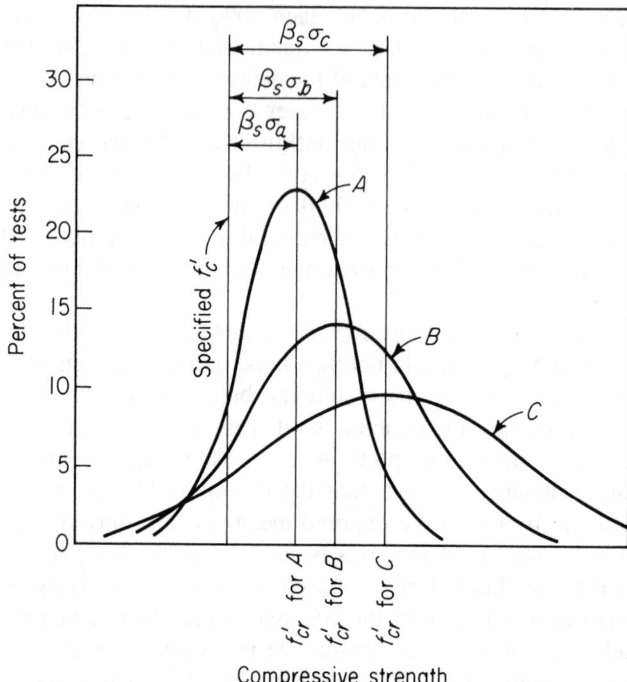

FIGURE 2.2
Frequency curves and average strengths for various degrees of control of concretes with specified design strength f_c'. *(Adapted from Ref. 2.13.)*

Equation (2.1) provides a probability of 1 in 100 that averages of three consecutive tests will be below the specified strength f_c', and Eq. (2.2) provides a probability of 1 in 100 that an individual test will be more than 500 psi below the specified strength f_c'. If no adequate record of concrete plant performance is available, the average strength must exceed f_c' by at least 1000 psi for f_c' of 3000 psi, by at least 1200 psi for f_c' between 3000 and 5000 psi, and by 1400 psi for f_c' over 5000 psi, according to the ACI Code.

It is seen that this method of control recognizes the fact that occasional deficient batches are inevitable. The requirements ensure (1) a small probability that such strength deficiencies as are bound to occur will be large enough to represent a serious danger and (2) an equally small probability that a sizable portion of the structure, as represented by three consecutive strength tests, will be made of below-par concrete.

In spite of scientific advances, building in general, and concrete making in particular, retain some elements of an art; they depend on many skills and imponderables. It is the task of systematic *inspection* to ensure close correspondence between plans and specifications and the finished structure. Inspection during construction should be carried out by a competent engineer, preferably the one who produced the design or one who is responsible to the design engineer. The inspector's main functions in regard to materials quality control are sampling, examination, and field testing of materials, control of concrete proportioning, inspection of batching, mixing, conveying, placing, compacting, and curing, and supervision of the preparation of specimens for laboratory tests. In addition, the inspector must inspect foundations, formwork, placing of reinforcing steel, and other pertinent features of the general progress of work; keep records of all the inspected items; and prepare periodic reports. The importance of thorough inspection to the correctness and adequate quality of the finished structure cannot be emphasized too strongly.

This brief account of concrete technology represents the merest outline of an important subject. Anyone in practice who is actually responsible for any of the phases of producing and placing concrete must be familiar with the details in much greater depth.

2.7 PROPERTIES IN COMPRESSION

a. Short-Time Loading

Performance of a structure under load depends to a large degree on the stress-strain relationship of the material from which it is made, under the type of stress to which the material is subjected in the structure. Since concrete is used mostly in compression, its compressive stress-strain curve is of primary interest. Such a curve is obtained by appropriate strain measurements in cylinder tests (Sec. 2.6) or on the compression side in beams. Figure 2.3 shows a typical set of such curves for normal density concrete, obtained from uniaxial compressive tests performed at normal, moderate testing speeds on concretes 28 days old. Figure 2.4 shows corresponding curves for lightweight concretes having a density of 100 pcf.

All the curves have somewhat similar character. They consist of an initial relatively straight elastic portion in which stress and strain are closely proportional, then begin to curve to the horizontal, reaching the maximum stress, i.e., the compressive strength, at a strain that ranges from about 0.002 to 0.003 for normal density concretes, and from about 0.003 to 0.0035 for lightweight concretes (Refs. 2.14 and 2.15), the larger values in each case corresponding to the higher strengths. All curves show a descending branch after the peak stress is reached; however the characteristics of the curves after peak stress are highly dependent upon the method of testing. If special procedures are followed in testing to ensure a constant strain rate while cylinder resistance is decreasing, long stable descending branches can be obtained (Ref. 2.16). In the absence of special devices, unloading past the point of peak stress may be rapid, particularly for the higher-strength concretes, which are generally more brittle than low-strength concrete.

In present practice, the specified compressive strength f_c' is commonly in the range from 3000 to 5000 psi for normal density cast-in-place concrete, and up to about 6000 psi for precast prestressed concrete members. Lightweight concrete strengths are somewhat below these values generally. The high-strength concretes, with f_c' to about 12,000 psi, are used with increasing frequency, particularly for heavily loaded columns in high-rise concrete buildings and for long-span bridges (mostly prestressed) where a significant reduction in dead load may be realized by minimizing member cross-section dimensions. (See Sec. 2.11).

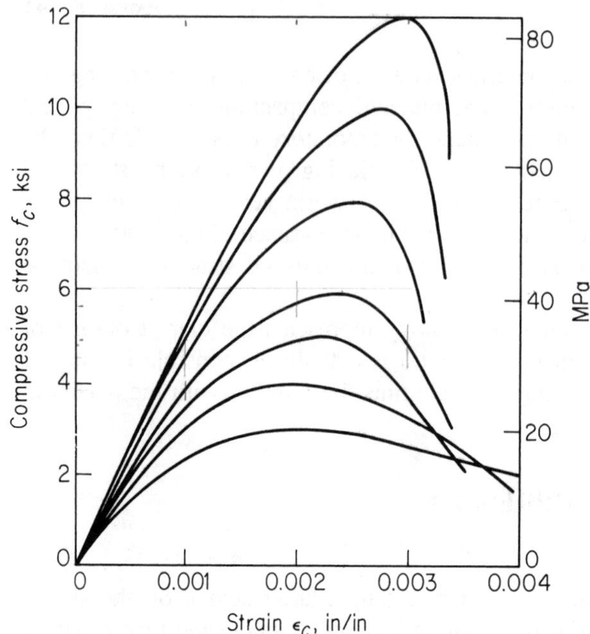

FIGURE 2.3
Typical compressive stress-strain curves for normal density concrete with w_c = 145 pcf. *(Adapted from Refs. 2.14 and 2.15.)*

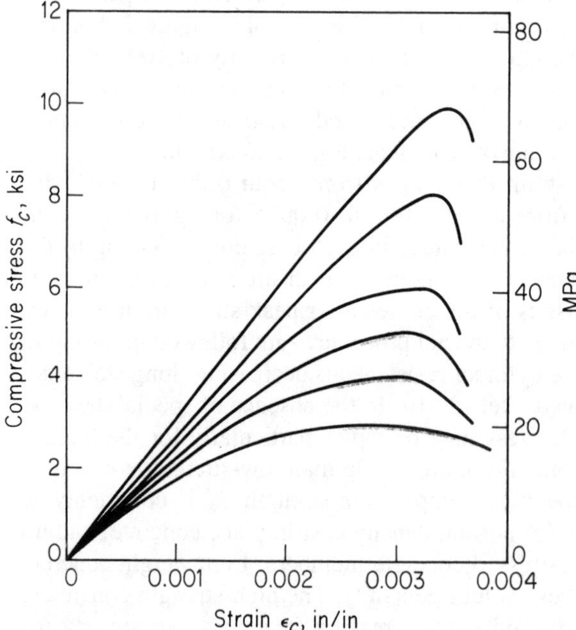

FIGURE 2.4
Typical compressive stress-strain curves for lightweight concrete with w_c = 100 pcf. *(Adapted from Refs. 2.14 and 2.15.)*

The *modulus of elasticity* E_c (in psi units), i.e., the slope of the initial straight portion of the stress-strain curve, is seen to be larger the higher the strength of the concrete. For concretes in the strength range to about 6000 psi, it can be computed with reasonable accuracy from the empirical equation found in the ACI Code:

$$E_c = 33w_c^{1.5}\sqrt{f_c'} \tag{2.3}$$

where w_c is the unit weight of the hardened concrete in pcf and f_c' its strength in psi. Equation (2.3) was obtained by testing structural concretes with values of w_c from 90 to 155 pcf. For normal sand-and-stone concretes, with $w_c = 145$ pcf, E_c may be taken as

$$E_c = 57{,}000\sqrt{f_c'} \tag{2.4}$$

For compressive strengths in the range from 6000 to 12,000 psi, the ACI Code equation overestimates E_c for both normal weight and lightweight material by as much as 20 percent. Based on recent research at Cornell University (Refs. 2.14 and 2.15), the following equation is recommended for normal density concretes with f_c' in the range of 3000 to 12,000 psi, and for lightweight concretes from 3000 to 9000 psi:

$$E_c = (40{,}000\sqrt{f_c'} + 1{,}000{,}000)\left(\frac{w_c}{145}\right)^{1.5} \tag{2.5}$$

where terms and units are as defined above for the ACI Code equations.

Information on concrete strength properties such as those discussed is usually obtained through tests made 28 days after pouring. However, cement continues to hydrate, and consequently concrete continues to harden, long after this age, at a decreasing rate. Figure 2.5 shows a typical curve of the gain of concrete strength with age for concrete made using Type I (normal) cement and also Type III (high early strength) cement, each curve normalized with respect to the 28-day compressive strength. High early strength cements produce more rapid strength gain at early ages, although the rate of strength gain at later ages is generally less. Concretes using Type III cement are often used in precasting plants, and often the strength f_c' is specified at 7 days, rather than 28 days.

It should be noted that the shape of the stress-strain curve for various concretes of the same cylinder strength, and even for the same concrete under various conditions of loading, varies considerably. An example of this is shown in Fig. 2.6, where different specimens of the same concrete are loaded at different rates of strain, from one corresponding to a relatively fast loading (0.001 in./in. per min) to one corresponding to an extremely slow application of load (0.001 in./in. per 100 days). It is seen that the descending branch of the curve, indicative of internal disintegration of the material, is much more pronounced at fast than at slow rates of loading. It is also seen that the peaks of the curves, i.e., the maximum strengths reached, are somewhat smaller at slower rates of strain.

When compressed in one direction, concrete, like other materials, expands in the direction transverse to that of the applied stress. The ratio of the transverse

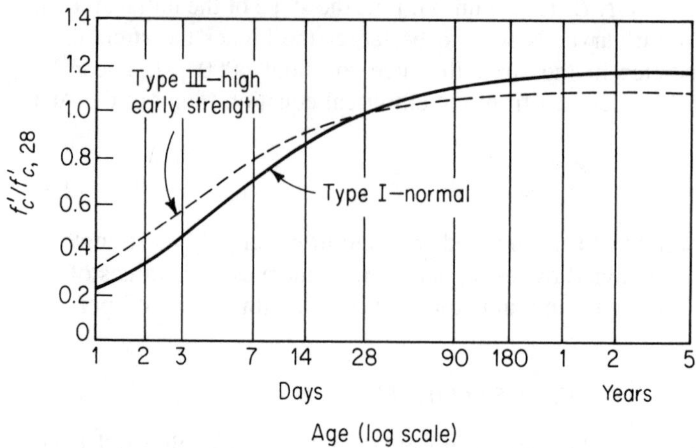

FIGURE 2.5
Effect of age on compressive strength f'_c for moist-cured concrete. *(Adapted from Ref. 2.17.)*

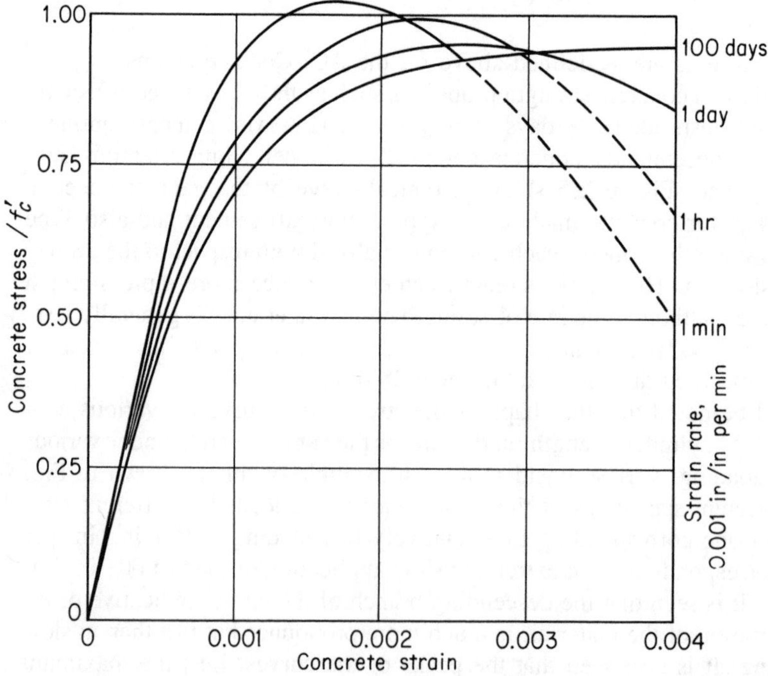

FIGURE 2.6
Stress-strain curves at various strain rates, concentric compression. *(Adapted from Ref. 2.18.)*

to the longitudinal strain in known as *Poisson's ratio* and depends somewhat on strength, composition, and other factors. At stresses lower than about $0.7f_c'$, Poisson's ratio for concrete falls within the limits of 0.15 to 0.20.

b. Long-Time Loading

In some engineering materials, such as steel, strength and the stress-strain relationships are independent of rate and duration of loading, at least within the usual ranges of rate of stress, temperature, and other variables. In contrast, Fig. 2.6 illustrates the fact that the influence of time, in this case of rate of loading, on the behavior of concrete under load is pronounced. The main reason for this is that concrete creeps under load, while steel does not exhibit creep under conditions prevailing in buildings, bridges, and similar structures.

Creep is the property of continuing to deform over considerable lengths of time at constant stress or load. The nature of the creep process is shown schematically in Fig. 2.7. This particular concrete was loaded after 28 days with resulting instantaneous strain ϵ_{inst}. The load was then maintained for 230 days, during which time creep was seen to have increased the total deformation to almost 3 times its instantaneous value. If the load were maintained, the deformation would follow the solid curve. If the load is removed, as shown by the dashed curve, most of the elastic instantaneous strain ϵ_{inst} is recovered, and some creep recovery is seen to occur. If the concrete is reloaded at some later date, instantaneous and creep deformations develop again, as shown.

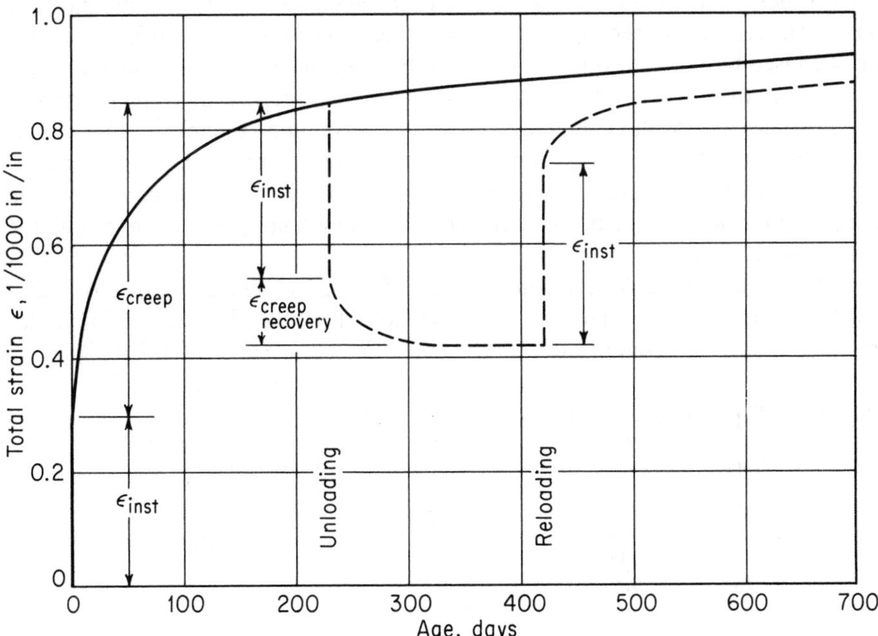

FIGURE 2.7
Typical creep curve (concrete loaded to 600 psi at age 28 days).

Creep deformations for a given concrete are practically proportional to the magnitude of the applied stress; at any given stress, high-strength concretes show less creep than lower-strength concretes. As seen in Fig. 2.7, with elapsing time, creep proceeds at a decreasing rate and ceases after 2 to 5 years at a final value which, depending on concrete strength and other factors, attains about 1.2 to 3 times the magnitude of the instantaneous strain. If, instead of being applied quickly and thereafter kept constant, the load is increased slowly and gradually, as is the case in many structures during and after construction, instantaneous and creep deformations proceed simultaneously. The effect is that shown in Fig. 2.6; i.e., the previously discussed difference in the shape of the stress-strain curve for various rates of loading is chiefly the result of the creep deformation of concrete.

For stresses not exceeding about half the cylinder strength, creep strains are directly proportional to stress. Because initial elastic stains are also proportional to stress in this range, this permits definition of the *creep coefficient:*

$$C_{cu} = \frac{\epsilon_{cu}}{\epsilon_{ci}} \tag{2.6}$$

where ϵ_{cu} is the final asymptotic value of the additional creep strain and ϵ_{ci} is the initial, instantaneous strain when the load is first applied. Creep may also be expressed in terms of the *specific creep* δ_{cu}, defined as the additional time-dependent strain per psi stress. It can easily be shown that

$$C_{cu} = E_c \delta_{cu} \tag{2.7}$$

In addition to the stress level, creep depends on the average ambient relative humidity, being more than twice as large for 50 percent as for 100 percent humidity (Ref. 2.4). This is so because part of the reduction in volume under sustained load is caused by outward migration of free pore water, which evaporates into the surrounding atmosphere. Other factors of importance include the type of cement and aggregate, age of the concrete when first loaded, and concrete strength (Ref. 2.19). The creep coefficient for high-strength concrete is much less than for low-strength concrete. However, sustained load stresses are apt to be higher so that the creep deformation may be as great for high-strength concrete, even though the creep coefficient is less.

The values of Table 2.1, quoted from Ref. 2.20 and extended for high-strength concrete based on recent research at Cornell University, are typical values for average humidity conditions, for concretes loaded at the age of 7 days.

Table 2.1 Typical creep parameters

Compressive strength		Specific creep δ_{cu}		Creep coefficient C_{cu}
psi	MPa	10^{-6} per psi	10^{-6} per MPa	
3000	21	1.00	145	3.1
4000	28	0.80	116	2.9
6000	41	0.55	80	2.4
8000	55	0.40	58	2.0
10,000	69	0.28	41	1.6
12,000	83	0.22	33	1.4

To illustrate, if the concrete in a column with $f_c' = 4000$ psi is subject to a long-time load that causes sustained stress of 1200 psi, then after several years under load the final value of the creep strain will be about $1200 \times 0.80 \times 10^{-6} = 0.00096$ in./in. Thus, if the column were 20 ft long, creep would shorten it by about $\frac{1}{4}$ in.

The creep coefficient at any time, C_{ct}, can be related to the ultimate creep coefficient C_{cu}. In Ref. 2.17, Branson suggests the equation:

$$C_{ct} = \frac{t^{0.60}}{10 + t^{0.60}} C_{cu} \tag{2.8}$$

where t = time in days after loading.

In many special situations, e.g., slender members or frames, or in prestressed construction, the designer must take account of the combined effects of creep and shrinkage (Sec. 2.10). In such cases, rather than relying on the sample values of Table 2.1, more accurate information on creep parameters should be obtained, such as from Refs. 2.17 or 2.20.

Sustained loads affect not only the deformation but also the strength of concrete. The cylinder strength f_c' is determined at normal rates of test loading (about 35 psi per sec). Tests by Rüsch (Ref. 2.18) and at Cornell University (Refs. 2.21 and 2.22) have shown that for concentrically loaded unreinforced concrete prisms and cylinders, the *strength under sustained load* is significantly smaller than f_c', of the order of 75 to 85 percent of f_c' for loads maintained for a year or more. Thus, a member subjected to a sustained overload causing compressive stress of, say, 85 percent of f_c' may fail after a period of time, even through the load is not increased.

c. Fatigue

When concrete is subject to fluctuating rather than sustained loading, its *fatigue strength*, as for all other materials, is considerably smaller than its static strength. When plain concrete in compression is stressed cyclically from zero to maximum stress, its fatigue limit is from 50 to 60 percent of the static compressive strength, for 2,000,000 cycles. A reasonable estimate can be made for other stress ranges using the modified Goodman diagram (see Ref. 2.20). For other types of applied stress, such as flexural compressive stress in reinforced concrete beams or flexural tension in unreinforced beams or on the tension side of reinforced beams, the fatigue limit likewise appears to be about 55 percent of the corresponding static strength. These figures, however, are for general guidance only. It is known that the fatigue strength of concrete depends not only on its static strength but also on moisture condition, age, and rate of loading (see Ref. 2.23).

2.8 TENSION STRENGTH

While concrete is best employed in a manner that utilizes its favorable compressive strength, its strength in tension is also of consequence in various connections. The conditions under which cracks form and propagate on the tension side of reinforced concrete flexural members depend strongly on the tension strength. Concrete tensile stresses also occur as a result of shear, torsion, and other actions, and in most cases member behavior changes upon cracking. As a result, it is important to be able to predict, with reasonable accuracy, the tensile strength of concrete.

There are considerable experimental difficulties in determining the true tensile strength of concrete. In *direct tension* tests, minor misalignments and stress concentrations in the gripping devices are apt to mar the results. For many years, tension strength has been measured in terms of the *modulus of rupture* f_r, the computed flexural tensile stress at which a test beam of plain concrete fractures. Because this nominal stress is computed on the assumption that concrete is an elastic material, and because this bending stress is localized at the outermost surface, it is apt to be larger than the strength of concrete in uniform axial tension. It is thus a measure of, but not identical with, the real axial tension strength.

More recently the result of the so-called *split-cylinder test* has established itself as a measure of the tensile strength of concrete. A 6 × 12 in. concrete cylinder, the same as is used for compressive tests, is inserted in a compression testing machine in the horizontal position, so that compression is applied uniformly along two opposite generators. Pads are inserted between the compression platens of the machine and the cylinder in order to equalize and distribute the pressure. It can be shown that in an elastic cylinder so loaded, a nearly uniform tensile stress of magnitude $2P/\pi dL$ exists at right angles to the plane of load application. Correspondingly, such cylinders, when tested, split into two halves along that plane, at a stress f_{ct} that can be computed from the above expression. P is the applied compressive load at failure, and d and L are the diameter and length of the cylinder respectively. Because of local stress conditions at the load lines and the presence of stresses at right angles to the aforementioned tension stresses, the results of the split-cylinder tests likewise are not identical with (but are believed to be a good measure of) the true axial tensile strength. The results of all types of tensile tests show considerably more scatter than those of compression tests.

Tensile strength, however determined, does not correlate well with the compressive strength f'_c. It appears that for sand-and-gravel concrete, the tensile strength depends primarily on the strength of bond between hardened cement paste and aggregate, whereas for lightweight concretes it depends largely on the tensile strength of the porous aggregate. The compressive strength, on the other hand, is much less determined by these particular characteristics.

Better correlation is found between the various measures of tensile strength and the square root of the compressive strength. The direct tensile strength, for example, ranges from about 3 to $5\sqrt{f'_c}$ for normal density concretes, and from about 2 to $3\sqrt{f'_c}$ for all-lightweight concrete. Typical ranges of values for direct tensile strength, split-cylinder strength, and modulus of rupture are summarized in Table 2.2. In these expressions, f'_c is expressed in psi units, and the resulting tensile strengths are obtained in psi.

These approximate expressions show that tension and compression strengths are by no means proportional, and that any increase in compression strength, such as that achieved by lowering the water-cement ratio, is accompanied by a much smaller percentage increase in tensile strength.

Table 2.2 Approximate range of tensile strengths of concrete

	Normal weight concrete, psi	Lightweight concrete, psi
Direct tensile strength f_t'	3 to $5\sqrt{f_c'}$	2 to $3\sqrt{f_c'}$
Split-cylinder strength f_{ct}	6 to $8\sqrt{f_c'}$	4 to $6\sqrt{f_c'}$
Modulus of rupture f_r	8 to $12\sqrt{f_c'}$	6 to $8\sqrt{f_c'}$

The ACI Code contains the recommendation that the modulus of rupture f_r' be taken to equal $7.5\sqrt{f_c'}$ for normal weight concrete, and that this value be multiplied by 0.85 for "sand-lightweight" and 0.75 for "all-lightweight" concretes, giving values of $6.4\sqrt{f_c'}$ and $5.6\sqrt{f_c'}$ respectively for those materials.

2.9 STRENGTH UNDER COMBINED STRESS

In many structural situations, concrete is subjected simultaneously to various stresses acting in various directions. For instance, in beams much of the concrete is subject simultaneously to compression and shear stresses, and in slabs and footings to compression in two perpendicular directions plus shear. By methods well known from the study of engineering mechanics, any state of combined stress, no matter how complex, can be reduced to three principal stresses acting at right angles to each other on an appropriately oriented elementary cube in the material. Any or all of the principal stresses can be either tension or compression. If any one of them is zero, a state of *biaxial* stress is said to exist; if two of them are zero, the state of stress is *uniaxial*, either simple compression or simple tension. In most cases only the uniaxial strength properties of a material are known from simple tests, such as the cylinder strength f_c' and the tensile strength f_t'. For predicting the strengths of structures in which concrete is subject to biaxial of triaxial stress, it would be desirable to be able to calculate the strength of concrete in such states of stress, knowing from tests only either f_c' or f_c' and f_t'.

In spite of extensive and continuing research, no general theory of the strength of concrete under combined stress has yet emerged. Modifications of various strength theories, such as maximum stress, maximum strain, the Mohr-Coulomb, and the octahedral shear stress theories, all of which are discussed in structural mechanics texts, have been adapted with varying partial success to concrete (Refs. 2.24 to 2.28). Recent research indicates that the nonlinear fracture mechanics approach may be used successfully in studying tensile crack propagation (Ref. 2.29). At present none of these theories has been generally accepted, and many have obvious internal contradictions. The main difficulty in developing an adequate general strength theory lies in the highly nonhomogeneous nature of concrete, and in the degree to which its behavior at high stresses and at fracture is influenced by microcracking and other discontinuity phenomena (Ref. 2.30).

However, the strength of concrete has been well established by tests, at least for the biaxial stress state (Refs. 2.31 and 2.32). Results may be presented in the form of an interaction diagram such as Fig. 2.8, which shows the strength in direction 1 as a function of the stress applied in direction 2. All stresses are normalized in terms of the uniaxial compressive strength f_c'. It is seen that in the quadrant representing biaxial compression a strength increase as great as about 20 percent over the uniaxial compressive strength is attained, the amount of increase depending upon the ratio of f_2 to f_1. In the biaxial tension quadrant, the strength in direction 1 is almost independent of stress in direction 2. When tension in direction 2 is combined with compression in direction 1, the compressive strength is reduced almost linearly, and vice versa. For example, lateral compression of about half the uniaxial compressive strength will reduce the tensile strength by almost half compared with its uniaxial value. This fact is of great importance in predicting diagonal tension cracking in deep beams or shear walls, for example.

Experimental investigations into the triaxial strength of concrete have been few, due mainly to the practical difficulty of applying load in three directions simultaneously without introducing significant restraint from the loading equipment (Ref. 2.33). From information now available the following tentative conclusions can be drawn relative to the triaxial strength of concrete: (1) in a state of equal triaxial compression, concrete strength may be an order of magnitude larger than the

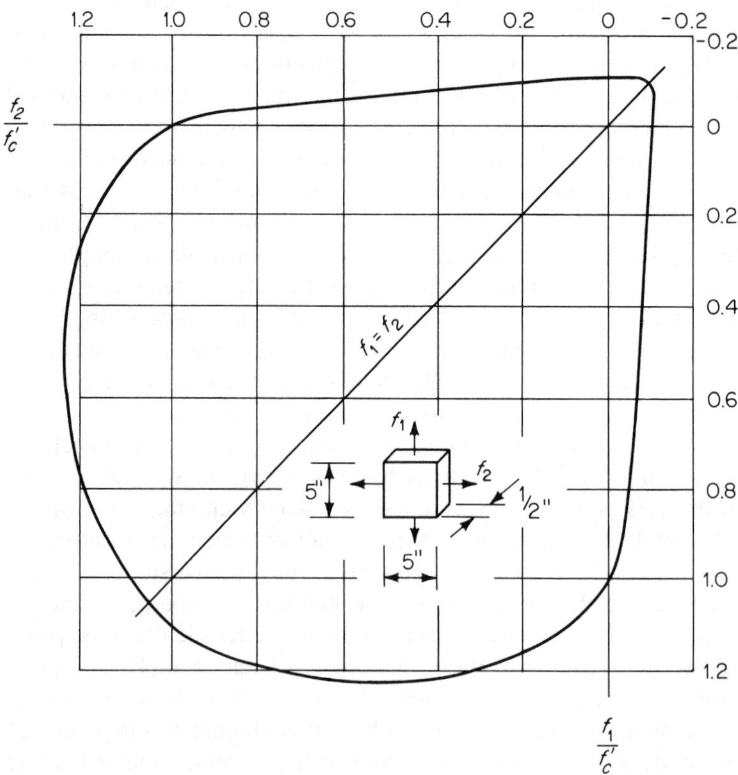

FIGURE 2.8
Strength of concrete in biaxial stress. *(Adapted from Ref. 2.32.)*

uniaxial compressive strength; (2) for equal biaxial compression combined with a smaller value of compression in the third direction, a strength increase greater than 20 percent can be expected; and (3) for stress states including compression combined with tension in at least one other direction, the intermediate principal stress is of little consequence, and the compressive strength can be predicted safely based on Fig. 2.8.

In fact, the strength of concrete under combined stress cannot yet be calculated rationally and, equally important, in many situations in concrete structures it is nearly impossible to calculate all the acting stresses and their directions; these are two of the main reasons for continued reliance on tests. Because of this, the design of reinforced concrete structures continues to be based more on extensive experimental information than on consistent analytical theory, particularly in the many situations where combined stresses occur.

2.10 SHRINKAGE AND TEMPERATURE EFFECTS

The deformations discussed in Sec. 2.7 were induced by stresses caused by external loads. Influences of a different nature cause concrete, even when free of any external loading, to undergo deformations and volume changes. The most important of these are shrinkage and the effects of temperature variations.

a. Shrinkage

As discussed in Secs. 2.2 and 2.4, any workable concrete mix contains more water than is needed for hydration. If the concrete is exposed to air, the larger part of this free water evaporates in time, the rate and completeness of drying depending on ambient temperature and humidity conditions. As the concrete dries, it shrinks in volume, probably due to the capillary tension that develops in the water remaining in the concrete. Conversely, if dry concrete is immersed in water, it expands, regaining much of the volume loss from prior shrinkage. Shrinkage, which continues at a decreasing rate for several months, depending on the configuration of the member, is a detrimental property of concrete in several respects. When not adequately controlled, it will cause unsightly and often deleterious cracks, as in slabs, walls, etc. In structures that are statically indeterminate (and most concrete structures are), it can cause large and harmful stresses. In prestressed concrete it leads to partial loss of initial prestress. For these reasons it is essential that shrinkage be minimized and controlled.

As is clear from the nature of the process, the chief factor that determines the amount of final shrinkage is the unit water content of the fresh concrete. This is clearly illustrated in Fig. 2.9, which shows the amount of shrinkage in units of 0.001 in./in. for varying amounts of mixing water. The same aggregates were used for all tests, but in addition to and independently of water content, the amount of cement was also varied, from 4 to 11 sacks per cubic yard of concrete. This very large variation of cement content had only very minor effects on the amount of shrinkage, compared with the effect of water content; this is evident from the narrowness of the band that comprises the test results for the

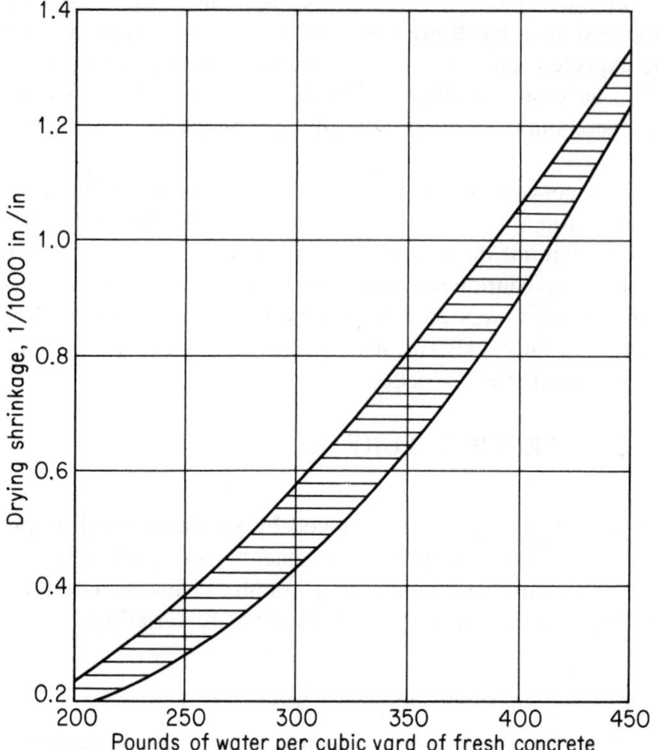

FIGURE 2.9
Effect of water content on drying shrinkage. *(From Ref. 2.4.)*

widely varying cement contents. It is evident from this that the chief means of reducing shrinkage is to reduce the water content of the fresh concrete to the minimum compatible with the required workability. In addition, prolonged and careful curing is beneficial for shrinkage control.

Values of final shrinkage for ordinary concretes are generally of the order of 400×10^{-6} to 800×10^{-6} in./in., depending on the initial water content, ambient temperature and humidity conditions, and the nature of the aggregate. Highly absorptive aggregates, such as some sandstones and slates, result in shrinkage values 2 and more times those obtained with less absorptive materials, such as granites and some limestones. Some lightweight aggregates, in view of their great porosity, easily result in much larger shrinkage values than ordinary concretes.

For some purposes, such as predicting the time-dependent loss of force in prestressed concrete beams, it is important to estimate the amount of shrinkage as a function of time. Long-term studies (Ref. 2.17) show that, for moist-cured concrete at any time t after the initial 7 days, shrinkage can be predicted satisfactorily by the equation

$$\epsilon_{sh,t} = \frac{t}{35 + t}\epsilon_{sh,u} \tag{2.9}$$

where $\epsilon_{sh,t}$ is the unit shrinkage strain at time t in days and $\epsilon_{sh,u}$ is the ultimate value after a long period of time. Equation (2.9) pertains to "standard" conditions, defined in Ref. 2.17 to exist for humidity not in excess of 40 percent and for an average thickness of member of 6 in., and it applies both for normal weight and lightweight concretes. Modification factors are to be applied for nonstandard conditions, and separate equations are given for steam-cured members.

b. Effect of Temperature Change

Like most other materials, concrete expands with increasing temperature and contracts with decreasing temperature. The effects of such volume changes are similar to those caused by shrinkage, i.e., temperature contraction can lead to objectionable cracking, particularly when superimposed on shrinkage. In indeterminate structures, deformations due to temperature changes can cause large and occasionally harmful stresses.

The coefficient of thermal expansion and contraction varies somewhat, depending upon the type of aggregate and richness of the mix. It is generally within the range of 4×10^{-6} to 7×10^{-6} in./in. per °F. A value of 5.5×10^{-6} is generally accepted as satisfactory for calculating stresses and deformations caused by temperature changes (Ref. 2.4).

2.11 HIGH-STRENGTH CONCRETE

In recent years there has been a rapid growth of interest in high-strength concrete. While the exact definition is arbitrary, the term generally refers to concrete having uniaxial compressive strength in the range of 6000 to 12,000 psi or higher. Such concretes can be made using carefully selected but widely available cements, sands, and stone; certain admixtures including water-reducing agents, fly ash, and microsilica; plus very careful quality control during production. In addition to higher strength in compression, most other engineering properties are improved. Modulus of elasticity is higher, as is tensile strength, and creep coefficient is lower. High-strength concretes are more durable and resist corrosion and abrasion better than normal concretes.

The most common application of high-strength concrete has been in the columns of tall concrete buildings, where normal concrete would result in unacceptably large cross sections, with loss of valuable floor space. It has been shown that the use of the more expensive high-strength concrete mixes in columns not only saves floor area, but also is more economical than increasing the amount of steel reinforcement. Concrete of up to 12,000 psi was specified for the lower-story columns of 311 South Wacker Drive in Chicago (see Fig. 2.10), presently the world's tallest concrete frame building, with total height of 946 ft.

For bridges, too, there are significant advantages through smaller cross sections, and the resulting reduction in dead load permits longer spans. The higher elastic modulus and lower creep coefficient result in reduced initial and long-term deflections, and in the case of prestressed bridges, initial and time-dependent losses of prestressing force are less. The East Huntington Bridge shown in Fig.

FIGURE 2.10
311 South Wacker Drive, Chicago, presently the world's tallest concrete building. High-strength concrete with $f_c' = 12{,}000$ psi was used in the lower-story columns. (*Courtesy of Portland Cement Association.*)

1.10 was built using 8000 psi concrete, and other recent bridges have used strength as high as 12,000 psi.

Other recent applications of high-strength concrete include offshore structures, parking garages, bridge deck overlays, dam spillways, warehouses, and heavy industrial slabs (Ref. 2.34).

An essential requirement for high-strength concrete is a low water-cement ratio. For normal concretes this usually falls in the range of about 0.40 to 0.60 by weight, while for high-strength mixes it may be as low as 0.25. In order to permit proper placement of what would otherwise be a zero slump mix, high range water-reducing admixtures, or "superplasticizers," are essential, and may increase slumps to as much as 6 or 8 inches.

Other additives often include fly ash and microsilica, or "silica fume." Fly ash, a by-product of the combustion of coal in power plants, is very finely divided and reacts with calcium hydroxide in the presence of moisture to form a cementitious material. It is used as a replacement of a portion of the Portland cement, usually up to about 8 to 12 percent by weight. Silica fume, also a by-product, results from the production of certain silicon alloys, mainly ferrochromium and ferromanganese. It is extremely finely divided and is highly cementitious. It, too, is used as a Portland cement replacement, to about 10 or 12 percent by weight. The addition of silica fume contributes mainly to strength gain at early ages, from 3 to 28 days, while fly ash builds strength mainly at ages later than 28 days. Because of the replacement of Portland cement with these additives, it is useful and customary to define water content for high-strength concretes in terms of the *water-cementitious* ratio, rather than the usual water-cement ratio.

Much research in recent years has been devoted to establishing the fundamental and engineering properties of high-strength concretes, as well as the engineering characteristics of structural members made using the new material (Refs. 2.35 to 2.37). A large body of information is now available, permitting the engineer to use high-strength concrete with confidence when its advantages justify the higher cost. Compressive strength curves shown in Figs. 2.3 and 2.4 illustrate important differences compared with normal concrete, including higher elastic modulus and more extended range of linear elastic response, but somewhat reduced ultimate strain capacity. Creep coefficients are significantly reduced, as indicated by Table 2.1. Strength under sustained load is a higher fraction of standard cylinder strength (Refs. 2.21 and 2.22), and new information will soon be published relating to durability and abrasion resistance. As experience is gained in practical applications, and as design codes are gradually updated to recognize the special properties of the higher-strength concretes now available, much broader use of high-strength concrete can be expected.

2.12 REINFORCING STEELS FOR CONCRETE

The useful strength of ordinary reinforcing steels in tension as well as compression, i.e., the yield strength, is about 15 times the compressive strength of common structural concrete, and well over 100 times its tensile strength. On the other hand, steel is a high-cost material compared with concrete. It follows that the two materials are best used in combination if the concrete is made to resist the compressive stresses and the steel the tensile stresses. Thus, in reinforced concrete beams, the concrete resists the compressive force, longitudinal steel reinforcing bars are located close to the tension face to resist the tension force, and usually additional steel bars are so disposed that they resist the inclined tension stresses that are caused by the shear force in the beams. However, reinforcement is also used for resisting compressive forces primarily where it is desired to reduce the cross-sectional dimensions of compression members, as in the lower-floor columns of multistory buildings. Even if no such necessity exists, a minimum amount of reinforcement is placed in all compression members to safeguard them against the effects of small accidental bending moments that might crack and even fail an unreinforced member.

For most effective reinforcing action, it is essential that steel and concrete deform together, i.e., that there be a sufficiently strong *bond* between the two materials to ensure that no relative movements of the steel bars and the surrounding concrete occur. This bond is provided by the relatively large *chemical adhesion* that develops at the steel-concrete interface, by the *natural roughness* of the mill scale of hot-rolled reinforcing bars, and by the closely spaced rib-shaped *surface deformations* with which reinforcing bars are furnished in order to provide a high degree of interlocking of the two materials.

Additional features that make for the satisfactory joint performances of steel and concrete are the following:

1. The *thermal expansion coefficients* of the two materials, about 6.5×10^{-6} for steel vs. an average of 5.5×10^{-6} for concrete, are sufficiently close to forestall cracking and other undesirable effects of differential thermal deformations.
2. While the *corrosion resistance* of bare steel is poor, the concrete that surrounds the steel reinforcement provides excellent corrosion protection, minimizing corrosion problems and corresponding maintenance costs.
3. The *fire resistance* of unprotected steel is impaired by its high thermal conductivity and by the fact that its strength decreases sizably at high temperatures. Conversely, the thermal conductivity of concrete is relatively low. Thus, damage caused by even prolonged fire exposure, if any, is generally limited to the outer layer of concrete, and a moderate amount of concrete cover provides sufficient thermal insulation for the embedded reinforcement.

Steel is used in two different ways in concrete structures: as reinforcing steel and as prestressing steel. Reinforcing steel is placed in the forms prior to casting of the concrete. Stresses in the steel, as in the hardened concrete, are caused only by the loads on the structure, except for possible parasitic stresses from shrinkage or similar causes. In contrast, in prestressed concrete structures large tension forces are applied to the reinforcement prior to letting it act jointly with the concrete in resisting external loads. The steels for these two uses are very different and will be discussed separately.

2.13 REINFORCING BARS

The most common type of reinforcing steel (as distinct from prestressing steel) is in the form of round bars, often called *rebars*, available in a large range of diameters from about $\frac{3}{8}$ to $1\frac{3}{8}$ in. for ordinary applications and in two heavy bar sizes of about $1\frac{3}{4}$ and $2\frac{1}{4}$ in. These bars are furnished with surface deformations for the purpose of increasing resistance to slip between steel and concrete. Minimum requirements for these deformations (spacing, projection, etc.) have been developed in experimental research. Different bar producers use different patterns, all of which satisfy these requirements. Figure 2.11 shows a variety of current types of deformations.

FIGURE 2.11
Types of deformed reinforcing bars (rebars).

Bar sizes are designated by numbers, Nos. 3 to 11 being commonly used and Nos. 14 and 18 representing the two special large-sized bars previously mentioned. Designation by number, instead of by diameter, has been introduced because the surface deformations make it impossible to define a single easily measured value of the diameter. The numbers are so arranged that the unit in the number designation corresponds closely to the number of $\frac{1}{8}$ in. of diameter size. A No. 5 bar, for example, has a nominal diameter of $\frac{5}{8}$ in. Table A.1 of App. A gives areas, perimeters, and weights of standard bars. Table A.2 to A.4 give similar information for groups of bars.

a. Grades and Strengths

In reinforced concrete a long-time trend is evident toward the use of higher-strength materials, both steel and concrete. Reinforcing bars with 40 ksi yield stress, almost standard 20 years ago, have largely been replaced by bars with 60 ksi yield stress, both because they are more economical and because their use tends to reduce congestion of steel in the forms. Table 2.3 lists all presently available reinforcing steels, their grade designation, the ASTM specifications that define their properties (including deformations) in detail, and their two main minimum specified strength values. Grades 40 and 50 bars, shown in Table 2.3, may not be generally available in the larger bar diameters.

Table 2.3 Standardized reinforcing and prestressing steels

Product	ASTM specification	Grade or type	Minimum yield strength		Minimum tensile strength	
			ksi	MPa	ksi	MPa
Reinforcing bars	A615	40	40	276	70	483
		60	60	414	90	620
	A616	50	50	345	80	552
		60	60	414	90	620
	A617	40	40	276	70	483
		60	60	414	90	620
	A706	60	60 (78 max.)	414 (538 max.)	80	552
Bar mats	A184[a]					
Wire, smooth	A82		70	483	80	552
deformed	A496		75	517	85	586
Welded wire fabric, smooth	A185		65	448	75	517
deformed	A497		70	483	80	552
Prestress bar	A722	Type I	127.5	880	150	1034
		Type II	120	827	150	1034
Prestress wire	A421		188–200	1296–1330	235–250	1620–1725
Prestress strand	A416	250	212.5	1465	250	1725
		270	229.5	1580	270	1860

[a] Same as reinforcing bars.

Source: From Ref. 2.38.

Welding of rebars in making splices, or for convenience in fabricating reinforcing cages for placement in the forms, may result in metallurgical changes that reduce both strength and ductility, and special restrictions must be placed both on the type of steel used and the welding procedures. The provisions of ASTM A706 relate specifically to welding.

The ACI Code permits reinforcing steels up to $f_y = 80$ ksi. Such high-strength steels usually yield gradually but have no yield plateau (see Fig. 2.13). In this situation it is required that at the specified minimum yield strength the total strain shall not exceed 0.0035. This is necessary to make current design methods, which were developed for sharp-yielding steels with a yield plateau, applicable to such higher-strength steels. Under special circumstances steel in this higher-strength range has its place, e.g., in lower-story columns of high-rise buildings.

In order for bars of various grades and sizes to be easily distinguished, which is necessary to avoid accidental use of lower-strength or smaller-size bars than called for in the design, all deformed bars are furnished with rolled-in markings. These identify the producing mill (usually an initial), the bar size number (3 to 18), the type of steel (*S* for billet, a rail sign for rail steel, *A* for axle, and *W* for low-alloy, corresponding to ASTM Specifications A615, A616, A617, and A706 respectively), and an additional marking for identifying the higher-strength steels.

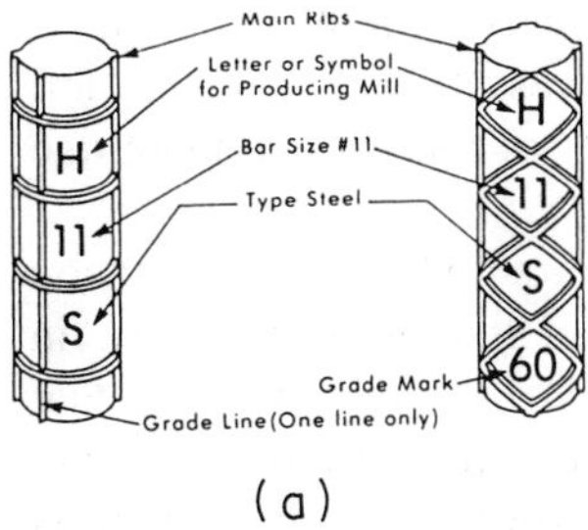

(a)

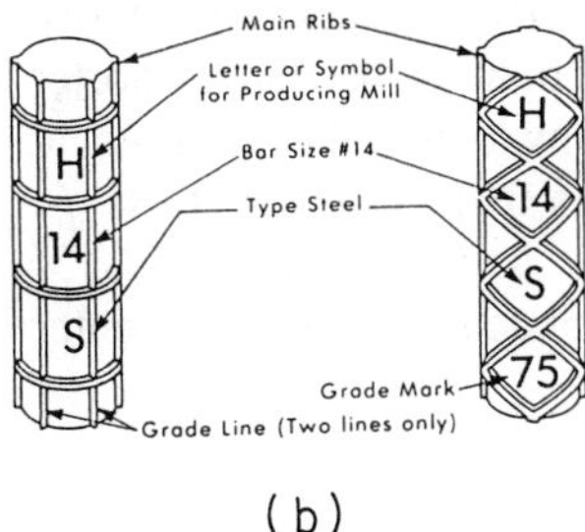

(b)

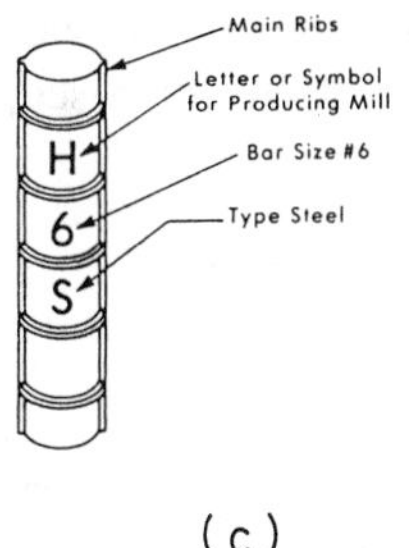

(c)

FIGURE 2.12
Marking system for reinforcing bars meeting ASTM Specifications A615, A616, A617, and A706: (*a*) Grades 60 and A706; (*b*) Grade 75; (*c*) Grades 40 and 50. (*Adapted from Ref. 2.39.*)

Grade 60 bars have either one longitudinal line or the number 60; Grade 75 bars have either two longitudinal lines or the number 75. The identification marks are shown in Fig. 2.12.

b. Stress-Strain Curves

The two chief numerical characteristics that determine the character of bar reinforcement are its *yield point* (generally identical in tension and compression) and its *modulus of elasticity* E_s. The latter is practically the same for all reinforcing steels (but not for prestressing steels) and is taken as $E_s = 29{,}000{,}000$ psi.

In addition, however, the shape of the stress-strain curve, and particularly of its initial portion, has significant influence on the performance of reinforced concrete members. The typical stress-strain curves of American reinforcing steels are shown in Fig. 2.13. The complete stress-strain curves are shown in the left part of the figure; the right part gives the initial portions of the curves magnified 10 times.

Low-carbon steels, typified by the Grade 40 curve, show an elastic portion followed by a *yield plateau*, i.e., a horizontal portion of the curve where strain continues to increase at constant stress. For such steels the yield point is that stress at which the yield plateau establishes itself. With further strains the stress begins to increase again, though at a slower rate, a process that is known as *strain-hardening*. The curve flattens out when the *tensile strength* is reached; it then turns down until fracture occurs. Higher-strength carbon steels, e.g., those with 60 ksi yield stress or higher, either have a yield plateau of much shorter length or enter strain-hardening immediately without any continued yielding at constant stress. In the latter case the ACI Code specifies that the yield stress f_y be the stress corresponding to a strain of 0.0035 in./in., as shown in Fig. 2.13. Low alloy, high-strength steels rarely show any yield plateau and usually enter strain-hardening immediately upon beginning to yield.

c. Fatigue Strength

In highway bridges and some other situations both steel and concrete are subject to large numbers of stress fluctuations. Under such conditions, steel, just like concrete (Sec. 2.7c), is subject to *fatigue*. In metal fatigue, one or more microscopic

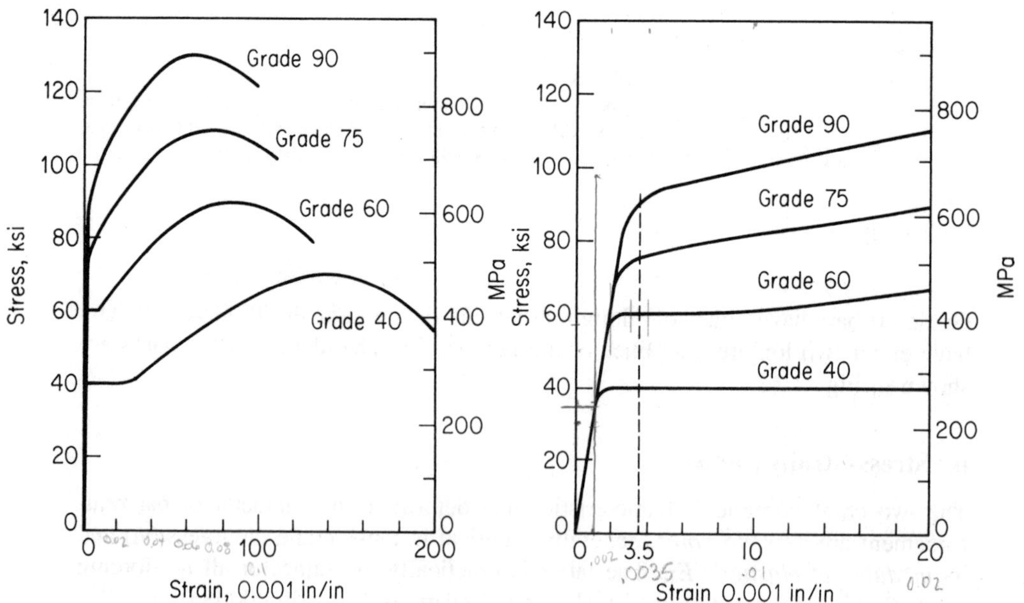

FIGURE 2.13
Typical stress-strain curves for reinforcing bars.

cracks form after cyclic stress has been applied a significant number of times. These fatigue cracks occur at points of stress concentrations or other discontinuities and gradually increase with increasing numbers of stress fluctuations. This reduces the remaining uncracked cross-sectional area of the bar until it becomes too small to resist the applied force. At this point the bar fails in a sudden, brittle manner.

For reinforcing bars it has been found (Refs. 2.23 and 2.40) that the fatigue strength, i.e., the stress at which a given stress fluctuation between f_{max} and f_{min} can be applied 2 million times or more without causing failure, is practically independent of the grade of steel. It has also been found that the stress range, i.e., the algebraic difference between maximum and minimum stress, $f_1 = f_{max} - f_{min}$, that can be sustained without fatigue failure depends on f_{min}. Further, in deformed bars the degree of stress concentration at the location where the rib joins the main cylindrical body of the bar tends to reduce the safe stress range. This stress concentration depends on the ratio r/h, where r is the base radius of the deformation and h its height. The radius r is the transition radius from the surface of the bar to that of the deformation; it is a fairly uncertain quantity that changes with roll wear as bars are being rolled.

On the basis of extensive tests (Ref. 2.40) the following formula has been developed for design:

$$f_r = 21 - 0.33 f_{min} + 8\frac{r}{h} \tag{2.10}$$

where f_r = safe stress range, ksi
f_{min} = minimum stress; positive if tension, negative if compression
r/h = ratio of base radius to height of rolled-on deformation (in the common situation where r/h is not known, a value of 0.3 may be used)

Where bars are exposed to fatigue regimes, stress concentrations such as welds or sharp bends should be avoided since they may impair fatigue strength.

d. Coated Reinforcing Bars

Galvanized or epoxy-coated reinforcing bars are often specified in order to minimize corrosion of reinforcement and consequent spalling of concrete under severe environmental conditions, such as in bridge decks or parking garages subject to de-icing chemicals, port and marine structures, and wastewater treatment plants (Refs. 2.39, 2.41, 2.42).

ASTM A767, "Standard Specification for Zinc-Coated (Galvanized) Steel Bars for Concrete Reinforcement," includes requirements for the zinc coating material, the galvanizing process, the class or weight of coating, finish and adherence of coating, and the method of fabrication. Rebars are usually galvanized after cutting and bending. Supplementary requirements pertain to coating of sheared ends and repair of damaged coating if bars are fabricated after galvanizing.

Epoxy-coated bars, presently more widely used than galvanized bars, are governed by ASTM A775, "Standard Specification for Epoxy-Coated Reinforcing

Steel Bars," which includes requirements for the coating material, surface preparation prior to coating, method of application, and limits on coating thickness. Typically, the coating is applied to straight bars in a production-line operation, and the bars are cut and bent after coating. Cut ends and small spots of damaged coating are suitably repaired after fabrication. Extra care is required in the field to ensure that the coating is not damaged during shipment and placing and that repairs are made if necessary.

2.14 WELDED WIRE FABRIC

Apart from single reinforcing bars, *welded wire fabric* is often used for reinforcing slabs and other surfaces, such as shells, and for shear reinforcement in thin beam webs, particularly in prestressed beams. Welded wire reinforcement consists of sets of longitudinal and transverse cold-drawn steel wires at right angles to each other and welded together at all points of intersection. The size and spacing of wires may be the same in both directions or may be different, depending on the requirements of the design.

The conventional notation used to describe the type and size of welded wire fabric is in a period of transition and two systems are in use. Previously fabric was referred to using the longitudinal and transverse wire spacing and the wire gage; for example, 6 × 6-4 × 4, meaning wire spacings at 6 in. each way using 4-gage wire each way. The new designations show spacing of wires in the same way, but the wire gage has been replaced by an identification W or D (for smooth or deformed wire) and the cross-sectional wire area in hundredths of a square inch. For example, 4 × 4-*W*5 × W5 indicates wire spacings at 4 in. each way with smooth wire having a cross-sectional area of 0.050 in^2 in each direction. Sizes and spacings of wires for common types of welded wire fabric and cross-sectional areas of steel per foot, as well as weight per 100 ft^2, are shown in Table A.12 of App. A.

ASTM Specifications A185 and A497 pertain to smooth and deformed welded wire fabric respectively, as shown in Table 2.3. Because the yield stresses shown are specified at a strain of 0.005, the ACI Code requires that f_y be taken equal to 60 ksi unless the stress at a strain of 0.0035 is used.

2.15 PRESTRESSING STEELS

Prestressing steel is used in three forms: round wires, stranded cable, and alloy steel bars. Prestressing wire ranges in diameter from 0.192 to 0.276 in. It is made by cold-drawing high-carbon steel after which the wire is stress-relieved by heat treatment to produce the prescribed mechanical properties. Wires are normally bundled in groups of up to about 50 individual wires to produce prestressing tendons of the required strength. Stranded cable, more common than wire in U.S. practice, is fabricated with six wires wound around a seventh of slightly larger diameter. The pitch of the spiral winding is between 12 and 16 times the nominal diameter of the strand. Strand diameters range from 0.250 to 0.600 in. Alloy steel bars for prestressing are available in diameters from 0.625 to 1.375 in., and as both plain round bars and deformed bars.

Specific requirements for prestressing steels are found in ASTM A421, "Standard Specification for Uncoated Stress-Relieved Steel Wire for Prestressed Concrete," ASTM A416, "Standard Specification for Steel Strand, Uncoated Seven-Wire Stress-Relieved for Prestressed Concrete," and ASTM A722, "Standard Specification for Uncoated High-Strength Steel Bar for Prestressing Concrete." Table A.15 of Appendix A provides design information for U.S. prestressing steels.

a. Grades and Strengths

The tensile strengths of prestressing steels range from about 2.5 to 6 times the yield strengths of commonly used rebars. The grade designations correspond to the minimum specified ultimate tensile strength in ksi. For the widely used seven-wire strand, two grades are available: Grade 250 (f_{pu} = 250 ksi) and Grade 270. The higher-strength Grade 270 strand is gradually displacing the slightly lower-strength strand. For alloy steel bars, two grades are used: the regular Grade 145 is most common, but special Grade 160 bars may be ordered. Round wires may be obtained in Grades 235, 240, and 250, depending on diameter.

b. Stress-Strain Curves

Figure 2.14 shows stress-strain curves for prestressing wires, strand, and alloy bars of various grades. For comparison, the stress-strain curve for a Grade 60 rebar is also shown. It is seen that, in contrast to reinforcing bars, prestressing steels do not show a sharp yield point or yield plateau; i.e., they do not yield at constant or nearly constant stress. Yielding develops gradually, and in the inelastic range the curve continues to rise smoothly until the tensile strength is reached. Because well-defined yielding is not observed in these steels, the yield strength is somewhat arbitrarily defined as the stress at a total elongation of 1 percent for strand and wire and at 0.7 percent for alloy steel bars. Figure 2.14 shows that the yield strengths so defined represent a good limit below which stress and strain are fairly proportional, and above which strain increases much more rapidly with increasing stress. It is also seen that the spread between tensile strength and yield strength is smaller in prestressing steels than in reinforcing steels. It may further be noted that prestressing steels have significantly less ductility.

While the modulus of elasticity E_s for bar reinforcement can be taken as 29,000,000 psi, the effective modulus of prestressing steel varies, depending on the type of steel (e.g., strand vs. wire or bars) and type of use, and is best determined by test or supplied by the manufacturer. For unbonded strand (i.e., strand not embedded in concrete), the modulus may be as low as 26,000,000 psi. For bonded strand, E_s is usually about 27,000,000 psi, while for smooth round wires E_s is about 29,000,000 psi, the same as for rebars. The elastic modulus of alloy steel bars is usually taken as E_s = 27,000,000 psi.

c. Relaxation

When prestressing steel is stressed to the levels that are customary during initial tensioning and at service loads, it exhibits a property known as *relaxation*. Relaxation is defined as the loss of stress in stressed material held at constant

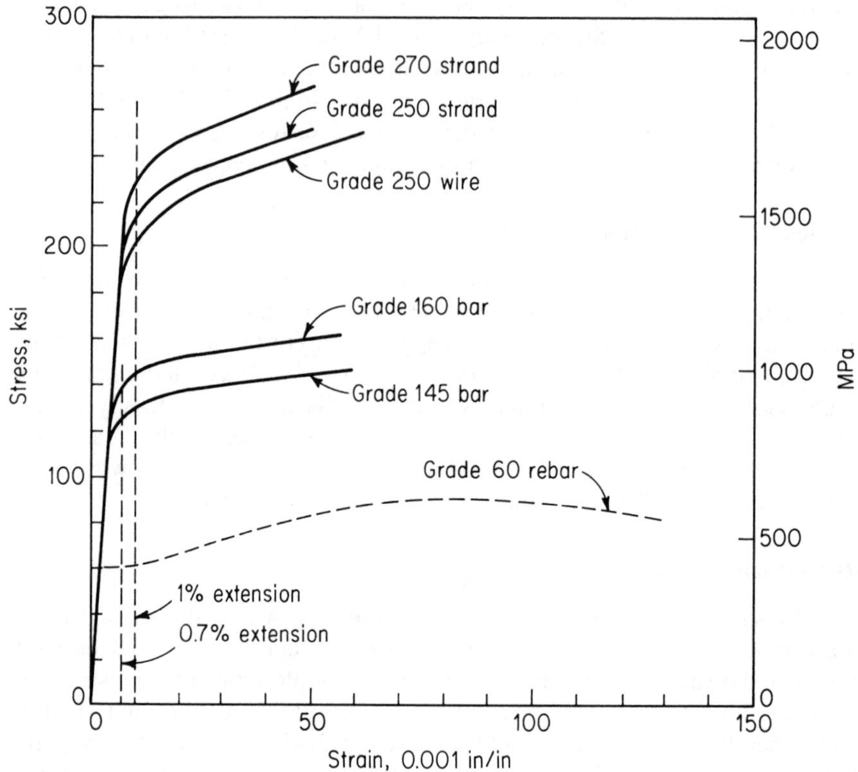

FIGURE 2.14
Typical stress-strain curves for prestressing steels.

length. (The same basic phenomenon is known as creep when defined in terms of change in strain of a material under constant stress.) To be specific, if a length of prestressing steel is stressed to a sizable fraction of its yield strength f_{py} (say 80 to 90 percent) and held at a constant strain between fixed points such as the ends of a beam, the steel stress f_p will gradually decrease from its initial value f_{pi}. In prestressed concrete members this stress relaxation is important because it modifies the internal stresses in the concrete and changes the deflections of the beam some time after initial prestress was applied.

The amount of relaxation varies, depending on the type and grade of steel, the time under load, and the initial stress level. A satisfactory estimate for ordinary strand and wires can be obtained from Eq. (2.11), which was derived from more than 400 relaxation tests of up to 9 years' duration:

$$\frac{f_p}{f_{pi}} = 1 - \frac{\log t}{10}\left(\frac{f_{pi}}{f_{py}} - 0.55\right) \tag{2.11}$$

where f_p is the final stress after t hours, f_{pi} is the initial stress, and f_{py} is the nominal yield stress (Ref. 2.43). In Eq. (2.11), log t is to the base 10, and f_{pi}/f_{py} not less than 0.55; below that value essentially no relaxation occurs.

The tests on which Eq. (2.11) is based were carried out on round, stress-relieved wires and are equally applicable to stress-relieved strand. In the absence of other information, results may be applied to alloy steel bars as well.

Special low-relaxation strand is available and its use is becoming common. According to ASTM A416, such steel shall exhibit relaxation after 1000 hours of not more than 2.5 percent when initially stressed to 70 percent of specified tensile strength and not more than 3.5 percent when loaded to 80 percent of tensile strength.

REFERENCES

2.1. "Selection and Use of Aggregates for Concrete," ACI Committee 621, *J. ACI*, vol. 58, no. 5, 1961, pp. 513–542.

2.2. "Guide for Structural Lightweight Aggregate Concrete," ACI Committee 213, *ACI Manual of Concrete Practice*, Part 1, 1988.

2.3. T. T. C. Hsu and F. O. Slate, "Tensile Bond Strength between Aggregate and Cement Paste or Mortar," *J. ACI*, vol. 60, no. 4, 1963, pp. 465–486.

2.4. G. E. Troxell, H. E. Davis, and J. W. Kelly, *Composition and Properties of Concrete*, 2d ed., McGraw-Hill, New York, 1968.

2.5. "Standard Practice for Selecting Proportions for Normal, Heavyweight, and Mass Concrete," ACI Committee 211, *ACI Manual of Concrete Practice*, Part 1, 1988.

2.6. "Standard Practice for Selecting Proportions for Structural Lightweight Concrete," ACI Committee 211, *ACI Manual of Concrete Practice*, Part 1, 1988.

2.7. "Guide for Use of Admixtures in Concrete," ACI Committee 212, *ACI Manual of Concrete Practice*, Part I, 1988.

2.8. "Developments in the Use of Superplasticizers," Special Publication SP-68, American Concrete Institute, Detroit, 1981.

2.9. *Design and Control of Concrete Mixtures*, 13th ed., Portland Cement Association, Skokie, Illinois, 1988.

2.10. "Standard Practice for Consolidation of Concrete," ACI Committee 309, *ACI Manual of Concrete Practice*, Part 2, 1988.

2.11. "Guide for Measuring, Mixing, Transporting, and Placing Concrete," ACI Committee 304, *ACI Manual of Concrete Practice*, Part 2, 1988.

2.12. "Cold Weather Concreting," ACI Committee 306, *ACI Manual of Concrete Practice*, Part 2, 1988.

2.13. "Recommended Practice for Evaluation of Strength Test Results of Concrete," ACI Committee 214, *ACI Manual of Concrete Practice*, Part 2, 1988.

2.14. R. L. Carrasquillo, A. H. Nilson, and F. O. Slate, "Properties of High Strength Concrete Subject to Short Term Loads," *J. ACI*, vol. 78, no. 3, 1981, pp. 171–178.

2.15. F. O. Slate, A. H. Nilson, and S. Martinez, "Mechanical Properties of High-Strength Lightweight Concrete," *J. ACI*, vol. 83, no. 4, 1986, pp. 606–613.

2.16. P. T. Wang, S. P. Shah, and A. E. Naaman, "Stress-Strain Curves of Normal and Lightweight Concrete in Compression," *J. ACI*, vol 75, no. 11, 1978, pp. 603–611.

2.17. D. E. Branson, *Deformation of Concrete Structures*, McGraw-Hill, New York, 1977.

2.18. H. Rüsch, "Researches Toward a General Flexural Theory for Structural Concrete," *J. ACI*, vol. 32, no. 1, 1960, pp. 1–28.

2.19. P. K. Mehta, *Concrete: Structure, Properties, and Materials,* Prentice Hall, 1986.

2.20. A. M. Neville, *Properties of Concrete*, 3d ed., Pittman, Marshfield, MA, 1981.

2.21. M. M. Smadi, F. O. Slate, and A. H. Nilson, "High-, Medium-, and Low-Strength Concretes Subject to Sustained Overloads," *J. ACI*, vol. 82, no. 5, 1985, pp. 657–664.

2.22. M. M. Smadi, F. O. Slate, and A. H. Nilson, "Shrinkage and Creep of High-, Medium, and Low-Strength Concretes, Including Overloads," *ACI Mater. J.*, vol. 84, no. 3, 1987, pp. 224–234.

2.23. "Fatigue of Concrete Structures," Special Publication SP-75, American Concrete Institute, Detroit, 1982.

2.24. S. Timoshenko, *Strength of Materials,* Part II, 3rd ed., Van Nostrand, Princeton, NJ, 1956.
2.25. D. McHenry and J. Karni, "Strength of Concrete under Combined Tensile and Compressive Stress," *J. ACI*, vol. 54, no. 10, 1958, pp. 829–840.
2.26. B. Bresler and K.S. Pister, "Strength of Concrete under Combined Stress," *J. ACI*, vol. 55, no. 3, 1958, pp. 321–345.
2.27. H. J. Cowan, "The Strength of Plain, Reinforced, and Prestressed Concrete under the Action of Combined Stresses," *Mag. Concr. Res.*, vol. 5, no. 14, 1953, pp. 75–86.
2.28. N. J. Carino and F. O. Slate, "Limiting Tensile Strain Criterion for Failure of Concrete," *J. ACI*, vol. 73, no. 3, March 1976, pp. 160–165.
2.29. *Fracture Mechanics: Application to Concrete*, Special Publication SP-118, American Concrete Institute, Detroit, 1989, 308 pp.
2.30. T. T. C. Hsu, F. O. Slate, G. M. Sturman, and G. Winter, "Microcracking of Plain Concrete and the Shape of the Stress-Strain Curve," *J. ACI*, vol. 60, no. 2, 1963, pp. 209–224.
2.31. H. Kupfer, H. K. Hilsdorf, and H. Rüsch, "Behavior of Concrete under Biaxial Stresses," *J. ACI*, vol. 66, no. 8, 1969, pp. 656–666.
2.32. M. E. Tasuji, F. O. Slate, and A. H. Nilson, "Stress-Strain Response and Fracture of Concrete in Biaxial Loading," *J. ACI*, vol. 75, no. 7, 1978, pp. 306–312.
2.33. K. H. Gerstle, D. H. Linse, et al., "Strength of Concrete under Multiaxial Stress States," *Proc. Douglas McHenry International Symposium on Concrete and Concrete Structures,* ACI Special Publication SP-55, American Concrete Institute, 1978, pp. 103–131.
2.34. H. G. Russell, S. H. Gebler, and D. Whiting, "High-Strength Concrete: Weighing the Benefits," *Civil Engineering*, New York, vol. 59, no. 11, 1989, pp. 59–61.
2.35. A. H. Nilson, "High-Strength Concrete—An Overview of Cornell Research," *Proceedings of Symposium on Utilization of High-Strength Concrete*, Stavanger, Norway, 1987, pp. 27–38.
2.36. A. H. Nilson, "Properties and Performance of High-Strength Concrete," *Proceedings of IABSE Symposium on Concrete Structures for the Future,"* Paris-Versailles, 1987, 389–394.
2.37. A. H. Nilson, "Design Implications of Current Research on High-Strength Concrete," *High Strength Concrete*, Special Publication SP-87, American Concrete Institute, Detroit, 1985, pp. 85–118.
2.38. "Steel Reinforcement Properties and Availability," ACI Committee 439, *J. ACI*, vol. 74, no. 10, 1977, pp. 481–492.
2.39. *Manual of Standard Practice*, 24th ed., Concrete Reinforcing Steel Institute, Schaumburg, IL, 1986.
2.40. W. G. Corley, J. M. Hanson, and T. Helgason, "Design of Reinforced Concrete for Fatigue," *J. of Struct. Div.*, ASCE, vol. 104, no. ST6, 1978, pp. 921–932.
2.41. "Epoxy-Coated Reinforcing Bars," *Engineering Data Report No. 14*, Concrete Reinforcing Steel Institute, Schaumburg, IL, 1982.
2.42. "Suggested Specifications Provisions for Epoxy-Coated Reinforcing Bars," *Engineering Data Report No. 19*, Concrete Reinforcing Steel Institute, Schaumburg, IL, 1984.
2.43. W. G. Corley, M. A. Sozen, and C. P. Siess, "Time-Dependent Deflections of Prestressed Concrete Beams," *Highway Research Board Bulletin No. 307*, 1961, pp. 1–25.

CHAPTER 3

FLEXURAL ANALYSIS AND DESIGN OF BEAMS

3.1 INTRODUCTION

The fundamental assumptions upon which the analysis and design of reinforced concrete members are based were introduced in Sec. 1.8 of Chap. 1, and the application of those assumptions to the simple case of axial loading was developed in Sec. 1.9. The student should review Secs. 1.8 and 1.9 at this time. In developing methods for the analysis and design of beams in this chapter, the same assumptions apply, and identical concepts will be utilized. This chapter will include analysis and design for flexure, including the dimensioning of the concrete cross section and the selection and placement of reinforcing steel. Other important aspects of beam design including shear reinforcement, bond and anchorage of reinforcing bars, and the important questions of serviceability (e.g., limiting deflections and controlling concrete cracking) will be treated in Chaps. 4, 5, and 6.

3.2 BENDING OF HOMOGENEOUS BEAMS

Reinforced concrete beams are nonhomogeneous in that they are made of two entirely different materials. The methods used in the analysis of reinforced concrete beams are therefore different from those used in the design or investigation of beams composed entirely of steel, wood, or any other structural material. The fundamental principles involved are, however, essentially the same. Briefly, these principles are as follows.

At any cross section there exist internal forces which can be resolved into components normal and tangential to the section. Those components that are normal to the section are the *bending* stresses (tension on one side of the neutral axis and compression on the other). Their function is to resist the bending moment at the section. The tangential components are known as the *shear* stresses, and they resist the transverse or shear forces.

Fundamental assumptions relating to flexure and flexural shear are as follows:

1. A cross section that was plane before loading remains plane under load. This means that the unit strains in a beam above and below the neutral axis are proportional to the distance from that axis.
2. The bending stress f at any point depends on the strain at that point in a manner given by the stress-strain diagram of the material. If the beam is made of a homogeneous material whose stress-strain diagram in tension and compression is that of Fig. 3.1*a*, the following holds. If the maximum strain at the outer fibers is smaller than the strain ϵ_p up to which stress and strain are proportional for the given material, then the compression and tension stresses on either side of the axis are proportional to the distance from the axis, as shown in Fig. 3.1*b*. However, if the maximum strain at the outer fibers is larger than ϵ_p, this is no longer true. The situation which then obtains is shown in Fig. 3.1*c*; i.e., in the outer portions of the beam, where $\epsilon > \epsilon_p$, stresses and strains are no longer proportional. In these regions the magnitude of stress at any level, such as f_2 in Fig. 3.1*c*, depends on the strain ϵ_2 at that level in the manner given by the stress-strain diagram of the material. In other words, for a given strain

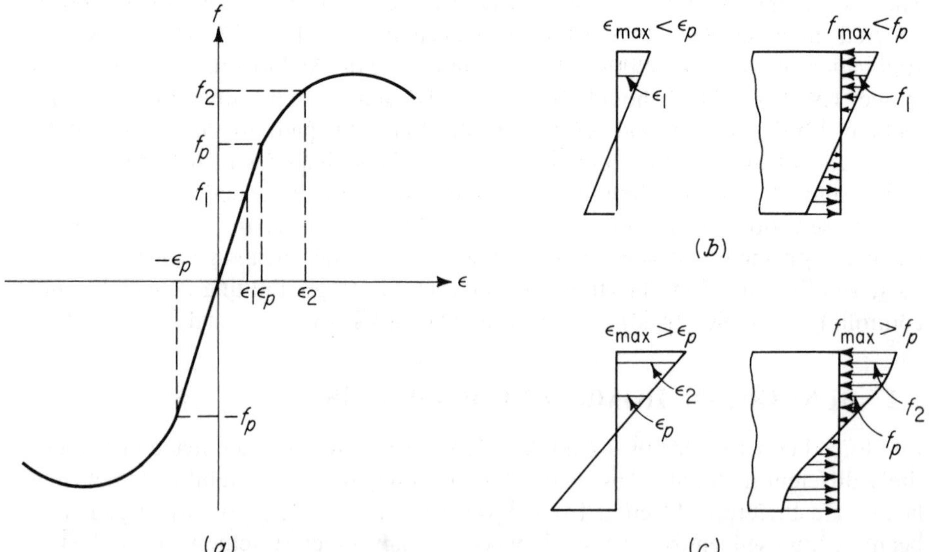

FIGURE 3.1
Elastic and inelastic stress distributions in homogeneous beams.

in the beam, the stress at a point is the same as that given by the stress-strain diagram for the same strain.

3. The distribution of the shear stresses v over the depth of the section depends on the shape of the cross section and of the stress-strain diagram. These shear stresses are largest at the neutral axis and equal to zero at the outer fibers. The shear stresses on horizontal and vertical planes through any point are equal.

4. Owing to the combined action of shear stresses (horizontal and vertical) and flexure stresses, at any point in a beam there are inclined stresses of tension and compression, the largest of which form an angle of 90° with each other. The intensity of the inclined maximum or principal stress at any point is given by

$$t = \frac{f}{2} \pm \sqrt{\frac{f^2}{4} + v^2} \tag{3.1}$$

where f = intensity of normal fiber stress
v = intensity of tangential shearing stress

The inclined stress makes an angle α with the horizontal such that $\tan 2\alpha = 2v/f$.

5. Since the horizontal and vertical shearing stresses are equal and the flexural stresses are zero at the neutral plane, the inclined tensile and compressive stresses at any point in that plane form an angle of 45° with the horizontal, the intensity of each being equal to the unit shear at the point.

6. When the stresses in the outer fibers are smaller than the proportional limit f_p, the beam behaves *elastically*, as shown in Fig. 3.1*b*. In this case the following pertains:

a. The neutral axis passes through the center of gravity of the cross section.

b. The intensity of the bending stress normal to the section increases directly with the distance from the neutral axis and is a maximum at the extreme fibers. The stress at any given point in the cross section is represented by the equation

$$f = \frac{My}{I} \tag{3.2}$$

where f = bending stress at a distance y from neutral axis
M = external bending moment at section
I = moment of inertia of cross section about neutral axis

The maximum bending stress occurs at the outer fibers and is equal to

$$f_{\max} = \frac{Mc}{I} = \frac{M}{S} \tag{3.3}$$

where c = distance from neutral axis to outer fiber
$S = I/c$ = section modulus of cross section

c. The shear stress (longitudinal equals transverse) v at any point in the cross section is given by

$$v = \frac{VQ}{Ib} \tag{3.4}$$

where V = total shear at section
Q = statical moment about neutral axis of that portion of cross section lying between a line through point in question parallel to neutral axis and nearest face (upper or lower) of beam
I = moment of inertia of cross section about neutral axis
b = width of beam at a given point

d. The intensity of shear along a vertical cross section in a rectangular beam varies as the ordinates of a parabola, the intensity being zero at the outer fibers of the beam and a maximum at the neutral axis. The maximum is $\frac{3}{202}V/ba$, since at the neutral axis $Q = ba^2/8$ and $I = ba^3/12$ in Eq. (3.4).

The remainder of this chapter deals only with bending stresses and their effects on reinforced concrete beams. Shear stresses and their effects are discussed separately in Chap. 4.

3.3 REINFORCED CONCRETE BEAM BEHAVIOR

Plain concrete beams are inefficient as flexural members because the tension strength in bending (modulus of rupture, see Sec. 2.8) is a small fraction of the compression strength. In consequence, such beams fail on the tension side at low loads long before the strength of the concrete on the compression side has been fully utilized. For this reason, steel reinforcing bars are placed on the tension side as close to the extreme tension fiber as is compatible with proper fire and corrosion protection of the steel. In such a reinforced concrete beam the tension caused by the bending moments is chiefly resisted by the steel reinforcement, while the concrete alone is usually capable of resisting the corresponding compression. Such joint action of the two materials is assured if relative slip is prevented. This is achieved by using deformed bars with their high bond strength at the steel-concrete interface (see Sec. 2.12) and, if necessary, by special anchorage of the ends of the bars. A simple example of such a beam, with the customary designations for the cross-sectional dimensions, is shown in Fig. 3.2. For simplicity, the discussion that follows will deal with beams of rectangular cross section, even though members of other shapes are very common in most concrete structures.

When the load on such a beam is gradually increased from zero to the magnitude that will cause the beam to fail, several different stages of behavior can be clearly distinguished. At low loads, as long as the maximum tension stress in the concrete is smaller than the modulus of rupture, the entire concrete is effective in resisting stress, in compression on one side and in tension on the other side of the neutral axis. In addition, the reinforcement, deforming the same

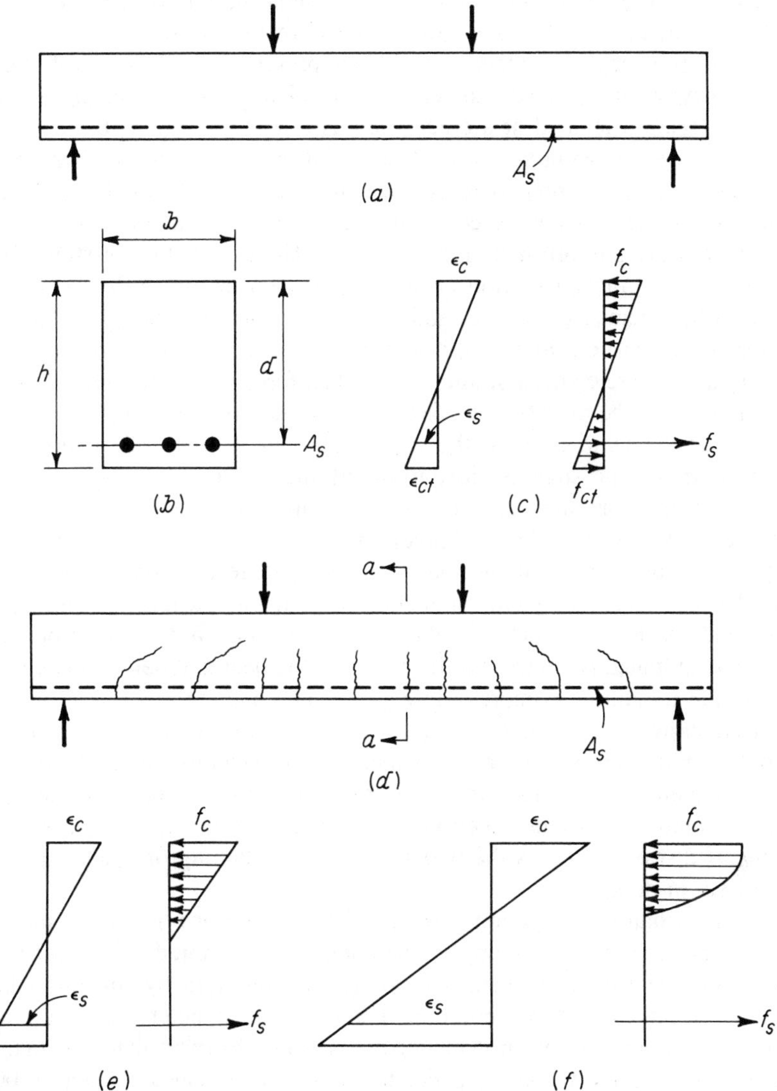

FIGURE 3.2
Behavior of reinforced concrete beam under increasing load.

amount as the adjacent concrete, is also subject to tension stresses. At this stage all stresses in the concrete are of small magnitude and are proportional to strains. The distribution of strains and stresses in concrete and steel over the depth of the section is as shown in Fig. 3.2*c*.

When the load is further increased, the tension strength of the concrete is soon reached, and at this stage tension cracks develop. These propagate quickly upward to or close to the level of the neutral plane, which in turn shifts upward with progressive cracking. The general shape and distribution of these tension cracks is shown in Fig. 3.2*d*. In well-designed beams the width of these cracks is so small (hairline cracks) that they are not objectionable from the viewpoint of either corrosion protection or appearance. Their presence, however, profoundly affects the behavior of the beam under load. Evidently, in a cracked section, i.e., in a cross section located at a crack such as *a-a* in Fig. 3.2*d*, the concrete does not transmit any tension stresses. Hence, just as in tension members (Sec. 1.9b), the steel is called upon to resist the entire tension. At moderate loads, if the concrete stresses do not exceed approximately $f_c'/2$, stresses and strains continue to be closely proportional (see Fig. 1.15). The distribution of strains and stresses at or near a cracked section is then that shown in Fig. 3.2*e*. When the load is still further increased, stresses and strains rise correspondingly and are no longer proportional. The ensuing nonlinear relation between stresses and strains is that given by the concrete stress-strain curve. Therefore, just as in homogeneous beams (see Fig. 3.1), the distribution of concrete stresses on the compression side of the beam is of the same shape as the stress-strain curve. Figure 3.2*f* shows the distribution of strains and stresses close to the ultimate load.

Eventually, the carrying capacity of the beam is reached. Failure can be caused in one of two ways. When relatively moderate amounts of reinforcement are employed, at some value of the load the steel will reach its yield point. At that stress the reinforcement yields suddenly and stretches a large amount (see Fig. 2.13), and the tension cracks in the concrete widen visibly and propagate upward, with simultaneous significant deflection of the beam. When this happens, the strains in the remaining compression zone of the concrete increase to such a degree that crushing of the concrete, the *secondary compression failure*, ensues at a load only slightly larger than that which caused the steel to yield. Effectively, therefore, attainment of the yield point in the steel determines the carrying capacity of moderately reinforced beams. Such yield failure is gradual and is preceded by visible signs of distress, such as the widening and lengthening of cracks and the marked increase in deflection.

On the other hand, if large amounts of reinforcement or normal amounts of steel of very high strength are employed, the compression strength of the concrete may be exhausted before the steel starts yielding. Concrete fails by crushing when strains become so large that they disrupt the integrity of the concrete. Exact criteria for this occurrence are not yet known, but it has been observed that rectangular beams fail in compression when the concrete strains reach values of about 0.003 to 0.004. Compression failure through crushing of the concrete is sudden, of an almost explosive nature, and occurs without warning. For this reason it is good practice to dimension beams in such a manner that should they be overloaded, failure would be initiated by yielding of the steel rather than by crushing of the concrete.

The analysis of stresses and strength in the different stages just described will be discussed in the next several sections.

a. Stresses Elastic and Section Uncracked

As long as the tensile stress in the concrete is smaller than the modulus of rupture, so that no tension cracks develop, the strain and stress distribution as shown in

Fig. 3.2*c* is essentially the same as in an elastic, homogeneous beam (Fig. 3.1*b*). The only difference is the presence of another material, the steel reinforcement. As shown in Sec. 1.9a, in the elastic range, for any given value of strain, the stress in the steel is n times that of the concrete [Eq. (1.6)]. In the same section it was shown that one can take account of this fact in calculations by replacing the actual steel-and-concrete cross section with a fictitious section thought of as consisting of concrete only. In this "transformed section" the actual area of the reinforcement is replaced with an equivalent concrete area equal to nA_s located at the level of the steel. The transformed, uncracked section pertaining to the beam of Fig. 3.2*b* is shown in Fig. 3.3.

Once the transformed section has been obtained, the usual methods of analysis of elastic homogeneous beams apply. That is, the section properties (location of neutral axis, moment of inertia, section modulus, etc.) are calculated in the usual manner, and, in particular, stresses are computed with Eqs. (3.2) to (3.4).

Example 3.1. A rectangular beam has the dimensions (see Fig. 3.2*b*) $b = 10$ in., $h = 25$ in., and $d = 23$ in., and is reinforced with three No. 8 bars so that $A_s = 2.35$ in^2. The concrete cylinder strength f'_c is 4000 psi, and the tensile strength in bending (modulus of rupture) is 475 psi. The yield point of the steel f_y is 60,000 psi, the stress-strain curves of the materials being those of Fig. 1.15. Determine the stresses caused by a bending moment $M = 45$ ft-kips.

Solution. With a value $n = E_s/E_c = 29{,}000{,}000/3{,}600{,}000 = 8$, one has to add to the rectangular outline an area $(n-1)A_s = 7 \times 2.35 = 16.45$ in^2, disposed as shown on Fig. 3.4, in order to obtain the uncracked, transformed section. Conventional calculations show that the location of the neutral axis of this section is given by $\bar{y} = 13.2$ in., and its moment of inertia about this axis is 14,710 in^4. For $M = 45$ ft-kips $= 540{,}000$ in-lb, the concrete compression stress at the top fiber is, from Eq. (3.3),

$$f_c = 540{,}000 \frac{13.2}{14{,}710} = 485 \text{ psi}$$

and, similarly, the concrete tension stress at the bottom fiber is

$$f_{ct} = 540{,}000 \frac{11.8}{14{,}710} = 433 \text{ psi}$$

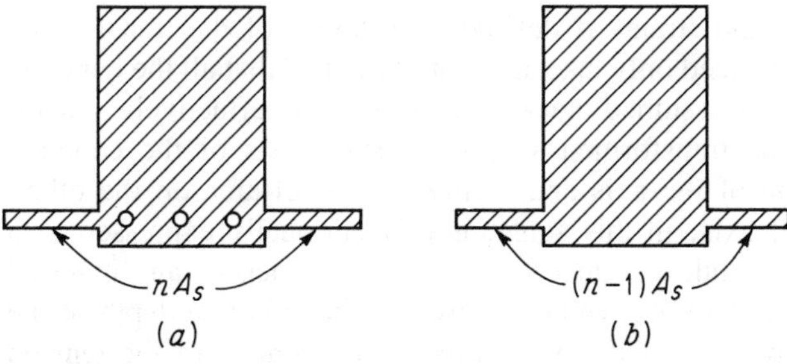

FIGURE 3.3
Uncracked transformed beam section.

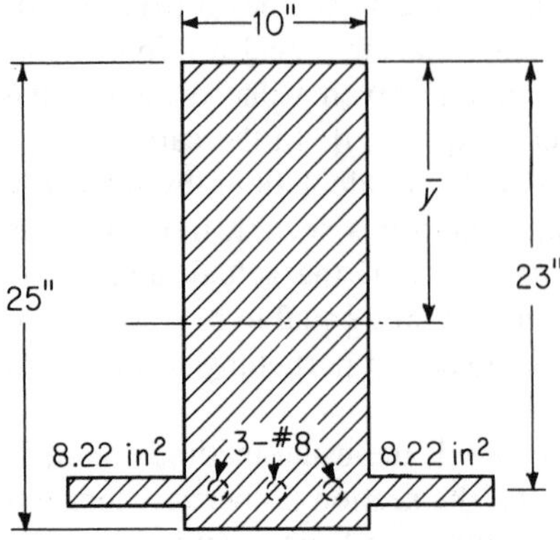

FIGURE 3.4
Transformed beam section of Example 3.1

Since this value is below the given tensile bending strength of the concrete, 475 psi, no tension cracks will form, and calculation by the uncracked, transformed section is justified. The stress in the steel, from Eqs. (1.6) and (3.2), is

$$f_s = n\frac{My}{I} = 8\left(540{,}000\frac{9.8}{14{,}710}\right) = 2880 \text{ psi}$$

By comparing f_c and f_s with the cylinder strength and the yield point respectively, it is seen that at this stage the actual stresses are quite small compared with the available strengths of the two materials.

b. Stresses Elastic and Section Cracked

When the tension stress f_{ct} exceeds the modulus of rupture, cracks form, as shown in Fig. 3.2*d*. If the concrete compression stress is less than approximately $\frac{1}{2}f'_c$ and the steel stress has not reached the yield point, both materials continue to behave elastically, or very nearly so. This situation generally obtains in structures under normal service conditions and loads, since at these loads the stresses are generally of the order of magnitude just discussed. At this stage, for simplicity and with little if any error, it is assumed that tension cracks have progressed all the way to the neutral axis and that sections plane before bending are plane in the bent member. The situation with regard to strain and stress distribution is then that shown in Fig. 3.2*e*.

To compute stresses, and strains if desired, the device of the transformed section can still be used. One need only take account of the fact that all the concrete which is stressed in tension is assumed cracked, and therefore effectively absent. As shown in Fig. 3.5*a*, the transformed section then consists of the concrete in compression on one side of the axis and n times the steel area on the other. The distance to the neutral axis, in this stage, is conventionally expressed as a fraction kd of the effective depth d. (Once the concrete is cracked, any material located below the steel is ineffective, which is why d is the effective depth of the beam.) To determine the location of the neutral axis, the moment of the tension area about the axis is set equal to the moment of the compression area, which gives

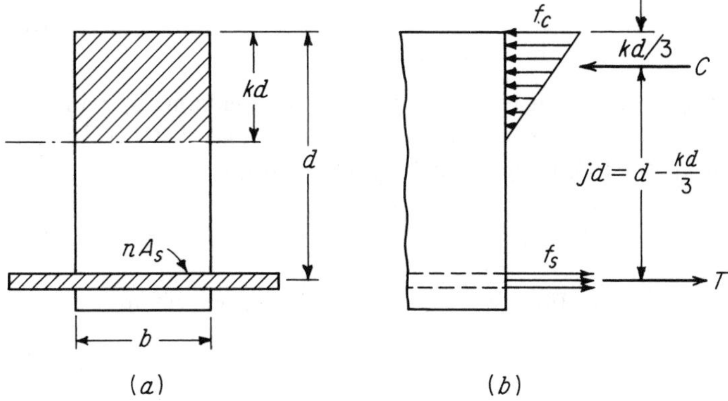

FIGURE 3.5
Cracked transformed section.

$$b\frac{(kd)^2}{2} - nA_s(d - kd) = 0 \tag{3.5}$$

Having obtained *kd* by solving this quadratic equation, one can determine the moment of inertia and other properties of the transformed section as in the preceding case. Alternatively, one can proceed from basic principles by accounting directly for the forces that act on the cross section. These are shown in Fig. 3.5*b*. The concrete stress, with maximum value f_c at the outer edge, is distributed linearly as shown. The entire steel area A_s is subject to the stress f_s. Correspondingly, the total compression force C and the total tension force T are

$$C = \frac{f_c}{2}bkd \qquad \text{and} \qquad T = A_s f_s \tag{3.6}$$

The requirement that these two forces be equal numerically has been taken care of by the manner in which the location of the neutral axis has been determined.

Equilibrium requires that the couple constituted by the two forces C and T be equal numerically to the external bending moment M. Hence, taking moments about C gives

$$M = Tjd = A_s f_s jd \tag{3.7}$$

where jd is the internal lever arm between C and T. From Eq. (3.7), the steel stress is

$$f_s = \frac{M}{A_s jd} \tag{3.8}$$

Conversely, taking moments about T gives

$$M = Cjd = \frac{f_c}{2}bkdjd = \frac{f_c}{2}kjbd^2 \tag{3.9}$$

from which the concrete stress is

$$f_c = \frac{2M}{kjbd^2} \tag{3.10}$$

In using Eqs. (3.6) through (3.10) it is convenient to have equations by which k and j may be found directly, in order to establish the neutral axis distance kd and the internal lever arm jd. First defining the *reinforcement ratio*

$$\rho = \frac{A_s}{bd} \tag{3.11}$$

then substituting $A_s = \rho bd$ into Eq. (3.5) and solving for k, one obtains

$$k = \sqrt{(\rho n)^2 + 2\rho n} - \rho n \tag{3.12}$$

From Fig. 3.5*b* it is seen that $jd = d - kd/3$, or

$$j = 1 - \frac{k}{3} \tag{3.13}$$

Values of k and j for elastic cracked section analysis, for common steel ratios and modular ratios, will be found in Table A.7 of App. A.

Example 3.2. The beam of Example 3.1 is subject to a bending moment $M = 90$ ft-kips (rather than 45 ft-kips as previously). Calculate the relevant properties and stresses.

Solution. If the section were to remain uncracked, the tension stress in the concrete would now be twice its previous value, that is, 866 psi. Since this exceeds by far the modulus of rupture of the given concrete (475 psi), cracks will have formed and the analysis must be adapted appropriately. Equation (3.5), with the known quantities b, n, and A_s inserted, gives the distance to the neutral axis $kd = 7.6$ in., or $k = 7.6/23 = 0.33$. From Eq. (3.13), $j = 1 - 0.33/3 = 0.89$. With these values the steel stress is obtained from Eq. (3.8) as $f_s = 22,400$ psi, and the maximum concrete stress from Eq. (3.10) as $f_c = 1390$ psi.

Comparing the results with the pertinent values for the same beam when subject to one-half the moment, as previously calculated, one notices that (1) the neutral plane has migrated upward so that its distance from the top fiber has changed from 13.2 to 7.6 in.; (2) even though the bending moment has only been doubled, the steel stress has increased from 2880 to 22,400 psi, or about 7.8 times, and the concrete compression stress has increased from 485 to 1390 psi, or 2.9 times; (3) the moment of inertia of the cracked transformed section is easily computed to be 5910 in^4, compared with 14,710 in^4 for the uncracked section. This affects the magnitude of the deflection, as discussed in Chap. 6. Thus it is seen how radical is the influence of the formation of tension cracks on the behavior of reinforced concrete beams.

c. Flexural Strength

It is of interest in structural practice to calculate those stresses and deformations that occur in a structure in service under design load. For reinforced concrete beams this can be done by the methods just presented, which assume elastic behavior of both materials. It is equally, if not more, important that the structural engineer be able to predict with satisfactory accuracy the ultimate strength of a structure or structural member. By making this strength larger by an appropriate amount than the largest loads that can be expected during the lifetime of the structure, an adequate margin of safety is assured. In the past, methods based on elastic analysis, like those just presented or variations thereof, have been used for this purpose. It is clear, however, that at or near the ultimate load, stresses are no longer proportional to strains. In regard to axial compression, this has been discussed in detail in Sec. 1.9, and in regard to bending, it has been pointed out that at high loads, close to the ultimate, the distribution of stresses and strains is that of Fig. 3.2*f* rather than the elastic distribution of Fig. 3.2*e*. More realistic methods of analysis, based on actual inelastic rather than assumed elastic behavior of the materials and on results of extremely extensive experimental research, have been developed to predict the ultimate strength. They are now used almost exclusively in structural design practice.

If the distribution of concrete compression stresses at or near ultimate load (Fig. 3.2*f*) had a well-defined and invariable shape—parabolic, trapezoidal, or otherwise—it would be possible to derive a completely rational theory of ultimate bending strength, just as the theory of elastic bending with its known triangular shape of stress distribution (Figs. 3.1*b* and 3.2*c* and *e*) is straightforward and rational. Actually, inspection of Figs. 2.3, 2.4, and 2.6, and of many more concrete stress-strain curves that have been published, shows that the geometrical shape of the stress distribution is quite varied and depends on a number of factors, such as the cylinder strength and the rate and duration of loading. For this and other reasons, a wholly rational flexural theory for reinforced concrete has not yet been developed (Refs. 3.1 to 3.3). Present methods of analysis, therefore, are based in part on known laws of mechanics and are supplemented, where needed, by extensive test information.

Let Fig. 3.6 represent the distribution of internal stresses and strains when the beam is about to fail. One desires a method to calculate that moment M_n (nominal ultimate moment) at which the beam will fail either by tension yielding of the steel or by crushing of the concrete in the outer compression fiber. For the first mode of failure the criterion is that the steel stress equal the yield point, $f_s = f_y$.

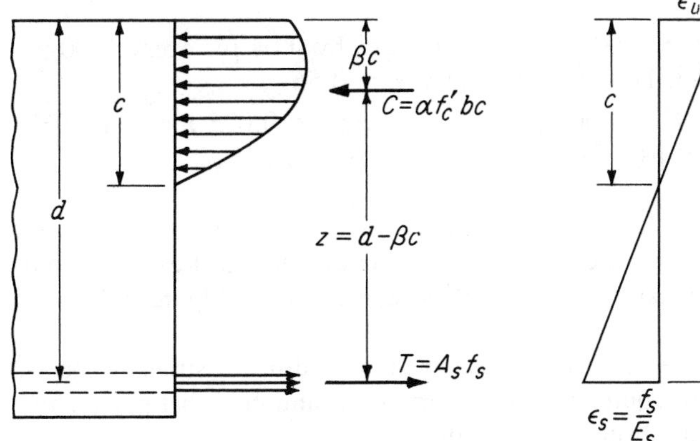

FIGURE 3.6
Stress distribution at ultimate load.

It has been mentioned before that an exact criterion for concrete compression failure is not yet known, but that for rectangular beams strains of 0.003 to 0.004 in/in have been measured immediately preceding failure. If one assumes, usually slightly conservatively, that the concrete is about to crush when the maximum strain reaches $\epsilon_u = 0.003$, comparison with a great many tests of beams and columns of a considerable variety of shapes and conditions of loading shows that a satisfactorily accurate and safe prediction of ultimate strength can be made (Ref. 3.4). In addition to these two criteria (yielding of the steel at a stress of f_y and crushing of the concrete at a strain of 0.003), it is not really necessary to know the exact shape of the concrete stress distribution in Fig. 3.6. What is necessary is to know, for a given distance c of the neutral axis, (1) the total resultant compression force C in the concrete and (2) its vertical location, i.e., its distance from the outer compression fiber.

In a rectangular beam the area that is in compression is bc, and the total compression force on this area can be expressed as $C = f_{av}bc$, where f_{av} is the average compression stress on the area bc. Evidently, the average compression stress that can be developed before failure occurs becomes larger the higher the cylinder strength f'_c of the particular concrete. Let

$$\alpha = \frac{f_{av}}{f'_c} \tag{3.14}$$

Then

$$C = \alpha f'_c bc \tag{3.15}$$

For a given distance c to the neutral axis, the location of C can be defined as some fraction β of this distance. Thus, as indicated in Fig. 3.6, for a concrete of given strength it is necessary to know only α and β in order to define completely the effect of the concrete compression stresses.

Extensive direct measurements, as well as indirect evaluation of numerous beam tests, have shown that the following values for α and β are satisfactorily accurate (see Ref. 3.5, where α is designated as k_1k_3 and β as k_2):

α equals 0.72 for $f'_c \leq 4000$ psi and decreases by 0.04 for every 1000 psi above 4000 up to 8000 psi. For $f'_c > 8000$ psi, $\alpha = 0.56$.

β equals 0.425 for $f'_c \leq 4000$ psi and decreases by 0.025 for every 1000 psi above 4000 up to 8000 psi. For $f'_c > 8000$ psi, $\beta = 0.325$.

The decrease in α and β for high strength concretes is related to the fact that such concretes are more brittle; i.e., they show a more sharply curved stress-strain plot with a smaller near-horizontal portion (see Figs. 2.3 and 2.4). Figure 3.7 shows these simple relations.

If this experimental information is accepted, the ultimate strength can be calculated from the laws of equilibrium and from the assumption that plane cross sections remain plane. Equilibrium requires that

$$C = T \qquad \text{or} \qquad \alpha f'_c bc = A_s f_s \tag{3.16}$$

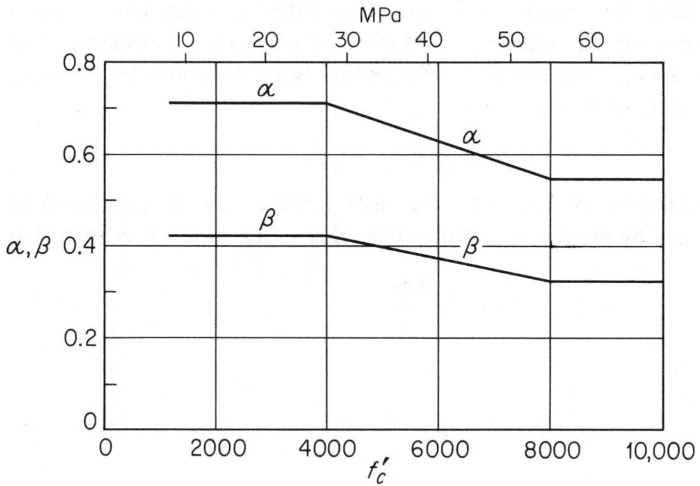

FIGURE 3.7
Variation of α and β with concrete strength f_c'.

Also, the bending moment, being the couple of the forces C and T, can be written as either

$$M = Tz = A_s f_s(d - \beta c) \tag{3.17}$$

or

$$M = Cz = \alpha f_c' bc(d - \beta c) \tag{3.18}$$

For *tension failure* by yielding of the steel, $f_s = f_y$. Substituting this value in Eq. (3.16), one obtains the distance to the neutral axis

$$c = \frac{A_s f_y}{\alpha f_c' b} \tag{3.19a}$$

Alternatively, using $A_s = \rho bd$, the neutral axis distance is

$$c = \frac{\rho f_y d}{\alpha f_c'} \tag{3.19b}$$

giving the distance to the neutral axis when tension failure occurs. The ultimate moment M_n is then obtained from Eq. (3.17) with the value for c just determined, and $f_s = f_y$; that is,

$$M_n = \rho f_y bd^2 \left(1 - \frac{\beta f_y \rho}{\alpha f_c'}\right) \tag{3.20a}$$

With the specific, experimentally obtained values for α and β given previously, this becomes

$$M_n = \rho f_y bd^2 \left(1 - 0.59\frac{\rho f_y}{f_c'}\right) \tag{3.20b}$$

On the other hand, for *compression failure* the criterion is that the compression strain in the concrete becomes $\epsilon_u = 0.003$, as previously discussed. The steel stress f_s, not having reached the yield point, is proportional to the steel strain ϵ_s; i.e., according to Hooke's law,

$$f_s = \epsilon_s E_s$$

From the strain distribution of Fig. 3.6, the steel strain ϵ_s can be expressed in terms of the distance c by evaluating similar triangles, after which it is seen that

$$f_s = \epsilon_u E_s \frac{d - c}{c} \tag{3.21}$$

Then, from Eq.(3.16),

$$\alpha f'_c bc = A_s \epsilon_u E_s \frac{d - c}{c} \tag{3.22}$$

and this quadratic may be solved for c, the only unknown for the given beam. With both c and f_s known, the ultimate moment of the beam, so heavily reinforced that failure occurs by crushing of the concrete, may be found from either Eq. (3.17) or Eq. (3.18).

It was noted that compression failure occurs explosively and without warning. For this reason it is good practice to keep the amount of reinforcement sufficiently small to ensure that, should the member be overstressed, it will give adequate warning before failing in a gradual manner by yielding of the steel rather than by crushing of the concrete. This can be done by keeping the reinforcement ratio $\rho = A_s/bd$ below a certain limiting value. This value, the so-called *balanced steel ratio* ρ_b, represents that amount of reinforcement necessary to make the beam fail by crushing of the concrete at the same load that causes the steel to yield. This means that the neutral axis must be so located that at the load at which the steel starts yielding, the concrete reaches its compressive strain limit ϵ_u. Correspondingly, setting $f_s = f_y$ in Eq. (3.21) and substituting the yield strain ϵ_y for f_y/E_s, one obtains the value of c defining the unique position of the neutral axis corresponding to simultaneous crushing of the concrete and initiation of yielding in the steel,

$$c = \frac{\epsilon_u}{\epsilon_u + \epsilon_y} d \tag{3.23}$$

Substituting that value of c into Eq. (3.16), with $A_s f_s = \rho b d f_y$, one obtains for the balanced steel ratio

$$\rho_b = \frac{\alpha f'_c}{f_y} \frac{\epsilon_u}{\epsilon_u + \epsilon_y} \tag{3.24}$$

In a well-designed member the actual steel ratio is kept well below the balanced ratio given by Eq. (3.24).

Example 3.3. Determine the ultimate moment M_n at which the beam of Examples 3.1 and 3.2 will fail.

Solution. For this beam the steel ratio $\rho = A_s/bd = 2.35/10 \times 23 = 0.0102$. The balanced steel ratio is found from Eq. (3.24) to be 0.0284. Since the amount of steel in the beam is less than that which would cause failure by crushing of the concrete, the beam will fail in tension by yielding of the steel. Its ultimate moment, from Eq. (3.20*b*), is

$$M_n = 0.0102 \times 60{,}000 \times 10 \times 23^2 \left(1 - 0.59 \frac{0.0102 \times 60{,}000}{4000}\right)$$

$$= 2{,}950{,}000 \text{ in-lb} = 246 \text{ ft-kips}$$

When the beam reaches its ultimate strength, the distance to its neutral axis, from Eq. (3.19*b*), is

$$c = \frac{0.0102 \times 60{,}000 \times 23}{0.72 \times 4000} = 4.89$$

It is interesting to compare the last result with those of Examples 3.1 and 3.2. In the previous calculations it was found that at low loads, when the concrete had not yet cracked in tension, the neutral axis was located at a distance of 13.2 in. from the compression edge; at higher loads, when the tension concrete was cracked but stresses were still sufficiently small to be elastic, this distance was 7.6 in. Immediately before the beam fails, as has just been shown, this distance has further decreased to 4.9 in. This migration of the neutral axis toward the compression edge as load is increased is a graphic illustration of the differences between the various stages of behavior through which a reinforced concrete beam passes as its load is increased from zero to that value that causes it to fail. The examples also illustrate the fact that ultimate moments cannot be determined accurately by elastic calculations.

3.4 DESIGN OF TENSION-REINFORCED RECTANGULAR BEAMS

For reasons that were explained in Chap. 1, the present design of reinforced concrete structures is based on the concept of providing sufficient strength to resist hypothetical overloads. The *nominal strength* of a proposed member is calculated, based on the best current knowledge of member and material behavior. That nominal strength is modified by a *strength reduction factor* ϕ, less than unity, to obtain the *design strength*. The *required strength*, should the hypothetical overload stage actually be realized, is found by applying *load factors* γ, greater than unity, to the loads actually expected. These expected *service loads* include the calculated dead load, the calculated or legally specified live load, and environmental loads such as those due to seismic action or temperature. Thus reinforced concrete members are proportioned so that, as shown in Eq. (1.5),

$$M_u \leq \phi M_n$$

$$P_u \leq \phi P_n$$

$$V_u \leq \phi V_n$$

where the subscripts n denote the nominal strengths in flexure, thrust, and shear respectively, and the subscripts u denote the factored load moment, thrust, and shear. The strength reduction factors ϕ normally differ, depending upon the type of strength to be calculated, the importance of the member in the structure, and other considerations discussed in detail in Chap. 1.

A member proportioned on the basis of adequate strength at a hypothetical overload stage must also perform in a satisfactory way under normal service load conditions. In specific terms, the deflection must be limited to an acceptable value, and concrete tensile cracks, which inevitably occur, must be of narrow width and well distributed throughout the tensile zone. It was formerly the practice to limit crack widths and deflections indirectly by limiting the stresses in the concrete and steel at the service load stage. The method known as *elastic design* or *service load design*, by which members are proportioned according to stress limits, with cracking and deflections controlled indirectly, is still permitted as an alternative according to the 1989 ACI Code (see ACI Code Appendix A). However, because of the limitations and inconsistencies associated with the service load design method, the *strength method* is preferred. After proportioning for adequate strength, deflections are calculated and compared against limiting values (or otherwise controlled), and crack widths limited by specific means. This approach to design, referred to in Europe, and to some extent in U.S. practice, as *limit states design*, is the main basis of the 1989 ACI Code, and it is the approach that will be followed in this and later chapters.

a. Equivalent Rectangular Stress Distribution

The method presented in Sec. 3.3*c* for calculating the ultimate flexural strength of reinforced concrete beams, derived from basic concepts of structural mechanics and pertinent experimental research information, also applies to situations other than the case of rectangular beams reinforced on the tension side. It can be used and gives valid answers for beams of other cross-sectional shapes, reinforced in other manners, and for members subject not only to simple bending but also to the simultaneous action of bending and axial force (compression or tension). However, the pertinent equations for these more complex cases become increasingly cumbersome and lengthy. What is more important, it becomes increasingly difficult for the designer to visualize the physical basis for the design methods and formulas; this could lead to a blind reliance on formulas, with a resulting lack of actual understanding. This is not only undesirable on general grounds but practically is more likely to lead to numerical errors in design work than when the designer at all times has a clear picture of the physical situation in the member being dimensioned or analyzed. Fortunately, it is possible, essentially by a conceptual trick, to formulate the strength analysis of reinforced concrete members in a different manner, which gives the same answers as the general analysis just developed but which is much more easily visualized and much more easily applied to cases of greater complexity than that of the simple rectangular beam. Its consistency is shown, and its application to more complex cases has been checked against the results of a vast number of tests on a great variety of types of members and conditions of loading (Ref. 3.4).

It was noted in the preceding section that the actual geometrical shape of the concrete compression-stress distribution varies considerably and that, in fact, one need not know this shape exactly, provided one does know two things: (1) the magnitude C of the resultant of the concrete compression stresses and (2) the location of this resultant. Information on these two quantities was obtained from the results of experimental research and expressed in the two parameters α and β.

Evidently, then, one can think of the actual complex stress distribution as replaced by a fictitious one of some simple geometric shape, provided that this fictitious distribution results in the same total compression force C applied at the same location as in the actual member when it is on the point of failure. Historically, a number of simplified, fictitious equivalent stress distributions has been proposed by investigators in various countries. The one generally accepted in this country, and increasingly abroad, was first proposed by C. S. Whitney and was subsequently elaborated and checked experimentally by others (see, for example, Ref. 3.4). The actual stress distribution immediately before failure and the fictitious equivalent distribution are shown in Fig. 3.8.

It is seen that the actual stress distribution is replaced by an equivalent one of simple rectangular outline. The intensity $\gamma f_c'$ of this equivalent constant stress and its depth $a = \beta_1 c$ are easily calculated from the two conditions that (1) the total compression force C and (2) its location, i.e., distance from the top fiber, must be the same in the equivalent rectangular as in the actual stress distribution. From Fig. 3.8*a* and *b* the first condition gives

$$C = \alpha f_c' cb = \gamma f_c' ab \qquad \text{from which} \qquad \gamma = \alpha \frac{c}{a}$$

With $a = \beta_1 c$ this gives $\gamma = \alpha/\beta_1$. The second condition simply requires that in the equivalent rectangular stress block the force C be located at the same distance βc from the top fiber as in the actual distribution. It follows that $\beta_1 = 2\beta$.

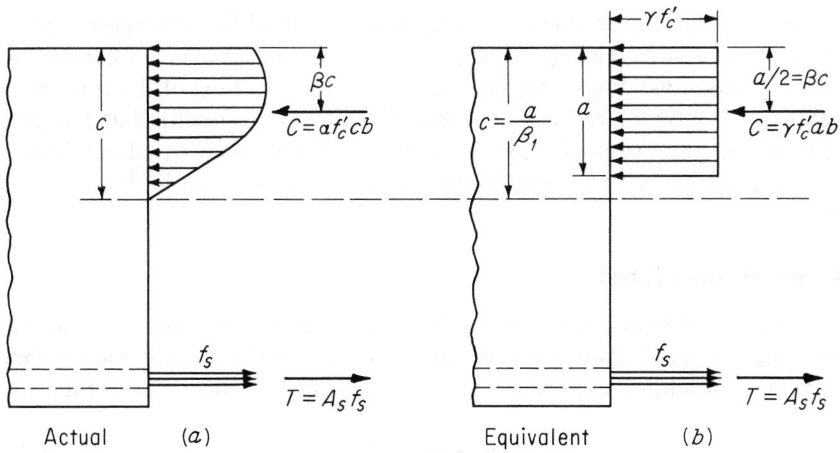

FIGURE 3.8
Actual and equivalent rectangular stress distributions at ultimate load.

Table 3.1 Concrete stress-block parameters

	f'_c, psi				
	≤ 4000	5000	6000	7000	≥ 8000
α	0.72	0.68	0.64	0.60	0.56
β	0.425	0.400	0.375	0.350	0.325
$\beta_1 = 2\beta$	0.85	0.80	0.75	0.70	0.65
$\gamma = \frac{\alpha}{\beta_1}$	0.85	0.85	0.85	0.86	0.86

To supply the details, the upper two lines of Table 3.1 present the experimental evidence of Fig. 3.7 in tabular form. The lower two lines give the just-derived parameters β_1 and γ for the rectangular stress block. It is seen that the stress intensity factor γ is essentially independent of f'_c and can be taken as 0.85 throughout. Hence, regardless of f'_c, the concrete compression force at failure in a rectangular beam of width b is

$$C = 0.85 f'_c ab \tag{3.25}$$

Also, for the common concretes with $f'_c \leq 4000$ psi, the depth of the rectangular stress block is $a = 0.85c$, c being the distance to the neutral axis. For higher strength concretes, this distance is $a = \beta_1 c$, with the β_1 values shown in Table 3.1. This is expressed in ACI Code 10.2.7.3 as follows: β_1 shall be taken as 0.85 for concrete strengths up to and including 4000 psi; for strengths above 4000 psi, β_1 shall be reduced continuously at a rate of 0.05 for each 1000 psi of strength in excess of 4000 psi, but β_1 shall not be taken less than 0.65. In mathematical terms the relationship between β_1 and f'_c can be expressed as

$$\beta_1 = 0.85 - 0.05\frac{f'_c - 4000}{1000} \quad \text{and} \quad 0.65 \leq \beta_1 \leq 0.85 \tag{3.26}$$

The equivalent rectangular stress distribution can be used for deriving the equations that have been developed in Sec. 3.3c. The failure criteria, of course, are the same as before: yielding of the steel at $f_s = f_y$ or crushing of the concrete at $\epsilon_u = 0.003$. Because the rectangular stress block is easily visualized and its geometric properties are extremely simple, many calculations are carried out directly without reference to formally derived equations, as will be seen in the following sections.

b. Balanced Steel Ratio

The balanced steel ratio can be established based on the condition that, at balanced failure, the steel strain is exactly equal to ϵ_y when the strain in the concrete simultaneously reaches the crushing strain of $\epsilon_u = 0.003$. Referring to Fig. 3.6,

$$c = \frac{\epsilon_u}{\epsilon_u + \epsilon_y} d \tag{3.27}$$

which is seen to be identical to Eq. (3.23). Then, from the equilibrium requirement that $C = T$

$$\rho_b f_y bd = 0.85 f_c' ab = 0.85 \beta_1 f_c' bc$$

from which

$$\rho_b = 0.85 \beta_1 \frac{f_c'}{f_y} \frac{\epsilon_u}{\epsilon_u + \epsilon_y} \tag{3.28a}$$

This is easily shown to be equivalent to Eq. (3.24). Substituting $\epsilon_u = 0.003$ and $E_s = 29{,}000{,}000$ psi in Eq. (3.28*a*) permits this expression for the balanced steel ratio to be rewritten in the alternative form

$$\rho_b = 0.85 \beta_1 \frac{f_c'}{f_y} \frac{87{,}000}{87{,}000 + f_y} \tag{3.28b}$$

c. Underreinforced Beams

It was pointed out in Sec. 3.3c that a compression failure in flexure, should it occur, gives little if any warning of distress, while a tension failure initiated by yielding of the steel is typically gradual. Distress is obvious from observing the large deflections and widening of concrete cracks associated with yielding of the steel reinforcement, and measures can be taken to avoid total collapse. In addition, most beams for which failure initiates by yielding possess substantial reserve strength because of strain-hardening of the rebars, which is not accounted for in the calculations of M_n.

Because of these differences in behavior, it is prudent to require that beams be designed such that failure, if it occurs, will be by yielding of the steel, not by crushing of the concrete. This can be done, theoretically, by requiring that the steel ratio ρ be less than the balanced ratio ρ_b given by Eq.(3.28*a*) or (3.28*b*).

In actual practice, the upper limit on ρ should be somewhat below ρ_b for the following reasons: (1) for a beam with ρ exactly equal to ρ_b, the compressive strain limit of the concrete would be reached, theoretically, at precisely the same moment that the steel reaches its yield stress, without significant yielding before failure, (2) material properties are never precisely known, (3) strain-hardening of the reinforcing steel, not accounted for in design, may lead to a brittle concrete compression failure even though ρ may be somewhat less than ρ_b, and (4) the actual steel area provided, considering standard rebar sizes, will always be equal to a larger than required, based on the selected steel ratio ρ, tending toward overreinforcement. For these reasons, ACI Code 10.3.3 specifies that

$$\rho_{\max} = 0.75 \rho_b \tag{3.29}$$

Thus, for all members designed according to the ACI Code, $f_s = f_y$ at failure, and the nominal flexural strength (referring to Fig. 3.9) is given by

$$M_n = A_s f_y \left(d - \frac{a}{2}\right) \tag{3.30}$$

where

$$a = \frac{A_s f_y}{0.85 f_c' b} \tag{3.31}$$

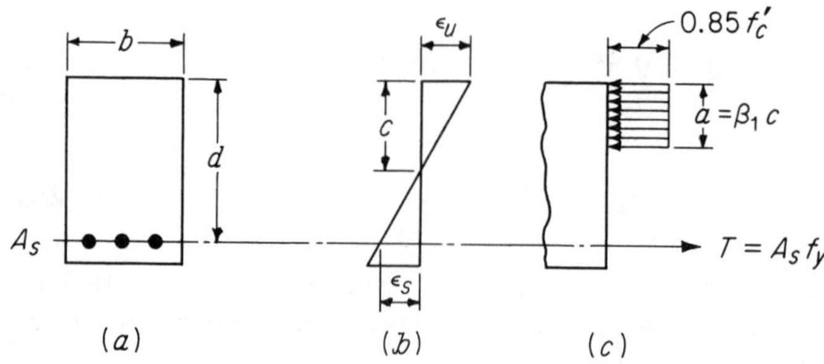

FIGURE 3.9
Singly reinforced rectangular beam.

Example 3.4. Using the equivalent rectangular stress distribution, calculate directly the ultimate strength of the previously analyzed beam of Example 3.3.

Solution. The distribution of stresses, internal forces, and strains is as shown in Fig. 3.9. The balanced steel ratio is calculated from Eq. (3.28*b*) as

$$\rho_b = 0.85 \times 0.85 \frac{4000}{60{,}000} \frac{87{,}000}{87{,}000 + 60{,}000} = 0.0284$$

and comparison with the actual steel ratio of 0.0102 confirms that the member is underreinforced and will fail by yielding of the steel. The depth of the equivalent stress block is found from the equilibrium condition that $C = T$. Hence $0.85 f_c' ab = A_s f_y$, or $a = 2.35 \times 60{,}000/0.85 \times 4000 \times 10 = 4.15$. The distance to the neutral axis, by definition of the rectangular stress block, is $c = a/\beta_1 = 4.15/0.85 = 4.89$. The ultimate moment is

$$M_n = A_s f_y \left(d - \frac{a}{2}\right) = 2.35 \times 60{,}000(23 - 2.07) = 2{,}950{,}000 \text{ in-lb} = 246 \text{ ft-kips}$$

The results of this simple and direct numerical analysis, based on the equivalent rectangular stress distribution, are identical with those previously determined from the general ultimate strength analysis of Sec. 3.3c.

It is convenient for everyday design to combine Eqs. (3.30) and (3.31) as follows. Noting that $A_s = \rho bd$, Eq. (3.31) can be rewritten as

$$a = \frac{\rho f_y d}{0.85 f_c'} \tag{3.32}$$

This is then substituted into Eq. (3.30) to obtain

$$M_n = \rho f_y bd^2 \left(1 - 0.59 \frac{\rho f_y}{f_c'}\right) \tag{3.33}$$

which is identical to Eq. (3.20*b*) derived in Sec. 3.3c. This basic equation can be simplified further as follows:

$$M_n = Rbd^2 \tag{3.34}$$

in which

$$R = \rho f_y \left(1 - 0.59 \frac{\rho f_y}{f'_c}\right) \tag{3.35}$$

The *flexural resistance factor* R depends only on the steel ratio and the strengths of the materials and is easily tabulated. Tables A.6a and A.6b of Appendix A give R values for ordinary combinations of steel and concrete and the full practical range of steel ratios.

In accordance with the safety provisions of the ACI Code, the nominal flexural strength M_n is reduced by imposing the strength reduction factor $\phi = 0.90$ for bending, to obtain the *design strength*:

$$\phi M_n = \phi A_s f_y \left(d - \frac{a}{2}\right) \tag{3.36}$$

or, alternatively,

$$\phi M_n = \phi \rho f_y b d^2 \left(1 - 0.59 \frac{\rho f_y}{f'_c}\right) \tag{3.37}$$

or

$$\phi M_n = \phi R b d^2 \tag{3.38}$$

d. Minimum Steel Ratio

Another mode of failure may occur in very lightly reinforced beams. If the flexural strength of the cracked section is less than the moment that produced cracking of the previously uncracked section, the beam will fail immediately and without warning of distress upon formation of the first flexural crack. To ensure against this type of failure, a *lower limit* can be established for the steel ratio by equating the cracking moment, computed from the concrete modulus of rupture (Sec. 2.8), to the strength of the cracked section. This results in a minimum steel ratio for rectangular beams of

$$\rho_{\min} = 1.8 \frac{\sqrt{f'_c}}{f_y} \tag{3.39a}$$

and for T beams with flange in compression of

$$\rho_{\min} = 2.7 \frac{\sqrt{f'_c}}{f_y} \tag{3.39b}$$

Further discussion of T beams will be found in Sec. 3.8.

ACI Code 10.5 imposes a minimum tensile steel ratio of

$$\rho_{\min} = \frac{200}{f_y} \tag{3.40}$$

at any section where positive reinforcement is required by analysis. (No minimum is imposed on tensile reinforcement in negative bending regions.) In T beams, the ratio ρ is to be computed using the width of the web. Alternatively, the area of reinforcement provided at every section, positive or negative, must be at least one-third greater than that required by analysis. For structural slabs of uniform thickness, however, the minimum area of steel is that required for shrinkage and temperature steel only (see Chapters 12 and 13), and the above minimum need not be applied.

It is easily confirmed that the ACI minimum steel ratio is conservative for rectangular beams for all but very high concrete strengths, and conservative for T beams with concrete strengths up to about 5000 psi. For other cases, use of Eq. (3.39*a*) or Eq. (3.39*b*) is recommended.

For convenient reference in design work, Table A.5 of App. A gives computed values of the balanced steel ratio ρ_b, the maximum steel ratio ρ_{max}, and the minimum steel ratio ρ_{min} (according to ACI Code) for common values of f_y and f'_c.

e. Examples of Rectangular Beam Review and Design

Flexural problems can be classified broadly as *review problems* or *design problems*. In review problems, the section dimensions, reinforcement, and material strengths are known, and the moment capacity is required. In the case of design problems, the required moment capacity is given, as are the material strengths, and it is required to find the section dimensions and reinforcement. Examples 3.5 and 3.6 that follow illustrate review and design, respectively.

Example 3.5 Flexural strength of a given member. A rectangular beam has width 12 in. and effective depth 17.5 in. It is reinforced with four No. 9 bars in one row. If $f_y = 60{,}000$ psi and $f'_c = 4000$ psi, what is the nominal flexural strength, and what is the maximum moment that can be utilized in design, according to the ACI Code?

Solution. From Table A.2 of App. A, the area of four No. 9 bars is 4.00 in^2. Thus the actual steel ratio is $\rho = 4.00/(12 \times 17.5) = 0.0190$. This is well below the balanced value from Eq. (3.28*b*) of

$$\rho_b = 0.85 \times 0.85 \left(\frac{4}{60}\right)\left(\frac{87}{147}\right) = 0.0285$$

so failure by tensile yielding would be obtained. For this underreinforced beam, from Eq. (3.31),

$$a = \frac{4.00 \times 60}{0.85 \times 4 \times 12} = 5.89 \text{ in.}$$

and, from Eq. (3.30),

$$M_n = 4.00 \times 60 \left(17.5 - \frac{5.89}{2}\right) = 3490 \text{ in-kips}$$

Applying the strength reduction factor for flexure, $\phi = 0.90$, the design strength should be taken as

$$\phi M_n = 0.90 \times 3490 = 3140 \text{ in-kips}$$

The ACI Code limits on the steel ratio,

$$\rho_{\max} = 0.75 \times 0.0285 = 0.0214$$

$$\rho_{\min} = \frac{200}{60{,}000} = 0.0033$$

are satisfied for this beam.

Example 3.6 Concrete dimensions and steel area to resist a given moment. Find the cross section of concrete and area of steel required for a simply supported rectangular beam with a span of 15 ft that is to carry a computed dead load of 1.27 kips/ft and a service live load of 2.44 kips/ft. Material strengths are $f_c' = 4000$ psi and $f_y = 60{,}000$ psi.

Solution. Load factors are first applied to the given service loads to obtain the factored load for which the beam is to be designed, and the corresponding moment:

$$w_u = 1.4 \times 1.27 + 1.7 \times 2.44 = 5.92 \text{ kips/ft}$$

$$M_u = \frac{1}{8} \times 5.92 \times 15^2 \times 12 = 2000 \text{ in-kips}$$

The concrete dimensions will depend upon the designer's choice of steel ratio. Selecting the maximum permissible steel ratio, $\rho_{\max} = 0.75\rho_b$, will result in the minimum concrete section. With $\rho_b = 0.0285$ as in Example 3.5,

$$\rho = 0.75 \times 0.0285 = 0.0214$$

Setting the required flexural strength equal to the design strength from Eq.(3.37), and substituting the selected values for ρ and material strengths,

$$M_u = \phi M_n$$

$$2000 = 0.90 \times 0.0214 \times 60bd^2\left(1 - 0.59\frac{0.0214 \times 60}{4}\right)$$

from which

$$bd^2 = 2130 \text{ in}^3$$

A beam width $b = 10$ in. and $d = 14.6$ in. will satisfy this requirement. The required steel area is found by applying the chosen steel ratio to the required concrete dimensions:

$$A_s = 0.0214 \times 10 \times 14.6 = 3.12 \text{ in}^2$$

Two No. 11 bars provide exactly this area.

Assuming 2.5 in. concrete cover from the centroid of the bars, the required total depth is $b = 17.1$ in. In actual practice, however, the concrete dimensions b and h are always rounded upward to the nearest inch, and often to the nearest multiple of 2 inches (see Sec. 3.5). The actual d is then found by subtracting the required concrete cover dimension from h. For the present example, $b = 10$ in. and $h = 18$ in. will be selected, resulting in effective depth $d = 15.5$ in. Improved economy then may be possible, refining the steel area based on the actual, larger, effective depth. One can obtain the revised steel requirement directly by solving Eq. (3.37) for ρ, with $\phi M_n = M_u$. A quicker solution can be obtained by iteration. First a reasonable value of a is assumed, and A_s is found from Eq. (3.36). From Eq. (3.31) a revised estimate of a is obtained, and A_s is revised. This method converges very rapidly. For example, assume $a = 5$ in. Then

$$A_s = \frac{2000}{0.90 \times 60(15.5 - 2.5)} = 2.85 \text{ in}^2$$

Checking the assumed a:

$$a = \frac{2.85 \times 60}{0.85 \times 4 \times 10} = 5.03 \text{ in.}$$

This is close enough to the assumed value that no further calculation is required. Nearly enough, the required steel of 2.85 in^2 could be provided using one No. 11 plus one No. 10 bar, but for simplicity of construction, two No. 11 bars will be used as before.

A somewhat larger beam cross section using less steel may be more economical, and will tend to reduce deflections. As an alternative solution, the beam will be redesigned with a lower reinforcement ratio of $\rho = 0.50\rho_b = 0.50 \times 0.0285 = 0.0143$. Setting the required strength equal to the design strength [Eq. (3.37)] as before:

$$2000 = 0.90 \times 0.0143 \times 60bd^2\left(1 - 0.59\frac{0.0143 \times 60}{4}\right)$$

and

$$bd^2 = 2960 \text{ in}^3$$

A beam with $b = 10$ in. and $d = 17.2$ in. will meet the requirement, for which

$$A_s = 0.0143 \times 10 \times 17.2 = 2.46 \text{ in}^2$$

Two No. 10 bars would be sufficient, providing an area of 2.53 in^2. If the total concrete depth is rounded upward to 20 in., a 17.5 in. effective depth results, reducing the required steel area to 2.41 in^2. Two No. 10 bars would still be the best choice.

It is apparent that an infinite number of solutions to the stated problem are possible, depending upon the steel ratio selected. That ratio may vary from an upper limit of $0.75\rho_b$ to a lower limit of $200/f_y$ for beams, according to the ACI Code. To compare the two solutions (using the theoretical dimensions, unrounded for the comparison, and assuming h is 2.5 inches greater than d in each case), increasing the concrete section area by 15 percent achieves a steel saving of 21 percent. The second solution would certainly be more economical and would be preferred, unless beam dimensions must be minimized for architectural or functional reasons.

There is a type of problem, occurring frequently, that does not fall strictly into either the review or design category. The concrete dimensions are given and are known to be adequate to carry the required moment, and it is necessary only to find the steel area. Typically, this is the situation at critical design sections of continuous beams, in which the concrete dimensions are often kept constant, although the steel reinforcement varies along the span according to the required flexural resistance. Dimensions b, d, and h are determined at the maximum moment section, usually at one of the supports. At other supports, and at midspan locations, where moments are usually smaller, the concrete dimensions are known to be adequate and only the tensile steel remains to be found. An identical situation was encountered in the design problem of Example 3.6, in which concrete dimensions were rounded upward from the minimum required values, and the required steel area was to be found. In either case, the iterative approach demonstrated in Example 3.6 is convenient.

Example 3.7 Determination of steel area. Using the same concrete dimensions as were used for the second solution of Example 3.6 ($b = 10$ in., $d = 17.5$ in., and $h = 20$ in.) and the same material strengths, find the steel area required to resist a moment M_u of 1600 in-kips.

Solution. Assume $a = 4.0$ in. Then

$$A_s = \frac{1600}{0.90 \times 60(17.5 - 2.0)} = 1.91 \text{ in}^2$$

Checking the assumed a:

$$a = \frac{1.91 \times 60}{0.85 \times 4 \times 10} = 3.37 \text{ in.}$$

Next assume $a = 3.2$ in. and recalculate A_s:

$$A_s = \frac{1600}{0.90 \times 60(17.5 - 1.6)} = 1.86 \text{ in}^2$$

No further iteration is required. Use $A_s = 1.86$ in^2. Two No. 9 bars will be used.

In solving these examples, the basic equations have been used, in order to develop familiarity with them. In actual practice, however, design aids such as Table A.5 of App. A, giving values of balanced, maximum, and minimum steel ratios, and Table A.6, providing values of flexural resistance factor R, are more convenient. The example problems will be repeated in Sec. 3.5 to demonstrate use of these aids.

f. Overreinforced Beams

According to the ACI Code, all beams are to be designed for the underreinforced condition, with the tensile steel ratio well below the balanced value and $f_s = f_y$ at failure. Occasionally, however, such as when reviewing the capacity of existing construction, it may be necessary to calculate the flexural strength of an overreinforced member, for which f_s is less than f_y at flexural failure.

In this case, the steel strain, in Fig. 3.9b, will be less than the yield strain, but can be expressed in terms of the concrete strain ϵ_u and the still-unknown distance c to the neutral axis:

$$\epsilon_s = \epsilon_u \frac{d - c}{c} \tag{3.41}$$

From the equilibrium requirement that $C = T$ one can write

$$0.85\beta_1 f_c' bc = \rho\epsilon_s E_s bd$$

Substituting the steel strain from Eq. (3.41) in the last equation, and defining $k_u = c/d$, one obtains a quadratic equation in k_u as follows:

$$k_u^2 + m\rho k_u - m\rho = 0$$

Here, $\rho = A_s/bd$ as usual and m is a material parameter given by

$$m = \frac{E_s\epsilon_u}{0.85\beta_1 f_c'} \tag{3.42}$$

Solving the quadratic equation for k_u:

$$k_u = \sqrt{m\rho + \left(\frac{m\rho}{2}\right)^2} - \frac{m\rho}{2} \tag{3.43}$$

The neutral axis depth for the overreinforced beam can then easily be found from $c = k_u d$, after which the stress-block depth $a = \beta_1 c$. With steel strain ϵ_s then computed from Eq.(3.41), and with $f_s = E_s\epsilon_s$, the nominal flexural strength is

$$M_n = A_s f_s \left(d - \frac{a}{2}\right) \tag{3.44}$$

3.5 DESIGN AIDS

Basic equations were developed in Sec. 3.4 for the analysis and design of reinforced concrete beams, and these were used directly in the examples. In practice, the design of beams and other reinforced concrete members is greatly facilitated by use of aids such as those in Appendix A of this text and in Refs. 3.6 through 3.8. Tables A.1, A.2, A.5 through A.8, and Graph A.1 of Appendix A relate directly to this chapter, and the student can scan this material to become familiar with the coverage. Other aids will be discussed, and their use demonstrated, in later chapters.

Eq. (3.38) of Sec. 3.4 gives the flexural design strength ϕM_n of a tension-reinforced rectangular beam having steel ratio at or below the balanced value. The flexural resistance factor R, from Eq. (3.35), is given in Table A.6a for lower steel ratios or Table A.6b for higher steel ratios. Alternatively, R can be obtained from Graph A.1. For *review* of the capacity of a section with known concrete dimensions b and d, having known steel ratio ρ, and with known materials strengths, the design strength ϕM_n can be obtained directly by Eq. (3.38).

For *design* purposes, where concrete dimensions and reinforcement are to be found and the factored load moment M_u is to be resisted, there are two possible approaches. One starts with selecting the optimum steel ratio, then calculating concrete dimensions, as follows:

1. Set the required strength M_u equal to the design strength ϕM_n from Eq. (3.38):

$$M_u = \phi R b d^2$$

2. With the aid of Table A.5, select an appropriate steel ratio between ρ_{max} and ρ_{min}. Often a ratio of about $0.50\rho_b$ will be an economical and practical choice.
3. From Table A.6, for the specified material strengths and selected steel ratio, find the flexural resistance factor R. Then

$$bd^2 = \frac{M_u}{\phi R}$$

4. Choose b and d to meet that requirement. Unless construction depth must be limited or other constraints exist, effective depth should be about 2 to 3 times the width.
5. Calculate the required steel area

$$A_s = \rho bd$$

   Then, referring to Table A.2, choose the size and number of bars, giving preference to the larger bar sizes to minimize placement costs.
6. Refer to Table A.8 to ensure that the selected beam width will provide room for the bars chosen, with adequate concrete cover and spacing. (These points will be discussed further in Sec. 3.6.)

The alternative approach starts with selecting concrete dimensions, after which the required reinforcement is found, as follows:

1. Select beam width b and effective depth d. Then calculate the required R:

$$R = \frac{M_u}{\phi bd^2}$$

2. From Table A.6, for specified material strengths, find the steel ratio ρ that will provide that R.
3. Calculate the required steel area

$$A_s = \rho bd$$

   and from Table A.2 select the size and number of bars.
4. Using Table A.8, confirm that the beam width is sufficient to contain the selected reinforcement.

Use of design aids to solve the example problems of Sec. 3.4 will be illustrated as follows.

Example 3.8 Flexural strength of a given member. Find the nominal flexural strength and design strength of the beam of Example 3.5, which has $b = 12$ in. and $d = 17.5$ in. and is reinforced with four No. 9 bars. Make use of the design aids of App. A. Material strengths are $f'_c = 4000$ psi and $f_y = 60{,}000$ psi.

Solution. From Table A.2, four No. 9 bars provide $A_s = 4.00$ in^2, and with $b = 12$ in. and $d = 17.5$ in., the steel ratio is $\rho = 4.00/(12 \times 17.5) = 0.0190$. According to Table A.8, this is below $\rho_{max} = 0.0214$ and above $\rho_{min} = 0.0033$. Then from Table

A.6b, with $f_c' = 4000$ psi, $f_y = 60{,}000$ psi, and $\rho = 0.019$, the value $R = 949$ psi is found. The nominal and design strengths are respectively

$$M_n = Rbd^2 = 949 \times 12 \times \frac{17.5^2}{1000} = 3490 \text{ in-kips}$$

$$\phi M_n = 0.90 \times 3490 = 3140 \text{ in-kips}$$

as before.

Example 3.9 Concrete dimensions and steel area to resist a given moment. Find the cross section of concrete and the area of steel required for the beam of Example 3.6, making use of the design aids of App. A. $M_u = 2000$ in-kips, $f_c' = 4000$ psi, and $f_y = 60{,}000$ psi. Use a steel ratio of one-half the balanced value.

Solution. From Table A.5, the balanced steel ratio is $\rho_b = 0.0285$. For economy, a value of $\rho = 0.50\rho_b = 0.0143$ will be used. For that value, by interpolation from Table A.6a, the required value of R is 750. Then

$$bd^2 = \frac{M_u}{\phi R} = \frac{2000 \times 1000}{0.90 \times 750} = 2960 \text{ in}^3$$

Concrete dimensions $b = 10$ in. and $d = 17.2$ in. will satisfy this, but the depth will be rounded upward to 17.5 in. to provide a total beam depth of 20.0 in. It follows that

$$R = \frac{M_u}{\phi bd^2} = \frac{2000 \times 1000}{0.90 \times 10 \times 17.5^2} = 726 \text{ psi}$$

and from Table A.6a, by interpolation, $\rho = 0.0138$. This leads to a steel requirement of $A_s = 0.0138 \times 10 \times 17.5 = 2.41 \text{ in}^2$ as before.

Example 3.10 Determination of steel area. Find the steel area required for the beam of Example 3.7, with concrete dimensions $b = 10$ in. and $d = 17.5$ in., known to be adequate to carry the factored load moment of 1600 in-lb. Material strengths are $f_c' = 4000$ psi and $f_y = 60{,}000$ psi.

Solution. Note that in such cases, in which the concrete dimensions are known to be adequate and only the reinforcement must be found, the iterative method used earlier is not required. The necessary flexural resistance factor is

$$R = \frac{M_u}{\phi bd^2} = \frac{1600 \times 1000}{0.90 \times 10 \times 17.5^2} = 580 \text{ psi}$$

According to Table A.6a, with the specified material strengths, this corresponds to a steel ratio of $\rho = 0.0107$, giving a steel area of

$$A_s = 0.0107 \times 10 \times 17.5 = 1.87 \text{ in}^2$$

as before (except for a small rounding difference). Two No. 9 bars will be used.

The tables and graphs of Appendix A give basic information and are used extensively throughout this text for illustrative purposes. The reader should be aware, however, of the greatly expanded versions of these tables, plus many other useful aids, that are found in Refs. 3.6 through 3.8 and elsewhere.

3.6 PRACTICAL CONSIDERATIONS IN THE DESIGN OF BEAMS

In order to focus attention initially on the basic aspects of flexural design, the preceding examples were carried out with only minimum regard for certain practical considerations that always influence the actual design of beams. These relate to optimal concrete proportion for beams, rounding of dimensions, standardization of dimensions, required cover for main and auxiliary reinforcement, and selection of bar combinations. Good judgment on the part of the design engineer is particularly important in translating from theoretical requirements to practical design. Several of the more important aspects are discussed here; much additional guidance is provided by the publications of ACI (Refs. 3.6 and 3.7) and CRSI (Refs. 3.8 to 3.10).

a. Concrete Protection for Reinforcement

To provide the steel with adequate concrete protection against fire and corrosion, the designer must maintain a certain minimum thickness of concrete cover outside of the outermost steel. The thickness required will vary, depending upon the type of member and conditions of exposure. According to ACI Code 7.7, for cast-in-place concrete, concrete protection at surfaces not exposed directly to the ground or weather should be not less than $\frac{3}{4}$ in. for slabs and walls and $1\frac{1}{2}$ in. for beams and columns. If the concrete surface is to be exposed to the weather or in contact with the ground, a protective covering of at least 2 in. is required ($1\frac{1}{2}$ in. for No. 5 bars and smaller), except that, if the concrete is poured in direct contact with the ground without the use of forms, a covering of at least 3 in. must be furnished.

In general, the centers of main flexural bars in beams should be placed $2\frac{1}{2}$ to 3 in. from the top or bottom surface of the beam, in order to furnish at least $1\frac{1}{2}$ in. of clear insulation for the bars and the stirrups (see Fig. 3.10). In slabs, 1 in. to the center of the bar is ordinarily sufficient to give the required $\frac{3}{4}$ in. insulation.

In order to simplify construction and thereby to reduce costs, the overall concrete dimensions of beams, b and h, are almost always rounded upward to the nearest inch, and often to the next multiple of 2 inches. As a result, the actual effective depth d, found by subtracting the sum of cover distance, stirrup diameter, and half the main rebar diameter from the total depth h, is seldom an even dimension. For slabs, the total depth is generally rounded upward to the nearest $\frac{1}{2}$ inch up to 6 in. in depth, and to the nearest inch above that thickness. The differences between h and d shown in Fig. 3.10 are not exact, but are satisfactory for design purposes for beams with No. 3 stirrups and No. 10 longitudinal bars or smaller, and for slabs using No. 4 bars or smaller. If larger bars are used for the main flexural reinforcement or for the stirrups, as is frequently the case, the corresponding dimensions are easily calculated.

Recognizing the closer tolerances that can be maintained under plant-control conditions, ACI Code 7.7.2 permits some reduction in concrete protection for reinforcement in precast concrete.

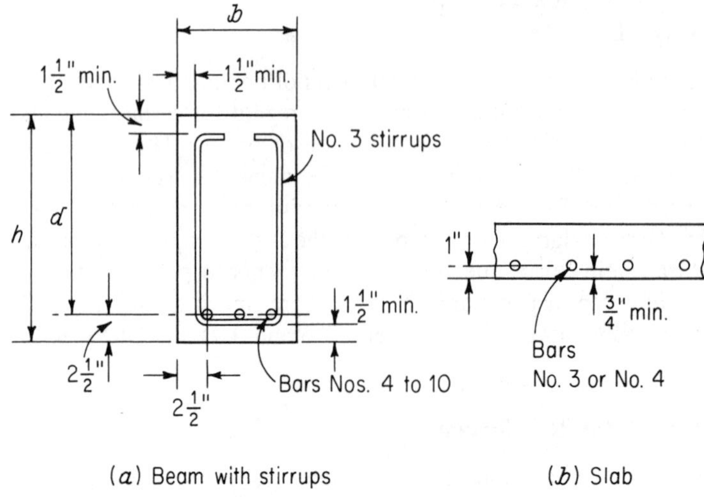

FIGURE 3.10
Requirements for concrete cover in beams and slabs.

b. Concrete Proportions

Reinforced concrete beams may be wide and shallow, or relatively narrow and deep. Consideration of maximum material economy often leads to proportions with effective depth d in the range from about 2 to 3 times the width b (or web width b_w for T beams). However, constraints may dictate other choices. For example, with one-way concrete joists supported by monolithic beams (see Chap. 20), use of beams and joists with the same total depth will permit use of a single flat-bottom form, resulting in fast, economical construction and permitting level ceilings. The beams will generally be wide and shallow, with heavier reinforcement than otherwise, but there may be an overall saving in construction cost. In other cases, it may be necessary to limit the total depth of floor or roof construction for architectural or other reasons. An advantage of reinforced concrete is its adaptability to such special needs.

c. Selection of Bars and Bar Spacing

As noted in Sec. 2.12, common reinforcing bar sizes range from No. 3 to No. 11, the bar number corresponding closely to the number of eighth-inches of bar diameter. The two larger sizes, No. 14 ($1\frac{3}{4}$ in. diameter) and No. 18 ($2\frac{1}{4}$ in. diameter) are used mainly in columns.

It is often desirable to mix bar sizes in order to meet steel area requirements more closely. In general, mixed bars should be of comparable diameter, for practical as well as theoretical reasons, and generally should be arranged symmetrically about the vertical centerline. Many designers limit the variation in diameter of bars in a single layer to two bar sizes, using, say, Nos. 10 and 8 bars together, but not Nos. 11 and 6. There is some practical advantage to minimizing the number of different bar sizes used for a given structure.

Normally, it is necessary to maintain a certain minimum distance between adjacent bars in order to ensure proper placement of concrete around them. Air pockets below the steel are to be avoided, and full surface contact between the bars and the concrete is desirable to optimize bond strength. ACI Code 7.6 specifies that the minimum clear distance between adjacent bars shall not be less than the nominal diameter of the bars, or 1 in. (For columns, these requirements are increased $1\frac{1}{2}$ bar diameters and $1\frac{1}{2}$ in.) Where beam reinforcement is placed in two or more layers, the clear distance between layers must not be less than 1 in., and the bars in the upper layer should be placed directly above those in the bottom layer.

The maximum number of bars that can be placed in a beam of given width is limited by bar diameter and spacing requirements and is also influenced by stirrup diameter, by concrete cover requirement, and by the maximum size of concrete aggregate specified. Table A.8 of App. A gives the maximum number of bars that can be placed in a single layer in beams, assuming $1\frac{1}{2}$ in. concrete cover and the use of No. 4 stirrups. There are also restrictions on the *minimum* number of bars that can be placed in a single layer, based on requirements for crack control (see Sec. 6.3). Table A.9 gives the minimum number of bars that will satisfy ACI Code requirements, which will be discussed in Chap. 6.

In large girders and columns, it is sometimes advantageous to "bundle" tensile or compressive reinforcement with two, three, or four bars in contact to provide for better deposition of concrete around and between adjacent bundles. These bars may be assumed to act as a unit, with not more than four bars in any bundle, provided that stirrups or ties enclose the bundle. No more than two bars should be bundled in one plane; typical bundle shapes are triangular, square, or L-shaped patterns. Individual bars in a bundle, cut off within the span of flexural members, should terminate at different points. ACI Code 7.6.6 requires at least 40 bar diameters stagger between points of cutoff. Where spacing limitations and minimum concrete cover requirements are based on bar diameter, a unit of bundled bars is to be treated as a single bar of such diameter as to provide the same bar area.

ACI Code 7.6.6 states that bars larger than No. 11 shall not be bundled in beams, although the AASHTO specification permits bundling of Nos. 14 and 18 bars in highway bridge girders. Careful attention must be paid to crack control if bars used as flexural reinforcement are bundled (see Sec. 6.3).

3.7 RECTANGULAR BEAMS WITH TENSION AND COMPRESSION REINFORCEMENT

If a beam cross section is limited because of architectural or other considerations, it may happen that the concrete cannot develop the compression force required to resist the given bending moment. In this case reinforcing is added in the compression zone, resulting in a so-called *doubly reinforced* beam, i.e., one with compression as well as tension reinforcement. The use of compression reinforcement has decreased markedly with the introduction and wider use of strength design methods, which account for the full strength potential of the concrete on the compressive side of the neutral axis. However, there are situations in which compressive reinforcement is used for reasons other than strength. It has been found that the inclusion of some compression steel will reduce the long-term deflections of members (see Sec. 6.5). In addition, in some cases, bars will be placed in the compression zone for minimum-moment loading (see Sec. 11.2) or as stirrup-support bars continuous throughout the beam span (see Chap. 4). It is often desirable to account for the presence of such reinforcement in flexural design.

a. Tension and Compression Steel Both at Yield Stress

If, in a doubly reinforced beam, the tensile steel ratio ρ is equal to or less than ρ_b, the strength of the beam may be approximated within acceptable limits by disregarding the compression bars. The strength of such a beam will be controlled by tensile yielding, and the lever arm of the resisting moment will ordinarily be but little affected by the presence of the compression bars.

If the tensile steel ratio is larger than ρ_b, a somewhat more elaborate analysis is required. In Fig. 3.11a, a rectangular beam cross section is shown with compression steel A_s' placed a distance d' from the compression face and with tensile steel A_s at effective depth d. It is assumed initially that both A_s' and A_s are stressed to f_y at failure. The total resisting moment can be thought of as the sum of two parts. The first part, M_{n1}, is provided by the couple consisting of the force in the compression steel A_s' and the force in an equal area of tension steel

$$M_{n1} = A_s' f_y (d - d') \tag{3.45a}$$

as shown in Fig. 3.11d. The second part, M_{n2}, is the contribution of the remaining tension steel $A_s - A_s'$ acting with the compression concrete:

$$M_{n2} = (A_s - A_s') f_y \left(d - \frac{a}{2}\right) \tag{3.45b}$$

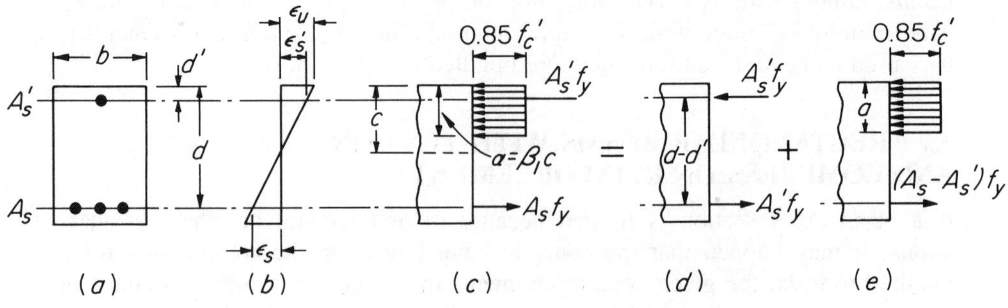

FIGURE 3.11
Doubly reinforced rectangular beam.

as shown in Fig. 3.11e, where the depth of the stress block is

$$a = \frac{(A_s - A_s')f_y}{0.85f_c'b} \tag{3.46a}$$

With the definitions $\rho = A_s/bd$ and $\rho' = A_s'/bd$, this can be written

$$a = \frac{(\rho - \rho')f_y d}{0.85f_c'} \tag{3.46b}$$

The total nominal resisting moment is then

$$M_n = M_{n1} + M_{n2} = A_s'f_y(d - d') + (A_s - A_s')f_y\left(d - \frac{a}{2}\right) \tag{3.47}$$

In accordance with the safety provisions of the ACI Code, this nominal capacity is reduced by the factor $\phi = 0.90$ to obtain the design strength.

It is highly desirable, for reasons given earlier, that failure, should it occur, be precipitated by tensile yielding rather than crushing of the concrete. This can be ensured by setting an *upper limit* on the tensile steel ratio. By setting the tensile steel strain in Fig. 3.11b equal to ϵ_y to establish the location of the neutral axis for the balanced failure condition and then summing horizontal forces shown in Fig. 3.11c (still assuming the compressive steel to be at the yield stress at failure), it is easily shown that the balance steel ratio $\overline{\rho}_b$ for a doubly reinforced beam is

$$\overline{\rho}_b = \rho_b + \rho' \tag{3.48}$$

where ρ_b is the balanced steel ratio for the corresponding singly reinforced beam and is calculated from Eq. (3.28a). To ensure the same margin against brittle concrete failure in a doubly reinforced beam as in a singly reinforced beam, according to ACI Code 10.3.3

$$\overline{\rho}_{\max} = 0.75\rho_b + \rho' \tag{3.49}$$

b. Compression Steel below Yield Stress

The preceding equations, through which the fundamental analysis of doubly reinforced beams is developed clearly and concisely, are valid *only* if the compression steel has yielded when the beam reaches its ultimate capacity. In many cases, such as for wide, shallow beams, beams with more than the usual concrete cover over the compression bars, or beams with relatively small amounts of tensile reinforcement, the compression bars will be below the yield stress at failure. It is necessary, therefore, to develop more generally applicable equations to account for the possibility that the compression reinforcement has not yielded when the doubly reinforced beam fails in flexure.

Whether or not the compression steel will have yielded at failure can be determined as follows. Referring to Fig. 3.11b, and taking as the limiting case $\epsilon_s' = \epsilon_y$, one obtains, from geometry,

$$\frac{c}{d'} = \frac{\epsilon_u}{\epsilon_u - \epsilon_y} \qquad \text{or} \qquad c = \frac{\epsilon_u}{\epsilon_u - \epsilon_y}d'$$

Summing forces in the horizontal direction (Fig. 3.11*c*) gives the *minimum* tensile steel ratio $\overline{\rho}_{cy}$ that will ensure yielding of the compression steel at failure:

$$\overline{\rho}_{cy} = 0.85\beta_1 \frac{f_c'}{f_y}\frac{d'}{d}\frac{\epsilon_u}{\epsilon_u - \epsilon_y} + \rho' \tag{3.50a}$$

Taking $\epsilon_u = 0.003$ as usual and $\epsilon_y = f_y/E_s$ with $E_s = 29{,}000{,}000$ psi, one obtains the alternative form

$$\overline{\rho}_{cy} = 0.85\beta_1 \frac{f_c'}{f_y}\frac{d'}{d}\frac{87{,}000}{87{,}000 - f_y} + \rho' \tag{3.50b}$$

If the *tensile steel ratio* is *less than* this limiting value, the neutral axis is sufficiently high that the compression steel stress at failure is less than the yield stress. In this case, it can easily be shown on the basis of Fig. 3.11*b* and *c* that the balanced steel ratio is

$$\overline{\rho}_b = \rho_b + \rho'\frac{f_s'}{f_y} \tag{3.51}$$

where

$$f_s' = E_s\epsilon_s' = E_s\left[\epsilon_u - \frac{d'}{d}(\epsilon_u + \epsilon_y)\right] \quad \text{and} \quad \le f_y \tag{3.52}$$

Hence, the maximum steel ratio permitted by ACI Code 10.3.3 is

$$\overline{\rho}_{\max} = 0.75\rho_b + \rho'\frac{f_s'}{f_y} \tag{3.53}$$

Thus Eqs. (3.51) and (3.53), with f_s' given by Eq. (3.52), are the generalized forms of Eqs. (3.48) and (3.49).

It should be emphasized that Eq. (3.52) for compression steel stress applies *only for a beam with exactly the balanced tensile steel ratio.*

If the tensile steel ratio is less than $\overline{\rho}_b$ as given by Eq. (3.51), and less than $\overline{\rho}_{cy}$ given by Eq. (3.50), then the tensile steel is at the yield stress at failure but the compression steel is not, and new equations must be developed for compression steel stress and flexural strength. The compression steel stress can be expressed in terms of the still-unknown neutral axis depth:

$$f_s' = \epsilon_u E_s \frac{c - d'}{c} \tag{3.54}$$

Consideration of horizontal force equilibrium (Fig. 3.11*c* with compression steel stress equal to f_s') then gives

$$A_s f_y = 0.85\beta_1 f_c' bc + A_s'\epsilon_u E_s \frac{c - d'}{c} \tag{3.55}$$

This is a quadratic equation in *c*, the only unknown, and is easily solved for *c*. The nominal flexural strength is found using the value of f_s' from Eq. (3.54), and

$a = \beta_1 c$ in the expression

$$M_n = 0.85 f'_c ab\left(d - \frac{a}{2}\right) + A'_s f'_s (d - d') \tag{3.56}$$

This nominal capacity is reduced by the factor $\phi = 0.90$ to obtain the design strength.

If compression bars are used in a flexural member, precautions must be taken to ensure that these bars will not buckle outward under load, spalling off the outer concrete. ACI Code 7.11.1 imposes the requirement that such bars be anchored in the same way that compression bars in columns are anchored by lateral ties (Sec. 8.2). Such ties must be used throughout the distance where the compression reinforcement is required.

c. Examples of Review and Design of Beams With Tension and Compression Steel

As was the case for beams with only tension reinforcement, doubly reinforced beam problems can be placed in one of two categories: review problems or design problems. For *review*, in which the concrete dimensions, reinforcement, and material strengths are given, one can find the flexural strength directly from the equations of Sec. 3.7a or Sec. 3.7b. First it must be confirmed that the tensile steel ratio is less than $\overline{\rho}_b$ given by Eq. (3.51), with compression steel stress from Eq. (3.52). Once it is established that the tensile steel has yielded, the tensile steel ratio defining compression steel yielding is calculated from Eq. (3.50*b*), and the actual tensile steel ratio is compared. If it is greater than $\overline{\rho}_{cy}$, then $f'_s = f_y$, and M_n is found from Eq. (3.47). If it is less than $\overline{\rho}_{cy}$, then $f'_s < f_y$. In this case, c is calculated by solving Eq. (3.55), f'_s comes from Eq. (3.54), and M_n is found from Eq. (3.56).

For the *design* case, in which the factored load moment M_u to be resisted is known and the section dimensions and reinforcement are to be found, a direct solution is impossible. The steel areas to be provided depend on the steel stresses, which are not known before the section is proportioned. It can be assumed that the compression steel stress is equal to the yield stress, but this must be confirmed; if it is not so, the design must be adjusted. The design procedure can be outlined as follows:

1. Calculate the maximum moment that can be resisted by the tension-reinforced section with $\rho = \rho_{\max} = 0.75\rho_b$. The corresponding tensile steel area is $A_s = \rho_{\max} bd$, and, as usual,

$$M_n = A_s f_y\left(d - \frac{a}{2}\right)$$

with

$$a = \frac{A_s f_y}{0.85 f'_c b}$$

2. Find the excess moment, if any, that must be resisted, and set $M_2 = M_n$, as calculated in step 1.

$$M_1 = \frac{M_u}{\phi} - M_2$$

A_s from step 1 is now defined as A_{s2}, i.e., that part of the tension steel area in the doubly reinforced beam that works with the compression force in the concrete. In Fig. 3.11*e*, $A_{s2} = (A_s - A_s')$.

3. Tentatively assume that $f_s' = f_y$. Then

$$A_s' = \frac{M_1}{f_y(d - d')}$$

4. Add an additional amount of tensile steel $A_{s1} = A_s'$. Thus the total tensile steel area A_s is A_{s2} from step 2 plus A_{s1}.

5. Review the doubly reinforced beam to see if $f_s' = f_y$; that is, check the tensile steel ratio against $\overline{\rho}_{cy}$.

6. If $\rho < \overline{\rho}_{cy}$, then the compression steel stress is less than f_y and the compression steel area must be increased in order to provide the needed force. This can be done as follows. The stress block depth is found from the requirement of horizontal equilibrium (Fig. 3.11*e*),

$$a = \frac{(A_s - A_s')f_y}{0.85 f_c' b}$$

and the neutral axis depth is $c = a/\beta_1$. From Eq. (3.54),

$$f_s' = \epsilon_u E_s \frac{c - d'}{c}$$

The revised compression steel area, acting at f_s', must provide the same force as the trial steel area that was assumed to act at f_y. Therefore,

$$A'_{s,\text{revised}} = A'_{s,\text{trial}} \frac{f_y}{f_s'}$$

The tensile steel area need not be revised, because it acts at f_y as assumed.

Example 3.11 Flexural strength of given member. A rectangular beam has a width of 12 in. and an effective depth to the centroid of the tension reinforcement of 18 in. The tension reinforcement consists of six No. 10 bars in two rows. Compression reinforcement consisting of two No. 9 bars is placed 2.5 in. from the compression face of the beam. If $f_y = 50{,}000$ psi and $f_c' = 5000$ psi, what is the design moment capacity of the beam?

Solution. The steel areas and ratios are

$$A_s = 7.59 \text{ in}^2 \qquad \rho = \frac{7.59}{12 \times 18} = 0.0352$$

$$A_s' = 2.00 \text{ in}^2 \qquad \rho' = \frac{2.00}{12 \times 18} = 0.0093$$

Check the beam first as a singly reinforced beam to see if the compression bars can be disregarded,

$$\rho_b = 0.85 \times 0.80 \times \frac{5}{50} \times \frac{87}{137} = 0.0432$$

$$\rho_{\max} = 0.75 \times 0.0432 = 0.0324$$

The actual $\rho = 0.0352$ is larger than $\rho_{\max}$, so the beam must be analyzed as doubly reinforced. From Eq. (3.50*b*),

$$\overline{\rho}_{cy} = \left(0.85 \times 0.80 \times \frac{5}{50} \times \frac{2.5}{18} \times \frac{87}{37}\right) + 0.0093 = 0.0315$$

The tensile steel ratio is greater than this, so the compression bars will yield when the beam fails. The balanced steel ratio and maximum steel ratio thus can be found from Eq. (3.48) and Eq. (3.49), respectively,

$$\overline{\rho}_b = 0.0432 + 0.0093 = 0.0525$$

$$\overline{\rho}_{\max} = (0.75 \times 0.0432) + 0.0093 = 0.0417$$

The actual tensile steel ratio is below the maximum value, as required. Then, from Eq. (3.46*a*),

$$a = \frac{5.59 \times 50}{0.85 \times 5 \times 12} = 5.48 \text{ in.}$$

and from Eq. (3.47),

$$M_n = 2.00 \times 50(18 - 2.5) + [5.59 \times 50(18 - 2.74)] = 5820 \text{ in-kips}$$

The design strength is

$$\phi M_n = 0.90 \times 5820 = 5240 \text{ in-kips}$$

Example 3.12 Design of doubly reinforced beam. A rectangular beam that must carry a service live load of 2.47 kips/ft and a calculated dead load of 1.05 kips/ft on an 18 ft simple span is limited in cross section for architectural reasons to 10 in. width and 20 in. total depth. If $f_y = 40,000$ psi and $f_c' = 3000$ psi, what steel area(s) must be provided?

Solution. The service loads are first increased by load factors to obtain the factored load of $1.4 \times 1.05 + 1.7 \times 2.47 = 5.66$ kips/ft. Then $M_u = 5.66 \times 18^2/8 = 229$ ft-kips $=$ 2750 in-kips. To satisfy spacing and cover requirements (see Sec. 3.6), assume that the tension steel centroid will be 4 in. above the bottom face of the beam and that compression steel, if required, will be placed 2.5 in. below the beam's top surface. Then $d = 16$ in. and $d' = 2.5$ in.

First check the capacity of the section if singly reinforced. Table A.5 shows $\rho_{\max}$ to be 0.0278, so $A_s = 0.0278 \times 10 \times 16 = 4.44 \text{ in}^2$. Then, with

$$a = \frac{4.44 \times 40}{0.85 \times 3 \times 10} = 6.96 \text{ in.}$$

the maximum nominal moment that can be developed is

$$M_n = 4.44 \times 40(16 - 3.48) = 2220 \text{ in-kips}$$

Alternatively, with $R = 869$ from Table A.6b, the nominal flexural strength is $M_n = 869 \times 10 \times 16^2/1000 = 2220$ in-kips. Since the corresponding design moment, $\phi M_n = 2000$ in-kips, is less than the required capacity, 2750 in-kips, compression steel is needed as well as additional tension steel. Assuming that $f'_s = f_y$ at failure, one has

$$M_1 = \frac{2750}{0.90} - 2220 = 836 \text{ in-kips}$$

$$A_{s1} = \frac{836}{40(16 - 2.5)} = 1.54 \text{ in}^2$$

giving the additional tension steel area required above that provided as the upper limit of the singly reinforced beam of the same concrete dimensions. This is also the compression steel requirement. Accordingly, the compression steel area will be

$$A'_s = 1.54 \text{ in}^2$$

and the tensile steel area is

$$A_s = 4.44 + 1.54 = 5.98 \text{ in}^2$$

The design should now be checked to confirm that the compressive bars would yield at failure as assumed. With $\rho' = 1.54/(10 \times 16) = 0.0096$, the limiting tensile steel ratio for yielding of the compression bars is found using Eq. (3.50*a*).

$$\bar{\rho}_{cy} = 0.85 \times 0.85 \frac{3}{40} \frac{2.5}{16} \frac{87}{47} + 0.0096 = 0.0253$$

The tentative steel ratio, $\rho = 5.98/(10 \times 16) = 0.0374$, is above the lower limit, ensuring that the compressive bars would yield as assumed.

Two No. 9 bars will be used for compression reinforcement, and six No. 9 bars will provide the required tensile steel area, as shown in Fig. 3.12. To place the tension bars within the 10 in. beam width, two rows of three bars each are used.

d. Tensile Steel below the Yield Stress

All doubly reinforced beams designed according to the ACI Code must be underreinforced, in the sense that the tensile steel ratio is limited to ensure yielding

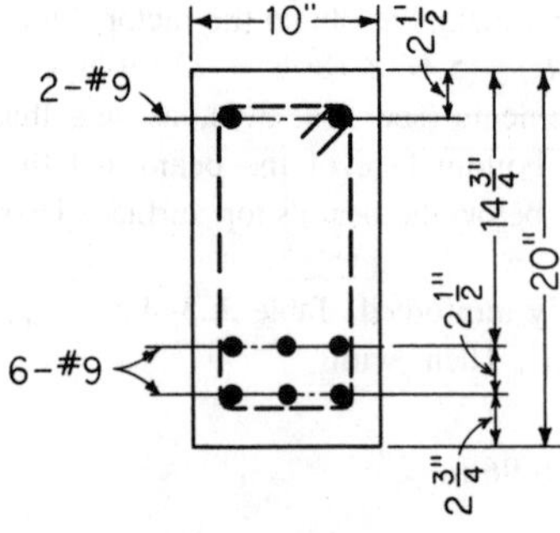

FIGURE 3.12
Doubly reinforced beam of Example 3.12.

at beam failure. Two cases were considered in Sec. 3.7a and 3.7b, respectively: (a) both tension steel and compression steel yield, and (b) tension steel yields but compression steel does not. Two other combinations may be encountered in reviewing capacity of existing beams: (c) tension steel does not yield, but compression steel does, and (d) neither tension steel nor compression steel yields. The last two cases are unusual, and in fact it would be difficult to place sufficient tension reinforcement to create such conditions, but it is possible. The solution in such cases is obtained as a simple extension of the treatment of Sec. 3.7b. An equation for horizontal equilibrium is written, in which both tension and compression steel stress are expressed in terms of the unknown neutral axis depth c. The resulting quadratic equation is solved for c, after which steel stresses can be calculated and the nominal flexural strength determined.

3.8 T BEAMS

With the exception of precast systems, reinforced concrete floors, roofs, decks, etc., are almost always monolithic. Forms are built for beam soffits and sides and for the underside of slabs, and the entire construction is poured at once, from the bottom of the deepest beam to the top of the slab. Beam stirrups and bent bars extend up into the slab. It is evident, therefore, that a part of the slab will act with the upper part of the beam to resist longitudinal compression. The resulting beam cross section is T-shaped rather than rectangular. The slab forms the beam flange, while the part of the beam projecting below the slab forms what is called the *web* or *stem*. The upper part of such a T beam is stressed laterally due to slab action in that direction. Although transverse compression at the level of the bottom of the slab may increase the longitudinal compressive strength by as much as 25 percent, transverse tension at the top surface reduces the longitudinal compression strength (see Sec. 2.9). Neither effect is usually taken into account in design.

a. Effective Flange Width

The next question to be resolved is that of the effective width of flange. In Fig. 3.13*a* it is evident that if the flange is but little wider than the stem width, the entire flange can be considered effective in resisting compression. For the floor system shown in Fig. 3.13*b*, however, it may be equally obvious that elements of the flange midway between the beam stems are less highly stressed in longitudinal

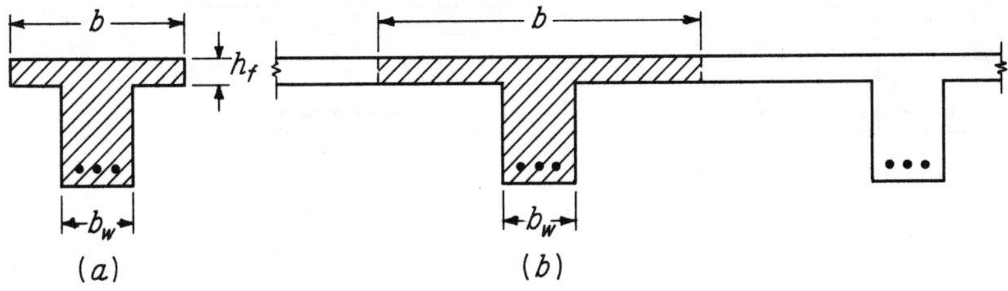

FIGURE 3.13
Effective flange width of T beams.

compression than those elements directly over the stem. This is so because of shearing deformation of the flange, which relieves the more remote elements of some compressive stress.

Although the actual longitudinal compression varies because of this effect, it is convenient in design to make use of an *effective flange width*, which may be smaller than the actual flange width but is considered to be uniformly stressed at the maximum value. This effective width has been found to depend primarily on the beam span and on the relative thickness of the slab.

The recommendations for effective width given in ACI Code 8.10 are as follows:

1. For symmetrical T beams the effective width b shall not exceed one-fourth the span length of the beam. The overhanging slab width on either side of the beam web shall not exceed 8 times the thickness of the slab nor go beyond one-half the clear distance to the next beam.
2. For beams having a slab on one side only, the effective overhanging slab width shall not exceed one-twelfth the span length of the beam, 6 times the slab thickness, or one-half the clear distance to the next beam.
3. For isolated beams in which the flange is used only for the purpose of providing additional compressive area, the flange thickness shall not be less than one-half the width of the web, and the total flange width shall not be more than 4 times the web width.

b. Strength Analysis

The neutral axis of a T beam may be either in the flange or in the web, depending upon the proportions of the cross section, the amount of tensile steel, and the strengths of the materials. If the calculated depth to the neutral axis is less than or equal to the slab thickness h_f, the beam can be analyzed as if it were a rectangular beam of width equal to b, the effective flange width. The reason for this is illustrated in Fig. 3.14a, which shows a T beam with neutral axis in the flange. The compressive area is indicated by the shaded portion of the figure. If the additional concrete indicated by areas 1 and 2 had been added when the beam

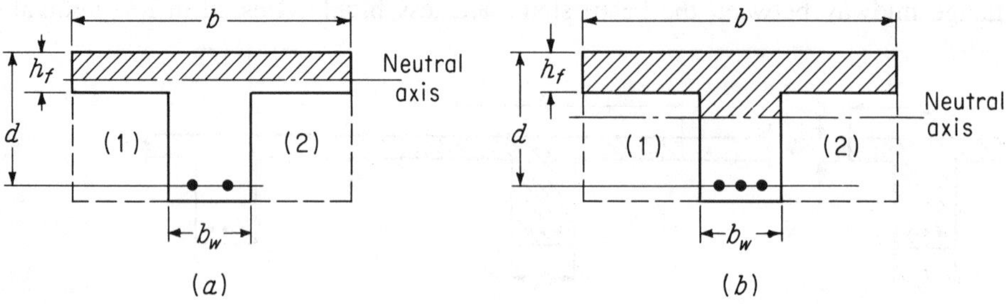

FIGURE 3.14
Effective cross sections of T beams.

was poured, the physical cross section would have been rectangular with a width b. No bending strength would have been added because areas 1 and 2 are entirely in the tension zone, and tension concrete is disregarded in flexural calculations. The original T beam and the rectangular beam are equal in flexural strength, and rectangular beam analysis for flexure applies.

When the neutral axis is in the web, as in Fig. 3.14*b*, the preceding argument is no longer valid. In this case, methods must be developed to account for the actual T-shaped compressive zone.

In treating T beams, it is convenient to adopt the same equivalent stress distribution that is used for beams of rectangular cross section. The rectangular stress block, having a uniform compressive-stress intensity $0.85f_c'$, was devised originally on the basis of tests of rectangular beams (see Sec. 3.4a), and its suitability for T beams may be questioned. However, extensive calculations based on actual stress-strain curves (reported in Ref. 3.11) indicate that its use for T beams, as well as for beams of circular or triangular cross section, introduces only minor error and is fully justified.

Accordingly, a T beam may be treated as a rectangular beam if the depth of the equivalent stress block is equal to or less than the flange thickness. Figure 3.15 shows a tensile-reinforced T beam with effective flange width b, web width b_w, effective depth to the steel centroid d, and flange thickness h_f. If for trial purposes the stress block is assumed to be completely within the flange,

$$a = \frac{A_s f_y}{0.85 f_c' b} = \frac{\rho f_y d}{0.85 f_c'} \tag{3.57}$$

where $\rho = A_s/bd$. If a is equal to or less than the flange thickness h_f, the member may be treated as a rectangular beam of width b and depth d. If a is greater than h_f, a T-beam analysis is required as follows.

It will be assumed that the strength of the T beam is controlled by yielding of the tensile steel. This will nearly always be the case because of the large compressive concrete area provided by the flange. In addition, an upper limit can be established for the steel ratio to ensure that this is so, as will be shown.

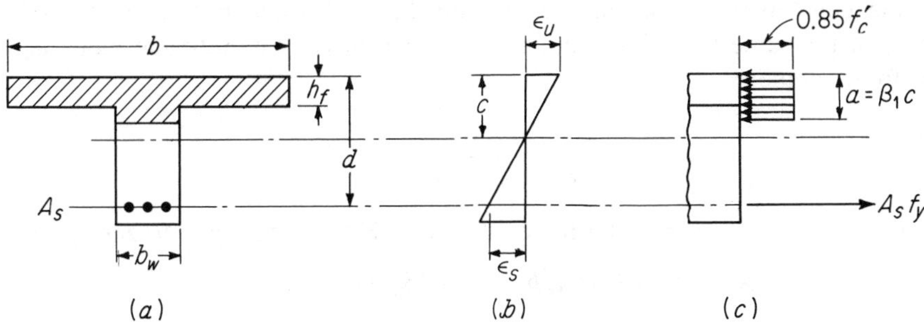

FIGURE 3.15
Strain and equivalent stress distributions for T beam.

As a computational device, it is convenient to divide the total tensile steel into two parts. The first part, A_{sf}, represents the steel area which, when stressed to f_y, is required to balance the longitudinal compressive force in the overhanging portions of the flange that are stressed uniformly at $0.85f'_c$. Thus,

$$A_{sf} = \frac{0.85f'_c(b - b_w)h_f}{f_y} \tag{3.58}$$

The force $A_{sf}f_y$ and the equal and opposite force $0.85f'_c(b - b_w)h_f$ act with a lever arm $d - h_f/2$ to provide the nominal resisting moment:

$$M_{n1} = A_{sf}f_y\left(d - \frac{h_f}{2}\right) \tag{3.59}$$

The remaining steel area, $A_s - A_{sf}$, at a stress f_y, is balanced by the compression in the rectangular portion of the beam. The depth of the equivalent rectangular stress block in this zone is found from horizontal equilibrium:

$$a = \frac{(A_s - A_{sf})f_y}{0.85f'_cb_w} \tag{3.60}$$

An additional moment M_{n2} is thus provided by the forces $(A_s - A_{sf})f_y$ and $0.85f'_cab_w$ acting at the lever arm $d - a/2$:

$$M_{n2} = (A_s - A_{sf})f_y\left(d - \frac{a}{2}\right) \tag{3.61}$$

and the total nominal resisting moment is the sum of the parts:

$$M_n = M_{n1} + M_{n2} = A_{sf}f_y\left(d - \frac{h_f}{2}\right) + (A_s - A_{sf})f_y\left(d - \frac{a}{2}\right) \tag{3.62}$$

This moment is reduced by the factor $\phi = 0.9$ in accordance with the safety provisions of the ACI Code to obtain the design strength.

As for rectangular beams, it is desired to ensure that the tensile steel will yield prior to sudden crushing of the compression concrete, as assumed in the preceding development. For balanced failure the steel strain in Fig. 3.15*b* will reach ϵ_y at the same time that the concrete strain reaches its ultimate value ϵ_u. Then, from geometry,

$$\frac{c}{d} = \frac{\epsilon_u}{\epsilon_u + \epsilon_y}$$

Setting the sum of the horizontal forces shown in Fig. 3.15*c* equal to zero gives

$$A_sf_y = 0.85\beta_1f'_cb_wc + 0.85f'_c(b - b_w)h_f$$

or

$$A_sf_y = 0.85\beta_1f'_cb_wc + A_{sf}f_y$$

Defining $\rho_w = A_s/b_w d$ and $\rho_f = A_{sf}/b_w d$ (i.e., expressing both total and partial steel areas in terms of the rectangular portion of the beam) results in the following balanced steel ratio ρ_{wb} for a T beam:

$$\rho_{wb} = 0.85\beta_1 \frac{f_c'}{f_y} \frac{\epsilon_u}{\epsilon_u + \epsilon_y} + \rho_f$$

The first term on the right-hand side of the last equation is simply the balanced steel ratio ρ_b for the rectangular portion of the beam, as comparison with Eq. (3.28*a*) will confirm. Thus the balanced steel ratio for a T beam is

$$\rho_{wb} = \rho_b + \rho_f \tag{3.63}$$

where all ratios are expressed in terms of the *rectangular portion* of the beam. To provide a margin against brittle failure of T beams, the ACI Code establishes that the steel ratio used shall not exceed

$$\rho_{w,\max} = 0.75(\rho_b + \rho_f) \tag{3.64}$$

The practical result of this restriction on tensile steel area is that the stress block of T beams will almost always be within the flange, except for unusual geometry or combinations of material strength. Consequently, rectangular beam equations may be applied in most cases.

The ACI Code restriction that the tensile steel ratio for beams must not be less than $\rho_{\min} = 200/f_y$ (see Sec. 3.4d) applies to T beams as well as rectangular beams. For T beams, the ratio ρ should be computed for this purpose based on the web width b_w.

c. Proportions of Cross Section

In the designing of T beams, as contrasted to reviewing the capacity of a given section, normally the slab dimensions and beam spacing will have been established by transverse flexural requirements. Consequently, the only additional section dimensions that must be determined from flexural considerations are the width and depth of the web and the tensile steel area.

If the stem dimensions were selected on the basis of concrete stress capacity in compression, they would be very small because of the large compression flange width furnished by the presence of the slab. Such a design would not represent the optimum solution because of the large tensile steel requirement resulting from the small effective depth, because of the excessive web reinforcement that would be required for shear, and because of large deflections associated with such a shallow member. It is better practice to select the proportions of the web (1) so as to keep an arbitrarily low web-steel ratio ρ_w or (2) so as to keep web-shear stress at desirably low limits or (3) for continuous T beams, on the basis of the flexural requirements at the supports, where the effective cross section is rectangular and of width b_w.

In addition to the main reinforcement calculated according to the preceding requirements, it is necessary to ensure the integrity of the compressive flange of T beams by providing steel in the flange in the direction transverse to the main span. In typical construction the slab steel serves this purpose. In other cases separate bars must be added to permit the overhanging flanges to carry, as cantilever beams, the loads directly applied. According to ACI Code 8.10.5, the spacing of such bars shall not exceed 5 times the thickness of the flange nor in any case exceed 18 in.

d. Examples of Review and Design of T Beams

For *reviewing* the capacity of a T beam with known concrete dimensions and tensile steel area, it is reasonable to start with the assumption that the stress block depth a does not exceed the flange thickness h_f. In that case, all ordinary rectangular beam equations (see Sec. 3.4) apply, with beam width taken equal to the effective width of the flange. If, upon checking that assumption, a proves to exceed h_f, then T beam analysis must be applied. Eqs. (3.58) through (3.62) can be utilized, in sequence, to obtain the nominal flexural strength, after which the design strength is easily calculated.

For *design*, the following sequence of calculations may be followed:

1. Establish flange thickness h_f based on flexural requirements of the slab, which normally spans transversely between parallel T beams.
2. Determine the effective flange width b according to ACI limits.
3. Choose web dimensions b_w and d based on either of the following:
   (*a*) negative bending requirements at the supports, if a continuous T beam
   (*b*) shear requirements, setting a reasonable upper-limit on the nominal unit shear stress v_u in the beam web (see Chapter 4)
4. With all concrete dimensions thus established, calculate a trial value of A_s, assuming that a does not exceed h_f, with beam width equal to flange width b. Use ordinary rectangular beam design methods.
5. For the trial A_s, check the depth of stress block a to confirm that it does not exceed h_f. If it should exceed that value, revise A_s using the T beam equations.
6. Check to ensure that $\rho_w \leq \rho_{w,\max}$. (This will almost invariably be the case.)
7. Check to ensure that $\rho_w \geq \rho_{w,\min} = 200/f_y$.

Example 3.13 Ultimate moment capacity of a given section. An isolated T beam is composed of a flange 28 in. wide and 6 in. deep poured monolithically with a web of 10 in. width which extends 24 in. below the bottom surface of the flange to produce a beam of 30 in. total depth. Tensile reinforcement consists of six No. 10 bars placed in two horizontal rows. The centroid of the bar group is 26 in. from the top of the beam. It has been determined that the concrete has a strength of 3000 psi and that the yield stress of the steel is 60,000 psi. What is the useful moment capacity of the beam?

Solution. It is easily confirmed that the flange dimensions are satisfactory according to the ACI Code for an isolated beam. The entire flange can be considered effective.

For six No. 10 bars, $A_s = 7.59$ in^2. First check the location of the neutral axis, on the assumption that rectangular beam equations may be applied,

$$\rho = \frac{7.59}{28 \times 26} = 0.0104$$

and from Eq. (3.57)

$$a = \frac{0.0104 \times 60 \times 26}{0.85 \times 3} = 6.37 \text{ in.}$$

This exceeds the flange thickness, and so a T beam analysis is required. From Eq. (3.58),

$$A_{sf} = 0.85 \times \frac{3}{60} \times 18 \times 6 = 4.59 \text{ in}^2$$

Hence

$$A_s - A_{sf} = 7.59 - 4.59 = 3.00 \text{ in}^2$$

The steel ratios are

$$\rho_w = \frac{7.59}{260} = 0.0292 \qquad \rho_f = \frac{4.59}{260} = 0.0177$$

while from Eq.(3.28*a*):

$$\rho_b = 0.85 \times 0.85 \times \frac{3}{60} \times \frac{87}{147} = 0.0214$$

According to the ACI Code, the maximum permitted tensile steel ratio is

$$\rho_{w,\max} = 0.75(0.0214 + 0.0177) = 0.0294$$

and comparison with ρ_w indicates that a ductile failure would be ensured. Then, from Eq. (3.59),

$$M_{n1} = 4.59 \times 60(26 - 3) = 6330 \text{ in-kips}$$

while from Eqs. (3.60) and (3.61)

$$a = \frac{3.00 \times 60}{0.85 \times 3 \times 10} = 7.07 \text{ in.}$$

$$M_{n2} = 3.00 \times 60(26 - 3.53) = 4050 \text{ in-kips}$$

When the ACI strength reduction factor is incorporated, the design strength is

$$\phi M_n = 0.90(6330 + 4050) = 9350 \text{ in-kips}$$

Example 3.14 Determination of steel area for a given moment. A floor system consists of a 3 in. concrete slab supported by continuous T beams of 24 ft span, 47 in. on centers. Web dimensions, as determined by negative-moment requirements at the supports, are $b_w = 11$ in. and $d = 20$ in. What tensile steel area is required at midspan to resist a moment of 6400 in-kips if $f_y = 60{,}000$ psi and $f_c' = 3000$ psi?

Solution. First determining the effective flange width,

$$16h_f + b_w = (16 \times 3) + 11 = 59 \text{ in.}$$

$$\frac{\text{Span}}{4} = 24 \times \frac{12}{4} = 72 \text{ in.}$$

$$\text{Centerline beam spacing} = 47 \text{ in.}$$

The centerline T beam spacing controls in this case, and $b = 47$ in. The concrete dimensions b_w and d are known to be adequate in this case, since they have been selected for the larger negative support moment applied to the effective rectangular section $b_w d$. The tensile steel at midspan is most conveniently found by trial. Assuming the stress-block depth equal to the flange thickness of 3 in., one gets

$$d - \frac{a}{2} = 20 - 1.50 = 18.50 \text{ in.}$$

Trial:

$$A_s = \frac{M_u}{\phi f_y(d - a/2)} = \frac{6400}{0.90 \times 60 \times 18.50} = 6.40 \text{ in}^2$$

$$\rho = \frac{A_s}{bd} = \frac{6.40}{47 \times 20} = 0.00681$$

$$a = \frac{\rho f_y d}{0.85 f_c'} = \frac{0.00681 \times 60 \times 20}{0.85 \times 3} = 3.20 \text{ in.}$$

Since a is greater than h_f, a T beam analysis is required.

$$A_{sf} = \frac{0.85 f_c'(b - b_w)h_f}{f_y} = \frac{0.85 \times 3 \times 36 \times 3}{60} = 4.58 \text{ in}^2$$

$$\phi M_{n1} = \phi A_{sf} f_y \left(d - \frac{h_f}{2}\right) = 0.90 \times 4.58 \times 60 \times 18.50 = 4570 \text{ in-kips}$$

$$\phi M_{n2} = M_u - \phi M_{n1} = 6400 - 4570 = 1830 \text{ in-kips}$$

Assume $a = 4.00$ in.:

$$A_s - A_{sf} = \frac{\phi M_{n2}}{\phi f_y(d - a/2)} = \frac{1830}{0.90 \times 60 \times 18.00} = 1.88 \text{ in}^2$$

Check:

$$a = \frac{(A_s - A_{sf})f_y}{0.85 f_c' b_w} = \frac{1.88 \times 60}{0.85 \times 3 \times 11} = 4.02 \text{ in.}$$

This is satisfactorily close to the assumed value of 4 in. Then

$$A_s = A_{sf} + (A_s - A_{sf}) = 4.58 + 1.88 = 6.46 \text{ in}^2$$

Checking to ensure that the maximum tensile steel ratio is not exceeded gives

$$\rho_w = \frac{6.46}{220} = 0.0294$$

$$\rho_f = \frac{4.58}{220} = 0.0208$$

$$\rho_b = 0.85 \times 0.85 \times \frac{3}{60} \times \frac{87}{147} = 0.0214$$

$$\rho_{w,\max} = 0.75(0.0214 + 0.0208) = 0.0316$$

indicating that the actual ρ_w is satisfactorily low.

The close agreement should be noted between the approximate tensile steel area of 6.40 in^2 found by assuming the stress-block depth equal to the flange thickness and the more exact value of 6.46 in^2 found by T beam analysis. The approximate solution would be satisfactory in most cases.

REFERENCES

3.1. H. Rusch, "Researches Toward a General Flexural Theory of Structural Concrete," *J. ACI*, vol. 32, no. 1, 1960, pp. 1–28.
3.2. L. B. Kriz, "Ultimate Strength Criteria for Reinforced Concrete," *J. Eng. Mech. Div. ASCE*, vol. 85, no. EM3, 1959, pp. 95–110.
3.3. L. B. Kriz and S. L. Lee, "Ultimate Strength of Overreinforced Beams," *Proc. ASCE*, vol. 86, no. EM3, 1960, pp. 95–106.
3.4. A. H. Mattock, L. B. Kriz, and E. Hogenstad, "Rectangular Concrete Stress Distribution in Ultimate Strength Design," *J. ACI*, vol. 32, no. 8, 1961, pp. 875–928.
3.5. P. H. Kaar, N. W. Hanson, and H. T. Capell, "Stress-Strain Curves and Stress Block Coefficients for High-Strength Concrete," *Proceedings Douglas McHenry Symposium*, ACI Special Publication SP-55, 1978.
3.6. *Design Handbook Vol. 1—Beams, One-Way Slabs, Brackets, Footings, and Pile Caps*, ACI Special Publication SP-17(84), American Concrete Institute, Detroit, 1984.
3.7. *ACI Detailing Manual*, ACI Special Publication SP-66(88), American Concrete Institute, Detroit, 1988.
3.8. *CRSI Handbook*, 6th ed., Concrete Reinforcing Steel Institute, Schaumburg, Illinois, 1984.
3.9. *Economical Concrete Construction*, Engineering Data Report No. 30, Concrete Reinforcing Steel Institute, Schaumburg, Illinois, 1988.
3.10. *Manual of Standard Practice*, 24th ed., Concrete Reinforcing Steel Institute, Schaumburg, Illinois, 1986.
3.11. C. W. Dolan, *Ultimate Capacity of Reinforced Concrete Sections Using a Continuous Stress-Strain Function*, MS Thesis, Cornell University, Ithaca, New York, June 1967.

PROBLEMS

3.1 A rectangular beam made using concrete with compressive strength $f'_c = 4000$ psi and steel with $f_y = 60{,}000$ psi has width $b = 24$ in., total depth $h = 18$ in., and effective depth $d = 15.5$ in. Concrete modulus of rupture $f_r = 475$ psi. The elastic modulus of the steel and concrete are, respectively, 29,000,000 psi and 3,600,000 psi. The tensile steel area is A_s = five No. 11 bars.

(*a*) Find the maximum service load moment that can be resisted without stressing the concrete higher than $0.45f'_c$ or the steel above $0.40f_y$,

(*b*) Determine the nominal flexural strength of the beam section, and calculate the overall factor of safety against flexural failure, (*c*) Determine whether this beam will show flexural cracking before reaching the service load calculated in (*a*).

3.2 A rectangular, tension-reinforced beam is to be designed for dead load of 500 lb/ft plus self weight and service live load of 1200 lb/ft, with a 22 ft simple span. Material strengths will be $f_y = 60$ ksi and $f_c' = 3$ ksi for steel and concrete respectively. The total beam depth must not exceed 16 in. Calculate the required beam width and tensile steel requirement, using a steel ratio of $0.5\rho_b$. Use ACI load factors and strength reduction factors. The effective depth may be assumed at 2.5 in. less than the total depth.

3.3 A beam of 20 ft simple span has cross-section dimensions $b = 10$ in., $d = 23$ in., and $h = 25$ in. (see Fig. 3.2*b* for notation). It carries a uniform service load of 2450 lb/ft in addition to its own weight.

(*a*) Check whether this beam, if reinforced with three No. 8 bars, is adequate to carry this load with a minimum factor of safety against failure of 1.85. If this requirement is not met, select a three-bar reinforcement of diameter or diameters adequate to provide this safety.

(*b*) Determine the maximum stress in the steel and in the concrete under service load, i.e., when the beam carries its own weight and the specified uniform load.

(*c*) Will the beam show hairline cracks on the tension side under service load? Material strengths are $f_c' = 4000$ psi and $f_y = 60{,}000$ psi. Assume a weight of 150 pcf for reinforced concrete.

3.4 A rectangular reinforced concrete beam has dimensions $b = 12$ in., $d = 21$ in., and $h = 24$ in., and is reinforced with three No. 10 bars. Material strengths are $f_y = 60{,}000$ psi and $f_c' = 4000$ psi.

(*a*) Find the moment that will produce the first cracking at the bottom surface of the beam, basing your calculation on I_g, the moment of inertia of the gross concrete section.

(*b*) Repeat the calculation using I_{ut}, the moment of inertia of the uncracked transformed section.

(*c*) Determine the maximum moment that can be carried without stressing the concrete beyond $0.45f_c'$ or the steel beyond $0.40f_y$.

(*d*) Find the nominal flexural strength and design strength of this beam.

3.5 A tensile-reinforced beam has $b = 10$ in. and $d = 20$ in. to the center of the bars, which are placed all in one row. If $f_y = 60{,}000$ psi and $f_c' = 4000$ psi, find the nominal flexural strength M_n for (*a*) A_s = two No. 8 bars, (*b*) A_s = two No. 10 bars, (*c*) A_s = three No. 10 bars.

3.6 A singly reinforced rectangular beam is to be designed, with effective depth approximately 1.5 times the width, to carry a service live load of 1500 lb/ft in addition to its own weight, on a 24 ft simple span. The ACI Code load factors are to be applied as usual. With $f_y = 40{,}000$ psi and $f_c' = 4000$ psi, determine the required concrete dimensions b, d, and h, and steel reinforcing bars (*a*) for $\rho = 0.40\rho_b$ and (*b*) for $\rho = \rho_{max}$. Include a sketch of each cross section drawn to scale. Allow for No. 3 stirrups. Comment on your results.

3.7 A four-span continuous beam of constant rectangular section is supported at A, B, C, D, and E. Factored moments resulting from analysis are

At supports, ft-kips	At midspan, ft-kips
$M_a = 92$	$M_{ab} = 105$
$M_b = 147$	$M_{bc} = 92$
$M_c = 134$	$M_{cd} = 92$
$M_d = 147$	$M_{de} = 105$
$M_e = 92$	

Determine the required concrete dimensions for this beam, using $d = 1.75b$, and find the required reinforcement for all critical moment sections. Use a maximum steel ratio of $\rho = 0.50\rho_b$, $f_y = 60{,}000$ psi, and $f'_c = 5000$ psi.

3.8 A two-span continuous concrete beam is to be supported by three masonry walls spaced 25 ft on centers. A service live load of 1.5 kips/ft is to be carried, in addition to the self-weight of the beam. A constant rectangular cross section is to be used, with $h = 2b$, but reinforcement is to be varied according to requirements. Find the required concrete dimensions and reinforcement at all critical sections. Allow for No. 3 stirrups. Include sketches, drawn to scale, of critical cross sections. Use $f'_c = 3000$ psi and $f_y = 40{,}000$ psi.

3.9 A rectangular concrete beam measures 12 in. wide and has an effective depth of 18 in. Compression steel consisting of two No. 8 bars is located 2.5 in. from the compression face of the beam. If $f'_c = 4000$ psi and $f_y = 60{,}000$ psi, what is the design moment capacity of the beam, according to the ACI Code, for the following alternative tensile steel areas: (*a*) A_s = three No. 10 bars in one layer, (*b*) A_s = four No. 10 bars in two layers, (*c*) A_s = six No. 10 bars in two layers? (*Note*: Check for yielding of compression steel in each case.)

3.10 A rectangular concrete beam of width $b = 24$ in. is limited by architectural considerations to a maximum total depth $h = 16$ in. It must carry a total factored load moment $M_u = 400$ ft-kips. Design the flexural reinforcement for this member, using compression steel if necessary. Allow 3 in. to the center of the bars from the compression or tension face of the beam. Material strengths are $f_y = 60{,}000$ psi and $f'_c = 4000$ psi. Select rebars to provide the needed areas, and show a sketch of your final design, including provision for No. 4 stirrups.

3.11 A rectangular beam with width $b = 24$ in., total depth $h = 14$ in., and effective depth to the tensile steel $d = 11.5$ in. is constructed using materials with strengths $f'_c = 4000$ psi and $f_y = 60{,}000$ psi. Tensile reinforcement consists of two No. 11 bars plus three No. 10 bars in one row. Compression reinforcement consisting of two No. 10 bars is placed at distance $d' = 2.5$ in. from the compression face. Calculate the nominal and design strengths of the beam (*a*) neglecting the compression reinforcement, (*b*) accounting for the compression reinforcement and assuming it acts at f_y, and (*c*) accounting for the compression reinforcement working at its actual stress f'_s, established by analysis.

3.12 A tensile-reinforced T beam is to be designed to carry a uniformly distributed load on a 20 ft simple span. The total moment to be carried is $M_u = 5780$ in-kips. Concrete dimensions, governed by web shear and clearance requirements, are $b = 20$ in., $b_w = 10$ in., $h_f = 5$ in., and $d = 20$ in. If $f_y = 60$ ksi and $f'_c = 4$ ksi, what tensile reinforcement is required at midspan? Select appropriate rebars to provide this area and check concrete cover limitations, assuming No. 3 stirrups. What total depth h is required?

3.13 A concrete floor system consists of parallel T beams spaced 10 ft on centers and spanning 32 ft between supports. The 6 in. thick slab is poured monolithically with T beam webs having width $b_w = 14$ in. and total depth, measured from the top of slab, of $h = 28$ in. The effective depth will be taken 3 in. less than the total depth. In addition to its own weight, each T beam must carry a superimposed dead load of 50 psf and service live load of 225 psf. Material strengths are $f_y = 60{,}000$ psi and $f'_c = 3000$ psi. Determine the required tensile steel area and select the needed rebars for a typical member.

3.14 A single precast T beam is to be used as a bridge over a small roadway. Concrete dimensions are $b = 48$ in., $b_w = 10$ in., $h_f = 5$ in., and $h = 25$ in. The effective depth $d = 20$ in. Concrete and steel strengths are 4000 psi and 60,000 psi, respectively. Using approximately one-half the maximum tensile reinforcement permitted by the ACI Code (select the actual size of bar and number to be used), determine the design moment capacity of the girder. If the beam is used on a 30 ft simple span, and if in addition to its own weight it must support railings, curbs, and suspended loads totaling 0.475 kips/ft, what uniform service live load limit should be posted?

CHAPTER 4

SHEAR AND DIAGONAL TENSION IN BEAMS

4.1 INTRODUCTION

The preceding chapter dealt with the flexural behavior and flexural strength of beams. Beams must also have an adequate safety margin against other types of failure, some of which may be more dangerous than flexural failure. This may be so because of greater uncertainty in predicting certain other modes of collapse, or because of the catastrophic nature of some other types of failure, should they occur.

Shear failure of reinforced concrete, more properly called *diagonal tension failure*, is one example. Shear failure is difficult to predict accurately. In spite of many decades of experimental research (Refs. 4.1 to 4.5) and the use of highly sophisticated analytical tools (Ref. 4.6), it is not yet fully understood. Furthermore, if a beam without properly designed shear reinforcement is overloaded to failure, shear collapse is likely to occur suddenly, with no advance warning of distress. This is in strong contrast with the nature of flexural failure. For typically underreinforced beams, flexural failure is initiated by gradual yielding of the tension steel, accompanied by obvious cracking of the concrete and large deflections, giving ample warning and providing the opportunity to take corrective measures. Because of these differences in behavior, reinforced concrete beams are generally provided with special *shear reinforcement* to ensure that flexural failure would occur before shear failure if the member should be severely overloaded.

Figure 4.1 shows a shear-critical beam tested at Cornell University under third-point loading. With no shear reinforcement provided, the member failed immediately upon formation of the critical crack in the high-shear region near the right support.

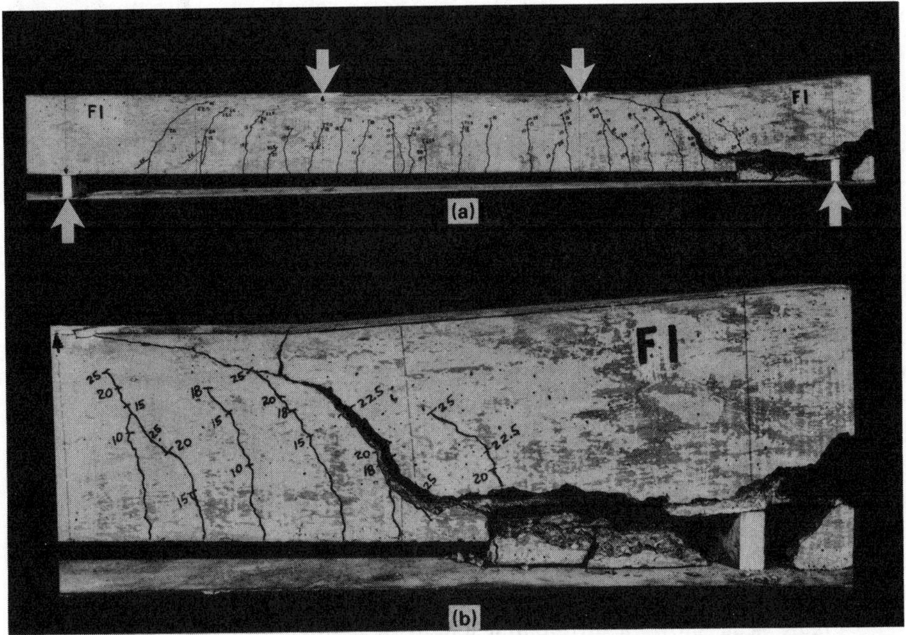

FIGURE 4.1
Shear failure of reinforced concrete beam: (*a*) overall view, (*b*) detail near right support.

It is important to realize that shear analysis and design is not really concerned with shear as such. The shear stresses in most beams are far below the direct shear strength of the concrete. The real concern is with *diagonal tension stress*, resulting from the combination of shear stress and longitudinal flexural stress. Most of this chapter deals with analysis and design for diagonal tension. Members without web reinforcement are studied first, in order to establish the location and orientation of cracks and the diagonal cracking load. Methods are then developed for the design of shear reinforcement, both in ordinary beams and in special types of members such as deep beams.

There are some circumstances, however, in which consideration of direct shear is appropriate. One example is in the design of composite members combining precast beams with a cast-in-place top slab. Horizontal shear stresses on the interface between components are important. The shear-friction theory, useful in this and other cases, will be presented following development of methods for the analysis and design of beams for diagonal tension.

4.2 DIAGONAL TENSION IN HOMOGENEOUS ELASTIC BEAMS

The stresses acting in homogeneous beams were briefly reviewed in Sec. 3.2. It was pointed out that when the material is elastic (stresses proportional to strains), shear stresses

$$v = \frac{VQ}{Ib} \tag{3.4}$$

act in any section in addition to the bending stresses

$$f = \frac{My}{I} \tag{3.2}$$

except for those locations at which the shear force V happens to be zero.

The role of shear stresses is easily visualized by the performance under load of the laminated beam of Fig. 4.2; it consists of two rectangular pieces bonded together along the contact surface. If the adhesive is strong enough, the member will deform as one single beam, as shown in Fig. 4.2*a*. On the other hand, if the adhesive is weak, the two pieces will separate and slide relative to each other, as shown in Fig. 4.2*b*. Evidently, then, when the adhesive is effective, there are forces or stresses acting in it which prevent this sliding or shearing. These horizontal shear stresses are shown in Fig. 4.2*c* as they act, separately, on the top and bottom pieces. The same stresses occur in horizontal planes in single-piece beams; they are different in intensity at different distances from the neutral axis.

Figure 4.2*d* shows a differential length of a single-piece rectangular beam acted upon by a shear force of magnitude V. Upward translation is prevented, i.e., vertical equilibrium is provided, by the vertical shear stresses v. Their average value is equal to the shear force divided by the cross-sectional area $v_{av} = V/ab$, but their intensity varies over the depth of the section. As is easily computed

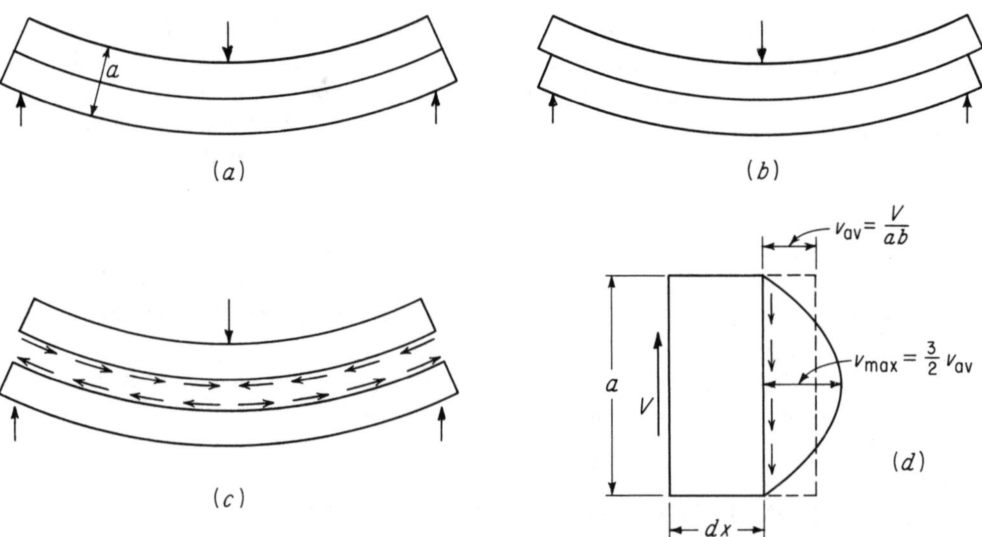

FIGURE 4.2
Shear in homogeneous rectangular beams.

from Eq. (3.4), the shear stress is zero at the outer fibers and has a maximum of $1.5v_{av}$ at the neutral axis, the variation being parabolic as shown. Other values and distributions are found for other shapes of the cross section, the shear stress always being zero at the outer fibers and of maximum value at the neutral axis. If a small square element located at the neutral axis of such a beam is isolated, as shown in Fig. 4.3*b*, the vertical shear stresses on it, equal and opposite on the two faces for reasons of equilibrium, act as shown. However, if these were the only stresses present, the element would not be in equilibrium; it would spin. Therefore, on the two horizontal faces there exist equilibrating horizontal shear stresses of the same magnitude. That is, at any point on the beam, the horizontal shear stresses of Fig. 4.3*b* are equal in magnitude to the vertical shear stresses of Fig. 4.2*d*.

It is proved in any strength-of-materials text that on an element cut at 45° these shear stresses combine in such a manner that their effect is as shown in Fig. 4.3*c*. That is, the action of the two pairs of shear stresses on the vertical and horizontal faces is the same as that of two pairs of normal stresses, one tension and one compression, acting on the 45° faces and of numerical value equal to that of the shear stresses. If an element of the beam is considered that is located neither at the neutral axis nor at the outer edges, its vertical faces are subject not only to the shear stresses but also to the familiar bending stresses whose magnitude is given by Eq. (3.2) (Fig. 4.3*d*). The six stresses that now act on the element

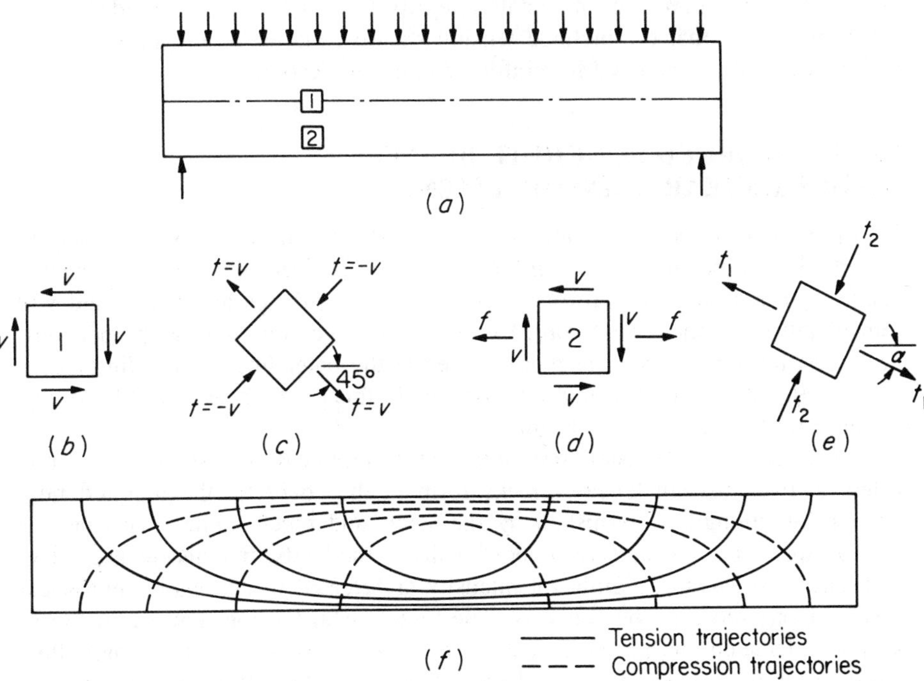

FIGURE 4.3
Stress trajectories in homogeneous rectangular beam.

can again be combined into a pair of inclined compression stresses and a pair of inclined tension stresses that act at right angles to each other. They are known as *principal stresses* (Fig. 4.3*e*). Their value, as mentioned in Sec. 3.2, is given by

$$t = \frac{f}{2} \pm \sqrt{\frac{f^2}{4} + v^2} \tag{3.1}$$

and their inclination α by $\tan 2\alpha = 2v/f$.

Since the magnitudes of the shear stresses v and the bending stresses f change both along the beam and vertically with distance from the neutral axis, the inclinations as well as the magnitudes of the resulting principal stresses t also vary from one place to another. Figure 4.3*f* shows the inclinations of these principal stresses for a rectangular beam uniformly loaded. That is, these stress trajectories are lines which, at any point, are drawn in that direction in which the particular principal stress, tension or compression, acts at that point. It is seen that at the neutral axis the principal stresses in a beam are always inclined at 45° to the axis. In the vicinity of the outer fibers they are horizontal near midspan.

An important point follows from this discussion. Tension stresses, which are of particular concern in view of the low tensile strength of the concrete, are not confined to the horizontal bending stresses f which are caused by bending alone. Tension stresses of various inclinations and magnitudes, resulting from shear alone (at the neutral axis) or from the combined action of shear and bending, exist in all parts of a beam and can impair its integrity if not adequately provided for. It is for this reason that the inclined tension stresses, known as *diagonal tension*, must be carefully considered in reinforced concrete design.

4.3 REINFORCED CONCRETE BEAMS WITHOUT SHEAR REINFORCEMENT

The discussion of shear in a homogeneous elastic beam applies very closely to a plain concrete beam *without* reinforcement. As the load is increased in such a beam, a tension crack will form where the tension stresses are largest and will immediately cause the beam to fail. Except for beams of very unusual proportions, the largest tension stresses are those caused at the outer fiber by bending alone, at the section of maximum bending moment. In this case, shear has little, if any, influence on the strength of a beam.

However, when tension reinforcement is provided, the situation is quite different. Even though tension cracks form in the concrete, the required flexural tension strength is furnished by the steel, and much higher loads can be carried. Shear stresses increase proportionally to the loads. In consequence, diagonal tension stresses of significant intensity are created in regions of high shear forces, chiefly close to the supports. The longitudinal tension reinforcement has been so calculated and placed that it is chiefly effective in resisting longitudinal tension near the tension face. It does not reinforce the tensionally weak concrete against the diagonal tension stresses that occur elsewhere, caused by shear alone or by the combined effect of shear and flexure. Eventually, these stresses attain magnitudes sufficient to open additional tension cracks in a direction perpendicular to the local tension stress. These are known as *diagonal* cracks, in distinction to the vertical flexural cracks. The latter occur in regions of large moments, the former in regions in which the shear forces are high. In beams in which no reinforcement is provided to counteract the formation of large diagonal tension cracks, their appearance has far-reaching and detrimental effects. For this reason methods of predicting the loads at which these cracks will form are desired.

a. Criteria for Formation of Diagonal Cracks

It is seen from Eq. (3.1) that the diagonal tension stresses t represent the combined effect of the shear stresses v and the bending stresses f. These in turn are, respectively, proportional to the shear force V and the bending moment M at the particular location in the beam [Eqs (3.2) and (3.4)]. Depending on configuration, support conditions, and load distribution, a given location in a beam may have a large moment combining with a small shear force, or the reverse, or large or small values for both shear and moment. Evidently, the relative values of M and V will affect the magnitude as well as the direction of the diagonal tension stresses. Figure 4.4 shows a few typical beams and their moment and shear diagrams and

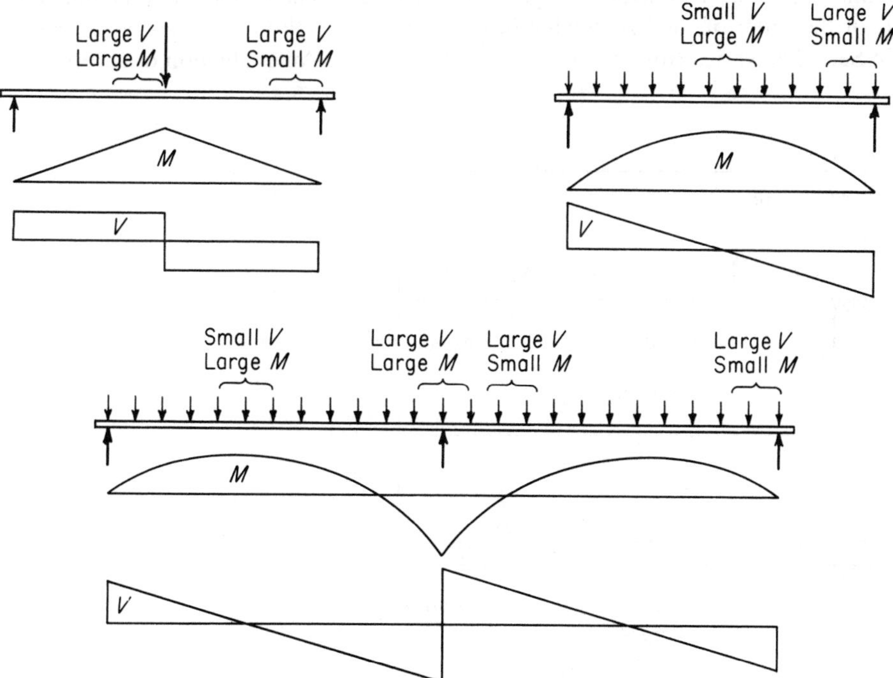

FIGURE 4.4
Typical locations of critical combinations of shear and moment.

draws attention to locations at which various combinations of high or low V and M occur.

At a location of large shear force V and small bending moment M, there will be little flexural cracking, if any, prior to the development of a diagonal tension crack. Consequently, the average shear stress prior to crack formation is

$$v = \frac{V}{bd} \tag{4.1}$$

The exact distribution of these shear stresses over the depth of the cross section is not known. It cannot be computed from Eq. (3.4) because this equation does not account for the influence of the reinforcement and because concrete is not an elastic homogeneous material. The value computed from Eq. (4.1) must therefore be regarded merely as a measure of the average intensity of shear stresses in the section. The maximum value, which occurs at the neutral axis, will exceed this average by an unknown but moderate amount.

If flexural stresses are negligibly small at the particular location, the diagonal tension stresses, as in Fig. 4.3b and c, are inclined at about 45° and are numerically equal to the shear stresses, with a maximum at the neutral axis. Consequently, diagonal cracks form mostly at or near the neutral axis and propogate from that location, as shown in Fig. 4.5a. These so-called *web-shear* cracks can be expected to form when the diagonal tension stress in the vicinity of the neutral axis becomes equal to the tension strength of the concrete. The former, as was indicated, is of the order of, and somewhat larger than, $v = V/bd$; the latter, as discussed in Sec. 2.8, varies from about $3\sqrt{f_c'}$ to about $5\sqrt{f_c'}$. An evaluation of a very large

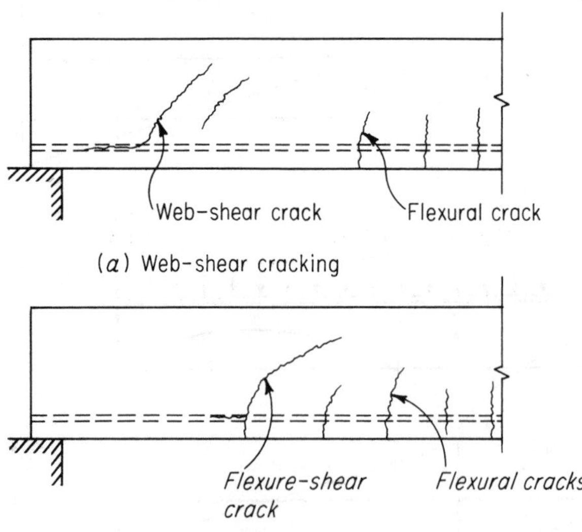

FIGURE 4.5
Diagonal tension cracking in reinforced concrete beams.

number of beam tests is in fair agreement with this reasoning (Ref. 4.1). It was found that in regions with large shear and small moment, diagonal tension cracks form at an average or nominal shear stress v_{cr} of about $3.5\sqrt{f'_c}$, that is,

$$v_{cr} = \frac{V_{cr}}{bd} = 3.5\sqrt{f'_c} \tag{4.2a}$$

where V_{cr} is that shear force at which the formation of the crack was observed.† Web-shear cracking is relatively rare and occurs chiefly near supports of deep, thin-webbed beams or at inflection points of continuous beams.

The situation is different when both the shear force and the bending moment have large values. At such locations, in a well-proportioned and reinforced beam, flexural tension cracks form first. Their width and length are well controlled and kept small by the presence of longitudinal reinforcement. However, when the diagonal tension stress at the upper end of one or more of these cracks exceeds the tensile strength of the concrete, the crack bends in a diagonal direction and continues to grow in length and width (see Fig. 4.5*b*). These cracks are known as *flexure-shear* cracks and are more common than web-shear cracks.

It is evident that at the instant at which a diagonal tension crack of this type develops, the average shear stress is larger than that given by Eq. (4.1). This is so because the preexisting tension crack has reduced the area of uncracked concrete which is available to resist shear to a value smaller that that of the uncracked area bd used in Eq. (4.1). The amount of this reduction will vary, depending on the unpredictable length of the preexisting flexural tension crack. Furthermore, the simultaneous bending stress f combines with the shear stress v to increase the diagonal tension stress t further [see Eq. (3.1)]. No way has been found to calculate reliable values of the diagonal tension stress under these conditions, and recourse must be had to test results.

A large number of beam tests have been evaluated for this purpose (Ref. 4.1). They show that in the presence of large moments (for which adequate longitudinal reinforcement has been provided) the nominal shear stress at which diagonal tension cracks form and propagate is conservatively given by

$$v_{cr} = \frac{V_{cr}}{bd} = 1.9\sqrt{f'_c} \tag{4.2b}$$

Comparison with Eq. (4.2*a*) shows that large bending moments can reduce the shear force at which diagonal cracks form to roughly one-half the value at which they would form if the moment were zero or nearly so. This is in qualitative agreement with the discussion just given.

It is evident, then, that the shear at which diagonal cracks develop depends on the ratio of shear force to bending moment, or, more precisely, on the ratio of

---

† Actually, diagonal tension cracks form at places where a compression stress acts in addition to and perpendicular to the diagonal tension stress, as shown in Fig. 4.3*d* and *e*. The crack therefore occurs at a location of biaxial stress rather than uniaxial tension. However, the effect of this simultaneous compression stress on the cracking strength appears to be small, in agreement with the information in Fig. 2.8.

shear stress v to bending stress f at the top of the flexural crack. Neither of these can be accurately calculated. It is clear, though, that $v = K_1(V/bd)$, where, by comparison with Eq. (4.1), constant K_1 depends chiefly on the depth of penetration of the flexural crack. On the other hand [see Eq. (3.10)], $f = K_2(M/bd^2)$, where K_2 also depends on crack configuration. Hence, the ratio

$$\frac{v}{f} = \frac{K_1}{K_2}\frac{Vd}{M}$$

must be expected to affect that load at which flexural cracks develop into flexure-shear cracks, the unknown quantity K_1/K_2 to be explored by tests. Equation (4.2*a*) gives the cracking shear for very large values of Vd/M, and Eq. (4.2*b*) for very small values. Moderate values of Vd/M result in magnitudes of v_{cr} intermediate between these extremes. Again, from evaluations of large numbers of tests (Ref. 4.1), it has been found that the nominal shear stress at which diagonal flexure-shear cracking develops is conservatively predicted from

$$v_{cr} = \frac{V_{cr}}{bd} = 1.9\sqrt{f_c'} + 2500\frac{\rho Vd}{M} \leq 3.5\sqrt{f_c'} \tag{4.2c}$$

where

$$V_{cr} = v_{cr}bd \tag{4.2d}$$

and $\rho = A_s/bd$, as before, and 2500 is an empirical constant in psi units. A graph of this relation and comparison with test data is given in Fig. 4.6.

Apart from the influence of Vd/M, it is seen from Eq. (4.2*c*) that increasing amounts of tension reinforcement, i.e., increasing values of the steel ratio ρ, have a beneficial effect in that they increase the shear at which diagonal cracks develop. This is so because larger amounts of longitudinal steel result in smaller and nar-

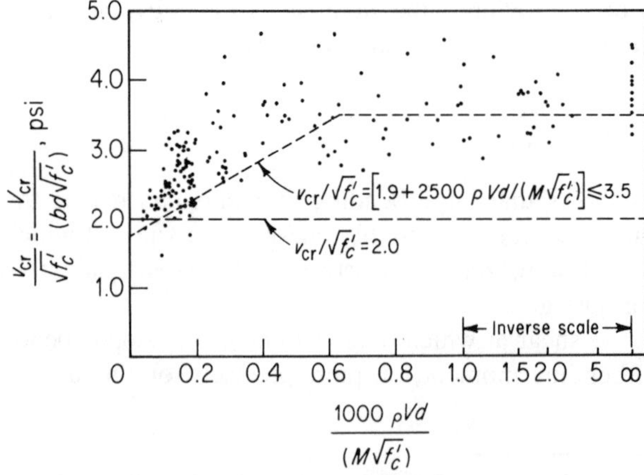

FIGURE 4.6
Correlation of Eq. (4.2*c*) with test results.

rower flexural tension cracks prior to the formation of diagonal cracking, leaving a larger area of uncracked concrete available to resist shear. [For more details on the development of Eq. (4.2*c*), see Ref. 4.1.]

More recent tests, summarized in Refs. 4.7 and 4.8, indicate that the stress at which flexure-shear cracks form in the presence of significant moments also depends on the tension-reinforcement ratio ρ in addition to the primary variable $\sqrt{f_c'}$ [see Eq. (4.2*b*)]. This is understandable because larger amounts of longitudinal steel more effectively counteract the widening of flexural cracks and their development into flexure-shear cracks. Correspondingly, the empirical equation for flexure-shear cracking, proposed in Ref. 4.7 to replace $v_{cr} = 1.9\sqrt{f_c'}$ in Eq. (4.2*b*), is

$$v_{cr} = (0.8 + 120\rho)\sqrt{f_c'} \tag{4.3}$$

but not greater than $2.3\sqrt{f_c'}$ and not less than $1.0\sqrt{f_c'}$. For steel ratios ρ smaller than about 0.009, Eq. (4.3) gives values progressively smaller than Eq. (4.2*b*), while for ρ larger than about 0.012 the reverse is true. Code changes proposed on this basis in Refs. 4.7 and 4.8 would eliminate Eq. (4.2*c*). They would take account of the effect of Vd/M only for unusually deep beams or when sizable concentrated loads are located close to supports.

b. Behavior of Diagonally Cracked Beams

In regard to flexural, as distinct from diagonal, tension, it was explained in Sec. 3.3 that cracks on the tension side of a beam are permitted to occur and are in no way detrimental to the strength of the member. One might expect a similar situation in regard to diagonal cracking caused chiefly by shear. The analogy, however, is not that simple. Flexural tension cracks are harmless only because adequate longitudinal reinforcement has been provided to resist the flexural tension stresses that the cracked concrete is no longer able to transmit. In contrast, the beams now being discussed, although furnished with the usual longitudinal reinforcement, are not equipped with any other reinforcement to offset the effects of diagonal cracking. This makes the diagonal cracks much more decisive in subsequent performance and strength of the beam than the flexural cracks.

Two types of behavior have been observed in the many tests on which present knowledge is based:

1. The diagonal crack, once formed, spreads either immediately or at only slightly higher load, traversing the entire beam from the tension reinforcement to the compression face, splitting it in two and failing the beam. This process is sudden and without warning and occurs chiefly in the shallower beams, i.e., beams with span-depth ratios of about 8 or more. Beams in this range of dimensions are very common. Complete absence of shear reinforcement would make them very vulnerable to accidental large overloads, which would result in catastrophic failures without warning. For this reason it is good practice to

provide a minimum amount of shear reinforcement even if calculation does not require it, because such reinforcement restrains growth of diagonal cracks, thereby increasing ductility and providing warning in advance of actual failure. Only in situations where an unusually large safety factor against inclined cracking is provided, i.e., where actual shear stresses are very small compared with v_{cr}, as in some slabs and most footings, is it permissible to omit shear reinforcement.

2. Alternatively, the diagonal crack, once formed, spreads toward and partially into the compression zone but stops short of penetrating to the compression face. In this case no sudden collapse occurs, and the failure load may be significantly higher than that at which the diagonal crack first formed. This behavior is chiefly observed in the deeper beams with smaller span-depth ratios and will be analyzed now.

Figure 4.7*a* shows a portion of a beam, arbitrarily loaded, in which a diagonal tension crack has formed. Consider the part of the beam to the left of the crack, shown in solid lines. There is an external upward shear force $V_{ext} = R_l - P_1$ acting on this portion.

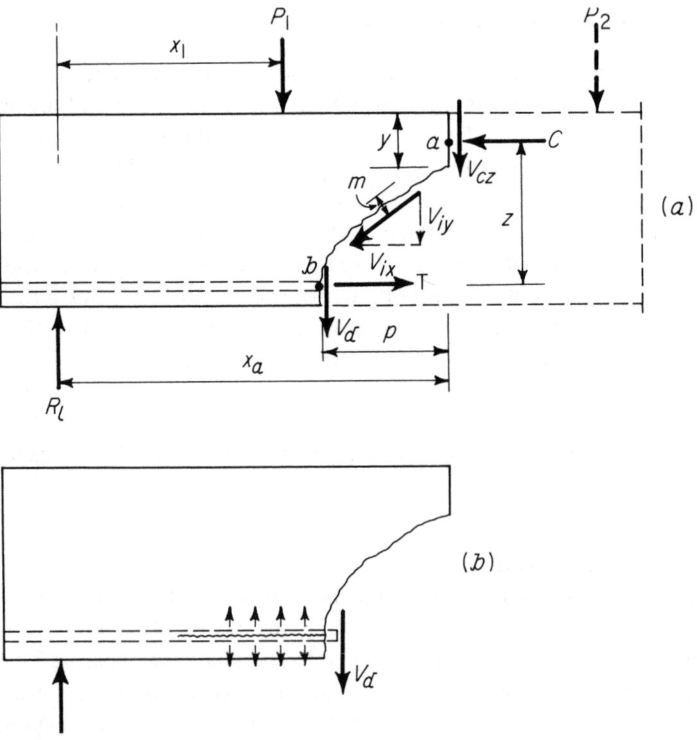

FIGURE 4.7
Forces at a diagonal crack in a beam without web reinforcement.

Once a crack is formed, no tension force perpendicular to the crack can be transmitted across it. However, as long as the crack is narrow, it can still transmit forces in its own plane through interlocking of the surface roughnesses. Sizable interlock forces V_i of this kind have in fact been measured, amounting to one-third and more of the total shear force. The components V_{ix} and V_{iy} of V_i are shown in Fig. 4.7*a*. The other internal vertical forces are those in the uncracked portion of the concrete, V_{cz}, and across the longitudinal steel, acting as a dowel, V_d. Thus, the internal shear force is

$$V_{\text{int}} = V_{cz} + V_d + V_{iy}$$

Equilibrium requires that $V_{\text{int}} = V_{\text{ext}}$ so that the part of the shear resisted by the uncracked concrete is

$$V_{cz} = V_{\text{ext}} - V_d - V_{iy} \tag{4.4}$$

In a beam provided with longitudinal reinforcement only, the portion of the shear force resisted by the steel in dowel action is usually quite small. In fact, the reinforcing bars on which the dowel force V_d acts are supported against vertical displacement chiefly by the thin concrete layer below. The bearing pressure caused by V_d creates, in this concrete, vertical tension stresses as shown in Fig. 4.7*b*. Because of these stresses, diagonal cracks often result in splitting of the concrete along the tension reinforcement, as shown. (See also Fig. 4.1.) This reduces the dowel force V_d and also permits the diagonal crack to widen. This, in turn, reduces the interface force V_i and frequently leads to immediate failure.

Next consider moments about points a at the intersection of V_{cz} and C; the external moment $M_{\text{ext},a}$ acts at a and happens to be $R_t x_a - P_1(x_a - x_1)$ for the loading shown. The internal moment is

$$M_{\text{int},a} = T_b z + V_d p - V_i m$$

Here p is the horizontal projection of the diagonal crack and m is the moment arm of the force V_i with respect to point a. The designation T_b for T is meant to emphasize that this force in the steel acts at point b rather than vertically below point a. Equilibrium requires that $M_{\text{int},a} = M_{\text{ext},a}$ so that the longitudinal tension in the steel at b is

$$T_b = \frac{M_{\text{ext},a} - V_d p + V_i m}{z} \tag{4.5}$$

Neglecting the forces V_d and V_i, which decrease with increasing crack opening, one has, with very little error,

$$T_b = \frac{M_{\text{ext},a}}{z} \tag{4.6}$$

The formation of the diagonal crack, then, is seen to produce the following redistribution of internal forces and stresses:

1. In the vertical section through point a, the average shear stress before crack formation was V_{ext}/bd. After crack formation, the shear force is resisted by

a combination of the dowel shear, the interface shear, and the shear force on the much smaller area *by* of the remaining uncracked concrete. As tension splitting develops along the longitudinal bars, V_d and V_i decrease; this, in turn, increases the shear force and the resulting shear stress on the remaining uncracked concrete area.

2. The diagonal crack, as described previously, usually rises above the neutral axis and traverses some part of the compression zone before it is arrested by the compression stresses. Consequently, the compression force C also acts on an area *by* smaller than that on which it acted before the crack was formed. Correspondingly, formation of the crack has increased the compression stresses in the remaining uncracked concrete.
3. Prior to diagonal cracking, the tension force in the steel at point *b* was caused by, and was proportional to, the bending moment in a vertical section through the same point *b*. As a consequence of the diagonal crack, however, Eq. (4.6) shows that the tension in the steel at *b* is now caused by, and is proportional to, the bending moment at *a*. Since the moment at *a* is evidently larger than that at *b*, formation of the crack has caused a sudden increase in the steel stress at *b*.

If the two materials are capable of resisting these increased stresses, equilibrium will establish itself after internal redistribution and further load can be applied before failure occurs. Such failure can then develop in various ways. For one, if only enough steel has been provided at *b* to resist the moment at that section, the increase of the steel force, in item 3, will cause the steel to yield because of the larger moment at *a*, thus failing the beam. If the beam is properly designed to prevent this occurrence, it is usually the concrete at the head of the crack that will eventually crush. This concrete is subject simultaneously to large compression and shear stresses, and this biaxial stress combination is conducive to earlier failure than would take place if either of these stresses were acting alone. Finally, if there is splitting along the reinforcement, it will cause the bond between steel and concrete to weaken to such a degree that the reinforcement may pull loose. This may either be the cause of failure of the beam or may occur simultaneously with crushing of the remaining uncracked concrete.

It was noted earlier that relatively deep beams will usually show continued and increasing resistance after formation of a critical diagonal tension crack, but relatively shallow beams will fail almost immediately upon formation of the crack. The amount of reserve strength, if any, was found to be erratic. In fact, in several test series in which two specimens as identical as one can make them were tested, one failed immediately upon formation of a diagonal crack, while the other reached equilibrium under the described redistribution and failed at a higher load.

For this reason, this reserve strength is discounted in modern design procedures. As previously mentioned, most beams are furnished with at least a minimum of web reinforcement. For those flexural members which are not, such as slabs, footings, and others, design is based on that shear force V_{cr} or shear stress v_{cr} at which formation of inclined cracks must be expected. Thus, Eq. (4.2*c*), or some equivalent of it, has become the design criterion for such members.

4.4 REINFORCED CONCRETE BEAMS WITH WEB REINFORCEMENT

Economy of design demands, in most cases, that a flexural member be capable of developing its full moment capacity rather than having its strength limited by premature shear failure. This is also desirable because structures, if overloaded, should not fail in the sudden and explosive manner characteristic of many shear failures, but should show adequate ductility and warning of impending distress. The latter, as pointed out earlier, is typical of flexural failure caused by yielding of the longitudinal bars, which is preceded by gradual excessively large deflections and noticeable widening of cracks. Therefore, if a fairly large safety margin relative to the available shear strength as given by Eq. (4.2*c*) or its equivalent does not exist, special shear reinforcement, known as *web reinforcement*, is used to increase this strength.

a. Types of Web Reinforcement

Typically, web reinforcement is provided in the form of vertical *stirrups*, spaced at varying intervals along the axis of the beam depending on requirements, as shown in Fig. 4.8*a*. Relatively small size bars are used, generally Nos. 3 to 5. Simple U-shaped bars similar to Fig. 4.8*b* are most common, although multiple-leg stirrups such as shown in Fig. 4.8*c* are sometimes necessary. Stirrups are formed to fit around the main longitudinal rebars as shown; small-diameter longitudinal bars are often included at the top of the stirrups to provide support during construction. Because of the relatively short length of stirrup embedded in the compression zone of a beam, in most cases special anchorage must be provided in the form of hooks or bends; detailed requirements for anchorage of stirrups will be discussed in Chap. 5.

Alternatively, shear reinforcement may be provided by bending up a part of the longitudinal steel where it is no longer needed to resist flexural tension, as suggested by Fig. 4.8*d*. In continuous beams, these bent-up bars may also provide all or part of the necessary reinforcement for negative moments. The requirements for longitudinal flexural reinforcement often conflict with those for diagonal tension, and because the saving in steel resulting from use of the capacity of bent bars as shear resistance is small, many designers prefer to include vertical stirrups to provide for all the shear requirement, counting on the bent part of the longitudinal bars, if bent bars are used, only to increase the overall safety against diagonal tension failure.

Welded wire mesh is sometimes used for shear reinforcement, particularly for small, light-loaded members with thin webs, and for certain types of precast, prestressed beams.

b. Behavior of Web-Reinforced Concrete Beams

Web reinforcement has no noticeable effect prior to the formation of diagonal cracks. In fact, measurements show that the web steel is practically free of stress

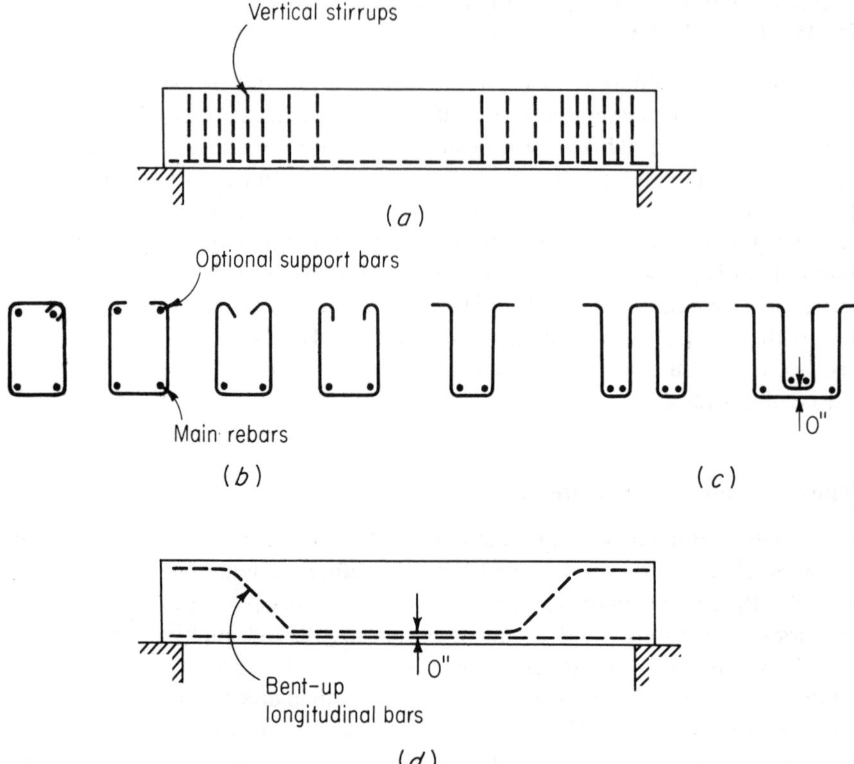

FIGURE 4.8
Types of web reinforcement.

prior to crack formation. After diagonal cracks have developed, web reinforcement augments the shear resistance of a beam in four separate ways:

1. Part of the shear force is resisted by the bars that traverse a particular crack. The mechanism of this added resistance is discussed below.
2. The presence of these same bars restricts the growth of diagonal cracks and reduces their penetration into the compression zone. This leaves more uncracked concrete available at the head of the crack for resisting the combined action of shear and compression, already discussed.
3. The stirrups also counteract the widening of the cracks, so that the two crack faces stay in close contact. This makes for a significant and reliable interface force V_i (see Fig. 4.7).
4. As seen in Fig. 4.8, the stirrups are arranged so that they tie the longitudinal reinforcement into the main bulk of the concrete. This provides some measure of restraint against the splitting of concrete along the longitudinal reinforcement, shown in Fig. 4.1 and Fig. 4.7*b*, and increases the share of the shear force resisted in dowel action.

From this it is clear that failure will be imminent when the stirrups start yielding. This not only exhausts their own resistance but also permits a wider crack opening with consequent reduction of the beneficial restraining effects, points 2 to 4, above.

It will be realized from this description that the behavior, once a crack is formed, is quite complex and dependent in its details on the particulars of crack configuration (length, inclination, and location of the main or critical crack). The latter, in turn, is quite erratic and has so far defied purely analytical prediction. For this reason, the concepts that underlie present design practice are not wholly rational. They are based partly on rational analysis, partly on test evidence, and partly on successful long-time experience with structures in which certain procedures for designing web reinforcement have resulted in satisfactory performance. Research in this field, both on plain and web-reinforced members, has furnished important experimental data for recent improvements in shear design and analysis. It has not yet resulted in a completely consistent rational analysis of shear behavior.

Beams with Vertical Stirrups. Since web reinforcement is ineffective in the uncracked beam, the magnitude of the shear force or stress that causes cracking to occur is the same as in a beam without web reinforcement and is given by Eq. (4.2*c*). Most frequently, web reinforcement consists of *vertical stirrups*; the forces acting on the portion of such a beam between the crack and the nearby support are shown in Fig. 4.9. They are the same as those of Fig. 4.7 except that each stirrup traversing the crack exerts a force $A_v f_v$ on the given portion of the beam. Here A_v is the cross-sectional area of the stirrup (in the case of the U-shaped stirrup of Fig. 4.8*b* it is twice the area of one bar) and f_v is the tension stress in the stirrup. Equilibrium in the vertical direction requires

$$V_{\text{ext}} = V_{cz} + V_d + V_{iy} + V_s \tag{a}$$

where $V_s = nA_v f_v$ is the vertical force in the stirrups, n being the number of stirrups traversing the crack. If s is the stirrup spacing and p the horizontal projection of the crack, as shown, then $n = p/s$.

The approximate distribution of the four components of the internal shear force with increasing external shear V_{ext} is shown schematically in Fig. 4.10. It is seen that after inclined cracking, the portion of the shear $V_s = nA_v f_v$ carried by the stirrups increases linearly, while the sum of the three other components,

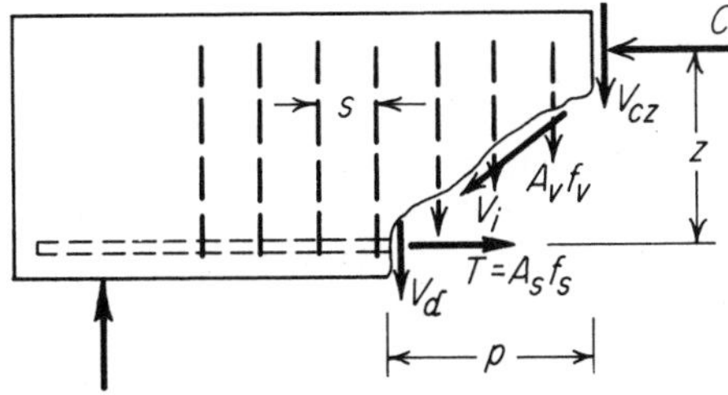

FIGURE 4.9
Forces at a diagonal crack in a beam with vertical stirrups.

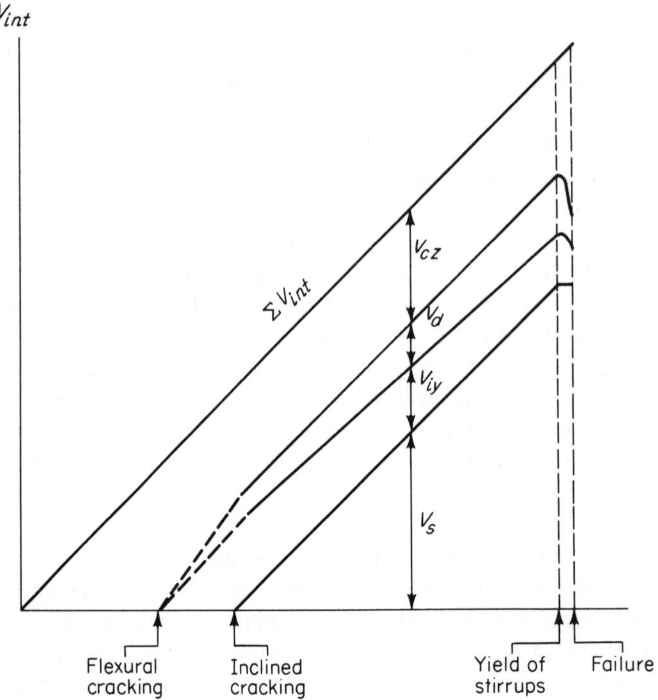

FIGURE 4.10
Redistribution of internal shear forces in a beam with stirrups. (*Adapted from Ref. 4.3.*)

$V_{cz} + V_d + V_{iy}$, stays nearly constant. When the stirrups yield, their contribution remains constant at the yield value $V_s = nA_v f_y$. However, because of widening of the inclined cracks and longitudinal splitting, V_{iy} and V_d fall off rapidly. This overloads the remaining uncracked concrete and very soon precipitates failure.

While total shear carried by the stirrups at yielding is known, the individual magnitudes of the three other components are not. Limited amounts of test evidence have led to the conservative assumption in present-day methods that just prior to failure of a web-reinforced beam, the sum of these three internal shear components is equal to the cracking shear V_{cr} as given by Eq. (4.2*d*). This sum is generally (somewhat loosely) referred to as the *contribution of the concrete* to the total shear resistance, and is denoted V_c. Thus $V_c = V_{cr}$ and

$$V_c = V_{cz} + V_d + V_{iy} \tag{b}$$

The number of stirrups n spaced a distance s apart was seen to depend on the length p of the horizontal projection of the diagonal crack. This length is assumed equal to the effective depth of the beam; thus $n = d/s$, implying a crack somewhat flatter than 45°. Then, at failure when $V_{ext} = V_n$, Eqs. (*a*) and (*b*) yield for the nominal ultimate shear strength

$$V_n = V_c + \frac{A_v f_y d}{s} \tag{4.7a}$$

where V_c is taken equal to the cracking shear V_{cr} given by Eq. (4.2*c*); that is,

$$V_c = \left(1.9\sqrt{f_c'} + 2500\frac{\rho V d}{M}\right) bd \leq 3.5\sqrt{f_c'}bd \tag{4.8}$$

Dividing both sides of Eq. (4.7*a*) by *bd*, one obtains the same relation expressed in terms of the nominal ultimate shear stress:

$$v_n = \frac{V_n}{bd} = v_c + \frac{A_v f_y}{bs} \tag{4.7b}$$

In Ref. 4.1 the results of 166 beam tests are compared with Eq. (4.7*b*). It is shown that the equation predicts the actual shear strength quite conservatively, the observed strength being on the average 45 percent larger than predicted; a very few of the individual test beams developed strength just slightly below that of Eq. (4.7*b*). At the same time the very considerable scatter, even though almost entirely on the conservative side, indicates that a deeper and more precise understanding of shear strength is still needed.

Beams with Inclined Bars. The function of *inclined web reinforcement* (Fig. 4.8*d*) can be discussed in very similar terms. Figure 4.11 again indicates the forces that act on the portion of the beam to one side of the diagonal crack that results in eventual failure. The crack with horizontal projection p and inclined length $i = p/(\cos\theta)$ is crossed by inclined bars horizontally spaced a distance s apart. The inclination of the bars is α and that of the crack θ, as shown. The distance between bars measured parallel to the direction of the crack is seen from the irregular triangle to be

$$a = \frac{s}{\sin\theta(\cot\theta + \cot\alpha)}$$

The number of bars crossing the crack, $n = i/a$, after some transformation, is obtained as

$$n = \frac{p}{s}(1 + \cot\alpha\tan\theta)$$

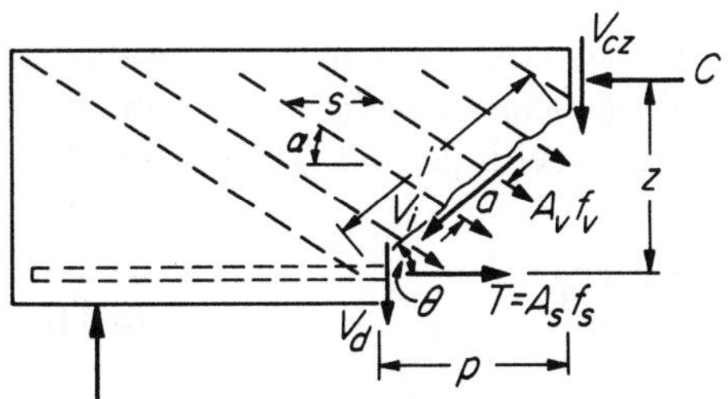

FIGURE 4.11
Forces at a diagonal crack in a beam with inclined web reinforcement.

The vertical component of the force in one bar or stirrup is $A_v f_v \sin \alpha$, so that the total vertical component of the forces in all bars which cross the crack is

$$V_s = nA_v f_v \sin\alpha = A_v f_v \frac{p}{s}(\sin\alpha + \cos\alpha \tan\theta)$$

As in the case of vertical stirrups, shear failure occurs when the stress in the web reinforcement reaches the yield point. Also, the same assumptions are made as in the case of stirrups, namely, that the horizontal projection of the diagonal crack is equal to the effective depth d, and that $V_{cz} + V_d + V_{iy}$ is equal to V_c. Lastly, the inclination θ of the diagonal crack, which varies somewhat depending on various influences, is generally assumed to be 45°. On this basis, the ultimate strength when failure is caused by shear is obtained as

$$V_n = V_c + \frac{A_v f_y d(\sin\alpha + \cos\alpha)}{s} \tag{4.9}$$

It is seen that Eq. (4.7*a*), developed for vertical stirrups, is only a special case, for $\alpha = 90°$, of the more general expression (4.9).

It should be noted that Eqs. (4.7) and (4.9) apply only if web reinforcement is so spaced that any conceivable diagonal crack is traversed by at least one stirrup or inclined bar. Otherwise web reinforcement would not contribute to the shear strength of the beam, because diagonal cracks which could form between widely spaced web reinforcement would fail the beam at the load at which it would fail if no web reinforcement were present. This imposes upper limits on the permissible spacing s to ensure that the web reinforcement is actually effective as calculated.

To summarize, at this time the nature and mechanism of diagonal tension failure are clearly understood qualitatively, but some of the quantitative assumptions that have been made in the preceding development cannot be proved by rational analysis. However, the calculated results are in acceptable and generally conservative agreement with a very large body of empirical data, and structures designed on this basis have proved satisfactory.

4.5 ACI CODE PROVISIONS FOR SHEAR DESIGN

According to ACI Code 11.1.1, the design of beams for shear is to be based on the relation

$$V_u \le \phi V_n \tag{4.10}$$

where V_u is the total shear force applied at a given section of the beam due to factored loads and $V_n = V_c + V_s$ is the nominal shear strength, equal to the sum of the contributions of the concrete and the web steel if present. Thus for vertical stirrups

$$V_u \le \phi V_c + \frac{\phi A_v f_y d}{s} \tag{4.11a}$$

and for inclined bars

$$V_u \le \phi V_c + \frac{\phi A_v f_y d(\sin\alpha + \cos\alpha)}{s} \tag{4.11b}$$

where all terms are as previously defined. The strength reduction factor ϕ is to be taken equal to 0.85 for shear. The slight additional conservatism, compared with the value of $\phi = 0.90$ for bending, reflects both the sudden nature of diagonal tension failure and the still-imperfect understanding of that failure mode.

For typical support conditions, where the reaction from the support surface or from a monolithic column introduces vertical compression at the end of the beam, sections located less than a distance d from the face of the support may be designed for the same shear V_u as that computed at a distance d. However, if concentrated loads act within that distance, or if the reaction causes vertical tension rather than compression (e.g., if the beam is supported at the lower end of a monolithic vertical element), the critical design section should be taken at the face of the support.

a. Shear Strength Provided by the Concrete

The nominal shear strength contribution of the concrete (including the contributions from aggregate interlock, dowel action of the main rebars, and that of the uncracked concrete) is basically the same as Eq. (4.8) with slight notational changes. To permit application of Eq. (4.8) to T beams having web width b_w, the rectangular beam width b is replaced by b_w with the understanding that for rectangular beams b is used for b_w. For T beams with a tapered web width, such as typical concrete joists, the average web width is used, unless the narrowest part of the web is in compression, in which case b_w is taken as the minimum width. Further, in Eq. (4.8), the shear V and moment M are designated V_u and M_u to emphasize that they are the values computed at factored loads. Thus, for members subject to shear and flexure, according to ACI Code 11.3.2, the concrete contribution to shear strength is

$$V_c = \left(1.9\sqrt{f'_c} + 2500\,\frac{\rho_w V_u d}{M_u}\right) b_w d \le 3.5\sqrt{f'_c}\, b_w d \tag{4.12a}$$

where ρ_w = longitudinal tensile steel ratio $A_s/b_w d$ or A_s/bd. With the section dimension b_w and d in inches and $V_u d$ and M_u in consistent units, V_c is expressed in pounds. In Eq. (4.12a), the quantity $V_u d/M_u$ is not to be taken greater than 1.0.

While Eq. (4.12a) is perfectly well suited to computerized design or for research, for manual calculations its use is tedious because ρ_w, V_u, and M_u generally change along the span, requiring that V_c be calculated at frequent intervals. For this reason, an alternative equation for V_c is permitted by ACI Code 11.3.1:

$$V_c = 2\sqrt{f'_c}\, b_w d \tag{4.12b}$$

Referring to Fig. 4.6, it is clear that Eq. (4.12b) is very conservative in regions where the shear-moment ratio is high, such as near the ends of simple spans or near the inflection points of continuous spans; however, because of its simplicity, it is generally used in practice.

The tests on which Eq. (4.12*a*) and Eq. (4.12*b*) are based used beams with concrete compressive strength mostly in the range of 3000 to 5000 psi. More recent experimental results, obtained at Cornell University and elsewhere (Refs. 4.9 to 4.11), have shown that in beams constructed using high strength concrete (see Sec. 2.11) with f'_c above 6000 psi, the concrete contribution to shear strength, V_c, is less than predicted by those equations. Differences become increasingly significant the higher the concrete strength. For this reason, ACI Code 11.1.2 places an upper limit of 100 psi on the value of $\sqrt{f'_c}$ to be used in Eqs. (4.12*a*) and (4.12*b*), *as well as in all other ACI Code shear provisions*. However, values of $\sqrt{f'_c}$ greater than 100 psi may be used in computing V_c if the usual minimum amount of web reinforcement is increased (see Sec. 4.5b).

Code provisions for computing V_c according to Eq. (4.12*a*) or (4.12*b*) apply to normal weight concrete. Lightweight aggregate concretes having density from 90 to 120 pcf are used increasingly, particularly for precast elements. Their tensile strength, of particular importance in shear and diagonal tension calculations, is known to be significantly less than that of normal weight concrete of the same compressive strength (see Table 2.2 and Ref. 4.12). It is advisable, when designing with lightweight concrete, to obtain an accurate estimate of the actual tensile strength of the material. The split-cylinder strength f_{ct} is not identical with the direct tensile strength, but it serves as a convenient and reliable measure.

For normal concrete the split-cylinder strength is often taken equal to $6.7\sqrt{f'_c}$. Accordingly, the ACI Code specifies that $f_{ct}/6.7$ shall be substituted for $\sqrt{f'_c}$ in all equations for V_c, with the further restriction that $f_{ct}/6.7$ shall not exceed $\sqrt{f'_c}$. If the split-cylinder strength is not available, values of V_c calculated using $\sqrt{f'_c}$ shall be multiplied by 0.75 for "all-lightweight" concrete and by 0.85 for "sand-lightweight" concrete. All other shear provisions remain unchanged.

b. Minimum Web Reinforcement

If V_u, the shear force at factored loads, is no larger than ϕV_c, calculated by Eq. (4.12*a*) or alternatively by Eq. (4.12*b*), then theoretically no web reinforcement is required. Even in such a case, however, ACI Code 11.5.5 requires provision of at least a minimum area of web reinforcement equal to

$$A_v = 50\,\frac{b_w s}{f_y} \tag{4.13}$$

where s = longitudinal spacing of web reinforcement, in.
f_y = yield strength of web steel, psi
A_v = total cross-sectional area of web steel within distance s, in^2

This provision holds unless V_u is one-half or less of the design shear strength ϕV_c provided by the concrete. Specific exceptions to this requirement for minimum web steel are made for slabs and footings, for concrete joist floor construction, and for beams with total depth not greater than 10 in., $2\frac{1}{2}$ times the thickness of the flange, or one-half the web width (whichever is greatest). These members are excluded because of their capacity to redistribute internal forces before diagonal tension failure, as confirmed both by tests and successful design experience.

For high-strength concrete beams, the usual limitation of 100 psi imposed on the value of $\sqrt{f_c'}$ used in calculating V_c by Eq. (4.12*a*) or Eq. (4.12*b*) is waived by ACI Code 11.1.2.1 if such beams are designed with minimum web reinforcement equal to $f_c'/5000$ times, but not more than three times, the amount required by Eq. (4.13). The concrete contribution to shear strength may be calculated based on the full concrete compressive strength. Tests described in Ref. 4.9 indicate that for beams with concrete strength above about 6000 psi, the concrete contribution V_c was significantly less than predicted by the ACI Code equations, although the steel contribution V_s was higher. The total nominal shear strength V_n was greater than predicted by ACI Code methods in all cases. The ACI Code provision modifying minimum web steel for high strength concrete beams, resulting in more web steel in most cases, is intended to enhance the post-cracking capacity, thus resulting in safe designs even though the concrete contribution to shear strength is overestimated.†

Example 4.1 Beam without web reinforcement. A rectangular beam is to be designed to carry a shear force V_u of 30 kips. No web reinforcement is to be used, and f_c' is 4000 psi. What is the minimum cross section if controlled by shear?

Solution. If no web reinforcement is to be used, the cross-sectional dimensions must be selected so that the applied shear V_u is no larger than one-half the design shear strength ϕV_c. The calculations will be based on Eq. (4.12*b*). Thus

$$V_u = \frac{1}{2}\phi(2\sqrt{f_c'}b_w d)$$

$$b_w d = \frac{30,000}{0.85\sqrt{4000}} = 560 \text{ in}^2$$

A beam with $b_w = 18$ in. and $d = 31$ in. is required. Alternately, if the minimum amount of web reinforcement given by Eq. (4.13) is used, the concrete shear resistance may be taken at its full value ϕV_c and it is easily confirmed that a beam with $b_w = 12$ in. and $d = 23.5$ in. will be sufficient.

c. Region in which Web Reinforcement Is Required

If the required shear strength V_u is greater than the design shear strength ϕV_c provided by the concrete in any portion of a beam, there is a theoretical requirement for web reinforcement. Elsewhere in the span, web steel at least equal to the

---

† The shortcomings of the ACI Code "$V_c + V_s$" approach to shear design, particularly the provisions relating to the concrete contribution V_c, have provided motivation for the development of more rational procedures, not yet implemented in the United States. An alternative method of shear design will be discussed in Sec. 4.8.

amount given by Eq. (4.13) must be provided, unless the calculated shear force is less than $\frac{1}{2}\phi V_c$.

The portion of any span through which web reinforcement is theoretically necessary can be found by drawing the shear diagram for the span and superimposing a plot of the shear strength of the concrete. Where the shear force V_u exceeds ϕV_c, shear reinforcement must provide for the excess. The additional length through which at least the minimum web steel is needed can be found by superimposing a plot of $\phi V_c/2$.

Example 4.2 Limits of web reinforcement. A simply supported rectangular beam 16 in. wide having an effective depth of 22 in. carries a total factored load of 7.9 kips/ft on a 20 ft clear span. It is reinforced with 9.86 in^2 of tensile steel, which continues uninterrupted into the supports. If $f_c' = 3000$ psi, throughout what part of the beam is web reinforcement required?

Solution. The maximum external shear force occurs at the ends of the span, where $V_u = 7.9 \times 20/2 = 79.0$ kips. At the critical section for shear, a distance d from the support, $V_u = 79.0 - 7.9 \times 1.83 = 64.5$ kips. The shear force varies linearly to zero at midspan. The variation of V_u is shown in Fig. 4.12*a*. Adopting Eq. (4.12*b*) gives

$$V_c = 2\sqrt{3000} \times 16 \times 22 = 38,600 \text{ lb}$$

Hence $\phi V_c = 0.85 \times 38.6 = 32.8$ kips. This value is superimposed on the shear diagram, and, from geometry, the point at which web reinforcement theoretically is no longer required is

$$10\left(\frac{79.0 - 32.8}{79.0}\right) = 5.86 \text{ ft}$$

from the support face. However, according to the ACI Code, at least a minimum amount of web reinforcement is required wherever the shear force exceeds $\phi V_c/2$, or 16.4 kips in this case. As seen from Fig. 4.12*a*, this applies to a distance

$$10\left(\frac{79.0 - 16.4}{79.0}\right) = 7.92 \text{ ft}$$

from the support face. To summarize, at least the minimum web steel must be provided within a distance of 7.92 ft from the supports, and within 5.86 ft the web steel must provide for the shear force corresponding to the shaded area.

If the alternative Eq. (4.12*a*) is used, the variation along the span of ρ_w, V_u, and M_u must be known so that V_c can be calculated. This is best done in tabular form, as shown in Table 4.1.

The applied ultimate shear V_u and the design shear capacity ϕV_c are plotted in Fig. 4.12*b*. From the graph it is found that stirrups are theoretically no longer required 5.70 ft from the support face. However, from the plot of $\phi V_c/2$ it is found that at least the minimum web steel is to be provided within a distance of 7.90 ft.

When Fig. 4.12*a* and *b* are compared, it is evident that the length over which web reinforcement is needed is nearly the same for this example whether Eq. (4.12*a*) or (4.12*b*) is used. However, the smaller shaded area of Fig. 4.12*b* indicates that substantially less web-steel area would be needed within that required distance if the more accurate Eq. (4.12*a*) were adopted.

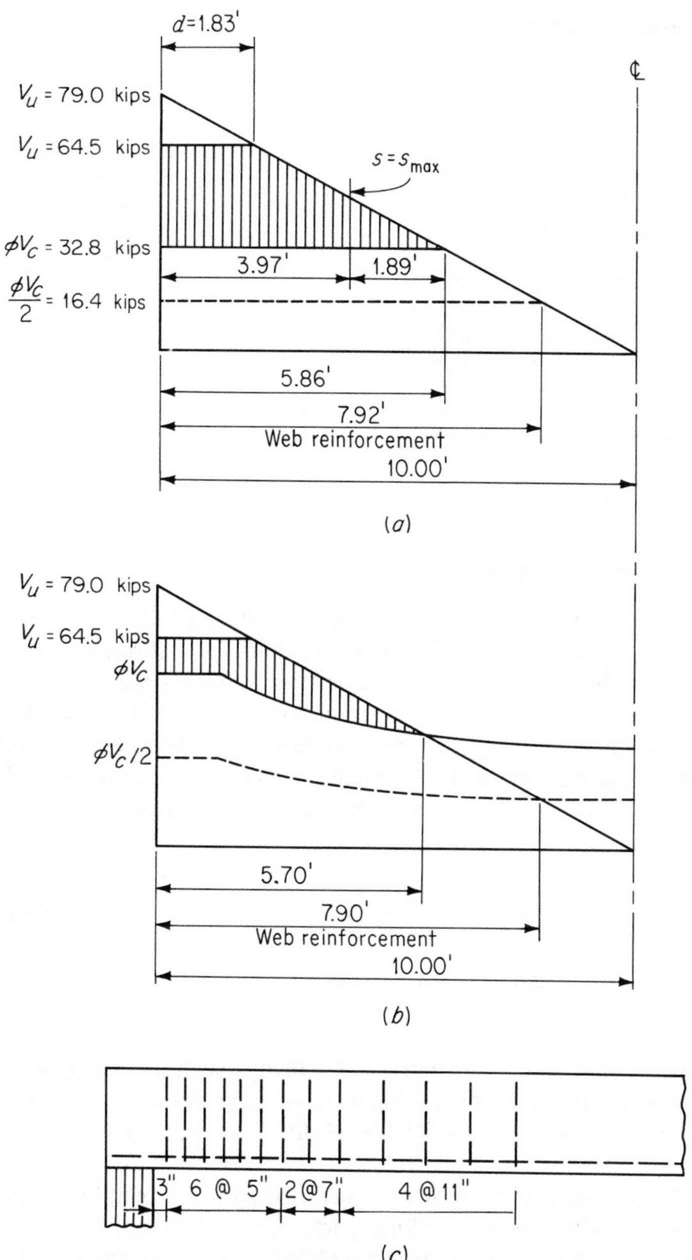

FIGURE 4.12
Shear design example.

Table 4.1 Shear design example

Distance from support, ft	M_u, ft-kips	V_u, kips	V_c[a]	ϕV_c
0	0	79.0	61.2	52.0
1	76	71.1	61.2	52.0
2	142	63.2	56.7	48.2
3	201	55.3	48.9	41.6
4	252	47.4	45.1	38.3
5	296	39.5	42.6	36.2
6	332	31.6	40.8	34.7
7	360	23.7	39.4	33.5
8	379	15.8	38.4	32.6
9	391	7.9	37.7	32.0
10	395	0	36.6	31.1

[a] $V_c = (1.9\sqrt{f_c'} + 2500\rho_w V_u d/M_u)b_w d \leq 3.5\sqrt{f_c'}b_w d$ and $V_u d/M_u \leq 1.0$

d. Design of Web Reinforcement

The design of web reinforcement, under the provisions of the ACI Code, is based on Eq. (4.11*a*) for vertical stirrups and Eq. (4.11*b*) for inclined stirrups or bent bars. In design, it is usually convenient to select a trial web-steel area A_v based on standard stirrup sizes (usually in the range from Nos. 3 to 5 for stirrups, and according to the longitudinal rebar size for bent-up bars), for which the required spacing s can be found. Equating the design strength ϕV_n to the required strength V_u and transposing Eqs. (4.11*a*) and (4.11*b*) accordingly, one finds that the required spacing of web reinforcement is

for vertical stirrups:
$$s = \frac{\phi A_v f_y d}{V_u - \phi V_c} \tag{4.14a}$$

for bent bars:
$$s = \frac{\phi A_v f_y d(\sin\alpha + \cos\alpha)}{V_u - \phi V_c} \tag{4.14b}$$

It should be emphasized that when conventional U stirrups such as in Fig. 4.8*b* are used, the web area A_v provided by each stirrup is *twice* the cross-sectional area of the bar; for stirrups such as those of Fig. 4.8*c*, A_v is four times the area of the bar used.

While the ACI Code requires only that the inclined part of a bent bar make an angle of at least 30° with the longitudinal part, usually bars are bent at a 45° angle. Only the center three-fourths of the inclined part of any bar is to be considered effective as web reinforcement.

It is undesirable to space vertical stirrups closer than about 4 in.; the size of the stirrups should be chosen to avoid a closer spacing. When vertical stirrups are required over a comparatively short distance, it is good practice to space them uniformly over the entire distance, the spacing being calculated for the point of greatest shear (minimum spacing). If the web reinforcement is required over a long distance, and if the shear varies materially throughout this distance, it is more

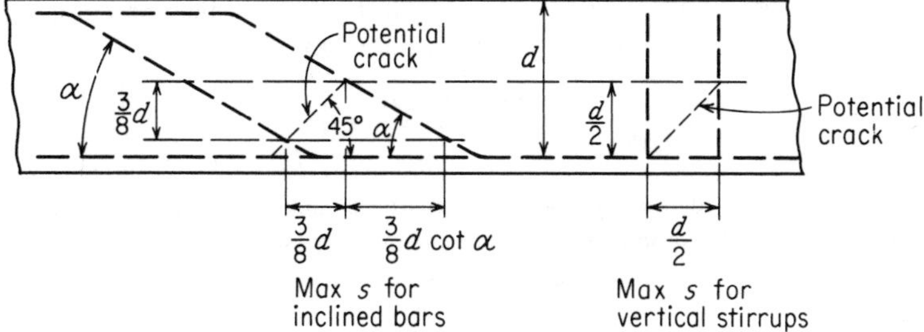

FIGURE 4.13
Maximum spacing of web reinforcement as governed by diagonal crack interception.

economical to compute the spacings required at several sections and to place the stirrups accordingly, in groups of varying spacing.

Where web reinforcement is needed, the Code requires it to be spaced so that every 45° line, representing a potential diagonal crack and extending from the middepth $d/2$ of the member to the longitudinal tension bars, is crossed by at least one line of web reinforcement; in addition, the Code specifies a maximum spacing of 24 in. When V_s exceeds $4\sqrt{f'_c}b_w d$, these maximum spacings are halved. These limitations are shown in Fig. 4.13 for both vertical stirrups and inclined bars, for situations in which the excess shear does not exceed the stated limit.

For design purposes, Eq. (4.13) giving the minimum web-steel area A_v is more conveniently inverted to permit calculation of maximum spacing s for the selected A_v. Thus, for the usual case of vertical stirrups, with $V_s \leq 4\sqrt{f'_c}b_w d$, the maximum spacing of stirrups is the smallest of

$$s_{\max} = \frac{A_v f_y}{50 b_w} \tag{4.15a}$$

$$s_{\max} = \frac{d}{2} \tag{4.15b}$$

$$s_{\max} = 24 \text{ in.} \tag{4.15c}$$

For longitudinal bars bent at 45°, Eq. (4.15*b*) is replaced by $s_{\max} = 3d/4$, as confirmed by Fig. 4.13.

To avoid excessive crack width in beam webs, the ACI Code limits the yield strength of the reinforcement to $f_y = 60{,}000$ psi or less. In no case, according to the ACI Code, is V_s to exceed $8\sqrt{f'_c}b_w d$ regardless of the amount of web steel used.

Example 4.3 Design of web reinforcement. Using vertical U stirrups, with $f_y = 40{,}000$ psi, design the web reinforcement for the beam of Example 4.2.

Solution. The solution will be based on the shear diagram in Fig. 4.12*a*. The stirrups must be designed to resist that part of the shear shown shaded. With No. 3 stirrups used for trial, the three maximum spacing criteria are first applied. For $\phi V_s = V_u - \phi V_c =$

31,700 lb, which is less than $4\phi\sqrt{f_c'}b_w d = 65{,}600$ lb, the maximum spacing must exceed neither $d/2 = 11$ in. nor 24 in. Also, from Eq. (4.15*a*),

$$s_{max} = \frac{A_v f_y}{50 b_w} = \frac{0.22 \times 40{,}000}{50 \times 16} = 11 \text{ in.}$$

The first and third criteria identically control in this case, and a maximum spacing of 11 in. is imposed. From the support to a distance d from the support, the excess shear $V_u - \phi V_c$ is 31,700 lb. In this region the required spacing is

$$s = \frac{\phi A_v f_y d}{V_u - \phi V_c} = \frac{0.85 \times 0.22 \times 40{,}000 \times 22}{31{,}700} = 5.2 \text{ in.}$$

This is neither so small that placement problems would result nor so large that maximum spacing criteria would control, and the choice of No. 3 stirrups is confirmed. Solving Eq. (4.14*a*) for the excess shear at which the maximum spacing can be used gives

$$V_u - \phi V_c = \frac{\phi A_v f_y d}{s} = \frac{0.85 \times 0.22 \times 40{,}000 \times 22}{11} = 14{,}960 \text{ lb}$$

With reference to Fig. 4.12*a*, this is attained at a distance x_1 from the point of zero excess shear, where $x_1 = 5.86 \times 14{,}960/46{,}200 = 1.89$ ft. This is 3.97 ft from the support face. With this information, a satisfactory spacing pattern can be selected. The first stirrup is usually placed at a distance $s/2$ from the support. The following spacing pattern is satisfactory:

1 space at	3 in. =	3 in.
6 spaces at	5 in. =	30 in.
2 spaces at	7 in. =	14 in.
4 spaces at	11 in. =	44 in.
	Total =	91 in. = 7 ft 7 in.

The resulting stirrup pattern is shown in Fig. 4.12*c*. As an alternative solution, it is possible to plot a curve showing required spacing as a function of distance from the support. Once the required spacing at some reference section, say at the support, is determined,

$$s_0 = \frac{0.85 \times 0.22 \times 40{,}000 \times 22}{79{,}000 - 32{,}800} = 3.55 \text{ in.}$$

it is easy to obtain the required spacings elsewhere. In Eq. (4.14*a*) only $V_u - \phi V_c$ changes with distance from the support. For uniform load this quantity is a linear function of distance from the point of zero excess shear, 5.86 ft from the support face. Hence, at 1 ft intervals,

$$s_1 = 3.55 \times 5.86/4.86 = 4.27 \text{ in.}$$

$$s_2 = 3.55 \times 5.86/3.86 = 5.39 \text{ in.}$$

$$s_3 = 3.55 \times 5.86/2.86 = 7.27 \text{ in.}$$

$$s_4 = 3.55 \times 5.86/1.86 = 11.18 \text{ in.}$$

$$s_5 = 3.55 \times 5.86/0.86 = 24.20 \text{ in.}$$

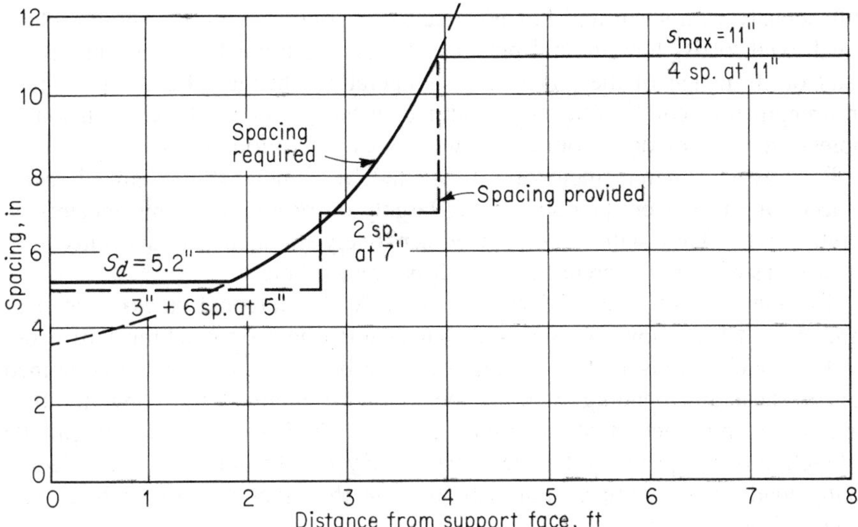

FIGURE 4.14
Required stirrup spacings for Example 4.3.

This is plotted in Fig. 4.14 together with the maximum spacing of 11 in., and a practical spacing pattern is selected. The spacing at a distance d from the support face is selected as the minimum requirement, in accordance with the ACI Code. The pattern of No. 3 U-shaped stirrups selected (shown on the graph) is identical with the previous solution. In many cases the experienced designer would find it unnecessary actually to plot the spacing diagram of Fig. 4.14 and would select a spacing pattern directly after calculating the required spacing at intervals along the beam.

If the web steel were to be designed on the basis of the excess-shear diagram of Fig. 4.12*b*, the second approach illustrated above would necessarily be selected, and spacings would be calculated at intervals along the span. In this particular case, the maximum permitted spacing of 11 in. is less than that required by excess shear anywhere between the point of zero excess shear and a distance d from the support. Consequently, the following spacing could be used:

$$
\begin{aligned}
1 \text{ space at } 5 \text{ in.} &= 5 \text{ in.} \\
8 \text{ spaces at } 11 \text{ in.} &= \underline{88 \text{ in.}} \\
\text{Total} &= 93 \text{ in.} = 7 \text{ ft } 9 \text{ in.}
\end{aligned}
$$

Nine No. 3 stirrups would be used, rather than the 13 previously calculated, in each half of the span. In designs for which the stress governs the stirrup spacing for both methods rather than the maximum spacing requirements, the saving obtained by the second solution would be even more significant.

Stop II

4.6 EFFECT OF AXIAL FORCES

The beams considered in the preceding sections were subjected to shear and flexure only. Reinforced concrete beams may also be subjected to axial forces, acting simultaneously with shear and flexure, due to a variety of causes. These include external axial loads, longitudinal prestressing, and restraint forces introduced as a result of shrinkage of the concrete or temperature changes. Beams may have their strength in shear significantly modified in the presence of axial tension or compression, as is evident from a review of Secs. 4.1 through 4.4.

Prestressed concrete members are treated by somewhat specialized methods, according to present practice, based largely on results of testing prestressed concrete beams. They will be considered separately in Chap. 21, and only non-prestressed reinforced concrete beams will be treated here.

The main effect of axial load is to modify the diagonal cracking load of the member. It was shown in Sec. 4.3 that diagonal tension cracking will occur when the principal tensile stress in the web of a beam, resulting from combined action of shear and bending, reaches the tensile strength of the concrete. It is clear that the introduction of longitudinal force, which modifies the magnitude and direction of the principal tensile stresses, may significantly alter the diagonal cracking load. Axial compression will increase the cracking load, while axial tension will decrease it.

For members carrying only flexural and shear loading, the shear force at which diagonal cracking would occur, V_{cr}, was predicted by Eq. (4.2*c*), based on a combination of theory and experimental evidence. Furthermore, for reasons that were explained in Sec. 4.4b, in beams with web reinforcement, the contribution of the concrete to shear strength V_c was taken equal to the diagonal cracking load V_{cr}. Thus, according to the ACI Code, the concrete contribution is calculated by Eq. (4.12*a*) or (4.12*b*). For members carrying flexural and shear loading plus axial loads, V_c can be calculated by suitable modifications of these equations as follows.

a. Axial Compression

In developing Eq. (4.2*c*) for V_{cr}, it was pointed out that the diagonal cracking load depends on the ratio of shear stress v to bending stress f at the top of the flexural crack. While these stresses were never actually determined, they were conveniently expressed as

$$v = K_1\left(\frac{V}{bd}\right) \tag{a}$$

and

$$f = K_2\left(\frac{M}{bd^2}\right) \tag{b}$$

Equation (*a*) relates the concrete shear stress at the top of the flexural crack to the average shear stress; Eq. (*b*) relates the flexural tension in the concrete at the top of the crack to the tension in the flexural steel, through the modular ratio $n = E_s/E_c$, as follows:

$$f = K_0\frac{f_s}{n} = K_0\frac{M}{nA_sjd}$$

or

$$f = K_0 \frac{M}{n\rho jbd^2} \tag{c}$$

where jd is the internal lever arm between C and T, and K_0 is an unknown constant. Thus the previous constant K_2 is equal to $K_0/n\rho j$.

Now consider a beam subject to axial compression N as well as M and V, as shown in Fig. 4.15*a*. In Fig. 4.15*b*, the external moment, shear, and thrust acting on the left side of a small element of the beam, having length dx, are equilibrated by the internal stress resultants T, C, and V acting on the right. It is convenient to replace the external loads M and N with the statically equivalent load N acting at eccentricity $e = M/N$ from the middepth, as shown in Fig. 4.15*c*. The lever arm of the eccentric force N with respect to the compressive resultant C is

$$e' = e + d - \frac{h}{2} - jd \tag{d}$$

The steel stress f_s can now be found taking moments about the point of application of C:

$$f_s = \frac{Ne'}{A_s jd}$$

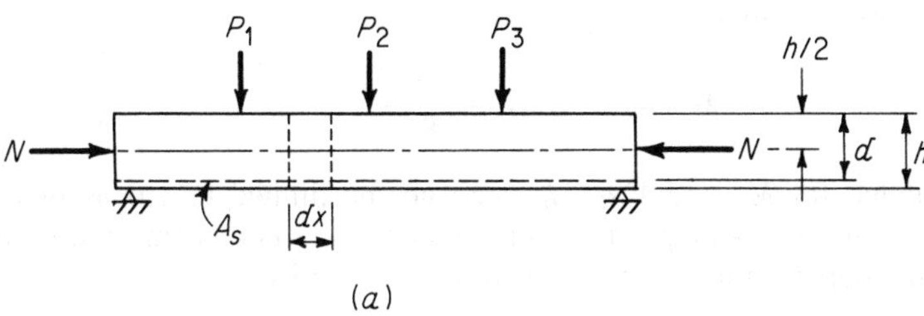

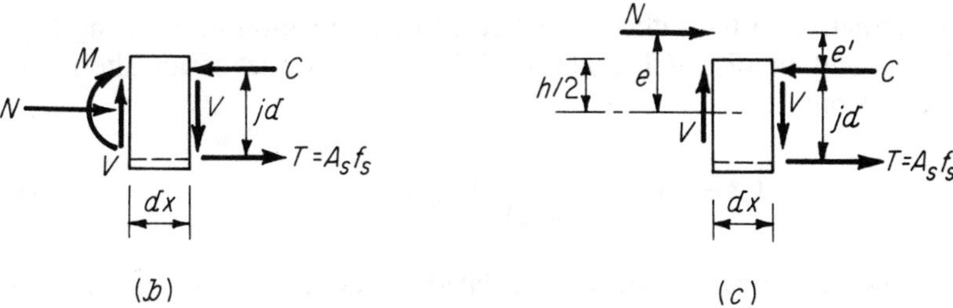

FIGURE 4.15
Beams subject to axial compression plus bending and shear loads.

from which

$$f_s = \frac{M + N(d - h/2 - jd)}{A_s jd}$$

Noting that j is very close to $\frac{7}{8}$ for loads up to that producing diagonal cracking, the term in parentheses in the last equation above can be written as $(d - 4h)/8$. Then with $f = K_0 f_s/n$ as before, one obtains for the concrete tensile stress at the head of the flexural crack:

$$f = K_0 \frac{M - N(4h - d)/8}{n\rho jbd^2} = K_2 \frac{M - N(4h - d)/8}{bd^2} \tag{e}$$

Comparing Eq. (*e*) with Eqs. (*c*) and (*b*) makes it clear that the previous derivation for flexural tension f holds for the present case including axial loads if a modified moment $M - N(4h - d)/8$ is substituted for M. It follows that Eq. (4.2*c*) can be used to calculate V_{cr} with the same substitution of modified for actual moment.

The ACI Code treatment is based on this development. The concrete contribution to shear strength V_c is taken equal to V_{cr} and is given by Eq. (4.12*a*) as before:

$$V_c = \left(1.9\sqrt{f_c'} + 2500\frac{\rho_w V_u d}{M_u}\right) b_w d \tag{4.12a}$$

except that the modified moment

$$M_m = M_u - N_u \frac{4h - d}{8} \tag{4.16}$$

is to be substituted for M_u and $V_u d/M_u$ need not be limited to 1.0 as before. The thrust N_u is to be taken positive for compression. For beams with axial compression, the upper limit of $3.5\sqrt{f_c'} b_w d$ is replaced by

$$V_c = 3.5\sqrt{f_c'} b_w d \sqrt{1 + \frac{N_u}{500A_g}} \tag{4.17}$$

where A_g is the gross area of the concrete and N_u/A_g is expressed in psi units.

As an alternative to the rather complicated determination of V_c using Eqs. (4.12*a*), (4.16), and (4.17) ACI Code 11.3.1.2 permits use of an alternative simplified expression:

$$V_c = 2\left(1 + \frac{N_u}{2000A_g}\right)\sqrt{f_c'} b_w d \tag{4.18}$$

Figure 4.16 shows a comparison of V_c calculated by the more complex and simplified expressions for beams with compression load. Equation (4.18) is seen to be generally quite conservative, particularly for higher values of N_u/A_g. However, because of its simplicity, it is widely used in practice.

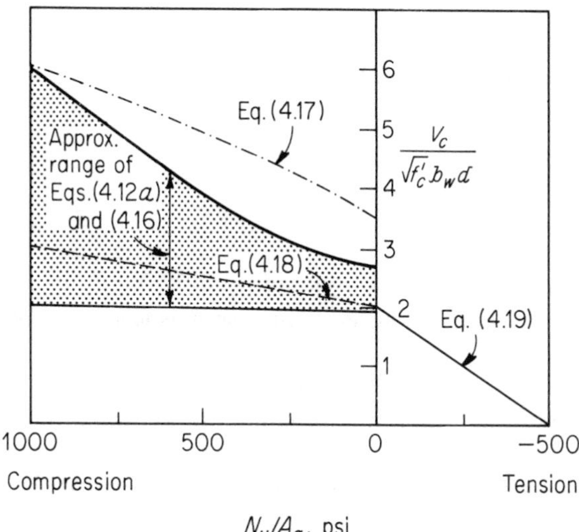

FIGURE 4.16
Comparison of equations for V_c for members subject to axial loads.

b. Axial Tension

The approach developed above for beams with axial compression does not correlate well with experimental evidence for beams subject to axial tension, and often predicts strengths V_c higher than actually measured. For this reason, the ACI Code provides that, for members carrying significant axial tension as well as bending and shear, the contribution of the concrete be taken as

$$V_c = 2\left(1 + \frac{N_u}{500A_g}\right)\sqrt{f'_c}b_w d \tag{4.19}$$

where N_u is negative for tension. As a simplifying alternative, the Code suggests that, for beams carrying axial tension, V_c be taken equal to zero and the shear reinforcement be required to carry the total shear. The variation of V_c with N_u/A_g for beams with tension is shown in Fig. 4.16 also.

Example 4.4 Effect of axial forces on V_c. A beam with dimensions $b = 12$ in., $d = 24$ in., and $h = 27$ in., with $f'_c = 4000$ psi, carries a single concentrated factored load of 100 kips at midspan. Find the maximum shear strength of the concrete, V_c, at the first critical section for shear at a distance d from the support (a) if no axial forces are present, (b) if axial compression of 60 kips acts, and (c) if axial tension of 60 kips acts. In each case, compute V_c by both the more complex and simplified expressions of the ACI Code. Neglect self-weight of the beam. At the section considered, tensile reinforcement consists of three No. 10 bars with a total area of 3.79 in^2.

Solution. At the critical section, $V_u = 50$ kips and $M_u = 50 \times 2 = 100$ ft-kips, while $\rho = 3.79/12 \times 24 = 0.013$.

(a) If $N_u = 0$, Eq. (4.12a) predicts

$$V_c = \left(1.9\sqrt{4000} + 2500\frac{0.013 \times 50 \times 2}{100}\right)12 \times \frac{24}{1000} = 44.0 \text{ kips}$$

not to exceed the value of

$$V_c = 3.5\sqrt{4000} \times 12 \times \frac{24}{1000} = 63.8 \text{ kips}$$

If the simplified Eq. (4.12*b*) is used,

$$V_c = 2\sqrt{4000} \times 12 \times \frac{24}{1000} = 36.4 \text{ kips}$$

which is about 17 percent below the more exact value of Eq. (4.12*a*).

(*b*) With a compression of 60 kips introduced, the modified moment is found from Eq. (4.16) to be

$$M_m = 100 - 60\,\frac{4 \times 27 - 24}{8 \times 12} = 47.5 \text{ ft-kips}$$

After introduction of that value into Eq. (4.12*a*) in place of M_u, the concrete shear strength is

$$V_c = \left(1.9\sqrt{4000} + 2500\frac{0.013 \times 50 \times 2}{47.5}\right) 12 \times \frac{24}{1000} = 54.3 \text{ kips}$$

and, according to Eq. (4.17), should not exceed

$$V_c = 63.8\sqrt{1 + \frac{60{,}000}{500 \times 12 \times 27}} = 74.6 \text{ kips}$$

If the simplified Eq. (4.18) is used,

$$V_c = 2\left(1 + \frac{60{,}000}{2000 \times 12 \times 27}\right)\sqrt{4000} \times 12 \times \frac{24}{1000} = 39.8 \text{ kips}$$

Comparing the results of the more exact calculation for (*a*) and (*b*), it is seen that the introduction of an axial compressive stress of $60{,}000/12 \times 27 = 185$ psi increases the concrete shear V_c by about 25 percent.

(*c*) With an axial tension of 60 kips acting, the reduced V_c is found from Eq. (4.19) to be

$$V_c = 2\left(1 - \frac{60{,}000}{500 \times 12 \times 27}\right)\sqrt{4000} \times 12 \times \frac{24}{1000} = 22.9 \text{ kips}$$

a reduction of almost 50 percent from the value for $N_u = 0$. The alternative of using Eq. (4.19) for this case, according to the ACI Code, would be to set $V_c = 0$.

In all cases above, the strength reduction factor $\phi = 0.85$ would be applied to V_c to obtain the design strength.

4.7 BEAMS WITH VARYING DEPTH

Reinforced concrete members having varying depth are frequently used in the form of haunched beams for bridges or portal frames, as in Fig. 4.17*a*, as precast roof girders such as shown in Fig. 4.17*b*, or as cantilever slabs. Generally the depth increases in the direction of increasing moments. For beams with varying

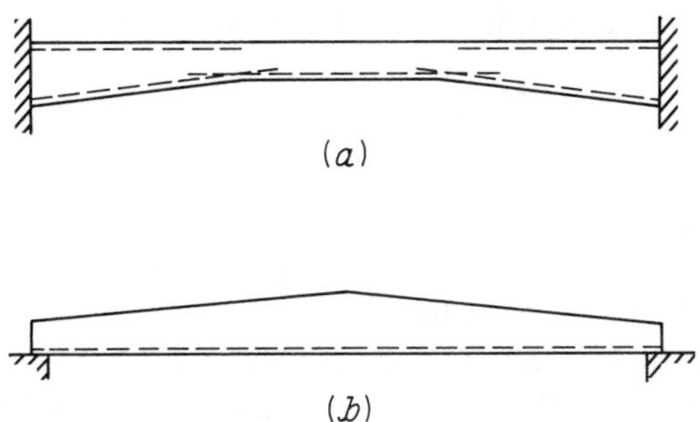

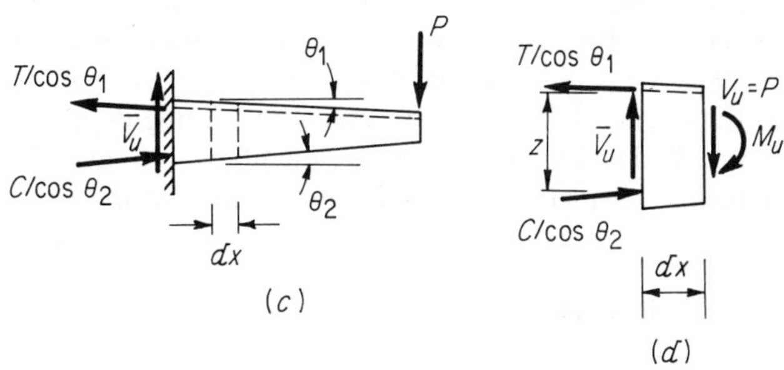

FIGURE 4.17
Effect of varying beam depth on shear.

depth, the inclination of the internal compressive and tensile stress resultants may significantly affect the shear for which the beam should be designed. In addition, the shear resistance of such members may differ from that of prismatic beams.

Figure 4.17*c* shows a cantilever beam, with fixed support at the left end, carrying a single concentrated load P at the right. The depth increases linearly in the direction of increasing moment. In such cases, the internal tension in the steel and the compressive stress resultant in the concrete are inclined, and introduce components transverse to the axis of the member. With reference to Fig. 4.17*d*, showing a short length dx of the beam, if the slope of the top surface is θ_1 and that of the bottom is θ_2, the net shear force $\overline{V}_u$ for which the beam should be designed is very nearly equal to

$$\overline{V}_u = V_u - T \tan \theta_1 - C \tan \theta_2$$

where V_u is the external shear force equal to the load P here, and $C = T = M_u/z$. The internal lever arm $z = (d - a/2)$ as usual. Thus, in a case for which the beam depth increases in the direction of increasing moment, the shear for which the member should be designed is approximately

$$\overline{V}_u = V_u - \frac{M_u}{z}(\tan \theta_1 + \tan \theta_2) \tag{4.20a}$$

For the infrequent case in which the member depth decreases in the direction of increasing moment, it is easily confirmed that the corresponding equation is

$$\overline{V}_u = V_u + \frac{M_u}{z}(\tan\theta_1 + \tan\theta_2) \tag{4.20b}$$

These equations are approximate because the direction of the internal forces is not exactly as assumed; however, the equations may be used without significant error provided the slope angles do not exceed about 30°.

There has been very little research studying the shear strength of beams having varying depth. Tests reported in Ref. 4.13 on simple span beams with haunches at slopes up to about 15° and with depths both increasing and decreasing in the direction of increasing moments indicate no appreciable change in the cracking load V_{cr} compared with that for prismatic members. Furthermore, the strength of the haunched beams, which contained vertical stirrups as web reinforcement, was not significantly decreased or increased, regardless of the direction of decreasing depth. Based on this information, it appears safe to design beams with varying depth for shear using equations for V_c and V_s developed for prismatic members, provided the actual depth d at the section under consideration is used in the calculations.

4.8 TRUSS MODELS FOR SHEAR ANALYSIS AND DESIGN

The ACI Code method of design for shear and diagonal tension in beams, presented in preceding sections of this chapter, is essentially empirical. While generally leading to safe designs, the ACI Code "$V_c + V_s$" approach lacks a proper physical model for the behavior of beams subject to shear combined with bending, and its shortcomings are now generally recognized. The "concrete contribution" V_c is generally considered to be some combination of force transfer by dowel action of the main steel, aggregate interlock along a diagonal crack, and shear in the uncracked concrete beyond the end of the crack. The values of each contribution are not identified. A rather vague rationalization is followed in adopting the diagonal cracking load of a member *without* web steel as the concrete contribution to the shear strength of an otherwise identical beam *with* web steel (see Sec. 4.4). Furthermore, Eq. (4.12*a*), used to predict that diagonal cracking load, is now recognized to overestimate concrete shear strength for beams with low steel ratios (Refs 4.7 and 4.8) and to overestimate the gain in shear strength resulting from use of high strength concrete (Refs 4.9 to 4.11). It has also been shown to overestimate the influence of $V_u d/M_u$ (Ref. 4.3). Additional research has shown that shear strength decreases, compared with the prediction of Eq. (4.12*a*), as member size increases (Ref. 4.14).

Ad hoc procedures are built into the ACI Code to adjust for certain of these deficiencies, but it follows that it is necessary to include equations, also empirically developed for the most part, for specific classes of members (e.g., deep beams vs. normal beams, beams with axial loads, prestressed vs. nonprestressed beams, etc.) with restrictions on the range of applicability of such equations.

Attention is increasingly being given to the development of design approaches based on rational behavioral models, generally applicable, rather than on empirical evidence alone. The best model for beams with web reinforcement appears to be the *truss model*. Originally introduced by Ritter (Ref. 4.15) and Morsch (Ref. 4.16) at the turn of the last century, a simplified version of truss model has long provided the basis for design of shear steel. However, the concept has been greatly extended by the recent work of Schlaich, Thurlimann, Marti, Collins, MacGregor, and others (Refs. 4.17 to 4.21). In its present form, the truss analogy is broadly applicable, not only to the design of beams of normal proportions for shear, but also for torsion, deep beams, beams with axial tension or compression, and brackets and corbels. It is effective for many other problems in reinforced concrete design, such as the design of "disturbed regions" near reactions or concentrated loads and the detailing of joint reinforcement (see Chap. 10).

Applied to normal beams subject to shear and bending, the truss model provides a clear insight into the flow of forces and provides an excellent basis for design and detailing of reinforcement. The essential features of the truss model may be understood with reference to Fig. 4.18*a*, which shows one-half the span of a simply supported, uniformly loaded beam. The combined action of flexure and shear produces the pattern of cracking shown. Reinforcement consists of the main flexural steel near the tension face and vertical stirrups distributed over the span.

The structural action can be represented by the truss of Fig. 4.18*b*, with the main steel providing the tension chord, the concrete top flange acting as the compression chord, the stirrups providing the vertical tension web members, and the concrete between inclined cracks acting as compression diagonals. The truss is formed by lumping all of the stirrups cut by section *a-a* into one vertical member and all of the diagonal concrete struts cut by section *b-b* into one compression diagonal.

Fig. 4.18*c* shows a more complicated but more realistic truss representation and illustrates the five basic components of a complete truss model: (a) struts, or concrete compression members uniaxially loaded, (b) ties, or steel tension members, (c) joints at the intersection of truss members, assumed to be pin-connected, (d) compression fans, which form at "disturbed" regions such as at the supports or under concentrated loads, transmitting the forces into the beam, and (e) diagonal compression fields, occurring where parallel compression struts transmit force from one stirrup to another. The slope angle θ of the diagonal compression struts is not necessarily 45 degrees, as is assumed by current ACI methods for stirrup design, but may range between about 25 and 65 degrees, depending to a large extent on the arrangement of reinforcement. It is assumed in developing the truss model that all of the stirrups reach the yield stress at failure. With the force in all the verticals known and equal to $A_v f_y$, the truss of Fig. 4.18*c* becomes statically determinate. Rational design equations can be based on the truss model for ordinary cases, but the model also permits the direct numerical solution for required reinforcement for special cases (see Refs. 4.18 and 4.19).

The Canadian National Standard for reinforced concrete (Ref. 4.22) includes a method of shear design that is essentially the same as the present ACI method but that also includes as an alternative the "general method" based on the truss model. Referred to as the *compression field theory* (Refs. 4.20 and 4.21), the method includes consideration of equilibrium and compatibility requirements and uses appropriate stress-strain relations for the reinforcement and the concrete. In its complete form, the diagonal compression field theory is capable of predicting not only the failure load but also the complete load-deformation response.

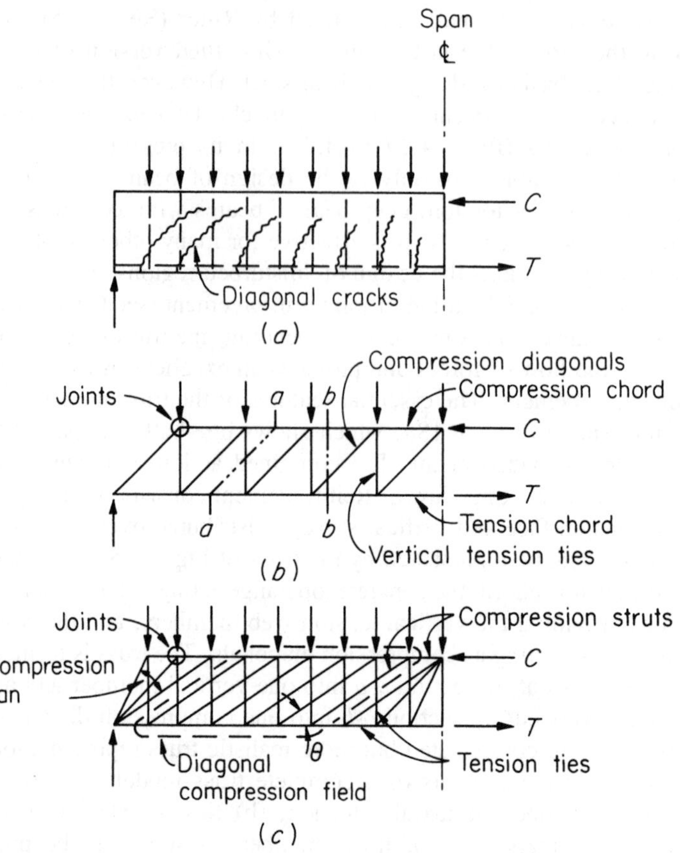

FIGURE 4.18
Truss model for beams with web reinforcement: (*a*) uniformly-loaded beam; (*b*) simple truss model; (*c*) more realistic model.

The basic elements of the compression field theory, applied to members carrying combined flexure and shear, will be clear from Fig. 4.19. Figure 4.19*a* shows a simple-span concrete beam, reinforced with longitudinal bars and transverse stirrups, and carrying a uniformly distributed loading along the top face. The light diagonal lines are an idealized representation of potential tensile cracking in the concrete.

Figure 4.19*b* illustrates that the net shear V at a section a distance x from the support is resisted by the vertical component of the diagonal compression force in the concrete struts. The horizontal component of the compression in the struts must be equilibrated by the total tension force ΔN in the longitudinal steel. Thus, with reference to Figs. 4.19*b* and *e*, the magnitude of the longitudinal tension resulting from shear is

$$\Delta N = \frac{V}{\tan\theta} \tag{4.21}$$

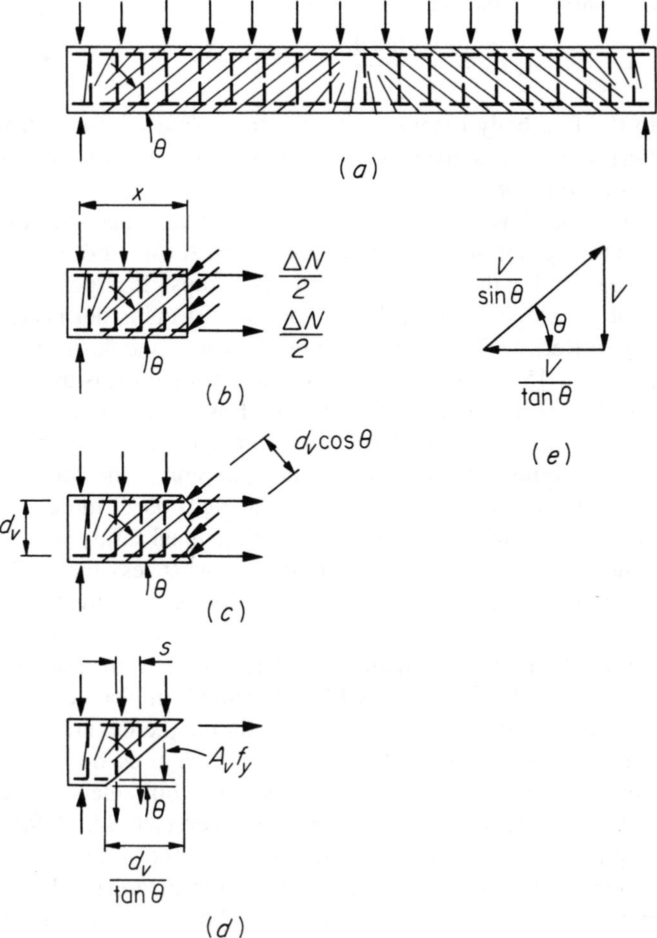

FIGURE 4.19
Basis of compression field theory for shear: (*a*) beam with shear and longitudinal steel; (*b*) tension in horizontal bars due to shear; (*c*) diagonal compression on beam web; (*d*) vertical tension in stirrups; (*e*) equilibrium diagram of forces due to shear. (*Adapted from Ref. 4.20.*)

where θ is the angle of inclination of the diagonal struts. These forces superimpose on the longitudinal forces owing to flexure, not shown in Fig. 4.19*b*.

The effective depth for shear calculations, according to this method, is taken at the distance between longitudinal force resultants, d_v. Thus, from Fig. 4.19*c*, the diagonal compressive stress in a web having width b_v is

$$f_d = \frac{V}{b_v d_v \sin\theta \cos\theta} \tag{4.22}$$

The tensile force in the vertical stirrups, each having area A_v and assumed to act at the yield stress f_y, can be found from the free body of Fig. 4.19*d*. With stirrups assumed to be at uniform spacing s,

$$A_v f_y = \frac{V s \tan\theta}{d_v} \tag{4.23}$$

Note, with reference to the free body diagram, that the transverse reinforcement within the length $d_v/\tan\theta$ can be designed to resist the lowest shear that occurs within this length, i.e., the shear at the right end.

In the ACI Code method developed in Sec. 4.4, it was assumed that the angle θ was 45°. With that assumption, and if d is substituted for d_v, Eq. (4.23) is identical to that used earlier for the design of vertical stirrups. It is generally recognized, however, that the slope angle of the compression struts is not necessarily 45°, and according to Ref. 4.22 that angle can be chosen by the designer to be anywhere in the range from 15 to 75°, provided the same value of θ is used in satisfying all requirements at a section. It is evident from Eqs. (4.21) and (4.23) that, if a lower slope angle is selected, less vertical reinforcement but more horizontal reinforcement will be required. In addition, the compression in the concrete diagonals will be increased. Conversely, if a higher slope angle is used, more vertical steel but less horizontal steel will be needed, and the diagonal thrust will be less. It is generally economical to use a slope angle θ somewhat less than 45°, with the limitation that the concrete diagonal struts must not be overstressed in compression.

The requirement for additional longitudinal steel resulting from diagonal compression is not recognized explicitly in the ACI Code method for beam design. However, the ACI Code does contain requirements for extending the flexural steel a distance d (or 12 bar diameters) beyond the point at which it is no longer required to resist flexure. One of the rationalizations for this requirement is the redistribution of forces that occurs in a diagonally cracked beam (see Sec. 5.9*b*).

The complete compression field theory includes development of an expression for the theoretical angle of inclination of the compression diagonals, based on compatibility of strains in the longitudinal, transverse, and diagonal directions, and incorporating the stress-strain relations applicable to the concrete and steel. Included also are limitations on the diagonal compression stress in the concrete struts, established recognizing the biaxial compression-tension state of stress in the concrete (see Sec. 2.9). Other considerations include control of diagonal crack widths, provision for minimum reinforcement, and requirements for steel detailing. A complete development can be found in Ref. 4.20, and codified design procedures are contained in Ref. 4.22.

In addition to providing a sound basis for the design of reinforcement for shear, the truss model gives important insights into detailing needs (Ref. 4.23). For example, it is clear from the truss model that stirrups must be capable of developing their full tensile strength throughout the entire stirrup height, not only at or near the middepth of the beam as assumed by ACI Code 12.13 (see also Sec. 5.5 of this text). ACI Code 12.13.2.2, in particular, permits anchorage of stirrups by embedment within the distance $d/2$ above or below middepth on the compression side of the member for a full development length, but not less than 24 times the stirrup bar diameter. This represents an unsafe detail and should not be used.

For wide beams, it is often the practice to use conventional U stirrups, with the vertical tension from the stirrups concentrated around the outermost bars. As shown by the truss model, diagonal compression struts can transmit forces only at the joints. Lack of stirrup legs in the interior of the beam web would force joints to form only at the exterior longitudinal bars, which would concentrate the diagonal compression in the outer faces of the beam and possibly result in premature failure. It is best to form a truss joint at each of the longitudinal bars, and multiple leg stirrups should always be used in wide beams (see Fig. 4.8*c*).

4.9 DEEP BEAMS

Some concrete members have depth much greater than normal, in relation to their span, while the thickness in the perpendicular direction is much smaller than either span or depth. The main loads and reactions act in the plane of the member, and a state of *plane stress* in the concrete is approximated. Members of this type are called *deep beams*. They can be defined, with reference to Fig. 4.20, as beams having a ratio of span to depth, l_n/h, of about 5 or less, or having a shear span a less than about twice the depth.

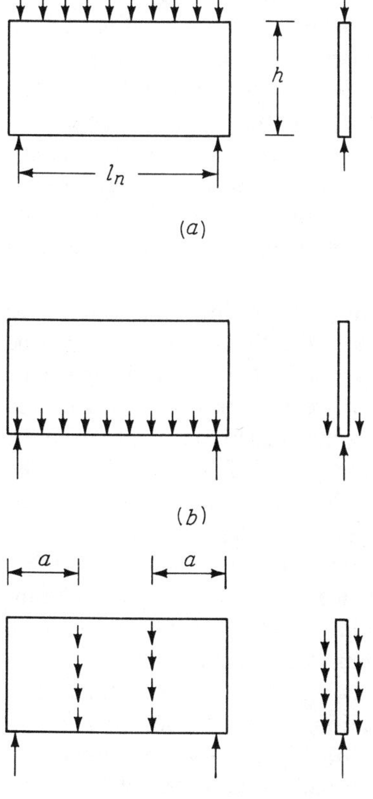

FIGURE 4.20
Placement of loads on deep beams: (*a*) loads applied along the compression edge; (*b*) loads suspended along the tension edge; (*c*) loads distributed through depth.

Examples of members of this type are found in transfer girders used in multistory buildings to provide column offsets, in foundation walls, walls of rectangular tanks and bins, floor diaphragms, and shear walls, as well as in folded plate roof structures. The behavior of deep beams is significantly different from that of beams of more normal proportions, requiring special consideration in analysis, design, and detailing of reinforcement.

Deep beams are usually loaded along the top edge, as in Fig. 4.20*a*, with reactions provided at the bottom. However, in some cases, e.g., the side walls of storage bins, the loads may be applied along the bottom edge, as in Fig. 4.20*b*. Loads may also be applied more or less uniformly throughout the depth, as in Fig. 4.20*c*, by other deep members framing in at right angles, and reactions may also be distributed through the depth. Deep beams may be simply supported or continuous.

Because of their proportions, they are likely to have strength controlled by shear. On the other hand, their shear strength is likely to be significantly greater than predicted by the usual equations. Special design methods account for these differences.

a. Behavior

Stresses in deep beams before cracking can be studied using the methods of two-dimensional elasticity, photoelasticity, or finite element analysis. Such studies confirm that the usual hypothesis, that plane sections before bending remain plane after bending, does not hold for deep beams. Significant warping of the cross sections occurs because of high shear stresses (Refs. 4.24 and 4.25). Consequently, flexural stresses are not linearly distributed, even in the elastic range, and the usual methods for calculating section properties and stresses cannot be applied.

On the other hand, elastic stress analysis is of limited interest because for such members deflections at service load are not likely to be troublesome. The main purpose of elastic analysis is to predict the location and orientation of flexural and shear cracks.

Of far greater importance is strength analysis to determine the ultimate-load capacity. Theory, confirmed by tests, indicates that the flexural strength can be predicted with sufficient accuracy using the same methods employed for beams of normal proportions. The equivalent rectangular stress block and the associated parameters can be employed without change. Thus the stress-block depth a can be found from Eq. (3.31) and the nominal flexural strength M_n from Eq. (3.30). The usual factor of $\phi = 0.90$ is applied to determine the design strength in bending. Although tests indicate that, because of biaxiality of compressive stress in the concrete compression zone, ultimate strains much larger than the usual $\epsilon_u = 0.003$ may be attained, this mainly affects the balanced steel ratio. Because it is seldom, if ever, feasible to use more than a small fraction of the balanced steel ratio in deep beams, the larger concrete strain capacity has little practical effect.

Shear strength of deep beams may be as much as 2 or 3 times greater than that predicted using conventional ACI Code equations developed for members of normal proportions, i.e., Eq. (4.12*a*) or Eq. (12.4*b*). In Sec. 4.4 it was explained that shear transfer in diagonally cracked beams is usually assumed to take place by four mechanisms: (a) direct transfer in the uncracked concrete compression zone,

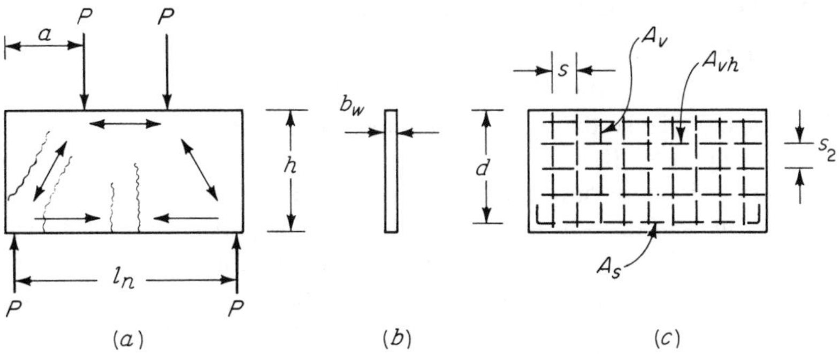

FIGURE 4.21
Deep beam carrying concentrated load: (*a*) loads, reactions, and internal forces; (*b*) cross section; (*c*) reinforcement.

(b) aggregate interlock, (c) dowel action of the main flexural reinforcement, and (d) direct tension in the web steel. For deep beams, however, a significant part of the load is transferred directly from the point of application to the supports by diagonal compression struts, as illustrated by Fig. 4.21*a*. Diagonal cracks that form roughly in a direction parallel to a line from the load to support isolate a compression strut, which acts with the horizontal compression in the concrete and the tension in the main reinforcement to equilibrate the loads. The geometry of this mechanism and the relative importance of each contribution to shear strength clearly depend on the proportions of the member as well as the placement of the loads and reactions. The parameter a/d, shown in Fig. 4.21*a*, is significant. For a deep beam with load distributed along the top edge, as in Fig. 4.20*a*, the parameter a/d may be replaced by the equivalent parameter M/Vd. The equivalence, for a beam with concentrated loads, as in Fig. 4.21*a*, is easily shown.

The reinforcement of deep beams differs from that of normal beams in ways dictated by the special features just described. The main flexural steel is placed near the tension edge, as usual, although because of the greater depth of the tension zone it may be advisable to distribute such steel over, say, the bottom third of the member (Ref. 4.24). Because the ultimate strength of deep beams depends upon strut-and-tie action, in which the main steel is fully stressed over nearly its entire length rather than only at the maximum moment section, special attention must be paid to the anchorage of such steel. Hooks or bends are normally used, even though deformed bars are specified.

Because of the orientation of the principal stresses in deep beams, when diagonal cracking occurs, it will be at a slope steeper than 45° in most cases. Consequently, while it is important to include vertical stirrups, they are apt to be less effective than horizontal web steel, placed as shown in Fig. 4.21*c* (Ref. 4.26). The horizontal bars are effective not only because they act more in the direction perpendicular to the diagonal crack, thus improving shear transfer by aggregate interlock, but also because they contribute to shear transfer by dowel action. Normal requirements for anchorage of web reinforcement apply to deep beams. Special attention must be paid to the importance of vertical web steel in functioning as hangers to carry loads suspended near the bottom edge of some deep beams, as in Fig. 4.20*b* and *c*.

b. ACI Code Provisions for Deep Beam Design

According to ACI Code 11.8, special provisions for shear are to be applied to beams for which l_n/d is less than 5 and which are loaded on one face and supported on the opposite face so that diagonal compression struts can form between the loads and the supports. If the loads are applied at the sides or bottom of a member, design provisions for ordinary beams apply. The special shear provisions apply to *simply supported* deep beams but not to *continuous* deep beams. In the latter case, the member is to be designed for shear according to the normal beam design procedures, or, as an alternative, the continuous deep beam can be designed according to the truss model concept (see Sec. 4.9c). The empirical nature of the deep beam provisions, derived for simple spans, requires this exclusion.

As usual, the design basis is that

$$V_u \leq \phi V_n \tag{4.10}$$

where $\phi = 0.85$ for shear, and

$$V_n = V_c + V_s \tag{4.24}$$

Regardless of the amount of reinforcement provided, the nominal strength V_n is not to be taken greater than the following:

For $l_n/d < 2$:
$$V_n = 8\sqrt{f'_c}\,b_w d \tag{4.25a}$$

For $2 \leq l_n/d \leq 5$:
$$V_n = \frac{2}{3}\left(10 + \frac{l_n}{d}\right)\sqrt{f'_c}\,b_w d \tag{4.25b}$$

The analogous ACI Code restriction for normal beams states that $V_s \leq 8\sqrt{f'_c}\,b_w d$; if the concrete contribution V_c is taken equal to $2\sqrt{f'_c}\,b_w d$ as usual, this amounts to an upper limit of $V_n = 10\sqrt{f'_c}\,b_w d$ for beams with l_n/d greater than 5. The variation of the maximum permissible V_n, as a function of l_n/d, is shown in Fig. 4.22.†

† The justification for the relatively small reduction, not greater than 20 percent, in the permissible V_n for deep beams, which in fact are stronger in shear than normal beams, is not clear.

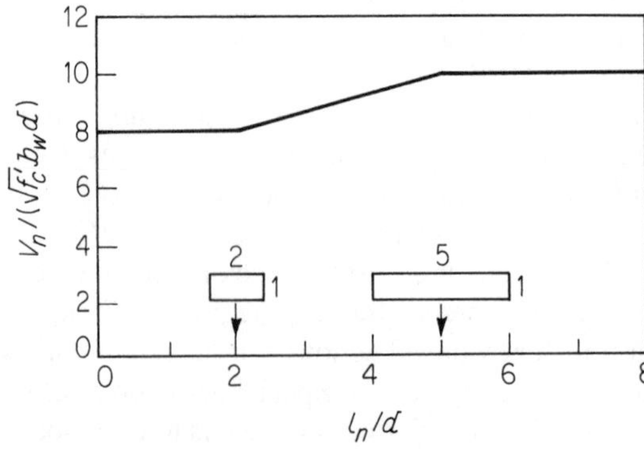

FIGURE 4.22
ACI Code limitation on total nominal shear strength V_n for deep beams.

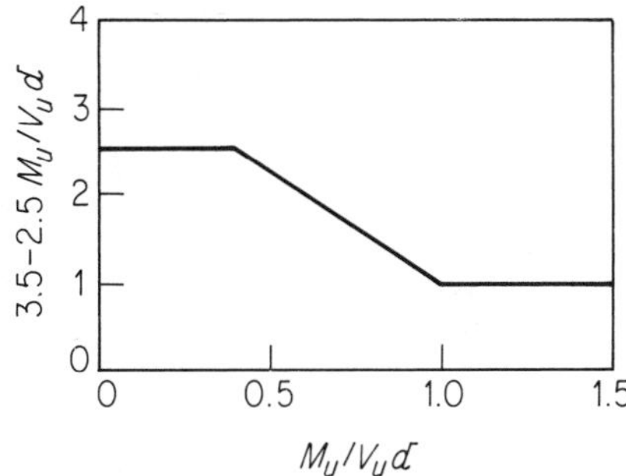

FIGURE 4.23
Shear strength multiplier for deep beams.

The critical section for shear is to be taken a distance $0.15l_n$ from the face of supports for uniformly distributed loads and $0.5a$ for beams with concentrated loads, but not to exceed a distance d from the support face in either case. Shear reinforcement required by calculation or other ACI Code provision at the critical section is to be used throughout the span.

Because of the strength increase attainable for deep beams owing to strut-and-tie action, the ACI Code provisions permit the usual value of the concrete shear strength V_c, calculated by Eq. (4.12*a*), to be increased by a multiplier that depends upon the ratio $M_u/V_u d$. For deep beams, the concrete contribution to shear strength can be computed from

$$V_c = \left(3.5 - 2.5\frac{M_u}{V_u d}\right)\left(1.9\sqrt{f_c'} + 2500\rho_w\frac{V_u d}{M_u}\right)b_w d \tag{4.26}$$

with the restrictions that the multiplier $(3.5 - 2.5M_u/V_u d)$ must not exceed 2.5 and that V_c must not be taken greater than $6\sqrt{f_c'}b_w d$. In Eq. (4.26), M_u and V_u are the moment and shear force, at factored loads, occurring simultaneously at the critical section.† Figure 4.23 shows the value of the multiplier in Eq. (4.26) as a function of the parameter $M_u/V_u d$, and illustrates the significant strength enhancement of V_c for deep members, for which $M_u/V_u d$ is typically low at the critical section for shear. For example, a deep beam with a span-depth ratio of 3, loaded at the third points, will have $M_u/V_u d = 0.5$ at the critical section, permitting an increase in V_c to 2.25 times the value for normal beams.

When the shear force V_u at factored loads exceeds the design shear strength of the concrete ϕV_c, shear reinforcement must be provided to carry the excess shear. The contribution of the web steel V_s is to be calculated from

$$V_s = \left[\frac{A_v}{s}\left(\frac{1 + l_n/d}{12}\right) + \frac{A_{vh}}{s_2}\left(\frac{11 - l_n/d}{12}\right)\right]f_y d \tag{4.27}$$

† The concrete shear strength may also be computed by the approximate equation $V_c = 2\sqrt{f_c'}b_w d$, according to the ACI Code, although the use of this alternative expression would be highly conservative and uneconomical.

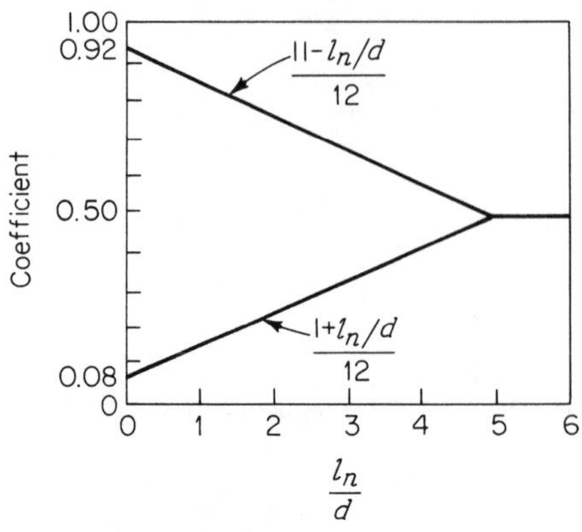

FIGURE 4.24
Effectiveness coefficients for vertical and horizontal web reinforcement in deep beams.

in which A_v is the area of shear reinforcement perpendicular to the main flexural steel within a distance s and A_{vh} is the area of shear reinforcement parallel to the main flexural steel within a distance s_2 (see Fig. 4.21c).

Combining Eqs. (4.10), (4.24), and (4.27) and rearranging terms, one obtains for the required shear reinforcement for deep beams the expression

$$\frac{A_v}{s}\left(\frac{1 + l_n/d}{12}\right) + \frac{A_{vh}}{s_2}\left(\frac{11 - l_n/d}{12}\right) = \frac{V_u - \phi V_c}{\phi f_y d} \tag{4.28}$$

The relative amounts of horizontal and vertical web steel that are used, based on Eq. (4.28), may vary within the following restrictions: the area A_v must not be less than $0.0015 b_w s$, and s must not exceed $d/5$ or 18 in. The area A_{vh} must not be less than $0.0025 b_w s_2$, and s_2 must not exceed $d/3$ or 18 in.

For design purposes, it is useful to note that the coefficients in parenthesis in Eq. (4.28) are weighting factors for the relative effectiveness of the vertical and horizontal web steel. The values of these factors are plotted in Fig. 4.24 as a function of the parameter l_n/d. It is seen that, for very deep beams with small l_n/d, the horizontal steel A_{vh} is dominantly effective, and the addition of vertical web steel A_v will have little effect in increasing strength. As the ratio l_n/d increases, the effectiveness of the vertical steel tends to increase, until at $l_n/d = 5$ (the limit of deep beams according to the ACI Code definition), vertical and horizontal steel are taken to be equally effective. Thus, for very deep beams it is more efficient to add web steel, if needed, in the form of horizontal bars, while satisfying the minimum requirements for vertical steel.†

---

† Publications resulting from the work of ACI Committee 426 on Shear and Diagonal Tension have drawn attention to certain inconsistencies in the use of Eq. (4.27) particularly for beams in the transition range from deep to normal proportions. It is to be hoped that they will be resolved in future editions of the Code (Refs. 4.8 and 4.27).

Example 4.5. A transfer girder is to carry two columns, each with factored load of 1200 kips, at the third points of its 36 ft span. A distributed factored load of 3.96 kips/ft will be applied along its top edge. The general arrangement is shown in Fig. 4.25*a*. A beam width of 2 ft and total depth of 12 ft are tentatively selected according to geometric requirements. Design the beam for the given loads, plus self-weight, using $f_y = 60,000$ psi and $f_c' = 4000$ psi.

Solution. For the trial beam dimensions, the self-weight is $2 \times 12 \times 150/1000 = 3.6$ kips/ft. Applying the usual load factor of 1.4 and adding the superimposed load, provide for a total distributed load of $1.4 \times 3.6 + 3.96 = 9.0$ kips/ft in addition to the column loads. The resulting shear and moment diagrams are shown in Fig. 4.25*b* and *c*.

Allowing for multiple layers of large bars as the main tensile reinforcement, take the effective depth to be 8 in. less than the total depth, or 136 in. Thus $l_n/d = (36 - 1.5) \times 12/136 = 3.04$, less than the limiting value of 5, confirming that deep-beam provisions apply.

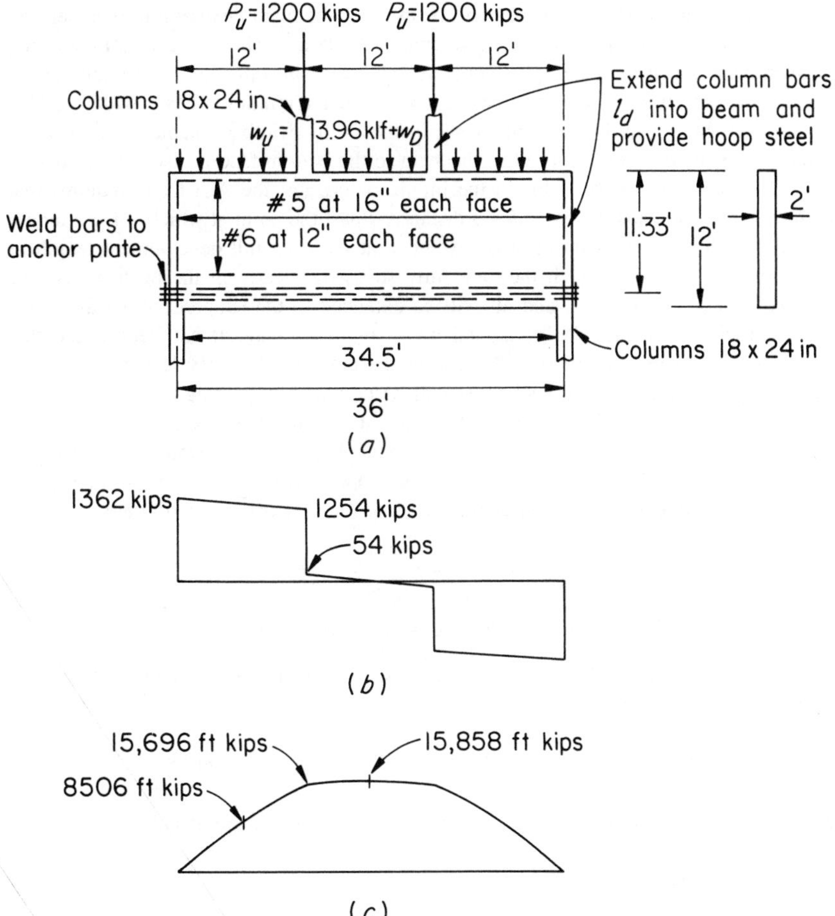

FIGURE 4.25
Deep beam example: (*a*) beam dimensions and reinforcement; (*b*) shear diagram; (*c*) moment diagram.

The main flexural reinforcement will be designed using ordinary beam equations. Assuming the stress-block depth $a = 20$ in for trial gives

$$A_s = \frac{M_u}{\phi f_y(d - a/2)} = \frac{15{,}858 \times 12}{0.9 \times 60(136 - 10)} = 27.96 \text{ in}^2$$

Checking the initial assumption gives

$$a = \frac{A_s f_y}{0.85 f_c' b} = \frac{27.96 \times 60}{0.85 \times 4 \times 24} = 20.6 \text{ in.}$$

sufficiently close to the assumed value for no revision to be necessary. For the materials used, the balanced steel ratio is

$$\rho_b = 0.85\beta_1 \frac{f_c'}{f_y} \frac{87}{87 + f_y} = 0.85 \times 0.85 \times \frac{4}{60} \times \frac{87}{147} = 0.0285$$

and, according to the ACI Code, the maximum steel ratio permitted is $0.75 \times 0.0285 = 0.0214$. The actual steel ratio is $27.96/(24 \times 136) = 0.0086$, well below the permitted maximum. A total of 18 No. 11 bars will be used for the main flexural reinforcement, in three layers of six bars each, providing an area of 28.08 in^2. Allowing for $1\frac{3}{8}$ in. spacing between bars of that diameter, plus concrete cover of 3 in. outside of the main bars to provide for web steel and concrete protection, one has a minimum beam width of 21.1 in.; the bars can easily be accommodated in the 24 in. width available. The assumed distance of 8 in. from the centroid of the steel to the bottom face is also satisfactory.

To allow for possible shift in the moment diagram, the ACI Code requires that the tensile reinforcement be extended a distance d past the point at which it is no longer needed. In addition, the bars must be extended at least the full development length of 59 in. past the point at which they are fully stressed. Clearly in the present case the main bars cannot be cut off, and all will be extended to the supports. Note that there is not sufficient clearance for horizontal hooks or bends here. It has been found that vertical hooks create a plane of weakness in the concrete that may lead to premature failure. Consequently, special anchorage will be provided by passing all bars through a steel anchor plate and welding them to the plate at the outside surface.

The critical section for shear, according to the ACI Code provisions, is at a distance of $0.5a = 0.50(12 - 0.75) = 5.63$ ft from the support face, or 6.38 ft from the support centerline. The moment and shear at factored loads at that critical section are

$$M_u = 1362 \times 6.38 - 9 \times \frac{6.38^2}{2} = 8506 \text{ ft-kips}$$

$$V_u = 1362 - 9 \times 6.38 = 1305 \text{ kips}$$

According to the ACI Code provisions, the upper limit of the nominal shear strength is determined by Eq. (4.25*b*):

$$V_{n,\max} = \frac{2}{3}(10 + 3.04)\sqrt{4000} \times 24 \times \frac{136}{1000} = 1794 \text{ kips}$$

from which $\phi V_{n,\max} = 0.85 \times 1794 = 1525$ kips, well above the actual value of shear V_u.

Next the ratio $M_u/V_u d$ is calculated:

$$\frac{M_u}{V_u d} = \frac{8506 \times 12}{1305 \times 136} = 0.575$$

from which

$$3.5 - 2.5\frac{M_u}{V_u d} = 3.5 - 2.5 \times 0.575 = 2.06$$

This is confirmed to be below the upper limit value of 2.5. The concrete contribution to shear strength can now be found from Eq. (4.26):

$$V_c = 2.06\left(1.9\sqrt{4000} + 2500 \times \frac{0.0086}{0.575}\right) \times 24 \times \frac{136}{1000} = 1059 \text{ kips}$$

This is below the limiting value of $6\sqrt{4000} \times 24 \times 136/1000 = 1239$ kips; the design is satisfactory to this point.

The required web reinforcement will now be determined from Eq. (4.28):

$$\frac{A_v}{s}\left(\frac{1 + 3.04}{12}\right) + \frac{A_{vh}}{s_2}\left(\frac{11 - 3.04}{12}\right) = \frac{1305 - 0.85 \times 1059}{0.85 \times 60 \times 136}$$

$$0.337\frac{A_v}{s} + 0.663\frac{A_{vh}}{s_2} = 0.058$$

It is clear from the last result that the horizontal steel A_{vh} is more effectively utilized than the vertical bars A_v. Consequently, only the minimum required A_v will be provided. For these bars, the maximum spacing is not to exceed

$$s_{\max} = \frac{136}{5} = 27 \text{ in.}$$

or 18 in., which controls in this case. If one selects No. 5 bars for trial, with area $A_v = 0.31 \times 2 = 0.62$ in^2, the required spacing is based on the minimum requirement of $A_v = 0.0015 b_w s$:

$$s = \frac{0.62}{0.0015 \times 24} = 17.22 \text{ in.}$$

This is below, and close to, the maximum permitted spacing of 18 in. No. 5 vertical bars at 16 in. spacing each face will be used.

The spacing of the horizontal web bars must not exceed

$$s_{2,\max} = \frac{136}{3} = 45 \text{ in.}$$

or 18 in., which controls. The required area is found based on Eq. (4.28):

$$0.663\frac{A_{vh}}{s_2} = 0.058 - 0.337 \times \frac{0.62}{16} = 0.045$$

With No. 6 bars, each face provides an area $A_{vh} = 2 \times 0.44 = 0.88$ in^2, requiring a spacing of

$$s_2 = 0.663\frac{0.88}{0.045} = 12.96 \text{ in.}$$

No. 6 horizontal bars at 12 in. spacing each face will be used.

The arrangement of the web steel is summarized in Fig. 4.25*a*. The vertical bars will be detailed in the form of closed hoops with anchorage provided by bending around the horizontal bars top and bottom. At the ends of the horizontal web bars, 180° hooks will be used.

In order to ensure proper transmission of column loads and end reactions into the transfer girder, vertical bars from all columns will be extended into the girder by the full development length. To ensure against local failure at the heavy load concentrations, horizontal column ties will be provided throughout the length of the extended bars.

c. Truss Models for Deep Beam Design

Numerous deficiencies, inconsistencies, and arbitrary limitations will be noted in reviewing the ACI Code method for the design of deep beams. While beams designed according to ACI Code provisions generally have performed satisfactorily, such difficulties stem from attempting to adapt the empirical "$V_c + V_s$" approach, with its "concrete contribution" from dowel action, aggregate interlock, and direct concrete shear transfer, to members for which the dominant mechanism for force transfer from load to reaction is the direct compression strut. An alternative approach to deep beam design is based on the *truss model*, introduced in Sec. 4.8 for the design of normal beams for shear. It is particularly well-suited for the design of deep beams (Refs. 4.19 and 4.28). It is perfectly general and can easily be applied to beams with any span-depth ratio, to continuous as well as simple spans, and to deep beams with loads and reactions applied in any practical arrangement.

ACI Code 11.8.3 permits use of "methods that satisfy equilibrium and strength requirements" as an alternative to using the equations summarized in Sec. 4.9b, and ACI Commentary R11.8.3 relates this statement to the truss model approach through two basic references. Thus a more rational method for deep beam design is endorsed by ACI Code, although no specific guidance is given.

Basic features of deep beam analysis based on the truss model will be illustrated by the center-loaded simple span beam shown in Fig. 4.26. Elastic analysis of the uncracked member shows the direction of principal stresses, with dashed lines for compression trajectories and solid lines for tension. As load is increased, cracks can be expected to form perpendicular to the tension trajectories (i.e., locally parallel to the dashed lines). The largest tensile stresses act along the bottom edge of the beam, while the maximum compressive stresses act more or less parallel to lines from load to supports.

Fig. 4.26*b* shows a simplified representation of the internal force flow, where compression trajectories, now represented as compression struts, are shown as dashed lines, and tension trajectories, now simplified as tension ties, are shown as solid lines. The main tension tie is along the bottom of the beam between the supports, but because of the divergence of the thrust lines under the load, secondary tension ties are required perpendicular to the diagonal lines between load and support points. Representation of the force flow can be further simplified by the truss model shown in Fig. 4.26*c*, although this model does not provide an explanation for the inclined diagonal cracking that would occur.

The five basic features of the complete truss model, introduced in Sec. 4.8, included (a) compression struts, (b) tension ties, (c) joints, or nodes, (d)

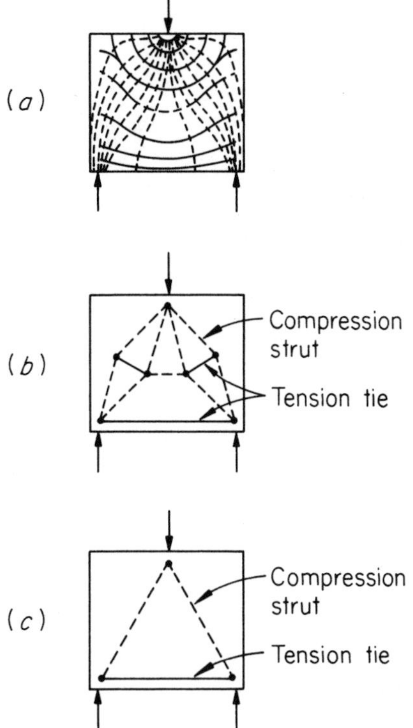

FIGURE 4.26
Center-loaded deep beam: (*a*) stress trajectories; (*b*) truss model; (*c*) simplified truss model. (*Adapted from Ref. 4.19.*)

compression fans, and (e) diagonal compressive fields. Use of these to develop truss models for deep beams will be illustrated for the third-point-loaded beams shown in Figs. 4.27*a* and 4.27*b*.

For the beam without stirrups, Fig. 4.27*a*, the concentrated loads are carried by two compression struts, shown shaded, along lines approximately between load points and reactions. The outward thrust at the bottom of the struts is equilibrated by a properly anchored tension tie. A horizontal compression strut between loads equilibrates the inward thrust at the top of the diagonal struts.

Joints, or nodes, occur at the intersection of the lines of action of the struts, ties, and loads, as shown by the darker shaded areas. These regions are loaded in biaxial compression, generally equal in the two principal directions. Care must be taken in developing the geometry of the truss model so that the centroids of the members and the lines of action of forces and reactions meet at any given joint.

The truss model of Fig. 4.27*a* can fail (a) by yielding of the tie, (b) by crushing of one of the struts, or (c) by crushing in a nodal region. All three types of failure have been observed in tests. If failure is to occur, the ductile tension failure would be preferred, so the beam should be proportioned with the strength of the tie governing.

The addition of stirrups will modify the internal flow of forces in the truss model. This is illustrated by the deep beam of Fig. 4.27*b*, otherwise identical to that of Fig. 4.27*a*, but with vertical, equally spaced stirrups between load and

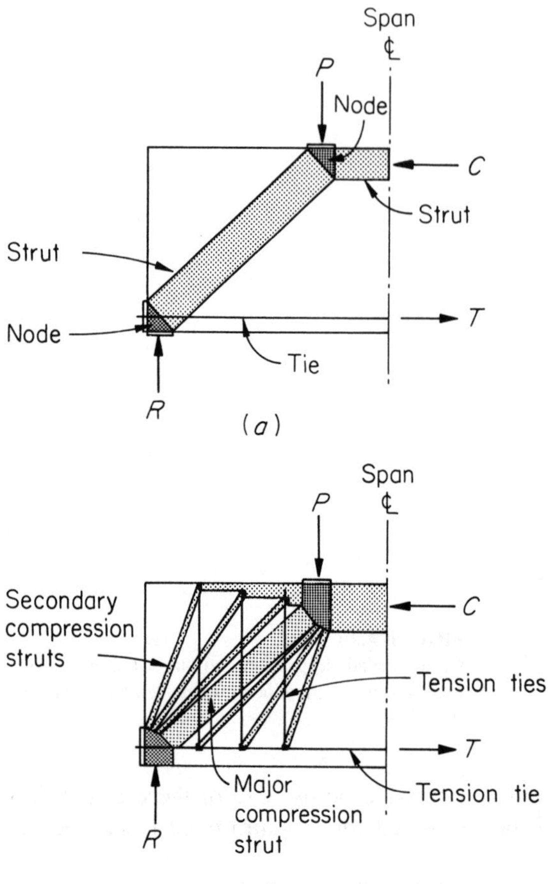

FIGURE 4.27
Truss models of deep beams: (*a*) beam without stirrups; (*b*) beam with stirrups

reaction. It is seen that the model is the sum of two trusses. The main force flow is through a major compression diagonal between load and support. The second truss utilizes the stirrups as vertical tension members, equilibrating the thrusts in the secondary struts in compression fans over the support and under the load. Although it appears to be considerably more complex than the truss in Fig. 4.27*a*, the truss becomes statically determinate with the assumption that the stress in each stirrup is at the yield, and the sharing of the load between the major compression strut and the minor struts can be determined. Note that from joint equilibrium along the bottom chord of the truss, the tension force in the tie is reduced stepwise at each stirrup, in contrast to the constant force in the tie for the beam of Fig. 4.27*a*. Similarly, the horizontal compression thrust along the top of the beam decreases stepwise along the span from the maximum value at midspan.

Perhaps the most difficult aspect of the truss model analysis is the layout of the truss itself. The following procedure is suggested in Ref. 4.18: (a) draw the truss model to scale, (b) visualize the force flow using consistent equilib-

Table 4.2 Effective concrete compressive strength

Structural member	f_{ce}
Truss node	
Joints bounded by compressive struts and bearing areas	$0.85f'_c$
Joints anchoring one tension tie	$0.65f'_c$
Joints anchoring tension ties in more than one direction	$0.50f'_c$
Isolated compression struts in deep beams or disturbed regions	$0.50f'_c$
Severely cracked webs of slender beams	$0.25f'_c$ to $0.45f'_c$

Source: from Ref. 4.19.

rium considerations, and (c) ensure with a careful detailing that the truss member forces can be developed and transferred at the nodes. The designer must take into account the necessary dimensions of the concrete struts and the possible placing and detailing of the reinforcement. An iterative procedure is normally required. Starting with an initial assumed truss, the designer can determine the forces and necessary strut widths, then modify the truss geometry and repeat the procedure until a satisfactory solution is obtained.

The size of the compression struts is chosen so that the concrete capacity at the struts and at the nodes is not exceeded at ultimate load. The effective concrete strength f_{ce} is less than the uniaxial compressive strength f'_c mainly because of the presence of lateral tensile stress (see Sec. 2.9 and Fig. 2.8 of Chap. 2) or cracking along the compressive stress trajectories and because of nonuniform strain conditions in the concrete. Values for f_{ce} given in Table 4.2 are suggested in Ref. 4.19. Ref. 4.18 suggests that an average value of $0.6f'_c$ be used.

Further information on the development of truss models and their use in the design of deep beams will be found in Refs. 4.18, 4.19, and 4.28.

4.10 SHEAR-FRICTION DESIGN METHOD

Generally, in reinforced concrete design, shear is used merely as a convenient measure of diagonal tension, which is the real concern. In contrast, there are circumstances such that direct shear may cause failure of reinforced concrete members. Such situations occur commonly in precast concrete structures, particularly in the vicinity of connections, as well as in composite construction combining cast-in-place concrete with either precast concrete or structural steel elements. Potential failure planes can be established for such cases along which direct shear stresses are high, and failure to provide adequate reinforcement across such planes may produce disastrous results.

The necessary reinforcement may be determined on the basis of the *shear-friction method* of design (Refs. 4.29 to 4.33). The basic approach is to assume that the concrete may crack in an unfavorable manner, or that slip may occur along a predetermined plane of weakness. Reinforcement must be provided crossing the potential or actual crack or shear plane to prevent direct shear failure.

The shear-friction theory is very simple, and the behavior is easily visualized. Figure 4.28*a* shows a cracked block of concrete, with the crack crossed by reinforcement. A shear force V_n acts parallel to the crack, and the resulting tendency for the upper block to slip relative to the lower is resisted largely by friction along the concrete interface at the crack. Since the crack surface is naturally rough and irregular, the effective coefficient of friction may be quite high. In addition, the irregular surface will cause the two blocks of concrete to separate slightly, as shown in Fig. 4.28*b*.

If reinforcement is present normal to the crack, then slippage and subsequent separation of the concrete will stress the steel in tension. Tests have confirmed that well-anchored steel will be stressed to its yield strength when shear failure is obtained. (Ref. 4.31). The resulting tensile force sets up an equal and opposite pressure between the concrete faces on either side of the crack. It is clear from the free body of Fig. 4.28*c* that the maximum value of this interface pressure is $A_{vf} f_y$, where A_{vf} is the total area of steel crossing the crack and f_y is its yield strength.

The concrete resistance to sliding may be expressed in terms of the normal force times a coefficient of friction μ. By setting the summation of horizontal forces equal to zero

$$V_n = \mu A_{vf} f_y \tag{4.29}$$

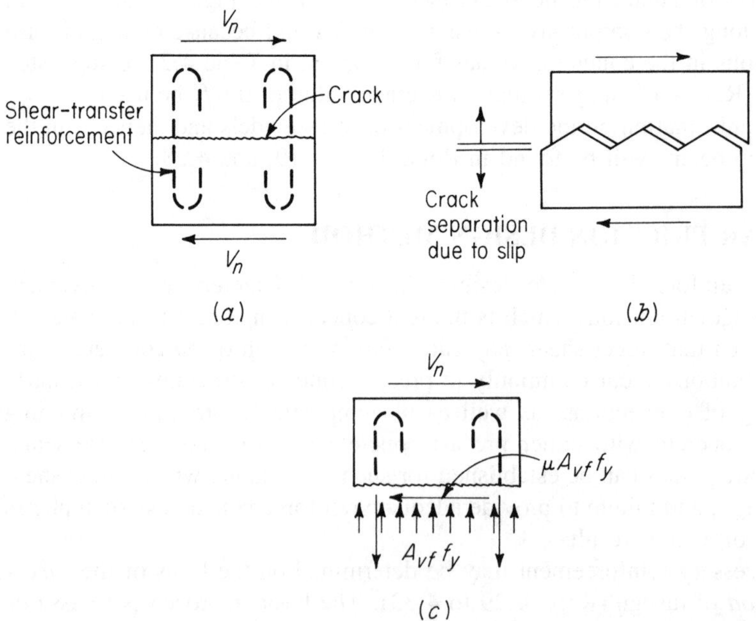

FIGURE 4.28
Basis of shear-friction design method: (*a*) applied shear; (*b*) enlarged representation of crack surface; (*c*) free body sketch of concrete above crack.

Defining the steel ratio $\rho = A_{vf}/A_c$, where A_c in this case is the area of the cracked surface, allows Eq. (4.29) to be rewritten in terms of the nominal shear stress v_n:

$$v_n = \mu \rho f_y \tag{4.30}$$

The relative movement of the concrete on opposite sides of the crack also subjects the individual reinforcing bars to shearing action, and the dowel resistance of the bars to this shearing action contributes to shear resistance. However, it is customary to neglect the dowel effect for simplicity in design and to compensate for this by using an artificially high value of the friction coefficient.

Based on early tests, μ may be taken equal to 1.4 for cracks in monolithic concrete, but V_n should not be assumed greater than $0.2f'_cA_c$ or $800A_c$ lb (Ref. 4.29).

In Fig. 4.29 the shear-transfer strength predicted by Eq. (4.30) is compared with experimental values obtained in more recent tests conducted at the University of Washington (Ref. 4.31). It is evident that Eq. (4.30) gives a conservative estimate of shear strength. It is also clear that strength considerably in excess of the upper limit of 800 psi can be developed if appropriate reinforcement is provided. It has been proposed (Ref. 4.31) that a modified form of Eq. (4.30) be adopted when ρf_y exceeds 600 psi, as follows:

$$v_n = \mu \rho f_y \left(\frac{300}{\rho f_y} + 0.5 \right) \tag{4.31}$$

The strengths predicted by Eq. (4.31) (indicated by the dashed line in Fig. 4.29) appear to give a satisfactory correlation with experimental results for concrete strengths greater than 2500 psi. Pending further data, it is recommended that an upper limit of $v_n = 1300$ psi be imposed for Eq. (4.31).

The provisions of ACI Code 11.7 are based on Eq. (4.29). The design strength is to be taken equal to ϕV_n where $\phi = 0.85$ for shear-friction design, and V_n is not to exceed the smaller of $0.2f'_cA_c$ or $800A_c$ lb. Recommendations for friction factor μ are as follows:

Concrete placed monolithically	1.4λ
Concrete placed against hardened concrete with surface intentionally roughened	1.0λ
Concrete placed against hardened concrete not intentionally roughened	0.6λ
Concrete anchored to as-rolled structural steel by headed studs or rebars	0.7λ

where $\lambda = 1.0$ for normal weight concrete, 0.85 for "sand-lightweight" concrete, and 0.75 for "all-lightweight" concrete. The yield strength of the reinforcement is not to exceed 60,000 psi. Direct tension across the shear plane, if present, is to be provided for by additional reinforcement, and permanent net compression across the shear plane may be taken as additive to the force in the shear-friction reinforcement $A_{vf}f_y$ when calculating the required A_{vf}.

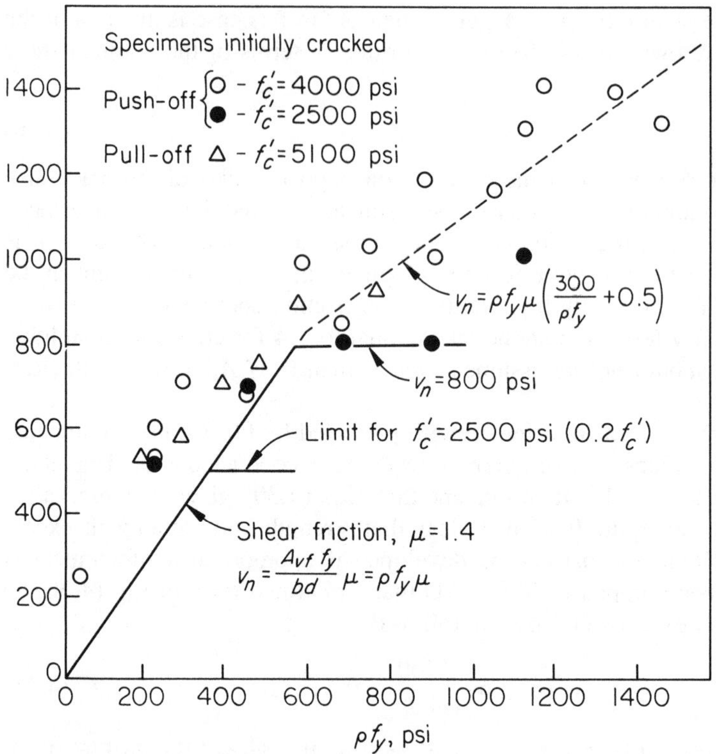

FIGURE 4.29
Calculated vs. experimental shear-transfer strength for initially cracked specimens. (*From Ref. 4.31*)

When shear is transferred between concrete newly placed against hardened concrete, the surface roughness is an important variable; an intentionally roughened surface is defined to have a full amplitude of approximately $\frac{1}{4}$ in. In any case, the old surface must be clean and free of laitance. When shear is to be transferred between as-rolled steel and concrete, the steel must be clean and without paint, according to ACI Code 11.7.

If V_u is the shear force to be resisted at factored loads, then with $V_u = \phi V_n$, the required steel area is found by transposition of Eq. (4.29):

$$A_{vf} = \frac{V_u}{\phi \mu f_y} \tag{4.32}$$

In some cases the shear-friction reinforcement may not cross the shear plane at 90° as described in the preceding paragraphs. If the shear-friction reinforcement is inclined to the shear plane so that the shear force is applied in the direction to increase tension in the steel, as in Fig. 4.30*a*, then the component of that tension parallel to the shear plane, shown in Fig. 4.30*b*, contributes to the resistance to slip. Then the shear strength may be computed from

$$V_n = A_{vf} f_y (\mu \sin \alpha_f + \cos \alpha_f) \tag{4.33}$$

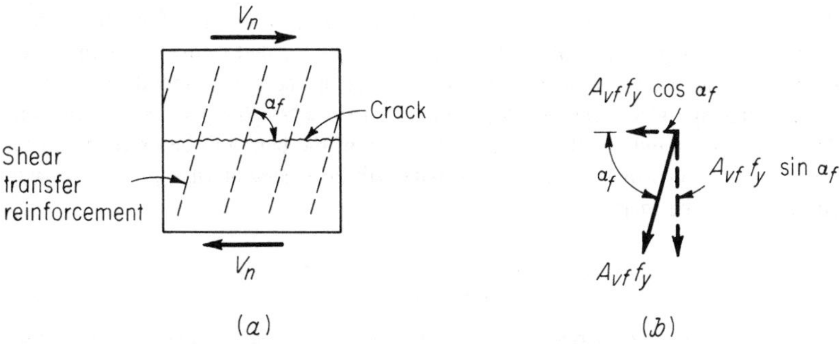

FIGURE 4.30
Shear-friction reinforcement inclined with respect to crack face.

in lieu of Eq. (4.29). Here α_f is the angle between the shear-friction reinforcement and the shear plane. If α_f is larger than 90°, i.e., if the inclination of the steel is such that the tension in the bars tends to be reduced by the shear force, then the assumption that the steel stress equals f_y is not valid, and a better arrangement of bars should be made.

Certain precautions should be observed in applying the shear-friction method of design. Reinforcement, of whatever type, should be well anchored to develop the yield strength of the steel, by the full development length or by hooks or bends, in the case of rebars, or by proper heads and welding, in the case of studs joining concrete to structural steel. The concrete should be well confined, and the liberal use of hoops has been recommended (Ref. 4.29). Care must be taken to consider all possible failure planes and to provide sufficient well-anchored steel across these planes.

Example 4.6 Design of beam bearing detail. A precast beam must be designed to resist a support reaction, at factored loads, of $V_u = 125$ kips applied to a 3×3 steel angle as shown in Fig. 4.31. In lieu of a calculated value, a horizontal force T_u, owing to restrained volume change, will be assumed at 20 percent of the vertical reaction, or 25 kips. Determine the required auxilliary reinforcement, using steel of yield strength $f_y = 60{,}000$ psi. Concrete design strength $f_c' = 5000$ psi.

Solution. A potential crack will be assumed at 20°, initiating at a point 4 in. from the end of the beam, as shown in Fig. 4.31*a*. The total required steel A_{vf} is the sum of that required to resist the effects of V_u and T_u. Equation (4.32) is modified accordingly:

$$A_{vf} = \frac{V_u \cos 20^\circ + T_u \sin 20^\circ}{\phi \mu f_y}$$

$$= \frac{125 \times 0.940 + 25 \times 0.340}{0.85 \times 1.4 \times 60}$$

$$= 1.76 \text{ in}^2$$

The net compression normal to the potential crack would be no less than V_u sin 20 − T_u cos 20 = 19 kips. This could be counted upon to reduce the required shear friction steel, according to ACI Code, but it will be discounted conservatively here. Four No. 6 bars will be used, providing an area of 1.77 in^2. They will be welded to the 3 × 3 angle and will extend into the beam a sufficient distance to develop the yield strength of the bars. According to the ACI Code, the development length for a No. 6 bar is 18 in. However, considering the uncertainty of the exact crack location, the bars will be extended 24 in. into the beam as shown in Fig. 4.31*a*. The bars will be placed at an angle of 15° with the bottom face of the member.

$$A_c = 16\left(\frac{4}{\sin 20°}\right) = 187 \text{ in}^2$$

Thus, according to the ACI Code, the maximum nominal shear strength of the surface is not to exceed $V_n = 0.2f_c'A_c = 187$ kips or $V_n = 800A_c = 150$ kips. The maximum design strength to be used is $\phi V_n = 0.85 \times 150 = 128$ kips. The applied shear on the interface at factored loads is

$$V_n = 125\cos 20° + 25\sin 20° = 126 \text{ kips}$$

and so the design is judged satisfactory to this point.

A second possible crack must be considered, as shown in Fig. 4.31*b*, resulting from the tendency of the entire anchorage weldment to pull horizontally out of the beam.

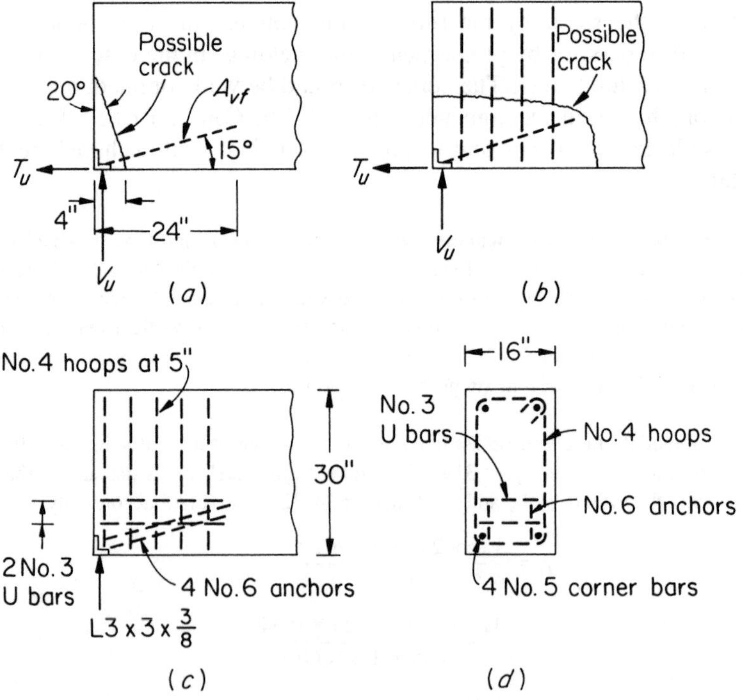

FIGURE 4.31
Design of beam bearing shoe: (*a*) diagonal crack; (*b*) horizontal crack; (*c*) reinforcement; (*d*) cross section.

The required steel area A_{sh} and the concrete shear stress will be calculated based on the development of the full yield tension in the bars A_{vf}. (Note that the factor ϕ need not be used here because it has already been introduced in computing A_{vf}.)

$$A_{sh} = \frac{A_{vf} f_y \cos 15^\circ}{\mu f_y}$$

$$= \frac{1.76 \times 0.966}{1.4}$$

$$= 1.21 \text{ in}^2$$

Four No. 4 hoops will be used, providing an area of 1.57 in^2.

The maximum shear force that can be transferred, according to the ACI Code limits, will be based conservatively on a horizontal plane 24 in. long. No strength reduction factor need be included in the calculation of this maximum value because it was already introduced in determining the steel area A_{vf} by which the shear force is applied. Accordingly,

$$V_n \leq 800 \times 16 \times 24 = 308 \text{ kips}$$

The maximum shear force that could be applied in the given instance is

$$V_n = 1.76 \times 60 \cos 15^\circ = 102 \text{ kips}$$

well below the specified maximum.

Additional confinement steel is recommended in Ref. 4.33, in the amount $V_u/8f_y$, to be placed in the form of hoops or hairpin bars confining the concrete near the bottom and end faces of a precast beam at the reaction. In the present case,

$$A_{cv} = A_{ch} = \frac{125}{8 \times 60}$$

$$= 0.26 \text{ in}^2$$

One additional No. 4 hoop will be placed as near as possible to the end face, and two U-shaped No. 3 bars will be added parallel to the bottom of the beam as shown in Fig. 4.31*c* and *d*. Also shown in Fig. 4.31*d* are four No. 5 corner bars that will provide anchorage for the hoop steel.

REFERENCES

4.1. "Shear and Diagonal Tension," pt. 2, ACI-ASCE Committee 326, *J. ACI*, vol. 59, no. 2, 1962, pp. 277–333.

4.2. B. Bresler and J. G. MacGregor, "Review of Concrete Beams Failing in Shear," *Proc. ASCE*, vol. 93, no. ST1, 1967, pp. 343–372.

4.3. "The Shear Strength of Reinforced Concrete Members," ASCE-ACI Task Committee 426, *Proc. ASCE*, vol. 99, no. ST6, 1973, pp. 1091–1187 (with extensive bibliography).

4.4. "The Shear Strength of Reinforced Concrete Members—Slabs," ASCE-ACI Task Committee 426, *Proc. ASCE*, vol. 100, no. ST8, 1974, pp. 1543–1591.

4.5. *Shear in Reinforced Concrete*, Vols. 1 and 2, Special Publication SP-42, American Concrete Institute, Detroit, 1974.

4.6. A. H. Nilson (ed.), *Finite Element Analysis of Reinforced Concrete*, American Society of Civil Engineers, New York, 1982.

4.7. J. G. MacGregor and P. Gergely, "Suggested Revisions to ACI Building Code Clauses Dealing with Shear in Beams," *J. ACI*, vol. 74, no. 10, 1977, pp. 493–500.

4.8. "Suggested Revisions to Shear Provisions for Building Codes," ACI-ASCE Committee 426, *J. ACI*, vol. 74, no. 9, 1977, pp. 458–469 (also see ACI Publ. 426. 1R-77, 84 pp., 1977).

4.9. A. H. Elzanaty, A. H. Nilson, and F. O. Slate, "Shear Capacity of Reinforced Concrete Beams Using High-Strength Concrete," *J. ACI*, vol. 83, no. 2, 1986, pp. 290–296.

4.10. A. G. Mphonde and G. C. Franz, "Shear Tests of High- and Low-Strength Concrete Beams Without Stirrups," *J. ACI*, vol. 81, no. 4, 1984, pp. 350–357.

4.11. S. H. Ahmad, A. R. Khaloo, and A. Proveda, "Shear Capacity of Reinforced High-Strength Concrete Beams," *J. ACI*, vol. 83, no. 2, 1986, pp. 297–305.

4.12. S. Martinez, A. H. Nilson, and F. O. Slate, "Short-Term Mechanical Properties of High-Strength Lightweight Concrete," Research Report No. 82-9, Department of Structural Engineering, Cornell University, August 1982.

4.13. S. Y. Debaiky and E. I. Elmiema, "Behavior and Strength of Reinforced Concrete Haunched Beams in Shear," *J. ACI*, vol. 79, no. 3, 1982, pp. 184–194.

4.14. G. N. J. Kani, "How Safe Are Our Large Reinforced Concrete Beams," *J. ACI*, vol. 64, no. 3, 1967, pp. 128–141.

4.15. W. Ritter, "Die Bauweise Hennebique" (The Hennebique System), Schweizerische Bauzeitung, XXXIII, no. 7, 1899.

4.16. E. Morsch, *Der Eisenbetonbau, seine Theorie and Anwendung* (Reinforced Concrete Theory and Application), Verlag Konrad Wittwer, Stuttgart, 1912.

4.17. J. Schlaich, K. Shafer, and M. Jennewein, "Toward a Consistent Design of Structural Concrete," *J. Prestressed Concr. Inst.*, vol. 32, no. 3, 1987, pp. 74–150.

4.18. P. Marti, "Truss Models in Detailing," *Conc. Int.*, vol. 7, no. 12, 1985, pp. 66–73.

4.19. J. G. MacGregor, *Reinforced Concrete*, Prentice Hall, Englewood Cliffs, NJ, 1988.

4.20. M. P. Collins and D. Mitchell, "Shear and Torsion Design of Prestressed and Nonprestressed Concrete Beams," *J. Prestressed Concr. Inst.*, PCI, vol. 25, no. 5, 1980, pp. 32–100.

4.21. F. J. Vecchio and M. P. Collins, "Predicting the Response of Reinforced Concrete Beams Subjected to Shear Using Modified Compression Field Theory," *ACI Struct. J.*, vol. 85, no. 3, 1988, pp. 258– 268.

4.22. *Design of Concrete Structures for Buildings*, CAN3-A23.3-M84, Canadian Standards Associations, Toronto, 1984.

4.23. N. S. Anderson and J. A. Ramirez, "Detailing of Stirrup Reinforcement," *ACI Struc. J.*, vol 86, no. 5, 1989, pp. 507–515.

4.24. L. Chow, H. D. Conway, and G. Winter, "Stresses in Deep Beams," *Trans. ASCE*, vol. 118, 1953, p. 686.

4.25. H. A. P. dePaiva and C. P. Siess, "Strength and Behavior of Deep Beams in Shear," *J. Struct. Div. ASCE*, vol. 91, no. ST5, 1965, p. 19.

4.26. R. A. Crist, "Shear Behavior of Deep Reinforced Concrete Beams," *Proc. Symp. Effects Repeated Loading Mat. Structural Elements, Mexico City*, vol. 4, 1966, RILEM, Paris.

4.27. J. G. MacGregor and N. M. Hawkins, "Suggested Revisions to ACI Building Code Clauses Dealing With Shear Friction and Shear in Deep Beams and Corbels," *J. ACI*, vol. 74, no. 11, 1977, pp. 537–545.

4.28. D. M. Rogowsky and J. G. MacGregor, "Design of Reinforced Concrete Deep Beams," Concrete International, vol. 8, no. 8, 1986, pp. 49–58.

4.29. P. W. Birkeland and H. W. Birkeland, "Connections in Precast Concrete Construction," *J. ACI*, vol. 63, no. 3, 1966, pp. 345–368.

4.30. R. F. Mast, "Auxiliary Reinforcement in Precast Concrete Connections," *J. Structural Division, ASCE*, vol. 94, no. ST6, June 1968, pp. 1485–1504.

4.31. A. H. Mattock and N. M. Hawkins, "Shear Transfer in Reinforced Concrete—Recent Research," *J. Prestressed Concr. Inst.*, vol. 17, no. 2, 1972, pp. 55–75.

4.32. A. H. Mattock, "Shear Transfer in Concrete Having Reinforcement at an Angle to the Shear Plane," Special Publication SP-42, American Concrete Institute, Detroit, 1974.

4.33. *PCI Design Handbook*, 3rd ed., Prestressed Concrete Institute, Chicago, 1985.

PROBLEMS

4.1 A beam is to be designed for factored loads causing a maximum shear of 50.0 kips, using concrete with $f_c' = 4000$ psi. Proceeding on the basis that the concrete dimensions will be determined by diagonal tension, select the appropriate width and effective depth (*a*) for a beam in which no web reinforcement is to be used, (*b*) for a beam in which only the minimum web reinforcement is provided, as given by Eq. (4.13), and (*c*) for a beam in which web reinforcement provides shear strength $V_s = 2V_c$. Follow the ACI Code requirements, and let $d = 2b$ in each case. Calculations may be based on the more approximate value of V_c given by Eq. (4.12*b*).

4.2 A rectangular beam having $b = 12$ in. and $d = 22$ in. spans 20 ft face-to-face of simple supports. It is reinforced for flexure with three No. 11 bars that continue uninterrupted to the ends of the span. It is to carry service dead load $D = 1.63$ kips/ft (including self-weight) and service live load $L = 3.26$ kips/ft, both uniformly distributed along the span. Design the shear reinforcement, using No. 3 vertical U stirrups. The more approximate Eq. (4.12*b*) for V_c may be used. Material strengths are $f_c' = 4000$ psi and $f_y = 60{,}000$ psi.

4.3 Redesign the shear reinforcement for the beam of Prob. 4.2, basing V_c on the more accurate Eq. (4.12*a*). Comment on your results, with respect to design time and probable construction cost difference.

4.4 A beam of 11 in. width and effective depth of 16 in. carries a factored uniformly distributed load of 6.0 kips/ft, including its own weight, in addition to a central, concentrated factored load of 14 kips. It spans 18 ft, and restraining end moments at full factored load are 155 ft-kips at each support. It is reinforced with three No. 9 bars for both positive and negative bending. If $f_c' = 4000$ psi, through what part of the beam is web reinforcement theoretically required (*a*) if Eq. (4.12*b*) is used, (*b*) if Eq. (4.12*a*) is used? Comment.

4.5 What effect would an additional clockwise moment of 200 ft-kips at the right support have on the requirement for shear reinforcement determined in part (*a*) of Prob. 4.4?

4.6 Design the web reinforcement for the beam of Prob. 4.4, with V_c determined by the more approximate ACI equation, using No. 3 vertical stirrups with $f_y = 60{,}000$ psi.

4.7 Design the web reinforcement for the beam of Prob. 4.5, with V_c determined by the more approximate ACI equation, using No. 3 vertical stirrups with $f_y = 60{,}000$ psi.

4.8 The beam of Prob. 4.2 will be subjected to a factored axial compression load of 150 kips on the 12×25 in gross cross section, in addition to the loads described earlier. What is the effect on concrete shear strength V_c (*a*) by the more accurate ACI equation, (*b*) by the more approximate ACI equation?

4.9 The beam of Prob. 4.2 will be subjected to a factored axial tension load of 75 kips on the 12×25 in gross cross section, in addition to the loads described earlier. What is the effect on concrete shear strength V_c (*a*) by the more accurate ACI equation, (*b*) by the more conservative ACI approach?

4.10 For a deep beam having a span-depth ratio of 3, loaded at its upper third points and supported at its lower corners (*a*) sketch the stress trajectories indicating the principal stress directions for the uncracked beam, and based on these, show the probable crack pattern; (*b*) identify and explain the means by which the beam resists shear after cracking; (*c*) explain how the web reinforcement differs from that for a beam of normal proportions; (*d*) list the possible modes of failure if the beam should be overloaded.

4.11 A deep transfer girder is to be 11 ft in overall depth and will span 22 ft between column supports. In addition to its own weight, it will pick up a uniformly distributed factored load of 3.8 kips/ft applied along its top edge from an adjacent floor, and will carry a column delivering a concentrated factored load of 1000 kips from floors above, applied at midspan. Girder width must not be greater than 16 in., but may be less. Design the beam for the loads given, finding the required width b, reinforcement A_s, A_v, and A_{vh}, based on the ACI Code provisions. Comment on the need for special detailing of reinforcement and sketch your recommendations. Specified material strengths are $f'_c = 5000$ psi and $f_y = 60{,}000$ psi.

4.12 A precast concrete beam having cross-section dimensions $b = 10$ in. and $h = 24$ in. is designed to act in a composite sense with a cast-in-place top slab having depth $h_f = 5$ in. and width 48 in. At factored loads, the maximum compressive stress in the flange at midspan is 2400 psi; at the supports of the 28 ft simple span the flange force must be zero. Vertical U stirrups provided for flexural shear will be extended into the slab and suitably anchored to provide also for transfer of the flange force by shear-friction. Find the minimum number of No. 4 stirrups that must be provided, based on shear-friction requirements. Concrete in both precast and cast-in-place parts will have $f'_c = 4000$ psi and $f_y = 60{,}000$ psi. The top surface of the precast web will be intentionally roughened according to the ACI Code definition.

4.13 Redesign the beam-end reinforcement of Example 4.6, given that a roller support will be provided so that $T_u = 0$.

CHAPTER
5

BOND, ANCHORAGE, AND DEVELOPMENT LENGTH

5.1 FUNDAMENTALS OF FLEXURAL BOND

If the reinforced concrete beam of Fig. 5.1*a* were constructed using plain round reinforcing bars, and, furthermore, if those bars were to be greased or otherwise lubricated before the concrete were poured, the beam would be very little stronger than if it were built of plain concrete, without reinforcement. If a load were applied, as in Fig. 5.1*b*, the bars would tend to maintain their original length as the beam deflects. The bars would slip longitudinally with respect to the adjacent concrete, which would experience tensile strain due to flexure. Proposition 2 of Sec. 1.8, the assumption that the strain in an embedded reinforcing bar is the same as that in the surrounding concrete, would not be valid. In order for reinforced concrete to behave as intended, it is essential that *bond forces* be developed on the interface between concrete and steel, such as to prevent significant slip from occurring at that interface.

Figure 5.1*c* shows the bond forces that act on the concrete at the interface as a result of bending, while Fig. 5.1*d* shows the equal and opposite bond forces acting on the reinforcement. It is through the action of these interface bond forces that the slip indicated in Fig. 5.1*b* is prevented.

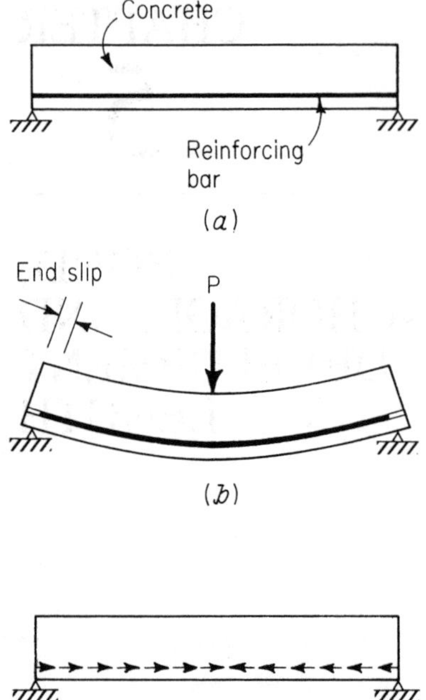

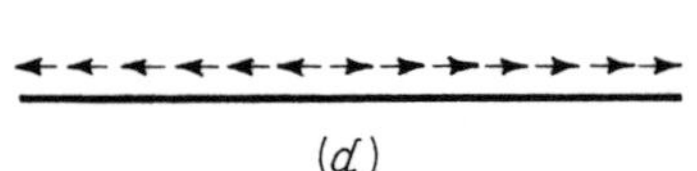

FIGURE 5.1
Bond stresses due to flexure: (*a*) beam before loading; (*b*) unrestrained slip between concrete and steel; (*c*) bond forces acting on concrete; (*d*) bond forces acting on steel.

Some years ago, when plain bars without surface deformations were used, initial bond strength was provided only by the relatively weak chemical adhesion and mechanical friction between steel and concrete. Once adhesion and static friction were overcome at larger loads, small amounts of slip led to interlocking of the natural roughness of the bar with the concrete. However, this natural bond strength is so low that in beams reinforced with plain bars the bond between steel and concrete was frequently broken. Such a beam will collapse as the bar is pulled through the concrete. To prevent this, end anchorage was provided, chiefly in the form of hooks, as in Fig. 5.2. If the anchorage is adequate, such a beam will not collapse, even if the bond is broken over the entire length between anchorages. This is so because the member acts as a tied arch, as shown in Fig. 5.2, the uncracked concrete shown shaded representing the arch and the anchored bars the tie rod. In this case, over the length in which the bond is broken, bond stresses are zero. This means that over the entire unbonded length the force in the steel is constant and equal to $T = M_{\max}/z$. In consequence, the total steel elongation in such beams is larger than in beams in which bond is preserved, resulting in larger deflections and greater widths of cracks.

To improve this situation, deformed bars are now universally used in the United States and many other countries (see Sec. 2.12). With such bars, the

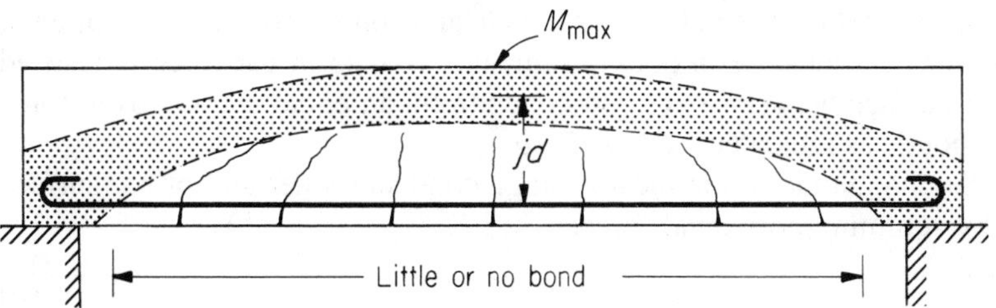

FIGURE 5.2
Tied-arch action in a beam with little or no bond.

shoulders of the projecting ribs bear on the surrounding concrete and result in greatly increased bond strength. It is then possible in most cases to dispense with special anchorage devices such as hooks. In addition, crack widths as well as deflections are reduced.

a. Bond Stress Based on Simple Cracked Section Analysis

In a short piece of a beam of length dx, such as shown in Fig. 5.3a, the moment at one end will generally differ from that at the other end by a small amount dM. If this piece is isolated, and if one assumes that, after cracking, the concrete does not resist any tension stresses, the internal forces are those shown in Fig. 5.3a. The change in bending moment dM produces a change in the bar force

$$dT = \frac{dM}{jd} \tag{a}$$

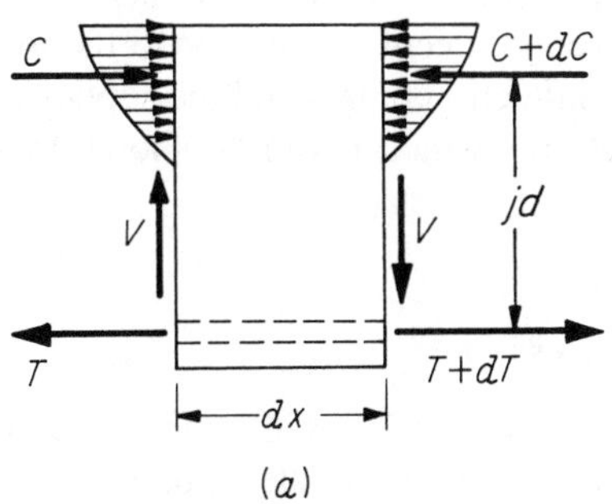

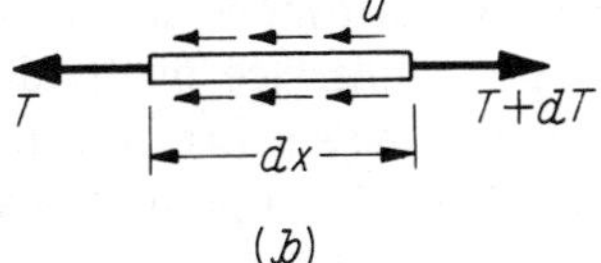

FIGURE 5.3
Forces and stresses acting on elemental length of beam: (a) free body sketch of reinforced concrete element; (b) free body sketch of steel element.

where jd is the internal lever arm between tensile and compressive force resultants. Since the bar or bars must be in equilibrium, this change in bar force is resisted at the contact surface between steel and concrete by an equal and opposite force produced by bond, as indicated by Fig. 5.3*b*.

If u is the magnitude of the local average bond stress per unit of bar surface area, then, by summing horizontal forces

$$u\Sigma_0 dx = dT \tag{b}$$

where Σ_0 is the sum of the perimeters of all the bars. Thus

$$u = \frac{dT}{\Sigma_0 dx} \tag{5.1}$$

indicating that the local bond stress is proportional to the rate of change of bar force along the span. Alternatively, substituting Eq. (*a*) in Eq. (5.1), the unit bond stress can be written as

$$u = \frac{dM}{\Sigma_0 jd dx} \tag{b}$$

from which

$$u = \frac{V}{\Sigma_0 jd} \tag{5.2}$$

Equation (5.2) is the conventional "elastic cracked section equation" for flexural bond stress, and it indicates that the unit bond stress is proportional to the shear at a particular section, i.e., to the rate of change of bending moment.

Note that Eq. (5.2) applies to the *tension* bars in a concrete zone that is assumed to be fully cracked, with the concrete resisting no tension. It applies, therefore, to the tensile bars in simple spans, or, in continuous spans, either to the bottom bars in the positive bending region between inflection points or to the top bars in the negative bending region between the inflection points and the supports. It does not apply to compression reinforcement, for which it can be shown that the flexural bond stresses are very low.

b. Actual Distribution of Flexural Bond Stress

The actual distribution of bond stress along deformed reinforcing bars is more complex than that represented by Eq. (5.2), and Eq. (5.1) provides a better basis for understanding beam behavior. Figure 5.4 shows a beam segment subject to pure bending. The concrete fails to resist tensile stresses only where the actual crack is located; there the steel tension is maximum and has the value predicted by simple theory: $T = M/jd$. Between cracks, the concrete *does* resist moderate amounts of tension, introduced by bond stresses acting along the interface in the direction shown in Fig. 5.4*a*. This reduces the tensile force in the steel, as illustrated by Fig. 5.4*c*. From Eq. (5.1), it is clear that the bond stress u is proportional to

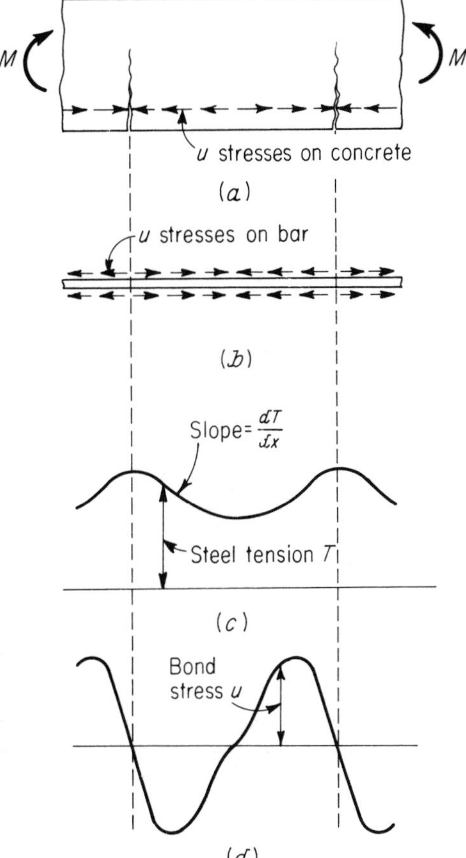

FIGURE 5.4
Variation of steel force and bond stress in reinforced concrete member subject to pure bending: (*a*) cracked concrete segment; (*b*) bond stresses acting on reinforcing bar; (*c*) variation of tensile force in steel; (*d*) variation of bond stress along steel.

the rate of change of bar force, and thus will vary as shown in Fig. 5.4*d*; bond stresses are highest where the slope of the steel force curve is greatest, and are zero where the slope is zero. Very high local bond stresses adjacent to cracks have been measured in tests at Cornell University (Refs. 5.1 and 5.2). They are so high that inevitably some slip occurs between concrete and steel adjacent to each crack.

Beams are seldom subject to pure bending moment; they generally carry transverse loads producing shear and moment that vary along the span. Figure 5.5*a* shows a beam carrying a distributed load. The cracking indicated is typical. The steel force T predicted by simple cracked section analysis is proportional to the moment diagram and is as shown by the dashed line in Fig. 5.5*b*. However, the actual value of T is less than that predicted by the simple analysis everywhere except at the actual crack locations. The actual variation of T is shown by the solid line of Fig. 5.5*b*. In Fig. 5.5*c*, the bond stresses predicted by the simplified theory are shown by the dashed line, and the actual variation shown by the solid line. Note that the value of u is equal to that given by Eq. (5.2)

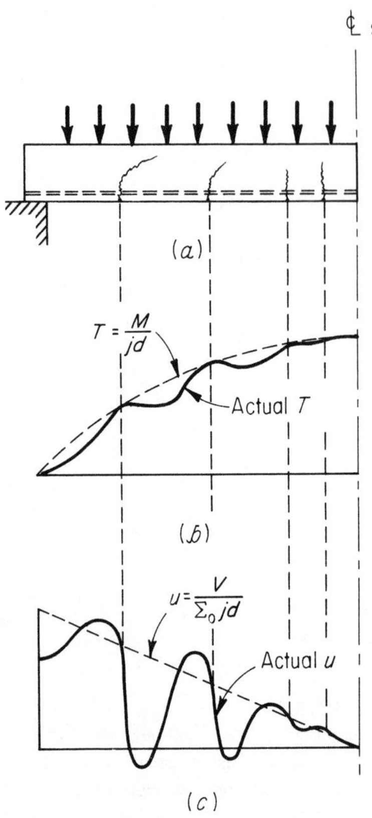

FIGURE 5.5
Effect of flexural cracks on bond stresses in beam: (*a*) beam with flexural cracks; (*b*) variation of tensile force T in steel along span; (*c*) variation of bond stress u along span.

only at those locations where the slope of the steel force diagram equals that of the simple theory. Elsewhere, if the slope is greater than assumed, the bond stress is greater; if the slope is less, bond stress is less. Just to the left of the cracks, for the present example, bond stresses are much higher than predicted by Eq. (5.2), and in all probability will result in local bond failure. Just to the right of the cracks, bond stresses are much lower than predicted, and in fact are generally negative very close to the crack; i.e., the bond forces act in the reverse direction.

It is evident that actual bond stresses in beams bear very little relation to those predicted by Eq. (5.2) except in the general sense that they are highest in the regions of high shear.

5.2 ULTIMATE BOND STRENGTH AND DEVELOPMENT LENGTH

When bond failure ensues, it results generally in splitting of the concrete along the bars, either in vertical planes as in Fig. 5.6*a* or in horizontal planes as in

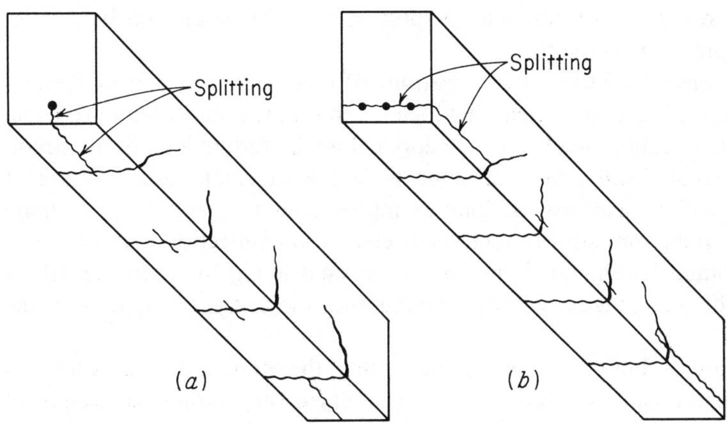

FIGURE 5.6
Splitting of concrete along reinforcement.

Fig. 5.6*b*. Such splitting comes chiefly from wedging action when the ribs of the deformed bars bear against the concrete (Refs. 5.3 and 5.4). The horizontal type of splitting of Fig. 5.6*b* frequently begins at a diagonal crack. In this case, as discussed in connection with Fig. 4.7*b* and shown in Fig. 4.1, the dowel action increases the tendency toward splitting. This indicates that shear and bond failures are often intricately interrelated.

When splitting has spread all the way to the end of an unanchored bar, complete bond failure occurs. That is, sliding of the steel relative to the concrete leads to immediate collapse of the beam.

For modern deformed bars, tests at the University of Texas and at the United States Bureau of Standards (Refs. 5.5 to 5.7) indicate that splitting occurs when the total bond force $U = u\Sigma_0$ per inch of length of bar, which is transmitted from steel to concrete, reaches a critical value. This ultimate bond force, in pounds per linear inch of bar, is largely independent of bar size or perimeter. The concept of a wedging action is in reasonable conformity with this finding, since the effects of a wedge of given shape depend more on the force with which it is driven than on its size.

It was found that the ultimate average bond force per inch of length of bar is approximately

$$U_n = 35\sqrt{f_c'} \tag{5.3}$$

Bond forces of this magnitude were observed to result in failure by splitting, in large and excessive slip, or in other prohibitive deformations. These tests were made mostly on beams of the type in Fig. 5.6*a*, a single bar causing vertical splitting. For several bars in one layer, at typical spacings found in beams, the ultimate bond force is about 80 percent of that of Eq. (5.3). The fact that bond resistance was found to correlate better with $\sqrt{f_c'}$ than with f_c' agrees with the concept that the resistance of concrete to splitting depends chiefly on its tensile, rather than compressive, strength.

If one considers the large local variations of bond stress caused by flexural and diagonal cracks, it becomes clear that local bond failures immediately adjacent to cracks will often occur at loads considerably below the failure load of the beam. These result in small local slips and some widening of cracks and increase of deflections, but will be harmless as long as failure does not propagate all along the bar, with resultant total slip. In fact, as discussed in connection with Fig. 5.2, when end anchorage is reliable, bond can be severed along the entire length of bar, excluding the anchorages, without endangering the carrying capacity of the beam.

In the beam of Fig. 5.7, the moment, and therefore the steel stress, is evidently zero at the supports and maximal at point *a* (neglecting the weight of the beam). If one designates the steel stress at *a* as f_s, the total tension force at that point in a bar of area A_b is $T_s = A_b f_s$, while at the end of the bar it is zero. Evidently, this force has been transferred from the concrete into the bar over the length l by bond stresses on the surface. Therefore, the average bond force per unit length over the length l is

$$U = \frac{A_b f_s}{l} \tag{d}$$

If this unit bond force U is smaller than the ultimate value U_n of Eq. (5.3), no splitting or other complete failure will occur over the length l.

To put it differently, the minimum length that is necessary to develop, by bond, a given bar force $A_b f_s$ is

$$l_d = \frac{A_b f_s}{U_n} \tag{e}$$

This length l_d is known as the *development length* of the bar. In particular, in order to ensure that a bar is securely anchored by bond to develop its maximum usable strength (the yield stress), this development length must be approximately

$$l_d = \frac{A_b f_y}{U_n} = \frac{0.028 A_b f_y}{\sqrt{f_c'}} \tag{5.4}$$

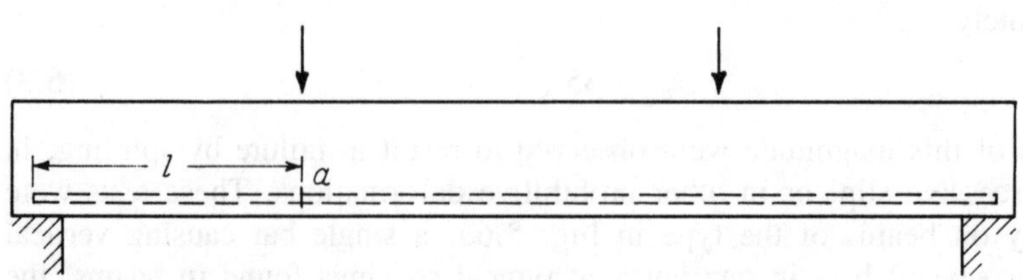

FIGURE 5.7
Development length.

for single bars, or bars that are well-spaced laterally. For bars at the relatively close spacings typical for beams, dividing by 0.8, one has

$$l_d = \frac{0.035A_b f_y}{\sqrt{f'_c}} \tag{5.5}$$

Therefore in the beam of Fig. 5.7, if the actual length l is equal to or larger than the development length l_d no premature bond failure will occur. That is, the beam will fail in bending or shear rather than by bond failure. This will be so even if in the immediate vicinity of cracks local slips may have occurred over small regions along the beam.

From this discussion it is seen that the main requirement for safety against bond failure is this: the length of the bar, from any point of given steel stress (f_s, or at most f_y) to its nearby free end must be at least equal to its development length, as given by Eq. (5.4) or Eq. (5.5). If this requirement is satisfied, the magnitude of the nominal flexural bond stress along the beam, as given by Eq. (5.2), is of only secondary importance, since the integrity of the member is ensured even in the face of possible minor local bond failures. However, if the actual available length is inadequate for full development, special anchorage, such as by hooks, must be provided to ensure adequate strength.

The actual local intensity of the unit bond force along the embedded length of a bar varies, depending on the crack pattern, on the distance from the section of maximum steel stress and on other factors. This was illustrated by Figs. 5.4 and 5.5. The ultimate value of Eq. (5.3) was obtained, from tests with a variety of embedded lengths l, by dividing the bar force by this length. This quantity, therefore, is merely the average value over the embedded length at bond failure, regardless of local variations.

It is seen that U_n is largely independent of the diameter of the bar. This means that the bond force per inch is substantially the same for large as for small bars. On the other hand, for a given steel stress f_s, the bar force $A_b f_s = f_s \pi d_b^2/4$ is proportional to the square of the diameter. If this is substituted in Eq. (*e*), the result is $l_d = f_s \pi d_b^2/4U_n$, showing that the necessary development length increases with the square of the bar diameter. Therefore, a beam reinforced with a larger number of small bars requires a smaller development length than a beam reinforced with a small number of large bars of the same total area. This demonstrates the superiority of small bars in developing bond strength.

Recent research (Refs. 5.7 to 5.10) takes explicit account of three factors that influence bond strength or required embedment length, i.e., concrete cover over the bar, spacing of bars, and lateral reinforcement such as stirrups. From Fig. 5.6*a* it is apparent that if the bottom concrete cover is increased, more concrete tensile strength can be developed which will delay vertical splitting. Similarly, Fig. 5.6*b* illustrates that if the bar spacing is increased (e.g., if only two instead of three bars are used), again more concrete per bar would be available to resist horizontal splitting. Finally, if stirrups of the types shown in Fig. 4.8 are used, their tensile yield strength in addition to the tensile strength of the concrete will be available to resist both vertical and horizontal splitting. Hence, any of these factors increases bond strength or reduces required embedment length.

5.3 ACI CODE PROVISIONS FOR DEVELOPMENT OF TENSION REINFORCEMENT

The basic approach to flexural bond in present design practice derives from the principles set forth in the preceding section. The fundamental requirement is that the calculated force in the reinforcement at each section of a reinforced concrete member must be developed on each side of that section by adequate embedment length, hooks, mechanical anchorage, or a combination of these, to ensure against pullout. Local high bond stresses, such as are known to exist adjacent to cracks in beams, are not considered to be significant. Even though some local slip may occur, that slip will not be progressive or general, and it will not lead to collapse of the member, if suitable anchorage is provided. Generally, the force to be developed is calculated based on the yield stress in the reinforcement, i.e., the bar strength is to be fully developed.

a. Basic Development Length in Tension

On the basis of experimental evidence, the development length for bars of common sizes, at spacings usually found in beams, is given by Eq. (5.5) of Sec. 5.2. Rather than introducing a ϕ factor for bond, as is done for flexure and shear, the experimental length is increased by approximately 15 percent to ensure safety of the member. This results in the ACI Code equation for basic development length in tension

$$l_{db} = \frac{0.04\, A_b f_y}{\sqrt{f'_c}} \quad \text{for No. 11 bars and smaller and deformed wire} \tag{5.6a}$$

according to ACI Code 12.2.2, where A_b is the cross-sectional area of the individual bar. Note that in this, and in all other ACI Code equations relating to development length and splices of reinforcement, *values of* $\sqrt{f'_c}$ *are not to be taken larger than 100 psi* because of the lack of experimental evidence on bond strengths obtainable with concretes having compressive strength in excess of 10,000 psi.

Tests indicate that Eq. (5.6*a*) tends to overestimate the required development length for the large-diameter Nos. 14 and 18 bars; accordingly, a 6 percent reduction is permitted for No. 14 bars and a 22 percent reduction for No. 18 bars. The adjustments result in the following:

$$l_{db} = \frac{0.085 f_y}{\sqrt{f'_c}} \quad \text{for No. 14 bars} \tag{5.6b}$$

$$l_{db} = \frac{0.125 f_y}{\sqrt{f'_c}} \quad \text{for No. 18 bars} \tag{5.6c}$$

b. Modification Factors to Account for Bar Spacing, Cover, and Transverse Reinforcement

It was pointed out in Sec. 5.2 that development length is influenced by a number of factors, including spacing of the bars, concrete cover, and the presence or absence of lateral reinforcement such as stirrups or spirals. Development length is further influenced by the depth of concrete deposited below the bar, by the type of concrete (i.e., lightweight vs. normal weight), and the presence of epoxy coating, sometimes specified for corrosion protection. Furthermore, if the amount of steel provided is greater than that required, a reduction in development length is permissible. These influences are accounted for in the ACI Code by modification factors that are applied to the basic development length l_{db} to obtain the required length l_d to be used in design. In no case is l_d to be less than 12 in., according to ACI Code 12.2.2. Provisions relating to cover, spacing, and confinement steel are somewhat complex, but they reflect research reported in Refs. 5.7 through 5.10. They will be summarized in this section. Other modification factors will be developed in Sec. 5.3c.

In beams, bars are typically spaced about one or two bar diameters apart. For such cases, it was pointed out in Sec. 5.2 that bond failure generally involves splitting of the concrete in the plane of the reinforcement, as shown in Fig. 5.6*b*. On the other hand, for slabs, footings, and certain other types of members, bar spacings are generally much greater, and the required development length is reduced. Cover distance, measured from the surface of the bar to the concrete face, and measured either in the plane of the bars or perpendicular to that plane, also influences splitting. Small cover distances correlate with the need for increased development length. On the other hand, it was pointed out in Sec. 5.2 that lateral reinforcement, such as that provided by stirrups or continuous spirals, improves the resistance of tensile bars to pullout by resisting both vertical and horizontal splitting of the concrete. The effectiveness of such transverse reinforcement depends on its cross-sectional area and its spacing along the length of the bars being developed; and the amount of transverse steel required, if used, is a function of the size and number of bars to be developed.

These various interactive effects are treated in detail in ACI Code 12.2.3, which introduces the modification factors summarized in part A of Table 5.1.

c. Other Modification Factors for Tension Reinforcement

Superimposed on the modification factors just described are those pertaining to top reinforcement, lightweight aggregate concrete, epoxy-coated bars, and excess reinforcement. These factors are as follows.

Top Reinforcement. If bars are placed in the beam forms during construction such that a substantial depth of concrete is placed below those bars, there is a tendency for excess water, often used in the mix for workability, and for entrapped air to rise toward the top of the concrete during vibration. Air and water tend to

Table 5.1 Modification factors to be applied to basic development length to obtain required development length l_d in tension

A. Modification factors for bar spacing, cover, and transverse steel:

1. Bars satisfying any one of conditions (*a*), (*b*), (*c*), *or* (*d*) as described in the following 1.0

   (*a*) Bars in beams or columns with minimum cover as required for concrete protection for reinforcement (see Sec. 3.6a), *and* with transverse reinforcement satisfying tie requirements for compression members or minimum stirrup requirements along the development length for beams, *and* with clear spacing of not less than $3d_b$

   (*b*) Bars in beams or columns with minimum cover as required for concrete protection for reinforcement *and* enclosed within transverse reinforcement A_{tr} along the development length of minimum area

$$A_{tr} \geq \frac{d_b s N}{40}$$

   where A_{tr} = total cross-sectional area of transverse steel within a spacing s and perpendicular to the plane of bars being spliced or developed, in^2

   d_b = diameter of the bar being developed, in.

   s = spacing of stirrups or ties, in.

   N = number of bars in layer being spliced or developed

   (*c*) Bars in the inner layer of slab or wall reinforcement and with clear spacing not less than $3d_b$

   (*d*) Any bars with cover not less than $2d_b$ and with clear spacing not less than $3d_b$

2. Bars with cover of d_b or less *or* with clear spacing of $2d_b$ or less 2.0
3. Bars not included in 1 or 2 .... 1.4
4. For No. 11 bars and smaller with clear spacing not less than $5d_b$ and with cover from face of member to edge bar, measured in the plane of the bars, not less than $2.5d_b$, the factors of 1, 2, or 3 may be multiplied by 0.8
5. For reinforcement enclosed within spiral reinforcement not less than $\frac{1}{4}$ in. diameter and not more than 4 in. pitch, within No. 4 or larger circular ties spaced at not more than 4 in. on center, or within No. 4 or larger ties or stirrups spaced not more than 4 in. on center and arranged such that alternate bars shall have support provided by the corner of a tie or hoop with an included angle of not more than 135 degrees, the factors of 1, 2, or 3 may be multiplied by 0.75
6. The basic development length, multiplied by the applicable factor of 1, 2, or 3, with modifiers of 4, 5, or both, shall not be less than $0.03d_b f_y/\sqrt{f'_c}$.

B. Other modification factors:

The basic development length l_{db} as modified by the factors of part *A* shall also be multiplied by applicable factor or factors as follows:

1. Top reinforcement, i.e., horizontal reinforcement so placed that more than 12 in. of fresh concrete is cast in the member below the development length or splice 1.3
2. Lightweight aggregate concrete 1.3
   or when f_{ct} is specified .... $6.7\sqrt{f'_c}/f_{ct}$
   .... but not less than 1.0
3. Epoxy-coated reinforcement

   (*a*) Bars with cover less than $3d_b$ or clear spacing between bars less than $6d_b$ 1.5

   (*b*) All other conditions 1.2

   (*c*) The product of the factor for top reinforcement and for epoxy-coated reinforcement need not be taken greater than 1.7
4. Excess reinforcement

   Development length may be reduced where reinforcement in a flexural member is in excess of that required by analysis, except where anchorage or development for f_y is specifically required or the reinforcement is designed for a region of high seismic risk, by the factor (A_s required/A_s provided)

accumulate to some extent on the underside of the bars. Tests have shown a significant loss in bond strength for bars with more than 12 in. of fresh concrete cast beneath them, and accordingly the development length is increased by the factor 1.3, according to ACI Code 12.2.4, for such cases. Note that this definition may require that bars relatively near the bottom of a deep member be treated as top bars.

Lightweight Aggregate Concrete. The term $\sqrt{f'_c}$ found in the denominator of the several expressions for basic development length l_{db} reflects the influence of the tensile strength of the concrete. For normal concretes the split-cylinder tensile strength f_{ct} is generally taken to be $f_{ct} = 6.7\sqrt{f'_c}$. Accordingly, if the split-cylinder strength f_{ct} is known for a particular lightweight concrete, then $\sqrt{f'_c}$ in the equations should be replaced by $f_{ct}/6.7$ (i.e., the development length should be modified by the factor $6.7\sqrt{f'_c}/f_{ct}$, with the ACI Code restriction that this modification factor must not be less than 1.0).

Epoxy-Coated Reinforcement. Epoxy-coated bars are used increasingly in projects where the work will be subjected to corrosive environmental conditions or de-icing chemicals, such as for highway bridge decks and parking garages. Studies have shown that bond strength is reduced because the epoxy coating prevents adhesion between the concrete and the bar, and for typical cases a factor of 1.5 is imposed on development length. If cover and spacing is large, the effect of the epoxy coating is not so pronounced, and the factor is reduced to 1.2. The beneficial effect of transverse steel in preventing splitting failure for small spacing and cover has been demonstrated, justifying the superposition of the modification factors for transverse steel and epoxy coating. Further, since the bond strength of epoxy-coated bars is already reduced due to lack of adhesion, an upper limit of 1.7 is established for the product of top reinforcement and epoxy-coating factors.

Excess Reinforcement. Not infrequently, tensile reinforcement somewhat in excess of the calculated requirement will be provided, e.g., as a result of upward rounding of A_s when bars are selected or when minimum steel area requirements govern. In this case, according to ACI Code, the required development length may be reduced by the ratio of steel area required to steel area actually provided. This modification factor is to be applied only where anchorage or development for the full yield strength of the bar is not required by other provisions of the ACI Code.

These various additional modification factors, specified in ACI Code 12.2.4. and 12.2.5, are summarized in part B of Table 5.1.

The basic development lengths l_{db} for reinforcing bars are easily tabulated for bars of given size and strength and for particular values of f'_c, and these are given in Table A.11 of Appendix A. The basic development length must be modified by the applicable factor or factors of Table 5.1, and minimum length requirements must be imposed. The total sequence of steps in determining development length l_d to be used in design is as follows, with reference to the paragraphs of Table 5.1:

1. Calculate basic development length l_{db} from Eqs. (5.6*a*), (5.6*b*), or (5.6*c*).
2. Multiply the basic development length by the applicable factor for spacing, cover, and transverse steel (paragraphs A.1 through A.3).
3. Multiply by the factors for wide bar spacing and confinement steel, if applicable (paragraphs A.4 and A.5).
4. Impose a minimum development length (paragraph A.6).
5. Multiply by the factor for top reinforcement, if applicable (paragraph B.1).
6. Multiply by factor for lightweight aggregate concrete, if applicable (paragraph B.2).
7. Multiply by the factor for epoxy-coated bars, if applicable (paragraph B.3).
8. Multiply by the ratio for excessive reinforcement, if applicable (paragraph B.4).
9. Check the minimum l_d of 12 in.

The complexity of the requirements for development length, new with the 1989 ACI Code, requires repetitive and time-consuming calculations for a typical project, and it is evident that a short computer program for development length would be a useful tool. It may be noted that many beams will meet the requirements along the development length of (*a*) or (*b*) of paragraph A.1 of Table 5.1, and a modification factor of 1.0 can be used, multiplied further by the factor of 1.3 for top bars if applicable. Transverse reinforcement provided for shear can be used to satisfy the requirement for A_{tr}. For most slabs, the factor 0.8 of paragraph A.4 applies.

Example 5.1 Development length in tension. Figure 5.8 shows a beam-column joint in a continuous building frame. Based on frame analysis, the negative steel required at

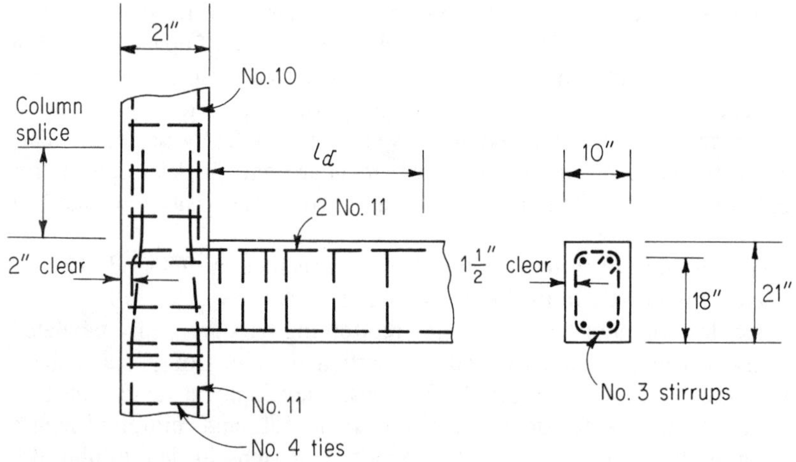

FIGURE 5.8
Bar details at beam-column joint, for bar development examples.

the end of the beam is 2.90 in^2.; two No. 11 bars are used, providing $A_s = 3.12$ in^2. Beam dimensions are $b = 10$ in., $d = 18$ in., and $h = 21$ in. The design will include No. 3 stirrups spaced four at 3 in., followed by a constant 5 in. spacing in the region of the support, with 1.5 in. clear cover. Normal density concrete is to be used, with $f_c' = 4000$ psi, and rebars have $f_y = 60{,}000$ psi. Find the minimum distance l_d at which the negative bars can be cut off, based on development of the required steel area at the face of the column.

Solution. According to the ACI Code, the basic development length for No. 11 deformed bars in tension is given by Eq. (5.6*a*):

$$l_{db} = \frac{0.04 \times 1.56 \times 60{,}000}{\sqrt{4000}} = 59 \text{ in.}$$

Checking for lateral spacing of the No. 11 bars, determines that the clear distance between the bars is $[10 - (2 \times 1.5) - (2 \times 0.375) - (2 \times 1.41)] = 3.43$ in., or 2.43 times the bar diameter d_b. This does not meet the minimum of $3d_b$ of (*a*), (*c*), and (*d*) of Table 5.1 paragraph A.1. The transverse steel provided by the stirrups is $2 \times 0.11 = 0.22$ in^2. With the requirement of (*b*) of paragraph A.1,

$$A_{tr} \geq \frac{1.410 \times 5 \times 2}{40} = 0.35 \text{ in}^2.$$

It is seen that this requirement is not met either, so a modification factor larger than 1.0 must be imposed. However, cover and spacing distances both exceed the requirements of paragraph A.2, so the 1.4 factor of paragraph A.3 can be used. This results in l_d, based on cover, spacing, and transverse steel, of $1.4 \times 59 = 83$ in. The minimum of

$$\frac{0.03 \times 1.41 \times 60{,}000}{\sqrt{4000}} = 40 \text{ in.}$$

is checked next, but it does not control here.

Next it is noted that more than 12 in. of concrete will be cast below the No. 11 bars, so they must be classed as top bars, requiring an additional factor of 1.3. However, a reduction can be made because the steel provided exceeds that required; the applicable factor is $2.90/3.12 = 0.93$. Thus the total required development length, combining all applicable factors, is

$$59 \times 1.4 \times 1.3 \times 0.93 = 100 \text{ in.}$$

Note that the final cutoff point selected will depend also on the shape of the moment diagram (see Sec. 5.9).

5.4 ANCHORAGE OF TENSION BARS BY HOOKS

a. Standard Dimensions

In the event that the desired tensile stress in a bar cannot be developed by bond alone, it is necessary to provide special anchorage at the ends of the bar, usually by means of a 90° or a 180° hook. The dimensions and bend radii for such hooks have been standardized in ACI Code 7.1 as follows (see Fig. 5.9):

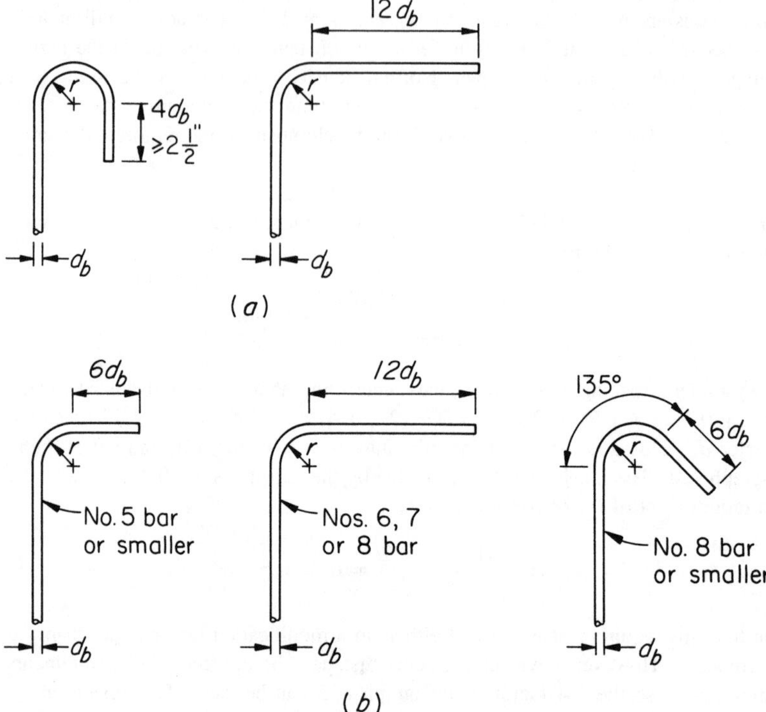

FIGURE 5.9
Standard bar hooks: (*a*) main reinforcement; (*b*) stirrups and ties.

1. A 180° bend plus an extension of at least 4 bar diameters, but not less than $2\frac{1}{2}$ in. at the free end of the bar, or
2. A 90° bend plus an extension of at least 12 bar diameters at the free end of the bar, or
3. For stirrup and tie anchorage only:
   a. For No. 5 bars and smaller, a 90° bend plus an extension of at least 6 bar diameters at the free end of the bar, or
   b. For Nos. 6, 7, and 8 bars, a 90° bend plus an extension of at least 12 bar diameters at the free end of the bar, or
   c. For No. 8 bars and smaller, a 135° bend plus an extension of at least 6 bar diameters at the free end of the bar.

The minimum diameter of bend, measured on the inside of the bar, for standard hooks other than for stirrups or ties in sizes Nos. 3 through 5, should be not less than the values shown in Table 5.2. For stirrup and tie hooks, for bar sizes No. 5 and smaller, the inside diameter of bend should not be less than 4 bar diameters, according to the ACI Code.

Table 5.2 Minimum diameters of bend for standard hooks

Bar size	Minimum diameter
Nos. 3 through 8	6 bar diameters
Nos. 9,10, and 11	8 bar diameters
Nos. 14 and 18	10 bar diameters

When welded wire fabric (smooth or deformed wires) is used for stirrups or ties, the inside diameter of bend should not be less than 4 wire diameters for deformed wire larger than D6 and 2 wire diameters for all other wires. Bends with inside diameter of less than 8 wire diameters should not be less than 4 wire diameters from the nearest welded intersection.

b. Development Length and Modification Factors for Hooked Bars

Hooked bars resist pullout by the combined actions of bond along the straight length of bar leading to the hook, and anchorage provided by the hook. Tests indicate that the main cause of failure of hooked bars in tension is splitting of the concrete in the plane of the hook. This splitting is due to the very high stresses in the concrete inside of the hook; these stresses are influenced mainly by the bar diameter d_b for a given tensile force, and the radius of bar bend. Resistance to splitting has been found to depend on the concrete cover for the hooked bar, measured laterally from the edge of the member to the bar perpendicular to the plane of the hook, and measured to the top (or bottom) of the member from the point where the hook starts, parallel to the plane of the hook. If these distances must be small, the strength of the anchorage can be substantially increased by providing confinement steel in the form of closed stirrups or ties.

ACI Code 12.5 provisions for hooked bars in tension are based on research summarized in Refs. 5.9 and 5.10. The Code requirements account for the combined contribution of bond along the straight bar leading to the hook, plus the hooked anchorage. A total development length l_{dh} is defined as shown in Fig. 5.10 and is measured from the critical section to the farthest point on the bar, parallel to the straight part of the bar. For standard hooks, as shown in Fig. 5.9, the basic development length for bars having $f_y = 60{,}000$ psi is

$$l_{hb} = \frac{1200 d_b}{\sqrt{f'_c}} \tag{5.7}$$

This basic development length is to be multiplied by certain applicable modifying factors summarized in Table 5.3 to determine the development length l_{dh} to be used in design. These factors are combined as appropriate; e.g., if side cover of at least $2\frac{1}{2}$ in. is provided for a 180° hook, and if, in addition, ties are provided, the basic development length is multiplied by the product of 0.7 and 0.8. In any case, the length l_{dh} is not to be less than 8 bar diameters and not less than 6 in.

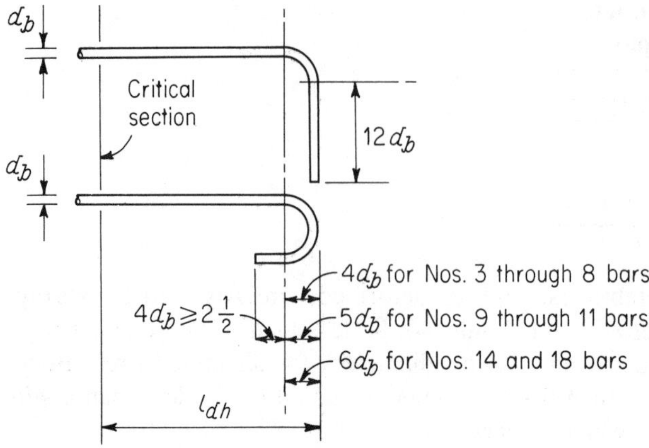

FIGURE 5.10
Bar details for development of standard hooks.

Lateral confinement steel is essential if the full bar strength must be developed with minimum concrete confinement, such as when hooks may be required at the ends of a simply supported beam (Fig. 5.11) or where a beam in a continuous structure frames into an end column and does not extend past the column (Ref. 5.11). According to ACI Code 12.5.4, for bars hooked at the discontinuous ends of members with both side cover and top or bottom cover less than $2\frac{1}{2}$ in., as shown in Fig. 5.11, hooks *must* be enclosed with closed stirrups or ties along the full development length. The spacing of the confinement steel must not exceed 3

Table 5.3 Development lengths for hooked deformed bars in tension

A.	Basic development length l_{hb} for hooked bars with $f_y = 60{,}000$ psi	$\dfrac{1200d_b}{\sqrt{f'_c}}$
B.	Modification factors to be applied to l_{hb}	
	Bars with f_y other than 60,000 psi	$\dfrac{f_y}{60{,}000}$
	For No. 11 bars and smaller, side cover not less than $2\frac{1}{2}$ in. and for 90° hook, cover on bar extension beyond hook not less than 2 in.	0.7
	For No. 11 bars and smaller, hook enclosed vertically or horizontally within ties or stirrup-ties spaced along the full development length l_{dh} with spacing not greater than $3d_b$	0.8
	Reinforcement in excess of that required by analysis	$\dfrac{A_s \text{ required}}{A_s \text{ provided}}$
	Lightweight aggregate concrete	1.3

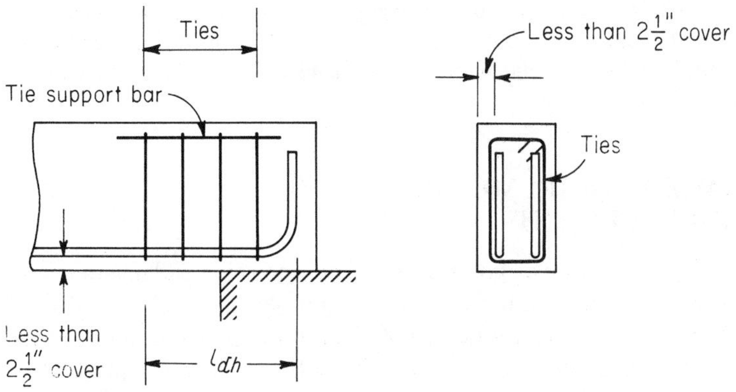

FIGURE 5.11
Lateral reinforcement requirements at discontinuous ends of members with small cover distances.

times the diameter of the hooked bar. In such cases, the factor 0.8 of Table 5.3 does not apply.

c. Mechanical Anchorage

For some special cases, e.g., at the ends of main flexural reinforcement in deep beams, there is not room for hooks or the necessary confinement steel, and special mechanical anchorage devices may be used. These may consist of welded plates or manufactured devices, the adequacy of which must be established by tests. Development of reinforcement, when such devices are employed, may consist of the combined contributions of bond along the length of the bar leading to the critical section, plus that of the mechanical anchorage; that is to say, the total resistance is the sum of the parts.

Example 5.2 Development of hooked bars in tension. Referring to the beam-column joint shown in Fig. 5.8. the No. 11 negative bars are to be extended into the column and terminated in a standard 90° hook, keeping 2 in. clear to the outside face of the column. The column width in the direction of beam width is 16 in. Find the minimum length of embedment of the hook past the column face, and specify hook details.

Solution. The basic development length for hooked bars, measured from the critical section along the bar to the far side of the vertical hook, is given by Eq. (5.7):

$$l_{hb} = \frac{1200 \times 1.41}{\sqrt{4000}} = 27 \text{ in.}$$

In this case, side cover for the No. 11 bars exceeds 2.5 in. and cover beyond the bent bar is adequate, so a modifying factor of 0.7 can be applied. The only other factor applicable is for excess reinforcement, which is 0.93 as for Example 5.1. Accordingly, the minimum development length for the hooked bars is

$$l_{dh} = 27 \times 0.7 \times 0.93 = 18 \text{ in.}$$

With $21 - 2 = 19$ in. available, the required length is contained within the column. The hook will be bent to a minimum diameter of $8 \times 1.41 = 11.28$ in.; a 6 in. bend radius will be specified. The bar will continue for 12 bar diameters, or 17 in. past the end of the bend in the vertical direction.

5.5 ANCHORAGE REQUIREMENTS FOR WEB REINFORCEMENT

Stirrups should be carried as close as possible to the compression and tension faces of a beam, and special attention must be given to proper anchorage. The truss model (see Sec. 4.8 and Fig. 4.18) for design of shear reinforcement indicates the development of diagonal compressive struts, the thrust from which is equilibrated, near the top and bottom of the beam, by the tension web members (i.e., the stirrups). Thus, at the factored load stage, the tensile strength of the stirrups must be developed for almost their full height. Clearly, it is impossible to do this by development length. For this reason, stirrups normally are provided with 90° or 135° hooks at their upper end (see Fig. 5.9*b* for standard hook details) and, at their lower end, are bent 90° to pass around the longitudinal reinforcement. In simple spans, or in the positive bending region of continuous spans, where no top bars are required for flexure, stirrup support bars should be used. These are usually about the same diameter as the stirrups themselves, and they not only provide improved anchorage of the hooks but also facilitate fabrication of the reinforcement cage, holding the stirrups in position during pouring of the concrete.

ACI Code 12.13 includes special provisions for anchorage of web reinforcement. The ends of single-leg, simple-U, or multiple-U stirrups are to be anchored by one of the following means:

1. For No. 5 bars and smaller, and for Nos. 6, 7, and 8 bars with f_y of 40,000 psi or less, a standard hook around longitudinal reinforcement, as shown in Fig. 5.12*a*.
2. For Nos. 6, 7, and 8 stirrups with f_y greater than 40,000 psi, a standard hook around a longitudinal bar, plus an embedment between midheight of the member and the outside end of the hook equal to or greater than $0.014 d_b f_y / \sqrt{f'_c}$ in., as shown in Fig. 5.12*b*.

ACI Code 12.13 specifies further that, between anchored ends, each bend in the continuous portion of a simple-U or multiple-U stirrup shall enclose a longitudinal bar, as in Fig. 5.12*c*. Longitudinal bars bent to act as shear reinforcement, if extended into a region of tension, shall be continuous with longitudinal reinforcement and, if extended into a region of compression, shall be anchored beyond middepth $d/2$ as specified for development length. Pairs of U-stirrups or ties so placed as to form a closed unit shall be considered properly spliced when length of laps are $1.3 l_d$, as in Fig. 5.12*d*.

Other provisions are contained in the ACI Code relating to the use of welded wire fabric, which is sometimes used for web reinforcement in precast and prestressed concrete beams.

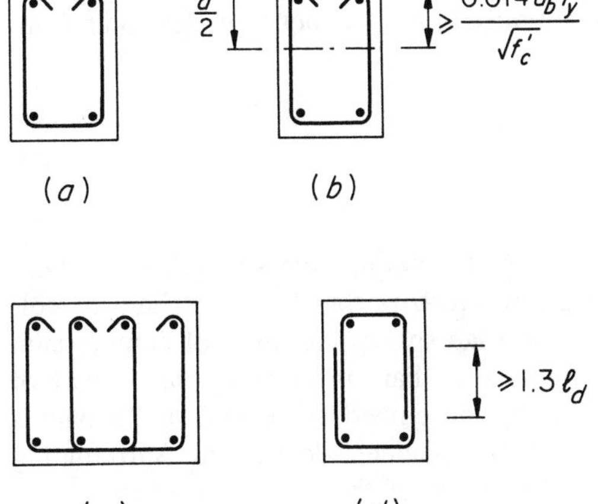

FIGURE 5.12
ACI requirements for stirrup anchorage: (*a*) No. 5 stirrups and smaller, and Nos. 6, 7, and 8 stirrups with fy not exceeding 40,000 psi; (*b*) Nos. 6, 7, and 8 stirrups with f_y exceeding 40,000 psi; (*c*) wide beam with multiple-leg U stirrups; (*d*) pairs of U stirrups forming a closed unit. See Fig. 5.9 for optional standard hook details.

5.6 WELDED WIRE FABRIC

Tensile reinforcement consisting of welded wire fabric, with either deformed or smooth wires, is commonly used in one-way and two-way slabs and certain other types of members (see Sec. 2.14). For *deformed* wire fabric, some of the development is assigned to the welded cross wires and some to the embedded length of the deformed wire. According to ACI Code 12.7, the basic development length for deformed wire fabric, measured from the critical section to the end of the wire, with at least one cross wire within the development length not less than 2 in. from the critical section, is

$$l_{db} = \frac{0.03 d_b (f_y - 20{,}000)}{\sqrt{f_c'}} \tag{5.8a}$$

but not less than

$$l_{db} = 0.20 \frac{A_w f_y}{s_w \sqrt{f_c'}} \tag{5.8b}$$

where d_b and A_w are the diameter and cross-sectional area of the individual wire, with units in inches and square inches respectively, and s_w in inch units is the lateral spacing of the wires to be developed. The development length l_d to be used in design is to be calculated as the product of the basic development length l_{db} and the applicable modification factors of Table 5.1, but is not to be less than 8 in any case.

For *smooth* wire fabric, ASTM specifications require that welds be stronger than for deformed wire fabric, and development is considered to be provided by embedment of two cross wires, with the closer wire not less than 2 in. from the critical section. However, the basic development length measured from the critical section to the outermost wire is not to be less than

$$l_{db} = 0.27 \frac{A_w f_y}{s_w \sqrt{f_c'}} \tag{5.9}$$

according to ACI Code 12.8. Modification factors pertaining to excess reinforcement and lightweight concrete may be applied, but l_d is not to be less than 6 in. for smooth wire fabric.

5.7 DEVELOPMENT OF BARS IN COMPRESSION

Reinforcement may be required to develop its compressive strength by embedment under various circumstances; e.g., where bars transfer their share of column loads to a supporting footing or where lap splices are made of compression bars in column (see Sec. 5.11). In the case of bars in compression, a part of the total force is transferred by bond along the embedded length, and a part is transferred by end bearing of the bars on the concrete. Because the surrounding concrete is relatively free of cracks and because of the beneficial effect of end bearing, shorter basic development lengths are permissible for compression bars than for tension bars. If lateral confinement steel is present, such as spiral column reinforcement or special spiral steel around an individual bar, the required development length is further reduced. Hooks such as are shown in Fig. 5.9 are *not* effective in transferring compression from bars to concrete, and, if present for other reasons, should be disregarded in determining required embedment length.

The ACI Code equation for basic development length in compression (ACI Code 12.3) is

$$l_{db} = \frac{0.02 d_b f_y}{\sqrt{f_c'}} \tag{5.10}$$

but not less than $0.0003 d_b f_y$. Modification factors summarized in part *B* of Table 5.4, as applicable, are applied to the basic development length in compression to obtain the development length l_d to be used in design. In no case is l_d to be less than 8 in., according to the ACI Code. Basic and modified compressive development lengths are given in Table A.12 of App. A.

Table 5.4 Development lengths for deformed bars in compression

A.	Basic development length l_{db}	$\frac{0.02 d_b f_y}{\sqrt{f_c'}}$ but $\geq 0.0003 d_b f_y$
B.	Modification factors to be applied to l_{db}	
	Reinforcement in excess of that required by analysis	$\frac{A_s \text{ required}}{A_s \text{ provided}}$
	Reinforcement enclosed within spiral reinforcement not less than $\frac{1}{4}$ in. diameter and not more than 4 in. pitch or within No. 4 ties spaced at not more than 4 in. on centers	0.75

5.8 BUNDLED BARS

It was pointed out in Sec. 3.6c that it is sometimes advantageous to "bundle" tensile reinforcement in large beams, with two, three, or four bars in contact, in order to provide for improved placement of concrete around and between bundles of bars. Bar bundles are typically triangular or L shaped for three bars, and square for four. When bars are cut off in a bundled group, the cutoff points must be staggered at least 40 diameters. The development length of individual bars within a bundle, for both tension and compression, is that of the individual bar increased by 20 percent for a three-bar bundle and 33 percent for a four-bar bundle, to account for the probable deficiency of bond at the inside of the bar group.

5.9 BAR CUTOFF AND BEND POINTS IN BEAMS

Chapter 3 dealt with moments, flexural stresses, concrete dimensions, and longitudinal bar areas at the critical moment sections of beams. These critical moment sections are generally at the face of the supports (negative bending) and near the middle of the span (positive bending). Occasionally, haunched members having variable depth or width are used so that the concrete flexural capacity will agree more closely with the variation of bending moment along a span or series of spans. Usually, however, prismatic beams with constant concrete cross-section dimensions are used to simplify formwork and thus to reduce cost.

The steel requirement, on the other hand, is easily varied in accordance with requirements for flexure, and it is common practice either to cut off bars where they are no longer needed to resist stress or, sometimes in the case of continuous beams, to bend up the bottom steel (usually at 45°) so that it provides tensile reinforcement at the top of the beam over the supports.

a. Theoretical Points of Cutoff or Bend

The tensile force to be resisted by the reinforcement at any cross section is

$$T = A_s f_s = \frac{M}{z}$$

where M is the value of bending moment at that section and z is the internal lever arm of the resisting moment. The lever arm z varies only within narrow limits and is never less than the value at the maximum-moment section. Consequently, the tensile force can be taken with good accuracy directly proportional to the bending moment. Since it is desirable to design so that the steel everywhere in the beam is as nearly fully stressed as possible, it follows that the required steel area is very nearly proportional to the bending moment.

To illustrate, the moment diagram for a uniformly loaded simple-span beam shown in Fig. 5.13*a* can be used as a steel-requirement diagram. At the maximum-moment section, 100 percent of the tensile steel is required (0 percent can be discontinued or bent), while at the supports, 0 percent of the steel is theoretically required (100 percent can be discontinued or bent). The percentage of bars that could be discontinued elsewhere along the span is obtainable directly from the moment diagram, drawn to scale. To facilitate the determination of cutoff or bend points for simple spans, Graph A.2 of App. A has been prepared. It represents a half-moment diagram for a uniformly loaded simple span.

To determine cutoff or bend points for continuous beams, the moment diagrams resulting from loading for maximum span moment and maximum support moment are drawn. A moment envelope results, which defines the range of values of moment at any section. Cutoff or bend points can be found from the appropriate moment curve as for simple spans. Figure 5.13*b* illustrates, for example, a continuous beam with moment envelope resulting from alternate loadings to produce maximum

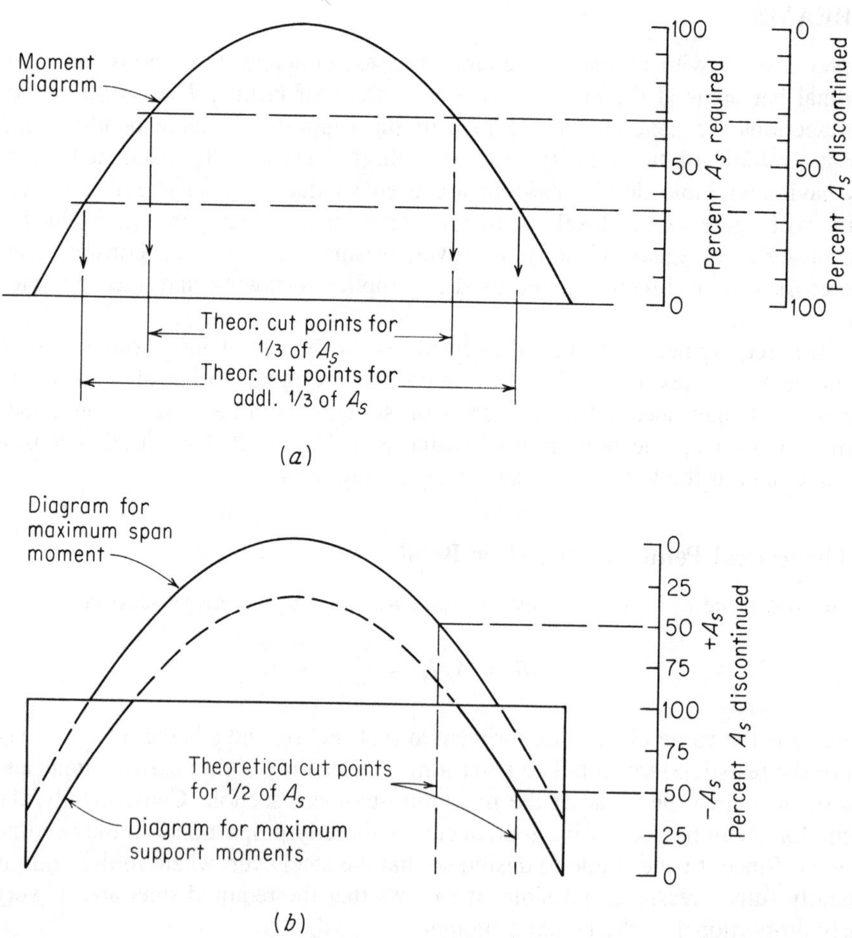

FIGURE 5.13
Bar cutoff points from moment diagrams.

span and maximum support moments. The locations of the points at which 50 percent of the bottom and top steel may theoretically be discontinued are shown.

According to ACI Code 8.3, uniformly loaded, continuous reinforced concrete beams of fairly regular span may be designed using moment coefficients (see Table 11.1). These coefficients, analogous to the numerical constant in the expression $\frac{1}{8}wL^2$ for simple-beam bending moment, give a conservative approximation of span and support moments for continuous beams. When such coefficients are used in design, bend points may conveniently be found from Graph A.3 of App. A. Moment curves corresponding to the various span- and support-moment coefficients are given at the top and bottom of the chart respectively.

Alternatively, if moments are found by frame analysis rather than from ACI moment coefficients, the location along the span where bending moment reduces to any particular value (e.g., as determined by the bar group after some bars are cut off), or to zero, is easily computed by statics.

b. Practical Considerations and ACI Code Requirements

Actually, in no case should the tensile steel be discontinued exactly at the theoretically described points. As described in Sec. 4.4 and shown in Fig. 4.9, when diagonal tension cracks form, an internal redistribution of forces occurs in a beam. Prior to cracking, the steel tensile force at any point is proportional to the moment at a vertical section passing through the point. However, after the crack has formed, the tensile force in the steel at the crack is governed by the moment at a section nearer midspan, which may be much larger. Furthermore, the actual moment diagram may differ from that used as a design basis, due to approximation of the real loads, approximations in the analysis, or the superimposed effect of settlement or lateral loads. In recognition of these facts, ACI Code 12.10 requires that every bar should be continued at least a distance equal to the effective depth of the beam or 12 bar diameters (whichever is larger) beyond the point at which it is theoretically no longer required to resist stress.

In addition, it is necessary that the calculated stress in the steel at each section be developed by adequate embedded length or end anchorage, or a combination of the two. For the usual case, with no special end anchorage, this means that the full development length l_d must be provided beyond critical sections at which peak stress exists in the bars. These critical sections are located at points of maximum moment and at points where adjacent terminated reinforcement is no longer needed to resist bending.†

† The ACI Code is ambiguous as to whether or not the extension length d or $12d_b$ is to be added to the required development length l_d. The Code Commentary presents the view that these requirements need not be superimposed, and Fig. 5.14 has been prepared on that basis. However, the argument just presented regarding possible shifts in moment curves or steel stress distribution curves leads to the conclusion that these requirements should be superimposed. In such cases, each bar should be continued a distance l_d plus the greater of d or $12d_b$ beyond the peak stress location.

Further reflecting the possible change in peak-stress location, ACI Code 12.11 requires that at least one-third of the positive-moment steel (one-fourth in continuous spans) must be continued uninterrupted along the same face of the beam a distance at least 6 in. into the support. When a flexural member is a part of a primary lateral load resisting system, positive moment reinforcement required to be extended into the support must be anchored to develop the yield strength of the bars at the face of support, to account for the possibility of reversal of moment at the supports. According to ACI Code 12.12, at least one-third of the total reinforcement provided for negative moment at the support must be extended beyond the extreme position of the point of inflection a distance not less than one-sixteenth the clear span, or d, or $12d_b$, whichever is greatest.

Requirements for bar-cutoff or bend-point locations are summarized in Fig. 5.14. If negative bars L are to be cut off, they must extend a full development

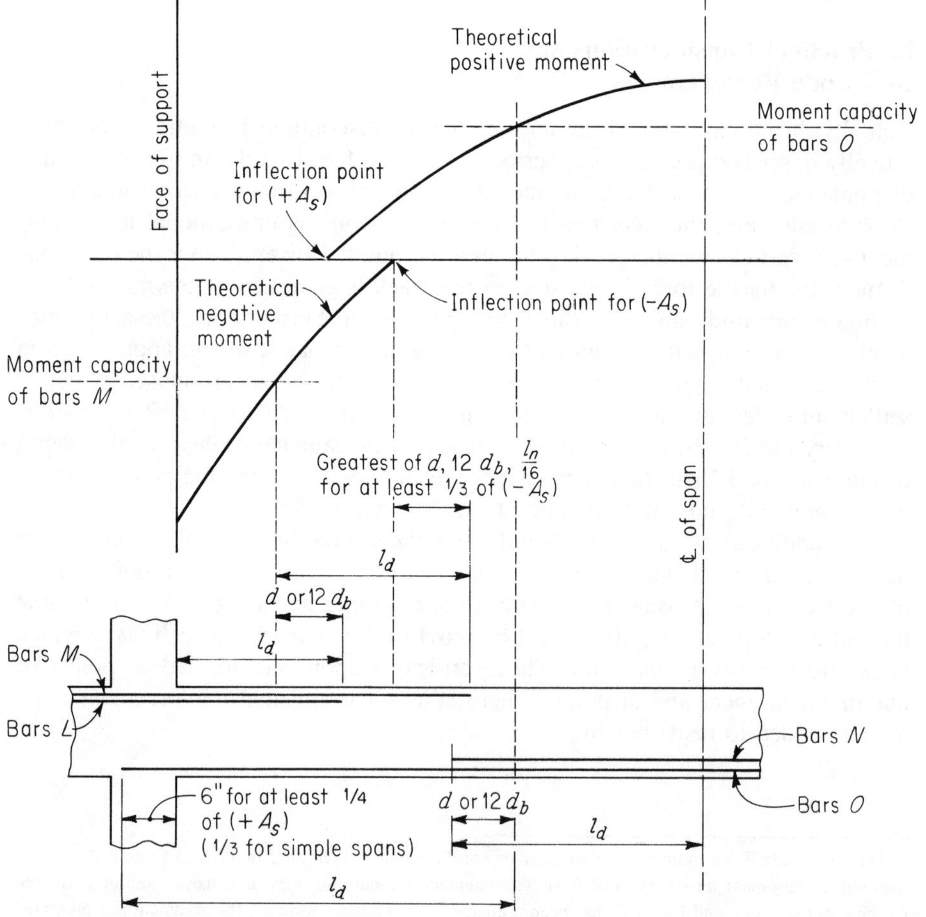

FIGURE 5.14
Bar cutoff requirements of the ACI Code.

length l_d beyond the face of the support. In addition, they must extend a distance d or $12d_b$ beyond the theoretical point of cutoff defined by the moment diagram. The remaining negative bars M (at least one-third of the total negative area) must extend at least l_d beyond the theoretical point of cutoff of bars L and in addition must extend d, $12d_b$, or $l_n/16$ (whichever is greatest) past the point of inflection of the negative-moment diagram.

If the positive bars N are to be cut off, they must project l_d past the point of theoretical maximum moment, as well as d or $12d_b$ beyond the cutoff point from the positive-moment diagram. The remaining positive bars O must extend l_d past the theoretical point of cutoff of bars N and must extend at least 6 in. into the face of the support.

When bars are cut off in a tension zone, there is a tendency toward the formation of premature flexural and diagonal tension cracks in the vicinity of the cut end. This may result in a reduction of shear capacity and a loss in overall ductility of the beam. ACI Code 12.10 requires special precautions, specifying that no flexural bar shall be terminated in a tension zone unless *one* of the following conditions is satisfied:

1. The shear is not over two-thirds that normally permitted, including allowance for shear reinforcement, if any.
2. Stirrups in excess of those normally required are provided over a distance along each terminated bar from the point of cutoff equal to $\frac{3}{4}d$. These "binder" stirrups shall provide an area A_v such that $A_v f_y / b_w s$ is not less than 60 psi. In addition, the stirrup spacing shall not exceed $d/8\beta_b$, where β_b is the ratio of the area of bars cut off to the total area of bars at the section.
3. The continuing bars, if No. 11 or smaller, provide twice the area required for flexure at that point, and the shear does not exceed three-quarters of that permitted.

As an alternative to cutting off the steel, tension bars may be anchored by bending them across the web and making them continuous with the reinforcement on the opposite face. Although this leads to some complication in detailing and placing the steel, thus adding to construction cost, some engineers prefer the arrangement because added insurance is provided against the spread of diagonal tension cracks. In some cases, particularly for relatively deep beams in which a large percentage of the total bottom steel is to be bent, it may be impossible to locate the bend-up point for bottom bars far enough from the support for the same bars to meet the requirements for top steel. The theoretical points of bend should be checked carefully for both bottom and top steel.

Because the determination of cutoff or bend points may be rather tedious, particularly for frames that have been analyzed by elastic methods rather than by moment coefficients, many designers specify that bars be cut off or bent at more or less arbitrarily defined points that experience has proven to be safe. For nearly equal spans, uniformly loaded, in which not more than about one-half the tensile steel is to be cut off or bent, the locations shown in Fig. 5.15 are satisfactory.

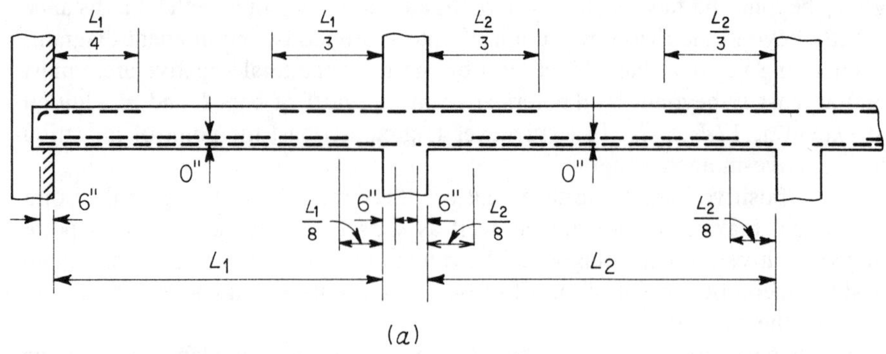

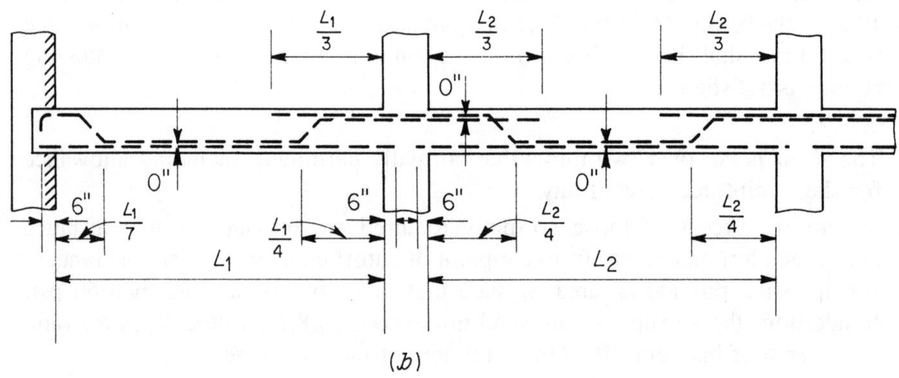

FIGURE 5.15
Standard cutoff or bend points for bars in approximately equal spans with uniformly distributed loads.

Note, in Fig. 5.15, that the beam at the exterior support at the left is shown to be simply supported. If the beam is monolithic with exterior columns or with a concrete wall at that end, details for a typical interior span could be used for the end span as well.

c. Special Requirements near the Point of Zero Moment

While the basic requirement for flexural tensile reinforcement is that a full development length l_d be provided beyond the point where the bar is assumed fully stressed to f_y, this requirement may *not* be sufficient to ensure safety against bond distress. Figure 5.16 shows the moment and shear diagram representative of a uniformly loaded continuous beam. Positive bars provided to resist the maximum moment at c are required to have a full development length beyond the point c, measured in the direction of decreasing moment. Thus l_d in the limiting case could be exactly equal to the distance from point c to the point of inflection.

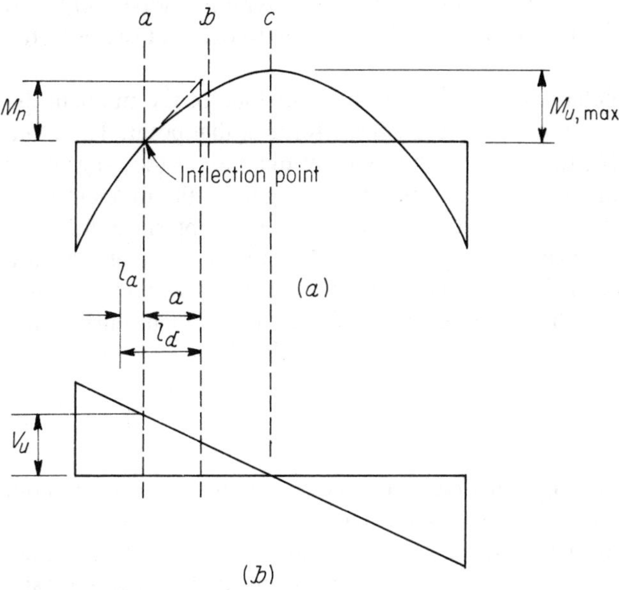

FIGURE 5.16
Development length requirement at point of inflection.

However, if that requirement were exactly met, then at point b, halfway from c to the point of inflection, those bars would have only half their development length remaining, whereas the moment would be three-quarters of that at point c, and three-quarters of the bar force must yet be developed. This situation arises whenever the moments over the development length are greater than those corresponding to a linear reduction to zero. Therefore the problem is a concern in the positive-moment region of continuous uniformly loaded spans, but not in the negative-moment region.

It was shown in Sec. 5.2 that the bond force U per unit length along the tensile reinforcement in a beam is $U = dT/dx$, where dT is the change in bar tension in the length dx. Since $dT = dM/z$, this can be written

$$U = \frac{dM}{z\,dx} \tag{a}$$

that is, the bond force per unit length of bar, generated by bending, is proportional to the slope of the moment diagram. In reference to Fig. 5.16a, the maximum bond force U in the positive moment region would therefore be at the point of inflection, and U would gradually diminish along the beam toward point c. Clearly, a conservative approach in evaluating adequacy in bond for those bars that are continued as far as the point of inflection (not necessarily the full A_s provided for M_u at point c) would be to require that the pullout resistance, which is assumed to increase linearly along the bar from its end, would be governed by the maximum rate of moment increase, i.e., the maximum slope dM/dx of the moment diagram, which for positive bending is seen to occur at the inflection point.

From elementary mechanics, it is known that the slope of the moment diagram at any point is equal to the value of the shear force at that point. Therefore, with reference to Fig. 5.16, the slope of the moment diagram at the point of inflection is V_u. A dashed line may therefore be drawn tangent to the moment curve at the point of inflection having the slope equal to the value of shear force V_u. Then if M_n is the nominal flexural strength provided by those bars that extend to the point of inflection, and if the moment diagram were conservatively assumed to vary linearly along the dashed line tangent to the actual moment curve, from the basic relation that $M_n/a = V_u$, a distance a is established:

$$a = \frac{M_n}{V_u} \tag{b}$$

If the bars in question were fully stressed at a distance a to the right of the point of inflection, and if the moments diminished linearly to the point of inflection, as suggested by the dashed line, then pullout would not occur if the development length l_d did not exceed the distance a. The actual moments are less than indicated by the dashed line, so the requirement is on the safe side.

If the bars extend past the point of inflection toward the support, as is always required, then the extension can be counted as contributing toward satisfying the requirement for embedded length. Arbitrarily, according to ACI Code 12.11, a length past the point of inflection not greater than the larger of the beam depth d or 12 times the bar diameter d_b may be counted toward satisfying the requirement. Thus, the requirement for tensile bars at the point of inflection is that

$$l_d \le \frac{M_n}{V_u} + l_a \tag{5.11}$$

where

M_n = nominal flexural strength assuming all reinforcement at section to be stressed to f_y
V_u = factored shear force at section
l_a = embedded length of bar past point of zero moment, but not to exceed the greater of d or $12d_b$

A corresponding situation occurs near the supports of simple spans carrying uniform loads, and similar requirements must be imposed. However, because of the beneficial effect of vertical compression in the concrete at the end of a simply supported span, which tends to prevent splitting and bond failure along the bars, the value M_n/V_u may be increased 30 percent for such cases, according to ACI Code 12.11. Thus, at the ends of a simply supported span, the requirement for tension reinforcement is

$$l_d \le 1.3\frac{M_n}{V_u} + l_a \tag{5.12}$$

The consequence of these special requirements at the point of zero moment is that, in some cases, smaller bar sizes must be used to obtain smaller l_d, even though requirements for development past the point of maximum stress are met.

It may be evident from review of Sec. 5.9b and Sec. 5.9c that the determination of cutoff or bend points in flexural members is complicated and can be extremely time-consuming in design. It is important to keep the matter in perspective and to recognize that the overall cost of construction will be increased very little if some rebars are slightly longer than absolutely necessary, according to calculation, or as dictated by ACI Code provisions. In addition, simplicity in construction is a desired goal, and can, in itself, produce compensating cost savings. Accordingly, many engineers in practice continue *all* positive reinforcement into the face of the supports the required 6 in. and extend *all* negative reinforcement the required distance past the points of inflection, rather than using staggered cutoff points.

d. Structural Integrity Provisions

Experience with structures that have been subjected to damage to a major supporting element, such as a column, owing to accident or abnormal loading has indicated that total collapse can be prevented through relatively minor changes in bar detailing. If some reinforcement, properly confined, is carried continuously through a support, then even if that support is damaged or destroyed, catenary action of the beams can prevent total collapse. In general, if beams have bottom and top steel meeting or exceeding the requirements summarized in Secs. 5.9b and 5.9c, and if binding steel is provided in the form of *closed stirrups*, then that catenary action can usually be ensured. According to ACI Code 7.13.2, if closed stirrups are *not* provided, then at least one-quarter of the positive reinforcement required at midspan must be continuous or must be spliced at the support with a class A tension splice (see Sec. 5.11a). At noncontinuous supports, such bars must be terminated with a standard hook (see Sec. 5.4a).

Special requirements pertain to beams around the perimeter of a structure. According to ACI Code 7.13.2, in perimeter beams, at least one-sixth of the negative reinforcement required at the supports and at least one-quarter of the positive steel required at midspan must be made continuous around the perimeter and tied with closed stirrups. Normally, top steel would be spliced at midspan and bottom steel spliced near the supports with class A tension splices.

It may be noted that these provisions require very little additional steel in the structure. At least one-quarter of the bottom bars must be extended 6 in. into the supports by other ACI Code provisions; the structural integrity provisions merely require that these bars be made continuous or spliced. Similarly, other ACI Code provisions require that at least one-third of the negative bars be extended a certain minimum distance past the point of inflection; the structural integrity provisions for perimeter beams require only that half of those bars be further extended and spliced at midspan. The use of closed stirrups has become almost standard practice in any case, because of the need to support negative rebars during concrete placement.

5.10 INTEGRATED BEAM DESIGN EXAMPLE

In this and in the preceding chapters, the several aspects of the design of reinforced concrete beams have been studied more or less separately: first the flexural design, then design for shear, and finally for bond and anchorage. The following example is presented to show how the various requirements for beams, which are often in some respects conflicting, are satisfied in the overall design of a representative member.

Example 5.3 Integrated design of T beam. A floor system consists of single span T beams 8 ft on centers, supported by 12 in. masonry walls spaced at 25 ft between inside faces. The general arrangement is shown in Fig. 5.17*a*. A 5 in. monolithic slab carries a uniformly distributed service live load of 165 psf. The T beams, in addition to the slab load and their own weight, must carry two 16,000 lb equipment loads applied 3 ft from the span centerline as shown. A complete design is to be provided for the T beams, using concrete of 4000 psi strength and bars with 60,000 psi yield stress.

Solution. According to the ACI Code, the span length is to be taken as the clear span plus the beam depth, but need not exceed the distance between the centers of supports. The latter provision controls in this case, and the effective span is 26 ft. Estimating the beam web dimensions to be 12 × 24 in., the calculated and factored dead loads are

Slab: $$\frac{5}{12} \times 150 \times 7 = 440 \text{ plf}$$

Beam: $$\frac{12 \times 24}{144} 150 = 300$$

$$w_d = 740 \text{ plf}$$

$$1.4 w_d = 1040 \text{ plf}$$

The uniformly distributed live load is

$$w_l = 165 \times 8 = 1310 \text{ plf}$$

$$1.7 w_l = 2230 \text{ plf}$$

Live load overload factors are applied to the two concentrated loads to obtain P_u = 16,000 × 1.7 = 27,000 lb. Factored loads are summarized in Fig. 5.17*b*.

In lieu of other controlling criteria, the beam web dimensions will be selected on the basis of shear. The left and right reactions under factored load are 27.0 + 3.27 × 13 = 69.5 kips. With the effective beam depth estimated to be 20 in., the maximum shear which need be considered in design is 69.5 − 3.27(0.50 + 1.67) = 62.4 kips. Although the ACI Code permits V_s as high as $8\sqrt{f_c'}b_w d$, this would require very heavy web reinforcement. A lower limit of $4\sqrt{f_c'}b_w d$ will be adopted. With $V_c = 2\sqrt{f_c'}b_w d$ this results in a maximum $V_n = 6\sqrt{f_c'}b_w d$. Then $b_w d = V_u/(6\phi\sqrt{f_c'}) = 62{,}400/(6 \times 0.85\sqrt{4000}) = 194 \text{ in}^2$. Cross-sectional dimensions b_w = 12 in. and d = 18 in. are selected, providing a total beam depth of 22 in. The assumed dead load of the beam need not be revised.

According to the Code, the effective flange width b is the smallest of the three quantities

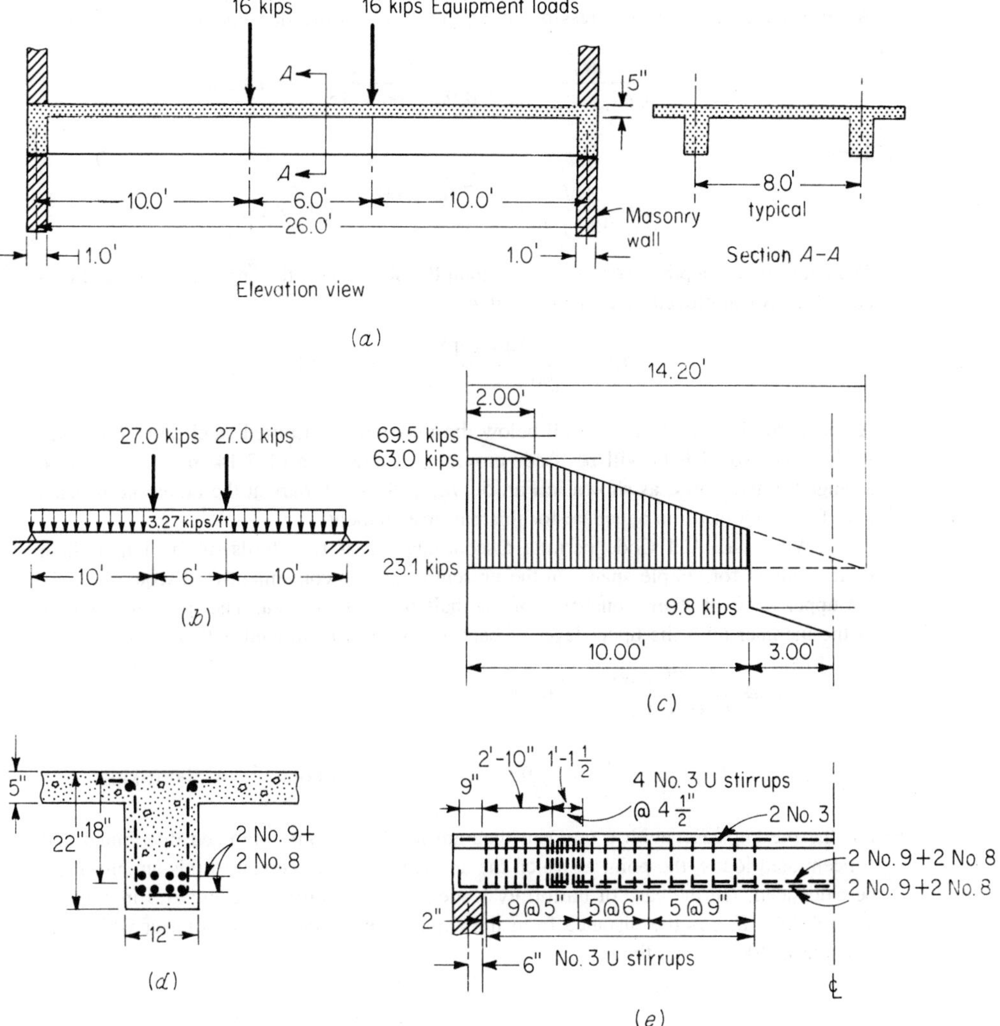

FIGURE 5.17
T beam design example.

$$\frac{L}{4} = \frac{26 \times 12}{4} = 78 \text{ in.}$$

$$16h_f + b_w = 80 + 12 = 92 \text{ in.}$$

$$\text{Centerline spacing} = 96 \text{ in.}$$

The first controls in this case. The maximum moment is at midspan, where

$$M_u = \frac{1}{8} \times 3.27 \times 26^2 + 27.0 \times 10 = 546 \text{ ft-kips}$$

Assuming for trial that the stress-block depth will equal the slab thickness leads to

$$A_s = \frac{M_u}{\phi f_y(d - a/2)} = \frac{546 \times 12}{0.90 \times 60 \times 15.5} = 7.82 \text{ in}^2$$

Then

$$a = \frac{A_s f_y}{0.85 f'_c b} = \frac{7.82 \times 60}{0.85 \times 4 \times 78} = 1.78 \text{ in.}$$

The stress-block depth is seen to be less than the slab depth; rectangular beam equations are valid. An improved determination of A_s is

$$A_s = \frac{546 \times 12}{0.90 \times 60 \times 17.11} = 7.10 \text{ in}^2$$

A check confirms that this is well below the maximum permitted steel ratio. Four No. 9 plus four No. 8 bars will be used, providing a total area of 7.14 in^2. They will be arranged in two rows, as shown in Fig. 5.17*d*, with No. 9 bars at the outer end of each row. Beam width b_w is adequate for this bar arrangement.

While the ACI Code permits discontinuation of two-thirds of the longitudinal reinforcement for simple spans, in the present case it is convenient to discontinue only the upper layer of steel, consisting of one-half of the total area. The moment capacity of the member after the upper layer of bars has been discontinued is then found:

$$a = \frac{3.57 \times 60}{0.85 \times 4 \times 78} = 0.81 \text{ in.}$$

$$\phi M_n = \phi A_s f_y \left(d - \frac{a}{2}\right) = 0.90 \times 3.57 \times 60 \times 18.66 \times \frac{1}{12} = 300 \text{ ft-kips}$$

For the present case, with a moment diagram resulting from combined distributed and concentrated loads, the point at which the applied moment is equal to this amount must be calculated. (In the case of uniformly loaded beams, Graphs A.2 and A.3 in App. A are helpful.) If x is the distance from the support centerline to the point at which the moment is 300 ft-kips, then

$$69.5x - \frac{3.27x^2}{2} = 300$$

$$x = 4.80$$

The upper bars must be continued at least $d = 1.50$ ft or $12d_b = 1.13$ ft beyond this theoretical point of cutoff. In addition, the full development length l_d must be provided past the maximum-moment section at which the stress in the bars to be cut is assumed to be f_y. Because of the heavy concentrated loads near the midspan, the point of peak stress will be assumed to be at the concentrated load rather than at midspan. The basic development length is found from Table A.11 to be 38 in. for the No. 9 bars, or 3.17 ft. This must be increased 100 percent, to 6.34 ft, according to the modification factor of Table 5.1*A*, because of the close bar spacing and lack of transverse steel meeting the stated requirements. Thus the bars will be continued at least $3.00 + 6.34 = 9.34$ ft past the midspan point, but in addition must continue to a point $4.80 - 1.50 = 3.30$ ft from the support centerline. The second requirement controls, and the upper bars will be terminated, as shown in Fig. 5.17*e*, 2.80 ft from the support face. The lower bars extend to a point 3 in. from the end of the beam, providing 5.55 ft embedment past the critical section. This is slightly less than the 6.34 ft required by ACI Code, and the deficiency will be made up by providing standard 90° bends at the ends of the bottom bars as shown in Fig. 5.17*e*. (Note that a simple design, using very little extra steel, would result from extending all eight positive bars in to the support.)

Checking by Eq. (5.12) to ensure that the continued steel is of sufficiently small diameter determines that

$$l_a \leq 1.3 \frac{333 \times 12}{69.5} + 3 = 78 \text{ in.}$$

The actual l_d of 76 in. meets this restriction.

Since the cut bars are located in the tension zone, special binding stirrups will be used to control cracking; these will be selected after the normal shear reinforcement has been determined.

The shear diagram resulting from application of factored loads is shown in Fig. 5.17*c*. The shear contribution of the concrete is

$$\phi V_c = 0.85 \times 2\sqrt{4000} \times 12 \times 18 = 23{,}000 \text{ lb}$$

Thus web reinforcement must be provided for that part of the shear diagram shown shaded.

No. 3 stirrups will be selected. The maximum spacings must not exceed $d/2 = 9$ in., 24 in., or $A_v f_y / 50 b_w = 0.22 \times 60{,}000/50 \times 12 = 22$ in. The first criterion controls here. For reference, from Eq. (4.14*a*) the hypothetical stirrup spacing at the support is

$$s_0 = \frac{0.85 \times 0.22 \times 60 \times 18}{69.5 - 23.1} = 4.35 \text{ in.}$$

and at 2 ft intervals along the span,

$$s_2 = 5.06 \text{ in.}$$
$$s_4 = 6.05 \text{ in.}$$
$$s_6 = 7.53 \text{ in.}$$
$$s_8 = 9.96 \text{ in.}$$
$$s_{10} = 14.70 \text{ in.}$$

The spacing need not be closer than that required 2.00 ft from the support centerline. In addition, stirrups are not required past the point of application of concentrated load, since beyond that point the shear is less than half of ϕV_c. The final spacing of vertical stirrups selected is

1 space at 2 in. = 2 in.
9 spaces at 5 in. = 45 in.
5 spaces at 6 in. = 30 in.
5 spaces at 9 in. = 45 in.
Total = 122 in. = 10 ft 2 in.

For convenience during construction, two No. 3 longitudinal bars will be added to fix the top of the stirrups.

In addition to the shear reinforcement just specified, it is necessary to provide extra web reinforcement over a distance equal to $\frac{3}{4}d$, or 13.5 in., from the cut ends of the discontinued steel. The spacing of this extra web reinforcement must not exceed $d/8\beta_b = 18/(8 \times \frac{1}{2}) = 4.5$ in. In addition, the area of added steel within the distance s must not be less than $60b_w s/f_y = 60 \times 12 \times 4.5/60{,}000 = 0.054$ in^2. For convenience, No. 3 stirrups will be used for this purpose also, providing an area of 0.22 in^2 in the distance s. The placement of the four extra stirrups is shown in Fig. 5.17*e*.

5.11 BAR SPLICES

In general, reinforcing bars are stocked by suppliers in lengths of 60 ft for bars from No. 5 to No. 18, and in 20 or 40 ft lengths for smaller sizes. For this reason, and because it is often more convenient to work with shorter bar lengths, it is frequently necessary to splice bars in the field. Splices in reinforcement at points of maximum stress should be avoided, and when splices are used they should be staggered, although neither condition is practical, for example, in compression splices in columns.

Splices for No. 11 bars and smaller are usually made simply by lapping the bars a sufficient distance to transfer stress by bond from one bar to the other. The lapped bars are usually placed in contact and lightly wired so that they stay in position as the concrete is poured. Alternatively, splicing may be accomplished by welding or by sleeves or mechanical devices that provide "full positive connection" between the bars. ACI Code 12.14.2 prohibits use of lapped splices for bars larger than No. 11. For bars that will carry only compression, it is possible to transfer load by end bearing of square cut ends, if the bars are accurately held in position by a sleeve or other device.

Lap splices of bundled bars are based on the lap splice length required for individual bars within the bundle but must be increased in length by 20 percent for three-bar bundles and by 33 percent for four-bar bundles because of the reduced effective perimeter. Individual bar splices within a bundle should not overlap, and entire bundles must not be lap spliced.

According to ACI Code 12.14.3, welded splices must be butted and welded so that the connection will develop in tension at least 125 percent of the specified yield strength of the bar. The same requirement applies to full mechanical connections. This ensures that an overloaded spliced bar would fail by ductile yielding in the region away from the splice, rather than at the splice where brittle failure is likely. Connections not meeting this requirement may used at points of less than maximum stress.

a. Lap Splices in Tension

The required length of lap for tension splices, established by test, may be stated in terms of the development length l_d. In the process of calculating l_d from the basic development length l_{db}, the usual modification factors are applied (see Table 5.1), except that the reduction factor for excess reinforcement should not be applied because that factor is already accounted for in the splice specification.

Two different classifications of lap splices are established corresponding to the minimum length of lap required: a Class A splice requires a lap of $1.0l_d$, and a class B splice requires a lap of $1.3l_d$. In either case, a minimum length of 12 in. applies. Lap splices, in general, must be class B splices, according to ACI Code 12.15.2, except that class A splices are allowed when the area of reinforcement provided is at least twice that required by analysis over the entire length of the splice *and* when one-half or less of the total reinforcement is spliced within the required lap length. The effect of these requirements is to encourage designers to locate splices away from regions of maximum stress, to a location where the actual steel area is at least twice that required by analysis, and to stagger splices.

b. Compression Splices

Reinforcing bars in compression are spliced mainly in columns, where bars are normally terminated just above each floor or every other floor. This is done partly for construction convenience, to avoid handling and supporting very long column bars, but it is also done to permit column steel area to be reduced in steps, as loads become lighter at higher floors.

Compression bars may be spliced by lapping, by direct end bearing, or by welding or mechanical devices that provide positive connection. The minimum length of lap for compression splices is set according to ACI Code 12.16:

For bars with $f_y \le 60{,}000$ psi	$0.0005 f_y d_b$
For bars with $f_y > 60{,}000$ psi	$(0.0009 f_y - 24) d_b$

but not less than 12 in. For f'_c less than 3000 psi, the required lap is increased by one-third. When bars of different size are lap spliced in compression, the splice length is to be the larger of the development length of the larger bar and the splice length of the smaller bar. In exception to the usual restriction on lap splices for large diameter bars, No. 14 and No. 18 bars *may* be lap spliced to No. 11 and smaller bars.

Direct end bearing of the bars has been found by test and experience to be an effective means for transmitting compression. In such a case, the bars must be held in proper alignment by a suitable device. The bar ends must terminate in flat surfaces within 1.5° of a right angle, and the bars must be fitted within 3° of full bearing after assembly, according to ACI Code 12.16.4. Ties, closed stirrups, or spirals must be used.

c. Column Splices

Lap splices, butt-welded splices, mechanical connections, or end-bearing splices may be used in columns, with certain restrictions. Reinforcing bars in columns may be subjected to compression or tension, or, for different load combinations, both tension and compression. Accordingly, column splices must conform in some cases to the requirements for compression splices only or tension splices only or to requirements for both. ACI Code 12.17 requires that a minimum tension capacity be provided in each face of all columns, even where analysis indicates compression only. Ordinary compressive lap splices provide sufficient tensile resistance, but end-bearing splices may require additional bars for tension, unless the splices are staggered.

For lap splices, where the bar stress due to factored loads is compression, column lap splices must conform to the requirements presented in Sec. 5.11b for compression splices. Where the stress is tension and does not exceed $0.5f_y$, lap splices must be Class B if more than half the bars are spliced at any section, or Class A if half or fewer are spliced and alternate lap splices are staggered by l_d. If the stress is tension and exceeds $0.5f_y$, then lap splices must be Class B, according to ACI Code.

If lateral ties are used throughout the splice length having an area of at least $0.0015hs$, where s is the spacing of ties and h is the overall thickness of the member, the required splice length may be multiplied by 0.83 but must not be less than 12 in. If spiral reinforcement confines the splice, the length required may be multiplied by 0.75 but again must not be less than 12 in.

End-bearing splices, as described above, may be used for column bars stressed in compression, if the splices are staggered or additional bars are provided at splice locations. The continuing bars in each face must have a tensile strength of not less than $0.25f_y$ times the area of reinforcement in that face.

Example 5.4 Compression splice of column reinforcement. In reference to Fig. 5.8, four No. 11 column bars from the floor below are to be lap spliced with four No. 10 column bars from above, and the splice is to be made just above a construction joint at floor level. The column, measuring 12 in. × 21 in. in cross section, will be subject to compression only for all load combinations. Lateral reinforcement consists of No. 4 ties at 16 in. spacing. All vertical bars may be assumed to be fully stressed. Calculate the required splice length. Material strengths are $f_y = 60{,}000$ psi and $f'_c = 4000$ psi.

Solution. The length of the splice must be the larger of the development length of the No. 11 bars and the splice length of the No. 10 bars. For the No. 11 bars, the development length, from Eq. (5.10), is

$$l_{db} = \frac{0.02 \times 1.41 \times 60{,}000}{\sqrt{4000}} = 27 \text{ in.}$$

but is not to be less than $0.0003 \times 1.41 \times 60{,}000 = 25$ in. The first criterion controls. No modification factors apply. For the No. 10 bars, the compression splice length is $0.0005 \times 60{,}000 \times 1.27 = 38$ in. In the check for use of the modification factor for tied columns, the critical column dimension is 21 in., and the required effective tie area is thus $0.0015 \times 21 \times 16 = 0.50$ in^2. The No. 4 ties provide an area of only $0.20 \times 2 = 0.40$ in^2, so the reduction factor of 0.83 cannot be applied to the splice length. Thus the compression splice length of 38 in., which exceeds the development length of 27 in. for the No. 11 bars, controls here, and a lap splice of 38 in. is required. Note that if the spacing of the ties at the splice were reduced to 12.8 in. or less (say 12 in.), the required lap would be reduced to $38 \times 0.83 = 32$ in. This would save steel, and, although placement cost would increase slightly, would probably represent the more economical design.

REFERENCES

5.1. R. M. Mains, "Measurement of the Distribution of Tensile and Bond Stresses along Reinforcing Bars," *J. ACI*, vol. 23, no. 3, 1951, pp. 225-252.
5.2. A. H. Nilson, "Internal Measurement of Bond Slip," *J. ACI*, vol. 69, no. 7, 1972, pp. 439–441.
5.3. Y. Goto, "Cracks Formed in Concrete around Deformed Tension Bars," *J. ACI*, vol. 68, no. 4, 1971, pp. 244–251.
5.4. L. A. Lutz and P. Gergely, "Mechanics of Bond and Slip of Deformed Bars in Concrete," *J. ACI*, vol. 64, no. 11, 1967, pp. 711–721.
5.5. P. M. Ferguson and J. N. Thompson, "Development Length of High Strength Reinforcing Bars in Bond," *J. ACI*, vol. 59, no. 7, July 1962, pp. 887–922.
5.6. R. G. Mathey and D. Watstein, "Investigation of Bond in Beam and Pullout Specimens with High-Strength Deformed Bars," *J. ACI*, vol. 32, no. 9, 1961, pp. 1071–1090.
5.7. "Bond Stress—The State of the Art," ACI Committee 408, *J. ACI*, vol. 63, no. 11, 1966, pp. 1161–1190.
5.8. P. M. Ferguson, "Small Bar Spacing or Cover—A Bond Problem for the Designer," *J. ACI*. vol. 74, no. 9, 1977, pp. 435–439.
5.9. "Suggested Development, Splice, and Standard Hook Provisions for Deformed Bars in Tension," ACI Committee 408, *Conc. Int.*, vol. 1, no. 7, 1979, pp. 44–46.
5.10. J. O. Jirsa, L. A. Lutz, and P. Gergely, "Rationale for Suggested Development, Splice, and Standard Hook Provisions for Deformed Bars in Tension," *Conc. Int.*, vol. 1, no. 7, 1979, pp. 47–61.
5.11. J. L. G. Marques and J. O. Jirsa, "A Study of Hooked Bar Anchorages in Beam-Column Joints," *J. ACI*, vol. 72, no. 5, 1975, pp. 198–209.

PROBLEMS

5.1 The short beam of Fig. P5.1 cantilevers from a supporting column at the left. It must carry calculated dead load of 1.5 kips/ft including its own weight, and service live load of 3.0 kips/ft. Tensile flexural reinforcement consists of two No. 11 bars at 21 in. effective depth.

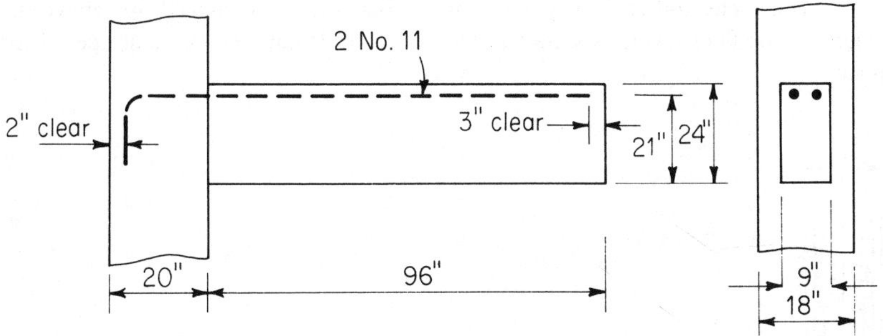

FIGURE P5.1

(*a*) If materials for the beam provide $f_y = 60{,}000$ psi and $f'_c = 3000$ psi, check to see if proper development length can be provided in the beam for the No. 11 bars.

(*b*) If the column material strengths are $f_y = 60{,}000$ psi and $f'_c = 5000$ psi, check to see if adequate embedment can be provided within the column for the No. 11 bars. If hooks are required, specify detailed dimensions.

5.2 The beam of Fig. P5.2 is simply supported with a clear span of 24.75 ft and is to carry a distributed dead load of 0.54 kips/ft including its own weight, and live load of 1.08 kips/ft, unfactored, in service. The reinforcement consists of three No. 10 bars at 16 in. effective depth, one of which is to be discontinued where no longer needed. Material strengths specified are $f_y = 60{,}000$ psi and $f_c' = 4000$ psi.

(*a*) Calculate the point where the center bar can be discontinued.

(*b*) Check to be sure that adequate embedded length is provided for continued and discontinued bars.

(*c*) Check special requirements at the support, where $M_u = 0$.

(*d*) If No. 3 bars are used for lateral reinforcement, specify special reinforcing details in the vicinity where the No. 10 bar is cut off.

(*e*) Comment on the practical aspects of the proposed design. Would you recommend cutting of the steel as suggested? Could two bars be discontinued rather than one?

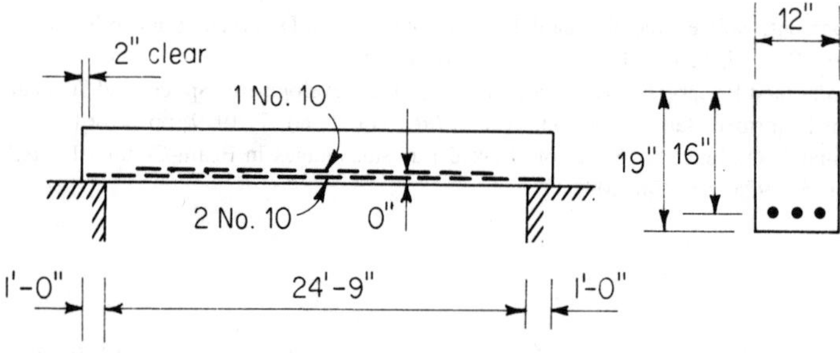

FIGURE P5.2

5.3 Figure P5.3 shows the column reinforcement for a 16 in. diameter concrete column, with $f_y = 60{,}000$ psi and $f_c' = 5000$ psi. Analysis of the building frame indicates a required $A_s = 7.10$ in^2 in the lower column and 5.60 in^2 in the upper column. Spiral reinforcement consists of a $\frac{3}{8}$ in. diameter rod at 2 in. pitch. Column bars are to be spliced just above the construction joint at the floor level, as shown in the sketch. Calculate the minimum permitted length of splice.

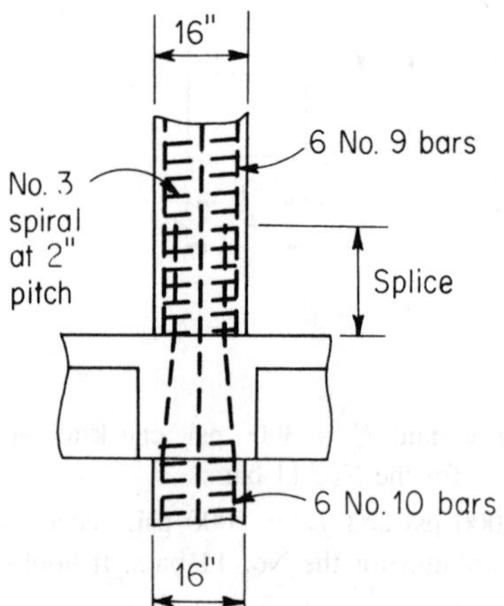

FIGURE P5.3

5.4 The short cantilever shown in Fig. P5.4 carries a heavy concentrated load 6 in. from its outer end. Flexural analysis indicates that three No. 8 bars are required, suitably anchored in the supporting wall and extending to a point no closer than 2 in. from the free end. The bars will be fully stressed to f_y at the fixed support. Investigate the need for hooks and lateral confinement steel at the right end of the member. Material strengths are $f_y = 60{,}000$ psi and $f_c' = 4000$ psi. If hooks and lateral steel are required, show details in a sketch.

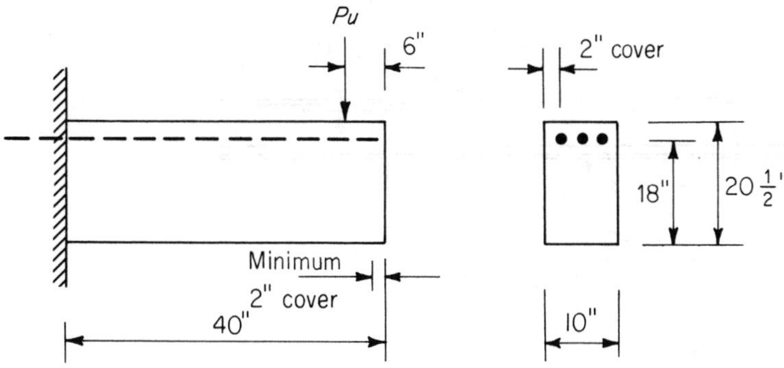

FIGURE P5.4

5.5 A continuous-strip wall footing is shown in cross section in Fig. P5.5. It is proposed that tensile reinforcement be provided using No. 8 bars at 16 in. spacing along the length of the wall, to provide a bar area of 0.59 in^2/ft. The bars have strength $f_y = 60{,}000$ psi and the footing concrete has $f_c' = 3000$ psi. The critical section for bending is assumed to be at the face of the supported wall, and the effective depth to the tensile steel is 12 in. Check to ensure that sufficient development length is available for the No. 8 bars, and if hooks are required, sketch details of the hooks giving dimensions.

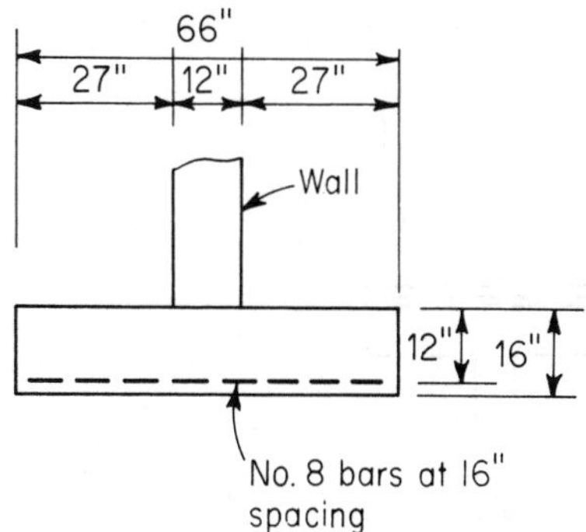

FIGURE P5.5

Note: If hooks are required for the No. 8 bars, prepare an alternate design using bars having the same area per foot but of smaller diameter such that hooks could be eliminated; use the largest bar size possible to minimize the cost of steel placement.

5.6 The continuous beam shown in Fig. P5.6 has been designed to carry a service dead load of 2 kips/ft including self-weight, and service live load of 3 kips/ft. Flexural design has been based on ACI moment coefficients of $\frac{1}{11}$ and $\frac{1}{16}$ at the face of support and midspan respectively, resulting in a concrete section with $b = 14$ in. and $d = 22$ in. Negative reinforcement at the support face is provided by four No. 10 bars, which will be cut off in pairs where no longer required by the ACI Code. Positive bars consist of four No. 8 bars, which will also be cut off in pairs. Specify the exact point of cutoff for all negative and positive steel. Specify also any supplementary web reinforcement that may be required. Check for satisfaction of ACI Code requirements at the point of inflection and suggest modifications of reinforcement if appropriate. Material strengths are $f_y = 60{,}000$ psi and $f_c' = 4000$ psi.

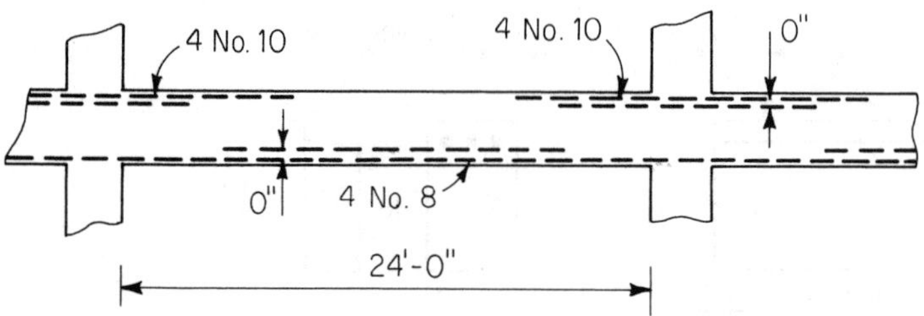

FIGURE P5.6

5.7 Figure P5.7 shows a deep transfer girder that carries two heavy column loads at its outer ends from a high-rise concrete building. Ground-floor columns must be offset 8 ft as shown. The loading produces an essentially constant moment (neglect self-weight of girder) calling for a concrete section with $b = 22$ in. and $d = 50$ in., with main tensile reinforcement at the top of the girder comprised of twelve No. 11 bars in three layers of four bars each. The maximum available bar length is 60 ft, so tensile splices must be provided. Design and detail all splices, following ACI Code provisions. Splices will be staggered, with no more than four bars spliced at any section. Also, investigate the need for special anchorage at the outer ends of main reinforcement, and specify details of special anchorage if required. Material strengths are $f_y = 60{,}000$ psi and $f_c' = 5000$ psi.

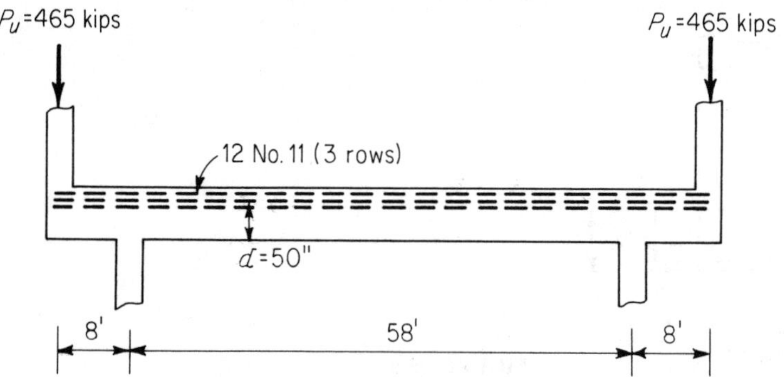

FIGURE P5.7

II Start

CHAPTER 6

SERVICEABILITY

6.1 INTRODUCTION

Chapters 3, 4, and 5 have dealt mainly with the strength design of reinforced concrete beams. Methods have been developed to ensure that beams will have a proper safety margin against failure in flexure or shear, or due to inadequate bond and anchorage of the reinforcement. The member has been assumed to be at a hypothetical overload state for this purpose.

It is also important that member performance in normal service be satisfactory, when loads are those actually expected to act, i.e., when load factors are 1.0. This is not guaranteed simply by providing adequate strength. Service load deflections under full load may be excessively large, or long-term deflections due to sustained loads may cause damage. Tension cracks in beams may be wide enough to be visually disturbing, or may even permit serious corrosion of reinforcing bars. These and other questions, such as vibration or fatigue, require consideration.

Serviceability studies are carried out based on elastic theory, with stresses in both concrete and steel assumed to be proportional to strain. The concrete on the tension side of the neutral axis may be assumed uncracked, partially cracked, or fully cracked, depending on the loads and material strengths (see Sec. 3.3).

In the past, questions of serviceability were dealt with indirectly, by limiting the stresses in concrete and steel at service loads to the rather conservative values that had resulted in satisfactory performance. Now, with strength design methods in general use that permit more slender members through more accurate assessment of capacity, and with higher strength materials further contributing to the trend toward smaller member sizes, such indirect methods will no longer do. The current approach is to investigate service load cracking and deflections specifically, after proportioning members based on strength requirements. ACI Code provisions reflect this change in thinking.

In this chapter, methods will be developed to ensure that the cracks associated with flexure of reinforced concrete beams are narrow and well distributed, and that short- and long-term deflections at loads up to the full service load are not objectionably large.

6.2 CRACKING IN FLEXURAL MEMBERS

All reinforced concrete beams crack, generally starting at loads well below service level, and possibly even prior to loading due to restrained shrinkage. Flexural cracking due to loads is not only inevitable, but actually necessary for the reinforcement to be used effectively. Prior to the formation of flexural cracks, the steel stress is no more than n times the stress in the adjacent concrete, where n is the modular ratio, E_s/E_c. For materials common in current practice, n is approximately 8. Thus, when the concrete is close to its modulus of rupture of about 500 psi, the steel stress will be only $8 \times 500 = 4000$ psi, far too low to be very effective as reinforcement. At normal service loads, steel stresses eight or nine times that value can be expected.

In a well-designed beam, flexural cracks are fine, so-called "hairline" cracks, almost invisible to the casual observer, and they permit little if any corrosion of the reinforcement. As loads are gradually increased above the cracking load, both the number and width of cracks increase, and at service load level a maximum width of crack of about 0.01 in. is typical. If loads are further increased, crack widths increase further, although the number of cracks is more or less stable.

Cracking of concrete is a random process, highly variable and influenced by many factors. Because of the complexity of the problem, present methods for predicting crack widths are based primarily on test observations. Most equations that have been developed predict the probable maximum crack width, which usually means that about 90 percent of the crack widths in the member are below the calculated value. However, isolated cracks exceeding twice the computed width can sometimes occur (Ref. 6.1).

a. Variables Affecting Width of Cracks

In the discussion of the importance of a good bond between steel and concrete in Sec. 5.1, it was pointed out that if proper end anchorage is provided a beam will not fail prematurely, even though the bond is destroyed along the entire span. However, crack widths will be greater than for an otherwise identical beam in which good resistance to slip is provided along the length of the span. In general, beams with smooth round bars will display a relatively small number of rather wide cracks in service, while beams with good slip resistance ensured by proper surface deformations on the bars will show a larger number of very fine, almost invisible cracks. Because of this improvement, reinforcing bars in current practice are always provided with surface deformations, the maximum spacing and minimum height of which are established by ASTM Specifications A615, A616, and A617.

A second variable of importance is the stress in the reinforcement. Studies by Gergely and Lutz (Ref. 6.2) and others have confirmed that crack width is proportional to f_s^n, where f_s is the steel stress and n is an exponent that varies in the range from about 1.0 to 1.4. For steel stresses in the range of practical interest, say from 20 to 36 ksi, n may be taken equal to 1.0. The steel stress is easily computed based on elastic cracked-section analysis (Sec. 3.3b). Alternatively, f_s may be taken equal to $0.60f_y$ according to ACI Code 10.6.4.

Experiments by Broms (Ref. 6.3) and others have showed that both crack spacing and crack width are related to the concrete cover distance d_c, measured from the center of the bar to the face of the concrete. In general, increasing the cover increases the spacing of cracks and also increases crack width. Furthermore, the distribution of the reinforcement in the tension zone of the beam is important. Generally, to control cracking, it is better to use a larger number of smaller-diameter bars to provide the required A_s than to use the minimum number of larger bars, and the bars should be well distributed over the tensile zone of the concrete.

b. The Gergely-Lutz Equation for Crack Width

Based on their research at Cornell University (Ref. 6.2), which involved the statistical analysis of a large amount of experimental data, Gergely and Lutz proposed the following equation for predicting the maximum width of crack at the tension face of a beam:

$$w = 0.076\beta f_s \sqrt[3]{d_c A} \tag{6.1}$$

in which w is the maximum width of crack, in thousandth inches, and f_s is the steel stress at the load for which the crack width is to be determined, measured in ksi. The geometric parameters are shown in Fig. 6.1 and are as follows:

d_c = thickness of concrete cover measured from tension face to center of bar closest to that face, in.

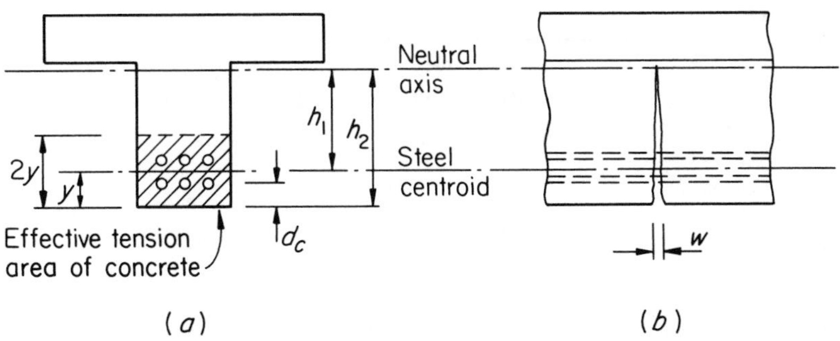

FIGURE 6.1
Geometric basis of crack width calculations.

β = ratio of distances from tension face and from steel centroid to neutral axis, equal to h_2/h_1

A = concrete area surrounding one bar, equal to total effective tension area of concrete surrounding reinforcement and having same centroid, divided by number of bars, in^2

Equation (6.1), which applies only to beams in which deformed bars are used, includes all the factors just named as having an important influence on the width of cracks: steel stress, concrete cover, and the distribution of the reinforcement in the concrete tensile zone. In addition, the factor β is added, to account for the increase in crack width with distance from the neutral axis (see Fig. 6.1*b*).

c. Permissible Crack Widths

The acceptable width of flexural cracks in service depends mostly on the conditions of exposure and should be established in view of the possibility of corrosion of the reinforcement. The recommendations of ACI Committee 224 (Ref. 6.1) are summarized in Table 6.1. However, good engineering judgment must be used in setting limiting values in particular cases. It should be recognized that, because of the random nature of cracking, individual cracks significantly wider than predicted by Eq. (6.1) are likely. The designer should also keep in mind, however, that Eq. (6.1) predicts the crack width at the surface of the member, whereas the width of crack is known to be less at the steel-concrete interface (Ref. 6.3). Increasing the concrete cover, even though it increases the surface crack width, may be beneficial in avoiding corrosion.

d. Cyclic and Sustained Load Effects

Both cyclic and sustained loading account for increasing crack width. While there is a large amount of scatter in test data, results of fatigue tests and sustained loading tests indicate that a doubling of crack width can be expected with time

Table 6.1 Tolerable crack widths for reinforced concrete

	Tolerable crack width	
Exposure condition	**in.**	**mm**
Dry air or protective membrane	0.016	0.41
Humidity, moist air, soil	0.012	0.30
Deicing chemicals	0.007	0.18
Seawater and seawater spray; wetting and drying	0.006	0.15
Water-retaining structures, excluding nonpressure pipes	0.004	0.10

Source: From Ref. 6.1.

(Ref. 6.1). Under most conditions the spacing of cracks does not change with time at constant levels of sustained stress or cyclic stress range.

6.3 ACI CODE PROVISIONS FOR CRACK CONTROL

In view of the random nature of cracking and the wide scatter of crack width measurements, even under laboratory conditions, excessive precision in calculating the crack width is not justified. Accordingly, Eq. (6.1) can be simplified for typical beams by adopting a representative value of $\beta = 1.2$. Then a parameter z may be defined as follows:

$$z = f_s \sqrt[3]{d_c A} \tag{6.2}$$

in which

$$z = \frac{w}{0.076 \times 1.2} = \frac{w}{0.091}$$

Control of the maximum crack width can be thus obtained by setting an upper limit on the parameter z. ACI Code 10.6.4 specifies that z shall not exceed 175 for interior exposure and 145 for exterior exposure. These limits correspond to maximum crack widths of 0.016 and 0.013 in, respectively. In addition, ACI Code 10.6.3 specifies that tension reinforcement shall be well distributed in the zone of maximum concrete tension. ACI Code 9.4 states that designs shall not be based on a yield strength f_y in excess of 80,000 psi.

When bars having different diameters are used together, as is often advantageous in practice, the concrete tension area per bar should be computed using an equivalent number of bars, found by dividing the total area of reinforcement by the area of the largest bar used, according to the ACI Code. Thus, if a total steel area of 3.27 in^2 were provided by one No. 10 bar plus two No. 9 bars, the equivalent number of bars to be used in calculating A would be $3.27/1.27 = 2.6$. When bundled bars are used, Lutz (Ref. 6.4) has recommended that each bundle be counted as the equivalent of 1.4 bars in computing A, recognizing that the bundled bars have an increased bond perimeter compared to a single bar having the same area as the bundle.

Equations (6.1) and (6.2) can also be used for one-way slabs. For typical slabs, however, where effective depth is less than for beams and concrete cover below the bars may be about 1 in., a representative value of β is about 1.35 rather than 1.2 as for beams. For a given limiting crack width, therefore, limiting z values should be multiplied by the ratio 1.2/1.35. Thus, for slabs z should not exceed 155 for interior exposure and 130 for exterior exposure, corresponding to crack widths of 0.016 in. and 0.013 in. respectively.

When concrete T-beam flanges are in tension, as in the negative-moment region of continuous T beams, concentration of the reinforcement over the web may result in excessive crack width in the overhanging slab, even though cracks directly over the web are fine and well distributed. To prevent this, the tensile

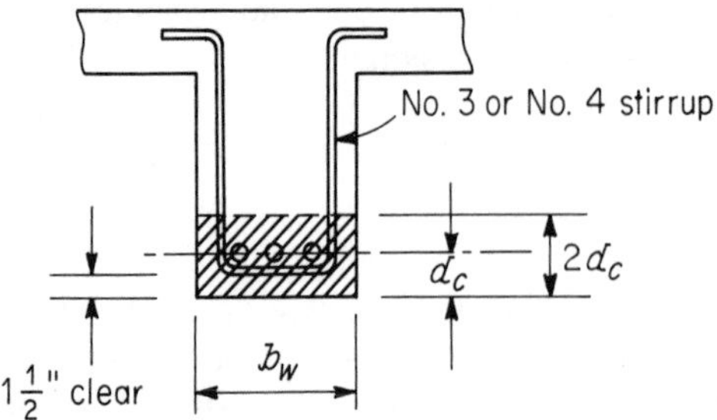

FIGURE 6.2
Minimum number of bars in beam stem.

reinforcement should be distributed over the width of the flange, rather than concentrated. However, because of shear lag, the outer bars in such a distribution would be considerably less highly stressed than those directly over the web, producing an uneconomical design. As a reasonable compromise, ACI Code 10.6.6 requires that the tension reinforcement in such cases be distributed over the effective flange width or a width equal to one-tenth the span, whichever is smaller. If the effective flange width exceeds one-tenth the span, some longitudinal reinforcement must be provided in the outer portions of the flange. The amount of such additional reinforcement is left to the discretion of the designer; it should at least be the equivalent of temperature reinforcement for the slab (see Sec. 12.3), and is often taken as twice that amount.

For beams with relatively deep webs, some reinforcement should be placed near the vertical faces of the web to control the width of cracks in the concrete tension zone above the level of the main rebars. Without such steel, crack widths in the web wider than those at the level of the main bars have been observed. According to ACI Code 10.6.7, if the depth of the web exceeds 3 ft, longitudinal "skin" reinforcement must be uniformly distributed along both side faces of the member for a distance $d/2$ nearest the flexural tension steel. The area of skin reinforcement, A_{sk}, per foot of height on each side face, must not be less than $0.012(d - 30)$ in^2 per ft. The maximum spacing must not exceed $d/6$ or 12 in. The *total* area of longitudinal skin reinforcement in both faces need not exceed one-half the area of the required flexural tensile reinforcement. The contribution of the skin steel to flexural strength is usually disregarded, although it may be included in the strength calculations if a strain compatibility analysis is used to establish the stress in the skin steel at the flexural failure load.

For beams with main flexural reinforcement in one layer in the web, a convenient design aid can be developed, based on Eq. (6.2), that permits tabulation of the minimum number of bars that will satisfy ACI Code requirements for crack control. With reference to Fig. 6.2, the total tensile area of concrete is equal to $2d_c b_w$. Thus the tensile area per bar is

$$A = \frac{2d_c b_w}{m} \qquad (a)$$

where m is the number of bars in the single layer of reinforcement. Then from Eq. (6.2),

$$\left(\frac{z}{f_s}\right)^3 = \frac{2d_c^2 b_w}{m} \tag{b}$$

from which

$$m = \frac{2d_c^2 b_w}{(z/f_s)^3} \tag{c}$$

For the typical Grade 60 bars, f_s may be taken as $0.6 \times 60 = 36$ ksi, and for bars of diameter d_b with 1.5 in. cover below the stirrups, which generally are Nos. 3 or 4 bars,

$$d_c = 2.0 + 0.5d_b \tag{d}$$

That value is then substituted in Eq. (*c*) to obtain

$$m = \frac{2(2.0 + 0.5d_b)^2 b_w}{(z/36)^3} \tag{6.3}$$

giving the minimum number of bars that can be used to satisfy crack control requirements, in terms of bar diameter d_b and beam width b_w, for the imposed limit on z. Table A.9 of App. A gives values of m, the minimum number of bars that will satisfy ACI Code requirements for crack control for both interior and exterior exposure.

Example 6.1 Check for satisfactory crack widths. The T beam of Fig. 6.3 carries a service load moment of 5850 in-kips. Estimate the maximum width of crack to be expected at the bottom surface of the member at full service load and determine if reinforcing details, with respect to cracking, are satisfactory for exterior exposure according to the ACI Code.

Solution. The total tensile steel area provided by the six No. 10 bars is 7.59 in^2. The steel stress at service load can be estimated closely by taking the internal lever arm equal to the distance $d - t/2$:

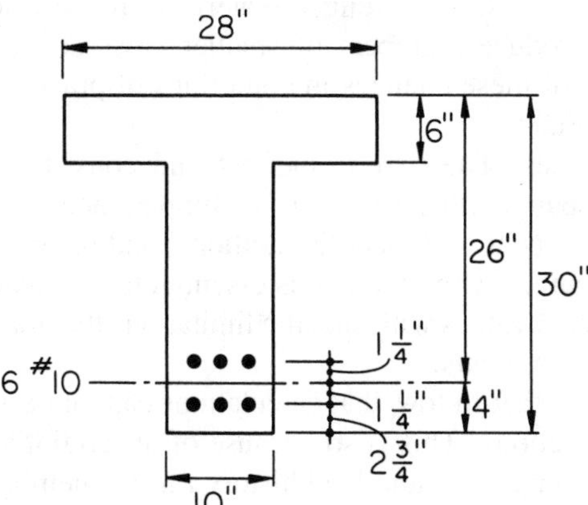

FIGURE 6.3
T beam for crack width determination in Example 6.1.

$$f_s = \frac{M_s}{A_s(d - t/2)} = \frac{5850}{7.59 \times 23} = 33.6 \text{ ksi}$$

(Alternatively, the ACI Code permits using $f_s = 0.60 f_y$ giving 36.0 ksi). The distance from the steel centroid to the tensile face of the beam is 4 in.; hence, the total effective concrete area for purposes of cracking calculations is $4 \times 2 \times 10 = 80$ in^2 and

$$A = \frac{80}{6} = 13.30 \text{ in}^2$$

Then, by Eq. (6.1), with $d_c = 2\frac{3}{4}$ in.,

$$w = 0.076 \times 1.2 \times 33.6 \sqrt[3]{2.75 \times 13.30}$$

$$= 10 \text{ thousandth in.} = 0.010 \text{ in.}$$

Alternatively, by Eq. (6.2),

$$z = 33.6 \sqrt[3]{2.75 \times 13.30} = 112$$

well below the limit of 145 imposed by the ACI Code for exterior construction. If the results had been unfavorable, a redesign using a larger number of smaller-diameter bars, thus reducing the value of A, would have been indicated.

6.4 CONTROL OF DEFLECTIONS

In addition to limitations on cracking, described in the preceding articles of this chapter, it is usually necessary to impose certain controls on deflections of beams in order to ensure serviceability. Excessive deflections can lead to cracking of supported walls and partitions, ill-fitting doors and windows, poor roof drainage, misalignment of sensitive machinery and equipment, or visually offensive sag. It is important, therefore, to maintain control of deflections, in one way or another, so that members designed mainly for strength at prescribed overloads will also perform well in normal service.

In the past, deflection control was achieved indirectly, by limiting service load stresses in concrete and steel to conservatively low values. The resulting members were generally larger, and consequently stiffer, than those designed by current methods based on strength. In addition, stronger materials are now in general use, and this, too, tends to produce members of smaller cross section that are less stiff than before. Because of these changes in conditions of practice, deflection control is increasingly important.

There are presently two approaches. The first is indirect and consists in setting suitable upper limits on the span-depth ratio. This is simple, and it is satisfactory in many cases where spans, loads and load distributions, and member sizes and proportions fall in the usual ranges. Otherwise, it is essential to calculate deflections and to compare those predicted values with specific limitations that may be imposed by codes or by special requirements.

It will become clear, in the sections that follow, that calculations can, at best, provide a guide to probable actual deflections. This is so because of uncertainties regarding material properties, effects of cracking, and load history for the member under consideration. Extreme precision in the calculations, therefore, is never justified, because highly accurate results are unlikely. However, it is generally sufficient to know, for example, that the deflection under load will be about $\frac{1}{2}$ in. rather than 2 in., while it is relatively unimportant to know whether it will actually be $\frac{5}{8}$ in. rather than $\frac{1}{2}$ in.

The deflections of concern are generally those that occur during the normal service life of the member. In service, a member sustains the full dead load, plus some fraction or all of the specified service live load. Safety provisions of the ACI Code and similar design specifications ensure that, under loads up to the full service load, stresses in both steel and concrete remain within the elastic ranges. Consequently, deflections that occur at once upon application of load, the so-called *immediate deflections*, can be calculated based on the properties either of the uncracked elastic member, the cracked elastic member, or some combination of these (see Sec. 3.3).

It was pointed out in Secs. 2.7 and 2.10, however, that in addition to concrete deformations that occur immediately when load is applied, there are other deformations that take place gradually over an extended period of time. These time-dependent deformations are chiefly due to concrete creep and shrinkage. As a result of these influences, reinforced concrete members continue to deflect with the passage of time. Long-term deflections continue over a period of several years, and may eventually be two or more times the initial elastic deflections. Clearly, methods for predicting both instantaneous and time-dependent deflections are essential.

6.5 IMMEDIATE DEFLECTIONS

Elastic deflections can be expressed in the general form

$$\Delta = \frac{f(\text{loads, spans, supports})}{EI}$$

where EI is the flexural rigidity and f(loads, spans, supports) is a function of the particular load, span, and support arrangement. For instance, the deflection of a uniformly loaded simple beam is $5wl^4/384EI$, so that $f = 5wl^4/384$. Similar deflection equations have been tabulated or can easily be computed for many other loadings and span arrangements, simple, fixed, or continuous, and the corresponding f functions can be determined. The particular problem in reinforced concrete structures is therefore the determination of the appropriate flexural rigidity EI for a member consisting of two materials with properties and behavior as widely different as steel and concrete.

If the maximum moment in a flexural member is so small that the tension stress in the concrete does not exceed the modulus of rupture f_r, no flexural tension cracks will occur. The full, uncracked section is then available for resisting stress and providing rigidity. This stage of loading has been analyzed in Sec. 3.3a. In agreement with this analysis, the effective moment of inertia for this low range of loads is that of the uncracked, transformed section I_{ut}, and E is the modulus of concrete E_c as given by Eq. (2.3). Correspondingly, for this load range,

$$\Delta_{iu} = \frac{f}{E_c I_{ut}} \qquad (a)$$

At higher loads, flexural tension cracks are formed. In addition, if shear stresses exceed v_{cr} [see Eq. (4.2*c*)] and web reinforcement is employed to resist them, diagonal cracks can exist at service loads. In the region of flexural cracks the position of the neutral axis varies: directly at each crack it is located at the level calculated for the cracked, transformed section (see Sec. 3.3b); midway between cracks it dips to a location closer to that calculated for the uncracked transformed section. Correspondingly, flexural-tension cracking causes the effective moment of inertia to be that of the cracked transformed section in the immediate neighborhood of flexural-tension cracks, and closer to that of the uncracked transformed section midway between cracks, with a gradual transition between these extremes.

It is seen that the value of the local moment of inertia varies in those portions of the beam in which the bending moment exceeds the cracking moment of the section

$$M_{cr} = \frac{f_r I_{ut}}{y_t} \qquad (6.4)$$

where y_t is the distance from the neutral axis to the tension face and f_r is the modulus of rupture. The exact variation of I depends on the shape of the moment diagram and on the crack pattern, and is difficult to determine. This makes an exact deflection calculation impossible.

However, extensively documented studies (Ref. 6.5) have shown that deflections Δ_{ic} occurring in a beam after the maximum moment M_a has reached and exceeded the cracking moment M_{cr} can be calculated by using an effective moment of inertia I_e; that is,

$$\Delta_{ic} = \frac{f}{E_c I_e} \qquad (b)$$

where

$$I_e = \left(\frac{M_{cr}}{M_a}\right)^3 I_{ut} + \left[1 - \left(\frac{M_{cr}}{M_a}\right)^3\right] I_{cr} \quad \text{and} \quad \le I_{ut} \qquad (6.5)$$

where I_{cr} is the moment of inertia of the cracked transformed section.

In Fig. 6.4, the effective moment of inertia, given by Eq. (6.5), is plotted as a function of the ratio M_a/M_{cr} (the reciprocal of the moment ratio used in the equation). It is seen that, for values of maximum moment M_a less than the cracking moment M_{cr}, that is, M_a/M_{cr} less than 1.0, $I_e = I_{ut}$. With increasing values of M_a, I_e approaches I_{cr}, and for values of M_a/M_{cr} of 3 or more, I_e is almost the same as I_{cr}. Typical values of M_a/M_{cr} at full service load range from about 1.5 to 3.

Figure 6.5 shows the growth of deflections with increasing moment for a simple-span beam, and illustrates the use of Eq. (6.5). For moments no larger

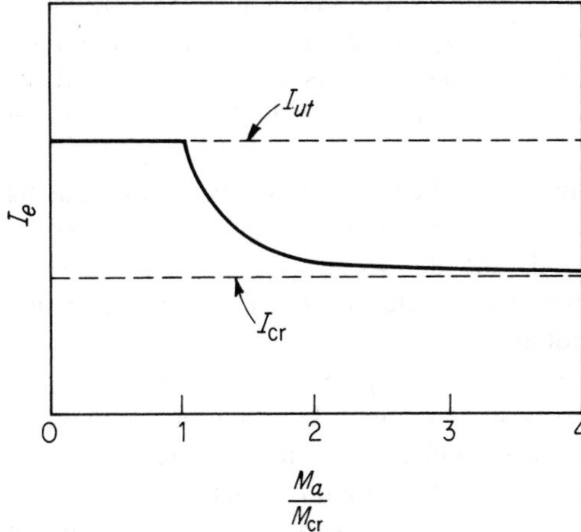

FIGURE 6.4
Variation of I_e with moment ratio.

than M_{cr}, deflections are practically proportional to moments and the deflection at which cracking begins is obtained from Eq. (*a*) with $M = M_{cr}$. At larger moments, the effective moment of inertia I_e becomes progressively smaller, according to Eq. (6.5), and deflections are found by Eq. (*b*) for the load level of interest. The moment M_2 might correspond to the full service load, for example, while the

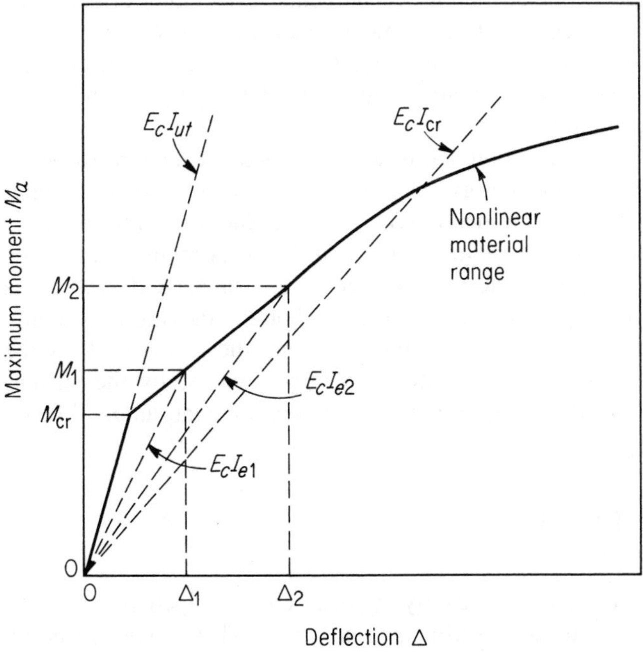

FIGURE 6.5
Deflection of a reinforced concrete beam.

moment M_1 would represent the dead load moment for a typical case. A moment-deflection curve corresponding to the line $E_c I_{cr}$ represents an upper bound for deflections, consistent with Fig. 6.4, except that at loads somewhat beyond the service load, the nonlinear response of steel or concrete or both causes a further nonlinear increase in deflections.

Note that to calculate the increment of deflection due to live load, causing a moment increase $M_2 - M_1$, a two-step computation is required: the first for deflection Δ_2 due to live and dead load, and the second for deflection Δ_1 due to dead load alone, each with the appropriate value of I_e. Then the deflection increment due to live load is found, equal to $\Delta_2 - \Delta_1$.

Most reinforced concrete spans are continuous, not simply supported. The concepts just introduced for simple spans can be applied, but the moment diagram for a given span will include both negative and positive regions, reflecting the rotational restraint provided at the ends of the spans by continuous frame action. The effective moment of inertia for a continuous span can be found by a simple averaging procedure, according to ACI Code, that will be described in Sec. 6.7c.

A fundamental problem for continuous spans is that, although the deflections are based on the moment diagram, that moment diagram depends, in turn, on the flexural rigidity *EI* for each member of the frame. The flexural rigidity depends on the extent of cracking, as has been demonstrated. Cracking, in turn, depends on the moments, which are to be found. The circular nature of the problem is evident.

One could use an iterative procedure, initially basing the frame analysis on uncracked concrete members, determining the moments, calculating effective *EI* terms for all members, then recalculating moments, adjusting the *EI* values, etc. The process could be continued for as many iterations as needed, until changes are not significant. However, such an approach would be expensive and time-consuming, even with computer use.

Usually, a very approximate approach is adopted. Member flexural rigidities for the frame analysis are based simply on properties of uncracked rectangular concrete cross sections. This can be defended noting that the moments in a continuous frame depend only on the *relative* values of *EI* in its members, not the *absolute* values. Hence, if a consistent assumption, i.e., uncracked section, is used for all members, the results should be valid. Although cracking is certainly more prevalent in beams than in columns, thus reducing the relative *EI* for the beams, this is compensated to a large extent, in typical cases, by the stiffening effect of the flanges in the positive bending regions of continuous T beam construction.

6.6 DEFLECTIONS DUE TO LONG-TERM LOADS

Initial deflections are increased significantly if loads are sustained over a long period of time, due to the effects of shrinkage and creep. These two effects are usually combined in deflection calculation. Creep generally dominates, but for some types of members, shrinkage deflections are large and should be considered separately (see Sec. 6.8).

It was pointed out in Sec. 2.7 that creep deformations of concrete are directly proportional to the compressive stress up to and beyond the usual service load range. They increase asymptotically with time and, for the same stress, are larger for low-strength than for high-strength concretes. The ratio of additional time-dependent strain to initial elastic strain is given by the creep coefficient C_{cu} (see Table 2.1).

For a reinforced concrete beam, the long-term deformation is much more complicated than for an axially loaded cylinder, because while the concrete creeps under sustained load, the steel does not. The situation in a reinforced concrete beam is illustrated by Fig. 6.6. Under sustained load, the initial strain ϵ_i at the top face of the beam increases, due to creep, by the amount ϵ_t, while the strain ϵ_s in the steel is essentially unchanged. Because the rotation of the strain distribution diagram is therefore about a point at the level of the steel, rather than about the cracked elastic neutral axis, the neutral axis moves down as a result of creep, and

$$\frac{\phi_t}{\phi_i} < \frac{\epsilon_t}{\epsilon_i} \tag{a}$$

demonstrating that the usual creep coefficients could not be applied to initial curvatures to obtain creep curvatures (hence deflections).

The situation is further complicated. Due to the lowering of the neutral axis associated with creep (see Fig. 6.6*b*) and the resulting increase in compression area, the compressive stress required to produce a given resultant C to equilibrate $T = A_s f_s$ is less than before, in contrast to the situation in a creep test of a compressed cylinder, because the beam creep occurs at a gradually diminishing stress. On the other hand, with the new lower neutral axis, the internal lever arm between compressive and tensile resultant forces is less, calling for an increase in both resultants for a constant moment. This, in turn, will require a small increase in stress, and hence strain, in the steel; thus ϵ_s is not constant as assumed originally.

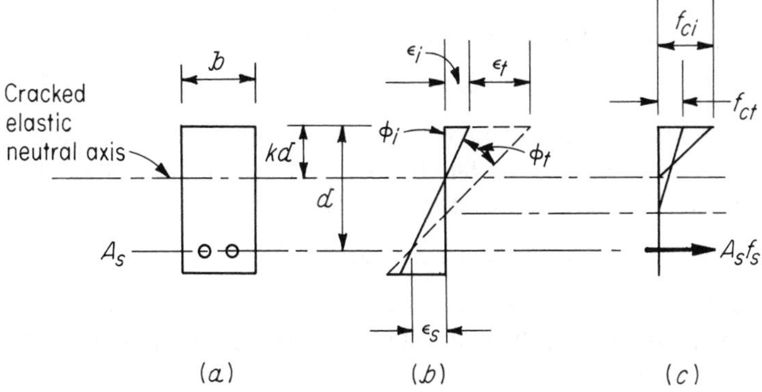

FIGURE 6.6
Effect of concrete creep on curvature: (*a*) beam cross section; (*b*) strains; (*c*) stresses and forces. (*Adapted from Ref. 6.6.*)

Because of such complexities, it is necessary in practice to calculate additional, time-dependent deflections of beams due to creep (and shrinkage) using a simplified, empirical approach by which the initial elastic deflections are multiplied by a factor λ to obtain the additional long-time deflections. Values of λ for use in design are based on long-term deflection data for reinforced concrete beams (Refs. 6.6 to 6.9). Thus

$$\Delta_t = \lambda \Delta_i \tag{6.6}$$

where Δ_t is the *additional* long-term deflection due to the combined effect of creep and shrinkage, and Δ_i is the initial elastic deflection calculated by the methods of Sec. 6.5.

The coefficient λ depends on the duration of the sustained load. It also depends on whether the beam has only reinforcement A_s on the tension side, or whether additional longitudinal reinforcement A_s' is provided on the compression side. In the latter case, the long-term deflections are much reduced. This is so because when no compression reinforcement is provided, the compression concrete is subject to unrestrained creep and shrinkage. On the other hand, since steel is not subject to creep, if additional bars are located close to the compression face, they will resist and thereby reduce the amount of creep and shrinkage and the corresponding deflection (Ref. 6.9). Compression steel may be included for this reason alone. Specific values of λ, used to account for the influence of creep and compression reinforcement, will be given in Sec. 6.7.

If a beam carries a certain sustained load W (e.g., the dead load plus the average traffic load on a bridge) and is subject to a short-term heavy live load P (e.g., the weight of an unusually heavy vehicle), the maximum total deflection under this combined loading is obtained as follows:

1. Calculate the instantaneous deflection Δ_{iw} caused by the sustained load W by methods given in Sec. 6.5.
2. Calculate the additional long-term deflection caused by W; i.e.,

$$\Delta_{tw} = \lambda \Delta_{iw}$$

3. Then the total deflection caused by the sustained part of the load is

$$\Delta_w = \Delta_{iw} + \Delta_{tw}$$

4. In calculating the additional instantaneous deflection caused by the short-term load P, account must be taken of the fact that the load-deflection relation after cracking is nonlinear, as illustrated by Fig. 6.5. Hence

$$\Delta_{ip} = \Delta_{i(w+p)} - \Delta_{iw}$$

   where $\Delta_{i(w+p)}$ is the total instantaneous deflection that would be obtained if W and P were applied simultaneously, calculated by using I_e determined for the moment caused by $W + P$.
5. Then the total deflection under the sustained load plus heavy short-term load is

$$\Delta = \Delta_w + \Delta_{ip}$$

In calculations of deflections, careful attention must be paid to the load history, i.e., the time-sequence in which loads are applied, as well as to the magnitude of the loads. The short-term peak load on the bridge girder just described might be applied early in the life of the member, before time-dependent deflections had taken place. Similarly, for buildings, heavy loads such as stacked material are often placed during construction. These temporary loads may be equal to, or even greater than, the design live load. The state of cracking will correspond to the *maximum load* that was carried, and the sustained load deflection, on which the long-term effects are based, would correspond to that cracked condition. I_e for the maximum load reached should be used to recalculate the sustained load deflection before calculating long-term effects.

This will be illustrated referring to Fig. 6.7, showing the load-deflection plot for a building girder that is designed to carry a specified dead and live load. Assume first that the dead and live loads increase monotonically. As the full dead load W_d is applied, the load deflection curve follows the path 0-1, and the dead load deflection, Δ_d, is found using I_{e1} calculated from Eq. (6.5), with $M_a = M_d$. The time-dependent effect of the dead load would be $\lambda\Delta_d$. As live load is then applied, path 1-2 would be followed. Live load deflection, Δ_l, would be found in two steps, as described in Sec. 6.5, first finding Δ_{d+l} based on I_{e2}, with M_a in Eq. (6.5) equal to M_{d+l}, then subtracting dead load deflection Δ_d.

If, on the other hand, short-term construction loads were applied, then removed, the deflection path 1-2-3 would be followed. Then, under dead load only,

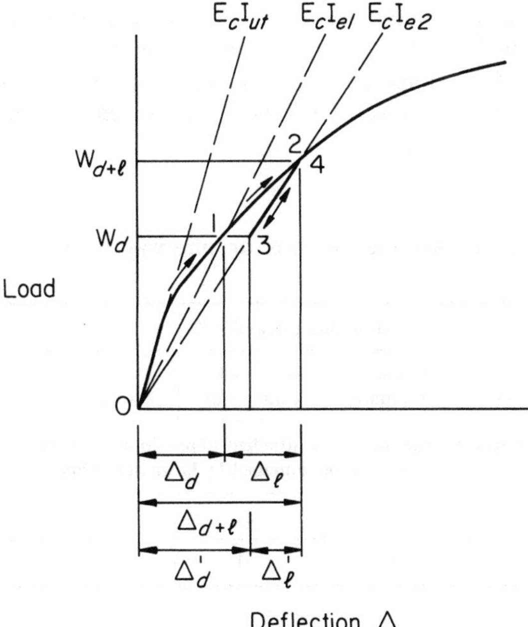

FIGURE 6.7
Effect of load history on deflection of a building girder.

the resulting deflection would then be Δ'_d. Note that this deflection can be found in one step using W_d, but with I_{e2} corresponding to the maximum load reached. The long-term deflection now would be $\lambda\Delta'_d$, significantly *larger* than before.

On the other hand, should the full design *live* load then be applied, the deflection would follow path 3-4, and the live load deflection would be *less* than for the first case. It, too, can be calculated by a simple one-step calculation using W_l alone, in this case, and with moment of inertia equal to I_{e2}.

Clearly, in calculating deflections, the engineer must anticipate, as nearly as possible, both the magnitude and time-sequence of the loadings. Although long-term deflections are often calculated assuming monotonic loading, with both immediate and long-term effects of dead load occurring before application of live load, in many cases this is not realistic.

6.7 ACI CODE PROVISIONS FOR CONTROL OF DEFLECTIONS

a. Minimum Depth-Span Ratios

As pointed out in Sec. 6.4, two approaches to deflection control are in current use, both acceptable under the provisions of the ACI Code, within prescribed limits. The simpler of these is to impose restrictions on the minimum member depth h, relative to the span l, to ensure that the beam will be sufficiently stiff that deflections are unlikely to cause problems in service. Deflections are greatly influenced by support conditions (e.g., a simply supported uniformly loaded beam will deflect 5 times as much as an otherwise identical beam with fixed supports), so minimum depths must vary depending on conditions of restraint at the ends of the spans.

According to ACI Code 9.5.2, the minimum depths of Table 6.2 apply to one-way construction *not* supporting or attached to partitions or other construction likely to be damaged by large deflections, unless computation of deflections indicates a lesser depth can be used without adverse effects. Values given in Table

Table 6.2 Minimum depth of nonprestressed beams or one-way slabs unless deflections are computed

	Minimum thickness, h			
Member	**Simply supported**	**One end continuous**	**Both ends continuous**	**Cantilever**
	Members not supporting or attached to partitions or other construction likely to be damaged by large deflections			
Solid one-way slabs	$l/20$	$l/24$	$l/28$	$l/10$
Beams or ribbed one-way slabs	$l/16$	$l/18.5$	$l/21$	$l/8$

6.2 are to be used directly for normal weight concrete with $w_c = 145$ pcf and reinforcement with $f_y = 60{,}000$ psi. For members using lightweight concrete with density in the range from 90 to 120 pcf, the values of Table 6.2 should be multiplied by $(1.65 - 0.005w_c) \geq 1.09$. For yield strengths other than 60,000 psi, the values should be multiplied by $(0.4 + f_y/100{,}000)$.

b. Calculation of Immediate Deflections

When there is need to use member depths shallower than are permitted by Table 6.2, or when members support construction that is likely to be damaged by large deflections, or for prestressed members, deflections must be calculated and compared with limiting values (see Sec. 6.7e). The calculation of deflections, when required, proceeds along the lines described in Secs. 6.5 and 6.6. For design purposes, the moment of the uncracked transformed section, I_{ut}, can be replaced by that of the gross concrete section, I_g, neglecting reinforcement, without serious error. With this simplification, Eqs. (6.4) and (6.5) are replaced by the following:

$$M_{\text{cr}} = \frac{f_r I_g}{y_t} \tag{6.7}$$

and

$$I_e = \left(\frac{M_{\text{cr}}}{M_a}\right)^3 I_g + \left[1 - \left(\frac{M_{\text{cr}}}{M_a}\right)^3\right] I_{\text{cr}} \quad \text{and} \quad \leq I_g \tag{6.8}$$

The modulus of rupture for normal weight concrete is to be taken equal to

$$f_r = 7.5\sqrt{f_c'} \tag{6.9a}$$

For lightweight concrete, the modulus of rupture may not be known, but the split-cylinder strength f_{ct} is often specified and determined by tests. For normal weight concretes, the split-cylinder strength is generally assumed to be $f_{ct} = 6.7\sqrt{f_c'}$. Accordingly, in Eq. (6.9*a*), $f_{ct}/6.7$ can be substituted for $\sqrt{f_c'}$ for the purpose of calculating the modulus of rupture. Then for lightweight concrete, if f_{ct} is known,

$$f_r = 7.5\frac{f_{ct}}{6.7} = 1.12 f_{ct} \tag{6.9b}$$

where $f_{ct}/6.7$ is not to exceed $\sqrt{f_c'}$ according to ACI Code 9.5.2. In lieu of test information on tensile strength, f_r can be calculated by Eq. (6.9*a*) multiplied by 0.75 for "all-lightweight" concrete and 0.85 for "sand-lightweight" concrete.

c. Continuous Spans

For continuous spans, ACI Code 9.5.2 calls for a simple average of values obtained from Eq. (6.8) for the critical positive- and negative-moment sections, i.e.,

$$I_e = 0.50 I_{em} + 0.25(I_{e1} + I_{e2}) \tag{6.10a}$$

where I_{em} is the effective moment of inertia for the midspan section and I_{e1} and I_{e2} those for the negative-moment sections at the respective beam ends, each calculated from Eq. (6.8) using the applicable value of M_a. It is shown in Ref. 6.10 that a somewhat improved result can be had for continuous prismatic members using a weighted average for beams with both ends continuous of

$$I_e = 0.70I_{em} + 0.15(I_{e1} + I_{e2}) \tag{6.10b}$$

and for beams with one end continuous and the other simply supported of

$$I_e = 0.85I_{em} + 0.15I_{e1} \tag{6.10c}$$

where I_{e1} is the effective moment of inertia at the continuous end. The ACI Code, as an option, also permits use of I_e for continuous prismatic beams to be taken equal to the value obtained from Eq. (6.8) at midspan; for cantilevers, I_e calculated at the support section may be used.

After I_e is found, deflections may be computed using the moment-area method (Ref. 6.11), with due regard for rotations of the tangent to the elastic curve at the supports. In general, in computing the maximum deflection, the loading producing the maximum positive moment may be used, and the midspan deflection may normally be used as an acceptable approximation of the maximum deflection. Coefficients for deflection calculation such as derived by Branson in Ref. 6.5 are helpful. For members where supports may be considered fully fixed or hinged, handbook equations for deflections may be used.

d. Long-Term Deflection Multipliers

On the basis of empirical studies (Refs. 6.5, 6.7, and 6.9), ACI Code 9.5.2 specifies that *additional* long-term deflections Δ_t due to the combined effects of creep and shrinkage shall be calculated by multiplying the immediate deflection Δ_i by the factor

$$\lambda = \frac{\xi}{1 + 50\rho'} \tag{6.11}$$

where $\rho' = A'_s/bd$ and ξ is a time-dependent coefficient that varies as shown in Fig. 6.8. In Eq. (6.11) the quantity $1/(1 + 50\rho')$ is a reduction factor that is essentially a section property, reflecting the beneficial effect of compression reinforcement A'_s in reducing long-term deflections, whereas ξ is a material property depending on creep and shrinkage characteristics. For simple and continuous spans, the value of ρ' used in Eq. (6.11) should be that at the midspan section, according to the ACI Code, or that at the support for cantilevers. Equation (6.11) and the values of ξ given by Fig. 6.8 apply to both normal weight and lightweight concrete beams. The additional, time-dependent deflections are thus found using values of λ from Eq. (6.11) in Eq. (6.6).

Values of ξ given in the ACI Code and Commentary are satisfactory for ordinary beams and one-way slabs, but may result in underestimation of time-dependent deflections of two-way slabs, for which Branson has suggested a five-year value of $\xi = 3.0$ (Ref. 6.5).

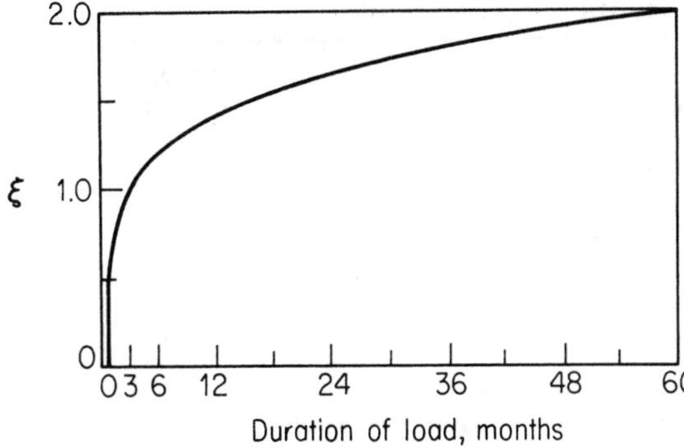

FIGURE 6.8
Time variation of ξ for long-term deflections.

Recent research at Cornell University indicates that Eq. (6.11) does not properly reflect the reduced creep that is characteristic of the higher strength concretes that are now often used (Ref. 6.12). As indicated in Table 2.1 of Chap. 2, the creep coefficient for high-strength concrete may be as low as $\frac{1}{2}$ the value for normal concrete. Clearly, the long-term deflection of high-strength concrete beams under sustained load, expressed as a ratio of immediate elastic deflection, correspondingly will be less. This suggests a lower value of the *material modifier*, ξ, in Eq. (6.11) and Fig. 6.8. On the other hand, in high-strength concrete beams, the influence of compression steel in reducing creep deflections is less pronounced, requiring an adjustment in the *section modifier*, $1/(1 + 50\rho')$, in that equation.

Based on long-term tests conducted at Cornell University, plus correlation with results of six other experimental programs, the following modified form of Eq. (6.11) is recommended:

$$\lambda = \frac{\mu\xi}{1 + 50\mu\rho'} \tag{6.12}$$

in which

$$\mu = 1.4 - f_c'/10{,}000$$
$$0.4 \le \mu \le 1.0 \tag{6.13}$$

The proposed equation gives results identical to Eq. (6.11) for concrete strengths of 4000 psi and below, and much improved predictions for concrete strengths between 4000 psi and 12,000 psi.

e. Permissible Deflections

To ensure satisfactory performance in service, ACI Code 9.5.2 imposes certain limits on deflections calculated according to the procedures just described. These limits are given in Table 6.3. Limits depend on whether or not the member supports or is attached to other nonstructural elements, and whether or not those nonstructural elements are likely to be damaged by large deflections. When long-term deflections are computed, that part of the deflection that occurs before attachment

Table 6.3 Maximum allowable computed deflections

<table>
<tr><th>Type of member</th><th>Deflection to be considered</th><th>Deflection limitation</th></tr>
<tr><td>Flat roofs not supporting or attached to nonstructural elements likely to be damaged by large deflections</td><td>Immediate deflection due to the live load L</td><td>$\frac{l}{180}$</td></tr>
<tr><td>Floors not supporting or attached to nonstructural elements likely to be damaged by large deflections</td><td>Immediate deflection due to the live load L</td><td>$\frac{l}{360}$</td></tr>
<tr><td>Roof or floor construction supporting or attached to nonstructural elements likely to be damaged by large deflections</td><td rowspan="2">That part of the total deflection which occurs after attachment of the nonstructural elements, the sum of the long-time deflection due to all sustained loads, and the immediate deflection due to any additional live load</td><td>$\frac{l}{480}$</td></tr>
<tr><td>Roof or floor construction supporting or attached to nonstructural elements not likely to be damaged by large deflections</td><td>$\frac{l}{240}$</td></tr>
</table>

of the nonstructural elements may be deducted; information from Fig. 6.8 is useful for this purpose. The last two limits of Table 6.3 may be exceeded under certain conditions, according to the ACI Code.

Example 6.2 Deflection calculation. The beam shown in Fig. 6.9 is a part of the floor system of an apartment house and is designed to carry calculated dead load w_d of 1.1 kips/ft and a service live load w_l of 2.2 kips/ft. Of the total live load, 20 percent is sustained in nature, while 80 percent will be applied only intermittently over the life of the structure. Under full dead and live load, the moment diagram is as shown in Fig. 6.9*c*. The beam will support nonstructural partitions that would be damaged if large deflections were to occur. They will be installed shortly after construction shoring is removed and dead loads take effect, but before significant creep occurs. Calculate that part of the total deflection that would adversely affect the partitions, i.e., the sum of long-time deflection due to dead and partial live load plus the immediate deflection due to the nonsustained part of the live load. Material strengths are $f'_c = 2500$ psi and $f_y = 40$ ksi.

Solution. For the specified materials, $E_c = 57{,}000\sqrt{2500} = 2.85 \times 10^6$ psi, and with $E_s = 29 \times 10^6$ psi, the modular ratio $n = 10$. The modulus of rupture $f_r = 7.5\sqrt{2500} = 375$ psi. The effective moment of inertia will be calculated for the moment diagram shown in Fig. 6.9*c* corresponding to the full service load, on the basis that the extent of cracking will be governed by the full service load, even though that load is intermittent. In the positive-moment region, the centroidal axis of the uncracked T section of Fig. 6.9*b* is found, by taking moments about the top surface, to be at 7.66 in. depth, and $I_g = 33{,}160$ in^4. By similar means, the centroidal axis of the cracked transformed T section shown in Fig. 6.9*d* is located 4.14 in. below the top of the slab and $I_{cr} = 13{,}180$ in^4. The cracking moment is then found by means of Eq. (6.7):

$$M_{cr} = 375 \times \frac{33{,}160}{16.84} \times \frac{1}{12{,}000} = 62 \text{ ft-kips}$$

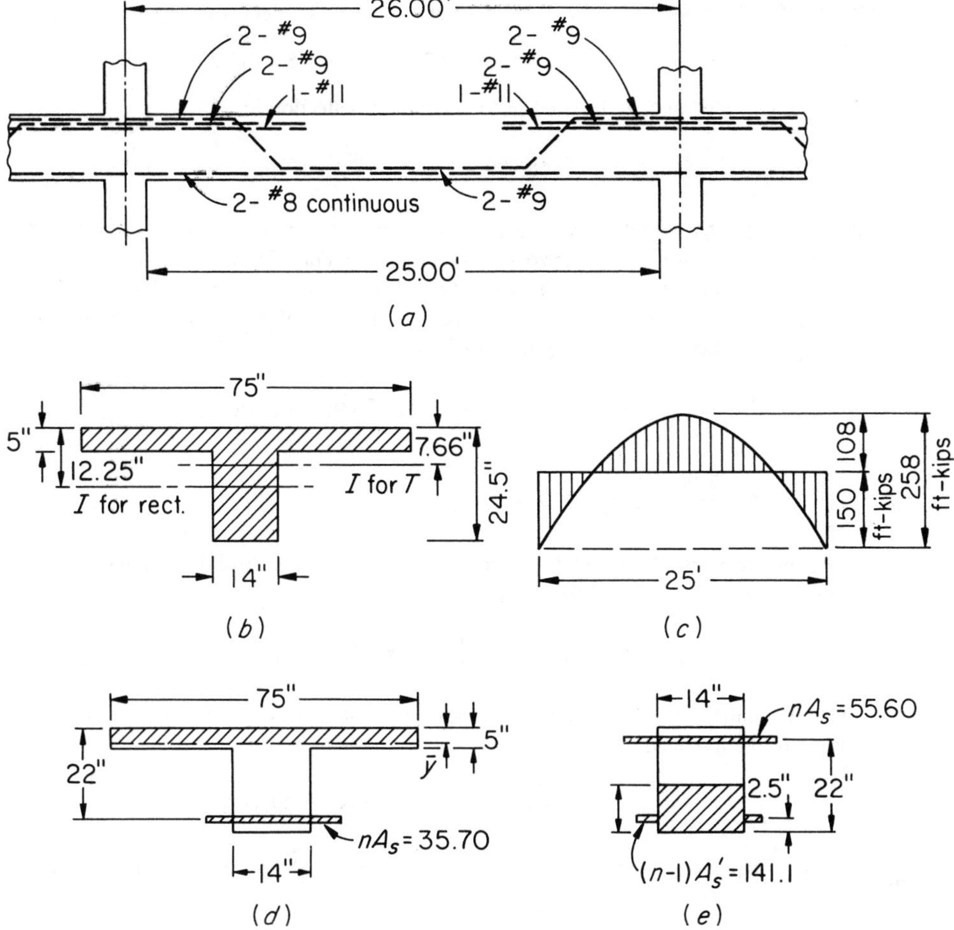

FIGURE 6.9
Continuous T beam for deflection calculations in Example 6.2.

With $M_{\text{cr}}/M_a = 62/108 = 0.574$, the effective moment of inertia in the positive bending region is found from Eq. (6.8) to be

$$I_e = 0.574^3 \times 33{,}160 + (1 - 0.574^3) \times 13{,}180 = 16{,}970 \text{ in}^4$$

In the negative bending region, the gross moment of inertia will be based on the rectangular section shown dotted in Fig. 6.9*b*. For this area, the centroid is 12.25 in. from the top surface and $I_g = 17{,}200 \text{ in}^4$. For the cracked transformed section shown in Fig. 6.9*e*, the centroidal axis is found, taking moments about the bottom surface, to be 9.33 in. from that level, and $I_{\text{cr}} = 13{,}368 \text{ in}^4$. Then

$$M_{\text{cr}} = 375 \times \frac{17{,}200}{12.25} \times \frac{1}{12{,}000} = 44 \text{ ft-kips}$$

giving $M_{cr}/M_a = 44/150 = 0.293$. Thus, for the negative-moment regions,

$$I_e = 0.293^3 \times 17{,}200 + (1 - 0.293^3) \times 13{,}368 = 13{,}530 \text{ in}^4$$

The average value of I_e to be used in calculation of deflection is

$$I_{e,\text{av}} = \frac{1}{2}(16{,}970 + 13{,}530) = 15{,}250 \text{ in}^4$$

It is next necessary to find the sustained-load deflection multiplier given by Eq. (6.11) and Fig. 6.8. For the positive bending zone, with no compression reinforcement, $\lambda_{\text{pos}} = 2.00$.

For convenient reference, the deflection of the member under full dead plus live load of 3.3 kips/ft, corresponding to the moment diagram of Fig. 6.9*c*, will be found. Making use of the moment-area principles,

$$\Delta_{d+l} = \frac{1}{EI}\left[(\frac{2}{3} \times 258 \times 12.5 \times \frac{5}{8} \times 12.5) - (150 \times 12.5 \times 6.25)\right] = \frac{5100}{EI}$$

$$= \frac{5100 \times 1728}{2850 \times 15{,}250} = 0.203 \text{ in.}$$

Using this figure as a basis, the time-dependent portion of dead load deflection (the only part of the total that would affect the partitions) is

$$\Delta_d = 0.203 \times \frac{1.1}{3.3} \times 2.00 = 0.135 \text{ in.}$$

while the sum of the immediate and time-dependent deflection due to the sustained portion of the live load is

$$\Delta_{0.20l} = 0.203 \times \frac{2.2}{3.3} \times 0.20 \times 3.00 = 0.081 \text{ in.}$$

and the instantaneous deflection due to application of the short-term portion of the live load is

$$\Delta_{0.80l} = 0.203 \times \frac{2.2}{3.3} \times 0.80 = 0.108 \text{ in.}$$

Thus the total deflection that would adversely affect the partitions, from the time they are installed until all long-time and subsequent instantaneous deflections have occurred, is

$$\Delta = 0.135 + 0.081 + 0.108 = 0.324 \text{ in.}$$

For comparison, the limitation imposed by the ACI Code in such circumstances is $l/480 = 26 \times 12/480 = 0.650$ in., indicating that the stiffness of the proposed member is sufficient.

It may be noted that relatively little error would have been introduced in the above solution if the cracked-section moment of inertia had been used for both positive and negative sections rather than I_e. Significant savings in computational effort would have resulted. If M_{cr}/M_a is less than $\frac{1}{3}$, use of I_{cr} would almost always be acceptable. It should be noted further that computation of the moment of inertia for both uncracked and cracked sections is greatly facilitated by design aids like those included in Ref. 6.13.

6.8 DEFLECTIONS DUE TO SHRINKAGE AND TEMPERATURE CHANGES

Concrete shrinkage will produce compressive stress in the longitudinal reinforcement in beams and slabs and equilibrating tensile stress in the concrete. If, as usual, the reinforcement is not symmetrically placed with respect to the concrete centroid, then shrinkage will produce curvature and corresponding deflection. The deflections will be in the same direction as those produced by the loads, if the reinforcement is mainly on the side of the member subject to flexural tension.

Shrinkage deflection is not usually calculated separately, but is combined with creep deflection, according to ACI Code procedures (see Sec. 6.7d). However, there are circumstances where a separate and more accurate estimation of shrinkage deflection may be necessary, particularly for thin, lightly loaded slabs. Compression steel, while it has only a small effect in reducing immediate elastic deflections, contributes significantly in reducing deflections due to shrinkage (as well as creep), and is sometimes added for this reason.

Curvatures due to shrinkage of concrete in an unsymmetrically reinforced concrete member can be found by the fictitious tensile force method (Ref. 6.5). Figure 6.10*a* shows the member cross section, with compression steel area A_s' and tensile steel area A_s, at depths d' and d respectively from the top surface. In Fig. 6.10*b*, the concrete and steel are imagined to be temporarily separated, so that the concrete can assume its free shrinkage strain ϵ_{sh}. Then a fictitious compressive force $T_{sh} = (A_s + A_s')\epsilon_{sh}E_s$ is applied to the steel, at the centroid of all the bars, a distance e below the concrete centroid, such that the steel shortening will exactly equal the free shrinkage strain of the concrete. The equilibrating tension force T_{sh} is then applied to the recombined section, as in Fig. 6.10*c*. This produces a moment $T_{sh}e$, and the corresponding shrinkage curvature is

$$\phi_{sh} = \frac{T_{sh}e}{EI}$$

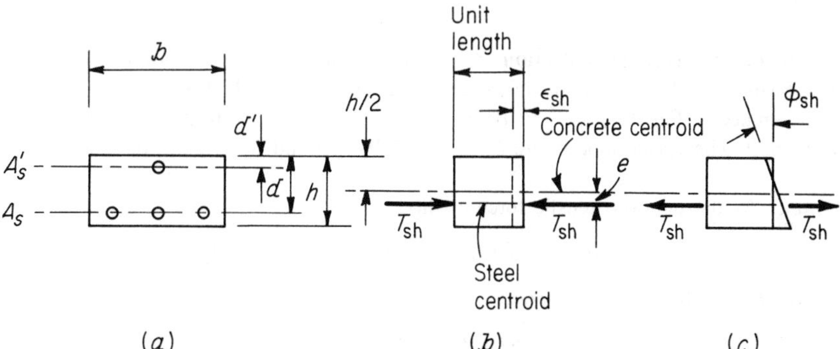

FIGURE 6.10
Shrinkage curvature of a reinforced concrete beam or slab: (*a*) cross section; (*b*) free shrinkage strain; (*c*) shrinkage curvature.

The effects of concrete cracking and creep complicate the analysis, but comparisons with experimental data (Ref. 6.5) indicate that good results can be obtained using e_g and I_g for the uncracked gross concrete section and by using a reduced modulus E_{ct} equal to $\frac{1}{2}E_c$ to account for creep. Thus

$$\phi_{sh} = \frac{2T_{sh}e_g}{E_cI_g} \tag{6.14}$$

where E_c is the usual value of concrete modulus given by Eq. (2.3).

Empirical methods are also used, in place of the fictitious tensile force method, to calculate shrinkage curvatures. These methods are based on the simple but reasonable proposition that the shrinkage curvature is a direct function of the free shrinkage and steel percentage, and an inverse function of the section depth (Ref. 6.5). Branson suggests that for steel percentage $(p - p') \leq 3$ percent (where $p = 100A_s/bd$ and $p' = 100A_s'/bd$),

$$\phi_{sh} = 0.7\frac{\epsilon_{sh}}{h}(p - p')^{1/3}\left(\frac{p - p'}{p}\right)^{1/2} \tag{6.15a}$$

and for $(p - p') > 3$ percent,

$$\phi_{sh} = \frac{\epsilon_{sh}}{h} \tag{6.15b}$$

With shrinkage curvature calculated by either method, the corresponding member deflection can be determined by any convenient means such as the moment-area or conjugate-beam method. If steel percentages and eccentricities are constant along the span, the deflection ϵ_{sh} resulting from the shrinkage curvature can be determined from

$$\Delta_{sh} = K_{sh}\phi_{sh}l^2 \tag{6.16}$$

where K_{sh} is a coefficient equal to 0.500 for cantilevers, 0.125 for simple spans, 0.065 for interior spans of continuous beams, and 0.090 for end spans of continuous beams (Ref. 6.5).

Example 6.3 Shrinkage deflection. Calculate the midspan deflection of a simply supported beam of 20 ft span due to shrinkage of the concrete for which $\epsilon_{sh} = 780 \times 10^{-6}$. With reference to Fig. 6.10*a*, $b = 10$ in., $d = 17.5$ in., $h = 20$ in., $A_s = 3.00$ in^2, and $A_s' = 0$. The elastic moduli are $E_c = 3.6 \times 10^6$ psi and $E_s = 29 \times 10^6$ psi.

Solution. By the fictitious tensile force method,

$$T_{sh} = 3.00 \times 780 \times 10^{-6} \times 29 \times 10^6 = 67{,}900 \text{ lb}$$

and from Eq. (6.14) with $I_g = 6670$,

$$\phi_{sh} = \frac{2 \times 67{,}900 \times 7.5}{3.6 \times 10^6 \times 6670} = 42.4 \times 10^{-6}$$

while from Eq. (6.16) with $K_{sh} = 0.125$ for the simple span,

$$\Delta_{sh} = 0.125 \times 42.4 \times 10^{-6} \times 240^2 = 0.305 \text{ in.}$$

Alternatively, by Branson's approximate Eq. (6.15*a*) with $p = 100 \times 3/175 = 1.7$ percent and $p' = 0$,

$$\phi_{sh} = \frac{0.7 \times 780 \times 10^{-6}}{20}(1.7)^{1/3} = 32.5 \times 10^{-6}$$

compared with 42.4×10^{-6} obtained by the equivalent tensile force method. Considering the uncertainties such as the effects of cracking and creep, the approximate approach can usually be considered satisfactory.

Deflections will be produced as a result of differential temperatures varying from top to bottom of a member also. Such variation will result in a strain variation with member depth that may usually be assumed to be linear. For such cases, the deflection due to differential temperature can be calculated using Eq. (6.16) in which ϕ_{sh} is replaced by $\alpha \Delta T/h$, where the thermal coefficient α for concrete may be taken as 5.5×10^{-6} per °F and ΔT is the temperature differential in degrees Fahrenheit from one side to the other. The presence of the reinforcement has little influence on curvatures and deflections resulting from differential temperatures, because the thermal coefficient for the steel is very close to that for concrete.

6.9 MOMENT VS. CURVATURE FOR REINFORCED CONCRETE SECTIONS

Although it is not needed explicitly in ordinary design and is not a part of ACI Code procedures, the relation between moment applied to a given beam section and the resulting curvature, through the full range of loading to failure, is important in several contexts. It is basic to the study of member ductility, understanding the development of plastic hinges, and accounting for the redistribution of elastic moments that occurs in most reinforced concrete structures before collapse (see Sec. 11.9).

It will be recalled, with reference to Fig. 6.11, that curvature is defined as the angle change per unit length at any given location along the axis of a member subjected to bending loads:

$$\psi = \frac{1}{r} \tag{6.17}$$

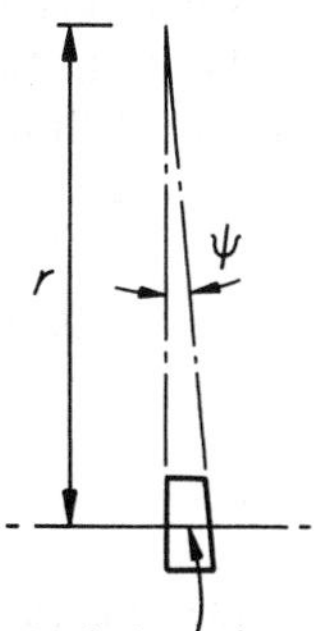

FIGURE 6.11
Unit curvature resulting from bending of beam section.

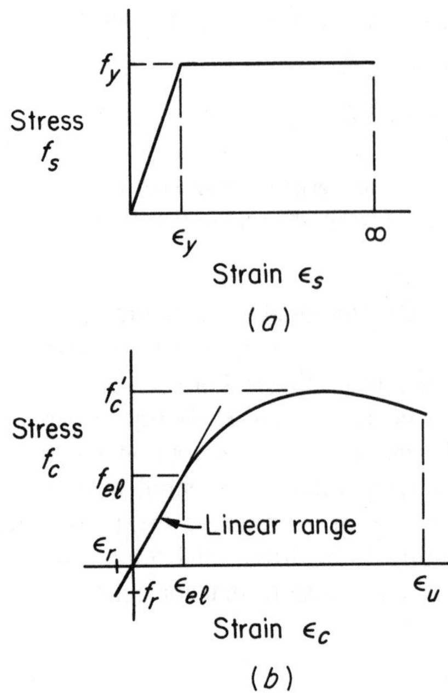

FIGURE 6.12
Idealized stress-strain curves: (*a*) steel; (*b*) concrete.

where ψ = unit curvature and r = radius of curvature. With the stress-strain relationships for steel and concrete, represented in idealized form in Fig. 6.12*a* and Fig. 6.12*b*, respectively, and the usual assumptions regarding perfect bond and plane sections, it is possible to calculate the relation between moment and curvature for a typical underreinforced concrete beam section, subject to flexural cracking, as follows.

Figure 6.13*a* shows the transformed cross section of a rectangular, tensile-reinforced beam in the uncracked elastic stage of loading, with steel represented by the equivalent concrete area nA_s, i.e. with area $(n - 1)A_s$ added outside of the rectangular concrete section.† The neutral axis, a distance c_1 below the top surface of the beam, is easily found (see Sec. 3.3a). In the limit case, the concrete stress at the tension face is just equal to the modulus of rupture, f_r, and the strain is $\epsilon_r = f_r/E_c$. The steel is well below yield at this stage, which can be confirmed by computing, from the strain diagram, the steel strain, $\epsilon_s = \epsilon_{cs}$, where ϵ_{cs} is the concrete strain at the level of the steel. It is easily confirmed, also, that the maximum concrete compressive stress will be well below the proportional limit. The curvature is seen, in Fig. 6.13*b*, to be

$$\psi_{cr} = \frac{\epsilon_1}{c_1} = \frac{\epsilon_r}{c_2} \tag{6.18}$$

† Note that compression reinforcement, or multiple layers of tension reinforcement, can easily be included in the analysis with no essential complication.

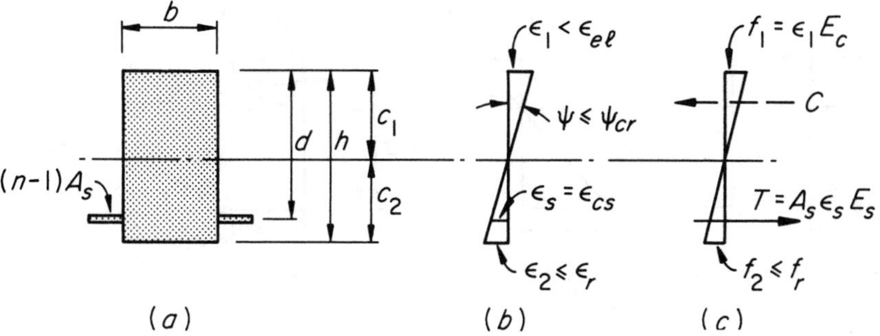

FIGURE 6.13
Uncracked beam in the elastic range of loading: (*a*) transformed cross section; (*b*) strains; (*c*) stresses and forces.

and the corresponding moment is

$$M_{cr} = \frac{f_r I_{ut}}{c_2} \tag{6.19}$$

where I_{ut} is the moment of inertia of the uncracked transformed section. Eq. (6.18) and Eq. (6.19) provide the information needed to plot point 1 of the moment-curvature graph of Fig. 6.16*a*.

When tensile cracking occurs at the section, the stiffness is immediately reduced, and curvature increases to point 2 in Fig. 6.16 with no increase in moment. The analysis now is based on the cracked, transformed section of Fig. 6.14*a*, with steel represented by the transformed area nA_s and tension concrete deleted. The cracked, elastic neutral axis distance $c_1 = kd$ is easily found by the usual methods (see Sec. 3.3b). In the limit case, the concrete strain just reaches the proportional limit, as shown in Fig. 6.14*b*, and, typically, the steel is still below the yield strain. The curvature is easily computed by

$$\psi_{el} = \frac{\epsilon_1}{c_1} = \frac{\epsilon_{el}}{c_1} \tag{6.20}$$

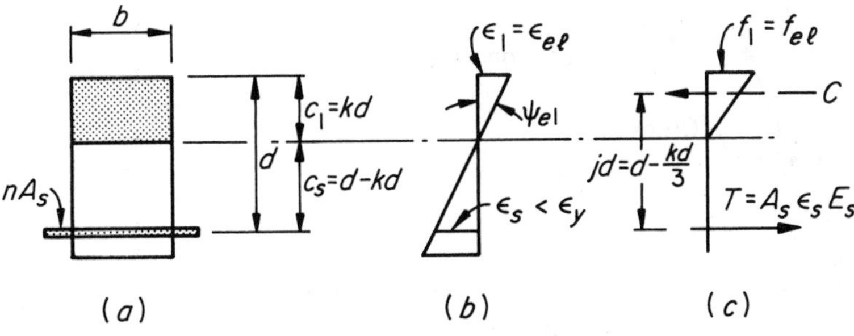

FIGURE 6.14
Cracked beam in the elastic range of material response: (*a*) transformed cross section; (*b*) strains; (*c*) stresses and forces.

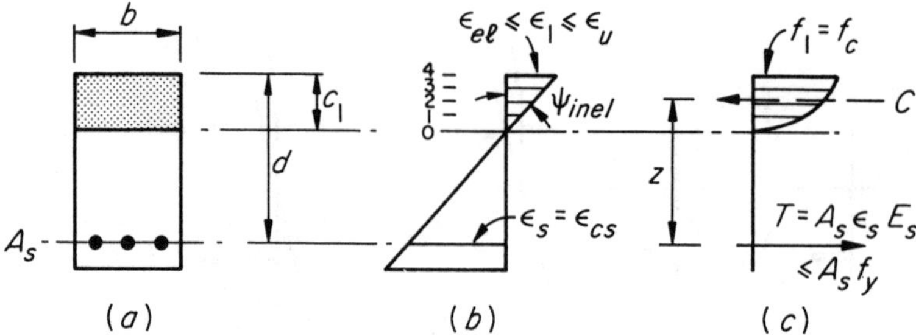

FIGURE 6.15
Cracked beam with concrete in the inelastic range of loading: (*a*) cross section; (*b*) strains; (*c*) stresses and forces.

and the corresponding moment is

$$M_{el} = \frac{1}{2} f_{el} kjbd^2 \tag{6.21}$$

as was derived in Sec. 3.3b. This provides point 3 in Fig. 6.16. The curvature at point 2 can now be found from the ratio M_{cr}/M_{el}.

Next, the cracked, inelastic stage of loading is shown in Fig. 6.15. Here the concrete is well into the inelastic range, although the steel has not yet yielded. The neutral axis depth, c_1, is less than the elastic kd and is changing with increasing load as the shape of the concrete stress distribution changes and the steel stress changes.

It is now convenient to adopt a numerical representation of the concrete compressive stress distribution, to find both the total concrete compressive force

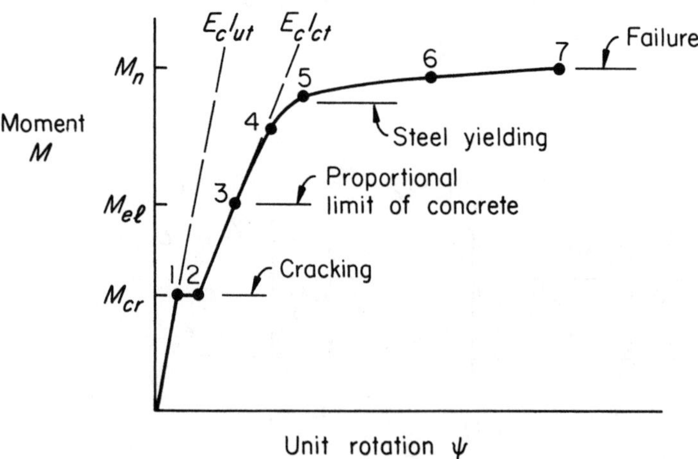

FIGURE 6.16
Moment-curvature relation for tensile-reinforced beam.

C and the location of its centroid, for any arbitrarily selected value of maximum concrete strain ϵ_1 in this range. The compressive strain diagram is divided into an arbitrary number of steps (e.g., four, in Fig. 6.15*b*), and the corresponding compressive stresses for each strain read from the stress-strain curve of Fig. 6.12*b*. The stepwise representation of the actual continuous stress block is integrated numerically to find C, and its point of application is located taking moments of the concrete forces about the top of the section. The basic equilibrium requirement, $C = T$, then can be used to find the correct location of the neutral axis, for the particular compressive strain selected, following an iterative procedure.

The entire process can be summarized as follows:

1. Select any top face concrete strain ϵ_1 in the inelastic range, i.e., between ϵ_{el} and ϵ_u.
2. Assume the neutral axis depth, a distance c_1 below the top face.
3. From the strain diagram geometry, determine $\epsilon_s = \epsilon_{cs}$.
4. Compute $f_s = \epsilon_s E_s$, but $\leq f_y$, and $T = A_s f_s$.
5. Determine C by integrating numerically under the concrete stress distribution curve.
6. Check to see if $C = T$. If not, the neutral axis must be adjusted upward or downward, for the particular concrete strain that was selected in Step 1, until equilibrium is satisfied. This determines the correct value of c_1.

Curvature can then be found from

$$\psi_{\text{inel}} = \frac{\epsilon_1}{c_1} \tag{6.22}$$

The internal lever arm, z, from the centroid of the concrete stress distribution to the tensile resultant, Fig. 6.15*c*, is calculated, after which

$$M_{\text{inel}} = Cz = Tz \tag{6.23}$$

The sequence of steps 1 through 6 is then repeated for newly selected values of concrete strain ϵ_1. The end result will be a series of points, such as 4, 5, 6, and 7 in Fig. 6.16. The limit of the moment-curvature plot is reached when the concrete top face strain equals ϵ_u, corresponding to point 7. The steel would be well past the yield strain at this loading, and at the yield stress.

It is important to be aware of the difference between a moment-unit curvature plot, such as Fig. 6.16, and a moment-rotation diagram for the hinging region of a reinforced concrete beam. The hinging region normally includes a number of discrete cracks, but between those cracks, the uncracked concrete reduces the steel strain, leading to what is termed the "tension stiffening" effect. The result is that the total rotation at the hinge is much less than would be calculated by multiplying the curvature per unit length at the cracked section by the observed or assumed length of the hinging region. Furthermore, the sharp increase in unit curvature shown in Fig. 6.16 at cracking would not be seen on the moment-rotation plot, only a small, but progressive, reduction of the slope of the diagram.

REFERENCES

6.1. "Control of Cracking in Concrete Structures," ACI Committee 224, *Concr. Int.*, vol. 2, no. 10, 1980, pp. 35–76.

6.2. P. Gergely and L. A. Lutz, "Maximum Crack Width in Reinforced Concrete Flexural Members," in *Causes, Mechanisms, and Control of Cracking in Concrete*, ACI Special Publication SP-20, American Concrete Institute, 1968, pp. 1–17.

6.3. B. B. Broms, "Crack Width and Crack Spacing in Reinforced Concrete Members," *J. ACI*, vol. 62, no. 10, 1965, pp. 1237–1256.

6.4. L. A. Lutz, "Crack Control Factor for Bundled Bars and for Bars of Different Sizes," *J. ACI*, vol. 71, no. 1, 1974, pp. 9–10.

6.5. D. E. Branson, *Deformation of Concrete Structures*, McGraw-Hill, New York, 1977.

6.6. "Deflections of Reinforced Concrete Flexural Members," ACI Committee 435, *J. ACI*, vol. 63, no. 6, 1966, pp. 637–674.

6.7. W. W. Yu and G. Winter, "Instantaneous and Long-Time Deflections of Reinforced Concrete Beams under Working Loads," *J. ACI*, vol. 57, no. 1, 1960, pp. 29–50.

6.8. "Prediction of Creep, Shrinkage and Temperature Effects in Concrete Structures," in *Designing for the Effects of Creep, Shrinkage, and Temperature in Concrete Structures*, ACI Committee 209, Special Publication SP-27, American Concrete Institute, Detroit, 1971.

6.9. D. E. Branson, "Compression Steel Effect on Long-Time Deflections," *J. ACI*, vol. 68, no. 8, 1971, pp. 555–559.

6.10. "Proposed Revisions by Committee 435 to ACI Building Code and Commentary Provisions on Deflections," ACI Committee 435, *J. ACI*, vol. 75, no. 6, June 1978, pp. 229–238.

6.11. C. H. Norris, J. B. Wilbur, and S. Utku, *Elementary Structural Analysis*, 3d ed., McGraw-Hill, New York, 1976.

6.12. K. Paulson, A. H. Nilson, and K. C. Hover, *Immediate and Long-Term Deflection of High-Strength Concrete Beams*, Research Report No. 89-3, Department of Structural Engineering, Cornell University, 1989.

6.13. *CRSI Handbook*, 6th ed., Concrete Reinforcing Steel Institute, Chicago, 1984.

PROBLEMS

6.1 A rectangular beam of width $b = 12$ in., effective depth $d = 20.5$ in., and total depth $h = 23$ in. spans 18.5 ft between simple supports. It will carry a computed dead load of 1.27 kips/ft including self-weight, plus a service live load of 2.44 kips/ft. Reinforcement consists of four No. 8 bars in one row. Material strengths are $f_y = 60{,}000$ psi and $f'_c = 4000$ psi.

(*a*) Compute the stress in the steel at full service load, and using the Gergely-Lutz equation estimate the maximum width of crack.

(*b*) Assuming exterior exposure to moist air, confirm the suitability of the proposed design.

6.2 For the beam of Prob. 6.1:

(*a*) Compute the ACI z value using $f_s = 0.60 f_y$ as permitted by the ACI Code.

(*b*) Compare against ACI Code limitations to determine if the design is satisfactory with respect to cracking.

(*c*) Compare with indication of Table A.8 in App. A.

6.3 To save steel-handling costs, an alternative design is proposed for the beam of Prob. 6.1, using two No. 11 bars to provide approximately the same steel area as the originally proposed four No. 8 bars. Check to determine if the redesigned beam is satisfactory with respect to cracking (*a*) by the Gergely-Lutz equation and Table 6.1, and (*b*) according to the ACI Code. What modification could you suggest that would minimize the number of bars to reduce cost, yet satisfy requirements of crack control?

6.4 For the beam of Prob. 6.1:

(*a*) Calculate the increment of deflection resulting from the first application of the short-term live load.

(*b*) Find the creep portion of the sustained load deflection plus the immediate deflection due to live load.

(*c*) Compare your results with the limitations imposed by the ACI Code, as summarized in Table 6.3.

Assume that the beam is a part of a floor system and supports cinder block partitions susceptible to cracking if deflections are excessive.

6.5 A beam having $b = 12$ in., $d = 21.5$ in., and $h = 24$ in. is reinforced with three No. 11 bars. Material strengths are $f_y = 60{,}000$ psi and $f_c' = 4000$ psi. It is used on a 28 ft simple span to carry a total service load of 2430 lb/ft. For this member, the sustained loads include self-weight of the beam plus additional superimposed dead load of 510 lb/ft, plus 400 lb/ft representing that part of the live load that acts more or less continuously, such as furniture, equipment, and time-average occupancy load. The remaining 1220 lb/ft live load consists of short-duration loads, such as the brief peak load in the corridors of an office building at the end of a working day.

(*a*) Find the increment of deflection under sustained loads due to creep.

(*b*) Find the additional deflection increment due to the intermittent part of the live load.

In your calculations, you may assume that the peak load is applied almost immediately after the building is placed in service, then reapplied intermittently. Compare with ACI Code limits from Table 6.3. Assume that, for this long-span floor beam, construction details are provided that will avoid damage to supported elements due to deflections. If ACI Code limitations are not met, what changes would you recommend to improve the design?

6.6 A reinforced concrete beam is continuous over two equal 22 ft spans, simply supported at the two exterior supports, and fully continuous at the interior support. Concrete cross-section dimensions are $b = 10$ in., $h = 22$ in., and $d = 19.5$ in. for both positive and negative bending regions. Positive reinforcement in each span consists of one No. 10 bar and one No. 8 bar, and negative reinforcement at the interior support is made up of three No. 10 bars. No compression steel is used. Material strengths are $f_y = 60{,}000$ psi and $f_c' = 5000$ psi. The beam will carry a service live load, applied early in the life of the member, of 1800 lb/ft distributed uniformly over both spans; 20 percent of this load will be sustained more or less permanently, while the rest is intermittent. The total service dead load is 1000 lb/ft including self-weight. Find: (*a*) the immediate deflection when shores are removed and the full dead load is applied, (*b*) the long-term deflection under sustained load, (*c*) the increment of deflection when the short-term part of the live load is applied.

Compare with ACI Code deflection limits; piping and brittle conduits are carried that would be damaged by large deflections. Note that midspan deflection may be used as a close approximation of maximum deflection.

6.7 Recalculate the deflections of Prob. 6.6 based on the assumption that 20 percent of the live load represents the normal service condition of loading and is sustained more or less continuously, while the remaining 80 percent is a short-term peak loading that would probably not be applied until most creep deflections have occurred. Compare with your earlier results.

6.8 The tensile-reinforced rectangular beam shown in Fig. P6.8*a* is made using steel with $f_y = 60{,}000$ psi and $E_s = 29{,}000{,}000$ psi. A perfectly plastic response after yielding can be assumed. The concrete used has the stress-strain curve shown in Fig. P6.8*b*, with limit of elastic response at a strain of 0.0005, maximum stress at 0.0020, and ultimate strain of 0.0030.

The concrete elastic modulus is $E_e = 3{,}600{,}000$ psi, and modulus of rupture is $f_r = 475$ psi. Based on this information, plot a curve relating applied moment to unit curvature at a section subjected to flexural cracking. Label points corresponding to first cracking, limit of concrete elastic response, first yielding of steel, and ultimate flexural strength.

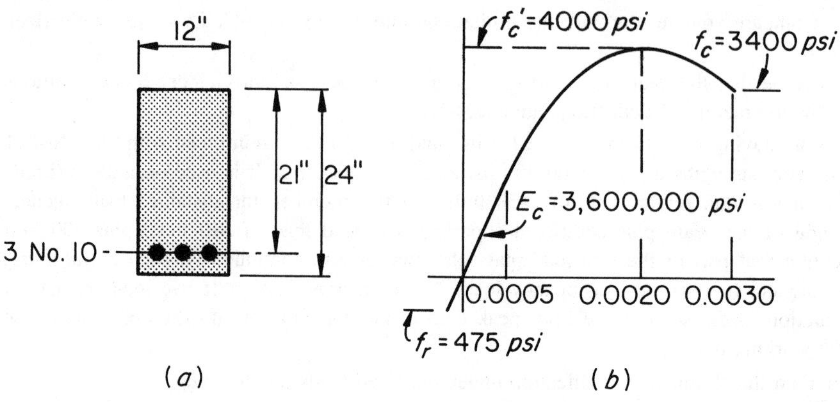

FIGURE P6.8

CHAPTER 7

ANALYSIS AND DESIGN FOR TORSION

7.1 INTRODUCTION

Reinforced concrete members are commonly subjected to bending moments, transverse shears associated with those bending moments, and, in the case of columns, to axial forces often combined with bending and shear. In addition, torsional forces may act, tending to twist a member about its longitudinal axis. Such torsional forces seldom act alone, but are almost always concurrent with bending moment and transverse shear, and sometimes with axial force as well.

For many years, torsion was regarded as a secondary effect, and was not considered explicitly in design, its influence being absorbed in the overall factor of safety of rather conservatively designed structures. In recent years, however, it has become necessary to take account of torsional effects in member design in many cases and to provide reinforcement to increase torsional strength. There are two reasons for this change. First, improved methods of analysis and design, such as the strength design approach now favored, have permitted a somewhat lower overall factor of safety through more accurate appraisal of load capacity, and have led to somewhat smaller member sizes. Second, there is increasing use of structural members for which torsion is a central feature of behavior, examples including curved bridge girders, eccentrically loaded box beams, and helical stairway slabs. Consequently, there has been a great increase, since the 1960s, in research activity relating to torsion in reinforced concrete, and practical design rules have been formulated. While further changes can be expected, procedures are included in design specifications such as the ACI Code that appear to be both safe and reasonably economical.

It is useful in considering torsion to distinguish between primary and secondary torsion in reinforced concrete structures. *Primary torsion*, sometimes called *equilibrium torsion* or *statically determinate torsion*, exists when the external load has no alternative but must be supported by torsion. For such cases, the torsion required to maintain static equilibrium can be uniquely determined. An example is the cantilevered slab of Fig. 7.1*a*. Loads applied to the slab surface cause twisting moments m_t to act along the length of the supporting beam. These are equilibrated

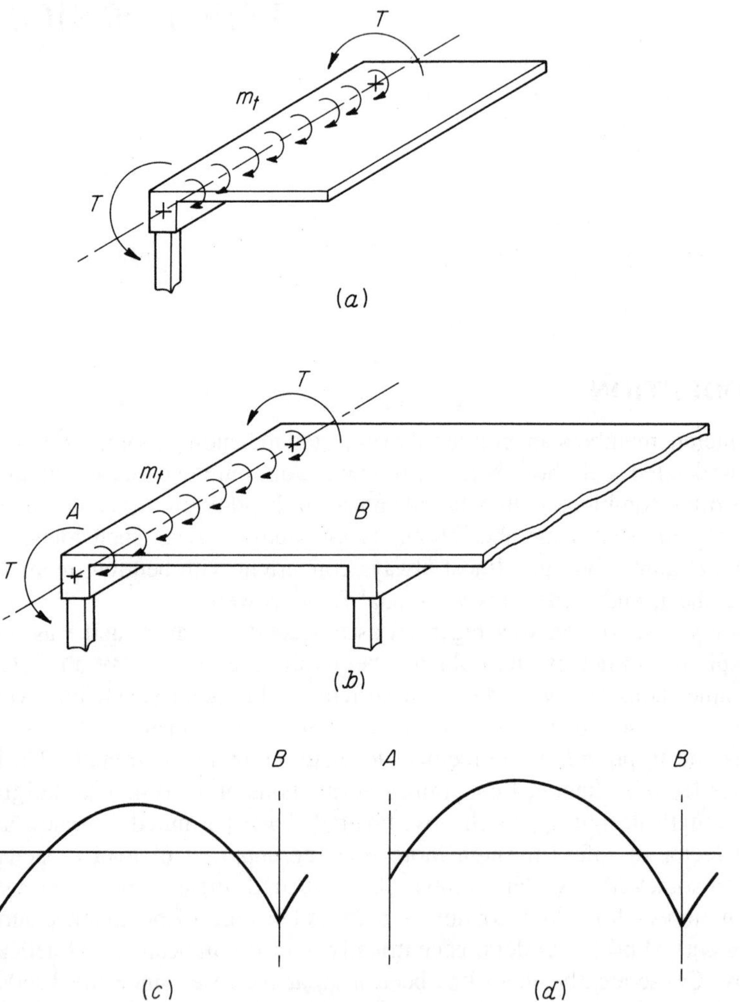

FIGURE 7.1
Torsional effects in reinforced concrete: (*a*) primary or equilibrium torsion at a cantilevered slab; (*b*) secondary or compatibility torsion at an edge beam; (*c*) slab moments if edge beam is stiff torsionally; (*d*) slab moments if edge beam is flexible torsionally.

FIGURE 7.2
Curved continuous beam bridge designed for torsion effects.

by the resisting torque T provided at the columns. Without the torsional moments, the structure will collapse.

In contrast to this condition, *secondary torsion*, also called *compatibility torsion* or *statically indeterminate torsion*, arises from the requirements of continuity, i.e., compatibility of deformation between adjacent parts of a structure. For this case, the torsional moments cannot be found based on static equilibrium alone. Disregard of continuity in the design will often lead to extensive cracking, but generally will not cause collapse. An internal readjustment of forces is usually possible and an alternative equilibrium of forces found. An example of secondary torsion is found in the spandrel or edge beam supporting a monolithic concrete slab, shown in Fig. 7.1*b*. If the spandrel beam is torsionally stiff and suitably reinforced, and if the columns can provide the necessary resisting torque T, then the slab moments will approximate those for a rigid exterior support as shown in Fig. 7.1*c*. However, if the beam has little torsional stiffness and inadequate torsional reinforcement, cracking will occur to further reduce its torsional stiffness, and the slab moments will approximate those for a hinged edge, as shown in Fig. 7.1*d*. If the slab is designed to resist the altered moment diagram, collapse will not occur.†

While current techniques for analysis permit the realistic evaluation of torsional moments for statically indeterminate conditions as well as determinate, designers often neglect secondary torsional effects when torsional stresses are low and alternative equilibrium states are possible. This is permitted according to the ACI Code and many other design specifications. On the other hand, when torsional strength is an essential feature of the design, such as for the bridge shown in Fig. 7.2, special analysis and special torsional reinforcement is required, as described in the remainder of this chapter.

† This phenomenon has been described by H. W. Birkeland as "the wisdom of the structure" (*J. ACI*, vol. 75, no. 4, 1978, pp. 105–111), i.e., the structure will strive to act as the designer intended. The concept should not be carried past certain limits.

7.2 TORSION IN PLAIN CONCRETE MEMBERS

Figure 7.3 shows a portion of a prismatic member subjected to equal and opposite torques T at the ends. If the material is elastic, St. Venant's torsion theory indicates that torsional shear stresses are distributed over the cross section, as shown in Fig. 7.3*b*. The largest shear stresses occur at the middle of the wide faces and are equal to

$$\tau_{\max} = \frac{T}{\alpha x^2 y} \tag{7.1}$$

where α is a shape factor approximately equal to $\frac{1}{4}$ and x and y are, respectively, the shorter and longer sides of the rectangle, as shown. If the material is inelastic, the stress distribution is similar, as shown by dashed lines, and the maximum shear stress is still given by Eq. (7.1) except that α assumes a larger value.

Shear stresses in pairs act on an element at or near the wide surface, as shown in Fig. 7.3*a*. As shown in strength of materials texts, this state of stress corresponds to equal tension and compression stresses on the faces of an element at 45° to the direction of shear. These inclined tension stresses are of the same kind as those caused by transverse shear, discussed in Sec. 4.2. However, in the case of torsion, since the torsional shear stresses are of opposite signs in the two halves of the member (Fig. 7.3*b*), the corresponding diagonal tension stresses in the two halves are at right angles to each other (Fig. 7.3*a*).

When the diagonal tension stresses exceed the tension resistance of the concrete, a crack forms at some accidentally weaker location and spreads immediately

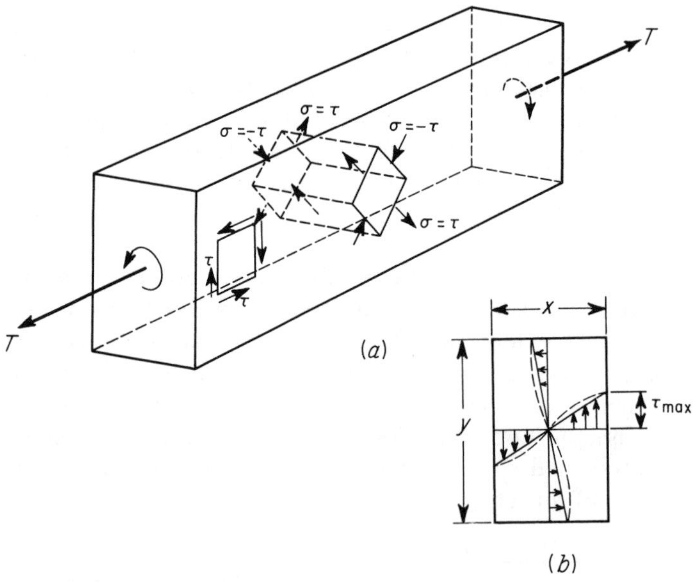

FIGURE 7.3
Stresses caused by torsion.

across the beam, as shown in Fig. 7.4. Observation shows that the tension crack (on the near face on Fig. 7.4*a*) forms at practically 45°, that is, perpendicular to the diagonal tension stresses. The cracks on the two narrow faces, where diagonal tension stresses are smaller, are of more indefinite inclination, as shown, and the fracture line on the far face connects the cracks at the short faces. This completes the formation of an entire fracture surface across the beam that fails the member.

For purposes of analysis this somewhat warped fracture surface can be replaced by a plane section inclined at 45° to the axis, as in Fig. 7.4*b*. Test observation shows (Ref. 7.1) that on such a plane, failure is more nearly by bending than by twisting. As shown in Fig. 7.4*b* and *c*, the applied torque T can be resolved

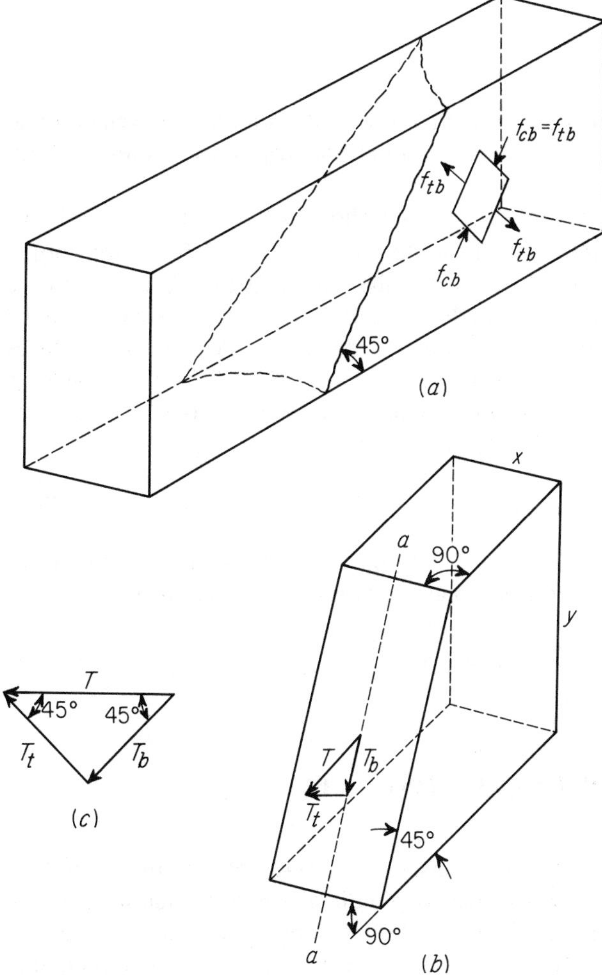

FIGURE 7.4
Torsional crack in plain concrete member.

into a component T_b that causes bending about the axis a-a of the failure plane and a component T_t that causes twisting. It is seen that

$$T_b = T \cos 45^\circ$$

The section modulus of the failure plane about a-a is

$$Z = \frac{x^2 y \csc 45^\circ}{6}$$

Then the maximum bending (tension) stress in the concrete is

$$f_{tb} = \frac{T_b}{Z} = T \sin 45^\circ \cos 45^\circ \frac{6}{x^2 y}$$

or

$$f_{tb} = \frac{3T}{x^2 y} \tag{7.2}$$

It is seen that the tension stress so calculated is identical with the St. Venant shear stress $\tau_{\max}$ [Eq. (7.1)] or with the corresponding diagonal tension stress σ (Fig. 7.3) for $\alpha = \frac{1}{3}$.

If f_{tb} were the only stress acting, cracking should occur when $f_{tb} = f_r$, the modulus of rupture of concrete, which can be taken as $f_r = 7.5\sqrt{f_c'}$ for normal density concrete. However, at right angles to the tension stress f_{tb} there exists a compression stress f_{cb} of equal magnitude (see Fig. 7.4*a*). For this state of biaxial stress, tests show that the presence of equal perpendicular compression reduces the tension strength of concrete by about 15 percent (see Fig. 2.8). Consequently, a crack forms and the member fails approximately when $f_{tb} = 0.85 f_r = 6\sqrt{f_c'}$. Let this value of f_{tb} be designed as the cracking stress

$$f_{\text{cr}} = \tau_{\text{cr}} = 6\sqrt{f_c'} \tag{7.3}$$

Then, upon substitution of f_{cr} for f_{tb} in Eq. (7.2), one obtains the magnitude of the torque that will crack and fail a plain rectangular concrete member:

$$T_{\text{cr}} = 6\sqrt{f_c'}\frac{x^2 y}{3} \tag{7.4}$$

7.3 TORSION IN REINFORCED CONCRETE MEMBERS

To resist torsion, reinforcement must consist of closely spaced stirrups and of longitudinal bars. Tests have shown that longitudinal bars alone hardly increase the torsional strength, test results showing an improvement of at most 15 percent (Ref 7.2). This is understandable because the only way in which longitudinal steel alone can contribute to torsional strength is by dowel action, which is particularly weak and unreliable if longitudinal splitting along bars is not restrained by transverse reinforcement. Thus, the torsional strength of members reinforced only with longitudinal steel is satisfactorily, and somewhat conservatively, predicted by Eqs. (7.3) and (7.4).

When members are adequately reinforced, as in Fig. 7.5*a*, the concrete cracks at a torque equal to or only somewhat larger than in an unreinforced member [Eq. (7.4)]. The cracks form a spiral pattern, as shown for one single crack in Fig. 7.5*b*. In actuality, a great number of such spiral cracks develop at close spacing. Upon cracking, the torsional resistance of the concrete drops to about half of that of the uncracked member, the remainder being now resisted by reinforcement. This redistribution of internal resistance is reflected in the torque-twist curve (Fig. 7.6), which at the cracking torque shows continued twist at constant torque until the reinforcement has picked up the portion of the torque no longer carried by concrete. Any further increase of applied torque must then be carried by the reinforcement. Failure occurs when somewhere along the member the concrete crushes along a line such as *a-d* in Fig. 7.5. In a well-designed member such crushing occurs only after the stirrups have started to yield.

The torsional strength can be analyzed by considering the equilibrium of the internal forces that are transmitted across the potential failure surface, shown shaded in Fig. 7.5. This surface is seen to be bounded by a 45° tension crack

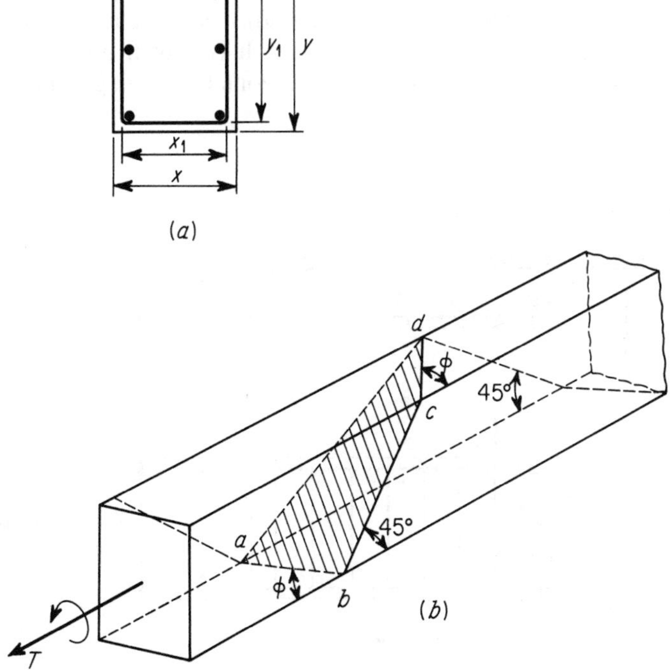

FIGURE 7.5
Torsional crack in reinforced concrete member.

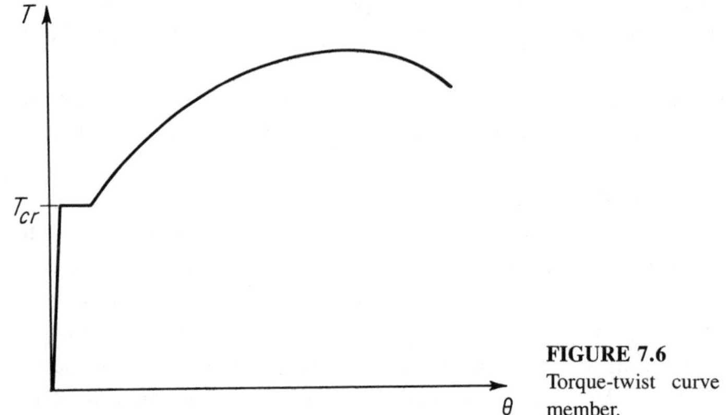

FIGURE 7.6
Torque-twist curve in reinforced concrete member.

across one wider face, two cracks across the narrower faces of inclination ϕ, an angle generally between 45 and 90°, and the zone of concrete crushing along line a-d. The failure is basically flexural, as for plain beams, with a concrete compression zone developing adjacent to a-d (Ref. 7.2).

Figure 7.7 shows the partially cracked failure surface, including the compression zone of concrete (shaded) and the horizontal and vertical stirrup forces S_h and S_v of all the stirrups intersecting the failure surface, except for those located in the compression zone. The number of horizontal stirrup legs, top or bottom, crossing the surface is seen to be $n_h = (x_1 \cot \phi)/s$, and the number of vertical legs opposite the compression zone, $n_v = y_1/s$. It is known from tests that at failure the vertical stirrup legs yield while the horizontal legs are generally

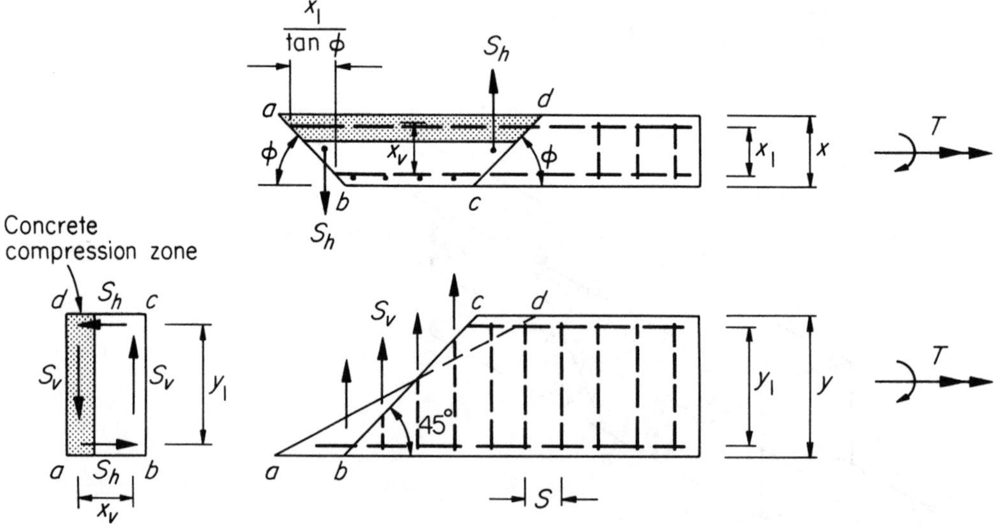

FIGURE 7.7
Failure surface in skewed bending.

not stressed to yielding. Correspondingly, the twisting couple produced by the horizontal stirrup forces is

$$T_h = n_h S_h y_1 = \frac{x_1 \cot \phi}{s} A_t f_{sh} y_1 = \kappa_h \frac{x_1 y_1}{s} A_t f_y \qquad (a)$$

where A_t = area of one stirrup leg
f_{sh} = tension stress in horizontal stirrup leg
f_y = yield point
κ_h = $\cot \phi (f_{sh}/f_y)$

To analyze the torque produced by the vertical stirrup forces near the front face, one must note, first, that the equilibrating forces near the rear face, in the compression zone, are fairly indeterminate. They consist, at the least, of a shear force S_c and a compression force P_c in the concrete and forces in the stirrup legs located in that zone. It is clear, however, that because of equilibrium, all these forces must have a resultant R equal and opposite to the sum of the vertical stirrup forces S_v. Correspondingly, the torque produced by the vertical stirrup forces can be written

$$T_v = n_v S_v x_v = \frac{y_1}{s} A_t f_y \kappa_v x_1 = \kappa_v \frac{x_1 y_1}{s} A_t f_y \qquad (b)$$

where x_v is the lever arm of the internal forces S_v and R, and $\kappa_v = x_v / x_1$.

It is seen that Eqs. (*a*) and (*b*) are identical except for κ_h and κ_v. At the present stage of knowledge, neither of these constants can be determined analytically, and recourse must be had to extensive tests. When one designates $\alpha_t = \kappa_h + \kappa_v$, the total torque contributed by the stirrups, $T_s = T_v + T_h$, is seen to be

$$T_s = \alpha_t \frac{x_1 y_1}{s} A_t f_y \qquad (7.5)$$

Tests (Ref. 7.2) have shown that α_t depends primarily on the ratio of cross-sectional dimensions and can be taken as

$$\alpha_t = 0.66 + 0.33 \frac{y_1}{x_1} \leq 1.50 \qquad (7.6)$$

It was mentioned previously that after cracking, the torque T_0 contributed by the concrete compression zone is about half of the cracking torque T_{cr} [Eq. (7.4)]. Taking the fraction conservatively to be 40 percent leads to

$$T_0 = 2.4 \sqrt{f'_c} \frac{x^2 y}{3} = 0.8 \sqrt{f'_c} x^2 y \qquad (7.7)$$

The total nominal torsion strength is then $T_n = T_0 + T_s$, or

$$T_n = 0.8 \sqrt{f'_c} x^2 y + \alpha_t A_t \frac{x_1 y_1}{s} f_y \qquad (7.8)$$

From the derivation of T_n it is evident that this nominal torsional strength will be developed only if the stirrups are spaced close enough for any failure surface to intersect an adequate number of stirrups. For this reason maximum spacing limits must be set for the stirrups (see Sec. 7.6).

The role of the longitudinal reinforcement in providing torsional strength is not yet clearly understood, but it is known that T_n can be developed only if adequate longitudinal reinforcement is provided. Its chief functions are the following:

1. It anchors the stirrups, particularly at the corners, which enables them to develop their full yield strength.
2. It provides at least some resisting torque because of the dowel forces which develop where the bars cross torsional cracks.
3. It has been observed that, after cracking, members subject to torsion tend to lengthen as the spiral cracks widen and become more pronounced. Longitudinal reinforcement counteracts this tendency and controls crack width.

Tests indicate that for Eq. (7.8) to be valid the total volume of longitudinal steel in a unit length of the member should be between 0.7 and 1.5 times the total volume of stirrups in that same length. It is customary to design torsional members so that these two volumes are equal. It is easily verified that this is so if the total area of longitudinal reinforcement is

$$A_l = 2A_t \frac{x_1 + y_1}{s} \tag{7.9}$$

7.4 TORSION PLUS SHEAR IN MEMBERS WITHOUT STIRRUPS

It was mentioned earlier that reinforced concrete members designed to carry torsion alone are quite unusual. The prevalent situation is that of a beam subject to the usual flexural moments and shear forces, which, in addition, must also resist torsional moments. In an uncracked member, shear forces as well as torques produce shear stresses. It must be expected, therefore, that simultaneously applied flexural shear forces and torques interact in a manner that will reduce the strength of the member compared with what it would be if shear or torsion were acting alone.

No satisfactory theories of this complex interaction have yet been developed, so that reliance must be placed on the extensive experimental investigations of this situation (Ref. 7.3). These test results are best represented in terms of an interaction equation or curve. Let V_0 and T_0, according to Eqs. (4.12*b*) and (7.7), be the cracking shear and torque of the member when subject, respectively, to flexural shear or to torsion alone. It will be recalled that for all practical purposes failure occurs at these same values almost immediately following cracking, so that for members without web reinforcement $V_0 = 2\sqrt{f_c'}bd$ and $T_0 = .08\sqrt{f_c'}x^2y$ adequately represent their ultimate strengths in the two modes. Further, let V_c and T_c, respectively, represent the shear capacity and the torsion capacity under combined loading, i.e., when the member is subject to simultaneous flexural shear and torsion. Then the numerous test results are well, and somewhat conservatively, represented by the circular-interaction equation

$$\left(\frac{V_c}{V_0}\right)^2 + \left(\frac{T_c}{T_0}\right)^2 = 1 \tag{7.10}$$

A graphical representation of this equation is shown in Fig. 7.8. It is seen that this interaction curve is quite favorable; i.e., the two modes do not interfere with each other very strongly. For instance, if a member carries a torque $T_0/2$, that is, one half of its pure torsion capacity, the curve shows that it can simultaneously carry about $0.85V_0$, that is, only 15 percent less than it could carry if no torsion were present at all.

It will be noted that the value of T_0 from Eq. (7.7) rather than Eq. (7.4) is used. The latter represents the torsion strength of previously uncracked concrete, while the former gives the strength after prior cracking. Under the combined action of bending, shear, and torsion, cracking may occur long before torsional failure. While this is not certain in all or even the majority of cases, the ACI Code provisions for torsion plus shear are based on the more conservative Eq. (7.7) rather than Eq. (7.4).

7.5 TORSION PLUS SHEAR IN MEMBERS WITH STIRRUPS

In Sec. 7.3 above, it has been shown that in members subject only to torsion and reinforced with stirrups and longitudinal bars, the nominal torsion strength T_n is furnished in part by the torsional strength of the concrete compression zone T_0 and in part by the torsional resistance of the steel T_s [see Eqs. (7.5), (7.7), and (7.8)].

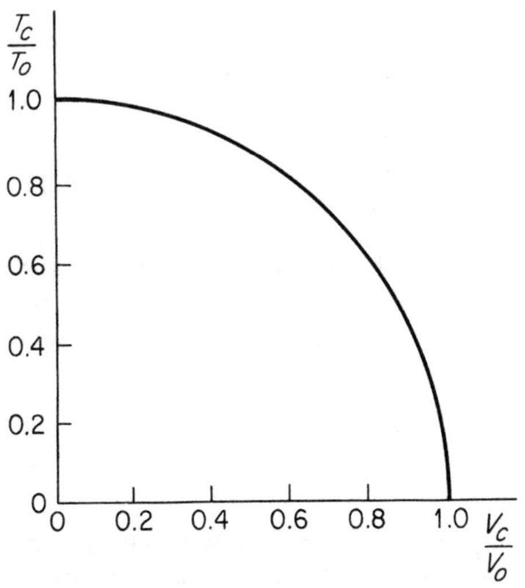

FIGURE 7.8
Interaction curve for combined torsion and flexural shear.

Experimental evidence on the strength of such reinforced members when subject to simultaneous torsion and shear is fairly sparse at this time (Ref. 7.4). Pending more extensive research, both experimental and analytical, the following approach appears reasonable and correlates satisfactorily and conservatively with the limited test evidence (Refs. 7.5 to 7.7):

1. In members with stirrups the portion of the total torsion carried by the concrete is determined by the same type of interaction equation [Eq. (7.10)] as in members without stirrups.
2. To carry the excess torque, over and above that resisted by the concrete, the same amount of reinforcement must be provided in members subject to torsion plus shear as would be required for purely torsional members. This torsional reinforcement is to be added to that required in the same member for carrying bending moments and flexural shears.

For a member subject to the factored external shear force V_u and the factored external torque T_u, it is now necessary to determine those portions V_c and T_c carried by the concrete. Solving Eq. (7.10) separately for each of these two quantities, one has

$$V_c = \frac{V_0}{\sqrt{1 + (V_0/T_0)^2(T_c/V_c)^2}} \qquad (a)$$

$$T_c = \frac{T_0}{\sqrt{1 + (T_0/V_0)^2(V_c/T_c)^2}} \qquad (b)$$

Using for V_0 and T_0 the values explained in Sec. 7.4 above, one obtains

$$\frac{T_0}{V_0} = 0.4\frac{x^2y}{bd} = \frac{0.4}{C_t} \qquad (c)$$

where

$$C_t = \frac{bd}{x^2y} \qquad (7.11)$$

It may be assumed with sufficient accuracy that the ratio of the shares of the applied shears and torques taken by the concrete is equal to the ratio of the applied, i.e., factored, shears and torques themselves. That is, it is assumed, with reasonable experimental justification, that

$$\frac{V_c}{T_c} = \frac{V_u}{T_u} \qquad (d)$$

If the values for V_0 and T_0 given in Sec. 7.4 above and the ratios given by Eqs. (*c*) and (*d*) are substituted into Eqs. (*a*) and (*b*), one obtains the shear force V_c

and the torsional moment T_c carried by the concrete in the case of simultaneous torsion and flexural shear:

$$V_c = \frac{2\sqrt{f'_c}bd}{\sqrt{1 + \left(2.5C_t \frac{T_u}{V_u}\right)^2}} \tag{7.12}$$

and

$$T_c = \frac{0.8\sqrt{f'_c}x^2y}{\sqrt{1 + \left(\frac{0.4V_u}{C_tT_u}\right)^2}} \tag{7.13}$$

It should be noted that the same unit of length, generally inches, must be used for T_u as for C_t in the denominator parentheses of Eqs. (7.12) and (7.13).

The nominal strength T_n is then the sum of the part taken by concrete, T_c [Eq. (7.13)], and by the reinforcement, T_s [Eq. (7.5)]. That is,

$$T_n = T_c + T_s \tag{7.14}$$

7.6 ACI CODE PROVISIONS FOR TORSION DESIGN

The basic principles upon which ACI Code design provisions are based have been presented in the preceding sections. ACI Code 11.6.5 safety provisions require that

$$T_u \leq \phi T_n = \phi(T_c + T_s) \tag{7.15}$$

where T_u = required torsional strength at factored loads
T_n = nominal torsional strength of member
T_c = torsional strength contributed by concrete
T_s = torsional strength contributed by steel reinforcement

The strength reduction factor $\phi = 0.85$ for torsion. T_c is based on Eq. (7.13) and T_s on Eq. (7.5).

a. T Beams and Box Sections

The previous development treated only rectangular members. For T beams, the torsional strength may be taken conservatively as the sum of the torsional strengths of the web and projecting flanges. Accordingly, the term x^2y in Eqs. (7.13), (7.11), and elsewhere above is replaced by Σx^2y, where x and y are, respectively, the smaller and larger side dimensions of each of the component rectangles of the cross section, partitioned in such a way as shown in Fig. 7.9, for example. The partitioning may be done, for members without torsion reinforcement, in such a way as to maximize the value of Σx^2y. For members with torsional reinforcement, the partitioning should be consistent with the reinforcement; i.e., if the closed

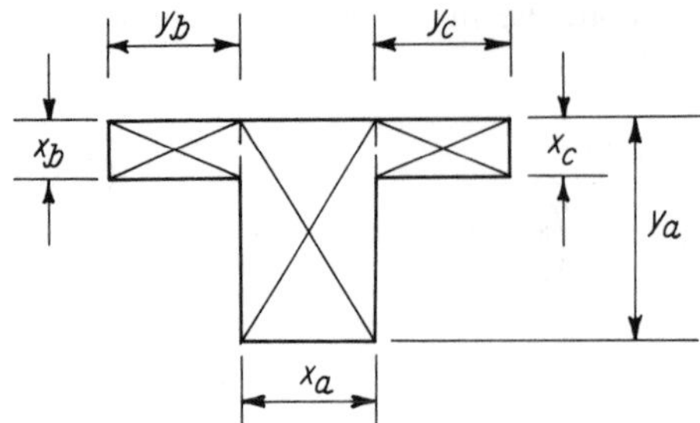

FIGURE 7.9
Basis of torsional section properties.

stirrups are located in the stem and extend close to the top of the slab, as usual, the partitioning would be as shown in Fig. 7.9. Alternately, closed stirrups may in some cases extend across the entire top flange, and the partitioning should be done accordingly. Tests indicate that the effective width of the projecting flange should not be taken greater than 3 times the flange thickness, for torsion.

Rectangular box section beams may be taken as equal to a solid section, provided the wall thickness h is at least $x/4$. Box beams with thinner walls may fail in a brittle way if subject to torsion, compared with the more ductile failure typical of solid beams, and the ratio of cracking torque to ultimate torque is less. Accordingly, ACI Code 11.6.1 provides that box beams with wall thickness less than $x/4$ but greater than $x/10$ may be treated as solid sections, except that $\Sigma x^2 y$ must be multiplied by $4h/x$ where h is the wall thickness. Box beams with wall thickness less than $x/10$ require special investigation of wall stiffness and stability. In all box sections, fillets should be provided at interior corners. Specific recommendations for fillets are found in the ACI Code Commentary.

b. Minimal Torsion

If the factored torsional moment T_u does not exceed $1.5\phi\sqrt{f'_c}\Sigma x^2 y/3$, torsional effects may be neglected, according to ACI Code 11.6.1. This lower limit is 25 percent of the torsional strength of the unreinforced concrete member, given by Eq. (7.4), reduced by the factor ϕ, as usual, for design purposes. The presence of torsional moment at or below that limit would not significantly reduce the flexural shear strength of the member (see Fig. 7.8).

c. Equilibrium vs. Compatibility Torsion

A distinction is made in the ACI Code between equilibrium (primary) torsion and compatibility (secondary) torsion. For the first condition, described earlier with reference to Fig. 7.1*a*, the supporting member *must* be designed to provide the torsional resistance required by static equilibrium. For the secondary torsion resulting from compatibility requirements, shown in Fig. 7.1*b*, it is assumed that cracking will result in a redistribution of internal forces, and according to ACI Code 11.6.3 the maximum factored torsional moment T_u may be reduced to $4\phi\sqrt{f'_c}\Sigma x^2 y/3$. The design moments and shears in the supported slab should be adjusted accordingly, and may be assumed to be uniformly distributed along the edge. The reduced value of T_u permitted by the ACI Code is intended to approximate the torsional cracking strength of the supporting beam, for combined torsional and flexural loading. The large rotations that occur at essentially constant torsional load would result in significant redistribution of internal forces, justifying use of the reduced value for design of the torsional member and the supported elements.

d. Reinforcement for Torsion

If the torsional moment exceeds that which can be carried by the concrete, reinforcement must be provided to carry the excess. It follows from Eq. (7.15) that the torque to be resisted by the reinforcement is

$$T_s = \frac{T_u - \phi T_c}{\phi} \tag{7.16}$$

From this, by Eq. (7.5), the required cross-sectional area of one stirrup leg for torsion is

$$A_t = \frac{(T_u - \phi T_c)s}{\alpha_t \phi f_y x_1 y_1} \tag{7.17}$$

where α_t is given by Eq. (7.6).

Even if torsion reinforcement is provided, in order to ensure ductile rather than brittle behavior, should failure occur, the torsional moment strength T_s must not be taken greater than $4T_c$, according to ACI Code 11.6.9.

Torsional reinforcement is composed of bars of two types: transverse stirrups and additional longitudinal reinforcement. Both must be provided, for reasons discussed in Sec. 7.3.

The transverse stirrups used for torsional reinforcement must be of a closed form, because principal tensile stress results on each of the four faces of a beam in torsion. U-shaped stirrups commonly used for transverse shear reinforcement are not suitable for torsional reinforcement. On the other hand, one-piece closed stirrups would make field assembly of beam reinforcement difficult, and for practical reasons torsional stirrups are generally two-piece stirrup-ties as shown in Fig. 7.10. A U-shaped stirrup is combined with a horizontal top-bar, suitably anchored.

Anchorage of the stirrups and top bars requires special attention. Tests have shown that in a member subject to torsional loading, a simple 90° hook will lead to premature failure due to spalling of the concrete outside of the hook. To avoid this, a 135° hook should be used for both the stirrup and top tie, as shown in Fig. 7.10*c*. An exception is permissible when lateral confinement is provided by a flange, as in Fig. 7.10*b* and at the right side of Fig. 7.10*a*, where a 90° hook is satisfactory. For beams where the width is greater than about 24 in., multiple stirrups are often used, as in Fig. 7.10*d*; in this case only the outer two legs are proportioned for torsion plus shear, and the interior legs are proportioned for vertical shear only. If flanges are included in the computation of torsional strength of T- or L-shaped beams, torsional stirrups should be provided in the flanges, as shown in Fig. 7.10*e*. U-shaped bars may be used but a full development length is needed interior to the face of the web.

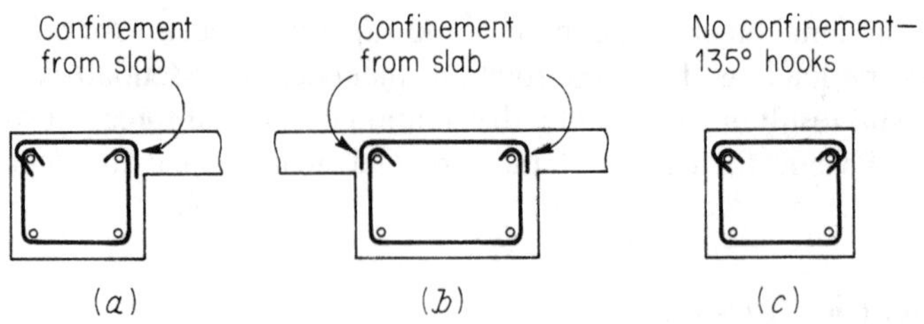

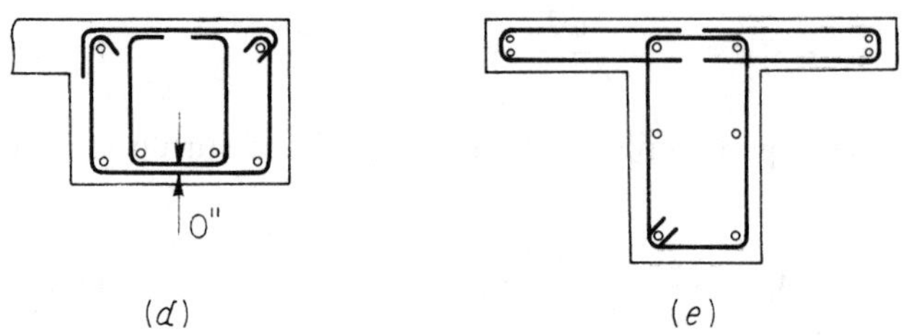

FIGURE 7.10
Stirrup-ties and longitudinal reinforcement for torsion: (*a*) spandrel beam with flanges on one side; (*b*) interior beam; (*c*) isolated rectangular section; (*d*) wide spandrel beam; (*e*) T beam with torsional reinforcement in flanges.

The required spacing of closed stirrups, with area A_t selected for trial based on standard bar sizes, can be found by suitable transposition of Eq. (7.17).

In order to control spiral cracking properly, the maximum spacing of torsional stirrups should not exceed $(x_1 + y_1)/4$ or 12 in., whichever is smaller. In addition, for members requiring both shear and torsion reinforcement, the minimum area of closed stirrups must be such that

$$(A_v + 2A_t) \geq 50\frac{b_w s}{f_y} \tag{7.18}$$

according to ACI Code 11.5.5.

The longitudinal bar area required to resist torsion is given by Eq. (7.9). In addition, according to ACI Code 11.6.9 provisions, it must not be less than

$$A_l = \left(\frac{400x\,s}{f_y}\frac{T_u}{T_u + V_u/3C_t} - 2A_t\right)\frac{x_1 + y_1}{s} \tag{7.19}$$

The value of A_l computed by Eq. (7.19) need not exceed that obtained by substituting $50b_w s/f_y$ for $2A_t$. The spacing of the longitudinal bars should not exceed 12 in., and they should be well distributed around the perimeter of the cross section to control cracking. The bars must not be less than No. 3 in size, and at least one bar must be placed in each corner of the stirrups. Careful attention must be paid to the anchorage of longitudinal torsional reinforcement so that it is able to develop its yield strength at the face of the supporting columns, where torsional moments are often maximum.

Reinforcement required for torsion may be combined with that required for other forces, provided that the area furnished is the sum of the individually required areas and that the most restrictive requirements for spacing and placement are met. According to ACI Code 11.6.7, torsional reinforcement must be provided at least a distance $d + b$ beyond the point theoretically required. Sections located less than a distance d from the face of a support may be designed for the same torsional stress as that computed at a distance d, recognizing the beneficial effects of support compression.

The effect of axial tension on torsional resistance has not been studied experimentally. Conservatively, the designer may provide torsional reinforcement to carry the total torque, disregarding the contribution of the concrete. On the basis that the effect of axial force on torsional strength should be similar to its effect on shear strength, the ACI Code permits Eqs. (7.12) and (7.13) to be used, but the resulting values of V_c and T_c must be reduced by the factor $1 + N_u/500A_g$, where N_u is the axial force, negative for tension.

Example 7.1 Design for torsion with shear. The 28 ft span beam shown in Fig. 7.11*a* and *b* carries a monolithic slab cantilevering 6 ft past the beam centerline, as shown in the section. The resulting L beam supports a live load of 50 psf uniformly distributed over its upper surface. The effective depth to the flexural steel centroid is 21.5 in., and the distance from the beam surfaces to the centroid of stirrup steel is $1\frac{3}{4}$ in. Material strengths are $f_c' = 4000$ psi and $f_y = 60{,}000$ psi. Design the torsional and shear reinforcement for the beam.

Solution. Applying ACI load factors gives the slab load as

$$\begin{aligned} 1.4w_d &= 75 \times 5.5 \times 1.4 = 580 \text{ lb/ft} \\ 1.7w_l &= 50 \times 1.7 \times 5.5 = \underline{470 \text{ lb/ft}} \\ &\text{Total} = 1050 \text{ lb/ft at 3.25 ft eccentricity} \end{aligned}$$

while the beam carries directly

$$\begin{aligned} 1.4w_d &= 300 \times 1.4 = 420 \text{ lb/ft} \\ 1.7w_l &= 50 \times 1.7 = \underline{85 \text{ lb/ft}} \\ &\text{Total} = 505 \text{ lb/ft} \end{aligned}$$

Thus, the uniformly distributed load on the beam is 1555 lb/ft, acting together with a uniformly distributed torque of $1050 \times 3.25 = 3410$ ft-lb/ft. At the face of the column the design shear force is $V_u = 1.555 \times 28/2 = 21.8$ kips. At the same location, the design torsional moment of $T_u = 3.410 \times 28/2 = 47.7$ ft-kips.

The variation of V_u and T_u with distance from the face of the supporting column is given by Fig. 7.11*c* and *d* respectively. The values of V_u and T_u at the critical design

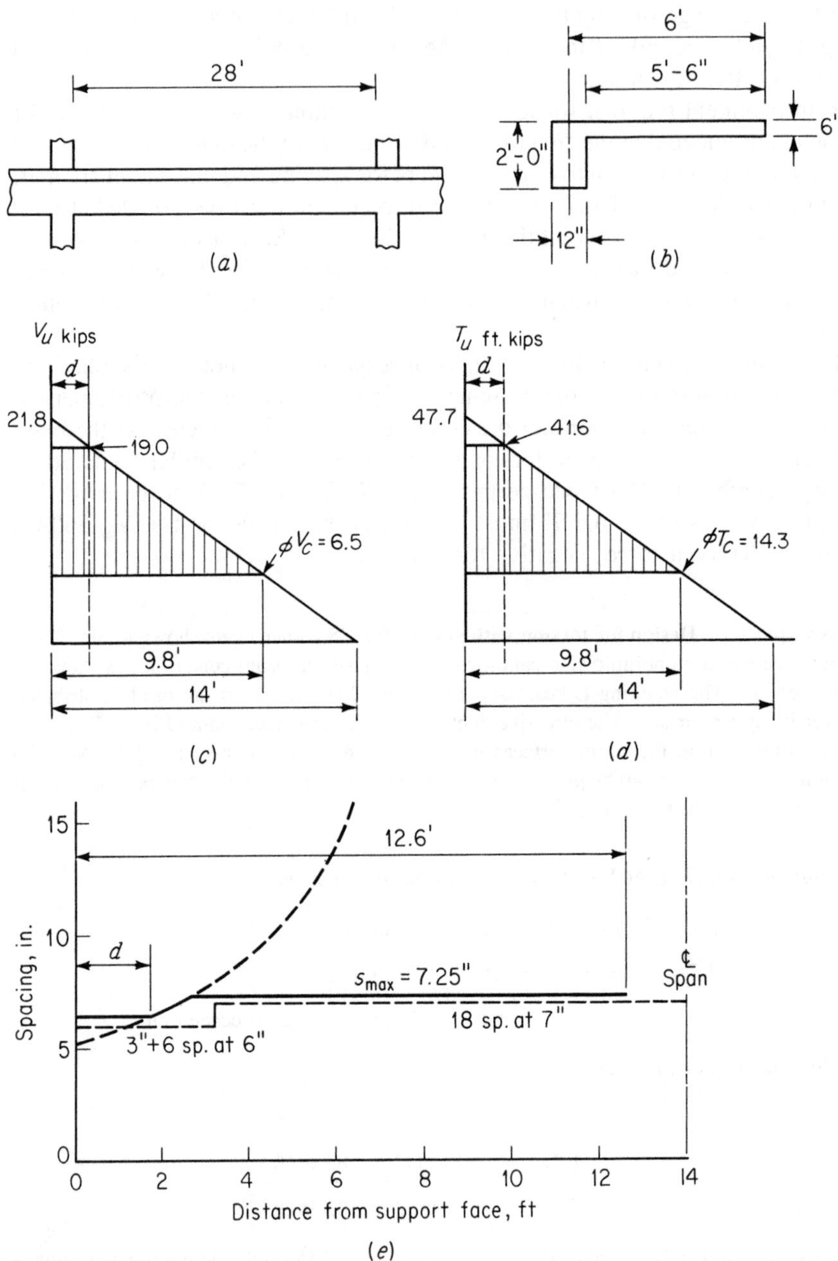

FIGURE 7.11
Shear and torsion design example.

section, a distance d from the column face, are

$$V_u = 21.8 \times \frac{12.21}{14} = 19.0 \text{ kips}$$

$$T_u = 47.7 \times \frac{12.21}{14} = 41.6 \text{ ft-kips}$$

For reference, for the present beam, $\Sigma x^2 y = 12^2 \times 24 + 6^2 \times 18 = 4098\text{in}^3$. According to the ACI Code, torsion may be neglected if $T_u \le 0.5 \times 0.85\sqrt{4000} \times 4098/12{,}000 = 9.2$ ft-kips. Torsion must clearly be considered in the present case. From Eq. (7.11), $C_t = (12 \times 21.5)/4098 = 0.0630$. Then from Eq. (7.12) the shear strength contributed by the concrete is

$$V_c = \frac{2\sqrt{4000}(12 \times 21.5)}{\sqrt{1 + \left(\dfrac{2.5 \times 0.0630 \times 41.6 \times 12}{19.0}\right)^2}} \times \frac{1}{1{,}000} = 7.7 \text{ kips}$$

and $\phi V_c = 0.85 \times 7.7 = 6.5$ kips, while from Eq. (7.13) the torsional strength contributed by the concrete is

$$T_c = \frac{0.8\sqrt{4000}(4098)}{\sqrt{1 + \left(\dfrac{0.4 \times 19.0}{0.0630 \times 41.6 \times 12}\right)^2}} \times \frac{1}{12{,}000} = 16.8 \text{ ft-kips}$$

resulting in $\phi T_c = 0.85 \times 16.8 = 14.3$ ft-kips. Thus, from Eq. (7.16), the torsion to be resisted by the reinforcement is

$$T_s = \frac{41.6 - 14.3}{0.85} = 32.1 \text{ ft-kips}$$

This is less than the upper limit of $4T_c = 67.2$ ft-kips, and so the beam section is satisfactory if properly reinforced.

Note that for this very common case of uniformly distributed transverse load coupled with uniformly distributed applied torque both T_u and V_u decrease linearly to zero at midspan. It follows that T_u/V_u remains constant at all sections along the span, and in turn T_c and V_c are constant, as shown by Fig. 7.11*c* and *d*. Furthermore, as noted in Sec. 7.5, $V_c/T_c = V_u/T_u$. This can be restated as $\phi V_c/V_u = \phi T_c/T_u$; that is, the proportion of the shear taken by the concrete is the same as the proportion of the torsion taken by the concrete. As a result, the point at which shear reinforcement is no longer theoretically required is the same as the point at which torsion reinforcement is no longer theoretically required. In the present case, this point is 9.8 ft from the support face.

With $1\frac{3}{4}$ in. cover to the center of the stirrup bar from all faces, $x_1 = 12 - 3.5 = 8.5$ in. and $y_1 = 24.0 - 3.5 = 20.5$ in. The coefficient $\alpha_t = 0.66 + 0.33 \times 20.5/8.5 = 1.46$. Computing the web reinforcement required for torsion at the column face (for reference only), at that location, from Eq. (7.17),

$$A_t = \frac{(T_u - \phi T_c)s}{\alpha_t \phi f_y x_1 y_1}$$

$$= \frac{(47.7 - 14.3) \times 12{,}000s}{1.46 \times 0.85 \times 60{,}000 \times 8.5 \times 20.5} = 0.0309s$$

for one leg of a closed vertical stirrup at a spacing s, or $0.0618s$ for two legs.

From Eq. (4.14*a*), the web reinforcement required for transverse shear, again computed at the column face, is

$$A_v = \frac{(V_u - \phi V_c)s}{\phi f_y d} = \frac{(21.8 - 6.5) \times 1000s}{0.85 \times 60{,}000 \times 21.5} = 0.0140s$$

to be provided in the two vertical legs. Thus the total area to be provided by the two vertical legs, for combined shear and torsion reinforcement at spacing s, at the face of the support is

$$2A_t + A_v = (0.0618 + 0.0140)s = 0.0758s$$

No. 4 closed stirrups will provide a total area in the two legs of 0.40 in^2; hence, the reference spacing at the column face would be $s_0 = 0.40/0.0758 = 5.28$ in. Since coefficients of s above are proportional to distance from the point of zero excess shear or zero excess torsion, 9.8 ft from the column face, the required spacings at d and at 2 ft intervals along the span can be found by proportion:

$$s_0 = 5.28 \text{ in.}$$

$$s_d = 5.28 \times 9.8/8.0 = 6.47 \text{ in.}$$

$$s_2 = 5.28 \times 9.8/7.8 = 6.62 \text{ in.}$$

$$s_4 = 5.28 \times 9.8/5.8 = 8.91 \text{ in.}$$

$$s_6 = 5.28 \times 9.8/3.8 = 13.6 \text{ in.}$$

$$s_8 = 5.28 \times 9.8/1.8 = 28.7 \text{ in.}$$

These values of s are plotted in Fig. 7.11*e*. ACI Code provisions for maximum spacing should now be checked. For torsion reinforcement the maximum spacing is the smaller of

$$\frac{x_1 + y_1}{4} = \frac{8.5 + 20.5}{4} = 7.25 \text{ in.}$$

or 12 in., while by the shear provisions the maximum spacing is $d/2 = 10.75$ in. The most restrictive provision is the first, and the maximum spacing of 7.25 in. is plotted in Fig. 7.11*e*. In addition, it is noted that the stirrups need not be spaced more closely than the requirement at a distance d from the column face. The resulting spacing requirements are shown by the solid line in the figure. These requirements are met in a practical way by No. 4 closed stirrups, the first 3 in. from the column face, followed by 6 at 6 in. spacing and 18 at 7 in. spacing. While stirrups could be discontinued $(d + b)$, or 2.8 ft, past the point of zero excess shear or torsion, the resulting 12.6 ft distance is sufficiently close to the half-span for reinforcement to be carried throughout. The minimum web steel provided, 0.40 in^2, satisfies the ACI Code minimum of $50b_w s/f_y = 50 \times 12 \times 7/60{,}000 = 0.07$ in^2.

The longitudinal steel required for torsion at a distance d from the column face is next computed. At that location

$$A_t = 0.0309 \times 5.28 \times \frac{8.01}{9.8} = 0.134 \text{ in}^2$$

and from Eq. (7.9),

$$A_l = 2 \times 0.134 \times \frac{8.5 + 20.5}{6.47} = 1.20 \text{ in}^2$$

with the total not to be less than

$$A_l = \left[\frac{400 \times 12 \times 6.47}{60{,}000} \times \frac{41{,}600 \times 12}{41{,}600 \times 12 + [19{,}000/(3 \times 0.0630)]} - (2 \times 0.134)\right]\frac{29}{6.47}$$

$$= 0.74 \text{ in}^2$$

According to the ACI Code, the spacing must not exceed 12 in. Reinforcement will be placed at the top, middepth, and bottom of the member, each level to provide not less than $1.20/3 = 0.40$ in^2. Two No. 4 bars will be used at the middepth, while reinforcement to be placed for flexure will be increased by 0.40 in^2 at the top and bottom of the member.

While A_l reduces in direct proportion to A_t and hence becomes zero at 9.8 ft from the column, for simplicity of construction the bars will be carried throughout the length of the member. Adequate embedment will be provided past the face of the column to fully develop f_y in the bars at that location.

7.7 COMPRESSION FIELD THEORY FOR TORSION DESIGN

In Sec. 4.8 it was pointed out that an alternative method of design for combined shear and flexure in beams has been developed, known as the *compression field theory*. It provides a simple, easily visualized physical model for the behavior of a reinforced concrete member subjected to shear and flexure, and leads to more rational design relationships with less dependence upon empirical adjustment than more traditional methods (Refs. 7.8 to 7.10).

The compression field theory is also applicable to reinforced concrete members subjected to torsion, or combinations of torsion, shear, and flexure, and offers the same advantages for these more complicated cases. The approach has been adopted as an acceptable alternative to torsion design in the Canadian National Standard, and provides the basis for European design methods as well (Refs. 7.11 and 7.12).

The model used for development of the theory is similar to that employed in developing the compression field theory for shear (see Sec. 4.8 and Fig. 4.19). The basic features will be described with reference to Fig. 7.12*a*, which shows a hollow box-section beam, reinforced with longitudinal bars at each corner and with transverse steel tension ties in the form of closed-hoop stirrups. The free body is subjected to a torque T at the far end, which is equilibrated by internal stresses and forces at the near end.

The light diagonal lines represent real or potential cracks in the concrete walls, each forming a more or less continuous spiral around the member; these cracks are perpendicular to the direction of principal tension.

It is assumed that the concrete resists no tension. Thus the model consists of parallel longitudinal reinforcing bars, transverse steel reinforcement in the form of closed hoops, and diagonal concrete compression struts, each defined by diagonal cracks on either side.

The torsion T is resisted by the tangential components of the diagonal compression in the concrete, which produce a shear flow q around the perimeter of

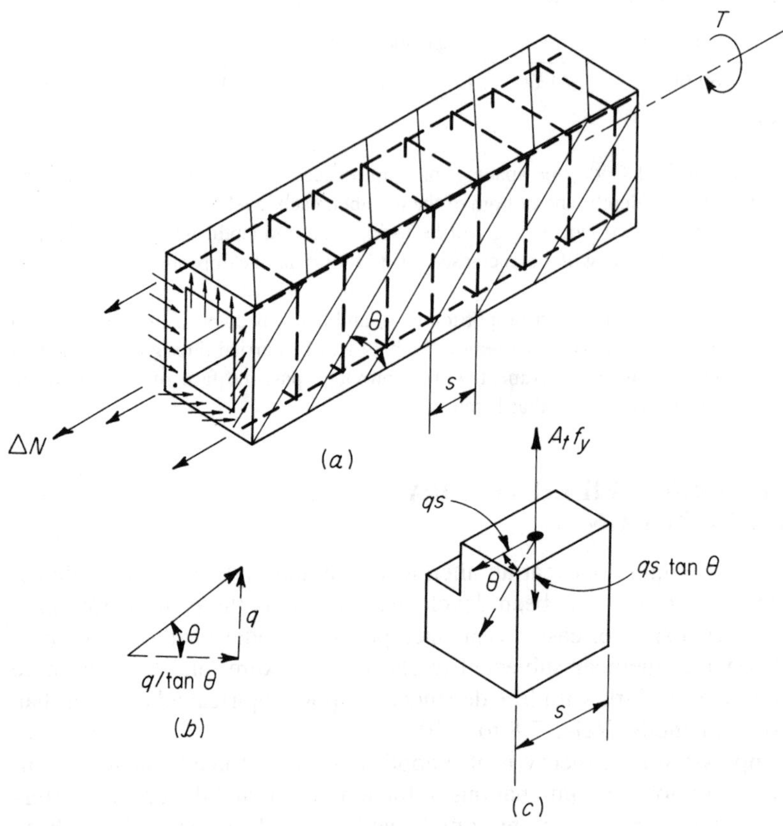

FIGURE 7.12
Basis of compression field theory for torsion: (*a*) free body of reinforced concrete hollow box-section member subject to torque T; (*b*) components of diagonal compressive stress in concrete; (*c*) stresses and forces acting at the corner element. (*Adapted from Ref. 7.11.*)

the cross section at the near end. This shear flow is related to the applied torque T by the equilibrium equation:

$$T = 2A_0 q \tag{7.20}$$

where A_0 is the area enclosed by the shear flow path and q is the shear per unit length around the perimeter.

The diagonal compression in the concrete struts produces a horizontal component of thrust (see Fig. 7.12*b*) which must be equilibrated by the total tension force ΔN in the longitudinal steel, given by

$$\Delta N = \frac{q p_0}{\tan \theta} = \frac{T}{2A_0} \frac{p_0}{\tan \theta} \tag{7.21}$$

in which p_0 is the perimeter of the shear flow path and θ is the slope angle of the compression diagonals. Thus the compression field theory provides a rational explanation for tension in the longitudinal reinforcement, while by the skewed bending theory, its role is not well defined (see Sec. 7.3). Such longitudinal forces are superimposed in those resulting from flexure, if present.

The outward thrust of the compression diagonals must be equilibrated by tension in the transverse steel, as shown in Fig. 7.12*c*. A shear stress of q per unit length on the transverse section of the member must be accompanied by an equal longitudinal unit shear stress, according to basic principles of mechanics. Thus the horizontal shear force acting on the top surface of the corner element of Fig. 7.12*c* is qs, where s is the spacing of the transverse steel. This force is introduced by the diagonal compression in the concrete struts, acting at a slope angle θ, so the outward component of thrust on the corner element is easily found. Vertical equilibrium then requires that

$$A_t f_y = qs \tan\theta = \frac{Ts}{2A_0} \tan\theta \tag{7.22}$$

According to the Canadian National Standard, the angle θ can be chosen by the designer to be anywhere in the range from 15 to 75°, just as for shear design. This has the same implications regarding the relative amounts of longitudinal and transverse reinforcement as described in Sec. 4.8 for shear. Also, just as for shear, in the case of torsion the assumption of too flat an angle will produce excessive compression stress in the diagonal concrete struts. This, in effect, sets the lower limit on the slope that can be assumed.

A full analysis requires consideration of strain compatibility in the walls of the torsional member and of appropriate stress-strain relations in the concrete and steel, as well as satisfaction of the equilibrium requirements described here. From such an analysis, the full behavioral response of members in torsion or combined loading can be predicted. Details will be found in Refs. 7.11 and 7.12.

While the 1989 ACI Code is based on the skewed bending theory of torsion, as described in earlier sections of this chapter, it appears likely that the compression field theory will be incorporated in future revisions, at least as an alternative method of design.

REFERENCES

7.1. T. T. C. Hsu, "Torsion of Structural Concrete—Plain Concrete Rectangular Sections," in *Torsion of Structural Concrete*, ACI Special Publication SP-18, 1968, pp. 203–238.

7.2. T. T. C. Hsu, "Torsion of Structural Concrete—Behavior of Reinforced Concrete Rectangular Members," in *Torsion of Structural Concrete*, ACI Special Publication SP-18, 1968, pp. 261–306.

7.3. U. Ersoy and P. M. Ferguson, "Concrete Beams Subject to Combined Torsion and Shear—Experimental Trends," in *Torsion of Structural Concrete*, ACI Special Publication SP-18, 1981, pp. 441–460.

7.4. D. L. Osborn, B. Mayoglou, and A. H. Mattock, "Strength of Reinforced Concrete Beams with Web Reinforcement in Combined Torsion, Shear, and Bending," *J. ACI*, vol. 66, no. 1, 1969, pp. 31–41.

7.5. "Tentative Recommendations for the Design of Reinforced Concrete Members to Resist Torsion," ACI Committee 438, *J. ACI*, vol. 66 no. 1, 1969, pp. 1–8.

7.6. T. T. C. Hsu and E. L. Kemp, "Background and Practical Application of Tentative Design Criteria for Torsion," *J. ACI*, vol. 66, no. 1, 1969, pp. 12–23.

7.7. P. Lampert and B. Thurlimann, "Ultimate Strength and Design of Reinforced Concrete Beams in Torsion and Bending," Int. Assoc. Bridge and Struct. Eng. Publ. 31-I, Zurich, 1971, pp. 107–131.

7.8. D. Mitchell and M. P. Collins, "Diagonal Compression Field Theory—A Rational Model for Structural Concrete in Pure Torsion," *J. ACI*, vol. 71, no. 8, 1974, pp. 396–408.
7.9. A. H. Mattock, "How to Design for Torsion," in *Torsion of Structural Concrete*, ACI Special Publication SP-18, 1968, pp. 469–495.
7.10. D. Mitchell and M. P. Collins, "Detailing for Torsion," *J. ACI*, vol. 73, no. 9, 1976, pp. 506–511.
7.11. M. P. Collins and D. Mitchell, "Shear and Torsion Design of Prestressed and Non-Prestressed Concrete Beams,"*J. Prestressed Concr. Inst.*, vol. 25, no. 5, 1980, pp. 32–100.
7.12. "Design of Concrete Structures for Buildings," CAN3-A23.3-M84, Canadian Standards Association, Toronto, 1984.

PROBLEMS

7.1 A beam of rectangular cross section having $b = 22$ in. and $h = 15$ in. is to carry a total factored load of 2000 lb/ft uniformly distributed over its 26 ft span, and in addition will be subjected to a uniformly distributed torsion of 3500 ft-lb/ft at factored loads. Closed stirrup-ties will be used to provide for flexural shear and torsion, placed with the stirrup steel centroid 1.75 in. from each concrete face. The corresponding flexural effective depth will be approximately 12.5 in. Design the transverse reinforcement for this beam and calculate the increment of longitudinal steel area needed to provide for torsion, using $f_c' = 4000$ psi and $f_y = 60{,}000$ psi.

7.2 Architectural and clearance requirements call for the use of a transfer girder shown in Fig. P7.2, spanning 20 ft between supporting column faces. The girder must carry from above a concentrated column load of 20 kips at midspan, applied with eccentricity 2 ft from the girder centerline. (Load factors are already included, as is an allowance for girder self-weight.) The member is to have dimensions $x = 10$ in., $y = 20$ in., $x_1 = 6$ in., $y_1 = 16$ in., and $d = 17$ in. Supporting columns provide full torsional rigidity; flexural rigidity at the ends of the span may be assumed to develop 40 percent of the maximum moment that would be obtained if the girder were simply supported. Design both transverse and longitudinal steel for the beam. Material strengths are $f_y = 40{,}000$ psi and $f_c' = 4000$ psi.

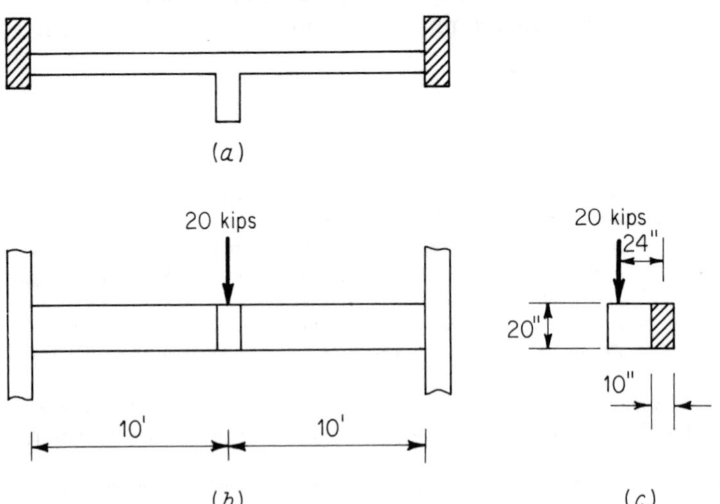

FIGURE P7.2
Transfer girder: (*a*) top view; (*b*) front view; (*c*) side view.

7.3 The beam shown in cross section in Fig. P7.3 is a typical interior member of a continuous building frame, with span 30 ft between support faces. At factored loads it will carry a uniformly distributed vertical load of 1700 lb/ft, acting simultaneously with a uniformly distributed torsion of 3750 ft-lb/ft. Transverse reinforcement for shear and torsion will consist of No. 4 stirrup-ties as shown, with 1.5 in. clear to all concrete faces. The effective depth to flexural steel may be taken equal to 22.5 in. for both negative and positive bending regions. Design the transverse reinforcement for shear and torsion, and calculate the longitudinal steel to be added to the flexural requirements to provide for torsion. Torsional reinforcement will be provided in the web only, not the flanges, so flanges should be neglected in the torsional computations. Material strengths are f_y = 60,000 psi and f_c' = 4000 psi.

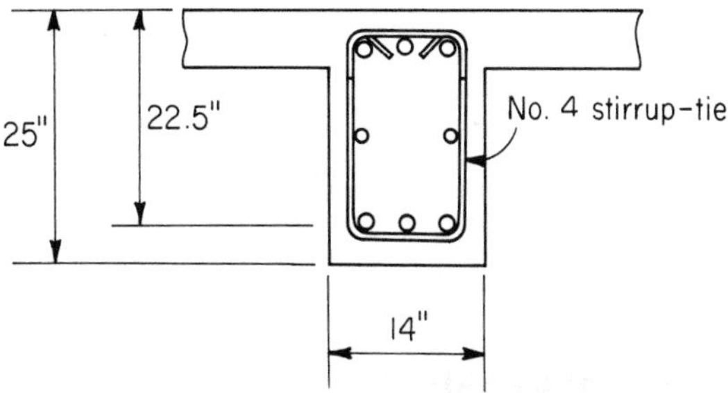

FIGURE P7.3

7.4 The single-span T-beam bridge described in Prob. 3.13 of Chap. 3 is reinforced for flexure with four No. 10 bars in two layers, which continue uninterrupted into the supports, permitting a service live load of 1.50 kips/ft to be carried, in addition to the dead load of 0.93 kips/ft, including self-weight. Assume now that only half that live load acts, but that it is applied over half the width of the member only, entirely to the right of the section centerline. Design the transverse reinforcement for shear and torsion, and calculate the modified longitudinal steel needed for this eccentric load condition. Torsional reinforcement may be provided in the slab if needed, as well as in the web. Stirrup-ties will be No. 3 or No. 4 bars, with 1.5 in. clear to all concrete faces. Supports provide no restraint against flexural rotations, but full restraint against twist. Show a sketch of your final design detailing all reinforcement. Material strengths are as given for Prob. 3.13.

II start

CHAPTER 8

SHORT COLUMNS

8.1 INTRODUCTION: AXIAL COMPRESSION

Columns are defined as members that carry loads chiefly in compression. Usually columns carry bending moments as well, about one or both axes of the cross section, and the bending action may produce tensile forces over a part of the cross section. Even in such cases, columns are generally referred to as compression members, because the compression forces dominate their behavior. In addition to the most common type of compression member, i.e., vertical elements in structures, compression members include arch ribs, rigid frame members inclined or otherwise, compression elements in trusses, shells or portions thereof that carry axial compression, and other forms. In this chapter the term *column* will be used interchangeably with the term *compression member*, for brevity and in conformity with general usage.

Three types of reinforced concrete compression members are in use:

1. Members reinforced with longitudinal bars and lateral ties.
2. Members reinforced with longitudinal bars and continuous spirals.
3. Composite compression members reinforced longitudinally with structural steel shapes, pipe, or tubing, with or without additional longitudinal bars, and various types of lateral reinforcement.

Types 1 and 2 are by far the most common, and most of the discussion of this chapter will refer to them.

The main reinforcement in columns is longitudinal, parallel to the direction of the load, and consists of bars arranged in a square, rectangular, or circular pattern, as was shown in Fig. 1.14. The ratio of longitudinal steel area A_{st} to gross concrete cross section A_g is in the range from 0.01 to 0.08, according to

ACI Code 10.9.1. The lower limit is necessary to ensure resistance to bending moments not accounted for in the analysis and to reduce the effects of creep and shrinkage of the concrete under sustained compression. Ratios higher than 0.08 not only are uneconomical, but also would cause difficulty owing to congestion of the reinforcement, particularly where the steel must be spliced. Generally, the larger diameter bars are used to reduce placement costs and to avoid unnecessary congestion. The special large-diameter No. 14 and No. 18 bars are produced mainly for use in columns. According to ACI Code 10.9.2, a minimum of four longitudinal bars is required when the bars are enclosed by spaced rectangular or circular ties, and a minimum of six bars must be used when the longitudinal bars are enclosed by a continuous spiral.

Columns may be divided into two broad categories: *short columns*, for which the strength is governed by the strength of the materials and the geometry of the cross section, and *slender columns*, for which the strength may be significantly reduced by lateral deflections. A number of years ago, an ACI-ASCE survey indicated that 90 percent of columns braced against sidesway and 40 percent of unbraced columns could be designed as short columns. Effective lateral bracing, which prevents relative lateral movement of the two ends of a column, is commonly provided by shear walls, elevator and stairwell shafts, diagonal bracing, or a combination of these. Although slender columns are more common now because of the wider use of high-strength materials and improved methods of dimensioning members, it is still true that most columns in ordinary practice can be considered short columns. Only short columns will be discussed in this chapter; the effects of slenderness in reducing column strength will be covered in Chapter 9.

The behavior of short, axially loaded compression members was discussed in Sec. 1.9 in introducing the basic aspects of reinforced concrete. It is suggested that that material be reviewed at this point. In that discussion, it was demonstrated that, for lower loads for which both materials remain in their elastic range of response, the steel carries a relatively small portion of the total load. The steel stress f_s is equal to n times the concrete stress:

$$f_s = nf_c \tag{8.1}$$

where $n = E_s/E_c$ is the modular ratio. In this range the axial load P is given by

$$P = f_c[A_g + (n - 1)A_{st}] \tag{8.2}$$

where the term in square brackets is the area of the transformed section. Equations (8.2) and (8.1) can be used to find concrete and steel stresses respectively, for given loads, provided both materials remain elastic. Example 1.1 demonstrated the use of these equations.

In Sec. 1.9 it was further shown that the nominal ultimate strength of an axially loaded column can be found, recognizing the nonlinear response of both materials, by

$$P_n = 0.85f_c'A_c + A_{st}f_y \tag{8.3a}$$

or

$$P_n = 0.85f_c'(A_g - A_{st}) + A_{st}f_y \tag{8.3b}$$

i.e., by summing the strength contributions of the two components of the column. At this stage, the steel carries a significantly larger fraction of the load than was the case at lower total load.

The calculation of nominal ultimate strength of an axially loaded column was demonstrated in Sec. 1.9.

According to ACI Code 10.3.5, the useful *design strength* of an axially loaded column is to be found based on Eq. (8.3*b*) with the introduction of certain strength reduction factors. The ACI factors are lower for columns than for beams, reflecting their greater importance in a structure. A beam failure would normally affect only a local region, whereas a column failure could result in collapse of the entire structure. In addition, these factors reflect differences in the behavior of tied columns and spirally reinforced columns that will be discussed in Sec. 8.2. A basic ϕ factor of 0.75 is used for spirally reinforced columns, and 0.70 for tied columns, vs. $\phi = 0.90$ for beams.

A further limitation on column strength is imposed by ACI Code 10.3.5 in order to allow for accidental eccentricities of loading not considered in the analysis. This could be included by specifying a certain minimum eccentricity to be used (as was done in earlier editions of the Code), or, more directly, by imposing an upper limit of capacity less than the calculated design strength. This upper limit is taken as 0.85 times the design strength for spirally reinforced columns, and 0.80 times the calculated strength for tied columns. Thus, according to ACI Code 10.3.5, for spirally reinforced columns

$$\phi P_{n(\max)} = 0.85\phi\left[0.85 f_c'(A_g - A_{st}) + f_y A_{st}\right] \tag{8.4a}$$

with $\phi = 0.75$. For tied columns

$$\phi P_{n(\max)} = 0.80\phi\left[0.85 f_c'(A_g - A_{st}) + f_y A_{st}\right] \tag{8.4b}$$

with $\phi = 0.70$.

8.2 LATERAL TIES AND SPIRALS

Figure 1.14 shows cross sections of the simplest types of columns, spirally reinforced or provided with lateral ties. Other cross sections frequently found in buildings and bridges are shown in Fig. 8.1. In general, in members with large axial forces and small moments, longitudinal bars are spaced more or less uniformly around the perimeter (Fig. 8.1*a* to *d*). When bending moments are large, much of the longitudinal steel is concentrated at the faces of largest compression or tension, i.e., at maximum distances from the axis of bending (Fig 8.1*e* to *h*). Specific recommended patterns for many combinations and arrangements of bars are found in Refs. 8.1 and 8.2. In heavily loaded columns with large steel percentages, the result of a large number of bars, each of them positioned and held individually by ties, is steel congestion in the forms and difficulties in placing the concrete. In such cases, bundled bars are frequently employed. Bundles consist of three or four bars tied in direct contact, wired, or otherwise fastened together. These are usually placed in the corners. Tests have shown that adequately bundled bars act as one unit; i.e., they are detailed as if a bundle constituted a single round bar of area equal to the sum of the bundled bars.

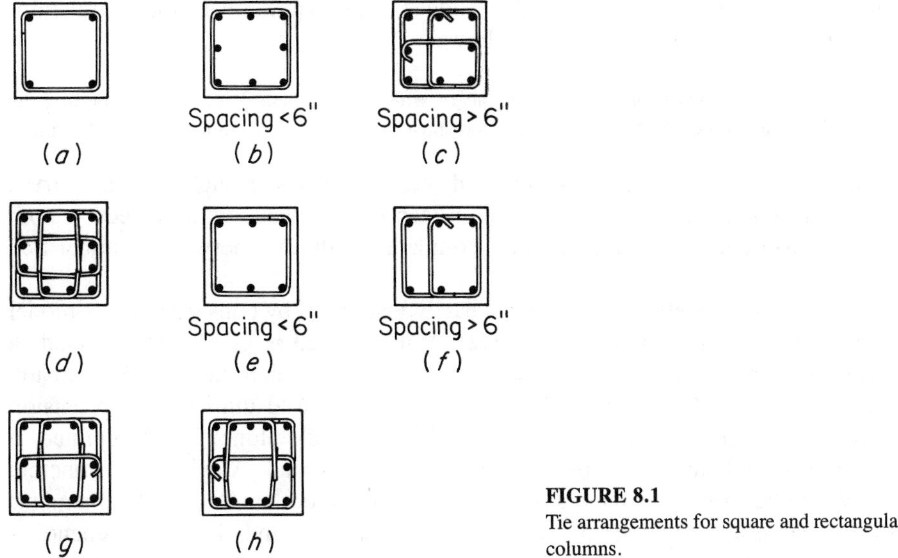

FIGURE 8.1
Tie arrangements for square and rectangular columns.

Lateral reinforcement, in the form of individual relatively widely spaced ties or a continuous closely spaced spiral, serves several functions. For one, such reinforcement is needed to hold the longitudinal bars in position in the forms while the concrete is being placed. For this purpose, longitudinal and transverse steel are wired together to form cages, which are then moved into the forms and properly positioned before placing the concrete. For another, transverse reinforcement is needed to prevent the highly stressed, slender longitudinal bars from buckling outward by bursting the thin concrete cover.

Closely spaced spirals evidently serve these two functions. Ties, which can be arranged and spaced in various ways, must be so designed that these two requirements are met. This means that the spacing must be sufficiently small to prevent buckling between ties and that in any tie plane a sufficient number of ties must be provided to position and hold all bars. On the other hand, in columns with many longitudinal bars, if the column section is crisscrossed by too many ties, they interfere with the placement of concrete in the forms. To achieve adequate tying yet hold the number of ties to a minimum, ACI Code 7.10.5 gives the following rules for tie arrangement:

> All bars of tied columns shall be enclosed by *lateral ties*, at least No. 3 in size for longitudinal bars up to No. 10, and at least No. 4 in size for Nos. 11, 14, and 18 and bundled longitudinal bars. The spacing of the ties shall not exceed 16 diameters of longitudinal bars, 48 diameters of tie bars, nor the least dimension of the column. The ties shall be so arranged that every corner and alternate longitudinal bar shall have lateral support provided by the corner of a tie having an included angle of not more than 135°, and no bar shall be farther than 6 in. clear on either side from such a laterally supported bar. Deformed wire or welded wire fabric of equivalent area may be used instead of ties. Where the bars are located around the periphery of a circle, complete circular ties may be used.

For spirally reinforced columns ACI Code 7.10.4 requirements for lateral reinforcement may be summarized as follows:

> Spirals shall consist of a continuous bar or wire not less than $\frac{3}{8}$ in. in diameter, and the clear spacing between turns of the spiral must not exceed 3 in. nor be less than 1 in.

In addition, a minimum ratio of spiral steel is imposed such that the structural performance of the column is significantly improved, with respect to both ultimate load and the type of failure, compared with an otherwise identical tied column.

The structural effect of a spiral is easily visualized by considering as a model a steel drum filled with sand (Fig. 8.2). When a load is placed on the sand, a lateral pressure is exerted by the sand on the drum, which causes hoop tension in the steel wall. The load on the sand can be increased until the hoop tension becomes large enough to burst the drum. The sand pile alone, if not confined in the drum, would have been able to support hardly any load. A cylindrical concrete column, to be sure, does have a definite strength without any lateral confinement. As it is being loaded, it shortens longitudinally and expands laterally, depending on Poisson's ratio. A closely spaced spiral confining the column counteracts the expansion, as did the steel drum in the model. This causes hoop tension in the spiral, while the carrying capacity of the confined concrete in the core is greatly increased. Failure occurs only when the spiral steel yields, which greatly reduces its confining effect, or when it fractures.

A tied column fails at the load given by Eq. (8.3*a* or *b*). At this load the concrete fails by crushing and shearing outward along inclined planes, and the longitudinal steel by buckling outward between ties (Fig 8.3). In a spiral-reinforced column, when the same load is reached, the longitudinal steel and the concrete within the core are prevented from outward failing by the spiral. The concrete

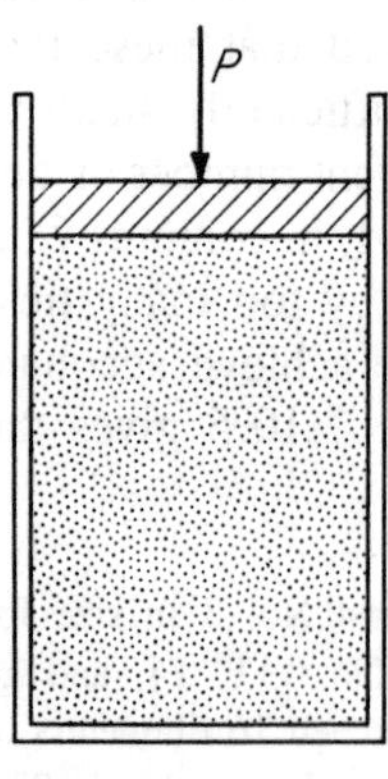

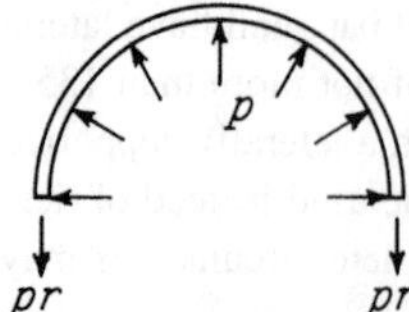

FIGURE 8.2
Model for action of a spiral.

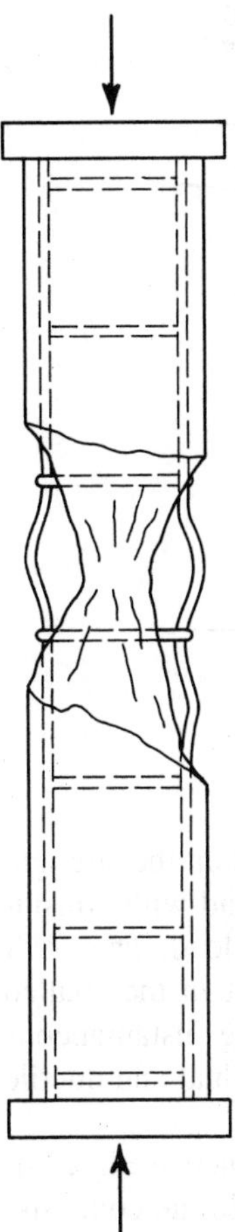

FIGURE 8.3
Failure of a tied column.

in the outer shell, however, not being so confined, does fail; i.e., the outer shell spalls off when the load P_n is reached. It is at this stage that the confining action of the spiral has a significant effect, and if sizable spiral steel is provided, the load that will ultimately fail the column by causing the spiral steel to yield or fracture can be much larger than that at which the shell spalled off. Furthermore, the axial strain limit when the column fails will be much greater than otherwise; the toughness of the column has been much increased.

In contrast to the practice in some foreign countries, it is reasoned in the United States that any excess capacity beyond the spalling load of the shell is wasted because the member, although not actually failed, would no longer be considered serviceable. For this reason the ACI Code provides a minimum spiral reinforcement of such an amount that its contribution to the carrying capacity is just slightly larger than that of the concrete in the shell. The situation is best understood from Fig. 8.4, which compares the performance of a tied column with that of a spiral column whose spalling load is equal to the ultimate load of the tied column. The failure of the tied column is abrupt and complete. This is true, to almost the same degree, of a spiral column with

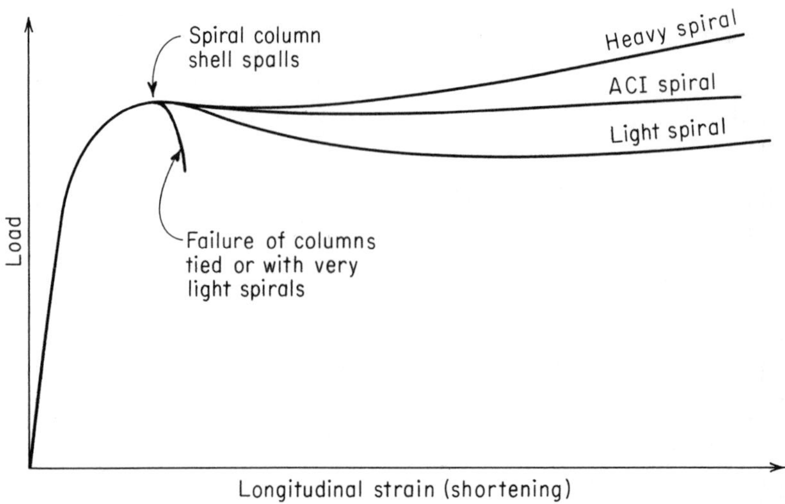

FIGURE 8.4
Behavior of spirally reinforced and tied columns.

a spiral so light that its strength contribution is considerably less than the strength lost in the spalled shell. With a heavy spiral the reverse is true, and with considerable prior deformation the spalled column would fail at a higher load. The "ACI spiral," its strength contribution about compensating for that lost in the spalled shell, hardly increases the ultimate load. However, by preventing instantaneous crushing of concrete and buckling of steel, it produces a more gradual and ductile failure, i.e., a tougher column.

It has been found experimentally (Refs. 8.3 to 8.5) that the increase in compressive strength of the core concrete in a column provided through the confining effect of spiral steel is closely represented by the equation

$$f_c^* - 0.85f_c' = 4.0f_2' \qquad (a)$$

where f_c^* = compressive strength of spirally confined core concrete
$0.85f_c'$ = compressive strength of concrete if unconfined
f_2' = lateral confinement stress in core concrete produced by spiral

The confinement stress f_2' is calculated assuming that the spiral steel reaches its yield stress f_y when the column eventually fails. With reference to Fig. 8.5, a hoop tension analysis of an idealized model of a short segment of column confined by one turn of lateral steel shows that

$$f_2' = \frac{2A_{sp}f_y}{d_c s} \qquad (b)$$

where A_{sp} = cross-sectional area of spiral wire
f_y = yield strength of spiral steel
d_c = outside diameter of spiral
s = spacing or pitch of spiral wire

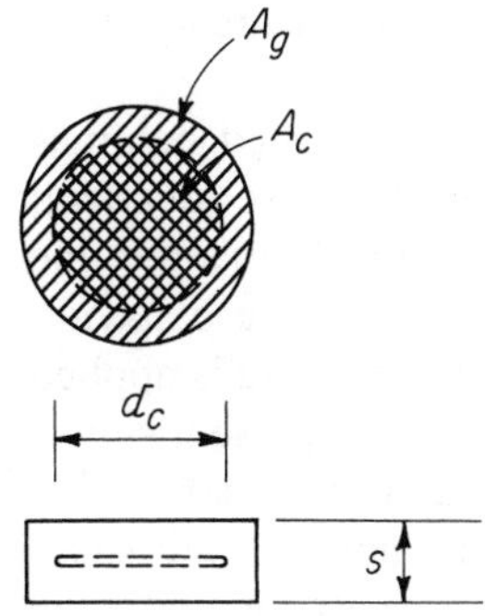

FIGURE 8.5
Confinement of core concrete due to hoop tension.

A volumetric ratio is defined as the ratio of the volume of spiral steel to the volume of core concrete:

$$\rho_s = \frac{2\pi d_c A_{sp}}{2} \frac{4}{\pi d_c^2 s}$$

from which

$$A_{sp} = \frac{\rho_s d_c s}{4} \tag{c}$$

Substituting the value of A_{sp} from Eq. (c) into Eq. (b) results in

$$f_2' = \frac{\rho_s f_y}{2} \tag{d}$$

To find the right amount of spiral steel one calculates

$$\text{Strength contribution of the shell} = 0.85 f_c'(A_g - A_c) \tag{e}$$

where A_g and A_c are, respectively, the gross and core concrete areas. Then substituting the confinement stress from Eq. (d) into Eq. (a) and multiplying by the core concrete area,

$$\text{Strength provided by the spiral} = 2\rho_s f_y A_c \tag{f}$$

The basis for the design of the spiral is that the strength gain provided by the spiral should be at least equal to that lost when the shell spalls, so combining Eqs. (e) and (f),

$$0.85 f_c'(A_g - A_c) = 2\rho_s f_y A_c$$

from which

$$\rho_s = 0.425\left(\frac{A_g}{A_c} - 1\right)\frac{f_c'}{f_y} \tag{g}$$

According to the ACI Code, this result is rounded upward slightly, and ACI Code 10.9.3 states that the ratio of spiral reinforcement shall not be less than

$$\rho_s = 0.45\left(\frac{A_g}{A_c} - 1\right)\frac{f_c'}{f_y} \tag{8.5}$$

It is further stipulated in the ACI Code that f_y must not be taken greater than 60,000 psi.

It follows from this development that two concentrically loaded columns designed to the ACI Code, one tied and one with spiral but otherwise identical, will fail at about the same load, the former in a sudden and brittle manner, the latter gradually with prior spalling of the shell and with more ductile behavior. This advantage of the spiral column is much less pronounced if the load is applied with significant eccentricity or when bending from other sources is present simultaneously with axial load. For this reason, while the ACI Code permits somewhat larger design loads on spiral than on tied columns when the moments are small or zero ($\phi = 0.75$ for spirally reinforced columns vs. $\phi = 0.70$ for tied), the difference is not large, and it is even further reduced for large eccentricities, for which ϕ approaches 0.90 for both.

The design of spiral reinforcement according to the ACI Code provisions is easily reduced to tabular form, as in Table A.14 of App. A.

8.3 COMPRESSION PLUS BENDING OF RECTANGULAR COLUMNS

Members that are axially, i.e., concentrically, compressed occur rarely, if ever, in buildings and other structures. Components such as columns and arches chiefly carry loads in compression, but simultaneous bending is almost always present. Bending moments are caused by continuity, i.e., by the fact that building columns are parts of monolithic frames in which the support moments of the girders are partly resisted by the abutting columns, by transverse loads such as wind forces, by loads carried eccentrically on column brackets, or in arches when the arch axis does not coincide with the pressure line. Even when design calculations show a member to be loaded purely axially, inevitable imperfections of construction will introduce eccentricities and consequent bending in the member as built. For this reason members that must be designed for simultaneous compression and bending are very frequent in almost all types of concrete structures.

When a member is subjected to combined axial compression P and moment M, such as in Fig. 8.6a, it is usually convenient to replace the axial load and moment with an equal load P applied at eccentricity $e = M/P$, as in Fig. 8.6b. The two loadings are statically equivalent. All columns may then be classified in

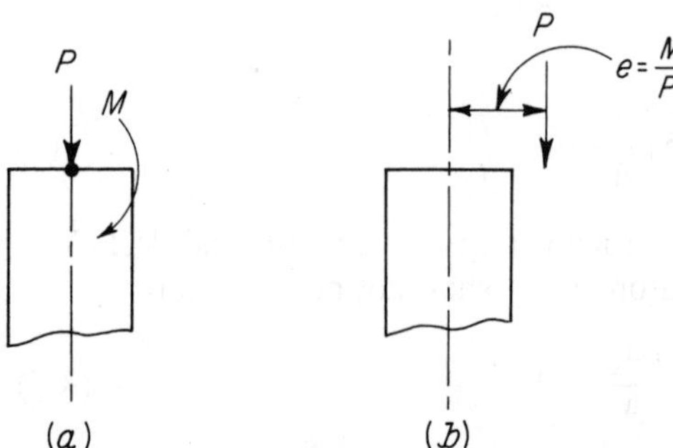

FIGURE 8.6
Equivalent eccentricity of column load.

terms of the equivalent eccentricity. Those having relatively small e are generally characterized by compression over the entire concrete section, and if overloaded will fail by crushing of the concrete accompanied by yielding of the steel in compression on the more heavily loaded side. Columns with large eccentricity are subject to tension over at least a part of the section, and if overloaded may fail due to tensile yielding of the steel on the side farthest from the load.

For columns, load stages below the ultimate are generally not important. Cracking of concrete, even for columns with large eccentricity, is usually not a serious problem, and lateral deflections at service load levels are seldom, if ever, a factor. Design of columns is therefore based on the factored overload stage, for which the required strength must not exceed the design strength as usual, i.e.,

$$\phi M_n \geq M_u \tag{8.6a}$$

$$\phi P_n \geq P_u \tag{8.6b}$$

8.4 STRAIN COMPATIBILITY ANALYSIS AND INTERACTION DIAGRAMS

Figure 8.7a shows a member loaded parallel to its axis by a compressive force P_n, at an eccentricity e measured from the centerline. The distribution of strains at a section a-a along its length, at incipient failure, is shown in Fig. 8.7b. With plane sections assumed to remain plane, concrete strains vary linearly with distance from the neutral axis, which is located a distance c from the more heavily loaded side

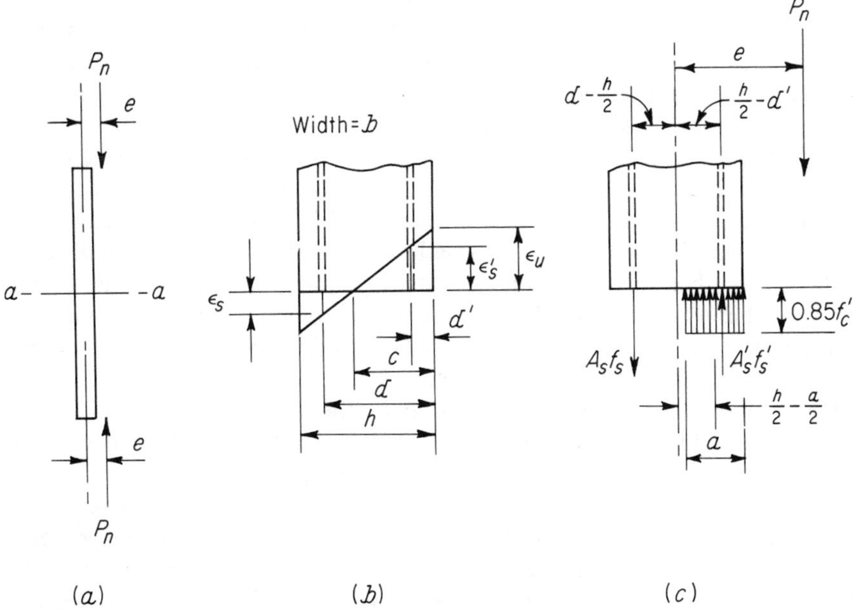

FIGURE 8.7
Column subject to eccentric compression: (a) loaded column; (b) strain distribution at section a-a; (c) stresses and forces at nominal ultimate strength.

of the member. With full compatibility of deformations, the steel strains at any location are the same as the strains in the adjacent concrete; thus if the ultimate concrete strain is ϵ_u, the strain in the bars nearest the load is ϵ_s' while that in the tension bars at the far side is ϵ_s. Compression steel having area A_s' and tension steel with area A_s are located at distances d' and d respectively from the compression face.

The corresponding stresses and forces are shown in Fig. 8.7*c*. Just as for simple bending, the actual concrete compressive stress distribution is replaced by an equivalent rectangular distribution having depth $a = \beta_1 c$. A large number of tests on columns of a variety of shapes have shown that the ultimate strengths computed on this basis are in satisfactory agreement with test results (Ref. 8.6).

Equilibrium between external and internal axial forces shown in Fig. 8.7*c* requires that

$$P_n = 0.85 f_c' ab + A_s' f_s' - A_s f_s \tag{8.7}$$

Also, the moment about the centerline of the section of the internal stresses and forces must be equal and opposite to the moment of the external force P_n, so that

$$M_n = P_n e = 0.85 f_c' ab \left(\frac{h}{2} - \frac{a}{2} \right) + A_s' f_s' \left(\frac{h}{2} - d' \right) + A_s f_s \left(d - \frac{h}{2} \right) \tag{8.8}$$

These are the two basic equilibrium relations for rectangular eccentrically compressed members.

The fact that the presence of the compression reinforcement A_s' has displaced a corresponding amount of concrete of area A_s' is neglected in writing these equations. If necessary, particularly for large steel ratios, one can account for this very simply. Evidently, in the above equations a nonexistent concrete compression force of amount $A_s'(0.85 f_c')$ has been included as acting in the displaced concrete at the level of the compression steel. This excess force can be removed in both equations by multiplying A_s' by $f_s' - 0.85 f_c'$ rather than by f_s'.

For large eccentricities, failure is initiated by yielding of the tension steel A_s. Hence, for this case, $f_s = f_y$. When the concrete reaches its ultimate strain ϵ_u, the compression steel may or may not have yielded; this must be determined based on compatibility of strains. For small eccentricities the concrete will reach its limit strain ϵ_u before the tension steel starts yielding; in fact, the bars on the side of the column farther from the load may be in compression, not tension. For small eccentricities, too, the analysis must be based on compatibility of strains between the steel and the adjacent concrete.

For a given eccentricity determined from the frame analysis (i.e., $e = M_u/P_u$) it is possible to solve Eqs. (8.7) and (8.8) for the load P_n and moment M_n that would result in failure as follows. In both equations, f_s', f_s, and a can be expressed in terms of a single unknown c, the distance to the neutral axis. This is easily done based on the geometry of the strain diagram, with ϵ_u taken equal to 0.003 as usual, and using the stress-strain curve of the reinforcement. The result is that the two equations contain only two unknowns, P_n and c, and can be solved for those values simultaneously. However, to do so in practice would be complicated algebraically, particularly because of the need to incorporate the limit f_y on both f_s' and f_s.

A better approach, providing the basis for practical design, is to construct a *strength interaction diagram* defining the failure load and failure moment for a given column for the full range of eccentricities from zero to infinity. For any eccentricity, there is a unique pair of values of P_n and M_n that will produce the state of incipient failure. That pair of values can be plotted as a point on a graph relating P_n and M_n, such as shown in Fig. 8.8. A series of such calculations, each corresponding to a different eccentricity, will result in a curve having a shape typically as shown in Fig. 8.8. On such a diagram, any radial line represents a particular eccentricity $e = M/P$. For that eccentricity, gradually increasing the load will define a load path as shown, and when that load path reaches the limit curve, failure will result. Note that the vertical axis corresponds to $e = 0$, and P_0 is the capacity of the column if concentrically loaded, as given by Eq. (8.3*b*). The horizontal axis corresponds to an infinite value of e, i.e., pure bending at moment capacity M_0. Small eccentricities will produce failure governed by concrete compression, while large eccentricities give a failure triggered by yielding of the tension steel.

For a given column, selected for trial, the interaction diagram is most easily constructed by selecting successive choices of neutral axis distance c, from infinity (axial load with eccentricity 0) to a very small value found by trial to give $P_n = 0$ (pure bending). For each selected value of c, the steel strains and stresses and the concrete force are easily calculated as follows. For the tension steel,

$$\epsilon_s = \epsilon_u \frac{d - c}{c} \tag{8.9}$$

$$f_s = \epsilon_u E_s \frac{d - c}{c} \quad \text{and} \quad \leq f_y \tag{8.10}$$

while for the compression steel,

$$\epsilon_s' = \epsilon_u \frac{c - d'}{c} \tag{8.11}$$

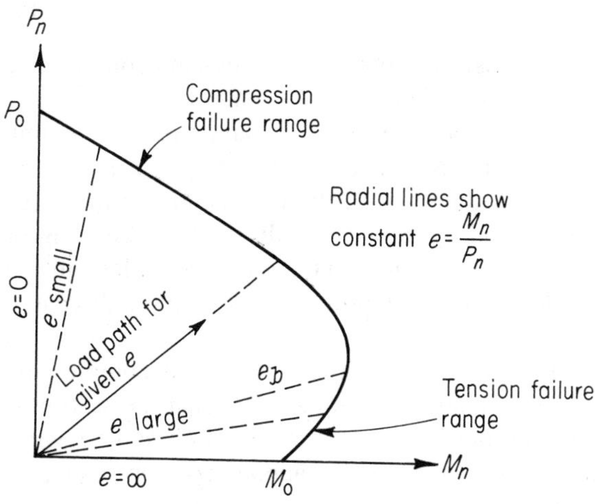

FIGURE 8.8
Interaction diagram for nominal column strength in combined bending and axial load.

$$f_s' = \epsilon_u E_s \frac{c - d'}{c} \qquad \text{and} \qquad \leq f_y \tag{8.12}$$

The concrete stress block has depth

$$a = \beta_1 c \qquad \text{and} \leq h \tag{8.13}$$

and consequently the concrete compressive resultant is

$$C = 0.85 f_c' ab \tag{8.14}$$

The axial force P_n and moment M_n corresponding to the selected neutral axis location can then be calculated from Eqs. (8.7) and (8.8) respectively, and thus a single point on the strength interaction diagram is established. The calculations are then repeated for successive choices of neutral axis to establish the curve defining the strength limits, such as Fig. 8.8. The calculations, of a repetitive nature, are easily programmed for the computer.

8.5 BALANCED FAILURE

As already noted, the failure line is divided into a compression failure range and a tension failure range. It is useful to define what is termed a *balanced failure mode* and corresponding eccentricity e_b with the load P_b and moment M_b acting in combination, to produce a failure with the concrete reaching its limit strain ϵ_u at precisely the same instant that the tensile steel on the far side of the column reaches yield strain. This point on the interaction diagram is the dividing point between compression failure (small eccentricities) and tension failure (large eccentricities).

The values of P_b and M_b are easily computed with reference to Fig. 8.7. For balanced failure,

$$c = c_b = d \frac{\epsilon_u}{\epsilon_u + \epsilon_y} \tag{8.15}$$

and

$$a = a_b = \beta_1 c_b \tag{8.16}$$

Equations (8.9) through (8.14) are then used to obtain steel stress and compressive resultant, after which P_b and M_b are found from Eqs. (8.7) and (8.8.)

It is to be noted that, in contrast to beam design, one cannot restrict column designs such that yielding failure rather than crushing failure would always be the result of overloading. The type of failure for a column depends on the value of eccentricity e, which in turn is defined by the load analysis of the building or other structure. However, the balanced failure point on the interaction diagram is a useful point of reference in connection with safety provisions, as will be discussed further in Sec. 8.9.

It is important to observe, in Fig. 8.8, that in the region of compression failure the larger the axial load P_n, the smaller the moment M_n that the section is able to sustain before failing. However, in the region of tension failure the reverse is true; the larger the axial load, the larger the simultaneous moment capacity.

This is easily understood. In the compression failure region failure occurs through overstraining of the concrete. The larger the concrete compressive strain caused by the axial load alone, the smaller the margin of additional strain available for the added compression caused by bending. On the other hand, in the tension failure region, yielding of the steel initiates failure. If the member is loaded in simple bending to the point at which yielding begins in the tension steel, and if an axial compression load is then added, the steel compressive stresses caused by this load will superimpose on the previous tensile stresses. This reduces the total steel stress to a value below its yield strength. Consequently, an additional moment can now be sustained of such magnitude that the combination of the steel stress from the axial load and the increased moment again reaches the yield strength.

The typical shape of a column interaction diagram shown in Fig. 8.8 has important design implications. In the range of tension failure, a *reduction of thrust* may produce failure for a given moment. In carrying out a frame analysis, the designer must consider all combinations of loading that may occur, including that which would produce minimum axial load paired with a given moment. Only that amount of compression that is certain to be present should be utilized in calculating the capacity of a column subject to a given moment.

Example 8.1 Column strength interaction diagram. A 12 × 20 in. column is reinforced with four No. 9 bars of area 1.0 in.^{2} each, one in each corner as shown in Fig. 8.9*a*. The concrete cylinder strength is $f_c' = 3500$ psi and the steel yield strength is 50 ksi. Determine (*a*) the load P_b, moment M_b, and corresponding eccentricity e_b for balanced failure; (*b*) the load and moment for a representative point in the tension failure region of the interaction curve; (*c*) the load and moment for a representative point in the compression failure region; (*d*) the axial load strength for zero eccentricity. Then (*e*) sketch the strength interaction diagram for this column. Finally, (*f*) design the lateral reinforcement, based on ACI Code provisions.

Solution

(*a*) The neutral axis for the balanced failure condition is easily found from Eq. (8.15) with $\epsilon_u = 0.003$ and $\epsilon_y = 50/29{,}000 = 0.0017$:

$$c_b = 17.5 \times \frac{0.003}{0.0047} = 11.1 \text{ in.}$$

giving a stress-block depth $a = 0.85 \times 11.1 = 9.44$ in. For the balanced failure condition, by definition, $f_s = f_y$. The compressive steel stress is found from Eq. (8.12):

$$f_s' = 0.003 \times 29{,}000 \frac{11.1 - 2.5}{11.1} = 67.4 \text{ ksi} \qquad \text{but} \qquad \leq 50 \text{ ksi}$$

confirming that the compression steel, too, is at the yield. The concrete compressive resultant is

$$C = 0.85 \times 3.5 \times 9.44 \times 12 = 337 \text{ kips}$$

The balanced load P_b is then found from Eq. (8.7) to be

$$P_b = 337 + 2.0 \times 50 - 2.0 \times 50 = 337 \text{ kips}$$

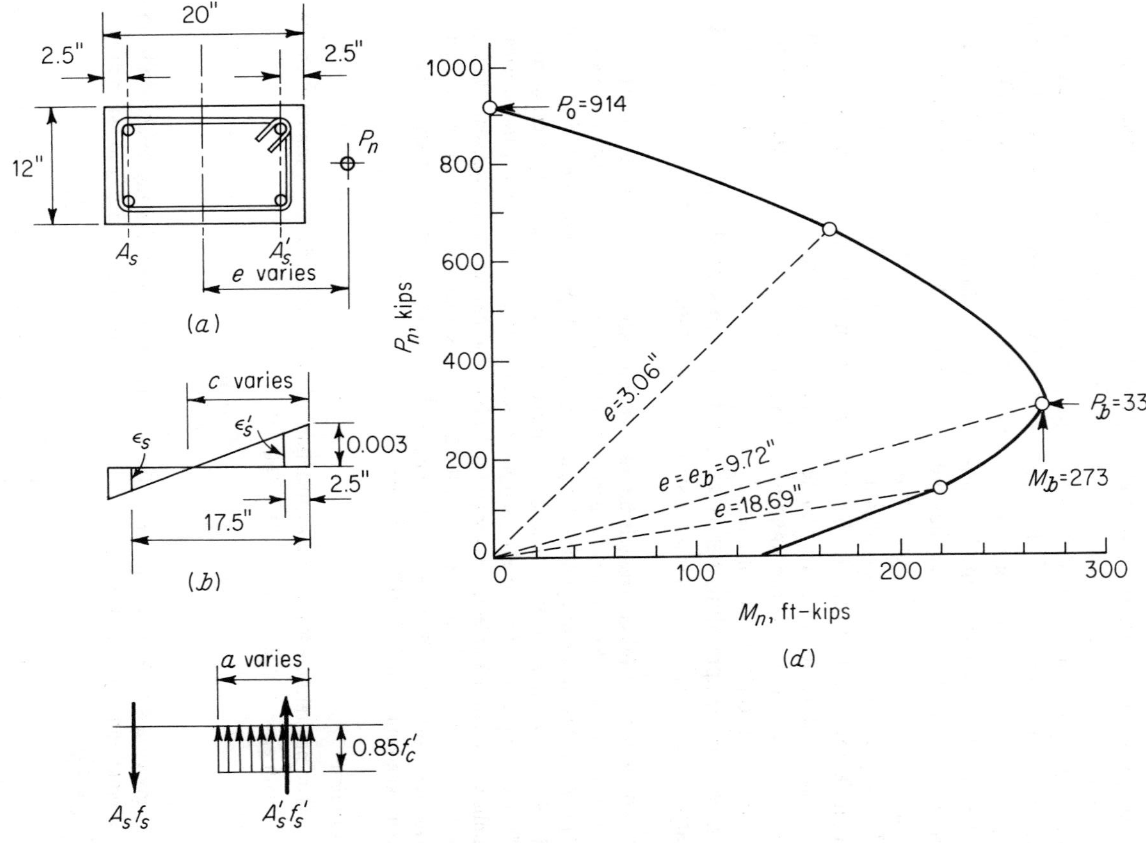

FIGURE 8.9
Column interaction diagram for Example 8.1: (*a*) cross section; (*b*) strain distribution; (*c*) stresses and forces; (*d*) strength interaction diagram.

and the balanced moment from Eq. (8.8) is

$$M_b = 337(10 - 4.72) + 2.0 \times 50(10 - 2.5) + 2.0 \times 50(17.5 - 10)$$

$$= 3280 \text{ in-kips} = 273 \text{ ft-kips}$$

The corresponding eccentricity of load is $e_b = 9.72$ in.

(*b*) Any choice of c *smaller than* $c_b = 11.1$ in. will give a point in the tension failure region of the interaction curve, with eccentricity larger than e_b. For example, choose $c = 5.0$ in. By definition, $f_s = f_y$. The compressive steel stress is found to be

$$f_s' = 0.003 \times 29{,}000 \frac{5.0 - 2.5}{5.0} = 43.5 \text{ ksi}$$

With the stress-block depth $a = 0.85 \times 5.0 = 4.25$, the compressive resultant is $C = 0.85 \times 3.5 \times 4.25 \times 12 = 152$ kips. Then from Eq. (8.7) the thrust is

$$P_n = 152 + 2.0 \times 43.5 - 2.0 \times 50 = 139 \text{ kips}$$

and the moment capacity from Eq. (8.8) is

$$M_n = 152(10 - 2.12) + 2.0 \times 43.5(10 - 2.5) + 2.0 \times 50(17.5 - 10)$$

$$= 2598 \text{ in-kips} = 217 \text{ ft-kips}$$

giving eccentricity $e = 2598/139 = 18.69$ in., well above the balanced value.

(*c*) Now selecting a c value *larger than* c_b to demonstrate a compression failure point on the interaction curve, choose $c = 18.0$ in., for which $a = 0.85 \times 18.0 = 15.3$ in. The compressive concrete resultant is $C = 0.85 \times 3.5 \times 15.3 \times 12 = 546$ kips. From Eq. (8.10) the stress in the steel at the left side of the column is

$$f_s = 0.003 \times 29{,}000 \frac{17.5 - 18.0}{18.0} = -2 \text{ ksi}$$

Note that the negative value of f_s indicates correctly that A_s is in compression if c is greater than d, as in the present case. The compressive steel stress is found from Eq. (8.12) to be

$$f_s' = 0.003 \times 29{,}000 \frac{18.0 - 2.5}{18.0} = 75 \text{ ksi} \qquad \text{but} \qquad \leq 50 \text{ ksi}$$

Then the column capacity is

$$P_n = 546 + 2.0 \times 50 + 2.0 \times 2 = 650 \text{ kips}$$

$$M_n = 546(10 - 7.65) + 2.0 \times 50(10 - 2.5) - 2.0 \times 2(17.5 - 10)$$

$$= 2000 \text{ in-kips} = 167 \text{ ft-kips}$$

(*d*) The axial strength of the column if concentrically loaded corresponds to $c = \infty$ and $e = 0$. For this case,

$$P_n = 0.85 \times 3.5 \times 12 \times 20 + 4.0 \times 50 = 914 \text{ kips}$$

Note that, for this as well as the preceding calculations, subtraction of the concrete displaced by the steel has been neglected. For comparison, if the deduction were made in the last calculation:

$$P_n = 0.85 \times 3.5(12 \times 20 - 4) + (4.0 \times 50) = 902 \text{ kips}$$

The error in neglecting this deduction is only 1 percent in this case; the difference generally can be neglected except perhaps for columns with steel ratios close to the maximum of 8 percent.

From the calculations just completed, plus similar repetitive calculations that will not be given here, the strength interaction curve of Fig. 8.9*d* is constructed. Note the characteristic shape, described earlier, the location of the balanced failure point as well as the "small eccentricity" and "large eccentricity" points just found, and the axial load capacity.

(*e*) The design of the lateral ties will be carried out following the ACI Code restrictions. For the minimum permitted tie diameter of $\frac{3}{8}$ in., used with No. 9 longitudinal bars having a diameter of $1\frac{1}{8}$ in., in a column the least dimension of which is 12 in., the tie spacing is not to exceed:

$$48 \times \frac{3}{8} = 18 \text{ in.}$$

$$16 \times \frac{9}{8} = 18 \text{ in.}$$

$$b = 12 \text{ in.}$$

The last restriction controls in this case, and No. 3 ties will be used at 12 in. spacing, detailed as shown in Fig. 8.9*a*. Note that the permitted spacing as controlled by the first and second criteria, 18 in., must be reduced because of the 12 in. column dimension, indicating that a saving in tie steel could be realized using a smaller tie diameter; however, this would not meet the ACI Code restriction on the minimum tie diameter in this case.

8.6 DISTRIBUTED REINFORCEMENT

When large bending moments are present, it is most economical to concentrate all or most of the steel along the outer faces parallel to the axis of bending. Such arrangements are shown in Fig. 8.1*e* to *h*. On the other hand, with small eccentricities so that the axial compression is prevalent, and when a small cross section is desired, it is often advantageous to place the steel more uniformly around the perimeter, as in Fig. 8.1*a* to *d*. In this case, special attention must be paid to the intermediate bars, i.e., those that are not placed along the two faces that are most highly stressed. This is so because when the ultimate load is reached, the stresses in these intermediate bars are usually below the yield point, even though the bars along one or both extreme faces may be yielding. This situation can be analyzed by a simple and obvious extension of the previous analysis based on compatibility of strains. A strength interaction diagram may be constructed just as before. A sequence of choices of neutral axis location results in a set of paired values of P_n and M_n, each corresponding to a particular eccentricity of load.

Example 8.2 Analysis of eccentric column with distributed reinforcement. The column of Fig. 8.10*a* is reinforced with ten No. 11 bars distributed around the perimeter as shown. Load P_n will be applied with eccentricity e about the strong axis. Material strengths are $f'_c = 6000$ psi and $f_y = 75$ ksi. Find the load and moment corresponding to a failure point with neutral axis $c = 18$ in. from the right face.

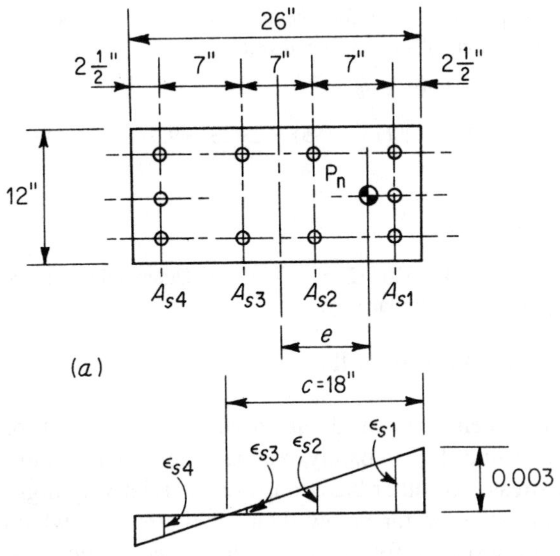

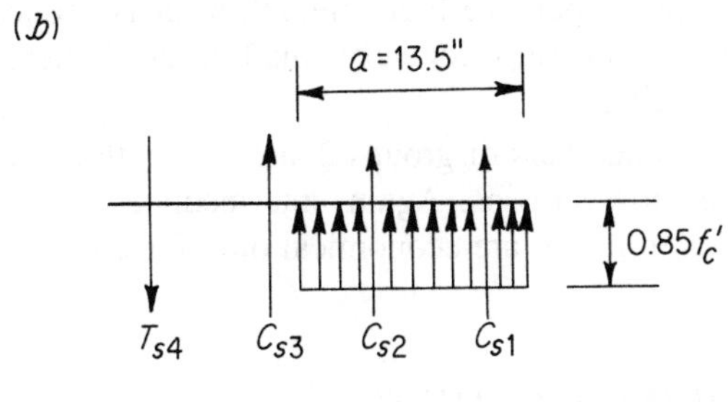

FIGURE 8.10
Column of Example 8.2: (*a*) cross section; (*b*) strain distribution; (*c*) stresses and forces.

Solution. When the concrete reaches its limit strain of 0.003, the strain distribution is that shown in Fig. 8.10*b*, the strains at the locations of the four bar groups are found from similar triangles, after which the stresses are found multiplying strains by $E_s = 29{,}000$ ksi applying the limit value f_y:

$$\begin{aligned}
\epsilon_{s1} &= 0.00258 & f_{s1} &= 75.0 \text{ ksi compression} \\
\epsilon_{s2} &= 0.00142 & f_{s2} &= 41.2 \text{ ksi compression} \\
\epsilon_{s3} &= 0.00025 & f_{s3} &= 7.3 \text{ ksi compression} \\
\epsilon_{s4} &= 0.00091 & f_{s4} &= 26.4 \text{ ksi tension}
\end{aligned}$$

For $f_c' = 6000$ psi, $\beta_1 = 0.75$ and the depth of the equivalent rectangular stress block is $a = 0.75 \times 18 = 13.5$ in. The concrete compressive resultant is $C = 0.85 \times 6.0 \times 13.5 \times 12 = 826$ kips, and the respective steel forces in Fig. 8.10*c* are:

$$\begin{aligned}
C_{s1} &= 4.68 \times 75.0 = 351 \text{ kips} \\
C_{s2} &= 3.12 \times 41.2 = 129 \text{ kips} \\
C_{s3} &= 3.12 \times 7.3 = 23 \text{ kips} \\
T_{s4} &= 4.68 \times 26.4 = 124 \text{ kips}
\end{aligned}$$

The thrust and moment that would produce failure for a neutral axis 18 in. from the right face are found by the obvious extensions of Eq. (8.7) and (8.8):

$$P_n = 826 + 351 + 129 + 23 - 124 = 1205 \text{ kips}$$

$$\begin{aligned} M_n &= 826(13 - 6.75) + 351(13 - 2.5) + 129(13 - 9.5) - 23(13 - 9.5) + 124(13 - 2.5) \\ &= 10{,}520 \text{ in-kips} \\ &= 877 \text{ ft-kips} \end{aligned}$$

The corresponding eccentricity is $e = 10,520/1205 = 8.73$ in. Other points on the interaction diagram can be computed in a similar way.

Two general conclusions can be made from this example:

1. Even with the relatively small eccentricity of about one-third of the depth of the section, only the bars of group 1 just barely reached their yield strain, and consequently their yield stress. All other bar groups of the relatively high-strength steel that was used are stressed far below their yield strength, which would also have been true for group 1 for a slightly larger eccentricity. It follows that the use of the more expensive high-strength steel is economical in symmetrically reinforced columns only for very small eccentricities, e.g., in the lower stories of tall buildings.
2. The contribution of the intermediate bars of groups 2 and 3 to both P_n and M_n is quite small because of their low stresses. Again, intermediate bars, except as they are needed to hold ties in place, are economical only for columns with very small eccentricities.

8.7 UNSYMMETRICAL REINFORCEMENT

Most reinforced concrete columns are symmetrically reinforced about the axis of bending. However, for some cases, such as the columns of rigid portal frames in which the moments are uniaxial and the eccentricity large, it is more economical to use an unsymmetrical pattern of bars, with most of the bars on the tension side such as shown in Fig. 8.11. Such columns can be analyzed by the same strain compatibility approach as described above. However, for an unsymmetrically reinforced column to be loaded concentrically the load must pass through a point known as the *plastic centroid.* The plastic centroid is defined as the point of application of the resultant force for the column cross section (including concrete and steel forces) if the column is compressed uniformly to the failure strain $\epsilon_u = 0.003$ over its entire cross section. Eccentricity of the applied load must be measured with respect to the plastic centroid, because only then will $e = 0$ correspond to the axial thrust with no moment. The location of the plastic centroid for the column of Fig. 8.11 is the resultant of the three internal forces to be accounted for. Its distance from the left face is

$$x = \frac{0.85 f_c' bh^2/2 + A_s f_y d + A_s' f_y d'}{0.85 f_c' bh + A_s f_y + A_s' f_y} \tag{8.17}$$

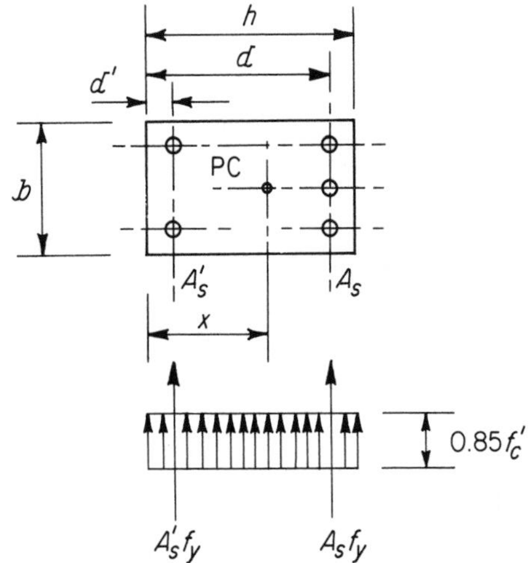

FIGURE 8.11
Plastic centroid of an unsymmetrically reinforced column.

Clearly, in a symmetrically reinforced cross section, the plastic centroid and the geometric center coincide.

8.8 CIRCULAR COLUMNS

It was mentioned in Sec. 8.2 that when load eccentricities are small, spirally reinforced columns show greater toughness, i.e., greater ductility, than tied columns, although this difference fades out as the eccentricity is increased. For this reason, as discussed in Sec. 8.2, the ACI Code provides a more favorable reduction factor $\phi = 0.75$ for spiral columns, compared with $\phi = 0.70$ for tied columns. Also, the maximum stipulated design load for entirely or nearly axially loaded members is larger for spiral-reinforced members than for comparable tied members (see Sec. 8.9). It follows that spirally reinforced columns permit a somewhat more economical utilization of the materials, particularly for small calculated eccentricities. Further advantages lie in the facts that spirals are available prefabricated, which may save labor in assembling column cages, and that the circular shape is frequently desired by the architect.

Figure 8.12 shows the cross section of a spiral-reinforced column. From six to ten and more longitudinal bars of equal size are provided for longitudinal reinforcement, depending on column diameter. The strain distribution at the instant at which the ultimate load is reached is shown in Fig. 8.12*b*. Bar groups 2 and 3 are seen to be strained to much smaller values than groups 1 and 4. The stresses in the four bar groups are easily found. For any of the bars with strains in excess of yield strain $\epsilon_y = f_y/E_s$, the stress at failure is evidently the yield stress of the bar. For bars with smaller strains, the stress is found from $f_s = \epsilon_s E_s$.

One then has the internal forces shown in Fig. 8.12*c*. They must be in force and moment equilibrium with the nominal strength P_n. It will be noted that the situation is analogous to that discussed in Secs. 8.4 to 8.6 for rectangular columns. Calculations can be carried out exactly as in Example 8.1 except that for circular

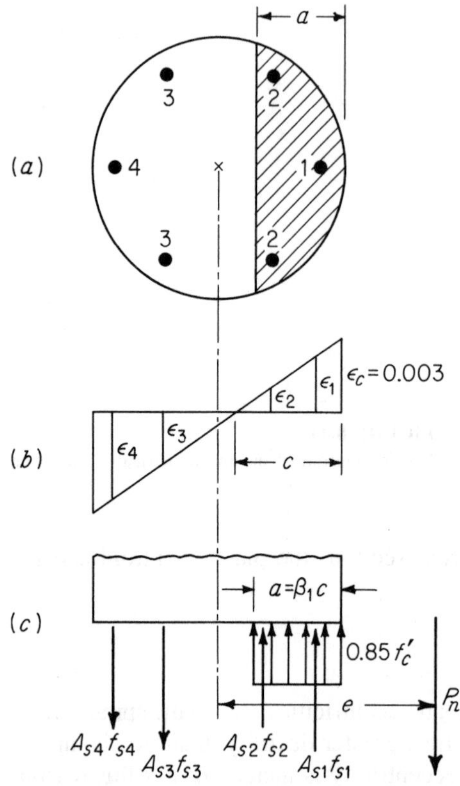

FIGURE 8.12
Circular column with compression plus bending.

columns the concrete compression zone subject to the equivalent rectangular stress distribution has the shape of a segment of a circle shown shaded in Fig. 8.12*a*.

Although the shape of the compression zone and the strain variation in the different groups of bars make longhand calculations awkward, no new principles are involved and computer solutions are easily developed.

Design or analysis of spirally reinforced columns is usually carried out by means of design aids, such as Graphs A.13 to A.16 of App. A. More elaborate and detailed tables and graphs are available, e.g., in Ref. 8.7. In developing such design aids, the entire steel area is assumed to be arranged in a uniform, concentric ring, rather than being concentrated in the actual bar locations; this simplifies calculations without noticeably affecting results.

It should be noted that in order to qualify for the more favorable safety provisions for spiral columns, the steel ratio of the spiral must be at least equal to that given by Eq. (8.5) for reasons discussed in Sec. 8.2.

8.9 ACI CODE SAFETY PROVISIONS

For columns, as for all members designed according to the ACI Code, adequate safety margins are established by applying overload factors to the service loads and strength reduction factors to the nominal ultimate strengths. Thus, for columns, $\phi P_n \geq P_u$ and $\phi M_n \geq M_u$ are the basic safety criteria. For members subject to axial compression or compression plus flexure, the ACI code provides basic reduction factors:

$$\phi = 0.70 \text{ for tied columns}$$

$$\phi = 0.75 \text{ for spirally reinforced columns}$$

The spread between these two values reflects the added safety furnished by the greater toughness of spirally reinforced columns.

There are various reasons why the ϕ values for columns are considerably lower than those for flexure or shear (0.90 and 0.85 respectively). One is that the strength of underreinforced flexural members is not much affected by variations in concrete strength, since it depends primarily on the yield strength of the steel, while the strength of axially loaded members depends strongly on the concrete compressive strength. Because the cylinder strength of concrete under site conditions is less closely controlled than the yield strength of mill-produced steel, a larger occasional strength deficiency must be allowed for. This is particularly true for columns, in which concrete, being placed from the top down in the long, narrow form, is more subject to segregation than in horizontally cast beams. Moreover, electrical and other conduits are frequently located in building columns; this reduces their effective cross sections, often to an extent unknown to the designer, even though this is poor practice and restricted by the ACI Code. Finally, the consequences of a column failure, say in a lower story, would be more catastrophic than that of a single beam in a floor system in the same building.

In addition to the basic ϕ factors for tied and spirally reinforced columns, stated above, the ACI Code includes special provisions in the ranges of very high and very low eccentricities.

In the tension failure range with eccentricities from e_b to infinity (pure bending), tension governs—the more so the smaller the axial force. Finally, when the axial force is zero, the member becomes an ordinary beam subject to pure bending. The ACI Code prescribes that beams must be designed so that they are underreinforced; tension governs and $\phi = 0.90$. It follows that some transition is in order from the ϕ values of 0.70 or 0.75 for columns, when compression governs, to $\phi = 0.90$ for beams without axial load, when tension governs. ACI Code 9.3.2 therefore stipulates that ϕ may be increased linearly to 0.90 as ϕP_n decreases from $0.10 f'_c A_g$ or ϕP_b, whichever is smaller, to zero. The value $0.10 f'_c A_g$ is an approximation that avoids the calculation of ϕP_b, which is usually somewhat larger.

At the other extreme, for columns with very small or zero calculated eccentricities, the ACI Code recognizes that accidental construction misalignments and other unforeseen factors may produce actual eccentricities in excess of these small design values. Also, the concrete strength under high, sustained axial loads may be somewhat smaller than the short-term cylinder strength. Therefore, regardless of the magnitude of the calculated eccentricity, ACI Code 10.3.5 limits the maximum design strength to $0.80\phi P_0$ for tied columns (with $\phi = 0.70$) and to $0.85\phi P_0$ for spirally reinforced columns (with $\phi = 0.75$). Here P_0 is the nominal strength of the axially loaded column with zero eccentricity [see Eq. (8.4)].

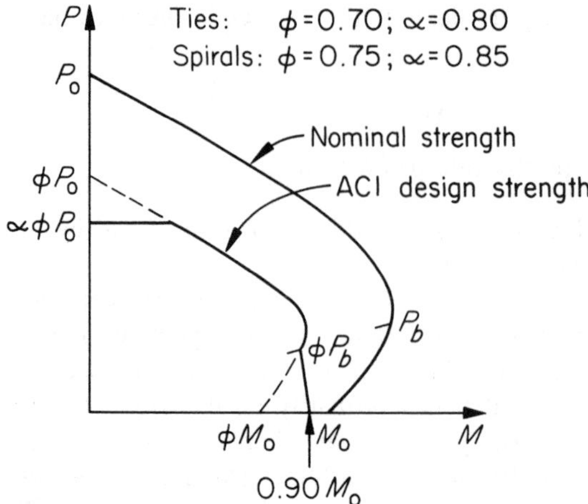

FIGURE 8.13
ACI safety provisions superimposed on column strength interaction diagram.

The effects of the safety provisions of the ACI Code are shown in Fig. 8.13. The solid curve labeled "nominal strength" is the same as Fig. 8.8 and represents the actual carrying capacity, as nearly as can be predicted. The smooth curve shown partially dashed, then solid, then dashed, represents the basic design strength obtained by reducing the nominal strengths P_n and M_n, for each eccentricity, by $\phi = 0.70$ for tied columns and $\phi = 0.75$ for spiral columns. The horizontal cutoff at $\alpha\phi P_0$ represents the maximum design load stipulated in the ACI Code for small eccentricities, i.e., large axial loads, as just discussed. At the other end, for large eccentricities, i.e., small axial loads, the ACI Code permits a linear transition of ϕ from 0.70 or 0.75, applicable at the lower of ϕP_b or $0.10 f_c' A_g$, to 0.90 at $P = 0$. This transition is shown by the solid line at the lower right end of the design strength curve.†

8.10 DESIGN AIDS

The design of eccentrically loaded columns using the strain compatibility method of analysis described requires that a trial column be selected. The trial column is then investigated to determine if it is adequate to carry any combination of Pu and M_u that may act on it should the structure be overloaded, i.e., to see if P_u and M_u from the analysis of the structure, when plotted on a strength interaction diagram such as Fig. 8.13, fall within the region bounded by the curve labeled "ACI design strength." Furthermore, economical design requires that the controlling combination of P_u and M_u be close to the limit curve. If these conditions are not met, a new column must be selected for trial.

† While the general intent of the ACI Code safety provisions relating to eccentric columns is clear and fundamentally sound, the end result is a set of strangely shaped column design charts following no discernible physical law, as is demonstrated by Graphs A.5 to A.16 of App. A. Improved column safety provisions, resulting in a smooth design curve appropriately related to the strength curve, would be simpler to use and more rational as well.

While a simple computer program can be written, based on the strain compatibility analysis, to calculate points on the design strength curve, and even to plot the curve, for any trial column, in practice design aids are used such as are available in handbooks and special volumes published by the American Concrete Institute (Ref. 8.7) and the Concrete Reinforcing Steel Institute (Ref. 8.8). They cover the most frequent practical cases, such as symmetrically reinforced rectangular and square columns and circular spirally reinforced columns.

Graphs A.5 through A.16 of App. A. are representative of the column design charts found in Ref. 8.7, in this case for concrete with $f'_c = 4000$ psi and steel of yield strength $f_y = 60$ ksi, for varying cover distances. In Ref. 8.7 are included such charts for a broad range of material strength. Graphs A.5 through A.8 are drawn for rectangular columns with reinforcement distributed around the column perimeter; Graphs A.9 through A.12 are for rectangular columns with reinforcement along two opposite faces. Circular columns with bars in a circular pattern are shown in Graphs A.13 through A.16.

The graphs are seen to consist of strength interaction curves of the type shown in Fig. 8.12 and labeled "ACI design strength," i.e., the ACI safety provisions are incorporated. However, instead of plotting ϕP_n vs. ϕM_n, corresponding parameters have been used to make the charts more generally applicable, i.e., load is plotted as $\phi P_n/A_g$ while moment is expressed as $(\phi P_n/A_g)(e/h)$. Families of curves are drawn for various values of $\rho_g = A_{st}/A_g$. They are used in most cases in conjunction with the family of radial lines representing different eccentricity ratios e/h.

Charts such as these permit the direct design of eccentrically loaded columns throughout the common range of strength and geometric variables. They may be used in one of two ways as follows. For a given factored load P_u and equivalent eccentricity $e = M_u/P_u$:

1. (*a*) Select trial cross section dimensions b and h (refer to Fig. 8.7).
(*b*) Calculate the ratio γ based on required cover distances to the bar centroids, and select the corresponding column design chart.
(*c*) Calculate P_u/A_g and $M_u/A_g h$, where $A_g = bh$.
(*d*) From the graph, for the values found in (*c*), read the required steel ratio ρ_g.
(*e*) Calculate the total steel area $A_{st} = \rho_g bh$.

2. (*a*) Select the steel ratio ρ_g.
(*b*) Choose a trial value of h and calculate e/h and γ.
(*c*) From the corresponding graph, read P_u/A_g and calculate the required A_g.
(*d*) Calculate $b = A_g/h$.
(*e*) Revise the trial value of h if necessary to obtain a well-proportioned section.
(*f*) Calculate the total steel area $A_{st} = \rho_g bh$.

Use of the column design charts will be illustrated in Examples 8.3 and 8.4.

Other design aids pertaining to lateral ties and spirals, as well as recommendations for standard practice, will be found in Refs. 8.7 through 8.9.

Example 8.3 Selection of reinforcement for column of given size. In a two-story structure an exterior column is to be designed for a service dead load of 142 kips, maximum live load of 213 kips, dead load moment of 83 ft-kips, and live load moment of 124 ft-kips. The minimum live load compatible with the full live load moment is 106 kips, obtained when no live load is placed on the roof but a full live load is placed on the second floor. Architectural considerations require that a rectangular column be used, with dimensions $b = 16$ in. and $h = 20$ in.
(*a*) Find the required column reinforcement for the condition that the full live load acts.
(*b*) Check to ensure that the column is adequate for the condition of no live load on the roof.
Material strengths are $f_c' = 4000$ psi and $f_y = 60{,}000$ psi.

Solution

(*a*) The column will be designed initially for full load, then checked for adequacy when live load is partially removed. According to the ACI safety provisions, the column must be designed for a factored load $P_u = 1.4 \times 142 + 1.7 \times 213 = 561$ kips and a factored moment $M_u = 1.4 \times 83 + 1.7 \times 124 = 327$ ft-kips. A column 16 × 20 in. is specified, and reinforcement distributed around the column perimeter will be used. Bar cover is estimated to be 2.5 in. from the column face to the steel centerline for each bar. The column parameters (assuming bending about the strong axis) are

$$\frac{P_u}{A_g} = \frac{561}{320} = 1.75 \text{ ksi}$$

$$\frac{M_u}{A_g h} = \frac{327 \times 12}{320 \times 20} = 0.61 \text{ ksi}$$

With 2.5 in. cover, the parameter $\gamma = (20 - 5)/20 = 0.75$. For this column geometry and material strengths, Graph A.7 of App. A applies. From that figure, with $\phi P_n/A_g = P_u/A_g = 1.75$ and $\phi M_n/A_g h = M_u/A_g h = 0.61$, $\rho_g = 0.039$. Thus the required reinforcement is $A_{st} = 0.039 \times 320 = 12.48$ in^2. Ten No. 10 bars will be used, four on each long face plus one intermediate bar on each short face, providing area $A_{st} = 12.66$ in^2. The steel ratio is well within the permissible range from 0.01 to 0.08 according to the ACI Code.

(*b*) With the roof live load absent, the column will carry a factored load $P_u = 1.4 \times 142 + 1.7 \times 106 = 379$ kips and factored moment $M_u = 327$ ft-kips, as before. Thus the column parameters for this condition are

$$\frac{P_u}{A_g} = \frac{379}{320} = 1.18 \text{ ksi}$$

$$\frac{M_u}{A_g h} = \frac{327 \times 12}{320 \times 20} = 0.61 \text{ ksi}$$

and $\gamma = 0.75$ as before. From Graph A.7 it is found that a steel ratio of $\rho_g = 0.032$ is sufficient for this condition, less than that required in part (*a*), so no modification is required.

Selecting No. 3 ties for trial, the maximum tie spacing must not exceed $48 \times 0.375 = 18$ in., $16 \times 1.27 = 20.3$ in., or 16 in. Spacing is controlled by the least column dimension here, and No. 3 ties will be used at 16 in. spacing, in the pattern shown in Fig. 8.1*b*.

Example 8.4 Selection of column size for a given reinforcement ratio. A column is to be designed to carry a factored load $P_u = 518$ kips and factored moment $M_u = 530$

ft-kips. Materials having strengths $f_y = 60{,}000$ psi and $f'_c = 4000$ psi are specified. Cost studies for the particular location indicate that a steel ratio ρ_g of about 0.03 is optimum. Find the required dimensions b and h of the column. Bending will be about the strong axis, and an arrangement of steel with bars concentrated in two layers, adjacent to the outer faces of the column and parallel to the axis of bending, will be used.

Solution. It is convenient to select a trial column dimension h, perpendicular to the axis of bending; a value of $h = 24$ in. will be selected, and assuming a concrete cover of 3 in. to the bar centers, the parameter $\gamma = 0.75$. Graph A.11 of App. A applies. For the stated loads the eccentricity is $e = 530 \times 12/518 = 12.3$ in., and $e/h = 12.3/24 = 0.51$. From Graph A.11 with $e/h = 0.51$ and $\rho_g = 0.03$, $\phi P_n/A_g = P_u/A_g = 1.35$. For the trial dimension $h = 24$ in., the required column width is

$$b = \frac{P_u}{1.35h} = \frac{518}{1.35 \times 24} = 15.98 \text{ in.}$$

A column 16 × 24 in. will be used, for which the required steel area is $A_{st} = 0.03 \times 16 \times 24 = 11.52 \text{ in}^2$. Eight No. 11 bars will be used, providing $A_{st} = 12.50$, arranged in two layers of four bars each, similar to the sketch shown in Graph A.11.

8.11 BIAXIAL BENDING

The methods discussed in the preceding sections permit rectangular or square columns to be designed if bending is present about only one of the principal axes. There are situations, by no means exceptional, in which axial compression is accompanied by simultaneous bending about both principal axes of the section. Such is the case, for instance, in corner columns of tier buildings where beams and girders frame into the columns in the directions of both walls and transfer their end moments into the columns in two perpendicular planes. Similar loading may occur at interior columns, particularly if the column layout is irregular.

The situation with respect to strength of biaxially loaded columns is shown in Fig. 8.14. Let X and Y denote the directions of the principal axes of the cross section. In Fig. 8.14*a* the section is shown subject to bending about the Y axis only, with load eccentricity e_x measured in the X direction. The corresponding strength interaction curve is shown as Case (*a*) in the three-dimensional sketch of Fig. 8.14*d* and is drawn in the plane defined by the axes P_n and M_{ny}. Such a curve can be established by the usual methods for uniaxial bending. Similarly, Fig. 8.14*d* shows bending about the X axis only, with eccentricity e_y measured in the Y direction. The corresponding interaction curve is shown as Case (*b*) in the plane of P_n and M_{nx} in Fig. 8.14*d*. For Case (*c*), which combines X and Y axis bending, the orientation of the resultant eccentricity is defined by the angle λ:

$$\lambda = \arctan \frac{e_x}{e_y} = \arctan \frac{M_{ny}}{M_{nx}}$$

Bending for this case is about an axis defined by the angle θ with respect to the X axis. The angle λ in Fig. 8.14*c* establishes a plane in Fig. 8.14*d*, passing through the vertical P_n axis and making an angle λ with the M_{nx} axis, as shown. In that plane, column strength is defined by the interaction curve labeled Case (*c*).

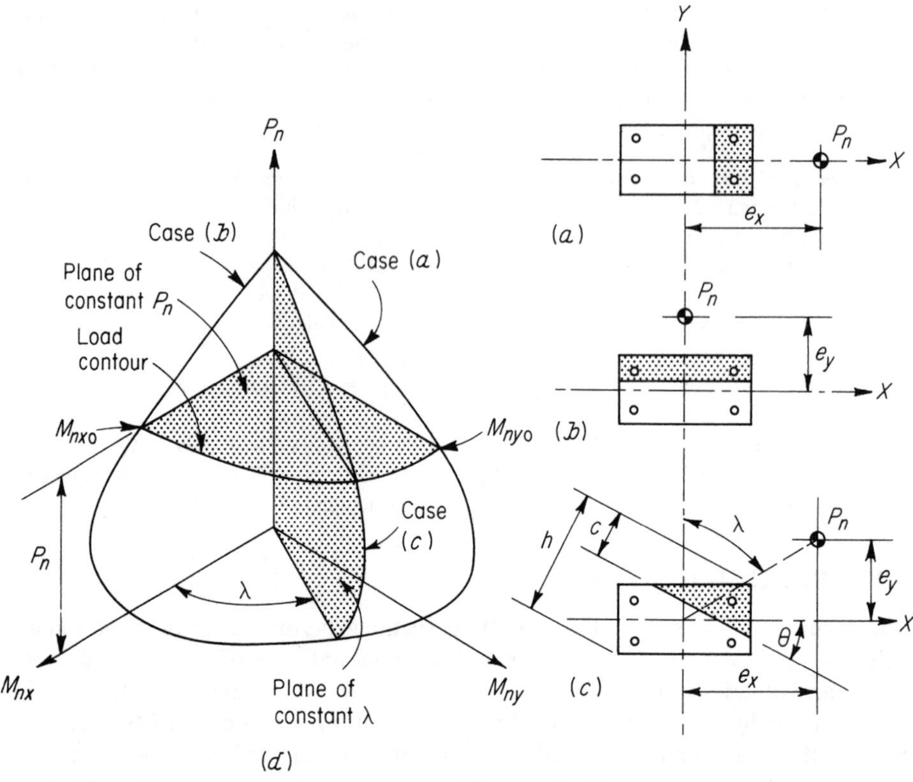

FIGURE 8.14
Interaction diagram for compression plus biaxial bending: (*a*) uniaxial bending about *Y* axis; (*b*) uniaxial bending about *X* axis; (*c*) biaxial bending about diagonal axis; (*d*) interaction surface.

For other values of λ similar curves are obtained to define a *failure surface* for axial load plus biaxial bending, such as shown in Fig. 8.14*d*. The surface is exactly analogous to the *failure line* for axial load plus uniaxial bending. Any combination of P_u, M_{ux}, and M_{uy} falling inside the surface can be applied safely, but any point falling outside the surface would represent failure. Note that the failure surface can be described either by a set of curves defined by radial planes passing through the P_n axis, such as shown by Case (*c*), or by a set of curves defined by horizontal plane intersections, each for a constant P_n, defining load contours.

Constructing such an interaction surface for a given column would appear to be an obvious extension of uniaxial bending analysis. In Fig. 8.14*c*, for a selected value of θ, successive choices of neutral axis distance *c* could be taken. For each, using strain compatibility and stress-strain relations to establish bar forces and the concrete compressive resultant, then using the equilibrium equations to find P_n, M_{nx}, and M_{ny}, one can determine a single point on the interaction surface. Repetitive calculations, easily done by computer, then establish sufficient points to define the surface. The triangular or trapezoidal compression zone, such as shown in Fig. 8.14*c*, is a complication, and in general the strain in each reinforcing bar will be different, but these features can be incorporated.

The main difficulty, however, is that the neutral axis will not, in general, be perpendicular to the resultant eccentricity, drawn from the column center to the load P_n. For each successive choice of neutral axis, there are unique values of P_n, M_{nx}, and M_{ny}, and only for special cases will the ratio of M_{ny}/M_{nx} be such that the eccentricity is perpendicular to the neutral axis chosen for the calculation. The result is that, for successive choices of c for any given θ, the value of λ in Fig. 8.14*c* and *d* will vary. Points on the failure surface established in this way will wander up the failure surface for increasing P_n, not representing a plane intersection, as shown for Case (*c*) in Fig. 8.14*d*.

In practice, the factored load P_u and the factored moments M_{ux} and M_{uy} to be resisted are known from the frame analysis of the structure. Therefore the actual value of $\lambda = \text{arc tan } (M_{uy}/M_{ux})$ is established, and one needs only the curve of Case (*c*), Fig. 8.14*d*, to test the adequacy of the trial column. An iterative computer method to establish the interaction line for the particular value of λ that applies will be described in Sec. 8.14.

Alternatively, simple approximate methods are widely used. These will be described in Secs. 8.12 and 8.13.

8.12 LOAD CONTOUR METHOD

The load contour method is based on representing the failure surface of Fig. 8.14*d* by a family of curves corresponding to constant values of P_n (Ref. 8.10). The general form of these curves can be approximated by a nondimensional interaction equation:

$$\left(\frac{M_{nx}}{M_{nx0}}\right)^{\alpha 1} + \left(\frac{M_{ny}}{M_{ny0}}\right)^{\alpha 2} = 1.0 \tag{8.18}$$

where

$$\begin{aligned} M_{nx} &= P_n e_y \\ M_{nx0} &= M_{nx} \qquad \text{when } M_{ny} = 0 \\ M_{ny} &= P_n e_x \\ M_{ny0} &= M_{ny} \qquad \text{when } M_{nx} = 0 \end{aligned}$$

and α_1 and α_2 are exponents depending on column dimensions, amount and distribution of steel reinforcement, stress-strain characteristics of steel and concrete, amount of concrete cover, and size of lateral ties or spiral. When $\alpha_1 = \alpha_2 = \alpha$, the shapes of such interaction contours are as shown in Fig. 8.15 for specific α values.

Introduction of the ACI ϕ factors for reducing nominal axial and flexural strengths to design strengths presents no difficulty. With the appropriate ϕ factors applied to P_n, M_{nx}, and M_{ny}, a new failure surface is defined, similar to the original but inside it. With the introduction of ϕ factors, and with $\alpha_1 = \alpha_2 = \alpha$, Eq. (12.18) becomes

$$\left(\frac{\phi M_{nx}}{\phi M_{nx0}}\right)^{\alpha} + \left(\frac{\phi M_{ny}}{\phi M_{ny0}}\right)^{\alpha} = 1.0 \tag{8.19}$$

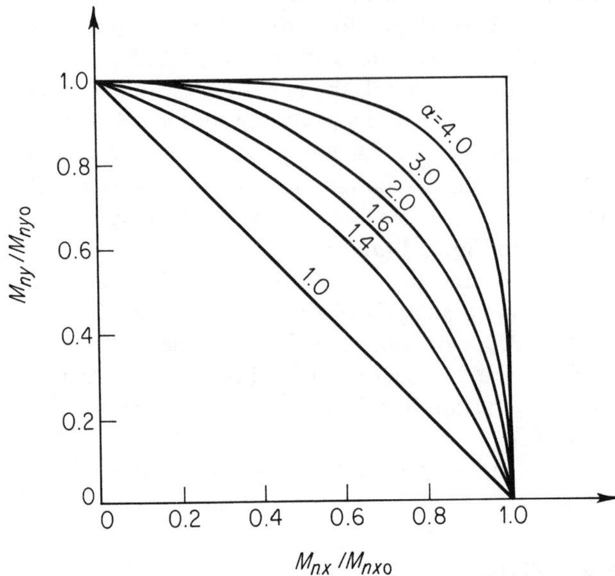

FIGURE 8.15
Interaction contours at constant P_n for varying α. (*Adapted from Ref. 8.10.*)

Obviously the ϕ factors cancel in Eq. (12.19), so Fig. 8.15 can also be used to describe the load contours for the surface of *design strength*, with the coordinates relabeled according to Eq. (8.19).

Calculations reported by Bresler in Ref. 8.10 indicate that α falls in the range from 1.15 to 1.55 for square and rectangular columns. Values near the lower end of that range are the more conservative. Methods and design aids permitting a more defined estimation of α will be found in Ref. 8.7.

In practice, the values of P_u, M_{ux}, and M_{uy} are known from the analysis of the structure. For a trial column section, the values of ϕM_{nx0} and ϕM_{ny0} corresponding to the load P_u can easily be found by the usual methods for uniaxial bending. Then replacing ϕM_{nx} with M_{ux} and ϕM_{ny} with M_{uy} in Eq. (8.19), or alternatively by plotting M_{ux} and M_{uy} in Fig. 8.15, it can be confirmed that a particular combination of factored moments falls within the load contour (safe design) or outside the contour (failure), and the design modified if necessary.

An approximate approach to the load contour method, in which the curved load contour is represented by a bilinear approximation, will be found in Ref. 8.13. It leads to a method of *trial design* in which the biaxial bending moments are represented by an equivalent uniaxial bending moment. Design charts based on this approximate approach will be found in the ACI *Column Design Handbook* (Ref. 8.7). Trial designs arrived at in this way should be checked for adequacy by the load contour method, described above, or by the method of reciprocal loads which follows.

8.13 RECIPROCAL LOAD METHOD

A simple, approximate design method developed by Bresler (Ref. 8.10) has been satisfactorily verified by comparison with results of extensive tests and accurate calculations (Ref. 8.14). It is noted that the column interaction surface of Fig. 8.14*d* can, alternatively, be plotted as a function of the axial load P_n and eccentricities $e_x = M_{ny}/P_n$ and $e_y = M_{nx}/P_n$, as is shown in Fig. 8.16*a*. The surface S_1 of Fig. 8.16*a* can be transformed into an equivalent failure surface S_2, as shown in Fig. 8.16*b*, where e_x and e_y are plotted against $1/P_n$ rather than P_n. Thus $e_x = e_y = 0$ corresponds to the inverse of the capacity of the column if it were concentrically loaded, P_0, and this is plotted as point *C*. For $e_y = 0$ and any given value of e_x, there is a load P_{ny0} (corresponding to moment M_{ny0}) that would result in failure. The reciprocal of this load is plotted as point *A*. Similarly, for $e_x = 0$ and any given value of e_y, there is a certain load P_{nx0} (corresponding to moment M_{nx0}) that would cause failure, the reciprocal of which is point *B*. The values of P_{nx0} and P_{ny0} are easily established, for known eccentricities of loading applied to a given column, using the methods already established for uniaxial bending, or using design charts for uniaxial bending.

An oblique plane S_2' is defined by the three points: *A, B*, and *C*. This plane is used as an approximation of the actual failure surface S_2. Note that, for any point on the surface S_2 (i.e., for any given combination of e_x and e_y), there is a corresponding plane S_2'. Thus the approximation of the true failure surface S_2

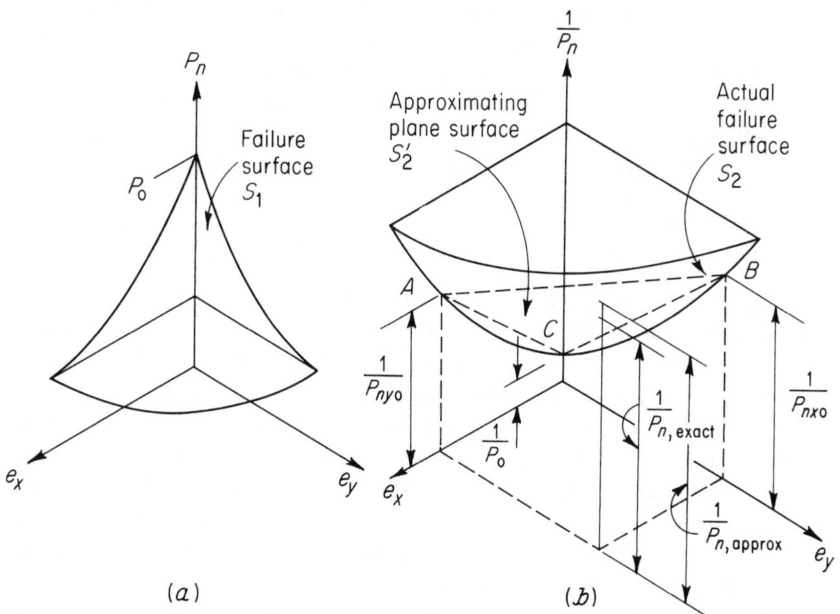

FIGURE 8.16
Interaction surfaces for the reciprocal load method.

involves an infinite number of planes S_2' determined by particular pairs of values of e_x and e_y, i.e., by particular points A, B, and C.

The vertical ordinate $1/P_{n,\text{exact}}$ to the true failure surface will always be conservatively estimated by the distance $1/P_{n,\text{approx}}$ to the oblique plane ABC (extended), because of the concave upward eggshell shape of the true failure surface. In other words, $1/P_{n,\text{approx}}$ is always greater than $1/P_{n,\text{exact}}$, which means that $P_{n,\text{approx}}$ is always less than $P_{n,\text{exact}}$.

Bresler's reciprocal load equation derives from the geometry of the approximating plane. It can be shown that

$$\frac{1}{P_n} = \frac{1}{P_{nx0}} + \frac{1}{P_{ny0}} - \frac{1}{P_0} \tag{8.20}$$

where

P_n = approximate value of ultimate load in biaxial bending with eccentricities e_x and e_y

P_{ny0} = ultimate load when only eccentricity e_x is present ($e_y = 0$)

P_{nx0} = ultimate load when only eccentricity e_y is present ($e_x = 0$)

P_0 = ultimate load for concentrically loaded column

Equation (8.20) has been found to be acceptably accurate for design purposes provided $P_n \geq 0.10P_0$. It is not reliable where biaxial bending is prevalent and accompanied by an axial force smaller than $P_0/10$. In the case of such strongly prevalent bending, failure is initiated by yielding of the steel in tension, and the situation corresponds to the lowest tenth of the interaction diagram of Fig. 8.14*d*. In this range it is conservative and accurate enough to neglect the axial force entirely and to calculate the section for biaxial bending only.

The introduction of the ACI strength reduction factors does not change the preceding development in any fundamental way, as long as the ϕ factor is constant for all terms, and for design purposes the Bresler equation can be rewritten as

$$\frac{1}{\phi P_n} = \frac{1}{\phi P_{nx0}} + \frac{1}{\phi P_{ny0}} - \frac{1}{\phi P_0} \tag{8.21}$$

In the range for which the Bresler method is applicable, above $0.10P_0$, ϕ is constant, except that for very small eccentricities the ACI Code imposes an upper limit on the maximum design strength that has the effect of flattening the upper part of the column strength interaction curve (see Sec. 8.9 and Graphs A.5 through A.16 of App. A). When using the Bresler method for biaxial bending, it is necessary to use the uniaxial strength curve *without* the horizontal cutoff (as shown by dotted lines in the graphs of App. A) in obtaining values for use in Eq. (8.21). The value of ϕP_n obtained in this way should then be subject to the restriction, as for uniaxial bending, that it must not exceed $0.80\phi P_0$ for tied columns and $0.85\phi P_0$ for spirally reinforced columns.

In a typical design situation, given the size and reinforcement of the trial column and the load eccentricities e_y and e_x, one finds by computation or from design charts the ultimate loads ϕP_{nx0} and ϕP_{ny0} for uniaxial bending around the

X and Y axes respectively, and the ultimate load ϕP_0 for concentric loading. Then $1/\phi P_n$ is computed from Eq. (8.21) and, from that, ϕP_n is calculated. The design requirement is that the factored load P_u must not exceed ϕP_n, as modified by the horizontal cutoff mentioned above, if applicable.

Example 8.5 Design of column for biaxial bending. The 12 × 20 in. column shown in Fig. 8.17 is reinforced with eight No. 9 bars arranged around the column perimeter, providing an area $A_{st} = 8.00 \text{ in}^2$. A factored load P_u of 275 kips is to be applied with eccentricities $e_y = 3$ in. and $e_x = 6$ in., as shown. Material strengths are $f'_c = 4$ ksi and $f_y = 60$ ksi. Check the adequacy of the trial design (*a*) using the reciprocal load method and (*b*) using the load contour method.

Solution

(*a*) By the reciprocal load method, first considering bending about the Y axis, $\gamma = 15/20 = 0.75$, and $e/h = 6/20 = 0.30$. With the reinforcement ratio of $A_{st}/bh = 8.00/240 = 0.033$, Graph A.7 of App. A indicates

$$\frac{\phi P_{ny0}}{A_g} = 1.75 \qquad \phi P_{ny0} = 1.75 \times 240 = 420 \text{ kips}$$

$$\frac{\phi P_0}{A_g} = 3.65 \qquad \phi P_0 = 3.65 \times 240 = 876 \text{ kips}$$

Then for bending about the X axis, $\gamma = 7/12 = 0.58$ (say 0.60), and $e/h = 3/12 = 0.25$. Graph A.6 of App. A gives

$$\frac{\phi P_{nx0}}{A_g} = 1.80 \qquad \phi P_{nx0} = 1.80 \times 240 = 432 \text{ kips}$$

$$\frac{\phi P_0}{A_g} = 3.65 \qquad \phi P_0 = 3.65 \times 240 = 876 \text{ kips}$$

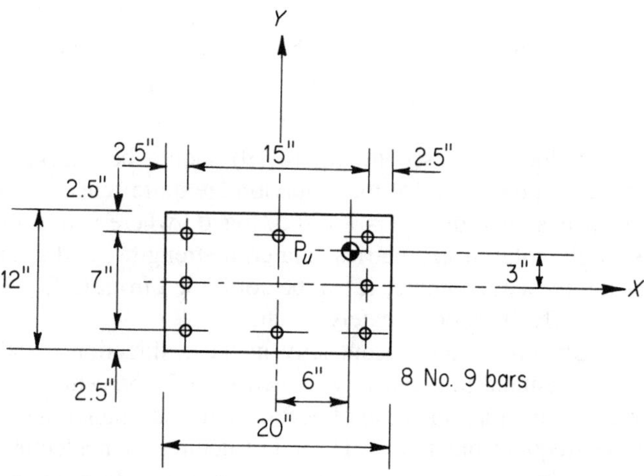

FIGURE 8.17
Column cross section for Example 8.5.

Substituting these values in Eq. (8.21) results in

$$\frac{1}{\phi P_n} = \frac{1}{432} + \frac{1}{420} - \frac{1}{876} = 0.00356$$

from which $\phi P_n = 281$ kips. Thus, according to the Bresler method, the design load of $P_u = 275$ kips can be applied safely.

(*b*) By the load contour method, for Y axis bending with $P_u = \phi P_n = 275$ kips, and $\phi P_n / A_g = 275/240 = 1.15$, Graph A.7 of App. A indicates that

$$\frac{\phi M_{ny0}}{A_g h} = 0.62$$

Hence $\phi M_{ny0} = 0.62 \times 240 \times 20 = 2980$ in-kips. Then for X axis bending, with $\phi P_n / A_g = 1.15$, as before, from Graph A.6,

$$\frac{\phi M_{nx0}}{A_g h} = 0.53$$

So $\phi M_{nx0} = 0.53 \times 240 \times 12 = 1530$ in-kips. The factored load moments about the Y and X axes respectively are

$$M_{uy} = 275 \times 6 = 1650 \text{ in-kips}$$

$$M_{ux} = 275 \times 3 = 830 \text{ in-kips}$$

Adequacy of the trial design will now be checked using Eq. (8.19) with an exponent α conservatively taken equal to 1.15. Then with $\phi M_{nx} = M_{ux}$ and $\phi M_{ny} = M_{uy}$, that equation indicates

$$\left(\frac{830}{1530}\right)^{1.15} + \left(\frac{1650}{2980}\right)^{1.15} = 0.495 + 0.507 = 1.002$$

This is close enough to 1.0 that the design would be considered safe by the load contour method also.

In actual practice, the value of α used in Eq. (8.19) should be checked, for the specific column, because predictions of that equation are quite sensitive to changes in α. In Ref. 8.13 it is shown that $\alpha = \log 0.5/\log \beta$, where values of β can be tabulated for specific column geometries, material strengths, and load ranges (see Ref. 8.7). For the present example, it can be confirmed from Ref. 8.7 that $\beta = 0.56$ and hence $\alpha = 1.19$, approximately as chosen.

One observes that, in Example 8.5*a*, eccentricity in the Y direction of 50 percent that in the X direction resulted in a capacity reduction of 33 percent, i.e., from 420 to 281 kips. For cases in which the ratio of eccentricities is smaller, there is some justification for the frequent practice in framed structures of neglecting the bending moments in the direction of the smaller eccentricity. *In general, biaxial bending should be taken into account when the estimated eccentricity ratio approaches or exceeds 0.2.*

8.14 COMPUTER ANALYSIS FOR BIAXIAL BENDING OF COLUMNS

Although the load contour method and the reciprocal load method are widely used in practice, each has serious shortcomings. With the load contour method, selection of the appropriate value of the exponent α is made difficult by a number of factors relating to column shape and bar distribution. For many cases the usual assumption that $\alpha_1 = \alpha_2$ is a poor approximation. Design aids are available, but they introduce further approximations, e.g., the use of a bilinear representation of the load contour. The reciprocal load method is very simple to use, but the representation of the curved failure surface by an approximating plane is not reliable in the range of large eccentricities, where failure is initiated by steel yielding.

With the general availability of desktop computers, it is better to use simpler methods to obtain faster, and more exact, solutions to the biaxial column problem. Such a method is that developed by Ehsani (Ref. 8.15). A column strength interaction line is established for a trial column, exactly analogous to the curve for axial load plus uniaxial bending, as described in Secs. 8.3 to 8.7. However, the curve is generated for the particular value of the eccentricity angle that applies, as determined by the ratio of M_{uy}/M_{ux} from the structural frame analysis (see curve (*c*) of Fig. 8.14*d*). This is done by taking successive choices of neutral axis distance, measured in this case along one face of the column from the most heavily compressed corner, from very small (large eccentricity) to very large (small eccentricity), then calculating the axial force P_n and moments M_{nx} and M_{ny}. For each neutral axis distance, iteration is performed with successive values of the orientation angle θ, Fig. 8.14*c*, until $\lambda = \arctan M_{ny}/M_{nx}$ is in agreement with the value of $\lambda = \arctan M_{uy}/M_{ux}$ from the structural frame analysis. Thus one point on the curve (*c*) of Fig. 8.14*d* is established. The sequence of calculations is repeated: another choice of neutral axis distance is made, a value of θ is selected, the axial force and moments are calculated, λ is found, and the value of θ is iterated until λ is correct. Thus the next point is established, and so on, until the complete strength interaction curve for that particular value of λ is achieved. ACI Code safety provisions may then be imposed in the usual way, and the adequacy of the proposed design tested, for the known load and moments, against the design strength curve for the trial column.

The method is obviously impractical for manual calculation, but the iterative steps are easily and quickly performed even by desktop computers, which may also provide a graphical presentation of results. Full details will be found in Ref. 8.15.

8.15 BAR SPLICING IN COLUMNS

The main vertical reinforcement in columns is usually spliced just above each floor, or sometimes at alternate floors. This permits the column steel area to be reduced progressively at the higher levels in a building, where loads are smaller, and in addition avoids handling and supporting very long column bars. Column steel may be spliced by lapping, by butt welding, by various types of patented mechanical connections, or by direct end bearing, using special devices to ensure proper alignment of bars.

Special attention must be given to the problem of bar congestion at splices. Lapping the bars, for example, effectively doubles the steel area in the column cross section at the level of the splice and can result in problems placing concrete and meeting the ACI Code requirement for minimum lateral spacing of bars ($1.5d_b$ or 1.5 in.). To avoid difficulty, column steel percentages are sometimes limited in practice to not more than about 4 percent, or the bars are extended two stories and staggered splices are used.

The most common method of splicing column steel is the simple lapped bar splice, with the bars in contact throughout the lapped length. It is standard practice to offset the lower bars, as shown in Fig. 8.18, to permit the proper positioning of the upper bars. In order to prevent outward buckling of the bars at the bottom bend point of such an offset, with spalling of the concrete cover, it is necessary to provide special lateral reinforcement in the form of extra ties. According to ACI Code 7.8.1, the slope of the inclined part of an offset bar must not exceed 1 in 6, and lateral steel must be provided to resist $1\frac{1}{2}$ times the horizontal component of the computed force in the inclined part of the offset bar. This special reinforcement must be placed not more than 6 in. from the point of bend, as shown in Fig. 8.18. Elsewhere in the column, above and below the floor, the usual spacing requirements described in Sec. 8.2 apply, except that ties must be located not more than one-half the normal spacing above the floor. Where beams frame from four directions into a joint, as shown in Fig. 8.18, ties may be terminated not more than 3 in. below the lowest reinforcement in the shallowest of such beams, according to ACI Code 7.10.5. If beams are not present on

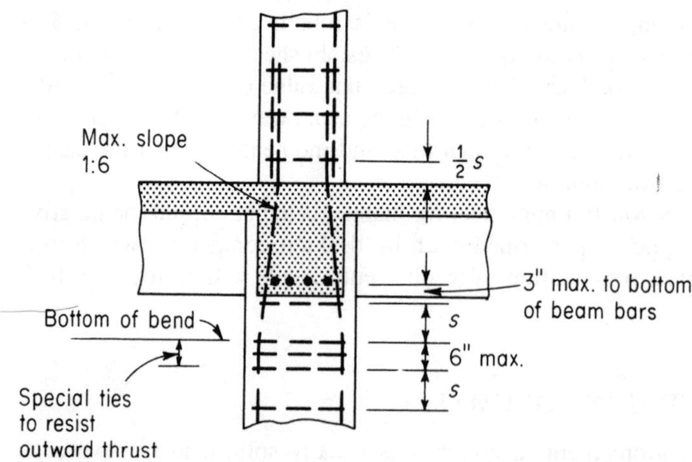

FIGURE 8.18
Splice details at typical interior column.

four sides, such as for exterior columns, ties must be placed vertically at the usual spacing through the depth of the joint to a level not more than one-half the usual spacing below the lowest reinforcement in the slab.

Analogous requirements are found in ACI Code 7.10.4 and are illustrated in Ref. 8.9 for spirally reinforced columns.

Column splices are mainly compression splices, although load combinations producing moderate to large eccentricity require that splices transmit tension as well. ACI Code 12.17 permits splicing by lapping, butt welding, mechanical connectors, or end bearing if the calculated factored load stress ranges from f_y in compression to $0.5f_y$ in tension, and it requires that the total tensile strength provided in each face of the column by splices alone or by splices in combination with continuing unspliced bars acting at f_y be at least twice the calculated tension at that face. If the factored load stress exceeds $0.5f_y$ in tension, the lap splices must develop the full strength f_y in tension, or full welded splices or mechanical connectors must be used. In any case, the Code requires that, at sections where splices are located, a minimum tensile strength be provided in each face of the column equal to the capacity of one-quarter of the vertical reinforcement acting at f_y.

Requirements for both compression and tension lap splices were discussed in Sec. 5.11, and the design of a compression splice in a typical column was illustrated in Example 5.4.

REFERENCES

8.1. *ACI Detailing Manual,* ACI Publication SP-66, Detroit, 1980.

8.2. *CRSI Handbook,* 6th ed., Concrete Reinforcing Steel Institute, Chicago, 1984.

8.3. F. E. Richart, A. Brandtzaeg, and R. L. Brown, "A Study of the Failure of Concrete under Combined Compressive Stresses," Univ. Ill. Eng. Exp. Stn. Bull. 185, 1928.

8.4. F. E. Richart, A. Brandtzaeg, and R. L. Brown, "The Failure of Plain and Spirally Reinforced Concrete in Compression," Univ. Ill. Eng. Exp. Stn. Bull. 190, 1929.

8.5. S. Martinez, A. H. Nilson, and F. O. Slate, "Spirally Reinforced High Strength Concrete Columns," *J. ACI,* vol. 81, no. 5, 1984, pp. 431–442.

8.6. A. H. Mattock, L. B. Kriz, and E. Hognestad, "Rectangular Concrete Stress Distribution in Ultimate Strength Design," *J. ACI,* vol. 32, no. 8, 1961, pp. 875–928.

8.7. *Design Handbook,* Vol. 2, *Columns,* ACI Publication SP-17A(78), Detroit, 1978.

8.8. *CRSI Design Handbook,* 6th ed., Concrete Reinforcing Steel Institute, Chicago, 1984.

8.9. *ACI Detailing Manual,* ACI Publication SP-66, Detroit, 1980.

8.10. B. Bresler, "Design Criteria for Reinforced Columns under Axial Load and Biaxial Bending," *J. ACI,* vol. 32, no. 5, 1960, pp. 481–490.

8.11. R. W. Furlong, "Ultimate Strength of Square Columns under Biaxially Eccentric Loads," *J. ACI,* vol. 32, no. 9, 1961, pp. 1129–1140.

8.12. F. N. Pannell, "Failure Surfaces for Members in Compression and Biaxial Bending," *J. ACI,* vol. 60, no. 1, 1963, pp. 129–140.

8.13. A. L. Parme, J. M. Nieves, and A. Gouwens, "Capacity of Reinforced Concrete Rectangular Members Subject to Biaxial Bending," *J. ACI,* vol. 63, no. 9, 1966, pp. 911–923.

8.14. L. N. Ramamurthy, "Investigation of the Ultimate Strength of Square and Rectangular Columns under Biaxially Eccentric Loads," in *Symp. Reinforced Concrete Columns,* ACI Publication SP-13, 1966, pp. 263–298.

8.15. M. R. Ehsani, "CAD for Columns," *Concr. Int.* vol. 8, no. 9, 1986, pp. 43–47.

PROBLEMS

8.1 A 16 in. square column is reinforced with four No. 14 bars, one in each corner, with cover distances 3 in. to the steel center in each direction. Material strengths are $f_c' = 5000$ psi and $f_y = 60{,}000$ psi. Construct the interaction diagram relating axial strength P_n and flexural strength M_n. Bending will be about an axis parallel to one face. Calculate the coordinates for P_o, P_b, and at least three other representative points on the curve.

8.2 Plot the design strength curve relating ϕP_n and ϕM_n for the column of Prob. 8.1. Design and detail the tie steel required by the ACI Code. Is the column a good choice to resist a load $P_u = 540$ kips applied with an eccentricity $e = 4.44$ in?

8.3 The short column shown in Fig. P8.3 will be subjected to an eccentric load causing uniaxial bending about the Y axis. Material strengths are $f_y = 60$ ksi and $f_c' = 6$ ksi.

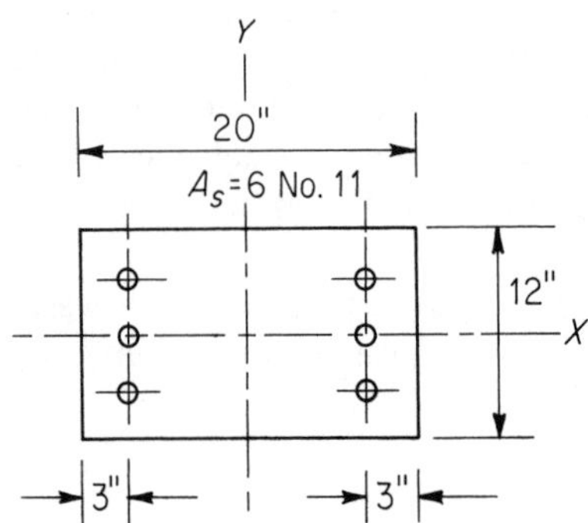

FIGURE P8.3

(*a*) Construct the strength interaction curve for this column, calculating no fewer than five points, including those corresponding to pure bending, pure axial thrust, and balanced failure.

(*b*) Show on the same drawing the design strength curve obtained through introduction of the ACI ϕ factors.

(*c*) Design the lateral reinforcement for the column, giving key dimensions for ties.

8.4 The column shown in Fig. P8.4 is subjected to axial load and bending moment causing bending about an axis parallel to that of the rows of bars. What moment would cause the column to fail if the axial load applied simultaneously was 500 kips? Material strengths are $f_c' = 4000$ psi and $f_y = 60$ ksi.

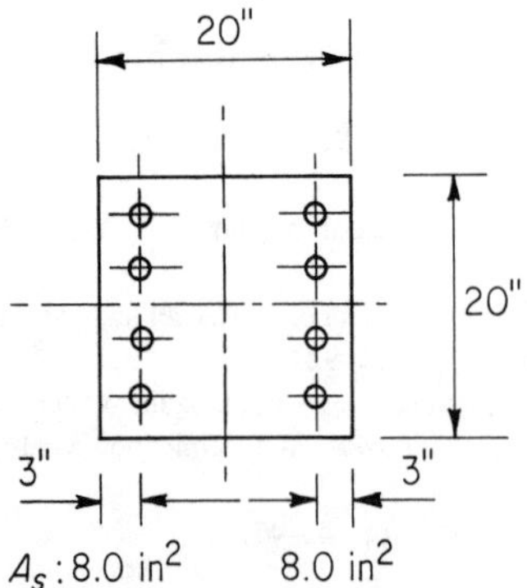

FIGURE P8.4

8.5 What is the strength M_n of the column of Prob. 8.4 if it were loaded in pure bending (axial force = 0) about one principal axis?

8.6 Construct the interaction diagram relating P_n to M_n for the building column shown in Fig. P8.6. Bending will be about the axis *a-a*. Calculate specific coordinates for concentric loading ($e = 0$), for P_b, and at least three other points, well chosen, on the curve. Material strengths are $f'_c = 8000$ psi and $f_y = 60{,}000$ psi. Show also the curve of design strength relating ϕP_n and ϕM_n.

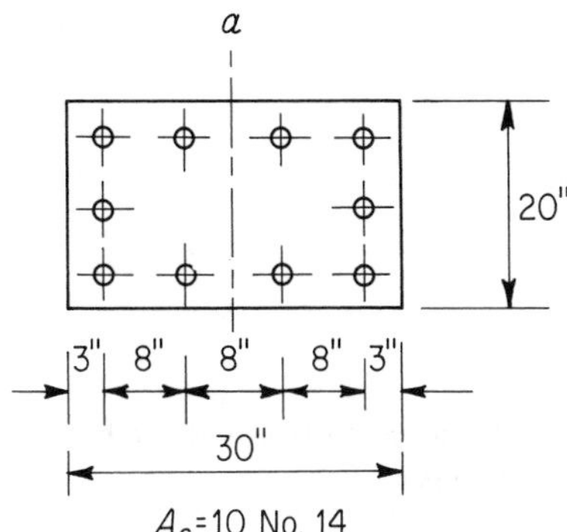

FIGURE P8.6

8.7 A short rectangular reinforced concrete column shown in Fig. P8.7 is to be a part of a long-span rigid frame and will be subjected to high bending moments combined with relatively low axial loads, causing bending about the strong axis. Because of the high eccentricity, steel is placed unsymmetrically as shown, with three No. 14 bars near the tension face and two No. 11 bars near the compression face. Material strengths are $f'_c = 6$ ksi and $f_y = 60$ ksi. Construct the complete strength interaction diagram plotting P_n vs. M_n, relating eccentricities to the plastic centroid of the column (not the geometric center). On the same diagram, show the curve relating ϕP_n to ϕM_n.

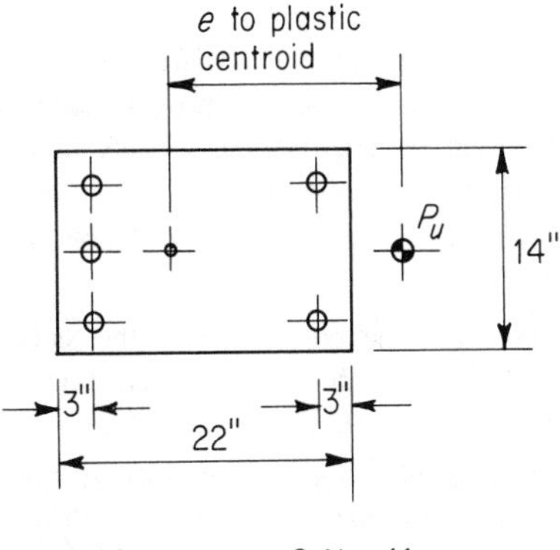

FIGURE P8.7

8.8 Construct the strength interaction diagram and design strength curves for the square column of Fig. P8.8, given that the column will be subjected to biaxial bending with equal eccentricities about both principal axes. Take $f_y = 60$ ksi and $f_c' = 4$ ksi.

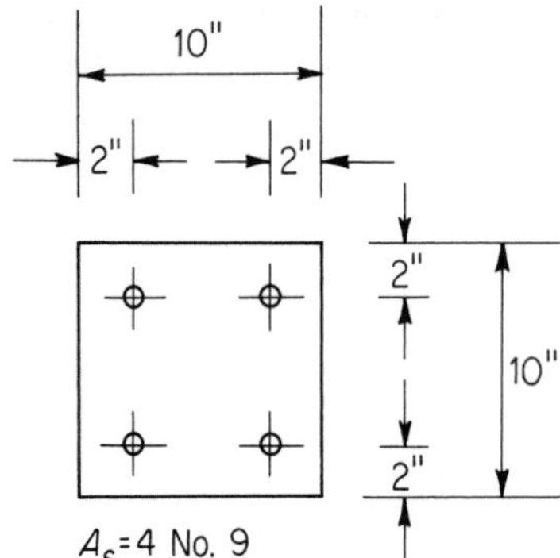

FIGURE P8.8

8.9 The square column shown in Fig. P8.9 is a corner column subject to axial load and biaxial bending. Material strengths are $f_y = 60{,}000$ psi and $f_c' = 6000$ psi.

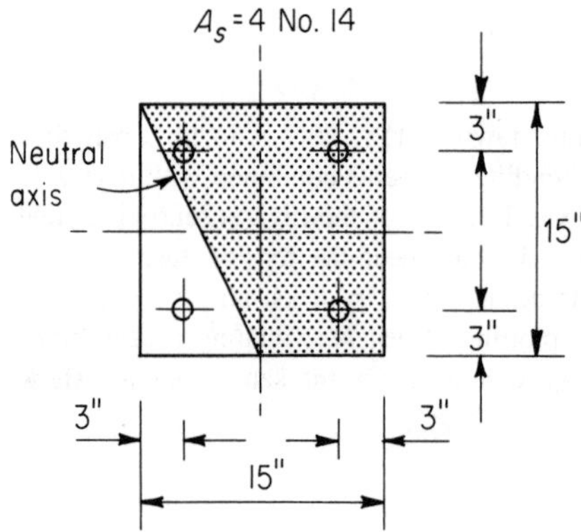

FIGURE P8.9

(*a*) Find the unique combination of $P_n, M_{nx},$ and M_{ny} and that will produce incipient failure with the neutral axis located as in the figure. The compressive zone is shown shaded. Note that the actual neutral axis is shown, not the equivalent rectangular stress block limit; however, the rectangular stress block may be used as the basis of calculations.

(*b*) Find the angle between the neutral axis and the eccentricity axis, the latter defined as the line from the column center to the point of load.

8.10 For the axial load P_n found in Prob. 8.9, and for the same column, with the same eccentricity ratio e_y/e_x, find the values of M_{nx} and M_{ny} that would produce incipient failure using the load contour method. Compare with the results of Prob. 8.9. Take $\alpha = 1.30$.

8.11 For the eccentricities e_x and e_y found in Prob. 8.9, find the value of axial load P_n that would produce incipient failure using the reciprocal load (Bresler) method. Compare with the results of Prob. 8.9 and 8.10.

CHAPTER

9

SLENDER COLUMNS

9.1 INTRODUCTION

The material presented in Chapter 8 pertained to concentrically or eccentrically loaded *short columns*, for which the strength is governed entirely by the strength of the materials and the geometry of the cross section. Most columns in present-day practice fall in that category. However, with the increasing use of high strength materials and improved methods of dimensioning members, it is now possible, for a given value of axial load, with or without simultaneous bending, to design a much smaller cross section than in the past. This clearly makes for more slender members. It is because of this, together with the use of more innovative structural concepts, that rational and reliable design procedures for slender columns have become increasingly important.

A column is said to be *slender* if its cross-sectional dimensions are small compared with its length. The degree of slenderness is generally expressed in terms of the slenderness ratio l/r, where l is the unsupported length of the member and r is the radius of gyration of its cross section, equal to $\sqrt{I/A}$. For square or circular members, the value of r is the same about either axis; for other shapes r is smallest about the minor principal axis, and it is generally this value that must be used in determining the slenderness ratio of free-standing column.

It has long been known that a member of great slenderness will collapse under a smaller compression load than a stocky member of the same cross-sectional dimensions. When a stocky member, say with $l/r = 10$ (e.g., a square column of length equal to about 3 times its cross-section dimension h), is loaded in axial compression, it will fail at the load given by Eq. (8.3), because at that load both concrete and steel are stressed to their maximum carrying capacity and give way, respectively, by crushing and by yielding. If a member with the same cross section has a slenderness ratio $l/r = 100$ (e.g., a square column hinged at both ends and of length equal to about 30 times its section dimension), it may fail under an axial load of one-half or less that given by Eq. (8.3). In this case, collapse is caused by buckling, i.e., by sudden lateral displacement of the member between its ends, with consequent overstressing of steel and concrete by the bending stresses that superimpose on the axial compression stresses.

Most columns in practice are subjected to bending moments as well as axial loads, as was made clear in Chapter 8. These moments produce lateral deflection of a member between its ends and may also result in relative lateral displacement of joints. Associated with these lateral displacements are *secondary moments* that add to the primary moments and that may become very large for slender columns, leading to failure. A practical definition of a slender column is one for which there is a significant reduction in axial load capacity because of these secondary moments. In the development of ACI Code column provisions, for example, any reduction greater than about 5 percent was considered significant, requiring consideration of slenderness effects.

The ACI Code and Commentary contain detailed provisions governing the design of slender columns. ACI Code 10.11 presents an approximate method for accounting for slenderness through use of *moment magnification factors*. Provisions are quite similar to those used for steel columns designed under the American Institute of Steel Construction (AISC) Specification. Alternatively, in ACI Code 10.10 a more fundamental approach is endorsed, in which the effect of lateral displacements is accounted for directly in the frame analysis. Because of the increasing complexity of the moment magnification approach, as it has been refined in recent years, with its many detailed requirements, and because of the general availability of computers in the design office, there is increasing interest in "second order analysis" as suggested in ACI Code 10.10, in which the effect of lateral displacements is computed directly.

As noted, most columns in practice continue to be short columns. Simple expressions are included in the ACI Code to determine whether the slenderness effect must be considered. These will be presented in Sec. 9.4 following the development of background information in Sec. 9.2 and Sec. 9.3 relating to column buckling and slenderness effects.

9.2 CONCENTRICALLY LOADED COLUMNS

The basic information on the behavior of straight, concentrically loaded slender columns was developed by Euler more than 200 years ago. In generalized form, it states that such a member will fail by buckling at the critical load

$$P_c = \frac{\pi^2 E_t I}{(kl)^2} \tag{9.1}$$

It is seen that the buckling load decreases rapidly with increasing *slenderness ratio* kl/r (Ref. 9.1).

For the simplest case of a column hinged at both ends and made of elastic material, E_t simply becomes Young's modulus and kl is equal to the actual length

l of the column. At the load given by Eq. (9.1) the originally straight member buckles into a half sine wave, as in Fig. 9.1*a*. In this bent configuration, bending moments Py act at any section such as a; y is the deflection at that section. These deflections continue to increase until the bending stress caused by the increasing moment, together with the original compression stress, overstresses and fails the member.

If the stress-strain curve of a short piece of the given member is of the shape of Fig. 9.2*a*, as it would be for reinforced concrete columns, E_t is equal to Young's modulus, provided the buckling stress P_c/A is below the proportional limit f_p. If it is larger than f_p, buckling occurs in the inelastic range. In this case, in Eq. (9.1), E_t is the tangent modulus, i.e., the slope of the tangent to the stress-strain curve. As the stress increases, E_t decreases. A plot of the buckling load vs. the slenderness ratio, a so-called column curve, therefore has the shape given in Fig. 9.2*b*, which shows the reduction in buckling strength with increasing slenderness. For very stocky columns one finds that the value of the buckling load, calculated from Eq. (9.1), exceeds the direct crushing strength of the stocky column, given by Eq. (8.3). This is also shown in Fig. 9.2*b*. Correspondingly, there is a limiting slenderness ratio $(kl/r)_{\text{lim}}$. For values smaller than this, failure occurs by simple crushing, regardless of kl/r; for values larger than $(kl/r)_{\text{lim}}$, failure occurs by buckling, the buckling load or stress decreasing for greater slenderness.

If a member is fixed against rotation at both ends, it buckles in the shape of Fig. 9.1*b*, with inflection points (i.p.) as shown. The portion between the inflection points is in precisely the same situation as the hinge-ended column of Fig. 9.1*a*, and thus the *effective length kl* of the fixed-fixed column, i.e., the distance between inflection points, is seen to be $kl = l/2$. Equation (9.1) shows that an elastic column fixed at both ends will carry 4 times as much load as when hinged.

Columns in real structures are rarely either hinged or fixed but have ends partially restrained against rotation by abutting members. This is schematically shown in Fig. 9.1*c*, from which it is seen that for such members the effective length kl, i.e., the distance between inflection points, has a value between l and $l/2$. The precise value depends on the degree of end restraint, i.e., on the ratio of the rigidity EI/l of the column to the sum of rigidities EI/l of the restraining members at both ends.

In the columns of Fig. 9.1*a* to *c* it was assumed that one end was prevented from moving laterally relative to the other end, by horizontal bracing or otherwise. In this case it is seen that the effective length kl is always smaller than (or at most it is equal to) the real length l.

If a column is fixed at one end and entirely free at the other (cantilever column or flagpole), it buckles as shown in Fig. 9.1*d*. That is, the upper end moves laterally with respect to the lower, a kind of deformation known as *sidesway*. It buckles into a quarter of a sine wave and is therefore analogous to the upper half of the hinged column of Fig. 9.1*a*. The inflection points, one at the end of the actual column and the other at the imaginary extension of the sine wave, are a distance $2l$ apart, so that the effective length is $kl = 2l$.

If the column is rotationally fixed at both ends but one end can move laterally with respect to the other, it buckles as shown in Fig. 9.1*e*, with an effective length

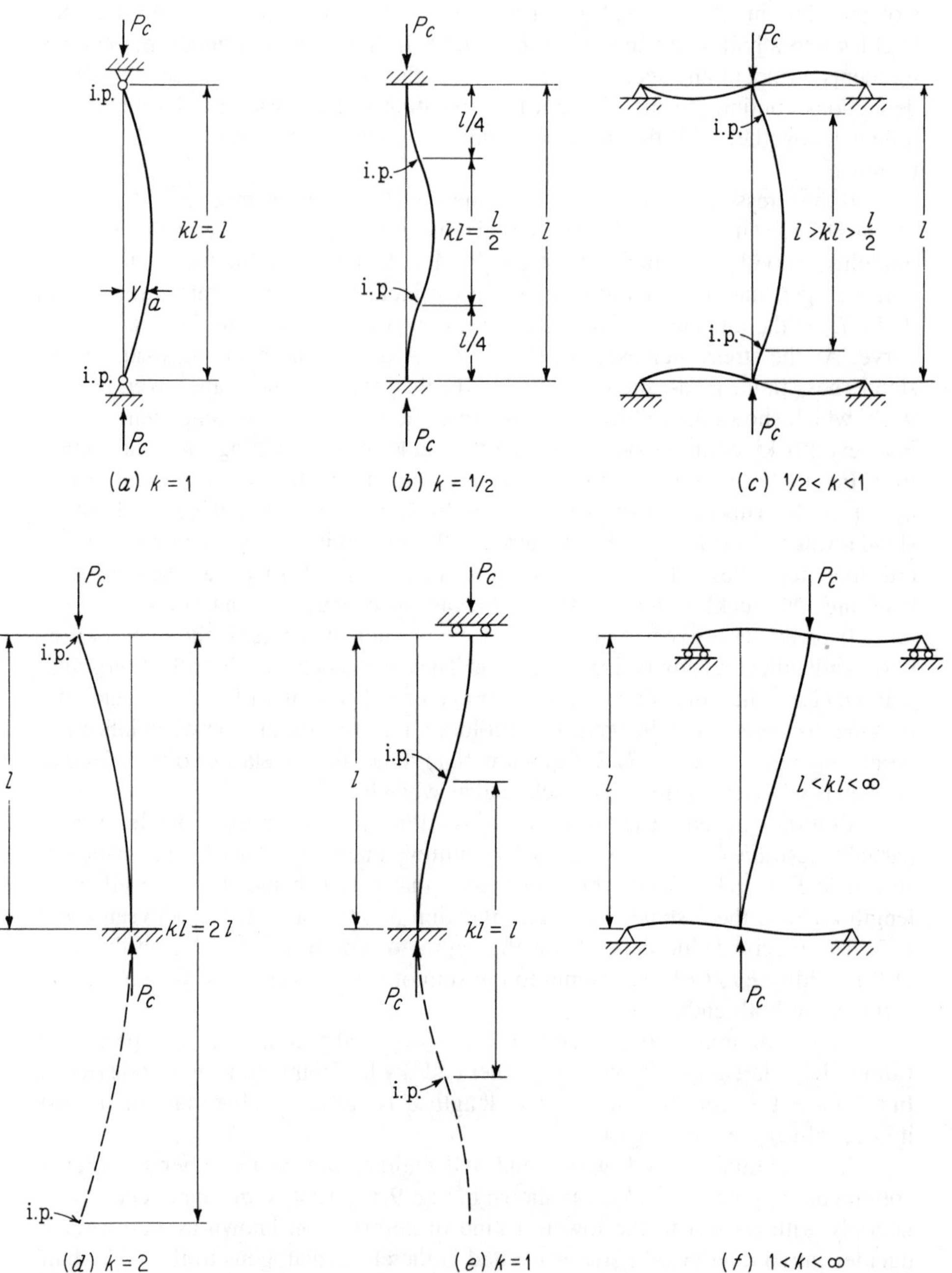

FIGURE 9.1
Buckling and effective length of axially loaded columns.

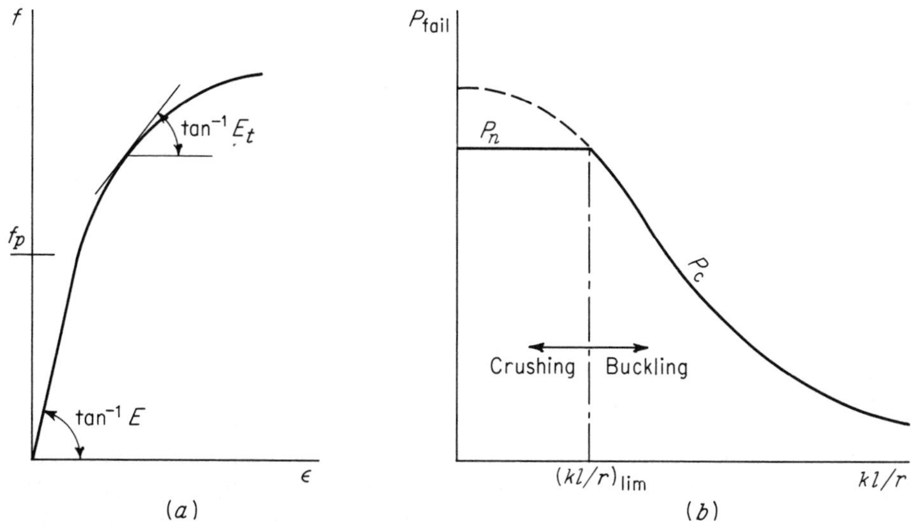

FIGURE 9.2
Effect of slenderness on strength of axially loaded columns.

$kl = l$. If one compares this column, fixed at both ends but free to sidesway, with a fixed-fixed column that is braced against sidesway (Fig. 9.1*b*), one sees that the effective length of the former is twice that of the latter. By Eq. (9.1), this means that the buckling strength of an elastic fixed-fixed column that is free to sidesway is only one-quarter that of the same column when braced against sidesway. This is an illustration of the general fact that *compression members free to buckle in a sidesway mode are always considerably weaker than when braced against sidesway.*

Again, the ends of columns in actual structures are rarely either hinged, fixed, or entirely free but are usually restrained by abutting members. If sidesway is not prevented, buckling occurs as in Fig. 9.1*f*, and the effective length, as before, depends on the degree of restraint. If the cross beams are very rigid compared with the column, the case of Fig. 9.1*e* is approached and *kl* is only slightly larger than *l*. On the other hand, if the restraining members are extremely flexible, a hinged condition is approached at both ends. Evidently, a column hinged at both ends and free to sidesway is unstable. It will simply topple, being unable to carry any load whatever.

In reinforced concrete structures one is rarely concerned with single members but rather with rigid frames of various configurations. The manner in which the described relationships affect the buckling behavior of frames is illustrated by the simple portal frame of Fig. 9.3, with loads applied concentrically to the columns. If sidesway is prevented, as indicated schematically by the brace of Fig. 9.3*a*, the buckling configuration will be as shown. The buckled shape of the column corresponds to that of Fig. 9.1*c*, except that the lower end is hinged. It is seen that the effective length *kl* is smaller than *l*. On the other hand, if no sidesway

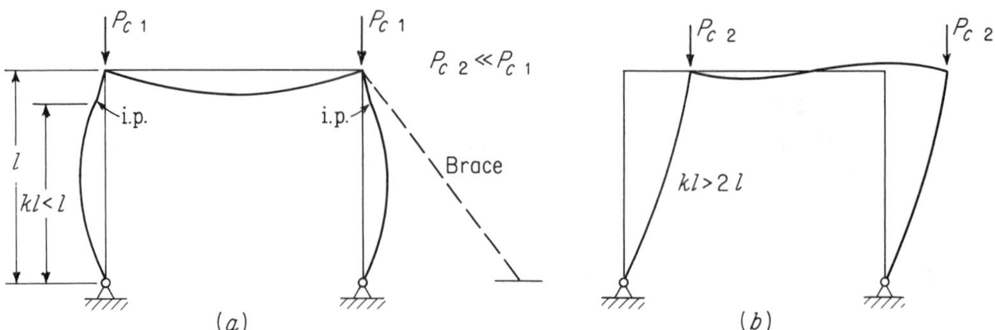

FIGURE 9.3
Rigid-frame buckling: (*a*) laterally braced; (*b*) unbraced.

bracing is provided to an otherwise identical frame, buckling occurs as in Fig. 9.3*b*. The column is in a situation similar to that of Fig. 9.1*d*, upside down, except that the upper end is not fixed but only partially restrained by the girder. It is seen that the effective length kl exceeds $2l$ by an amount depending on the degree of restraint. The buckling strength depends on kl/r in the manner shown in Fig. 9.2*b*. In consequence, even though they are dimensionally identical, the unbraced frame will buckle at a radically smaller load than the braced frame.

In summary, the following can be noted:

1. The strength of concentrically loaded columns decreases with increasing slenderness ratio kl/r.
2. In columns that are *braced against sidesway* or that are parts of frames braced against sidesway, the effective length kl, i.e., the distance between inflection points, falls between $l/2$ and l, depending on the degree of end restraint.
3. The effective lengths of columns that are *not braced against sidesway* or that are parts of frames not so braced are always larger than l, the more so the smaller the end restraint. In consequence, the buckling load of a frame not braced against sidesway is always substantially smaller than that of the same frame when braced.

9.3 COMPRESSION PLUS BENDING

Most reinforced concrete compression members are also subject to simultaneous flexure, caused by transverse loads or by end moments owing to continuity. The behavior of members subject to such combined loading also depends greatly on their slenderness.

Figure 9.4*a* shows such a member, axially loaded by P and bent by equal end moments M_e. If no axial load were present, the moment M_0 in the member would be constant throughout and equal to the end moments M_e. This is shown in Fig. 9.4*b*. In this situation, i.e., in simple bending without axial compression, the member deflects as shown by the dashed curve of Fig. 9.4*a*, where y_0 represents the deflection at any point caused by bending only. When P is applied,

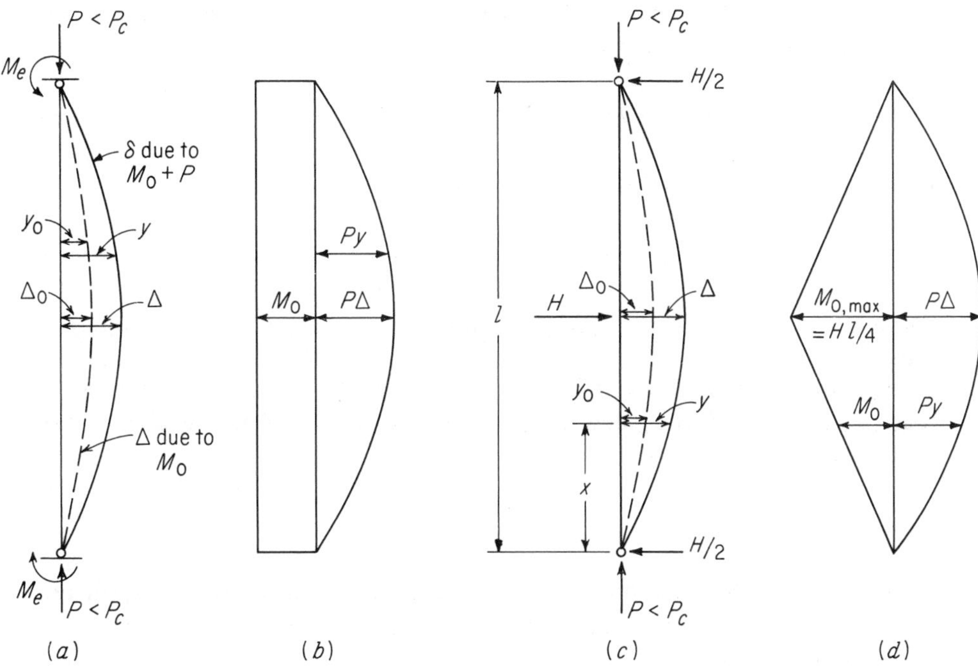

FIGURE 9.4
Moments in slender members with compression plus bending, bent in single curvature.

the moment at any point increases by an amount equal to P times its lever arm. The increased moments cause additional deflections, so that the deflection curve under the simultaneous action of P and M_0 is the solid curve of Fig. 9.4*a*. At any point, then, the total moment is now

$$M = M_0 + Py \tag{9.2}$$

i.e., the total moment consists of the moment M_0 that acts in the presence of P and the additional moment caused by P, equal to P times the deflection. This is one illustration of the so-called $P\Delta$ effect.

A similar situation is shown in Fig. 9.4*c*, where bending is caused by the transverse load H. When P is absent, the moment at any point x is $M_0 = Hx/2$, with a maximum value at midspan equal to $Hl/4$. The corresponding M_0 diagram is shown in Fig. 9.4*d*. When P is applied, additional moments Py are caused again, distributed as shown, and the total moment at any point in the member consists of the same two parts as in Eq. (9.2).

The deflections y of elastic columns of the type shown in Fig. 9.4 can be calculated from the deflections y_0, that is, from the deflections of the corresponding beam without axial load, using the following expression (see, for example, Ref. 1).

$$y = y_0 \frac{1}{1 - P/P_c} \tag{9.3}$$

Let Δ be the deflection at the point of maximum moment M_{max}, as in Fig. 9.4. Then, from Eqs. (9.2) and (9.3),

$$M_{max} = M_0 + P\Delta = M_0 + P\Delta_0 \frac{1}{1 - P/P_c} \tag{9.4}$$

It can be shown (Ref. 2) that Eq. (9.4) can be written

$$M_{max} = M_0 \frac{1 + \psi P/P_c}{1 - P/P_c} \tag{9.5}$$

where ψ is a coefficient that depends on the type of loading and varies between about ± 0.20 for most practical cases. Considering that P/P_c is always significantly smaller than 1, it is seen that the second term in the numerator of Eq. (9.5) becomes quite small compared with 1. Neglecting this term, one obtains the simplified design equation

$$M_{max} = M_0 \frac{1}{1 - P/P_c} \tag{9.6}$$

where $1/(1 - P/P_c)$ is known as the *moment magnification factor*, which reflects the amount by which the moment M_0 is magnified by the presence of a simultaneous axial force P.

Since P_c decreases with increasing slenderness ratio, it is seen from Eq. (9.6) that the moment M in the member increases with slenderness ratio kl/r. The situation is shown schematically in Fig. 9.5. It indicates that, for a given transverse loading (i.e., a given value of M_0), an axial force P causes a larger additional moment in a slender member than in a stocky member.

In the two members in Fig. 9.4, the largest moment caused by P, namely $P\Delta$, adds directly to the maximum value of M_0; for example,

$$M_0 = \frac{Hl}{4}$$

in Fig. 9.4*d*. As P increases, the maximum moment at midspan increases at a rate faster than that of P in the manner given by Eqs. (9.2) and (9.6) and shown in Fig. 9.6. The member will fail when the simultaneous values of P and M become

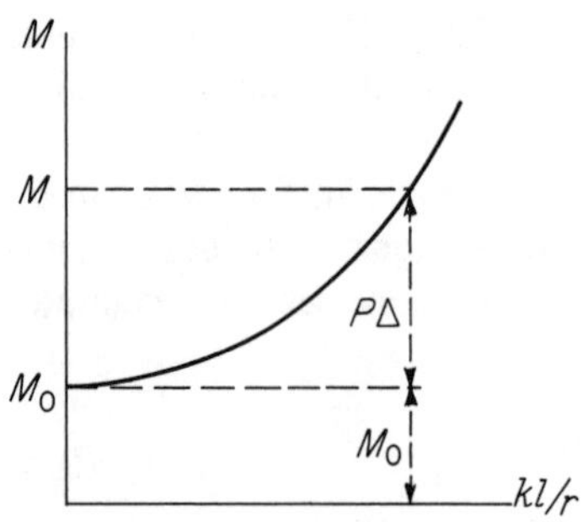

FIGURE 9.5
Effect of slenderness on column moments.

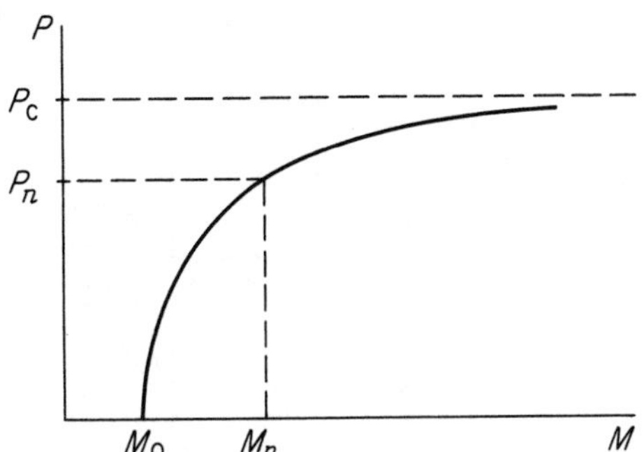

FIGURE 9.6
Effect of axial load on column moments.

equal to P_n and M_n, the ultimate strength of the cross section at the location of maximum moment.

This direct addition of the maximum moment caused by P to the maximum moment caused by the transverse load, clearly the most unfavorable situation, does not result for all types of deformations. For instance, the member of Fig. 9.7*a*, with equal and opposite end moments, has the M_0 diagram of Fig. 9.7*b*. The deflections caused by M_0 alone are again magnified when an axial load P is applied. In this case, these deflections under simultaneous bending and compression can be approximated by (Ref. 9.1) as

$$y = y_0 \frac{1}{1 - P/4P_c} \tag{9.7}$$

By comparison with Eq. (9.3) it is seen that the deflection magnification here is much smaller.

The additional moments Py caused by the axial load are distributed as shown in Fig. 9.7*c*. Although the M_0 moments are largest at the ends, the Py moments are seen to be largest some distance from the ends. Depending on their relative magnitudes, the total moments $M = M_0 + Py$ are distributed as in either Fig. 9.7*d* or *e*. In the former case, the maximum moment continues to act at the end and to be equal to M_e; the presence of the axial force, then, does not result in any increase in the maximum moment. Alternatively, in the case of Fig. 9.7*e*, the maximum moment is located at some distance from the end; at that location M_0 is significantly smaller than its maximum value M_e, and for this reason the added moment Py increases the maximum moment to a value only moderately greater than M_e.

Comparing Figs. 9.4 and 9.7, one can generalize as follows. The moment M_0 will be magnified most strongly when the location where M_0 is largest coincides with that where the deflection y_0 is largest. This results for members bent into single curvature by symmetrical loads or equal end moments. If the two end moments of Fig. 9.4*a* are unequal but of the same sign, i.e., producing single curvature, M_0 will still be strongly magnified, though not quite so much as for equal end moments. On the other hand, as evident from Fig. 9.7, there will be little or possibly no magnification if the end moments are of opposite sign and produce an inflection point along the member.

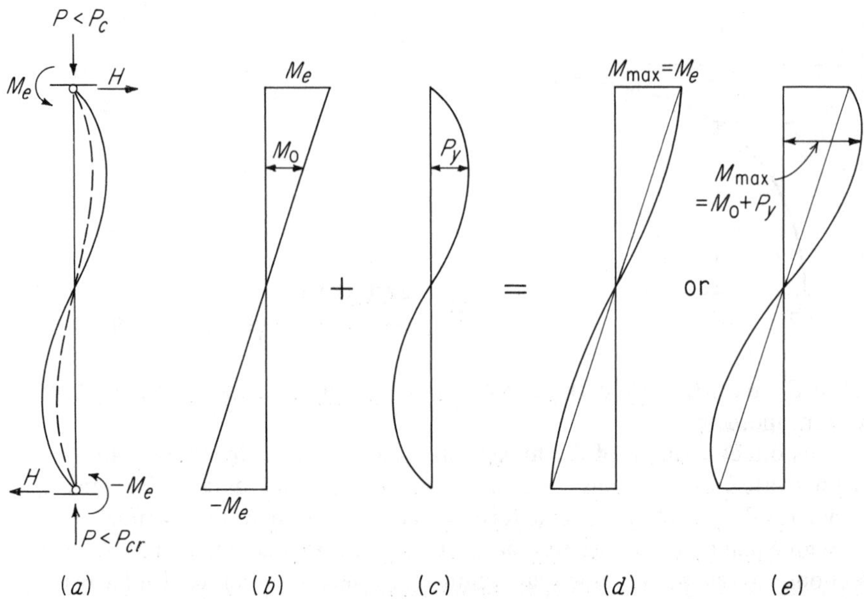

FIGURE 9.7
Moments in slender members with compression plus bending, bent in double curvature.

It can be shown (Ref. 9.2) that the way in which moment magnification depends on the relative magnitude of the two end moments (as in Figs. 9.4*a* and 9.7) can be expressed by a modification of Eq. (9.6):

$$M_{\max} = M_0 \frac{C_m}{1 - P/P_c} \tag{9.8}$$

where

$$C_m = 0.6 + 0.4 \frac{M_{1b}}{M_{2b}} \geq 0.4 \tag{9.9}$$

Here M_{1b} is the numerically smaller and M_{2b} the numerically larger of the two end moments; hence, by definition, $M_0 = M_{2b}$. The fraction M_{1b}/M_{2b} is defined as positive if the end moments produce single curvature and negative if they produce double curvature. It is seen that when $M_{1b} = M_{2b}$, as in Fig. 9.4*a*, $C_m = 1$, so that Eq. (9.8) becomes Eq. (9.6) as it should. It is to be noted that Eq. (9.9) *applies only to members braced against sidesway.* As will become apparent from the discussion that follows, for members not braced against sidesway, maximum moment magnification usually occurs, that is, $C_m = 1$.

Members that are braced against sidesway include columns that are parts of structures in which sidesway is prevented in one of various ways: by walls sufficiently strong and rigid in their own planes effectively to prevent horizontal displacement; by special bracing in vertical planes; in buildings by designing the utility core to resist horizontal loads and furnish bracing to the frames; or by bracing the frame against some other essentially immovable support.

If no such bracing is provided, *sidesway can occur only for the entire frame simultaneously*, not for individual columns in the frame. If this is the case, the combined effect of bending and axial load is somewhat different from that in braced columns. For illustration, consider the simple portal frame of Fig. 9.8*a* subject to a horizontal load H, such as a wind load, and compression forces P, such as from gravity loads. The moments M_0 caused by H alone, in the absence of P, are shown in Fig. 9.8*b*; the corresponding deformation of the frame is given in dashed curves. When P is added, horizontal moments are caused that result in the magnified deformations shown in solid curves and in the moment diagram of Fig. 9.8*c*. It is seen that the maximum values of M_0, both positive and negative, and the maximum values of the additional moments M_P of the same sign occur at the same locations, namely, at the ends of the columns. They are therefore fully additive, leading to a large moment magnification. In contrast, if the frame of Fig. 9.8 is laterally braced and vertically loaded, Fig. 9.9 shows that the maximum values of the two different moments occur in different locations; the moment magnification, if any, is therefore much smaller, as correctly expressed by C_m.

It should be noted that the moments that cause a frame to sidesway need not be caused by horizontal loads as in Fig. 9.8. Asymmetries, either of frame configuration or vertical loading or both, also result in sidesway displacements. In this case the presence of axial column loads results in the same deflection and moment magnification as before.

In summary, it can be stated as follows:

1. In flexural members the presence of axial compression causes additional deflections and additional moments Py. Other things being equal, the additional moments increase with increasing slenderness ratio kl/r.

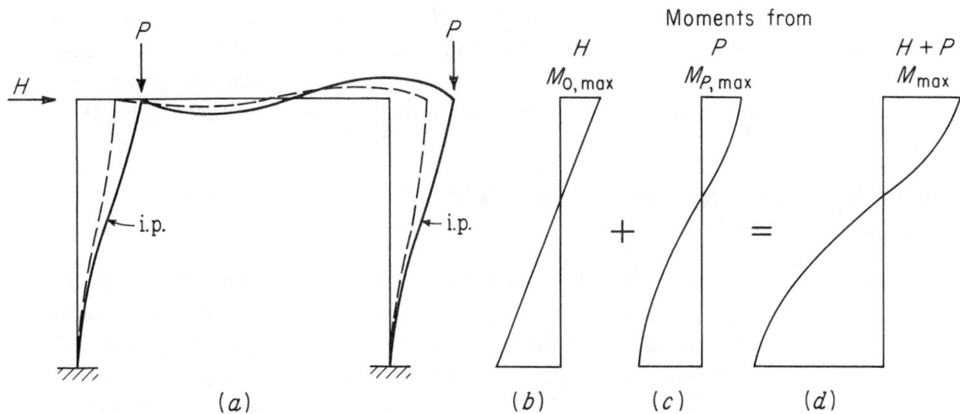

FIGURE 9.8
Fixed portal frame, laterally unbraced.

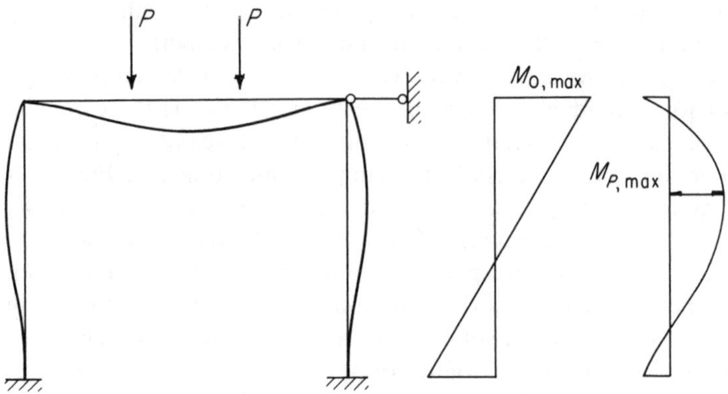

FIGURE 9.9
Fixed portal frame, laterally braced.

2. In *members braced against sidesway* and bent in single curvature, the maxima of both types of moments, M_0 and Py, occur at the same or at nearby locations and are fully additive; this leads to large moment magnifications. If the M_0 moments result in double curvature (i.e., in the occurrence of an inflection point), the opposite is true and less or no moment magnification occurs.
3. In *members of frames not braced against sidesway*, the maximum moments of both kinds, M_0 and Py, almost always occur at the same locations, the ends of the columns; they are fully additive, regardless of the presence or absence of an inflection point. Here, too, other things being equal, the additional deflections and the corresponding moments increase with increasing kl/r.

This discussion is a simplified presentation of a fairly complex subject. The provisions of the ACI Code regarding slender columns are based on the behavior and the corresponding equations which have just been presented. They take account, in an approximate manner, of the additional complexities that arise from the fact that concrete is not an elastic material, that tension cracking changes the moment of inertia of a member, and that under sustained load, creep increases the short-time deflections and, thereby, the moments caused by these deflections.

9.4 ACI CRITERIA FOR NEGLECT OF SLENDERNESS

The procedure of designing slender columns is inevitably fairly lengthy, particularly because it involves a trial and error process. At the same time, studies of existing structures have shown that most columns in actual buildings are sufficiently stocky that slenderness effects reduce their capacity by only a few percent. As stated in Chapter 8, an ACI-ASCE survey indicated that 90 percent of columns braced against sway, and 40 percent of unbraced columns, could be designed as short columns, i.e., they could develop essentially the full cross-sectional strength with little or no reduction resulting from slenderness (Ref. 3). Furthermore, lateral bracing is usually provided by shear walls, elevator shafts, stairwells, or other elements for which resistance to lateral deflection is much greater than for the columns of the building frame. It may be concluded that in most cases in reinforced concrete buildings, slenderness effects may be neglected.

To permit the designer to dispense with the complicated analysis required for slender column design for these ordinary cases, ACI Code 10.11.4 provides limits of slenderness below which the effects of slenderness are insignificant and may be neglected. These limits are adjusted to result in a maximum unaccounted reduction of column capacity of no more than 5 percent. The Code provisions are as follows:

1. For compression members braced against sidesway, the effects of slenderness may be neglected when kl_u/r is less than $34 - 12M_{1b}/M_{2b}$.
2. For compression members not braced against sidesway, the effects of slenderness may be neglected when kl_u/r is less than 22.

In these provisions, k is the effective length factor (see Sec. 9.2); l_u is the unsupported length taken as the clear distance between floor slabs, beams, or other members providing lateral support; M_{1b} is the smaller factored end moment due to loads that result in no appreciable sidesway, positive if the member is bent in single curvature and negative if bent in double curvature; and M_{2b} is the larger factored end moment due to loads that result in no appreciable sidesway, always positive.

The radius of gyration r for rectangular columns may be taken as $0.30h$, where h is the overall cross-sectional dimension in the direction in which stability is being considered. For circular members, it may be taken as 0.25 times the diameter. For other shapes, r may be computed for the gross concrete section.

Further guidance is provided in ACI Commentary 10.11.4, which notes that in most *braced frames* it is sufficiently accurate to use estimated values of effective length factor k in the above criteria for slenderness, as follows:

1. For columns hinged at both ends, k of 1.00 should be used.
2. For stocky columns restrained by flat slab floors, k ranges from about 0.95 to 1.00 and conservatively may be taken as 1.00.
3. For columns in beam-column frames, k ranges from about 0.75 to 0.90 and conservatively may be taken as 0.90.

9.5 ACI CRITERIA FOR BRACED VS. UNBRACED FRAMES

It is clear from the discussion of Sec. 9.3 that there are important differences in the behavior of slender columns in braced frames and corresponding columns in unbraced frames. ACI Code provisions and Commentary guidelines for the approximate design of slender columns reflect this, and there are separate provisions in each relating to the important parameters in braced vs. unbraced frames, including moment magnification factors and effective length factors.

In actual structures, a frame is seldom either completely braced or completely unbraced. It is necessary to determine in advance if bracing provided by shear walls, elevator and utility shafts, stairwells, or other elements is adequate to restrain the frame against significant sway effects. ACI Commentary 10.11.2 provides guidance in this respect, with two alternative methods.

In applying the moment magnifier method for slender column design, a compression member braced against sidesway is a member within a story in which horizontal displacements do not significantly affect the moments in the structure. When the *stability index*

$$Q = \frac{\Sigma P_u \Delta_u}{H_u h_s} \tag{9.10}$$

for a story is not greater than 0.04, the secondary moments should not exceed 5 percent of the first-order moments, and the structure can be considered braced (Ref. 9.3). In Eq. (9.10), H_u is the total factored lateral force acting within the story; Δ_u is the lateral deflection due to H_u (computed by first-order analysis neglecting sway effects) at the top of the story relative to the bottom of the story; ΣP_u is the sum of the total factored axial loads on all the columns of the story; and h_s is the height of the story, center to center of floors or roof.

An alternative, more approximate method is also suggested in ACI Commentary 10.11.2. A compression member may be assumed braced if it is located in a story in which the bracing elements (shear walls, etc.) have a total stiffness resisting lateral movement of the story at least 6 times the sum of the lateral stiffnesses of all the columns within the story.

9.6 ACI MOMENT MAGNIFIER METHOD FOR BRACED FRAMES

A slender reinforced concrete column reaches the limit of its strength when the combination of P and M at the most highly stressed section will make that section fail. In general, P is essentially constant along the length of the member. This means that the column approaches failure when at the most highly stressed section the axial force P combines with a moment $M = M_{\max}$, as given by Eq. (9.8), so that this combination becomes equal to P_n and M_n, which will cause the section to fail. This is easily visualized by means of Fig. 9.10.

For a column of given cross section, Fig. 9.10 presents a typical interaction curve. For simplicity, suppose that the column is bent in single curvature with equal eccentricities at both ends. For this eccentricity the strength of the cross section is given by point A on the interaction curve. If the column is stocky enough for the moment magnification to be negligibly small, then $P_{n,\text{stocky}}$ at point A represents the member strength of the column under the simultaneous moment $M_{n,\text{stocky}} = e_0 P_{n,\text{stocky}}$.

On the other hand, if the same column is sufficiently slender, significant moment magnification will occur with increasing P. Then the moment at the most

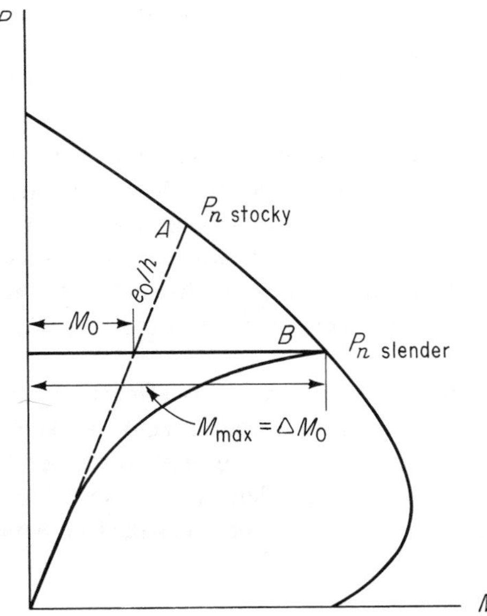

FIGURE 9.10
Effect of slenderness on carrying capacity.

highly stressed section is $M_{\max}$, as given by Eq. (9.8), with $C_m = 1$ because of equal end eccentricities. The solid curve in Fig. 9.10 shows the nonlinear increase of $M_{\max}$ as P increases. The point where this curve intersects the interaction curve, i.e., point B, defines the member strength $P_{n,\text{slender}}$ of the slender column, combined with the simultaneously applied end moments $M_0 = e_0 P_{n,\text{slender}}$. If end moments are unequal, the factor C_m is to be included, as discussed in Sec. 9.3.

ACI Code 10.11.5 specifies that axial loads and end moments in columns may be determined by a conventional elastic frame analysis (see Chap. 11). The member is then to be designed for that axial load and a simultaneous magnified column moment.

For a moment-resisting frame that is effectively braced against sidesway, the ACI Code equation for magnified moment, acting with the factored axial load P_u, can be written as follows:

$$M_c = \delta_b M_{2b} \tag{9.11}$$

where the moment magnification factor is

$$\delta_b = \frac{C_m}{1 - P_u/\phi P_c} \geq 1 \tag{9.12}$$

In Eqs. (9.11) and (9.12), the subscript b denotes a braced frame. The value of ϕ is the same as usual for strength design of columns and is equal to 0.70 and 0.75, respectively, for tied and spirally reinforced columns in the compression range of failure. The critical load P_c, in accordance with Eq. (9.1), is given as

$$P_c = \frac{\pi^2 EI}{(kl_u)^2} \tag{9.13}$$

where l_u is defined as the unsupported length of the compression member.

In Eq. (9.12), the value of C_m is as previously given by Eq. (9.9):

$$C_m = 0.6 + 0.4\frac{M_{1b}}{M_{2b}} \geq 0.4 \tag{9.9}$$

for members braced against sidesway and without transverse loads between supports. Here M_{2b} is the larger of the two end moments and M_{1b}/M_{2b} is positive when the end moments produce single curvature and negative when they produce double curvature. The variation of C_m with M_{1b}/M_{2b} is shown in Fig. 9.11. In Eq. (9.12), when the calculated value of δ_b is smaller than 1, this indicates that the larger of the two end moments, M_{2b}, is the largest moment in the column, a situation depicted in Fig. 9.7*d*.

In this way the ACI code provides for the capacity-reducing effects of slenderness in braced frames by means of the moment magnification factor δ_b. However, it is well known that for columns with no or very small applied moments, i.e., axially or nearly axially loaded columns, increasing slenderness also reduces the column strength. For this situation, ACI Code 10.11.5.4 provides the following guidelines:

> If computations show that there is no moment at both ends of a braced compression member or that computed end eccentricities are less than $(0.6 + 0.03h)$ in., M_{2b} shall be based on a minimum eccentricity of $(0.6 + 0.03h)$ in. about each principal axis separately. (Here h is the thickness of the section in the direction perpendicular to the pertinent principal axis.) The ratio M_{1b}/M_{2b} for calculating C_m shall be determined by either of the following:
>
> (*a*) When computed eccentricities are present, but less than $(0.6 + 0.03h)$ in., computed end moments may be used to evaluate M_{1b}/M_{2b} in calculating C_m.
>
> (*b*) If computations show that there is essentially no moment at both ends of the member, the ratio M_{1b}/M_{2b} shall be taken equal to 1.

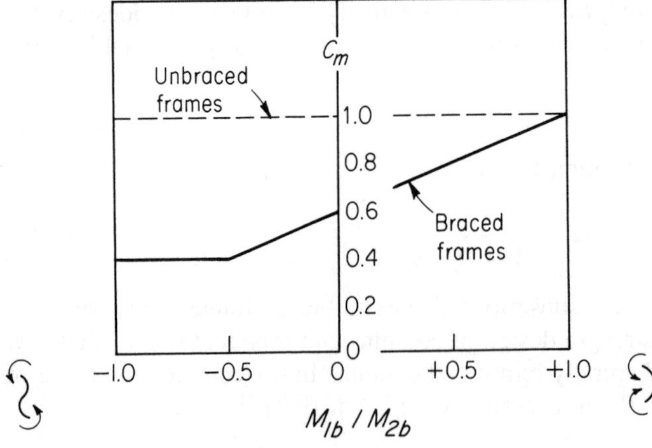

FIGURE 9.11
Values of C_m for slender columns in braced and unbraced frames.

In homogeneous elastic members, such as steel columns, the rigidity *EI* is easily obtained from Young's modulus and the usual moment of inertia. Reinforced concrete columns, however, are nonhomogeneous, since they consist of both steel and concrete. While steel is substantially elastic, concrete is not, and is, in addition, subject to creep and to cracking if tension occurs on the convex side of the column. All these factors affect the effective rigidity *EI* of a reinforced concrete member. It is possible by computer methods to calculate fairly realistic effective rigidities, taking account of these factors. Even these calculations are no more accurate than the assumptions on which they are based. On the basis of elaborate studies, both analytical and experimental (Ref. 9.4), the ACI Code permits *EI* to be determined by either

$$EI = \frac{E_c I_g/5 + E_s I_{se}}{1 + \beta_d} \tag{9.14}$$

or by the simpler expression

$$EI = \frac{E_c I_g/2.5}{1 + \beta_d} \tag{9.15}$$

where E_c = modulus of elasticity of concrete, psi
I_g = moment of inertia of gross section of column, in^4
E_s = modulus of elasticity of steel = 29,000,000 psi
I_{se} = moment of inertia of reinforcement about centroidal axis of member cross section, in^4
β_d = ratio of maximum factored axial dead load to maximum factored axial total load

The factor β_d accounts approximately for the effect of creep. That is, the larger the sustained dead loads the larger are the creep deformations and corresponding curvatures. Consequently, the larger the sustained loads relative to the temporary loads the smaller the effective rigidity, as correctly reflected in Eqs. (9.14) and (9.15). However, of the two materials, only concrete is subject to creep, while reinforcing steel as ordinarily used is not. It therefore appears that in the more accurate Eq. (9.14) the creep parameter $1 + \beta_d$ should be applied only to the term $E_c I_g/5$, which refers to concrete, but not to $E_s I_{se}$. This would improve economy of design, particularly for more heavily reinforced columns (large ρ ratios) and for most composite columns, i.e., columns reinforced with structural steel shapes.

Both Eqs. (9.14) and (9.15) are conservative lower limits for large numbers of investigated actual members (Ref. 9.3). The simpler but more conservative Eq. (9.15) is not unreasonable for lightly reinforced members, but greatly underestimates the effect of reinforcement for more heavily reinforced members, i.e., for the range of higher ρ values. Equation (9.14) is more reliable for the entire range of ρ and definitely preferable for medium and high ρ values (Ref. 9.5).

An accurate determination of the effective length factor k is essential in connection with Eqs. (9.11) and (9.13). In Sec. 9.2 it was shown that for frames braced against sidesway, k varies from $\frac{1}{2}$ to 1, while for laterally unbraced frames,

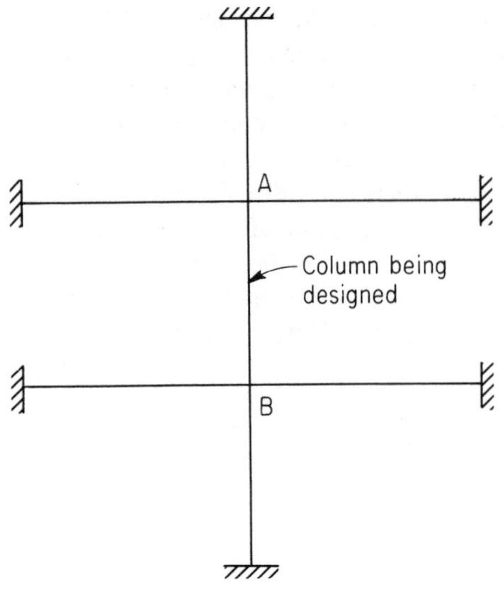

FIGURE 9.12
Section of rigid frame including column to be designed.

it varies from 1 to ∞, depending on the degree of rotational restraint at both ends. This was illustrated in Fig. 9.1. For rigid frames, it is seen that this degree of rotational restraint depends on whether the rigidities of the beams framing into the column at top and bottom are large or small compared with the rigidity of the column itself. An approximate but generally satisfactory way of determining k is by means of *alignment charts* based on isolating the given column plus all members framing into it at top and bottom, as shown in Fig. 9.12. The degree of end restraint at each end is $\psi = \Sigma(EI/l \text{ of columns})/\Sigma(EI/l \text{ of floor members})$. Only floor members that are in a plane at either end of the column are to be included. The value of k can then be read directly from the charts of Fig. 9.13 as illustrated by the dashed lines.†

It is seen that k must be known before a column in a frame can be dimensioned. Yet k depends on the rigidity EI/l of the members to be dimensioned, as well as on that of the abutting members. Thus, the dimensioning process necessarily involves iteration; i.e., one assumes member sizes, calculates rigidities and corresponding k values, and then calculates more accurate member sizes on the basis of these k values until assumed and final member sizes coincide or are satisfactorily close. The question of determining the correct rigidities EI/l for both beams and columns is not a simple one. In the near-failure state for which dimensioning is done, the beams will be cracked on the tension side and the columns may or may not show tension cracking, depending on the eccentricity

† Alternative to the use of charts are equations for the determination of effective length factors k, developed in Refs. 9.6, 9.7, and 9.8 and given in ACI Commentary 10.11.2. The equations are more convenient in developing computer solutions.

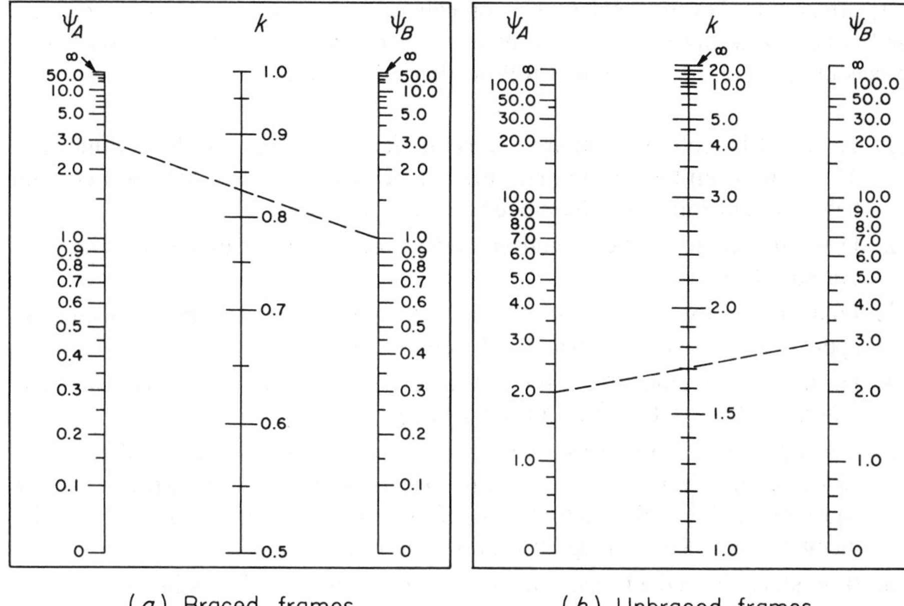

FIGURE 9.13
Alignment charts for effective length factors k.

ratio e/h. ACI Code 10.11.2 specifies that in determining k, due consideration shall be given to the effects on the rigidities of reinforcement and of cracking.

As a first approximation in using Fig. 9.13, when only a rough estimate of member sizes has been made, it is sufficiently accurate to use one-half of the gross moment of inertia for the beams and the gross moment of inertia for the columns. This reflects the fact that the reduction of rigidity because of cracking is generally larger for beams than for columns. ACI Commentary 10.11.2 notes that this basis will result in reasonable member sizes for columns with kl_u/r less than 60. The moment of inertia of T beams can be estimated as two times that of the rectangular cross section of width b_w and depth h; thus, the effects of cracking and flange contribution in T beams tend to cancel.

Once tentative member sizes have been established on this basis and a preliminary slenderness analysis made, a more accurate determination may be called for. Refinement is essential if kl_u/r is 60 or greater. Comparisons of a computer study with results of laboratory tests on simple frames have shown that the following procedure leads to satisfactory results. Calculate EI/l for the beams based on the moment of inertia of the cracked transformed section of the beams, taking the average between support and span sections if these sections are different, and calculate EI/l for the columns from Eq. (9.14), taking the creep factor $\beta_d = 0$ (Ref. 9.9).

Thus, the overall design sequence for slender columns is necessarily an iterative procedure because member sizes and reinforcement, unknown at the outset, affect such key parameters as moments of inertia, effective length factors, and critical buckling load. An outline of the separate steps in the analysis/design procedure for braced frames will follow along these lines:

1. Select a trial column section to carry the factored axial load P_u and moment $M_u = M_2$ from the elastic first-order frame analysis, assuming short column behavior and following the procedures of Chapter 8.
2. Determine if the frame should be considered braced or unbraced, using the criteria of Sec. 9.5.
3. Find the unsupported length l_u and the effective length factor k using the approximate values suggested at the end of Sec. 9.4.
4. For the trial column, check for consideration of slenderness effects using the criteria of Sec. 9.4 with estimated k values.
5. If slenderness is tentatively found to be important, refine the calculation of k based on the alignment chart of Fig. 9.13a, with member stiffnesses EI/l and rotational restraint factors ψ based on gross moments of inertia and trial member sizes. Recheck against the slenderness criteria.
6. If moments from the frame analysis are small, check to determine if minimum eccentricities and moments control.
7. Calculate the equivalent uniform moment factor C_m from Eq. (9.9).
8. Calculate β_d, EI from Eq. (9.14) or Eq. (9.15), and P_c from Eq. (9.13) for the trial column.
9. Calculate the moment magnification factor δ_b from Eq. (9.12) and magnified moment M_c from Eq. (9.11).
10. Check the adequacy of the column to resist axial load and magnified moment, using column design charts of Appendix A in the usual way. Revise the column section and reinforcement if necessary.
11. If results appear to justify it, refine the calculations for k, ψ, and P_c further, based on moments of inertia of the cracked beam and the column, in each case accounting for reinforcement. Determine the revised moment magnification factor and adjust the column design if necessary.

Example 9.1 Design of a Slender Column in a Braced Frame. Fig. 9.14 shows an elevation view of a multistory concrete frame building, with 48 in. wide × 12 in. deep beams on all column lines, carrying two-way slab floors and roof. The clear height of the columns is 13 ft. Interior columns are tentatively dimensioned at 18 × 18 in., and exterior columns at 16 × 16 in. The frame is effectively braced against sway by stair and elevator shafts having concrete walls monolithic with the floors, located in the building corners (not shown in the figure). The structure will be subjected to vertical dead and live loads. Trial calculations by first order analysis indicate that the pattern of live loading indicated in Fig. 9.14, with full load distribution on roof and upper floors and a checkerboard pattern adjacent to Column C3, produces maximum moments with single curvature in that column, at nearly maximum axial load. Dead loads act on all spans. Service load

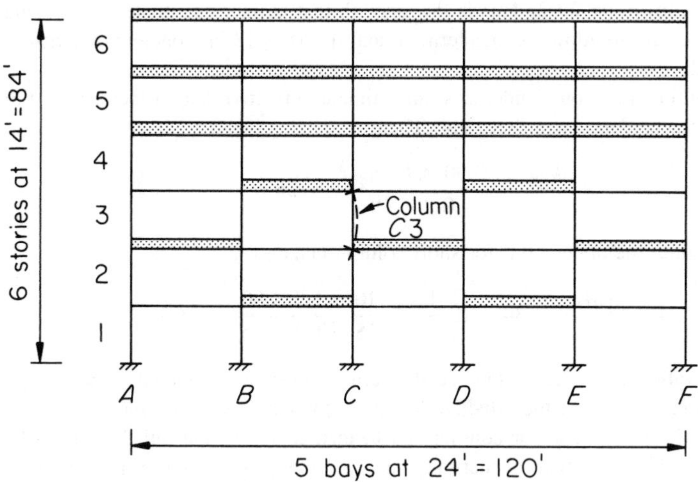

FIGURE 9.14
Concrete building frame of Example 9.1.

values of dead and live load axial force and moments for the typical interior column C3 are as follows:

Dead load	*Live load*
$P = 230$ kips	$P = 173$ kips
$M_2 = 18$ ft-kips	$M_2 = 95$ ft-kips
$M_1 = 10$ ft-kips	$M_1 = 90$ ft-kips

Design column C3, using the ACI moment magnifier method. Use $f_c' = 4000$ psi and $f_y = 60,000$ psi.

Solution. The column will first be designed as a short column, assuming no slenderness effect. With the application of the usual load factors,

$$P_u = 1.4 \times 230 + 1.7 \times 173 = 616 \text{ kips}$$

$$M_u = 1.4 \times 18 + 1.7 \times 95 = 187 \text{ ft-kips}$$

For an 18 × 18 in. column, with 1.5 in. clear to the outside steel, No. 3 stirrups, and (assumed) No. 10 longitudinal steel:

$$\gamma = (18.00 - 2 \times 1.50 - 2 \times 0.38 - 1.27)/18 = 0.72$$

Graph A.7 for $\gamma = 0.75$, with bars arranged around the column perimeter, will be used. Then

$$\frac{P_u}{A_g} = \frac{616}{324} = 1.90 \text{ ksi}$$

$$\frac{M_u}{A_g h} = \frac{187 \times 12}{324 \times 18} = 0.38 \text{ ksi}$$

and from the graph, $\rho_g = 0.02$. This is low enough that an increase in steel area could be made, if necessary to allow for slenderness, and the 18 × 18 in. concrete dimensions will be retained.

For an initial check on slenderness, an estimated effective length factor k = 0.9 will be used, in accordance with Sec. 9.4. Then

$$\frac{kl_u}{r} = \frac{0.90 \times 13 \times 12}{0.3 \times 18} = 26$$

For a braced frame, the upper limit for short column behavior is

$$34 - 12\frac{M_{1b}}{M_{2b}} = 34 - 12\,\frac{1.4 \times 10 + 1.7 \times 90}{1.4 \times 18 + 1.7 \times 95} = 23.3$$

The calculated value of 26 exceeds this, so slenderness must be considered in the design. A more refined calculation of the effective length factor k is thus called for.

Because E_c is the same for column and beams, it will be cancelled in the stiffness calculations. If the moment of inertia is based on the gross section of the columns, $I_g = 18 \times 18^3/12 = 8750$ in^4, so for the columns, $I_g/l_c = 8750/(14 \times 12) = 52.1$ in. Calculations for the beams will be based on the gross section of the web, assuming that the cracking effects and flange contributions will cancel. Thus $I_g = 48 \times 12^3/12 = 6910$ in^4, and $I_g/l_c = 6910/(24 \times 12) = 24.0$ in^3. Rotational restraint factors at the top and bottom of Column C3 are the same and are

$$\psi_a = \psi_b = \frac{52.1 + 52.1}{24.0 + 24.0} = 2.17$$

From Fig. 9.13*a* for the braced frame, the value of k is 0.87, rather than 0.90 as used previously. Consequently,

$$\frac{kl_u}{r} = \frac{0.87 \times 13 \times 12}{0.3 \times 18} = 25.1$$

This is still above the limit value of 23.3, confirming that slenderness must be considered.

A check will now be made of minimum moment. According to the ACI Code, the eccentricity must not be less than $(0.6 + 0.03 \times 18) = 1.14$ in., corresponding to a moment in this case of $616 \times (1.14/12) = 58$ ft-kips. It is seen that this does not control.

The coefficient C_m can now be found from Eq. (9.9) with $M_{1b} = 1.4 \times 10 + 1.7 \times 90 = 167$ ft-kips and $M_{2b} = 1.4 \times 18 + 1.7 \times 95 = 187$ ft-kips:

$$C_m = 0.6 + 0.4\,\frac{167}{187} = 0.96$$

Next the factor β_d will be found based on the ratio of factored dead to total axial loads:

$$\beta_d = \frac{1.4 \times 230}{1.4 \times 230 + 1.7 \times 173} = 0.52$$

For a relatively low column steel ratio, one estimated to be in the range of 0.02 to 0.03, the more approximate Eq. (9.15) for *EI* will be used, and

$$EI = \frac{3.60 \times 10^6 \times 8750}{2.5(1 + 0.52)} = 8.29 \times 10^9 \text{ in}^2\text{-lb}$$

The critical buckling load is found from Eq. (9.13) to be

$$P_c = \frac{\pi^2 EI}{(kl_u)^2} = \frac{\pi^2 \times 8.29 \times 10^9}{(0.87 \times 13 \times 12)^2} = 4.44 \times 10^6 \text{ lb}$$

The moment magnification factor can now be found from Eq. (9.12), with $\phi = 0.70$:

$$\delta_b = \frac{C_m}{1 - P_u/\phi P_c} = \frac{0.96}{1 - 616/0.7 \times 4440} = 1.20$$

Thus the required axial strength of the column is $P_u = 616$ kips (as before) and the magnified design moment is $M_c = \delta_b M_{2b} = 1.20 \times 187 = 224$ ft-kips. With reference again to the column design chart A.7 with

$$\frac{P_u}{A_g} = \frac{616}{324} = 1.90 \text{ ksi}$$

$$\frac{M_u}{A_g h} = \frac{224 \times 12}{324 \times 18} = 0.46 \text{ ksi}$$

it is seen that the required steel ratio is increased from 0.020 to 0.028 because of slenderness. The steel area now required is

$$A_{st} = 0.028 \times 324 = 9.07 \text{ in}^2$$

which can closely be provided using four No. 10 and four No. 9 bars, arranged as shown in Fig. 9.15. No. 3 ties will be used, at a spacing not to exceed the least dimension of the column (18 in.), 48 tie diameters (18 in.), or 16 bar diameters (18 in.). Single ties at 18 in. spacing, as shown in the figure, will meet requirements of the ACI Code.

Further refinements in the calculation of effective length factor k could be made, after the reinforcement in both the columns and beams has been selected, using the moment of inertia of the cracked transformed section of the T beams, and using Eq. (9.14) for the columns, with $\beta_d = 0$. The critical buckling load could also be refined using Eq. (9.14). These adjustments are not justified here because the column slenderness is barely above the upper limit for short column behavior, and the moment magnification is not great.

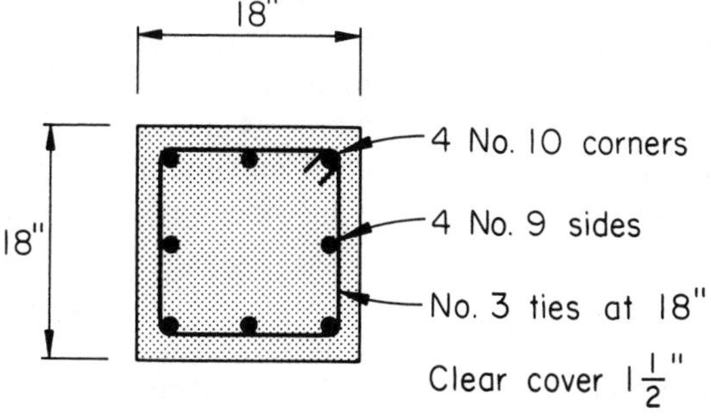

FIGURE 9.15
Cross section of column C3, Example 9.1.

9.7 ACI MOMENT MAGNIFIER METHOD FOR UNBRACED FRAMES

The important differences in behavior between columns braced against sidesway and columns for which sidesway is possible were discussed in Sec. 9.2 and Sec. 9.3. The critical load for a column, P_c, depends on the effective length kl_u, and although the effective length factor k falls between 0.5 and 1.0 for braced columns, it is between 1.0 and ∞ for columns that are unbraced (see Fig. 9.1 and Fig. 9.13). Consequently, an unbraced column will buckle at a much smaller load than an otherwise identical braced column.

Columns subject to sidesway do not normally stand alone but are part of a structural system including floors and roof. A floor or roof is normally very stiff in its own plane. Consequently, all columns at a given story level in a structure are subject to essentially identical sway displacements; i.e., sidesway of a particular story can occur only by simultaneous lateral motion of all columns of that story. Clearly, all columns at a given level must be considered together in evaluating slenderness effects relating to sidesway.

On the other hand, it is also possible for a single column in an unbraced frame to buckle individually under the gravity loads, the ends of the column being held against relative lateral movement by other, stiffer columns at the same floor level. This possibility, resulting in magnification of non-sway moments due to gravity loads, must also be considered in the analysis and design of slender columns in unbraced frames.

The ACI moment magnifier approach can still be used, but in frames subject to sidesway, two moment magnifiers are applicable, one relating to the individual column considered as a part of a braced frame and the other relating to the sway behavior of all of the columns in the given story. Thus, it is necessary, according to the ACI moment magnifier method, to separate the loads acting on a structure into two categories: loads that result in no appreciable sidesway, and loads that result in appreciable sidesway. Clearly two separate frame analyses are required, one for loads of each type. In general, gravity loads acting on reasonably symmetrical frames produce little sway, and the effects of gravity loads may therefore be placed in the first category. ACI Commentary 10.11.5.1 makes specific reference to tests and analysis confirming this, noting that sway magnification of the gravity moment by the sway multiplier is unwarranted.

The maximum magnified moments caused by sway loading occur at the ends of the column, but those due to gravity loads may occur somewhere in the midheight region of a column, the exact location of the latter varying depending on the end moments. The conservative approximation is made in ACI Code procedures that, although the magnified gravity moments and magnified sway moments do not occur at the same location, the actual maximum moment cannot exceed their sum. Consequently, for cases involving sidesway, equation Eq. (9.11) is replaced by

$$M_c = \delta_b M_{2b} + \delta_s M_{2s} \tag{9.16}$$

in which M_{2b} = the larger factored end moment due to loads that result in no appreciable sidesway (gravity loads, in the usual case)

M_{2s} = the larger factored end moment due to loads that result in appreciable sidesway (horizontal loads in the usual case)

δ_b = moment magnification factor for frames braced against sidesway, to reflect effects of curvature between the ends of the member

δ_s = moment magnification factor for frames not braced against sidesway, to reflect lateral drift resulting from lateral (and sometimes gravity) loads

The moment magnifier for the braced frame slenderness magnification for the column is calculated just as described in Sec. 9.6; that for the sway frame magnification is a common moment magnifier applicable to *all* columns in the story and is based on $\Sigma P_u / \Sigma P_c$, where ΣP_u is the total axial load on all the columns and ΣP_c is the total critical buckling load. The summations cover *all* columns in the story, including any short columns. The coefficient C_m is found as usual for the braced frame multiplier:

$$C_m = 0.6 + 0.4 \frac{M_{1b}}{M_{2b}} \geq 0.4 \tag{9.9}$$

but C_m for the sway frame multiplier is always equal to 1.0. Thus,

$$\delta_b = \frac{C_m}{1 - P_u / \phi P_c} \geq 1 \tag{9.17}$$

$$\delta_s = \frac{1}{1 - \Sigma P_u / \phi \Sigma P_c} \geq 1 \tag{9.18}$$

in which P_c is found from Eq. (9.13). In calculation of the moment magnifier δ_b in Eq. (9.17), P_c should be based on effective length factors k for *braced* frames. In calculating δ_s in Eq. (9.18), effective length factors k for *unbraced frames* should be used in finding P_c values.

When gravity loads are applied to an unbraced frame, either unsymmetrical load arrangements or lack of symmetry of the frame itself will normally produce some sway. Unless the lateral deflection of a floor due to such effects is appreciable, the minor effect of this sway component can be neglected, and the corresponding moments considered as nonsway moments to be magnified by the *braced frame magnifier*, even though the frame is, in fact, unbraced. According to ACI Commentary 10.11.5.1, sway effects can be neglected if the first-order analysis indicates that

$$\frac{\Delta}{l_u} \leq \frac{1}{1500}$$

where Δ is computed considering only the nonlateral loads on the unbraced frame.

The sequence of design steps for slender columns in braced frames will be similar to that outlined in Sec. 9.6 for unbraced frames, except for the use of two separate moment magnification factors according to Eq. (9.16). This, in turn, requires that the loads be separated into gravity loads, which are assumed to produce no sway, and horizontal loads producing sway and requires that separate frame analyses be made to determine the effects of each. It will be noted that, according to ACI Code 9.2 (see also Table 1.2 of Chapter 1), if wind effects W are included in the design, three possible factored load combinations are to be applied:

$$U = 1.4D + 1.7L$$
$$U = 0.75(1.4D + 1.7L + 1.7W) \text{ with } L \text{ equal to both its full value and 0}$$
$$U = 0.9D + 1.3W$$

Similar provisions are included for cases where earthquake loads are to be considered. This represents a significant complication in the sway frame analysis; however, the factored loads can be separated into gravity effects and sway effects, as required, and a separate analysis can be performed for each.

For unbraced frames, just as for braced frames, it may be necessary to account for the $P\Delta$ effect even though computed end moments are very small. ACI Code 10.11.5.5 provides guidance as follows:

> If computations show that there is no moment at both ends of a compression member not braced against sidesway or that computed end eccentricities are less than $(0.6 + 0.03h)$ in., M_{2s} in Eq (9.16) shall be based on a minimum eccentricity of $(0.6 + 0.03h)$ in. about each principal axis separately.

This provision for minimum sway moment would not be applicable for the first load combination above, $1.4D + 1.7L$, unless the lateral deflection due to gravity dead and live service loads is appreciable, with $\Delta/l_u > 1/1500$. In that case, minimum moments should be calculated at the stated eccentricity and included in the analysis, with the corresponding sway magnifier.

It is pointed out in Ref. 9.10 that designing for at least the minimum eccentricity of loading in sway frame is roughly equivalent to designing for an accidental lean of the building.

It is important to realize that, for frames not braced against sidesway, *the beams must be designed for the total magnified end moments of the compression members at the joint.* Even though the columns may be very rigid, if plastic hinges were to form in the restraining beams adjacent to the joints, the effective column length would be greatly increased and the critical column load much reduced.

Example 9.2 Design of Slender Columns in Unbraced Frame. Consider now that the concrete building frame of Example 9.1 is *unbraced* in the plane of the sketch, without the stairwells or elevator shafts described earlier. In the absence of such bracing, column sizes will be increased tentatively to 20×20 in. for interior columns and 18×18 in. for exterior columns. Beam sizes and clear heights are the same as for Example 9.1. The building will be subjected to gravity dead and live loads and horizontal wind loads. Elastic first-order analysis of the frame at service loads (all load factors $=$ 1.0) gives the following results at the third story:

	Cols. A3 and F3	Cols. B3 and E3	Cols. C3 and D3
P_{dead}	115 kips	230 kips	230 kips
P_{live}	90 kips	173 kips	173 kips
P_{wind}	±30 kips	±18 kips	±6 kips
$M_{2,\text{dead}}$			18 ft-kips
$M_{2,\text{live}}$			95 ft-kips
$M_{2,\text{wind}}$			±89 ft-kips
$M_{1,\text{dead}}$			10 ft-kips
$M_{1,\text{live}}$			90 ft-kips
$M_{1,\text{wind}}$			±89 ft-kips

Column C3 is to be designed for the critical loading condition, using $f'_c = 4000$ psi and $f_y = 60,000$ psi as before.

Solution. The column size and reinforcement must satisfy requirements for each of the three load combinations noted above.

(a) Gravity loads only. Considering first the gravity loads acting alone, assume that lateral sway is negligible because of frame symmetry, even though partial live loadings would be unsymmetrical and might produce some slight drift. Then $U = 1.4D + 1.7L$, and with reference to the load table, for column C3,

$$P_u = 1.4 \times 230 + 1.7 \times 173 = 616 \text{ kips}$$

$$M_u = 1.4 \times 18 + 1.7 \times 95 = 187 \text{ ft-kips}$$

Short column behavior will be assumed initially. For a 20 × 20 in. column, with 1.5 in. clear to the outside steel, No. 3 ties, and (assumed) No. 11 main bars,

$$\gamma = (20.00 - 2 \times 1.50 - 2 \times 0.38 - 1.41)/20 = 0.74$$

Graph A.7 for $\gamma = 0.75$, with bars arranged around the column perimeter, will be used. The needed parameters are

$$\frac{P_u}{A_g} = \frac{616}{400} = 1.54 \text{ ksi}$$

$$\frac{M_u}{A_g h} = \frac{187 \times 12}{400 \times 20} = 0.28 \text{ ksi}$$

From Graph A.7, it appears that the required steel ratio is less than the ACI Code minimum of 0.01 and that the column size can be reduced. However, anticipating results of the other load cases, tentatively retain the 20 × 20 in. column size. Now checking for slenderness, and basing moments of inertia on the gross concrete sections, find the following:

$$\text{Columns: } I_g = 20 \times 20^3/12 = 13,300 \text{ in}^4$$

$$I_g/l_c = 13,300/(14 \times 12) = 79.2 \text{ in}^3$$

$$\text{Beams: } I_g = 48 \times 12^3/12 = 6910 \text{ in}^4$$

$$I_g/l_c = 6910/(24 \times 12) = 24.0 \text{ in}^3$$

The rotational restraint factors at the top and bottom are thus

$$\psi_a = \psi_b = \frac{79.2 + 79.2}{24.0 + 24.0} = 3.30$$

and, with reference to the alignment charts of Fig. 9.13*a* and Fig. 9.13*b*, effective length factors are $k = 0.89$ for a braced frame and (for later reference) $k = 1.90$ for an unbraced frame. For the present load case with negligible sway, the single column is assumed braced, so

$$\frac{kl_u}{r} = \frac{0.89 \times 13 \times 12}{0.3 \times 20} = 23.1$$

The limit value for short column behavior for nonsway effects is

$$34 - 12\,\frac{1.4 \times 10 + 1.7 \times 90}{1.4 \times 18 + 1.7 \times 95} = 23.3$$

which is larger than the calculated 23.1, indicating that the member can be designed as a short column for the case with gravity loads acting alone. For this load case, a 20 × 20 in. column with steel ratio equal to the minimum of 0.01 will be adequate.

(b) Gravity plus wind load. Two load combinations must be considered when wind effects are included: $U = 0.75(1.4D + 1.7L + 1.7W)$ and $U = 0.9D + 1.3W$. The combination $U = 0.9D + 1.3W$ is unlikely to control here, because for column C3 uplift from wind is very small compared with dead load. It will not be considered further. Then, first assuming short column behavior with no slenderness effects, check the column for the full factored dead loads and moments:

$$P_u = 0.75(1.4 \times 230 + 1.7 \times 173 + 1.7 \times 6) = 469\,\text{kips}$$

$$M_u = 0.75(1.4 \times 18 + 1.7 \times 95 + 1.7 \times 89) = 254\,\text{ft-kips}$$

The column parameters are thus

$$\frac{P_u}{A_g} = \frac{469}{400} = 1.17\text{ ksi}$$

$$\frac{M_u}{A_g h} = \frac{254 \times 12}{400 \times 20} = 0.38\text{ ksi}$$

Graph A.7 indicates the proposed 20 × 20 in. column with 1 percent steel is just satisfactory for this combined load case, if slenderness is not a factor. However, the column must be checked for slenderness when considered as a part of an unbraced frame. With the value of $k = 1.90$, found earlier from the unbraced alignment chart,

$$\frac{kl_u}{r} = \frac{1.90 \times 13 \times 12}{0.3 \times 20} = 49.4$$

This is much above the limit value of 22 for short column behavior in an unbraced frame, so slenderness must be considered here. The loads must therefore be separated into gravity loads and sway loads, and the appropriate magnification factor must be computed and applied for each case. The factored end moment resulting from the nonsway loads on column C3 is

$$M_{2b} = 0.75(1.4 \times 18 + 1.7 \times 95) = 140\,\text{ft-kips}$$

The nonsway multiplier, δ_b, is found as follows:

$$\beta_d = \frac{0.75(1.4 \times 230)}{0.75(1.4 \times 230 + 1.7 \times 173 + 1.7 \times 6)} = 0.51$$

$$EI = \frac{3.6 \times 10^6 \times 13300}{2.5(1 + 0.51)} = 12.68 \times 10^9 \text{ in}^2\text{-lb}$$

$$P_c = \frac{\pi^2 EI}{(kl_u)^2} = \frac{\pi^2 \times 12.68 \times 10^9}{(0.89 \times 13 \times 12)^2} = 6.49 \times 10^6 \text{ lb}$$

$$C_m = 0.6 + 0.4 \frac{167}{187} = 0.96$$

$$\delta_b = \frac{C_m}{1 - P_u/\phi P_c} = \frac{0.96}{1 - 469/(0.7 \times 6490)} = 1.07$$

Next, the sway effects will be considered, for which

$$M_{2s} = 0.75(1.7 \times 89) = 113 \text{ ft-kips}$$

The sway moment multiplier will be computed next. This requires calculation of the sum of all factored axial forces on the columns of the floor under consideration, and the sum of all critical buckling loads. In reference to the load table,

Columns A3 and F3: $P_u = 0.75(1.4 \times 115 + 1.7 \times 90 + 1.7 \times 30) = 274$ kips

Columns B3 and E3: $P_u = 0.75(1.4 \times 230 + 1.7 \times 173 + 1.7 \times 18) = 485$ kips

Columns C3 and D3: $P_u = 0.75(1.4 \times 230 + 1.7 \times 173 + 1.7 \times 6) = 469$ kips

Thus, $\Sigma P_u = 2(274 + 485 + 469) = 2456$ kips. Critical loads P_c must be calculated for each of the columns as follows. For columns A3 and E3,

Columns: $I_g = 18 \times 18^3/12 = 8750 \text{ in}^4$. and $I_g/l_c = 8750/(14 \times 12)$
$= 52.1 \text{ in}^3$

Beam: $I_g = 6910 \text{ in}^4$. and $I_g/l_c = 6910/(24 \times 12) = 24.0 \text{ in}^3$.

Rotational restraint factors for this case, with two columns and one beam framing into the joint, are

$$\psi_a = \psi_b = \frac{52.1 + 52.1}{24.0} = 4.34$$

which, with reference to the alignment chart for unbraced frames, gives $k = 2.08$. Next β_d is found:

$$\beta_d = \frac{0.75(1.4 \times 115)}{0.75(1.4 \times 115 + 1.7 \times 90 + 1.7 \times 30)} = 0.44$$

after which

$$EI = \frac{3.6 \times 10^6 \times 8750}{2.5(1 + 0.44)} = 8.75 \times 10^9 \text{ in}^2\text{-lb}$$

Then the critical load is

$$P_c = \frac{\pi^2 \times 8.75 \times 10^9}{(2.08 \times 13 \times 12)^2} = 0.82 \times 10^6 \text{ lb}$$

For columns B3 and E3, from the earlier calculations for column C3, $k = 1.90$ for the sway loading case. For these columns

$$\beta_d = \frac{0.75(1.4 \times 230)}{0.75(1.4 \times 230 + 1.7 \times 173 + 1.7 \times 18)} = 0.50$$

$$EI = \frac{3.6 \times 10^6 \times 13300}{2.5(1 + 0.50)} = 12.77 \times 10^9 \text{ in}^2\text{-lb}$$

$$P_c = \frac{\pi^2 \times 12.77 \times 10^9}{(1.90 \times 13 \times 12)^2} = 1.43 \times 10^6\text{-lb}$$

Finally, for columns C3 and D3, with $k = 1.90$, $\beta_d = 0.51$, and $EI = 12.68 \times 10^9$ in^2-lb,

$$P_c = \frac{\pi^2 \times 12.68 \times 10^9}{(1.90 \times 13 \times 12)^2} = 1.42 \times 10^6 \text{ lb}$$

Thus, for all the columns at this level of the structure,

$$\Sigma P_c = 2(820 + 1430 + 1420) = 7340 \text{ kips}$$

and finally, the sway moment magnification factor is

$$\delta_s = \frac{1}{1 - 2456/0.7 \times 7340} = 1.92$$

The total magnified moment is thus

$$M_c = \delta_b M_{2b} + \delta_s M_{2s} = 1.07 \times 140 + 1.92 \times 113 = 367 \text{ ft-kips}$$

combined with factored axial load of $P_u = 469$ kips. In reference to Graph A.7 with column parameters

$$\frac{P_u}{A_g} = \frac{469}{400} = 1.17 \text{ ksi}$$

$$\frac{M_u}{A_g h} = \frac{367 \times 12}{400 \times 20} = 0.55 \text{ ksi}$$

it is seen that $\rho_g = 0.025$. This is substantially above the steel ratio needed if slenderness were not a factor, but is still within the practical and economical range. The required steel area of

$$A_{st} = 0.025 \times 400 = 10.00 \text{ in}^2$$

will be provided using eight No. 10 bars, arranged as shown in Fig. 9.16. Spacing of No. 3 ties must not exceed the least dimension of the column, 48 ties diameters, or 16 main bar diameters. The second criterion controls, and No. 3 ties at 18 in spacing will be used in the pattern shown in Fig. 9.16.

Further refinement is possible in this design by using cracked transformed section properties for the beams and I for the columns accounting for the reinforcement, thus modifying the rotational restraint factors, the effective lengths, the critical buckling loads, and eventually the moment magnification factors. This would add little to the example and will not be carried out here.

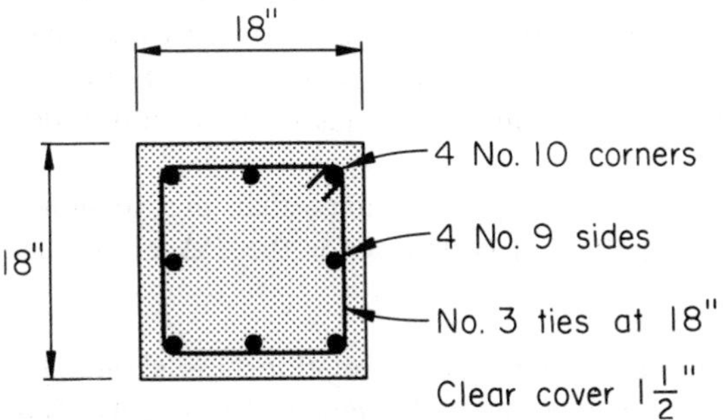

FIGURE 9.16
Cross section of column C3, Example 9.2.

9.8 SECOND ORDER ANALYSIS FOR SLENDERNESS EFFECTS

It may be evident from the preceding examples that, although the moment magnifier method of ACI Code 10.11 works well enough for braced frames, its application to unbraced frames is complicated, with many opportunities for error. With the general availability of computers in design offices, and because of the increasing complexity of the moment magnifier method as it is further refined with each edition of the ACI Code, there is increasing interest in a rational second-order frame analysis, or $P\Delta$ analysis, in which the effect of lateral deflection on moments, axial forces, and, in turn, lateral deflections is computed directly. The resulting moments and deflections include the effects of slenderness, and so the problem is strictly nonlinear.

With reference to Eq. (9.16) for magnified moments in unbraced frames, the total magnified moment is the sum of nonsway effects and sway effects. If a second order analysis is used to account for the moments associated with lateral deflection of the joints of a frame, Eq. (9.16) is replaced by

$$M_c = \delta_b M_{2b} + M_{\text{second order}} \tag{9.19}$$

Here, the second term of Eq. (9.16) is replaced by the results of a direct calculation of sway effects. The first term still must be included, and usually nonsway effects resulting from column curvature between the joints would be accounted for by the usual moment magnifier method.

The second order analysis is encouraged in ACI Code 10.10.1 and ACI Commentary 10.10.1. A second order analysis is *required* by ACI Code 10.11.4.3 for all compression members with kl_u/r greater than 100. According to the Code, "Such analysis shall take into account the influence of axial loads and variable moment of inertia on member stiffness and fixed-end moments, the effect of deflections on moments and forces, and the effects of duration of loads." ACI Code 10.10.3 notes that the detailed requirements of ACI Code 10.11 (the moment magnifier method) need not be applied if slenderness effects in compression members are evaluated in accordance with a rational analysis including all of the effects just listed.

A rational second order analysis will give a better approximation to actual moments and forces than will the moment magnifier method. Differences are particularly significant for irregular frames, frames subject to significant sway forces, or lightly braced frames. There may be important economies in the resulting design.

ACI Commentary 10.10.1 states five considerations that are regarded as minimum for an adequate frame analysis accounting for the $P\Delta$ effect, as follows:

1. Realistic moment-curvature or moment end-rotation relationships should be used to provide accurate values of deflections and secondary moments. Because column design and stability considerations are considered at the ultimate limit state, the stiffness used in an elastic analysis should be representative of this state. In lieu of more precise values, it is satisfactory to take EI as $E_c I_g(0.2 + 1.2\rho_g E_s/E_c)$ in computing column stiffnesses and $0.5E_c I_g$ when computing the beam stiffnesses.
2. The effect of foundation rotations on the lateral deformations should be considered.
3. It is necessary to consider the effect of axial loads on the stiffnesses and carryover factors for very slender columns ($l_u/r > 45$).
4. In frames subject to sustained lateral loads, as in the case of buildings resisting a horizontal reaction from an arch or unbalanced horizontal earth forces, and in frames in which unbalanced dead loads give rise to differential shortening of the two sides of a building resulting in lateral deflections, the effects of creep should be considered.
5. The maximum moment in the compression members should be determined considering the effects of the lateral deflections of the frame and the deflections of the compression member itself.

Only recently have practical methods been developed for performing a full second-order analysis (Refs. 9.3, 9.10, 9.11, 9.12, and 9.13). General purpose programs are commercially available that perform a full nonlinear analysis including sway effects. However, existing linear first order analysis programs can be modified to produce acceptable results, and ordinary desktop computers can be used for this purpose. This requires an iterative approach, which can be summarized as follows.

Fig. 9.17a shows a simple frame subject to lateral loads H and vertical loads P. The lateral deflection Δ is calculated by ordinary first-order analysis. As the frame is displaced laterally, the column end moments must equilibrate the lateral loads and a moment equal to $(\Sigma P)\Delta$:

$$\Sigma(M_{\text{top}} + M_{\text{bot}}) = Hl_c + \Sigma P\Delta \tag{9.20}$$

where Δ is the lateral deflection of the top of the frame with respect to the bottom, and ΣP is the sum of the vertical forces acting. The moment $\Sigma P\Delta$ in a given story can be represented by equivalent shear forces $(\Sigma P)\Delta/l_c$, where l_c is the story height, as shown in Fig. 9.17b. These shears give an overturning moment equal to that of the loads P acting at a displacement Δ.

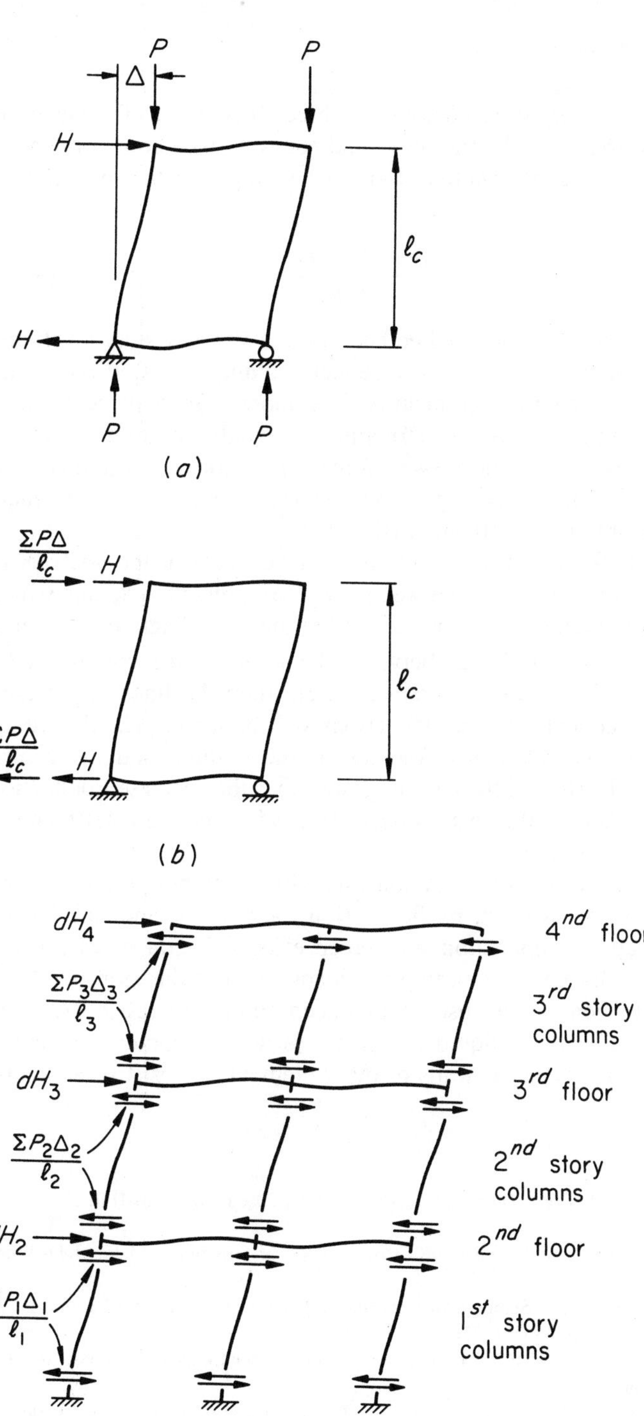

FIGURE 9.17
Basis for iterative $P\Delta$ analysis: (*a*) vertical and lateral loads on rectangular frame; (*b*) real lateral forces H and fictitious sway forces dH; (*c*) three-story frame subject to sway forces. (*Adapted from Ref. 9.10.*)

Fig. 9.17*c* shows the story shears acting in a three-story frame. The algebraic sum of the story shears from the columns above and below a given floor correspond in effect to a sway force dH acting on that floor. For example, at the second floor the sway force is

$$dH_2 = \frac{\Sigma P_1 \Delta_1}{l_1} - \frac{\Sigma P_2 \Delta_2}{l_2} \tag{9.21}$$

The sway forces must be added to the applied lateral force H at any story level, and the structure is then reanalyzed, giving new deflections and increased moments. If the lateral deflections increase significantly (say more than 5 percent), new dH sway forces are computed, and the structure is reanalyzed for the sum of the applied lateral forces and the new sway forces. Iteration is continued until changes are insignificant. Generally one or two cycles of iteration are adequate for structures of reasonable lateral stiffness (Ref. 9.3).

It is noted in Ref. 9.10 that a correction must be made in the analysis to account for the differences in shape between the $P\Delta$ moment diagram which has the same shape as the deflected column, and the moment diagram associated with the $P\Delta/l$ forces, which is linear between the joints at the column ends. The area of the actual $P\Delta$ moment diagram is larger than the linear equivalent representation, and consequently lateral deflections will be larger. The difference will vary depending on the relative stiffnesses of the column and the beams framing into the joints. In Ref. 9.10 it is suggested that the increased deflection can be accounted for by taking the sway forces dH as 15 percent greater than the calculated value for each iteration.

The accuracy of the results of a $P\Delta$ analysis will be strongly influenced by the values of member stiffness used, by foundation rotations, if any, and by the effects of concrete creep. In connection with creep effects, lateral loads causing significant sway are usually wind or earthquake loads of short duration, so creep effects are minimal. In general the use of unbraced frames to resist *sustained* lateral loads, e.g., from earth or liquid pressures, is not recommended, and it would be preferable to include shear walls or other elements to resist these loads.

REFERENCES

9.1. S. P. Timoshenko and J. M. Gere, *Theory of Elastic Stability*, 3d ed., McGraw-Hill, New York, 1969.

9.2. B. G. Johnson (ed.), *Guide to Stability Design Criteria for Metal Structures*, 3d ed., John Wiley and Sons, New York, 1976.

9.3. J. G. MacGregor and S. E. Hage, "Stability Analysis and Design of Concrete Frames," *Proc. ASCE*, vol. 103, no. ST 10, 1977, pp. 1953–1977.

9.4. J. G. MacGregor, J. E. Breen, and E. O. Pfrang, "Design of Slender Concrete Columns," *J. ACI*, vol. 67, no. 1, 1970, pp. 6–28.

9.5. J. G. MacGregor, V. H. Oelhafen, and S. E. Hage, "A Reexamination of the *EI* Value for Slender Columns," *Reinforced Concrete Columns*, American Concrete Institute, Detroit, 1975, pp. 1–40.

9.6. *Code of Practice for the Structural Use of Concrete*, Part 1, "Design Materials and Workmanship," (CP110: Part 1, 1972), British Standards Institution, London, 1972.

9.7. W. B. Cranston, "Analysis and Design of Reinforced Concrete Columns," *Research Report No. 20*, Paper 41.020, Cement and Concrete Association, London, 1972.

9.8. R. W. Furlong, "Column Slenderness and Charts for Design," *J. ACI*, vol. 68, no. 1, 1971, pp. 9–18.
9.9. J. E. Breen, J. G. MacGregor, and E. O. Pfrang, "Determination of Effective Length Factors For Slender Concrete Columns," *J. ACI*, vol. 69, no. 11, 1972, pp. 669–672.
9.10. J. G. MacGregor, *Reinforced Concrete*, Prentice Hall, Englewood Cliffs, NJ, 1988.
9.11. B. R. Wood, D. Beaulieu, and P. F. Adams, "Column Design by *P*-Delta Method," *Proc. ASCE*, vol. 102, no. ST2, 1976, pp. 411–427.
9.12. B. R. Wood, D. Beaulieu, and P. F. Adams, "Further Aspects of Design by *P*-Delta Model," *Proc. ASCE*, vol. 102, no. ST3, 1976, pp. 487–500.
9.13. R. W. Furlong, "Rational Analysis of Multistory Concrete Structures," *Conc. Int.* vol. 3, no. 6, 1981, pp. 29–35.

PROBLEMS

9.1 The 15 × 15 in. column shown in Fig. P9.1 must extend from footing level to the second floor of a braced frame structure with an unsupported length of 20.5 ft. Exterior exposure requires 2 in. clear cover for the outermost steel. Analysis indicates the critical loading results in the following service loads: (*a*) from dead loads, $P = 150$ kips, $M_{\text{top}} = 29$ ft-kips, $M_{\text{bot}} = 14.5$ ft-kips; (*b*) from live loads, $P = 90$ kips, $M_{\text{top}} = 50$ ft-kips, $M_{\text{bot}} = 25$ ft-kips, with the column bent in double curvature as shown. The effective length factor k determined from preliminary calculations is 0.90. Material strengths are $f_c' = 4000$ psi and $f_y = 60,000$ psi. Using the ACI moment-magnifier method, determine whether the column is adequate to resist these loads.

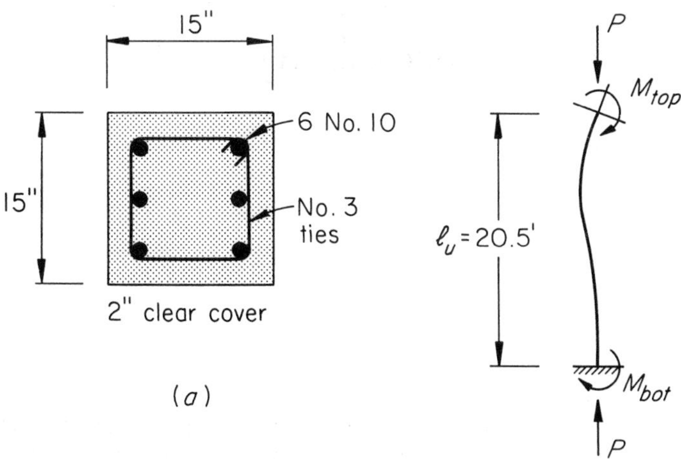

FIGURE P9.1

9.2 The structure shown in Fig. P9.2*a* requires tall slender columns at the left side. It is fully braced by shear walls on the right. All columns are 16 × 16 in., as in Fig. P9.2*b*, and all beams are 24 × 18 in. with 6 in. monolithic floor slab, as in Fig. P9.2*c*. Trial calculations call for column reinforcement as shown and for beam steel an area of 6.00 in^2 for both negative and positive bending, with effective depth 15.5 in. Alternate load analysis indicates the critical condition with column *AB* bent in single curvature, and service load thrust and moments as follows: from dead load, $P = 139$ kips, $M_{top} = 61$ ft-kips, $M_{bot} = 41$ ft-kips; from live load, $P = 93$ kips, $M_{top} = 41$ ft-kips, $M_{bot} = 27$ ft-kips. Material strengths are $f'_c = 4000$ psi and $f_y = 60,000$ psi. Is the proposed column, reinforced as shown, satisfactory for this load condition?

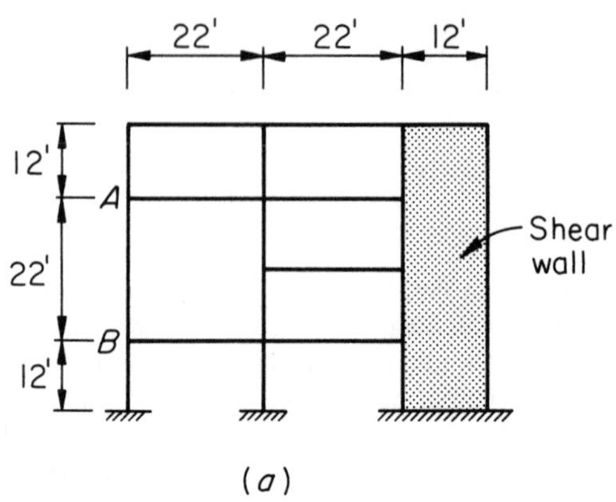

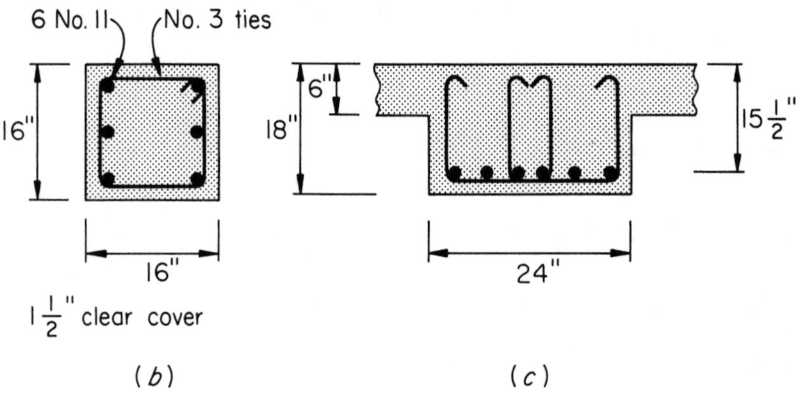

FIGURE P9.2

9.3 Refine the calculations of Problem 9.2, accounting for the reinforcement in the beams and columns in the calculation of *EI* as recommended in Sec. 9.6. Assume reinforcement will be approximately as given in Problem 9.2. Comment on your results.

9.4 An interior column in a braced frame has unbraced length of 20 ft and carries the following service load forces and moments: (*a*) from dead loads, $P = 180$ kips, $M_{top} = 28$ ft-kips, $M_{bot} = 28$ ft-kips; (*b*) from live loads, $P = 220$ kips, $M_{top} = 112$ ft-kips, $M_{bot} = 112$ ft-kips, with the column bent in single curvature. Rotational restraint factors at the top and bottom may be taken equal to 1.0. Design a square-tied column to resist these loads, with reinforcement ratio of about 0.02. Use $f'_c = 4000$ psi and $f_y = 60,000$ psi.

9.5 In reference to the braced frame of Example 9.1, shown in Fig. 9.14, the exterior column A3 carries the following service loads for the critical loading pattern:

Dead load	**Live load**
P = 153 kips	P = 87 kips
M = 75 ft-kips	M = 80 ft-kips
M = 65 ft-kips	M = 70 ft-kips

with the column bent in single curvature. Find the required reinforcement for the column, considering slenderness effects, using the ACI moment magnifier method. Trial column dimensions of 16 × 16 in may be modified if appropriate. Stiffness calculations may be based on gross concrete sections, modified for flange effect and cracking in the beams as suggested in Sec. 9.6. Material strengths are as given for Example 9.1.

CHAPTER

10

DESIGN OF REINFORCEMENT AT JOINTS

10.1 INTRODUCTION

Most reinforced concrete failures occur not because of any inadequacies in analysis of the structure or in design of the members but because of inadequate attention to the detailing of reinforcement. Most often, the problem is at the connections of main structural elements (Ref. 10.1).

There is an unfortunate tendency in modern structural practice for the engineer to rely upon a detailer, employed by the reinforcing bar fabricator, to provide the joint design. Certainly, in many cases, standard details such as those found in the ACI Detailing Manual (Ref. 10.2) can be followed, but only the design engineer, with the complete results of analysis of the structure at hand, can make this judgment. In many other cases, special requirements for force transfer require that joint details be fully specified on the engineering drawings, including bend configurations and cutoff points for main bars and provision of supplementary reinforcement.

The basic requirement at joints is that all of the forces existing at the ends of the members must be transmitted through the joint to the supporting members. Complex stress states exist at the junction of beams and columns, for example, that must be recognized in designing the reinforcement. Sharp discontinuities occur in the direction of internal forces, and it is essential to place reinforcing bars, properly anchored, to resist the resulting tension. Some frequently used connection details, when tested, have been found to provide as little as 30 percent of the strength required (Refs. 10.1 and 10.3).

In recent years, important research has been directed toward establishing a better basis for joint design (Refs. 10.4 and 10.5). Full-scale tests of beam-column joints have led to improved design methods such as those described in "Recommendations for Design of Beam-Column Joints in Monolithic Reinforced Concrete Structures," reported by ACI Committee 352 (Ref. 10.6). Although they are not a part of the ACI Code, such recommendations provide a basis for the safe design of beam-column joints both for ordinary construction and for buildings subject to seismic forces. Other tests have given valuable insights into the behavior of beam-girder joints, wall junctions, and other joint configurations, thus providing a sound basis for design.

Practicality of the joint design should not be overlooked. Beam reinforcement entering a beam-column joint must clear the vertical column bars, and timely consideration of this fact in selecting member widths and bar size and spacing can avoid costly delays in the field. Similarly, beam steel and girder steel, intersecting at right angles at a typical beam-girder-column joint, cannot be in the same horizontal plane as they enter the joint. Figure 10.1 shows an extreme illustration of the congestion of reinforcing bars at such an intersection. Concrete placement in such a region is difficult at best.

Most of this chapter treats the design of joint regions for typical continuous-frame monolithic structures that are designed according to the strength requirements of the ACI Code for gravity loads or normal wind loads. Joints connecting members that must sustain strength under reversals of deformation into the inelastic range, as for earthquake loads or very high winds, represent a separate

FIGURE 10.1
Steel congestion at beam-girder-column joint.

category and will not be treated here (Ref. 10.6). Brackets and corbels, although they are most often a part of precast buildings rather than monolithic construction, have features in common with monolithic joints, and these will be covered in this chapter.

10.2 BEAM-COLUMN JOINTS

A *beam-column joint* is defined as the portion of a column within the depth of the beams that frame into it. Formerly, the design of monolithic joints was limited to providing adequate anchorage for the reinforcement. However, the increasing use of high-strength concrete, resulting in smaller member cross sections, and the use of larger diameter reinforcing bars now require that more attention be given to joint design and detailing. Although very little guidance is provided by the ACI Code, the ACI Committee 352 report "Recommendations for Design of Beam-Column Joints in Monolithic Structures" (Ref. 10.6) provides a basis for the design of joints in both ordinary structures and structures required to resist heavy cyclic loading into the inelastic range.

a. Classification of Joints

Reference 10.6 classifies structural joints into two categories. A *Type 1* joint connects members in an ordinary structure designed on the basis of strength, according to the main body of the ACI Code, to resist gravity and normal wind load. A *Type 2* joint connects members designed to have sustained strength under deformation reversals into the inelastic range, such as members in a structure designed for earthquake motions, very high winds, or blast effects. Only Type 1 joints will be considered in this chapter.

Figure 10.2 shows a typical *interior joint* in a monolithic reinforced concrete frame, with beams 1 and 2 framing into opposite faces of the column and beams

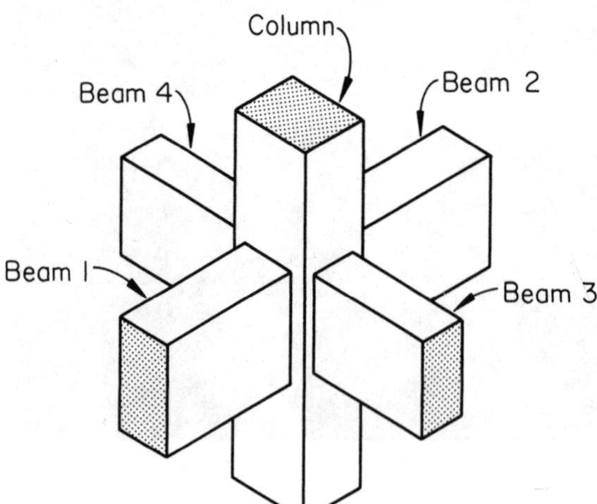

FIGURE 10.2
Typical monolithic interior beam-column joint.

3 and 4 framing into the column faces in the perpendicular direction. An *exterior joint* would include beams 1, 2, and 3, or in some cases only beams 1 and 2. A *corner joint* would include only beams 1 and 3, or occasionally only a single beam 1. A joint may have beams framing into it from two perpendicular directions as shown, but for purposes of analysis and design each direction can be considered separately.

b. Joint Loads and Resulting Forces

The joint region must be designed to resist forces that the beams and column transfer to the joint, including axial loads, bending, torsion, and shear. Figure 10.3*a* shows joint loads acting on the free body of a typical joint of a frame subject to gravity loads, with moments M_1 and M_2 acting on opposite faces, in the opposing sense. In general these moments will be unequal, with their difference equilibrated by the sum of the column moments M_3 and M_4. Fig. 10.3*b* shows the resulting forces to be transmitted through the joint. Similarly, Fig. 10.4*a* shows the loads on a joint in a structure subjected to sidesway loading. The corresponding joint forces are shown in Fig. 10.4*b*. Only for very heavy lateral loading, such as from seismic forces, would the moments acting on opposite faces of the joint act in the same sense, as shown here, producing very high horizontal shears within the joint.

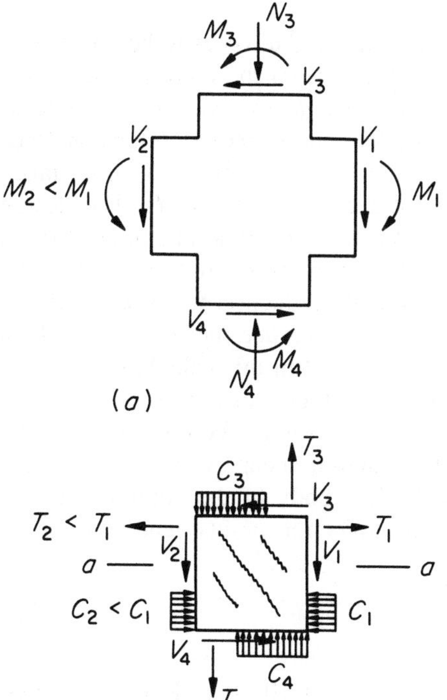

FIGURE 10.3
Joint loads and forces resulting from gravity loads: (*a*) forces and moments on the free body of a joint region; (*b*) resulting internal forces.

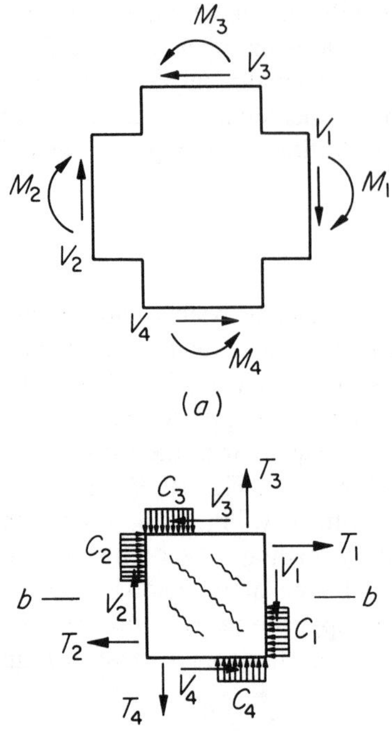

FIGURE 10.4
Joint loads and forces resulting from lateral loads: (*a*) forces and moments on the free body of a joint; (*b*) resulting internal forces.

According to Committee 352 recommendations, the forces to be considered in designing joint regions are not those determined from the conventional frame analysis; rather, they are calculated based on the *nominal strengths of the members*. Where a typical underreinforced beam meets the column face, the tension force from the negative moment reinforcement at the top of the beam is to be taken as $T = A_s f_y$, and the compression force at the face is $C = T$. The design moment applied at the joint face is that corresponding to these maximum forces, $M_u = M_n = A_s f_y(d - a/2)$, rather than that from the overall analysis of the frame. Note that the inclusion of the usual strength reduction factor ϕ would be unconservative in the present case because it would reduce the forces for which the joint is to be designed; it is therefore not included in this calculation.

With the moment applied to each joint face found in this way, the corresponding column forces for joint design are those forces required to keep the connection in equilibrium. To illustrate, the column shears V_3 and V_4 of Figs. 10.3*a* and 10.4*a* are calculated based on the free body of the column between inflection points, as shown in Fig. 10.5. The inflection points generally can be assumed at column midheight as shown.

c. Shear Strength of a Joint

A joint subject to the forces shown in Figs. 10.3*b* or 10.4*b* will develop a pattern of diagonal cracking owing to the diagonal tensile stresses that result from the

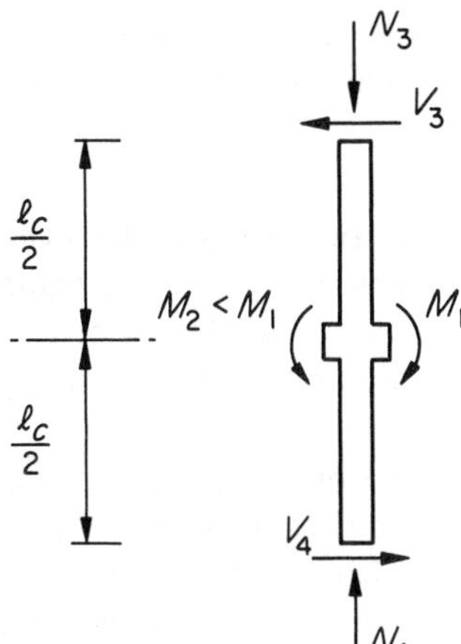

FIGURE 10.5
Free body diagram of an interior column and joint region.

normal forces and shears, as indicated by those figures. The approach used by Committee 352 is to limit the shear force on a horizontal plane through the joint to a value established by tests. The design basis is

$$V_u \leq \phi V_n \tag{10.1}$$

where V_u is the applied shear force, V_n is the nominal shear strength of the joint, and ϕ is taken equal to 0.85.

The shear force V_u is to be calculated on a horizontal plane at midheight of the joint, such as plane *a-a* of Fig. 10.3*b* or plane *b-b* of Fig. 10.4*b*, by summing horizontal forces acting on the joint above that plane. For example, in Fig. 10.3*b* the joint shear on plane *a-a* is

$$V_u = T_1 - T_2 - V_3$$

and in Fig. 10.4*b*, the joint shear on plane *b-b* is

$$V_u = T_1 + C_2 - V_3$$
$$= T_1 + T_2 - V_3$$

The nominal shear strength V_n is given by the equation

$$V_n = \gamma \sqrt{f'_c} b_j h \tag{10.2}$$

where b_j is the effective joint width in inches, h is the thickness in inches of the column in the direction of the load being considered, and $\sqrt{f'_c}$ is expressed in psi units. The value of f'_c used in Eq. (10.2) is not to be taken greater than 6000 psi, even though the actual strength may be larger, because of the lack of research information on connections using high-strength concrete.

The coefficient γ in Eq. (10.2) depends on the confinement of the joint provided by the beams framing into it, as follows:

Interior joint	$\gamma = 24$
Exterior joint	$\gamma = 20$
Corner joint	$\gamma = 15$

The definitions of interior, exterior, and corner joints were discussed in Sec. 10.2a and shown in Fig. 10.2. However, there are restrictions to be applied for purposes of determining γ as follows:

1. An *interior joint* has beams framing into all four sides of the joint. However, to be classified as an interior joint, the beams should cover at least $\frac{3}{4}$ the width of the column, and the total depth of the shallowest beam should not be less than $\frac{3}{4}$ the total depth of the deepest beam. Interior joints that do not satisfy this requirement should be classified as *exterior joints*.
2. An *exterior joint* has at least two beams framing into opposite sides of the joint. However, to be classified as an exterior joint, the widths of the beams on the two opposite faces of the joint should cover at least $\frac{3}{4}$ the width of the column, and the depths of these two beams should be not less than $\frac{3}{4}$ the total depth of the deepest beam framing into the joint. Joints that do not satisfy this requirement should be classified as *corner joints*.

For joints with beams framing in from two perpendicular directions, as for a typical interior joint, the horizontal shear should be checked independently in each direction. Although such a joint is designed to resist shear in two directions, only one classification is made for the joint in this case (i.e., only one value of γ is selected based on the joint classification, and that value is used to compute V_n when checking the design shear capacity in each direction).

According to Committee 352 recommendations, the effective joint width b_j to be used in Eq. (10.2) depends on the transverse width of the beams that frame into the column as well as the transverse width of the column. With regard to the beam width b_b, if there is a single beam framing into the column in the load direction, then b_b is the width of that beam. If there are two beams in the direction of shear, one framing into each column face, then b_b is the average of the two beam widths. In reference to Figure 10.6*a*, when the beam width is less than the column width, the effective joint width is the average of the beam width and column width, but it should not exceed the beam width plus one-half the column depth h on each side of the beam. That is,

$$b_j = \frac{b_b + b_c}{2} \quad \text{and} \quad b_j \le b_b + h \tag{10.3}$$

If the beam frames flush with one face of the column, as is common for exterior joints, the same criteria result in an effective joint width of

$$b_j = \frac{b_b + b_c}{2} \quad \text{and} \quad b_j \le b_b + \frac{h}{2} \tag{10.4}$$

as shown in Fig. 10.6*b*. If the beam width b_b *exceeds* the column width (permitted for Type 1 joints only), the effective joint width b_j is equal to the column width b_c, as shown in Fig. 10.6*c*.

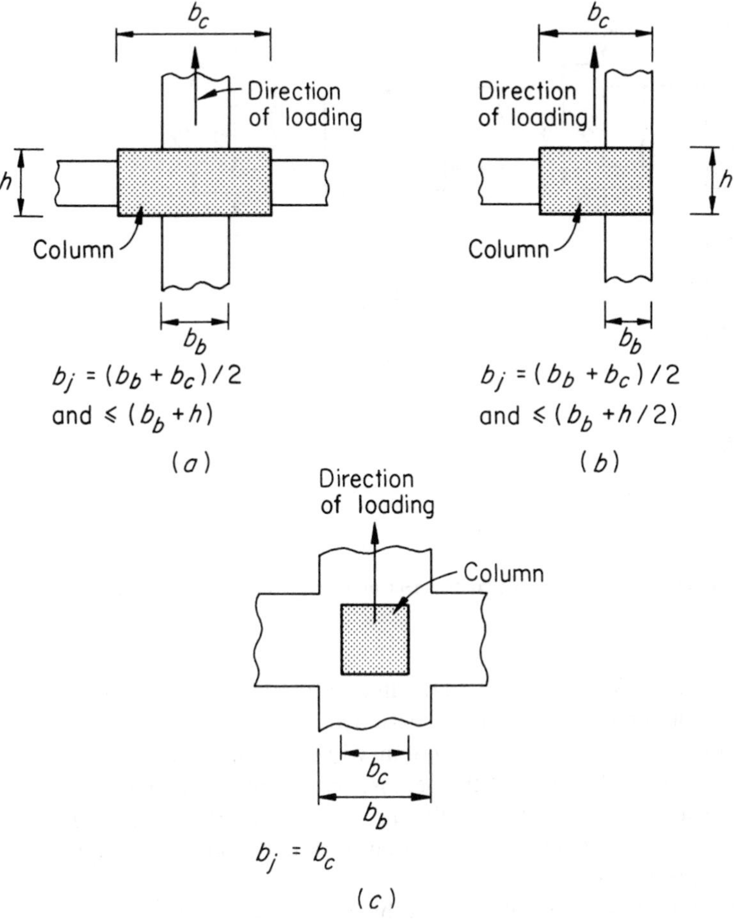

FIGURE 10.6
Determination of effective joint width b: (*a*) interior joint; (*b*) exterior or corner joint.

d. Confinement and Transverse Joint Reinforcement

The successful performance of a beam-column joint depends strongly on the lateral confinement of the joint. Confinement has two benefits: (a) the core concrete is strengthened and its strain capacity improved, and (b) the vertical column bars are prevented from buckling outward. Confinement can be provided either by the beams that frame into the joint or by special column ties provided within the joint region.

Confinement by beams is illustrated in Fig. 10.7. According to Committee 352 recommendations, if beams frame into four sides of the joint, as in Fig. 10.7*a*, adequate confinement is provided if each beam width is at least $\frac{3}{4}$ the width of the intersected column face and if no more than 4 in. of column face is exposed on

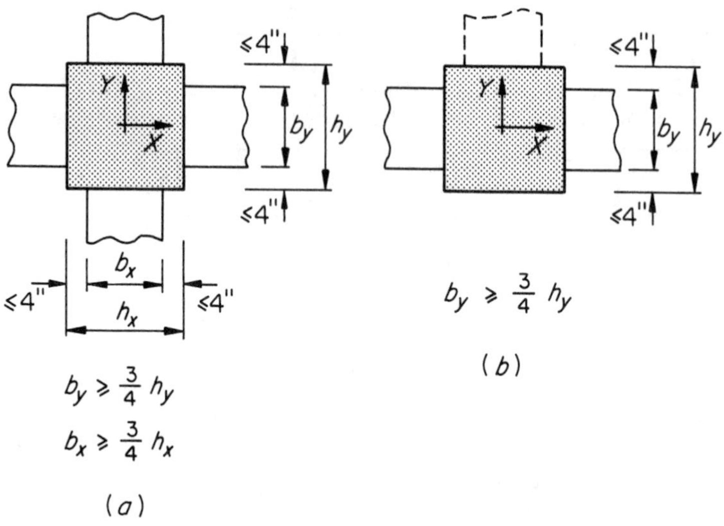

FIGURE 10.7
Confinement of joint concrete by beams: (*a*) confinement in *X* and *Y* directions; (*b*) confinement in *X* direction only.

either side of the beam. Where beams frame into only two sides of the joint, as in Fig. 10.7*b*, adequate confinement can be assumed *in the direction of the beams* if the beam widths are at least $\frac{3}{4}$ the column width and no more than 4 in. of concrete is exposed on either side of the beams. In the other direction, transverse reinforcement must be provided for confinement. The presence of a third beam, but not a fourth, in the perpendicular direction does not modify the requirement for transverse reinforcement.

If adequate confinement is not provided by beams according to these criteria, then transverse reinforcement must be provided. If confinement steel is needed, it must meet all the usual requirements for column ties (see Sec. 8.2 of Chap. 8). In addition, there must be at least two layers of ties between the top and bottom flexural steel in the beams at the joint, and the vertical center-to-center spacing of these ties must not exceed 12 in. If the beam-column joint is part of the primary system for resisting nonseismic lateral loads, this maximum spacing is reduced to 6 in.

e. Anchorage and Development of Beam Reinforcement

For interior joints, normally the flexural reinforcement in a beam entering one face of the joint is continued through the joint to become the flexural steel for the beam entering the opposite face. Therefore, for loadings associated with Type 1 joints, pullout is unlikely, and no special recommendations are made. However, for exterior or corner joints, where one or more of the beams do not continue beyond the joint, a problem of bar anchorage exists. The critical section for development of the yield strength of the beam steel is at the face of the column. Column dimensions seldom permit development of the steel entering the joint by straight embedment alone, and hooks are usually needed for the negative beam reinforcement. Ninety degree hooks are used, with the hook extending toward and beyond the middepth of the joint. If the bottom bars entering the joint need to develop their strength $A_s f_y$ at the face of the joint, as they do if the beam is a part of a primary lateral load-resisting system, they should have 90° hooks also, in this case turned upward to extend toward the middepth of the joint. Requirements for development of hooked bars given in Chapter 5 are applicable in both cases, including modifying factors for concrete cover and for enclosure with ties or stirrups.

Example 10.1 Design of exterior Type 1 joint. The exterior joint shown in Fig. 10.8 is a part of a continuous, monolithic, reinforced concrete frame designed to resist gravity

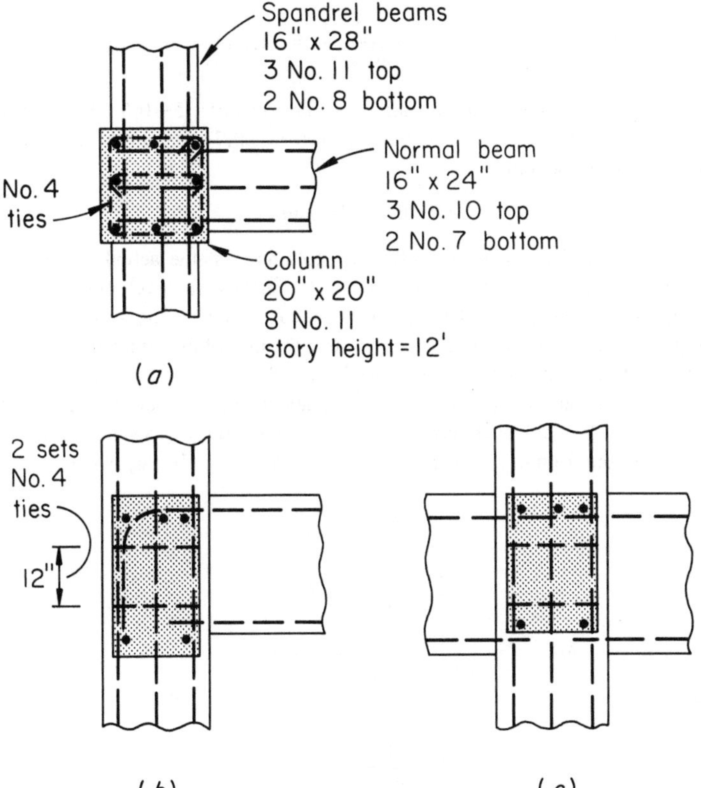

FIGURE 10.8
Exterior beam-column joint for Example 10.1: (*a*) plan view; (*b*) cross section through spandrel beam; (*c*) cross section through normal beam. Note that beam stirrups and column ties outside of the joint are not shown.

loads only. Member section dimensions $b \times h$ and reinforcements are as shown. The frame story height is 12 ft. Material strengths are $f'_c = 4000$ psi and $f_y = 60{,}000$ psi. Design the joint, following the recommendations of the Committee 352 report.

Solution. First the joint geometry must be carefully laid out, to be sure that beam bars and column bars do not interfere with one another and that placement and vibration of the concrete are practical. In this case, bar layout is much simplified by making the column 4 in. wider than the beams. Column steel is placed with the usual 1.5 in. of concrete outside the No. 4 ties. Beam top and bottom bars are placed just inside the outer column bars. The slight offset of the center top beam bars to avoid the center column bars is of no concern. Top bars of the spandrel beams are placed just under the top normal beam bars, except for the outer spandrel bar, which is above the hook shown in Fig. 10.8*b*. Bottom bars enter the joint at different levels without interference.

No anchorage problems exist for the spandrel top reinforcement, which is continuous through the joint. However, the normal beam top steel must be provided with hooks to develop its yield strength at the face of the column. Referring to Table 5.3 of Chap. 5, the basic development length for No. 10 hooked bars is

$$l_{hb} = \frac{1200 d_b}{\sqrt{f'_c}} = \frac{1200 \times 1.27}{\sqrt{4000}} = 24.1 \text{ in.}$$

Being inside the column bars, the beam top bars have side cover of $1.5 + 0.5 + 1.4 = 3.4$ inches. This exceeds 2.5 inches, so a modification factor of 0.7 is applicable, and the required hook development length is

$$l_{dh} = 24.1 \times 0.7 = 16.9 \text{ in.}$$

If the hooked bars are carried down just inside the column ties, the actual embedded length is $20.0 - 1.5 - 0.5 = 18.0$ inches, exceeding 16.9 inches, so development is ensured. None of the beams are a part of the primary, lateral load–resisting system of the frame, so the bottom bars simply can be carried 6 in. into the face of the joint and stopped.

Next the shear strength of the joint must be checked. In the direction of the spandrel beams, moments applied to the joint will be about the same and acting in the opposite sense, so very little joint shear is expected in that direction. However, the normal beam will subject the joint to horizontal shears. In reference to Fig. 10.9*a*, which shows

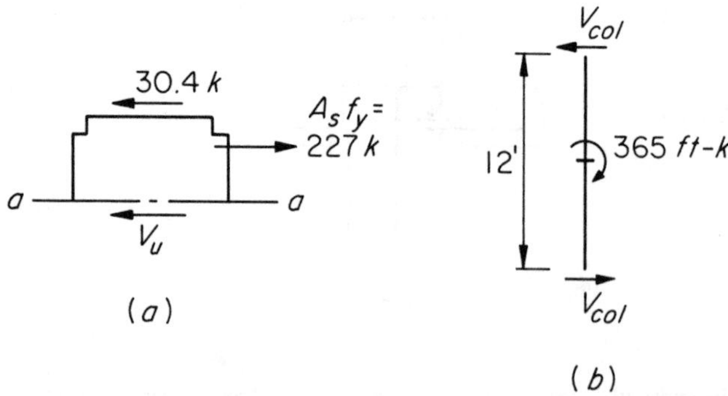

FIGURE 10.9
Basis of column shear for Example 10.1: (*a*) horizontal forces on joint free body sketch; (*b*) free body sketch of column between inflection points.

a free body sketch of the top half of the joint, the maximum force from the beam top steel is

$$A_s f_y = 3.79 \times 60 = 227 \text{ kips}$$

The joint moment is calculated based on this tensile force. The normal beam effective depth is $d = 24.0 - 1.5 - 0.5 - 1.27/2 = 21.4$ in. and with stress block depth $a = A_s f_y / 0.85 f_c' b_w = 227/(0.85 \times 4 \times 16) = 4.17$ in., the design moment is

$$M_u = M_n = A_s f_y \left(d - \frac{a}{2}\right) = \frac{227}{12}\left(21.4 - \frac{4.17}{2}\right) = 365 \text{ ft-kips}$$

Column shears corresponding to this joint moment are found based on the free body of the column between assumed midheight inflection points, as shown in Fig. 10.9*b*: $V_{col} = 365/12 = 30.4$ kips. Then summing horizontal forces on the joint above the middepth plane *a-a*, the joint shear in the direction of the normal beam is

$$V_u = 227 - 30.4 = 197 \text{ kips}$$

For purposes of calculating the joint shear strength, the joint can be classified as exterior, because the 16 in. width of the spandrel beams exceeds $\frac{3}{4}$ the column width of 15 in., and the spandrels are the deepest beams framing into the joint. Thus, $\gamma = 20$. The effective joint width is

$$b_j = \frac{b_b + b_c}{2} = \frac{16 + 20}{2} = 18 \text{ in.}$$

but not to exceed $b_b + h = 16 + 20 = 36$ inches, which does not control. Then the nominal and design shear strengths of the joint are respectively

$$V_n = \gamma \sqrt{f_c'} b_j h = 20\sqrt{4000} \times 18 \times \frac{20}{1000} = 455 \text{ kips}$$

$$\phi V_n = 0.85 \times 455 = 387 \text{ kips}$$

The applied shear $V_u = 197$ kips does not exceed the design strength, so shear is satisfactory.

Confinement is provided in the direction of the spandrel beams by the beams themselves because the spandrel width of 16 in. exceeds $\frac{3}{4}$ the column width and no more than 4 in. of column face is exposed on either side. However, in the direction of the normal beam, confinement must be provided by column ties within the joint. Two sets of No. 4 ties will be provided, as shown in Figs. 10.8*a* and 10.8*b*. The clear distance between column bars is 5.89 in. here, less than 6 in., so strictly speaking the single-leg cross tie is not required. However, it will improve the joint confinement, guard against outward buckling of the central No. 11 column bar, and add little to the cost of construction, so it will be specified as shown in Fig. 10.8*a*. Note that a 90° hook at one end, rather than the 135° bend shown, would meet ACI Code tie anchorage requirements and would facilitate steel fabrication.

Example 10.2 Design of interior Type 1 joint. Figure 10.10 shows a proposed interior joint of a reinforced concrete building, with beam and column dimensions and reinforcement as indicated. The building frame is to carry gravity loads and normal wind loads. Design and detail the joint reinforcement.

Solution. Because the joint is to be a part of the primary, lateral load–resisting system, beam bottom bars as well as top bars are carried straight through the joint for anchorage.

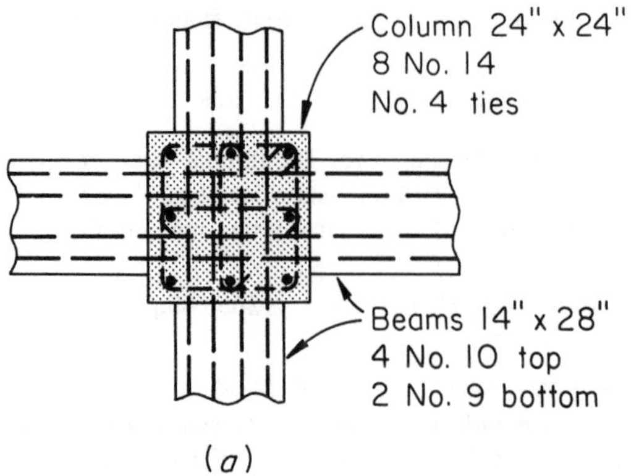

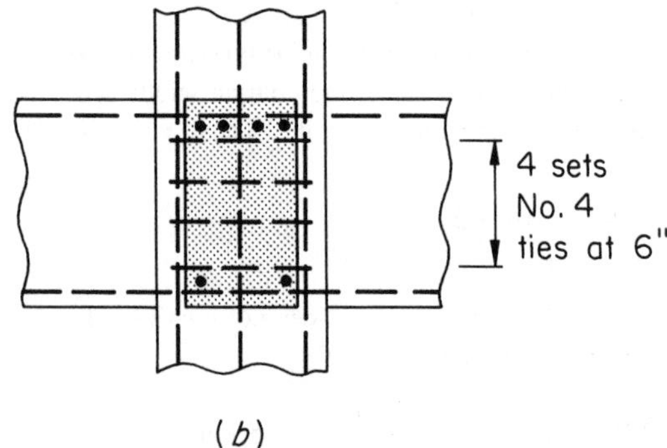

FIGURE 10.10
Interior beam-column joint for Example 10.2: (*a*) plan view; (*b*) section through beam.

In such cases, it is usually convenient to lap splice the bottom steel near the point of inflection of the beams.

In Figs. 10.10*a* and 10.10*b*, top and bottom beam bars entering the joint in one direction must pass, respectively, under and over the corresponding bars in the perpendicular direction. It will be assumed that this has been recognized by adjusting the effective depths in designing the beams. Because the column is 10 in. wider than the beams, the outer beam bars can be passed inside the corner column bars without interference. Four bars are used for the beam top steel in order to avoid interference with the center column bar.

Even the combination of normal wind loading with gravity loads should not produce large unbalanced moment on opposite faces of this interior column, and it can be safely assumed that joint shear will not be critical. However, confinement of the joint region by the beams is considered inadequate because (a) the beam width of 14 in. is less than $\frac{3}{4}$ the column width of 24 in., and (b) the exposed column face outside the beam is $(24 - 14)/2 = 5$ in., which exceeds the 4 in. limit. Consequently, transverse column ties must be added within the joint for confinement. For the 24 in. square column, the spacing between the vertical bars exceeds 6 in., so it is necessary, according to ACI Code, to provide ties to support the intermediate bars as well as the corner bars. Three ties are used per set, as shown in Fig. 10.10*a*. Since the joint is a part of the lateral load–resisting system, the maximum vertical spacing of these tie sets is 6 in. Four sets within the joint, as indicated in Fig. 10.10*b*, are adequate to satisfy this requirement.

f. Wide-Beam Joints

In multistory buildings, in order to reduce the construction depth of each floor and so to reduce the overall building height, wide shallow beams are sometimes used. Joint design in cases where the beams are wider than the column introduces some important concepts not addressed in the Committee 352 report, although most of the report's provisions can be applied. It is important to equilibrate all of the forces applied to the joint. The tension from the top bars in the usual case, with beam width no greater than the column, will be equilibrated by the horizontal component of a diagonal compression strut within the joint. The diagonal compression at the ends of the strut, in turn, is equilibrated by the beam compression and the thrust from the column. (See Sec. 10.3 for a more complete description of the strut-and-tie model.) If the outer bars of the normal beam pass *outside* of the column, as they might in wide-beam designs, the diagonal strut also will be outside of the column, with no equilibrating vertical compression at its upper and lower ends. The outer parts of the beam would tend to shear off, resulting in premature failure.

Two possibilities exist to overcome this problem. The first requires that all of the beam top steel be placed within the width of the column, and preferably inside the outer column bars. Second, if the normal beam bars are carried outside the joint, vertical stirrups can be provided through the joint region to carry the vertical component of thrust from the compression strut.

In extreme but not unusual cases, very wide beams are used, several times wider than the column, with beam depth only about two times the slab depth. In such cases, a safe basis for joint design is to treat the wide beam as a slab and follow the recommendations for slab-column connections contained in Chapter 13.

Example 10.3 Design of exterior Type 1 joint with wide beams. Figure 10.11 shows a typical exterior joint in the floor of a wide-beam structure, designed to resist gravity loads. Here the beams in each direction are 8 in. wider than the corresponding column dimension. Check the proposed joint geometry and shear strength, and design the transverse joint reinforcement. Material strengths are $f_c' = 4000$ psi and $f_y = 60{,}000$ psi. Story height is 12 ft.

Solution. For the present case, all normal beam top steel is passed inside the core of the joint, terminating in 90° hooks at the outside of the column. Top steel in the spandrel beams is continuous through the joint but is also carried inside the joint core. Bottom beam bars, in each case, can be spread across the width of the beam, and they are carried only 6 in. into the joint for both the spandrels and the normal beam because the joint is not a part of the primary, lateral load–resisting system. Beam stirrups outside of the joint, not shown in Fig. 10.11, would be carried outside of the outer bottom bars and bent up. They would require small-diameter horizontal bars inside the hooks for proper anchorage at the upper ends of their vertical legs.

Checking the required development length of the No. 10 top bars of the normal beam gives

$$l_{hb} = \frac{1200 d_b}{\sqrt{f_c'}} = \frac{1200 \times 1.27}{\sqrt{4000}} = 24.1 \text{ in.}$$

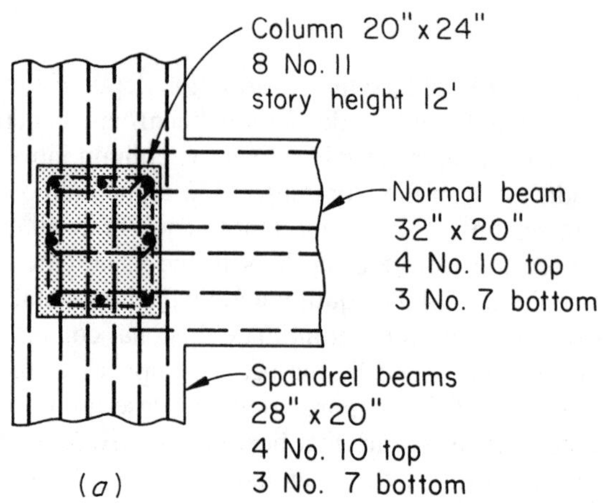

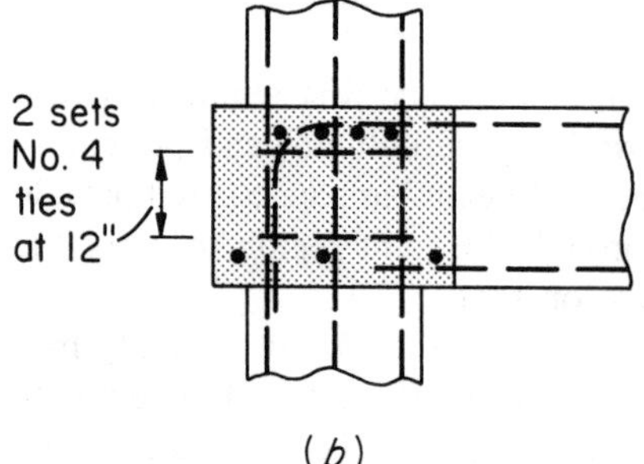

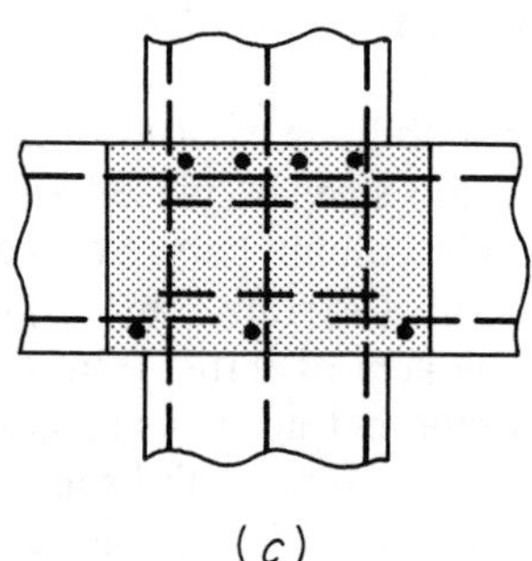

FIGURE 10.11
Exterior beam-column joint for Example 10.3: (*a*) plan view; (*b*) section through spandrel beam; (*c*) section through normal beam.

With lateral cover well in excess of 2.5 in., a modification factor of 0.7 is applicable, and the necessary hook development length is

$$l_{dh} = 24.1 \times 0.7 = 16.9 \text{ in.}$$

If the hooks are carried down in the plane of the outer column bars, the available embedment is $20.0 - 1.5 - 0.5 = 18.0$ in., exceeding the minimum required embedment.

Moments from the spandrels on either side of the joint will be about equal, so no joint shear problem exists in that direction. In the direction of the normal beam, shear must be checked. The tensile force applied by the top bars is $A_s f_y = 5.06 \times 60 = 304$ kips. The depth of the beam compressive stress block is $a = A_s f_y / 0.85 f'_c b_w = 304/(0.85 \times 4 \times 32) = 2.79$ inches, and the corresponding moment is

$$M_u = M_n = A_s f_y \left(d - \frac{a}{2}\right) = \frac{304}{12}\left(17.6 - \frac{2.79}{2}\right) = 411 \text{ ft-kips}$$

Column shears are based on a free body corresponding to that of Fig. 10.9*b*, and are equal to $V_{\text{col}} = 411/12 = 34.3$ kips. Thus, the joint shear at middepth is $V_u = 304 - 34.3 = 270$ kips.

The spandrel beams provide full-width joint confinement in their direction, and the joint can be classed as exterior, so $\gamma = 20$. In the perpendicular direction, when the beam width exceeds the column width, the joint width b_j is to be taken equal to the column width (24 in. in the present case). The nominal and design shear strengths are respectively

$$V_n = \gamma \sqrt{f'_c} b_j h = 20 \sqrt{4000} \times 24 \times 20 = 607 \text{ kips}$$

$$\phi V_n = 0.85 \times 607 = 516 \text{ kips}$$

Because the design strength is well above the applied shear of 270 kips, the shear requirement is met.

Transverse confinement steel must be provided in the direction of the normal beam, between the top and bottom bars of the normal beam, with spacing not to exceed 12 in. Two sets of No. 4 column ties will be used, as shown in Fig. 10.11. In addition to the hoop around the outside bars, a single-leg cross tie is required for the middle column bars because the clear distance between column bars exceeds 6 in.

10.3 STRUT-AND-TIE MODEL FOR JOINT BEHAVIOR

Although the Committee 352 report (Ref. 10.6) is an important contribution to the safe design of joints of certain standard configurations, the recommendations are based mainly on results of tests. Consequently they must be restricted to joints whose geometry closely matches that of the tested joints. This leads to many seemingly arbitrary geometric limitations, and little guidance is provided for the design of joints that may not meet these limitations. An illustration of this is the wide-beam joint discussed in Sec. 10.2f. Such joints are mentioned only very briefly in the report.

Good physical models are available for many aspects of reinforced concrete behavior—for example, for predicting the flexural strength of a beam or the strength of an eccentrically loaded column—but no clear physical model is evident in the Committee 352 recommendations for the behavior of a joint. For this reason, among others, increasing attention is being given to the so-called *strut-and-tie model* as a basis for the design of "discontinuity regions" or "disturbed regions" such as joints (Refs. 10.7, 10.8, 10.9 and 10.10).

The essential features of a strut-and-tie model of joint behavior may be understood with reference to Fig. 10.12, which shows a joint of a frame subject to lateral loading, with clockwise moments from the beams equilibrated by counterclockwise moments from the columns. The line of action of the horizontal forces C_1 and T_2 intersects that of the vertical forces C_3 and T_4 at a *nodal zone*, where the resultant force is equilibrated by a diagonal *compression strut* within the joint. At the lower end of the strut, the diagonal compression equilibrates the resultant of the horizontal forces T_1 and C_2 and the vertical forces T_3 and C_4. The tension bars must be well anchored by extension into and through the joint, or in the case of discontinuous bars (such as the top beam steel in an exterior joint), by hooks. The concrete within the nodal zone is subjected to a biaxial or, in many cases, a triaxial state of stress.

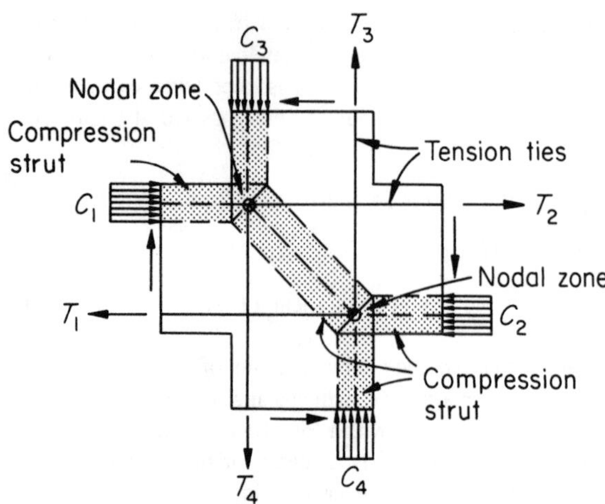

FIGURE 10.12
Strut-and-tie model for behavior of a beam-column joint.

With this simple model, the flow of forces in a joint is easily visualized, satisfaction of the requirements of equilibrium is confirmed, and the need for proper anchorage of bars is emphasized. In a complete strut-and-tie model analysis, through proper attention to deformations within the joint, serviceability is ensured through control of cracking.

According to the strut-and-tie model, the main function of the column ties required within the joint region by conventional design procedures, in addition to preventing outward buckling of the vertical column bars, is to confine the concrete in the compression strut, thereby improving both its strength and ductility, and to control the cracking that may occur owing to diagonal tension perpendicular to the axis of the compression strut. The main load is carried by the uniaxially loaded struts and ties.

The strut-and-tie model not only provides valuable insights into the behavior of ordinary beam-column joints but also represents an important tool for the design of joints that fall outside of the limited range of those considered in Ref. 10.6. Further development of joint design methods may well incorporate this approach. In the sections of this chapter that follow, a number of types of joints will be considered that occur commonly in reinforced concrete structures, for which the strut-and-tie models provide essential aid in developing proper bar details.

10.4 BEAM-TO-GIRDER JOINTS

Commonly in concrete construction, secondary floor beams are supported by primary girders, as shown in Figs. 10.13*a* and 10.13*b*. It is often assumed that the reaction from the floor beam is more or less uniformly distributed through the depth of the interface between beam and girder. This incorrect assumption is perhaps encouraged by the ACI Code "$V_c + V_s$" approach to shear design, which makes use of a nominal average shear stress in the concrete, $v_c = V_c/b_w d$, suggesting a uniform distribution of shear stress through the beam web.

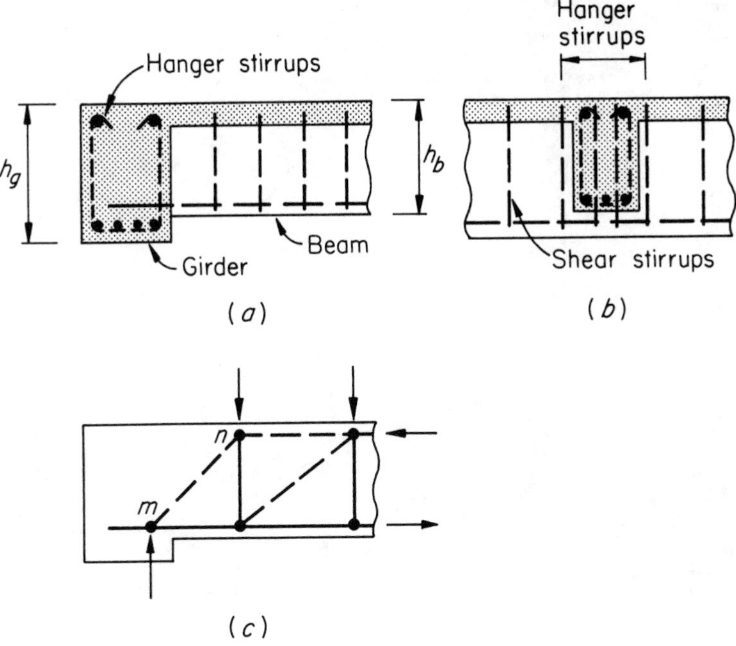

FIGURE 10.13
Main girder supporting secondary beam: (*a*) cross section through girder showing hanger stirrups; (*b*) cross section through beam; (*c*) truss model showing transfer of beam load to girder at load near ultimate.

The actual behavior of a diagonally cracked beam, as indicated by tests, is quite different, and the flow of forces can be represented in somewhat simplified form by the truss model of the beam shown in Fig. 10.13*c* (Ref. 10.11). The main reaction is delivered from beam to girder by a diagonal compression strut *mn*, which applies its thrust near the bottom of the carrying girder. Failure to provide for this thrust may result in splitting off the concrete at the bottom of the girder followed by collapse of the beam. A graphic example of lack of support for diagonal compression at the junction of a beam and its supporting girder is shown in Fig. 10.14.

Proper detailing of steel in the region of such a joint requires the use of well-anchored "hanger" stirrups in the girder, as shown in Figs. 10.13*a* and 10.13*b* to provide for the downward thrust of the compression strut at the end of the beam (Ref. 10.12). These stirrups serve as tension ties to transmit the reaction of the beam to the compression zone of the girder, where it can be equilibrated by diagonal compression struts in the girder. The hanger stirrups, which are required *in addition* to the normal girder stirrups required for shear, can be designed based on equilibrating part or all of the reaction from the beam, with the hanger stirrups assumed to be stressed to their yield stress f_y at the factored load stage. If the beam and girder are the same depth, the hangers should take the full reaction. However, if the beam depth is much less than that of the girder, hangers may

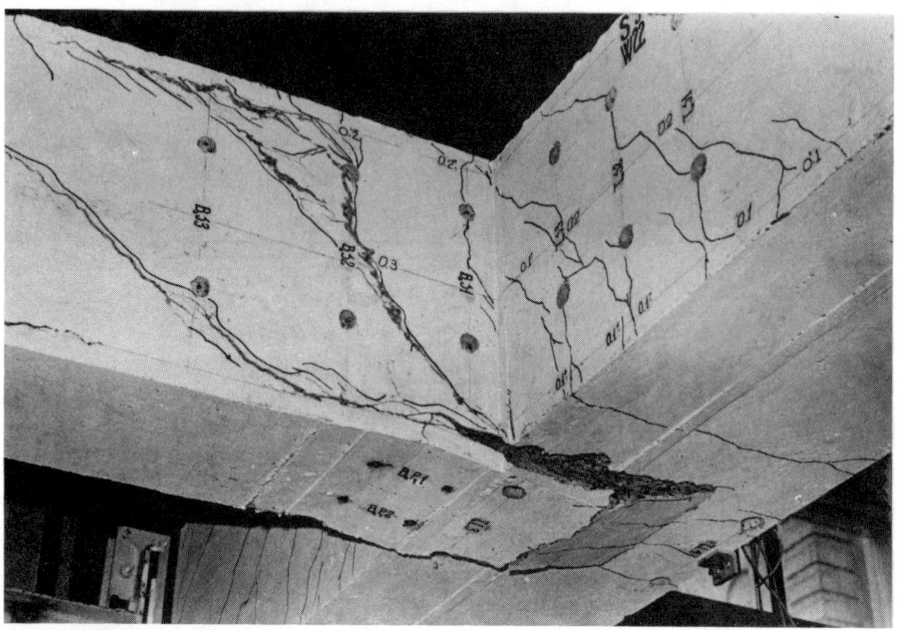

FIGURE 10.14
Failure owing to lack of support for diagonal compression in beam-girder joint. (*Courtesy M. P. Collins, University of Toronto.*)

prove unnecessary. It is suggested in Ref. 10.12 that hanger stirrups be placed to resist a downward force of V_s^*, where

$$V_s^* = \frac{h_b}{h_g} V \tag{10.5}$$

Here h_b is the depth of the beam, h_g is the depth of the carrying girder, as indicated by Fig. 10.13, and V is the end reaction received from the beam.

Hangers will also be unnecessary if the factored beam shear is less than ϕV_c (as is usually the case for one-way joists, for example), because in such a case diagonal cracks would not form in the supported member. The predictions of the truss model would not be valid, and the reaction would be more nearly uniform through the depth.

The hanger stirrups should pass around the main flexural reinforcement of the girder, as shown in Fig. 10.13. If the beam and girder have the same depth, the main flexural bars in the girder should pass below those entering the connection from the beam in order to provide the best possible reaction platform for the diagonal compression strut.

10.5 LEDGE GIRDERS

Frequently in precast concrete construction, an L or inverted-T girder is used to provide a seat, or ledge, to support precast beams framing into the carrying girder from the perpendicular direction. Typical ledge girder cross sections are shown in Fig. 10.15. The end reaction of the beams introduces a heavy concentrated load near the bottom of such girders, requiring special reinforcement in the projecting ledge and in the girder web.

The design of such reinforcement is facilitated through use of a strut-and-tie model, as illustrated in Fig. 10.16. The downward reaction of the supported beam creates a compression fan in the ledge that distributes the reaction along a length greater than that of the bearing plate, as shown in Fig. 10.16*b*. The horizontal components of the fan are equilibrated by a compression strut along the lower flange of the girder.

In the cross section view of Fig. 10.16*a*, the downward thrust under the bearing plate is equilibrated by a diagonal compression strut, with the outward thrust at the top of the strut causing tension in the upper horizontal leg of closed hoop stirrups in the lower part of the girder. In many cases, a short structural steel angle is used just under the bearing plate, and the main tie at the top of the ledge]is welded to the angle to ensure positive anchorage. At the bottom of the diagonal strut, the horizontal component of thrust is equilibrated by the opposing thrust from the other side, and the vertical component causes tension in stirrups that extend to the top of the girder. These stirrups are used in addition to those required for girder shear. Proper anchorage at the nodes is ensured by passing longitudinal bars inside the bends of both sets of stirrups.

Design of the reinforcing bars at the factored load stage is based on development of the yield stress f_y in the steel. A strength reduction factor of $\phi = 0.90$ is used for direct tension, according to provisions of the ACI Code. The tension force to be resisted by the bars can be found by simple statics, for the geometry defined by the strut-and-tie model. A carefully drawn sketch is often the best basis for developing such a model (Refs. 10.7, 10.8, 10.9, and 10.10).

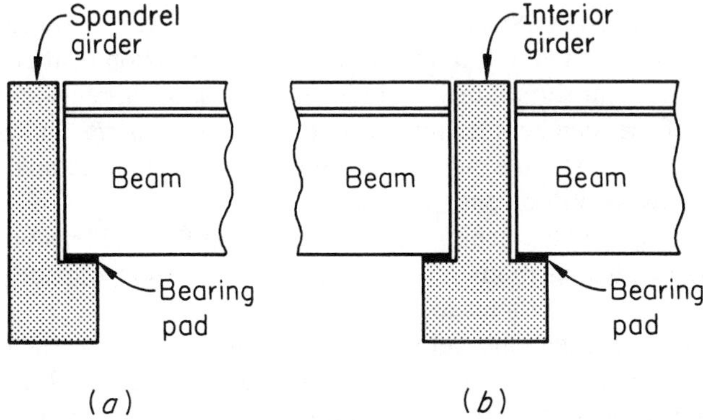

FIGURE 10.15
Ledge girders carrying precast T beams: (*a*) L girder providing exterior support for T beam; (*b*) inverted T girder carrying two T beam reactions.

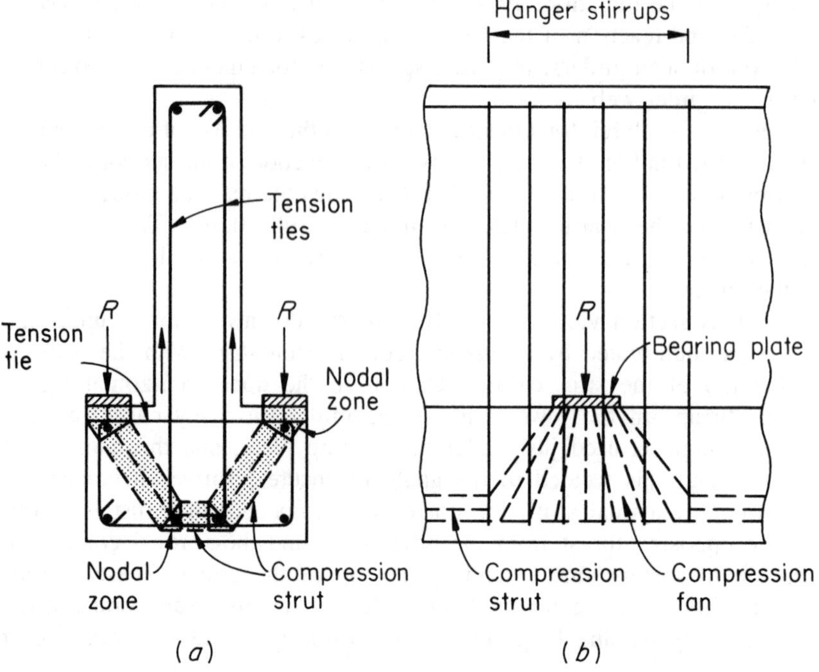

FIGURE 10.16
Strut-and-tie model for behavior of inverted-T ledge girder: (*a*) girder cross section; (*b*) side elevation.

10.6 CORNERS AND T JOINTS

In many common types of reinforced concrete structures, moments and other forces must be transmitted around corners. Some examples, shown in Fig. 10.17, include gable frames, retaining walls, liquid storage tanks, and large box culverts. Reinforcement detailing at the corners is rarely obvious. A comprehensive experimental study of such joints by Nilsson and Losberg (Ref. 10.3) showed that many commonly used joint details will transmit only a small fraction of their assumed strength. Ideally, the joint should resist a moment at least as large as the calculated failure moment of the members framing into it (i.e., the *joint efficiency* should be at least 100 percent). Test have shown that, for common rebar details, joint efficiency may be as low as 30 percent.

Corner joints may be subjected to opening moments, causing flexural tension on the inside of the joint, or closing moments, causing tension on the outside. Generally, the first case is the more difficult to detail properly.

Consider, for example, a corner joint subjected to opening moments, such as an exterior corner of the liquid storage tank shown in the plan view in Fig. 10.17*d*. Figure 10.18*a* shows the system of forces acting on such a corner. The rebar pattern shown is *not* recommended. Formation of crack 1, radiating inward from the corner, is perhaps obvious. Crack 2, which may lead to splitting off the outside corner, may not be so obvious. However, the resultant of the two

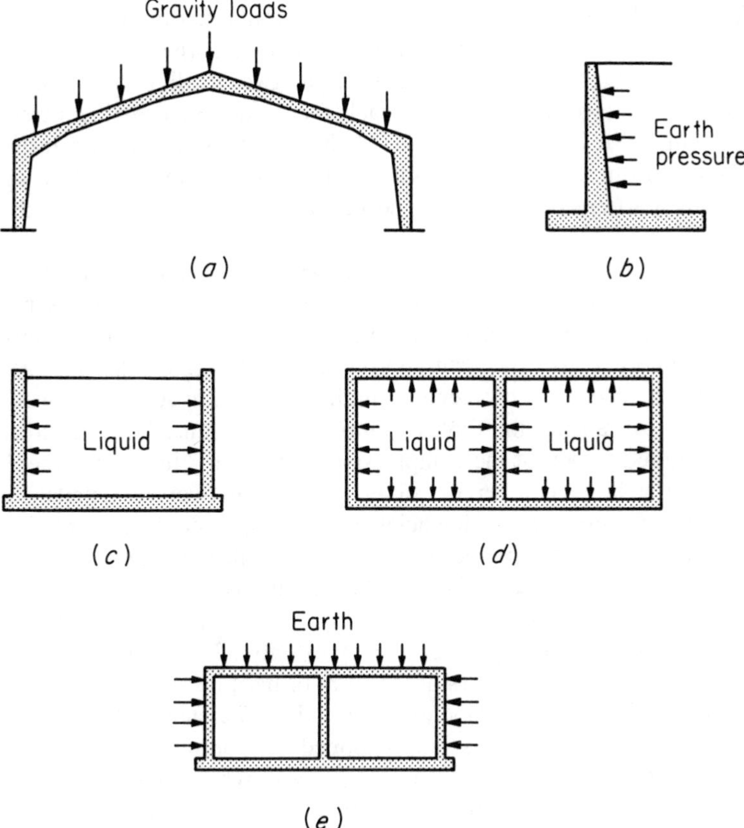

FIGURE 10.17
Structures with corners subject to opening or closing moments: (*a*) gable frame; (*b*) earth-retaining wall; (*c*) liquid storage tank; (*d*) plan view of multicell liquid storage tank; (*e*) large box culvert.

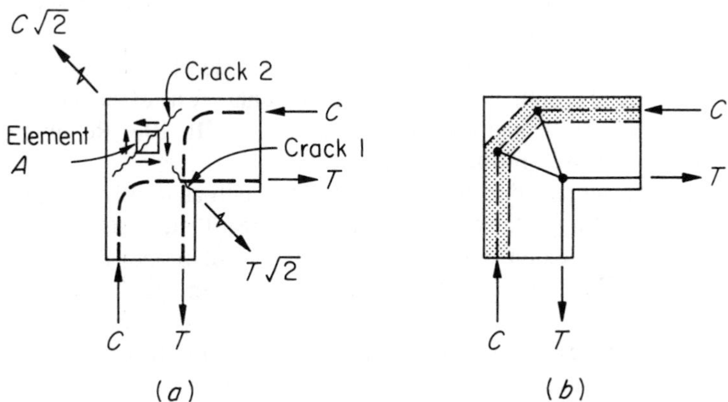

FIGURE 10.18
Corner joint subject to opening moments: (*a*) cracking in an improperly designed joint; (*b*) strut-and-tie model of joint behavior.

compressive forces C, having a magnitude $C\sqrt{2}$, is equilibrated by the resultant tension $T\sqrt{2}$. These two forces, one applied near the outer corner and one near the inner corner, require high tensile stress between the two, leading to formation of crack 2 as shown. The same conclusion is reached considering a small concrete element A in the corner. It is subjected to the shearing forces shown as a result of the forces C and T from the entering members. The resultant of these shearing stresses is 45° principal tension across the corner, confirming formation of the crack 2.

One may, at first, be tempted to add an L-shaped bar around the outside of the corner in an attempt to confine the outer concrete. Such a bar would serve no purpose, however, because the bar would be in compression and may actually assist in pushing the corner off. The strut-and-tie model of Fig. 10.18*b* provides valuable insight into the needed reinforcement, indicating that, in addition to well-anchored tensile bars to transmit the forces T into the joint, some form of radial reinforcement is required to permit the compressive forces C to "turn the corner."

Test results for a large number of joints with alternative bar details are reported in Ref. 10.3. Comparative efficiencies for some specific details, relating the maximum moment transmitted by the corner joint to the flexural capacity of the entering members, are summarized in Fig. 10.19. In all cases the reinforcement ratio of the entering members is 0.75 percent. Fig. 10.19*a* is a simple detail, probably often used, but it provides joint efficiency of only 32 percent. The detail of Fig. 10.19*b*, reinforced with bent bars in the form of hairpins with the plane of the hooks parallel to the inside face of the joint, provides efficiency of 68 percent. In Fig. 10.19*c* the main reinforcement is simply looped and continued out the other leg of the joint, resulting in an efficiency of 77 percent. The somewhat similar detail of Fig. 10.19*d*, in which the bars entering the joint are terminated with separate loops, gives an efficiency of 87 percent. The best performance results from the detail of Fig. 10.19*e*, the same as Fig. 10.19*d* except for the addition of a diagonal bar. This improves joint efficiency to 115 percent, so that the joint is actually stronger than the design strength of the members framing into it. It was determined experimentally that the area of the diagonal bar should be about one-half that of the main reinforcement.

The joints between the vertical wall and horizontal base slab of retaining walls (see Fig. 10.17*b*) are also subjected to opening moments. Tests of such joints confirm the benefit of placing a diagonal bar similar to Fig. 10.19*e*. Retaining wall bar details will be discussed further in Chapter 19.

T joints also may be subjected to bending moments, such as if only one cell of the multiple-cell liquid storage tank of Fig. 10.17*d* were filled. Tests of such joints, reported in Ref. 10.3, again indicate the importance of proper detailing. The rebar arrangement of Fig. 10.20*a*, which is sometimes seen, permits a joint efficiency of only 24 to 40 percent, but the simple rearrangement shown in Fig. 10.20*b* improves the efficiency to between 82 and 110 percent. In both cases, efficiency depends upon the main steel ratio in the entering members, with highest efficiency corresponding to the lowest tensile steel ratio.

Joints subjected to closing moments, with main reinforcement passing around the corner close to the outside face, cause few detailing problems because

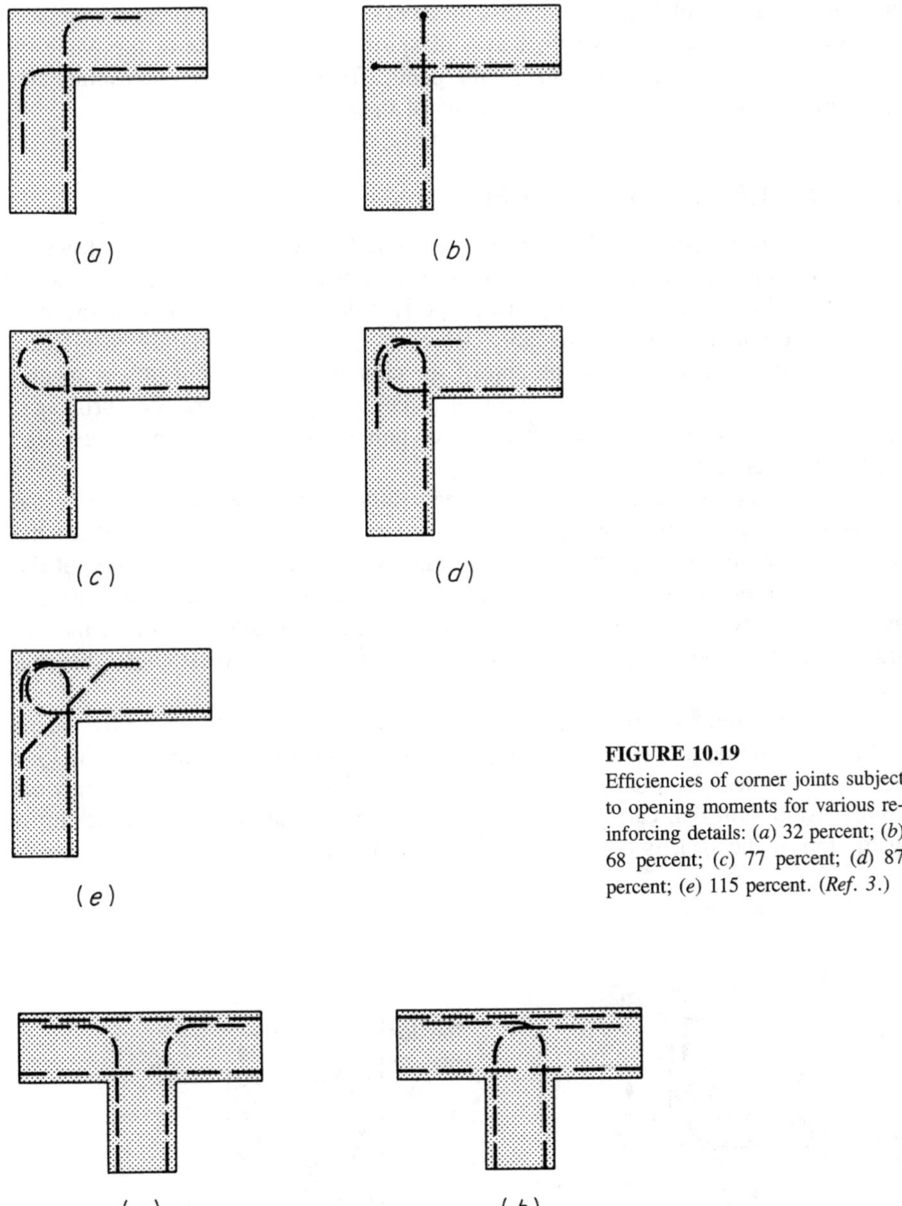

FIGURE 10.19
Efficiencies of corner joints subject to opening moments for various reinforcing details: (*a*) 32 percent; (*b*) 68 percent; (*c*) 77 percent; (*d*) 87 percent; (*e*) 115 percent. (*Ref. 3.*)

(a) (b)

FIGURE 10.20
Comparative efficiencies of T joints subject to bending moment: (*a*) 24 to 40 percent depending on steel ratio; (*b*) 82 to 110 percent depending on steel ratio.

the main tension steel from the entering members can be carried around the outside of the corner. There is, however, a risk of splitting the concrete in the plane of the bend, or concrete crushing inside the bend. The efficiency of such joints can be improved by increasing the bend radius of the bar.

10.7 BRACKETS AND CORBELS

Brackets such as shown in Fig. 10.21*a* are widely used in precast construction for supporting precast beams at the columns. When they project from a wall, rather than from a column, they are properly called corbels, although the two terms are often used interchangeably. Brackets are designed mainly to provide for the vertical reaction V_u at the end of the supported beam, but unless special precautions are taken to avoid horizontal forces caused by restrained shrinkage, creep (in the case of prestressed beams), or temperature change, they must also resist a horizontal force N_{uc}.

Steel bearing plates or angles are generally used at the top surface of the brackets, as shown, to provide a uniform contact surface and to distribute the reaction. A corresponding steel bearing plate or angle is usually provided at the lower corner of the supported member. If the two plates are welded together, as is commonly done, horizontal forces clearly must be allowed for in the design. However, they may be avoided by use of Teflon or elastomeric bearing pads.

The structural performance of a bracket can be visualized easily by means of the strut-and-tie model shown in Fig. 10.21*b*. The downward thrust of the load V_u is equilibrated by the vertical component of the reaction from the diagonal compression strut that carries the load down into the column. The outward thrust at the top of the strut is balanced by the tension in the horizontal tie bars across

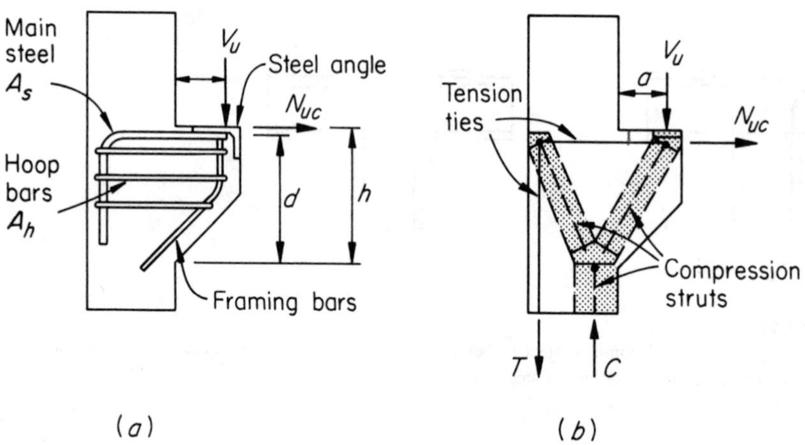

FIGURE 10.21
Typical reinforced concrete bracket: (*a*) loads and reinforcement; (*b*) strut-and-tie model for internal forces.

the top of the bracket; these also take the tension, if any, imparted by the horizontal force N_{uc}. At the left end of the horizontal tie, the tension is equilibrated by the horizontal component of thrust from the second compression strut shown. The vertical component of this thrust requires the tensile force shown acting downward at the left side of the supporting column.

The steel required, according to the strut-and-tie model, is shown in Fig. 12.21*a*. The main bars A_s must be carefully anchored because they need to develop their full yield strength f_y directly under the load V_u, and for this reason they are usually welded to the underside of the bearing angle. A 90° hook is provided for anchorage at the left side. Closed hoop bars with area A_h confine the concrete in the two compression struts and resist a tendency for splitting in a direction parallel to the thrust. The framing bars shown are usually of about the same diameter as the stirrups and serve mainly to improve the stirrup anchorage at the outer face of the bracket.

The bracket may also be considered as a very short cantilevered beam, with flexural tension at the column face resisted by the top bars A_s. Either concept will result in about the same area of main reinforcement.

A second possible mode of failure is by direct shear along a plane more or less flush with the vertical face of the main part of the column. Shear-friction reinforcement crossing such a crack (see Sec. 4.10) would include the area A_s previously placed in the top tie and the area A_h from the hoops below it. Other failure modes include flexural tension failure, with yielding of the top bars followed by crushing of the concrete at the bottom of the bracket; crushing of the concrete under the bearing angle (particularly if end rotation of the supported beam causes the force V_u to be applied too close to the outer corner of the bracket); and direct tension failure, if the horizontal force N_{uc} is larger than anticipated.

The provisions of ACI Code 11.9 for the design of brackets and corbels have been developed mainly based on tests (Refs. 10.13, 10.14, and 10.15) and relate to the flexural model of bracket behavior. They apply to brackets and corbels with a shear span ratio a/d of 1.0 or less (see Fig. 10.21*a*). The distance d is measured at the column face, and the depth at the outside edge of the bearing area must not be less than $0.5d$. The usual design basis is employed, i.e., $M_u \leq \phi M_n$ and $V_u \leq \phi V_n$, and for brackets and corbels (for which shear dominates the design), ϕ is to be taken equal to 0.85 for all strength calculations, including flexure and direct tension as well as shear.

The section at the face of the supporting column must simultaneously resist the shear V_u, the moment $M_u = V_u a + N_{uc}(h - d)$, and the horizontal tension N_{uc}. Unless special precautions are taken, a horizontal tension not less than 20 percent of the vertical reaction must be assumed to act. This tensile force is to be regarded as live load, and a load factor of 1.7 should be applied.

An amount of steel A_f to resist the moment M_u can be found by the usual methods for flexural design. Thus,

$$A_f = \frac{M_u}{\phi f_y (d - a/2)} \tag{10.6}$$

where $a = A_f f_y / 0.85 f_c' b$. An additional area of steel A_n must be provided to resist the tensile component of force:

$$A_n = \frac{N_{uc}}{\phi f_y} \tag{10.7}$$

The total area required *for flexure and direct tension* at the top of the bracket is thus

$$A_s \geq A_f + A_n \tag{10.8}$$

Design for shear is based on the shear-friction method of Sec. 4.10, and the total shear-friction reinforcement A_{vf} is found by

$$A_{vf} = \frac{V_u}{\phi \mu f_y} \tag{10.9}$$

where the friction factor μ for monolithic construction is 1.40 for normal weight concrete, 1.19 for "sand-lightweight" concrete, and 1.05 for "all-lightweight" concrete. The usual limitations that $V_n = V_u/\phi$ must not exceed the smaller of $0.2 f_c' A_c$ or $800 A_c$ apply to the critical section at the support face. (For brackets and corbels, A_c is to be taken equal to the area bd.) Then, according to ACI Code 11.9, the total area required *for shear plus direct tension* at the top of the bracket is

$$A_s \geq \frac{2}{3} A_{vf} + A_n \tag{10.10}$$

with the remaining part of A_{vf} placed in form of closed hoops having area A_h in the lower part of the bracket, as shown in Fig. 10.21*a*.

Thus, the total steel area A_s required at the top of the bracket is equal to the larger of the values given by Eq. (10.8) or Eq. (10.10). An additional restriction, that A_s must not be less than $0.04(f_c'/f_y)bd$, is intended to avoid the possibility of sudden failure upon formation of a flexural tensile crack at the top of the bracket.

According to the ACI Code, closed hoop stirrups having area A_h (see Fig. 10.21*a*) not less than $0.5(A_s - A_n)$ must be provided and be uniformly distributed within two-thirds of the effective depth adjacent to and parallel to A_s. This requirement is more clearly stated as follows:

$$A_h \geq 0.5 A_f \qquad \text{and} \qquad \geq \frac{1}{3} A_{vf} \tag{10.11}$$

Example 10.4 Design of column bracket. A column bracket having the general features shown in Fig. 10.22 is to be designed to carry the end reaction from a long-span precast girder. Vertical reactions from service dead and live loads are 25 kips and 51 kips, respectively, applied 5.5 in. from the column face. A steel bearing plate will be provided for the girder, which will rest directly on a $5 \times 3 \times \frac{3}{8}$ in. steel angle anchored at the outer corner of the bracket. Bracket reinforcement will include main steel A_s welded to the underside of the steel angle, closed hoop stirrups having total area A_h distributed appropriately through the bracket depth, and framing bars in a vertical plane near the outer face. Select appropriate concrete dimensions, and design and detail all reinforcement. Material strengths are $f_c' = 5000$ psi and $f_y = 60{,}000$ psi.

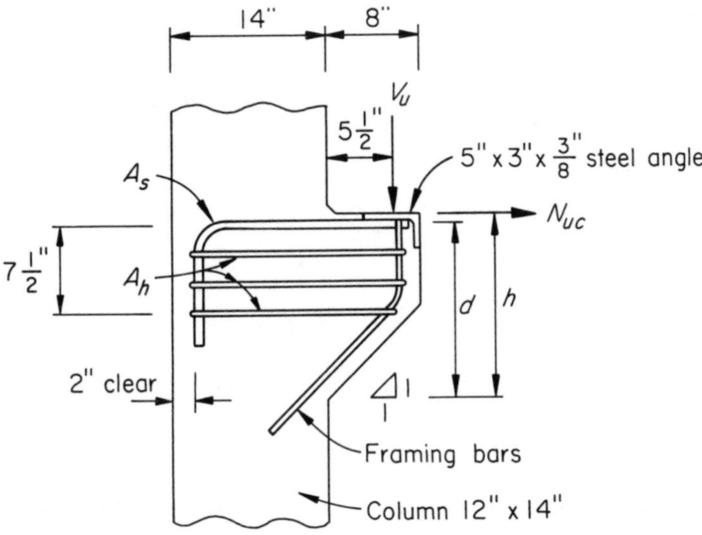

FIGURE 10.22
Column bracket design example.

Solution. The vertical factored load to be carried is

$$V_u = 1.4 \times 25 + 1.7 \times 51 = 122 \text{ kips}$$

In the absence of a roller or low-friction support pad, a horizontal tensile force of

$$N_{uc} = 0.20 \times 122 = 24 \text{ kips}$$

will be included. According to the shear friction provisions of the ACI Code, the nominal shear strength V_n must not exceed $0.2f_c'bd$ or $800bd$. With $f_c' = 5000$ psi, the second limit controls. Then, with $V_u = \phi V_n$ and with the column width $b = 12$ in.,

$$122 = 0.85 \times 0.800 \times 12d$$

from which $d = 14.95$ in. Estimating 1 in. from the center of the main steel to the top surface of the bracket, a total depth $h = 16$ in. will be selected, with d approximately equal to 15 in., the exact value depending on the bar diameter chosen for A_s. If a 45° slope is used, as indicated in Fig. 10.22, the bracket depth at the outside of the bearing area will be 8 in. This is not less than $0.5d = 7.5$ in., as required. For the bracket geometry selected, $a/d = 5.5/15 = 0.37$. This does not exceed the 1.0 limit imposed by the ACI Code.

The total shear friction steel is found from Eq. (10.9):

$$A_{vf} = \frac{V_u}{\phi \mu f_y} = \frac{122}{0.85 \times 1.4 \times 60} = 1.71 \text{ in}^2$$

The bending moment to be resisted is

$$M_u = V_u a + N_{uc}(h - d)$$

$$= 122 \times 5.5 + 24 \times 1 = 695 \text{ kips}$$

The depth of the flexural compression stress block will be estimated to be 2 in., so, from Eq. (10.6),

$$A_f = \frac{M_u}{\phi f_y(d - a/2)} = \frac{695}{0.85 \times 60(15 - 1.0)} = 0.97 \text{ in}^2$$

Checking the stress block depth gives

$$a = \frac{A_f f_y}{0.85 f_c' b} = \frac{0.97 \times 60}{0.85 \times 5 \times 12} = 1.14 \text{ in.}$$

so the revised steel area is

$$A_f = \frac{695}{0.85 \times 60(15 - 0.57)} = 0.94 \text{ in}^2$$

The tensile force of 24 kips requires an additional steel area, from Eq. (10.7), of

$$A_n = \frac{N_{uc}}{\phi f_y} = \frac{24}{0.85 \times 60} = 0.47 \text{ in}^2$$

Thus, from Eq. (10.8) and Eq. (10.10), respectively, the total steel area at the top of the bracket must not be less than

$$A_s \geq A_f + A_n = 0.94 + 0.47 = 1.41 \text{ in}^2$$

or not less than

$$A_s \geq \frac{2}{3}A_{vf} + A_n = \frac{2}{3} \times 1.71 + 0.47 = 1.61 \text{ in}^2$$

The second requirement controls here. The minimum steel requirement of

$$A_{s,\min} = 0.04\frac{f_c'}{f_y}bd = 0.04 \times \frac{5}{60} \times 12 \times 15 = 0.60 \text{ in}^2$$

is seen not to control. A total of three No. 7 bars, providing $A_s = 1.80 \text{ in}^2$, will be used.

Closed hoop steel having a total area A_h not less than $0.5(A_s - A_n)$ must be provided. Thus,

$$A_h \geq 0.5A_f = 0.5 \times 0.94 = 0.47 \text{ in}^2$$

and

$$A_h \geq 0.5 \times \frac{2}{3}A_{vf} = \frac{1}{3} \times 1.71 = 0.57 \text{ in}^2$$

The second requirement controls. Three No. 3 closed hoops will be provided, giving total area $A_h = 0.66 \text{ in}^2$. These must be placed within $\frac{2}{3}$ of the effective depth of the main steel. A spacing of 2.5 in. will be satisfactory, as indicated in Fig. 10.22. A pair of No. 3 framing bars will be added at the inside corner of the hoops to improve anchorage, as shown.

Anchorage of the No. 7 bars will be provided at the right end by welding to the underside of the steel angle and at the left end by a standard 90° bend (see Fig. 5.10). The basic development length (Table 5.3) is

$$l_{hb} = \frac{1200 d_b}{\sqrt{f_c'}} = \frac{1200 \times 0.875}{\sqrt{5000}} = 14.8 \text{ in.}$$

Two modification factors apply here. The first is 0.7, provided at least 2 in. cover is maintained at the end of the hook, and the second is (required A_s)/(provided A_s) =

$1.61/1.80 = 0.89$. Thus, the required development length past the face of the column is

$$l_{dh} = 14.8 \times 0.7 \times 0.89 = 9.22 \text{ in.}$$

This requirement is easily met. The hook extension will be $12d_b = 12 \times 0.875 = 10.5$ inches. For the hoop bars, a standard 135° hook, as shown in Fig. 5.9*b*, will be used.

REFERENCES

10.1. "Reinforced Concrete Design Includes Approval of Details," CRSI Engineering Practice Committee, *Concr. Int.*, vol. 10, no. 1, 1988, pp. 21–22.

10.2. *ACI Detailing Manual*, Special Publication SP-66(88), American Concrete Institute, Detroit, 1988.

10.3. I. H. E. Nilsson and A. Losberg, "Reinforced Concrete Corners and Joints Subjected to Bending Moment," *J. Struct. Div.*, ASCE, vol. 102, no. ST6, 1976, pp. 1229–1254.

10.4. D. F. Meinheit and J. O. Jirsa, "Shear Strength of Reinforced Concrete Beam-Column Connections," *J. Struct. Div.*, ASCE, vol. 107, no. ST11, 1981, pp. 2227–2244.

10.5. J. G. L. Marques and J. O. Jirsa, "A Study of Hooked Bar Anchorages in Beam-Column Joints," *J. ACI*, vol. 72, no. 5, 1975, pp. 198–209.

10.6. "Recommendations for Design of Beam-Column Joints in Monolithic Reinforced Concrete Structures," Reported by ACI Committee 352, *ACI Struct. J.*, vol. 82, no. 3, 1985, pp. 266–283.

10.7. P. Marti, "Basic Tools of Reinforced Concrete Beam Design," *J. ACI*, vol. 82, no. 1, 1985, pp. 46–56.

10.8. P. Marti, "Truss Models in Detailing," *Concr. Int.*, vol. 7, no. 12, 1985, pp. 66–73.

10.9. J. Schlaich, K. Schafer, and M. Jennewein, "Toward a Consistent Design of Structural Concrete," *J. Prestressed Concr. Inst.*, vol. 32, no. 3, 1987, pp. 74–150.

10.10. W. D. Cook and D. Mitchell, "Studies of Disturbed Regions Near Discontinuities in Reinforced Concrete Members," *ACI Struct. J.*, vol. 85, no. 2, 1988, pp. 206–216.

10.11. N. S. Anderson and J. A. Ramirez, "Detailing of Stirrup Reinforcement," *ACI Struct. J.*, vol. 86, no. 5, 1989, pp. 507–515.

10.12. R. Park and T. Pauley, *Reinforced Concrete Structures*, John Wiley, New York, 1975.

10.13. L. B. Kriz and C. H. Raths, "Connections in Precast Concrete Structures—Strength of Corbels," *J. Prestressed Concr. Inst.*, vol. 10, no. 1, 1965, pp. 16–47.

10.14. A. H. Mattock, K. C. Chen, and K. Soongswang, "The Behavior of Reinforced Concrete Corbels," *J. Prestressed Concr. Inst.*, vol. 21, no. 2, 1976, pp. 52–77.

10.15. A. H. Mattock, "Design Proposals for Reinforced Concrete Corbels," *J. Prestressed Concr. Inst.*, vol. 21, no. 3, 1976, pp. 18–24.

PROBLEMS

10.1 An interior Type 1 joint, which is to be considered a part of the primary lateral load-resisting system, is to be designed. The 16 in. square column, with main steel consisting of four No. 11 bars, is intersected by two 12 × 18 in. beams in the *X* direction, reinforced with three No. 10 top bars and three No. 8 bottom bars. In the *Y* direction there are two 12 × 22 in. girders, reinforced with three No. 11 top bars and three No. 9 bottom bars. Concrete cover is 2.5 in. to the center of the bars, except for the top steel in the girders, which is carried just under the top steel of the beams. Design and detail the joint, using $f_c' = 4000$ psi and $f_y = 60{,}000$ psi. Specify placement of all bars and cutoff points.

10.2 A typical exterior joint of the building of Problem 10.1 is identical to the interior joint except that the 12 × 18 in. beam occurs on one side of the column only; the girders frame into two opposite faces, as before. All reinforcement is the same as for the joint of Problem 10.1. Design and detail the joint, specifying bar placement, cutoff points, and details such as bar hook dimensions.

10.3 A beam-to-girder joint similar to Fig. 10.13 is to be designed. The beam, with top steel consisting of four No. 11 bars and two No. 9 bottom bars, delivers a total factored-load reaction of 110 kips to the girder. Detail the connection, following the recommendations of Sec. 10.4, indicating cutoff points and the dimensions of main-bar hooks if required, and selecting suitable hanger stirrups to be added to the normal shear reinforcement of the girder. Use $f_c' = 4000$ psi and $f_y = 60{,}000$ psi.

10.4 The precast columns of a proposed parking garage will incorporate symmetrical brackets to carry the end reactions of short girders that, in turn, carry long-span precast, prestressed double T floor units. The girder reactions will be applied 6 in. from the column face, as shown in Fig. P10.4, and a total width of bracket of 9 in. must be provided for proper bearing. Column width in the perpendicular direction is 20 in. Service load reactions applied at the top face of the brackets are 45 kips dead load and 36 kips live load. Select all unspecified concrete dimensions and design and detail the reinforcement. A corner angle is suggested at the outer top edge of the bracket. Column material strengths are $f_c' = 6000$ psi and $f_y = 60{,}000$ psi.

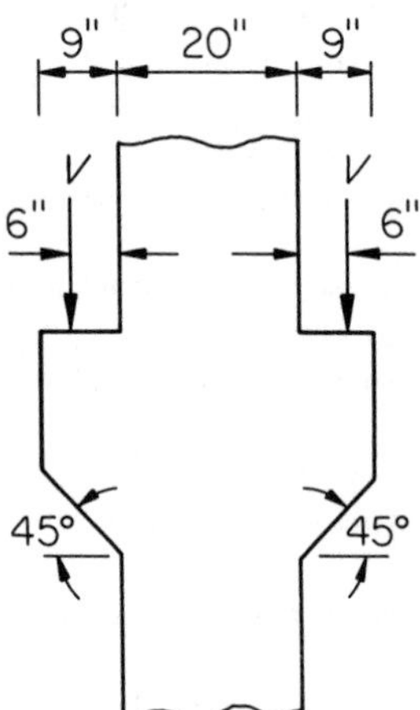

FIGURE P10.4

CHAPTER 11

ANALYSIS OF INDETERMINATE BEAMS AND FRAMES

11.1 CONTINUITY

The individual members that compose a steel or timber structure are fabricated or cut separately and joined together by rivets, bolts, welds, or nails. Unless the joints are specially designed for rigidity, they are too flexible to transfer moments of significant magnitude from one member to another. In contrast, in reinforced concrete structures, as much of the concrete as is practical is poured in one single operation. Reinforcing steel is not terminated at the ends of a member but is extended through the joints into adjacent members. At construction joints, special care is taken to bond the new concrete to the old by carefully cleaning the latter, by extending the reinforcement through the joint, and by other means. As a result, reinforced concrete structures usually represent monolithic, or continuous, units. A load applied at one location causes deformation and stress at all other locations. Even in precast concrete construction, which resembles steel construction in that individual members are brought to the job site and joined in the field, connections are often designed to provide for the transfer of moment as well as shear and thrust, producing at least partial continuity.

The effect of continuity is most simply illustrated by a continuous beam, such as shown in Fig. 11.1*a*. With simple spans, such as provided in many types of steel construction, only the loaded member *CD* would deform, and all other members of the structure would remain straight. But with continuity from one member to the next through the support regions, as in a reinforced concrete structure, the distortion caused by a load on one single span is seen to spread

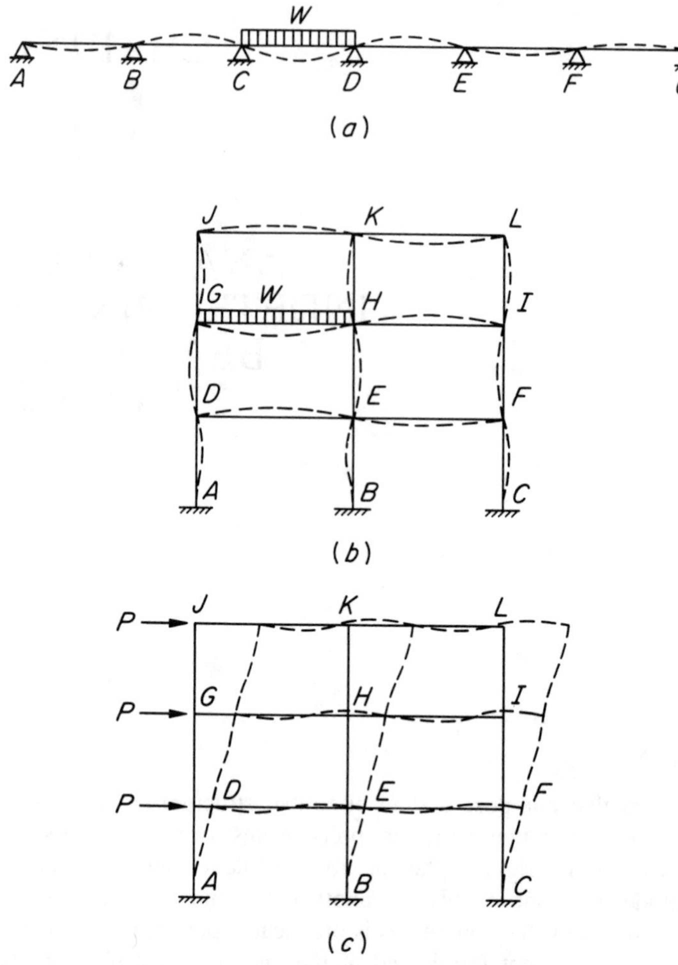

FIGURE 11.1
Deflected shape of continuous beams and frames.

to all other spans, although the magnitude of deformation decreases with increasing distance from the loaded member. All members of the six-span structure are subject to curvature, and thus also to bending moment, as a result of loading span *CD*.

Similarly, for the rigid-jointed frame of Fig. 11.1*b*, the distortion caused by a load on the single member *GH* spreads to all beams and all columns, although, as before, the effect decreases with increasing distance from the load. All members are subject to bending moment, even though they may carry no transverse load.

If horizontal forces, such as forces caused by wind or seismic action, act on a frame, it deforms as illustrated by Fig. 11.1*c*. Here, too, all members of the frame distort, even though the forces act only on the left side; the amount of distortion is seen to be the same for all corresponding members, regardless of their distance from the points of loading, in contrast to the case of vertical loading. A member such as *EH*, even though it carries no transverse load, will experience deformations and associated bending moment.

In *statically determinate structures*, such as simple-span beams, the deflected shape and the moments and shears depend only on the type and magnitude of the loads and the dimensions of the member. In contrast, inspection of the *statically indeterminate structures* of Fig. 11.1 shows that the deflection curve of any member depends not only on the loads but also on the joint rotations, whose magnitudes in turn depend on the distortion of adjacent, rigidly connected members. For a rigid joint such as joint *H* of the frame of Fig. 11.1*b* or Fig. 11.1*c*, all the rotations at the near ends of all members framing into that joint must be the same. For a correct design of continuous beams and frames, it is evidently necessary to determine moments, shears, and thrusts considering the effect of continuity at the joints.

The determination of these internal forces in continuously reinforced concrete structures is usually based on *elastic analysis* of the structure at factored loads with methods that will be described in Secs. 11.2 through 11.6. For checking the results of more exact analysis, the approximate methods of Sec. 11.7 are useful. For many structures, a full elastic analysis is not justified, and the ACI coefficient method of analysis described in Sec. 11.8 provides an adequate basis for design moments and shears.

Before failure, reinforced concrete sections are usually capable of considerable inelastic rotation at nearly constant moment, as was described in Sec. 6.9. This permits a *redistribution of elastic moments* and provides the basis for *plastic analysis* of beams, frames, and slabs. Plastic analysis of beams and frames will be developed in Sec. 11.9 for beams and frames and in Chapters 14 and 15 for slabs.

11.2 PLACEMENT OF LOADS

The individual members of a structural frame must be designed for the worst combination of loads that can reasonably be expected to occur during its useful life. Internal moments, shears, and thrusts are brought about by the combined effect of dead and live loads. While the former are constant, live loads such as floor loads from human occupancy can be placed in various ways, some of which will result in larger effects than others.

In Fig. 11.2*a* only span *CD* is loaded by live load. The distortions of the various frame members are seen to be largest in, and immediately adjacent to, the loaded span and to decrease rapidly with increasing distance from the load. Since bending moments are proportional to curvatures, the moments in more remote members are correspondingly smaller than those in, or close to, the loaded span. However, the loading of Fig. 11.2*a* does not produce the maximum possible positive moment in *CD*. In fact, if additional live load were placed on span *AB*, this span would bend down, *BC* would bend up, and *CD* itself would bend down in the same manner, although to a lesser degree, as it is bent by its own load. Hence, the positive moment in *CD* is increased if *AB*, and by the same reasoning *EF*, are loaded simultaneously. By expanding the same reasoning to the other

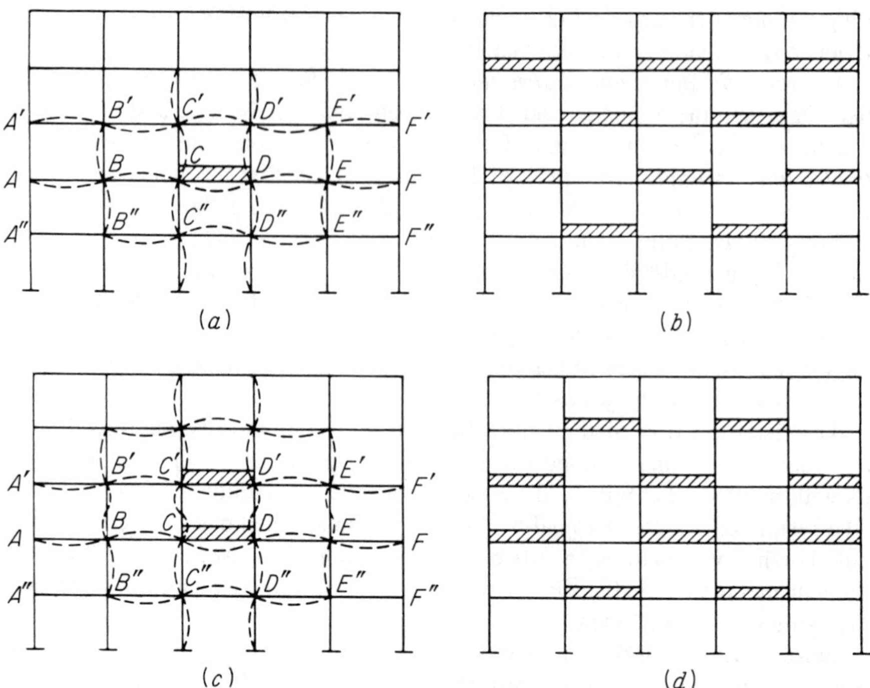

FIGURE 11.2
Alternate live loadings for maximum effects.

members of the frame, one can easily see that the checkerboard pattern of live load of Fig. 11.2*b* produces the largest possible positive moments not only in *CD* but in all loaded spans. Hence, two such checkerboard patterns are required to obtain the maximum positive moments in all spans.

In addition to maximum span moments, it is often necessary to investigate minimum span moments. Dead load, acting as it does on all spans, usually produces only positive span moments. However, live load, placed as in Fig. 11.2*a*, and even more so in Fig. 11.2*b*, is seen to bend the unloaded spans upward, i.e., to produce negative moments in the span. If these negative live load moments are larger than the generally positive dead load moments, a given girder, depending on load position, may be subject at one time to positive span moments and at another to negative span moments. It must be designed to withstand both types of moments; i.e., it must be furnished with tensile steel at both top and bottom. Thus, the loading of Fig. 11.2*b*, in addition to giving maximum span moments in the loaded spans, gives minimum span moments in the unloaded spans.

Maximum negative moments at the supports of the girders are obtained, on the other hand, if loads are placed on the two spans adjacent to the particular support and in a corresponding pattern on the more remote girders. A separate loading scheme of this type is then required for each support for which maximum negative moments are to be computed.

In each column, the largest moments occur at the top or bottom. While the loading of Fig. 11.2*c* results in large moments at the ends of columns CC' and DD', the reader can easily be convinced that these moments are further augmented if additional loads are placed as shown in Fig. 11.2*d*.

It is seen from this brief discussion that in order to calculate the maximum possible moments at all critical points of a frame, live load must be placed in a great variety of different schemes. In most practical cases, however, consideration of the relative magnitude of effects will permit limitation of analysis to a small number of significant cases.

11.3 SIMPLIFICATIONS IN FRAME ANALYSIS

Considering the complexity of many practical building frames and the need to account for the possibility of alternative loadings, there is evidently a need to simplify. This can be done by means of certain approximations that allow the determination of moments with reasonable accuracy while reducing substantially the amount of computation.

Numerous trial computations have shown that, for building frames with a reasonably regular outline, not involving unusual asymmetry of loading or shape, the influence of sidesway caused by vertical loads can be neglected. In that case, moments due to vertical loads are determined with sufficient accuracy by dividing the entire frame into simpler subframes. Each of these consists of one continuous beam, plus the top and bottom columns framing into that particular beam. Placing the live loads on the beam in the most unfavorable manner permits sufficiently accurate determination of all beam moments, as well as the moments at the top ends of the bottom columns and the bottom ends of the top columns. For this partial structure, the far ends of the columns are considered fixed, except for such first-floor or basement columns where soil and foundation conditions dictate the assumption of hinged ends. Such an approach is explicitly permitted by ACI Code 8.9, which specifies the following for floor and roof members:

1. The live load may be considered to be applied only to the floor or roof under consideration, and the far ends of columns built integrally with the structure may be considered fixed.
2. The arrangement of live load may be limited to combinations of (*a*) factored dead load on all spans with full factored live load on two adjacent spans, and (*b*) factored dead load on all spans with full factored live load on alternate spans.

When investigating the maximum negative moment at any joint, negligible error will result if the joints second removed in each direction are considered to be completely fixed. Similarly, in determining maximum or minimum span moments, the joints at the far ends of the adjacent spans may be considered fixed. Thus, individual portions of a frame of many members may be investigated separately.

In regard to columns, ACI Code 8.8 indicates:

1. Columns shall be designed to resist the axial forces from factored loads on all floors or roof and the maximum moment from factored loads on a single adjacent span of the floor or roof under consideration. The loading condition giving the maximum ratio of moment to axial load shall also be considered.
2. In frames or continuous construction, consideration shall be given to the effect of unbalanced floor or roof loads on both exterior and interior columns and of eccentric loading due to other causes.
3. In computing moments in columns due to gravity loading, the far ends of columns built integrally with the structure may be considered fixed.
4. The resistance to moments at any floor or roof level shall be provided by distributing the moment between columns immediately above and below the given floor in proportion to the relative column stiffness and conditions of restraint.

11.4 METHODS FOR ELASTIC ANALYSIS

Many methods have been developed over the years for the elastic analysis of continuous beams and frames. The so-called classical methods (Ref. 11.1), such as application of the theorem of three moments, the method of least work (Castigliano's second theorem), and the general method of consistent deformation, will prove useful only in the analysis of continuous beams having few spans or of very simple frames. For the more complicated cases generally encountered in practice, such methods prove exceedingly tedious, and alternative approaches are preferred.

For many years moment distribution (Ref. 11.1) provided the basic analytical tool for the analysis of indeterminate concrete beams and frames, originally with the aid of the slide rule and later with hand-held programmable calculators. For relatively small problems, moment distribution may still provide the most rapid results, and it is often used in current practice. However, with the widespread availability of computers, manual methods have been replaced largely by matrix analysis, which provides rapid solutions with a high degree of accuracy (Refs. 11.2 to 11.6).

Approximate methods of analysis, based either on careful sketches of the shape of the deformed structure under load or on moment coefficients, still provide a means for rapid estimation of internal forces and moments (Ref. 11.7). Such estimates are useful in preliminary design and in checking more exact solutions for gross errors that might result from input errors. In structures of minor importance, approximations may provide the basis for final design.

In view of the number of excellent texts now available that treat methods of analysis (e.g. Refs. 11.1 to 11.7), the present discussion will be confined to an evaluation of the usefulness of several of the more important of these, with particular reference to the analysis of reinforced concrete structures. Certain idealizations and approximations that facilitate the solution in practical cases will be described in more detail.

a. Slope Deflection

The method of slope deflection was developed independently by Bendixen in Germany in 1914 and by Maney in the United States in 1915. The method entails writing two equations for each member of a continuous frame, one at each end, expressing the end moment as the sum of four contributions: (1) the restraining moment associated with an assumed fixed-end condition for the loaded span, (2) the moment associated with rotation of the tangent to the elastic curve at the near end of the member, (3) the moment associated with rotation of the tangent at the far end of the member, and (4) the moment associated with translation of one end of the member with respect to the other. These equations are related through application of requirements of equilibrium and compatibility at the joints. A set of simultaneous, linear algebraic equations results for the entire structure, in which the structural displacements are the unknowns. Solution for these displacements permits the calculation of all internal forces and moments (Ref. 11.1).

The method is well suited to solving continuous beams, provided there are not very many spans. Its usefulness is extended through modifications that take advantage of symmetry and antisymmetry and of hinged-end support conditions where they exist. However, for multistory and multibay frames in which there are a large number of members and joints, and that will, in general, involve translation as well as rotation of these joints, the effort required to solve the correspondingly large number of simultaneous equations will be prohibitive. Other methods of analysis are more attractive.

b. Moment Distribution

In 1932, Hardy Cross developed the method of moment distribution to solve problems in frame analysis that involve many unknown joint displacements and rotations. For the next three decades, moment distribution provided the standard means in engineering offices for the analysis of indeterminate frames. Even now, it serves as the basic analytical tool when computer facilities are not available.

The moment distribution method (Ref. 11.1) can be regarded as an iterative solution of the slope-deflection equations. The fixed-end moments for each member are modified in a series of cycles, each converging on the precise final result, to account for rotation and translation of the joints. The resulting series can be terminated whenever one reaches the degree of accuracy required. After member end moments are obtained, all member stress resultants can be obtained by use of the laws of statics.

It has been found by comparative analyses that, except in unusual cases, building-frame moments found by modifying fixed-end moments by only two cycles of moment distribution will be sufficiently accurate for design purposes (Ref. 11.8).

c. Matrix Analysis

The introduction of matrix methods of analysis to the structural engineering profession in the early 1950s, coupled with the increasing availability of computers, has produced changes in practice that can only be described as revolutionary. Use of matrix theory makes it possible to reduce the detailed numerical operations required in the analysis of an indeterminate structure to systematic processes of matrix manipulation that can be performed automatically and rapidly by computer. Such methods permit the rapid solution of problems involving large numbers of unknowns. As a consequence, less reliance is placed on special techniques limited to certain types of problems, and powerful methods of general applicability have emerged, such as the matrix displacement method (Refs. 11.2 to 11.6). By such means, an "exact" determination of moments, shears, and thrusts throughout an entire building frame can be obtained quickly and at small expense. Account can be taken of such factors as rotational restraint provided by members perpendicular to the plane of a frame. A large number of alternative loadings may be considered. Highly refined analyses are possible at lower cost than for approximate analyses previously employed.

Some engineers prefer to write their own programs for structural analysis particularly suited to their own needs, using programming languages such as BASIC, FORTRAN, or PASCAL. However, most will make use of readily available general-purpose programs—many suitable for the microcomputers now found in most offices—that may be used for a broad range of problems. Input, including loads, material properties, structural geometry, and member dimensions, is provided by the user, often in an interactive mode. Output includes joint displacements and rotations and moments, shears, and thrusts at critical sections throughout the structure. For more complicated problems such as shearwall-frame interaction under lateral loading, more powerful programs are necessary. Their use will often require mainframe computers, either in-house or accessed through workstations and telephone lines.

11.5 IDEALIZATION OF THE STRUCTURE

It is seldom possible for the engineer to analyze an actual complex redundant structure. Almost without exception, certain idealizations must be made in devising an analytical model, so that the analysis will be practically possible. Thus, three-dimensional members are represented by straight lines, generally coincident with the actual centroidal axis. Supports are idealized as rollers, hinges, or rigid joints. Loads actually distributed over a finite area are assumed to be point loads. In three-dimensional framed structures, analysis is commonly confined to plane frames, each of which is assumed to act independently, even though they are actually linked and interact.

In the idealization of reinforced concrete frames, certain questions require special comment. The most important of these pertain to effective span lengths, effective moments of inertia, and conditions of support.

a. Effective Span Length

In elastic frame analysis, a structure is usually represented by a simple line diagram, based dimensionally on the centerline distances between columns and between floor beams. Actually, the depths of beams and the widths of columns (in the plane of the frame) amount to sizable fractions of the respective lengths of these members; their clear lengths are therefore considerably smaller than their centerline distances between joints.

It is evident that the usual assumption in frame analysis that the members are prismatic, with constant moment of inertia between centerlines, is not strictly correct. A beam intersecting a column may be prismatic up to the column face, but from that point to the column centerline it has a greatly increased depth, with a moment of inertia that could be considered infinite compared with that of the remainder of the span. A similar variation in width and moment of inertia is obtained for the columns. Thus, to be strictly correct, the actual variation in member depth should be considered in the analysis. Qualitatively, this would increase beam support moments somewhat and decrease span moments. In addition, it is apparent that the critical section for design for negative bending would be at the face of the support, and not at the centerline, since for all practical purposes an unlimited effective depth is obtained in the beam across the width of the support.

It will be observed that, in the case of the columns, the moment gradient is not very steep, so that the difference between centerline moment and the moment at the top or bottom face of the beam is small and can in most cases be disregarded. However, the slope of the moment diagram for the beam is usually quite steep in the region of the support, and there will be a substantial difference between the support centerline moment and face moment. If the former were used in proportioning the member, an unnecessarily large section would result. It is desirable, then, to reduce support moments found by elastic analysis to account for the finite width of the supports.

In Fig. 11.3, the change in moment between the support centerline and the support face will be equal to the area under the shear diagram between those two points. For knife-edge supports, this shear area is seen to be very nearly equal to $VaL/2$. Actually, however, the reaction is distributed in some unknown way across the width of the support. This will have the effect of modifying the shear diagram as shown by the dashed line; it has been proposed that the reduced area be taken as equal to $VaL/3$. The fact that the reaction is distributed will modify the moment diagram as well as the shear diagram, causing a slight rounding of the negative moment peak, as shown in the figure, and the reduction of $VaL/3$ is properly applied to the moment diagram after the peak has been rounded. This will give nearly the same face moment as would be obtained by deducting the amount $VaL/2$ from the peak moment.

Another effect is present, however: the modification of the moment diagram due to the increased moment of inertia of the beam at the column. This effect is similar to that of a haunch, and it will mean slightly increased negative moment and slightly decreased positive moment. For ordinary values of the ratio a, this shift in the moment curve will be of the order of $VaL/6$. Thus, it is convenient simply to deduct the amount $VaL/3$ from the unrounded peak moment obtained from elastic analysis. This allows for (1) the actual rounding of the shear diagram and the negative moment peak due to the distributed reaction and (2) the downward

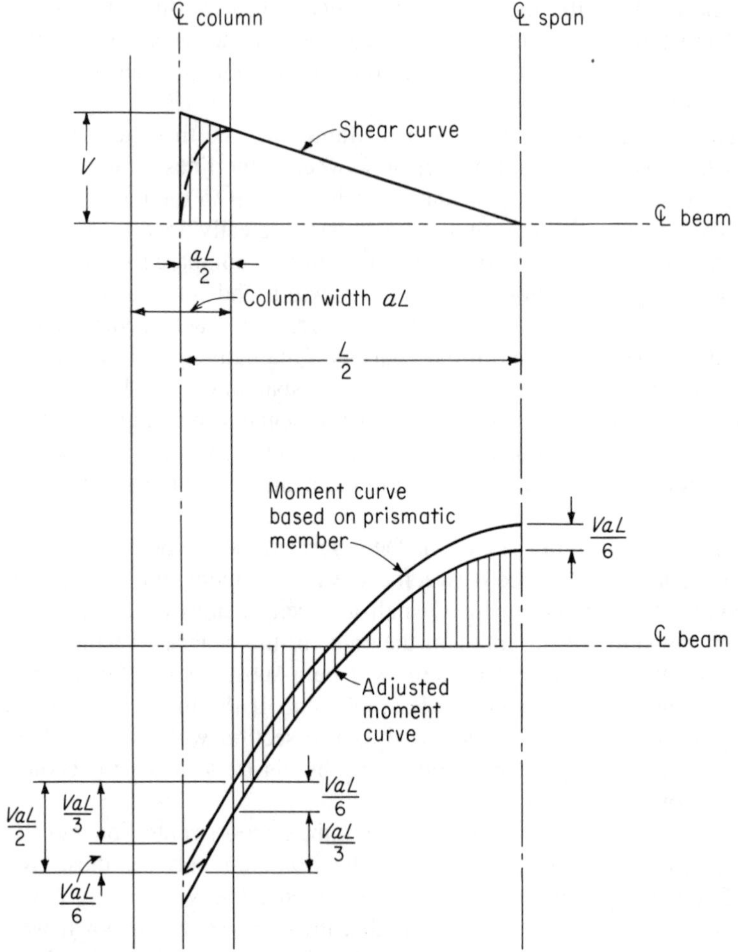

FIGURE 11.3
Reduction of negative and positive moments in a frame.

shift of the moment curve due to the haunch effect at the supports. The consistent reduction in positive moment of $VaL/6$ is illustrated in Fig. 11.3.

In connection with moment reductions, it should be noted that there are certain conditions of support for which no reduction in negative moment is justified. For example, when a continuous beam is carried by a girder of approximately the same depth, the negative moment in the beam at the centerline of the girder should be used in designing the negative reinforcing steel.

b. Moments of Inertia

Selection of reasonable values for moments of inertia of beams and columns for use in the frame analysis is far from a simple matter. The design of beams and columns is based on cracked section theory, i.e., on the supposition that tension concrete is ineffective. It might seem, therefore, that moments of inertia to be used should be determined in the same manner, i.e., based on the cracked transformed section, in this way accounting for the effects of cracking and presence of reinforcement. Things are not this simple, unfortunately.

Consider first the influence of cracking. For typical members, the moment of inertia of a cracked beam section is about one-half that of the uncracked gross concrete section. However, the extent of cracking depends on the magnitude of the moments relative to the cracking moment. In beams, no flexural cracks would be found near the inflection points. Columns, typically, are mostly uncracked, except for those having relatively large eccentricity of loading. A fundamental question, too, is the load level to consider for the analysis. Elements that are subject to cracking will have more extensive cracks near ultimate load than at service load. Compression members will be unaffected in this respect. Thus, the relative stiffness depends on load level.

A further complication results from the fact that the effective cross section of beams varies along a span. In the positive bending region, a beam usually has a T section. For typical T beams, with flange width about 4 to 6 times web width and flange thickness from 0.2 to 0.4 times the total depth, the gross moment of inertia will be about 2 times that of the rectangular web with width b_w and depth h. However, in the negative bending region near the supports, the bottom of the section is in compression. The T flange is cracked, and the effective cross section is therefore rectangular.

The amount and arrangement of reinforcement is also influential. In beams, if bottom bars are continued through the supports, as is often done, this steel acts as compression reinforcement and stiffens the section. In columns, steel ratios are generally much higher than in beams, adding to the stiffness.

Given these complications, it is clear that some simplifications are necessary. It is helpful to note that, in most cases, it is only the *ratio* of member stiffnesses that influences the final result, not the absolute value of the stiffnesses. The stiffness ratios may be but little affected by different assumptions in computing moment of inertia if there is consistency for all members.

In practice, it is generally sufficiently accurate to base stiffness calculations for frame analysis on the gross concrete cross section of the columns. In continuous T beams, cracking will reduce the moment of inertia to about one-half that of the uncracked section. Thus, the effect of the flanges and the effect of cracking may nearly cancel in the positive bending region, and the moment of inertia for the beams can be based simply on the rectangular area $b_w h$. In the negative moment regions there are no flanges; however, if bottom bars continue through the supports to serve as compression steel, the added stiffness tends to compensate for lack of compression flange. Thus, for beams, generally a constant moment of inertia can be used. This is, at least, consistent with ACI Commentary 10.11.2, which suggests that, in calculation of effective length of columns, a value of $0.5I_g$ can be used for flexural members to account for the effects of cracking and reinforcement on relative stiffness and that I_g be used for compression members.

c. Conditions of Support

For purposes of analysis, many structures can be divided into a number of two-dimensional frames. Even for such cases, however, there are situations in which it is impossible to predict with accuracy what the conditions of restraint might be at the ends of a span; yet moments are frequently affected to a considerable degree by the choice made. In many other cases, it is necessary to recognize that structures may be three-dimensional. The rotational restraint at a joint may be influenced or even governed by the characteristics of members framing into that joint at right angles. Adjacent members or frames parallel to the one under primary consideration may likewise influence its performance.

If floor beams are cast monolithically with reinforced concrete walls (frequently the case when first-floor beams are carried on foundation walls), the moment of inertia of the wall about an axis parallel to its face may be so large that the beam end could be considered completely fixed for all practical purposes. If the wall is relatively thin or the beam particularly massive, the moment of inertia of each should be calculated, that of the wall being equal to $bt^3/12$, where t is the wall thickness and b the wall width tributary to one beam.

If the outer ends of concrete beams rest on masonry walls, as is sometimes the case, an assumption of zero rotational restraint (i.e., hinged support) is probably closest to the actual case.

For columns supported on relatively small footings, which in turn rest on compressible soil, a hinged end is generally assumed, since such soils offer but little resistance to rotation of the footing. If, on the other hand, the footings rest on solid rock, or if a cluster of piles is used with their upper portion encased by a concrete cap, the effect is to provide almost complete fixity for the supported column, and this should be assumed in the analysis. Columns supported by a continuous foundation mat should likewise be assumed fixed at their lower ends.

If members framing into a joint in a direction perpendicular to the plane of the frame under analysis have sufficient torsional stiffness, and if their far ends be considered fixed or nearly so, their effect on joint rigidity should be included in the computations. The torsional stiffness of a member of length L is given by the expression GJ/L, where G is the shear modulus of elasticity of concrete (approximately to $E_c/2.2$) and J is the torsional stiffness factor of the member. For beams of rectangular cross section or of sections made up of rectangular elements, J can be taken equal to $\Sigma(hb^3/3 - b^4/5)$, in which h and b are the cross-sectional dimensions of each rectangular element, b being the lesser dimension in each case. In moment distribution, when the effect of torsional rigidity is included, it is important that the absolute flexural stiffness $4EI/L$ be used rather than relative I/L values.

A common situation in beam-and-girder floors and concrete joist floors is illustrated in Fig. 11.4. The sketch shows a beam-and-girder floor system in which longitudinal beams are placed at the third points of each bay, supported by transverse girders, in addition to the longitudinal beams supported directly by the columns. If the transverse girders are quite stiff, it is apparent that the flexural stiffness of all beams in the width w should be balanced against the stiffness

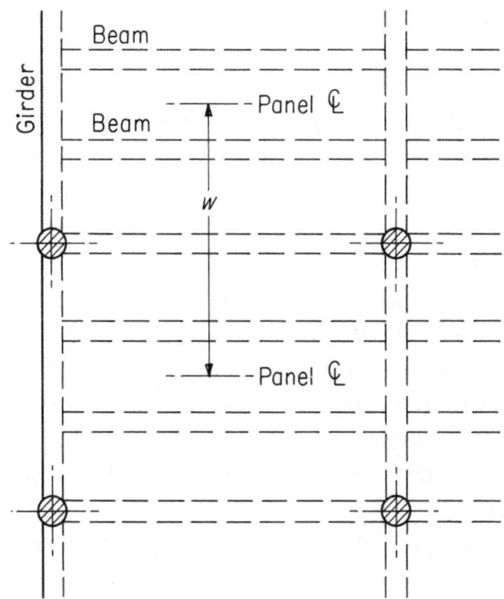

FIGURE 11.4
Slab, beam, and girder floor system.

of one set of columns in the longitudinal bent. If, on the other hand, the girders have little torsional stiffness, there would be ample justification for making two separate longitudinal analyses, one for the beams supported directly by the columns, in which the rotational resistance of the columns would be considered, and a second for the beams framing into the girders, in which case hinged supports would be assumed. Probably, it would generally be sufficiently accurate to consider the girders stiff torsionally and to add directly the stiffness of all beams tributary to a single column. This has the added advantage that all longitudinal beams will have the same cross-sectional dimensions and the same reinforcing steel, which will greatly facilitate construction. Plastic redistribution of loads upon overloading would generally ensure nearly equal restraint moments on all beams before collapse as assumed in design. Torsional moments should not be neglected in designing the girders.

11.6 PRELIMINARY DESIGN

In making an elastic analysis of a structural framework, it is necessary to know at the outset the cross-sectional dimensions of the members, so that moments of inertia and stiffnesses can be calculated. Yet the determination of these same cross-sectional dimensions is the precise purpose of the elastic analysis. Obviously, a preliminary estimate of member sizes must be one of the first steps in the analysis. Subsequently, with the results of the analysis at hand, members are proportioned, and the resulting dimensions compared with those previously assumed. If necessary, the assumed section properties are modified, and the analysis is repeated. Since the procedure may become quite laborious, it is obviously advantageous to make the best possible original estimate of member sizes, in the hope of avoiding repetition of the analysis.

In this connection, it is worth repeating that in the ordinary frame analysis, one is concerned with relative stiffnesses only, not the absolute stiffnesses. If, in the original estimate of member sizes, the stiffnesses of all beams and columns are overestimated or underestimated by about the same amount, correction of these estimated sizes after the first analysis will have little or no effect. Consequently, no revision of the analysis would be required. If, on the other hand, a nonuniform error in estimation is made, and relative stiffnesses differ from assumed values by more than about 30 percent, a new analysis should be made.

The experienced designer can estimate beam and column sizes with surprising accuracy. Those with little or no experience must rely on trial calculations or arbitrary rules, modified to suit particular situations. In building frames, beam sizes are usually governed by the negative moments and the shears at the supports, where their effective section is rectangular. Moments can be approximated by the fixed-end moments for the particular span, or by using the ACI moment coefficients (see Sec. 11.8). In most cases, shears will not differ greatly from simple beam shears. Alternatively, many designers prefer to estimate the depth of beams at about $\frac{3}{4}$ in. per foot of span, with the width about one-half the depth. Obviously, these dimensions are subject to modification, depending on the type and magnitude of the loads, methods of design, and material strength.

Column sizes are governed primarily by axial loads, which can be estimated quickly, although the presence of moments in the columns is cause for some increase of the area as determined by axial loads. For interior columns, in which unbalanced moments will not be large, a 10 percent increase may be sufficient, while for exterior columns, particularly for upper stories, an increase of 50 percent in area may be appropriate. In deciding on these estimated increases, the following factors should be considered. Moments are larger in exterior than in interior columns, since in the latter dead load moments from adjacent spans will largely balance, in contrast to the case in exterior columns. In addition, the influence of moments, compared with that of axial loads, is larger in upper-floor than in lower-floor columns, because the moments are usually of about the same magnitude, while the axial loads are larger in the latter than in the former. Judicious consideration of factors such as these will enable a designer to produce a reasonably accurate preliminary design, which in most cases will permit a satisfactory analysis to be made on the first trial.

11.7 APPROXIMATE ANALYSIS

In spite of the development of refined methods for the analysis of beams and frames, increasing attention is being paid to various approximate methods of analysis (Ref. 11.7). There are several reasons for this. Prior to performing a complete analysis of an indeterminate structure, it is necessary to estimate the proportions of its members in order to know their relative stiffness, upon which the analysis depends. These dimensions can be obtained on the basis of approximate analysis. Also, even with the availability of computers, most engineers find it desirable to make

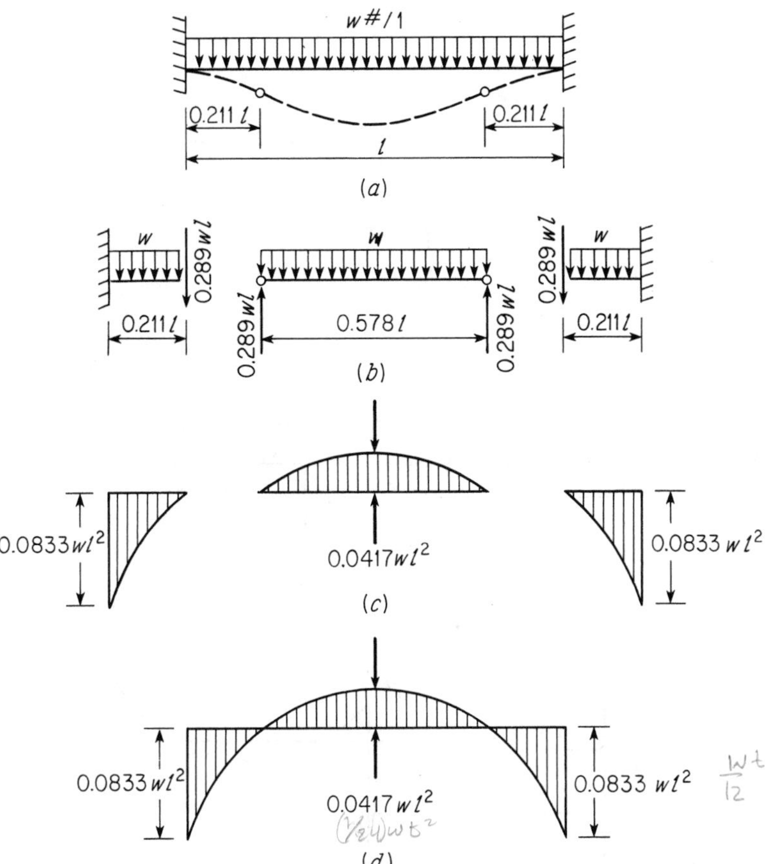

FIGURE 11.5
Analysis of fixed-end beam by locating inflection points.

a rough check of results, using approximate means, to detect gross errors. Further, for structures of minor importance, it is often satisfactory to design on the basis of results obtained by rough calculation. For these reasons, many engineers at some stage in the design process estimate the values of moments, shears, and thrusts at critical locations, using approximate sketches of the structure deflected by its loads.

Provided that points of inflection (locations in members at which the bending moment is zero and there is a reversal of curvature of the elastic curve) can be located accurately, the stress resultants for a framed structure can usually be found on the basis of static equilibrium alone. Each portion of the structure must be in equilibrium under the application of its external loads and the internal stress resultants.

For the fixed-end beam of Fig. 11.5*a*, for example, the points of inflection under uniformly distributed load are known to be located $0.211l$ from the ends of the span. Since the moment at these points is zero, imaginary hinges can be placed

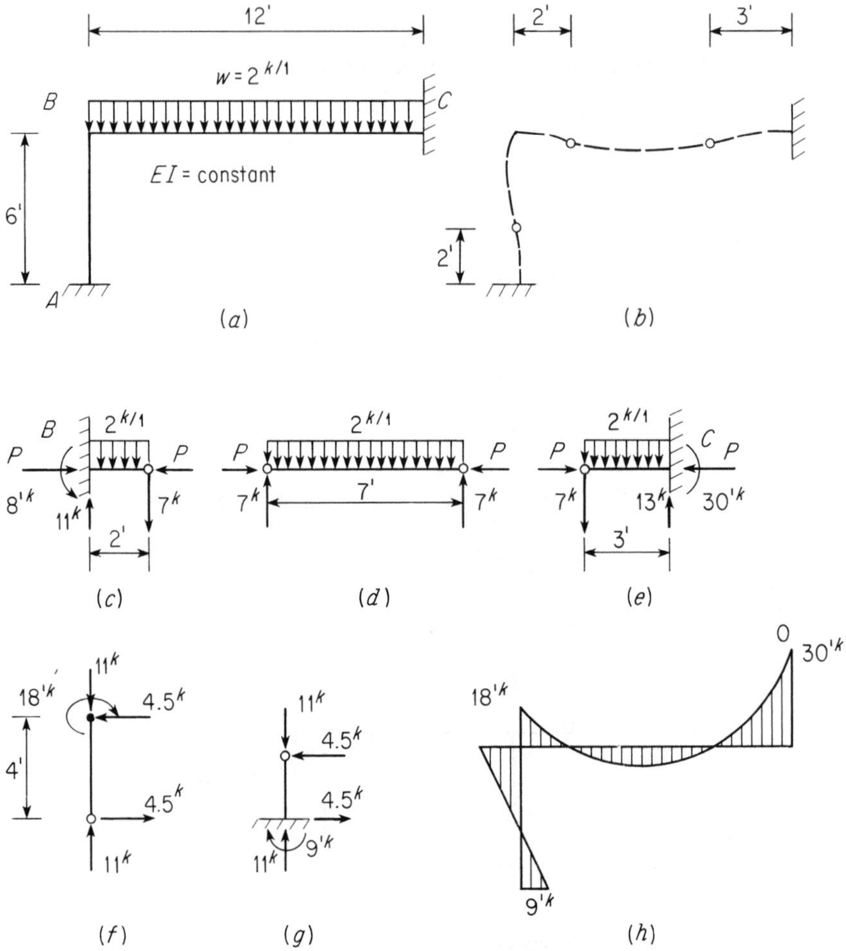

FIGURE 11.6
Approximate analysis of rigid frame.

there without modifying the member behavior. The individual segments between hinges can be analyzed by statics, as shown in Fig. 11.5*b*. Starting with the center segment, shears equal to $0.289wl$ must act at the hinges. These, together with the transverse load, produce a midspan moment of $0.0417wl^2$. Proceeding next to the outer segments, a downward load is applied at the hinge representing the shear from the center segment. This, together with the applied load, produces support moments of $0.0833wl^2$. Note that, for this example, since the correct position of the inflection points was known at the start, the resulting moment diagram of Fig. 11.5*c* agrees exactly with the true moment diagram for a fixed-end beam shown in Fig. 11.5*d*. In more practical cases inflection points must be estimated, and the results obtained will only approximate the true values.

The use of approximate analysis in determining stress resultants in frames is illustrated by Fig. 11.6. Figure 11.6*a* shows the geometry and loading of a two-member rigid frame. In Fig. 11.6*b* an exaggerated sketch of the probable deflected shape is given, together with the estimated location of points of inflection. On this basis, the central portion of the girder is analyzed by statics, as shown in Fig. 11.6*d*, to obtain girder shears at the inflection points of 7 kips, acting with a thrust P (still not determined). Similarly, the requirements of statics applied to the outer segments of the girder in Fig. 11.6*c* and *e* give vertical shears of 11 and 13 kips at B and C respectively, and end moments of 18 and 30 ft-kips at the same locations. Proceeding then to the upper segment of the column, shown in Fig. 11.6*f*, with known thrust of 11 kips and top moment of 18 ft-kips acting, a horizontal shear of 4.5 kips at the inflection point is required for equilibrium. Finally, static analysis of the lower part of the column indicates a requirement of 9 ft-kips moment at A, as shown in Fig. 11.6*g*. The value of P equal to 4.5 kips is obtained by summing horizontal forces at joint B.

The moment diagram resulting from approximate analysis is shown in Fig. 11.6*h*. For comparison, an exact analysis of the frame indicates member end moments of 8 ft-kips at A, 16 ft-kips at B, and 28 ft-kips at C. The results of the approximate analysis would be satisfactory for design in many cases; if a more exact analysis is to be made, a valuable check is available on the magnitude of results.

A specialization of the approximate method described, known as the *portal method*, is commonly used to estimate the effects of sidesway due to lateral forces acting on multistory building frames. For such frames, it is usual to assume that horizontal loads are applied at the joints only. If this is true, moments in all members vary linearly and, except in hinged members, have opposite signs close to the midpoint of each member.

For a simple rectangular portal frame having three members, the shear forces are the same in both legs and are each equal to half the external horizontal load. If one of the legs is more rigid than the other, it would require a larger horizontal force to displace it horizontally the same amount as the more flexible leg. Consequently, the portion of the total shear resisted by the stiffer column is larger than that of the more flexible column.

In multistory building frames, moments and forces in the girders and columns of each individual story are distributed in substantially the same manner as just discussed for single-story frames. The portal method of computing approximate moments, shears, and axial forces from horizontal loads is therefore based on the following three simple propositions:

1. The total horizontal shear in all columns of a given story is equal and opposite to the sum of all horizontal loads acting above that story.
2. The horizontal shear is the same in both exterior columns; the horizontal shear in each interior column is twice that in an exterior column.
3. The inflection points of all members, columns and girders, are located midway between joints.

Although the last of these propositions is commonly applied to all columns, including those of the bottom floor, the authors prefer to deal with the latter separately, depending on conditions of foundation. If the actual conditions are such as practically to prevent rotation (foundation on rock, massive pile foundations, etc.), the inflection points of the bottom columns are above midpoint and may be assumed to be at a distance $2h/3$ from the bottom. If little resistance is offered to rotation, e.g., for relatively small footings on compressible soil, the inflection point is located closer to the bottom and may be assumed to be at a distance $h/3$ from the bottom, or even lower. (In ideal hinges, the inflection point is at the hinge, i.e., at the very bottom.) Since shears and corresponding moments are largest in the bottom story, a judicious evaluation of foundation conditions as they affect the location of inflection points is of considerable importance.

The first of the three cited propositions follows from the requirement that horizontal forces be in equilibrium at any level. The second takes account of the fact that in building frames interior columns are generally more rigid than exterior ones because (1) the larger axial loads require larger cross section and (2) exterior columns are restrained from joint rotation only by one abutting girder, while interior columns are so restrained by two such members. The third proposition is very nearly true because, except for the top and bottom columns and, to a minor degree, for the exterior girders, each member in a building frame is restrained about equally at both ends. For this reason it deflects under horizontal loads in an antisymmetrical manner, with the inflection point at midlength.

The actual computations in this method are extremely simple. Once column shears are determined from propositions 1 and 2 and inflection points located from proposition 3, all moments, shears, and forces are simply computed by statics. The process is illustrated in Fig. 11.7*a*.

Consider joints C and D. The total shear in the second story is $3 + 6 = 9$ kips. According to proposition 2, the shear in each exterior column is $9/6 = 1.5$ kips, and in each interior column $2 \times 1.5 = 3.0$ kips. The shears in the other floors, obtained in the same manner, act at the hinges as shown. Consider the equilibrium of the rigid structure between hinges a, b, and c; the column moments, 3.0 and 9.0 respectively, are obtained directly by multiplying the shears by their lever arms, 6 ft. The girder moment at C, to produce equilibrium, is equal and opposite to the sum of the column moments. The shear in the girder is obtained by recognizing that its moment (i.e., shear times half the girder span) must be equal to the girder moment at C. Hence, this shear is $12.0/10 = 1.2$ kips. The moment at end D is equal to that at C, since the inflection point is at midspan. At D, column moments are computed in the same manner from the known column shears and lever arms. The sum of the two girder moments, to produce equilibrium, must be equal and opposite to the sum of the two column moments, from which the girder moment to the right of C is $18.0 + 6.0 - 12.0 = 12.0$. Axial forces in the columns also follow from statics. Thus, for the rigid body aEd, a vertical shear of 0.3 kip is seen to act upward at d. To equilibrate it, a tensile force of -0.3 kip is required in the column CE. In the rigid body abc an upward shear of 1.2 kips at b is added to the previous upward tension of 0.3 kip at a. To equilibrate these two forces, a tension force of -1.5 kips is required in column AC. If the equilibrium of all other partial structures between hinges is considered in a similar manner, all moments, forces, and shears are rapidly determined.

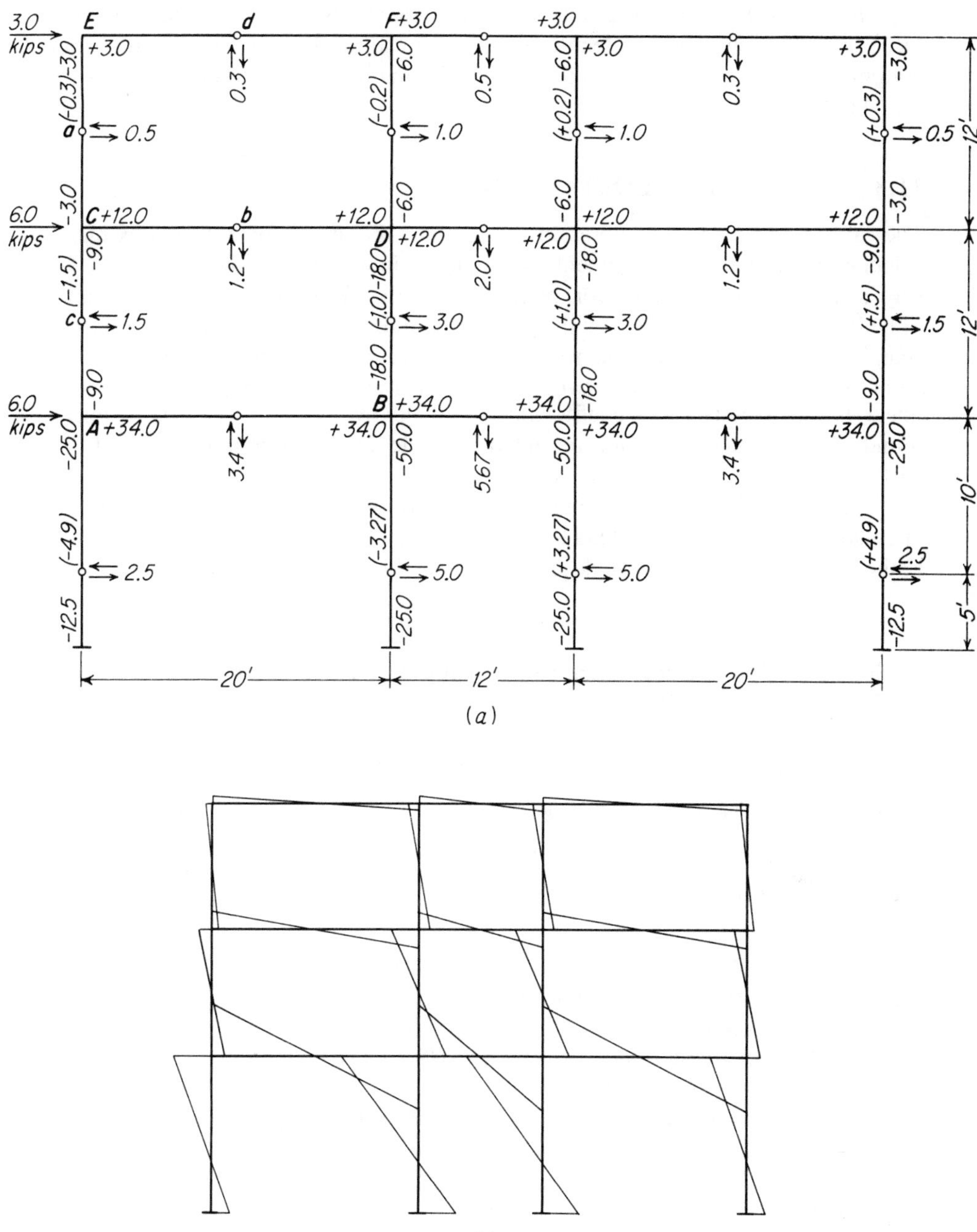

FIGURE 11.7
Portal method for determining moments from wind load in a building frame: (*a*) moments, shears, and thrusts; (*b*) variations of moments.

In the present case relatively flexible foundations were assumed, and the location of the lowermost inflection points was estimated to be at $h/3$ from the bottom. The general character of the resulting moment distribution is shown in Fig. 11.7*b*.

11.8 ACI MOMENT COEFFICIENTS

The ACI Code includes expressions that may be used for the approximate calculation of maximum moments and shears in continuous beams and one-way slabs. The expressions for moment take the form of a coefficient multiplied by $w_u l_n^2$, where w_u is the total factored load per unit length on the span and l_n is the clear span from face to face of supports for positive moment, or the average of the two adjacent clear spans for negative moment. Shear is taken equal to a coefficient multiplied by $w_u l_n/2$. The coefficients, found in ACI Code 8.3.3, are reprinted in Table 11.1 and summarized in Fig. 11.8.

The ACI moment coefficients were derived by elastic analysis, considering alternative placement of live load to yield maximum negative or positive moments at the critical sections, as was described in Sec. 11.2. They are applicable within

Table 11.1 Moment and shear values using ACI coefficients†

Positive moment	
End spans	
If discontinuous end is unrestrained	$\frac{1}{11} w_u l_n^2$
If discontinuous end is integral with the support	$\frac{1}{14} w_u l_n^2$
Interior spans	$\frac{1}{16} w_u l_n^2$
Negative moment at exterior face of first interior support	
Two spans	$\frac{1}{9} w_u l_n^2$
More than two spans	$\frac{1}{10} w_u l_n^2$
Negative moment at other faces of interior supports	$\frac{1}{11} w_u l_n^2$
Negative moment at face of all supports for (1) slabs with spans not exceeding 10 ft and (2) beams and girders where ratio of sum of column stiffness to beam stiffness exceeds 8 at each end of the span	$\frac{1}{12} w_u l_n^2$
Negative moment at interior faces of exterior supports for members built integrally with their supports	
Where the support is a spandrel beam or girder	$\frac{1}{24} w_u l_n^2$
Where the support is a column	$\frac{1}{16} w_u l_n^2$
Shear in end members at first interior support	$1.15 \frac{w_u l_n}{2}$
Shear at all other supports	$\frac{w_u l_n}{2}$

† w_u = total factored load per unit length of beam or per unit area of slab.
l_n = clear span for positive moment and shear and the average of the two adjacent clear spans for negative moment.

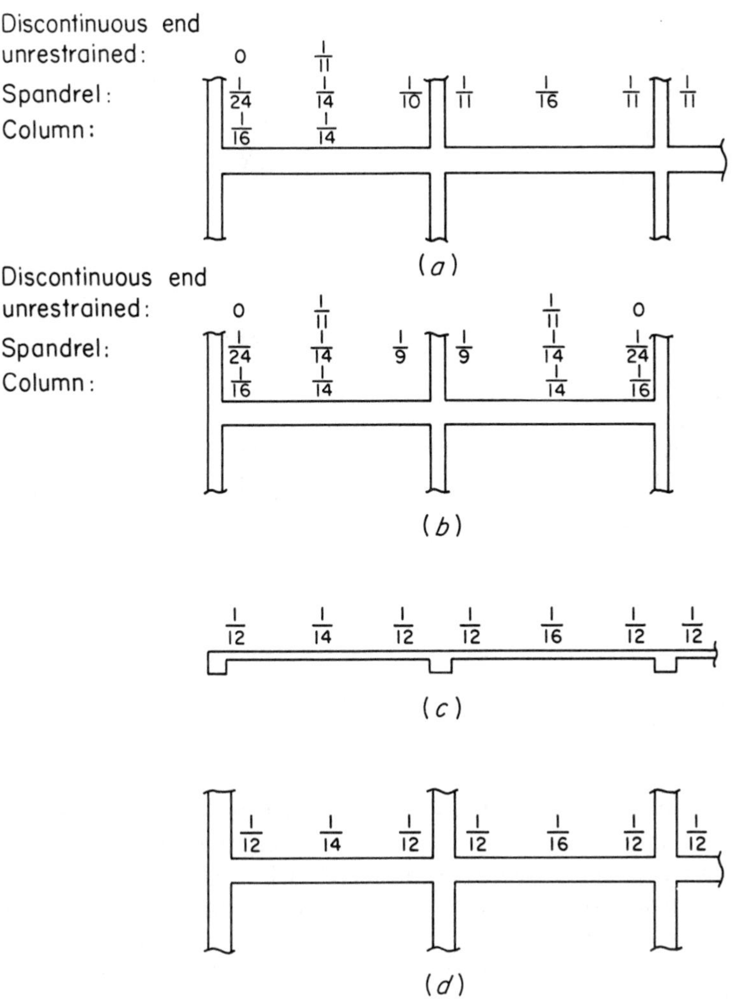

FIGURE 11.8
Summary of ACI moment coefficients: (*a*) beams with more than two spans; (*b*) beams with two spans only; (*c*) slabs with spans not exceeding 10 ft; (*d*) beams in which the sum of column stiffnesses exceeds 8 times the sum of beam stiffnesses at each end of the span.

the following limitations:

1. There are two or more spans.
2. Spans are approximately equal, with the longer of two adjacent spans not greater than the shorter by more than 20 percent.
3. Loads are uniformly distributed.
4. The unit live load does not exceed 3 times the unit dead load.
5. Members are prismatic.

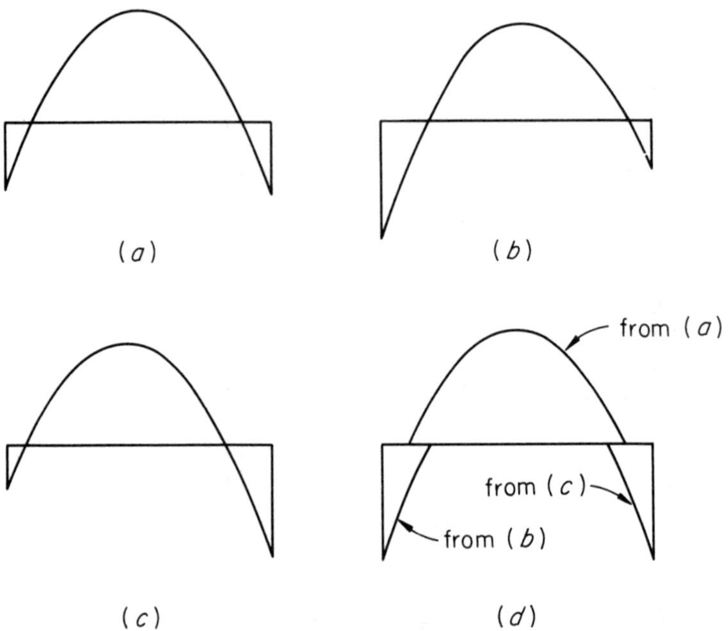

FIGURE 11.9
Maximum moment diagrams and moment envelope for a continuous beam: (*a*) maximum positive moment; (*b*) maximum negative moment at left end; (*c*) maximum negative moment at right end; (*d*) composite moment envelope.

Because alternative loading patterns are considered, the result of applying the Code moment coefficients is not a single moment diagram for a given span but an envelope of maximum moments, as illustrated by Fig. 11.9 for one span in a continuous frame. For maximum positive moment, that span would carry dead and live loads, while adjacent spans would carry dead load only, producing the diagram of Fig. 11.9*a*. For maximum negative moment at the left support, dead and live loads would be placed on the given span and that to the left, while the adjacent span on the right would carry only dead load, with result shown in Fig. 11.9*b*. Fig. 11.9*c* shows the corresponding results for maximum moment at the right support. The moment envelope of Fig. 11.9*d*, which provides the basis for design of the span, is a composite moment diagram formed from the controlling portions of those just developed. It is seen that there is a range of positions for the points of inflection resulting from alternate loadings. The extreme locations, required to determine bar cutoff points, can be found with the aid of Graph A.3 of Appendix A. In the region of the inflection point, it is evident from Fig. 11.9*d* that there may be a reversal of moments for alternative load patterns. However, within the stated limits for use of the coefficients, there should be no reversal of moments at the critical design sections near midspan or at the support faces.

Comparison of the moments found using the ACI coefficients with those calculated by more exact analysis will usually indicate that the coefficient moments are quite conservative. Actual elastic moments may be considerably smaller. Consequently, in many reinforced concrete structures, significant economy can be achieved by making a more precise analysis. This is mandatory for beams and slabs with spans differing by more than 20 percent, sustaining loads that are not uniformly distributed, or carrying live loads greater than 3 times the dead load.

Because the load patterns in a continuous frame that produce critical moments in the columns are different from those for maximum negative moments in the beams, column moments must be found separately. According to ACI Code 8.8, columns must be designed to resist the axial load from factored dead and live loads on all floors above and on the roof plus the maximum moment from factored loads on a single adjacent span of the floor or roof under consideration. In addition, because of the characteristic shape of the column strength interaction diagram (see Chapter 8) it is necessary to consider the case that gives the maximum ratio of moment to axial load. In multistory structures, this results from a checkerboard loading pattern (see Fig. 11.2*d*), which gives maximum column moments but at a less-than-maximum axial force. As a simplification, in computing moments owing to gravity loads, the far ends of the columns may be considered fixed. The moment found at a column-beam joint for a given loading is to be assigned to the column above and the column below in proportion to the relative column stiffness and conditions of restraint.

The shears at the ends of spans in a continuous frame are modified from the value of $w_u l_n/2$ for a simply supported beam because of the usually unbalanced end moments. For interior spans, within the limits of the ACI coefficient method, this effect will seldom exceed about 8 percent, and it may be neglected, as suggested in Table 11.1. However, for end spans, at the face of the first interior support, the additional shear is significant, and a 15 percent increase above the simple beam shear is indicated in Table 11.1. The corresponding reduction in shear at the face of the exterior support is conservatively neglected.

11.9 LIMIT ANALYSIS

a. Introduction

Presently, most reinforced concrete structures are designed for moments, shears, and thrusts found by elastic theory with methods such as those described in Secs. 11.1 through 11.8 of this chapter. On the other hand, the actual proportioning of members is done by strength methods, with the recognition that inelastic section and member response would result upon overloading. Factored loads are used in the elastic analysis to find moments in a continuous beam, for example, after which the critical beam sections are designed with the knowledge that the steel would be well into the yield range and the concrete stress distribution very nonlinear before final collapse. Clearly this is an inconsistent approach to the total analysis-design process, although it can be shown to be both safe and conservative. A beam or frame so analyzed and designed will not fail at a load lower than the value calculated in this way.†

On the other hand, it is known that a continuous beam or frame normally will not fail when the ultimate moment capacity of just one critical section is reached.

† See the discussion of upper and lower bound theorems of the theory of plasticity, Sec. 14.2, for an elaboration on this point.

A *plastic hinge* will form at that section, permitting large rotation to occur at essentially constant resisting moment and thus transferring load to other locations along the span where the limiting resistance has not yet been reached. Normally in a continuous beam or frame, excess capacity will exist at those other locations because they would have been reinforced for moments resulting from different load distributions selected to produce maximum moments at those other locations.

As loading is further increased, additional plastic hinges may form at other locations along the span and eventually result in collapse of the structure, but only after a significant *redistribution of moments* has occurred. The ratio of negative to positive moments found from elastic analysis is no longer correct, for example, and the true ratio after redistribution depends upon the flexural strengths actually provided at the hinging sections.

Recognition of redistribution of moments can be important because it permits a more realistic appraisal of the actual load-carrying capacity of a structure, thus leading to improved economy. In addition, it permits the designer to modify, within limits, the moment diagrams for which members are to be designed. Certain sections can be deliberately underreinforced if moment resistance at adjacent critical sections is increased correspondingly. Adjustment of design moments in this way enables the designer to reduce the congestion of reinforcement that often occurs in high-moment areas, such as at the beam-column joints.

The formation of plastic hinges is well established by tests such as that pictured in Fig. 11.10, which was carried out in the George Winter Laboratory at Cornell University. The three-span continuous beam illustrates the inelastic response typical of heavily overloaded members. It was reinforced in such a way that plastic hinges would form at the interior support sections before the limit capacity of sections elsewhere was reached. The beam continued to carry increasing load well beyond the load that produced first yielding at the supports. The extreme deflections and sharp changes in slope of the member axis that are seen here were obtained only slightly before final collapse.

FIGURE 11.10
Three-span continuous beam after the formation of plastic hinges at the interior supports.

The *inconsistency* of the present approach to the total analysis-design process, the possibility of utilizing the *reserve strength* of concrete structures resulting from moment redistribution, and the opportunity to *reduce steel congestion* in critical regions have motivated considerable interest in limit analysis for reinforced concrete based on the concepts just described. For beams and frames, ACI Code 8.4 permits limited redistribution of moments, depending upon the tensile steel ratio. For slabs, which generally use very low steel ratios and consequently have great ductility, plastic design methods are especially suitable.

b. Plastic Hinges and Collapse Mechanisms

If a short segment of a reinforced concrete beam is subjected to a bending moment, curvature of the beam axis will result, and there will be a corresponding rotation of one face of the segment with respect to the other. It is convenient to express this in terms of an angular change per unit length of the member. The relation between moment and angle change per unit length of beam, or curvature, at a reinforced concrete beam section subject to tensile cracking was developed in Section 6.9 of Chapter 6. Methods were presented there by which the theoretical moment-curvature graph might be drawn for a given beam cross section, as in Fig. 6.16.

The actual moment-curvature relationship measured in beam tests differs somewhat from that shown in Fig. 6.16, mainly because, from tests, curvatures are calculated from average strains measured over a finite gage length, usually about equal to the effective depth of the beam. In particular, the sharp increase in curvature upon concrete cracking shown in Fig. 6.16 is not often seen because the crack occurs at only one discrete location along the gage length. Elsewhere, the uncracked concrete shares in resisting flexural tension, resulting in what is known as *tension stiffening*. This tends to reduce curvature. Furthermore, the exact shape of the moment-curvature relation depends strongly upon the steel ratio as well as upon the exact stress-strain curves for the concrete and steel.

Fig. 11.11 shows a somewhat simplified moment-curvature diagram for an actual concrete beam section having tensile steel ratio of about one-half the balanced value. The diagram is linear up to the cracking moment M_{cr}, after which a nearly straight line of somewhat flatter slope is obtained. At the moment that initiates yielding, M_y, the unit rotation starts to increase disproportionately. Further increase in applied moment causes extensive inelastic rotation until eventually the compressive strain limit of the concrete is reached at the ultimate rotation ψ_u. The resisting moment at ultimate capacity is often somewhat above the calculated flexural strength M_n, due largely to strain hardening of the reinforcement.

The effect of inelastic concrete response prior to steel yielding is small for typically underreinforced sections, as is indicated in Fig. 6.16, and the yield moment can be calculated based on the elastic concrete stress distribution shown in Fig. 11.11*b*:

$$M_y = A_s f_y \left(d - \frac{kd}{3} \right) \tag{11.1}$$

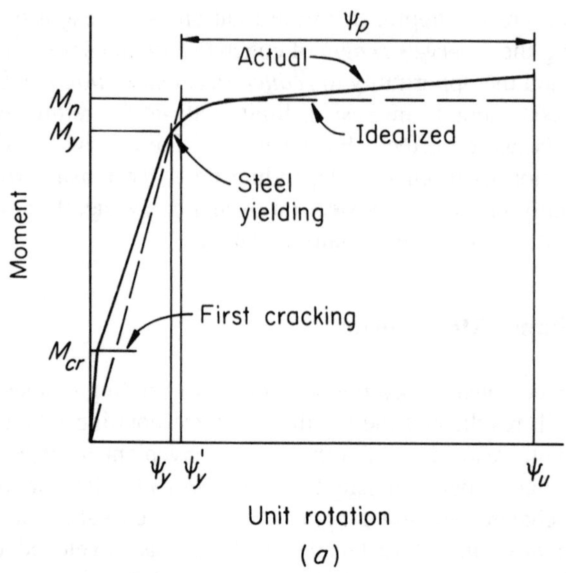

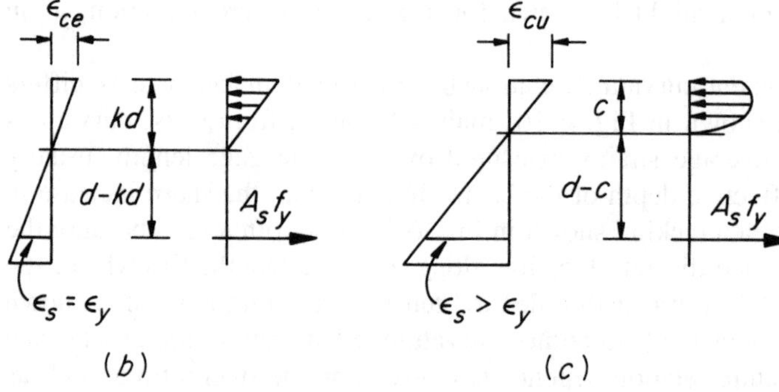

FIGURE 11.11
Plastic hinge characteristics in a reinforced concrete member: (*a*) typical moment-rotation diagram; (*b*) strains and stresses at start of yielding; (*c*) strains and stresses at incipient failure.

where kd is the distance from the compression face to the cracked elastic neutral axis (see Sec. 3.3b). The ultimate moment capacity M_n, based on Fig. 11.11*c*, is calculated by the usual expression

$$M_n = A_s f_y \left(d - \frac{a}{2}\right) = A_s f_y \left(d - \frac{\beta_1 c}{2}\right) \tag{11.2}$$

For purposes of limit analysis, the $M - \psi$ curve is usually idealized, as shown by the dashed line in Fig. 11.11*a*. The slope of the elastic portion of the curve is obtained with satisfactory accuracy using the moment of inertia of the cracked transformed section. After the calculated ultimate moment M_n is reached, continued plastic rotation is assumed to occur with no change in applied moment. The elastic curve of the beam will show an abrupt change in slope at such a section. The beam behaves as if there were a hinge at that point. However, the hinge will not be "friction-free," but will have a constant resistance to rotation.

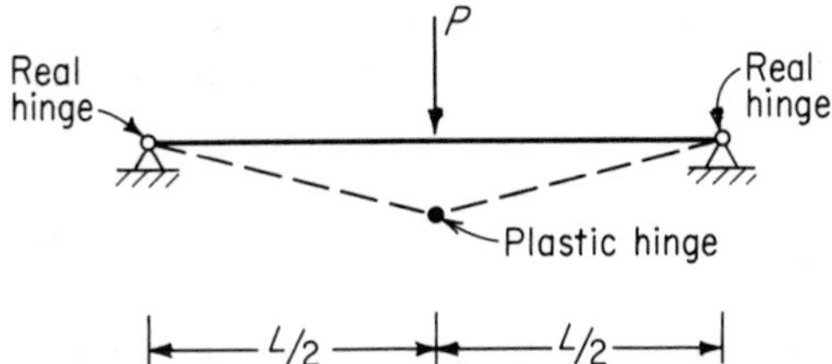

FIGURE 11.12
Statically determinate member after the formation of plastic hinge.

If such a plastic hinge forms in a determinate structure, as shown in Fig. 11.12, uncontrolled deflection takes place, and the structure will collapse. The resulting system is referred to as a *mechanism*, an analogy to linkage systems in mechanics. Generalizing, one can say that a statically determinate system requires the formation of only one plastic hinge in order to become a mechanism.

This is not so for indeterminate structures. In this case, stability may be maintained even though hinges have formed at several cross sections. The formation of such hinges in indeterminate structures permits a redistribution of moments within the beam or frame. It will be assumed for simplicity that the indeterminate beam of Fig. 11.13*a* is symmetrically reinforced, so that the negative bending

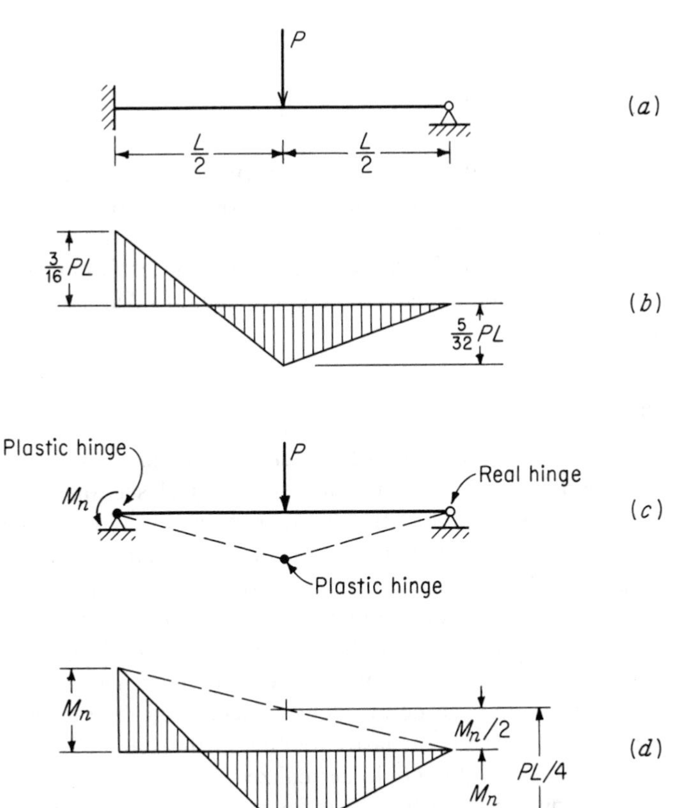

FIGURE 11.13
Indeterminate beam with plastic hinges at support and midspan.

capacity is the same as the positive. Let the load P be increased gradually until the elastic moment at the fixed support, $\frac{3}{16}PL$, is just equal to the plastic moment capacity of the section M_n. This load is

$$P = P_{\text{el}} = \frac{16}{3}\frac{M_n}{L} = 5.33\frac{M_n}{L} \tag{a}$$

At this load, the positive moment under the load is $\frac{5}{32}PL$, as shown in Fig. 11.13*b*. The beam still responds elastically everywhere but at the left support. At that point the actual fixed support can be replaced for purposes of analysis with a plastic hinge offering a known resisting moment M_n. Because a redundant reaction has been replaced by a known moment, the beam is now determinate.

The load can be increased further until the moment under the load also becomes equal to M_n, at which load the second hinge forms. The structure is converted into a mechanism, as shown in Fig. 11.13*c*, and collapse occurs. The moment diagram at collapse load is shown in Fig. 11.13*d*.

The magnitude of load causing collapse is easily calculated from the geometry of Fig. 11.13*d*:

$$M_n + \frac{M_n}{2} = \frac{PL}{4}$$

from which

$$P = P_n = \frac{6M_n}{L} \tag{b}$$

By comparison of Eqs. (*b*) and (*a*), it is evident that an increase in P of 12.5 percent is possible, beyond that load which caused the formation of the first plastic hinge, before the beam will actually collapse. Due to the formation of plastic hinges, a redistribution of moments has occurred such that, at failure, the ratio between the positive moment and negative moment is equal to that assumed in reinforcing the structure.

c. Rotation Requirement

It may be evident that there is a direct relation between the amount of redistribution desired and the amount of inelastic rotation at the critical sections of a beam required to produce the desired redistribution. In general, the greater the modification of the elastic-moment ratio, the greater the required rotation capacity to accomplish that change. To illustrate, if the beam of Fig. 11.13*a* had been reinforced according to the elastic-moment diagram of Fig. 11.13*b*, no inelastic-rotation capacity at all would be required. The beam would, at least in theory, yield simultaneously at the left support and at midspan. On the other hand, if the reinforcement at the left support had been deliberately reduced (and the midspan reinforcement correspondingly increased), inelastic rotation at the support would be required before the strength at midspan could be realized.

The amount of rotation required at plastic hinges for any assumed moment diagram can be found by considering the requirements of compatibility. The member

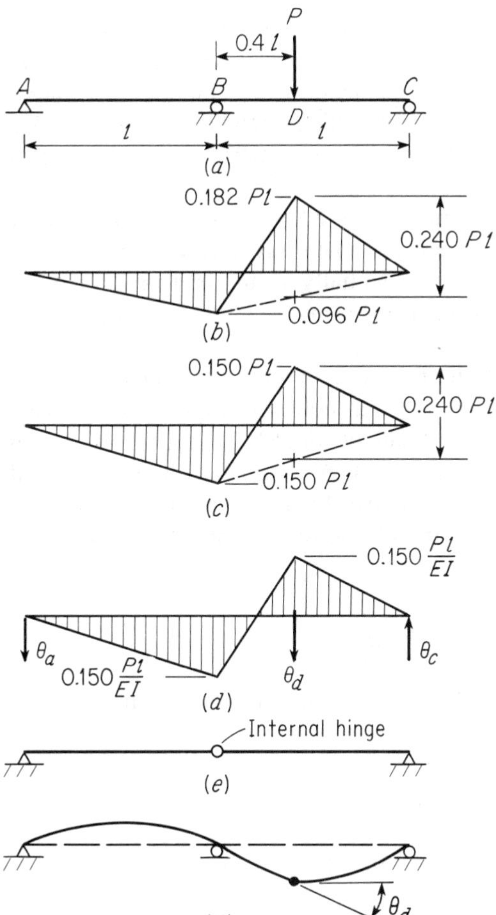

FIGURE 11.14
Moment redistribution in two-span beam: (*a*) a loaded beam; (*b*) elastic moments; (*c*) modified moments; (*d*) M/EI loads; (*e*) a conjugate beam; (*f*) deflection curve.

must be bent, under the combined effects of elastic moment and plastic hinges, so that the correct boundary conditions are satisfied at the supports. Usually, zero support deflection is to be maintained. The moment-area and conjugate-beam principles are useful in quantitative determination of rotation requirements (Ref. 11.9). In deflection calculations, it is convenient to assume that plastic hinging occurs at a point, rather than being distributed over a finite *hinging length*, as is actually the case. Consequently, in loading the conjugate beam with unit rotations, plastic hinges are represented as concentrated loads.

Calculation of rotation requirements will be illustrated by the two-span continuous beam of Fig. 11.14*a*. The elastic-moment diagram resulting from a single concentrated load is shown in Fig. 11.14*b*. The moment at support *B* is $0.096Pl$, while that under the load is $0.182Pl$. If the deflection of the beam at support *C* were calculated using the unit rotations equal to M/EI, based on this elastic-moment diagram, a zero result would be obtained.

Figure 11.14*c* shows an alternative, statically admissible moment diagram that was obtained by arbitrarily increasing the support moment from $0.096Pl$ to $0.150Pl$. If the beam deflection at C were calculated using this moment diagram as a basis, a nonzero value would be obtained. This indicates the necessity for inelastic rotation at one or more points to maintain geometric compatibility at the right support.

If the beam were reinforced according to Fig. 11.14*c*, increasing loads would produce the first plastic hinge at D, where the beam has been deliberately made understrength. Continued loading would eventually result in formation of the second plastic hinge at B, creating a mechanism and leading to collapse of the structure.

Limit analysis requires calculation of rotation at all plastic hinges up to, but not including, the last hinge that triggers actual collapse. Figure 11.14*d* shows the M/EI load to be imposed on the conjugate beam of Fig. 11.14*e*. Also shown is the concentrated angle change θ_d, which is to be evaluated. Starting with the left span, taking moments of the M/EI loads about the internal hinge of the conjugate beam at B, one obtains the left reaction of the conjugate beam (equal to the slope of the real beam):

$$\theta_a = 0.025 \frac{Pl^2}{EI}$$

With that reaction known, moments are taken about the support C of the conjugate beam and set equal to zero to obtain

$$\theta_d = 0.060 \frac{Pl^2}{EI}$$

This represents the necessary discontinuity in the slope of the elastic curve shown in Fig. 11.14*f* to restore the beam to zero deflection at the right support. The beam must be capable of developing at least that amount of plastic rotation if the modified moment diagram assumed in Fig. 11.14*c* is to be valid.

d. Rotation Capacity

The capacity of concrete structures to absorb inelastic rotations at plastic-hinge locations is not unlimited. The designer adopting full limit analysis in concrete must calculate not only the amount of rotation required at critical sections to achieve the assumed degree of moment redistribution but also the rotation capacity of the members at those sections to ensure that it is adequate.

Curvature at initiation of yielding is easily calculated from the elastic strain distribution shown in Fig. 11.11*b*.

$$\psi_y = \frac{\epsilon_y}{d(1-k)} \tag{11.3}$$

in which the ratio k establishing the depth of the elastic neutral axis is found from Eq. (3.12). The ultimate unit rotation can be obtained from the geometry of Fig.

11.11c:

$$\psi_u = \frac{\epsilon_{cu}}{c} \tag{11.4}$$

Although it is customary in flexural strength analysis to adopt $\epsilon_{cu} = 0.003$, for purposes of limit analysis a more refined value is needed. Extensive experimental studies (Refs. 11.10 and 11.11) indicate that the ultimate strain capacity of concrete is strongly influenced by the beam width b, by the moment gradient, and by the presence of additional reinforcement in the form of compression steel and "binding" steel (i.e., web reinforcement). The last parameter is conveniently introduced by means of a steel ratio ρ'', defined as the ratio of the volume of one stirrup plus its tributary compressive steel volume to the concrete volume tributary to one stirrup. On the basis of empirical studies, the ultimate flexural strain at a plastic hinge is

$$\epsilon_{cu} = 0.003 + 0.02\frac{b}{z} + \left(\frac{\rho'' f_y}{14.5}\right)^2 \tag{11.5}$$

where z is the distance between points of maximum and zero moment. Based on Eqs. (11.3) to (11.5), the inelastic unit rotation for the idealized relation shown in Fig. 11.11a is

$$\psi_p = \psi_u - \psi_y \frac{M_n}{M_y} \tag{11.6}$$

This plastic rotation is not confined to one cross section but is distributed over a finite length referred to as the *hinging length*. The experimental studies upon which Eq. (11.5) is based measured strains and rotations in a length equal to the effective depth d of the test members. Consequently, ϵ_{cu} is an *average* value of ultimate strain over a finite length, and ψ_p, given by Eq. (11.6), is an *average* value of unit rotation. The total inelastic rotation θ_p can be found by multiplying the average unit rotation by the hinging length:

$$\theta_p = \left(\psi_u - \psi_y \frac{M_n}{M_y}\right) l_p \tag{11.7}$$

On the basis of current evidence, it appears that the hinging length l_p in support regions, on either side of the support, can be approximated by the expression

$$l_p = 0.5d + 0.05z \tag{11.8}$$

in which z is the distance from the point of maximum moment to the nearest point of zero moment.

e. Moment Redistribution under the ACI Code

Full utilization of the plastic capacity of reinforced concrete beams and frames requires an extensive analysis of all possible mechanisms and an investigation of rotation requirements and capacities at all proposed hinge locations. The increase of design time may not be justified by the gains obtained. On the other hand, a restricted amount of redistribution of elastic moments can safely be made without complete analysis, yet may be sufficient to obtain most of the advantages of limit analysis.

A limited amount of redistribution is permitted by ACI Code 8.4, depending upon a rough measure of available ductility, without explicit calculation of rotation requirements and capacities. The ratio ρ/ρ_b, or in the case of doubly reinforced members, $(\rho - \rho')/\rho_b$, is used as an indicator of rotation capacity, where ρ_b is the balanced steel ratio, given by Eq. (3.28). For singly reinforced members with $\rho = \rho_b$, experiments indicate almost no plastic rotation capacity, since the concrete strain is nearly equal to ϵ_{cu} when steel yielding is initiated. Similarly, in a doubly reinforced member, when $\rho - \rho' = \rho_b$, little rotation will occur after yielding before the concrete crushes. However, when ρ' or $\rho - \rho'$ is low, extensive rotation is usually possible. Accordingly, ACI Code 8.4 provides as follows:

> Except where approximate values for moments are used, the negative moments calculated by elastic theory at the supports of continuous flexural members for any assumed loading arrangement may be increased or decreased by not more than $20[1 - (\rho - \rho')/\rho_b]$ percent. These modified negative moments shall be used for calculation of the moments at sections within the spans. Such an adjustment shall be made only when the section at which the moment is reduced is so designed that ρ or $\rho - \rho'$ is equal to or less than $0.50\rho_b$.

Redistribution for steel ratios above $0.50\rho_b$ is conservatively prohibited. ACI Code provisions are shown graphically in Fig. 11.15.

To demonstrate the advantage of moment redistribution when alternative loadings are involved, consider the concrete beam of Fig. 11.16. A three-span continuous beam is shown, with dead load of 1 kip/ft and live load of 2 kips/ft. To obtain maximum moments at all critical design sections, it is necessary to consider three alternative loadings. Case *a*, with live and dead load over exterior spans and dead load only over interior span, will produce the maximum positive moment in the exterior spans. Case *b*, with dead load on exterior spans and dead and live load on the interior span, will produce the maximum positive moment in the

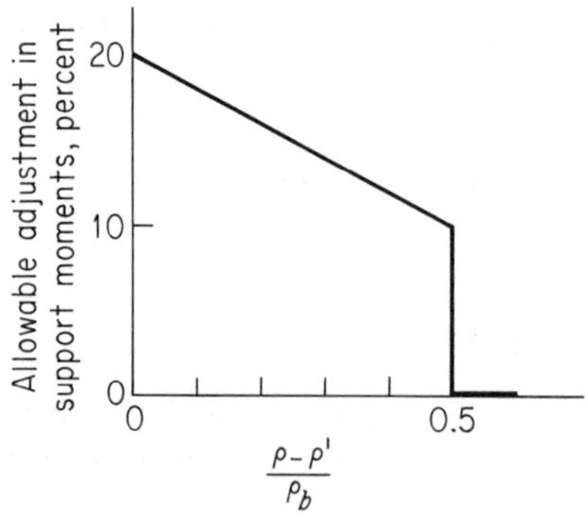

FIGURE 11.15
Allowable moment redistribution under the ACI Code.

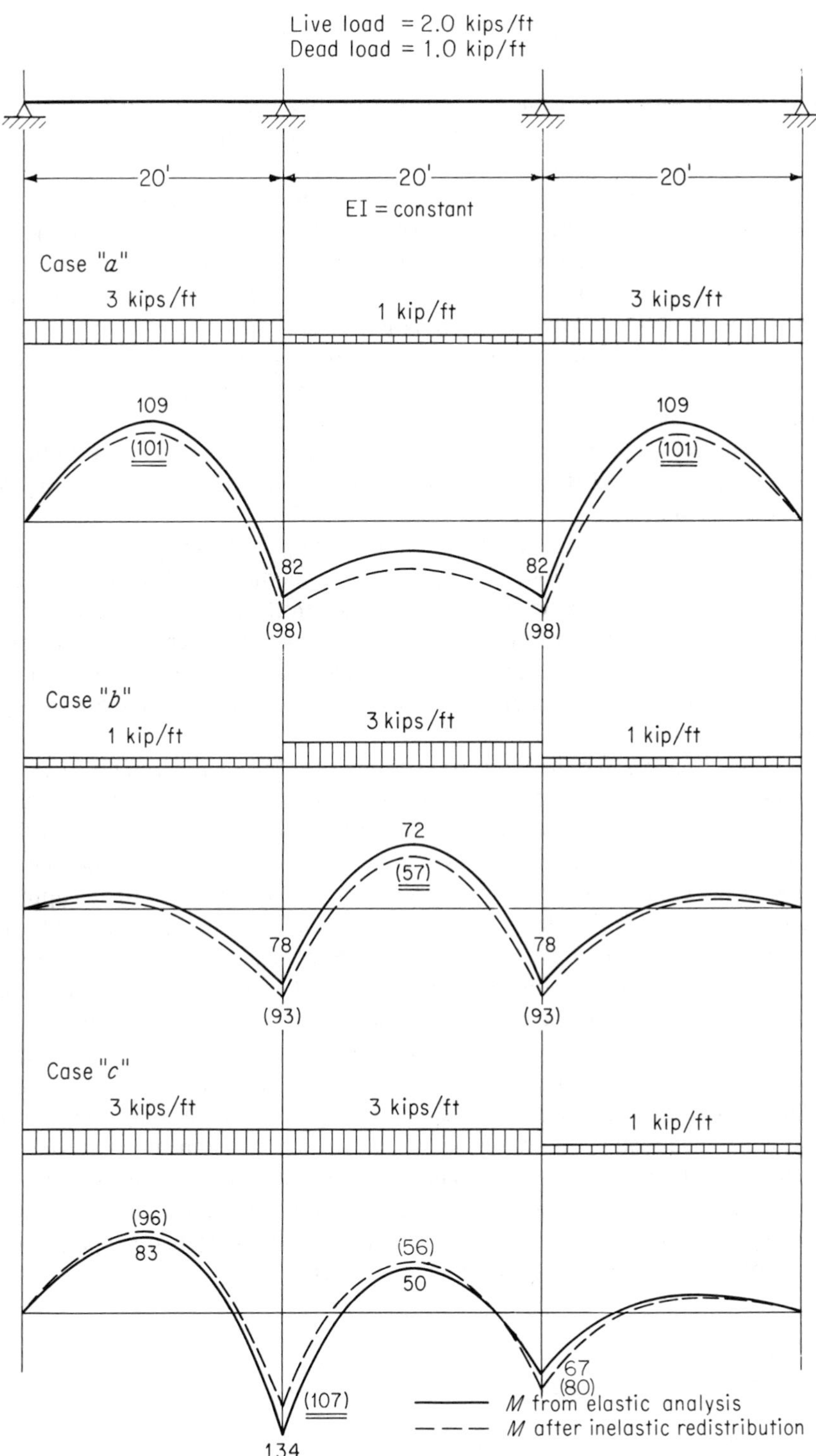

FIGURE 11.16
Redistribution of moments in a three-span continuous beam.

interior span. The maximum negative moment over the interior support is obtained by placing dead and live load on the two adjacent spans and dead load only on the far exterior span, as shown in case *c*.

It will be assumed for simplicity that a 20 percent adjustment of support moments is permitted throughout, provided span moments are modified accordingly. An overall reduction in design moments through the entire three-span beam may be possible. Case *a*, for example, produces an elastic maximum span moment in the exterior spans of 109 ft-kips. Corresponding to this is an elastic negative moment of 82 ft-kips at the interior support. Adjusting the support moment upward by 20 percent, one obtains a negative moment of 98 ft-kips, which results in a downward adjustment of the span moment to 101 ft-kips.

Now consider case *b*. By a similar redistribution of moments, a reduced middle-span moment of 57 ft-kips is obtained through an increase of the support moment from 78 to 93 ft-kips.

The moment obtained at the first interior support for loading case *c* can be adjusted in the reverse direction; i.e., the support moment is decreased by 20 percent to 107 ft-kips. To avoid increasing the controlling span moment of the interior span, the right interior support moment is adjusted upward by 20 percent to 80 ft-kips. The positive moments in the left exterior span and in the interior span corresponding to these modified support moments are 96 and 56 ft-kips respectively.

It will be observed that the reduction obtained for the span moments in cases *a* and *b* were achieved at the expense of increasing the moment at the first interior support. However, the increased support moment in each case was less than the moment for which that support would have to be designed based on the loading *c*, which produced the maximum support moment. Similarly, the reduction in support moment in case *c* was taken at the expense of an increase in span moments in the two adjacent spans. However, in each case the increased span moments were less than the maximum span moments obtained for other loading conditions. The final design moments at all critical sections are underlined in Fig. 11.16. It can be seen, then, that the net result is a reduction in design moments over the entire beam. This modification of moments does not mean a reduction in safety factor below that implied in code safety provisions; rather, it means a reduction of the *excess* strength that would otherwise be present in the structure because of the actual redistribution of moments that would occur before failure. It reflects the fact that the maximum design moments are obtained from alternative load patterns, which could not exist concurrently. The end result is a more realistic appraisal of the actual collapse load of the indeterminate structure.

REFERENCES

11.1. C. H. Norris, J. B. Wilbur, and S. Utku, *Elementary Structural Analysis*, 3d ed., McGraw-Hill, New York, 1976.

11.2. W. McGuire and R. H. Gallagher, *Matrix Structural Analysis*, Wiley, New York, 1978.

11.3. M. D. Vanderbilt, *Matrix Structural Analysis*, Quantum, New York, 1974.

11.4. W. Weaver and J. M. Gere, *Matrix Analysis of Framed Structures*, 2d ed., D. Van Nostrand, New York, 1980.

11.5. C. K. Wang, *Structural Analysis on Microcomputers*, MacMillan, New York, 1986.
11.6. J. F. Fleming, *Structural Engineering Analysis on Personal Computers*, McGraw-Hill, New York, 1986.
11.7. J. R. Benjamin, *Statically Indeterminate Structures*, McGraw-Hill, New York, 1959.
11.8. *Continuity in Concrete Building Frames*, 4th ed., Portland Cement Association, Chicago, 1959.
11.9. G. C. Ernst, "A Brief for Limit Design," *Trans. ASCE*, vol. 121, 1956, pp. 605–632.
11.10. A. H. Mattock, "Rotation Capacity of Hinging Regions in Reinforced Concrete Frames," *Proc. Int. Symp. Flexural Mech. Reinforced Concrete*, ACI Publication SP-12, 1964.
11.11. J. S. Ford, D. C. Chang, and J. E. Breen, "Design Implications from Tests of Unbraced Multi-panel Concrete Frames," *Concr. Int.*, vol. 3, no. 3, 1981, pp. 37–47.

PROBLEMS

11.1 A concrete beam with $b = 12$ in., $h = 26.5$ in., and $d = 24$ in., having a span of 24 ft, can be considered fully fixed at the left support and supported vertically but with no rotational restraint (e.g., roller) at the right end. It is reinforced for positive bending with a combination of bars giving $A_s = 2.45$ in^2, and for negative bending at the left support with $A_s = 2.88$ in^2. Positive bars are carried 6 in. into the face of the left support, according to the ACI Code requirements, but lack the embedded length to be considered effective as compression steel. No. 3 closed hoop stirrups are provided at 9 in. spacing over the full span. The factored load consists of a single concentrated force of 63.3 kips at midspan. Self-weight of the beam may be neglected in the calculations. Calculate the rotation requirement at the first plastic hinge to form (*a*) if the beam is reinforced according to the description above, (*b*) if, in order to reduce bar congestion at the left support, that steel area is reduced by 12.5 percent, with appropriate increase in the positive steel area, and (*c*) if the steel area at the left support is reduced by 25 percent, compared with the original description, with appropriate increase in the positive steel area. Calculate also the rotation capacity of the critical section, for comparison with the requirements of (*a*), (*b*), and (*c*). Comment on your results and compare with the approach to moment redistribution presented in the ACI Code. Material strengths are $f_y = 60$ ksi and $f_c' = 4$ ksi.

11.2 A 12-span continuous reinforced concrete T beam is to carry a calculated dead load of 900 lb/ft including self-weight, plus service live load of 1400 lb/ft on uniform spans measuring 26.5 ft between centers of supporting columns (25 ft clear spans). The slab thickness is 6 in. and the effective flange width is 75 in. Web proportions are $b_w = 0.6d$, and the maximum steel ratio will be set at $0.50\rho_b$. All columns will be 18 in. square. Material strengths are $f_c' = 4000$ psi and $f_y = 60{,}000$ psi.

(*a*) Find the design moments for the exterior and first interior span based on the ACI Code moment coefficients of Table 11.1.

(*b*) Find the design moments in the exterior and first interior span by elastic frame analysis, assuming the floor-to-floor height to be 10 ft. Note that alternative live loadings should be considered (see Sec. 11.2) and that moments can be reduced to account for the support width (see Sec. 11.5a). Compare your results with those obtained using the ACI moment coefficients.

(*c*) Adjust the design negative and positive moments taking maximum advantage of the redistribution provisions of the ACI Code.

(*d*) Design the exterior and first interior spans for flexure and shear, finding concrete dimensions and bar requirements, basing your design on the modified moments.

11.3 A continuous reinforced concrete frame consists of a two-span rectangular beam *ABC*, with center-to-center spans *AB* and *BC* of 24 ft. Columns measuring 14 in. square are provided at *A*, *B*, and *C*. The columns may be considered fully fixed at the floors above and below for purposes of analysis. The beam will carry a service live load of 1200 plf and a calculated dead load of 1000 plf, including self-weight. Floor-to-floor height is 12 ft. Material strengths are $f_y = 60{,}000$ psi and $f'_c = 4000$ psi.

(*a*) Carry out an elastic analysis of the two-span frame, considering alternate live loadings to maximize the bending moment at all critical sections. Design the beams, using maximum steel ratio of $0.50\rho_b$ and $d = 2b$. Find the required concrete section and required steel areas at positive and negative bending sections. Select the rebars. Cutoff points can be determined according to Fig. 5.15*a*. Note that negative design moments are at the face of supports, not support centerlines.

(*b*) Take maximum advantage of the redistribution provisions of ACI Code 8.4 (see sec. 11.9e) to reduce design moments at all critical sections, and redesign the steel for the beams. Keep the concrete section unchanged. Select rebars and determine cutoff points.

(*c*) Comment on your two designs with regard to the amount of steel required and the possible congestion of steel at the critical bending sections. You may assume that the shear reinforcement is unchanged in the redesigned beam.

CHAPTER

12

EDGE-SUPPORTED SLABS

12.1 TYPES OF SLABS

In reinforced concrete construction slabs are used to provide flat, useful surfaces. A reinforced concrete slab is a broad, flat plate, usually horizontal, with top and bottom surfaces parallel or nearly so. It may be supported by reinforced concrete beams (and is usually poured monolithically with such beams), by masonry or reinforced concrete walls, by structural steel members, directly by columns, or continuously by the ground.

Slabs may be supported on two opposite sides only, as in Fig. 12.1*a*, in which case the structural action of the slab is essentially *one-way,* the loads being carried by the slab in the direction perpendicular to the supporting beams. There may be beams on all four sides, as in Fig. 12.1*b*, so that *two-way* slab action is obtained. Intermediate beams, as shown in Fig. 12.1*c*, may be provided. If the ratio of length to width of one slab panel is larger than about 2, most of the load is carried in the short direction to the supporting beams and one-way action is obtained in effect, even though supports are provided on all sides.

Concrete slabs may in some cases by carried directly by columns, as in Fig. 12.1*d*, without the use of beams or girders. Such slabs are described as *flat plates* and are commonly used where spans are not large and loads not particularly heavy. *Flat slab* construction, shown in Fig. 12.1*e*, is also beamless but incorporates a thickened slab region in the vicinity of the column and often employs flared column tops. Both are devices to reduce stresses due to shear and negative bending around the columns. They are referred to as *drop panels* and *column capitals* respectively. Closely related to the flat plate slab is the two-way joist or *grid slab,* shown in Fig. 12.1*f*. To reduce the dead load of solid-slab construction, voids are formed

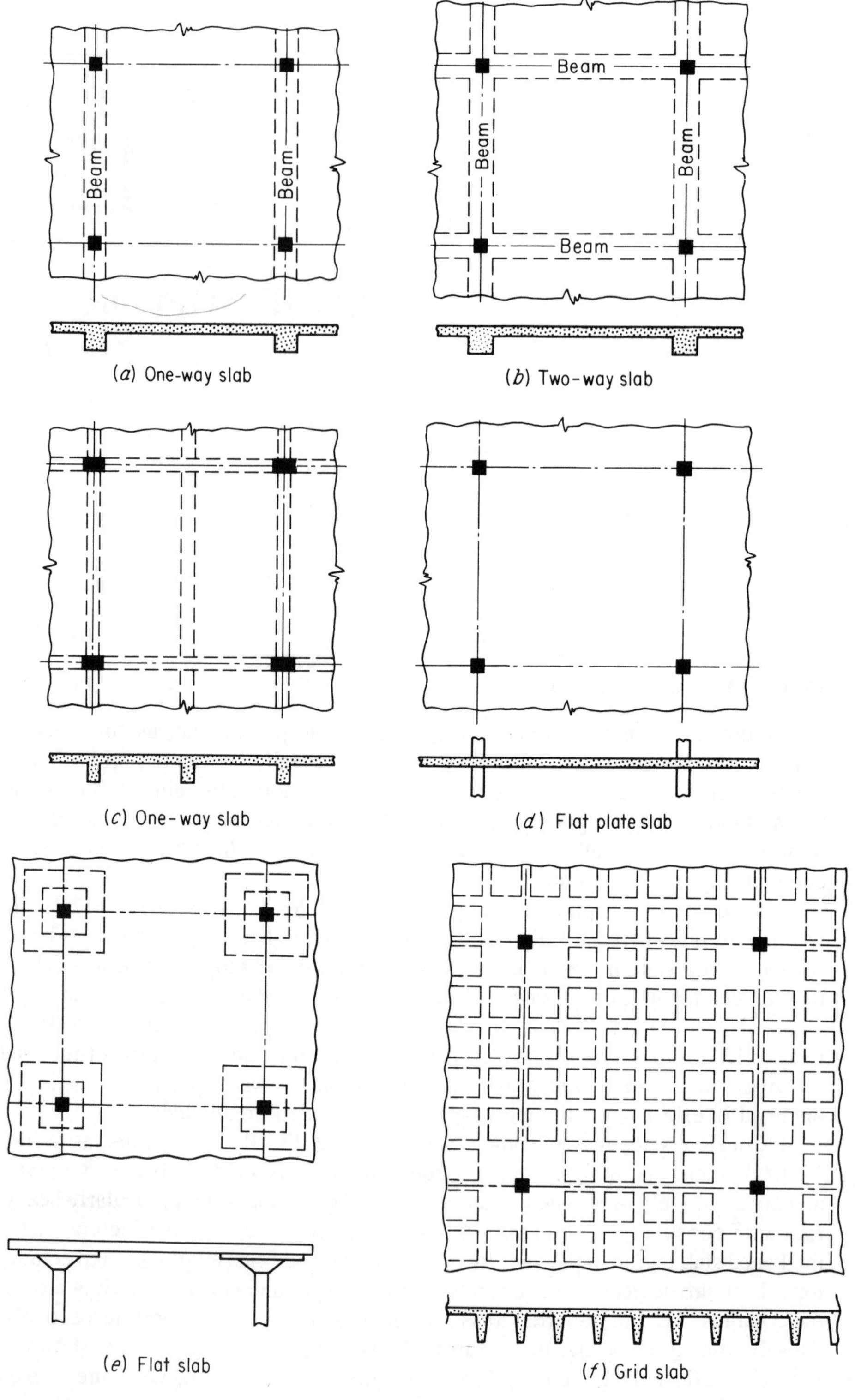

FIGURE 12.1
Types of structural slabs.

in a rectilinear pattern through use of metal or fiberglass form inserts. A two-way ribbed construction results. Usually inserts are omitted near the columns, so a solid slab is formed to resist moments and shears better in these areas.

In addition to the column-supported types of construction shown in Fig. 12.1, many slabs are supported continuously on the ground, as for highways, airport runways, and warehouse floors. In such cases, a well-compacted layer of crushed stone or gravel is usually provided to ensure uniform support and to allow for proper subgrade drainage.

Reinforcing steel for slabs is primarily parallel to the slab surfaces. Straight-bar reinforcement is generally used, although in continuous slabs bottom bars are sometimes bent up to provide for negative reinforcement over the supports. Welded wire fabric is commonly employed for slabs on the ground. Bar or rod mats are available for the heavier reinforcement sometimes needed in highway slabs and airport runways. Slabs may also be prestressed using high tensile strength strands.

Reinforced concrete slabs of the types shown in Fig. 12.1 are usually designed for loads assumed to be uniformly distributed over one entire slab panel, bounded by supporting beams or column centerlines. Minor concentrated loads can be accommodated through two-way action of the reinforcement (two-way flexural steel for two-way slab systems or one-way flexural steel plus lateral distribution steel for one-way systems). Heavy concentrated loads generally require supporting beams.

One-way and two-way edge-supported slabs such as shown in Fig. 12.1*a, b,* and *c* will be discussed in this chapter. Two-way beamless systems such as shown in Fig. 12.1*d, e,* and *f,* as well as two-way edge-supported slabs (Fig. 12.1*b*), will be treated in Chap. 13, and ground-supported slabs such as used for highways and airports or for industrial and warehouse floors will be discussed in Chap. 16. Special methods based on limit analysis at the overload state, applicable to all types of slabs, will be presented in Chapters 14 and 15.

12.2 DESIGN OF ONE-WAY SLABS

The structural action of a one-way slab may be visualized in terms of the deformed shape of the loaded surface. Figure 12.2 shows a rectangular slab, simply supported along its two opposite long edges and free of any support along the two opposite short edges. If a uniformly distributed load is applied to the surface, the deflected shape will be as shown by the solid lines. Curvatures, and consequently bending moments, are the same in all strips s spanning in the short direction between supported edges, whereas there is no curvature, hence no bending moment, in the long strips l parallel to the supported edges. The surface is cylindrical.

For purposes of analysis and design, a unit strip of such a slab cut out at right angles to the supporting beams, as in Fig. 12.3, may be considered as a rectangular beam of unit width, with a depth h equal to the thickness of the slab and a span l_a equal to the distance between supported edges. This strip can then be analyzed by the methods that were used for rectangular beams, the bending moment being computed for the strip of unit width. The load per unit area on the slab becomes the load per unit length on the slab strip. Since all the load on the slab must

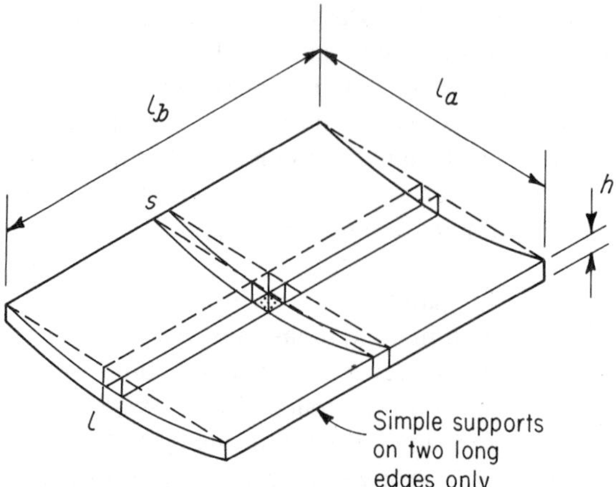

FIGURE 12.2
Deflected shape of uniformly loaded one-way slab.

be transmitted to the two supporting beams, it follows that all the reinforcement should be placed at right angles to these beams, with the exception of any bars that may be placed in the other direction to control shrinkage and temperature cracking. A one-way slab thus consists of a set of rectangular beams side by side.

This simplified analysis, which assumes Poisson's ratio to be zero, is slightly conservative. Actually, flexural compression in the concrete in the direction of l_a will result in lateral expansion in the direction of l_b unless the compressed concrete is restrained. In a one-way slab, this lateral expansion is resisted by adjacent slab strips, which tend to expand also. The result is a slight strengthening and stiffening in the span direction, but this effect is small and can be disregarded.

The ratio of steel in a slab can be determined by dividing the sectional area of one bar by the area of concrete between two successive bars, the latter area being the product of the depth to the center of the bars and the distance between them, center to center. The ratio of steel can also be determined by dividing the average area of steel per foot of width by the effective area of concrete in a 1 ft strip. The average area of steel per foot of width is equal to the area of one bar times the average number of bars in a 1 ft strip (12 divided by the spacing in inches), and the effective area of concrete in a 1 ft (or 12 in.) strip is equal to 12 times the effective depth d.

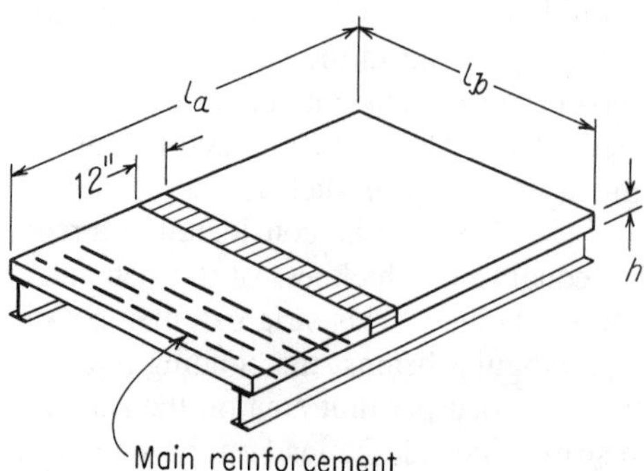

FIGURE 12.3
Unit-strip basis for flexural design.

To illustrate the latter method of obtaining the steel ratio ρ, assume a 5 in. slab with an effective depth of 4 in., with No. 4 bars spaced $4\frac{1}{2}$ in. center to center. The average number of bars in a 12 in. strip of slab is $12/4.5 = 2.7$ bars, and the average steel area in a 12 in. strip is $2.7 \times 0.20 = 0.54$ in^2. Hence $\rho = 0.54/(12 \times 4) = 0.0112$. By the other method

$$\rho = \frac{0.20}{4.5 \times 4} = 0.0112$$

The spacing of bars that is necessary to furnish a given area of steel per foot of width is obtained by dividing the number of bars required to furnish this area into 12. For example, to furnish an average area of 0.46 in^2/ft, with No. 4 bars, requires $0.46/0.20 = 2.3$ bars per foot; the bars must be spaced not more than $12/2.3 = 5.2$ in. center to center. The determination of slab steel areas for various combinations of bars and spacings is facilitated by Table A.4 of App. A.

Design moments and shears in one-way slabs can be found either by elastic analysis or through the use of the same coefficients as used for beams (see Chap. 11). If the slab rests freely on its supports, the span length may be taken equal to the clear span plus the depth of the slab but need not exceed the distance between centers of supports, according to ACI Code 8.7.1. In general, center-to-center distances should be used in continuous slab analysis, but a reduction is allowed in negative moments to account for support width (see Chap. 11). For slabs with clear spans not more than 10 ft that are built integrally with their supports, ACI Code 8.7.4 permits analysis as a continuous slab on knife-edge supports with spans equal to the clear spans and the width of the beams otherwise neglected. If moment and shear coefficients are used, computations should be based on clear spans.

One-way slabs are normally designed with tensile steel ratios well below the maximum permissible value of $0.75\rho_b$. Typical steel ratios range from about 0.004 to 0.008. This is partially for reasons of economy, because the saving in steel associated with increasing the effective depth more than compensates for the cost of the additional concrete, and partially because very thin slabs with high steel ratios would be likely to permit large deflections. Thus flexural design may start with selecting a relatively low steel ratio, say about $0.20\rho_b$, setting $M_u = \phi M_n$ in Eq. (3.37), and solving for the required effective depth d, given that $b = 12$ in. for the unit strip. Alternatively, Table A.6 or Graph A.1 of App. A may be used. Table A.10 is also useful. The required steel area per 12 in. strip, $A_s = \rho b d$, is then easily found.

ACI Code 9.5.2 specifies the minimum thickness in Table 12.1 for nonprestressed slabs of normal weight concrete ($w_c = 145$ pcf) using Grade 60

Table 12.1 Minimum thickness h of nonprestressed one-way slabs

Simply supported	$l/20$
One end continuous	$l/24$
Both ends continuous	$l/28$
Cantilever	$l/10$

reinforcement, provided that the slab is not supporting or attached to construction that is likely to be damaged by large deflections. Lesser thicknesses may be used if calculation of deflections indicates no adverse effects. For concretes having unit weight w_c in the range from 90 to 120 pcf, the tabulated values should be multiplied by $(1.65 - 0.005w_c)$, but not less than 1.09. For reinforcement having a yield stress f_y other than 60,000 psi, the tabulated values should be multiplied by $(0.4 + f_y/100{,}000)$. Slab deflections may be calculated, if required, by the same methods as for beams (see Sec. 6.7).

Shear will seldom control the design of one-way slabs, particularly if low tensile steel ratios are used. It will be found that the shear capacity of the concrete, ϕV_c, will almost without exception be well above the required shear strength V_u at factored loads.

The total slab thickness h is usually rounded to the next higher $\frac{1}{4}$ in. for slabs up to 6 in. thickness, and to the next higher $\frac{1}{2}$ in. for thicker slabs. The concrete protection below the reinforcement should follow the requirements of the ACI Code, calling for $\frac{3}{4}$ in. below the bottom of the steel (see Fig. 3.10*b*). In a typical slab, 1 in. below the center of the steel may be assumed. The lateral spacing of the bars, except those used only to control shrinkage and temperature cracks (see Sec. 12.3), should not exceed 3 times the thickness h or 18 in., whichever is less, according to ACI Code 7.6.5. Generally, bar size should be selected so that the actual spacing is not less than about 1.5 times the slab thickness, to avoid excessive cost for bar fabrication and handling. Also, to reduce cost, straight bars are usually used for slab reinforcement, cut off where permitted as described for beams in Sec. 5.9.

12.3 TEMPERATURE AND SHRINKAGE REINFORCEMENT

Concrete shrinks as the cement paste hardens, as was pointed out in Sec. 2.10. It is advisable to minimize such shrinkage by using concretes with the smallest possible amounts of water and cement compatible with other requirements, such as strength and workability, and by thorough moist-curing of sufficient duration. However, no matter what precautions are taken, a certain amount of shrinkage is usually unavoidable. If a slab of moderate dimensions rests freely on its supports, it can contract to accommodate the shortening of its length produced by shrinkage. Usually, however, slabs and other members are joined rigidly to other parts of the structure and cannot contract freely. This results in tension stresses known as *shrinkage stresses*. A decrease in temperature relative to that at which the slab was poured, particularly in outdoor structures such as bridges, may have an effect similar to shrinkage. That is, the slab tends to contract and if restrained from doing so becomes subject to tensile stresses.

Since concrete is weak in tension, these temperature and shrinkage stresses are likely to result in cracking. Cracks of this nature are not detrimental, provided their size is limited to what are known as *hairline cracks*. This can be achieved by placing reinforcement in the slab to counteract contraction and distribute the cracks uniformly. As the concrete tends to shrink, such reinforcement resists the contraction and consequently becomes subject to compression. The total shrinkage

Table 12.2 Minimum ratios of temperature and shrinkage reinforcement in slabs

Slabs where Grade 40 or 50 deformed bars are used	0.0020
Slabs where Grade 60 deformed bars or welded wire fabric (smooth or deformed) are used	0.0018
Slabs where reinforcement with yield strength exceeding 60,000 psi measured at yield strain of 0.35 percent is used	$\dfrac{0.0018 \times 60,000}{f_y}$

in a slab so reinforced is less than that in one without reinforcement; in addition, whatever cracks do occur will be of smaller width and more evenly distributed by virtue of the reinforcement.

In one-way slabs the reinforcement provided for resisting the bending moments has the desired effect of reducing shrinkage and distributing cracks. However, as contraction takes place equally in all directions, it is necessary to provide special reinforcement for shrinkage and temperature contraction in the direction perpendicular to the main reinforcement. This added steel is known as *temperature* or *shrinkage reinforcement,* or *distribution steel.*

Reinforcement for shrinkage and temperature stresses normal to the principal reinforcement should be provided in a structural slab in which the principal reinforcement extends in one direction only. ACI Code 7.12 specifies the minimum ratios of reinforcement area to gross concrete area shown in Table 12.2 but in no case shall such reinforcing bars be placed farther apart than 5 times the slab thickness or more than 18 in. In no case is the steel ratio to be less than 0.0014.

The steel required by the ACI Code for shrinkage and temperature crack control also represents the minimum permissible reinforcement in the span direction of one-way slabs; the usual minimum ratio $200/f_y$ does not apply.

Example 12.1 One-way slab design. A reinforced concrete slab is built integrally with its supports and consists of two equal spans, each with a clear span of 15 ft. The service live load is 100 psf, and 4000 psi concrete is specified for use with steel of yield stress equal to 60,000 psi. Design the slab, following the provisions of the ACI Code.

Solution. The thickness of the slab is first estimated, based on the minimum thickness of Table 12.1: $l/28 = 15 \times 12/28 = 6.43$ in. A trial thickness of 6.50 in. will be used, for which the weight is $150 \times 6.50/12 = 81$ psf. The specified live load and computed dead load are multiplied by the ACI load factors:

$$\begin{aligned} \text{Dead load} &= 81 \times 1.4 = 113 \text{ psf} \\ \text{Live load} &= 100 \times 1.7 = \underline{170} \text{ psf} \\ \text{Total} &= 283 \text{ psf} \end{aligned}$$

For this case, design moments at critical sections may be found using the ACI moment coefficients (see Table 11.1):

$$\begin{aligned} &\text{At interior support:} && -M = \tfrac{1}{9} \times 0.283 \times 15^2 = 7.06 \text{ ft-kips} \\ &\text{At midspan:} && +M = \tfrac{1}{14} \times 0.283 \times 15^2 = 4.53 \text{ ft-kips} \\ &\text{At exterior support:} && -M = \tfrac{1}{24} \times 0.283 \times 15^2 = 2.65 \text{ ft-kips} \end{aligned}$$

The maximum steel ratio permitted by the ACI Code is, according to Eq. (3.29):

$$0.75\rho_b = 0.75 \times 0.85^2 \times \tfrac{4}{60} \times \tfrac{87}{147} = 0.021$$

If that maximum value of ρ were actually used, the minimum required effective depth, controlled by negative moment at the interior support, would be found from Eq. (3.37) to be

$$d^2 = \frac{M_u}{\phi \rho f_y b(1 - 0.59\rho f_y / f'_c)}$$

$$= \frac{7.06 \times 12}{0.90 \times 0.021 \times 60 \times 12(1 - 0.59 \times 0.21 \times \frac{60}{4})} = 7.67 \text{ in}^2$$

$$d = 2.77 \text{ in}\dagger$$

This is less than the effective depth of $6.50 - 1.00 = 5.50$ in. resulting from application of Code restrictions, and the latter figure will be adopted. At the interior support, if the stress-block depth $a = 1.00$ in., the area of steel required per foot of width in the top of the slab is [Eq. (3.36)]

$$A_s = \frac{M_u}{\phi f_y(d - a/2)} = \frac{7.06 \times 12}{0.90 \times 60 \times 5.00} = 0.31 \text{ in}^2$$

Checking the assumed depth a by Eq. (3.31), one gets

$$a = \frac{A_s f_y}{0.85 f'_c b} = \frac{0.31 \times 60}{0.85 \times 4 \times 12} = 0.46 \text{ in.}$$

A second trial will be made with $a = 0.46$ in. Then

$$A_s = \frac{7.06 \times 12}{0.90 \times 60 \times 5.27} = 0.30 \text{ in}^2$$

for which $a = 0.46 \times 0.30/0.31 = 0.45$ in. No further revision is necessary. At other critical-moment sections, it will be satisfactory to use the same lever arm to determine steel areas, and

At midspan: $$A_s = \frac{4.53 \times 12}{0.90 \times 60 \times 5.27} = 0.19 \text{ in}^2$$

At exterior support: $$A_s = \frac{2.65 \times 12}{0.90 \times 60 \times 5.27} = 0.11 \text{ in}^2$$

The minimum reinforcement is that required for control of shrinkage and temperature cracking. This is

$$A_s = 0.0018 \times 12 \times 6.50 = 0.14 \text{ in}^2$$

per 12 in. strip. This requires a small increase in the amount of steel used at the exterior support.

---

†This depth is more easily found using Graph A.1 of App. A. For $\rho = \rho_{max}$, $M_u/\phi b d^2 = 1050$, from which $d = 2.75$ in. Table A.6*a* may also be used.

The factored shear force at a distance d from the face of the interior support is

$$V_u = 1.15 \times \frac{283 \times 15}{2} - 283 \times \frac{5.50}{12} = 2310 \text{ lb}$$

By Eq. (4.12*b*), the nominal shear strength of the concrete slab is

$$V_n = V_c = 2\sqrt{f_c'}bd = 2\sqrt{4000} \times 12 \times 5.50 = 8350 \text{ lb}$$

Thus the design strength of the concrete slab, $\phi V_c = 0.85 \times 8350 = 7100$ lb, is well above the required strength in shear of $V_u = 2310$ lb.

The required tensile steel areas may be provided in a variety of ways, but whatever the selection, due consideration must be given to the actual placing of the steel during construction. The arrangement should be such that the steel can be placed rapidly with the minimum of labor costs even though some excess steel is necessary to achieve this end.

Two possible arrangements are shown in Fig. 12.4. In Fig. 12.4*a* bent bars are used, while in Fig. 12.4*b* all bars are straight.

In the arrangement of Fig. 12.4*a*, No. 4 bars at 10 in. furnish 0.24 in^2 of steel at midspan, slightly more than required. If two-thirds of these bars are bent upward for negative reinforcement over the interior support, the average spacing of such bent bars at the interior support will be $(10 + 20)/2 = 15$ in. Since an identical pattern of bars is bent upward from the other side of the support, the effective spacing of the No. 4 bars over the interior support is $7\frac{1}{2}$ in. This pattern closely satisfies the required steel area of 0.30 in^2 per foot width of slab over the support. The bars bent at the interior support

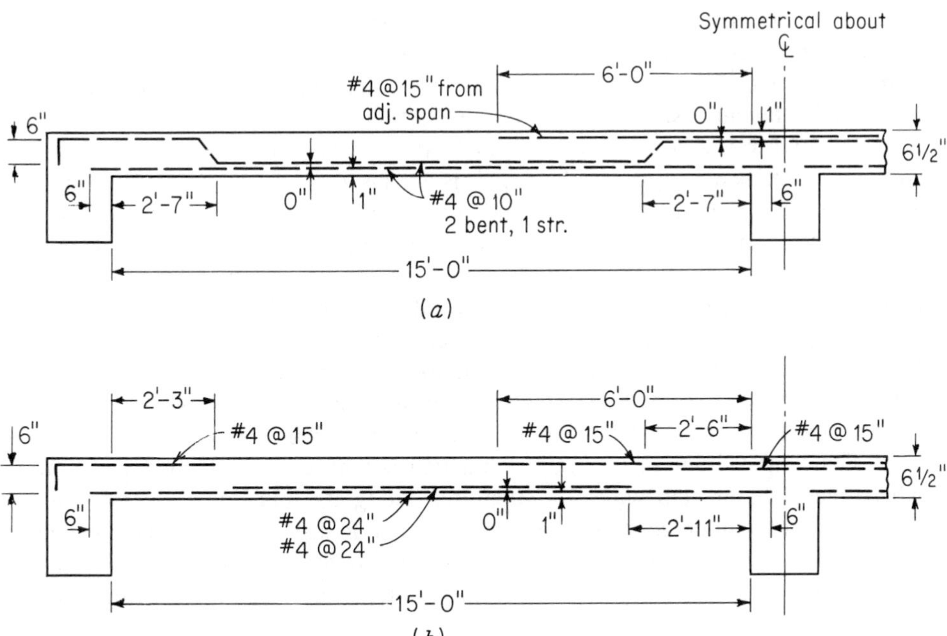

FIGURE 12.4
One-way slab design example.

will also be bent upward for negative reinforcement at the exterior support, providing reinforcement equivalent to No. 4 bars at 15 in., or 0.16 in^2 of steel.

Note that it is not necessary to achieve uniform spacing of reinforcement in slabs, and that the steel provided can be calculated safely on the basis of average spacing as in the example. Care should be taken to satisfy requirements for both minimum and maximum spacing of principal reinforcement, however.

The locations of bend and cutoff points shown in Fig. 12.4*a* were obtained using Graph A.3 of App. A as explained in Sec. 5.9 and Table A.11 (see also Fig. 5.14).

The arrangement of Fig. 12.4*b* uses only straight bars. Although it is satisfactory according to the ACI Code (since the shear stress does not exceed two-thirds of that permitted), cutting off the shorter positive and negative bars as shown leads to an undesirable condition at the ends of those bars, where there will be concentrations of stress in the concrete. The design would be improved if the negative bars were cut off at 3 ft from the face of the interior support rather than 2 ft 6 in. as shown, and if the positive steel were cut off at 2 ft 2 in. rather than at 2 ft 11 in. This would result in an overlap of approximately $2d$ of the cut positive and negative bars. Figure 5.15*a* suggests a somewhat simpler arrangement that would also prove satisfactory.

The required area of steel to be placed normal to the main reinforcement for purposes of temperature and shrinkage crack control is 0.14 in^2. This will be provided by No. 4 bars at 16 in. spacing, placed directly on top of the main reinforcement in the positive-moment region and below the main steel in the negative-moment zone.

12.4 BEHAVIOR OF TWO-WAY EDGE-SUPPORTED SLABS

The slabs discussed in Secs. 12.2 and 12.3 deform under load into a cylindrical surface. The main structural action is one-way in such cases, in the direction normal to supports on two opposite edges of a rectangular panel. In many cases, however, rectangular slabs are of such proportions and are supported in such a way that two-way action results. When loaded, such slabs bend into a dished surface rather than a cylindrical one. This means that at any point the slab is curved in both principal directions, and since bending moments are proportional to curvatures, moments also exist in both directions. To resist these moments, the slab must be reinforced in both directions, by a least two layers of bars perpendicular, respectively, to two pairs of edges. The slab must be designed to take a proportionate share of the load in each direction.

Types of reinforced concrete construction that are characterized by two-way action include slabs supported by walls or beams on all sides (Fig. 12.1*b*), flat plates (Fig. 12.1*d*), flat slabs (Fig. 12.1*e*), and grid slabs (Fig. 12.1*f*).

The simplest type of two-way slab action is that represented by Fig. 12.1*b*, where the slab, or slab panel, is supported along its four edges by relatively deep, stiff, monolithic concrete beams or by walls or steel girders. If the concrete edge beams are shallow or are omitted altogether, as for flat plates and flat slabs, deformation of the floor system along the column lines significantly alters the distribution of moments in the slab panel itself (Ref. 12.1). Two-way systems of this type are considered separately, in Chap. 13. The present discussion pertains to the former type, in which edge supports are stiff enough to be considered unyielding.

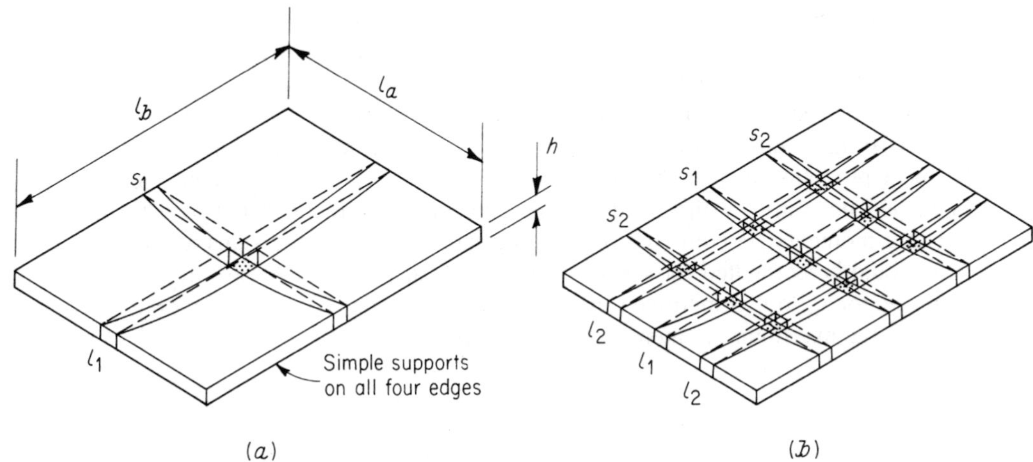

FIGURE 12.5
Two-way slab on simple edge supports: (*a*) bending of center strips of slab; (*b*) grid model of slab.

Such a slab is shown in Fig. 12.5*a*. To visualize its flexural performance it is convenient to think of it as consisting of two sets of parallel strips, in each of the two directions, intersecting each other. Evidently, part of the load is carried by one set and transmitted to one pair of edge supports, and the remainder by the other.

Figure 12.5*a* shows the two center strips of a rectangular plate with short span l_a and long span l_b. If the uniform load is w per square foot of slab, each of the two strips acts approximately like a simple beam uniformly loaded by its share of w. Because these imaginary strips actually are part of the same monolithic slab, their deflections at the intersection point must be the same. Equating the center deflections of the short and long strips gives

$$\frac{5w_a l_a^4}{384EI} = \frac{5w_b l_b^4}{384EI} \tag{a}$$

where w_a is the share of the load w carried in the short direction and w_b is the share of the load w carried in the long direction. Consequently,

$$\frac{w_a}{w_b} = \frac{l_b^4}{l_a^4} \tag{b}$$

One sees that the larger share of the load is carried in the short direction, the ratio of the two portions of the total load being inversely proportional to the fourth power of the ratio of the spans.

This result is approximate because the actual behavior of a slab is more complex than that of the two intersecting strips. An understanding of the behavior of the slab itself can be gained from Fig. 12.5*b*, which shows a slab model consisting of two sets of three strips each. It is seen that the two central strips s_1 and l_1 bend in a manner similar to that of Fig. 12.5*a*. The outer strips s_2 and l_2, however, are not only bent but also twisted. Consider, for instance, one of the intersections of s_2 with l_2. It is seen that at the intersection the exterior edge of strip l_2 is at a higher elevation than the interior edge, while at the nearby end of strip l_2 both edges are at the same elevation; the strip is twisted. This twisting results in torsional stresses and torsional moments that are seen to be most pronounced near the corners. Consequently, the total load on the slab is carried not only by the bending moments in two directions but also by the twisting moments. For this reason bending moments in elastic slabs are smaller than would be computed for sets of unconnected strips loaded by w_a and w_b. For instance, for a simply supported square slab, $w_a = w_b = w/2$. If only bending were present, the maximum moment in each strip would be

$$\frac{(w/2)l^2}{8} = 0.0625wl^2 \qquad (c)$$

The exact theory of bending of elastic plates shows that, actually, the maximum moment in such a square slab is only $0.048wl^2$, so that in this case the twisting moments relieve the bending moments by about 25 percent.

The largest moment occurs where the curvature is sharpest. Figure 12.5*b* shows this to be the case at midspan of the short strip s_1. Suppose the load is increased until this location is overstressed, so that the steel at the middle of strip s_1 is yielding. If the strip were an isolated beam, it would now fail. Considering the slab as a whole, however, one sees that no immediate failure will occur. The neighboring strips (those parallel as well as those perpendicular to s_1), being actually monolithic with it, will take over that share of any additional load which strip s_1 can no longer carry until they in turn start yielding. This inelastic redistribution will continue until in a rather large area in the central portion of the slab all the steel in both directions is yielding. Only then will the entire slab fail. From this reasoning, which is confirmed by tests, it follows that slabs need not be designed for the absolute maximum moment in each of the two directions (such as $0.048wl^2$ in the example of the previous paragraph), but only for a smaller average moment in each of the two directions in the central portion of the slab. For instance, one of the several analytical methods in general use permits the above square slab to be designed for a moment of $0.036wl^2$. By comparison with the actual elastic maximum moment $0.048wl^2$, it is seen that, owing to inelastic redistribution, a moment reduction of 25 percent is provided.

The largest moment in the slab occurs at midspan of the short strip s_1 of Fig. 12.5*b*. It is evident that the curvature, and hence the moment, in the short strip s_2 is less than at the corresponding location of strip s_1. Consequently, a variation of short-span moment occurs in the long direction of the span. This variation is shown qualitatively in Fig. 12.6. The short-span moment diagram in Fig. 12.6*a* is valid only along the center strip at 1-1. Elsewhere the maximum-moment value is less, as shown. Other moment ordinates are reduced proportionately. Similarly, the long-span moment diagram in Fig. 12.6*b* applies only at the longitudinal centerline of the slab; elsewhere ordinates are reduced according to the variation

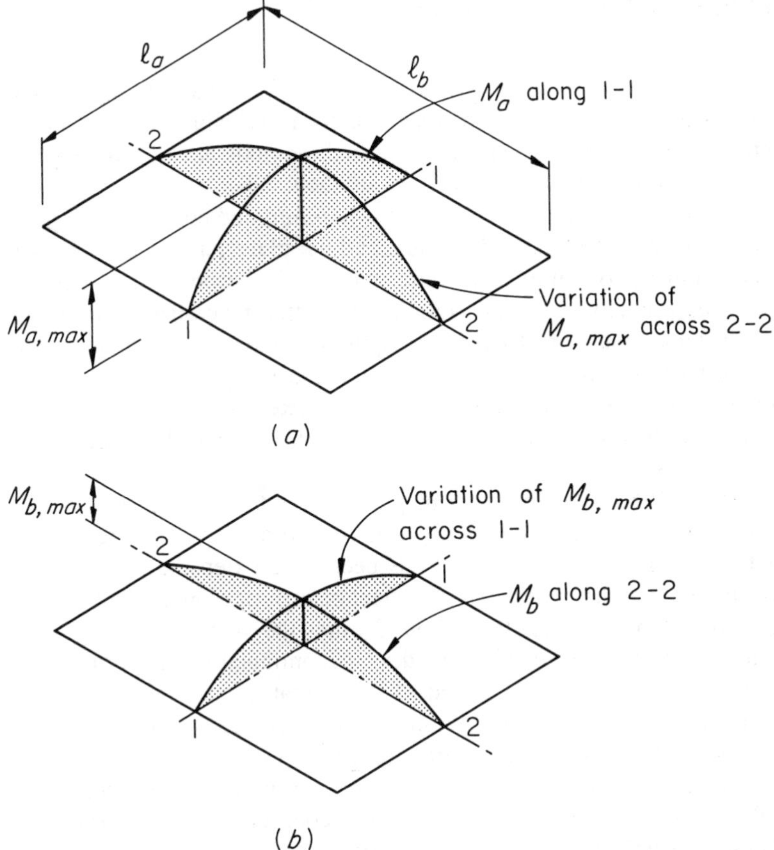

FIGURE 12.6
Moments and moment variations in a uniformly loaded slab with simple supports on four sides.

shown. These variations in maximum moment across the width and length of a rectangular slab are accounted for in an approximate way in most practical design methods by designing for a reduced moment in the outer quarters of the slab span in each direction.

It should be noted that only slabs with side ratios less than about 2 need be treated as two-way slabs. From Eq. (*b*) above it is seen that for a slab of this proportion the share of the load carried in the long direction is only of the order of one-sixteenth that in the short direction. Such a slab acts almost as if it were spanning in the short direction only. Consequently, rectangular slab panels with an aspect ratio of 2 or more may be reinforced for one-way action, with the main steel perpendicular to the long edges. Shrinkage and temperature steel should be provided in the long direction, of course, and auxiliary reinforcement should be provided over, and perpendicular to, the short support beams and at the slab corners to control cracking (see Sec. 12.5).

12.5 ANALYSIS BY THE COEFFICIENT METHOD

The precise determination of moments in two-way slabs with various conditions of continuity at the supported edges is mathematically formidable and not suited to design practice. For this reason, various simplified methods have been adopted for determining moments, shears, and reactions of such slabs.

According to the 1989 ACI Code, all two-way reinforced concrete slab systems, including edge-supported slabs, flat slabs, and flat plates, are to be analyzed and designed according to one unified method, which will be presented in detail in the following chapter. However, the complexity of the generalized approach, particularly for systems that do not meet the requirements permitting analysis by the "direct design method" of the present code, has led many engineers to continue to use the design method of the 1963 ACI Code (Ref. 12.2) for the special case of two-way slabs supported on four sides of each slab panel by relatively deep, stiff, edge beams.

Method 3 of the 1963 ACI Code will be presented in this chapter. Originally developed by Marcus (Ref. 12.3) and widely used in Europe, it was introduced in the United States by Rogers (Ref. 12.4). It has been used extensively here since 1963 for slabs supported at the edges by walls, steel beams, or monolithic concrete beams having a total depth not less than about 3 times the slab thickness. While it was not a part of the 1977 or later ACI Codes, its continued use is permissible under the current code provision (ACI Code 13.3.1) that a slab system may be designed by any procedure satisfying conditions of equilibrium and geometric compatibility, if it is shown that the design strength at every section is at least equal to the required strength, and that serviceability requirements are met.

The method makes use of tables of moment coefficients for a variety of conditions. These coefficients are based on elastic analysis but also account for inelastic redistribution. In consequence, the design moment in either direction is smaller by an appropriate amount than the elastic maximum moment in that direction. The moments in the middle strips in the two directions are computed from

$$M_a = C_a w l_a^2 \tag{12.1}$$

and

$$M_b = C_b w l_b^2 \tag{12.2}$$

where C_a, C_b = tabulated moment coefficients
w = uniform load, psf
l_a, l_b = length of clear span in short and long directions respectively

The method provides that each panel be divided in both directions into a middle strip whose width is one-half that of the panel and two edge or column strips of one-quarter of the panel width (see Fig. 12.7). As discussed before and shown in Fig. 12.6*a* and *b*, the moments in both directions are larger in the center portion of the slab than in regions close to the edges. Correspondingly, it is provided that the entire middle strip be designed for the full, tabulated design moment. In the edge strips this moment is assumed to decrease from its full value

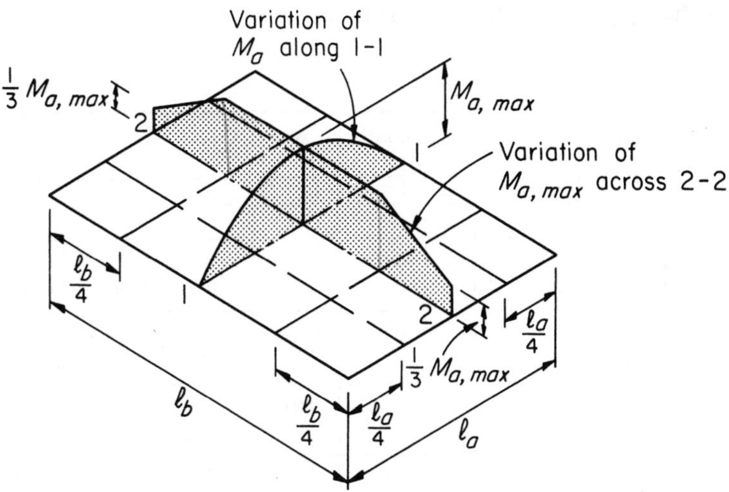

FIGURE 12.7
Variation of moments across the width of critical sections assumed for design.

at the edge of the middle strip to one-third of this value at the edge of the panel. This distribution is shown for the moments M_a in the short span direction in Fig. 12.7. The lateral variation of the long span moments M_b is similiar.

The discussion so far has been restricted to a single panel simply supported at all four edges. An actual situation is shown in Fig. 12.8, in which a system of beams supports a two-way slab. It is seen that some panels, such as A, have two discontinuous exterior edges, while the other edges are continuous with their

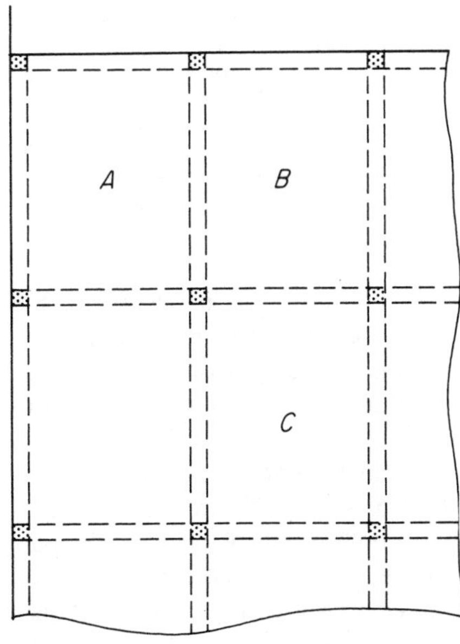

FIGURE 12.8
Plan of a typical two-way slab floor with beams on column lines.

Table 12.3 Coefficients for negative moments in slabs[a]

$$M_{a,\text{neg}} = C_{a,\text{neg}} w l_a^2$$
$$M_{b,\text{neg}} = C_{b,\text{neg}} w l_b^2$$
where w = total uniform dead plus live load

Ratio $m = \dfrac{l_a}{l_b}$		Case 1	Case 2	Case 3	Case 4	Case 5	Case 6	Case 7	Case 8	Case 9
1.00	$C_{a,\text{neg}}$		0.045		0.050	0.075	0.071		0.033	0.061
	$C_{b,\text{neg}}$		0.045	0.076	0.050			0.071	0.061	0.033
0.95	$C_{a,\text{neg}}$		0.050		0.055	0.079	0.075		0.038	0.065
	$C_{b,\text{neg}}$		0.041	0.072	0.045			0.067	0.056	0.029
0.90	$C_{a,\text{neg}}$		0.055		0.060	0.080	0.079		0.043	0.068
	$C_{b,\text{neg}}$		0.037	0.070	0.040			0.062	0.052	0.025
0.85	$C_{a,\text{neg}}$		0.060		0.066	0.082	0.083		0.049	0.072
	$C_{b,\text{neg}}$		0.031	0.065	0.034			0.057	0.046	0.021
0.80	$C_{a,\text{neg}}$		0.065		0.071	0.083	0.086		0.055	0.075
	$C_{b,\text{neg}}$		0.027	0.061	0.029			0.051	0.041	0.017
0.75	$C_{a,\text{neg}}$		0.069		0.076	0.085	0.088		0.061	0.078
	$C_{b,\text{neg}}$		0.022	0.056	0.024			0.044	0.036	0.014
0.70	$C_{a,\text{neg}}$		0.074		0.081	0.086	0.091		0.068	0.081
	$C_{b,\text{neg}}$		0.017	0.050	0.019			0.038	0.029	0.011
0.65	$C_{a,\text{neg}}$		0.077		0.085	0.087	0.093		0.074	0.083
	$C_{b,\text{neg}}$		0.014	0.043	0.015			0.031	0.024	0.008
0.60	$C_{a,\text{neg}}$		0.081		0.089	0.088	0.095		0.080	0.085
	$C_{b,\text{neg}}$		0.010	0.035	0.011			0.024	0.018	0.006
0.55	$C_{a,\text{neg}}$		0.084		0.092	0.089	0.096		0.085	0.086
	$C_{b,\text{neg}}$		0.007	0.028	0.008			0.019	0.014	0.005
0.50	$C_{a,\text{neg}}$		0.086		0.094	0.090	0.097		0.089	0.088
	$C_{b,\text{neg}}$		0.006	0.022	0.006			0.014	0.010	0.003

[a] A crosshatched edge indicates that the slab continues across, or is fixed at, the support; an unmarked edge indicates a support at which torsional resistance is negligible.

neighbors. Panel B has one edge discontinuous and three continuous edges, the interior panel C has all edges continuous, and so on. At a continuous edge in a slab, moments are negative, just as at interior supports of continuous beams. Also, the magnitude of the positive moments depends on the conditions of continuity at all four edges.

Correspondingly, Table 12.3 gives moment coefficients C, for *negative moments at continuous edges*. The details of the tables are self-explanatory. Maximum negative edge moments are obtained when both panels adjacent to the particular edge carry full dead and live load. Hence the moment is computed for this total load. *Negative moments at discontinuous edges* are assumed equal to one-third of the positive moments for the same direction. One must provide for such

Table 12.4 Coefficients for dead load positive moments in slabs[a]

$$M_{a,\text{pos},dl} = C_{a,dl} w l_a^2$$
$$M_{b,\text{pos},dl} = C_{b,dl} w l_b^2$$
where w = total uniform dead load

Ratio $m = \frac{l_a}{l_b}$		Case 1	Case 2	Case 3	Case 4	Case 5	Case 6	Case 7	Case 8	Case 9
1.00	$C_{a,dl}$	0.036	0.018	0.018	0.027	0.027	0.033	0.027	0.020	0.023
	$C_{b,dl}$	0.036	0.018	0.027	0.027	0.018	0.027	0.033	0.023	0.020
0.95	$C_{a,dl}$	0.040	0.020	0.021	0.030	0.028	0.036	0.031	0.022	0.024
	$C_{b,dl}$	0.033	0.016	0.025	0.024	0.015	0.024	0.031	0.021	0.017
0.90	$C_{a,dl}$	0.045	0.022	0.025	0.033	0.029	0.039	0.035	0.025	0.026
	$C_{b,dl}$	0.029	0.014	0.024	0.022	0.013	0.021	0.028	0.019	0.015
0.85	$C_{a,dl}$	0.050	0.024	0.029	0.036	0.031	0.042	0.040	0.029	0.028
	$C_{b,dl}$	0.026	0.012	0.022	0.019	0.011	0.017	0.025	0.017	0.013
0.80	$C_{a,dl}$	0.056	0.026	0.034	0.039	0.032	0.045	0.045	0.032	0.029
	$C_{b,dl}$	0.023	0.011	0.020	0.016	0.009	0.015	0.022	0.015	0.010
0.75	$C_{a,dl}$	0.061	0.028	0.040	0.043	0.033	0.048	0.051	0.036	0.031
	$C_{b,dl}$	0.019	0.009	0.018	0.013	0.007	0.012	0.020	0.013	0.007
0.70	$C_{a,dl}$	0.068	0.030	0.046	0.046	0.035	0.051	0.058	0.040	0.033
	$C_{b,dl}$	0.016	0.007	0.016	0.011	0.005	0.009	0.017	0.011	0.006
0.65	$C_{a,dl}$	0.074	0.032	0.054	0.050	0.036	0.054	0.065	0.044	0.034
	$C_{b,dl}$	0.013	0.006	0.014	0.009	0.004	0.007	0.014	0.009	0.005
0.60	$C_{a,dl}$	0.081	0.034	0.062	0.053	0.037	0.056	0.073	0.048	0.036
	$C_{b,dl}$	0.010	0.004	0.011	0.007	0.003	0.006	0.012	0.007	0.004
0.55	$C_{a,dl}$	0.088	0.035	0.071	0.056	0.038	0.058	0.081	0.052	0.037
	$C_{b,dl}$	0.008	0.003	0.009	0.005	0.002	0.004	0.009	0.005	0.003
0.50	$C_{a,dl}$	0.095	0.037	0.080	0.059	0.039	0.061	0.089	0.056	0.038
	$C_{b,dl}$	0.006	0.002	0.007	0.004	0.001	0.003	0.007	0.004	0.002

[a] A crosshatched edge indicates that the slab continues across, or is fixed at, the support; an unmarked edge indicates a support at which torsional resistance is negligible.

moments because some degree of restraint is generally provided at discontinuous edges by the torsional rigidity of the edge beam or by the supporting wall.

For *positive moments* there will be little, if any, rotation at the continuous edges if *dead load* alone is acting, because the loads on both adjacent panels tend to produce opposite rotations which cancel, or nearly so. For this condition, the continuous edges can be regarded as fixed, and the appropriate coefficients for the dead load positive moments are given in Table 12.4. On the other hand, the maximum *live load positive moments* are obtained when live load is placed only on the particular panel and not on any of the adjacent panels. In this case, some rotation will occur at all continuous edges. As an approximation it is assumed that there is 50 percent restraint for calculating these live load moments. The

Table 12.5 Coefficients for live load positive moments in slabs[a]

$M_{a,\text{pos},ll} = C_{a,ll} w l_a^2$
$M_{b,\text{pos},ll} = C_{b,ll} w l_b^2$ where w = total uniform live load

Ratio		Case 1	Case 2	Case 3	Case 4	Case 5	Case 6	Case 7	Case 8	Case 9
$m = \frac{l_a}{l_b}$		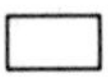		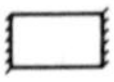	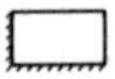		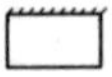	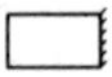	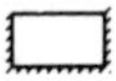	
1.00	$C_{a,ll}$	0.036	0.027	0.027	0.032	0.032	0.035	0.032	0.028	0.030
	$C_{b,ll}$	0.036	0.027	0.032	0.032	0.027	0.032	0.035	0.030	0.028
0.95	$C_{a,ll}$	0.040	0.030	0.031	0.035	0.034	0.038	0.036	0.031	0.032
	$C_{b,ll}$	0.033	0.025	0.029	0.029	0.024	0.029	0.032	0.027	0.025
0.90	$C_{a,ll}$	0.045	0.034	0.035	0.039	0.037	0.042	0.040	0.035	0.036
	$C_{b,ll}$	0.029	0.022	0.027	0.026	0.021	0.025	0.029	0.024	0.022
0.85	$C_{a,ll}$	0.050	0.037	0.040	0.043	0.041	0.046	0.045	0.040	0.039
	$C_{b,ll}$	0.026	0.019	0.024	0.023	0.019	0.022	0.026	0.022	0.020
0.80	$C_{a,ll}$	0.056	0.041	0.045	0.048	0.044	0.051	0.051	0.044	0.042
	$C_{b,ll}$	0.023	0.017	0.022	0.020	0.016	0.019	0.023	0.019	0.017
0.75	$C_{a,ll}$	0.061	0.045	0.051	0.052	0.047	0.055	0.056	0.049	0.046
	$C_{b,ll}$	0.019	0.014	0.019	0.016	0.013	0.016	0.020	0.016	0.013
0.70	$C_{a,ll}$	0.068	0.049	0.057	0.057	0.051	0.060	0.063	0.054	0.050
	$C_{b,ll}$	0.016	0.012	0.016	0.014	0.011	0.013	0.017	0.014	0.011
0.65	$C_{a,ll}$	0.074	0.053	0.064	0.062	0.055	0.064	0.070	0.059	0.054
	$C_{b,ll}$	0.013	0.010	0.014	0.011	0.009	0.010	0.014	0.011	0.009
0.60	$C_{a,ll}$	0.081	0.058	0.071	0.067	0.059	0.068	0.077	0.065	0.059
	$C_{b,ll}$	0.010	0.007	0.011	0.009	0.007	0.008	0.011	0.009	0.007
0.55	$C_{a,ll}$	0.088	0.062	0.080	0.072	0.063	0.073	0.085	0.070	0.063
	$C_{b,ll}$	0.008	0.006	0.009	0.007	0.005	0.006	0.009	0.007	0.006
0.50	$C_{a,ll}$	0.095	0.066	0.088	0.077	0.067	0.078	0.092	0.076	0.067
	$C_{b,ll}$	0.006	0.004	0.007	0.005	0.004	0.005	0.007	0.005	0.004

[a] A crosshatched edge indicates that the slab continues across, or is fixed at, the support; an unmarked edge indicates a support at which torsional resistance is negligible.

corresponding coefficients are given in Table 12.5. Finally, for computing shear in the slab and loads on the supporting beams, Table 12.6 gives the fractions of the total load W that are transmitted in the two directions.

12.6 REINFORCEMENT FOR TWO-WAY EDGE-SUPPORTED SLABS

Consistent with the assumptions of the analysis of two-way edge-supported slabs, the main flexural reinforcement is placed in an orthogonal pattern, with rebars parallel and perpendicular to the supported edges. As the positive steel is placed in two layers, the effective depth d for the upper layer is smaller than that for

Table 12.6 Ratio of load W in l_a and l_b directions for shear in slab and load on supports[a]

Ratio $m = \frac{l_a}{l_b}$		Case 1	Case 2	Case 3	Case 4	Case 5	Case 6	Case 7	Case 8	Case 9
1.00	W_a	0.50	0.50	0.17	0.50	0.83	0.71	0.29	0.33	0.67
	W_b	0.50	0.50	0.83	0.50	0.17	0.29	0.71	0.67	0.33
0.95	W_a	0.55	0.55	0.20	0.55	0.86	0.75	0.33	0.38	0.71
	W_b	0.45	0.45	0.80	0.45	0.14	0.25	0.67	0.62	0.29
0.90	W_a	0.60	0.60	0.23	0.60	0.88	0.79	0.38	0.43	0.75
	W_b	0.40	0.40	0.77	0.40	0.12	0.21	0.62	0.57	0.25
0.85	W_a	0.66	0.66	0.28	0.66	0.90	0.83	0.43	0.49	0.79
	W_b	0.34	0.34	0.72	0.34	0.10	0.17	0.57	0.51	0.21
0.80	W_a	0.71	0.71	0.33	0.71	0.92	0.86	0.49	0.55	0.83
	W_b	0.29	0.29	0.67	0.29	0.08	0.14	0.51	0.45	0.17
0.75	W_a	0.76	0.76	0.39	0.76	0.94	0.88	0.56	0.61	0.86
	W_b	0.24	0.24	0.61	0.24	0.06	0.12	0.44	0.39	0.14
0.70	W_a	0.81	0.81	0.45	0.81	0.95	0.91	0.62	0.68	0.89
	W_b	0.19	0.19	0.55	0.19	0.05	0.09	0.38	0.32	0.11
0.65	W_a	0.85	0.85	0.53	0.85	0.96	0.93	0.69	0.74	0.92
	W_b	0.15	0.15	0.47	0.15	0.04	0.07	0.31	0.26	0.08
0.60	W_a	0.89	0.89	0.61	0.89	0.97	0.95	0.76	0.80	0.94
	W_b	0.11	0.11	0.39	0.11	0.03	0.05	0.24	0.20	0.06
0.55	W_a	0.92	0.92	0.69	0.92	0.98	0.96	0.81	0.85	0.95
	W_b	0.08	0.08	0.31	0.08	0.02	0.04	0.19	0.15	0.05
0.50	W_a	0.94	0.94	0.76	0.94	0.99	0.97	0.86	0.89	0.97
	W_b	0.06	0.06	0.24	0.06	0.01	0.03	0.14	0.11	0.03

[a] A crosshatched edge indicates that the slab continues across, or is fixed at, the support; an unmarked edge indicates a support at which torsional resistance is negligible.

the lower layer by one bar diameter. Because the moments in the long direction are the smaller ones, it is economical to place the steel in that direction on top of the bars in the short direction. The stacking problem does not exist for negative reinforcement perpendicular to the supporting edge beams except at the corners where moments are small.

Either straight bars, cut off where no longer required, or bent bars may be used for two-way slabs, but economy of bar fabrication and placement will generally favor all straight bars. The precise locations of inflection points (or lines of inflection) are not easily determined, because they depend upon the side ratio, the ratio of live to dead load, and continuity conditions at the edges. The standard cutoff and bend points for beams, summarized in Fig. 5.15, may be used for edge-supported slabs as well.

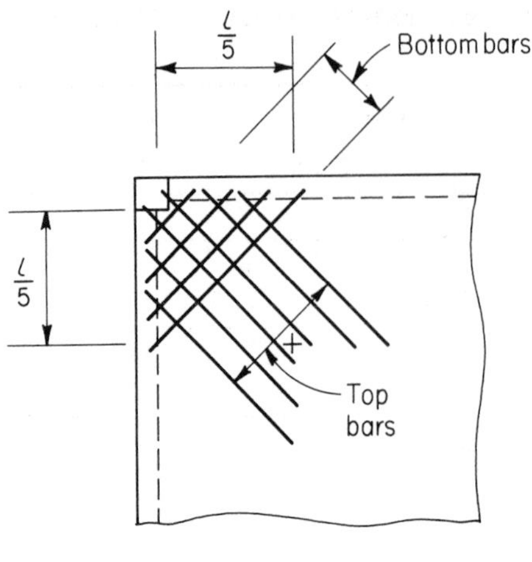

FIGURE 12.9
Special reinforcement at exterior corners of a beam-supported two-way slab.

According to ACI Code 13.4, the minimum reinforcement in each direction for two-way slabs is that required for shrinkage and temperature crack control, as given in Table 12.2. For two-way systems the spacing of flexural reinforcement at critical sections must not exceed 2 times the slab thickness h.

The twisting moments discussed in Sec. 12.4 are usually of consequence only at exterior corners of a two-way slab system, where they tend to crack the slab at the bottom along the panel diagonal, and at the top perpendicular to the panel diagonal. Special reinforcement should be provided at exterior corners in both the bottom and top of the slab, for a distance in each direction from the corner equal to one-fifth the longer span of the corner panel, as shown in Fig. 12.9. The reinforcement at the top of the slab should be parallel to the diagonal from the corner, while that at the bottom should be perpendicular to the diagonal. Alternatively, either layer of steel may be placed in two bands parallel to the sides of the slab. The positive and negative reinforcement, in any case, should be of a size and spacing equivalent to that required for the maximum positive moment in the panel, according to ACI Code 13.4.6.

Example 12.2 Design of two-way edge-supported slab. A monolithic reinforced concrete floor is to be composed of rectangular bays measuring 21 × 26 ft, as shown in Fig. 12.10. Beams of width 12 in. and depth 24 in. are provided on all column lines; thus the clear-span dimensions for the two-way slab panels are 20 × 25 ft. The floor is to be designed to carry a service live load of 137 psf uniformly distributed over its surface, in addition to its own weight, using concrete of strength $f_c' = 3000$ psi and reinforcement having $f_y = 60{,}000$ psi. Find the required slab thickness and reinforcement for the corner panel shown.

Solution. The minimum thickness for slabs of this type is often taken equal to 1/180 times the panel perimeter:

$$h = 2(20 + 25) \times \tfrac{12}{180} = 6 \text{ in.}$$

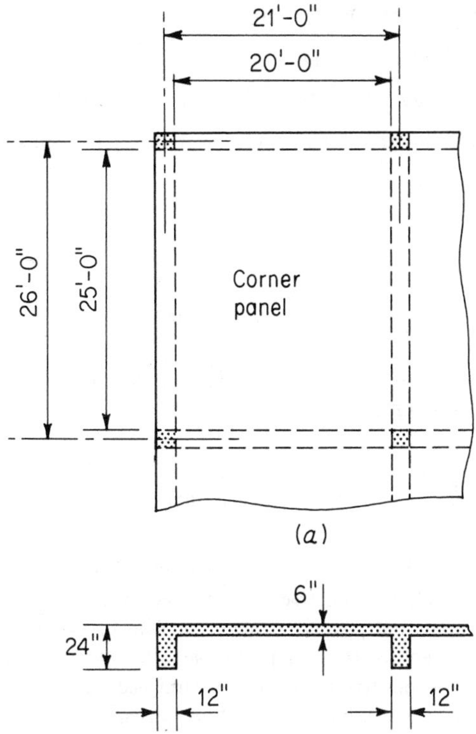

FIGURE 12.10
Two-way edge-supported slab example: (*a*) partial floor plan; (*b*) typical cross section.

This will be selected for a trial depth. The corresponding dead load is $\frac{1}{2} \times 150 = 75$ psf. Thus, the factored loads on which the design is to be based are

$$\begin{aligned} \text{Live load} &= 1.7 \times 137 = 233 \text{ psf} \\ \text{Dead load} &= 1.4 \times \ \ 75 = \underline{105} \text{ psf} \\ \text{Total load} &= 338 \text{ psf} \end{aligned}$$

With the ratio of panel sides $m = l_a/l_b = 20/25 = 0.8$, the moment calculations for the slab *middle strips* are as follows.

Negative moments at continuous edges (Table 12.3)

$$M_{a,\text{neg}} = 0.071 \times 338 \times 20^2 = 9600 \text{ ft-lb} = 115{,}000 \text{ in-lb}$$

$$M_{b,\text{neg}} = 0.029 \times 338 \times 25^2 = 6130 \text{ ft-lb} = 73{,}400 \text{ in-lb}$$

Positive moments (Tables 12.4 and 12.5)

$$\begin{aligned} M_{a,\text{pos},dl} &= 0.039 \times 105 \times 20^2 = 1638 \text{ ft-lb} = 19{,}700 \text{ in-lb} \\ M_{a,\text{pos},ll} &= 0.048 \times 233 \times 20^2 = 4470 \text{ ft-lb} = \underline{53{,}700} \text{ in-lb} \\ M_{a,\text{pos,tot}} &= 73{,}400 \text{ in-lb} \\ M_{b,\text{pos},dl} &= 0.016 \times 105 \times 25^2 = 1050 \text{ ft-lb} = 12{,}600 \text{ in-lb} \\ M_{b,\text{pos},ll} &= 0.020 \times 233 \times 25^2 = 2910 \text{ ft-lb} = \underline{35{,}000} \text{ in-lb} \\ M_{b,\text{pos,tot}} &= 47{,}600 \text{ in-lb} \end{aligned}$$

Negative moments at discontinuous edges ($\frac{1}{3}$ × positive moments)

$$M_{a,\text{neg}} = \tfrac{1}{3}\,(73{,}400) = 24{,}500 \text{ in-lb}$$

$$M_{b,\text{neg}} = \tfrac{1}{3}\,(47{,}600) = 15{,}900 \text{ in-lb}$$

The required reinforcement in the *middle strips* will be selected with the help of Graph A.1 of App. A.

Short direction

1. Midspan

$$\frac{M_u}{\phi bd^2} = \frac{73{,}400}{0.90 \times 12 \times 5^2} = 272 \qquad \rho = 0.0048$$

$A_s = 0.0048 \times 12 \times 5 = 0.288$ in²/ft. From Table A.4 of App. A, No. 4 bars at 7 in. spacing are selected, giving $A_s = 0.34$ in²/ft.

2. Continuous edge

$$\frac{M_u}{\phi bd^2} = \frac{115{,}000}{0.90 \times 12 \times 5^2} = 426 \qquad \rho = 0.0078\dagger$$

$A_s = 0.0078 \times 12 \times 5 = 0.468$ in²/ft. If two of every three positive bars are bent up, and likewise for the adjacent panel, the negative-moment steel area furnished at the continuous edge will be $\frac{4}{3}$ times the positive-moment steel in the span, or $A_s = \frac{4}{3} \times 0.34 = 0.453$ in²/ft. It is seen that this is 3 percent less than the required amount of 0.468. On the other hand, the positive-moment steel furnished, 0.34 in²/ft, represents about 15 percent more than the required amount. As discussed in Sec. 11.9*e*, the ACI Code permits a certain amount of inelastic redistribution, within strictly specified limits. In the case at hand, the negative steel furnished suffices for only 97 percent of the calculated moment, but the positive steel permits about 115 percent of the calculated moment to be resisted. This more than satisfies the conditions for inelastic moment redistribution set by the ACI Code. This situation illustrates how such moment redistribution can be utilized to obtain a simpler and more economical distribution of steel.

3. Discontinuous edge. The negative moment at the discontinuous edge is one-third the positive moment in the span; it would be adequate to bend up every third bar from the bottom to provide negative-moment steel at the discontinuous edge. However, this would result in a 21 in. spacing, which is larger than the maximum spacing of $2h = 12$ in. permitted by the ACI Code. Hence, for the discontinuous edge, two of every three bars will be bent up from the bottom steel.

Long direction

1. Midspan

$$\frac{M_u}{\phi bd^2} = \frac{47{,}600}{0.90 \times 12 \times 4.5^2} = 218 \qquad \rho = 0.0038$$

---

†Note that this value of ρ, which is the maximum required anywhere in the slab, is about half the permitted maximum value of $0.75\rho_b = 0.0160$, indicating that a thinner slab might be used. However, use of the minimum possible thickness would require an increase in the tensile steel area and would be less economical for this reason. In addition, a thinner slab may produce undesirably large deflections. The trial depth of 6 in. will be retained for the final design.

(The positive-moment steel in the long direction is placed on top of that for the short direction. This is the reason for using $d = 4.5$ in. for the positive-moment steel in the long direction and $d = 5$ in. in all other locations.) $A_s = 0.0038 \times 12 \times 4.5 = 0.205$ in^2/ft. From Table A.4 of App. A, No. 3 bars at 6 in. spacing are selected, giving $A_s = 0.22$ in^2/ft.

2. Continuous edge

$$\frac{M_u}{\phi bd^2} = \frac{73{,}400}{0.90 \times 12 \times 5^2} = 272 \qquad \rho = 0.0048$$

$A_s = 0.0048 \times 12 \times 5 = 0.288$ in^2/ft. Again bending up two of every three bottom bars from both panels adjacent to the continuous edge, one has, at that edge, $A_s = \frac{4}{3} \times 0.22 = 0.29$ in^2/ft.

3. Discontinuous edge. For the reasons discussed in connection with the short direction, two out of every three bottom bars will, likewise, be bent up at this edge.

The preceding steel selections refer to the *middle strips* in both directions. For the *column strips*, the moments are assumed to decrease linearly from the full calculated value at the inner edge of the column strip to one-third of this value at the edge of the supporting beam. To simplify steel placement, a uniform spacing will be used in the column strips. The average moments in the column strips being two-thirds of the corresponding moments in the middle strips, adequate column steel will be furnished if the spacing of this steel is $\frac{3}{2}$ times that in the middle strip. Maximum spacing limitations should be checked.

Bend points for rebars will be located as suggested in Fig. 5.15, i.e., $l/4$ from the face of the supporting beam at the continuous ends and $l/7$ from the beam face at the discontinuous ends. The corresponding distances from the beam face to bend point are 5 ft 0 in. and 2 ft 10 in. for the short-direction positive bars, and 6 ft 3 in. and 3 ft 7 in. for the long-direction positive bars, at the continuous end and discontinuous end, respectively. Negative bars carried over from the adjacent panels will be cut off at $l/3$ from the support face, at 6 ft 8 in. for the short-direction negative bars and 8 ft 4 in. for the long-direction negative bars. At the exterior edges, negative bars will be extended as far as possible into the supporting beams, then bent downward in a 90° hook to provide anchorage.

At the exterior corner of the panel, No. 4 bars at 8 in. spacing will be used, parallel to the slab diagonal at the top, and perpendicular to the diagonal at the bottom, according to Fig. 12.9. This will provide an area of 0.29 in^2/ft each way, equal to that required for the maximum positive bending moment in the panel. This reinforcement will be carried to a point $25/5 = 5$ ft from the corner, with lengths varying as indicated in Fig. 12.9.

The reactions of the slab are calculated from Table 12.6, which indicates that 71 percent of the load is transmitted in the short direction and 29 percent in the long direction. The total load on the panel being $20 \times 25 \times 338 = 169{,}000$ lb, the load per foot on the long beam is $(0.71 \times 169{,}000)/(2 \times 25) = 2400$ lb/ft and on the short beam is $(0.29 \times 169{,}000)/(2 \times 20) = 1220$ lb/ft. The shear to be transmitted by the slab to these beams is numerically equal to these beam loads, reduced to a critical section a distance d from the beam face. The shear strength of the slab is

$$\phi V_c = 0.85 \times 2\sqrt{3000} \times 12 \times 5 = 5590 \text{ lb}$$

well above the required shear strength at factored loads.

12.7 DEFLECTION CONTROL

Edge-supported slabs are typically thin relative to their span, and may show large deflections even though strength requirements are met, unless certain limitations are imposed in the design to prevent this. The simplest approach to deflection control is to impose a minimum thickness-span ratio. The 1963 ACI Code, in which the coefficient method of analysis described in the preceding section was introduced, provided that the slab thickness should not be less than 3.5 in. and not less than the total panel perimeter divided by 180. These limitations have given satisfactory results for edge-supported slabs. The more general method of two-way system analysis found in more recent editions of the ACI Code contains three equations governing minimum slab thickness. These equations account for the relative stiffness of slab and edge beams, the ratio of long to short panel side dimensions, and conditions of restraint along the edges. These more complex equations will be presented in Chap. 13.

Alternative to the use of minimum-depth equations, the deflection at the center of a slab panel can be calculated and results compared against limitations such as those of ACI Code 9.5. These limitations, summarized in Table 6.3, apply to two-way floor systems as well as to beams.

The calculation of deflections for slabs is complicated by many factors, such as the varying rotational restraint at the edges, the influence of alternative loading arrangements, varying ratio of side lengths, and the effects of cracking, as well as the time-dependent influences of shrinkage and creep. The deflection of an edge-supported slab can be estimated with reasonable accuracy, however, based on the moment coefficients used in the flexural analysis (see Sec. 12.5). The deflection components of concern are usually the long-term deflections due to sustained loads and the immediate deflection due to live load. Table 6.3 gives upper limits for these deflection components in terms of the span l. For slabs, it is not clear from the ACI Code or Commentary whether the longer or shorter span is to be used as the basis, but it is conservative (and reasonable when considering possible damage to supported elements) to base calculated limits on the shorter span.

It should be realized that the moment coefficient analysis yields *maximum values* of positive and negative moments at the critical sections of edge-supported slabs. The coefficients of Tables 12.3, 12.4, and 12.5 were derived considering alternate loading conditions, including full dead and live loads on both panels adjacent to particular edges for maximum negative moments, and including, for positive-moment calculation, two load patterns: dead load on both panels and live load on only the panel for which maximum positive moment was to be found. It would be incorrect to assume in deflection calculations that the maximum moments found in this way could act simultaneously, because only one load arrangement can exist at any one time.

Maximum live load deflection, for example, will normally be obtained when the live load acts on the given panel, but not on adjacent panels, i.e., in a checkerboard pattern. Therefore, live load deflection should be based on the maximum positive moments found using Table 12.5 corresponding to that loading arrangement, together with statically consistent negative moments at the supported edges.

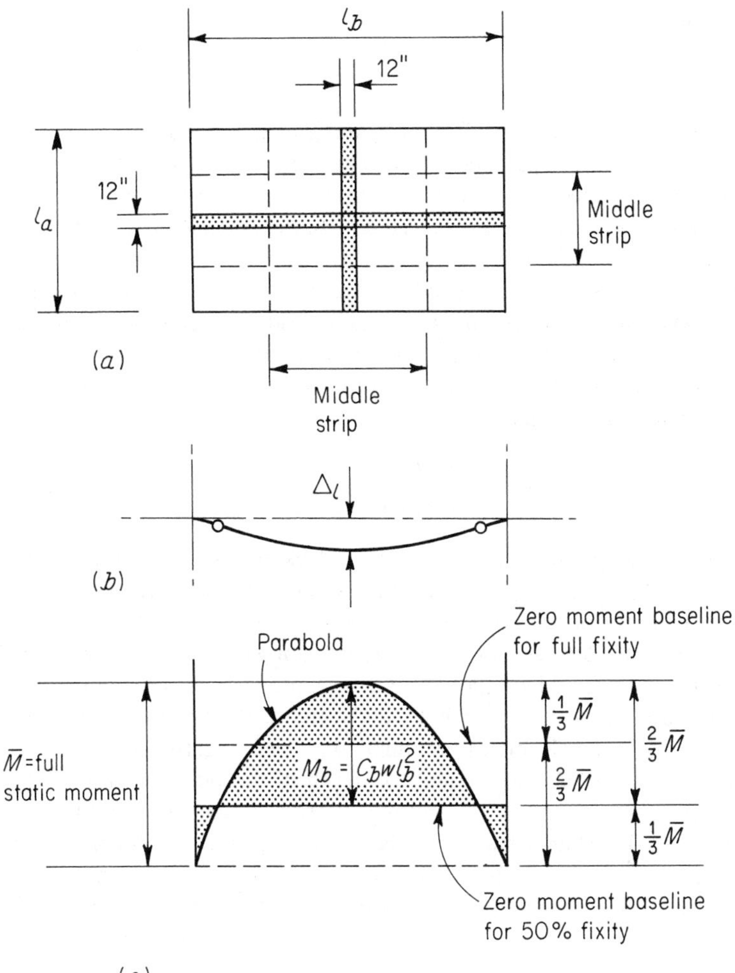

FIGURE 12.11
Live load deflection analysis: (*a*) plan of slab; (*b*) deflection curve for unit strip; (*c*) diagram for maximum positive live load moment.

This will be illustrated for the slab shown in Fig. 12.11*a*, considering the middle strip of unit width in the long direction of the panel. The variation of moment for a uniformly distributed load is parabolic, and the sum of the positive and average negative moments must, according to statics, be

$$\overline{M} = \tfrac{1}{8} w_b l_b^2 \tag{a}$$

where w_b is the fractional part of the load transmitted in the long direction of the panel (see Fig. 12.11*c*). If full fixity were obtained at the supports, the negative moment would be

$$M_{\text{neg}} = \tfrac{1}{12} w_b l_b^2 = \tfrac{2}{3}\overline{M} \tag{b}$$

and the positive moment would be

$$M_{\text{pos}} = \tfrac{1}{24} w_b l_b^2 = \tfrac{1}{3}\overline{M} \qquad (c)$$

It has been noted earlier that the coefficients for maximum live load positive moments were derived assuming 50 percent, not 100 percent, fixity. Accordingly, the zero-moment baseline associated with the maximum positive moment M_b obtained using Table 12.5 is as shown in Fig. 12.11c, and the statically consistent negative moments are one-half the positive moment M_b.

Deflection calculations are thus based on the parabolic moment diagram, with maximum ordinate M_b at midspan and negative end moments one-half that value. The midspan live load deflection, Δ_l, of the slab strip shown in Fig. 12.11b can easily be found based on the moment diagram of Fig. 12.11c. For the slab shown, with both edges continuous,

$$\Delta_l = \frac{3}{32} \frac{M_b l_b^2}{E_c I_{\text{eff}}} \qquad (12.3)$$

where M_b is the live load positive moment obtained using the appropriate coefficient of Table 12.5, E_c is the elastic modulus of the concrete, and I_{eff} is the effective moment of inertia of the concrete cross section of unit width.

Equation (12.3) was derived for a typical interior slab panel, with equal restraining moments at each end of the slab strip. Similar equations can easily be derived for the cases where one or both ends are discontinuous. However, according to the coefficient method of moment analysis, negative moments at discontinuous slab edges are assumed equal to one-third the positive moment in the same direction, so it is clear that the resulting deflection equation would differ very little from Eq. (12.3). That equation can be used for panel strips with one or both ends discontinuous, but monolithic with supporting beams, with very little error. For the special case where edges are completely free of restraint, as if, for example, the slab were supported by masonry walls, the midspan live load deflection is

$$\Delta_l = \frac{5}{48} \frac{M_b l_b^2}{E_c I_{\text{eff}}} \qquad (12.4)$$

The dead load deflection should be based on the moment diagram found using maximum dead load positive moment based on Table 12.4, which assumes all panels loaded. It should be recalled that continuous edges were regarded as fully fixed in deriving the coefficients of Table 12.4. Accordingly, it can easily be shown that the midspan dead load deflection Δ_d, for the case with both ends continuous, is

$$\Delta_d = \frac{1}{16} \frac{M_b l_b^2}{E_c I_{\text{eff}}} \qquad (12.5)$$

where M_b is, in this case, the dead load positive moment obtained using the coefficients of Table 12.4. For the special case where both ends are free of restraint, the midspan dead load deflection can be found from

$$\Delta_d = \frac{5}{48} \frac{M_b l_b^2}{E_c I_{\text{eff}}} \tag{12.6}$$

While the deflections discussed above have been with reference to a unit strip spanning in the longer direction of the panel of Fig. 12.11*a*, calculations may also be based on the strip in the shorter direction. The resulting deflection at the center of the panel should be the same in either case, although small differences can be expected because of the approximate nature of the calculations. A reasonable procedure is to calculate the deflection each way and then to average the results.

The ACI Code suggests that the moment of inertia I_{eff} be used, as given by Eq. (6.5), to reflect the influence of concrete cracking in reducing stiffness. For two-way edge-supported slabs, the amount of cracking at service load is usually not extensive, and little error is introduced if calculations are based on the gross moment of inertia I_g of the unit strip.

Slab deflections calculated according to the equations just given are the initial elastic deflections produced immediately upon application of the loads. For sustained loads such as from dead loads, the ACI Code recommends that the increase in deflection with time can be found based on Eq. (6.11), with a time-dependent multiplier ξ that varies according to Fig. 6.8 and with an ultimate value of 2.0. Experience has indicated that a value of 2.0 often underestimates time-dependent deflections of slabs, probably because slabs have much lower ratios of thickness to span than beams, which provided the basis for long-term multipliers. Branson suggests in Ref. 12.5 that $\xi = 3.0$ be used for slabs.

Example 12.3 Deflection of two-way edge-supported slab. The floor slab of Example 12.2 will support rigid partitions and other nonstructural elements that would be damaged by large deflections. These elements will be installed 3 months after construction shoring is removed and dead load imposed. Calculate the increment of dead load and service live load deflection that would affect the superimposed elements, and compare with ACI Code limit values.

Solution. Deflection calculations will be based on the moment analysis of Example 12.2. However, those moments were based on load factors of 1.4 applied to dead loads and 1.7 to live loads, and moments must be reduced accordingly to obtain service load moments. The modulus of elasticity, from Eq. (2.4), is $E_c = 57{,}000\sqrt{3000} = 3.12 \times 10^6$ psi. The moment of inertia will be taken as that of the gross concrete section, and for a 12 in. strip is $I_g = 12 \times 6^3/12 = 216\ \text{in}^4$.

The immediate deflection at midpanel due to dead load will be found by Eq. (12.5), first based on the long-direction strip, then the short-direction strip, and the results averaged if they differ. In the long direction, from Example 12.2, the dead load positive moment is 12,600 in-lb at factored loads, or 12,600/1.4 = 9000 in-lb at service loads. Thus

$$\Delta_d = \frac{9000\ (25 \times 12)^2}{16 \times 3.12 \times 10^6 \times 216} = 0.08 \text{ in.}$$

For comparison, in the short direction the service load moment due to dead load is 19,700/1.4 = 14,100 in-lb and the corresponding deflection at midpanel is

$$\Delta_d = \frac{14{,}100\ (20 \times 12)^2}{16 \times 3.12 \times 10^6 \times 216} = 0.08 \text{ in.}$$

just as before.

The time-dependent increment of deflection will be calculated based on a 5-year multiplier $\xi = 3.0$, but the ACI Code time variation shown in Fig. 6.8 is used. That figure indicates that one-half the time-dependent deflection would have occurred at 3 months. Only the remaining half would occur after installation of the partitions and other elements. Thus the fractional part of the time-dependent dead load deflections that may cause damage is $0.08 \times 3 \times \frac{1}{2} = 0.12$ in.

Live load deflection will be calculated from Eq. (12.3). In the long-span direction the live load positive moment at service load is 35,000/1.7 = 20,600 in-lb and the deflection at midpanel is

$$\Delta_l = \frac{3 \times 20{,}600\ (25 \times 12)^2}{32 \times 3.12 \times 10^6 \times 216} = 0.26 \text{ in.}$$

As a check, in the short-span direction the live load positive moment is 53,700/1.7 = 31,600 in-lb and the deflection is

$$\Delta_l = \frac{3 \times 31{,}600\ (20 \times 12)^2}{32 \times 3.12 \times 10^6 \times 216} = 0.25 \text{ in.}$$

the same as before for all practical purposes.

The deflection causing potential damage is the sum of the incremental time-dependent dead load deflection occurring after 3 months and the immediate deflection due to live load, i.e.,

$$\Delta = 0.12 + 0.26 = 0.38 \text{ in.}$$

According to the ACI Code limits of Table 6.3, the maximum allowable deflection for the stated conditions is $20 \times 12/480 = 0.50$ in., so on the basis of deflections the design can be considered satisfactory.

12.8 OTHER CONSIDERATIONS

Not infrequently in practice, openings must be provided in slabs, such as for heating and ventilating ducts or for piping. Two-way slabs with edge beams are less sensitive to holes than are beamless flat plate slabs, for example, and for openings of moderate size it is usually sufficient to provide extra reinforcement in each direction, adjacent to the opening on each side, equivalent to the interrupted steel and extending a full development length past the opening in all directions. Forty-five degree bars at the corners should be specified to control cracking of the slab at the corners of the opening. Larger openings require special consideration, and may require strengthening of the slab with beams to restore continuity. Concentrated loads, too, may require supporting beams.

Openings in slabs will be discussed in more detail in Chap. 13, and special methods of analysis that can account for both openings in slabs and concentrated loads will be developed in Chapters 14 and 15.

REFERENCES

12.1. W. L. Gamble, "Moments in Beam-Supported Slabs," *J. ACI,* vol. 69, no. 3, 1972, pp. 149–157.

12.2. "Building Code Requirements for Reinforced Concrete," ACI Publication 318-63, American Concrete Institute, Detroit, 1963.

12.3. H. Marcus, *Die Vereinfachte Beredhnung Biegsamer Platten,* Julius Springer, Berlin, 1929.

12.4. P. Rogers, "Two-Way Reinforced Concrete Slabs," *J. ACI,* vol. 41, no. 1, 1944, pp. 21–36.

12.5. D. E. Branson, *Deformation of Concrete Structures,* McGraw-Hill, New York, 1977.

PROBLEMS

12.1 A footbridge is to be built, consisting of a one-way solid slab spanning 16 ft between masonry abutments, as shown in Fig. P12.1. A service live load of 100 psf must be carried. In addition, a 2000 lb concentrated load, assumed to be uniformly distributed across the bridge width, may act at any location on the span. A 2 in. asphalt wearing surface will be used, weighing 20 psf. Precast concrete curbs are attached so as to be nonstructural. Prepare a design for the slab, using material strengths f_y = 60,000 psi and f' = 4000 psi, and summarize your results in the form of a sketch showing all concrete dimensions and reinforcement.

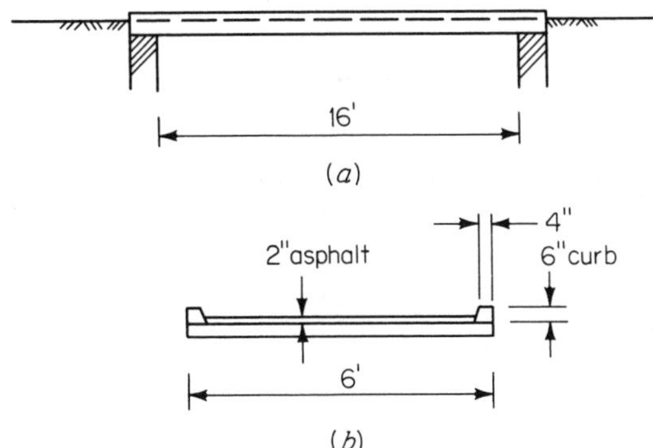

FIGURE P12.1

12.2 A reinforced concrete building floor system consists of a continuous one-way slab built monolithically with its supporting beams, as shown in cross section in Fig. P12.2. Service live load will be 125 psf. Dead loads include a 10 psf allowance for nonstructural lightweight concrete floor fill and surface, and a 10 psf allowance for suspended loads, plus the self-weight of the floor. Using ACI coefficients, calculate the design moments and shears and design the slab, using a maximum tensile steel ratio of 0.006. Use all straight bar reinforcement. One-half

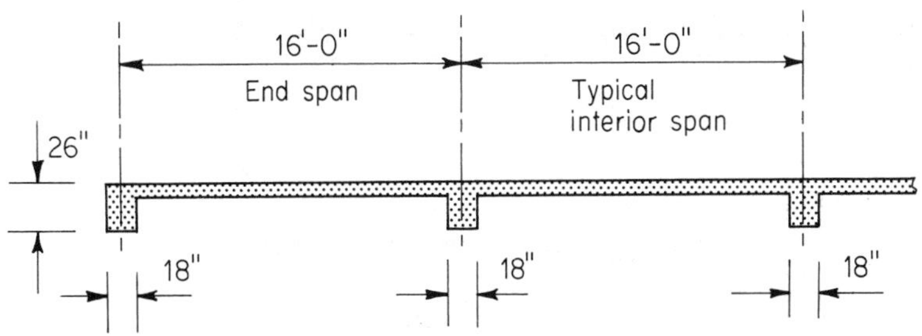

FIGURE P12.2

of the positive moment bars will be discontinued where no longer required; the other half will be continued into the supporting beams as specified by the ACI Code. All negative steel will be discontinued at the same distance from the support face in each case. Summarize your design with a sketch showing concrete dimensions, and size, spacing, and cutoff points for all rebars. Material strengths are $f_y = 60{,}000$ psi and $f_c' = 3000$ psi.

12.3 For the one-way slab floor of Prob. 12.2, calculate the immediate and long-term deflection due to dead loads. Assume that all dead loads are applied when the construction shoring is removed. Also determine the deflection due to application of the full service live load. Assuming that sensitive equipment will be installed 6 months after the shoring is removed, calculate the relevant deflection components and compare the total with maximum values recommended in the ACI Code.

12.4 A two-way concrete slab roof is to be designed to cover a transformer vault. The outside dimensions of the vault are 17×20 ft and supporting walls are 8 in. brick. A service live load of 80 psf distributed uniformly over the roof surface will be assumed, and a dead load allowance of 10 psf added to the self-weight of the slab. Design the roof as a two-way edge-supported slab, using $f' = 4000$ psi and $f_y = 50{,}000$ psi.

12.5 A parking garage is to be designed using a two-way slab supported by 16×26 in. monolithic beams on the column lines, as shown in Fig. P12.5. Live loading of 100 psf is

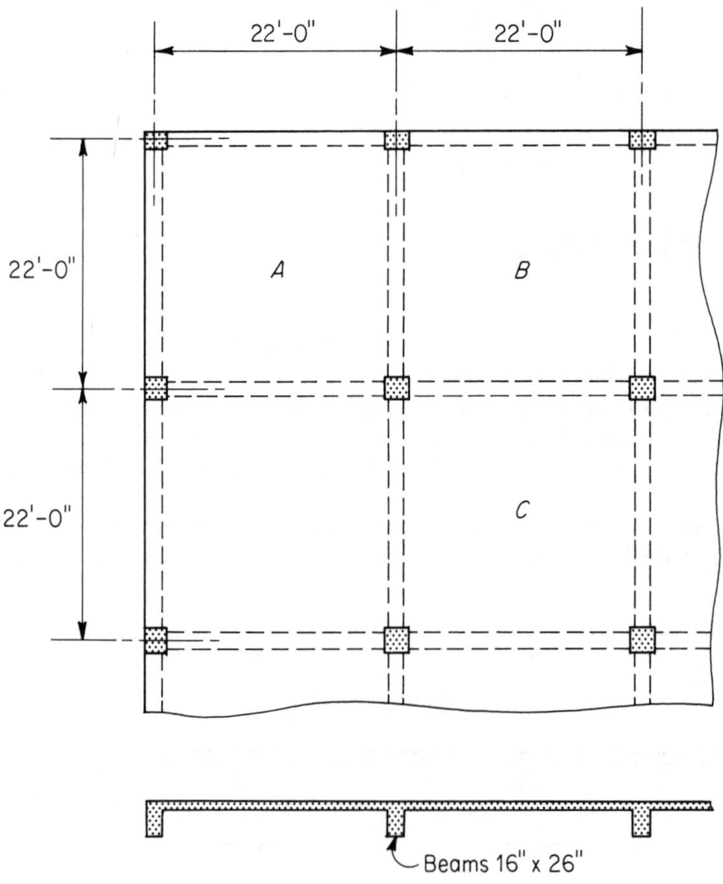

FIGURE P12.5

specified. Find the required slab thickness, using a steel ratio of approximately 0.005 maximum, and design the reinforcement for a typical corner panel *A*, edge panel *B*, and interior panel *C*. Detail the reinforcement, showing size, spacing, and length of rebars. All straight bars will be used. Material strengths will be $f_y = 60{,}000$ psi and $f_c' = 4000$ psi.

12.6 For the typical interior panel *C* of the parking garage of Prob. 12.5, (*a*) compute the immediate and long-term deflections due to dead load, and (*b*) compute the deflection due to the full service live load. Compare with ACI Code maximum permissible values, given that there are no elements attached that would be damaged by large deflections.

CHAPTER 13

TWO-WAY COLUMN-SUPPORTED SLABS

13.1 INTRODUCTION

When two-way slabs are supported by relatively shallow, flexible beams (Fig. 12.1*b*), or if column-line beams are omitted altogether, as for flat plates (Fig. 12.1*d*), flat slabs (Fig. 12.1*e*), or two-way joist systems (Fig. 12.1*f*), a number of new considerations are introduced. Figure 13.1*a* shows a portion of a floor system in which a rectangular slab panel is supported by relatively shallow beams on four sides. The beams, in turn, are carried by columns at the intersection of their centerlines. If a surface load w is applied, that load is shared between imaginary slab strips l_a in the short direction and l_b in the long direction, as described earlier in Sec. 12.4. Note that the portion of the load that is carried by the long strips l_b is delivered to the beams $B1$ spanning in the short direction of the panel. The portion carried by the beams $B1$ plus that carried directly in the short direction by the slab strips l_a, sums up to 100 percent of the load applied to the panel. Similarly, the short-direction slab strips l_a deliver a part of the load to long-direction beams $B2$. That load, plus the load carried directly in the long direction by the slab, includes 100 percent of the applied load. It is clearly a requirement of statics that, for column-supported construction, *100 percent of the applied load must be carried in each direction*, jointly by the slab and its supporting beams (Ref. 13.1).

A similar situation is obtained in the flat plate floor shown in Fig. 13.1*b*. In this case beams are omitted. However, broad strips of the slab centered on

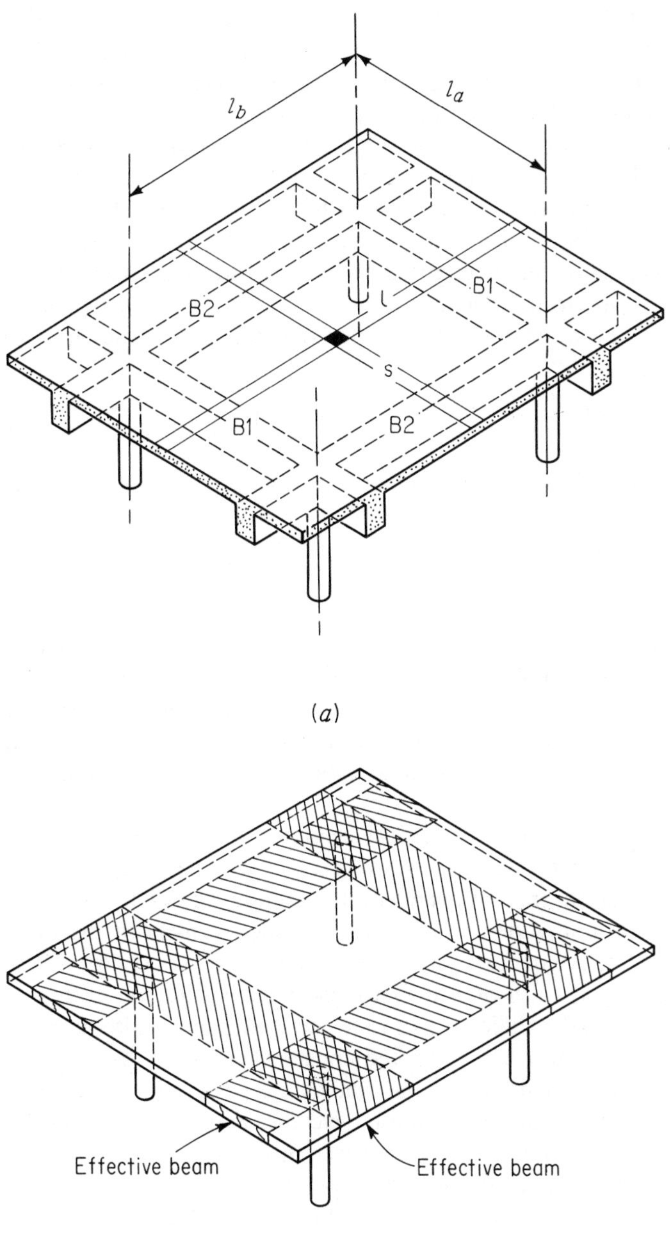

FIGURE 13.1
Column-supported two-way slabs: (*a*) two-way slab with beams; (*b*) two-way slab without beams.

the column lines in each direction serve the same function as the beams of Fig. 13.1*a*; for this case, also, the full load must be carried in each direction. The presence of dropped panels or column capitals in the double-hatched zone near the columns does not modify this requirement of statics.

Figure 13.2*a* shows a flat plate floor supported by columns at *A, B, C,* and *D*. Figure 13.2*b* shows the moment diagram for the direction of span l_1. In this direction, the slab may be considered as a broad, flat beam of width l_2. Accordingly, the load per foot of span is wl_2. In any span of a continuous beam, the sum of the midspan positive moment and the average of the negative moments at adjacent supports is equal to the midspan positive moment of a corresponding simply supported beam. In terms of the slab, this requirement of statics may be written

$$\frac{1}{2}(M_{ab} + M_{cd}) + M_{ef} = \frac{1}{8}wl_2l_1^2 \qquad (a)$$

A similar requirement exists in the perpendicular direction, leading to the relation

$$\frac{1}{2}(M_{ac} + M_{bd}) + M_{gh} = \frac{1}{8}wl_1l_2^2 \qquad (b)$$

These results disclose nothing about the relative magnitudes of the support moments and span moments. The proportion of the total static moment that exists at each critical section can be found from an elastic analysis that considers the relative span lengths in adjacent panels, the loading pattern, and the relative stiffness of the supporting beams, if any, and that of the columns. Alternatively, empirical methods that have been found to be reliable under restricted conditions may be adopted.

The moments across the width of critical sections such as *AB* or *EF* are not constant but vary as shown qualitatively in Fig. 13.2*c*. The exact variation depends on the presence or absence of beams on the column lines, the existence of dropped panels and column capitals, as well as on the intensity of the load. For design purposes it is convenient to divide each panel as shown in Fig. 13.2*c* into column strips, having a width of one-fourth the panel width, on each side of the column centerlines, and middle strips in the one-half panel width between two column strips. Moments may be considered constant within the bounds of a middle strip or column strip, as shown, unless beams are present on the column lines. In the latter case, while the beam must have the same curvature as the adjacent slab strip, the beam moment will be larger in proportion to its greater stiffness, producing a discontinuity in the moment-variation curve at the lateral face of the beam. Since the total moment must be the same as before, according to statics, the slab moments must be correspondingly less.

Chapter 13 of the ACI Code deals in a unified way with all such two-way systems. Its provisions apply to slabs supported by beams and to flat slabs and flat plates, as well as to two-way joist slabs. While permitting design "by any procedure satisfying the conditions of equilibrium and geometrical compatibility," specific reference is made to two alternative approaches: a semiempirical *direct design method* and an approximate elastic analysis known as the *equivalent frame method*.

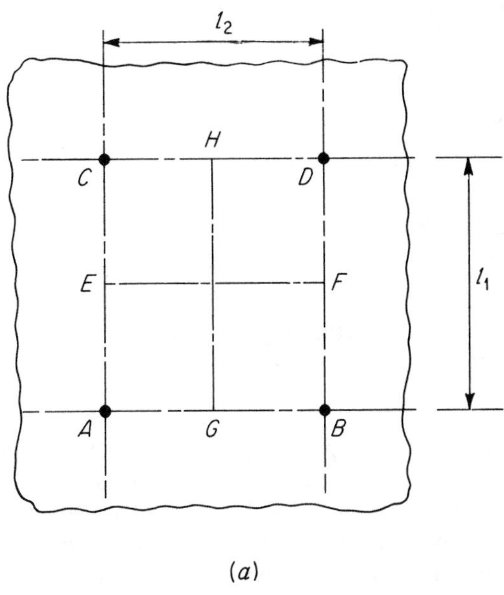

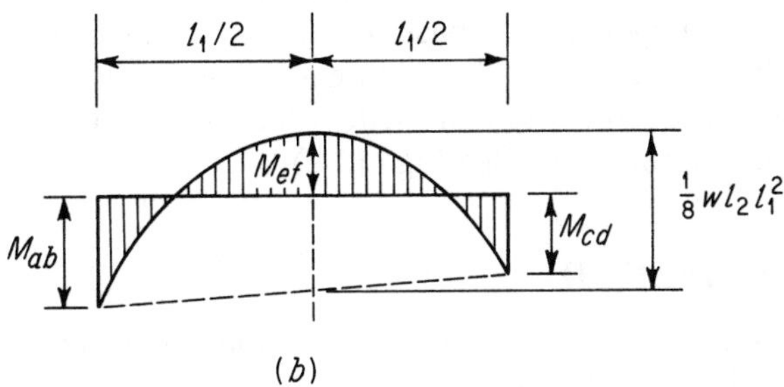

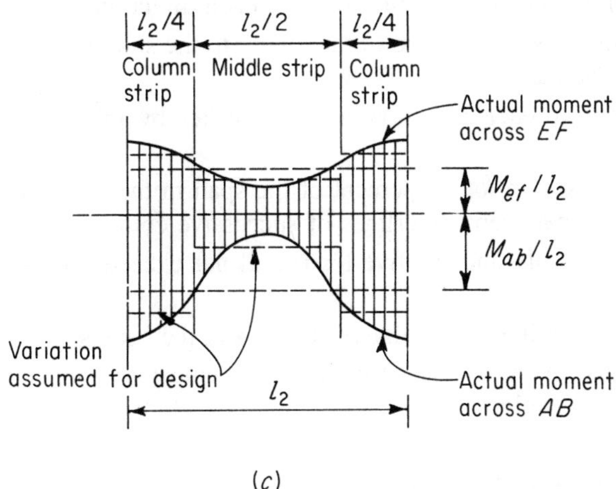

FIGURE 13.2
Moment variation in column-supported two-way slabs: (*a*) critical moment sections; (*b*) moment variation along a span; (*c*) moment variation across the width of critical sections.

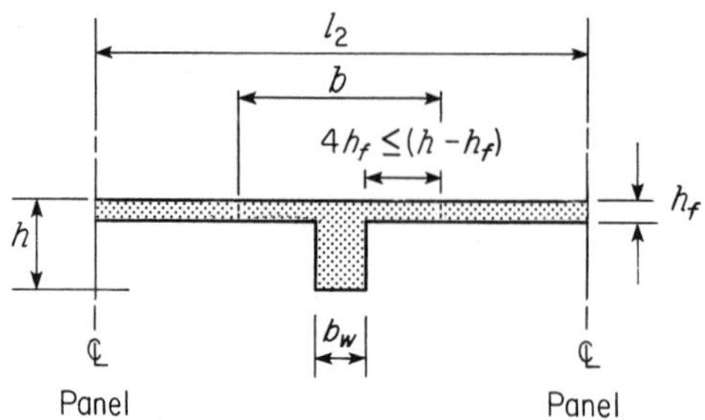

FIGURE 13.3
Cross section of a slab and effective beam.

In either case a typical panel is divided, for purposes of design, into *column strips* and *middle strips*. A column strip is defined as a strip of slab having a width on each side of the column centerline equal to one-fourth the smaller of the panel dimensions l_1 and l_2. Such a strip includes column-line beams, if present. A middle strip is a design strip bounded by two column strips. In all cases, l_1 is defined as the span in the direction of the moment analysis and l_2 as the span in the lateral direction. Spans are measured to column centerlines, except where otherwise noted. In the case of monolithic construction, beams are defined to include that part of the slab on each side of the beam extending a distance equal to the projection of the beam above or below the slab (whichever is greater) but not greater than 4 times the slab thickness (see Fig. 13.3).

13.2 DIRECT DESIGN METHOD

Moments in two-way slabs can be found using a semiempirical direct design method subject to the following restrictions:

1. There shall be a minimum of three continuous spans in each direction.
2. The panels shall be rectangular, with the ratio of the longer to the shorter spans within a panel not greater than 2.
3. The successive span lengths in each direction shall not differ by more than one-third the longer span.
4. Columns may be offset a maximum of 10 percent of the span in the direction of the offset from either axis between centerlines of successive columns.
5. Loads shall be due to gravity only and the live load shall not exceed 3 times the dead load.
6. If beams are used on the column lines, the relative stiffness of the beams in the two perpendicular directions, given by the ratio $\alpha_1 l_2^2/\alpha_2 l_1^2$, must be between 0.2 and 5.0 (see below for definitions).

a. Total Static Moment at Factored Loads

For purposes of calculating the total static moment M_0 in a panel, the clear span l_n in the direction of moments is used. The clear span is defined to extend from face to face of the columns, capitals, brackets, or walls but is not to be less than $0.65l_1$. The total factored moment in a span, for a strip bounded laterally by the centerline of the panel on each side of the centerline of supports, is

$$M_0 = \frac{w_u l_2 l_n^2}{8} \tag{13.1}$$

b. Assignment of Moments to Critical Sections

For interior spans, the total static moment is apportioned between the critical positive and negative bending sections according to the following ratios:

$$\text{Negative factored moment: Neg } M_u = 0.65M_0 \tag{13.2a}$$

$$\text{Positive factored moment: Pos } M_u = 0.35M_0 \tag{13.2b}$$

as illustrated by Fig. 13.4. The critical section for negative bending is taken at the face of rectangular supports, or at the face of an equivalent square support having the same cross-sectional area as a round support.

In the case of end spans, the apportionment of the total static moment among the three critical moment sections (interior negative, positive, and exterior negative, as illustrated by Fig. 13.4) depends upon the flexural restraint provided for the slab by the exterior column or the exterior wall, as the case may be, and depends also upon the presence or absence of beams on the column lines. ACI Code 13.6.3 specifies five alternative sets of moment distribution coefficients for end spans, as shown in Table 13.1 and as illustrated in Fig. 13.5.

In case (*a*), the exterior edge has no moment restraint, such as would be the condition with a masonry wall, which provides vertical support but no rotational

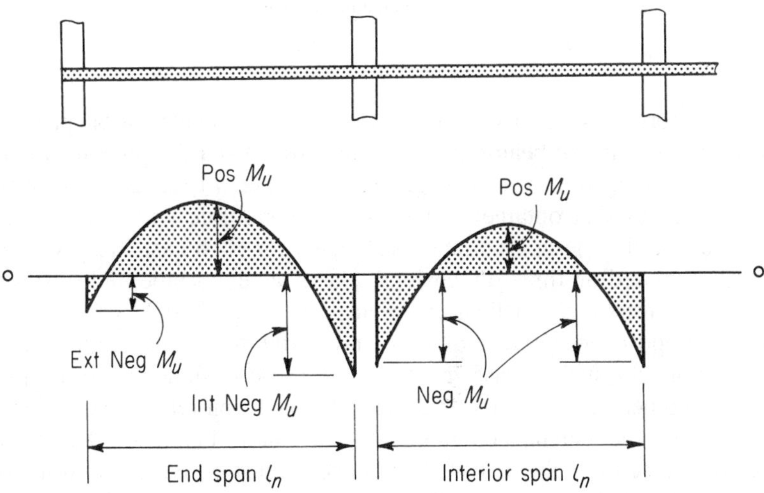

FIGURE 13.4
Distribution of total static moment M_0 to critical sections for positive and negative bending.

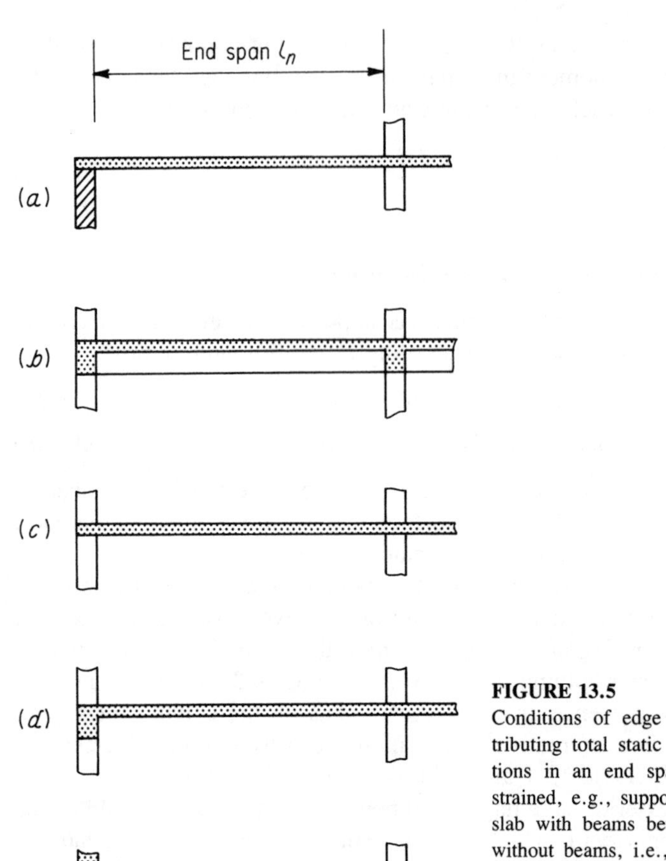

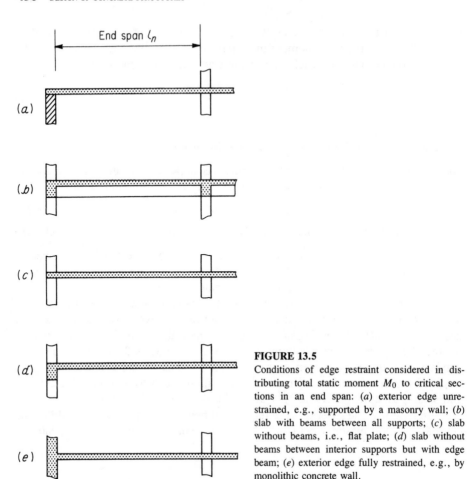

FIGURE 13.5
Conditions of edge restraint considered in distributing total static moment M_0 to critical sections in an end span: (*a*) exterior edge unrestrained, e.g., supported by a masonry wall; (*b*) slab with beams between all supports; (*c*) slab without beams, i.e., flat plate; (*d*) slab without beams between interior supports but with edge beam; (*e*) exterior edge fully restrained, e.g., by monolithic concrete wall.

restraint. Case (*b*) represents a two-way slab with beams on all sides of the panels. Case (*c*) is a flat plate, with no beams at all, while case (*d*) is a flat plate in which a beam is provided along the exterior edge. Finally, case (*e*) represents a fully restrained edge, such as that obtained if the slab is monolithic with a very stiff reinforced concrete wall. The appropriate coefficients for each case are given in Table 13.1, and are based on three-dimensional elastic analysis modified to some extent in the light of tests and practical experience (Refs. 13.2 to 13.8).

At interior supports, negative moments may differ for spans framing into the common support. In such a case the slab should be designed to resist the larger of the two moments, unless a special analysis based on relative stiffnesses is made to distribute the unbalanced moment (see Chap. 11). Edge beams if they are used, or the edge of the slab if they are not, must be designed to resist in torsion their share of the exterior negative moment indicated by Table 13.1 (see Chap. 7).

Table 13.1 Distribution factors applied to static moment M_0 for positive and negative moments in end span

	(a)	(b)	(c)	(d)	(e)
			Slab without beams between interior supports		
	Exterior edge unrestrained	Slab with beams between all supports	Without edge beam	With edge beam	Exterior edge fully restrained
Interior negative moment	0.75	0.70	0.70	0.70	0.65
Positive moment	0.63	0.57	0.52	0.50	0.35
Exterior negative moment	0	0.16	0.26	0.30	0.65

c. Lateral Distribution of Moments

Having distributed the moment M_0 to the positive- and negative-moment sections as just described, the designer still must distribute these design moments across the width of the critical sections. For design purposes, as discussed in Sec. 13.1, it is convenient to consider the moments constant within the bounds of a middle strip or column strip unless there is a beam present on the column line. In the latter case, because of its greater stiffness, the beam will tend to take a larger share of the column-strip moment than the adjacent slab. The distribution of total negative or positive moment between slab middle strips, slab column strips, and beams depends upon the ratio l_2/l_1, the relative stiffness of the beam and the slab, and the degree of torsional restraint provided by the edge beam.

A convenient parameter defining the relative stiffness of the beam and slab spanning in either direction is

$$\alpha = \frac{E_{cb} I_b}{E_{cs} I_s} \tag{13.3}$$

in which E_{cb} and E_{cs} are the moduli of elasticity of the beam and slab concrete (usually the same) and I_b and I_s are the moments of inertia of the effective beam and the slab. Subscripted parameters α_1 and α_2 are used to identify α computed for the directions of l_1 and l_2 respectively.

The flexural stiffnesses of the beam and slab may be based on the gross concrete section, neglecting reinforcement and possible cracking, and (in applying the direct method of analysis) variations due to column capitals and drop panels may be neglected. For the beam, if present, I_b is based on the effective cross section defined as in Fig. 13.3. For the slab, I_s is taken equal to $bh^3/12$, where b in this case is the width between panel centerlines on each side of the beam.

The relative restraint provided by the torsional resistance of the effective transverse edge beam is reflected by the parameter β_t, defined by

$$\beta_t = \frac{E_{cb}C}{2E_{cs}I_s} \tag{13.4}$$

where I_s, as before, is calculated for the slab spanning in direction l_1 and having width bounded by panel centerlines in the l_2 direction. The constant C pertains to the torsional rigidity of the effective transverse beam, which is defined according to ACI Code 13.7.5 as the largest of the following:

1. A portion of the slab having a width equal to that of the column or capital in the direction in which moments are taken.
2. The portion of the slab specified in 1 plus that part of any transverse beam above and below the slab.
3. The transverse beam defined as in Fig. 13.3.

The constant C is calculated by dividing the section into its component rectangles, each having smaller dimension x and larger dimension y, and summing the contributions of all the parts by means of the equation

$$C = \Sigma\left(1 - 0.63\frac{x}{y}\right)\frac{x^3y}{3} \tag{13.5}$$

The subdivision can be done in such a way as to maximize C.

With these parameters defined, the ACI Code direct design method distributes the negative and positive moments between column strips and middle strips, assigning to the column strips the percentages of positive and negative moments shown in Table 13.2. Linear interpolations are to be made between the values shown.

Implementation of these provisions is facilitated by the interpolation charts of Graph A.4 of App. A. Interior negative- and positive-moment percentages can be read directly from the chart for known values of l_2/l_1 and $\alpha_1 l_2/l_1$. For exterior negative moment, the parameter β_t requires an additional interpolation, facilitated by the auxiliary diagram on the right side of the chart. To illustrate its use for $l_2/l_1 = 1.55$ and $\alpha_1 l_2/l_1 = 0.6$, the dotted line indicates moment percentages of 100 for $\beta_t = 0$ and 65 for $\beta_t = 2.5$. Projecting to the right as indicated by the arrow to find the appropriate vertical scale of 2.5 divisions for an intermediate value of β_t, say 1.0, then upward and finally to the left, one reads the corresponding percentage of 86 on the main chart.

The column-line beam spanning in the direction l_1 is to be proportioned to resist 85 percent of the column-strip moment if $\alpha_1 l_2/l_1$ is equal to or greater than 1.0. For values between one and zero the proportion to be resisted by the beam may be obtained by linear interpolation. Concentrated loads applied directly to such a beam should be accounted for separately.

Table 13.2 Column-strip moment, percent of total moment at critical section

		l_2/l_1		
		0.5	**1.0**	**2.0**
Interior negative moment				
$\alpha_1 l_2/l_1 = 0$		75	75	75
$\alpha_1 l_2/l_1 \geq 1.0$		90	75	45
Exterior negative moment				
$\alpha_1 l_2/l_1 = 0$	$\beta_t = 0$	100	100	100
	$\beta_t \geq 2.5$	75	75	75
$\alpha_1 l_2/l_1 \geq 1.0$	$\beta_t = 0$	100	100	100
	$\beta_t \geq 2.5$	90	75	45
Positive moment				
$\alpha_1 l_2/l_1 = 0$		60	60	60
$\alpha_1 l_2/l_1 \geq 1.0$		90	75	45

The portion of the moment not resisted by the column strip is proportionately assigned to the adjacent half-middle strips. Each middle strip is designed to resist the sum of the moments assigned to its two half-middle strips. A middle strip adjacent and parallel to a wall is designed for twice the moment assigned to the half-middle strip corresponding to the first row of interior supports.

d. Shear in Slab Systems with Beams

Special attention must be given to providing the proper resistance to shear, as well as to moment, when designing by the direct method. According to ACI Code 13.6.8, beams with $\alpha_1 l_2/l_1$ equal to or greater than one must be proportioned to resist the shear caused by loads on a tributary area defined as shown in Fig. 13.6. For values of $\alpha_1 l_2/l_1$ between one and zero, the proportion of load carried by beam shear is to be found by linear interpolation. The remaining fraction of the load on the shaded area is assumed to be transmitted directly by the slab to the columns at the four corners of the panel, and the shear stress in the slab computed accordingly (see Sec. 13.6).

e. Design of Columns

Columns in two-way construction must be designed to resist the moments found from analysis of the slab-beam system. The column supporting an edge beam must provide a resisting moment equal to the moment applied from the edge of the slab (see Table 13.1). At interior locations, slab negative moments are found, assuming that dead and full live loads act. For the column design, a more severe loading results from partial removal of the live load. Accordingly, ACI Code 13.6.9 requires that interior columns resist a moment

$$M = 0.07[(w_d + 0.5w_l)l_2 l_n^2 - w_d' l_2'(l_n')^2] \tag{13.6}$$

In Eq. (13.6) the primed quantities refer to the shorter of the two adjacent spans (assumed to carry dead load only), and the unprimed quantities refer to the longer span (assumed to carry dead load and half live load). In all cases, the moment is distributed to the upper and lower columns in proportion to their relative flexural stiffness.

When β_a, the ratio of dead to live load (without load factors), is less than 2, *one* of the following two conditions must be satisfied:

1. The sum of the flexural stiffnesses of the columns above and below the slab must be such that $\alpha_c = \Sigma K_c/\Sigma(K_s + K_b)$ is not less than the minimum value given by Table 13.3.
2. If α_c for the columns is less than α_{min} specified in Table 13.3, the positive design moment in the panels supported by these columns must be multiplied by the coefficient δ_s:

$$\delta_s = 1 + \frac{2 - \beta_a}{4 + \beta_a}\left(1 - \frac{\alpha_c}{\alpha_{min}}\right) \tag{13.7}$$

In the equation for α_c just given, and in Table 13.3, the parameters K_c, K_s, and K_b are respectively the flexural stiffnesses of the column, slab, and column-line beam. Values may be calculated according to the usual equations of structural

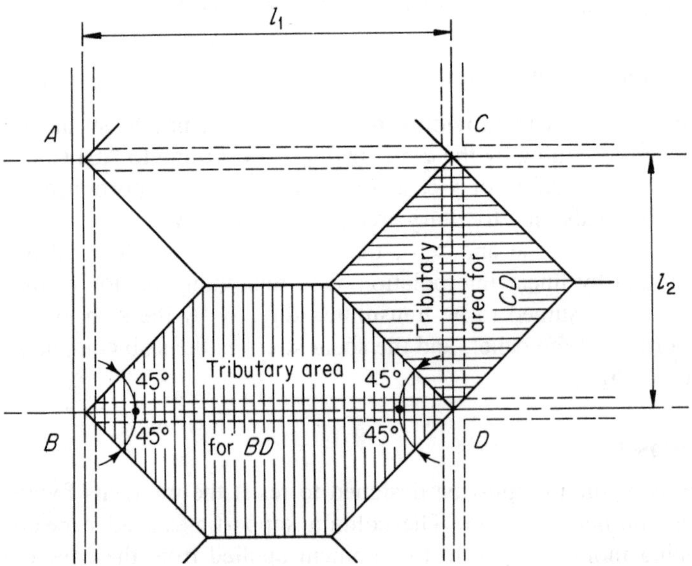

FIGURE 13.6
Tributary areas for shear calculation.

Table 13.3 $\alpha_{\min}$ = **minimum values**[a] *of* $\alpha_c = \Sigma K_c / \Sigma (K_s + K_b)$

β_a	Aspect ratio l_2/l_1	Relative beam stiffness $\alpha = \dfrac{E_{cb}I_b}{E_{cs}I_s}$				
		0	0.5	1.0	2.0	4.0
2.0	0.5–2.0	0	0	0	0	0
1.0	0.5	0.6	0	0	0	0
	0.8	0.7	0	0	0	0
	1.0	0.7	0.1	0	0	0
	1.25	0.8	0.4	0	0	0
	2.0	1.2	0.5	0.2	0	0
0.5	0.5	1.3	0.3	0	0	0
	0.8	1.5	0.5	0.2	0	0
	1.0	1.6	0.6	0.2	0	0
	1.25	1.9	1.0	0.5	0	0
	2.0	4.9	1.6	0.8	0.3	0
0.33	0.5	1.8	0.5	0.1	0	0
	0.8	2.0	0.9	0.3	0	0
	1.0	2.3	0.9	0.4	0	0
	1.25	2.8	1.5	0.8	0.2	0
	2.0	13.0	2.6	1.2	0.5	0.3

[a] See the ACI Code and Ref. 13.9.

mechanics, based on the gross concrete sections, and for members having uniform cross section along their length, $K = 4E_cI_g/l$. The value of I_g for the beam is based on the effective cross section illustrated in Fig. 13.3, and that of the slab is equal to $bh^3/12$, where b is equal to the transverse panel dimension l_2. For both beam and slab, the span l to be used in computing K is the span l_1 in the direction of the moment analysis. For the column, the span is the vertical length floor to floor.

13.3 FLEXURAL REINFORCEMENT

Consistent with the assumptions made in analysis, flexural reinforcement in two-way slab systems is placed in an orthogonal grid, with bars parallel to the sides of the panels. Bar diameters and spacings may be found as described in Sec. 12.2. Straight bars are generally used throughout, although in some cases positive-moment steel is bent up where no longer needed, in order to provide for part or all of the negative requirement. To provide for local concentrated loads, as well as to ensure that tensile cracks are narrow and well distributed, a maximum bar spacing at critical sections of 2 times the total slab thickness is specified by ACI Code 13.4 for two-way slabs. At least the minimum steel required for temperature and shrinkage crack control (see Sec. 12.3) must be provided. For protection of the steel against damage from fire or corrosion, at least $\frac{3}{4}$ in concrete cover must be maintained.

Because of the stacking that results when bars are placed in perpendicular layers, the inner steel will have an effective depth 1 bar diameter less than the outer steel. For flat plates and flat slabs, the stacking problem relates to middle-strip positive steel and column-strip negative bars. In two-way slabs with beams on the column lines, stacking occurs for the middle-strip positive steel, and in the column strips is important mainly for the column-line beams, because slab moments are usually very small in the region where column strips intersect.

In the discussion of the stacking problem for two-way slabs supported by walls or stiff edge beams, in Sec. 12.6 it was pointed out that, because curvatures and moments in the short direction are greater than in the long direction of a rectangular panel, short-direction bars are normally placed closer to the top or bottom surface of the slab, with the larger effective depth d, and long-direction bars are placed inside these, with the smaller d. For two-way beamless flat plates, or slabs with relatively flexible edge beams, things are not so simple.

Consider a rectangular interior panel of a flat plate floor. If the slab column strips provided unyielding supports for the middle strips spanning in the perpendicular direction (as was assumed for the slabs of Chap. 12), the short-direction middle-strip curvatures and moments would be the larger. In fact, the column strips deflect downward under load, and this softening of the effective support greatly reduces curvatures and moments in the supported middle strip.

For the entire panel, including both middle strips and column strips in each direction, the moments in the long direction will be larger than those in the short direction, as is easily confirmed by calculating the static moment M_0 in each direction for a rectangular panel. Noting that the apportioning of M_0 first to negative- and positive-moment sections, and then laterally to column and middle strips, is done by applying exactly the same ratios in each direction to the corresponding section, it is clear that the middle-strip positive moments (for example) are larger in the long direction than the short direction, exactly the opposite of the situation for the slab with stiff edge beams. In the column strips, positive and negative moments are larger in the long than in the short direction. On this basis, the designer is led to place the long-direction negative and positive bars, in both middle and column strips, closer to the top or bottom surface of the slab, respectively, with the larger effective depth.

If column-line beams are added, and if their stiffness is progressively increased for comparative purposes, it will be found that the short-direction slab moments gradually become dominant, as was true for the slabs of Chap. 12, although the long-direction beams carry larger moments than the short-direction beams. This will be clear from a careful study of Table 13.2.

The situation is further complicated by the influence of the ratio of short to long side dimensions of a panel, and by the influence of varying conditions of edge restraint (e.g., corner vs. typical exterior vs. interior panel). The best guide in specifying steel placement order in areas where stacking occurs is the relative magnitudes of design moments obtained from analysis for a particular case, with maximum d provided for the bars resisting the largest moment. No firm rules can be given. For square slab panels, many designers calculate the required steel area based on the average effective depth, thus obtaining the same bar size and spacing in each direction. This is slightly conservative for the outer layer, and slightly unconservative for the inner steel. Redistribution of loads and moments before failure would provide for the resulting differences in capacities in the two directions.

Rebar cutoff points could be calculated from moment envelopes if available; however, when the direct design method is used, moment envelopes and lines of inflection are not found explicitly. In such a case (and often when the equivalent frame method of Sec. 13.5 is used as well), standard bar cutoff points from Fig. 13.7 are used, as recommended in the ACI Code.

ACI Code 13.4.8.5 requires that at least two of the column strip bottom bars in each direction be continuous or spliced at the support with class A splices (see Sec. 5.11a) or anchored within the support. These bars are intended to preserve some measure of structural integrity through catenary action by bridging the damaged area in the event of local punching shear failure of the slab (see Sec. 13.6a). The bars must pass through the column, according to the ACI Code, and must be placed within the column core (i.e., within the main column steel cage).

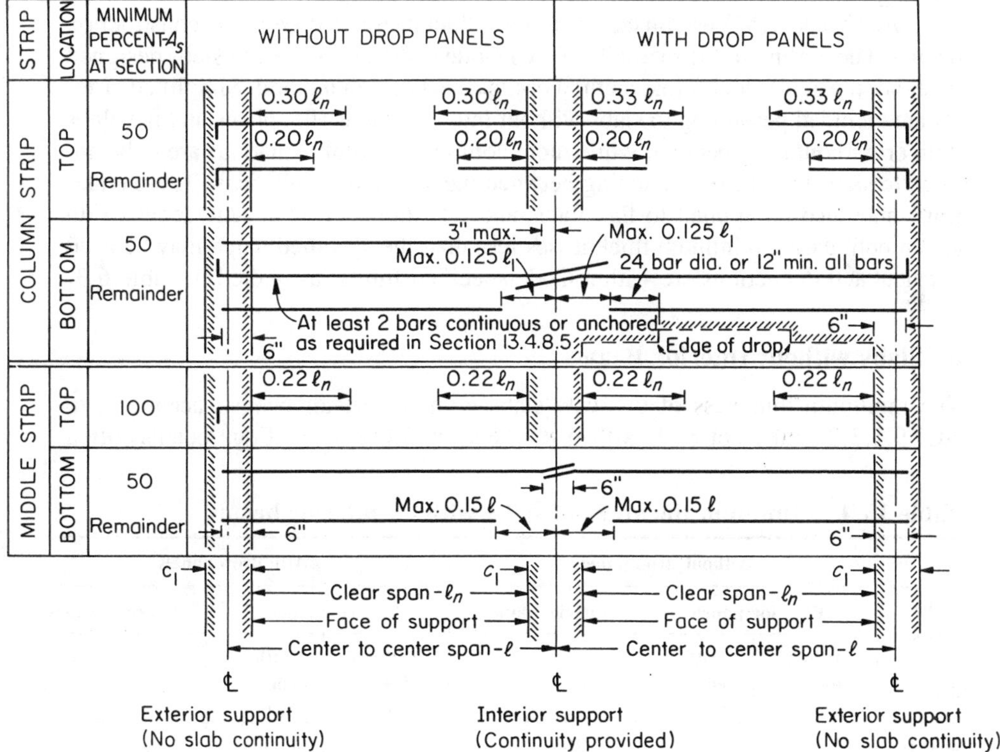

FIGURE 13.7
Minimum length of slab reinforcement in a slab without beams.

The need for special reinforcement at the exterior corners of two-way beam-supported slabs was described in Sec. 12.6, and typical corner reinforcement is shown in Fig. 12.9. According to the ACI Code, such reinforcement is required for slabs with beams between supporting columns if the value of α given by Eq. (13.3) is greater than 1.0.

13.4 DEPTH LIMITATIONS OF THE ACI CODE

To ensure that slab deflections in service will not be troublesome, the best approach is to compute deflections for the total load or load component of interest and to compare the computed deflections with limiting values. In recent years, methods have been developed that are both simple and acceptably accurate for predicting deflections of two-way slabs. A method for calculating the deflection of two-way column-supported slabs will be found in Sec. 13.9.

Alternatively, deflection control can be achieved indirectly by adhering to more or less arbitrary limitations on minimum slab thickness, limitations developed from review of test data and study of the observed deflections of actual structures. As a result of efforts to improve the accuracy and generality of the limiting equations, they have become increasingly complex.

ACI Code 9.5.3 establishes minimum thicknesses for two-way construction designed according to the methods of ACI Code Chapter 13, i.e., to slabs designed by either the equivalent frame method or the direct design method. Simplified criteria are included pertaining to slabs without interior beams (flat plates and flat slabs with or without edge beams), while more complicated limit equations are to be applied to slabs with beams spanning between the supports on all sides. These equations may also be applied to flat plates and flat slabs if it is advantageous to do so. In both cases, minimum thicknesses less than the specified value may be used if calculated deflections are within Code-specified limits, as quoted in Table 6.3.

a. Slabs without Interior Beams

The minimum thickness of two-way slabs without interior beams, according to ACI 9.5.3.2, must not be less than provided by Table 13.4. Edge beams, often

Table 13.4 Minimum thickness of slabs without interior beams

Yield stress f_y psi	Without drop panels			With drop panels		
	Exterior panels		Interior panels	Exterior panels		Interior panels
	Without edge beams	With edge beams[a]		Without edge beams	With edge beams[a]	
40,000	$l_n/33$	$l_n/36$	$l_n/36$	$l_n/36$	$l_n/40$	$l_n/40$
60,000	$l_n/30$	$l_n/33$	$l_n/33$	$l_n/33$	$l_n/36$	$l_n/36$

[a] Slabs with beams along exterior edges. The value of α for the edge beam shall not be less than 0.8.

provided even for two-way slabs otherwise without beams, in order to improve moment and shear transfer at the exterior supports, permit a reduction in minimum thickness of about 10 percent in exterior panels. In all cases, the minimum thickness of slabs without interior beams must not be less than the following:

For slabs without drop panels	5 in.
For slabs with drop panels	4 in.

b. Slabs with or without Beams on All Sides

According to ACI Code 9.5.3.3, slabs with or without beams spanning between the supports on all sides of each panel shall have minimum thickness not less than

$$h = \frac{l_n(0.8 + f_y/200,000)}{36 + 5\beta[\alpha_m - 0.12(1 + 1/\beta)]} \tag{13.8a}$$

and not less than

$$h = \frac{l_n(0.8 + f_y/200,000)}{36 + 9\beta} \tag{13.8b}$$

However, the thickness need not be more than

$$h = \frac{l_n(0.8 + f_y/200,000)}{36} \tag{13.8c}$$

where l_n = clear span in long direction, in.

α_m = average value of α for all beams on edges of a panel [see Eq. (13.3)]

β = ratio of clear span in long direction to clear span in short direction

For slabs without beams but with drop panels of length not less than one-sixth the span length in each direction and with a projection not less than $h/4$ below the slab, the thickness required by Eqs. (13.8) may be reduced by 10 percent. Unless an edge beam with a stiffness such that α is at least 0.80 is provided at all discontinuous edges, the minimum thickness required by Eqs. 13.8 must be increased by 10 percent in the panel having the discontinuous edge.

The values of h obtained from Eqs. 13.8, modified as in the preceding paragraph, must not be less than

For $\alpha_m < 2.0$	5 in.
For $\alpha_m \geq 2.0$	3.5 in.

Note that the provisions of Eqs. 13.8, as modified if appropriate, may also be used as an option for slabs without interior beams, instead of the values of Table 13.4. In all cases, slab thickness less than the stated minimum may be used if it can be shown by calculation that deflections will not exceed the limit values of Table 6.3.

Eqs. (13.8*a*, *b*, and *c*) can be restated in the general form

$$h = \frac{l_n(0.8 + f_y/200{,}000)}{F} \tag{13.8d}$$

where F is the value of the denominator in each case. Figure 13.8 shows the value of F as a function of α_m, for comparative purposes, for three panel aspect ratios β:

1. Square panel, with $\beta = 1.0$
2. Rectangular panel, with $\beta = 1.5$
3. Rectangular panel, with $\beta = 2.0$, the upper limit of applicability of Eqs. (13.8).

This illustrates the range through which each equation controls for these combinations of α_m and β. Note that the *lower value* of F from Eq. (13.8*a*) or (13.8*b*) controls (i.e., it requires the larger h), but no value lower than that from Eq. (13.8*c*) need be considered. It is clear that, for beamless flat plates, with $\alpha_m = 0$, Eq. (13.8*c*) will always govern, while for typical two-way slabs with edge beams pro-

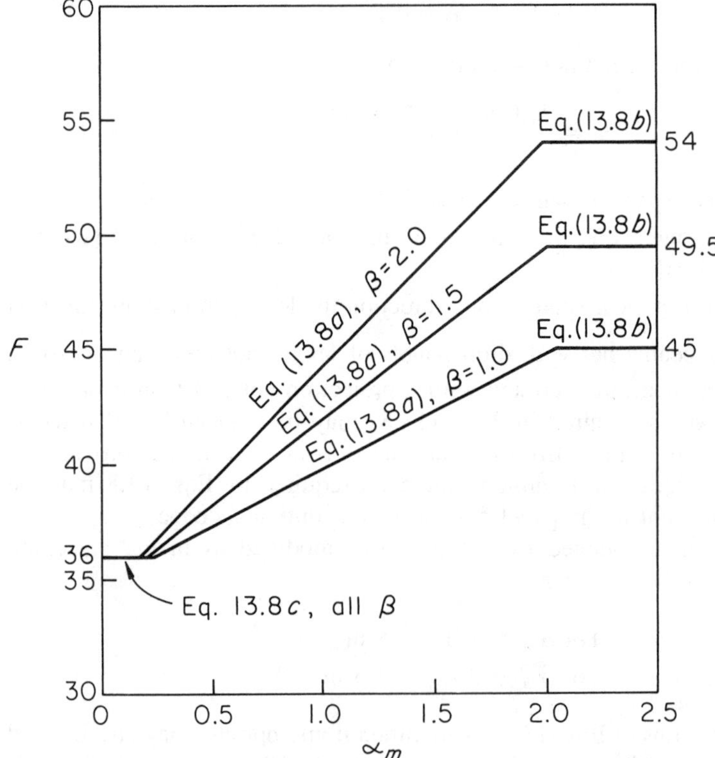

FIGURE 13.8
Parameter F governing minimum thickness of two-way slabs; minimum thickness $h = l_n(0.8 + f_y/200{,}000)/F$.

viding α_m of 2 or greater, Eq. (13.8*b*) controls. Equation (13.8*a*) provides a transition for slabs with shallow column-like beams having α_m in the range of 0 to 2.

Example 13.1. Design of two-way slab with edge beams.† A two-way reinforced concrete building floor system is composed of slab panels measuring 20 × 25 ft in plan, supported by shallow column-line beams cast monolithically with the slab, as shown in Fig. 13.9. Using concrete with $f_c' = 4000$ psi and steel having $f_y = 60{,}000$ psi, design a typical exterior panel to carry a service live load of 125 psf in addition to the self-weight of the floor.

Solution. The floor system satisfies all limitations stated in Sec. 13.2, and the ACI direct design method will be used. For illustrative purposes, only a typical exterior panel, as shown in Fig. 13.9, will be designed. The depth limitations of Sec. 13.4 will be used as a guide to the desirable slab thickness. To use Eqs. (13.8) a trial value of $h = 7$ will be introduced, and beam dimensions 14 × 20 in. will be assumed, as shown in Fig. 13.9. The effective flange projection beyond the face of the beam webs is the lesser of $4h_f$ or

† The design of a two-way slab without beams, i.e., a flat plate floor system, which may also be done by the direct design method if the restrictions of Sec. 13.2 are met, will be illustrated by an example in Sec. 13.5.

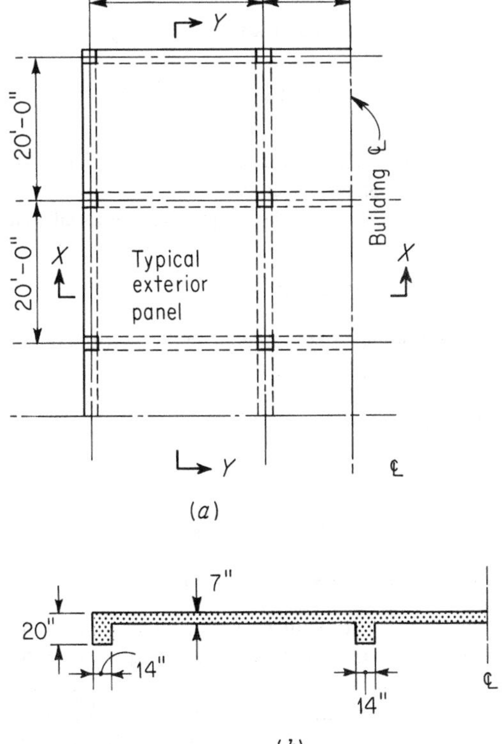

FIGURE 13.9
Two-way slab floor with beams on column lines: (*a*) partial floor plan; (*b*) section *X-X* (section *Y-Y* similar).

$h - h_f$, and in the present case is 13 in. The moment of inertia of the T beams will be estimated as multiples of that of the rectangular portion as follows:

For the edge beams: $I = \frac{1}{12} \times 14 \times 20^3 \times 1.5 = 14{,}000 \text{ in}^4$

For the interior beams: $I = \frac{1}{12} \times 14 \times 20^3 \times 2 = 18{,}600 \text{ in}^4$

For the slab strips:

For the 13.1 ft width: $I = \frac{1}{12} \times 13.1 \times 12 \times 7^3 = 4500 \text{ in}^4$

For the 20 ft width: $I = \frac{1}{12} \times 20 \times 12 \times 7^3 = 6900 \text{ in}^4$

For the 25 ft width: $I = \frac{1}{12} \times 25 \times 12 \times 7^3 = 8600 \text{ in}^4$

Thus, for the edge beam $\alpha = 14{,}000/4500 = 3.1$, for the two 25 ft long beams $\alpha = 18{,}600/6900 = 2.7$, and for the 20 ft long beam $\alpha = 18{,}600/8600 = 2.2$, producing an average value $\alpha_m = 2.7$. The ratio of long to short clear spans is $\beta = 23.8/18.8 = 1.27$. Then the minimum thickness given by Eq. (13.8*a*) is

$$h = \frac{286(0.8 + 60/200)}{36 + 5 \times 1.27(2.7 - 0.12 \times 1.79)} = 6.08 \text{ in.}$$

However, the minimum thickness is not to be less than that given by Eq. (13.8*b*):

$$h = \frac{286(0.8 + 60/200)}{36 + 9 \times 1.27} = 6.63 \text{ in.}$$

and the limit need not be greater than the value from Eq. 13.8*c*:

$$h = \frac{286(0.8 + 60/200)}{36} = 8.74 \text{ in.}$$

The 3.5 in. limitation of Sec. 13.4 clearly does not control in this case, and the 7 in. depth tentatively adopted will provide the basis of further calculation.

For a 7 in. slab, the dead load is $\frac{7}{12} \times 150 = 88$ psf. Applying the usual load factors to obtain design load gives

$$w = 1.4 \times 88 + 1.7 \times 125 = 335 \text{ psf}$$

For the *short-span direction*, for the slab-beam strip centered on the interior column line, the total static design moment is

$$M_0 = \frac{1}{8} \times 0.335 \times 25 \times 18.8^2 = 371 \text{ ft-kips}$$

This is distributed as follows:

Negative design moment = 371 × 0.65 = 241 ft-kips

Positive design moment = 371 × 0.35 = 130 ft-kips

The column strip has a width of 2 × 20/4 = 10 ft. With $l_2/l_1 = 25/20 = 1.25$ and $\alpha_1 l_2/l_1 = 2.2 \times 25/20 = 2.75$, Graph A.4 of App. A indicates that 68 percent of the negative moment, or 163 ft-kips, is taken by the column strip, of which 85 percent, or 139 ft-kips, is taken by the beam and 24 ft-kips by the slab. The remaining 78 ft-kips is allotted to the slab middle strip. Graph A.4 also indicates that 68 percent of the positive moment, or 88 ft-kips, is taken by the column strip, of which 85 percent, or 75 ft-kips, is assigned to the beam and 13 ft-kips to the slab. The remaining 42 ft-kips is taken by the slab middle strip.

A similar analysis is performed for the short-span direction, for the slab-beam strip at the edge of the building, based on a total static design moment of

$$M_0 = \tfrac{1}{8} \times 0.335 \times 13.1 \times 18.8^2 = 194 \text{ ft-kips}$$

of which 65 percent is assigned to the negative and 35 percent to the positive bending sections as before. In this case, $\alpha_1 l_2/l_1 = 3.1 \times 25/20 = 3.9$. The distribution factor for column-strip moment, from Graph A.4, is 68 percent for positive and negative moments as before, and again 85 percent of the column-strip moments is assigned to the beams.

In summary, the short-direction moments, in ft-kips, are as follows:

	Beam moment	Column-strip slab moment	Middle-strip slab moment
Interior slab-beam strip—20 ft span			
Negative	139	24	78
Positive	75	13	42
Exterior slab-beam strip—20 ft span			
Negative	73	13	40
Positive	39	7	22

The total static design moment in the *long direction* of the exterior panel is

$$M_0 = \tfrac{1}{8} \times 0.335 \times 20 \times 23.8^2 = 475 \text{ ft-kips}$$

This will be apportioned to the negative and positive moment sections according to Table 13.1, and distributed laterally across the width of critical moment sections with the aid of Graph A.4. The moment ratios to be applied to obtain exterior negative, positive, and interior negative moments are, respectively, 0.16, 0.57, and 0.70. The torsional constant for the edge beam is found from Eq. (13.5) for a 14×20 in. rectangular shape with a 7×13 in. projecting flange:

$$C = \left(1 - 0.63 \times \frac{14}{20}\right) \frac{14^3 \times 20}{3} + \left(1 - 0.63 \times \frac{7}{13}\right) \frac{7^3 \times 13}{3} = 11{,}205$$

With $l_2/l_1 = 0.80, \alpha_1 l_2/l_1 = 2.7 \times 20/25 = 2.2$, and from Eq. (13.4), $\beta_t = 11,205/(2 \times 6900) = 0.82$, Graph A.4 indicates that the column strip will take 93 percent of the exterior negative moment, 81 percent of the positive moment, and 81 percent of the interior negative moment. As before, the column-line beam will account for 85 percent of the column-strip moment. The results of applying these moment ratios are as follows:

	Beam moment	Column-strip slab moment	Middle-strip slab moment
Exterior negative—25 ft span	60	11	5
Positive—25 ft span	187	33	51
Interior negative—25 ft span	229	40	63

It is convenient to tabulate the design of the slab reinforcement as in Table 13.5. In the 25 ft direction, the two half-column strips may be combined for purposes of calculation into one strip of 106 in. width. In the 20 ft direction the exterior half-column strip and the interior half-column strip will normally differ and are treated separately. Design moments from the previous distributions are summarized in column 3 of the table.

The short-direction positive steel will be placed first, followed by the long-direction positive bars. If $\frac{3}{4}$ in. clear distance below the steel is allowed and use of No. 4 bars is anticipated, the effective depth in the short direction will be 6 in., while that in the long direction will be 5.5 in. A similar situation obtains for the top steel.

After calculating the design moment per foot strip of slab (column 6), find the minimum effective slab depth required for flexure. For the material strengths to be used, the maximum permitted steel ratio is $0.75\rho_b = 0.0214$. For this ratio

$$d^2 = \frac{M_u}{\phi \rho f_y b(1 - 0.59\rho f_y/f_c')}$$

$$= \frac{M_u}{0.90 \times 0.0214 \times 60,000 \times 12(1 - 0.59 \times 0.0214 \times 60/4)} = \frac{M_u}{11,300}$$

Table 13.5 Design of slab reinforcement

(1)	(2) Location	(3) M_u, ft-kips	(4) b, in.	(5) d, in.	(6) $M_u \times 12/b$, ft-kips/ft	(7) ρ	(8) A_s, in^2	(9) Number of No. 4 bars
25 ft span Two half-column strips	Exterior negative	11	106	5.5	1.25	0.0023[a]	1.34	7
	Positive	33	106	5.5	3.74	0.0023	1.34	7
	Interior negative	40	106	5.5	4.53	0.0029	1.69	9
Middle strip	Exterior negative	5	120	5.5	0.50	0.0023[a]	1.52	9[b]
	Positive	51	120	5.5	5.10	0.0033	2.18	11
	Interior negative	63	120	5.5	6.30	0.0041	2.71	14
20 ft span Exterior half-column strip	Negative	13	53	6	2.94	0.0021[a]	0.67	4
	Positive	7	53	6	1.58	0.0021[a]	0.67	4
Middle strip	Negative	78	180	6	5.20	0.0028	3.03	16
	Positive	42	180	6	2.80	0.0021[a]	2.27	13[b]
Interior half-column strip	Negative	12	53	6	2.71	0.0021[a]	0.67	4
	Positive	6.5	53	6	1.47	0.0021[a]	0.67	4

[a] Steel ratio controlled by shrinkage and temperature requirements.

[b] Number of bars controlled by maximum spacing requirements.

Hence, $d = \sqrt{M_u/11{,}300}$. Thus, the following minimum effective depths are needed:

$$\text{In 25 ft direction:} \qquad d = \sqrt{6.30 \times \frac{12{,}000}{11{,}300}} = 2.59 \text{ in.}$$

$$\text{In 20 ft direction:} \qquad d = \sqrt{5.20 \times \frac{12{,}000}{11{,}300}} = 2.35 \text{ in.}$$

both well below the depth dictated by deflection requirements. An underreinforced slab results. The required steel ratios (column 7) are conveniently found from Table A.6 with $R = M_u/\phi bd^2$ or from Table A.10. Note that a minimum steel area equal to 0.0018 times the gross concrete area must be provided for control of temperature and shrinkage cracking. For a 12 in. slab strip the corresponding area is $0.0018 \times 7 \times 12 = 0.151$ in^2. Expressed in terms of minimum steel ratio for actual effective depths, this gives

$$\text{In 25 ft direction:} \qquad \rho_{\min} = \frac{0.151}{5.5 \times 12} = 0.0023$$

$$\text{In 20 ft direction:} \qquad \rho_{\min} = \frac{0.151}{6 \times 12} = 0.0021$$

This requirement controls at the locations indicated in Table 13.5.

The total steel area in each band is easily found from the steel ratio and is given in column 8. Finally, with the aid of Table A.2, the required number of bars is obtained. Note that in two locations, the number of bars used is dictated by the maximum spacing requirement of $2 \times 7 = 14$ in.

The shear capacity of the slab is checked on the basis of the tributary areas shown in Fig. 13.6. At a distance d from the face of the long beam,

$$V_u = 0.335\left(10 - \frac{14}{2 \times 12} - \frac{6}{12}\right) = 2.99 \text{ kips}$$

The design shear strength of the slab is

$$\phi V_c = 0.85 \times 2\sqrt{4000} \times 12 \times \frac{6}{1000}$$
$$= 7.74 \text{ kips}$$

well above the shear applied at factored loads.

Each beam must be designed for its share of the total static moment, as found in the above calculations, as well as the moment due to its own weight; this moment may be distributed to positive and negative bending sections, using the same ratios used for the static moments due to slab loads. Beam shear design should be based on the loads from the tributary areas shown in Fig. 13.6. Since no new concepts would be introduced, the design of the beams will not be presented here.

Since $0.85 \times 93 = 79$ percent of the exterior negative moment in the long direction is carried directly to the column by the column-line beam in this example, torsional stresses in the spandrel beam are very low and may be disregarded. In other circumstances, the spandrel beams would be designed for torsion following the methods of Chap. 7.

13.5 EQUIVALENT FRAME METHOD

a. Basis of Analysis

The direct design method for two-way slabs described in Sec. 13.2 is useful if each of the six restrictions on geometry and load is satisfied by the proposed structure.

Otherwise, a more general method is needed. One such method, proposed by Peabody in 1948 (Ref. 13.10), was incorporated in subsequent editions of the ACI Code as *design by elastic analysis*. The method was greatly expanded and refined based on research at the University of Illinois in the 1960s (Refs. 13.11 and 13.12), and it appears in Chapter 13 of the current ACI Code as the *equivalent frame method*.

It will be evident that the equivalent frame method was derived with the assumption that the analysis would be done by hand calculation using the moment distribution method (see Ch. 11). If analysis is done by computer using a standard frame analysis program, special modeling devices are necessary. This point will be discussed further in Sec. 13.5e.

By the equivalent frame method the structure is divided, for analysis, into continuous frames centered on the column lines and extending both longitudinally and transversely, as shown by the shaded strips in Fig. 13.10. Each frame is composed of a row of columns and a broad continuous beam. The beam, or slab beam, includes the portion of the slab bounded by panel centerlines on either side of the columns, together with column-line beams or drop panels if used. For vertical loading, each floor with its columns may be analyzed separately, the columns assumed fixed at the floors above and below. In calculating bending moment at a support, it is convenient and sufficiently accurate to assume that the

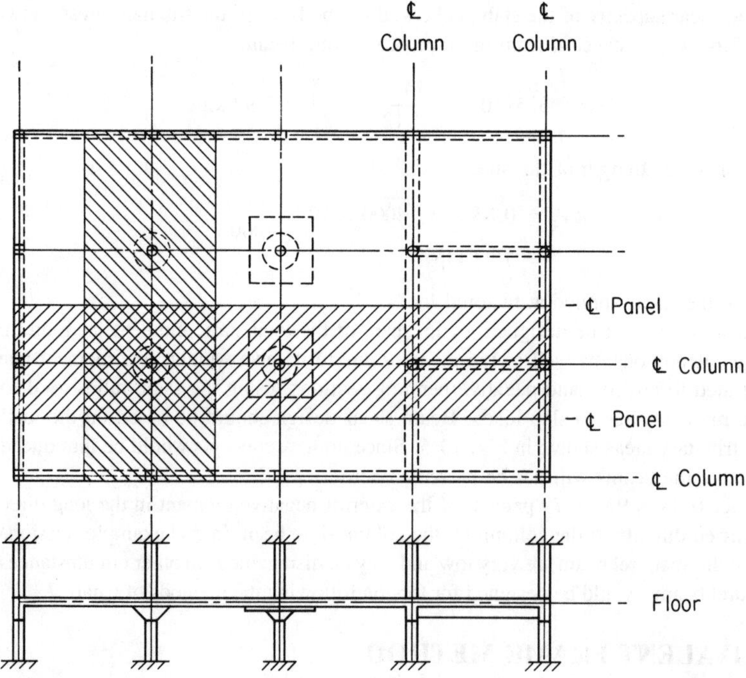

FIGURE 13.10
Building idealization for equivalent frame analysis.

continuous frame is completely fixed at the support two panels removed from the given support, provided the frame continues past that point.

b. Moment of Inertia of Slab Beam

Moments of inertia used for analysis may be based on the concrete cross section, neglecting reinforcement, but variations in cross section along the member axis should be accounted for.

For the beam strips, the first change from the midspan moment of inertia normally occurs at the edge of drop panels, if they are used. The next occurs at the edge of the column or column capital. While the stiffness of the slab strip could be considered infinite within the bounds of the column or capital, at locations close to the panel centerlines (at each edge of the slab strip) the stiffness is much less. According to ACI Code 13.7.3, from the center of the column to the face of the column or capital, the moment of inertia of the slab is taken equal to that at the face of the column or capital, divided by the quantity $(1 - c_2/l_2)^2$, where c_2 and l_2 are the size of the column or capital and the panel span respectively, both measured transverse to the direction in which moments are being determined.

Accounting for these changes in moments of inertia results in a member, for analysis, in which the moment of inertia varies in a stepwise manner. The stiffness factors, carryover factors, and uniform-load fixed-end moment factors needed for moment distribution analysis (see Chap. 11) are given in Table A.14*a* of App. A for a slab without drop panels, and in Table A.14*b* for a slab with drop panels of depth equal to 1.25 times the slab depth and of length equal to one-third the span length.

c. The Equivalent Column

In the equivalent frame method of analysis, the columns are considered to be attached to the continuous slab beam by torsional members transverse to the direction of the span for which moments are being found; the torsional member extends to the panel centerlines bounding each side of the slab beam under study. Torsional deformation of these transverse supporting members reduces the effective flexural stiffness provided by the actual column at the support. This effect is accounted for in the analysis by use of what is termed an *equivalent column* having stiffness less than that of the actual column.

The action of a column and the transverse torsional member is easily explained with reference to Fig. 13.11, which shows, for illustration, the column and transverse beam at the exterior support of a continuous slab-beam strip. From Fig. 13.11 it is clear that the rotational restraint provided at the end of the slab spanning in the direction l_1 is influenced not only by the flexural stiffness of the column but also by the torsional stiffness of the edge beam AC. With distributed torque m_t applied by the slab and resisting torque M_t provided by the column, the edge-beam sections at A and C will rotate to a greater degree than the section at B, owing to torsional deformation of the edge beam. To allow for this effect, the actual column and beam are replaced by an equivalent column, so defined that

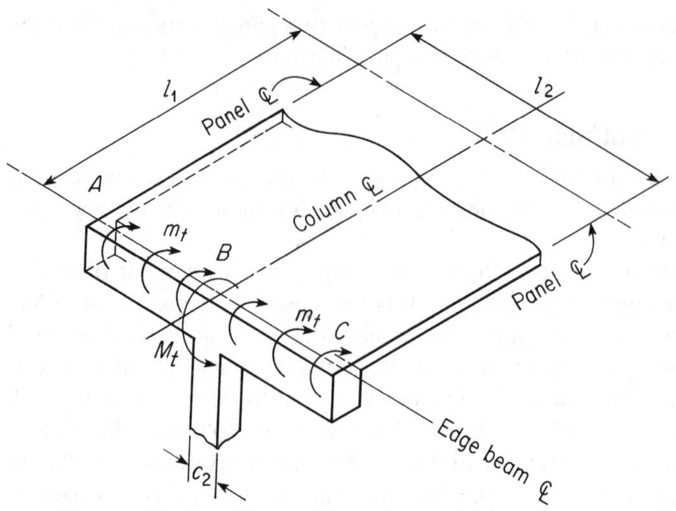

FIGURE 13.11
Torsion at a transverse supporting member illustrating the basis of the equivalent column.

the total flexibility (inverse of stiffness) of the equivalent column is the sum of the flexibilities of the actual column and beam. Thus

$$\frac{1}{K_{ec}} = \frac{1}{\Sigma K_c} + \frac{1}{K_t} \tag{13.9}$$

where K_{ec} = flexural stiffness of equivalent column
K_c = flexural stiffness of actual column
K_t = torsional stiffness of edge beam

all expressed in terms of moment per unit rotation. In computing K_c, the moment of inertia of the actual column is assumed to be infinite from the top of the slab to the bottom of the slab beam, and I_g is based on the gross concrete section elsewhere along the length. Stiffness and carryover factors for such a case are given in Table A.14*c*.

The effective cross section of the transverse torsional member, which may or may not include a beam web projecting below the slab, as shown in Fig. 13.11, is the same as defined earlier in Sec. 13.2c. The torsional constant C is calculated by Eq. (13.5) based on the effective cross section so determined. The torsional stiffness K_t can then be calculated by the expression

$$K_t = \sum \frac{9E_{cs}C}{l_2(1 - c_2/l_2)^3} \tag{13.10}$$

where E_{cs} = modulus of elasticity of slab concrete
c_2 = size of rectangular column, capital, or bracket in direction l_2
C = cross-sectional constant

The summation applies to the typical case in which there are edge beams on both sides of the column.

If a panel contains a beam parallel to the direction in which moments are being determined, the value of K_t obtained from Eq. (13.10) leads to values of K_{ec} that are too low. Accordingly, it is recommended that in such cases the value of K_t found by Eq. (13.10) be multiplied by the ratio of the moment of inertia of the slab with such a beam to the moment of inertia of the slab without it.

The concept of the equivalent column, illustrated with respect to an exterior column, is employed at *all* supporting columns for each continuous slab beam, according to the equivalent frame method.

d. Moment Analysis

With the effective stiffness of the slab-beam strip and the supports found as described, the analysis of the equivalent frame can proceed by moment-distribution (see Chap. 11).

In keeping with the requirements of statics (see Sec. 13.1), equivalent beam strips in each direction must each carry 100 percent of the load. If the live load does not exceed three-quarters of the dead load, maximum moment may be assumed to obtain at all critical sections when the full factored live load (plus factored dead load) is on the entire slab, according to ACI Code 13.7.6. Otherwise pattern loadings must be used to maximize positive and negative moments. Maximum positive moment is calculated with three-quarters factored live load on the panel and on alternate panels, while maximum negative moment at a support is calculated with three-quarters factored live load on the adjacent panels only. Use of three-quarters live load rather than the full value recognizes that maximum positive and negative moments cannot occur simultaneously (since they are found from different loadings) and that redistribution of moments to less highly stressed sections will take place before failure of the structure occurs. Factored moments must not be taken less than those corresponding to full factored live load on all panels, however.

Negative moments obtained from that analysis apply at the centerlines of supports. Since the support is not a knife-edge but a rather broad band of slab spanning in the transverse direction, some reduction in the negative design moment is proper (see also Sec. 11.5a). At interior supports, the critical section for negative bending, in both column and middle strips, may be taken at the face of the supporting column or capital, but in no case at a distance greater than $0.175l_1$ from the center of the column, according to ACI Code 13.7.7. To avoid excessive reduction of negative moment at the exterior supports (where the distance to the point of inflection is small) for the case where columns are provided with capitals, the critical section for negative bending in the direction perpendicular to an edge should be taken at a distance from the face of support not greater than one-half the projection of the capital beyond the face of the support.

With positive and negative design moments obtained as just described, it still remains to distribute these moments across the widths of the critical sections.

For design purposes, the total strip width is divided into column strip and adjacent half-middle strips, defined previously, and moments are assumed constant within the bounds of each. The distribution of moments to column and middle strips is done using the same percentages given in connection with the direct design method. These are summarized in Table 13.2 and by the interpolation charts of Graph A.4 of App. A.

The distribution of moments and shears to column-line beams, if present, is in accordance with the procedures of the direct design method also. Restriction 6 of Sec. 13.2, pertaining to the relative stiffness of column-line beams in the two directions, applies here also if these distribution ratios are used.

Example 13.2 Design of flat plate floor by equivalent frame method. An office building is planned using a flat plate floor system with the column layout as shown in Fig. 13.12. No beams, dropped panels, or column capitals are permitted. Specified live load is 100 psf and dead load will include the weight of the slab plus an allowance of 20 psf for finish floor plus suspended loads. The columns will be 18 in. square, and the floor-to-floor height of the structure will be 12 ft. Design the interior panel C, using material strengths $f_y = 60{,}000$ psi and $f_c' = 4000$ psi. Straight-bar reinforcement will be used.

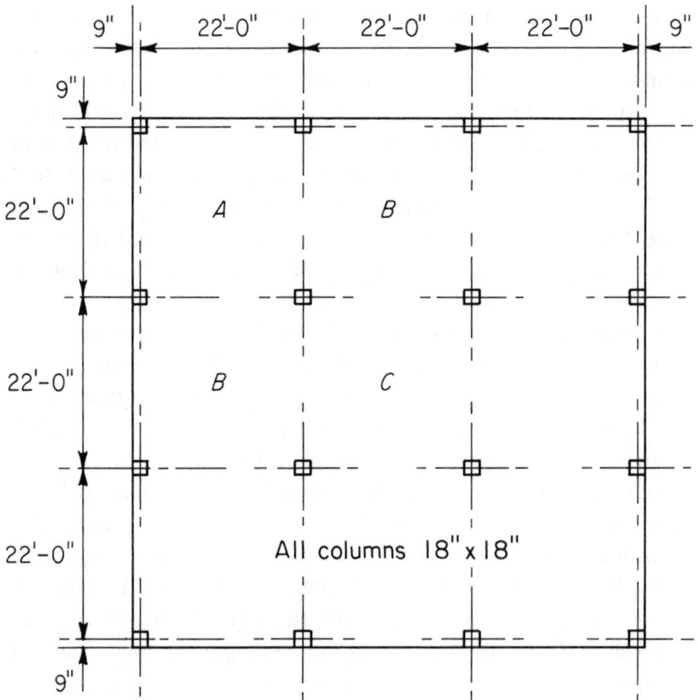

FIGURE 13.12
Two-way flat plate floor.

Solution. Minimum thickness h for a flat plate, according to the ACI Code, may be found by Eq. (13.8*c*).† For the present example, the minimum h is

$$h = \frac{20.5 \times 12(0.8 + 60/200)}{36} = 7.52 \text{ in.}$$

For slabs without edge beams, as for this example, the minimum thickness in all panels with discontinuous edges must be increased by 10 percent, resulting in a minimum $h = 8.27$ in. This will be rounded upward for practical reasons, with calculations based on a trial thickness of 8.5 in. for all panels. Thus the dead load of the slab is $150 \times 8.5/12 = 106$ psf, to which the superimposed dead load of 20 psf must be added. The factored design loads are

$$1.4w_d = 1.4(106 + 20) = 176 \text{ psf}$$
$$1.7w_l = 1.7 \times 100 = 170 \text{ psf}$$

The structure is identical in each direction, permitting the design for one direction to be used for both (an average effective depth to the tensile steel will be used in the calculations). While the restrictions of Sec. 13.2 are met and the direct design method of analysis is permissible, the equivalent frame method will be adopted to demonstrate its features. Moments will be found by the method of moment distribution.

For flat plate structures it is usually acceptable to calculate stiffnesses as if all members were prismatic, neglecting the increase in stiffness within the joint region, as it generally has negligible effect on design moments and shears. Then, for the slab spans,

$$K_s = \frac{4E_cI_c}{l}$$

$$= \frac{4E_c(264 \times 8.5^3)}{12 \times 264} = 205E_c$$

and the column stiffnesses are

$$K_c = \frac{4E_c(18 \times 18^3)}{12 \times 144} = 243E_c$$

Calculation of the equivalent column stiffness requires consideration of the torsional deformation of the transverse strip of slab that functions as the supporting beam. Applying the criteria of the ACI Code establishes that the effective torsional member has width 18 in. and depth 8.5 in. For this section, the torsional constant C from Eq. (13.5) is

$$C = \left(1 - 0.63 \times \frac{8.5}{18}\right) 8.5^3 \times \frac{18}{3} = 2590 \text{ in}^4$$

and the torsional stiffness, from Eq. (13.10), is

$$K_t = \frac{9E_c \times 2590}{264(1 - 1.5/22)^3} = 109E_c$$

† In many flat plate floors, the minimum slab thickness is controlled by requirements for shear transfer at the supporting columns, and h is determined either to avoid supplementary shear reinforcement or to limit the excess shear to a reasonable margin above that which can be carried by the concrete. Design for shear in flat plates and flat slabs will be treated in Sec. 13.6.

From Eq. (13.9), accounting for two columns and two torsional members at each joint,

$$\frac{1}{K_{ec}} = \frac{1}{2 \times 243E_c} + \frac{1}{2 \times 109E_c}$$

from which $K_{ec} = 151E_c$. Distribution factors at each joint are then calculated in the usual way.

For the present example, the ratio of service live load to dead load is $100/126 = 0.79$, and because this exceeds 0.75, according to the ACI Code maximum positive and negative moments must be found based on pattern loadings, with full factored dead load in place and three-quarters factored live load positioned to cause the maximum effect. In addition, the design moments must not be less than those produced by full factored live and dead loads on all panels. Thus three load cases must be considered: (*a*) full factored dead and live load, 346 psf, on all panels; (*b*) factored dead load of 176 psf on all spans plus three-quarters factored live load, 128 psf, on panel *C*; and (*c*) full factored dead load on all spans and three-quarters live load on first and second spans. Fixed end moments and final moments obtained from moment distribution are summarized in Table 13.6. The results indicate that load case (*a*) controls the slab design in the support region, while load case (*b*) controls at the midspan of panel *C*. Moment diagrams for the two controlling cases are shown in Fig. 13.13*a*. According to the ACI Code the critical

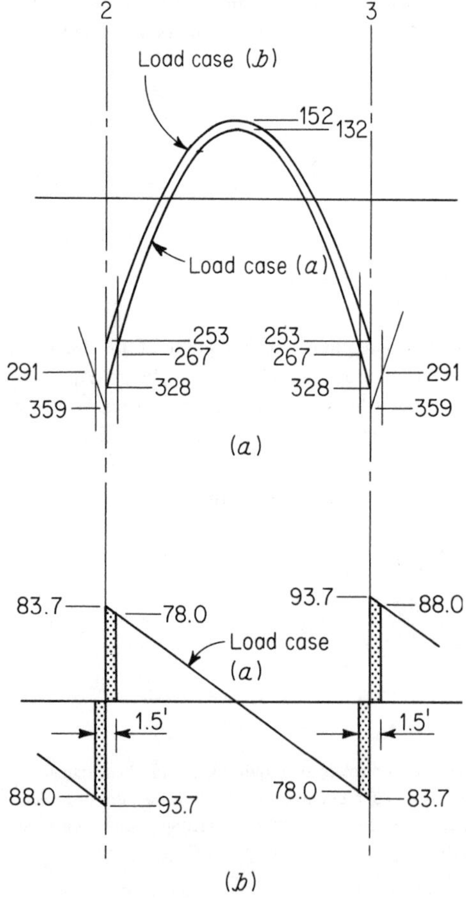

FIGURE 13.13
Design moments and shears for flat plate floor: (*a*) moments; (*b*) shears.

Table 13.6 Moments in flat plate floor

Panel	B		C		B	
Joint	**1**	**2**	**2**	**3**	**3**	**4**
(*a*) 346 psf all panels						
Fixed end moments	+307	−307	+307	−307	+307	−307
Final moments	+139	−359	+328	−328	+359	−139
Span moment in *C*			132			
(*b*) 176 psf panels *B* and 304 psf panel *C*						
Fixed end moments	+156	−156	+270	−270	+156	−156
Final moments	+ 59	−229	+253	−253	+229	− 59
Span moment in *C*			152			
(*c*) 304 psf panels *B* (left) and *C* and 176 psf panel *B* (right)						
Fixed end moments	+270	−270	+270	−270	+156	−156
Final moments	+120	−325	+306	−235	+220	− 62
Span moment in *C*			134			

section at interior supports may be taken at the face of supports, but not greater than $0.175l_1$ from the column centerline. The former criterion controls here, and the negative design moment is calculated by subtracting the area under the shear diagram between the centerline and face of support, for load case (*a*), from the negative moment at the support centerline. The shear diagram for load case (*a*) is given in Fig. 13.13*b*, with the adjusted design moments shown in Fig. 13.13*a*.

Because the effective depth for all panels will be the same, and because the negative steel for panel *C* will continue through the support region to become the negative steel for panels *B*, the larger negative moment found for the panels *B* will control. Accordingly, the design negative moment is 291 ft-kips and the design positive moment is 152 ft-kips.†

† When slab systems that meet the restrictions of the direct design method are designed by the equivalent frame method, according to ACI Code 13.7.7 the resulting design moments may be reduced proportionately so that the sum of the positive and average negative moments in a span are no greater than M_0 calculated for the direct design method according to Eq. (13.1). There is no theoretical basis for this. The reduction is less than 5 percent in the present example, and it will not be included in the design calculations.

Table 13.7 Design of flat plate reinforcement

(1)	(2) Location	(3) M_u, ft-kips	(4) b, in.	(5) d, in.	(6) $M_u \times 12/b$, ft-kips/ft	(7) ρ	(8) A_s, in^2	(9) No. and size of bars
Column strip	Negative	218	132	7	19.82	0.0081	7.48	17 No. 6
	Positive	91	132	7	8.27	0.0032	2.96	10 No. 5
Two half-middle strips	Negative	73	132	7	6.64	0.0026	2.40	8 No. 5
	Positive	61	132	7	5.55	0.0021	1.94	8 No. 5[a]

[a] Number of bars controlled by maximum spacing requirement.

Moments will be distributed laterally across the slab width according to Table 13.2, which indicates that 75 percent of the negative moment will be assigned to the column strip and 60 percent of the positive moment assigned to the column strip. The design of the slab reinforcement is summarized in Table 13.7.

Other important aspects of the design of flat plates include design for punching shear at the columns, which may require supplementary shear reinforcement, and transfer of unbalanced moments to the columns, which may require additional flexural bars in the negative bending region of the column strips or adjustment of spacing of negative steel. These considerations are of special importance at exterior columns and corner columns, such as shown in Fig. 13.12. Shear and moment transfer at the columns will be discussed in Secs. 13.6 and 13.7, respectively.

e. Equivalent Frame Analysis by Computer

It is clear that the equivalent frame method, as described in the ACI Code and the ACI Code Commentary, is oriented toward hand analysis using the method of moment distribution. Presently, most offices make use of computers, and frame analysis is done using general-purpose programs based on the direct stiffness method. Plane frame analysis programs can be used for slab analysis based on the concepts of the equivalent frame method, but the frame must be specially modeled. Variable moments of inertia along the axis of slab-beams and columns require nodal points (continuous joints) between sections where I is to be considered constant (i.e., in the slab at the junction of slab and drop panel, drop panel and capital, and in the columns at the bottom of the capitals). In addition, it is necessary to compute K_{ec} for each column, then to compute the equivalent value of the moment of inertia for the column.

Alternately, a three-dimensional frame analysis may be used, in which the torsional properties of the transverse supporting beams may be included directly. A third option is to make use of specially written computer programs, the most widely used being "ADOSS—Analysis and Design of Concrete Floor Systems," developed by the Portland Cement Association. Other such programs are available.

13.6 SHEAR DESIGN IN FLAT PLATES AND FLAT SLABS

When two-way slabs are supported directly by columns, as in flat slabs and flat plates, or when slabs carry concentrated loads, as in footings, shear near the columns is of critical importance. Tests of flat plate structures indicate that, in most practical cases, the capacity is governed by shear (Ref. 13.13).

a. Slabs without Special Shear Reinforcement

Two kinds of shear may be critical in the design of flat slabs, flat plates, or footings. The first is the familiar beam-type shear leading to diagonal tension failure. Applicable particularly to long narrow slabs or footings, this analysis considers the slab to act as a wide beam, spanning between supports provided by the perpendicular column strips. A potential diagonal crack extends in a plane across the entire width l_2 of the slab. The critical section is taken a distance d from the face of the column or capital. As for beams, the design shear strength ϕV_c must be at least equal to the required strength V_u at factored loads. The nominal shear strength V_c should be calculated by either Eq. (4.12*a*) or Eq. (4.12*b*), with $b_w = l_2$ in this case.

Alternatively, failure may occur by *punching shear*, with the potential diagonal crack following the surface of a truncated cone or pyramid around the column, capital, or drop panel, as shown in Fig. 13.14*a*. The failure surface extends from the bottom of the slab, at the support, diagonally upward to the top surface. The angle of inclination with the horizontal, θ (see Fig. 13.14*b*), depends upon the nature and amount of reinforcement in the slab. It may range between about 20 and 45°. The critical section for shear is taken perpendicular to the plane of the slab and a distance $d/2$ from the periphery of the support, as shown. The shear force V_u to be resisted can be calculated as the total factored load on the area bounded by panel centerlines around the column less the load applied within the area defined by the critical shear perimeter, unless significant moments must be transferred from the slab to the column (see Sec. 13.7).

At such a section, in addition to the shearing stresses and horizontal compressive stresses due to negative bending moment, vertical or somewhat inclined compressive stress is present, owing to the reaction of the column. The simultaneous presence of vertical and horizontal compression increases the shear strength of the concrete. For slabs supported by columns having a ratio of long to short sides not greater than 2, tests indicate that the nominal shear strength may be taken equal to

$$V_c = 4\sqrt{f_c'}\,b_0 d \tag{13.11a}$$

according to ACI Code 11.12.2, where b_0 = the perimeter along the critical section.

However, for slabs supported by very rectangular columns, the shear strength predicted by Eq. (13.11*a*) has been found to be unconservative. According to tests

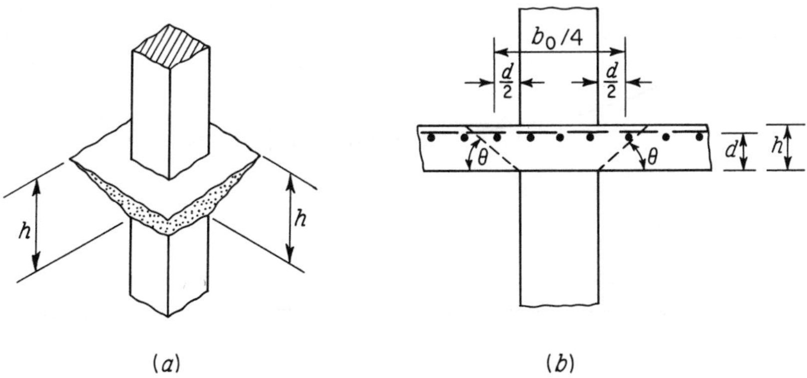

FIGURE 13.14
Failure surface defined by punching shear.

reported in Ref. 13.14, the value of V_c approaches $2\sqrt{f_c'}b_0d$ as β_c, the ratio of long to short sides of the column, becomes very large. Reflecting this test data, ACI Code 11.12.2 states further that V_c in punching shear shall not be taken greater than

$$V_c = \left(2 + \frac{4}{\beta_c}\right)\sqrt{f_c'}b_0d \tag{13.11b}$$

The variation of the shear strength coefficient, as governed by Eqs. (13.11*a*) and (13.11*b*) is shown in Fig. 13.15 as a function of β_c.

Further tests, reported in Ref. 13.15, have shown that the shear strength V_c decreases as the ratio of critical perimeter to slab depth, b_0/d, increases. Accordingly, ACI Code 11.12.2 states that V_c in punching shear must not be taken greater than

$$V_c = \left(\frac{\alpha_s d}{b_0} + 2\right)\sqrt{f_c'}b_0d \tag{13.11c}$$

where α_s is 40 for interior columns, 30 for edge columns, and 20 for corner columns, i.e., columns having critical sections with 4, 3, or 2 sides, respectively.

Thus, according to ACI Code, the punching shear strength of slabs and footings is to be taken as the smallest of the values of V_c given by Eqs. (13.11*a*), (13.11*b*), and (13.11*c*). The design strength is taken as ϕV_c as usual, where $\phi = 0.85$ for shear. The basic requirement is then $V_u \leq \phi V_c$.

For columns of nonrectangular cross section the ACI Code indicates that the perimeter b_0 must be of minimum length, but need not approach closer than $d/2$ to the perimeter of the reaction area. The manner of defining the critical perimeter b_0 and the ratio β_c for such irregular support configurations is illustrated in Fig. 13.16.

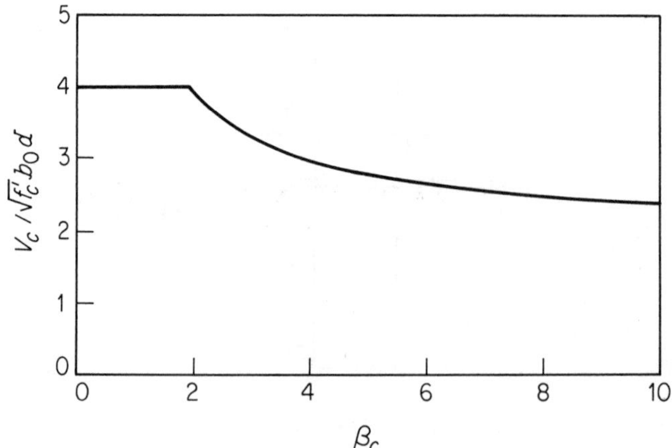

FIGURE 13.15
Shear strength coefficient for flat plates as a function of ratio β_c of long side to short side of support.

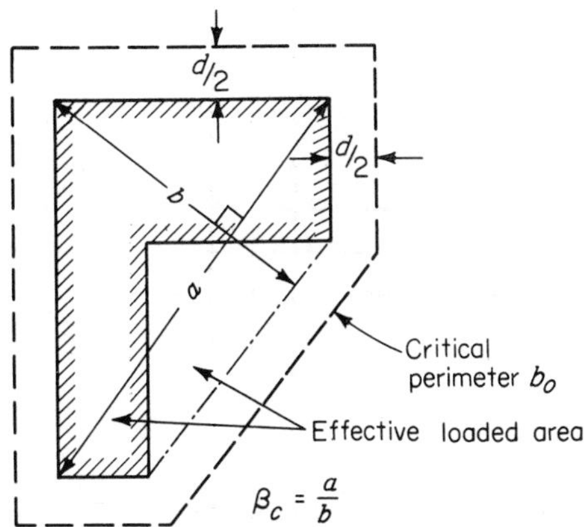

FIGURE 13.16
Punching shear for columns of irregular shape.

b. Types of Shear Reinforcement

Special shear reinforcement is often used at the supports for flat plates, and sometimes for flat slabs as well. It may take several forms. A few common types are shown in Fig. 13.17.

The *shearheads* shown in (*a*) and (*c*) consist of standard structural steel shapes embedded in the slab and projecting beyond the column. They serve to increase the effective perimeter b_0 of the critical section for shear. In addition, they may contribute to the negative bending resistance of the slab. The reinforcement shown in (*a*) is particularly suited for use with concrete columns. It consists of short lengths of I or wide-flange beams, cut and welded at the crossing point so that the arms are continuous through the column. Normal negative slab reinforcement passes over the top of the structural steel, while bottom bars are stopped short of the shearhead. Column bars pass vertically at the corners of the column without interference. The effectiveness of this type of shearhead has been well documented by testing at the laboratories of the Portland Cement Association (Ref. 13.16). The channel frame of (*c*) is very similar in its action, but is adapted for use with steel columns. The bent-bar arrangement of (*b*) is suited for use with concrete columns. The bars are usually bent at 45° across the potential diagonal tension crack, and extend along the bottom of the slab a distance sufficient to develop their strength by bond. The flanged collar of (*d*) is designed mainly for use with lift-slab construction (see Chap. 20). It consists of a flat bottom plate with vertical stiffening ribs. It may incorporate sockets for lifting rods, and usually is used in conjunction with shear pads welded directly to the column surfaces below the collar to transfer the vertical reaction.

Another type of shear reinforcement is illustrated in Fig. 13.17*e*, where vertical stirrups have been used in conjunction with supplementary horizontal bars radiating outward in two perpendicular directions from the support, to form what are termed *integral beams* contained entirely within the slab thickness. These beams act in the same general way as the shearheads of Fig. 13.17*a* and *c*. Adequate anchorage of the stirrups is difficult in slabs thinner than about 10 in.

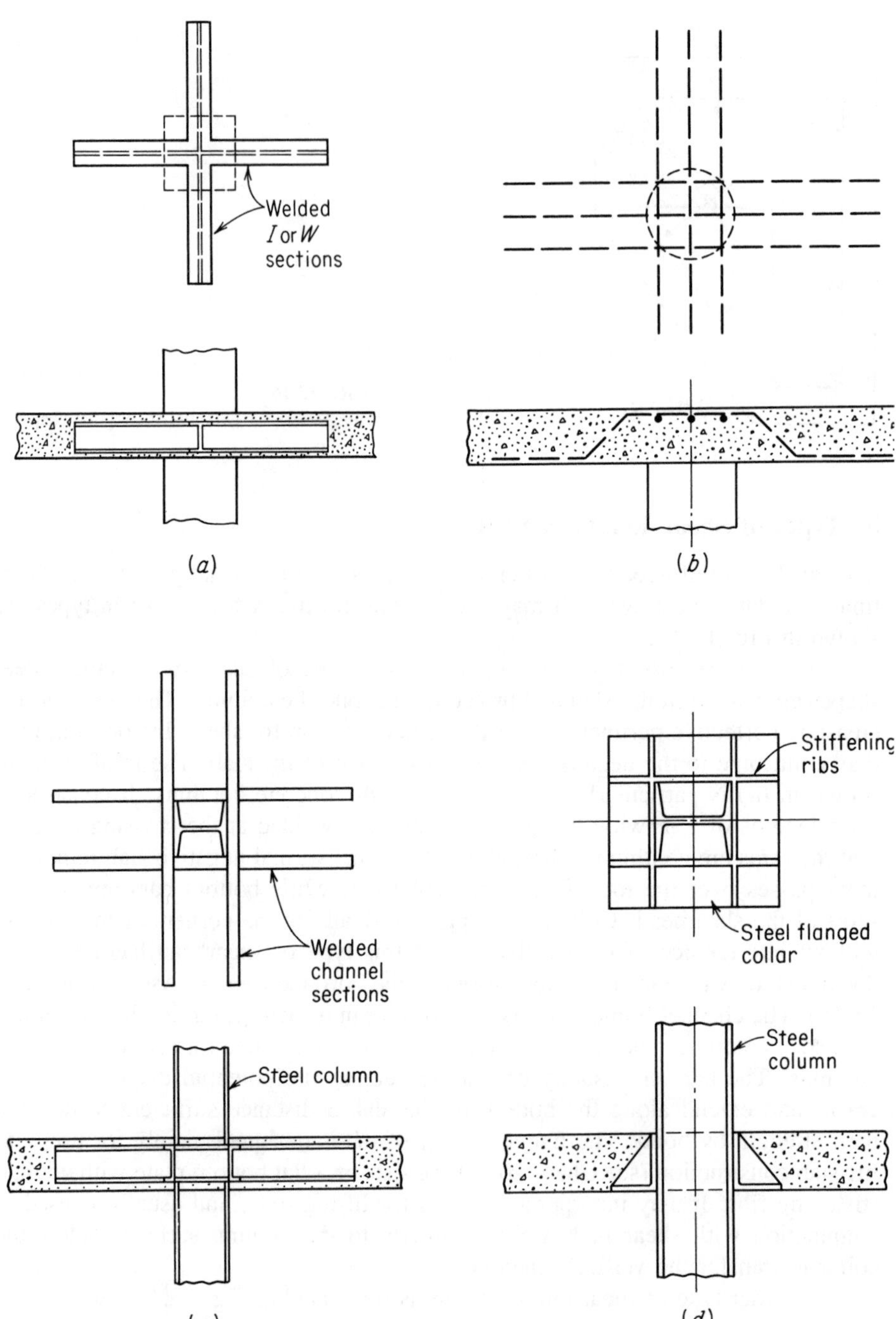

FIGURE 13.17
Shear reinforcement for flat plates.

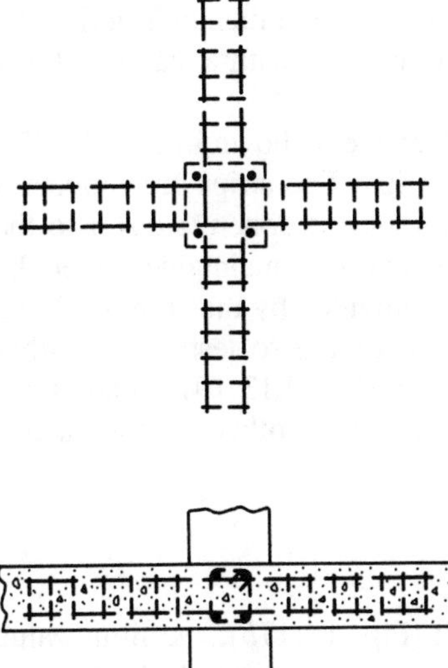

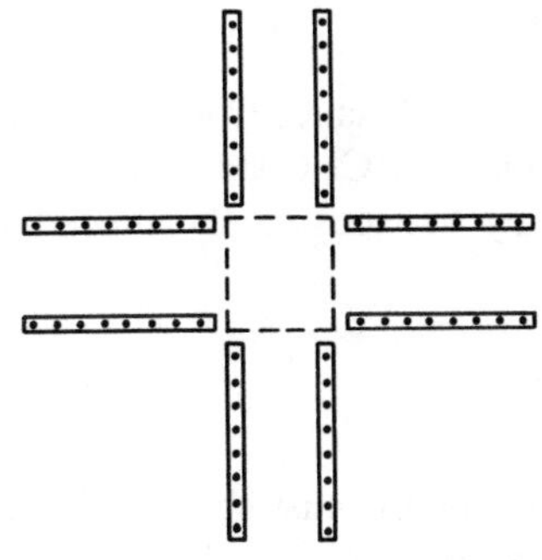

(e)

FIGURE 13.17
(*continued*)

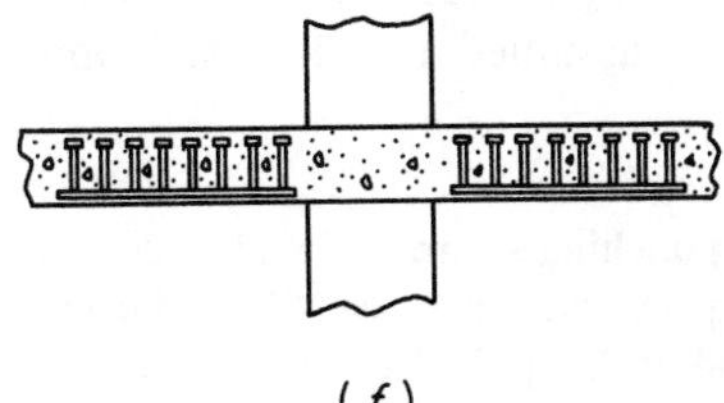

(f)

FIGURE 13.17
(*continued*)

In all cases, closed hoop stirrups should be used, with a large diameter horizontal bar at each bend point, and the stirrups must be terminated with a standard hook (Ref. 13.17).

A recent development is the shear stud reinforcement shown in Fig. 13.17*f*. This consists of large-head studs welded to steel strips. The strips are supported on wire chairs during construction to maintain the required concrete cover to the bottom of the slab below the strip, and the usual cover is maintained over the top of the head. Because of the positive anchorage provided by the stud head and the steel strip, these devices are more effective, according to tests, than either the bent bar or integral beam reinforcement (Refs. 13.18 and 13.19). In addition, they can be placed more easily, with less interference with other reinforcement, than other types of shear steel.

c. Design of Bent Bar Reinforcement

If shear reinforcement in the form of bars is used (Fig. 13.17*b*), the limit value of nominal shear strength V_n, calculated at the critical section $d/2$ from the support face, may be increased to $6\sqrt{f'_c}b_0 d$ according to ACI Code 11.12.3. Because of diagonal cracking, the shear resistance of the concrete, V_c, is reduced to $2\sqrt{f'_c} \times b_0 d$, and reinforcement must provide for the excess shear above ϕV_c. The total bar area A_v crossing the critical section at slope angle α is easily obtained by equating the vertical component of the steel force to the excess shear force to be accommodated:

$$\phi A_v f_y \sin\alpha = V_u - \phi V_c$$

Where inclined shear reinforcement is all bent at the same distance from a support, $V_s = A_v f_y \sin\alpha$ is not to exceed $3\sqrt{f'_c}b_0 d$, according to ACI Code 11.5.6. The required area of reinforcement for shear is found by transposing the preceding equation:

$$A_v = \frac{V_u - \phi V_c}{\phi f_y \sin\alpha} \tag{13.12}$$

Successive sections at increasing distances from the support must be investigated and reinforcement provided where V_u exceeds ϕV_c as given by Eq. (13.11). Only the center three-quarters of the inclined portion of the bent bars can be considered effective in resisting shear, and full development length must be provided past the location of peak stress in the steel, which is assumed to be at slab mid-depth $d/2$.

Example 13.3 Design of bar reinforcement for punching shear. A flat plate floor has thickness $h = 7\frac{1}{2}$ in. and is supported by 18 in. square columns spaced 20 ft on centers each way. The floor will carry a total factored load of 350 psf. Check the adequacy of the slab in resisting punching shear at a typical interior column, and provide shear reinforcement, if needed, using bent bars similar to Fig. 13.17*b*. An average effective depth $d = 6$ in. may be used. Material strengths are $f_y = 60{,}000$ psi and $f'_c = 4000$ psi.

Solution. The first critical section for punching shear is a distance $d/2 = 3$ in. from the column face, providing a shear perimeter $b_0 = 24 \times 4 = 96$ in. Based on the tributary area of loaded floor, the factored shear is

$$V_u = 350(20^2 - 2^2) = 139{,}000 \text{ lb}$$

and, if no shear reinforcement is used, the design strength of the slab, controlled by Eq. (13.11*a*), is

$$\phi V_c = 0.85 \times 4\sqrt{4000} \times 96 \times 6 = 124{,}000 \text{ lb}$$

confirming that shear reinforcement is required. Bars bent at 45° will be used in two directions, as shown in Fig. 13.18. When shear strength is provided by a combination of reinforcement and concrete, the concrete contribution is reduced to

$$\phi V_c = 0.85 \times 2\sqrt{4000} \times 96 \times 6 = 62{,}000 \text{ lb}$$

and so the shear V_s to be resisted by the reinforcement is

$$V_s = \frac{V_u - \phi V_c}{\phi} = \frac{139{,}000 - 62{,}000}{0.85} = 90{,}600 \text{ lb}$$

This is below the maximum permissible value of $3\sqrt{4000} \times 96 \times 6 = 109{,}000$ lb. The required bar area is then found from Eq. (13.12) to be

$$A_v = \frac{90{,}600}{60{,}000 \times 0.707} = 2.14 \text{ in}^2$$

A total of four bars will be used (two in each direction), and with eight legs crossing the critical section, the necessary area per bar is $2.14/8 = 0.27$ in^2. No. 5 bars will be used as shown in Fig. 13.18. The upper limit of $V_n = 6\sqrt{f_c'}b_0 d$ is automatically satisfied in this case, given the more stringent limit on V_s.

With bars bent at 45° and effective through the center three-fourths of the inclined length, the next critical section is approximately $\frac{3}{4}$ times the effective depth, or 4.5 in.,

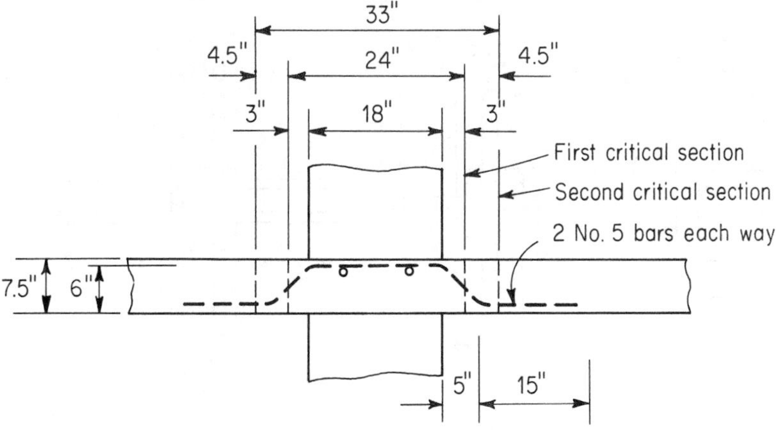

FIGURE 13.18
Bar reinforcement for punching shear in flat plate slab.

past the first, as shown, giving a shear perimeter of $33 \times 4 = 132$ in. The factored shear at that critical section is

$$V_u = 350(20^2 - 2.75^2) = 137{,}000 \text{ lb}$$

and the design capacity of the concrete is

$$\phi V_c = 0.85 \times 4\sqrt{4000} \times 132 \times 6 = 170{,}000 \text{ lb}$$

confirming that no additional bent bars are needed. The No. 5 bars will be extended along the bottom of the slab the full development length of 15 in., as shown in Fig. 13.18.

d. Design of Shearhead Reinforcement

If embedded structural steel shapes are used, as shown in Fig. 13.17*a* and *c*, the limiting value of V_n may be increased to $7\sqrt{f_c'}b_0d$. Such a shearhead, provided it is sufficiently stiff and strong, has the effect of moving the critical section out away from the column, as shown in Fig. 13.19. According to ACI Code 11.12.4, this critical section crosses each arm of the shearhead at a distance equal to three-quarters of the projection beyond the face of the support, and is defined so that the perimeter is a minimum. It need not approach closer than $d/2$ to the face of the support.

Moving the critical section out in this way provides the double benefit of increasing the effective perimeter b_0 and decreasing the total shear force for which the slab must be designed. The nominal shear V_n at the new critical section must not be taken greater than $4\sqrt{f_c'}b_0d$, according to ACI Code 11.12.4.

Tests reported in Ref. 13.16 indicate that throughout most of the length of a shearhead arm the shear is constant, and, further, that the part of the total shear

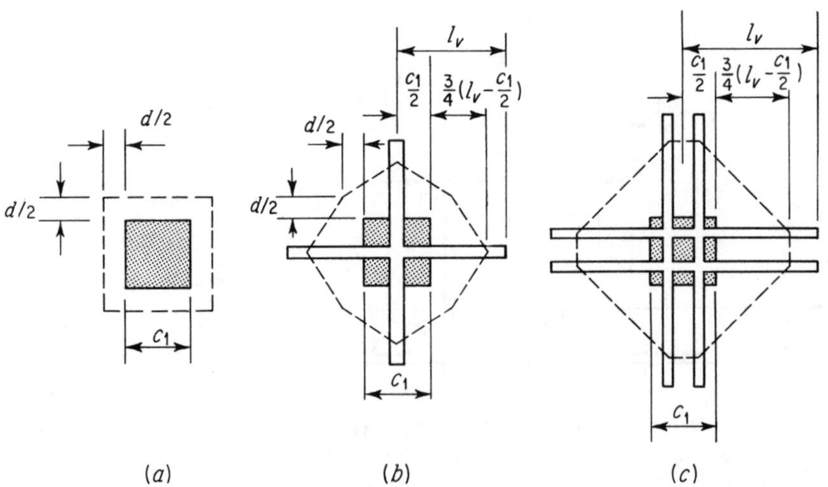

FIGURE 13.19
Critical sections for shear for flat plates: (*a*) no shearhead; (*b*) small shearhead; (*c*) large shearhead.

carried by the shearhead arm is proportional to α_v, its relative flexural stiffness, compared with that of the surrounding concrete section:

$$\alpha_v = \frac{E_s I_s}{E_c I_c} \tag{13.13}$$

The concrete section is taken with an effective width of $c_2 + d$, where c_2 is the width of the support measured perpendicular to the arm direction. Properties are calculated for the cracked, transformed section, including the shearhead. The observation that shear is essentially constant, at least up to the diagonal cracking load, implies that the reaction is concentrated largely at the end of the arm. Thus, if the total shear at the support is V and if the shearhead has η identical arms (generally $\eta = 4$ for shearheads at interior columns), the constant shear force in each arm is equal to $\alpha_v V/\eta$.

If the load is increased past that which causes diagonal cracking immediately around the column, tests indicate that the increased shear above the cracking shear V_c is carried mostly by the steel shearhead, and that the shear force in the projecting arm within a distance from the column face equal to h_v, the depth of the arm, assumes a nearly constant value greater than $\alpha_v V/\eta$. This increased value is very nearly equal to the total shear per arm, $V_u/\phi\eta$, minus the shear carried by the partially cracked concrete. The latter term is equal to $(V_c/\eta)(1 - \alpha_v)$; hence, the idealized shear diagram of Fig. 13.20*b* is obtained.

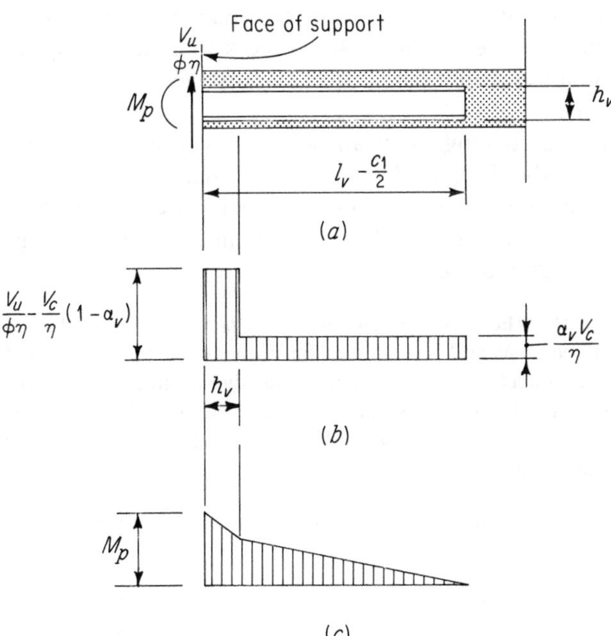

FIGURE 13.20
Stress resultants in shearhead arm: (*a*) shearhead arm; (*b*) shear; (*c*) moment.

The moment diagram of Fig. 13.20*c* is obtained by integration of the shear diagram. If V_c is equal to $V_n/2 = V_u/2\phi$, as tests indicate for shearheads of common proportions, it is easily confirmed that the plastic moment M_p at the face of the support, for which the shearhead arm must be proportioned, is

$$M_p = \frac{V_u}{2\phi\eta}\left[h_v + \alpha_v\left(l_v - \frac{c_1}{2}\right)\right] \tag{13.14}$$

in which the capacity-reduction factor ϕ is taken equal to 0.90 for bending.

According to ACI Code 11.12.4, the value of α_v must be at least equal to 0.15; more flexible shearheads have proved ineffective. The compression flange must not be more than $0.3d$ from the bottom surface of the slab, and the steel shapes used must not be deeper than 70 times the web thickness.

For flexural design of the slab, moments found at the support centerline by the equivalent frame method are reduced to moments at the support face, assumed to be the critical section for moment. By the direct method, support-face moments are calculated directly through use of clear-span distance. If shearheads are used, they have the effect of reducing the design moment in the column strips still further by increasing the effective support width. This reduction is proportional to the share of the load carried by the shearhead and to its size, and can be estimated conservatively (see Fig. 13.20*b* and *c*) by the expression

$$M_v = \frac{\phi\alpha_v V_u}{2\eta}\left(l_v - \frac{c_1}{2}\right) \tag{13.15}$$

where $\phi = 0.90$. According to ACI Code 11.12.4, the reduction may not be greater than 30 percent of the total design moment for the slab column strip, or greater than the change in column-strip moment over the distance l_v, or greater than M_p given by Eq. (13.14).

Limited test information pertaining to shearheads at a slab edge indicates that behavior may be substantially different due to torsional and other effects. If shearheads are to be used at an edge or corner column, special attention must be given to anchorage of the embedded steel within the column. The use of edge beams or a cantilevered slab edge may be preferred.

Example 13.4. Design of shearhead reinforcement. A flat plate slab $7\frac{1}{2}$ in. thick is supported by 10 in. square columns and is reinforced for negative bending with No. 5 bars 5 in. on centers in each direction, with an average effective depth d of 6 in. The concrete strength f_c' is 3000 psi. The slab must transfer an ultimate shear V_u of 113,000 lb to the column. What special slab reinforcement is required, if any, at the column to transfer the required ultimate shear?

Solution. The nominal shear strength at the critical section $d/2$ from the face of the column is found from Eq. (13.11*a*) to be

$$V_c = 4\sqrt{3000} \times 64 \times 6 = 84.1 \text{ kips}$$

and $\phi V_c = 0.85 \times 84.1 = 71.5$ kips. This is less than $V_u = 113$ kips, indicating that shear reinforcement is necessary. A shearhead similar to Fig. 13.17*a* will be used, fabricated from I-beam sections with $f_y = 36$ ksi. Maintaining $\frac{3}{4}$ in. clearance below such steel, bar clearance at the top of the slab permits use of an I beam of $4\frac{5}{8}$ in. depth;

a nominal 4 in. section will be used. With such reinforcement, the upper limit of shear V_n on the critical section is $7\sqrt{3000}(64 \times 6) = 147$ kips, and $\phi V_n = 0.85 \times 147 = 125$ kips, well above the value of V_u to be resisted. The required perimeter b_0 can be found by setting $V_u = \phi V_c$, where V_c is given by Eq. (13.11*a*):

$$b_0 = \frac{V_u}{4\phi\sqrt{f'_c}d} = \frac{113{,}000}{4 \times 0.85\sqrt{3000} \times 6} = 101 \text{ in.}$$

(Note that the actual shear force to be transferred at the critical section is slightly less than 113 kips because a part of the floor load is within the effective perimeter b_0; however, the difference is small except for very large shearheads.) The required projecting length l_v of the shearhead arm is found from geometry, with b_0 expressed in terms of l_v:

$$b_0 = 4\sqrt{2}\left[\frac{c_1}{2} + \frac{3}{4}\left(l_v - \frac{c_1}{2}\right)\right] = 101 \text{ in.}$$

from which $l_v = 22.2$ in. To determine the required plastic section modulus for the shear arm it is necessary to assume a trial value of the relative stiffness α_v. After selecting 0.25 for trial, the required moment capacity is found from Eq. (13.14):

$$M_p = \frac{113{,}000}{8 \times 0.90}[4 + 0.25(22.2 - 5)] = 130{,}000 \text{ in-lb}$$

A standard I beam S4 × 7.7, with yield stress of 36 ksi, provides 126,000 in-lb resistance and will tentatively be adopted. The $E_s I_s$ value provided by the beam is 174×10^6 in^2-lb. The effective cross section of the slab strip is shown in Fig. 13.21. Taking moments of the composite cracked section about the bottom surface to locate the neutral axis gives

$$y = \frac{8.90 \times 6 + 19.9 \times 2.75 + 8y^2}{8.90 + 19.9 + 16y}$$

from which $y = 2.29$ in. The moment of inertia of the composite section is

$$I_c = \tfrac{1}{3} \times 16 \times 2.29^3 + 8.90 \times 3.71^2 + 6 \times 9 + 19.9 \times 0.46^2 = 244 \text{ in}^4$$

the flexural stiffness of the effective composite slab strip is

$$E_c I_c = 3.1 \times 10^6 \times 244 = 756 \times 10^6 \text{ in}^2\text{-lb}$$

and, from Eq. (9.13),

$$\alpha_v = \frac{174}{756} = 0.23$$

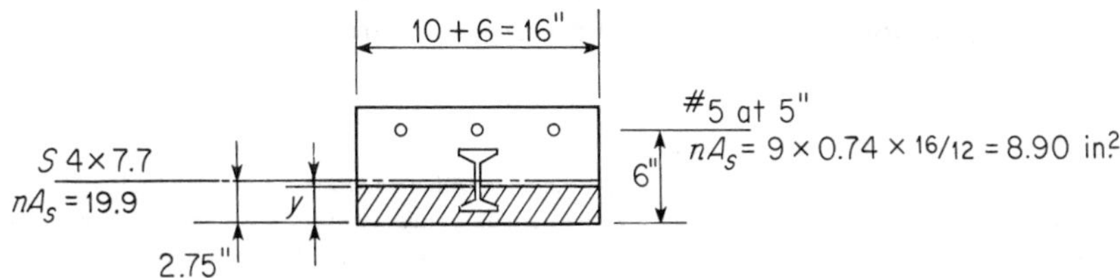

FIGURE 13.21
Effective section of slab.

This is greater than the specified minimum of 0.15 and close to the 0.25 value assumed earlier. The revised value of M_p is

$$M_p = \frac{113{,}000}{8 \times 0.90}[4 + 0.23(22.2 - 5)] = 122{,}000 \text{ in-lb}$$

The 4 in. beam is adequate. The calculated length l_v of 22.2 in. will be increased to 24 in. for practical reasons. The reduction in column-strip moment in the slab may be based on this actual length. From Eq.(13.15),

$$M_v = \frac{0.90 \times 0.23 \times 113{,}000}{8}(24 - 5) = 55{,}600 \text{ in-lb}$$

This value is less than M_p, as required by specification, and must also be less than 30 percent of the design negative moment in the column strip and less than the change in the column-strip moment in the distance l_v.

e. Design of Integral Beams with Vertical Stirrups

Steel shearheads of the type described in Sec. 13.6d have not been widely used, primarily because of their cost, but also because of difficulty in placing the slab flexural reinforcement to pass the structural steel sections and because of interference with the column steel. The bent bar shear reinforcement cages of Sec. 13.6c are less expensive, but also lead to troublesome congestion of reinforcement in the column-slab joint region. Shear reinforcement using vertical stirrups in *integral beams* as shown in Fig. 13.17*e* avoids much of this difficulty.

The first critical section for shear design in the slab is taken at $d/2$ from the column face, as usual, and the stirrups, if needed, are extended outward from the column in four directions for the typical interior case (three or two directions for exterior or corner columns, respectively), until the concrete alone can carry the shear, with $V_c = 4\sqrt{f'_c}b_0 d$ at the second critical section.† Within the region adjacent to the column, where shear resistance is provided by a combination of concrete and steel, the nominal shear strength V_n must not exceed $6\sqrt{f'_c}b_0 d$, according to ACI Code 11.12.3. In this region, the concrete contribution is reduced to $V_c = 2\sqrt{f'_c}b_0 d$. The second critical section crosses each integral beam at a distance $d/2$ measured outward from the last stirrup and is located so that its perimeter b_0 is a minimum (i.e., for the typical case, defined by 45 degree lines between the integral beams). The required spacing of the vertical stirrups is found by Eq. (4.14*a*) of Chap. 4.

The problem of anchorage of the shear reinforcement in shallow flat plates is critical, and closed hoop stirrups, terminating in standard hooks, always should be provided with interior corner bars to improve pullout resistance.

---

† Neither the ACI Code nor the ACI Commentary makes clear whether Eqs. (13.11*b*) and (13.11*c*) are to be applied at successive critical sections past the first, immediately adjacent to the column. The research on which these equations was based considered only the first critical section at the column. Except in extreme cases, the aspect ratio of the column, in Eq. (13.11*b*), seems less relevant with increasing distance from the column; however, the b_0/d ratio, in Eq. (13.11*c*), may be influential, and that equation might conservatively be applied.

Example 13.5 Design of an Integral Beam with Vertical Stirrups. The flat plate slab of 7.5 in. total thickness and 6 in. effective depth shown in Fig. 13.22 is carried by 12 in. square columns 15 ft on centers in each direction. A factored load of 135 kips must be transmitted from the slab to a typical interior column. Concrete and steel strengths used are respectively $f_c' = 4000$ psi and $f_y = 60,000$ psi. Determine if shear reinforcement is required for the slab, and if so, design integral beams with stirrups to carry the excess shear.

Solution. The design shear strength of the concrete alone at the critical section $d/2$ from the face of the column, by the controlling Eq. (13.11*a*), is

$$\phi V_c = 0.85 \times 4\sqrt{4000} \times 72 \times 6 = 92.9 \text{ kips}$$

This is less than $V_u = 135$ kips, indicating that shear reinforcement is required. In this case, the maximum design strength allowed by the ACI Code is

$$\phi V_n = 0.85 \times 6\sqrt{4000} \times 72 \times 6 = 139.4 \text{ kips}$$

satisfactorily above the actual V_u. When shear is resisted by combined action of concrete and bar reinforcement, the concrete contribution is reduced to

$$\phi V_c = 0.85 \times 2\sqrt{4000} \times 72 \times 6 = 46.4 \text{ kips}$$

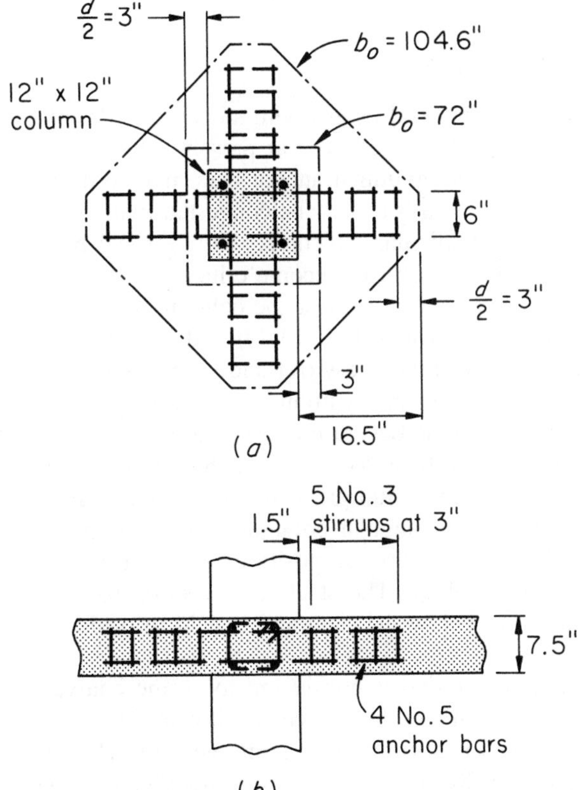

FIGURE 13.22
Vertical stirrup shear reinforcement for slab of Example 13.5.

For trial, No. 3 vertical closed hoop stirrups will be selected and arranged along four integral beams as shown in Fig. 13.22. Thus, the A_v provided is $4 \times 2 \times 0.11 = 0.88$ in^2 at the first critical section, a distance $d/2$ from the column face, and the required spacing can be found from Eq. (4.14*a*):

$$s = \frac{\phi A_v f_y d}{V_u - \phi V_c} = \frac{0.85 \times 0.88 \times 60 \times 6}{135 - 46.4} = 3.04 \text{ in.}$$

However, the maximum spacing of $d/2 = 3$ in. controls here, and No. 3 stirrups at a constant spacing of 3 in. will be used. In other cases, stirrup spacing might be increased with distance from the column, as excess shear is less, although this would complicate rebar placement and generally save little steel.

The required perimeter of the second critical section, at which the concrete alone can carry the shear, is found from the controlling Eq. (13.11*a*) as follows:

$$\phi V_c = 0.85 \times 4\sqrt{4000} \times b_0 \times 6 = 135{,}000 \text{ lb}$$

from which the minimum perimeter $b_0 = 104.6$ in. It is easily confirmed that this requires a minimum projection of the critical section past the face of the column of 14.25 in. Five stirrups at a constant 3 in. spacing will be sufficient, the first placed at $s/2 = 1.5$ in. from the column face, as indicated in Fig. 13.22. This provides a perimeter b_0 at the second critical section of $(16.5\sqrt{2} + 6) \times 4 = 117$ in., exceeding the requirement.

Four longitudinal No. 5 bars will be provided inside the corners of each closed hoop stirrup, as shown, to provide for proper anchorage of the shear reinforcement.

f. Design of Shear Stud Reinforcement

Slab shear reinforcement consisting of integral beams with stirrups, as described in Sec. 13.6e, is probably the most widely used type at present. However, the cage that is formed by the stirrups and longitudinal anchor bars may be difficult to install. Also, the slab-column joint region is somewhat congested, with top and bottom slab steel running in two perpendicular directions, with vertical bars in the column, and with the stirrups. Congestion can become critical when the slab has openings, which are frequently required, at or near the column faces.

Shear stud reinforcing strips, as shown in Fig. 13.17*f* and in Fig. 13.23*a* and *b*, have been widely used in West Germany, Switzerland, and more recently in Canada (Refs. 13.18 and 13.19). They are mentioned briefly in ACI Code Commentary R11.12.3, although no specific design provisions are included.

These devices are composed of vertical bars with anchor heads at their top, welded to a steel strip at the bottom. Multiple strips are arranged in two perpendicular directions for square and rectangular columns or usually in radial directions for circular columns. They are secured in position in the forms before the top and bottom flexural steel is in place. The steel strip rests on bar chairs to maintain the needed concrete cover below the steel and is held in position by nails through holes in the strip.

For design purposes, an individual stud is considered to be the equivalent of one vertical leg of one stirrup. Design can then proceed following the general procedures illustrated in Sec. 13.6e for stirrup shear reinforcement. However, based on extensive testing (Refs. 13.20 and 13.21), some modifications have been proposed. Ghali (Refs. 13.18 and 13.19) recommends the following:

FIGURE 13.23
Shear stud reinforcements for concrete slabs: (*a*) shear stud assembly; (*Courtesy of Amin Ghali and Walter H. Dilger.*)

FIGURE 13.23
Shear stud reinforcements for concrete slabs: (*b*) shear reinforcement installed in forms for prestressed concrete slab. (*Courtesy of Amin Ghali and Walter H. Dilger.*)

1. The upper limit for the nominal shear stress at $d/2$ from the column face is increased to $8\sqrt{f_c'}b_0 d$.
2. The allowable stud spacing is increased to between $2d/3$ and $3d/4$, depending on the maximum nominal shear stress at factored loads.
3. Within the shear-reinforced zone, the contribution of the concrete is increased to $3\sqrt{f_c'}b_0 d$.

In addition to the above, Ghali has recommended the following details:

(a) Top anchors are in the form of circular or square plates, the areas of which are at least 10 times the area of the stem.

(b) When the top anchor plates and the bottom strips are of uniform thickness, the thickness should be greater than or equal to one-half the stud diameter.

(c) If the top anchor plate is tapered, the thickness at the connection with the stem should be greater than or equal to $\frac{2}{3}$ the stud diameter.

(d) The width of the bottom strip should be greater than or equal to 2.5 times the stud diameter.

(e) Bottom anchor strips should be aligned with the column faces of square or rectangular columns.

(f) In the direction parallel to the column face, the distance between anchor strips should not exceed twice the effective depth of the slab.

(g) The minimum concrete cover above and below the stud strips should be as normally specified for slab bars, and the cover should not exceed the minimum plus $\frac{1}{2}$ the bar diameter of the flexural reinforcement.

13.7 TRANSFER OF MOMENTS AT COLUMNS

The analysis for punching shear in flat plates and flat slabs presented in Sec. 13.6 assumed that the shear force V_u was resisted by shearing stresses uniformly distributed around the perimeter b_0 of the critical section, a distance $d/2$ from the face of the supporting column. The nominal shear strength V_c was given by Eqs. (13.11*a, b,* or *c*).

If significant moments are to be transferred from the slab to the columns, as would result from unbalanced gravity loads on either side of a column or from horizontal loading due to wind or seismic effects, the shear stress on the critical section is no longer uniformly distributed.

The situation can be modeled as shown in Fig. 13.24*a*. Here V_u represents the total vertical reaction to be transferred to the column, and M_u represents the unbalanced moment to be transferred, both at factored loads. The vertical force V_u causes shear stress distributed more or less uniformly around the perimeter of the critical section as assumed earlier, represented by the inner pair of vertical arrows, acting downward. The unbalanced moment M_u causes additional loading on the joint, represented by the outer pair of vertical arrows, which add to the

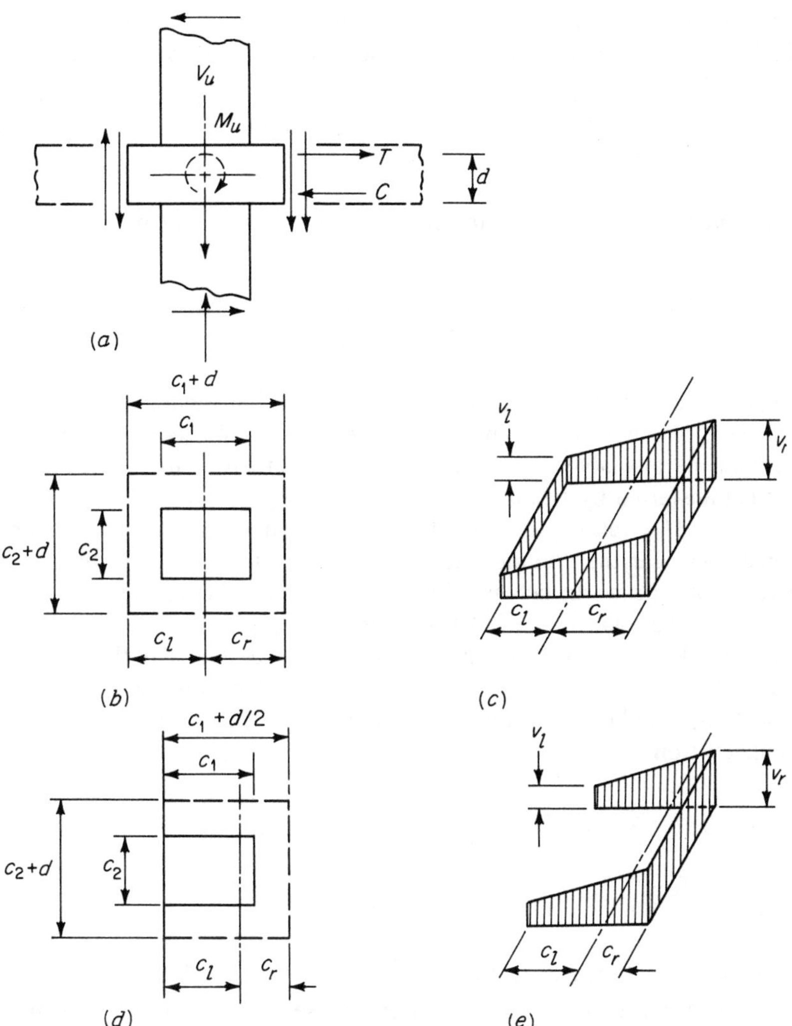

FIGURE 13.24
Transfer of moment from slab to column: (*a*) forces resulting from vertical load and unbalanced moment; (*b*) critical section for an interior column; (*c*) shear stress distribution for an interior column; (*d*) critical section for an edge column; (*e*) shear stress distribution for an edge column.

shear stresses otherwise present on the right side, in the sketch, and subtract on the left side.

Tests indicate that for square columns about 60 percent of the unbalanced moment is transferred by flexure (forces T and C in Fig. 13.24*a*) and about 40 percent by shear stresses on the faces of the critical section (Ref. 13.22). For rectangular columns, it is reasonable to suppose that the portion transferred by flexure increases as the width of the critical section that resists the moment increases, i.e., as $c_2 + d$ becomes larger relative to $c_1 + d$ in Fig. 13.24*b*. According to ACI Code 13.3.3, the moment considered to be transferred by flexure is

$$M_{ub} = \frac{1}{1 + \frac{2}{3}\sqrt{(c_1 + d)/(c_2 + d)}} M_u \tag{13.16a}$$

while that assumed to be transferred by shear, by ACI Code 11.12.6, is

$$M_{uv} = \left[1 - \frac{1}{1 + \frac{2}{3}\sqrt{(c_1 + d)/(c_2 + d)}}\right] M_u \tag{13.16b}$$

It is seen that for a square column these equations indicate that 60 percent of the unbalanced moment is transferred by flexure and 40 percent by shear, in accordance with the available data. If c_2 is very large relative to c_1, nearly all the moment is transferred by flexure.

The moment M_{ub} can be accommodated by concentrating the slab column-strip reinforcement near the column. According to ACI Code 13.3.3, this steel must be placed within a width between lines $1.5h$ on each side of the column or capital, where h is the total thickness of the slab or drop panel.

The moment M_{uv}, together with the vertical reaction delivered to the column, causes shear stresses assumed to vary linearly with distance from the centroid of the critical section, as indicated for an interior column by Fig. 13.24*c*. The stresses can be calculated from

$$v_l = \frac{V_u}{A_c} - \frac{M_{uv} c_l}{J_c} \tag{13.17a}$$

$$v_r = \frac{V_u}{A_c} + \frac{M_{uv} c_r}{J_c} \tag{13.17b}$$

where A_c = area of critical section $= 2d[(c_1 + d) + (c_2 + d)]$

c_l, c_r = distances from centroid of critical section to left and right face of section respectively

J_c = property of critical section analogous to polar moment of inertia

The quantity J_c is to be calculated from

$$J_c = \frac{2d(c_1 + d)^3}{12} + \frac{2(c_1 + d)d^3}{12} + 2d(c_2 + d)\left(\frac{c_1 + d}{2}\right)^2 \tag{13.18}$$

Note the implication, in the use of the parameter J_c in the form of a polar moment of inertia, that shear stresses indicated on the near and far faces of the critical section in Fig. 13.24*c* have horizontal as well as vertical components.

According to ACI Code 11.12.6, the maximum shear stress calculated by Eq. (13.17) must not exceed ϕv_n. For slabs without shear reinforcement, $\phi v_n = \phi V_c / b_0 d$, where V_c is the smallest value given by Eqs. (13.11*a*),

(13.11*b*), or (13.11*c*). For slabs with shear reinforcement other than shearheads, $\phi v_n = \phi(V_c + V_s)/b_0 d$, where V_c and V_s are as established in Secs. 13.6c, e, or f. Where shearhead reinforcement (see Sec. 13.6d) is used, the sum of the shear stresses due to vertical load on the second critical section, near the end of the shearhead arms, and the shear stresses resulting from moment transfer about the centroid of the first critical section $d/2$ from the support faces must not exceed $4\phi\sqrt{f_c'}$. In support of the last calculation, ACI Code Commentary R11.12.6.3 notes that tests indicate the first critical section is appropriate for calculation of stresses caused by transfer of moments even when shearheads are used. Even though the critical sections for direct shear transfer and shear due to moment transfer differ, they coincide or are in close proximity at the column corners where failures initiate, and it is conservative to take the maximum shear as the sum of the two components.

Equations similar to those above can be derived for the edge columns shown in Fig. 13.24*d* and *e* or for a corner column. Note that although the centroidal distances c_l and c_r are equal for the interior column, this is not true for the edge column of Fig. 13.24*d* or for a corner column.

The application of moment to a column from a slab or beam introduces shear to the column also, as is clear from Fig. 13.24*a*. This shear must be considered in the design of lateral column reinforcement.

As was pointed out in Sec. 13.6, most flat plate structures, if they are overloaded, fail in the region close to the column, where large shear and bending forces must be transferred. There has been much important new research aimed at developing improved design details for this region. The design engineer should consult Refs. 13.22 through 13.24 for additional specific information.

13.8 OPENINGS IN SLABS

Almost invariably, slab systems must include openings. These may be of substantial size, as required by stairways and elevator shafts, or they may be of smaller dimensions, such as those needed to accommodate heating, plumbing, and ventilating risers; floor and roof drains; and access hatches.

Relatively small openings usually are not detrimental in *beam-supported slabs*. As a general rule, the equivalent of the interrupted reinforcement should be added at the sides of the opening. Additional diagonal bars should be included at the corners to control the cracking that will almost inevitably occur there. The importance of small openings in *slabs supported directly by columns* (flat slabs and flat plates) depends upon the location of the opening with respect to the columns. From a structural point of view, they are best located away from the columns, preferably in the area common to the slab middle strips. Unfortunately, architectural and functional considerations usually cause them to be located close to the columns. In this case, the reduction in effective shear perimeter is the major concern, because such floors are usually shear-critical.

According to ACI Code 11.12.5, if the opening is close to the column (within 10 slab thicknesses or within the column strips), then that part of b_0 included within the radial lines projecting from the opening to the centroid of the column should be considered ineffective. If shearheads (see Sec. 13.6d) are used under such circumstances, the reduction in width of the critical section is found in the same way, except that only one-half the perimeter included within the radial lines need be deducted.

With regard to flexural requirements, the total amount of steel required by calculation must be provided regardless of openings. Any steel interrupted by holes should be matched with an equivalent amount of supplementary reinforcement on either side, properly lapped to transfer stress by bond. Concrete compression area to provide the required strength must be maintained; usually this would be restrictive only near the columns. According to ACI Code 13.5.2, openings of any size may be located in the area common to intersecting middle strips. In the area common to intersecting column strips, not more than one-eighth of the width of the column strip in either span can be interrupted by openings. In the area common to one middle strip and one column strip, not more than one-quarter of the reinforcement in either strip may be interrupted by the opening.

ACI Code 13.5.1 permits openings of *any* size if it can be shown by analysis that the strength of the slab is at least equal to that required and that all serviceability conditions, i.e., cracking and deflection limits, are met. The *strip method* of analysis and design for openings in slabs, by which specially reinforced integral beams, or *strong bands*, of depth equal to the slab depth are used to frame the openings, will be described in detail in Chapter 15. Very large openings should preferably be framed by beams or slab bands of increased depth to restore, as nearly as possible, the continuity of the slab. The beams must be designed to carry a portion of the floor load, in addition to loads applied directly by partition walls, elevator support beams, or stair slabs.

13.9 DEFLECTION CALCULATIONS

The deflection of a uniformly loaded flat plate, flat slab, or two-way slab supported by beams on column lines can be calculated by an equivalent frame method that corresponds with the method for moment analysis described in Sec. 13.5 (Ref. 13.25). The definition of column and middle strips, the longitudinal and transverse moment-distribution coefficients, and many other details are the same as for the moment analysis. Following the calculation of deflections by this means, they can be compared directly with limiting values like those of Table 6.3, which are applicable to slabs as well as to beams, according to the ACI Code.

A slab region bounded by column centerlines is shown in Fig. 13.25. While no column-line beams, drop panels, or column capitals are shown, the presence of any of these introduces no fundamental complication.

The deflection calculation considers the deformation of such a typical region in one direction at a time, after which the contributions from each direction are added to obtain the total deflection at any point of interest.

In reference to Fig. 13.25*a*, the slab is considered to act as a broad, shallow beam of width equal to the panel dimension l_y and having the span l_x. Initially the slab is considered to rest on unyielding support lines at $x = 0$ and $x = l_x$.

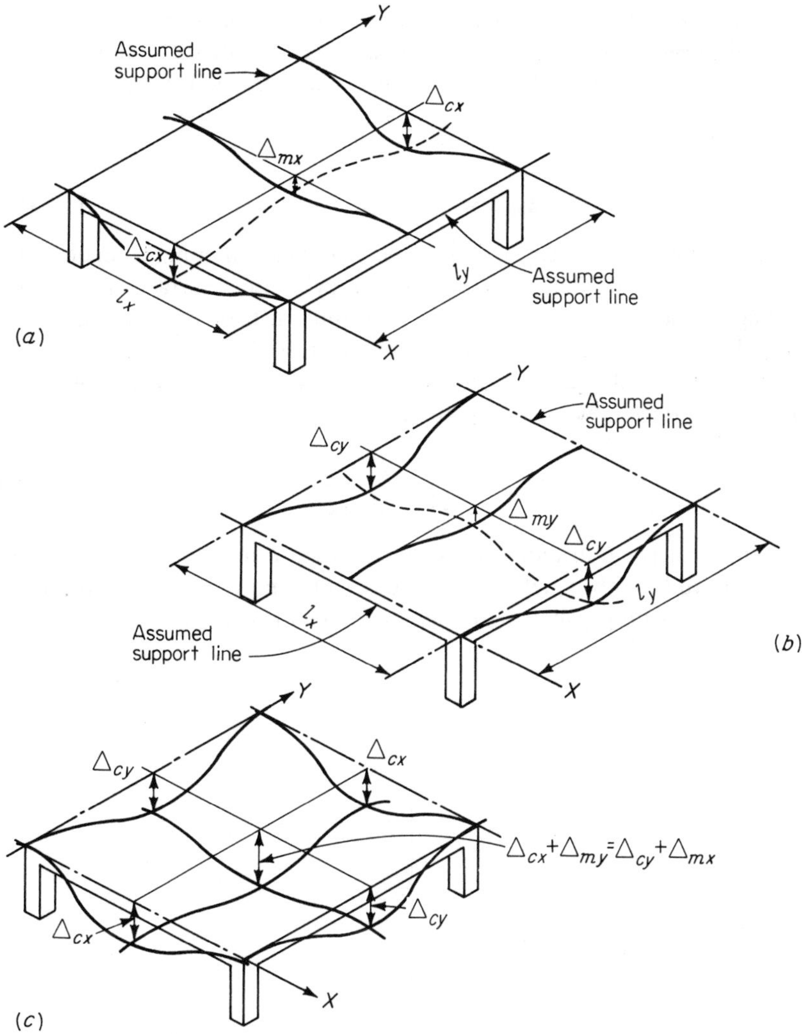

FIGURE 13.25
Basis of equivalent frame method for deflection analysis: (*a*) *X*-direction bending; (*b*) *Y*-direction bending; (*c*) combined bending.

Because of variation of moment as well as flexural rigidity across the width of the slab, all unit strips in the *X* direction will not deform identically. Typically the slab curvature in the middle-strip region will be less than that in the region of the column strips because the middle-strip moments are less. The result is as indicated in Fig. 13.25*a*.

Next the slab is analyzed for bending in the *Y* direction (Fig. 13.25*b*). Once again the effect of transverse variation of bending moment and flexural rigidity is seen.

The actual deformed shape of the panel is represented in Fig. 13.25*c*. The mid-panel deflection is the sum of the midspan deflection of the column strip in one direction and that of the middle strip in the other direction; i.e.,

$$\Delta_{\max} = \Delta_{cx} + \Delta_{my} \tag{13.19a}$$

or

$$\Delta_{\max} = \Delta_{cy} + \Delta_{mx} \tag{13.19b}$$

In calculations of the deformation of the slab panel in either direction, it is convenient first to assume that it deforms into a cylindrical surface, as it would if the bending moment at all sections were uniformly distributed across the panel width and if lateral bending of the panel were suppressed. The supports are considered to be fully fixed against both rotation and vertical displacement at this stage. Thus a *reference deflection* is computed:

$$\Delta_{f,\mathrm{ref}} = \frac{wl^4}{384E_cI_{\mathrm{frame}}} \tag{13.20}$$

where w is the load per foot along the span of length l and I_{frame} is the moment of inertia of the full-width panel (Fig. 13.26*a*) including the contribution of the column-line beam or drop panels and column capitals if present.

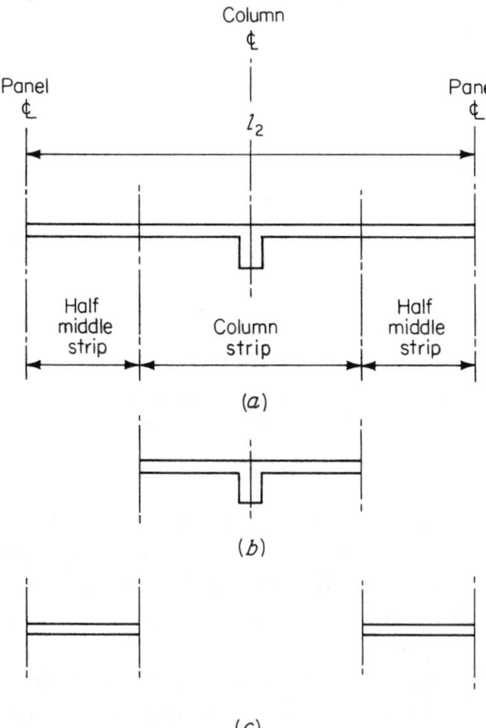

FIGURE 13.26
Effective cross sections for deflection calculations: (*a*) full-width frame; (*b*) column strip; (*c*) middle strips.

The effect of the actual moment variation across the width of the panel and the variation of stiffness due to beams, variable slab depth, etc., are accounted for by multiplying the reference deflection by the ratio of M/EI for the respective strips to that of the full-width frame:

$$\Delta_{f,\text{col}} = \Delta_{f,\text{ref}} \frac{M_{\text{col}}}{M_{\text{frame}}} \frac{E_c I_{\text{frame}}}{E_c I_{\text{col}}} \tag{13.21a}$$

$$\Delta_{f,\text{mid}} = \Delta_{f,\text{ref}} \frac{M_{\text{mid}}}{M_{\text{frame}}} \frac{E_c I_{\text{frame}}}{E_c I_{\text{mid}}} \tag{13.21b}$$

The subscripts relate the deflection Δ, the bending moment M, or the moment of inertia I to the full-width frame, column strip, or middle strip, as shown in Fig. 13.26*a, b,* and *c* respectively.

It may be noted that the moment ratios $M_{\text{col}}/M_{\text{frame}}$ and $M_{\text{mid}}/M_{\text{frame}}$ are identical to the lateral moment-distribution factors already found for the flexural analysis (see Table 13.2). A minor complication results from the fact that the lateral distribution of bending moments, according to the ACI Code, is not the same at the negative- and positive-moment sections. However, it appears consistent with the degree of accuracy usually required, as well as consistent with deflection methods endorsed elsewhere in the ACI Code, to use a simple average of lateral distribution coefficients for the negative and positive portions of each strip.

The presence of drop panels or column capitals in the column strip of a flat slab floor requires consideration of the variation of moment of inertia in the span direction (see Fig. 13.27). It is suggested in Ref. 13.26 that a weighted average moment of inertia be used in such cases:

$$I_{\text{av}} = 2\frac{l_c}{l} I_c + 2\frac{l_d}{l} I_d + \frac{l_s}{l} I_s \tag{13.22}$$

where I_c = moment of inertia of slab including both drop panel and capital
I_d = moment of inertia of slab with drop panel only
I_s = moment of inertia of slab alone

Span distances are defined in Fig. 13.27.

Next it is necessary to correct for the rotations of the equivalent frame at the supports, which until now were considered fully fixed. If the ends of the columns

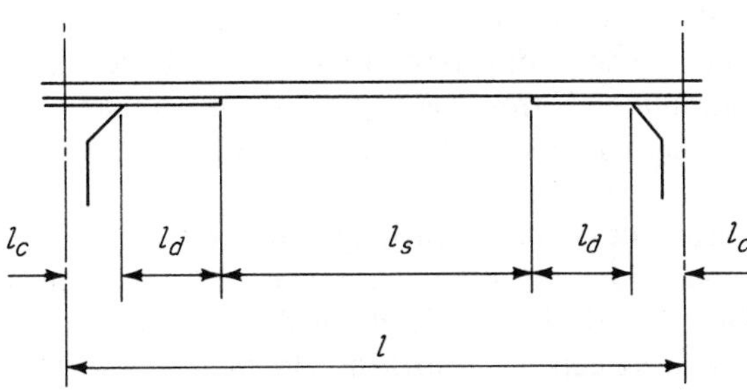

FIGURE 13.27
Flat slab span with variable moment of inertia.

are considered fixed at the floor above and floor below, as usual for frame analysis, the rotation of the column at the floor divided by the stiffness of the equivalent column

$$\theta = \frac{M_{\text{net}}}{K_{ec}} \tag{13.23}$$

where θ = angle change, radians
M_{net} = difference in floor moments to left and right of column
K_{ec} = stiffness of equivalent column (see Sec. 13.5c)

In some cases, the connection between the floor slab and column transmits negligible moment, as for lift slabs; thus $K_{ec} = 0$. The flexural analysis will indicate that the net moment is zero. The support rotation can be found in such cases by applying the moment-area theorems, taking moments of the M/EI area about the far end of the span, and dividing by the span length.

Once the rotation at each end is known, the associated midspan deflection of the equivalent frame can be calculated. It is easily confirmed that the midspan deflection of a member experiencing an end rotation of θ radians, the far end being fixed, is

$$\Delta_\theta = \frac{\theta l}{8} \tag{13.24}$$

Thus the total deflection at midspan of the column strip or middle strip is the sum of the three parts

$$\Delta_{\text{col}} = \Delta_{f,\text{col}} + \Delta_{\theta l} + \Delta_{\theta r} \tag{13.25a}$$

$$\Delta_{\text{mid}} = \Delta_{f,\text{mid}} + \Delta_{\theta l} + \Delta_{\theta r} \tag{13.25b}$$

where the subscripts l and r refer to the left and right ends of the span respectively.

The calculations described are repeated for the equivalent frame in the second direction of the structure, and the total deflection at midpanel is obtained by summing the column-strip deflection in one direction and the middle-strip deflection in the other, as indicated by Eqs. (13.19).

The midpanel deflection should be the same whether calculated by Eq. (13.19*a*) or Eq. (13.19*b*). Actually, a difference will usually be obtained because of the approximate nature of the calculations. For very rectangular panels, the main contribution to midpanel deflection is that of the long-direction column strip. Consequently, the midpanel deflection is best found by summing that of the long-direction column strip and the short-direction middle strip. However, for exterior panels, the important contribution is from the column strips perpendicular to the discontinuous edge, even though the long side of the panel may be parallel to that edge.

In slabs, as in beams, the effect of concrete cracking is to reduce the flexural stiffness. According to ACI code 9.5.3, the effective moment of inertia given by Eq. (6.8) is applicable to slabs as well as beams, although other values may be used if results are in reasonable agreement with tests. In most cases, two-way slabs will be essentially uncracked at service loads, and it is satisfactory to base deflection calculations on the uncracked moment of inertia I_g (see Ref. 13.25 for comparison with tests). In Ref. 13.27, Branson suggests the following refinements: (1) for slabs without beams, use I_g for all dead load deflections; for dead plus live load deflections use I_g for middle strips and I_e for column strips; (2) for slabs with beams use I_g for all dead load deflections; for dead plus live load deflections use I_g for column strips and I_e for middle strips. For continuous spans, I_e can be based on the midspan positive moment without serious error.

The deflections calculated using the procedure described are short-term deflections. Long-term slab deflections can be calculated by multiplying the short-term deflections by the factor λ of Eq. (6.11), as for beams. Because compression steel is seldom used in slabs, a multiplier of 2.0 results. Test evidence and experience with actual structures indicates that this may seriously underestimate long-term slab deflections, and multipliers for long-term deflection from 2.5 to 4.0 have been recommended (Refs. 13.27 to 13.29). A multiplier of 3.0 gives acceptable results in most cases.

It should be recognized that the prediction of slab deflections, both initial elastic and long-term, is complicated by the many uncertainties associated with actual building construction. Loading history, particularly during construction, has a profound effect on final deflections (Ref. 13.30). Construction loads can equal or exceed the service live load. Such loads may include the weight of stacked building material and usually include the weight of slabs above the one cast earlier, applied through shoring and reshoring to the lower slab. Because construction loads are applied to immature slabs, the immediate elastic deflections are large, and, upon removal of the construction loads, elastic recovery is less than the initial elastic deflection because E_c increases with age. Cracking resulting from construction loading does not disappear with removal of the temporary load and may result in live load deflections greater than expected. Creep during construction loading may be greater than expected because of the early age of the concrete when loaded. Shrinkage deflections of thin slabs are often of the same order of magnitude as the elastic deflections, and some cases must be calculated separately.

It is important to recognize that both initial and time-dependent slab deflections are subject to a high degree of variability. Calculated deflections are an estimate, at best, and considerable deviation from calculated values is to be expected in actual structures.

Example 13.6 Calculation of deflections. Find the deflections at the center of the typical exterior panel of the two-way floor designed in Example 13.1, due to dead and live loads. The live load may be considered a short-term load and will be distributed uniformly over all panels. The floor will support nonstructural elements that are likely to be damaged by large deflections. Take $E_c = 3.6 \times 10^6$ psi.

Solution. The elastic deflection due to the self-weight of 88 psf will be found, after which the additional long-term dead load deflection can be found by applying the factor $\lambda = 3.0$, and the short-term live load deflection due to 125 psf by direct proportion.

The effective concrete cross sections, upon which moment-of-inertia calculations will be based, are shown in Fig. 13.28 for the full-width frame, the column strip, and the middle strips, for the short-span and long-span directions. Note that the width of the

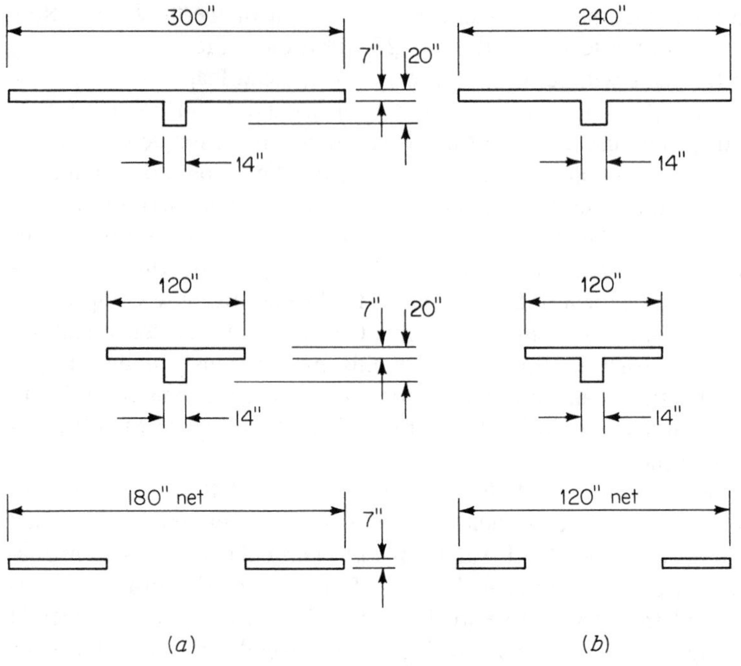

FIGURE 13.28
Cross-sectional dimensions for deflection example: (*a*) short-span direction frame, column strip, and middle strip; (*b*) long-span direction frame, column strip, and middle strip.

column strip in both directions is based on the shorter panel span, according to the ACI Code. The values of moment of inertia are as follows:

	Short direction	Long direction
I_{frame}	27,900 in^4	25,800 in^4
I_{col}	21,000 in^4	21,000 in^4
I_{mid}	5,150 in^4	3,430 in^4

First calculating the deflections of the floor in the *short-span direction* of the panel, from Eq. (13.20) the reference deflection is

$$\Delta_{f,\text{ref}} = \frac{88 \times 25(20 \times 12)^4}{12 \times 384 \times 3.6 \times 10^6 \times 27{,}900} = 0.016 \text{ in.}$$

(Note that the centerline span distance is used here, although clear span was used in the moment analysis to approximate the moment reduction due to support width, according to ACI Code procedures.) From the moment analysis in the short-span direction, it was concluded that 68 percent of the moment at both negative and positive sections was taken by the column strip and 32 percent by the middle strips. Accordingly, from Eqs. (13.21*a,b*),

$$\Delta_{f,\text{col}} = 0.016 \times 0.68 \times \frac{27{,}900}{21{,}000} = 0.014 \text{ in.}$$

$$\Delta_{f,\text{mid}} = 0.016 \times 0.32 \times \frac{27{,}900}{5150} = 0.028 \text{ in.}$$

For the panel under investigation, which is fully continuous over both supports in the short direction, it may be assumed that support rotations are negligible; consequently, $\Delta_{\theta l}$ and $\Delta_{\theta r} = 0$, and from Eqs. (13.25*a,b*),

$$\Delta_{\text{col}} = 0.014 \text{ in.}$$

$$\Delta_{\text{mid}} = 0.028 \text{ in.}$$

Now calculating the deformations in the *long direction* of the panel gives the reference deflection

$$\Delta_{f,\text{ref}} = \frac{88 \times 20(25 \times 12)^4}{12 \times 384 \times 3.6 \times 10^6 \times 25{,}800} = 0.033 \text{ in.}$$

From the moment analysis it was found that the column strip would take 93 percent of the exterior negative moment, 81 percent of the positive moment, and 81 percent of the interior negative moment. Thus the average lateral distribution factor for the column strip is

$$\left(\frac{93 + 81}{2} + 81\right)\frac{1}{2} = 0.84$$

or 84 percent, while the middle strips are assigned 16 percent. Then from Eqs. (13.21*a,b*),

$$\Delta_{f,\text{col}} = 0.033 \times 0.84 \times \frac{25{,}800}{21{,}000} = 0.034 \text{ in.}$$

$$\Delta_{f,\text{mid}} = 0.033 \times 0.16 \times \frac{25{,}800}{3430} = 0.040 \text{ in.}$$

While rotation at the interior column may be considered negligible, that at the exterior column cannot. For the dead load of the slab, the full static moment is

$$M_0 = \tfrac{1}{8} \times 0.088 \times 20 \times 25^2 = 137.5 \text{ ft-kips}$$

It was found that 16 percent of the static moment, or 22.0 ft-kips, should be assigned to the exterior support section. The resulting rotation is found from Eq. (13.23). It is easily confirmed that the stiffness of the equivalent column (see Sec. 13.5*c*) is $169 \times 3.6 \times 10^6$ in-lb/rad; hence

$$\theta = \frac{22{,}000 \times 12}{169 \times 3.6 \times 10^6} = 0.00043 \text{ rad}$$

From Eq. (13.24) the corresponding midpanel deflection component is

$$\Delta_{\theta l} = \frac{0.00043 \times 25 \times 12}{8} = 0.016 \text{ in.}$$

Thus from Eqs. (13.25*a,b*) the deflections of the column and middle strips in the long direction are

$$\Delta_{\text{col}} = 0.034 + 0.016 = 0.050 \text{ in.}$$

$$\Delta_{\text{mid}} = 0.040 + 0.016 = 0.056 \text{ in.}$$

and from Eq. (13.19*a*) the short-term midpanel deflection due to self-weight is

$$\Delta_{max} = 0.050 + 0.028 = 0.078 \text{ in.}$$

The long-term deflection due to dead load is $3.0 \times 0.078 = 0.234$ in., and the short-term live load deflection is $0.078 \times 125/88 = 0.111$ in.

The ACI limiting value for the present case is found to be 1/480 times the span, or $20 \times 12/480 = 0.500$ in., based on the sum of the long-time deflection due to sustained load and the immediate deflection due to live load. The sum of these deflection components in the present case is

$$\Delta_{max} = 0.234 + 0.111 = 0.345 \text{ in.}$$

well below the permissible value.

13.10 ANALYSIS FOR HORIZONTAL LOADS

Either the direct design method or the equivalent frame method, described in the preceding sections of this chapter, may be used for the analysis of two-way slab systems for gravity loads, according to ACI Code 13.3.1. However, the ACI Code provisions are not meant to apply to the analysis of buildings subject to lateral loads, such as loads caused by wind or earthquake. For lateral load analysis, the designer may select any method that is shown to satisfy equilibrium and geometric compatibility, and to give results that are in reasonable agreement with available test data. The results of the lateral load analysis may then be combined with those from the vertical load analysis, according to ACI Code 13.3.1.

Plane frame analysis, with the building assumed to consist of parallel frames each bounded laterally by the panel centerlines on either side of the column lines, has often been used in analyzing unbraced buildings for horizontal loads, as well as vertical. For vertical load analysis by the equivalent frame method, a single floor is usually studied as a substructure with attached columns assumed fully fixed at the floors above and below, but for horizontal frame analysis the equivalent frame includes all floors and columns, extending from the bottom to the top of the structure.

The main difficulty in equivalent frame analysis for horizontal loads lies in modeling the stiffness of the region at the beam-column (or slab-beam-column) connections. Transfer of forces in this region involves bending, torsion, shear, and thrust, and is further complicated by the effects of concrete cracking in reducing stiffness, and reinforcement in increasing it. Frame moments are greatly influenced by horizontal displacements at the floors, and a conservatively low value of stiffness should be used to ensure that a reasonable estimate of drift is included in the analysis.

While a completely satisfactory basis for modeling the beam-column joint stiffness has not been developed, at least two methods have been used in practice (Ref. 13.31). The first is based on an equivalent beam width αl_2, less than the actual width, to reduce the stiffness of the slab for purposes of analysis. Figure 13.29*a* shows a plate fixed at the far edge and supported by a column of width c_2 at the near side. If a rotation θ is imposed at the column, the plate rotation

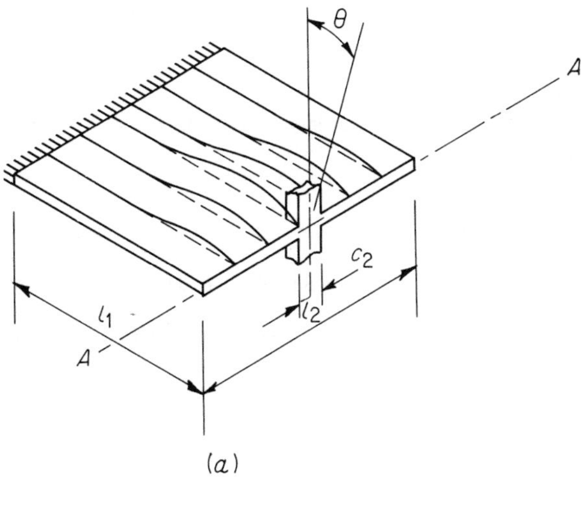

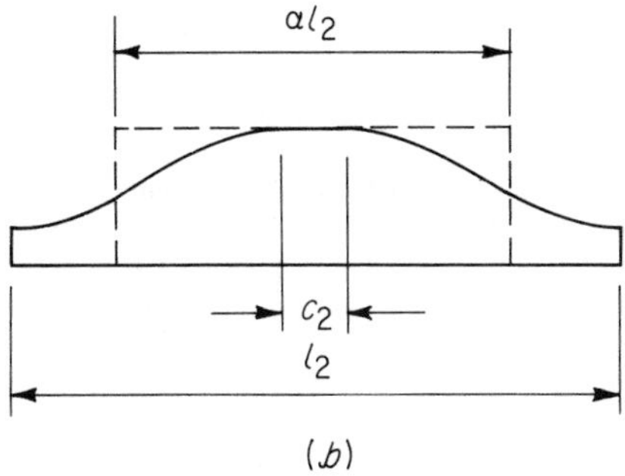

FIGURE 13.29
Equivalent beam width for horizontal load analysis.

along the axis A will vary as shown by Fig. 13.29a, from θ at the column to smaller values away from the column. An equivalent width factor α is obtained from the requirement that the stiffness of a prismatic beam of width αl_2 must equal the stiffness of the plate of width l_2. This equality is obtained if the areas under the two rotation diagrams of Fig. 13.29b are equal. Thus the frame analysis is based on a reduced slab (or slab-beam) stiffness found using αl_2 rather than l_2. Comparative studies indicate that, for flat plate floors, a value for α between 0.25 and 0.50 may be used (Ref. 13.31).

Alternatively, the beam-column stiffness can be modeled based on a transverse torsional member corresponding to that used in deriving the stiffness of the equivalent column for the vertical load analysis of two-way slabs by the equivalent frame method (see Sec. 13.5c). Rotational stiffness of the joint is a function of the flexural stiffness of the columns framing into the joint from above and below and the torsional stiffness of the transverse strip of slab or slab beam at the column. The equivalent column stiffness is found from Eq. (13.9) and the torsional stiffness from Eq. (13.10), as before.

REFERENCES

13.1. J. R. Nichols, "Statical Limitations upon the Steel Requirement in Reinforced Concrete Flat Slab Floors," *Trans. ASCE*, vol. 77, 1914, pp. 1670–1736.

13.2. D. S. Hatcher, M. A. Sozen, and C. P. Siess, "Test of a Reinforced Concrete Flat Plate," *Proc. ASCE*, vol. 91, no. ST5, 1965, pp. 205–231.

13.3. S. A. Guralnick and R. W. LaFraugh, "Laboratory Study of a Forty-Five Foot Square Flat Plate Structure," *J. ACI*, vol. 60, no. 9, 1963, pp. 1107–1185.

13.4. D. S. Hatcher, M. A. Sozen, and C. P. Siess, "Test of a Reinforced Concrete Flat Slab," *Proc. ASCE*, vol. 95, no. ST6, 1969, pp. 1051–1072.

13.5. J. O. Jirsa, M. A. Sozen, and C. P. Siess, "Test of a Flat Slab Reinforced with Welded Wire Fabric," *Proc. ASCE*, vol. 92, no. ST3, 1966, pp. 199–224.

13.6. W. L. Gamble, M. A. Sozen, and C. P. Siess, "Tests of a Two-Way Reinforced Concrete Floor Slab," *Proc. ASCE*, vol. 95, no. ST6, 1969, pp. 1073–1096.

13.7. M. D. Vanderbilt, M. A. Sozen, and C. P. Siess, "Test of a Modified Reinforced Concrete Two-Way Slab," *Proc. ASCE*, vol. 95, no. ST6, 1969, pp. 1097–1116.

13.8. W. L. Gamble, "Moments in Beam-Supported Slabs," *J. ACI*, vol. 69, no. 3, 1972, pp. 149–157.

13.9. J. O. Jirsa, M. A. Sozen, and C. P. Siess, "Pattern Loadings on Reinforced Concrete Floor Slabs," *Proc. ASCE*, vol. 95, no. ST6, 1969, pp. 1117–1137.

13.10. D. Peabody, Jr., "Continuous Frame Analysis of Flat Slabs," *J. Boston Society Civ. Eng.*, January 1948.

13.11. W. G. Corley, M. A. Sozen, and C. P. Siess, "The Equivalent Frame Method for Reinforced Concrete Slabs," Univ. Ill. Dept. Civ. Eng. Struct. Res. Series 218, June 1961.

13.12. W. G. Corley and J. O. Jirsa, "Equivalent Frame Analysis for Slab Design," *J. ACI*, vol. 67, no. 11, 1970, pp. 875–884.

13.13. "The Shear Strength of Reinforced Concrete Members—Slabs," ASCE-ACI Task Committee 426, *J. Struct. Div. ASCE*, vol. 100, no. ST8, 1974, pp. 1543–1591.

13.14. N. M. Hawkins, H. B. Fallsen, and R. C. Hinojosa, "Influence of Column Rectangularity on the Behavior of Flat Plate Structures," in *Cracking, Deflection, and Ultimate Load of Concrete Slab Systems*, ACI Publication SP-30, 1971, p. 127.

13.15. M. D. Vanderbilt, "Shear Strength of Continuous Plates," *J. Struct. Div. ASCE*, vol. 98, no. ST5, 1972, pp. 961–973.

13.16. W. G. Corley and N. M. Hawkins, "Shearhead Reinforcement for Slabs," *J. ACI*, vol. 65, no. 10, 1968, pp. 811–824.

13.17. N. W. Hawkins, D. Mitchell, and S. N. Hanna, "Effects of Shear Reinforcement on the Reversed Cyclic Loading Behavior of Flat Plate Structures," *Can. J. Civ. Eng.*, vol. 2, no. 4, 1975, pp. 572–582.

13.18. A. Ghali, "An Efficient Solution to Punching of Slabs," *Concr. Int.*, vol. 11, no. 6, 1989, pp. 50–54.

13.19. A. A. Elgabry and A. Ghali, "Design of Stud-Shear Reinforcement for Slabs," *ACI Struct. J.*, vol. 87, no. 3, 1990, pp. 350–361.

13.20. A. S. Mokhtar, A. Ghali, and W. Dilger, "Stud Shear Reinforcement for Flat Concrete Plates," *J. ACI*, vol. 82, no. 5, 1985, pp. 676–683.

13.21. A. A. Elgabry and A. Ghali, "Tests on Concrete Slab-Column Connections with Stud Shear Reinforcement Subjected to Shear-Moment Transfer," *ACI Struct. J.*, vol. 84, no. 5, 1987, pp. 433–442.

13.22. N. M. Hawkins, A. Bao, and J. Yamazaki, "Moment Transfer from Concrete Slabs to Columns," *ACI Struct. J.*, vol. 86, no. 6, 1989, pp. 705–716.

13.23. "Recommendations for Design of Slab-Column Connections in Monolithic Reinforced Concrete Structures," reported by ACI-ASCE Committee 352, *ACI Struct. J.*, vol. 85, no. 6, 1988, pp. 675–696.

13.24. J. P. Moehle, M. E. Kreger, and R. Leon, "Background to Recommendations for Design of Reinforced Concrete Slab-Column Connections," *ACI Struct. J.*, vol. 85, no. 6, 1988, pp. 636–644.

13.25. A. H. Nilson and D. B. Walters, "Deflection of Two-Way Floor Systems by the Equivalent Frame Method," *J. ACI*, vol. 72, no. 5, May 1975, pp. 210–218.

13.26. P. F. Rice, E. S. Hoffman, D. P. Gustafson and A. J. Gouwans, *Structural Design Guide to the ACI Building Code*, 3d ed., Van Nostrand Reinhold, New York, 1985.
13.27. D. E. Branson, *Deformation of Concrete Structures*, McGraw-Hill, New York, 1977.
13.28. P. J. Taylor and J. L. Heiman, "Long-Term Deflection of Reinforced Concrete Flat Slabs and Plates," *J. ACI*, vol. 74, no. 11, 1977, pp. 556–561.
13.29. C. J. Graham and A. Scanlon, "Long-Time Multipliers for Estimating Two-Way Slab Deflections," *J. ACI*, vol. 83, no. 6, 1986, pp. 899–908.
13.30. N. J. Gardner and A. Scanlon, "Long-Term Deflections of Two-Way Slabs," *Concr. Int.*, vol. 12, no. 1, 1990, pp. 63–67.
13.31. M. D. Vanderbilt and W. G. Corley, "Frame Analysis of Concrete Buildings," *Concr. Int.*, vol. 5, no. 12, 1983, pp. 33–43.

PROBLEMS

13.1 Redesign the corner floor panel of Example 12.2 of Chapter 12 using the ACI direct design method of Secs. 13.2 through 13.4. Summarize your design with a sketch showing concrete dimensions and the size, spacing, length, and placement of all slab reinforcement. The supporting beams need not be designed as a part of this problem. Compare your results with those of Example 12.2, and comment.

13.2 Design a typical interior panel of the floor system described in Example 12.2 of Chapter 12, using the moment coefficient method of Sec. 12.5. Summarize your design with a sketch showing concrete dimensions and the size, spacing, length, and placement of all slab reinforcement.

13.3 Redesign the typical interior panel of Problem 13.2 using the ACI direct design method of Secs. 13.2 through 13.4. The supporting beams need not be designed as a part of this problem. Compare your results with those of Problem 13.2, and comment.

13.4 Redesign the typical interior panel of Problem 13.2 using the ACI equivalent frame method of Sec. 13.5. Columns will be 16 in. square, and the floor-to-floor height will be 12 ft. Compare your results with those of Problem 13.2 (and Problem 13.3 if completed), and comment.

13.5 Redesign the typical exterior panel of the floor of Example 13.1 as a part of a flat plate structure, with no beams between interior columns but with beams provided along the outside edge to stiffen the slab. No dropped panels or column capitals are permitted, but shear reinforcement similar to Fig. 13.17*b* may be incorporated if necessary. Column size is 20 × 20 in., and the floor-to-floor height is 12 ft. Use either the direct design method or the equivalent frame method. Summarize your design by means of a sketch showing plan and typical cross sections.

13.6 A multistory commercial building is to be designed as a flat plate system with floors of uniform thickness having no beams or dropped panels. Columns are laid out on a uniform 20 ft spacing in each direction and have a 16 in. square section and a vertical dimension 10 ft from floor to floor. Specified service live load is 100 psf including partition allowance. Using the direct design method, design a typical interior panel, determining the required floor thickness, size and spacing of reinforcing bars, and bar details including cutoff points. To simplify construction, the reinforcement in each direction will be the same; use an average effective depth in the calculations. Use all straight bars. For moderate spans such as this, it has been determined that supplementary shear reinforcement would not be economical, although column capitals may be used if needed. Thus, slab thickness may be based on Eqs. (13.11*a*, *b*, and *c*), or column capital dimensions can be selected using those equations if slab thickness is based on the equations of Sec. 13.4. Material strengths are $f_y = 60,000$ psi and $f'_c = 4000$ psi.

13.7 Prepare alternative designs for shear reinforcement at the supports of the slab described in Example 13.4 (*a*) using bent bar reinforcement similar to Fig. 13.17*b*, and (*b*) using integral beams with vertical stirrups similar to Fig. 13.17*e*.

13.8 Prepare an alternative design for shear reinforcement at the supports of the slab described in Example 13.3, using a shearhead similar to Fig. 13.17*a*. As an alternative to shear reinforcement of any kind, calculate the smallest acceptable dimensions for a 45° column capital (see Fig. 12.1*e*) that would permit the concrete slab to resist the entire shear force. Dropped panels are not permitted.

13.9 Figure P13.9 shows a flat plate floor designed to carry a factored load of 350 psf. The total slab thickness $h = 7\frac{1}{2}$ in. and the average effective depth $d = 6$ in. Material strengths are $f_y = 60,000$ psi and $f_c' = 4000$ psi. The design for punching shear at a typical interior column *B*2 provided the basis for Example 13.3. At the exterior column *B*1, in order to provide a full perimeter b_0 the slab is cantilevered past the columns as shown. A total shear force $V_u = 105$ kips must be transmitted to the column, plus a bending moment $M_u = 120$ ft-kips about an axis parallel to the edge of the slab. Check for punching shear at column *B*1 and, if ACI Code restrictions are not met, suggest appropriate modifications in the proposed design. Edge beams are not permitted.

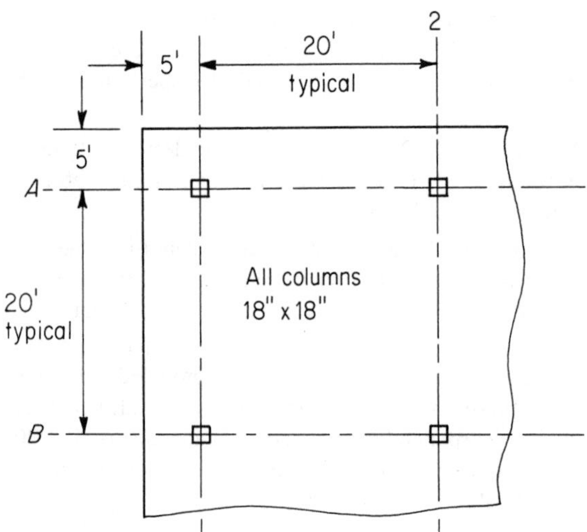

FIGURE P13.9

13.10 For the flat plate floor of Example 13.2, find the following deflection components at the center of panel *C*: (*a*) immediate deflection due to total dead load; (*b*) additional dead load deflection after a long period of time, due to total dead load; (*c*) immediate deflection due to three-quarters full live load. The moment of inertia of the cross concrete sections, I_g, may be used for all calculations. It may be assumed that maximum deflection will be obtained for the same loading pattern that would produce maximum positive moment in the panel. Check predicted deflection against ACI limitations, assuming that nonstructural attached elements would be damaged by excessive deflections.

CHAPTER
14

YIELD LINE ANALYSIS FOR SLABS

14.1 INTRODUCTION

Most concrete slabs are designed for moments found by the methods described in Chapters 12 and 13. These methods are based essentially upon elastic theory. On the other hand, reinforcement for slabs is calculated by strength methods that account for the actual inelastic behavior of members at the factored load stage. A corresponding contradiction exists in the process by which beams and frames are analyzed and designed, as was discussed in Sec. 11.9, and the concept of limit, or plastic, analysis of reinforced concrete was introduced. Limit analysis not only eliminates the inconsistency of combining elastic analysis with inelastic design but also accounts for the reserve strength characteristic of most reinforced concrete structures and permits, within limits, an arbitrary readjustment of moments found by elastic analysis to arrive at design moments that permit more practical reinforcing arrangements.

For slabs, there is still another good reason for interest in limit analysis. The elasticity-based methods of Chapter 12 and 13 are restricted in important ways. Slab panels must be square or rectangular. They must be supported along two opposite sides (one-way slabs), two pairs of opposite sides (two-way edge-supported slabs) or by a fairly regular array of columns (flat plates and related forms). Loads must be uniformly distributed, at least within the bounds of any single panel. There can be no large openings. But in practice, many slabs do not meet these restrictions. Answers are needed, for example, for round or triangular slabs, slabs with large openings, slabs supported on two or three edges only, and slabs carrying concentrated loads. Limit analysis provides a powerful and versatile tool for treating such problems.

It was evident from the discussion of Sec. 11.9 that full plastic analysis of a continuous reinforced concrete beam or frame would be tedious and time consuming because of the need to calculate the rotation requirement at all plastic hinges and to check rotation capacity at each hinge to ensure that it is adequate. Consequently, for beams and frames, the very simplified approach to plastic moment redistribution of ACI Code 8.4 is used. However, for slabs, which typically have tensile steel ratios much below the balanced value and consequently have large rotation capacity, it can be safely assumed that the necessary ductility is present. Practical methods for the plastic analysis of slabs are thus possible and have been developed. *Yield line theory,* presented in this chapter, is one of these. Although the ACI Code contains no specific provisions for limit or plastic analysis of slabs, ACI Code 1.4 permits use of "special systems of design or construction," the adequacy of which has been shown by successful use, analysis, or tests, and ACI Code Commentary 13.3.1 refers specifically to yield line analysis as an acceptable approach.

Yield line analysis for slabs was first proposed by Ingerslev (Ref. 14.1) and was greatly extended by Johansen (Refs. 14.2 and 14.3). Early publications were mainly in Danish, and it was not until Hognestad's English language summary (Ref. 14.4) of Johansen's work that the method received wide attention. Since that time, a number of important publications on the method have appeared (Refs. 14.5 through 14.15). A particularly useful and comprehensive treatment will be found in Ref. 14.15.

The *plastic hinge* was introduced in Sec. 11.9 as a location along a member in a continuous beam or frame at which, upon overloading, there would be large inelastic rotations at essentially a constant resisting moment. For slabs, the corresponding mechanism is the *yield line*. For the overloaded slab, the resisting moment per unit length measured along a yield line is constant as inelastic rotation occurs; the yield line serves as an axis of rotation for the slab segment.

Fig. 14.1*a* shows a simply supported, uniformly loaded reinforced concrete slab. It will be assumed to be underreinforced (as are almost all slabs), with $\rho < \rho_b$. The elastic moment diagram is shown in Fig. 14.1*b*. As the load is increased, when the applied moment becomes equal to the ultimate flexural capacity of the slab cross section, the tensile steel starts to yield along the transverse line of maximum moment.

Upon yielding, the curvature of the slab at the yielding section increases sharply, and deflection increases disproportionately. The elastic curvatures along the slab span are small compared with the curvature resulting from plastic deformation at the yield line, and it is acceptable to consider that the slab segments between the yield line and supports remain rigid, all the curvature occurring at the yield line as shown in Fig. 14.1*c*. The "hinge" that forms at the yield line rotates with essentially constant resistance, according to the relation shown earlier in Fig. 11.11*a*. The resistance per unit width of slab is the nominal flexural strength of the slab; that is, $m_p = m_n$, where m_n is calculated by the usual equations. For design purposes m_p would be taken equal to ϕm_n, with $\phi = 0.90$ as usual for flexure.

For a statically determinate slab like that of Fig. 14.1, the formation of one yield line results in collapse. A "mechanism" forms, i.e., the segments of the slab between the hinge and the supports are able to move without an increase in load.

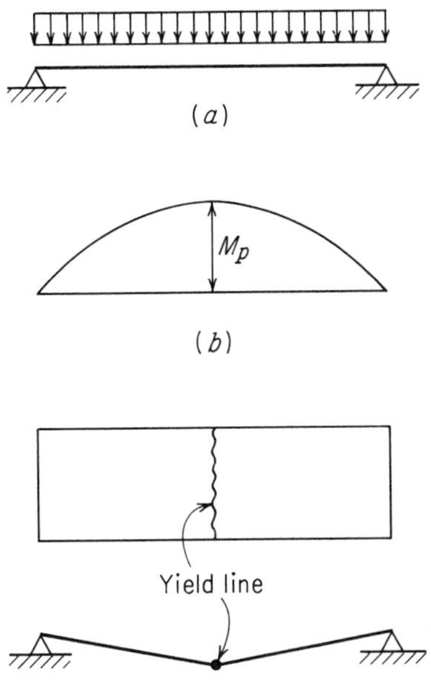

FIGURE 14.1
Simply supported, uniformly loaded one-way slab.

Indeterminate structures, however, can usually sustain their loads without collapse even after the formation of one or more yield lines. When it is loaded uniformly, the fixed-fixed slab of Fig. 14.2*a*, assumed here to be equally reinforced for positive and negative moments, will have an elastic distribution of moments as shown in Fig. 14.2*b*. As the load is gradually increased, the more highly stressed sections at the support start yielding. Rotations occur at the support line hinges, but restraining moments of constant value m_p continue to act. The load can be increased still further, until the moment at midspan becomes equal to the moment capacity there, and a third yield line forms as shown in Fig. 14.2*c*. The slab is now a mechanism, large deflections are permitted, and collapse takes place.

The moment diagram just before failure is shown in Fig. 14.2*d*. Note that the ratio of elastic positive to negative moments of 1:2 no longer holds. Due to inelastic deformation, the ratio of these moments just before collapse is 1:1 for this particular structure. Redistribution of moments was discussed earlier in Sec. 11.9, and it was pointed out that the moment ratios at the collapse stage depend upon the reinforcement provided, not upon the results of elastic analysis.

14.2 UPPER AND LOWER BOUND THEOREMS

Plastic analysis methods such as the yield line theory derive from the general theory of structural plasticity, which states that the ultimate collapse load of a structure lies between two limits, an upper bound and a lower bound of the true collapse load. These limits can be found by well-established methods. A full solution by the theory of plasticity would attempt to make the lower and upper bounds converge to a single correct solution.

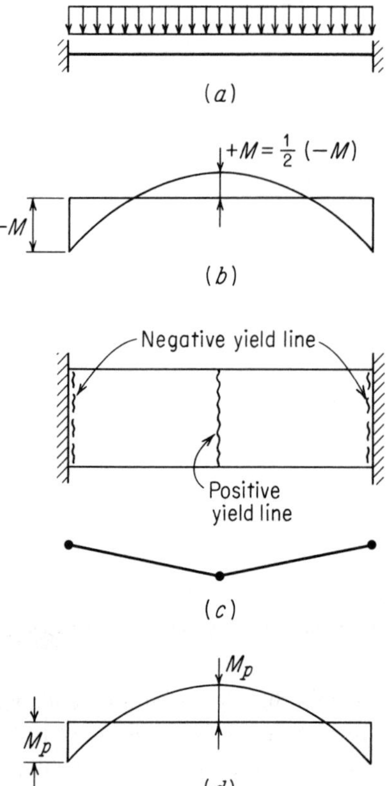

FIGURE 14.2
Fixed-end, uniformly loaded one-way slab.

The lower bound theorem and the upper bound theorem, when applied to slabs, can be stated as follows:

> **Lower bound theorem:** If, for a given external load, it is possible to find a distribution of moments that satisfies equilibrium requirements, with the moment not exceeding the yield moment at any location, and if the boundary conditions are satisfied, then the given load is a lower bound of the true carrying capacity.

> **Upper bound theorem:** If, for a small increment of displacement, the internal work done by the slab, assuming that the moment at every plastic hinge is equal to the yield moment and that boundary conditions are satisfied, is equal to the external work done by the given load for that same small increment of displacement, then that load is an upper bound of the true carrying capacity.

If the lower bound conditions are satisfied, the slab can certainly carry the given load, although a higher load may be carried if internal redistributions of moment occur. If the upper bound conditions are satisfied, a load greater than the given load will certainly cause failure, although a lower load may produce collapse if the selected failure mechanism is incorrect in any sense.

In practice, in the plastic analysis of structures, one works either with the lower bound theorem or the upper bound theorem, not both, and precautions are taken to ensure that the predicted failure load at least closely approaches the correct value.

The yield line method of analysis for slabs is an upper bound method, and consequently the failure load calculated for a slab with known flexural resistances may be higher than the true value. This is certainly a concern, as the designer would naturally prefer to be correct, or at least on the safe side. However, procedures can be incorporated in yield line analysis to help ensure that the calculated capacity is correct. Such procedures will be illustrated by the examples of Sec. 14.4 and 14.5.

14.3 RULES FOR YIELD LINES

The location and orientation of the yield line were evident for the simple slab of Fig. 14.1. Similarly, for the one-way indeterminate slab of Fig. 14.2 the yield lines were easily established. For other cases it is helpful to have a set of guidelines for drawing yield lines and locating axes of rotation. When a slab is on the verge of collapse because of the existence of a sufficient number of real or plastic hinges to form a mechanism, axes of rotation will be located along the lines of support or over point supports such as columns. The slab segments can be considered to rotate as rigid bodies in space about these axes of rotation. The yield line between any two adjacent slab segments is a straight line, being the intersection of two essentially plane surfaces. Because the yield line (as a line of intersection of two planes) contains all points common to these two planes, it must contain the point of intersection (if any) of the two axes of rotation, which is also common to the two planes. That is, the yield line (or yield line extended) must pass through the point of intersection of the axes of rotation of the two adjacent slab segments.

The terms *positive yield line* and *negative yield line* are used to distinguish between those associated with tension at the bottom and tension at the top of the slab, respectively.

Guidelines for establishing axes of rotation and yield lines are summarized as follows:

1. Yield lines are straight lines because they represent the intersection of two planes.
2. Yield lines represent axes of rotation.
3. The supported edges of the slab will also establish axes of rotation. If the edge is fixed, a negative yield line may form providing constant resistance to rotation. If the edge is simply supported, the axis of rotation provides zero restraint.
4. An axis of rotation will pass over any column support. Its orientation depends on other considerations.
5. Yield lines form under concentrated loads, radiating outward from the point of application.
6. A yield line between two slab segments must pass through the point of intersection of the axes of rotation of the adjacent slab segments.

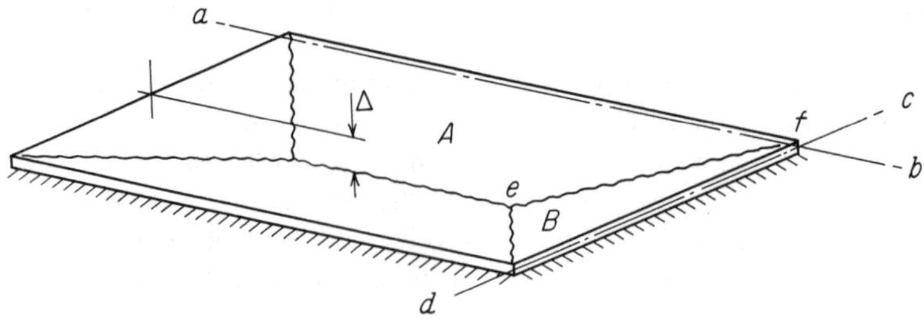

FIGURE 14.3
Two-way slab with simply supported edges.

In Fig. 14.3, which shows a slab simply supported along its four sides, rotation of slab segments A and B is about ab and cd, respectively. The yield line ef between these two segments is a straight line passing through f, the point of intersection of the axes of rotation.

Illustrations are given in Fig. 14.4 of the application of the guidelines to the establishment of yield line locations and failure mechanisms for a number of slabs with various support condition. Shown in (*a*) is a slab continuous over parallel supports. Axes of rotation are situated along the supports (negative yield lines) and near midspan, parallel to the supports (positive yield line). The particular location of the positive yield line in this case and the other cases of Fig. 14.4 depends upon the distribution of loading and the reinforcement of the slab. Methods for determining its location will be discussed later.

For the continuous slab on nonparallel supports, shown in (*b*), the midspan yield line (extended) must pass through the intersection of the axes of rotation over the supports. In (*c*) there are axes of rotation over all four simple supports. Positive yield lines form along the lines of intersection of the rotating segments of the slab. A rectangular two-way slab on simple supports is shown in (*d*). The diagonal yield lines must pass through the corners, while the central yield line is parallel to the two long sides (axes of rotation along opposite supports intersect at infinity in this case).

With this background, the reader should have no difficulty in applying the guidelines to the slabs of Fig. 14.4*e* to *g* to confirm the general pattern of yield lines shown. Many other examples will be found in Refs. 14.1 to 14.15.

Once the general pattern of yielding and rotation has been established by applying the guidelines just stated, the specific location and orientation of the axes of rotation and the failure load for the slab can be established by either of two methods. The first will be referred to as the *method of segment equilibrium* and will be presented in Sec. 14.4. It requires consideration of the equilibrium of the individual slab segments forming the collapse mechanism and leads to a set of simultaneous equations permitting solution for the unknown geometric parameters and for the relation between load capacity and resisting moments. The second, the *method of virtual work,* will be described in Sec. 14.5. This method is based on equating the internal work done at the plastic hinges with the external work done by the loads as the predefined failure mechanism is given a small virtual displacement.

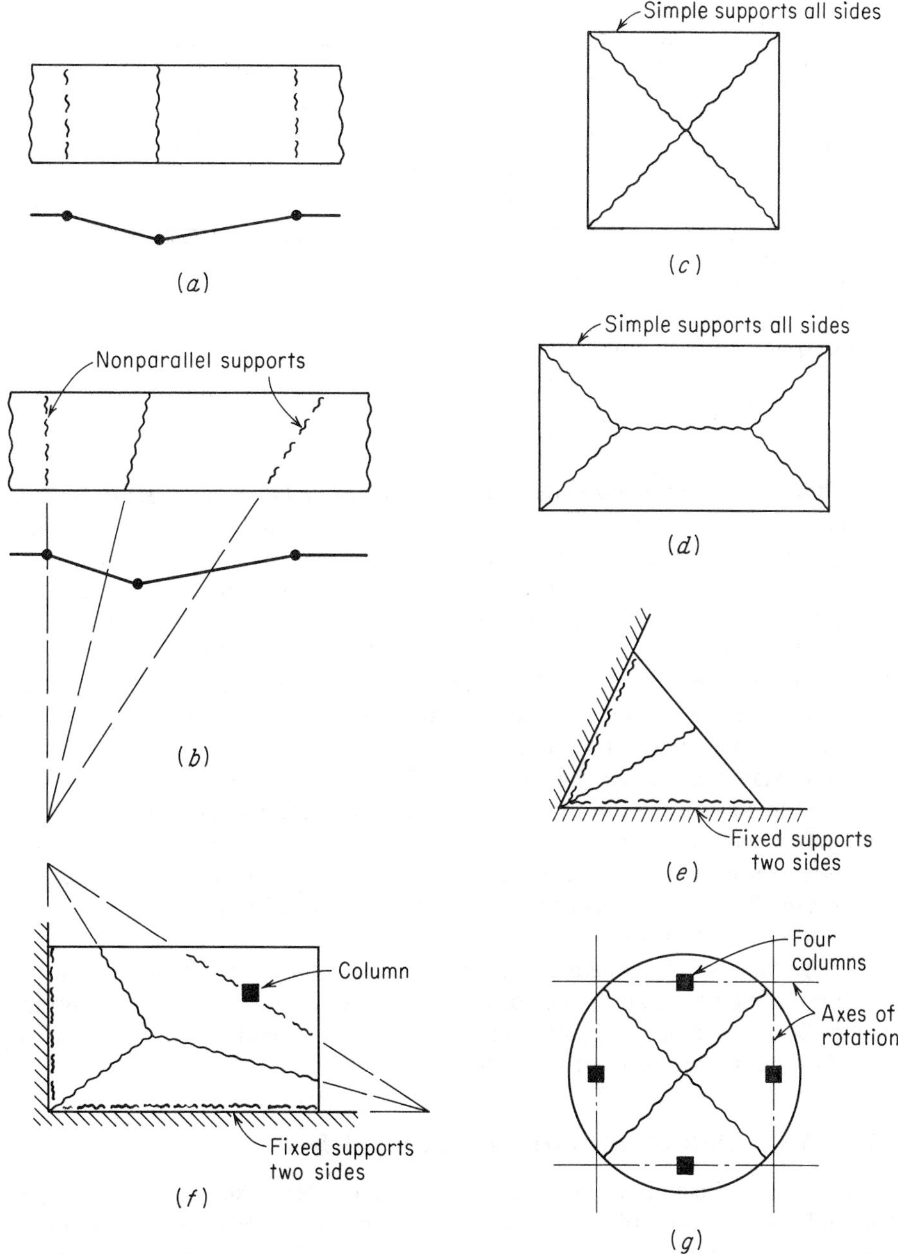

FIGURE 14.4
Typical yield line patterns.

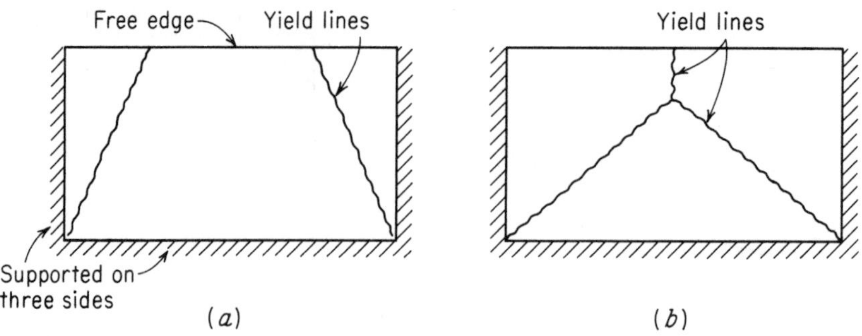

FIGURE 14.5
Alternative mechanisms for a slab supported on three sides.

It should be emphasized that *either method of yield line analysis is an upper bound approach* in the sense that the true collapse load will never by higher, but may be lower, than the load predicted. By either method, the solution has two essential parts: (a) establishing the correct failure pattern, and (b) finding the geometric parameters that define the exact location and orientation of the yield lines and solving for the relation between applied load and resisting moments. Either method can be developed in such a way as to lead to the correct solution for the mechanism chosen for study, but the true failure load will be found only if the correct mechanism has been selected.

For example, the rectangular slab of Fig. 14.5, supported along only three sides and free along the fourth, may fail by either of the two mechanisms shown. An analysis based on yield pattern *a* may indicate a slab capacity higher than one based on pattern *b*, or vice versa. It is necessary to investigate *all possible mechanisms* for any slab to confirm that the correct solution, giving the lowest failure load, has been found.

The method of segment equilibrium should not be confused with a true equilibrium method such as the strip method of Chapter 15. A true equilibrium method is a lower bound method of analysis—i.e., it will always give a *lower bound* of the true capacity of the slab.

14.4 ANALYSIS BY SEGMENT EQUILIBRIUM

Once the general pattern of yielding and rotation has been established by applying the guidelines of Sec. 14.3, the location and orientation of axes of rotation and the failure load for the slab can be established based on the equilibrium of the various segments of the slab. Each segment, studied as a free body, must be in equilibrium under the action of the applied loads, the moments along the yield lines, and the reactions or shear along the support lines. It is noted that because the yield moments are principal moments, twisting moments are zero along the yield lines, and in most cases the shearing forces are also zero. Only the unit moment m generally is considered in writing equilibrium equations.

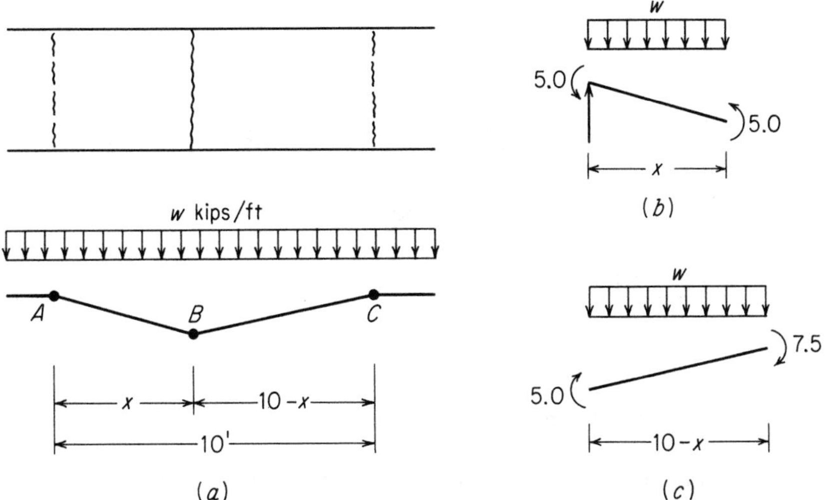

FIGURE 14.6
Analysis of a one-way slab by segment equilibrium equations.

Example 14.1 Segment equilibrium analysis of one-way slab. The method will be demonstrated first with respect to the one-way, uniformly loaded, continuous slab of Fig. 14.6*a*. The slab has a 10 ft span and is reinforced to provide a resistance to positive bending $\phi m_n = 5.0$ ft-kips/ft through the span. In addition, negative steel over the supports provides moment capacities of 5.0 ft-kips/ft at A and 7.5 ft-kips/ft at C. Determine the ultimate load capacity of the slab.

Solution. The number of equilibrium equations required will depend upon the number of unknowns. One unknown is always the relation between the resisting moments of the slab and the load. Other unknowns are needed to define the locations of yield lines. In the present instance, one additional equation will suffice to define the distance of the yield line from the supports. Taking the left segment of the slab as a free body and writing the equation for moment equilibrium about the left support line (see Fig. 14.6*b*) leads to

$$\frac{wx^2}{2} - 10.0 = 0 \tag{a}$$

Similarly, for the right slab segment,

$$\frac{w}{2}(10 - x)^2 - 12.5 = 0 \tag{b}$$

Solving Eqs. (*a*) and (*b*) simultaneously for w and x results in

$$w = 0.89 \text{ kips/ft}^2 \qquad x = 4.75 \text{ ft}$$

If a slab is reinforced identically in orthogonal directions, the ultimate resisting moment is the same in these two directions, as it is along any other line, regardless of its direction. Such a slab is said to be *isotropically* reinforced.

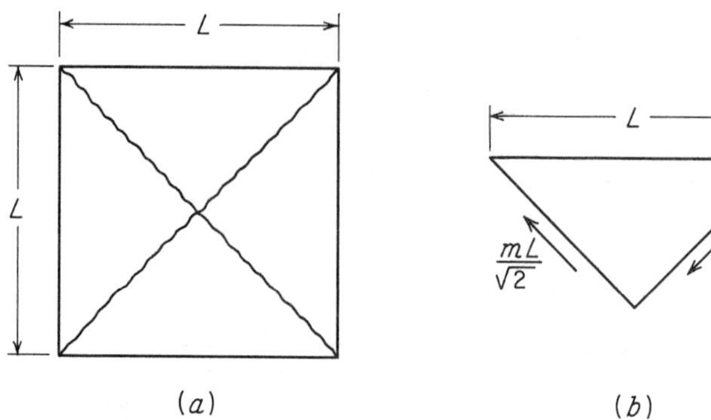

FIGURE 14.7
Analysis of a square two-way slab by segment equilibrium equations.

If, however, the ultimate strengths are different in two perpendicular directions, the slab is called *orthogonally anisotropic,* or simply *orthotropic*. Only isotropic slabs will be discussed in this section. Orthotropic reinforcement, which is very common in practice, will be discussed in Sec. 14.6.

It is convenient in yield line analysis to represent moments with vectors. The standard convention, in which the moment acts in a clockwise direction when viewed along the vector arrow, will be followed. Treatment of moments as vector quantities will be illustrated by the following example:

Example 14.2 Segment equilibrium analysis of square slab. A square slab is simply supported along all sides and is to be isotropically reinforced. Determine the resisting moment $m = \phi m_n$ per linear foot required just to sustain a uniformly distributed factored load of w psf.

Solution. Conditions of symmetry indicate the yield line pattern shown in Fig. 14.7*a*. Considering the moment equilibrium of any one of the identical slab segments about its support (see Fig. 14.7*b*), one obtains

$$\frac{wL^2}{4}\frac{L}{6} - 2\frac{mL}{\sqrt{2}}\frac{1}{\sqrt{2}} = 0$$

$$m = \frac{wL^2}{24}$$

In both examples just given, the resisting moment was constant along any particular yield line, i.e., the reinforcing bars were of constant diameter and equally spaced along a given yield line. On the other hand, it will be recalled that, by the elastic methods of slab analysis presented in Chaps. 12 and 13, reinforcing bars generally have a different spacing and may be of different diameter in middle strips compared with column or edge strips. A slab designed by elastic methods, leading to such variations, can easily be analyzed for strength by the yield line method. It is merely necessary to subdivide a yield line into its component parts, within any one of which the resisting moment per unit length of hinge is constant. Either the equilibrium equations of this section or the work equations of Sec. 14.5 can be modified in this way.

14.5 ANALYSIS BY VIRTUAL WORK

Alternative to the method of Sec. 14.4 is a method of analysis using the principle of virtual work. Since the moments and loads are in equilibrium when the yield line pattern has formed, an infinitesimal increase in load will cause the structure to deflect further. The external work done by the loads to cause a small arbitrary virtual deflection must equal the internal work done as the slab rotates at the yield lines to accommodate this deflection. The slab is therefore given a virtual displacement, and the corresponding rotations at the various yield lines can be calculated. By equating internal and external work, the relation between the applied loads and the ultimate resisting moments of the slab is obtained. Elastic rotations and deflections are not considered when writing the work equations, as they are very small compared with the plastic deformations.

a. External Work Done by Loads

An external load acting on a slab segment, as a small virtual displacement is imposed, does work equal to the product of its constant magnitude and the distance through which the point of application of the load moves. If the load is distributed over a length or an area, rather than concentrated, the work can be calculated as the product of the total load and the displacement of the point of application of its resultant.

Figure 14.8 illustrates the basis for external work calculation for several types of loads. If a square slab carrying a single concentrated load at its center (Fig. 14.8*a*) is given a virtual displacement defined by a unit value under the load, the external work is

$$W_e = P \times 1 \qquad (a)$$

If the slab of Fig. 14.8*b*, supported along three sides and free along the fourth, is loaded with a line load w per unit length along the free edge, and if that edge is given a virtual displacement having unit value along the central part, the external work is

$$W_e = (2wa) \times \frac{1}{2} + wb = w(a + b) \qquad (b)$$

When a distributed load w per unit area acts on a triangular segment defined by a hinge and yield lines such as Fig. 14.8*c*,

$$W_e = \frac{wab}{2} \times \frac{1}{3} = \frac{wab}{6} \qquad (c)$$

while for the rectangular slab segment of Fig. 14.8*d* carrying a distributed load w per unit area, the external work is

$$W_e = \frac{wab}{2} \qquad (d)$$

More complicated trapezoidal shapes may always be subdivided into component triangles and rectangles. The total external work is then calculated by summing

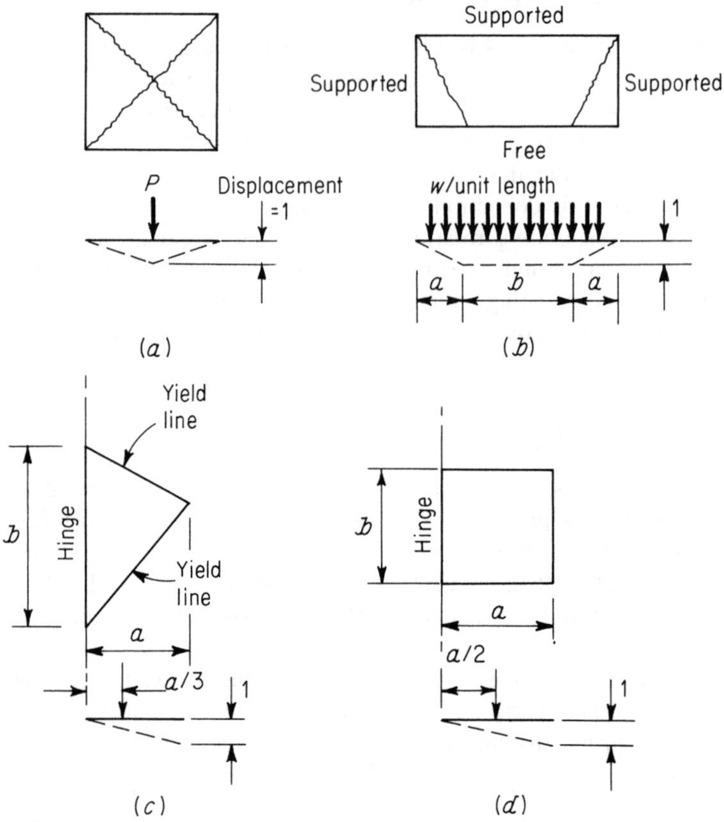

FIGURE 14.8
External work basis for various types of loads.

the work done by loads on the individual parts of the failure mechanism, with all displacements keyed to a unit value assigned somewhere in the system. There is no difficulty in combining the work done by concentrated loads, line loads, and distributed loads, if these act in combination.

b. Internal Work Done by Resisting Moments

The internal work done during the assigned virtual displacement is found by summing the products of yield moment m per unit length of hinge times the plastic rotation θ at the respective yield lines, consistent with the virtual displacement. If the resisting moment m is constant along a yield line of length l, and if a rotation θ is experienced, the internal work is

$$W_i = ml\theta \tag{e}$$

If the resisting moment varies, as would be the case if bar size or spacing is not constant along the yield line, the yield line is divided into n segments, within each one of which the moment is constant. The internal work is then

$$W_i = (m_1 l_1 + m_2 l_2 + \cdots + m_n l_n)\theta \tag{f}$$

For the entire system, the total internal work done is the sum of the contributions from all yield lines. Note that in all cases the internal work contributed is positive, regardless of the sign of m, because the rotation is in the same direction as the moment. External work, on the other hand, may be either positive or negative, depending on the direction of the displacement of the point of application of the force resultant.

Example 14.3 Virtual work analysis of one-way slab. Determine the load capacity of the one-way uniformly loaded continuous slab of Fig. 14.9, using the method of virtual work. The resisting moments of the slab are 5.0, 5.0, and 7.5 ft-kips/ft at A, B, and C respectively.

Solution. A unit deflection is given to the slab at B. Then the external work done by the load is the sum of the loads times their displacements and is equal to

$$\frac{wx}{2} + \frac{w}{2}(10 - x)$$

The rotations at the hinges are calculated in terms of the unit deflection (Fig. 14.9) and are

$$\theta_A = \theta_{B1} = \frac{1}{x} \qquad \theta_{B2} = \theta_C = \frac{1}{10 - x}$$

The internal work is the sum of the moments times their corresponding rotation angles:

$$5 \times \frac{1}{x} \times 2 + 5 \times \frac{1}{10 - x} + 7.5 \times \frac{1}{10 - x}$$

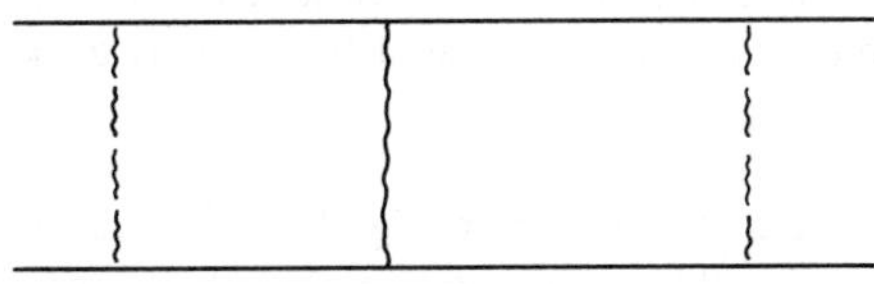

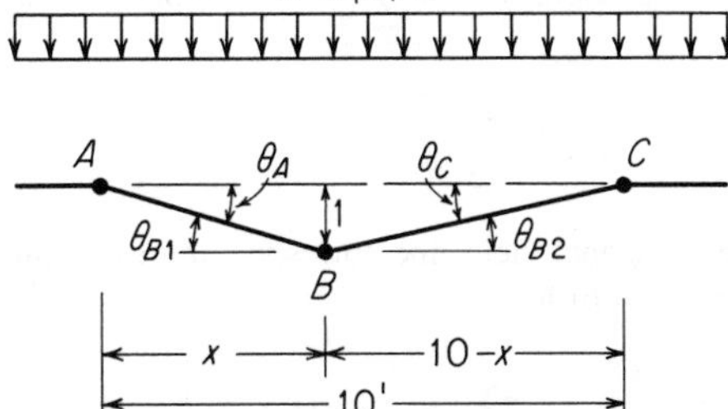

FIGURE 14.9
Virtual work analysis of one-way slab.

Equating the external and internal work gives

$$\frac{wx}{2} + 5w - \frac{wx}{2} = \frac{10}{x} + \frac{5}{10 - x} + \frac{7.5}{10 - x}$$

$$5w = \frac{10}{x} + \frac{25}{2(10 - x)}$$

$$w = \frac{2}{x} + \frac{5}{2(10 - x)}$$

To determine the minimum value of w, this expression is differentiated with respect to x and set equal to zero:

$$\frac{dw}{dx} = -\frac{2}{x^2} + \frac{5}{2(10 - x)^2} = 0$$

from which

$$x = 4.75 \text{ ft}$$

Substituting this value in the preceding expression for w, one obtains

$$w = 0.89 \text{ kips/ft}^2$$

as before.

In many cases, particularly those with yield lines established by several unknown dimensions (such as Fig. 14.4*f*), direct solution by virtual work would become quite tedious. The ordinary derivatives in the above example would be replaced by several partial derivatives, producing a set of equations to be solved simultaneously. In such cases it is often more convenient to select an arbitrary succession of possible yield line locations, solve the resulting mechanisms for the unknown load (or unknown moment), and determine the correct minimum load (or maximum moment) by trial.

Example 14.4 Virtual work analysis of rectangular slab. The two-way slab of Fig. 14.10 is simply supported on all four sides and carries a uniformly distributed ultimate load of w psf. Determine the ultimate moment resistance of the slab, which is to be isotropically reinforced.

Solution. Positive yield lines will form in the pattern shown in Fig. 14.10*a*, with the dimension a unknown. The correct dimension a will be such as to maximize the moment resistance required to support the load w. The values of a and m will be found by trial.

In Fig. 14.10*a* the length of the diagonal yield line is $\sqrt{25 + a^2}$. From similar triangles,

$$b = 5\frac{\sqrt{25 + a^2}}{a} \qquad c = a\frac{\sqrt{25 + a^2}}{5}$$

Then the rotation of the plastic hinge at the diagonal yield line corresponding to a unit deflection at the center of the slab (see Fig. 14.10*b*) is

$$\theta_1 = \frac{1}{b} + \frac{1}{c} = \frac{a}{5\sqrt{25 + a^2}} + \frac{5}{a\sqrt{25 + a^2}} = \frac{1}{\sqrt{25 + a^2}}\left(\frac{a}{5} + \frac{5}{a}\right)$$

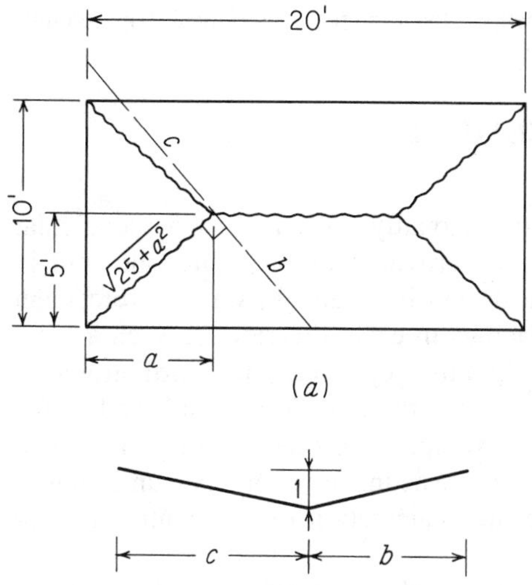

FIGURE 14.10
Virtual work analysis for rectangular two-way slab.

The rotation of the yield line parallel to the long edges of the slab is

$$\theta_2 = \frac{1}{5} + \frac{1}{5} = 0.40$$

For a first trial, let $a = 6$ ft. Then the length of the diagonal yield line is

$$\sqrt{25 + 36} = 7.81 \text{ ft.}$$

The rotation at the diagonal yield line is

$$\theta_1 = \frac{1}{7.81}\left(\frac{6}{5} + \frac{5}{6}\right) = 0.261$$

At the central yield line it is $\theta_2 = 0.40$. The internal work done as the incremental deflection is applied is

$$W_i = (m \times 7.81 \times 0.261 \times 4) + (m \times 8 \times 0.40) = 11.36m$$

The external work done during the same deflection is

$$W_e = (10 \times 6 \times \tfrac{1}{2}w \times \tfrac{1}{3} \times 2) + (8 \times 5w \times \tfrac{1}{2} \times 2) + (12 \times 5 \times \tfrac{1}{2}w \times \tfrac{1}{3} \times 2) = 80w$$

Equating W_i and W_e, one obtains

$$m = \frac{80w}{11.36} = 7.05w$$

Successive trials for different values of a result in the following data:

a	W_i	W_e	m
6.0	$11.36m$	$80.0w$	$7.05w$
6.5	$11.08m$	$78.4w$	$7.08w$
7.0	$10.87m$	$76.6w$	$7.04w$
7.5	$10.69m$	$75.0w$	$7.02w$

It is evident that the yield line pattern defined by $a = 6.5$ ft is critical. The required resisting moment for the given slab is $7.08w$.

14.6 ORTHOTROPIC REINFORCEMENT AND SKEWED YIELD LINES

Generally slab reinforcement is placed orthogonally, i.e., in two perpendicular directions. The same reinforcement may be provided in each direction, but in many practical cases, a more economical design is obtained using reinforcement having different bar areas or different spacings in each direction. In such a case, the reinforcement is said to be orthogonally anisotropic, or simply orthotropic.

Often yield lines will form at an angle with the directions established by the reinforcement; this was so in many of the examples considered earlier. For yield line analysis, it is necessary to calculate the resisting moment, per unit length, along such skewed yield lines. This requires calculation of the contribution to resistance from each of the two sets of bars.

Figure 14.11*a* shows an orthogonal grid of reinforcement, with angle α between the yield line and the X direction bars. Bars in the X direction are at spacing v and have moment resistance m_y per unit length about the Y axis, while bars in the Y direction are at spacing u and have moment resistance m_x per unit length about the X axis. The resisting moment per unit length for the bars in the Y and X directions will be determined separately, with reference to Fig. 14.11*b* and *c* respectively.

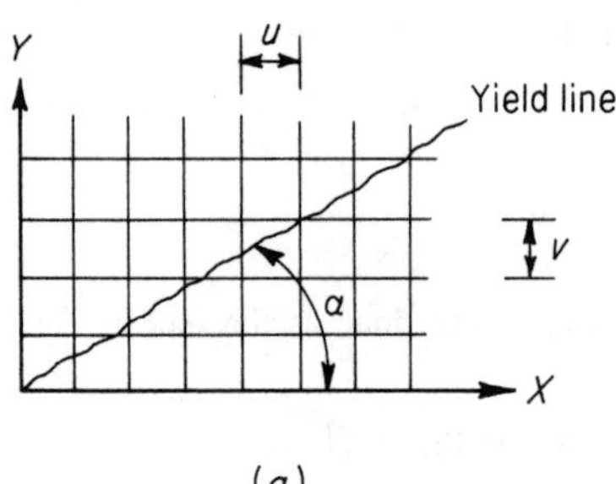

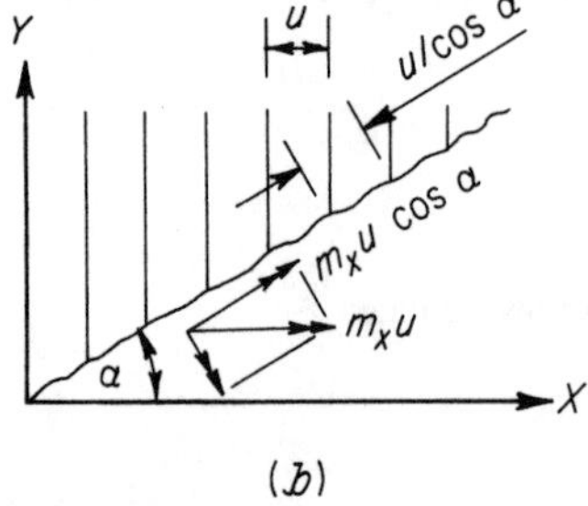

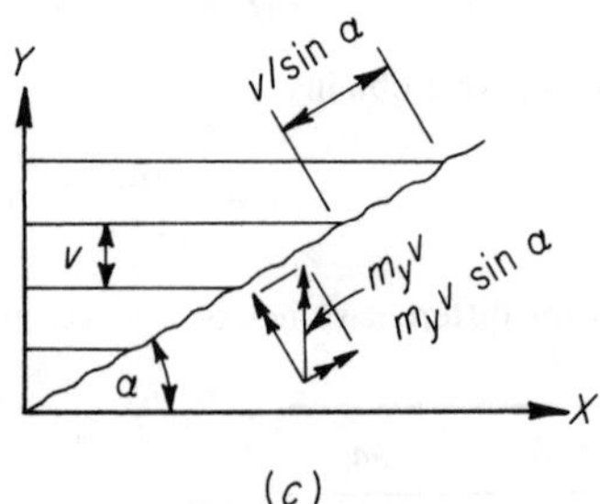

FIGURE 14.11
Yield line skewed with orthotropic reinforcement: (*a*) orthogonal grid and yield line; (*b*) Y direction bars; (*c*) X direction bars.

For the Y direction bars, the resisting moment *per bar* about the X axis is $m_x u$, and the component of that resistance about the α axis is $m_x u \cos \alpha$. The resisting moment per unit length along the α axis provided by the Y direction bars is therefore

$$m_{\alpha y} = \frac{m_x u \cos \alpha}{u/\cos \alpha} = m_x \cos^2 \alpha \qquad (a)$$

For the bars in the X direction, the resisting moment per bar about the Y axis is $m_y v$, and the component of that resistance about the α axis is $m_y v \sin \alpha$. Thus the resisting moment per unit length along the α axis provided by the X direction bars is

$$m_{\alpha x} = \frac{m_y v \sin \alpha}{v/\sin \alpha} = m_y \sin^2 \alpha \qquad (b)$$

Thus for the combined sets of bars, the resisting moment per unit length measured along the α axis is given by the sum of the resistances from Eqs. (a) and (b):

$$m_\alpha = m_x \cos^2 \alpha + m_y \sin^2 \alpha \qquad (14.1)$$

Note that, for the special case where $m_x = m_y = m$, with the same reinforcement provided in each direction,

$$m_\alpha = m(\cos^2 \alpha + \sin^2 \alpha) = m \qquad (14.2)$$

The slab is said to be *isotropically reinforced,* with the same resistance per unit length regardless of the orientation of the yield line.

The analysis just presented neglects any consideration of strain compatibility along the yield line, and assumes that the displacements at the level of the steel during yielding, which are essentially perpendicular to the yield line, are sufficient to produce yielding in both sets of bars. This is reasonably in accordance with test data except for values of α close to 0 to 90°. For such cases, it would be conservative to neglect the contribution of the bars nearly parallel to the yield line.

It has been shown that the analysis of an orthotropic slab can be simplified to that of a related isotropic slab, referred to as the *affine slab*, provided that the ratio of negative to positive reinforcement areas is the same in both directions. The horizontal dimensions and slab loads must be modified to permit this transformation. Details will be found in Refs. 14.1 to 14.5.

Example 14.5 Resisting moment along a skewed yield line. The balcony slab of Fig. 14.12 has fixed supports along two adjacent sides and is unsupported along the third side. It is reinforced for positive bending with No. 5 bars at 10 in. spacing and 5.5 in. effective depth, parallel to the free edge, and No. 4 bars at 10 in. spacing and 5.0 in. effective depth perpendicular to that edge. Concrete strength and steel yield stress are 4000 psi and 60,000 psi respectively. One possible failure mechanism includes a positive yield line at 30° with the long edge, as shown. Find the total resisting moment along the positive yield line provided by the two sets of bars.

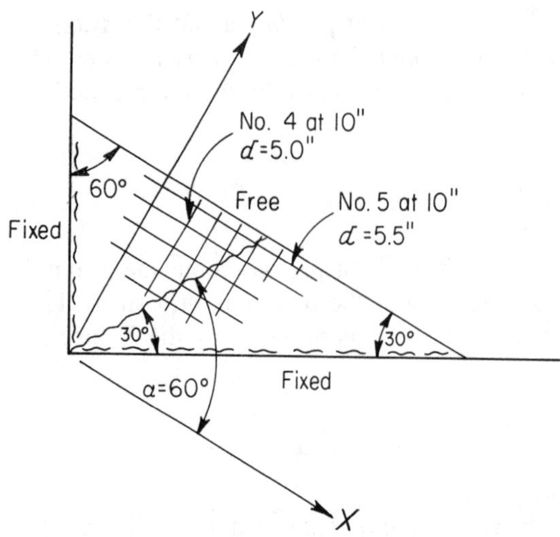

FIGURE 14.12
Skewed yield line example.

Solution. It is easily confirmed that the resisting moment about the X axis provided by the Y direction bars is $m_x = 5.21$ ft-kips/ft, and the resisting moment about the Y axis provided by the X direction bars is $m_y = 8.70$ ft-kips/ft (both with $\phi = 0.90$ included). The yield line makes an angle of 60° with the X axis bars. With cos $\alpha = 0.500$ and sin $\alpha = 0.866$, from Eq. (11.1) the resisting moment along the α axis is

$$m_\alpha = 5.21 \times 0.500^2 + 8.70 \times 0.866^2 = 7.83 \text{ ft-kips/ft}$$

14.7 SPECIAL CONDITIONS AT EDGES AND CORNERS

Certain simplifications were made in defining yield line patterns in some of the preceding examples, in the vicinity of edges and corners. In some cases, such as Fig. 14.4*b* and *f*, positive yield lines were shown intersecting an edge at an angle. Actually, at a free or simply supported edge, both bending and twisting moments should theoretically be zero. The principal stress directions are parallel and perpendicular to the edge, and consequently the yield lines should enter an edge perpendicular to it. Tests confirm that this is the case, but the yield lines generally turn only quite close to the edge, the distance t in Fig. 14.13 being small compared to the dimensions of the slab (Ref. 14.4).

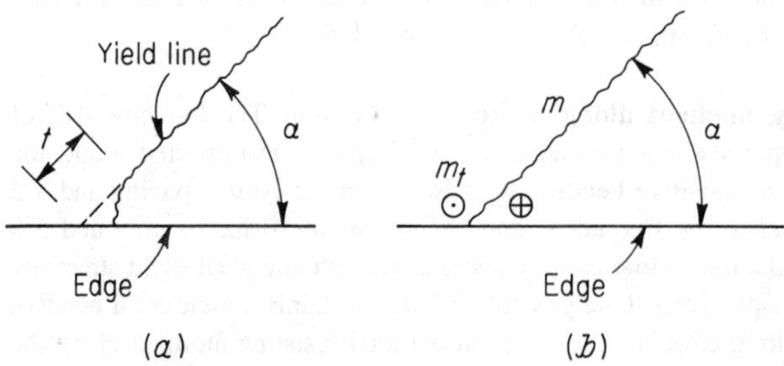

FIGURE 14.13
Conditions at edge of slab: (*a*) actual yield line; (*b*) simplified yield line.

Referring to Fig. 14.13, the actual yield line of *a* can be simplified by extending the yield line in a straight line to the edge, as in *b*, if a pair of concentrated shearing forces m_t is introduced at the corners of the slab segments. The force m_t acting downward at the acute corner (circled cross) and the force m_t acting upward at the obtuse corner (circled dot) together are the static equivalent of twisting moments and shearing forces near the edge. It is shown in Ref. 14.4 that the magnitude of the fictitious shearing forces m_t is given by the expression

$$m_t = m \cot \alpha \tag{14.3}$$

where m is the resisting moment per unit length along the yield line and α is the acute angle between simplified yield line and the edge of the slab.

It should be noted that, while the fictitious forces enter the solution by the equilibrium method, the virtual work solution is not affected because the net work done by the pair of equal and opposite forces moving through the identical virtual displacement is zero.

Also, in the preceding examples, it was assumed that yield lines enter the corners between the two intersecting sides. An alternative possibility is that the yield line forks before it reaches the corner, forming what is known as a *corner lever*, shown in Fig. 14.14*a*.

If the corner is not held down, the triangular element *abc* will pivot about the axis *ab* and lift off the supports. The development of such a corner lever is clearly shown in Fig. 14.15. The photograph shows a model reinforced concrete slab that was tested under uniformly distributed load. The edges were simply supported and were not restrained against upward movement. If the corner is held down, a similar situation obtains, except that the line *ab* becomes a yield line. If cracking at the corners of such a slab is to be controlled, top steel, more or less perpendicular to the line *ab* must be provided. The direction taken by the positive yield lines near the corner indicates the desirability of supplementary bottom-slab

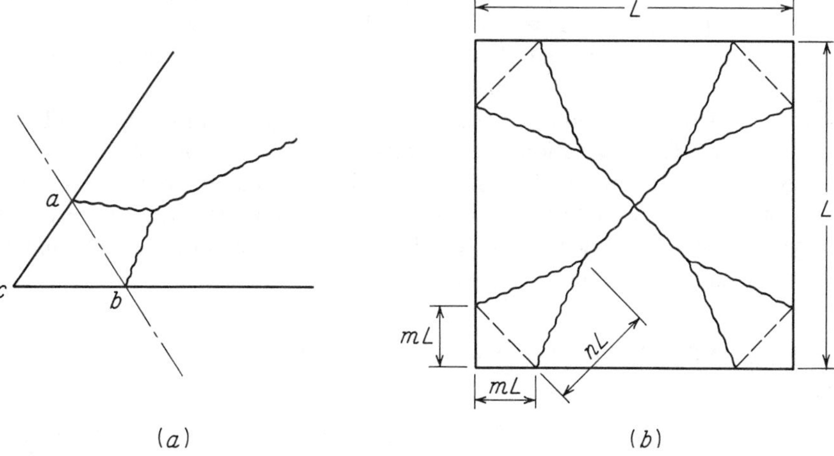

FIGURE 14.14
Corner conditions.

FIGURE 14.15
Development of corner levers in a simply supported, uniformly loaded slab.

reinforcement at the corners, placed approximately parallel to the line *ab* (see Sec. 12.6).

Although yield line patterns with corner levers are generally more critical than those without, they are often neglected in yield line analysis. The analysis becomes considerably more complicated if the possibility of corner levers is introduced, and the error made by neglecting them is usually small.

To illustrate, the uniformly loaded square slab of Example 14.2, when analyzed for the assumed yield pattern of Fig. 14.7, required an ultimate moment capacity of $wL^2/24$. The actual yield line pattern at failure is probably as shown in Fig. 14.14*b*. Since two additional parameters *m* and *n* have necessarily been introduced to define the yield line pattern, a total of three equations of equilibrium is now necessary. These equations are obtained by summing moments and vertical forces on the segments of the slab. Such an analysis results in a required resisting moment of $wL^2/22$, an increase of about 9 percent compared with the results of an analysis neglecting corner levers. The influence of such corner effects may be considerably larger when the corner angle is less than 90°.

14.8 FAN PATTERNS AT CONCENTRATED LOADS

If a concentrated load acts on a reinforced concrete slab at an interior location, away from any edge or corner, a negative yield line will form in a more-or-less circular pattern, as in Fig. 14.16*a*, with positive yield lines radiating outward

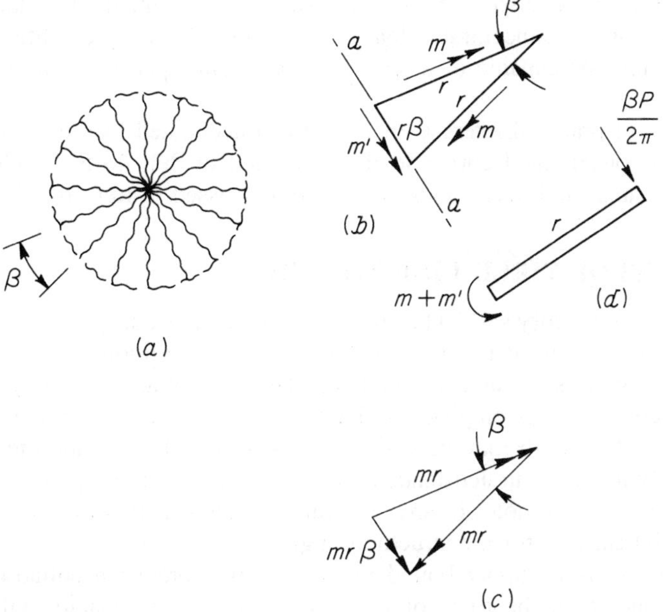

FIGURE 14.16
Yield fan geometry at concentrated load: (*a*) yield fan; (*b*) moment vectors acting on fan segment; (*c*) resultant of positive moment vectors; (*d*) edge view of fan segment.

from the load point. If the positive resisting moment per unit length is m and the negative resisting moment m', the moments per unit length acting along the edges of a single element of the fan, having a central angle β and radius r, are as shown in Fig. 14.16*b*. For small values of the angle β, the arc along the negative yield line can be represented as a straight line of length $r\beta$.

Figure 14.16*c* shows the moment resultant obtained by vector addition of the positive moments mr acting along the radial edges of the fan segment. The vector sum is equal to $mr\beta$, acting along the length $r\beta$, and the resultant positive moment, per unit length, is therefore m. This acts in the same direction as the negative moment m', as shown in Fig. 14.16*d*. Figure 14.16*d* also shows the fractional part of the total load P that acts on the fan segment.

Taking moments about the axis *a-a* gives

$$(m + m')r\beta - \frac{\beta P r}{2\pi} = 0$$

from which

$$P = 2\pi(m + m') \tag{14.4}$$

The collapse load P is seen to be independent of the fan radius r. With only a concentrated load acting, a complete fan of any radius could form with no change in collapse load.

It follows that Eq. (14.4) also gives the collapse load for a fixed-edge slab of any shape, carrying only a concentrated load P. The only necessary condition is that the boundary must be capable of a restraining moment equal to m' at all points.

Other load cases of practical interest, including a concentrated load near or at a free edge, and a concentrated corner load, are treated in Ref. 14.5. Loads distributed over small areas and load combinations are discussed in Ref. 14.12.

14.9 LIMITATIONS OF YIELD LINE THEORY

The usefulness of yield line theory should be apparent from the preceding sections. In general, elastic solutions are available only for restricted conditions, usually uniformly loaded rectangular slabs and slab systems. They do not account for the effects of inelastic action, except empirically. By yield line analysis, a rational determination of flexural strength may be had for slabs of any shape, supported in a variety of ways, with concentrated loads as well as distributed and partially distributed loads. The effects of holes of any size can be included. It is thus seen to be a powerful analytical tool for the structural engineer.

On the other hand, as an upper bound method, it will predict a collapse load that may be greater than the true collapse load. The actual capacity will be less than predicted if the selected mechanism is not the controlling one or if the specific locations of yield lines are not exactly correct. Most engineers would prefer an approach that would be in error, if at all, on the safe side. In this respect, the strip method of Chapter 15 is distinctly superior.

Beyond this, it should be evident that yield line theory provides, in essence, a method for determining the capacity of trial designs, arrived at by some other means, rather than for determining the amount and spacing of reinforcement. It is not, strictly speaking, a design method. To illustrate, yield line theory provides no inducement for the designer to place steel at anything other than a uniform lateral spacing along a yield line. It is necessary to consider the results of elastic analysis of a flat plate, for example, to recognize that reinforcement in that case should be placed in strong bands across the columns.

In applying yield line analysis to slabs, it must be remembered that the analysis is predicated upon available rotation capacity at the yield lines. If the slab reinforcement happens to correspond closely to the elastic distribution of moments in the slab, little rotation is required. If, on the other hand, there is a marked difference, it is possible that the required rotation will exceed the available rotation capacity, in which case the slab will fail prematurely. However, in general, because slabs are typically rather lightly reinforced, they will have adequate rotation capacity to attain the ultimate loads predicted by yield line analysis.

It should also be borne in mind that the yield line analysis focuses entirely on the flexural capacity of the slab. It is presumed that earlier failure will not occur due to shear or torsion and that cracking and deflections at service load will not be excessive. ACI Code 13.3.1 calls attention specifically to the need to meet "all serviceability conditions, including specified limits on deflections," and ACI Commentary 13.3.1 calls attention to the need for "evaluation of the stress conditions around the supports in relation to shear and torsion as well as flexure."

REFERENCES

14.1. A. Ingerslev, "The Strength of Rectangular Slabs," *J. Inst. Struct. Eng.*, London, vol. 1, no. 1, 1923, pp. 3–14.
14.2. K. W. Johansen, *Brutlinieteorier,* Jul. Gjellerups Forlag, Copenhagen, 1943 (see also *Yield Line Theory,* English translation, Cement and Concrete Association, London, 1962).
14.3. K. W. Johansen, *Pladeformler,* 2d ed., Polyteknisk Forening, Copenhagen, 1949 (see also *Yield Line Formulae for Slabs,* English translation, Cement and Concrete Association, London, 1972).
14.4. E. Hognestad, "Yield Line Theory for the Ultimate Flexural Strength of Reinforced Concrete Slabs," *J. ACI,* vol. 24, no. 7, 1953, pp. 637–656.
14.5. L. L. Jones and R. H. Wood, *Yield Line Analysis of Slabs,* American Elsevier, New York, 1967.
14.6. R. Taylor, D. R. H. Maher, and B. Hayes, "Effect of the Arrangement of Reinforcement on the Behavior of Reinforced Concrete Slabs," *Mag. Concr. Res.*, vol. 18, no. 55, 1966, pp. 85–94.
14.7. R. Lenschow and M. A. Sozen, "A Yield Criterion for Reinforced Concrete Slabs," *J. ACI,* vol. 64, no. 5, 1967, pp. 266–273.
14.8. S. H. Simmonds and A. Ghali, "Yield Line Design of Slabs," *J. Struct. Div.*, ASCE, vol. 102, no. ST1, 1976, pp. 109–123.
14.9. K. H. Chu and R. B. Singh, "Yield Analysis of Balcony Floor Slabs," *J. ACI,* vol. 63, no. 5, 1966, pp. 571–586.
14.10. E. C. Demsky and D. S. Hatcher, "Yield Line Analysis of Slabs Supported on Three Sides," *J. ACI,* vol. 66, no. 9, 1969, pp. 741–744.
14.11. A. Zaslavsky and C. H. Avraham, "Yield Line Design of Rectangular Reinforced Concrete Balconies," *J. ACI,* vol. 67, no. 1, 1970, pp. 53–56.
14.12. H. Gesund, "Limit Design of Slabs for Concentrated Loads," *J. Struct. Div.*, ASCE, vol. 107, no. ST9, 1981, pp. 1839–1856.
14.13. A. Zaslavsky, "Yield Line Analysis of Rectangular Slabs with Central Openings," *J. ACI,* vol. 64, no. 12, 1967, pp. 838–844.
14.14. S. Islam and R. Park, "Yield Line Analysis of Two-Way Reinforced Concrete Slabs with Openings," *J. Inst. Struct. Eng.*, vol. 49, no. 6, 1971, pp. 269–276.
14.15. R. Park and W. L. Gamble, *Reinforced Concrete Slabs,* John Wiley and Sons, New York, 1980.

PROBLEMS

14.1 A square slab measuring 10 ft on each side is simply supported on three sides and unsupported along the fourth. It is reinforced for positive bending with an isotropic mat of steel providing resistance ϕm_n of 7000 ft-lb/ft in each of the two principal directions. Determine the uniformly distributed load that would cause flexural failure, using the method of virtual work.

14.2 The triangular slab of Fig. P14.2 has fixed supports along the two perpendicular edges and is free of any support along the diagonal edge. Negative reinforcement perpendicular to the

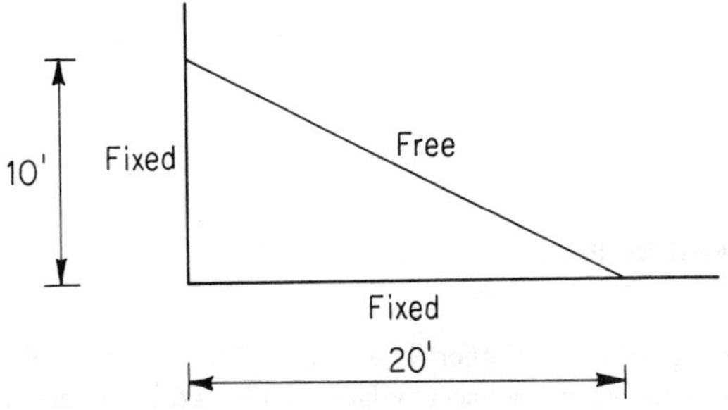

FIGURE P14.2

supported edges provides design strength $\phi m_n = 4$ ft-kips/ft. The slab is reinforced for positive bending by an orthogonal grid providing resistance $\phi m_n = 2.67$ ft-kips/ft in all directions. Find the total factored load w_u that will produce flexural failure. A virtual work solution is suggested.

14.3 The one-way reinforced concrete slab of Fig. P14.3 spans 20 ft. It is simply supported at its left edge, fully fixed at its right edge, and free of support along the two long sides. Reinforcement provides design strength $\phi m_n = 5$ ft-kips/ft in positive bending and $\phi m_n = 7.5$ ft-kips/ft in negative bending at the right edge. Using the equilibrium method, find the factored load w_u uniformly distributed over the surface that would cause flexural failure.

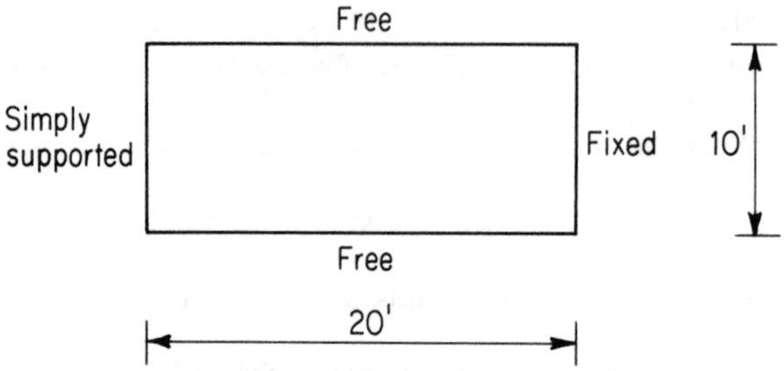

FIGURE P14.3

14.4 Solve Prob. 14.3 using the method of virtual work.

14.5 The triangular slab shown in Fig. P14.5 is to serve as weather protection over a loading dock. Support conditions are essentially fixed along *AB* and *BC*, and *AC* is a free edge. In addition to self-weight, a superimposed dead load of 15 psf and service live load of 40 psf must be provided for. Material strengths are $f'_c = 4000$ psi and $f_y = 60,000$ psi. Using yield line analysis, find the required slab thickness h and find the reinforcement required at critical sections. Neglect corner pivots. Use a maximum steel ratio of 0.005. Select bar sizes and spacings, and provide a sketch summarizing important aspects of the design. Make an approximate, conservative check of safety against shear failure for the design. Also include a conservative estimate of the deflection near the center of edge *AC* due to a full live load.

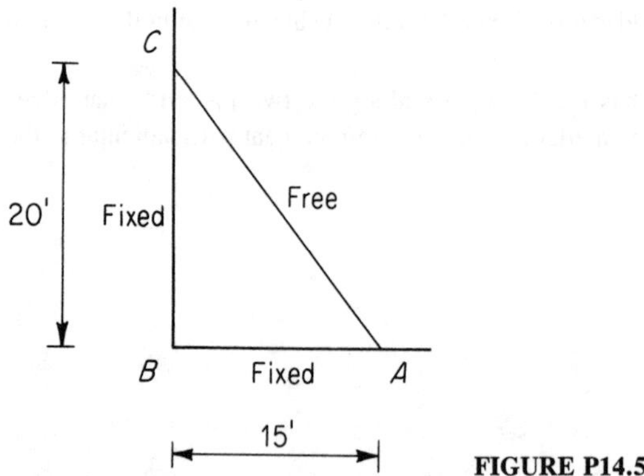

FIGURE P14.5

14.6 The square concrete slab shown in Fig. P14.6 is supported by monolithic concrete walls providing full vertical and rotational restraint along two adjacent edges, and by a 6 in. diameter steel pipe column, near the outer corner, that offers negligible rotational restraint. It is reinforced for positive bending by an orthogonal grid of bars parallel to the walls, providing design moment capacity $\phi m_n = 6.5$ ft-kips/ft in all directions. Negative reinforcement perpendicular to the walls, and negative bars at the outer corner parallel to the slab diagonal, provide $\phi m_n = 8.9$ ft-kips/ft. Neglecting corner pivots, find the total factored uniformly distributed load w_u that will initiate flexural failure. Solution by the method of virtual work is recommended, with collapse geometry established by successive trial. Yield line lengths and perpendicular distances are most easily found graphically. Include a check of the shear capacity of the slab, using approximate methods. The steel column is capped with a 12 × 12 in plate providing bearing.

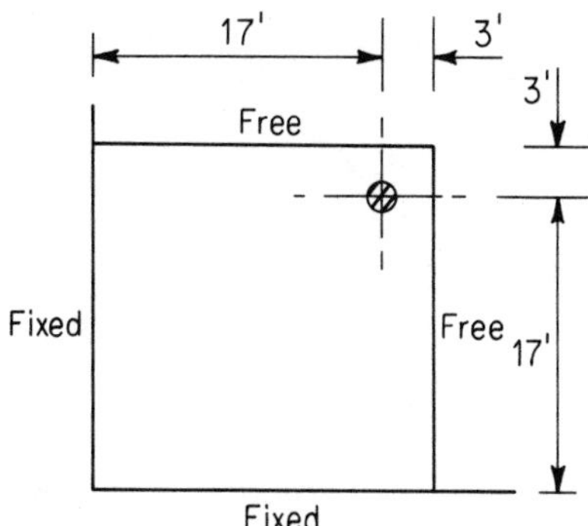

FIGURE P14.6

14.7 The square slab shown in Fig. P14.7 is supported by, and is monolithic with, a reinforced concrete wall along the edge *CD* that provides full fixity, and also is supported by a masonry wall along *AB* that provides a simply supported line. It is to carry a factored load w_u = 300 psf including its selfweight. Assuming a uniform 6 in. slab thickness, find the required reinforcement. Include a sketch summarizing details of your design, indicating placement and length of all reinforcing bars. Also check the shear capacity of the structure, making whatever assumptions appear reasonable and necessary. Use $f_c' = 4000$ psi and $f_y = 60{,}000$ psi.

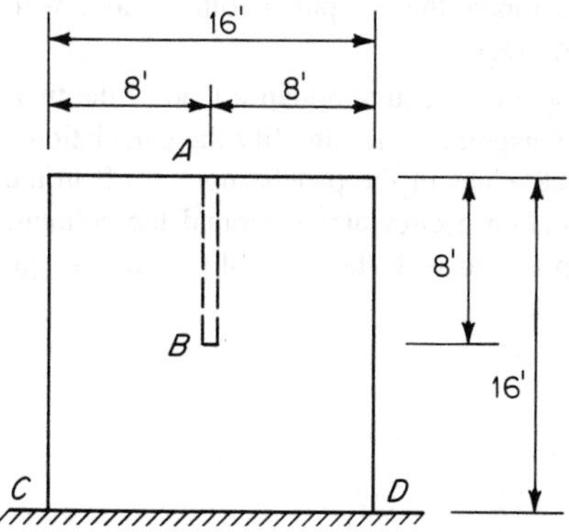

FIGURE P14.7

14.8 The slab of Fig. P14.8 is supported by three fixed edges but has no support along one long side. It has a uniform thickness of 7 in., resulting in effective depths in the long direction of 6.0. in. and in the short direction of 5.5 in. Bottom reinforcement consists of No. 4 bars at 14 in. centers in each direction, continued to the supports and the free edge. Top negative steel along the supported edges consists of No. 4 bars at 12 in. on centers, except that in a 2 ft wide "strong band" parallel and adjacent to the free edge, four No. 5 bars are used. All negative bars extend past the points of inflection, as required by ACI Code. Material strengths are $f_c' = 4000$ psi and $f_y = 60{,}000$ psi. Using the yield line method, determine the ultimate load w_u that can be carried.

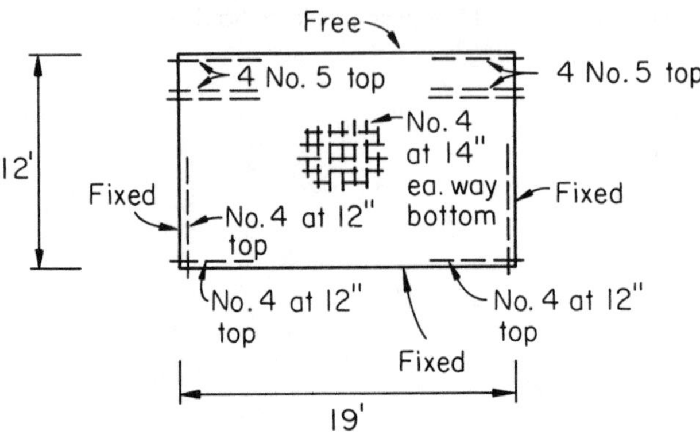

FIGURE P14.8

14.9 Using virtual work and yield line theory, compute the flexural collapse load of the one-way slab of Example 12.1 of Chapter 12. Assume that all straight bars are used, according to Fig. 12.4*b*. Compare the calculated collapse load with the original factored design load, and comment on differences.

14.10 Using virtual work and yield line theory, compute the flexural collapse load of the two-way edge-supported slab of Example 12.2 of Chapter 12. To simplify the calculations, assume that all straight bars are used and that all positive moment steel is carried into the face of the supporting beams, with no bars cut off in the span. Neglect corner effects. Note that the effect of continuity over an edge beam is to shift the positive moment yield line away from that edge toward the discontinuous edge at the opposite side. Compare the computed collapse load with the original factored design load, and comment on differences.

14.11 Using virtual work and yield line theory, compute the flexural collapse load of the two-way column-supported flat plate of Example 13.2 of Chapter 13. To simplify the calculations, assume that all positive moment bars are carried to the edges of the panels, not cut off in the span. Consider all possible failure mechanisms, including a circular fan around the column. Neglect corner effects. Compare the calculated collapse load with the original factored design load and comment on differences.

CHAPTER 15

STRIP METHOD FOR SLABS

15.1 INTRODUCTION

In Section 14.2 of Chapter 14, the upper and lower bound theorems of the theory of plasticity were presented, and it was pointed out that the yield line method of slab analysis was an *upper bound approach* to determining the ultimate flexural strength of slabs. An upper bound analysis, if in error, will be so on the unsafe side. The actual carrying capacity will be less than, or at best equal to, the capacity predicted, which is certainly a cause for concern in design. Also, when applying the yield line method, it is necessary to assume that the distribution of reinforcement is known over the whole slab. It follows that the yield line approach is a tool for *review* of the capacity of a given slab and can be used for *design* only in an iterative sense, for calculating the capacities of trial designs with varying reinforcement until a satisfactory arrangement is found.

These circumstances motivated Hillerborg to develop what is known as the *strip method* for slab design, his first results being published in Swedish in 1956 (Ref. 15.1). In contrast to yield line analysis, the strip method is a *lower bound approach*, based on satisfaction of equilibrium requirements everywhere in the slab. By the strip method (sometimes referred to as the *equilibrium theory*), a moment field is first determined that fulfills equilibrium requirements, after which the reinforcement of the slab at each point is designed for this moment field. If a distribution of moments can be found that satisfies both equilibrium and boundary conditions for a given external loading, and if the yield moment capacity of the slab is nowhere exceeded, then the given external loading will represent a lower bound of the true carrying capacity.

The strip method gives results on the safe side, which is certainly preferable in practice, and differences from the true carrying capacity will never impair safety. The strip method is a *design* method, by which the needed reinforcement can be calculated. It encourages the designer to vary the reinforcement in a logical way, leading to an economical arrangement of steel as well as a safe design. It is generally simple to use, even for slabs having holes or irregular boundaries.

In his original work in 1956, Hillerborg set forth the basic principles for edge-supported slabs and introduced the expression "strip method" (Ref. 15.1). He later expanded the method to include the practical design of slabs on columns and L-shaped slabs (Refs. 15.2 and 15.3). The first treatment of the subject in English was by Crawford (Ref. 15.4). Blakely, in 1964, translated the earlier Hillerborg work into English (Ref. 15.5). Important contributions, particularly regarding continuity conditions, have been made by Kemp (Refs. 15.6 and 15.7) and Wood and Armer (Refs. 15.8, 15.9, and 15.10). Load tests of slabs designed by the strip method were carried out by Armer (Ref. 15.11) and confirmed that the method produces safe and satisfactory designs. More recently, Hillerborg has written a book "for the practical designer, helping him in the simplest possible way to produce safe designs for most of the slabs that he will meet in practice, including slabs that are irregular in plan or that carry unevenly distributed loads" (Ref. 15.12). Subsequently, he published a paper in which he summarized what has become known as the "advanced strip method," pertaining to the design of slabs supported on columns, reentrant corners, or interior walls (Ref. 15.13). A useful summary of both the simple and advanced strip methods will be found in Ref. 15.14.

The strip method is appealing not only because it is safe, economical, and versatile over a broad range of applications, but also because it represents a formalization of the procedures followed instinctively by competent designers in placing reinforcement in the best possible position. In contrast with the yield line method, which provides no inducement to vary bar spacing, the strip method encourages use of strong bands of steel where needed, such as around openings or over columns, improving economy and reducing the likelihood of excessive cracking or large deflections under service loading.

15.2 BASIC PRINCIPLES

The governing equilibrium equation for a small slab element having sides dx and dy is

$$\frac{\partial^2 m_x}{\partial x^2} + \frac{\partial^2 m_y}{\partial y^2} - 2\frac{\partial^2 m_{xy}}{\partial x\,\partial y} = -w \tag{15.1}$$

where w is the external load per unit area; m_x and m_y are the bending moments per unit width in the X and Y directions, respectively; and m_{xy} is the twisting moment (Ref. 15.15). According to the lower bound theorem, any combination of m_x, m_y, and m_{xy} that satisfies the equilibrium equation at all points in the slab and that meets boundary conditions is a valid solution, provided that the reinforcement is placed to carry these moments.

The basis for the simple strip method is that the torsional moment is chosen equal to zero; no load is assumed to be resisted by the twisting strength of the slab. Therefore, if the reinforcement is parallel to the axes in a rectilinear coordinate system,

$$m_{xy} = 0$$

The equilibrium equation then reduces to

$$\frac{\partial^2 m_x}{\partial x^2} + \frac{\partial^2 m_y}{\partial y^2} = -w \tag{15.2}$$

This equation can be split conveniently into two parts, representing twistless beam strip action,

$$\frac{\partial^2 m_x}{\partial x^2} = -kw \tag{15.3a}$$

and

$$\frac{\partial^2 m_y}{\partial y^2} = -(1-k)w \tag{15.3b}$$

where the proportion of load taken by the strips is k in the X direction and $(1-k)$ in the Y direction. In many regions in slabs, the value of k will be either 0 or 1. With $k = 0$ all the load is dispersed by strips in the Y direction; with $k = 1$ all the load is carried in the X direction. In other regions, it may be reasonable to assume that the load is divided equally in the two directions (i.e., $k = 0.5$).

15.3 CHOICE OF LOAD DISTRIBUTION

Theoretically, the load w can be divided arbitrarily between the X and Y directions. Different divisions will, of course, lead to different patterns of reinforcement, and all will not be equally appropriate. The desired goal is to arrive at an arrangement of steel that is safe and economical and that will avoid problems at the service load level associated with excessive cracking or deflections. In general, the designer may be guided by his knowledge of the general distribution of elastic moments.

To see an example of the strip method and to illustrate the choices open to the designer, consider the square, simply supported slab shown in Fig. 15.1, with side length a and a uniformly distributed ultimate (factored) load w per unit area.

The simplest load distribution is obtained by setting $k = 0.5$ over the entire slab, as shown in Fig. 15.1. The load on all strips in each direction is then $w/2$, as illustrated by the load dispersion arrows of Fig. 15.1*a*. This gives maximum design moments

$$m_x = m_y = \frac{wa^2}{16} \tag{15.4}$$

over the whole slab, as shown in Fig. 15.1*c*, with uniform lateral distribution across the width of the critical section, as in Fig. 15.1*d*.

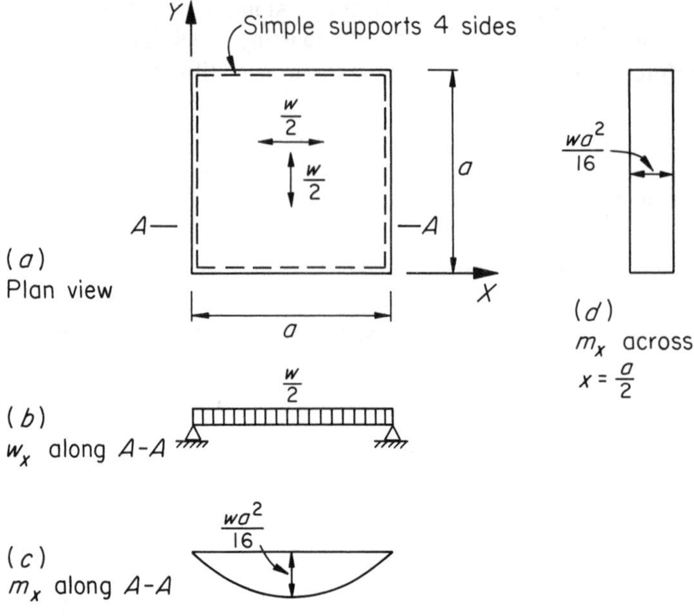

FIGURE 15.1
Square slab with load shared equally in two directions.

This would not represent an economical or serviceable solution because it is recognized that curvatures, hence moments, must be greater in the strips near the middle of the slab than near the edges in the direction parallel to the edge (see Fig. 12.5 of Chapter 12). If the slab were reinforced according to this solution, extensive redistribution of moments would be required, certainly accompanied by much cracking in the highly stressed regions near the middle of the slab.

An alternative, more reasonable distribution is shown in Fig. 15.2. Here the regions of different load dispersion, separated by the dash-dotted "discontinuity lines," follow the diagonals, and all of the load on any region is carried in the direction giving the shortest distance to the nearest support. The solution proceeds, giving k values of either 0 or 1, depending on the region, with load transmitted in the directions indicated by the arrows of Fig. 15.2*a*. For a strip *A-A* at a distance $y \leq a/2$ from the X axis, the design moment is

$$m_x = \frac{wy^2}{2} \tag{15.5}$$

The load acting on a strip *A-A* is shown in Fig. 15.2*b*, and the resulting diagram of moment m_x is given in Fig. 15.2*c*. The lateral variation of m_x across the width of the slab is as shown in Fig. 15.2*d*.

The lateral distribution of moments shown in Fig. 15.2*d* would theoretically require a continuously variable bar spacing, obviously an impracticality. One way of utilizing the distribution of Fig. 15.2, which is considerably more economical

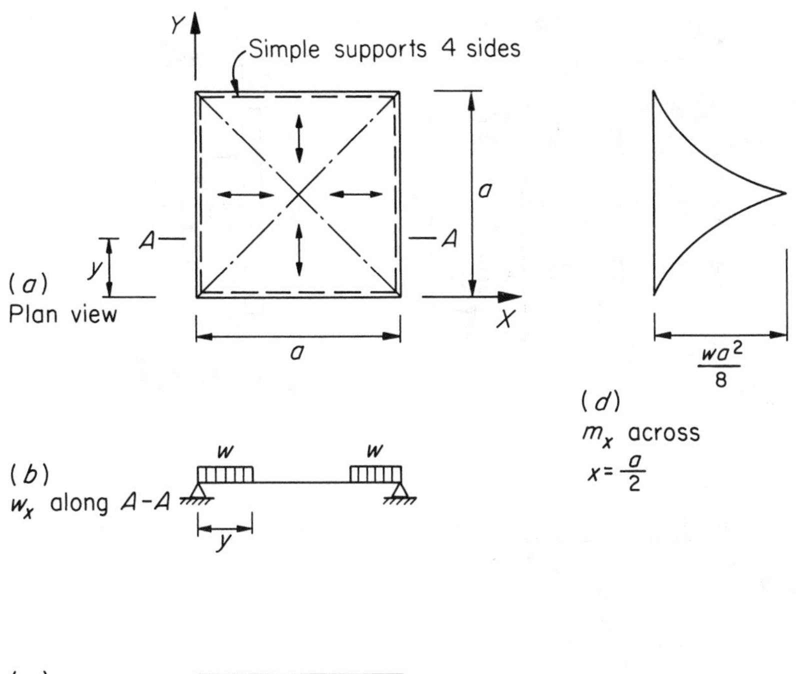

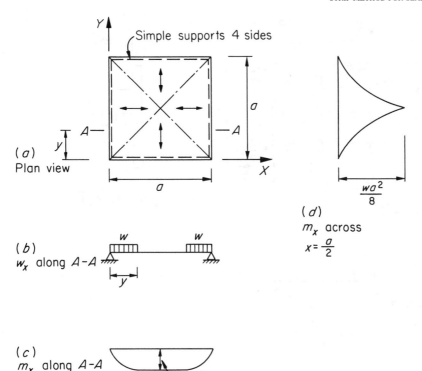

FIGURE 15.2
Square slab with load dispersion lines following diagonals.

than that of Fig. 15.1, would be to reinforce for the *average* moment over a certain width, approximating the actual lateral variation of Fig. 15.2*d* in a stepwise manner. Hillerborg notes that this is not strictly in accordance with the equilibrium theory and that the design is no longer certainly on the safe side, but other conservative assumptions, e.g., neglect of membrane strength in the slab and neglect of strain hardening of the reinforcement, would surely compensate for the slight reduction in safety margin.

A third alternative distribution is shown in Fig. 15.3. Here the division is made so that the load is carried to the nearest support, as before, but load near the diagonals has been divided, with one-half taken in each direction. Thus k is given values of 0 or 1 along the middle edges and a value of 0.5 in the corners and center of the slab, with load dispersion in the directions indicated by the arrows of Fig. 15.3*a*. Two different strip loadings are now identified. For an X direction strip along section A-A, the maximum moment is

$$m_x = \frac{w}{2} \times \frac{a}{4} \times \frac{a}{8} = \frac{wa^2}{64} \tag{15.6a}$$

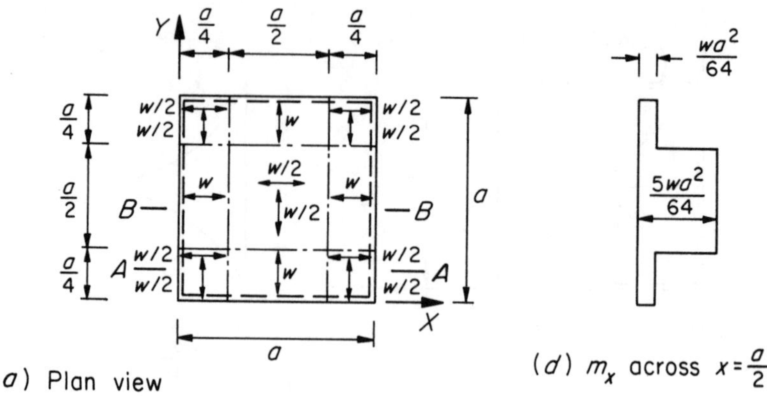

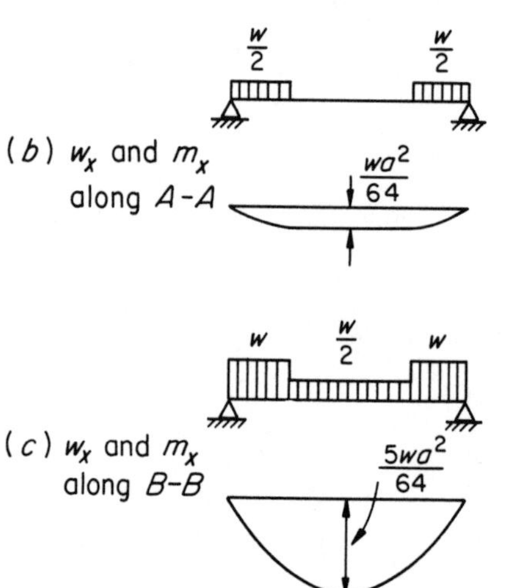

FIGURE 15.3
Square slab with load near diagonals shared equally in two directions.

and for a strip along section *B-B*, the maximum moment is

$$m_x = w \times \frac{a}{4} \times \frac{a}{8} + \frac{w}{2} \times \frac{a}{4} \times \frac{3a}{8} = \frac{5wa^2}{64} \qquad (15.6b)$$

This design leads to a practical arrangement of reinforcement, one with constant spacing through the center strip of width $a/2$ and a wider spacing through the outer strips, where the elastic curvatures and moments are known to be less. The averaging of moments necessitated in the second solution is avoided here, and the third solution is fully consistent with the equilibrium theory.

Comparing the three solutions just presented shows that the first would be unsatisfactory, as noted earlier, because it would require great redistribution of moments to achieve, possibly accompanied by excessive cracking and large deflections. The second, with discontinuity lines following the slab diagonals, has the advantage that the reinforcement more nearly matches the elastic distribution of moments, but it either leads to an impractical reinforcing pattern or requires an averaging of design moments in bands that involves a deviation from strict equilibrium theory. The third solution, with discontinuity lines parallel to the edges, does not require moment averaging and leads to a practical reinforcing arrangement, so it will often be preferred.

The three examples also illustrate the simple way in which moments in the slab can be found by the strip method, based on familiar beam analysis. It is important to note, too, that the load on the supporting beams is easily found because it can be computed from the end reactions of the slab beam strips in all cases. This information is not available from solutions such as those obtained by the yield line theory.

15.4 RECTANGULAR SLABS

With rectangular slabs, it is reasonable to assume that throughout most of the area the load will be carried in the short direction, consistent with elastic theory (see Section 12.4). In addition, it is important to take into account the fact that because of their length, longitudinal reinforcing bars will be more expensive than transverse bars of the same size and spacing. For a uniformly loaded rectangular slab on simple supports, Hillerborg presents one possible division, as shown in Fig. 15.4, with discontinuity lines originating from the slab corners at an angle depending on the ratio of short to long sides of the slab. All of the load in each region is assumed to be carried in the directions indicated by the arrows.

Instead of the solution of Fig. 15.4, which requires continuously varying reinforcement to be strictly correct, Hillerborg suggests that the load can be distributed as shown in Fig. 15.5, with discontinuity lines parallel to the sides of the slab. For such cases, it is reasonable to take edge bands of width equal to one-fourth the short span dimension. Here the load in the corners is divided equally in the X and Y directions as shown, while elsewhere all of the load is carried in the direction indicated by the arrows.

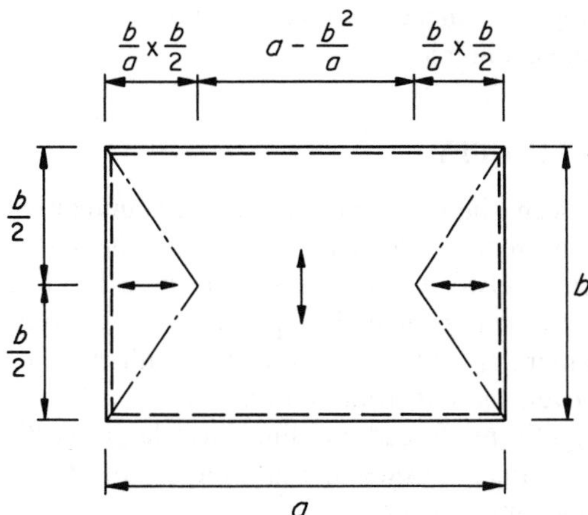

FIGURE 15.4
Rectangular slab with discontinuity lines originating at the corners.

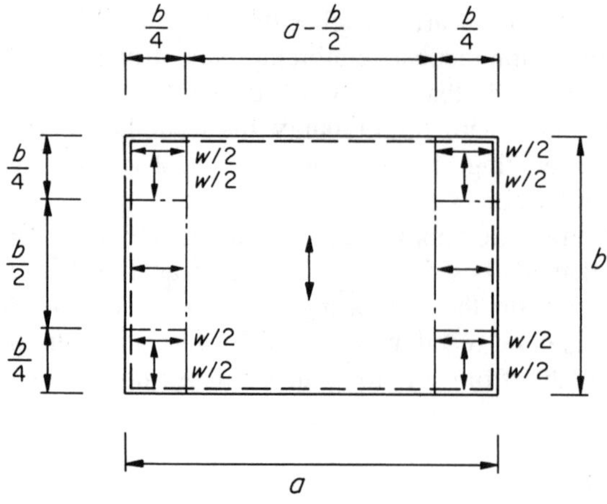

FIGURE 15.5
Discontinuity lines parallel to the sides for a rectangular slab.

The second, preferred arrangement, shown in Fig. 15.5, gives design moments as follows:

In the X direction:

Side strips:
$$m_x = \frac{w}{2} \times \frac{b}{4} \times \frac{b}{8} = \frac{wb^2}{64} \tag{15.7a}$$

Middle strips:
$$m_x = w \times \frac{b}{4} \times \frac{b}{8} = \frac{wb^2}{32} \tag{15.7b}$$

In the Y direction:

Side strips:
$$m_y = \frac{wb^2}{64} \tag{15.8a}$$

Middle strips:
$$m_y = \frac{wb^2}{8} \tag{15.8b}$$

This distribution, requiring no averaging of moments across band widths, is always on the safe side and is both simple and economical.

15.5 FIXED EDGES AND CONTINUITY

Designing by the strip method has been shown to provide a large amount of flexibility in assigning load to various regions of slabs. This same flexibility extends to the assignment of moments between negative and positive bending sections of slabs that are fixed or continuous over their supported edges. Some attention should be paid to elastic moment ratios to avoid problems with cracking and deflection at service loads. However, the redistribution that can be achieved in slabs, which are typically rather lightly reinforced and thus have large plastic rotation capacities when overloaded, permits considerable arbitrary readjustment of the ratio of negative to positive moments in a strip.

This is illustrated by Fig. 15.6, which shows a slab strip carrying loads only near the supports and unloaded in the central region, such as often occurs in designing by the strip method. It is convenient if the unloaded region is subject to a constant moment (and zero shear) because this simplifies the selection of positive reinforcement. The sum of the absolute values of positive span moment and negative end moment at the left or right end, shown as m_l and m_r in Fig. 15.6, depends only on the conditions at the respective end and is numerically equal to the negative moment if the strip carries the load as a cantilever. Thus, in determining moments for design, one calculates the "cantilever" moments, selects the span moment, and determines the corresponding support moments. Hillerborg notes that, as a general rule for fixed edges, the support moment should be about 1.5 to 2.5 times the span moment in the same strip. Higher values should be chosen for longitudinal strips that are largely unloaded, and in such cases a ratio of support to span moment of 3 to 4 may be used. However, little will be gained by using such a high ratio if the positive moment steel is controlled by minimum requirements of the Code.

For slab strips with one end fixed and one end simply supported, the dual goals of constant moment in the unloaded central region and a suitable ratio of negative to positive moments govern the location to be chosen for the discontinuity lines. Figure 15.7*a* shows a uniformly loaded rectangular slab having two adjacent edges fixed and the other two edges simply supported. Note that although the middle strips have the same width as those of Fig. 15.5, the discontinuity lines are shifted to account for the greater stiffness of the strips with fixed ends. Their location is defined by a coefficient α, with a value clearly less than 0.5 for the slab shown, its exact value yet to be determined. It will be seen that the selection of α relates directly to the ratio of negative to positive moments in the strips.

The moment curve of Fig. 15.7*b* is chosen so that moment is constant over the unloaded part, i.e., shearing force is zero. With constant moment, the positive steel can be fully stressed over most of the strip. The maximum positive moment in the *X* direction middle strip is then

$$m_{xf} = \frac{\alpha w b}{2} \times \frac{\alpha b}{4} = \alpha^2 \frac{w b^2}{8} \tag{15.9}$$

The cantilever moment at the left support is

$$m_x = (1 - \alpha)\frac{wb}{2}(1 - \alpha)\frac{b}{4} = (1 - \alpha)^2 \frac{wb^2}{8} \tag{15.10}$$

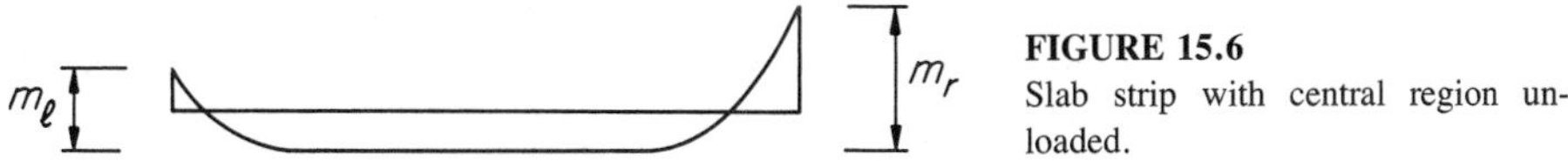

FIGURE 15.6
Slab strip with central region unloaded.

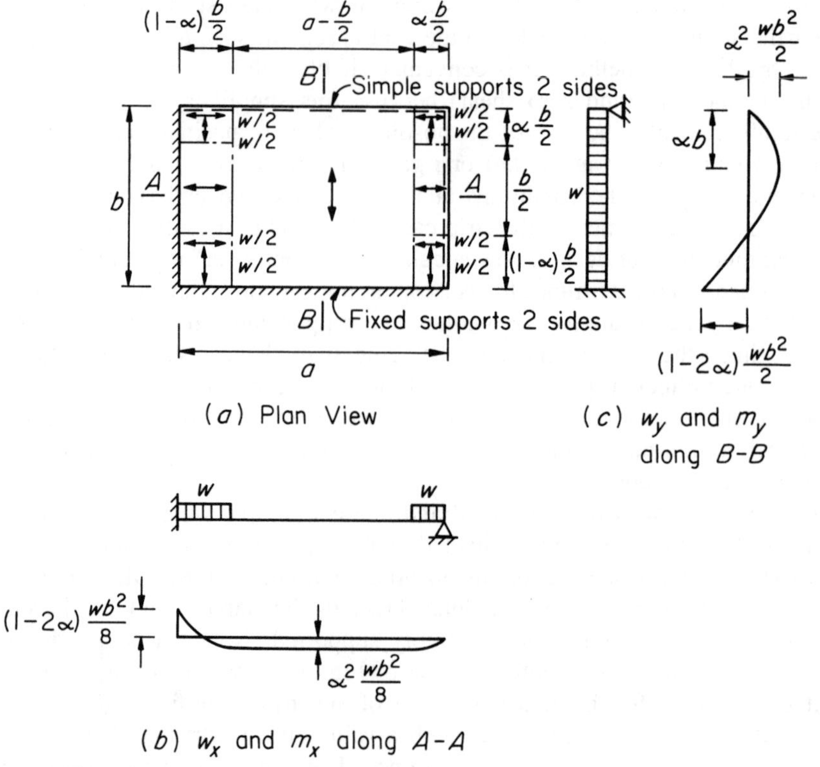

FIGURE 15.7
Rectangular slabs with two edges fixed and two edges simply supported.

and so the negative moment at the left support is

$$m_{xs} = (1-\alpha)^2 \frac{wb^2}{8} - \alpha^2 \frac{wb^2}{8} = (1-2\alpha)\frac{wb^2}{8} \tag{15.11}$$

For reference, the ratio of negative to positive moments in the X direction middle strip is

$$\frac{m_{xs}}{m_{xf}} = \frac{1-2\alpha}{\alpha^2} \tag{15.12}$$

The moments in the X direction edge strips are one-half of those in the middle strips because the load is half as great.

It is reasonable to choose the same ratio between support and span moments in the Y direction as in the X direction. Accordingly, the distance from the right support, Fig. 15.7c, to the maximum positive moment section is chosen as αb. It follows that the maximum positive moment is

$$m_{yf} = \alpha wb \times \frac{\alpha b}{2} = \alpha^2 \frac{wb^2}{2} \tag{15.13}$$

Applying the same methods as used for the X direction shows that the negative support moment in the Y direction middle strips is

$$m_{ys} = (1 - 2\alpha)\frac{wb^2}{2} \qquad (15.14)$$

It is easily confirmed that the moments in the Y direction edge strips are just one-eighth of those in Y direction middle strip.

With the above expressions, all the design moments for the slab can be found once a suitable value for α is chosen. From Eq. (15.12) it can be confirmed that values of α from 0.35 to 0.39 give corresponding ratios of negative to positive moments from 2.45 to 1.45, the range recommended by Hillerborg. For example, if it is decided that support moments are to be twice the span moments, the value of α should be 0.366, and the negative and positive moments in the central strip in the Y direction are respectively $0.134wb^2$ and $0.067wb^2$. In the middle strip in the X direction, moments are one-fourth those values; and in the edge strips in both directions, they are one-eighth of those values.

Example 15.1 Rectangular slab with fixed edges. Figure 15.8 shows a typical interior panel of a slab floor in which support is provided by beams on all column lines. Normally proportioned beams will be stiff enough, both flexurally and torsionally, that the slab can be assumed fully restrained on all sides. Clear spans for the slab, face to face of beams, are 25 ft and 20 ft as shown. The floor must carry a service live load of 130 psf, using concrete of strength $f_c' = 3000$ psi and steel with $f_y = 60,000$ psi. Find the moments at all critical sections, and determine the required slab thickness and reinforcement.

Solution. The minimum slab thickness required by ACI Code can be found from Eq. (13.8*b*) of Chapter 13, with $l_n = 25$ ft and $\beta = 1.25$:

$$h = \frac{25 \times 12(0.8 + 60/200)}{36 + 9 \times 1.25} = 6.98 \text{ in.}$$

A total depth of 6.75 in. will be selected, for which $w_d = 150 \times 6.75/12 = 84$ psf. Applying the usual factors of 1.4 and 1.7 to dead load and live load, respectively, determines that the total factored load for design is 338 psf. For strip analysis, discontinuity lines will be selected as shown in Fig. 15.8, with edge strips of width $b/4 = 20/4 = 5$ ft. In the corners, the load is divided equally in the two directions; elsewhere, 100 percent of the load is assigned to the direction indicated by the arrows. A ratio of support moment to span moment of 2.0 will be used. Calculation of moments then proceeds as follows:

X direction middle strip:

cantilever $\qquad m_x = \dfrac{wb^2}{32} = 338 \times \dfrac{400}{32} = 4225$ ft-lb/ft

negative $\qquad m_{xs} = 4225 \times \dfrac{2}{3} = 2817$

positive $\qquad m_{xf} = 4225 \times \dfrac{1}{3} = 1408$

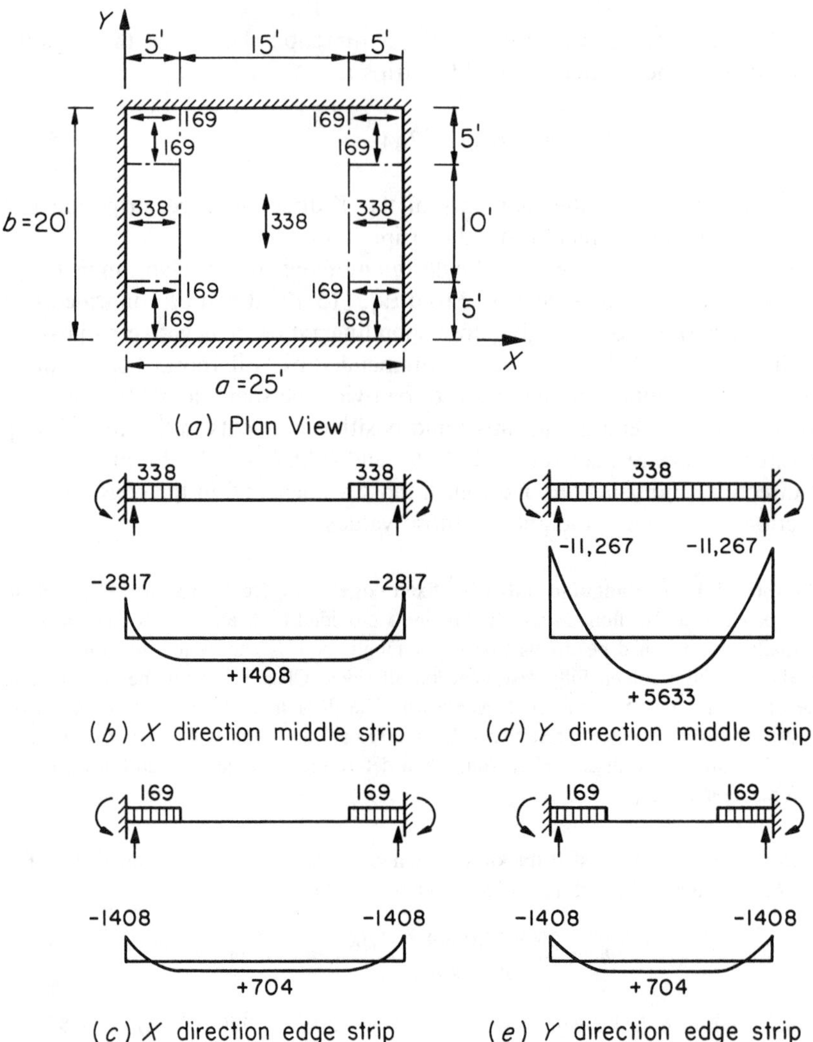

FIGURE 15.8
Design example: two-way slab with fixed edges.

X direction edge strips:

cantilever $$m_x = \frac{wb^2}{64} = 2113 \text{ ft-lb/ft}$$

negative $$m_{xs} = 2113 \times \frac{2}{3} = 1408$$

positive $$m_{xf} = 2113 \times \frac{1}{3} = 704$$

Y direction middle strip:

cantilever $$m_y = \frac{wb^2}{8} = 338 \times \frac{400}{8} = 16,900 \text{ ft-lb/ft}$$

negative $\quad m_{ys} = 16{,}900 \times \frac{2}{3} = 11,267$

positive $\quad m_{yf} = 16{,}900 \times \frac{1}{3} = 5633$

Y direction edge strips:

cantilever $\quad m_y = \frac{wb^2}{64} = 2113$ ft-lb/ft

negative $\quad m_{ys} = 2113 \times \frac{2}{3} = 1408$

positive $\quad m_{yf} = 2113 \times \frac{1}{3} = 704$

Strip loads and moment diagrams are as shown in Fig. 15.8. According to ACI Code 7.12, the minimum steel required for shrinkage and temperature crack control is $0.0018 \times 6.75 \times 12 = 0.146$ in^2/ft strip. With a total depth of 6.75 in., with $\frac{3}{4}$ in. concrete cover, and with estimated bar diameters of $\frac{1}{2}$ in., the effective depth of the slab in the short direction will be 5.75 in., and in the long direction, 5.25 in. Accordingly, the flexural steel ratio provided by the minimum steel acting at the smaller effective depth is

$$\rho_{\min} = \frac{0.146}{5.25 \times 12} = 0.0023$$

From Table A.6a of Appendix A, $R = 134$, and the flexural design strength is

$$\phi m_n = \phi R b d^2 = \frac{0.9 \times 134 \times 12 \times 5.25^2}{12} = 3324 \text{ ft-lb/ft}$$

Comparing this with the required moment resistance shows that the minimum steel will be adequate in the X direction in both middle and edge strips and in the Y direction edge strips. No. 3 bars at 9 in. spacing will provide the needed area. In the Y direction middle strip, for negative bending,

$$R = \frac{m_u}{\phi b d^2} = \frac{11,267 \times 12}{0.9 \times 12 \times 5.75^2} = 379$$

and from Table A.6a, the required steel ratio is 0.0069. The required steel is then

$$A_s = 0.0069 \times 12 \times 5.75 = 0.48 \text{ in}^2/\text{ft}$$

This will be provided with No. 5 bars at 8 in. on centers. For positive bending

$$R = \frac{5633 \times 12}{0.90 \times 12 \times 5.75^2} = 189$$

for which $\rho = 0.0033$, and the required positive steel area per strip is

$$A_s = 0.0033 \times 12 \times 5.75 = 0.23 \text{ in}^2/\text{ft}$$

to be provided by No. 4 bars on 10 in. centers. Note that slight adjustments downward and upward have been made in the steel required at negative and positive bending sections, as permitted by ACI Code 8.4, to arrive at practical bar spacings. Note also that all bar spacings are less than $2h = 2 \times 6.75 = 13.5$ in., as required by the Code, and that the steel ratios are well below the maximum permitted value of 0.0160.

Negative bar cutoff points can easily be calculated from the moment diagrams. For the X direction middle strip, the point of inflection a distance x from the left edge is found as follows:

$$1690x - 2817 - 338\frac{x^2}{2} = 0$$

$$x = 2.11 \text{ ft}$$

According to the Code, the negative bars must be continued at least d or $12d_b$ beyond that point, requiring a 6 in. extension in this case. Thus, the negative bars will be cut off 2.11 + 0.50 = 2.61 ft, say 2 ft 8 in., from the face of support. The same result is obtained for the X direction edge strips and the Y direction edge strips. For the Y direction middle strip, the distance $y = 4.21$ ft from face of support to inflection point is found in a similar manner. In this case, with No. 5 bars used, the required extension is 7.5 in., giving a total length past the face of supports of 4.21 + 0.63 = 4.84 ft, or 4 ft 10 in. All positive bars will be carried 6 in. into the face of the supporting beams.

15.6 UNSUPPORTED EDGES

The slabs considered in the preceding sections, uniformly loaded and with supports on four sides, could also have been designed by the methods of Chapter 12. The real power of the strip method becomes evident when dealing with nonstandard problems, such as slabs with an unsupported edge, slabs with holes, or slabs with reentrant corners (L-shaped slabs).

For a slab with one edge unsupported, for example, a reasonable basis for analysis by the simple strip method is that a strip along the unsupported edge takes a greater load per unit area than the actual unit load acting, i.e., that the strip along the unsupported edge acts as a support for the strips at right angles. Such strips have been referred to by Wood and Armer as "strong bands" (Ref. 15.8). A strong band is, in effect, an integral beam, usually having the same total depth as the remainder of the slab but containing a concentration of reinforcement. The strip may be made deeper than the rest of the slab to increase its carrying capacity, but this will not usually be necessary.

Figure 15.9*a* shows a rectangular slab carrying a uniformly distributed ultimate load w per unit area, with fixed edges along three sides and no support along one short side. Discontinuity lines are chosen as shown. The load on a unit middle strip in the X direction, shown in Fig. 15.9*b*, includes the downward load w in the region adjacent to the fixed left edge and the upward reaction kw in the region adjacent to the free edge. Summing moments about the left end, with moments positive clockwise and with the unknown support moment denoted m_{xs}, gives

$$m_{xs} + \frac{wb^2}{32} - \frac{kwb}{4}\left(a - \frac{b}{8}\right) = 0$$

from which

$$k = \frac{1 + 32m_{xs}/wb^2}{8(a/b) - 1} \qquad (15.15)$$

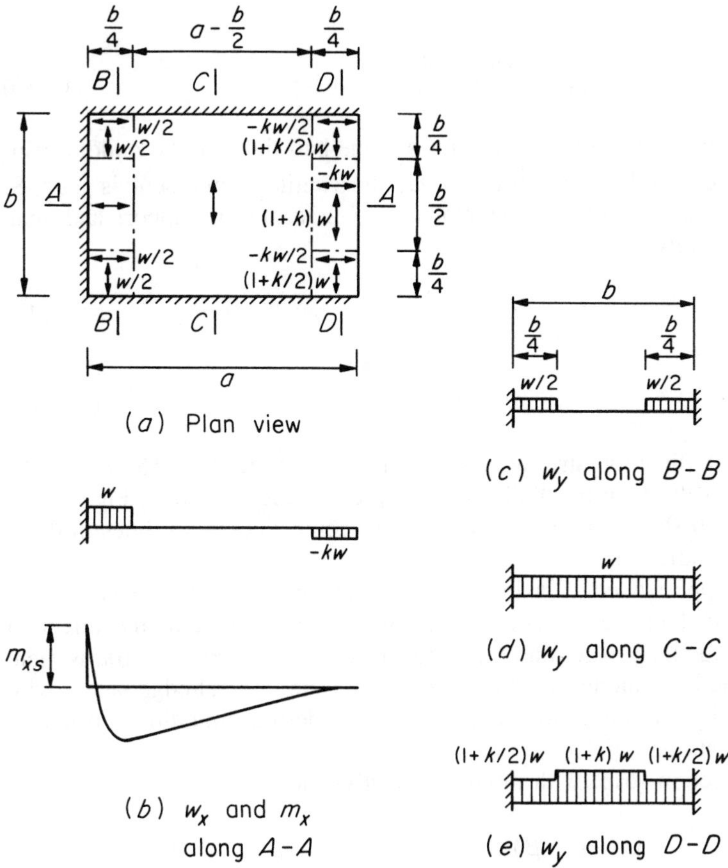

FIGURE 15.9
Slab with free edge along short side.

Thus, k can be calculated after the support moment is selected.

The appropriate value of m_{xs} to be used in Eq. (15.15) will depend on the shape of the slab. If a is large relative to b, the strong band in the Y direction at the edge will be relatively stiff, and the moment at the left support of the X direction strips will approach the elastic value for a propped cantilever. If the slab is nearly square, the deflection of the strong band would tend to increase the support moment; a value about half the free cantilever moment might be selected (Ref. 15.14).

Once m_{xs} is selected and k is known, it is easily shown that the maximum span moment occurs when

$$x = (1 - k)\frac{b}{4}$$

It has a value

$$m_{xf} = \frac{kwb^2}{32}\left(\frac{8a}{b} - 3 + k\right) \tag{15.16}$$

The moments in the X direction edge strips are one-half those in the middle strip. In the Y direction middle strip, Fig. 15.9d, the cantilever moment is $wb^2/8$. Adopting a ratio of support to span moment of 2 results in support and span moments, respectively, of

$$m_{ys} = \frac{wb^2}{12} \tag{15.17a}$$

$$m_{yf} = \frac{wb^2}{24} \tag{15.17b}$$

Moments in the Y direction strip adjacent to the fixed edge, Fig. 15.9c, will be one-eighth those values. In the Y direction strip along the free edge, Fig. 15.9e, moments can, with slight conservatism, be made equal to $(1 + k)$ times those in the Y direction middle strip.

If the unsupported edge is in the long-span direction, then a significant fraction of the load in the slab central region will be carried in the direction perpendicular to the long edges and the simple distribution of Fig. 15.10a is more suitable. A strong band along the free edge serves as an integral edge beam, with width βb normally chosen as low as possible considering limitations on tensile steel ratio in the strong band.

For a Y direction strip, with moments positive clockwise,

$$m_{ys} + \frac{1}{2}k_1 w(1 - \beta)^2 b^2 - k_2 w\beta b^2 (1 - \beta/2) = 0$$

from which

$$k_2 = \frac{k_1(1 - \beta)^2 + 2m_{ys}/wb^2}{\beta(2 - \beta)} \tag{15.18}$$

The value of k_1 may be selected so as to make use of the minimum steel in the X direction required by ACI Code 7.12. In choosing m_{ys} to be used in Eq. (15.18) for calculating k_2, one should again recognize that the deflection of the strong band along the free edge will tend to increase the Y direction moment at the supported edge above the propped cantilever value based on zero deflection. A value for m_{ys} of about half the free cantilever moment may be appropriate in typical cases. A high ratio of a/b will permit greater deflection of the free edge through the central region, tending to increase the support moment, and a low ratio will restrict deflection, reducing the support moment.

Example 15.2 Rectangular slab with long edge unsupported. The 12 ft × 19 ft slab shown in Fig. 15.11a, with three fixed edges and one long edge unsupported, must carry a uniformly distributed service live load of 110 psf. Concrete strength will be $f_c' = 4000$ psi, and rebars will have $f_y = 60,000$ psi. Select an appropriate slab

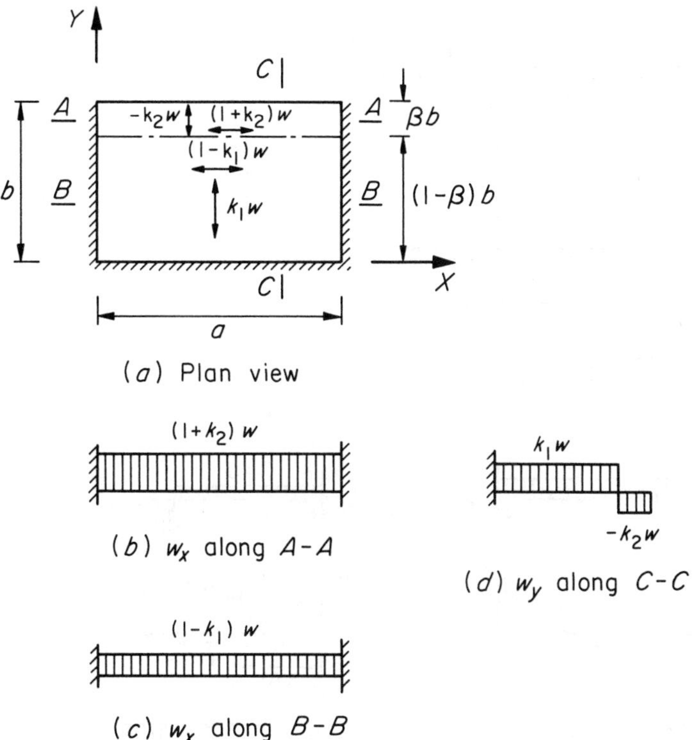

FIGURE 15.10
Slab with free edge in long-span direction.

thickness, determine all design moments, and select reinforcing bars and spacings for the slab.

Solution. The minimum thickness requirements of the ACI Code do not really apply to the type of slab considered here. However, Eq. (13.8*c*) of Chapter 13, which controls for beamless flat plates, can be applied conservatively because, although the present slab is beamless along the free edge, it has infinitely stiff supports on the other three edges. From Eq. (13.8*c*), with $l_n = 19$ ft,

$$h = \frac{19 \times 12(0.8 + 60/200)}{36} = 6.97 \text{ in.}$$

A total depth of 7 in. will be selected. The slab dead load is $150 \times \frac{7}{12} = 88$ psf, and the total factored design load is $1.4 \times 88 + 1.7 \times 110 = 310$ psf.

A strong band 2 ft wide will be provided for support along the free edge. In the main slab, a value $k_1 = 0.45$ will be selected, resulting in a slab load in the *Y* direction of $0.45 \times 310 = 140$ psf and in the *X* direction of $0.55 \times 310 = 170$ psf.

First, with regard to the *Y* direction slab strips, the negative moment at the supported edge will be chosen as one-half the free cantilever value, which in turn will be approximated based on 140 psf over an 11 ft distance from the support face to the center

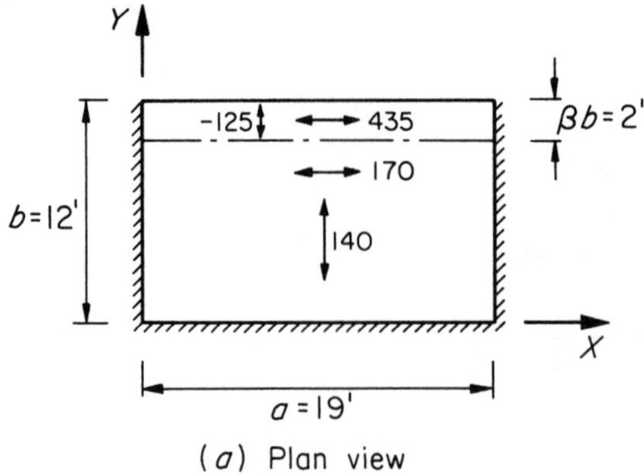

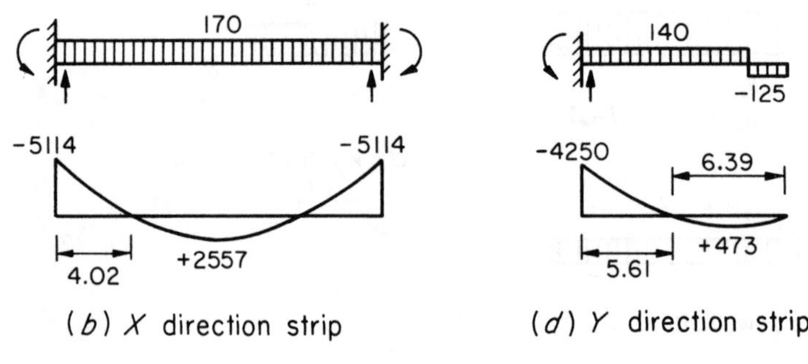

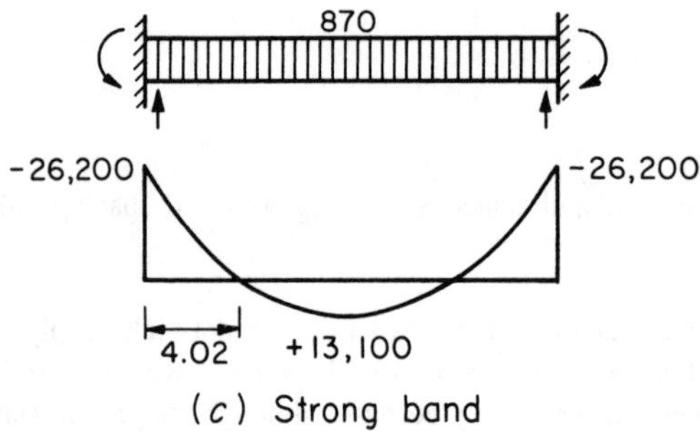

FIGURE 15.11
Design example: slab with long edge unsupported.

of the strong band. The restraining moment is thus

$$m_{ys} = \frac{1}{2} \times \frac{140 \times 11^2}{2} = 4235 \text{ ft-lb/ft}$$

Then, from Eq. (15.6)

$$k_2 = \frac{0.45(5/6)^2 - 2 \times 4235/(310 \times 144)}{(1/6)(2 - 1/6)} = 0.402$$

Thus, an uplift of 0.402 × 310 = 125 psf will be provided for the Y direction strips by the strong band, as shown in Fig. 15.11d. For this loading, the negative moment at the left support is

$$m_{ys} = 140 \times \frac{10^2}{2} - 125 \times 2 \times 11 = 4250 \text{ ft-lb/ft}$$

The difference from the original value of 4235 ft-lb/ft is caused by slight rounding errors introduced in the load terms. The statically consistent value of 4250 ft-lb/ft will be used for design. The maximum positive moment in the Y direction strips will be located at the point of zero shear. With y_1 as the distance of that point from the free edge to the zero shear location, and with reference to Fig. 15.11d,

$$125 \times 2 - 140(y_1 - 2) = 0$$

from which $y_1 = 3.79$ ft. The maximum positive moment, found at that location, is

$$m_{yf} = 125 \times 2(3.79 - 1) - 140 \times \frac{1.79^2}{2} = 473 \text{ ft-lb/ft}$$

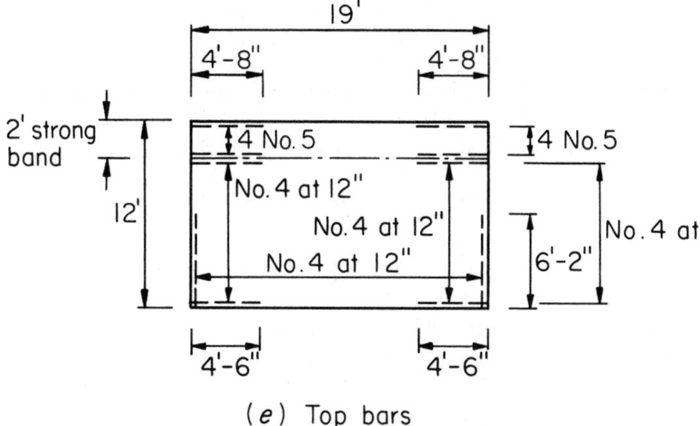

(*e*) Top bars

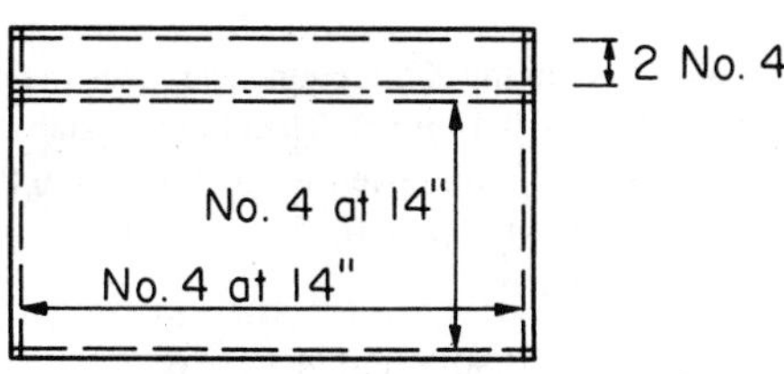

(*f*) Bottom bars

FIGURE 15.11
(*Continued*)

For later reference in cutting off bars, the point of inflection is located a distance y_2 from the free edge:

$$125 \times 2(y_2 - 1) - \frac{140}{2}(y_2 - 2)^2 = 0$$

resulting in $y_2 = 6.39$ ft.

For the X direction slab strips, the cantilever moment is

$$\text{cantilever } m_x = \frac{170 \times 19^2}{8} = 7671 \text{ ft-lb/ft}$$

A ratio of negative to positive moments of 2.0 will be chosen here, resulting in negative and positive moments, respectively, of

$$\text{negative} \quad m_{xs} = 7671 \times \frac{2}{3} = 5114 \text{ ft-lb/ft}$$

$$\text{positive} \quad m_{xf} = 7671 \times \frac{1}{3} = 2557 \text{ ft-lb/ft}$$

as shown in Fig. 15.11*b*.

The unit load on the strong band in the X direction is

$$(1 + k_2)w = (1 + 0.402) \times 310 = 435 \text{ psf}$$

so for the 2 ft wide band the load per foot is $2 \times 435 = 870$ plf, as indicated in Fig. 15.11*c*. The cantilever, negative, and positive strong band moments are respectively

$$\text{cantilever} \quad M_x = 870 \times 19^2/8 = 39{,}300 \text{ ft-lb}$$

$$\text{negative} \quad M_{xs} = 39{,}300 \times \frac{2}{3} = 26{,}200 \text{ ft-lb}$$

$$\text{positive} \quad M_{xf} = 39{,}300 \times \frac{1}{3} = 13{,}100 \text{ ft-lb}$$

With negative moment of $-26{,}200$ ft-lb and support reaction of $870 \times \frac{19}{2} = 8265$ lb, the point of inflection in the strong band is found as follows:

$$-26{,}200 + 8265x - \frac{870x^2}{2} = 0$$

giving $x = 4.02$ ft. The inflection point in the X direction slab strips will be at the same location.

In designing the slab steel in the X direction, one notes the minimum steel required by ACI Code is $0.0018 \times 7 \times 12 = 0.15$ in^2/ft. The effective slab depth in the X direction, assuming $\frac{1}{2}$ in. diameter bars with $\frac{3}{4}$ in. cover, is $7.0 - 1.0 = 6.0$ in. The corresponding flexural steel ratio in the X direction is $\rho = 0.15/(12 \times 6) = 0.0021$. From Table A.6a, $R = 124$, and the design strength is

$$\phi m_n = \phi R b d^2 = \frac{0.90 \times 124 \times 12 \times 6^2}{12} = 4018 \text{ ft-lb/ft}$$

It is seen that the minimum slab steel required by Code will provide for the positive bending moment of 2557 ft-lb/ft. The requirement of 0.15 in^2/ft could be met by No. 3 bars at 9 in. spacing, but to reduce placement costs, No. 4 bars at the maximum permitted spacing of $2h = 14$ in. will be selected, providing 0.17 in^2/ft. The X direction negative moment of 5114 ft-lb/ft requires

$$R = \frac{m_u}{\phi b d^2} = \frac{5114 \times 12}{0.90 \times 12 \times 6^2} = 158$$

and Table A.6a indicates that the required $\rho = 0.0027$. Thus, the negative bar requirement is $A_s = 0.0027 \times 12 \times 6 = 0.19$ in^2/ft. This will be provided by No. 4 bars at 12 in. spacing, continued $4.02 \times 12 + 6 = 54$ in., or 4 ft 6 in., from the support face.

In the Y direction, the effective depth will be one bar diameter less than in the X direction, or 5.5 in. Thus, the flexural steel ratio provided by the shrinkage and temperature steel is $\rho = 0.15/(12 \times 5.5) = 0.0023$. This results in $R = 135$, so the design strength is

$$\phi m_n = \frac{0.90 \times 135 \times 12 \times 5.5^2}{12} = 3675 \text{ ft-lb/ft}$$

well above the requirement for positive bending of 473 ft-lb/ft. No. 4 bars at 14 in. will be satisfactory for positive steel in this direction also. For the negative moment of 4250 ft-lb/ft,

$$R = \frac{4250 \times 12}{0.90 \times 12 \times 5.5^2} = 156$$

and from Table A.6a, the required $\rho = 0.0027$. The corresponding steel requirement is $0.0027 \times 12 \times 5.5 = 0.18$ in^2/ft. No. 4 bars at 12 in. will be used, and they will be extended $5.61 \times 12 + 6 = 74$ in., or 6 ft 2 in., past the support face.

In the strong band, the positive moment of 13,100 ft-lb requires

$$R = \frac{13{,}100 \times 12}{0.90 \times 24 \times 6^2} = 202$$

The corresponding steel ratio is 0.0035, and the required bar area is $0.0035 \times 24 \times 6 = 0.38$ in^2. This can be provided by two No. 4 bars, only slightly more steel than in a typical X direction slab strip. For the negative moment of 26,200 ft-lb,

$$R = \frac{26{,}200 \times 12}{0.90 \times 24 \times 6^2} = 404$$

resulting in $\rho = 0.0072$, and required steel $0.0072 \times 24 \times 6 = 1.04$ in^2. Four No. 5 bars, providing an area of 1.23 in^2, will be used, and they will be cut off $4.02 \times 12 + 7.5 = 56$ in., or 4 ft 8 in., from the support face.

The final arrangement of bar reinforcement is shown in Fig. 15.11*e*. Negative bar cutoff locations are as indicated, and development by embedded lengths into the supports will be provided. All positive bars in the slab and strong band will be carried 6 in. into the support faces.

A design problem commonly met in practice is that of a slab supported along three edges and unsupported along the fourth, with a distributed load that increases linearly from zero along the free edge to a maximum at the opposite supported edge. Examples include the wall of a rectangular tank subjected to liquid pressure and earth-retaining walls with buttresses or counterforts (see Sec. 19.1).

Fig. 15.12 shows such a slab, with load of intensity w_0 at the long, supported edge, reducing to zero at the free edge. In the main part of the slab, a constant load $k_2 w_0$ is carried in the X direction, as shown in Fig. 15.12c; thus, a constant load $k_2 w_0$ is deducted from the linear varying load in the Y direction, as shown in Fig. 15.12d. Along the free edge, a strong band of width βb is provided, carrying a load $k_1 w_0$, as in Fig. 15.12a, and so providing an uplift load equal to that amount at the end of the Y direction strip in Fig. 15.12d. The choice of k_1 and k_2 depends on the ratio of a/b. If this ratio is high, k_2 should be chosen with regard to the minimum slab reinforcement required by the ACI Code. The value of k_1 is then calculated by statics, based on a selected value of the restraining moment at the fixed edge, say one-half the free cantilever value. In many cases it will be convenient to let k_1 equal k_2. Then it is the support moment that follows from statics. The value of β is selected as low as possible considering the upper limit on tensile steel ratio in the strong band imposed by the Code for beams. The strong band is designed for a load of intensity $k_1 w_0$ distributed uniformly over its width βb.

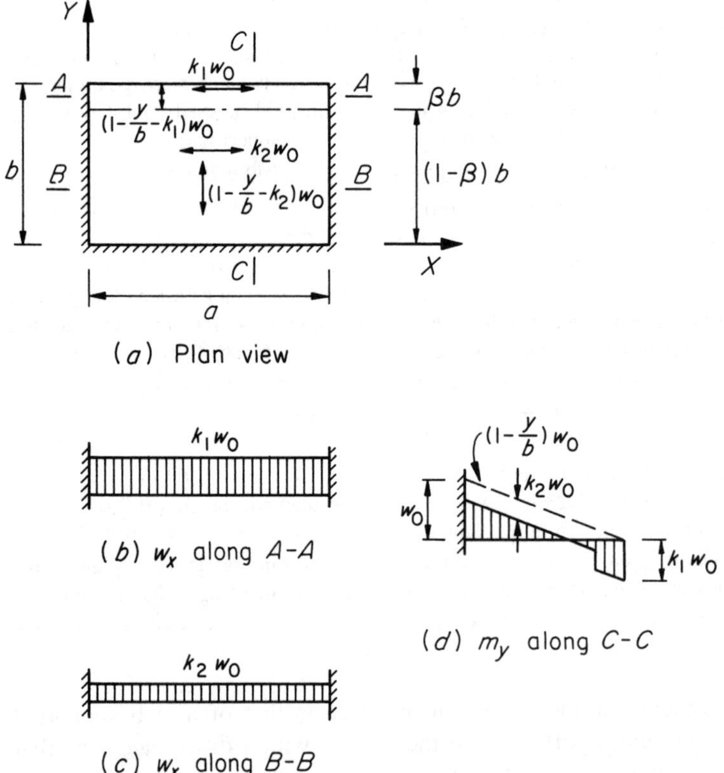

FIGURE 15.12
Slab with one free edge and linearly varying load.

15.7 SLABS WITH HOLES

Slabs with small openings can usually be designed as if there were no openings, replacing the interrupted steel with bands of rebars of equivalent area on either side of the opening in each direction (see Sec. 13.8). Slabs with larger openings must be treated more rigorously. The strip method offers a rational and safe basis for design in such cases. Integral load-carrying beams are provided along the edges of the opening, usually having the same depth as the remainder of the slab but with extra reinforcement, to pick up the load from the affected regions and transmit it to the supports. In general, these integral beams should be chosen so as to carry the loads most directly to the supported edges of the slab. The width of the strong bands should be selected so that the steel ratios are at or below the maximum permitted for beams by the ACI Code, thus insuring against over-reinforcement and compression failure.

Use of the strip method for analysis and design of a slab with a large central hole will be illustrated by the following example.

Example 15.3 Rectangular slab with central opening. Figure 15.13*a* shows a 16 ft × 28 ft slab with fixed supports along all four sides. A central opening 4 ft × 8 ft must be accommodated. Estimated slab thickness, from Eq. (13.8*b*) of chapter 13, is 7 in. The slab is to carry a uniformly distributed factored load of 300 psf, including self-weight. Devise an appropriate system of strong bands to reinforce the opening, and determine moments to be resisted at all critical sections of the slab.

Solution. The basic pattern of discontinuity lines and load dispersion will be selected according to Fig. 15.5. Edge strips are defined having width $\frac{16}{4}$ = 4 ft. In the corners, the load is equally divided in the two directions. In the central region, 100 percent of the load is assigned to the Y direction, while along the central part of the short edges, 100 percent of the load is carried in the X direction. Moments for this "basic case" without the hole will be calculated and later used as a guide in selecting moments for the actual slab with hole. A ratio of support to span moments of 2.0 will be used generally, as for the previous examples. Moments for the slab, neglecting the hole, would then be as follows:

X direction middle strips:

cantilever $$m_x = \frac{wb^2}{32} = 300 \times \frac{16^2}{32} = 2400 \text{ ft-lb/ft}$$

negative $$m_{xs} = 2400 \times \frac{2}{3} = 1600$$

positive $$m_{xf} = 2400 \times \frac{1}{3} = 800$$

X direction edge strips are $\frac{1}{2}$ middle strip values.

Y direction middle strips:

cantilever $$m_y = \frac{wb^2}{8} = 300 \times \frac{16^2}{8} = 9600 \text{ ft-lb/ft}$$

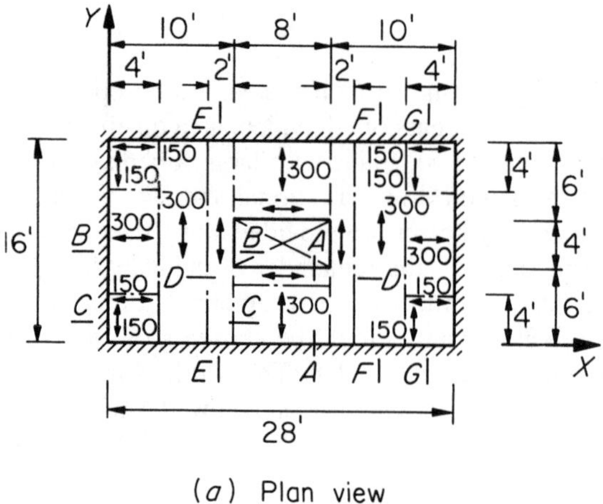

(a) Plan view

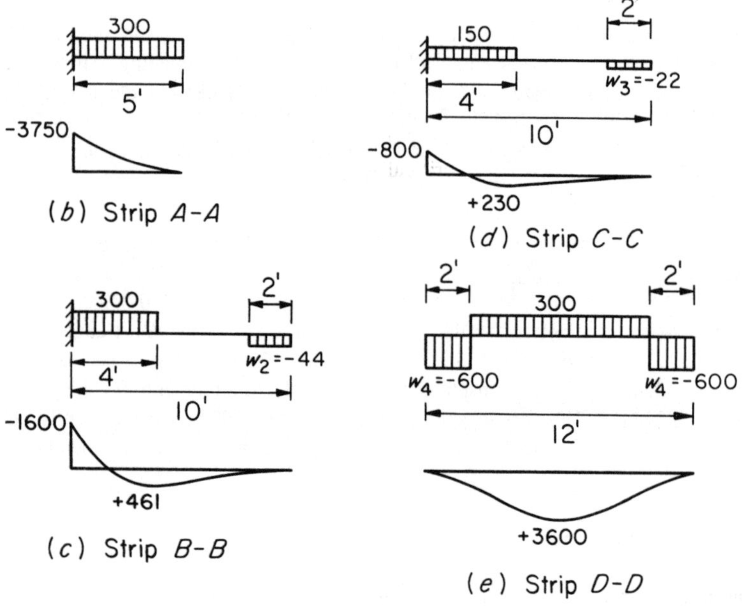

FIGURE 15.13
Design example: slab with central hole.

$$\text{negative} \qquad m_{ys} = 9600 \times \frac{2}{3} = 6400$$

$$\text{positive} \qquad m_{yf} = 9600 \times \frac{1}{3} = 3200$$

Y direction edge strips are $\frac{1}{8}$ middle strip values.

Because of the hole, certain strips lack support at one end. To support them, 1 ft wide strong bands will be provided in the X direction at the long edges of the hole

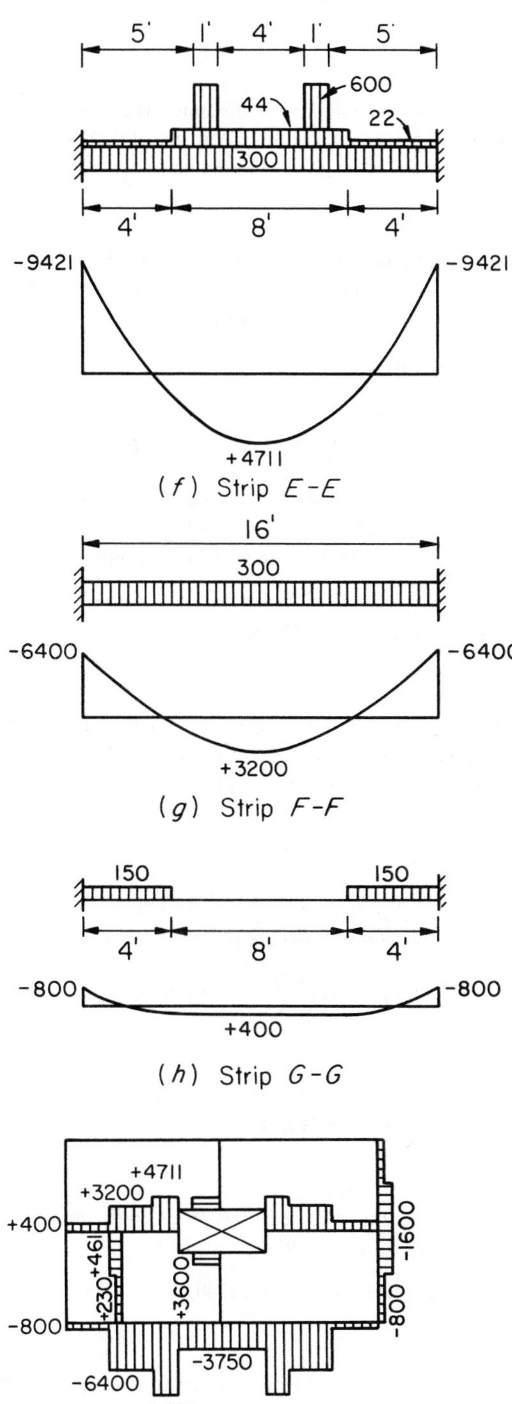

FIGURE 15.13
(*Continued*)

and 2 ft wide strong bands in the Y direction on each side of the hole. The Y direction bands will provide for the reactions of the X direction bands. With the distribution of loads shown in Fig. 15.13*a*, strip reactions and moments are found as follows:

Strip A-A

It may at first be assumed that propped cantilever action is obtained, with the restraint moment along the slab edge taken as 6400 ft-lb/ft, the same as for the basic case. Summing moments about the left end of the loaded strip then results in

$$w_1 = \frac{300 \times 6 \times 3 - 6400}{1 \times 5.5} = -182 \text{ psf}$$

The negative value indicates that the cantilever strips are serving as supports for strip *D-D*, and in turn for the strong bands in the Y direction, which is hardly a reasonable assumption. Instead, a discontinuity line will be assumed 5 ft from the support, as shown in Fig. 15.13*b*, terminating the cantilever and leaving the 1 ft strip *D-D* along the edge of the opening in the X direction to carry its own load. It follows that the support moment in the cantilever strip is

$$\text{negative } m_{ys} = 300 \times 5 \times 2.5 = 3750 \text{ ft-lb/ft}$$

Strip B-B

The restraint moment at the supported edge will be taken to be the same as the basic case, i.e., 1600 ft-lb/ft. Summing moments about the left end of the strip of Fig. 15.13*c* then results in an uplift reaction at the right end, to be provided by strip *E-E*, of

$$w_2 = \frac{300 \times 4 \times 2 - 1600}{2 \times 9} = 44 \text{ psf}$$

The left reaction is easily found to be 1112 lb, and the point of zero shear is 3.70 ft from the left support. The maximum positive moment, at that point, is

$$\text{positive } m_{xf} = 1112 \times 3.70 - 1600 - 300\frac{3.70^2}{2} = 461 \text{ ft-lb/ft}$$

Strip C-C

Negative and positive moments and the reaction to be provided by strip *E-E*, as shown in Fig. 15.13*d*, are all $\frac{1}{2}$ the corresponding values for strip *B-B*.

Strip D-D

The 1 ft wide strip carries 300 psf in the X direction with reactions provided by the strong bands *E-E*, as shown in Fig. 15.13*e*. The maximum positive moment is

$$m_{xf} = 600 \times 2 \times 5 - 300 \times 4 \times 2 = 3600 \text{ ft-lb/ft}$$

Strip E-E

In reference to Fig. 15.13*f*, the strong bands in the Y direction carry the directly applied load of 300 psf plus the 44 psf load from strip *B-B*, the 22 psf load from strip *C-C*, and the 600 psf end reaction from strip *D-D*. For strip *E-E* the cantilever, negative, and positive moments are

$$\begin{aligned}\text{cantilever} \quad m_y &= 300 \times 8 \times 4 + 22 \times 4 \times 2 + 44 \times 4 \times 6 + 600 \times 1 \times 5.5 \\ &= 14{,}132 \text{ ft-lb/ft}\end{aligned}$$

$$\text{negative} \qquad m_{ys} = 14{,}132 \times \frac{2}{3} = 9421$$

$$\text{positive} \qquad m_{yf} = 14{,}132 \times \frac{1}{3} = 4711$$

It should be emphasized that the loads shown are psf and would be multiplied by 2 to obtain loads per foot acting on the strong bands. Correspondingly, the moments just obtained are per foot width and must be multiplied by 2 to give the support and span moments for the 2 ft wide strong band.

Strip F-F

The moments for the *Y* direction middle strip of the basic case may be used without change; thus, in Fig. 15.13*g*,

$$\text{negative} \qquad m_{ys} = 6400 \text{ ft-lb/ft}$$

$$\text{positive} \qquad m_{yf} = 3200$$

Strip G-G

Moments for the *Y* direction edge strips of the basic case are used without change, resulting in

$$\text{negative} \qquad m_{ys} = 800 \text{ ft-lb/ft}$$

$$\text{positive} \qquad m_{yf} = 400$$

as shown in Fig. 15.13*h*.

The final distribution of moments across the negative and positive critical sections of the slab is shown in Fig. 15.13*i*. The selection of reinforcing bars and determination of cutoff points would follow the same methods as presented in Examples 15.1 and 15.2 and will not be given here. Reinforcing bar ratios needed in the strong bands are well below the maximum permitted for the 7 in. slab depth.

It should be noted that strips *B-B*, *C-C*, and *D-D* have been designed as if they were simply supported at the strong band *E-E*. To avoid undesirably wide cracks where these strips pass over the strong band, nominal negative reinforcement should be added in this region. Positive bars should be extended fully into the strong bands.

15.8 THE ADVANCED STRIP METHOD

The simple strip method described in the earlier sections of this chapter is not directly suitable for the design of slabs supported by columns (e.g., flat plates) or slabs supported at reentrant corners.† For such cases, Hillerborg introduced the advanced strip method (Refs. 15.2, 15.5, 15.12, and 15.13.).

† However, Wood and Armer, in Ref. 15.8, suggest that beamless slabs with column supports can be solved by the simple strip method through the use of strong bands between columns or between columns and exterior walls.

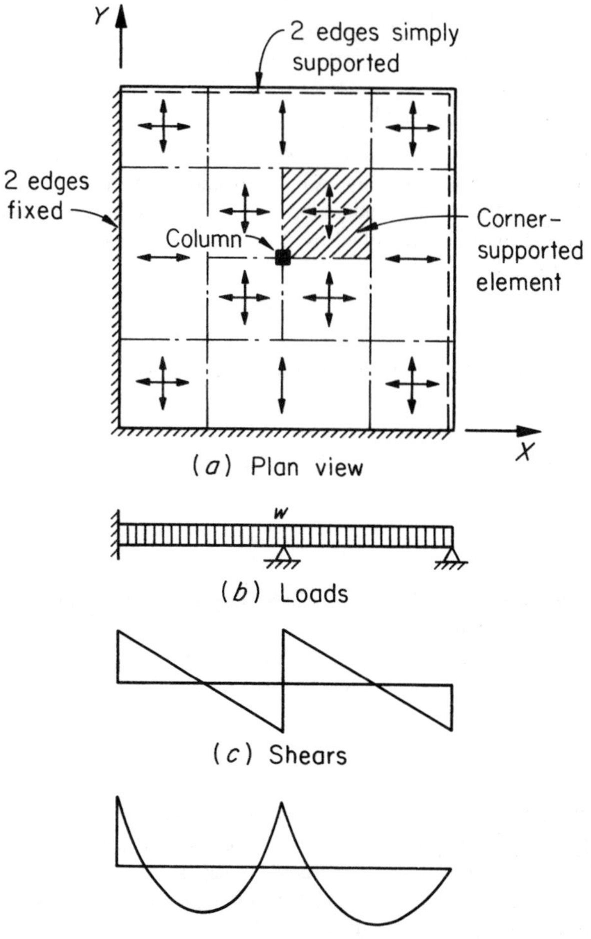

FIGURE 15.14
Slab with central supporting column.

Fundamental to the advanced strip method is the corner-supported element, such as that shown shaded in Fig. 15.14*a*. The corner-supported element is a rectangular region of the slab with the following properties:

1. The edges are parallel to the reinforcement directions.
2. It carries a uniform load w per unit area.
3. It is supported at only one corner.
4. No shear forces act along the edges.
5. No twisting moments act along the edges.
6. All bending moments acting along an edge have the same sign or are zero.
7. The bending moments along the edges are the design moments for the reinforcing bars.

A uniformly loaded strip in the X direction, shown in Fig. 15.14*b*, will thus have shear and moment diagrams as shown in Fig. 15.14*c* and Fig. 15.14*d*, respectively. Maximum moments are located at the lines of zero shear. The outer edges of the corner-supported element are defined at the lines of zero shear in both X and Y directions.

A typical corner-supported element, with an assumed distribution of moments along the edges, is shown in Fig. 15.15. It will be assumed that the bending moment is constant along each half of each edge. The vertical reaction is found by summing vertical forces:

$$R = wab \tag{15.19}$$

and moment equilibrium about the Y axis gives

$$m_{xfm} - m_{xsm} = \frac{wa^2}{2} \tag{15.20}$$

where m_{xfm} and m_{xsm} are the mean span and support moments per unit width, and beam sign convention is followed. Similarly,

$$m_{yfm} - m_{ysm} = \frac{wb^2}{2} \tag{15.21}$$

The last two equations are identical with the condition for a corresponding part of a simple strip—Eq. (15.20) spanning in the X direction and Eq. (15.21) in the Y direction—supported at the axis and carrying the load wb or wa per foot. So

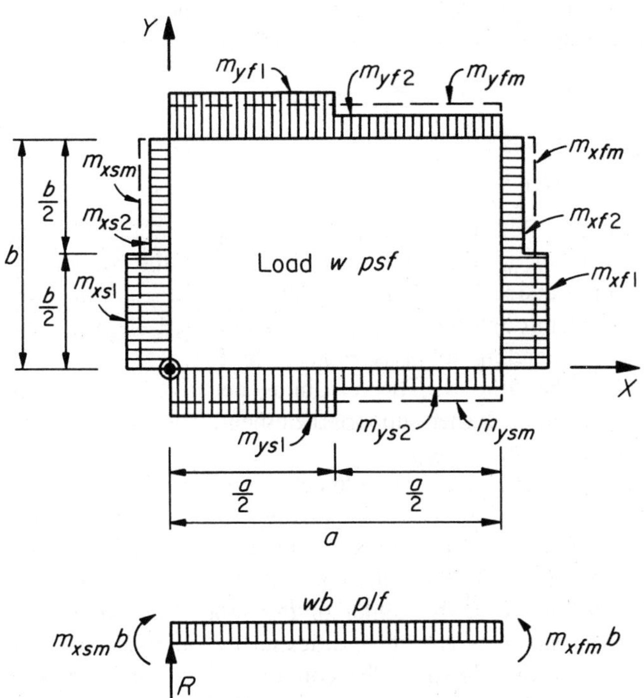

FIGURE 15.15
Corner-supported element.

if the corner-supported element forms a part of a strip, that part should carry 100 percent of the load w in each direction. (This requirement was discussed earlier in Chapter 13 and is simply a requirement of static equilibrium.)

The distribution of moments within the boundaries of a corner-supported element is complex. With the load on the element carried by a single vertical reaction at one corner, strong twisting moments must be present within the element; this contrasts with the assumptions of the simple strip method used previously.

The moment field within a corner-supported element and its edge moments have been explored in great detail in Ref. 15.12. It is essential that the edge moments, given in Fig. 15.15, are the design moments for the reinforcing bars (i.e., that nowhere within the element will a bar be subjected to a greater moment than at the edges). To meet this requirement, a limitation must be put on the moment distribution along the edges. Based on his studies (Ref. 15.12), Hillerborg has recommended the following restriction on edge moments:

$$m_{xf2} - m_{xs2} = \alpha \frac{wa^2}{2} \tag{15.22a}$$

with

$$0.25 \le \alpha \le 0.7 \tag{15.22b}$$

where m_{xf2} and m_{xs2} are the positive and negative X direction moments, respectively, in the outer half of the element, as shown in Fig. 15.15. The corresponding restriction applies in the Y direction. He notes further that for most practical applications, the edge moment distribution shown in Fig. 15.16 is appropriate, with

$$m_{xf1} = m_{xf2} = m_{xfm} \tag{15.23}$$

$$m_{xs2} = 0 \tag{15.24a}$$

$$m_{xs1} = 2m_{xsm} \tag{15.24b}$$

(Alternatively, it is suggested in Ref. 15.14 that negative support moments across the column line be taken at $1.5m_{xsm}$ in the half-element width by the column and at $0.5m_{xsm}$ in the remaining outside half-element width.) Positive reinforcement in the span should be carried through the whole corner-supported element. The negative reinforcement corresponding to $m_{xs1} - m_{xs2}$ of Fig. 15.15 must be extended at least $0.6a$ from the support. The remaining negative steel, if any, should be carried through the whole corner-supported element. The corresponding restrictions apply in the Y direction.

In practical applications, corner-supported elements are combined with each other and with parts of one-way strips, as shown in Fig. 15.14, to form a system of strips. In this system, each strip carries the total load w, as discussed earlier. In laying out the elements and strips, the concentrated corner support for the element may be assumed to be at the center of the supporting column, as shown in Fig. 15.14, unless supports are of significant size. In that case, the corner support may be taken at the corner of the column, and an ordinary simple strip may be included that spans between the column faces, along the edge of the corner-supported elements. Note in the figure that the corner regions of the slab

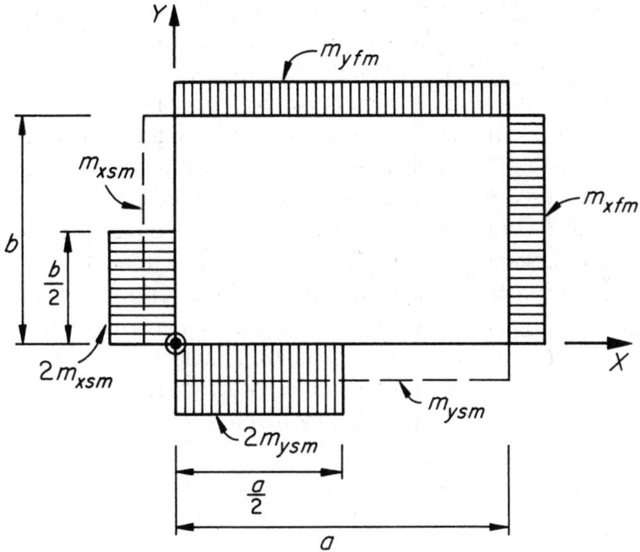

FIGURE 15.16
Recommended distribution of moments for typical corner-supported element.

are not included in the main strips that include the corner-supported elements. These may safely be designed for one-third of the corresponding moments in the main strips (Ref. 15.13.).

Example 15.4 Edge-supported flat plate with central column. Figure 15.17*a* illustrates a flat plate with overall dimensions 34 ft × 34 ft, with fixed supports along the left and lower edges in the sketch, hinged supports at the right and upper edges, and a single central column 16 in. square. It must carry a service live load of 30 psf over its entire surface plus its own weight and an additional superimposed dead load of 4 psf. Find the moments at all critical sections, and determine the required slab thickness and reinforcement. Material strengths are specified at $f_y = 60{,}000$ psi and $f_c' = 4000$ psi.

Solution. A trial slab depth will be chosen based on Eq. (13.8*c*), which governs for flat plates. It will be conservative for the present case, where continuous support is provided along the outer edges.

$$h = \frac{17 \times 12(0.8 + 60/200)}{36} = 6.23 \text{ in.}$$

A thickness of 6.5 in. will be selected tentatively, for which the self-weight is 150 × 6.5/12 = 81 psf. The total factored load to be carried is thus:

$$w_u = 1.4(81 + 4) + 1.7 \times 30 = 170 \text{ psf}$$

The average strip moments in the *X* direction in the central region caused by the load of 170 psf are found by elastic theory and are shown in Fig. 15.17*c*. The analysis in the *Y* direction is identical. The points of zero shear (and maximum moments) are

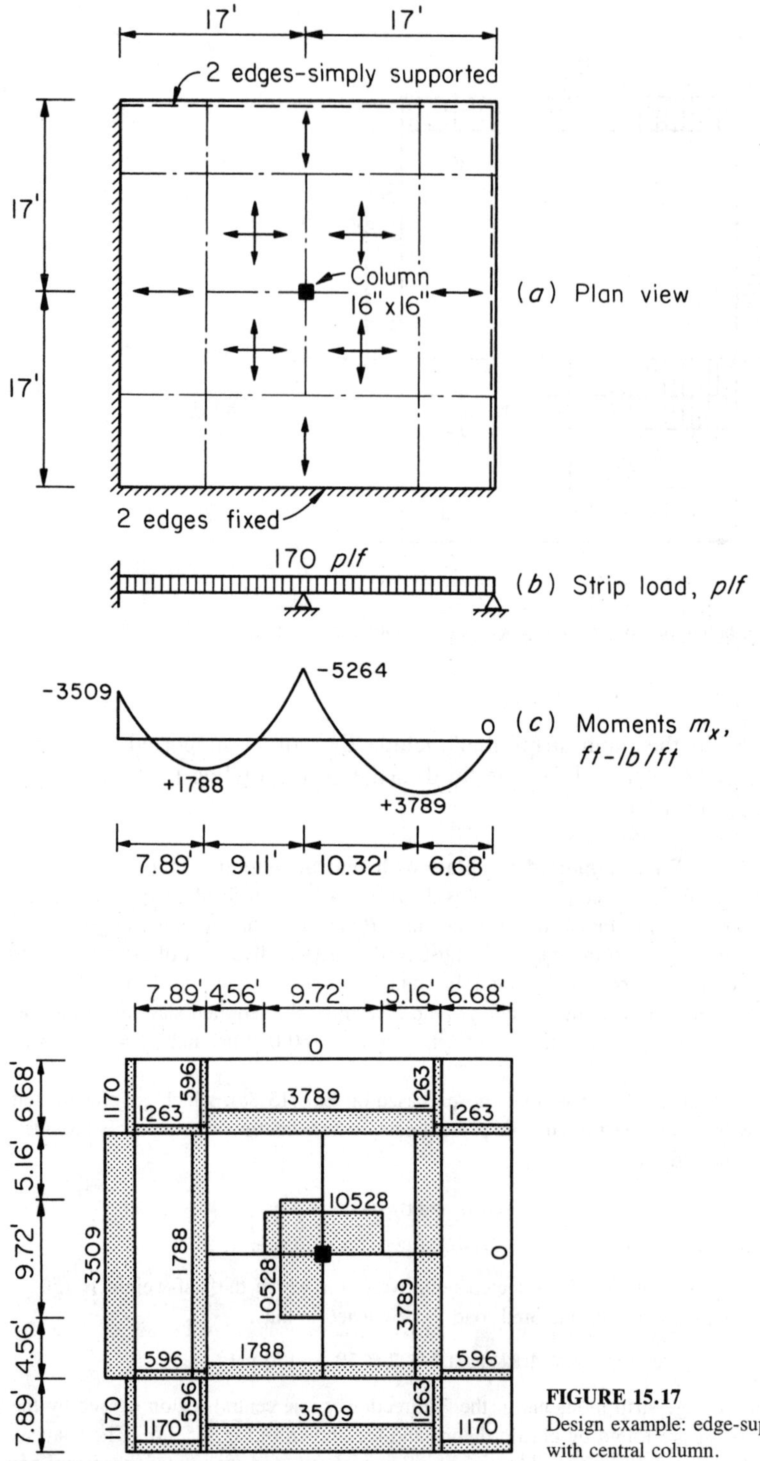

17'
17'
2 edges-simply supported
17'
17'
Column 16"x16"
(a) Plan view
2 edges fixed
170 plf
(b) Strip load, plf
-3509
-5264
0
+1788
+3789
(c) Moments m_x, ft-lb/ft
7.89'
9.11'
10.32'
6.68'
7.89' 4.56' 9.72' 5.16' 6.68'
0
1170
596
1263
3789
1263
1263
3509
1788
10528
10528
0
3789
596
1788
596
1170
596
3509
1263
1170
1170
6.68'
5.16'
9.72'
4.56'
7.89'

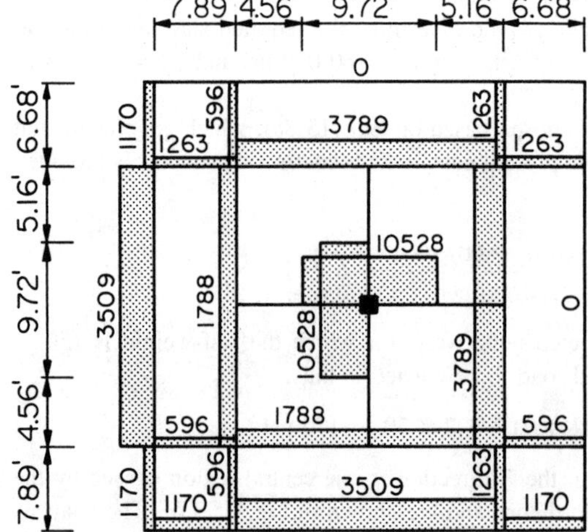

FIGURE 15.17
Design example: edge-supported flat plate with central column.

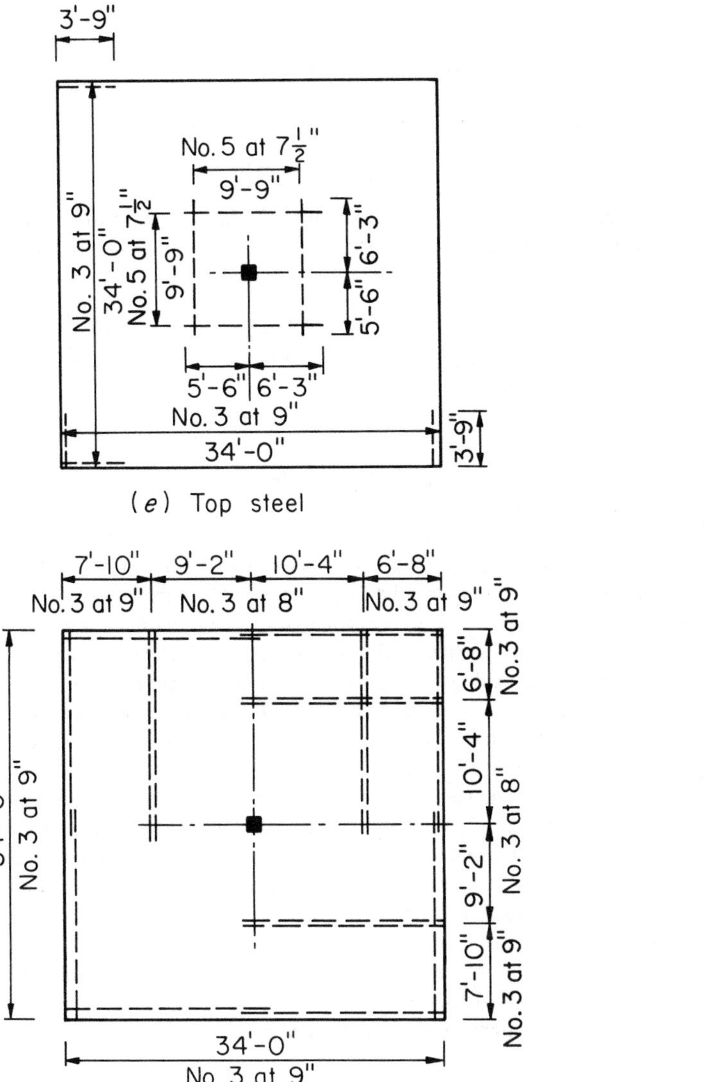

FIGURE 15.17
(Continued)

located 9.11 ft to the left of the column, and 10.32 ft to the right, as indicated. These dimensions determine the size of the four corner-supported elements.

Moments in the slab are then determined according to the preceding recommendations. At the fixed edge along the left side of the main strips, the moment m_{xs} is simply the moment per foot strip from the elastic analysis, 3509 ft-lb/ft. At the left edge of the corner-supported element in the left span,

$$m_{xf1} = m_{xf2} = m_{xfm} = 1788 \text{ ft-lb/ft}$$

Along the centerline of the slab, over the column, following the recommendations shown in Fig. 15.16,

$$m_{xs2} = 0$$

$$m_{xs1} = 2m_{xsm} = 10,528 \text{ ft-lb/ft}$$

At the right edge of the corner-supported element in the right span,

$$m_{xf1} = m_{xf2} = m_{xfm} = 3789 \text{ ft-lb/ft}$$

At the outer, hinge-supported edge, all moments are zero. Make a check of the α values, using Eq. (15.22*b*), and note from Eq. (15.20) that $wa^2/2 = m_{xfm} - m_{xsm}$. Thus, in the left span,

$$\alpha = \frac{m_{xf2} - m_{xs2}}{wa^2/2} = \frac{1788 - 0}{1788 + 5264} = 0.25$$

and in the right span,

$$\alpha = \frac{3789 - 0}{3789 + 5264} = 0.42$$

Because both values are within the range of 0.25 to 0.75, the proposed distribution of moments is satisfactory. If the first value had been below the lower limit of 0.25, the negative moment in the column half-strip might have been reduced from 10,528 ft-kips per ft, and the negative moment in the adjacent half-strip might have been increased above the 0 value used. Alternatively, the total negative moment over the column might have been somewhat decreased, with a corresponding increase in span moments.

Moments in the *Y* direction correspond throughout, and all results are summarized in Fig. 15.17*d*. Moments in the strips adjacent to the supported edges are set equal to one-third of those in the adjacent main strips.

With moments per ft strip known at all critical sections, the required reinforcement is easily found. With a $\frac{3}{4}$ in. concrete cover and $\frac{1}{2}$ in. bar diameter, in general the effective depth of the slab will be 5.5 in. Where bar stacking occurs—i.e., over the central column and near the intersection of the two fixed edges—an average effective depth equal to 5.25 in will be used. This will result in reinforcement identical in the two directions and will simplify construction.

For the 6.5 in. thick slab, minimum steel for shrinkage and temperature crack control is $0.0018 \times 6.5 \times 12 = 0.140$ in^2/ft strip, which will be provided by No. 3 bars at 9 in. spacing. The corresponding flexural steel ratio is

$$\rho_{\min} = \frac{0.140}{5.5 \times 12} = 0.0021$$

Interpolating from Table A.6a of Appendix A makes $R = 124$, and the design strength is

$$\phi m_n = \phi R b d^2 = 0.90 \times 124 \times 12 \times 5.5^2/12 = 3376 \text{ ft-lb/ft}$$

In comparison with the required strengths summarized in Fig. 15.17*d*, this will be adequate everywhere except for particular regions as follows:

Negative steel over column:

$$R = \frac{m_u}{\phi b d^2} = \frac{10,528 \times 12}{0.90 \times 12 \times 5.25^2} = 424$$

for which $\rho = 0.0076$ (from Table A.6a), and $A_s = 0.0076 \times 12 \times 5.25 = 0.48$ in^2/ft.

This will be provided using No. 5 bars at 7.5 in. spacing. They will be continued a distance $0.6 \times 9.11 = 5.47$ ft, say 5 ft 6 in., to the left of the column centerline, and $0.6 \times 10.32 = 6.19$ ft, say 6 ft 3 in., to the right.

Negative steel along fixed edges:

$$R = \frac{3509 \times 12}{0.90 \times 12 \times 5.50^2} = 129$$

for which $\rho = 0.0022$ and $A_s = 0.0022 \times 12 \times 5.5 = 0.15$ in^2/ft. No. 3 bars at 9 in. spacing will be adequate. The point of inflection for the slab in this region is easily found to be 3.30 ft from the fixed edge. The negative bars will be extended 5.5 in. beyond that point, resulting in a cutoff 45 in., or 3 ft 9 in., from the support face.

Positive steel in outer spans:

$$R = \frac{3789 \times 12}{0.90 \times 12 \times 5.50^2} = 139$$

resulting in $\rho = 0.0024$ and $A_s = 0.0024 \times 12 \times 5.5 = 0.16$ in^2/ft. No. 3 bars at 8 in. spacing will be used. In all cases, the maximum spacing of $2h = 13$ in. is satisfied. That maximum would preclude the economical use of larger diameter bars.

Bar size and spacing and cutoff points for the top and bottom steel are summarized in Figs. 15.17*e* and 15.17*f*, respectively.

Finally, the load carried by the central column is

$$P = 170 \times 19.43 \times 19.43 = 64,200 \text{ lb}$$

Investigating punching shear at a critical section taken $d/2$ from the face of the 16 in. column, with reference to Eq. (13.11*a*) and with $b_0 = 4 \times (16.00 + 5.25) = 85$ in., gives

$$\phi V_c = 4\phi\sqrt{f'_c}\, b_0 d = 4 \times 0.85\sqrt{4000} \times 85 \times 5.25 = 96,000 \text{ lb}$$

This is well above the applied shear of 64,200 lb, confirming that the slab thickness is adequate and that no shear reinforcement is required.

15.9 COMPARISONS OF METHODS FOR SLAB ANALYSIS AND DESIGN

The conventional methods of slab analysis and design, as described in Chapters 12 and 13 and as treated in Chapter 13 of the ACI Code, are limited to applications in which slab panels are supported on opposite sides or on all four sides by beams or walls or to the case of flat plates and related forms supported by a relatively regular array of columns. In all cases, slab panels must be square or rectangular, loads must be uniformly distributed within each panel, and slabs must be free of significant holes.

Both the yield line theory and the strip method offer the designer rational methods for slab analysis and design over a much broader range, including the following:

1. Boundaries of any shape, including rectangular, triangular, circular, and L-shaped boundaries with reentrant corners.

2. Supported or unsupported edges, skewed supports, column supports, or various combinations of these conditions.
3. Uniformly distributed loads, loads distributed over partial panel areas, linear varying distributed loads, line loads, and concentrated loads.
4. Slabs with significant holes.

The most important difference between the strip method and the yield line method is the fact that the strip method produces results that are always on the safe side, but yield line analysis may result in unsafe designs. A slab designed by the strip method may possibly carry a higher load than estimated, through internal force redistributions, before collapse; a slab analyzed by yield line procedures may fail at a lower load than anticipated if an incorrect mechanism has been selected as the basis or if the defining dimensions are incorrect.

Beyond this, it should be realized that the strip method is a tool for *design*, by which the slab thickness and reinforcing bar size and distribution may be selected to resist the specified loads. In contrast, the yield line theory offers only a means for *analyzing the capacity of a given slab*, with known reinforcement. According to the yield line approach, the design process is actually a matter of reviewing the capacities of a number of trial designs and alternative reinforcing patterns. All possible yield line patterns must be investigated and specific dimensions varied in order to be sure that the correct solution has been found. Except for simple cases, this is likely to be a time-consuming process.

Neither the strip method nor the yield line approach provides any information regarding cracking or deflections at service load. Both focus attention strictly on flexural strength. However, by the strip method, if care is taken at least to approximate the elastic distribution of moments, little difficulty should be experienced with excessive cracking. The methods for deflection prediction presented earlier in Sec. 12.7 of Chapter 12 and Sec. 13.9 of Chapter 13 can, without difficulty, be adapted for use with the strip method, because the concepts are fully compatible.

With regard to economy of reinforcement, it might be supposed that use of the strip method, which always leads to designs on the safe side, might result in more expensive structures than the yield line theory. Comparisons, however, indicate that, in most cases, this is not so (Refs. 15.8 and 15.12). Through proper use of the strip method, reinforcing bars are placed in a nonuniform way in the slab (e.g., in strong bands around openings) where they are used to best effect; yield line methods, on the other hand, often lead to uniform bar spacings, which may mean that individual bars are used inefficiently.

Many tests have been conducted on slabs designed by the strip method (Ref. 15.11; also, see the summary in Ref. 15.12). These tests included square slabs, rectangular slabs, slabs with both fixed and simply supported edges, slabs supported directly by columns, and slabs with large openings. The conclusions drawn determine that the strip method provides for safe design with respect to ultimate load and that at service load, behavior with respect to cracking and deflections is generally satisfactory. The method has been widely and successfully used in the Scandinavian countries since the 1960s and is now receiving increasing attention elsewhere.

REFERENCES

15.1. A. Hillerborg, "Equilibrium Theory for Reinforced Concrete Slabs," (in Swedish), *Betong*, vol. 41, no. 4, 1956, pp. 171–182.
15.2. A. Hillerborg, *Strip Method for Slabs on Columns, L-Shaped Plates, Etc.*, (in Swedish), Svenska Riksbyggen, Stockholm, 1959.
15.3. A. Hillerborg, "A Plastic Theory for the Design of Reinforced Concrete Slabs," *Proc. Sixth Congr.*, International Association for Bridge and Structural Engineering, Stockholm, 1960.
15.4. R. E. Crawford, *Limit Design of Reinforced Concrete Slabs*, thesis submitted to University of Illinois for the degree of Ph.D., Urbana, IL, 1962.
15.5. F. A. Blakey, *Strip Method for Slabs on Columns, L-Shaped Plates, Etc.*, (translation of Ref. 15.2), Commonwealth Scientific and Industrial Research Organization, Melbourne, 1964.
15.6. K. O. Kemp, "A Lower Bound Solution to the Collapse of an Orthotropically Reinforced Slab on Simple Supports," *Mag. Concr. Res.*, vol. 14, no. 41, 1962, pp. 79–84.
15.7. K. O. Kemp, "Continuity Conditions in the Strip Method of Slab Design," *Proc. Inst. Civ. Eng.*, vol. 45, 1970, p. 283 (supplement paper 7268s).
15.8. R. H. Wood and G. S. T. Armer, "The Theory of the Strip Method for the Design of Slabs," *Proc. Inst. Civ. Eng.*, vol. 41, 1968, pp. 285–311.
15.9. R. H. Wood, "The Reinforcement of Slabs in Accordance with a Predetermined Field of Moments," *Concrete*, vol. 2, no. 2, 1968, pp. 69–76.
15.10. G. S. T. Armer, "The Strip Method: A New Approach to the Design of Slabs," *Concrete*, vol. 2, no. 9, 1968, pp. 358–363.
15.11. G. S. T. Armer, "Ultimate Load Tests of Slabs Designed by the Strip Method," *Proc. Inst. Civ. Eng.*, vol. 41, 1968, pp. 313–331.
15.12. A. Hillerborg, *Strip Method of Design*, Viewpoint Publications, Cement and Concrete Association, Wexham Springs, Slough, England, 1975.
15.13. A. Hillerborg, "The Advanced Strip Method—a Simple Design Tool," *Mag. Concr. Res.*, vol. 34, no. 121, 1982, pp. 175–181.
15.14. R. Park and W. L. Gamble, *Reinforced Concrete Slabs* (Chapter 6), John Wiley, New York, 1980, pp. 205–273.
15.15. S. Timoshenko and S. Woinowsky-Krieger, *Theory of Plates and Shells*, 2d ed., McGraw-Hill, New York, 1959.

PROBLEMS

Note: For all the following problems, use material strengths $f_y = 60{,}000$ psi and $f_c' = 4000$ psi. All ACI Code requirements for minimum steel, maximum spacings, bar cutoff, and special corner reinforcement are applicable.

15.1. The square slab of Fig. P15.1 is simply supported by masonry walls along all four sides. It is to carry a service live load of 100 psf in addition to its self-weight. Specify a suitable load distribution; determine moments at all controlling sections; and select the slab thickness, reinforcing bars, and spacing.

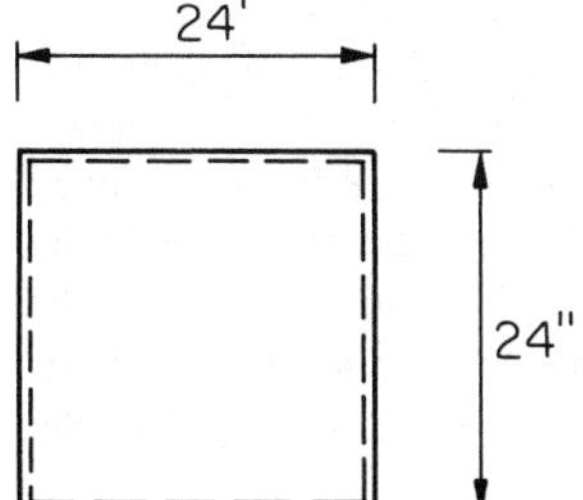

FIGURE P15.1

15.2. The rectangular slab shown in Fig. P15.2 is a typical interior panel of a large floor system having beams on all column lines. Columns and beam are sufficiently stiff that the slab can be considered fully restrained along all sides. A live load of 100 psf and a superimposed dead load of 30 psf must be carried in addition to the slab self-weight. Determine the required slab thickness, and specify all reinforcing bars and spacings. Cutoff points for negative bars should be specified; all positive steel may be carried into the supporting beams. Take support moments to be two times the span moments in the strips.

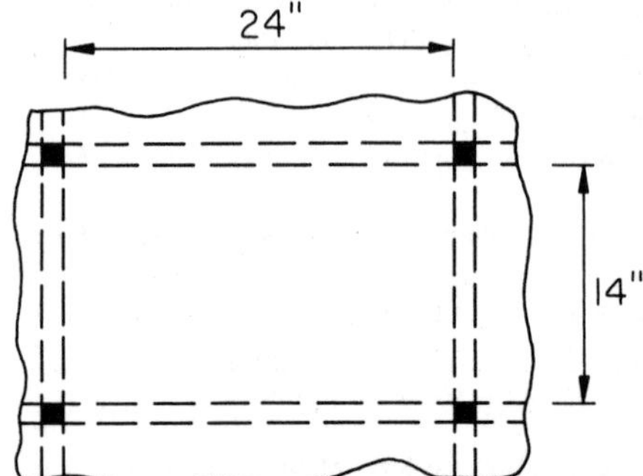

FIGURE P15.2

15.3. The slab of Figure P15.3 may be considered fully fixed along three edges, but it is without support along the fourth, long side. It must carry a uniformly distributed live load of 80 psf plus an external dead load of 40 psf. Specify a suitable slab depth, and determine reinforcement and cutoff points.

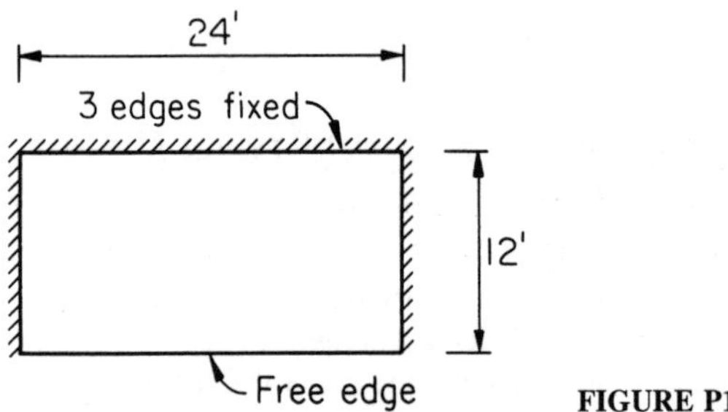

FIGURE P15.3

15.4. Figure P15.4 shows a counterfort retaining wall (see Sec. 19.1 of Chapter 19) consisting of a base slab and a main vertical wall of 16 ft height and constant thickness retaining the earth. Counterfort walls spaced at 19 ft on centers along the wall provide additional support for the main slab. Each section of the main slab, which is 16 ft high and 18 ft long, may be considered fully fixed at its base and also along its two vertical sides (because of full continuity and identical loadings on all such panels). The top of the main wall is without support. The horizontal earth pressure varies from 0 at the top of the wall to 587 psf at the top of the base slab. Determine a suitable thickness for the main wall, and select reinforcing bars and spacing.

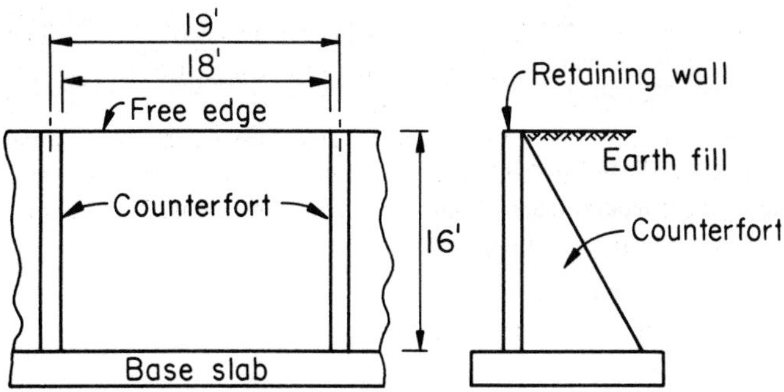

FIGURE P15.4

15.5. The triangular slab of Fig. P15.5, providing cover over a loading dock, is fully fixed along two adjacent sides and free of support along the diagonal edge. A uniform snow load of 60 psf is anticipated. Dead load of 10 psf will act, in addition to self-weight. Determine the required slab depth and specify all reinforcement. (Hint: The main bottom reinforcement should be parallel to the free edge, and the negative reinforcement should be perpendicular to the supported edges.)

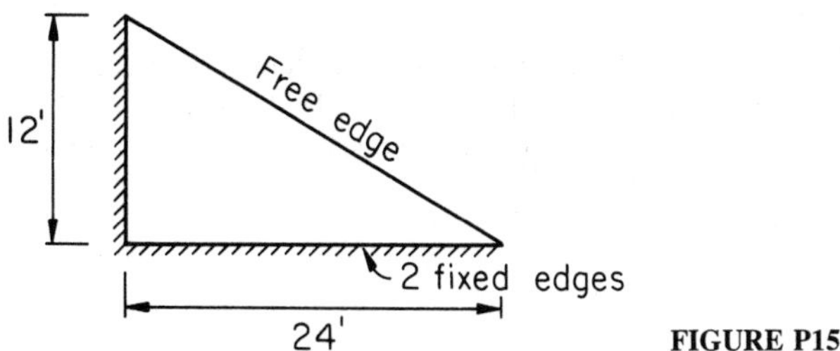

FIGURE P15.5

15.6. Figure P15.6 shows a rectangular slab with a large opening near one corner. It is simply supported along one long side and the adjacent short side, and the two edges adjacent to the opening are fully fixed. A factored load of 250 psf must be carried. Find the required slab thickness, and specify all reinforcement.

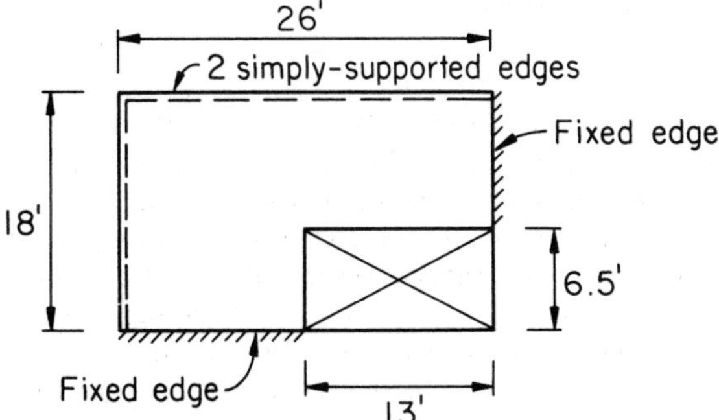

FIGURE P15.6

15.7. The roof deck slab of Fig. P15.7 is intended to carry a total factored load, including self-weight, of 165 psf. It will have fixed supports along the two long sides and one short side, but the fourth edge must be free of any support. Two 16 in. square columns will be located as shown.

(*a*) Determine an acceptable slab thickness.

(*b*) Select appropriate load dispersion lines.

(*c*) Determine moments at all critical sections.

(*d*) Specify bar sizes, spacings, and cutoff points.

(*e*) Check controlling sections in the slab for shear strength.

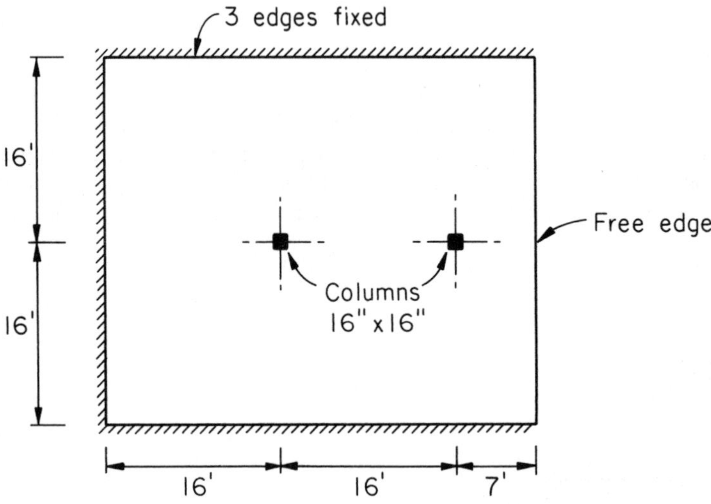

FIGURE P15.7

CHAPTER 16

SLABS ON GRADE

16.1 INTRODUCTION

Structural slabs are often supported directly by the natural ground surface, or by a prepared and compacted subbase and subgrade as shown in Fig. 16.1. Highway pavements, airport runways, and warehouse floors are common examples. The prepared subbase, consisting of crushed stone or gravel and generally from 8 to 16 in. thick, serves to provide more uniform support for the slab than if it were carried directly on the natural ground, and to improve the drainage of water from beneath the slab, particularly important in outdoor locations subject to freezing and thawing. The compacted subbase may be topped with sand, and often a waterproof membrane consisting of plastic film or asphalt-impregnated paper is used to avoid loss of fines from the freshly poured concrete and to reduce sliding friction under the slab. Reinforcement, not always used, may consist of two-way welded wire fabric or reinforcing bars.

If such a slab were loaded uniformly over its entire surface and supported by a perfectly uniform subbase, then stresses would be due mainly to concrete shrinkage and temperature changes. Such stresses alone may be sufficient to crack the slab, as will be shown. However, most foundation materials do not have ideally uniform properties, and most slabs are subjected to nonuniform loading as well. Truck wheels result in heavy concentrated loads on highway pavements. In

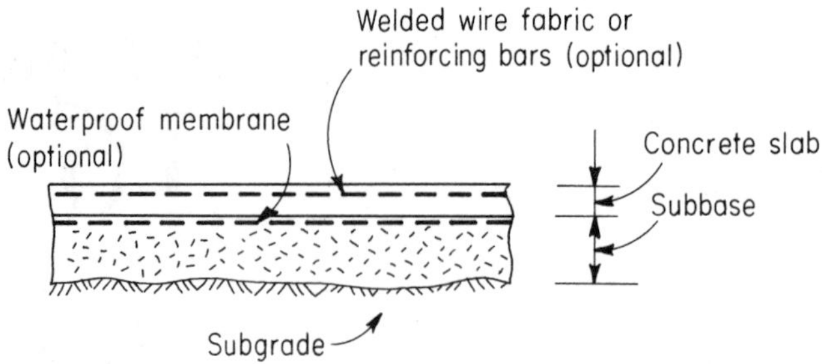

FIGURE 16.1
Typical slab on grade.

warehouses, the need to maintain clear aisles for access to stored materials often results in a checkerboard pattern of loading. The wheels of lift trucks used for materials handling cause heavy load concentrations.

Failures of ground-supported slabs are all too common. Unequal settlement, overloading, and restrained shrinkage and thermal displacements all tend to produce cracking. The passage of wheel loads over cracks or improperly made joints often leads to failure by progressive disintegration of the concrete. Slab failures, when they occur, are not spectacular and do not result in collapse in the usual sense, but the usefulness of the slab may be greatly impaired, and repairs are often costly. A remarkable example is provided in the U.S. interstate highway system; major sections are nearly impassable because of broken pavement, less than 25 years after construction.

Design methods for slabs on grade vary. There is a common theoretical basis that assumes highly idealized conditions, but results are modified in recognition of test data and practical experience. Generally the design is based on actual service loads (load factors equal to 1.0), and concrete stresses that are computed by elastic analysis are compared against specified limits. Steel reinforcement, if used, is placed mainly for crack control, although more modern methods of analysis and design account for its contribution in a structural sense.

In the following sections, a brief summary is given of methods of analysis and design of slabs on grade for shrinkage and temperature effects, and the effects of concentrated loads and loads on partial areas. Practical design methods for pavements and industrial slabs on grade are illustrated.

16.2 SHRINKAGE AND TEMPERATURE EFFECTS

Figure 16.2*a* shows a concrete pavement slab, unloaded except for its self-weight. Concrete shrinkage, or a decrease in temperature, tends to contract the slab, but this contraction is resisted by frictional drag between the slab and the subgrade. This causes tensile forces in the slab. If the total length of slab between contraction joints is l, then equilibrium of the horizontal forces shown in Fig. 16.2*b* for one

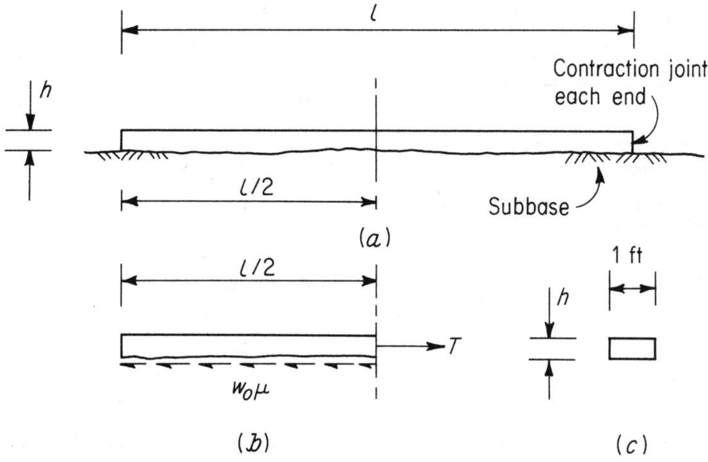

FIGURE 16.2
Subgrade drag effect: *(a)* Slab segment between joints; *(b)* Equilibrium of horizontal forces; *(c)* Cross section.

half the length, for a unit strip of slab, indicates that the tension force at a cross section at the midlength is

$$T = \frac{w_o \mu l}{2} \tag{16.1}$$

where w_o = self-weight of pavement slab, psf
μ = coefficient of friction between slab and subgrade
l = length between contraction or construction joints, ft
T = tensile force, lb per ft width of slab

The coefficient of friction varies widely, depending mainly on the roughness of the subgrade, and tests show a range between about 1.0 and 2.5 (Ref. 16.1). The coefficient may be less than 1.0 if plastic film is used between the slab and subgrade. For design of highway pavements, the *AASHTO Interim Guide* (Ref. 16.2) assumes a value of 1.5. If the slab has not cracked between contraction joints, the resulting concrete tensile stress, in psi units, is

$$f_t = \frac{T}{12h} \tag{16.2}$$

where h is the slab thickness in inches.

Commonly, there is a variation of temperature or concrete shrinkage through the slab depth. For example, the top surface, which is exposed to the air, will shrink more than the bottom, or a rapid drop in air temperature at night will affect the top surface more than the bottom. As a result, there is a tendency for the slab sections to curl upward, as shown in Fig. 16.3*a*. This is due to the difference in strains ϵ_{t1} and ϵ_{t2} at the top and bottom of the slab respectively, shown in Fig.

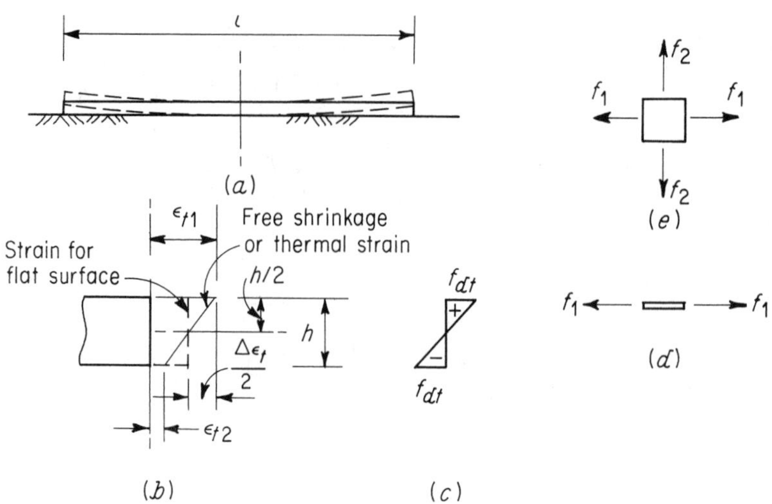

FIGURE 16.3
Effect of differential temperature or shrinkage: (*a*) warping of slab; (*b*) strains; (*c*) warping stresses; (*d*) and (*e*) biaxially-stressed element at top surface.

16.3*b*. However, in most cases the self-weight of the slab prevents this curling, enforcing a uniform distribution of strain as shown by the dotted line in Fig. 16.3*b*. This restraint induces warping stresses, as shown in Fig. 16.3*c*. Because this occurs in both horizontal directions, the result is a state of biaxial horizontal tension in a small element at the top surface (Fig. 16.3*d* and *e*).

For such a stress state, the tensile strain in direction 1 is

$$\epsilon_1 = \frac{f_1}{E_c} - \nu \frac{f_2}{E_c} \tag{a}$$

where E_c and ν are respectively the elastic modulus and Poisson's ratio for the concrete. In this case, $f_1 = f_2 = f_{dt}$, according to Fig. 16.3*c*. The flexural strain at the top surface is

$$\epsilon_1 = \frac{\epsilon_{t1} - \epsilon_{t2}}{2} = \frac{\Delta\epsilon_t}{2} \tag{b}$$

Substituting this value in Eq. (*a*) results in the expression for the concrete tensile stress at the surface of the slab due to variation of shrinkage or temperature between top and bottom:

$$f_{dt} = \frac{\Delta\epsilon_t E_c}{2(1 - \nu)} \tag{16.3}$$

where $\Delta\epsilon_t$ = difference in strains between top and bottom of slab due to shrinkage or temperature
E_c = modulus of elasticity of concrete, psi
f_{dt} = tensile stress due to warping tendency, psi

Field measurements reported in Ref. 16.1 show that for outdoor pavements with the top surface colder than the bottom, tensile stresses about equal to 50 psi can be expected at the top surface. When the top surface was warmer, tension of about 100 psi was reported at the bottom. For indoor slabs with uneven shrinkage, top tension of about 50 psi has been measured.

The warping stresses superimpose on the uniform tensile stresses calculated by Eq. (16.2).

It is common practice in the United States to place light reinforcement, usually in the form of welded wire fabric, in slabs on grade to control cracking associated with shrinkage and temperature. Equation (16.1) provides the basis for the design of such reinforcement, according to Ref. 16.2. Assuming that the slab has cracked and that all of the tensile force is resisted by the steel at an allowable stress of f_s, then $T = A_s f_s$ and

$$A_s = \frac{w_o \mu l}{2f_s} \tag{16.4}$$

per foot width of slab. An allowable stress of $f_s = 30,000$ psi is often used. By this approach, the sole function of the reinforcement is to limit the width of cracks, enhancing the transfer of vertical shears, such as from wheel loads, across the cracks by aggregate interlock. Warping effects are neglected.

16.3 WESTERGAARD ANALYSIS FOR CONCENTRATED LOADS

Highway pavements and airport runways, as well as many other slabs on grade, are subjected to heavy concentrated loads from truck wheels, aircraft landing gear, or other causes. Methods for the analysis of such slabs for the effects of concentrated loads are similar to those developed for beams on elastic foundations, and are based on the work of H. M. Westergaard (Refs. 16.3 to 16.5). The slab is assumed to be homogeneous, isotropic, and elastic; it is considered unreinforced and uncracked. Either of two assumptions is made relative to the subgrade. The first, that it behaves as a dense liquid, is equivalent to the assumption that the subgrade reaction is vertical, proportional to deflection, and at each point is independent of forces and displacements at other points. The stiffness of the soil is expressed in terms of the modulus of subgrade reaction k, usually in units of psi/in., or simply pci. The numerical value of k varies widely for different soil types and degrees of consolidation, and is generally based on tests. Typical values are given in Fig. 16.4.

The alternative assumption is that the subgrade behaves as a semi-infinite elastic solid, such that a concentrated load at one location on the surface causes vertical displacements everywhere, according to the Boussinesq equation. In this case there is interaction by all elements in the elastic solid in resisting forces and displacements.

Westergaard considered three separate cases, differentiated on the basis of location of the load with respect to the edges of the slab (see Fig. 16.5).

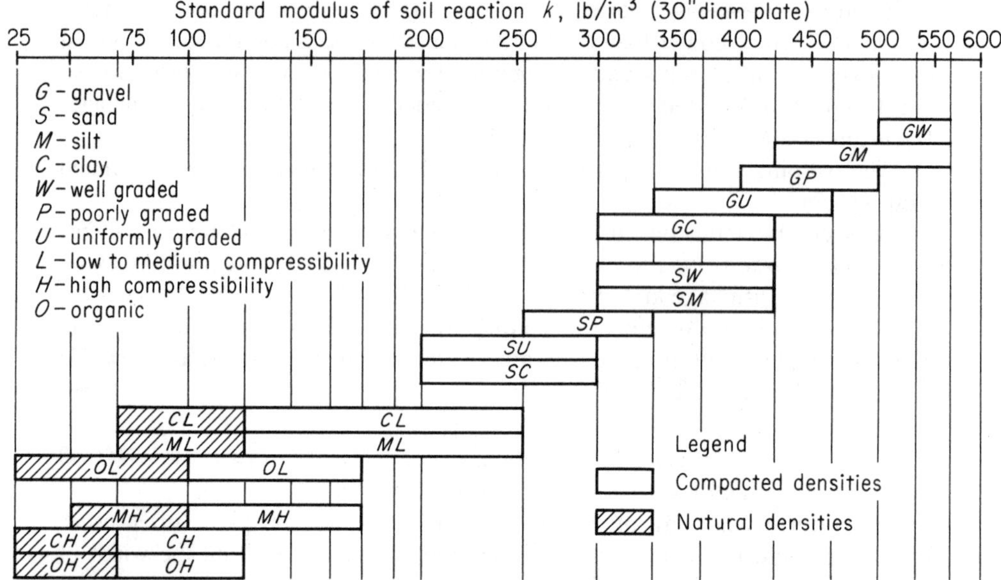

FIGURE 16.4
Soil classifications and modulus of subgrade reaction.

Case 1: Wheel load close to the corner of a large slab. With a load applied at the corner of a slab, the critical stress in the concrete is tension at the top surface of the slab. An approximate solution, due to A. T. Goldbeck, assumes a point load acting at the corner of the slab (see Fig. 16.6*a*). At small distances from the corner, the upward reaction of the soil has little effect, and the slab is considered to act as a cantilever. At a distance x from the corner, the bending moment is Px; it is assumed to be uniformly distributed across the width of the section of slab

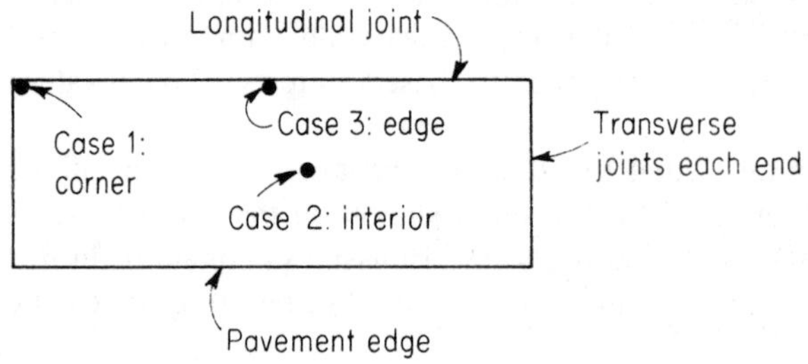

FIGURE 16.5
Concentrated loads on jointed highway pavement.

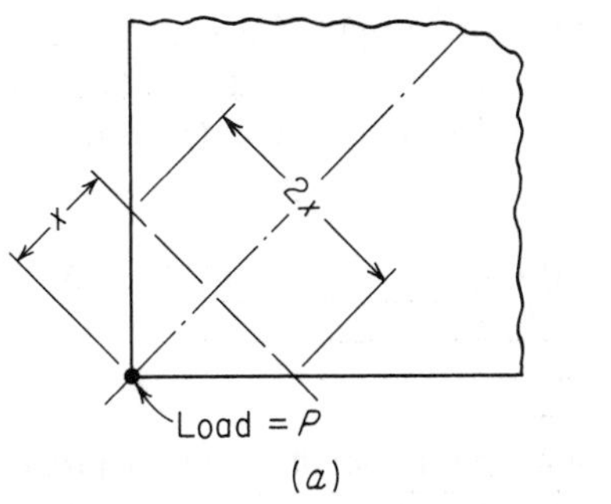

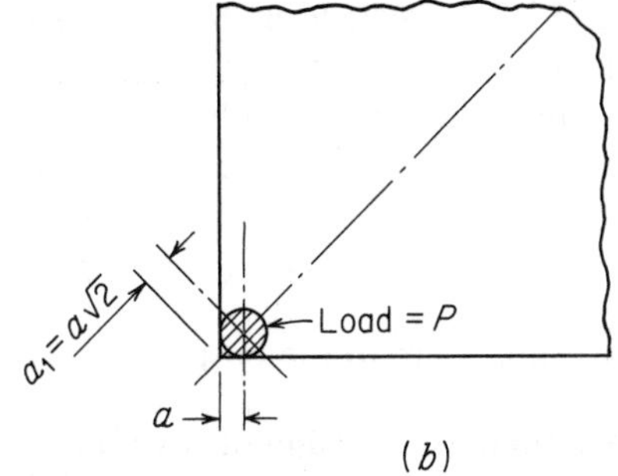

FIGURE 16.6
Corner load on ground-supported slab.

at right angles to the bisector of the corner angle. For a 90° corner, the width of this section is $2x$, and the bending moment per unit width of slab is

$$\frac{Px}{2x} = \frac{P}{2}$$

If h is the slab thickness (in.) and P the unfactored service load (lb), the concrete tensile stress (psi) at the top surface is

$$f_t = \frac{M}{S} = \frac{P/2}{h^2/6} = \frac{3P}{h^2} \tag{16.5}$$

Equation (16.5) will give reasonably close results only in the immediate vicinity of the slab corner and only if the load is applied over a small contact area.

In an analysis that considers the reaction of the subgrade and that considers the load to be applied over a contact area of radius a (see Fig. 16.6b), Westergaard derives the expression for critical tension at the top of the slab, occurring at a distance $2\sqrt{a_1 l}$ from the corner of the slab:

$$f_t = \frac{3P}{h^2}\left[1 - \left(\frac{a\sqrt{2}}{l}\right)^{0.6}\right] \tag{16.6}$$

in which l is the *radius of relative stiffness*, equal to

$$l = \sqrt[4]{\frac{E_c h^3}{12(1 - \nu^2)k}}$$

where E_c = elastic modulus of concrete, psi
ν = Poisson's ratio for concrete
k = modulus of subgrade reaction, pci

The value of l reflects the relative stiffness of the slab and the subgrade. It will be large for a stiff slab on a soft base and small for a flexible slab on a stiff base. Westergaard suggested that a typical value is 36 in. for highway pavements.

Case 2: Wheel load a considerable distance from the edges of a slab. When the load is applied some distance from the edges of the slab, the critical stress in the concrete will be tension at the bottom surface. This tension is greatest directly under the center of the loaded area and is given by the expression

$$f_b = 0.316\frac{P}{h^2}[\log h^3 - 4\log(\sqrt{1.6a^2 + h^2} - 0.675h) - \log k + 6.48] \quad (16.7)$$

where the logarithms are to the base 10.

Case 3: Wheel load at an edge of a slab but removed a considerable distance from a corner. When the load is applied at a point along an edge of the slab, the critical tensile stress is at the bottom of the concrete, directly under the load, and is equal to

$$f_b = 0.572\frac{P}{h^2}[\log h^3 - 4\log(\sqrt{1.6a^2 + h^2} - 0.675h) - \log k + 5.77] \quad (16.8)$$

In the event that the tensile stress in the slab computed by Eqs. (16.6) to (16.8) exceeds the allowable tensile stress on the concrete, it is necessary either to increase the thickness of the slab or to provide reinforcement. If reinforcement is used, it is generally assumed to act at an *allowable stress* equal to one half its yield strength but not greater than 24,000 psi. In such a case, sufficient steel is provided to resist all the concrete tension indicated by elastic analysis of the assumed homogeneous slab. Its centroid should be no closer to the neutral axis than that of the tension concrete stress block that it replaces.

Example 16.1: Design of an unreinforced concrete slab on grade. A warehouse floor, consisting of an unreinforced slab on ground, is to be designed to support a maximum load of 6000 lb, assumed to be applied over a contact area of 4 in. radius. Concrete of 4000 psi compressive strength is to be used, with an allowable tensile stress of 200 psi. The slab will be poured in 30 ft wide strips with construction joints at each edge, and contraction joints will be located every 30 ft in the long direction of the strip. It will be conservatively assumed that no shear transfer is possible across the joints. The load can be applied anywhere within each 900 ft^2 panel. The soil is a clay-gravel with a subgrade modulus of 300 pci.

Solution. It will be assumed that the critical load location is at the corner of the slab. With

$$E_c = 33w^{1.5}\sqrt{f_c'} = 33 \times 145^{1.5}\sqrt{4000} = 3,640,000 \text{ psi}$$

and assuming the slab thickness h to be 8 in., the radius of relative stiffness is

$$l = \sqrt[4]{\frac{Eh^3}{12(1-\nu^2)k}} = \sqrt[4]{\frac{3,640,000 \times 512}{12(1-0.15^2)300}} = 27 \text{ in.}$$

From Eq. (16.6),

$$h^2 = \frac{3P}{f_t}\left[1 - \left(\frac{a\sqrt{2}}{l}\right)^{0.6}\right] = \frac{3 \times 6000}{200}\left[1 - \left(\frac{4\sqrt{2}}{27}\right)^{0.6}\right] = 54.7\text{in}^2$$

$$h = 7.4 \text{ in.}$$

Tentatively, a total depth of $7\frac{1}{2}$ in. is selected. If the load is applied centrally within a panel, the tensile stress at the bottom of a $7\frac{1}{2}$ in. slab is, from Eq. (16.7),

$$f_b = \frac{0.316 \times 6000}{56.2}(\log 421 - 4 \log 3.96 - \log 300 + 6.48)$$

$$= 144 \text{ psi} < 200 \text{ psi}$$

If the load is applied along the edge of a slab panel, the critical tension at the bottom of the concrete is, from Eq. (16.8),

$$f_b = \frac{0.572 \times 6000}{56.2}(\log 421 - 4 \log 3.96 - \log 300 + 5.77)$$

$$= 215 \text{ psi} > 200 \text{ psi}$$

Since this exceeds the allowable tension by a small amount, a total slab depth of 8 in. will be used.

While the method of analysis and design just described has provided the basis for charts and diagrams in wide use (see Sec. 16.4) and, with modifications based on road tests, provides the basis for the design of rigid highway pavements according to AASHTO (see Appendix D of Ref. 16.2), its deficiencies and inconsistencies are obvious. The approach relies upon the flexural tensile strength of concrete, which is completely discounted in all other structural concrete design. The analysis is based on an uncracked slab, whereas cracking due to shrinkage and temperature effect, as well as loads, is unavoidable. The result of this approach is usually a relatively thick concrete slab, either without reinforcement or with a small percentage of steel generally designed on the basis of subgrade drag. Better methods have been proposed (see Sec. 16.7).

16.4 DESIGN AIDS FOR PAVEMENTS

The Westergaard equations permit calculation of stresses resulting from a single concentrated load on a pavement area defined by joints or edges. In practice, slab design must usually be based on a *pattern* of concentrated loads such as from the wheels of a heavy truck or an aircraft landing gear. For such cases, the determination of bending moments and stresses is greatly facilitated by influence charts developed in 1951 by Pickett and Ray (Refs. 16.6 and 16.7). Based on the Westergaard equations, these charts permit rapid calculation of bending moments resulting from loads at interior and edge locations.

Still more convenient for practical design are the charts published by the Portland Cement Association (Ref. 16.8), such as shown in Fig. 16.7. Based on the influence charts of Ref. 16.7, and in turn on the Westergaard equations, the PCA design charts permit rapid determination of flexural tensile stresses for specified truck axle loads. Comparative studies done at PCA indicate that the highest stresses occur, in almost all cases, when the wheels of a truck are at the end (transverse joint) of a section of pavement, the maximum tension being at the bottom of the slab acting in the direction parallel to the joint edge. Figure 16.7 is based on the application of loads, transmitted to the slab by two closely spaced

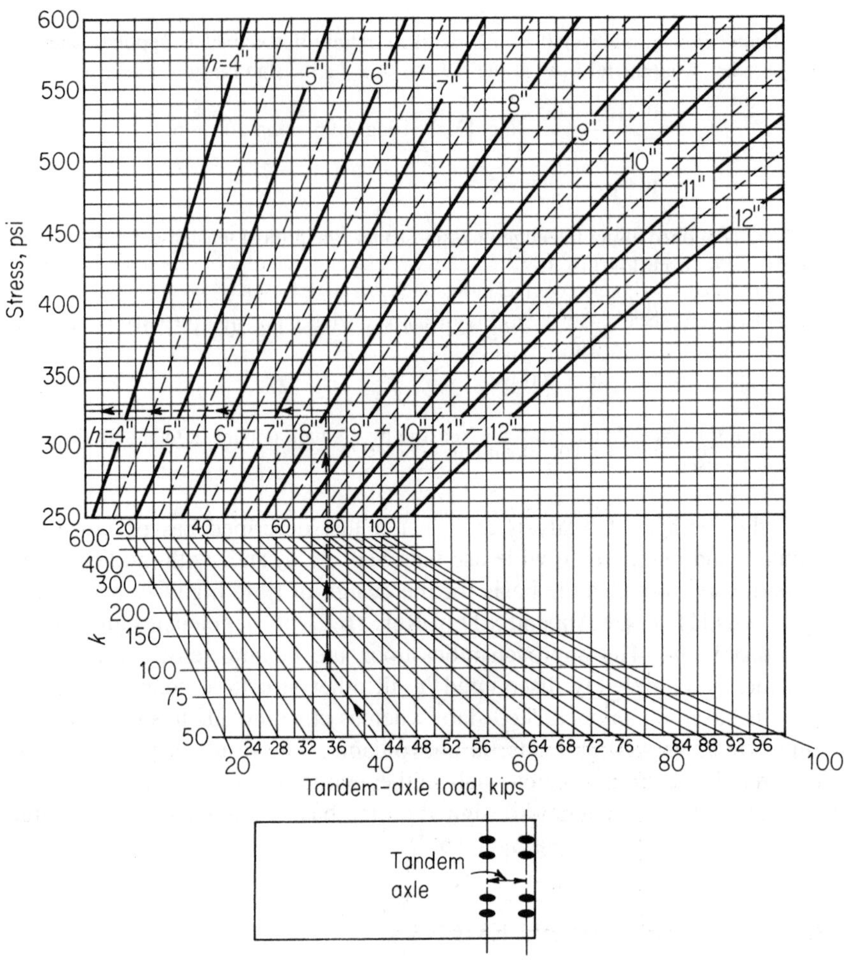

FIGURE 16.7
Design chart for tandem-axle load on ground-supported slab. *(From Ref. 16.8. Courtesy of Portland Cement Association.)*

axles with four tires per axle, with that wheel pattern located close to a transverse joint edge. To illustrate use of the chart, as indicated by the dashed arrowed lines, a tandem-axle load totaling 42 kips applied to an 8 in. thick slab resting on a subgrade having modulus k of 100 pci will result in a tensile stress at the bottom of the slab of 325 psi. According to the PCA method, allowable tensile stress is based on the required design life, number of load repetitions, and the modulus of rupture of the concrete, with an allowance for stresses due to shrinkage and temperature. Details and examples are found in Ref. 16.8.

Much useful information concerning pavement performance was gained as a result of the AASHTO road test carried out in Illinois from 1958 to 1962 (Ref. 16.9). Based on those results, for the particular combination of concrete strength f_r, elastic modulus E_c, and subgrade modulus k existing at the test site, an equation was developed relating loss of serviceability to pavement thickness, the equation having the form:

$$G_t = \beta(\log W_t - \log \rho) \tag{16.9}$$

where G_t = a function measuring loss of serviceability = log [0.333(4.5 $-p_t$)]
β = a function of design and load variables (see below)
p_t = a serviceability index
W_t = number of axle load applications by end of time period
ρ = a function of design and load variables (see below)

For standardized conditions, for which the single-axle load is taken equal to 18 kips, β and ρ are given by

$$\beta = 1.00 + \frac{1.624 \times 10^7}{(h + 1)^{8.46}} \tag{16.10}$$

$$\log \rho = 7.35 \log (h + 1) - 0.06 \tag{16.11}$$

where h = thickness of the slab in inch units. Combining Eqs. (16.9) through (16.11) and adopting a serviceability index p_t = 2.5 results in

$$\log W_{t,18} = 7.35 \log (h + 1) - 0.06 - \frac{0.1761}{\beta} \tag{16.12}$$

in which $W_{t,18}$ is the number of applications of an 18 kips single-axle load.

Equation (16.12) provides a basis for developing design charts. First, however, to account for conditions other than those that existed at the AASHTO road test, it is necessary to modify the basic equation. This is accomplished using what is known as the Spangler equation:

$$f_t = \frac{JP}{h^2}\left(1 - \frac{a\sqrt{2}}{l}\right) \tag{16.13}$$

where f_t = tensile stress in the concrete slab, psi
P = wheel load, lb
h = slab thickness, in.
a = radius of load contact area
l = radius of relative stiffness
J = load transfer coefficient, taken as = 3.2 for protected corner

Comparison of Eq. (16.13) with Eq. (16.6) based on Westergaard's earlier work will confirm that the two are identical except for a small difference in the coefficient and an increase in the exponent to 1.0.

The end result of the AASHTO development, details of which are found in Appendix D of Ref. 16.2, are presented in the form of the design chart of Fig. 16.8. Its use is illustrated by the dashed arrowed lines. For example, for 5×10^6 equivalent 18 kips single-axle load applications, with allowable flexural tensile stress in the slab of 500 psi, a diagonal line is drawn upward, intersecting

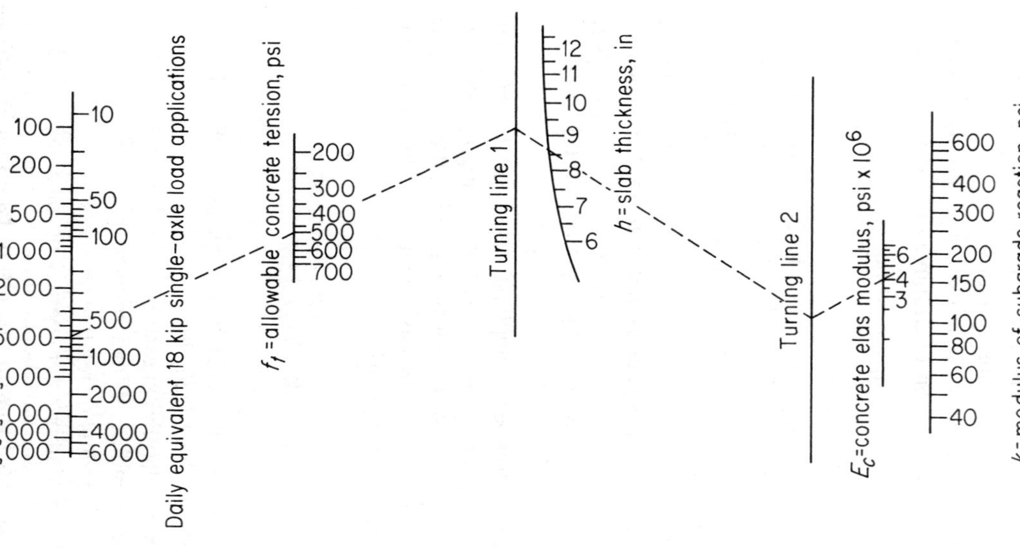

FIGURE 16.8
AASHTO design chart for rigid pavements. *(Adapted from Ref. 16.9.)*

the turning line 1. For $E_c = 4 \times 10^6$ psi and $k = 200$ psi, a second diagonal line is drawn downward, intersecting the turning line 2. A final line is drawn connecting the intersection points on the two turning lines, resulting in the slab thickness $h = 8.6$ in. This would be rounded upward to 9 in. in practice. The determination of equivalent load applications for load cases other than 18 kips is done by equations and charts contained in Ref. 16.2.

16.5 JOINTING PRACTICE FOR PAVEMENTS

Most major highways in the United States are designed as plain unreinforced concrete pavements or pavements with light mesh reinforcement to control temperature and shrinkage cracks. The design basis was described in Secs. 16.1 through 16.4. Jointing practice has varied widely over the years, and continues to vary from state to state. In the 1920s, most concrete pavements were built without joints, and they developed irregular wandering cracks that resulted in serious maintenance problems. Today, there appears to be general agreement that, for plain or mesh-reinforced pavements, both longitudinal and transverse joints are needed (Ref. 16.10).

Longitudinal joints are spaced at approximately 12 ft, coinciding with the standard width of traffic lane. Either keyways or tie bars are commonly used to transfer load, as shown in Fig. 16.9*a*. Typically, tie bars may be $\frac{5}{8}$ in. diameter, 30 in. long, and spaced at about 40 in. on centers.

Two types of transverse joints are used, contraction joints and expansion joints. Contraction joints are used to provide a uniform pattern of cracking when slab shortening due to shrinkage and temperature drop occurs. They may be sawed or hand-formed and should extend to about one quarter of the slab thickness, as shown in Fig. 16.9*b*. Spacing varies from 15 ft to about 50 ft, the present trend being toward spacings of 15 to 20 ft. Dowels are often used, of size and spacing as described for longitudinal joints. Normally the shrinkage and temperature mesh, if used, is continuous across the contraction joints.

Expansion joints are placed at larger spacing, about 100 ft, to provide for an increase in length of the pavement sections as temperatures rise. These are normally doweled, as shown in Fig. 16.9*c*, using plain round bars of about $1\frac{1}{8}$ in. diameter and 15 in. length, typically spaced at about 12 in. Some form of expansion cap is needed at the end of the dowel, and a premolded compressible seal is incorporated (Ref. 16.11).

In spite of improvements in jointing practice over the years, most plain or mesh-reinforced concrete pavements show serious cracking, particularly at the corners of slab sections, but also along the edges and ends of the panels. For this reason, more attention is being paid to continuously reinforced pavements or prestressed pavements without joints, as described in Secs. 16.6 and 16.8.

16.6 CONTINUOUSLY REINFORCED PAVEMENTS

It is generally recognized that most deterioration of pavements starts at the corners of slab segments, or at the transverse or longitudinal joints and edges, where

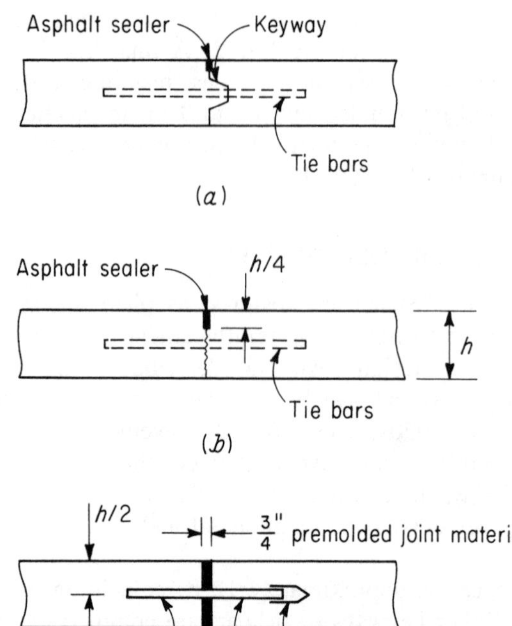

FIGURE 16.9
Common joint details for highway pavements: (*a*) longitudinal joint; (*b*) transverse contraction joint; (*c*) transverse expansion joint.

stresses and deflections due to wheel loads are greatest and where severe cracking frequently occurs. Intrusion of water and deicing chemicals, freezing and thawing, and erosion of the subgrade contribute further to the early and progressive breakup of the pavement. Furthermore, warping of slab sections between joints often leads to poor riding qualities. It is natural, therefore, to consider elimination of joints. This has led to development of continuously reinforced concrete pavements, in which longitudinal joints are often eliminated and transverse joints are spaced only at great distances. Continuously reinforced highways have been built in lengths up to 250 miles, with no joints except as required for construction convenience or for overpass structures (Ref. 16.12). Measurements indicate that regardless of total length, longitudinal movement will occur within only about 500 ft of the pavement ends.

Such pavements incorporate continuous, longitudinal, lap-spliced, deformed reinforcing bars in amounts from 0.5 to 1.0 percent of the gross concrete area, and about 0.05 to 0.1 percent transverse reinforcement. Figure 16.10 shows continuous slab reinforcement being placed with the aid of a traveling carriage.

Continuously reinforced pavements develop cracks due to shrinkage and temperature effects, but because of the relatively large steel ratio and good bond

FIGURE 16.10
Placing of bars for continuously reinforced highway pavement.

properties of the deformed bars, these cracks are narrow and permit full shear transfer between slab segments by dowel action and aggregate interlock. The spacing of these cracks is typically between about 3 and 10 ft, and crack widths are in the range from 0.003 to 0.030 in. They tend to close with rising temperature, and under most conditions are imperceptible. The thermal expansion of the slab is accommodated in part by compressive stresses in the concrete, but mainly by closing of the hairline cracks.

Slab thickness, reinforcement ratio, and bar placement for continuously reinforced highways is based mostly on experience, although design charts have been published that are similar to Fig. 16.8 (Refs. 16.13 to 16.17). For many years, thickness of continuously reinforced slabs was specified to be about three-quarters that for a conventional jointed pavement designed for the same conditions. A 1975 survey indicated that a typical interstate highway pavement designed for continuous reinforcement had a thickness of 8 in. and a steel ratio of about 0.006. Bars were typically No. 5 at 6 in. spacing, placed about one third of the slab depth from the top surface. Lap-splice lengths of 16 in. were representative (Ref. 16.18).

While continuously reinforced pavements have generally performed better in service than jointed pavements, there have been failures (Ref. 16.16). As a result, in 1978 the Federal Highway Administration recommended that continuously reinforced pavements should under most conditions have the same thickness as a jointed pavement designed to carry the same loads. Design charts such as Fig. 16.8, for jointed pavements, are therefore used to determine the thickness of continuously reinforced pavements as well (Ref. 16.13).

16.7 STRUCTURALLY REINFORCED PAVEMENTS

From a structural engineering viewpoint, it is not logical to design pavements as plain or very lightly reinforced slabs, relying on the flexural tensile strength of the concrete, which has a rather low and unpredictable value. For all other structural concrete members, it is recognized that it is not only inevitable that the concrete should crack, but necessary for it to do so in order for the steel reinforcement to act effectively. Thus the steel takes the tension that the concrete cannot resist efficiently; proper distribution of the reinforcement ensures that the cracks are narrow and well distributed, so that they will in no way interfere with the safety or function of the structure. Pavement slabs, too, can be designed from a structural engineering viewpoint, applying the same criteria used for structural concrete (Refs. 16.1 and 16.19).

Structurally reinforced pavements have generally been designed as continuously reinforced as well, without transverse or longitudinal joints. Thus, throughout most of the slab, the criteria design condition corresponds to Case 2 of Sec. 16.3; that of a concentrated wheel load applied some distance from the edges or ends of the slab. For this case, the deformation of the slab and variation of bending moment is as shown in Fig. 16.11. High tensile stresses are produced at the bottom of the pavement just under the load, calling for two-directional reinforcement near the bottom of the slab, while the smaller tensile stresses at the

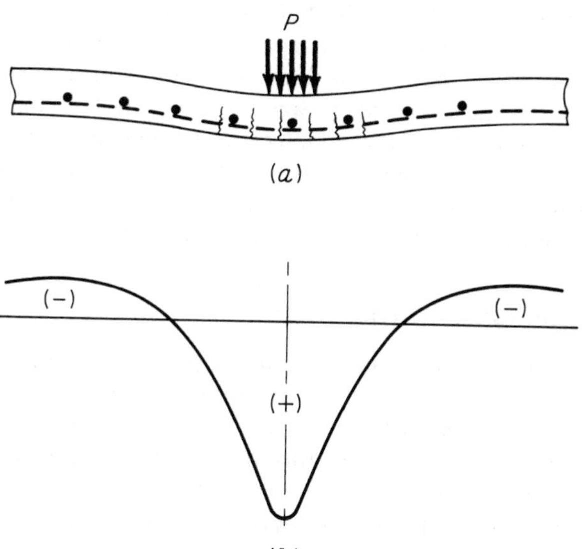

FIGURE 16.11
Action of a structurally reinforced concrete pavement: (*a*) deformation and cracking under concentrated wheel load; (*b*) bending moment variation. *(Adapted from Ref. 16.19.)*

top of the slab, in a circular area around the load, may be carried by the tensile strength of the plain concrete.

At the outside edges of such a pavement slab, and at transverse joints where they do occur, Case 1 or Case 3 of Sec. 16.3 applies. Extra reinforcement may be provided at the top of the slab, or at both top and bottom, to accommodate the local high bending moments.

Losberg, in Ref. 16.1 proposes a practical design method for structurally reinforced slabs, based on yield line theory (see Chap. 14), and cites several successful designs for runway pavements at military airfields in Sweden. The approach has not been widely used elsewhere.

16.8 PRESTRESSED CONCRETE PAVEMENTS

The possibility of prestressing continuous pavements should not be overlooked. In longitudinally prestressed concrete pavements, fluctuating stresses due to passing wheel loads remain in the compressive range. Cracking is eliminated and transverse joints are decreased in number or eliminated completely, resulting in longer pavement life and smoother riding characteristics, and generally permitting substantial reduction in slab thickness. Experience with experimental and prototype post-tensioned pavements has confirmed that they are practical and economically competitive (Refs. 16.20 to 16.22).

Techniques for longitudinal post-tensioning (see Chap. 21) vary, but some details of a project in Pensylvania are of interest (Ref. 16.23). Unbonded coated tendons of 0.68 in. diameter were used, placed 24 in. on centers, and located slightly below the middepth of the 6 in. thick slab. A slipform paver was modified so as to feed the 12 tendons into the slab as the paver passed. Polyethylene sheets were used on the base course under the slab so the slab could shorten during prestressing. The length between transverse joints was about 600 ft. Compared with a competing design using ordinary reinforced concrete, the thickness of the pavement was reduced from 9 in. to 6 in., and the amount of steel was reduced 90 percent by weight. Performance to date has been generally good, although some difficulties have been experienced at the welded anchorage assemblies.

16.9 INDUSTRIAL SLABS ON GRADE

Ground-supported concrete slabs are widely used for floors for commercial and industrial buildings. Often they are subjected to heavy wheel loads from materials handling equipment such as lift trucks, and, in addition, they frequently carry heavy loads from stored material, distributed nonuniformly over the surface. Design methods vary, ranging from purely empirical to those based on theoretical analysis such as the Westergaard equations. Actual service loads are used generally (i.e., load factors equal to 1.0), with the design criterion that the concrete tensile stress be kept a safe margin below the modulus of rupture. Industrial slabs are sometimes built without reinforcement of any kind, but generally a wire mesh or light bar reinforcement is recommended to control cracking. Slabs are usually provided with joints to provide for shrinkage or isolate machinery foundations or walls, or merely for construction convenience.

a. Loads Distributed over Partial Areas

In warehouses, heavy loads are often stacked about columns, leaving clear aisles midway between the column lines for access, as shown in Fig. 16.12. Live loads for warehouses are up to 250 psf according to ANSI A58.1 (Ref. 1.1), but actual loads are frequently much higher.

For more moderate column spacings, up to about 25 ft, construction joints are usually provided on the column lines only. For the typical load pattern shown in Fig. 16.12, high tensile stresses are produced at the top of the concrete slab along the aisle centerlines, and potential cracks are along the aisles. Such cracking would be particularly objectionable, because heavy wheel loads crossing the cracks would lead to rapid disintegration of the slab. Secondary cracking such as shown at the corner of one panel is likely.

In an analysis based on the loading and joint arrangement of Fig. 16.12, Rice derived an expression for the critical negative bending moment at the centerline of the aisle (Ref. 16.24):

$$M_c = \frac{w}{2\lambda^2} e^{-\lambda a} \sin \lambda a \tag{16.14}$$

where M_c = slab moment, in-lb/in.
$\lambda = \sqrt[4]{k/4E_cI_g}$, in^{-1}

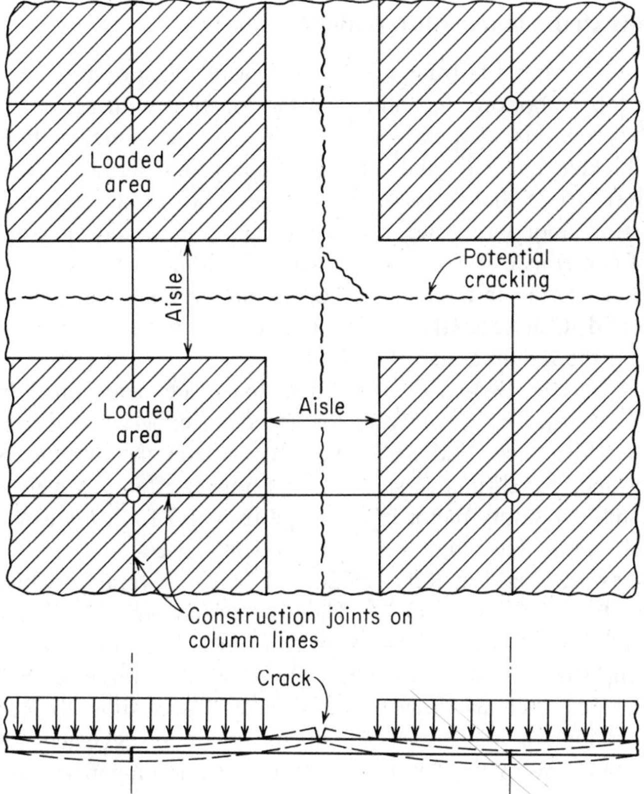

FIGURE 16.12
Warehouse slab loaded over partial area.

a = half-aisle width, in.
k = modulus of subgrade reaction, pci
w = uniform load, psi
e = base of natural logarithms

Recognizing that the width of the aisle cannot always be predicted exactly, Rice suggests that a *critical aisle width* be used. This width is such as to maximize Eq. (16.14) for bending moment. Tables are available (Ref. 16.24) that give critical aisle width and corresponding bending moment for various slab thicknesses and various values of subgrade modulus. More recent studies of partially loaded slabs using the finite element method of analysis have generally confirmed the results obtained by Rice for stresses at midaisle (Ref. 16.25).

For slabs jointed at intervals of about 25 ft or less, the stresses resulting from shrinkage are low, in the range of 25 psi, and are usually disregarded in the analysis. However, shrinkage and warping stresses can be calculated by the methods of Sec. 16.2 and superimposed on the stresses resulting from flexure.

b. Effect of Concentrated Loads from Vehicles

Wheel loads such as those from lift trucks produce flexural tensile stresses in industrial slabs, at the top or bottom depending on the location with respect to edges or corners of slab sections, that have been studied by the same methods used by Westergaard, Pickett and Ray, and others for analyzing highway pavements. It is shown in Ref. 16.25 that stresses under wheel loads on an otherwise unloaded area of industrial floor will control in many cases. It should be noted that the combination of centeraisle stresses due to pattern loading such as shown in Fig. 16.12, plus wheel loading in the aisle, produces slab stresses that tend to compensate rather than add. Consequently, the two load cases can be investigated separately in determining the required slab thickness.

Allowable flexural tensile stresses due to wheel loads are determined by dividing the modulus of rupture by a safety factor based on experience gained in studying pavement performance, and may take into account the number of repititions of load, allowance for shrinkage stresses, and impact effect, as well as the strength of the concrete. A safety factor in the range from 1.7 to 2.0 is usual, and values at the higher end of that range are most frequently used.

Useful design charts are found in Refs. 16.26 and 16.27, several of which are reproduced here, in Figs. 16.13 to 16.15. They are based on wheel loads applied at the interior of a slab, some distance away from a free edge or corner, and can also be applied at slab edges when the joints are provided with proper load transfer devices such as dowels and keyways, according to Ref. 16.26. Where such provision for load transfer is not made, it is suggested that the slab be thickened at the edge by approximately 50 percent, at a taper of about 1 in 10.

Figure 16.13 provides the basis for design of a slab to carry industrial trucks with axles equipped with single wheels. Use of the chart is indicated by the dashed arrows. First, an allowable working stress per 1000 lb of axle load is computed, by dividing the modulus of rupture by the selected safety factor and then dividing this result by the axle load in kips. For the specific effective contact area, known wheel spacing s, and subgrade modulus k, the required slab thickness may be read directly. For example, for a 25 kip axle load with two single wheels spaced at 37 in., effective contact area of 114 in^2, subgrade modulus $k = 100$ pci, modulus of rupture of 640 psi, and safety factor of 2, one calculates first the allowable working stress per kip of

$$f_{\text{all}} = \frac{640}{2 \times 25} = 12.8 \text{ psi/kip}$$

The horizontal arrowed line is followed to the right to the contact area of 114 in^2, then downward to the wheel spacing of 37 in., then to the right to the vertical axis corresponding to $k = 100$ pci, and the required thickness of 8 in. is read at the right scale of the chart.

For axles equipped with dual wheels, Fig. 16.14 is used to convert the dual-wheel axle load to an equivalent single-wheel axle load. For example, for dual-wheel spacing $s_d = 18$ in., with effective contact area 100 in^2 and trial slab thickness 10 in., an equivalent load factor of 0.775 is found. The equivalent

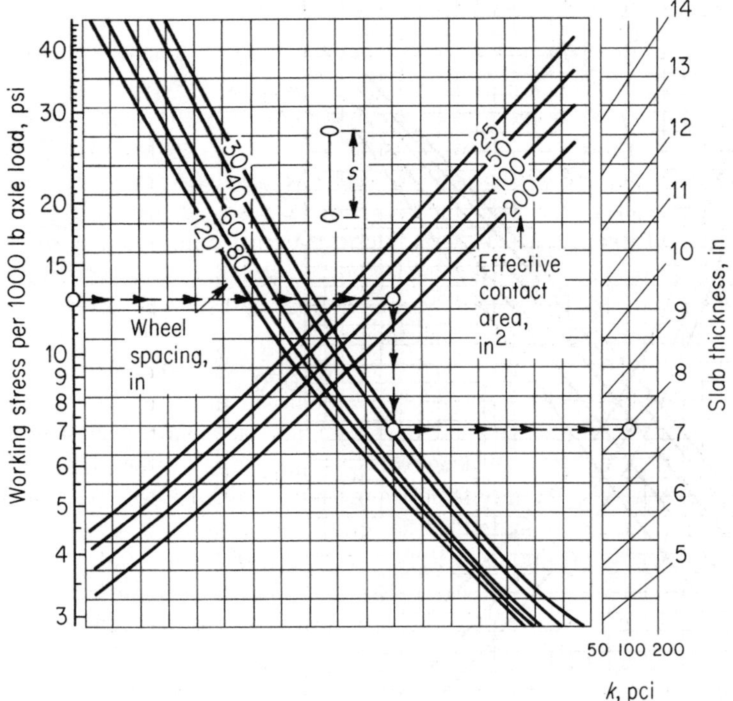

FIGURE 16.13
Industrial slab design chart for axles with single wheels. *(Adapted from Ref. 16.26.)*

single-axle load is thus 0.775 times the dual-wheel axle load. Figure 16.13 may then be used as before to determine the required slab thickness.

The load contact area is the area of slab in contact with one tire. If specific information is not available, the contact area can be estimated by dividing the wheel load by the inflation pressure. For example, for a 25 kip single-wheel axle load and an inflation pressure of 110 psi, the approximate contact area is $25{,}000/(2 \times 110) = 114 \text{ in}^2$. The *effective* contact pressure may be larger than that, depending on the thickness of the slab, and can be found with the aid of Fig. 16.15. For example, for an actual contact area of 62 in^2 and an estimated slab thickness of 10 in., the effective contact area is 71 in^2. In general, it is seen that thicker slabs result in greater increases in effective contact area, as could be expected. Figure 16.15 is based on the work of Westergaard.

c. Concentrated Loads from Storage Racks

Another loading that may be critical in storage warehouses is that from the legs of high storage racks, some of which are as tall as 100 ft. The design procedures are analogous to those for concentrated wheel loads. Details and design charts are found in Refs. 16.26 and 16.27.

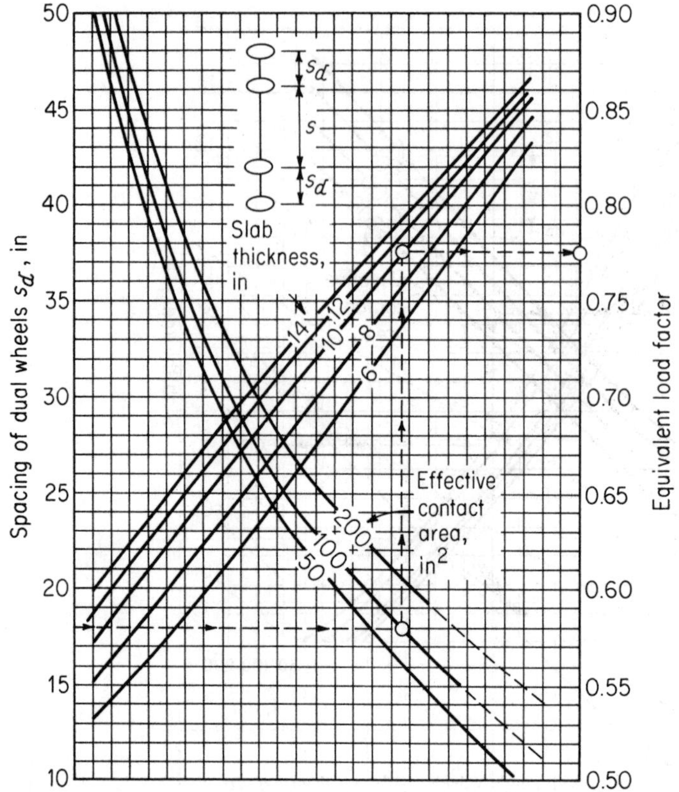

FIGURE 16.14
Industrial slab design chart for axles with dual wheels. *(Adapted from Ref. 16.26.)*

d. Jointing Practice for Industrial Slabs

Three types of joints are used in industrial slabs on grade: expansion joints, contraction joints, and construction joints. Expansion joints, sometimes called isolation joints, permit movement between the floor slab and fixed elements such as machinery bases or walls. They are also generally used around columns to avoid uncontrolled cracking of the slab over the column base. Typically a compressible material of about $\frac{3}{4}$ in. thickness is used, as shown in Fig. 16.16*a*, and a bituminous joint sealer placed after the slab is poured.

Contraction joints are used to ensure that shrinkage cracks, when they form, will be straight and in a regular pattern. Presently, they are usually saw-cut within one or two days after pouring the slab, to a depth of about one quarter of the total slab thickness, as shown in Fig. 16.16*b*. A liquid sealer is poured in the sawcut to prevent the accumulation of foreign material. Reinforcement, if used in the slab, is usually continued without interruption across contraction joints. They may be doweled as well, to improve load transfer.

Construction joints are required wherever operations must be discontinued and later resumed. A true construction joint permits neither vertical nor horizon-

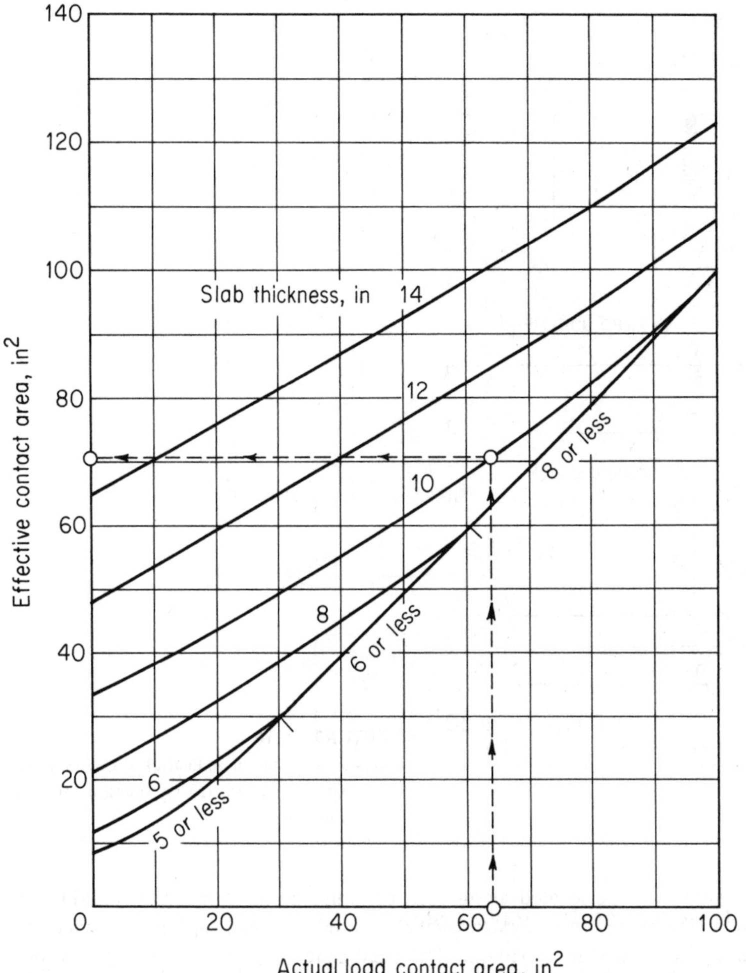

FIGURE 16.15
Effective load contact area vs. actual contact area. *(Adapted from Ref. 16.26.)*

tal movement between slab sections, and must usually be provided with tie bars and possibly a keyway as well, as shown in Fig. 16.16*c*. Because such joints are troublesome and expensive to construct, they are often avoided by terminating the work at expansion or contraction joints. A contraction joint used for this purpose should be provided with a shear key.

Other details and useful suggestions for joint layout are given in Ref. 16.27.

Example 16.2 Design of warehouse floor. A concrete slab on grade is to be designed for a warehouse having overall dimensions 400 × 1000 ft with columns on a uniform 20 × 20 ft grid. Material will be stored in a pattern similar to Fig. 16.12, with aisles 7 ft wide; the maximum service live load from stacked material will be 600 psf. Keyed

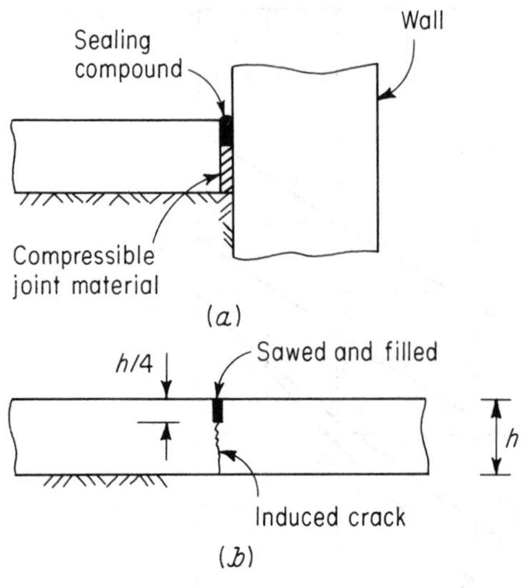

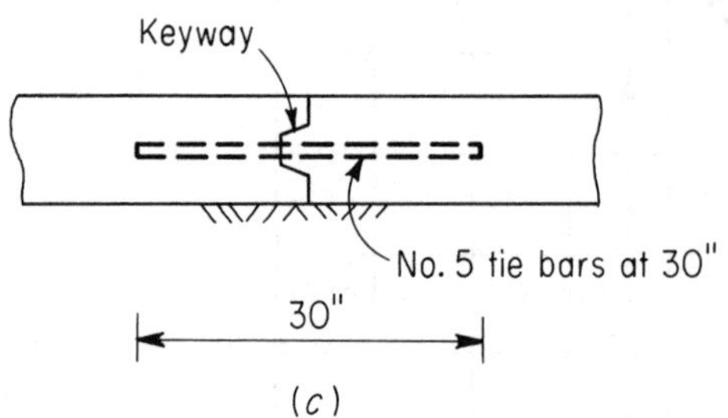

FIGURE 16.16
Types of joints in industrial slabs on grade: (*a*) expansion joint; (*b*) contraction joint; (*c*) construction joint.

contraction joints will be located along each column line in each direction. Materials handling will require the use of forklift trucks of 16,000 lb rated capacity, for which the single-axle load is 35,000 lb maximum. The pneumatic-tired vehicles have two wheels per axle, spaced at 60 in., and the tire contact area is 100 in^2. The slab concrete will have $f'_c = 4000$ psi, for which f_r may be taken as 475 psi. A safety factor of 1.7 with respect to flexural cracking is specified. The subgrade modulus $k = 100$ pci and the coefficient of friction between the slab and subbase is 1.5. For these conditions:

(*a*) Find the required slab thickness based on concentrated loads from the materials handling equipment.

(*b*) Check the tensile stresses in the slab resulting from the combination effect of pattern live loading and shrinkage. Shrinkage may be assumed to occur largely before placement of the live load. A strain gradient of 4×10^{-6} per inch of slab depth is anticipated, the exposed top surface drying more than the bottom of the slab. Revise the slab thickness found in (*a*) if necessary to provide the required safety factor against cracking.

Solution. Considering first the effect of the concentrated wheel loads, which would produce tension at the bottom of the slab at an interior location, the effective tire contact area is first found with the aid of Fig. 16.15. Assuming a slab depth of 10 in for trial, the chart indicates that for an actual loaded area of 100 in^2 the effective contact area is the same in this case. The working stress per 1000 lb axle load is

$$f_{all} = \frac{475}{1.7 \times 35} = 7.98 \text{ psi/kip}$$

Entering Fig. 16.13 with that value, proceeding to the right to the contact area of 100 in^2, then upward to a wheel spacing of 60 in., then to the right to the axis for $k = 100$ pci, the required thickness is found to be 10 in. With the use of keyed contraction joints, it is not considered necessary to thicken the slab at the joints.

Next the slab, tentatively with thickness 10 in, will be investigated for the combined effects of shrinkage and patterned distributed loading. The uniform component of shrinkage tension will be computed using the subgrade drag approach of Sec. 16.2. The assumption is made that the main part of the shrinkage will take place prior to live load application, with frictional resistance corresponding to the self-weight of the slab only. The tensile force per foot strip reaches a maximum value at the center of the aisles, midway between joints, and is computed from Eq. (16.1) based on the tentative slab thickness of 10 in. and weight of 125 psf:

$$T = 125 \times 1.5 \times 10 = 1875 \text{ lb}$$

producing concrete tensile stress of

$$f_t = \frac{1875}{12 \times 10} = 16 \text{ psi}$$

For the tentative slab thickness, the differential of free shrinkage strain between top and bottom is $\Delta\epsilon_t = 4 \times 10 \times 10^{-6} = 40 \times 10^{-6}$. With live loads and slab self-weight preventing warping, the warping tension at the top of the slab is found, from Eq. (16.3):

$$f_{dt} = \frac{40 \times 10^{-6} \times 3.60 \times 10^6}{2(1 - 0.20)} = 90 \text{ psi}$$

Thus the combined effect of subgrade drag and differential shrinkage produces concrete tension of $16 + 90 = 106$ psi tension at the top of the slab at midaisle, and $16 - 90 = -74$ psi compression at the bottom.

Moments due to the application of the pattern live loading of 600 psf (4.17 psi) are found using Eq. (16.14). For a 1 in. strip of slab, the slab moment of inertia is $I_g = 1 \times 10^3/12 = 83.3 \text{in}^4$ and the section modulus is $S = 83.3/5 = 16.7 \text{ in}^3$. The parameter λ is

$$\lambda = \sqrt[4]{\frac{100}{4 \times 3,600,000 \times 83.3}} = 0.017$$

and with a half-aisle width $a = 42$ in.,

$$\lambda a = 0.017 \times 42 = 0.714 \text{rad} = 40.9^\circ$$

From Eq. (16.14) the bending moment at the center of the aisle is

$$M_c = \frac{4.17}{2 \times 0.017^2 e^{0.714}} \sin 0.714 = 2310 \text{ in-lb/in.}$$

With the slab section modulus $S = 16.7$ in^3 per inch strip, this results in flexural tension at the top of the slab of

$$f_b = \frac{2310}{16.7} = 138 \text{ psi}$$

and compression at the bottom of the same amount.

The effects of subgrade drag, restraint of warping, and pattern live loading all produce maximum concrete tension at the top of the slab at the center of the aisle, and hence are additive. The total concrete tension is

$$f_t + f_{dt} + f_b = 16 + 90 + 138 = 244 \text{ psi}$$

This should be compared with the allowable flexural tension of $475/1.7 = 279$ psi. The 10 in. slab thickness is seen to be satisfactory, controlled by requirements for concentrated loads.

Considering the close spacing of joints and good margin of safety against cracking, accounting for combined effects, slab reinforcement appears unnecessary here, and a plain concrete floor 10 in. thick will be used.

REFERENCES

16.1. A. Losberg, "Pavements and Slabs on Grade with Structurally Active Reinforcement," *J. ACI,* vol. 75, no. 12, 1978, pp. 647–657.

16.2. *AASHTO Interim Guide for Design of Pavement Structures,* Subcommittee on Roadway Design, Amer. Assoc. State Highway and Trans. Officials, Washington, 1972 (revised 1981).

16.3. H. M. Westergaard, "Stresses in Concrete Pavements Computed by Theoretical Analysis," *Public Roads*, vol. 7, no. 2, 1926, pp. 25–35.

16.4. H. M. Westergaard, "Analysis of Stresses in Concrete Roads Caused by Variations in Temperature," *Public Roads*, vol. 8, no. 3, 1927, pp. 201–215.

16.5. H. M. Westergaard, "Analytical Tools for Judging Results of Structural Tests for Pavements," *Public Roads*, vol. 14, no. 10, 1933, pp. 185–188.

16.6. G. Pickett, M. E. Raville, W. C. Janes, and F. J. McCormick, "Deflections, Moments, and Reactive Pressures for Concrete Pavements," Bulletin 65, Kansas State College, 1951.

16.7. G. Pickett and G. K. Ray, "Influence Charts for Concrete Pavements," *Trans. ASCE*, vol. 116, 1951, pp. 49–73.

16.8. "Thickness Design for Concrete Pavements," Concrete Information Bulletin, Portland Cement Assoc., Chicago, 1966.

16.9. "The AASHO (ASSHTO) Road Test," Special Report No. 73, Highway Research Board, Washington, 1962.

16.10. G. K. Ray, "Thirty-five Years of Pavement Design and Performance," in *Proc. Second Int. Conf. on Concrete Pavement Design*, Purdue University, 1981.

16.11. W. A. Yrjanson, "Guide to Jointing Unreinforced Pavements," *Concrete Construction,* May 1983, pp. 401–404.

16.12. "Continuously Reinforced Concrete Pavement," Synthesis of Highway Practice No. 16, Highway (Transportation) Research Board, Washington, 1973.

16.13. *Design of Continuously Reinforced Concrete for Highways,* Concrete Reinforcing Steel Institute, Chicago, 1981.

16.14. *Construction of Continuously Reinforced Concrete Pavements,* Concrete Reinforcing Steel Institute, Chicago, 1983.

16.15. "A Design Procedure for Continuously Reinforced Concrete Pavements for Highways," ACI Committee 325 Report, *J. ACI,* vol. 69, no. 6, 1972, pp. 309–319.

16.16. "Failure and Repair of Continuously Reinforced Concrete Pavement," Synthesis of Highway Practice No. 60, Transportation Research Board, Washington, July 1979.

16.17. "Design of Continuously Reinforced Concrete Pavements for Airports," Reported by ACI Committee 325, *ACI Structural Journal,* vol. 86, no. 6, 1989, pp. 736–747.
16.18. *Continuously Reinforced Concrete Pavement—Design and Construction Practices of Various States,* Concrete Reinforcing Steel Institute, Chicago, January 1975.
16.19. A. Losberg, "Continuously Reinforced Pavements with a Very Thin Surface Layer," *Proc. First European Symp. on Concrete Pavements,* Paris, vol. 2, 1969, pp. 179–185.
16.20. M. Sargious and S. K. Wang, "Economical Design of Prestressed Concrete Pavements," *J. PCI,* vol. 16, no. 4, 1971, pp. 64–79.
16.21. M. Sargious and S. K. Wang, "Design of Prestressed Concrete Airfield Pavements under \Dual and Dual-Tandem Wheel Loading," *J. PCI,* vol. 16, no. 6, 1971, pp. 19–30.
16.22. J. A. Sindel, "A Design Procedure for Post-Tensioned Concrete Pavements," *Concrete International,* vol. 5, no. 2, 1983, pp. 51–57.
16.23. "Prestressed Concrete Pavements No Longer Seen as Experimental," *Engineering News-Record,* May 23, 1974, pp. 16–17.
16.24. P. Rice, "Design of Concrete Floors on Ground for Warehouse Loading," *J. ACI,* vol. 54, no. 2, 1957, pp. 105–114.
16.25. J. J. Panak and J. B. Rauhut, "Behavior and Design of Industrial Slabs on Grade," *J. ACI,* vol. 72, no. 5, 1975, pp. 219–224.
16.26. R. G. Packard, "Slab Thickness Design for Industrial Concrete Floors on Grade," Concrete Information Bulletin No. IS195.01D, Portland Cement Assoc., Skokie, Illinois, 1976.
16.27. R. E. Spears, "Concrete Floors on Ground," Concrete Information Bulletin No. EB075.01D, Portland Cement Assoc., Skokie, Illinois, 1978.

PROBLEMS

16.1 It is proposed to pave a pedestrian shopping mall using a 6 in. concrete slab having construction/contraction joints 25 ft on centers in both directions. The concrete specified has $f_c' = 3500$ psi, $w_c = 150$ pcf, and $\nu = 0.20$.

(*a*) What tensile stresses will result from concrete shrinkage, assumed to be uniform throughout the slab depth? The coefficient of friction between the slab and the compacted subbase may be taken as 1.5.

(*b*) What will be the effect of a temperature differential of 20°F between top and bottom of the slab, resulting from a rapid drop in temperature at night? The thermal coefficient may be assumed equal to 5.5×10^{-6} in/in per $°F$.

(*c*) Will the slab crack under the combined effects of shrinkage and temperature differential?

(*d*) Will the surface of the slab be flat or warped under the combined effects? Explain.

16.2 A concrete highway pavement is to be designed according to AASHTO recommendations, as summarized in Sec. 16.4, on the basis that the 12 ft wide traffic lanes are unreinforced plain concrete. It has been determined that the number of total equivalent 18 kip single-axle load applications will be 2 million over a 20 year period. The subbase and subgrade are a mixture of sand and gravel with an overall subgrade modulus $k = 300$ pci. Find the required slab thickness of the concrete which has $f_c' = 4000$ psi and an allowable flexural tensile stress of 250 psi. If contraction joints are sawed at 20 ft centers, what welded wire fabric would you specify to control shrinkage and temperature cracking? The coefficient of friction between slab and subgrade = 1.5, and an allowable tensile stress of 37.5 ksi is specified for the welded wire mesh. Sketch suitable details for transverse contraction and expansion joints, and for the longitudinal joint between traffic lanes.

16.3 Redesign the highway pavement of Prob. 16.2 as a continuously reinforced unjointed pavement, following the recommendations of Sec. 16.6. Note the current FHA position on slab thickness. Specify both longitudinal and transverse reinforcement and the position of each in the slab. If the thermal coefficient for reinforced concrete is 5.5×10^{-6} in/in per °F, and if the shrinkage of the reinforced concrete slab is 333×10^{-6} in/in, how many degrees temperature rise above the temperature at the time of casting will be required to completely close the cracks? If the crack spacing is 7 ft as an average, what crack width is to be expected prior to that temperature rise?

16.4 Redesign the warehouse floor slab of Example 16.2 as a continuously reinforced slab, constructed in 20 ft wide strips, with contraction/construction joints on the column lines in one direction only. Integrate the concepts of Secs. 16.6 and 16.9. Identify and justify any assumptions made. Sketch an appropriate detail for the longitudinal joints.

CHAPTER 17

COMPOSITE CONSTRUCTION

17.1 TYPES

While all reinforced concrete is composite construction in the sense that it combines two dissimilar materials in any single structural member, the term composite is usually applied to the combination of concrete with structural steel in the form of steel beams, columns, or floor deck panels. The structural steel, while capable of carrying loads if acting alone, can develop much greater strength and stiffness if it acts integrally with the concrete. Examples of composite construction include conventional hot-rolled steel beams with a concrete deck slab (Fig. 17.1*a*) for highway bridges or building floors, columns reinforced with structural steel shapes (Fig. 17.1*b*) used for tall buildings, and cold-formed sheet steel decking with an integral concrete slab on top (Fig. 17.1*c*) used for floors in steel frame office and apartment buildings. An essential feature of all these forms of composite construction is that slip between the steel and concrete is prevented, either by the natural bonding between the two materials or (more reliably) by interlocking devices welded or formed into the steel surface and embedded into the concrete as it is poured.

The most common form of composite construction, used for many years for highway bridges, results from the combined action of conventional steel beams with a cast-in-place concrete deck slab, as shown in Fig. 17.1*a*. Section 17.2 of this chapter pertains mainly to the analysis and design of composite highway bridges of this type, according to the AASHTO specification (Ref. 17.1).

The same type of composite construction is now used with increasing frequency in buildings. While the general principles are identical, details of analysis and design differ somewhat, due mainly to differences in the nature of the loading. Composite buildings are usually designed under the provisions of the AISC specification (Ref. 17.2); details are presented in Sec. 17.3).

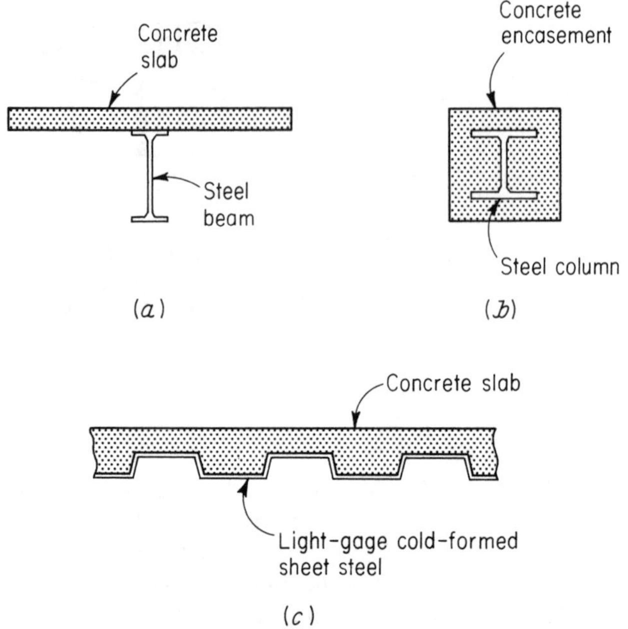

FIGURE 17.1
Types of composite construction: (*a*) concrete deck slab over conventional steel beam; (*b*) concrete-encased steel column; (*c*) concrete floor slab over light-gage cold-formed steel deck.

The combination of cold-formed steel deck panels with cast-in-place concrete floor or roof slab, shown in Fig. 17.1*c*, is widely used in the United States, usually in conjunction with building frames using conventional hot-rolled steel beam and column sections. Referred to as composite steel deck slabs, these floor systems are presently designed according to the recommendations of the ASCE Technical Council of Codes and Standards (Refs. 17.3 and 17.4); details will be found in Sec. 17.4 of this chapter.

Design criteria for composite steel-concrete columns are still being formulated. For this reason, and because proposed design methods are more closely tied to ordinary structural steel design than concrete design, composite columns will not be treated here.

17.2 COMPOSITE BEAM-AND-SLAB BRIDGES

A large percentage of all medium-span highway bridges in the United States, from about 60 to 120 ft span, are of composite construction such as illustrated in Figs. 17.1*a* and 17.2. Steel girders are placed parallel to one another in the span direction of the bridge, with lateral spacing typically from about 6 to 10 ft. Over these steel girders, and resting on their top flange, is a cast-in-place concrete slab that becomes a part of the roadway. The steel beams may be standard flanged W

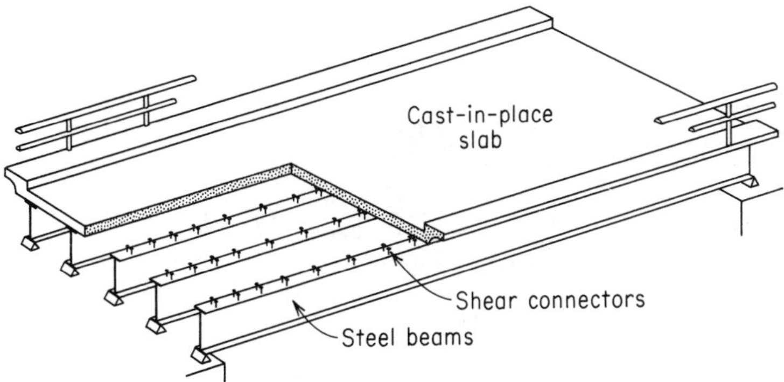

FIGURE 17.2
Composite structural steel and concrete slab highway bridge.

shapes, W sections plus cover plates (usually added only to the lower flange in the high moment regions), or built-up plate girders, fabricated by welding to form I-section beams. For longer spans, variable-depth steel members may be used. Such bridges may be continuous over one or more interior supports.

The concrete slab spans laterally, carrying its loads to the steel girders, and in this sense is a continuous one-way slab, with negative moments over the girders and positive moments between. Thus its principal reinforcement is in the direction transverse to the span direction of the bridge. In the direction of the girders, only shrinkage and distribution steel is used in the slab. The design of such a slab for bending in the transverse direction is carried out in the same way as for any continuous one-way slab (see Chap. 12).

Longitudinally, while the slab and the girders may be designed to act independently, if special precautions are taken to avoid slip at the interface, the slab becomes, in effect, a part of the compression flange of the girder, greatly increasing its strength and stiffness. Putting it another way, a smaller and lighter girder can be used to carry the specified loads, still satisfying requirements for strength and stiffness.

The need for special *shear connectors* is made clear from a study of Fig. 17.3; which shows a steel beam over which has been cast a concrete slab. As flexural loading is applied, compressive strains through the top half of the steel member, linearly varying with height, tend to cause shortening, and tensile strains through the bottom half tend to cause elongation, as shown in Fig. 17.3*b*. Similarly, if the slab were free to deform independently, the top surface would tend to shorten compared to the bottom surface, which would lengthen, as shown. Thus there is a tendency for slip to occur at the interface between the steel and concrete. The tendency to slip is a maximum near the supports of a simple span, where the external shear force (rate of change of moment) is greatest. It decreases to zero at the section of maximum moment (zero shear).

Actually such slip is prevented, in part by friction between the two surfaces, but mainly by special shear connectors such as shown in Fig. 17.3*a*. Studs are

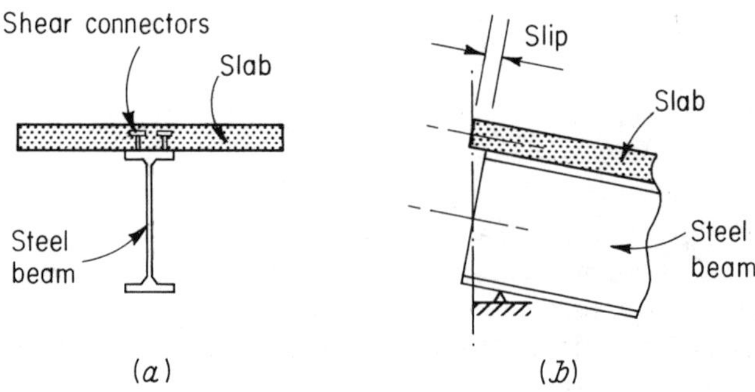

FIGURE 17.3
Composite beam subject to bending: (*a*) cross section; (*b*) slip at support region of simple span.

welded to the top flange of the steel beam and embedded in the concrete slab when it is poured. The presence of some such device to ensure that the steel and concrete act together in an integral sense in carrying the loads is an essential part of composite bridge construction.

a. Design Basis

In Sec. 1.5 of Chap. 1 it was pointed out that reinforced concrete structures may presently be designed by either of two alternative methods. The first, based on *strength criteria,* requires the use of load factors, greater than 1.0, by which expected service loads are increased to find design moments and forces at a hypothetical overload state. The members are then proportioned by equating their design strength, i.e., their nominal strength times a ϕ factor less than 1.0, to the required strength. This is the design approach favored in general for reinforced concrete, and which has been followed elsewhere in this text and in the ACI Code.

The alternative method of design is based on imposing limits on stresses in the materials when actual service loads act. These stresses are sufficiently below the material strengths to ensure the required degree of safety. Presently, the design of highway bridges under the AASHTO specification is based on such *service load design or allowable stress design*. Generally the materials will behave elastically up to the service load level, so familiar equations from elastic mechanics can be used, modified only to account for the nonhomogeneous nature of the construction, cracking of the concrete, and time-dependent effects.

Use of service load design for highway bridges is to some extent the result of historic precedent. However, because of the importance of fatigue in the design of bridges, there is reason to emphasize actual stresses in service. In addition, application of the load factor approach, basic to strength design, can become very complicated where many types of loads (e.g., dead, live, impact, centrifugal, wind, ice, etc.) may act in combination.

The details of service load design as used for highway bridge structures will become clear in the sections that follow.

b. Section Properties and Elastic Flexural Stresses

In analyzing or designing the beams of a composite bridge, it is necessary to consider the effects of both noncomposite and composite loads.

Noncomposite loads include the self-weight of the steel beams, the dead weight of formwork and fresh concrete of the deck, and any transverse bracing members. These loads cause flexural stresses in the steel beam, which acts alone at this stage in resisting them. Flexural stresses may be found by the usual methods of steel design and are based on the moment of inertia of the steel section, plus the cover plates if used.

After the slab has hardened, it becomes a part of the girder section, in effect, and subsequent loads cause bending stresses in the composite member. Composite loads include certain additional dead loads such as a roadway wearing surface, sidewalks, railings, piping, and lighting fixtures, and service live loads from traffic, snow, ice, etc. The composite flexural stresses, resulting from bending about a new neutral axis, superimpose on the preexisting flexural stresses in the steel beam, which were calculated for the original cross section.

In calculating the properties of the composite cross section, with the concrete slab serving as all or part of the compression zone, it is customary to account for the actual variable compressive stress across the width of the slab using a concept that was introduced earlier for reinforced concrete T beams. An *effective width* is defined that, if stressed uniformly to the maximum compression as found directly over the beam web, would give the same total compressive force as obtained for the actual, variable compression. According to the AASHTO specification, the effective width of the compression flange shall not exceed the following:

1. One fourth of the span length of the girder
2. The distance center to center of the girders
3. Twelve times the least thickness of the slab

For girders having a flange on one side only, the effective flange width shall not exceed one twelfth of the span length of the girder, nor six times the thickness of the slab, nor one half the distance center to center of the next girder, according to the AASHTO specification.

In computing the properties of the nonhomogeneous cross section, the concept of the *transformed cross section* is useful. While for reinforced concrete beams it is usual to transform the steel into fictitious tension concrete, in composite beams the reverse is convenient. The concrete compression slab is transformed into an equivalent area of steel, i.e., a steel area that provides the same total force, given the same strain, as the original concrete. The transformation is based on the modular ratio $n = E_s/E_c$, and is carried out as shown in Fig. 17.4. Figure 17.4*a* shows the steel beam and the slab, which has depth h_f and effective width b (determined according to the above criteria). If an equivalent steel area is substituted, having depth h_f and reduced width b/n, then the section may be treated as homogeneous, with the properties of steel. For a given flexural strain

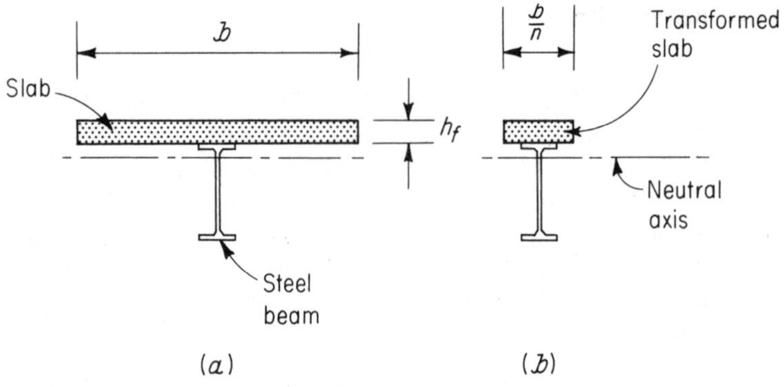

FIGURE 17.4
Section properties of composite section: (*a*) actual effective cross section; (*b*) transformed section.

distribution in the compression zone, the transformed steel will provide the same compression resultant, at the same level in the beam, as would the concrete in the actual section.

According to the AASHTO specification, the values of n listed in Table 17.1 may be used.

If part of the loads acting on a composite member are long term, such as dead loads, the effect of concrete creep is to diminish the effectiveness of the concrete part of the cross section in resisting stress. This can be accounted for in an approximate way, using a *reduced modulus of elasticity* for the concrete when computing properties of the transformed section to be used when calculating stresses due to those sustained loads. According to the AASHTO specification, the concrete modulus, when sustained loads are considered, is to be taken as one third of the usual value, or, put another way, the modular ratio n is to be taken as 3 times the usual value, when computing properties of the transformed section.

In composite design, the section must be found largely by trial. Although some tables are available giving the section properties for certain combinations of slab and steel beam (Ref. 17.5), and although methods have been developed to

Table 17.1 Values of $n = E_s/E_c$ for short-term loading, according to the AASHTO specification

f'_c	n
2000 to 2300 psi	11
2400 to 2800 psi	10
2900 to 3500 psi	9
3600 to 4500 psi	8
4600 to 5900 psi	7
6000 psi or more	6

reduce the labor of finding an appropriate section (Ref. 17.6), it is not practical to tabulate all desirable combinations of slab width and thickness, steel beam section, and cover plate size. Direct design of composite sections is therefore seldom possible, and the design process is actually a review of an assumed section, to determine if the stresses in the concrete and steel are within the allowable range. Stresses due to each type of loading are computed using the appropriate section properties, and these stresses are added algebraically to obtain the combined effect.

Under the AASHTO specification, maximum allowable bending stresses in ordinary A36 grade steel are 20,000 psi in tension and the same in compression. The allowable compressive stress in bending in the concrete is $0.40f'_c$.

Example 17.1 Section properties and flexural stresses in composite bridge girder. A 70 ft simple-span highway bridge will be comprised of W36 × 150 steel beams spaced laterally 6 ft 9 in. center to center, over which will be placed a concrete slab of 6.5 in. total depth. Shear connectors will be provided so that the two components act together in. resisting bending. Concrete strength is specified to be $f'_c = 3000$ psi. The maximum moments that will act on the cross section under consideration, with load factors of 1.0, are as follows: (*a*) noncomposite dead load moment on steel section alone = 320 ft-kips; (*b*) composite dead load moment = 120 ft-kips; (*c*) composite live load plus impact moment = 500 ft-kips. Find the flexural stresses at the top and bottom of the steel beam, and at the top of the concrete slab, resulting from the combined noncomposite and composite load (*a*) plus (*b*) plus (*c*).

Solution. According to the AASHTO specification, the effective width of the concrete compression flange is the smallest of:

$$\frac{L}{4} = 70 \times \frac{12}{4} = 210 \text{ in.}$$

$$\text{Center-to-center spacing} = 81 \text{ in.}$$

$$12 \times h = 12 \times 6.5 = 78 \text{ in.}$$

The last criterion controls, and the effective cross section is as shown in Fig. 17.5*a*. Next the concrete slab will be transformed to an equivalent width of steel, using $n = 9$ in. computing section properties for short-term loading, and $n = 3 \times 9 = 27$ for sustained (dead) loading. Thus the transformed slab width is $78/9 = 8.67$ in. for short-term loads, and $78/27 = 2.89$ in. for sustained loading, as shown in Fig. 17.5*b* and *c* respectively.

The depth, moment of inertia, cross-sectional area, and section modulus of the bare steel section are found from Ref. 17.2. With a section modulus of 504 in^3, the noncomposite moment of 320 ft-kips resulting from self-weight of the steel, wet concrete, etc., produces compressive and tensile flexural stresses at the top and bottom of the steel respectively of:

$$f_{ts} = 320 \times \frac{12{,}000}{504} = 7600 \text{ psi compression}$$

$$f_{bs} = 320 \times \frac{12{,}000}{504} = 7600 \text{ psi tension}$$

Calculation of section properties for the transformed section of Fig. 17.5*c*, with $n = 27$ for sustained loads, follows the usual methods of structural mechanics, and is

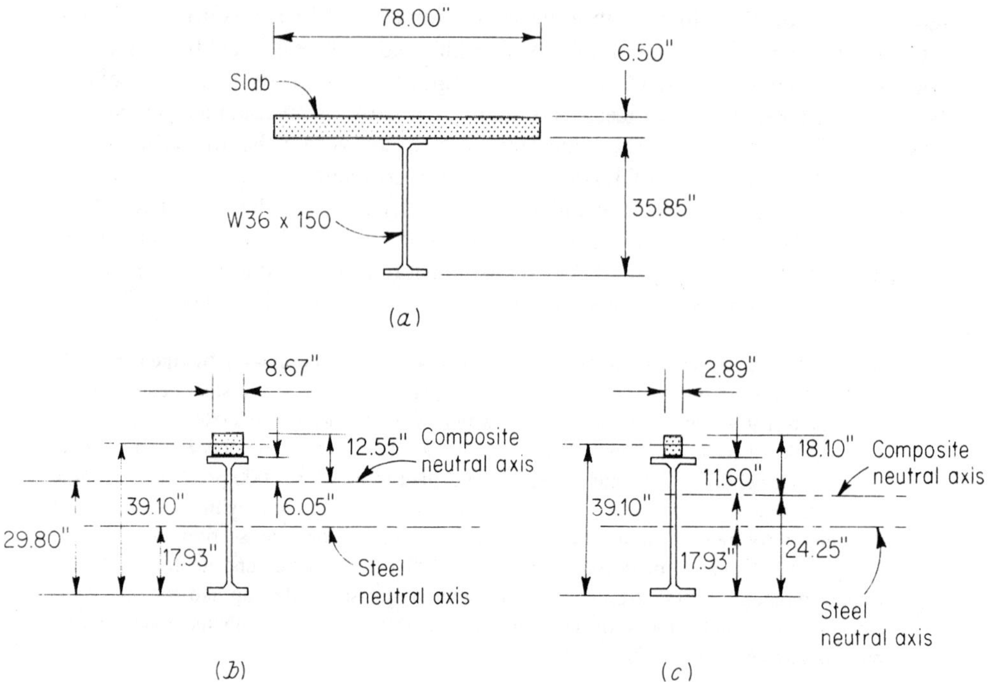

FIGURE 17.5
Composite cross section of Example 17.1: (*a*) effective section; (*b*) transformed section with $n = 9$; (*c*) transformed section with $n = 27$.

conveniently done in tabular form, as shown in Table 17.2. Taking moments of areas around the bottom of the section results in a composite neutral axis location $\bar{y} = 24.25$ in. above the bottom. The composite moment of inertia about that axis is $I = 15{,}016$ in^4. Dividing by the appropriate distances provides the section moduli with respect to the top of concrete, top of steel, and bottom of steel respectively:

$$S_{tc} = 830 \text{ in}^3$$

$$S_{ts} = 1290 \text{ in}^3$$

$$S_{bs} = 620 \text{ in}^3$$

Thus the stresses in the transformed section due to the composite sustained dead load moment of 120 ft-kips are:

$$f_{tc} = 120 \times \frac{12{,}000}{830} = 1700 \text{ psi compression}$$

$$f_{ts} = 120 \times \frac{12{,}000}{1290} = 1100 \text{ psi compression}$$

$$f_{bs} = 120 \times \frac{12{,}000}{620} = 2300 \text{ psi tension}$$

Table 17.2 Composite section properties of girder of Example 17.1

	I_0, in^4	A, in^2	y_b, in.	Ay_b, in^3	d to na, in.	Ad^2, in^4
Steel only	S_{bs} = 504 in^3			S_{ts} = 504 in^3		
Composite with n = 9						
Steel	9,040	44.2	17.93	793	11.87	6,230
Slab	198	56.4	39.10	2,205	9.30	4,880
	9,238	100.6		2,998		11,110
						9,238
						20,348
	$\bar{y}$ = 2,998/100.6 = 29.80 in. S_{tc} = 20,348/12.55 = 1,620 in^3 S_{ts} = 20,348/6.05 = 3,360 in^3 S_{bs} = 20,348/29.80 = 680 in^2					
Composite with n = 27						
Steel	9,040	44.2	17.93	793	6.32	1,760
Slab	66	18.8	39.10	735	14.85	4,150
	9,106	63.0		1,528		5,910
						9,106
						15,016
	y = 1,528/63.0 = 24.25 in. S_{tc} = 15,016/18.10 = 830 in^3 S_{ts} = 15,016/11.60 = 1,290 in^3 S_{bs} = 15,016/24.25 = 620 in^3					

It should be emphasized that the stress just computed at the level of the top of the slab is the stress in the *transformed steel,* and must be divided by $3n = 27$ to obtain the stress in the actual concrete of 60 psi compression.

Section properties applicable to short-term loads, using $n = 9$, follow in a similar way as shown in Table 17.2 and result in $I = 20{,}348$ in^4. Section moduli with respect to top of concrete, top of steel, and bottom of steel are respectively:

$$S_{tc} = 1620 \text{ in}^3$$

$$S_{ts} = 3360 \text{ in}^3$$

$$S_{bs} = 680 \text{ in}^3$$

so the stresses due to the composite, short-term, live load moment of 500 ft-kips are:

$$f_{tc} = 500 \times \frac{12{,}000}{1620} = 3700 \text{ psi compression}$$

$$f_{ts} = 500 \times \frac{12{,}000}{3360} = 1800 \text{ psi compression}$$

$$f_{bs} = 500 \times \frac{12{,}000}{680} = 8800 \text{ psi tension}$$

and the stress at the top of the concrete is 3700/9 = 410 psi compression.

Total stresses in the concrete and steel are easily found by summing the parts:

$$f_{tc} = 0 + 60 + 410 = 470 \text{ psi compression}$$

$$f_{ts} = 7600 + 1100 + 1800 = 10{,}500 \text{ psi compression}$$

$$f_{bs} = 7600 + 2300 + 8800 = 18{,}700 \text{ psi tension}$$

Checking these against allowable values, the concrete is well below the permitted $0.40 \times 3000 = 1200$ psi compression, and both compression and tension in the steel section are below the value of 20,000 psi permitted for Grade A36 steel, confirming that the proposed design is adequate.

c. Shear Connectors

Shear connectors are essential to the development of composite action. They must transfer the horizontal shear, with extremely small deformation, so that the structure behaves as a unit. They must also be capable of resisting any tendency for the slab to separate vertically from the steel flange due to buckling or other causes.

The two types of shear connector most widely used in U.S. practice are shown in Fig. 17.6. Short pieces of steel channel may be used, as in Fig. 17.6*a*, set transverse to the beam axis and welded to the beam along the heel and toe of the channel. According to specification, welds must be at least $\frac{3}{16}$ in. fillets. Stud shear connectors (Fig. 17.6*b*) are commonly available in diameters of $\frac{5}{8}$, $\frac{3}{4}$, and $\frac{7}{8}$ in. They are granular-flux-filled and are resistance-welded to the steel beam with special equipment. According to the AASHTO specification, the clear depth of concrete cover over the tops of all shear connectors must be not less than 2 in. The connectors must penetrate at least 2 in. above the bottom of the slab, and the clear distance between the edge of the girder flange and the edge of the connector must be not less than 1 in.

Connectors of both types are most economically installed in the fabricating shop. However, the advantages of this procedure may be offset by the possibility of damage during shipment and the difficulty in working along the top of the beams during erection.

According to the AASHTO specification, shear connectors must be designed for fatigue (Ref. 17.7), and then the number of connectors obtained checked to ensure that sufficient strength has been provided to develop the capacity of the member.

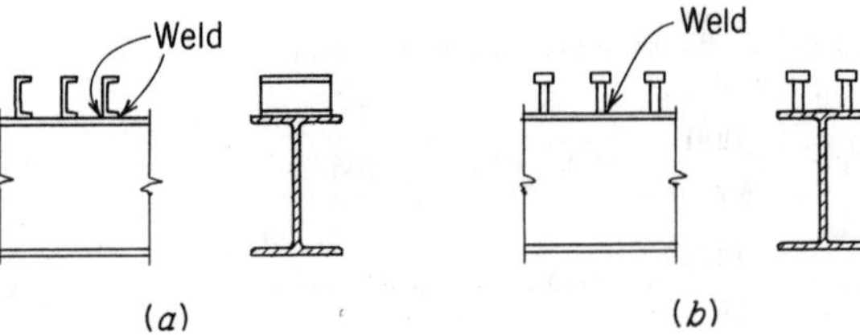

FIGURE 17.6
Typical shear connectors: (*a*) channels; (*b*) studs.

Design for fatigue. The horizontal shear per unit length along the interface between concrete slab and steel beam may be found by the usual equation of structural mechanics:

$$S = \frac{VQ}{I}$$

where S = horizontal shear, lb/in.
V = external shear force at section, lb
Q = static moment of transformed compression concrete area about neutral axis of entire section, in^3
I = moment of inertia of entire transformed section, in^4

In designing for fatigue, it is the variation of horizontal shear, as live load and impact load are applied, that is significant. Consequently, the preceding equation is restated in terms of the change in horizontal shear that occurs as live and impact loads are applied:

$$S_r = \frac{V_r Q}{I} \tag{17.1}$$

where S_r = range of horizontal shear, lb/in.
V_r = range of external shear force at section, due to live and impact loads, lb

and other terms are as already defined. The range of shear may be taken as the difference in the maximum and minimum shear envelopes, excluding dead loads.

The allowable range of horizontal shear Z_r per individual connector has been established by extensive fatigue testing (Ref. 17.7). The following values are accepted by AASHTO:

For channels:

$$Z_r = Bw \tag{17.2}$$

where Z_r = allowable range of horizontal shear per channel, lb
w = length of channel shear connector measured in a transverse direction on flange of girder, in.
B = 4000 for 100,000 cycles; 3000 for 500,000 cycles; 2400 for 2,000,000 cycles; 2100 for over 2,000,000 cycles

For welded studs:

$$Z_r = \alpha d^2 \tag{17.3}$$

where Z_r = allowable range of horizontal shear per stud, lb
d = diameter of stud, in. (height H must not be less than $4d$)
α = 13,000 for 100,000 cycles; 10,600 for 500,000 cycles; 7850 for 2,000,000 cycles; 5500 for over 2,000,000 cycles

The allowable range of horizontal shear of all the connectors at one transverse girder section, ΣZ_r, divided by the pitch p, or spacing of connector groups along the beam, must at least equal the range of horizontal shear, S_r. Thus the maximum pitch is

$$p = \frac{\Sigma Z_r}{S_r} \tag{17.4}$$

According to specification, the pitch is not to exceed 24 in.

Design for strength. According to specification, the number of connectors provided on the basis of fatigue must be checked to ensure that sufficient connectors have been provided to develop the strength of the composite girders. The number N of shear connectors needed between the point of maximum positive moment and the end supports is equal to

$$N = \frac{P}{\phi S_u} \tag{17.5}$$

where S_u = ultimate strength of one shear connector, lb
ϕ = capacity reduction factor, equal to 0.85
P = force to be developed, lb

The force P is equal to the smaller of the following:

$$P_1 = A_s f_y \tag{17.6a}$$

where A_s = total area of steel section, including cover plates, in^2
f_y = yield strength of steel beam, psi

or

$$P_2 = 0.85 f'_c bc \tag{17.6b}$$

where b = effective width of slab, in.
c = slab thickness, in.

The ultimate strength S_u of the shear connectors has been established by testing, and the following expressions are recommended:

For channels:

$$S_u = 550\left(h + \frac{t}{2}\right) w \sqrt{f'_c} \tag{17.7}$$

where h = average thickness of channel flange, in.
t = thickness of channel web, in.
w = length of channel measured in transverse direction on flange of girder, in.

For welded studs:

$$S_u = 0.4 d^2 \sqrt{f'_c E_c} \tag{17.8}$$

where d is the diameter of the stud, in inches (height H must not be less than $4d$).

An example of the application of these provisions for the design of shear connectors will be found in that part of Chap. 22 dealing with composite bridges.

d. Other Considerations

Shored construction. Normally in composite construction, a part of the dead loads must be carried by the bare steel beam. The noncomposite loads would include all loads applied prior to the time that the concrete slab has attained its strength. However, if temporary supports, or shoring, are used during construction, and if these supports are kept in place until the concrete has hardened (until it has attained 75 percent of its design strength, according to AASHTO), then all dead and live load stresses can be computed on the basis of composite action.

Flexural stresses in the steel are therefore lower, and it will generally be possible to use a smaller section, still satisfying requirements of strength and stiffness. This advantage must be weighed against possible objections to the presence of shores for the required period, as they are likely to interfere with construction activity or traffic flow beneath the bridge.

Continuous construction. In continuous spans, the positive bending region can be designed following the same procedures as for simple spans. In the negative bending region near the supports, the slab is on the tension side of the member, and due to inevitable tension cracking, it would have little influence on the stiffness and make no contribution to the strength. However, negative reinforcing bars can be added in the slab, in the direction parallel to the girder span. In this case, the presence of the reinforced slab can be counted upon both to stiffen and to strengthen the girders. Such bar reinforcement is analogous to a cover plate. It can be discontinued where no longer needed to resist flexure, just as bottom flange cover plates are discontinued where no longer needed in the positive bending region. If such negative bars are counted upon for composite action, shear connectors must be provided in the negative-moment region. These connectors are designed following the same procedures outlined for connectors in simple spans.

According to the AASHTO specification, in continuous composite spans, the minimum longitudinal reinforcement in the negative bending region must equal or exceed 1 percent of the gross area of the concrete slab, and two thirds of this requirement must be placed as closely as possible to the top of the slab, with due regard to the requirements for concrete cover.

Vertical shear. The intensity of unit shearing stress in a composite girder is computed based on the area of the web of the steel beam, as usual for steel design, neglecting the effects of the steel flanges and of the concrete slab. The shear is considered to be uniformly distributed through the depth of the web. For the typical A36 steel, the allowable shear stress computed in this way is 12,000 psi, according to the AASHTO specification.

17.3 COMPOSITE BEAM-AND-SLAB BUILDING FLOORS

Many buildings are constructed with structural steel columns, beams, and girders, plus a concrete slab that provides the floor surface. The slab can be, and often is, designed merely to carry the floor loads laterally to the beams, acting as a one-way continuous slab. As was true for bridges, however, significant advantages may be gained in buildings if the slab is also designed to act in a composite sense longitudinally, as an integral part of the beams, thereby providing greatly increased compression flange area. Flexural stresses are reduced and flexural stiffness increased. As a consequence, shallower and lighter beams can be used to meet requirements.

Composite construction in buildings is most efficient for longer spans, heavily loaded, with beams spaced laterally as widely apart as conditions permit. In all cases, the saving in steel weight resulting from composite action must be balanced against increased unit cost for fabrication and erection, compared with noncomposite members.

Composite floors and roofs in buildings are designed according to the provisions of the AISC specification (Ref. 17.2). While methods are similar to those used in designing bridges under the AASHTO specification, there are differences, mainly related to the differences in the nature and magnitude of the loads.

Two types of composite building construction are recognized in the AISC specification. *Fully encased beams,* such as shown in Fig. 17.7*a*, depend upon the natural bond between concrete and steel for interaction, without additional anchorage. *Beams with mechanical anchorage,* in the form of welded studs or channels, such as shown in Fig. 17.7*b*, do not have to be encased. For encased beams to be considered effective in a composite sense, according to specification, at least 2 in. of concrete cover must be provided on the sides of the beam and below it. The top of the beam must be at least 1.5 in. below the top of the slab, and at least 2 in. above the bottom. In addition, the encasement must have adequate mesh or other reinforcing steel throughout the depth of the beam and across the

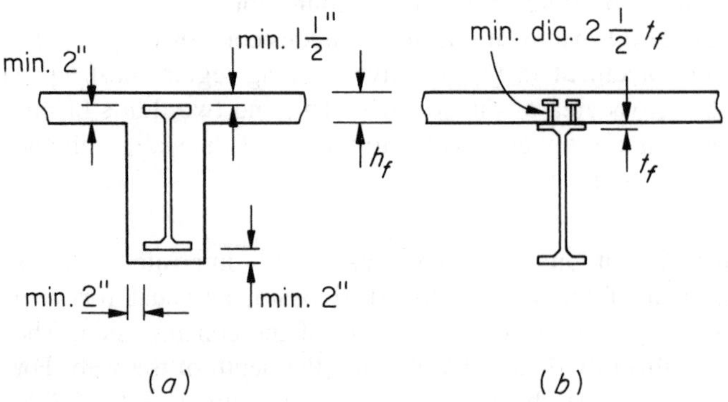

FIGURE 17.7
Composite construction in buildings: (*a*) beam totally encased; (*b*) slab on beam with shear connectors.

bottom to prevent spalling of the concrete. For beams with studs, in order to prevent tearing of a thin flange before the capacity of the stud is reached, the stud diameter must not be more than 2.5 times the thickness of the flange. This requirement does not apply if the stud is welded directly over the web.

The effective flange width for composite beams with a slab extending on both sides is defined under the AISC specification to be not more than one quarter the beam span, and the effective projection beyond the edge of the beam must not be taken greater than one half the clear distance to the next beam, nor more than 8 times the slab thickness. If a slab is present on one side only, the effective projection must not be taken more than one twelfth the beam span, nor 6 times the slab thickness, nor one half the clear distance.

As for composite bridges, design is based on allowable stresses at service loads, with all load factors set equal to 1.0. Flexural stresses are found based on the transformed section. The effective concrete flange width is transformed to an equivalent width of steel, dividing the concrete width by the modular ratio $n = E_s/E_c$. In stress calculations, there is no distinction made between dead and live loads (see below), so the same value of n is used for both sustained and short-term loads, in effect. However, when calculating deflections, it is recommended that the influence of creep be recognized by using a modular ratio 2 times the usual value when determining section properties to be used for dead load deflections.

a. Design Assumptions and Approximations

Unless temporary shores are used, *encased beams* with steel and concrete interconnected only by natural bond must be proportioned to support all dead loads applied prior to hardening of the concrete. The steel member is assumed to act unassisted. Dead loads and live loads applied after hardening of the concrete are resisted by composite action. The sum of noncomposite and composite stresses in the steel must not exceed 66 percent of the yield stress. For A36 steel the allowable bending stress is thus 24 ksi. Tensile stresses in the concrete are neglected in the calculations as usual.

Alternatively, for encased beams, the steel beam alone may be proportioned to resist, unassisted, the positive moment produced by *all* loads, live and dead, using an allowable bending stress of 76 percent of yield. This approximate allowance for the strengthening and stiffening effect of composite action eliminates the need to calculate composite section properties for encased sections, and is often adopted in engineering practice.

For *beams with shear connectors* the AISC specification permits, within certain limits, that stresses be calculated assuming that *all* the loads are resisted by composite action, including dead loads that are actually noncomposite, whether or not shoring is used during construction. Allowable stresses, so computed, are not to exceed 66 percent of yield. While it is difficult to explain this provision in terms of allowable stress design, the practice may be justified based on the concepts of strength design. At a hypothetical overload stage, the strength of the composite member would be calculated recognizing the nonlinear response of both materials at high loads. At such loads, elastic stress analysis is not pertinent and superposition of stresses is not valid. The order in which the loads are applied does not matter in comparing required strength and strength provided, and it does not matter whether the loads were initially applied to the noncomposite or composite member.

If the suggested procedure is followed, however, it is necessary to introduce certain limitations to ensure that bending stresses in the steel beam under service loading will be well below the yield stress, regardless of the ratio of composite to noncomposite loads. According to the AISC specification, for construction without temporary shoring, if stresses due to both dead and live loads are computed using the composite transformed section, the value of the transformed section modulus to be used must not exceed

$$S_{tr} = \left(1.35 + 0.35\frac{M_l}{M_d}\right)S_s \tag{17.9}$$

where S_{tr} = limiting value of section modulus of transformed composite section with respect to its bottom flange, in^3

S_s = section modulus of steel beam with respect to its bottom flange, in^3

M_l = moment caused by loads applied subsequent to the time when the concrete has reached 75 percent of its specified strength

M_d = moment caused by loads applied prior to this time

For the extreme case in which $M_l = 0$, the effect of Eq. (17.9) is to limit the stress in the tension flange to 89 percent of yield.

For calculating the concrete flexural compressive stress, the actual transformed section modulus is to be used, not the limiting value of Eq. (17.9), and for construction without shoring, the concrete stress is based on loads applied after the concrete has reached 75 percent of its specified strength. The allowable stress in the concrete is $0.45f_c'$.

The web and the end connections of the steel beam are designed to carry the total end shear and reaction for the composite beam.

b. Shear Connectors

According to elastic design concepts, the spacing of shear connectors should vary inversely with the horizontal shear stress along the top flange, and hence inversely with the applied shear load. However, tests have confirmed that beams with the total required number of shear connectors, uniformly spaced along the flange, develop the same ultimate strength. Only a slight amount of slip is required to redistribute elastic shear forces to less highly stressed connectors. Accordingly, the AISC specification permits connector design based simply on the concept that the total horizontal force must be transferred between the point of maximum moment and the point of zero moment, and does not require variable spacing. (Note the contrast with the AASHTO specification for bridges, which requires consideration not only of strength but also of fatigue under elastic conditions.) For buildings, the total force to be transferred for full composite action, between the points of maximum moment and zero moment, is to be taken as the smaller of

$$V_h = \frac{0.85 f_c' A_c}{2} \tag{17.10a}$$

or

$$V_h = \frac{A_s f_y}{2} \tag{17.10b}$$

where A_c = area of effective concrete flange, in^2
A_s = area of steel beam, in^2

The number of shear connectors, N_1, to be provided on each side of the point of maximum moment must not be less than the smaller value of V_h from Eq. (17.10) divided by q, the allowable value of resistance for one connector, as given in Table 17.3 (for normal density concrete), modified by the factors of Table 17.4 (for lightweight concrete). Note that, while the specification is worded in terms of allowable loads, it is based strictly on strength design concepts. The forces of Eq. (17.10) are computed based on the strength of each material, with that strength divided by 2 to permit use of *allowable* values of connector loads rather than ultimate *strength* values.

While, in general, shear connectors may be spaced uniformly, as just described, certain loading patterns may require a closer spacing over a part of the distance. Specifically, if concentrated loads act between the points of maximum moment and zero moment in the positive bending region, the moment at those concentrated loads may be only slightly less than the maximum moment at midspan. The number of shear connectors N_2 required between each end of the beam and the adjacent concentrated load would be only slightly less than the number N_1 required between each end and midspan, in that case. It can be shown that the number of connectors required between any such concentrated load and the nearest

Table 17.3 Allowable horizontal shear load q kips for one connector in normal density concrete slab

	Specified compressive strength of concrete, ksi		
Connector	**3.0**	**3.5**	**≥ 4.0**
$\frac{1}{2}$ in. diameter × 2 in. hooked or headed stud	5.1	5.5	5.9
$\frac{5}{8}$ in. diameter × 2.5 in. hooked or headed stud	8.0	8.6	9.2
$\frac{3}{4}$ in. diameter × 3 in. hooked or headed stud	11.5	12.5	13.3
$\frac{7}{8}$ in. diameter × 3.5 in. hooked or headed stud	15.6	16.8	18.0
Channel C3 × 4.1	4.3w	4.7w	5.0w
Channel C4 × 5.4	4.6w	5.0w	5.3w
Channel C5 × 6.7	4.9w	5.3w	5.6w

Notes: [a] Allowable horizontal loads may also be used for studs longer than shown.
[b] w = length of channel, in.

Table 17.4 Modification factors to be used for lightweight concrete slabs

Specified compressive strength of concrete, ksi	Air dry unit weight of concrete, pcf						
	90	95	100	105	110	115	120
≤ 4.0	0.73	0.76	0.78	0.81	0.83	0.86	0.88
≥ 5.0	0.82	0.85	0.87	0.91	0.93	0.96	0.99

point of zero moment is

$$N_2 = \frac{N_1(M\beta/M_{\max} - 1)}{\beta - 1} \tag{17.11}$$

where M = moment (less than maximum moment) at concentrated load
N_1 = number of connectors required between points of maximum moment and zero moment
β = S_{tr}/S_s or S_{eff}/S_s as applicable (see below)

In order to develop fully the strength of shear connectors, they must be placed with certain restrictions as to cover distance and spacing. According to the AISC specification, shear connectors must have at least 1 in. of lateral concrete cover. The minimum distance center to center of stud connectors along the beam axis is 6 diameters, and the minimum lateral spacing across the width of the steel beam flange is 4 diameters. In addition, the maximum center-to-center spacing along the beam must not exceed 8 times the total slab thickness.

c. Continuous Construction

Design of continuous composite building spans in the positive bending moment regions does not differ from that of simple spans. However, in the negative bending regions, the slab must incorporate bar reinforcement, parallel to the steel beam and within the slab effective width. Section properties are calculated including the bar reinforcement, but neglecting the concrete, which normally would be cracked well below service load. The usual ACI requirements for development length past the point of maximum stress, and extension past the point of inflection, are applied. In addition, shear connectors must be included between the interior support and each adjacent point of inflection to transfer the force from the reinforcing bars to the steel beam. The horizontal force to be transferred can be computed from

$$V_h = \frac{A_{sr} f_{yr}}{2} \tag{17.12}$$

where A_{sr} = total area of bar reinforcement at the interior support, in^2
f_{yr} = yield strength of bar reinforcement, ksi

The total tensile force at yield is divided by 2 in Eq. (17.12), to permit calculation of the required number of connectors based on allowable service loads, according to Tables 17.3 and 17.4; i.e., the design is based on strength concepts, but worded in terms of allowable loads in service, consistent with other provisions of the AISC specification. The total required number of connectors, V_h/q, may be uniformly distributed between the point of maximum moment and the point of inflection.

d. Partially Composite Beams

In some cases, it is neither feasible nor necessary to provide a sufficient number of shear connectors to satisfy requirements for full composite action. Research has shown that, for a given beam and concrete slab, the increase in bending strength intermediate between noncomposite and full composite action is dependent upon the number of shear connectors provided between these two limits (Ref. 17.8). For such intermediate cases, the horizontal shear V_h' that can be transferred is calculated as the product of the q value from Tables 17.3 and 17.4 multiplied by the number of connectors actually provided between the points of maximum and zero moment. This value must not be less than one quarter the smaller of the values given by Eqs. (17.10a). Then the effective section modulus with respect to the tension flange is determined from

$$S_{\text{eff}} = S_s + \sqrt{\frac{V_h'}{V_h}}(S_{tr} - S_s) \tag{17.13}$$

where V_h is defined by Eqs. (17.10a), V_h' is as stated in the paragraph above, and S_{tr} and S_s are the section moduli of the transformed section and bare steel section, as usual.

A similar equation applies in calculating moment of inertia for deflection calculations for beams that are only partially composite:

$$I_{\text{eff}} = I_s + \sqrt{\frac{V_h'}{V_h}}(I_{\text{tr}} - I_s) \tag{17.14}$$

with the obvious definition of terms.

Example 17.2 Composite building floor beam. Figure 17.8 shows a typical bay of a multistory building that is to be designed for composite construction. The slab thickness has been determined from transverse flexural requirements to be $h_f = 4.0$ in. Steel beams of A36 steel ($f_y = 36$ ksi) will be used, and the slab concrete has specified strength $f_c' = 4000$ psi. Construction requirements prevent use of temporary shoring. Service loads are as follows:

Floor live load	100 psf
Dead loads	
Estimated steel weight	5
4 in. concrete slab	50
Partition allowance	20
Suspended loads (ceiling, piping)	10

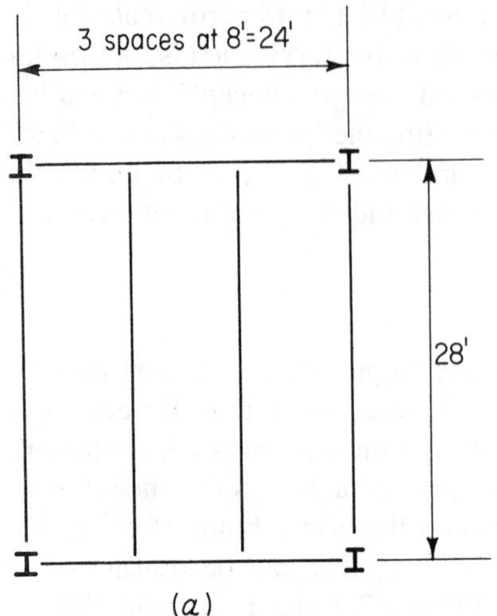

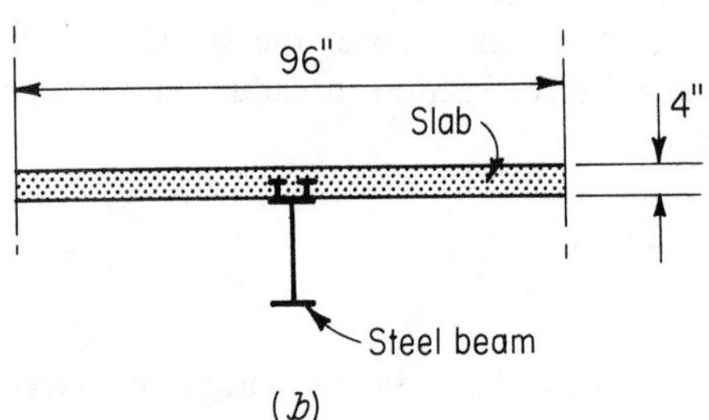

FIGURE 17.8
Composite floor of Example 17.2: (*a*) plan of typical bay; (*b*) beam cross section.

(*a*) Design the 28 ft spans for full composite action, according to the AISC specification. Assume simply supported spans.

(*b*) Calculate the flexural stresses in the section due to dead and live loads, based on bare steel, long-term composite, and short-term composite section properties, as applicable, and compare results with part (*a*).

(*c*) Calculate deflections due to the sustained dead loads and due to live load.

Solution. (*a*) According to AISC specifications, all loads may be considered to act on the composite section. Based on trial calculations, a W18 × 35 steel beam will be selected, with $A = 10.30\ \text{in}^2$, $I = 510\ \text{in}^4$, and $S_{ts} = S_{bs} = 57.6\ \text{in}^3$. Composite action will be ensured through the use of stud shear connectors. The effective flange width must not exceed:

$$28 \times \frac{12}{4} = 84\ \text{in.}$$

$$16 \times 4 + 6 = 70\ \text{in.}$$

$$\text{Center-to-center distance} = 96\ \text{in.}$$

The second criterion controls here. For the specified concrete, $E_c = 57{,}000\sqrt{4000} = 3{,}600{,}000$ psi, and hence the modular ratio $n = 29/3.6 = 8$. According to the specification, creep effects are not to be considered, and the transformed concrete flange width $b = 70/8 = 8.75$. The effective cross section is shown in Fig. 17.9*a*. Taking moments about the bottom face of the steel, the neutral axis distance from the bottom face of the composite section is found to be 17.23 in. as shown, and the composite moment of inertia is

$$I = 510 + 10.30(8.38)^2 + \frac{8.75(4.0)^3}{12} + 8.75 \times 4(2.47)^2$$

$$= 1490 \text{ in}^4$$

and the section moduli with respect to top of concrete, top of steel, and bottom of steel are

$$S_{tc} = \frac{1490}{4.47} = 333 \text{ in}^3$$

$$S_{ts} = \frac{1490}{0.47} = 3170 \text{ in}^3$$

$$S_{bs} = \frac{1490}{17.23} = 86.5 \text{ in}^3$$

The moments due to dead and live loads are respectively

$$M_d = \frac{(85 \times 8)28^2}{8} = 66.6 \text{ ft-kips}$$

$$M_l = \frac{(100 \times 8)28^2}{8} = 78.4 \text{ ft-kips}$$

Thus, according to Eq. (17.9), the maximum section modulus, with respect to the tension flange, that can be used in the stress calculations is

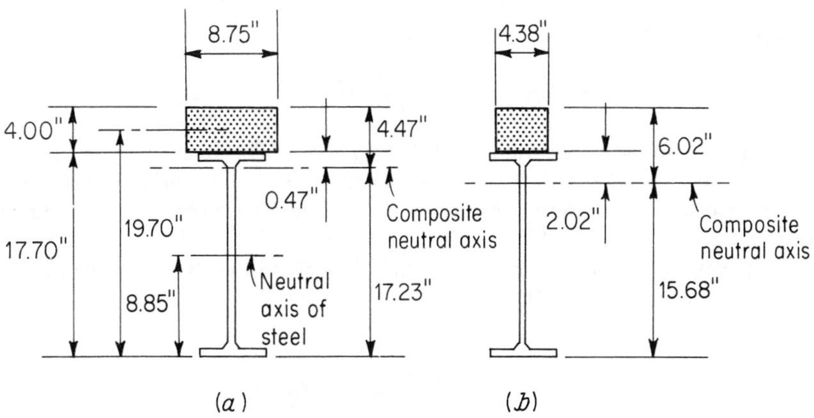

FIGURE 17.9
Transformed sections for Example 17.2: (*a*) short-term loading with $n = 8$; (*b*) sustained loading with $n = 16$.

$$S_{tr} = \left(1.35 + 0.35 \times \frac{78.4}{66.6}\right) 57.6 = 102 \text{ in}^3$$

The value calculated, $S_{bs} = 86.5$ in^3, is below this, as required. Then the stresses at the top of concrete, top of steel, and bottom of steel due to the total moment of 145.0 ft-kips are respectively

$$f_{tc} = 145 \times \frac{12{,}000}{333 \times 8} = 650 \text{ psi compression}$$

$$f_{ts} = 145 \times \frac{12{,}000}{3170} = 550 \text{ psi compression}$$

$$f_{bs} = 145 \times \frac{12{,}000}{86.5} = 20{,}100 \text{ psi tension}$$

The allowable stress limits of $0.45 \times 4000 = 1800$ psi in the concrete and $0.66 \times 36{,}000 = 24{,}000$ psi in the steel are seen to be satisfied.

The horizontal force to be transferred by the shear connectors is the lower value of the two given by Eq. (17.10):

$$V_h = 0.85 \times 4000 \times 4 \times \frac{70}{2} = 476{,}000 \text{ lb}$$

$$V_h = 10.30 \times \frac{36{,}000}{2} = 185{,}400 \text{ lb}$$

The second value is found to control here. Stud shear connectors having diameter $\frac{1}{2}$ in. and length 3 in. will be selected tentatively, used two per set across the 6 in. flange width. For the W18 × 35 beam, the flange thickness is 0.425 in., and hence the maximum diameter of stud that may be used is $2.5 \times 0.425 = 1.06$ in. This does not control. For $\frac{1}{2}$ in. studs, the minimum spacing along the flange is $6 \times \frac{1}{2} = 3$ in., the minimum transverse spacing is $4 \times \frac{1}{2} = 2$ in., and the maximum spacing along the flange is $8 \times 4 = 32$ in. From Table 17.3 the allowable load on one $\frac{1}{2}$ in. connector is $q = 5.9$ kips. Accordingly, the required number of connectors, from the maximum moment section to the support, in each direction, is

$$N_1 = \frac{185}{5.9} = 32$$

Sixteen sets of two studs each will be specified for each half-span, at a uniform spacing of $14 \times 12/16 = 10.5$ in. This is above the minimum of 3 in. and below the maximum of 32 in. previously determined. A lateral spacing across the flange of 3 in. will be specified.

(*b*) Flexural stresses will now be found, for comparison, based on the section properties that actually apply. The separate moments are

On bare steel:

$$M_{d1} = \frac{(55 \times 8)28^2}{8} = 43.1 \text{ ft-kips}$$

On composite beam:

$$M_{d2} = \frac{(30 \times 8)28^2}{8} = 23.5 \text{ ft-kips}$$

On composite beam:

$$M_l = \frac{(100 \times 8)28^2}{8} = 78.4 \text{ ft-kips}$$

Then the stresses at the top and bottom of the steel beam due to its self-weight plus the weight of the concrete slab are

$$S_{ts} = 43.1 \times \frac{12{,}000}{57.6} = 9000 \text{ psi compression}$$

$$S_{bs} = 9000 \text{ psi tension}$$

Calculation of stresses due to composite dead loads requires determination of transformed section properties using a reduced value of E_c, that is $n = 2 \times 8 = 16$. The transformed concrete slab has width $b = 70/16 = 4.38$ in. in this case, as shown in Fig. 17.9*b*. Calculations proceed as before, and result in a composite neutral axis 15.68 in. from the bottom surface of the beam, and $I = 1300 \text{ in}^4$. Thus the section moduli with respect to top of concrete, top of steel, and bottom of steel are respectively

$$S_{tc} = \frac{1300}{6.02} = 216 \text{ in}^3$$

$$S_{ts} = \frac{1300}{2.02} = 640 \text{ in}^3$$

$$S_{bs} = \frac{1300}{15.68} = 82.9 \text{ in}^3$$

and the corresponding stresses due to partitions and suspended loads are

$$f_{tc} = 23.5 \times \frac{12{,}000}{216 \times 16} = 80 \text{ psi compression}$$

$$f_{ts} = 23.5 \times \frac{12{,}000}{640} = 440 \text{ psi compression}$$

$$f_{bs} = 23.5 \times \frac{12{,}000}{82.9} = 3400 \text{ psi tension}$$

The section properties to be used for the live load moment of 78.4 ft-kips are the same as calculated in part (*a*), and the stresses are

$$f_{tc} = 78.4 \times \frac{12{,}000}{333 \times 8} = 350 \text{ psi compression}$$

$$f_{ts} = 78.4 \times \frac{12{,}000}{3170} = 300 \text{ psi compression}$$

$$f_{bs} = 78.4 \times \frac{12{,}000}{86.5} = 10{,}900 \text{ psi tension}$$

Superimposing stresses from the three calculations, the total stresses are

$$f_{tc} = 0 - 80 - 350 = 430 \text{ psi compression}$$

$$f_{ts} = -9000 - 400 - 300 = 9740 \text{ psi compression}$$

$$f_{bs} = +9000 + 3400 + 10{,}900 = 23{,}300 \text{ psi tension}$$

The concrete stress is well below the allowable value of 1800 psi and only about 65 percent of those found by the calculations in part (*a*). The steel stresses remain below the allowable value of 24,000 psi, but are about 15 percent higher than calculated earlier at the bottom face of the steel. Differences between the values found in part (*a*) by the AISC specification method and those such as found in part (*b*) considering the separate load effects will vary, depending on the ratio of live to dead load moments. The stress calculations of part (*b*) appear to be more realistic, but the procedures of part (*a*) can be rationalized in terms of strength design concepts.

(*c*) Three separate calculations of deflection will be made, the first for dead load of 55 × 8 = 440 lb/ft acting on the bare steel beam, the second for dead load of 30 × 8 = 240 lb/ft on the composite beam with $n = 16$ used for transformed section, and the third for live load of 100 × 8 = 800 lb/ft on the composite beam with $n = 8$ used. Thus

$$\Delta_{d1} = \frac{5 \times 440(28 \times 12)^4}{12 \times 384 \times 29{,}000{,}000 \times 510} = 0.41 \text{ in.}$$

$$\Delta_{d2} = \frac{5 \times 240(28 \times 12)^4}{12 \times 384 \times 29{,}000{,}000 \times 1300} = 0.09 \text{ in.}$$

$$\Delta_{l} = \frac{5 \times 800(28 \times 12)^4}{12 \times 384 \times 29{,}000{,}000 \times 1490} = 0.26 \text{ in.}$$

giving a total deflection of 0.76 in. The only limit imposed by AISC is that the live load deflection, for construction using plaster ceilings, must not exceed 1/360 times the span, or 0.93 in. in the present case. The actual live load deflection of 0.26 in. is well below this, as is typical for composite construction.

17.4 STEEL DECK REINFORCED COMPOSITE SLABS

Floor and roof slabs incorporating cold-formed steel deck panels, which serve both as form and reinforcement for the concrete placed over them, are widely used in buildings where the main framing is either of steel or composite construction. One of the many types of such slabs is shown in Fig. 17.10. There are many manufacturers of the steel deck used for composite slabs. Most have developed their own cross-section shapes and details. The steel sheet from which the panels are made ranges in thickness from about 0.024 to 0.060 in.

Such composite slabs have a number of advantages:

1. The steel deck, easily and quickly laid on the steel floor beams, serves as a working platform to support construction activity and to carry the freshly poured concrete. This eliminates the need for temporary falsework and forms.
2. The steel deck, with proper attention to details, can serve as the main tensile reinforcement for the slab.
3. If parts or all of the deck panels are formed into closed cells, as in Fig. 17.10, these cells can serve as ducts for electric and communication cables, or for heating and air conditioning ducts.

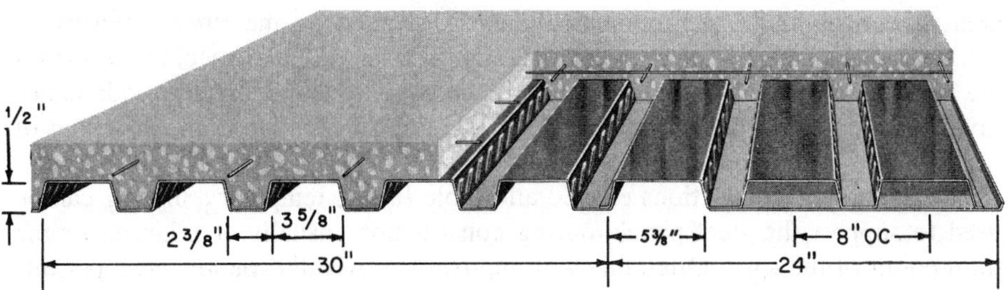

FIGURE 17.10
Composite concrete slab with cold-formed steel deck reinforcing. (*Courtesy of Inland-Ryerson Construction Products Co.*)

The designer must consider two stages in such construction. Prior to hardening of the freshly poured concrete, the steel deck panels, acting alone, must support their own weight, plus construction loads and the weight of the wet concrete. After the concrete has hardened, the steel and concrete act in a composite sense in resisting dead loads and live loads. In the first stage, it is customary to check stresses and deflections against allowable values, and in the second stage strength criteria are applied.

Guidance in the design of composite slabs is provided by a document presently available only in draft form: "Specifications for the Design and Construction of Composite Slabs" (Ref. 17.3; see also Ref. 17.4), prepared under sponsorship of the Technical Council of Codes and Standards of ASCE. Systems of this type are not included in the ACI Code for concrete building construction, or in the AISC specification for steel construction. However, noncomposite aspects of the design of composite slabs may be carried out according to the provisions of the AISI "Specification for the Design of Cold-formed Steel Structural Members" (Ref. 17.9).

a. Noncomposite Loads and Stresses

In designing for the noncomposite stage, with the steel deck acting alone, the loads W_2 to be carried, according to Ref. 17.3, include

1. Self-weight of the deck
2. Wet concrete
3. The more severe of: (*a*) a uniformly distributed load of 20 psf of surface or (*b*) a concentrated load of 150 lb acting on a 1 ft wide section of deck

The resulting flexural tensile stress in the deck must not exceed 60 percent of the yield stress, or 36 ksi, whichever is smaller. Allowable stresses in compression depend upon the geometry of the cross section as well as the properties of the steel, and limits are given in Ref. 17.9. The effective cross-section properties, including area, section moduli, and moment of inertia, depend on the stress intensity as well as the section geometry for construction of this type, and specific provisions are also found in Ref. 17.9. In practice, deck panel manufacturers provide tables giving section properties and allowable loads, greatly facilitating the checking of performance in the noncomposite stage.

If stresses or deflections exceed allowable limits, temporary shoring can be used to support the steel panels during construction, usually at midspan or the third points of the span. Often this will improve the overall economy, as it permits lighter deck sections to be used.

Deflections in the noncomposite stage are limited to 1/180 times the span, or $\frac{3}{4}$ in., whichever is smaller.

b. Basis of Strength Design

While stress and deflection limits must be satisfied in the noncomposite stage, the primary basis for the design of composite slabs is that adequate strength be provided to resist hypothetical overloads. Loads or load effects to be resisted are found by applying overload factors to service loads, as usual, and design strengths obtained by multiplying nominal strengths by strength reduction factors ϕ. Specifically, for the design of composite slabs,

$$V_u \leq \phi V_n \tag{17.15a}$$

$$M_u \leq \phi M_n \tag{17.15b}$$

where the usual notation is followed.

Here the load effect V_u and M_u are based on the factored loads:

$$W_u = 1.4(W_1 + W_3) + 1.7\text{LL} \tag{17.16}$$

where W_1 = weight of the steel deck and concrete, psf
W_3 = additional dead load exclusive of W_1, psf
LL = service live load, psf

The ϕ factors to be applied to the nominal strengths V_n and M_n to determine the design strengths vary depending on the type of resistance being calculated, according to the draft ASCE specification, just as is the case for ordinary reinforced concrete construction. Specific values will be given in the following sections.

c. Shear-Bond Strength

An essential feature of composite slab construction is that flexural bond stresses along the interface between steel and concrete must not cause horizontal slip between the two components. Just as was true for ordinary reinforcing bars, natural bond alone is generally not sufficient. For this reason, a variety of shear transfer devices is used. In most cases these consist of closely spaced embossments, such as shown in Fig. 17.10, or holes, welded buttons, or transverse wires. They act in a way similar to the ribs of deformed rebars. In addition, they counteract a tendency for deck and concrete to separate vertically. The embossments achieve this by being inclined to the horizontal so that they resist both horizontal and vertical forces. The other types of shear transfer devices act in a similar way.

Under load, a composite slab may fail in a mode that combines shear and bond failure, and hence is known as *shear-bond failure,* with certain features peculiar to this type of construction.

A shear-bond failure is shown in Fig. 17.11*a* as it occurs in a laboratory test under two-point loading. A major diagonal shear crack develops close to one of the loads. This causes loss of bond in the immediate vicinity of the crack, and this loss of bond generally propagates to the nearby end of the slab, causing bond failure over, the entire length l'. This results, in slip between concrete and steel, observable by the indicated end slip. From this it is seen that shear-bond failure is related to diagonal shear cracking, and this is reflected in the design approach to shear-bond.

On the basis of several hundred tests (Refs. 17.10 to 17.12) it has been found that the *shear-bond strength* of deck-reinforced composite slabs can be expressed in terms of an equation similar in form to Eq. (4.12*a*) for shear strength of

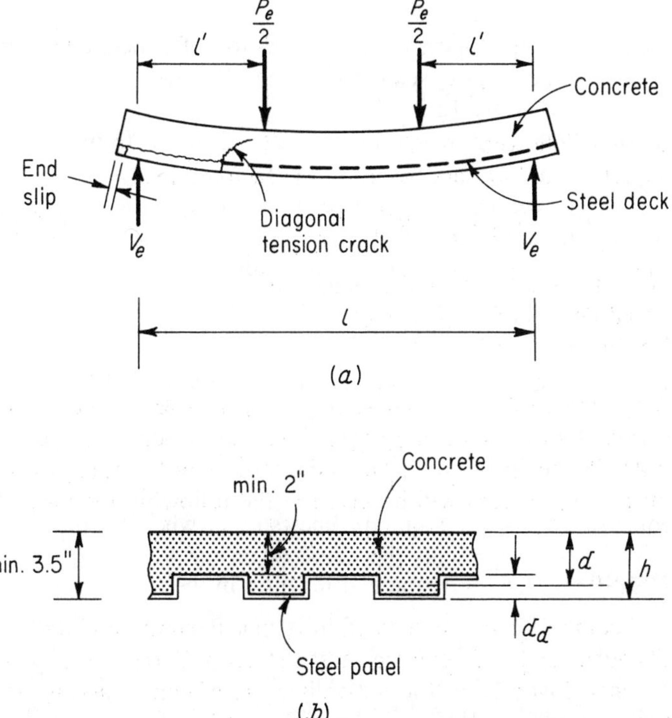

FIGURE 17.11
Shear-bond failure of composite slab: (*a*) diagonal cracking and bond failure at interface; (*b*) slab cross section.

conventionally reinforced flexural members:

$$V_n = \left(1.9\sqrt{f_c'} + 2500\rho_w \frac{V_u d}{M_u}\right) b_w d \tag{4.12a}$$

The constants 1.9 and 2500, of course, are not applicable to composite decks since they hold for ordinary bar-reinforced beams and slabs. For composite slabs, the values of these constants will depend on the peculiarities of the given deck, e.g., on cross-sectional shape, spacing and depth of ribs, sheet thickness, shape, spacing, and location of shear-transfer devices, etc. Thus, because of the considerable variety of decks being produced by the various manufacturers it is necessary to determine the appropriate constants for each type by tests.

These tests are standardized and are carried out on full-scale slab specimens loaded in bending by two-point loading; i.e., for a test span l two equal loads are applied, each at a distance l' from the nearby support, as shown in Fig. 17.11*a*. Thus, in the above equation for V_n, one has $M_u/V_u = l'$, the *shear span*. Denoting the unknown constants by k and m, the equation for shear-bond strength can be written as

$$V_n = \left(k\sqrt{f_c'} + m\frac{\rho d}{l'}\right) bd \tag{17.17}$$

where
- d = effective slab depth, i.e., distance from top of concrete to centroid of steel-deck cross section (see Fig. 17.11*b*)
- b = unit width, generally 12 in.
- l' = shear span, i.e., for two-point load test, distance from load to nearby support
- ρ = A_s/bd reinforcement ratio, where A_s = cross-sectional area of steel deck per unit width
- V_n = nominal shear-bond strength per unit width
- k, m = constants to be determined for each type of deck by standardized testing and evaluation procedure

For purposes of test evaluation, it is convenient to rewrite Eq. (17.17) as

$$\frac{V_e}{bd\sqrt{f_c'}} = k + m\frac{\rho d}{l'\sqrt{f_c'}} \tag{17.18}$$

In these terms, the equation represents a straight line, so that when $V_e/(bd\sqrt{f_c'})$ is plotted against $\rho d/(l'\sqrt{f_c'})$, k represents the intercept on the $V_e/(bd\sqrt{f_c'})$ axis and m is the slope of the line. Here V_e is the end reaction in the test at failure.

Such plotting of large numbers of tests has shown that the test points indeed follow an approximate straight line, with an experimental scattering of about ± 15 percent. For design purposes, the shear-bond strength is taken conservatively to be 15 percent less than the value predicted by the best-fit regression line. Thus, referring to Fig. 17.12, if the intercept and slope of the best-fit line are k_1 and m_1 respectively, for design purposes the values k and m are used, corresponding to a line 15 percent below the best-fit line.

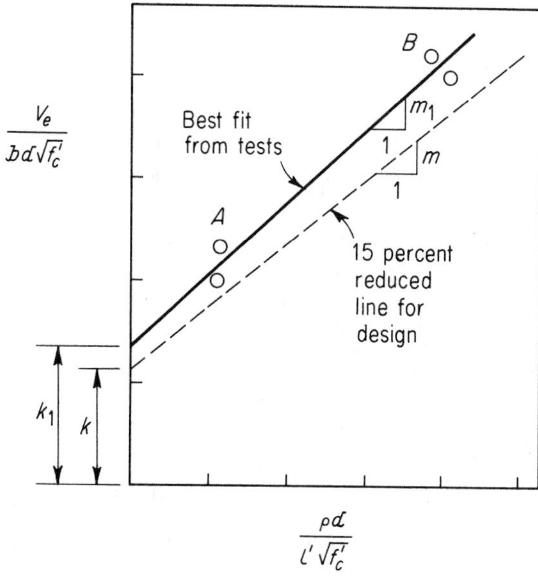

FIGURE 17.12
Typical shear-bond failure test results.

Each manufacturer of deck panels must establish shear-bond strength by tests in this way. These tests should cover the entire range of slab depths h and spans l for which the given deck is to be used. These tests are carried out with two-point loading in the manner of Fig. 17.11a, and plotted as in Fig. 17.12. In order to establish a reliable straight line, a minimum of two pairs of identical specimens must be tested to cover the range of shear spans l' and slab depths h. Thus, specimens for points A are made with small h and large l' those for points B with large h and small l.

d. Effect of Shoring

As long as the concrete placed on an unshored steel deck has not hardened, the entire weight W_1, consisting of the deck plus the wet concrete, is carried by the deck alone. Only the *additional* dead loads W_3 plus live loads LL applied after the concrete has hardened cause bond stresses. On the other hand, if the deck were to be temporarily supported until the concrete had hardened, then the shoring removed, *all* loads W_1, W_3, and LL cause bond stresses. This difference should be accounted for both in evaluating test results and in the design of the actual structure.

For the tests, this means the following. If the deck for the test was supported only at its ends when the concrete was placed, as in Fig. 17.13a, then W_1, the weight of the deck plus concrete, does not cause shear-bond stress. Hence it is only the applied testing machine load, P_e, that is resisted by shear-bond, and in Eq. (17.18) $V_e = P_e/2$, as is clear from Fig. 17.11a. However, on the other hand,

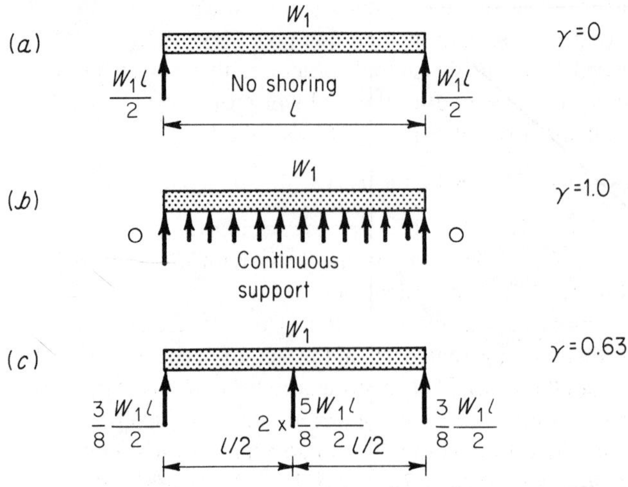

FIGURE 17.13
Effect of shoring on shears: (*a*) no shores during casting of slab; (*b*) full support over entire span; (*c*) shores at midspan.

if the deck were to be continuously supported, say, by resting on the laboratory floor when the concrete was poured, as in Fig. 17.13*b*, and then placed on the test supports after the concrete had hardened, the entire slab load W_1 plus the test load P_e would cause shear-bond stress. In this case, $V_e = P_e/2 + W_1 l/2$. If a single shore at midspan was used, as in Fig. 17.13*c*, then the effect of removing that shore is the same as introducing a downward concentrated load of $2(5/8)(W_1 l/2)$ at midspan, which would produce a reaction at each end of half that amount. For third-point shoring, a similar analysis is easily carried out. Thus, for the test,

$$V_e = \frac{P_e}{2} + \frac{\gamma W_1 l}{2} \tag{17.19}$$

where $\gamma = 1.0$ for complete support during casting
$\gamma = 0$ for no shoring during casting
$\gamma = 0.63$ for shoring at the center span point
$\gamma = 0.73$ for third-point shoring†

The constants k and m are then used by the design engineer to predict the shear-bond strength of the panels as used in the particular structure, using Eq. (17.17) with the values of f'_c, ρ, b, d, and l' that pertain. In situ, most slabs

† It is the practice of some manufacturers to establish the constants k and m based on $V_e = P_e/2$ regardless of shoring conditions during the test. While this is incorrect, the procedure is at least safe in that the actual shear force at failure was greater than that accounted for in determining k and m.

are designed to carry uniformly distributed loads, not concentrated loads. For distributed loads, as a good approximation, one can take $l' = l/4$.

In computing the factored loads in the actual structure, to be used in calculating V_u for checking shear-bond strength, the influence of shoring can be accounted for by using a modified form of Eq. (17.16) as follows:

$$W_u = 1.4(\gamma W_1 + W_3) + 1.7\text{LL} \tag{17.20}$$

where the values of γ are as just given for the various shoring conditions. The design requirement that $V_u \leq \phi V_n$ is then easily checked, with $\phi = 0.80$ for shear-bond strength, according to ASCE specification.

While it is apparently logical to account for the effect of shoring in the actual structure in this way, discounting that part of the actual service load that causes no shear-bond stress, both the AASHTO specification for composite bridges and the AISC specification for composite building beams take the position that *shear connectors* should be provided to develop the *full strength* of the composite member as governed by flexure (see Secs. 17.2*c* and 17.3*b*). Accordingly, the shear connectors are designed, under those specifications, for *total factored loads*, in effect, whether those loads were first applied in the noncomposite stage or the composite stage. Clearly this represents a more conservative basis for design.

e. Flexural Strength

The steel sheet thickness and depth of reinforcing decks is usually determined by the requirement that during construction and before concrete hardening the deck must carry its own weight and that of the concrete, plus such other construction loads as may occur. This means that even though the deck in the finished structure serves as reinforcement for the slab, its cross-sectional dimensions and thus its area A_s are generally governed by the temporary construction condition just described. In consequence, composite slabs may be overreinforced or underreinforced, depending on the particular combination of spans, loads, material strengths, and shoring conditions. It is not possible to avoid overreinforced slabs, in contrast to conventional slabs with the usual bar reinforcement.

For relatively shallow decks and deep slabs, i.e., when the slab thickness h is considerably larger than the depth d_d of the steel deck (see Fig. 17.11*b*), yielding is likely to spread over the entire depth d_d before the concrete compressive strain has reached the limiting value $\epsilon_u = 0.003$. Then the steel tensile force acts at the centroid of the steel deck section. In this case, the slab is *underreinforced* and the usual equation for the nominal flexural strength applies:

$$M_n = A_s f_y \left(d - \frac{a}{2}\right) \tag{17.21}$$

where $a = \dfrac{A_s fy}{0.85 f_c' b}$

d = effective depth from top of slab to steel centroid (see Fig. 17.11*b*)

b = width of unit strip, usually 12 in.

A_s = area of steel cross section, in^2

The design strength is ϕM_n calculated with $\phi = 0.90$ for underreinforced flexural failure.

An equation for the *balanced steel ratio* is easily derived, based on the condition that, for the balanced case, the tensile strain at the *top surface of the steel deck panel* reaches the yield strain at exactly the same instant of loading that the top surface of the concrete reaches its limit strain $\epsilon_u = 0.003$. The derivation is exactly analogous to that for the balanced steel ratio for ordinary bar-reinforced beams presented in Sec. 3.4*b*, and results in

$$\rho_b = 0.85\beta_1 \frac{f'_c}{f_y} \frac{\epsilon_u}{\epsilon_u + \epsilon_y} \frac{h - d_d}{d} \tag{17.22a}$$

With $\epsilon_u = 0.003$ and taking $E_s = 29.5 \times 10^6$ psi (slightly higher for cold-formed steel deck than for rebars) the last equation can also be written:

$$\rho_b = 0.85\beta_1 \frac{f'_c}{f_y} \frac{88{,}500}{88{,}500 + f_y} \frac{h - d_d}{d} \tag{17.22b}$$

Composite slabs with a steel ratio less than ρ_b are underreinforced, while those with a steel ratio above that limit are overreinforced with the steel stress less than f_y when the concrete strain limit is reached.

Calculation of the flexural strength of overreinforced slabs is complicated by the fact that the strains in the metal deck section vary with the depth depending upon the loading history, i.e., whether the steel deck was loaded in a noncomposite stage, whether it was shored and how, etc. For the case where *continuous shoring* is provided in the structure, such as shown in Fig. 17.13*b*, i.e., where *all* loads are composite, and with the approximation that the steel stress through the full depth of the deck is equal to its value at the centroid, a strain compatibility analysis leads to the following results.

Referring to a typical composite slab cross section as shown in Fig. 17.14*a*, with strains at flexural failure as shown in Fig. 17.14*b*,

$$\epsilon_s = \epsilon_u \frac{d - k_u d}{k_u d}$$

$$f_s = \epsilon_u E_s \frac{1 - k_u}{k_u}$$

Summing forces in the *X* direction in Fig. 17.14*c* and setting equal to zero results in

$$\rho b d\, \epsilon_u E_s \frac{1 - k_u}{k_u} = 0.85\beta_1 f'_c b k_u d \tag{a}$$

Defining the material parameter

$$\lambda = \frac{\epsilon_u E_s}{0.85\beta_1 f'_c} \tag{b}$$

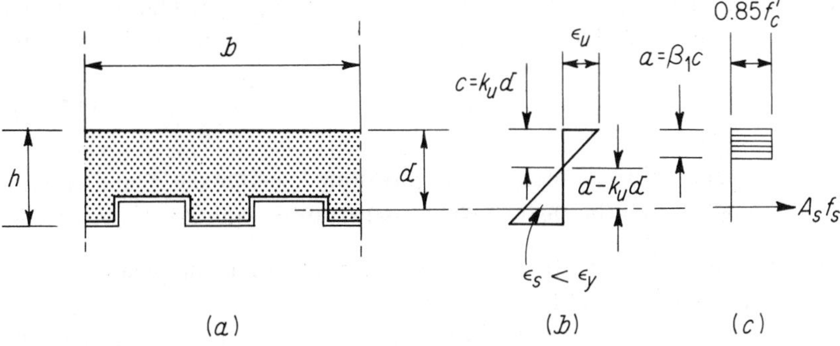

FIGURE 17.14
Flexural strain compatibility analysis for overreinforced composite slab: (*a*) cross section; (*b*) strain distribution; (*c*) stresses and forces.

Equation (*a*) can be written as a quadratic in k_u as follows:

$$k_u^2 + \rho\lambda k_u - \rho\lambda = 0 \tag{c}$$

from which

$$k_u = \sqrt{\rho\lambda + \left(\frac{\rho\lambda}{2}\right)^2} - \frac{\rho\lambda}{2} \tag{17.23}$$

Then $c = k_u d$ and $a = \beta_1 c$ as usual. The nominal flexural strength can then be found from the equation:

$$M_n = 0.85 f_c' ab \left(d - \frac{a}{2}\right) \tag{17.24}$$

and the design strength ϕM_n found by application of strength reduction $\phi = 0.75$ for the overreinforced case, according to the ASCE specification.

If continuous shoring is *not* provided during construction, prior strains in the steel panel due to construction loads affect the strain distribution in the composite section at failure, and the simple diagram of Fig. 17.14*b* is not strictly valid. However, neglect of these prior strains is fully consistent with assuming that the entire deck section acts at its centroidal stress f_s, as was done in the previous analysis, and would usually lead to only a small error.

For overreinforced composite slabs, the ASCE specification requires tests to confirm the computed capacity in any case.

f. Deflections, Shrinkage Reinforcement, Continuity

The ACI Code provisions for deflection calculations and limitations, including accounting for concrete creep as it affects deflections, can be applied to deck-reinforced composite slabs. The only difference noted in the ASCE specification refers to the effective moment of inertia used in deflection calculations. Tests show (Ref. 17.12) that a satisfactory value for the effective moment of inertia is

$$I_e = \frac{I_c + I_u}{2} \tag{17.25}$$

where I_c = moment of inertia of cracked transformed section, in^4
I_u = moment of inertia of uncracked transformed section, in^4

In either case, the moment of inertia of the steel deck about the composite neutral axis, cracked or uncracked, must be included.

Transverse shrinkage and temperature reinforcement must be provided in composite slabs, just as for other one-way slabs. It has been suggested that the quantity of such reinforcement can be reduced to about 60 percent of that required by the ACI Code for ordinary slabs because the deck itself, even though ribbed, provides some restraint against transverse shrinkage and temperature effects.

Composite slabs can be used either as simple spans, described above, or as continuous spans. In the case of simple spans, nominal negative bar reinforcement should be provided over the supports to control crack widths. For slabs that are designed as continuous over supports, the parts subject to negative bending require additional bars, placed near the top of the concrete slab, and these can be designed according to the usual methods for bar-reinforced slabs, accounting for the voids in the compression zone resulting from the particular panel cross-section shape used. The contribution of the steel deck on the compression side of the negative bending section is neglected.

Example 17.3 Shear-bond strength and flexural strength of composite slab. The composite slab of Fig. 17.15 consists of a steel panel of thickness 0.04 in. and cross-section shape shown, over which is cast a concrete slab having a total depth h = 5 in. It will be used as a part of the floor of an office building, with slabs spanning 10 ft between supporting steel beams. Mesh will be provided over the supports to control cracking, but spans will be designed as simply supported. Superimposed dead loads from flooring and partitions total 28 psf, and service live load is 50 psf. Material strengths are f'_c = 4000 psi and f_y = 60,000 psi. Confirm the adequacy of the proposed slab system (*a*) in flexure and (*b*) with respect to shear-bond strength. Manufacturer's tests indicate that the shear-bond constants are k = 0.40 and m = 3400.

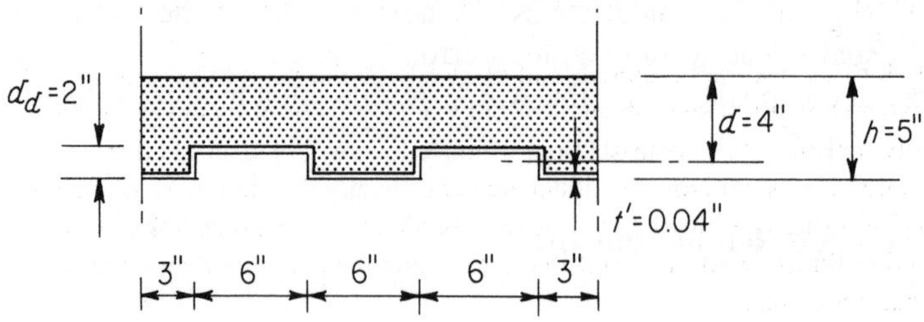

FIGURE 17.15
Composite slab of Example 17.3.

Solution. The composite slab will be checked for strength in bending and shear-bond at the overload stage applying load factors to *all* loads, whether those loads were originally introduced at the noncomposite or composite stages. (According to the draft ASCE specification, noncomposite dead loads can be excluded when checking shear-bond strength.) The self-weight of the composite slab will be based on the average thickness of 4 in. and neglects the small weight of the steel deck. The self-weight is therefore $W_0 = 150 \times 4/12 = 50$ psf, and the factored load to be considered is

$$W_u = 1.4(50 + 28) + 1.7 \times 50 = 194 \text{ psf}$$

and the required strengths in shear and flexure are

$$V_u = 194 \times \frac{10}{2} = 970 \text{ lb}$$

$$M_u = 194 \times \frac{100}{8} = 2430 \text{ ft-lb}$$

In checking flexural strength, it must first be determined whether the composite slab is underreinforced or overreinforced. From Eq. (17.22*b*) the balanced steel ratio is

$$\rho_b = 0.85 \times 0.85 \frac{4000}{60{,}000} \frac{88{,}500}{88{,}500 + 50{,}000} \frac{5-2}{4} = 0.023$$

The tensile steel area is found multiplying sheet thickness by perimeter per 12 in. width; thus $A_s = 0.04 \times 16 = 0.64$ in^2 and the actual steel ratio is $\rho = 0.64/(12 \times 4) = 0.013$. This is well below the balanced value, so yielding of the steel controls the flexural strength which can be computed based on Eq. (17.21). With $a = 0.64 \times 50/(0.85 \times 4 \times 12) = 0.78$ in., the nominal flexural strength is $M_n = 0.64 \times 50(4 - 0.78/2) = 116$ in-kips per 12 in. width or 9670 ft-lb. Hence the design strength is $\phi M_n = 0.90 \times 9670 = 8700$ ft-lb, greatly in excess of the required $M_u = 2430$ ft-lb.

Shear-bond strength will be computed based on Eq. (17.17). The effective shear span is $l' = 120/4 = 30$ in. With effective depth to the steel controid $d = 4$ in. and width $b = 12$ in., using the values of k and m from the manufacturer's test results, the nominal shear-bond strength is

$$V_n = \left(0.40\sqrt{4000} + \frac{3400 \times 0.013 \times 4}{30}\right) 12 \times 4 = 1500 \text{ lb}$$

and the design strength is $\phi V_n = 0.80 \times 1500 = 1200$ lb per 12 in. width of slab. This is above the required $V_u = 970$ lb, confirming that the proposed design is satisfactory with respect to shear-bond.

Note that, for the proposed factored load of 194 lb/ft, the composite slab is at 81 percent of its design strength in shear-bond, but only at 28 percent of its design moment capacity. Typically for this type of construction, ultimate load will be controlled by shear-bond, not by flexure. The panel cross-sectional area is generally dictated by requirements for carrying noncomposite dead load and construction load without exceeding allowable stresses or deflection limits, with the result that the A_s provided is more than required to resist factored load moments.

Conditions at the noncomposite stage will not be checked here, but would be investigated according to the provisions of the AISI specification (Ref. 17.9) for cold-formed steel members.

REFERENCES

17.1. "Standard Specifications for Highway Bridges," 14th ed., American Association of State Highway and Transportation Officials, Washington, 1989 (see also "Interim Specifications for Bridges," published annually by AASHTO).

17.2. "Specification for the Design, Fabrication, and Erection of Structural Steel for Buildings," in *Steel Construction Manual,* 8th ed., American Institute of Steel Construction, Chicago, 1980.

17.3. "Specifications for the Design and Construction of Composite Slabs," Technical Council of Codes and Standards, American Society of Civil Engineers, New York, June 1982.

17.4. "Commentary on Specifications for the Design and Construction of Composite Slabs," Technical Council of Codes and Standards, American Society of Civil Engineers, New York, September 1982.

17.5. I. M. Viest, R. S. Fountain, and R. C. Singleton, *Composite Construction in Steel and Concrete,* McGraw-Hill, New York, 1958.

17.6. R. S. Fountain and I. M. Viest, "Selection of the Cross Section for a Composite T Beam," *J. Struct. Div. ASCE,* vol 83, no. ST4, 1957, pp. 1–29.

17.7. R. G. Slutter and J. W. Fisher, "Fatigue Strength of Shear Connectors," Highway Research Record 147, Highway Research Board, Washington, 1966.

17.8. R. G. Slutter and G. C. Driscoll, "Flexural Strength of Steel-Concrete Composite Beams," *J. Struct. Div. ASCE,* vol. 91, no. ST2, 1965, pp. 71–99.

17.9. "Specification for the Design of Cold-Formed Steel Structural Members," 1980 ed., American Iron and Steel Institute, Washington, 1980.

17.10. R. M. Schuster, "Composite Steel Deck Concrete Floor Systems," *J. Struct. Div. ASCE,* vol. 102, no. ST5, 1976, pp. 899–918.

17.11. M. L. Porter, C. E. Ekberg, L. F. Greimann, and H. A. Elleby, "Shear-Bond Analysis of Steel Deck Reinforced Slabs," *J. Struct. Div. ASCE,* vol. 102, no. ST12, 1976, pp. 2255–2268.

17.12. M. L. Porter and C. E. Ekberg, "Design Recommendations for Steel Deck Floor Slabs,;; *J. Struct. Div. ASCE,* vol. 102., no. ST11, 1976, pp. 2121–2136.

CHAPTER 18

FOOTINGS AND FOUNDATIONS

18.1 TYPES AND FUNCTIONS

The substructure, or foundation, is the part of a structure that is usually placed below the surface of the ground and that transmits the load to the underlying soil or rock. All soils compress noticeably when loaded and cause the supported structure to settle. The two essential requirements in the design of foundations are that the total settlement of the structure be limited to a tolerably small amount and that differential settlement of the various parts of the structure be eliminated as nearly as possible. With respect to possible structural damage, the elimination of differential settlement, i.e., different amounts of settlement within the same structure, is even more important than limitations on uniform overall settlement.

To limit settlements as indicated, it is necessary (1) to transmit the load of the structure to a soil stratum of sufficient strength and (2) to spread the load over a sufficiently large area of that stratum to minimize bearing pressure. If adequate soil is not found immediately below the structure, it becomes necessary to use deep foundations such as piles or caissons to transmit the load to deeper, firmer layers. If satisfactory soil directly underlies the structure, it is merely necessary to spread the load, by footings or other means. Such substructures are known as *spread* foundations, and it is mainly this type that will be discussed. Information on the more special types of deep foundations can be found in texts on foundation engineering, e.g., Refs. 18.1 and 18.2.

18.2 SPREAD FOOTINGS

Spread footings can be classified as wall and column footings. The horizontal outlines of the most common types are given in Fig. 18.1. A wall footing is simply a strip of reinforced concrete, wider than the wall, that distributes its

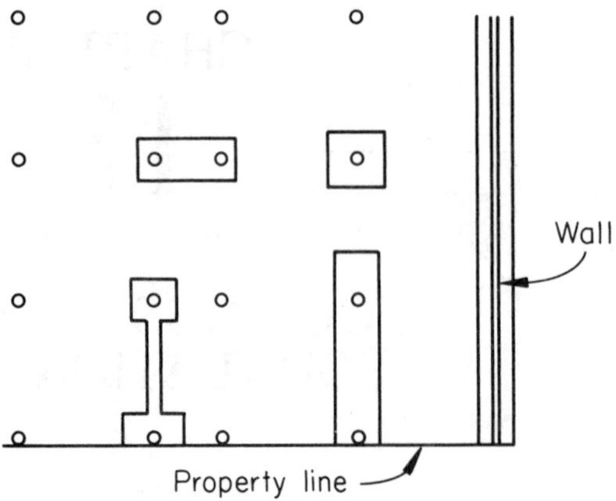

FIGURE 18.1
Types of spread footings.

pressure. Single-column footings are usually square, sometimes rectangular, and represent the simplest and most economical type. Their use under exterior columns meets with difficulties if property rights prevent the use of footings projecting beyond the exterior walls. In this case combined footings or strap footings are used that enable one to design a footing that will not project beyond the wall column. Combined footings under two or more columns are also used under closely spaced, heavily loaded interior columns where single footings, if they were provided, would completely or nearly merge.

Such individual or combined column footings are the most frequently used types of spread foundations on soils of reasonable bearing capacity. If the soil is weak and/or column loads are great, the required footing areas become so large as to be uneconomical. In this case, unless a deep foundation is called for by soil conditions, a mat or raft foundation is resorted to. This consists of a solid reinforced concrete slab which extends under the entire building and which consequently distributes the load of the structure over the maximum available area. Such a foundation, in view of its own rigidity, also minimizes differential settlement. It consists, in its simplest form, of a concrete slab reinforced in both directions. A form that provides more rigidity and at the same time is often more economical consists of an inverted beam-and-girder floor. Girders are located in the column lines in one direction, with beams in the other, mostly at closer intervals. If the columns are arranged in a square pattern, girders are equally spaced in both directions and the slab is provided with two-way reinforcement. Inverted flat slabs, with capitals at the bottoms of the columns, are also used for mat foundations.

18.3 DESIGN FACTORS

In ordinary construction the load on a wall or column is transmitted vertically to the footing, which in turn is supported by the upward pressure of the soil on which it rests. If the load is symmetrical with respect to the bearing area, the bearing pressure is assumed to be uniformly distributed (Fig. 18.2*a*). It is known that this is only approximately true. Under footings resting on coarse-grained soils the pressure is larger at the center of the footing and decreases toward the

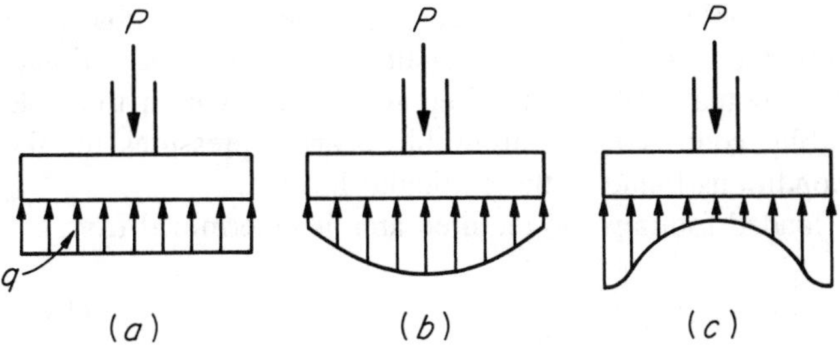

FIGURE 18.2
Bearing pressure distribution: (*a*) as assumed; (*b*) actual, for granular soils; (*c*) actual, for cohesive soils.

perimeter (Fig. 18.2*b*). This is so because the individual grains in such soils are somewhat mobile, so that the soil located close to the perimeter can shift very slightly outward in the direction of lower soil stresses. In contrast, in clay soils pressures are higher near the edge than at the center of the footing, since in such soils the load produces a shear resistance around the perimeter which adds to the upward pressure (Fig. 18.2*c*). It is customary to disregard these nonuniformities (1) because their numerical amount is uncertain and highly variable, depending on types of soil, and (2) because their influence on the magnitudes of bending moments and shearing forces in the footing is relatively small.

On compressible soils footings should be loaded concentrically to avoid tilting, which will result if bearing pressures are significantly larger under one side of the footing than under the opposite side. This means that single footings should be placed concentrically under the columns and wall footings concentrically under the walls and that for combined footings the centroid of the footings area should coincide with the resultant of the column loads. Eccentrically loaded footings can be used on highly compacted soils and on rock. It follows that one should count on rotational restraint of the column by a single footing only when such favorable soil conditions are present and when the footing is designed both for the column load and the restraining moment. Even then, less than full fixity should be assumed, except for footings on rock.

The accurate determination of stresses, particularly in single-column footings, is not practical, since they represent relatively massive blocks that cantilever from the column in all four directions. Under uniform upward pressure they deform in a bowl shape, a fact that would greatly complicate an accurate stress analysis. For this reason, present procedures for the design of such footings are based almost entirely on the results of two extensive experimental investigations, both carried out at the University of Illinois (Refs. 18.3 and 18.4). These tests have been reevaluated, particularly in the light of newer concepts of strength in shear and diagonal tension (Refs. 18.5 and 18.6).

18.4 LOADS, BEARING PRESSURES, AND FOOTING SIZE

Allowable bearing pressures are established from principles of soil mechanics, on the basis of load tests and other experimental determinations (see, for example,

Refs. 18.1 and 18.2). Allowable bearing pressures q_a under service loads are usually based on a safety factor of 2.5 to 3.0 against exceeding the ultimate bearing capacity of the particular soil and to keep settlements within tolerable limits. Many local building codes contain allowable bearing pressures for the types of soils and soil conditions found in the particular locality.

For concentrically loaded footings the required area is determined from

$$A_{\text{req}} = \frac{D+L}{q_a} \tag{18.1}$$

In addition, most codes permit a 33 percent increase in allowable pressure when the effects of wind W or earthquake E are included, in which case

$$A_{\text{req}} = \frac{D+L+W}{1.33q_a} \quad \text{or} \quad \frac{D+L+E}{1.33q_a} \tag{18.2}$$

It should be noted that footing sizes are determined for unfactored service loads and soil pressures, in contrast to the strength design of reinforced concrete members, which utilizes factored loads and factored nominal strengths. This is because for footing design safety is provided by the overall safety factors just mentioned, in contrast to the separate load and strength reduction factors for member dimensioning.

A_{req}, the required footing area, is the larger of those determined by Eq. (18.1) or (18.2). The loads in the numerators of Eqs. (18.1) and (18.2) must be calculated at the level of the base of the footing, i.e., at the contact plane between soil and footing. This means that the weight of the footing and surcharge (i.e., fill and possible liquid pressure on top of the footing) must be included. Wind loads and other lateral loads cause a tendency to overturning. In checking for overturning of a foundation, only those live loads that contribute to overturning should be included, and dead loads that stabilize against overturning should be multiplied by 0.9. A safety factor of at least 1.5 should be maintained against overturning, unless otherwise specified by the local building code (Ref. 18.6).

A footing is eccentrically loaded if the supported column is not concentric with the footing area or if the column transmits at its juncture with the footing, not only a vertical load but also a bending moment. In either case, the load effects at the footing base can be represented by the vertical load P and a bending moment M. The resulting bearing pressures are again assumed to be linearly distributed. As long as the resulting eccentricity $e = M/P$ does not exceed the kern distance k of the footing area, the usual flexure formula

$$q_{\substack{\max \\ \min}} = \frac{P}{A} \pm \frac{Mc}{I} \tag{18.3}$$

permits the determination of the bearing pressures at the two extreme edges, as shown in Fig. 18.3a. The footing area is found by trial and error from the condition $q_{\max} \leq q_a$. If the eccentricity falls outside the kern, Eq. (18.3) gives a negative value (tension) for q along one edge of the footing. Because no tension can be transmitted at the contact area between soil and footing, Eq. (18.3) is no

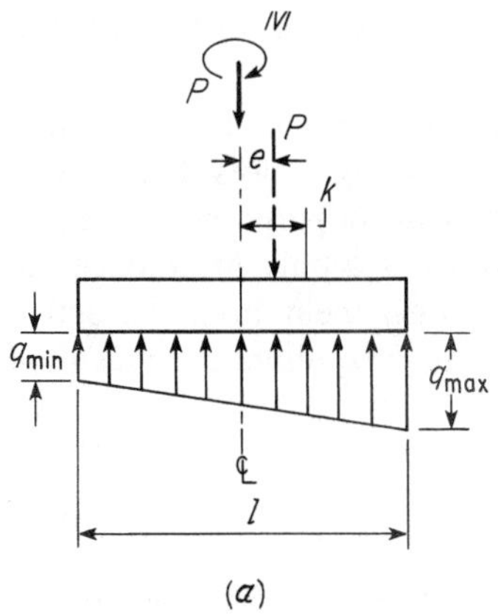

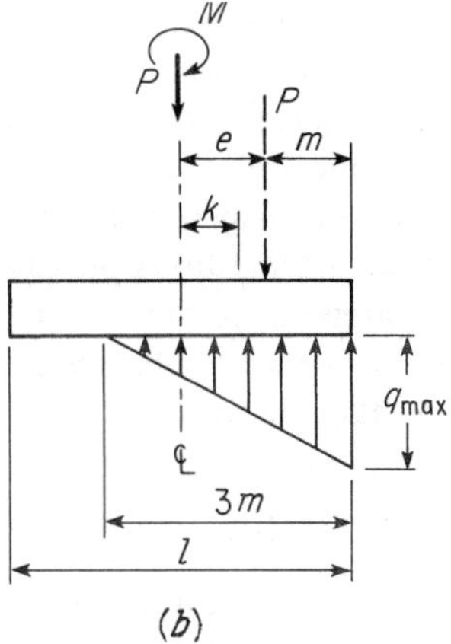

FIGURE 18.3
Assumed bearing pressures under eccentric footings.

longer valid and bearing pressures are distributed as in Fig. 18.3*b*. For rectangular footings of size $l \times b$ the maximum pressure can be found from

$$q_{\max} = \frac{2P}{3bm} \tag{18.4}$$

which, again, must be no larger than the allowable pressure q_a. For nonrectangular footing areas of various configurations, kern distances and other aids for calculating bearing pressures can be found in Refs. 18.1 and 18.6 and elsewhere.

Once the required footing area has been determined, the footing must then be designed to develop the necessary strength to resist all moments, shears, and other internal actions caused by the applied loads. For this purpose, the load factors of ACI Code 9.2 apply to footings as to all other structural components. Correspondingly, for strength design the footing is dimensioned for the effects of the following external loads (see Table 1.2):

$$U = 1.4D + 1.7L$$

or if wind effects are to be included,

$$U = 0.75(1.4D + 1.7L + 1.7W)$$

In seismic zones earthquake forces E must be considered, in which case $1.1E$ must be substituted for W. The requirement that

$$U = 0.9D + 1.3W$$

will hardly ever govern the strength design of a footing. However, lateral earth pressure H may on occasion affect footing design, in which case

$$U = 1.4D + 1.7L + 1.7H$$

For horizontal pressures F of liquids, such as groundwater, $1.4F$ must be substituted for $1.7H$ in the last equation. Vertical liquid pressures are to be added to dead loads; i.e., the load factor 1.4 applies to them.

These factored loads must be counteracted and equilibrated by corresponding bearing pressures of the soil. Consequently, once the footing area is determined, the bearing pressures are recalculated for the factored loads for purposes of strength computations. These are fictitious pressures that are needed only to produce the required ultimate strength of the footing. To distinguish them from the actual pressures q under service loads, the design pressures which equilibrate the factored loads U will be designated q_u.

18.5 WALL FOOTINGS

The simple principles of beam action apply to wall footings with only minor modifications. Figure 18.4 shows a wall footing with the forces acting on it. If bending moments were computed from these forces, the maximum moment would be found to occur at the middle of the width. Actually, the very large rigidity of the wall modifies this situation, and the tests cited in Sec. 18.3 show that for footings under concrete walls it is satisfactory to compute the moment at the face of the wall (section 1-1). Tension cracks in these tests formed at the locations shown in Fig. 18.4, i.e., under the face of the wall rather than in the middle. For footings supporting masonry walls, the maximum moment is computed midway between the middle and the face of the wall, because masonry is generally less rigid than concrete. The maximum bending moment in footings under concrete walls is therefore given by

$$M_u = \frac{1}{8} q_u (b - a)^2 \tag{18.5}$$

For determining shear stresses, the vertical shear force is computed on section 2-2, located, as in beams, at a distance d from the face of the wall. Thus,

$$V_u = q_u \left(\frac{b - a}{2} - d \right) \tag{18.6}$$

The calculation of development length is based on the section of maximum moment, i.e., section 1-1.

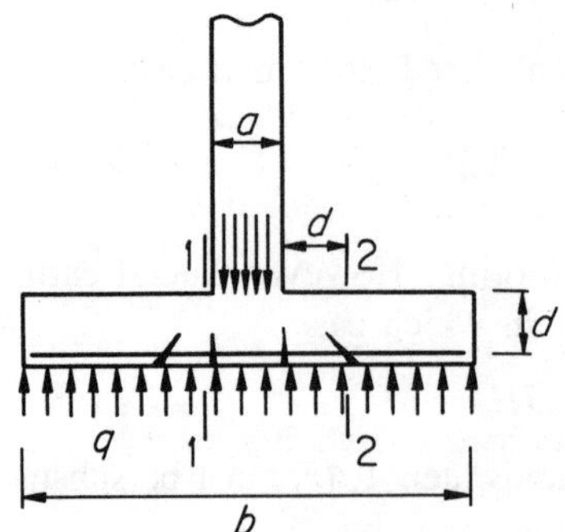

FIGURE 18.4
Wall footing.

Example 18.1 Design of wall footing. A 16 in. concrete wall supports a dead load $D = 14$ kips/ft and a live load $L = 10$ kips/ft. The allowable bearing pressure is $q_a = 4.5$ kips/ft^2 at the level of the bottom of the footing, which is 4 ft below grade. Design a footing for this wall using 3000 psi concrete and Grade 40 steel.

Solution. With a 12 in. thick footing, the footing weight per square foot is 150 psf, and the weight of the 3 ft fill on top of the footing is $3 \times 100 = 300$ psf. Consequently, the portion of the allowable bearing pressure that is available or effective for carrying the wall load is

$$q_e = 4500 - (150 + 300) = 4050 \text{ psf}$$

The required width of the footing is therefore $b = 24,000/4050 = 5.93$ ft. A 6 ft wide footing will be assumed.

The bearing pressure for strength design of the footing, caused by the factored loads, is

$$q_u = \frac{1.4 \times 14 + 1.7 \times 10}{6} \times 10^3 = 6100 \text{ psf}$$

From this, the required moment for strength design is

$$M_u = \frac{1}{8} \times 6100(6 - 1.33)^2 \times 12 = 199,500 \text{ in-lb/ft}$$

and assuming $d = 9$ in. the shear at section 2-2 is

$$V_u = 6100\left[\frac{1}{2}(6 - 1.33) - \frac{9}{12}\right] = 9700 \text{ lb/ft}$$

Shear usually governs the depth of footings, particularly since the use of shear reinforcements in footings is generally avoided as uneconomical. The design shear strength per foot [see Eq. (4.12*b*)] is

$$\phi V_c = \phi(2\sqrt{f'_c}bd) = 0.85(2\sqrt{3000} \times 12d) = 1117d \text{ lb/ft}$$

from which

$$d = \frac{9700}{1117} = 8.7 \text{ in.}$$

Since the ACI Code calls for a 3 in. clear cover of bars, a 12 in. thick footing will be selected, giving $d = 8.5$ in. This is sufficiently close to the assumed values, and the calculations need not be revised.

To determine the required steel area, $M_u/\phi bd^2 = 199{,}500/(0.9 \times 12 \times 8.5^2) = 256$ is used to enter Graph A.1 of App. A. For this value, the curve 40/3 gives the steel ratio $\rho = 0.0067$. The required steel area is then $A_s = 0.0067 \times 8.5 \times 12 = 0.68$ in^2/ft Number 7 bars, 10 in. on centers, furnish $A_s = 0.72$ in^2/ft. The required development length according to Table A.11 is 18 in. This length is to be furnished from section 1-1 outward. The length of each bar, if end cover is 3 in., is $72 - 6 = 66$ in., and the actual development length from section 1-1 to the nearby end is $\frac{1}{2}(66 - 16) = 25$ in., which is more than the required development length.

Longitudinal shrinkage and temperature reinforcement, according to ACI Code 7.12, must be at least $0.002 \times 12 \times 12 = 0.29$ in^2/ft. Number 4 bars on 8 in. centers will furnish 0.29 in^2/ft.

18.6 COLUMN FOOTINGS

In plan, single-column footings are usually square. Rectangular footings are used if space restrictions dictate this choice or if the supported columns are of strongly elongated rectangular cross section. In the simplest form, they consist of a single slab (Fig. 18.5*a*). Another type is that of Fig. 18.5*b*, where a pedestal or cap is interposed between the column and the footing slab; the pedestal provides for a more favorable transfer of load and in many cases is required in order to provide the necessary development length for dowels. This form is also known as a *stepped* footing. All parts of a stepped footing must be poured in a single pour, in order to provide monolithic action. Sometimes sloped footings like those in Fig. 18.5*c* are used. They require less concrete than stepped footings, but the additional labor necessary to produce the sloping surfaces (formwork, etc.) usually makes stepped footings more economical. In general, single-slab footings (Fig. 18.5*a*) are most economical for thicknesses up to 3 ft.

Single-column footings represent, as it were, cantilevers projecting out from the column in both directions and loaded upward by the soil pressure. Corresponding tension stresses are caused in both these directions at the bottom surface. Such footings are therefore reinforced by two layers of steel, perpendicular to each other and parallel to the edges.

The required bearing area is obtained by dividing the total load, including the weight of the footing, by the selected bearing pressure. Weights of footings, at this stage, must be estimated and usually amount to 4 to 8 percent of the column load, the former value applying to the stronger types of soils.

In computing bending moments and shears, only the upward pressure q_u that is caused by the factored column loads is considered. The weight of the footing proper does not cause moments or shears, just as, obviously, no moments or shears are present in a book lying flat on a table.

a. Shear

Once the required footing area A_{req} has been established from the allowable bearing pressure q_a and the most unfavorable combination of service loads, including weight of footing and overlying fill (and such surcharge as may be present), the thickness h of the footing must be determined. In single footings the effective depth d is mostly governed by shear. Since such footings are subject to two-way

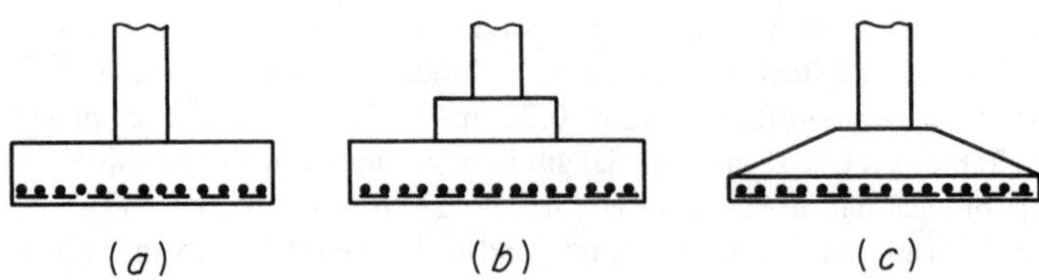

FIGURE 18.5
Types of single-column footings.

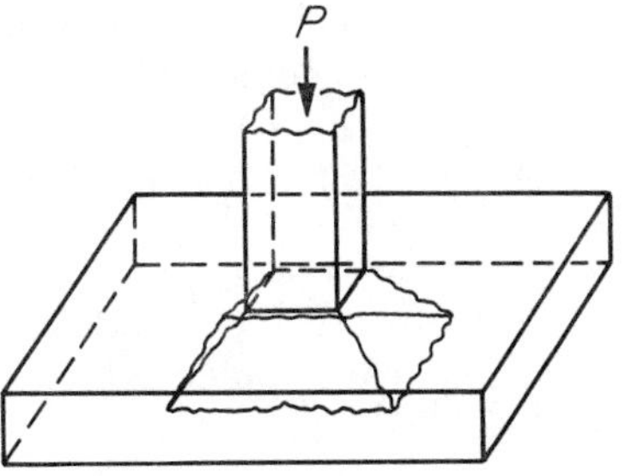

FIGURE 18.6
Punching-shear failure in single footing.

action, i.e., bend in both major directions, their performance in shear is much like that of flat slabs in the vicinity of columns (see Sec. 13.6). However, in contrast to two-way floor and roof slabs, it is generally not economical in footings to use shear reinforcement. For this reason only the design of footings in which all shear is carried by the concrete will be discussed here. For the rare cases where the thickness is restricted so that shear reinforcement must be used, the information in Sec. 13.6 about slabs applies also to footings.

Two different types of shear strength are distinguished in footings: two-way, or punching, shear and one-way, or beam, shear.

A column supported by the slab of Fig. 18.6 tends to punch through that slab because of the shear stresses that act in the footing around the perimeter of the column. At the same time the concentrated compression stresses from the column spread out into the footing so that the concrete adjacent to the column is in vertical or slightly inclined compression, in addition to shear. In consequence, if failure occurs, the fracture takes the form of the truncated pyramid shown in Fig. 18.6 (or of a truncated cone for a round column), with sides sloping outward at an angle approaching 45°. The average shear stress in the concrete that fails in this manner can be taken as that acting on vertical planes laid through the footing around the column on a perimeter a distance $d/2$ from the faces of the column (vertical section through $abcd$ on Fig. 18.7). The concrete subject to this shear stress v_{u1} is also in vertical compression from the stresses spreading out from the column, and in horizontal compression in both major directions because of the biaxial bending moments in the footing. This triaxility of stress increases the

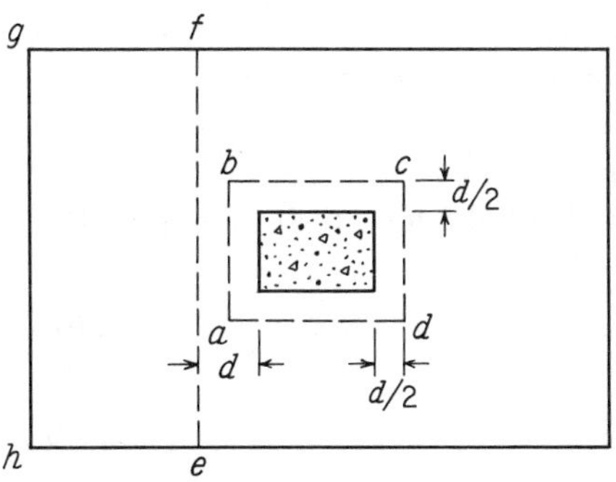

FIGURE 18.7
Critical sections for shear.

shear strength of the concrete. Tests of footings and of flat slabs have shown, correspondingly, that for punching-type failures the shear stress computed on the critical perimeter area is larger than in one-way action (e.g., beams).

As discussed in Sec. 13.6, the ACI Code equations (13.11*a,b,c*) give the nominal punching-shear strength on this perimeter:

$$V_c = 4\sqrt{f'_c}b_0d \tag{18.7a}$$

except for columns of very elongated cross section, for which

$$V_c = \left(2 + \frac{4}{\beta_c}\right)\sqrt{f'_c}b_0d \tag{18.7b}$$

For cases in which the ratio of critical perimeter to slab depth, b_0/d, is very large,

$$V_c = \left(\frac{\alpha_s d}{b_0} + 2\right)\sqrt{f'_c}b_0d \tag{18.7c}$$

Where b_0 is the perimeter *abcd* in Fig. 18.7; $\beta_c = a/b$ is the ratio of the long to short sides of the column cross section; and α_s is 40 for interior loading, 30 for edge loading, and 20 for corner loading of a footing. The punching-shear strength of the footing is to be taken as the smallest of the values given by Eqs. (18.7*a*), (18.7*b*), and (18.7*c*), and the design strength is ϕV_c, as usual, where $\phi = 0.85$ for shear.

The application of Eqs. (18.7) to punching shear in footings under columns of other than rectangular cross section is shown in Fig. 13.16. For such situations ACI Code 11.12.1 indicates that the perimeter b_o must be of minimum length but need not approach closer than $d/2$ to the perimeter of the actual loaded area. The manner of defining a and b for such irregular loaded areas is also shown in Fig. 13.16.

Shear failures can also occur, as in beams or one-way slabs, at a section a distance d from the face of the column, such as section *ef* of Fig. 18.7. Just as in beams and one-way slabs, the nominal shear strength is given by Eq. (4.12*a*), that is,

$$V_c = \left(1.9\sqrt{f'_c} + 2500\rho\frac{V_u d}{M_u}\right)bd \leq 3.5\sqrt{f'_c}bd \tag{18.8a}$$

where b = width of footing at distance d from face of column
= *ef* in Fig. 18.7
V_u = total factored shear force on that section
= q_u times footing area outside that section (area *efgh* in Fig. 18.7)
M_u = moment of V_u about *ef*

In footing design the simpler and somewhat more conservative Eq. (4.12*b*) is generally used, i.e.,

$$V_c = 2\sqrt{f'_c}bd \tag{18.8b}$$

The required depth of footing d is then calculated from the usual equation

$$V_u \leq \phi V_c \tag{18.9}$$

applied separately in connection with Eqs. (18.7) and (18.8). For Eq. (18.7), $V_u = V_{u1}$ is the total upward pressure caused by q_u on the area outside the perimeter *abcd* in Fig. 18.7. For Eq. (18.8), $V_u = V_{u2}$ is the total upward pressure on the area *efgh* outside the section *ef* in Fig. 18.7. The required depth is then the larger of those calculated from either Eq. (18.7) or (18.8). For shear, as usual, $\phi = 0.85$.

b. Bearing: Transfer of Forces at Base of Column

When a column rests on a footing or pedestal, it transfers its load to only a part of the total area of the supporting member. The adjacent footing concrete provides lateral support to the directly loaded part of the concrete. This causes triaxial compression stresses that increase the strength of the concrete that is loaded directly under the column. Based on tests, ACI Code 10.15.1 provides that when the supporting area is wider than the loaded area on all sides, the design bearing strength is

$$\phi P_n = 0.85\phi f_c' A_1 \sqrt{\frac{A_2}{A_1}} \leq 0.85\phi f_c' A_1 \times 2 \tag{18.10}$$

For bearing on concrete, $\phi = 0.70$, f_c' is the cylinder strength of the footing concrete, which frequently is less than that of the column, and A_1 is the loaded area. A_2 is the area of the lower base of the largest frustum of a pyramid, cone, or tapered wedge contained wholly within the support and having for its upper base the loaded area and having side slopes of 1 vertical to 2 horizontal. The meaning of this definition of A_2 may be clarified by Fig. 18.8. Note that, for the somewhat unusual case shown, where the top of the support is stepped, a step that is deeper or closer to the loaded area than that shown may result in reduction in the value of A_2. A footing for which the top surface is sloped away from the loaded area more steeply than 1 to 2 will result in a value of A_2 equal to A_1. In most usual cases, for which the top of the footing is flat and the sides are vertical, A_2 is simply the maximum area of the portion of the supporting surface that is geometrically similar to, and concentric with, the loaded area.

All axial forces and bending moments that act at the bottom section of a column must be transferred to the footing at the bearing surface by compression in the concrete and by reinforcement. With respect to the reinforcement, this may be done either by extending the column bars into the footing or by providing dowels that are embedded in the footing and project above it. In the latter case the column bars merely rest on the footing and in most cases are tied to dowels. This results in a simpler construction procedure than extending the column bars into the footing. To ensure the integrity of the junction between column and footing, ACI Code 15.8.2 requires that the minimum area of reinforcement that crosses the bearing surface (dowels or column bars) be 0.005 times the gross area of the supported

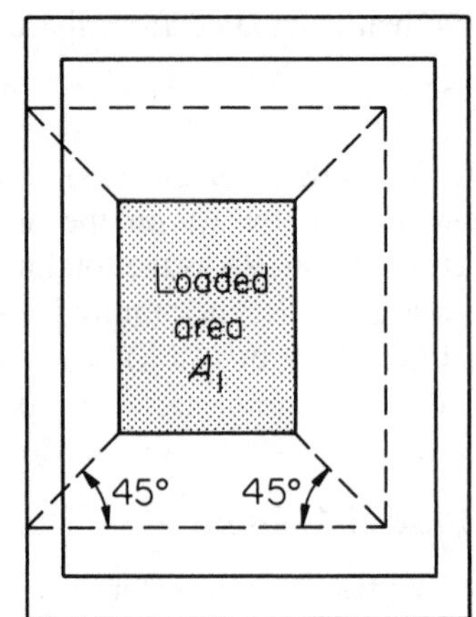

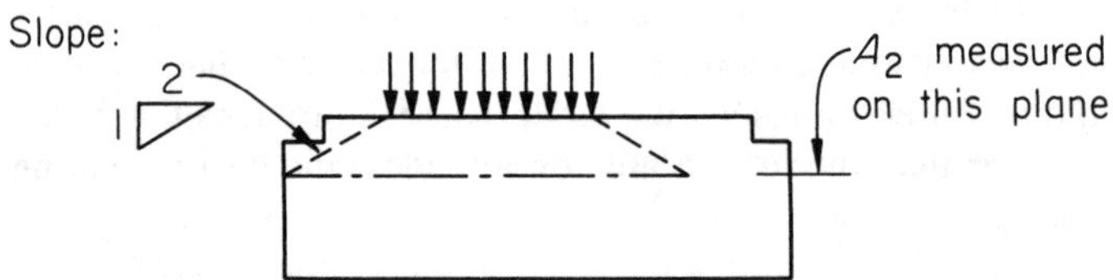

FIGURE 18.8
Definition of areas A_1 and A_2.

column. The length of the dowels or bars of diameter d_b must be sufficient on both sides of the bearing surface to provide the required development length for compression bars (see Sec. 5.7), that is, $l_d = 0.02 f_y d_b / \sqrt{f'_c}$ and $\geq 0.0003 f_y d_b$. In addition, if dowels are used, the lapped length must be at least that required for a lap splice in compression (see Sec. 5.11*b*), i.e., the length of lap must not be less than the usual development length in compression and must not be less than $0.0005 f_y d_b$. Where bars of different sizes are lap-spliced, the splice length should be the larger of the development length of the larger bar or the splice length of the smaller bar, according to the ACI Code.

The two largest bar sizes, Nos. 14 and 18, are frequently used in columns with large axial forces. Under normal circumstances the ACI Code specifically prohibits the lap splicing of these bars because tests have shown that welded splices or other positive connections are necessary to develop these heavy bars fully. However, a specific exception is made for dowels for Nos. 14 and 18 column bars. Relying on long-standing successful use, the ACI Code now permits these heavy bars to be spliced to dowels of lesser diameter (i.e., No. 11 or smaller) provided the dowels have a development length into the column corresponding to that of the column bar (i.e., Nos. 14 or 18, as the case may be) and into the footing as prescribed for the particular dowel size (i.e., No. 11 or smaller, as the case may be).

c. Bending Moments, Reinforcement, and Bond

If a vertical section is passed through a footing, the bending moment that is caused in the section by the net upward soil pressure (i.e., factored column load divided by bearing area) is obtained from simple statics. Figure 18.9 shows such a section *cd* located along the face of the column. The bending moment about *cd* is that caused by the upward pressure q_u on the area to one side of the section, i.e., the area *abcd*. The reinforcement perpendicular to that section, i.e., the bars running in the long direction, is calculated from this bending moment. Likewise, the moment about section *ef* is caused by the pressure q_u on the area *befg*, and the reinforcement in the short direction, i.e., perpendicular to *ef*, is calculated for this bending moment. In footings that support reinforced concrete columns, these critical sections for bending are located at the faces of the loaded area, as shown.

In footings supporting steel columns, the sections *ab* and *ef* are located not at the edge of the steel base plate but halfway between the edge of the column and that of the steel base plate, according to ACI Code 15.4.2.

In footings with *pedestals,* the width resisting compression in sections *cd* and *ef* is that of the pedestal; the corresponding depth is the sum of the thickness of pedestal and footing. Further sections parallel to *cd* and *ef* are passed at the edge of the pedestal, and the moments are determined in the same manner, to check the strength at locations in which the depth is that of the footing only.

For footings with relatively small pedestals, the latter are often discounted in moment and shear computation, and bending is checked at the face of the column, with width and depth equal to that of the footing proper.

In *square footings,* the reinforcement is uniformly distributed over the width of the footing in each of the two layers; i.e., the spacing of the bars is constant. The moments for which the two layers are designed are the same. However, the effective depth d for the upper layer is less by 1 bar diameter than that of the lower layer. Consequently, the required A_s is larger for the upper layer. Instead of using different spacings or different bar diameters in each of the two layers, it is customary to determine A_s based on average depth and to use the same arrangement of reinforcement for both layers.

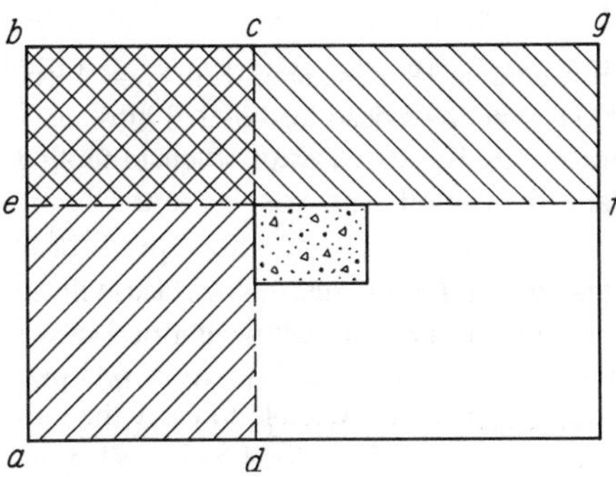

FIGURE 18.9
Critical sections for bending and bond.

In *rectangular footings,* the reinforcement *in the long direction* is again uniformly distributed over the pertinent (shorter) width. In locating the bars in the short direction, one has to consider that the support provided to the footing by the column is concentrated near the middle. Consequently, the curvature of the footing is sharpest, i.e., the moment per foot largest, immediately under the column, and it decreases in the long direction with increasing distance from the column. For this reason, a larger steel area per longitudinal foot is needed in the central portion than near the far ends of the footing. ACI Code 15.4.4 provides, therefore, the following:

> For reinforcement in the short direction, a portion of the total reinforcement [given by Eq. (18.11)] shall be distributed uniformly over a band width (centered on the centerline of the column or pedestal) equal to the length of the short side of the footing. The remainder of the reinforcement required in the short direction shall be distributed uniformly outside the center band width of the footing.

$$\frac{\text{Reinforcement in band width}}{\text{Total reinforcement in short direction}} = \frac{2}{\beta + 1} \tag{18.11}$$

In Eq. (18.11), β is the ratio of the long side to the short side of the footing.

Whether or not the minimum reinforcement ratio of $200/f_y$ prescribed by ACI Code 10.5 applies in footings separately to each of the two directions, or indeed at all, has caused much controversy. A strict reading of ACI Code 10.5.3 and Commentary R10.5.3 indicates that, as for structural slabs, only the minimum steel required for shrinkage and temperature controls. The reasoning is that in structural slabs, an overload would be distributed laterally and a sudden failure is therefore less likely. However, although that reasoning may apply to highly indeterminate building floors, the possibility for redistribution in a footing is much more limited. Because of this, and because of the importance of a footing to the safety of the structure, many engineers adhere to the minimum steel ratio of $200/f_y$ for footings.

The critical sections for development length of footing bars are the same as those for bending. Development length may also have to be checked at all vertical planes in which changes of section or of reinforcement occur, as at the edges of pedestals or where part of the reinforcement may be terminated.

Example 18.2 Design of a square footing. A column 18 in. square, with $f_c' = 4$ ksi, reinforced with eight No. 8 bars of $f_y = 50$ ksi, supports a dead load of 225 kips and a live load of 175 kips. The allowable soil pressure q_a is 5 ksf. Design a square footing with base 5 ft below grade, using $f_c' = 4$ ksi and $f_y = 50$ ksi.

Solution. Since the space between the bottom of the footing and the surface will be occupied partly by concrete and partly by soil (fill), an average unit weight of 125 pcf will be assumed. The pressure of this material at the 5 ft depth is $5 \times 125 = 625$ psf, leaving a bearing pressure of $q_e = 5000 - 625 = 4375$ psf available to carry the column service load. Hence, the required footing area $A_{\text{req}} = (225+175)/4.375 = 91.5\ \text{ft}^2$.

A base 9 ft 6 in. square is selected, furnishing a footing area of 90.3 ft^2, which differs from the required area by about 1 percent.

For strength design, the upward pressure caused by the factored column loads is $q_u = (1.4 \times 225 + 1.7 \times 175)/9.5^2 = 6.80$ ksf.

The footing depth in square footings is usually determined from the two-way or punching shear on the critical perimeter *abcd* of Fig. 18.10. Trial calculations suggest $d = 19$ in. Hence, the length of the critical perimeter is

$$b_0 = 4(18 + d) = 148 \text{ in.}$$

The shear force acting on this perimeter, being equal to the total upward pressure minus that acting within the perimeter *abcd*, is

$$V_{u1} = 6.8\left[9.5^2 - \left(\frac{37}{12}\right)^2\right] = 550 \text{ kips}$$

The corresponding nominal shear strength [Eq. (13.11*a*)] is

$$V_c = 4\sqrt{4000} \times 148 \times \frac{19}{1000} = 711 \text{ kips}$$

and

$$\phi V_c = 0.85 \times 711 = 604 \text{ kips}$$

Since the design strength exceeds the required strength V_{u1}, the depth $d = 19$ in. is adequate for punching shear. The selected value $d = 19$ in. will now be checked for one-way or beam shear on section *ef*. The factored shear force acting on that section is

$$V_{u2} = 6.8 \times 2.42 \times 9.5 = 156 \text{ kips}$$

and the nominal shear strength is

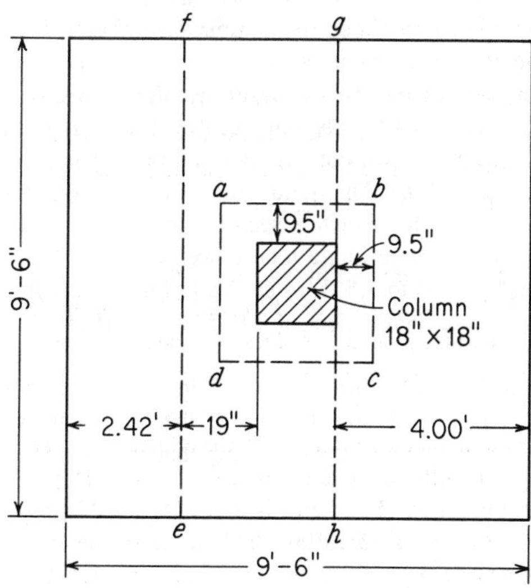

FIGURE 18.10
Critical sections for Example 18.2.

$$V_c = 2\sqrt{4000} \times 9.5 \times 12 \times \frac{19}{1000} = 274 \text{ kips}$$

The design shear strength $0.85 \times 274 = 233$ kips is larger than the required shear strength V_{u2}, so that $d = 19$ in. is also adequate for one-way shear.

The bending moment on section *gh* of Fig. 18.10 is

$$M_u = 6.8 \times 9.5 \frac{4.0^2}{2} 12 = 6200 \text{ in-kips}$$

The depth required for shear being greatly in excess of that required for bending, the steel ratio will be low and the corresponding depth of the rectangular stress block small. If $a = 2$ in., the required steel area is

$$A_s = \frac{6200}{0.9 \times 50(19 - 1)} = 7.65 \text{ in}^2$$

Checking the minimum steel ratio shows that $\rho_{\min} = 200/50{,}000 = 0.004$ gives the minimum reinforcement $A_s = \rho_{\min} bd = 0.004(12 \times 9.5)19 = 8.66$ in^2. This governs because it is larger than the 7.65 in^2 calculated for bending. Eleven No. 8 bars furnishing 8.64 in^2 will be used in each direction. The required development length beyond section *gh* is

$$l_d = 0.04 \times 0.79 \frac{50{,}000}{\sqrt{4000}} = 25.0 \text{ in.}$$

which is more than adequately met by the actual length of bars beyond section *gh*, namely, $48 - 3 = 45$ in.

Checking for transfer of forces at the base of the column shows that the footing concrete, which has the same f'_c as the column concrete and for which the strength is enhanced according to Eq. (18.10), is clearly capable of carrying that part of the column load transmitted by the column concrete. The force in the column carried by the steel will be transmitted to the footing using dowels to match the column bars. These must extend into the footing the full development length in compression, which is found from Table A.12 of App. A to be 16 in. for No. 8 bars. This is easily accommodated in a footing with $d = 19$ in. Above the top surface of the footing, the No. 8 dowels must extend into the column that same development length, but not less than the requirement for a lapped splice in compression (see Sec. 5.11*b*). The minimum lap splice length for the No. 8 bars is $0.0005 \times 1.0 \times 50{,}000 = 25$ in., which is seen to control here. Thus the bars will be carried 25 in. into the column, requiring a total dowel length of 41 in. This will be rounded upward for practical reasons to 3.5 ft, as shown in Fig. 18.11. It is easily confirmed that the minimum dowel steel requirement of $0.005 \times 18 \times 18 = 1.62$ in^2 does not control here.

For concrete in contact with ground, a minimum cover of 3 in. is required for corrosion protection. With $d = 19$ in., measured from the top of the footing to the center of the upper layer of bars, the total thickness of the footing that is required in order to provide 3 in. clear cover for the lower steel layer is

$$h = 19 + 1.5 \times 1 + 3 = 23.5 \text{ in.}$$

The footing, with 24 in. thickness, is shown in Fig. 18.11.

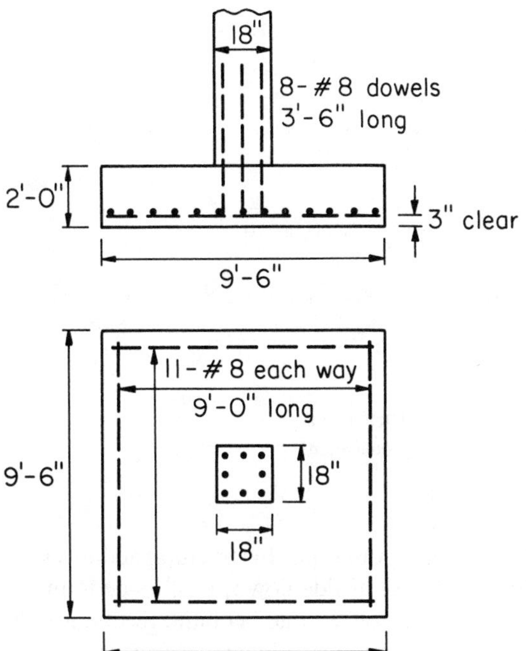

FIGURE 18.11
Footing of Example 18.2.

18.7 COMBINED FOOTINGS

Spread footings that support more than one column or wall are known as *combined footings*. They can be divided into two categories: those that support two columns and those that support more than two (generally large numbers of) columns.

Examples of the first type, i.e., two-column footings, are shown in Fig. 18.1. In buildings where the allowable soil pressure is large enough for single footings to be adequate for most columns, two-column footings are seen to become necessary in two situations: (1) if columns are so close to the property line that single-column footings cannot be made without projecting beyond that line, and (2) if some adjacent columns are so close to each other that their footings would merge. Both situations are shown in Fig. 18.1.

When the bearing capacity of the subsoil is low so that large bearing areas become necessary, individual footings are replaced by *continuous strip footings* that support more than two columns and usually all columns in a row. Sometimes such strips are arranged in both directions, in which case a *grid foundation* is obtained, as shown in Fig. 18.12. Such a foundation can be made to develop a much larger bearing area much more economically than can be done by single footings because the individual strips represent continuous beams whose moments are much smaller than the cantilever moments in large single footings that project far out from the column in all four directions.

For still lower bearing capacities, the strips are made to merge, resulting in a mat foundation, as shown in Fig. 18.13. That is, the foundation consists of

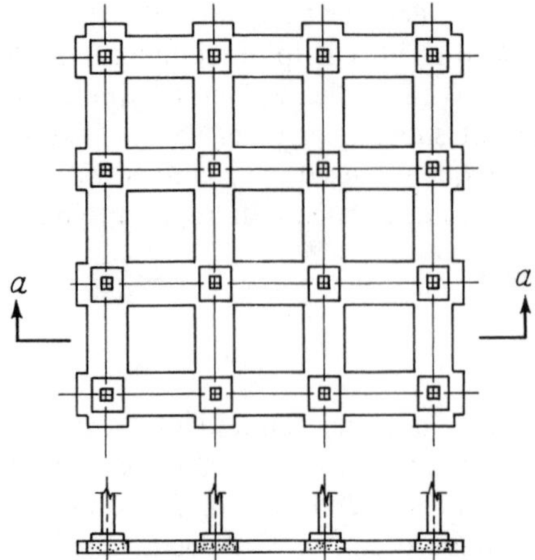

FIGURE 18.12
Grid foundation.

a solid reinforced concrete slab under the entire building. In structural action such a mat is very similar to a flat slab or a flat plate, upside down, i.e., loaded upward by the bearing pressure and downward by the concentrated column reactions. The mat foundation evidently develops the maximum available bearing area under the building. If the soil's capacity is so low that even this large bearing area is insufficient, some form of deep foundation, such as piles or caissons, must be used. These are discussed in texts on foundation design and fall outside the scope of the present volume.

Mat foundations may be designed with the column pedestals, as shown in Figs. 18.12 and 18.13, or without them, depending on whether or not they are necessary for shear strength and the development length of dowels.

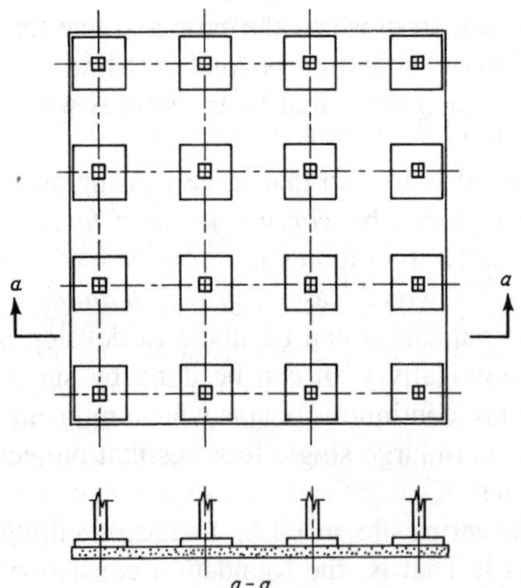

FIGURE 18.13
Mat foundation.

Apart from developing large bearing areas, another advantage of strip, grid, and mat foundations is that their continuity and rigidity help in reducing differential settlements of individual columns relative to each other, which may otherwise be caused by local variations in the quality of subsoil, or other causes. For this purpose, continuous spread foundations are frequently used in situations where the superstructure or the type of occupancy provides unusual sensitivity to differential settlement.

18.8 TWO-COLUMN FOOTINGS

It is desirable to design combined footings so that the centroid of the footing area coincides with the resultant of the two column loads. This produces uniform bearing pressure over the entire area and forestalls a tendency for the footings to tilt. In plan, such footings are rectangular, trapezoidal, or T shaped, the details of the shape being arranged to produce coincidence of centroid and resultant. The simple relationships of Fig. 18.14 facilitate the determination of the shape of the bearing area (from Ref. 18.6). In general, the distances m and n are given, the former being the distance from the center of the exterior column to the property line and the latter the distance from that column to the resultant of both column loads.

Another expedient that is used if a single footing cannot be centered under an exterior column is to place the exterior column footing eccentrically and to connect it with the nearest interior column footing by a beam or strap. This strap, being counterweighted by the interior column load, resists the tilting tendency of the eccentric exterior footings and equalizes the pressure under it. Such foundations are known as *strap, cantilever,* or *connected footings*.

The two examples that follow demonstrate some of the peculiarities of the design of two-column footings.

Example 18.3: Design of a combined footing supporting one exterior and one interior column. An exterior 24 × 18 in. column with D = 170 kips, L = 130 kips, and an interior 24 × 24 in. column with D = 250 kips, L = 200 kips are to be supported on a combined rectangular footing whose outer end cannot protrude beyond the outer face of the exterior column (see Fig. 18.1). The distance center to center of columns is 18 ft 0 in., and the allowable bearing pressure of the soil is 6000 psf. The bottom of the footing is 6 ft below grade, and a surcharge of 100 psf is specified on the surface. Design the footing for f'_c = 3000 psi, f_y = 60,000 psi.

Solution. The space between the bottom of the footing and the surface will be occupied partly by concrete (footing, concrete floor) and partly by backfill. An average unit weight of 125 pcf can be assumed. Hence, the effective portion of the allowable bearing pressure that is available for carrying the column loads is $q_e = q_a -$ (weight of fill and concrete + surcharge) = 6000 − (6 × 125 + 100) = 5150 psf. Then the required area A_{req} = sum of column loads/q_e = 750/5.15 = 145.5 ft^2. The resultant of the column loads is located from the center of the exterior column a distance 450 × 18/750 = 10.8 ft. Hence, the length of the footing must be 2(10.8 + 0.75) = 23.1 ft. A length of 23 ft 3 in. is selected. The required width is then 145.5/23.25 = 6.3 ft. A width of 6 ft 6 in. is selected (see Fig. 18.15).

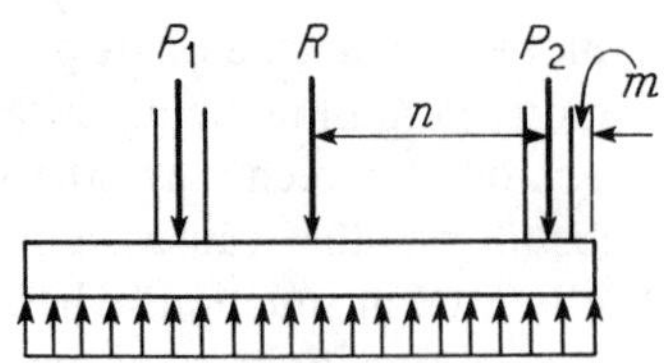

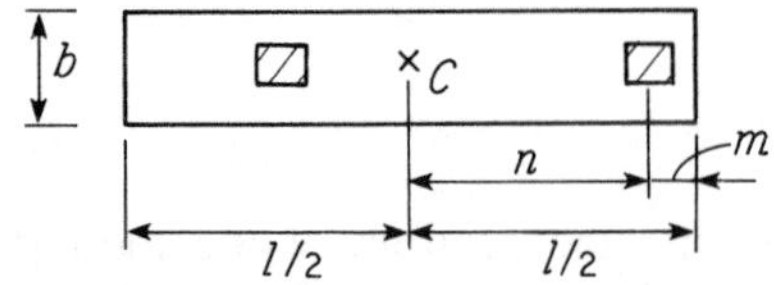

$$l = 2(m+n)$$

$$b = \frac{R}{q_e l}$$

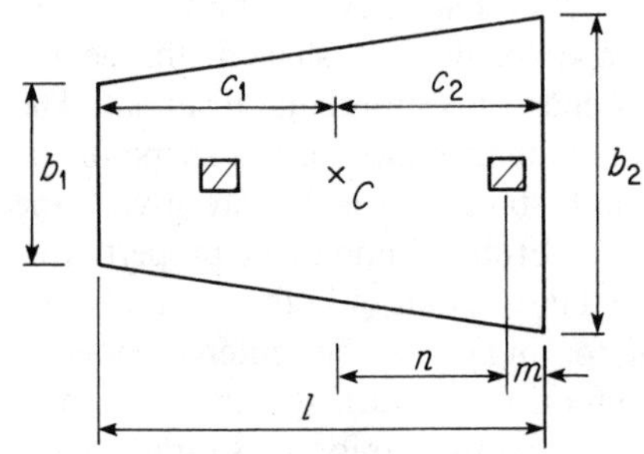

$$\frac{b_2}{b_1} = \frac{3(n+m)-l}{2l-3(n+m)}$$

$$(b_1+b_2) = \frac{2R}{q_e l}$$

$$c_1 = \frac{l(b_1+2b_2)}{3(b_1+b_2)}$$

$$c_2 = \frac{l(2b_1+b_2)}{3(b_1+b_2)}$$

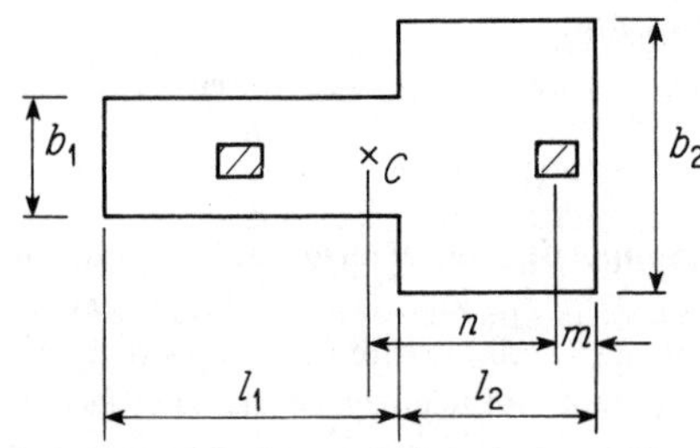

$$b_1 = \frac{R}{q_e} = \left[\frac{2(n+m)-l_2}{l_1(l_1+l_2)}\right]$$

$$b_2 = \frac{R}{l_2 q_e} - \frac{l_1 b_1}{l_2}$$

$$l_1 b_1 + l_2 b_2 = \frac{R}{q_e}$$

FIGURE 18.14
Two-column footings. (*Adapted from Ref. 18.6.*)

Longitudinally, the footing represents an upward-beam spanning between columns and cantilevering beyond the interior column. Since this beam is considerably wider than the columns, the column loads are distributed crosswise by transverse beams, one under each column. In the present relatively narrow and long footing it will be found that the required minimum depth for the transverse beams is smaller than is required for the footing in the longitudinal direction. These "beams," therefore, are not really distinct members but merely represent transverse strips in the main body of the footing, so reinforced that they are capable of resisting the transverse bending moments and the corresponding shears. It then becomes necessary to decide how large the effective width of this transverse beam can be assumed to be. Obviously, the strip directly under the column does not deflect independently and is strengthened by the adjacent parts of the

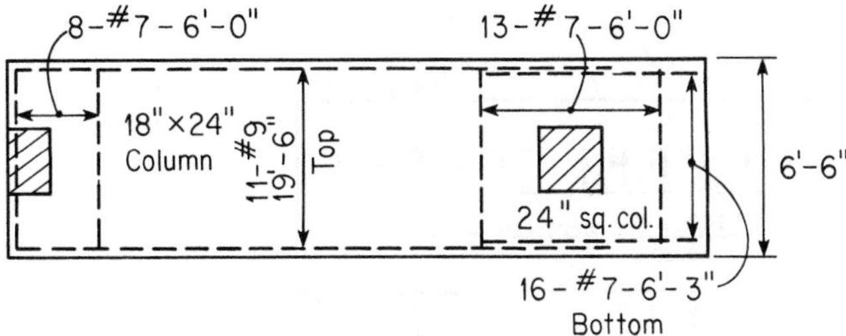

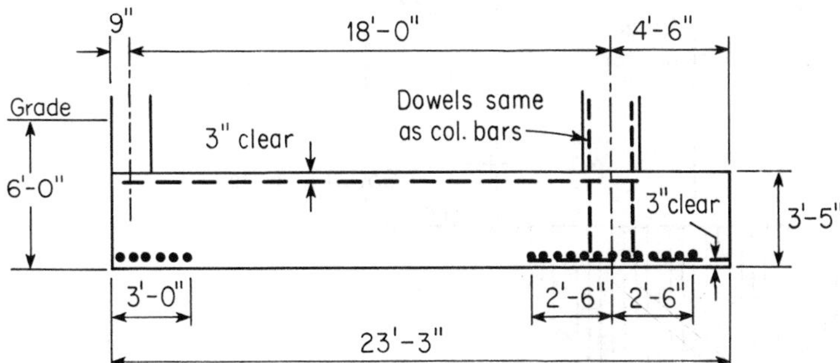

FIGURE 18.15
Combined footing of Example 18.3.

footing. The effective width of the transverse beams is therefore evidently larger than that of the column. In the absence of definite rules for this case, or of research results on which to base such rules, the authors recommend conservatively that the load be assumed to spread outward from the column into the footing at a slope of 2 vertical to 1 horizontal. This means that the effective width of the transverse beam is assumed to be equal to the width of the column plus $d/2$ on either side of the column, d being the effective depth of the footing.

Strength design in longitudinal direction

The net upward pressure caused by the factored column loads is

$$q_u = \frac{1.4(170 + 250) + 1.7(130 + 200)}{23.25 \times 6.5} = 7.60 \text{ kips/ft}^2$$

Then the net upward pressure per linear foot in the longitudinal direction is $7.60 \times 6.5 =$ 49.4 kips/ft. The maximum negative moment between the columns occurs at the section of zero shear. Let x be the distance from the outer edge of the exterior column to this section. Then (see Fig. 18.16)

$$V_u = 49{,}400x - 459{,}000 = 0$$

results in $x = 9.3$ ft. The moment at this section is

$$M_u = \left[49{,}400\frac{9.3^2}{2} - 459{,}000(9.30 - 0.75)\right]12 = -21{,}400{,}000 \text{ in-lb}$$

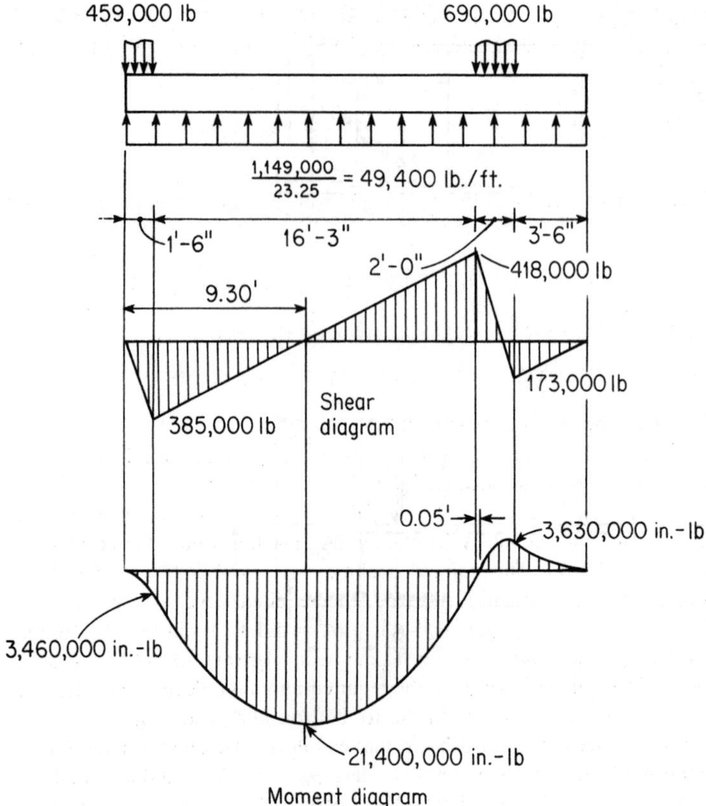

FIGURE 18.16
Moment and shear diagram for footing of Example 18.3.

The moment at the right edge of the interior column is

$$M_u = 49{,}400\frac{3.5^2}{2}12 = 3{,}630{,}000 \text{ in-lb}$$

and the details of the moment diagram are as shown in Fig. 18.16. Try $d = 37.5$ in.

From the shear diagram of Fig. 18.16 it is seen that the critical section for flexural shear is at a distance d to the left of the left face of the interior column. At that point the required shear strength is

$$V_u = 418{,}000 - \frac{37.5}{12}49{,}400 = 264{,}000 \text{ lb}$$

and the design shear strength

$$\phi V_c = 0.85 \times 2\sqrt{3000} \times 78 \times 37.5 = 272{,}000 \text{ lb} > V_u$$

indicating that $d = 37.5$ in. is adequate.

Additionally, as in single footings, punching shear should be checked on a perimeter section a distance $d/2$ around the column, on which the nominal shear stress

$v_c = 4\sqrt{3000} = 220$ psi. Of the two columns, the exterior one with a three-sided perimeter a distance $d/2$ from the column is more critical in regard to this punching shear. The perimeter is

$$b_0 = 2\left(1.5 + \frac{37.5/12}{2}\right) + \left(2.0 + \frac{37.5}{12}\right) = 11.24 \text{ ft}$$

and the shear force, being the column load minus the soil pressure within the perimeter, is

$$V_u = 459{,}000 - 3.06 \times 5.12(7600) = 340{,}000 \text{ lb}$$

On the other hand, the design shear strength on the perimeter section is

$$\phi V_c = 0.85 \times 220 \times 11.24 \times 12 \times 37.5 = 946{,}000 \text{ lb},$$

considerably larger than the required strength V_u.

With $d = 37.5$ in., and with 3.5 in. insulation from the center of the bars to the top surface of the footing, the total thickness is 41 in.

To determine the required steel area, $M_u/\phi bd^2 = 21{,}400{,}000/(0.9 \times 78 \times 37.5^2) = 217$ is used to enter Graph A.1*b* of App. A. For this value, the curve 60/3 gives the steel ratio $\rho = 0.0037$. The required steel area is $A_s = 0.0037 \times 37.5 \times 78 = 10.8$ in^2. Eleven No. 9 bars furnish 11.00 in^2. The required development length is $l_d = 1.4 \times 0.04 \times 1.00 \times 60{,}000/\sqrt{3000} = 61.3$ in. $= 5.1$ ft. From Fig. 18.16, the distance from the point of maximum moment to the nearer left end of the bars is seen to be $9.30 - \frac{3}{12} = 9.05$ ft, much larger than the required minimum development length. The selected reinforcement is therefore adequate for both bending and bond.

For the portion of the longitudinal beam that cantilevers beyond the interior column, the minimum required steel area controls. Here $\rho_{min} = 200/60{,}000 = 0.0033$. $A_s = 0.0033 \times 78 \times 37.5 = 9.75$ in^2. Sixteen No. 7 bars, with $A_s = 9.62$ in^2, are selected; their development length is computed similarly, but for bottom bars, is found adequate.

Design of transverse beam under interior column

The width of the transverse beam under the interior column can now be established as previously suggested and is $24 + 2(d/2) = 24 + 2 \times 18.75 = 61.5$ in. The net upward load per linear foot of the transverse beam is $690{,}000/6.5 = 106{,}000$ lb/ft. The moment at the edge of the interior column is

$$M_u = 106{,}000 \frac{2.25^2}{2} 12 = 3{,}220{,}000 \text{ in-lb}$$

Since the transverse bars are placed on top of the longitudinal bars (see Fig. 18.15), the actual value of d furnished is $37.5 - 1.0 = 36.5$ in. The minimum required steel area controls; i.e.,

$$A_s = \frac{200}{60{,}000} 61.5 \times 36.5 = 7.48 \text{ in}^2$$

Thirteen No. 7 bars are selected and placed within the 61.5 in. effective width of the transverse beam.

Punching shear at the perimeter a distance $d/2$ from the column has been checked before. The critical section for regular flexural shear, at a distance d from the face of the column, lies beyond the edge of the footing, and therefore no further check on shear is needed.

The design of the transverse beam under the exterior column is the same as the design of that under the interior column, except that the effective width is 36.75 in. The details of the calculations are not shown. It will be easily checked that eight No. 7 bars, placed within the 36.75 in. effective width, satisfy all requirements. Design details are shown in Fig. 18.15.

Example 18.4 Design of a strap footing. In a strap or connected footing, the exterior footing is placed eccentrically under its column so that it does not project beyond the property line. Such an eccentric position would result in a strongly uneven distribution of bearing pressure, which could lead to tilting or even tipping of the footing. To counteract this eccentricity, the footing is connected by a beam or strap to the nearest interior footing.

Both footings are so proportioned that under service load the pressure under each of them is uniform and the same under both footings. To achieve this, it is necessary, as in other combined footings, that the centroid of the combined area for the two footings coincide with the resultant of the column loads. The resulting forces are shown schematically in Fig. 18.17. They consist of the loads P_e and P_i of the exterior and interior columns, respectively, and of the net upward pressure q, which is uniform and equal under both footings. The resultants R_e and R_i of these upward pressures are also shown. Since the interior footing is concentric with the interior column, R_i and P_i are collinear. This is not the case for the exterior forces R_e and P_e where the resulting couple just balances the effect of the eccentricity of the column relative to the center of the footing. The strap proper is generally constructed so that it will not bear on the soil. This can be achieved by providing formwork not only for the sides but also for the bottom face and by withdrawing it before backfilling.

To illustrate this design, the columns of Example 18.3 will now be supported on a strap footing. Its general shape, plus dimensions as determined only subsequently by calculations, is seen in Fig. 18.18. With an allowable bearing pressure of $q_a = 6.0$ kips/ft^2 and a depth of 6 ft to the bottom of the footing as before, the bearing pressure available for carrying the external loads applied to the footing is $q_e = 5.15$ kips/ft^2, as in Example 18.3. These external loads, for the strap footing, consist of the column

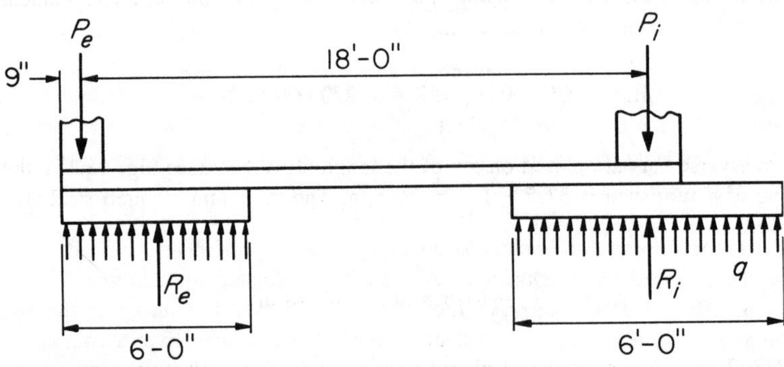

FIGURE 18.17
Forces and reactions on the strap footing of Example 18.4.

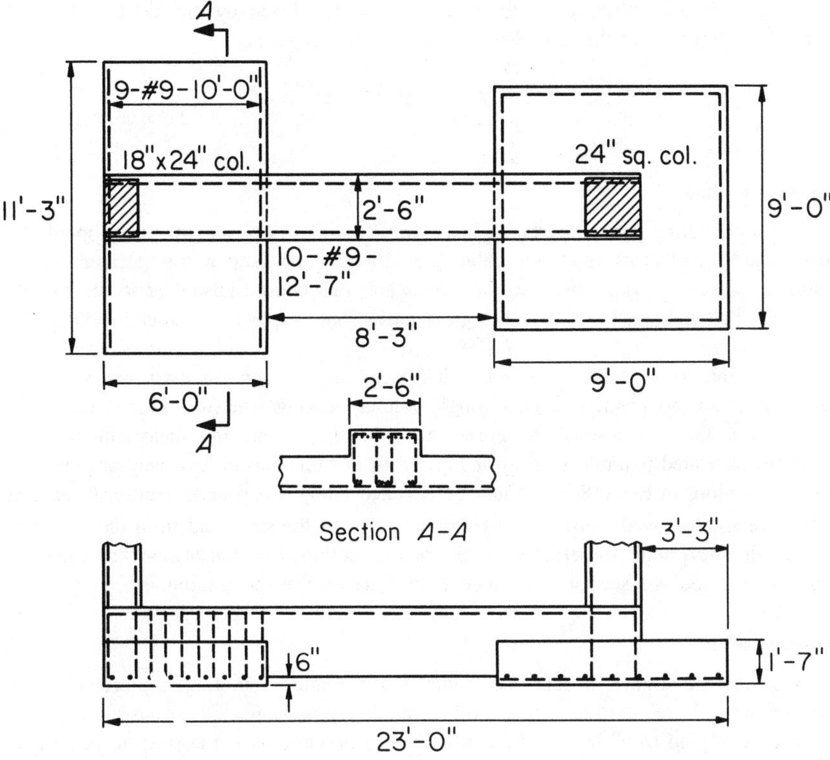

FIGURE 18.18
Strap footing of Example 18.4.

loads and of the weight plus fill and surcharge of that part of the strap which is located between the footings. (The portion of the strap located directly on top of the footing displaces a corresponding amount of fill and therefore is already accounted for in the determination of the available bearing pressure q.) If the bottom of the strap is 6 in. above the bottom of the footings to prevent bearing on soil, the total depth to grade is 5.5 ft. If the strap width is estimated to be 2.5 ft, its estimated weight plus fill and surcharge is $2.5 \times 5.5 \times 0.125 + 0.100 \times 2.5 = 2$ kips/ft. If the gap between footings is estimated to be 8 ft, the total weight of the strap is 16 kips. Hence, for purposes of determining the required footing area, 8 kips will be added to the dead load of each column. The required total area of both footings is then $(750 + 16)/5.15 = 149$ ft^2. The distance of the resultant of the two column loads plus the strap load from the axis of the exterior column, sufficiently accurately, is $458 \times 18/766 = 10.75$ ft, or 11.50 ft from the outer edge, almost identical with that calculated for Example 18.3. Trial calculations show that a rectangular footing 6 ft 0 in. $\times$ 11 ft 3 in. under the exterior column and a square footing 9×9 ft under the interior column have a combined area of 149 ft^2 and a distance from the outer edge to the centroid of the combined areas of $(6 \times 11.25 \times 3 + 9 \times 9 \times 18.75)/149 = 11.55$ ft, which is almost exactly equal to the previously calculated distance to the resultant of the external forces.

For *strength calculations*, the bearing pressure caused by the factored external loads, including that of the strap with its fill and surcharge, is

$$q_u = \frac{1.4(170 + 250 + 16) + 1.7(130 + 200)}{149} = 7.8 \text{ kips/ft}^2$$

Design of footings

The exterior footing performs exactly like a wall footing of 6 ft length. Even though the column is located at its edge, the balancing action of the strap was so arranged as to result in uniform bearing pressure, the downward load being transmitted to the footing uniformly by the strap. Hence, the design is carried out exactly as for a wall footing (see Sec. 18.5).

The interior footing, even though it merges in part with the strap, can safely be designed as an independent, square single-column footing (see Sec. 18.6). The main difference is that, because of the presence of the strap, punching shear cannot occur along the truncated pyramid surface of Fig. 18.6. For this reason, two-way or punching shear, according to Eq. (18.7), should be checked along a perimeter section located at a distance $d/2$ outward from the longitudinal edges of the strap and from the free face of the column, d being the effective depth of the footing. Flexural or one-way shear, as usual, is checked at a section a distance d from the face of the column.

Design of strap

Even though the strap is in fact monolithic with the interior footing, the effect on the strap of the soil pressure under this footing can safely be neglected because the footing has been designed to withstand the entire upward pressure as if the strap were absent. In contrast, the exterior footing having been designed as a wall footing which receives its load from the strap, the upward pressure from the wall footing becomes a load to be resisted by the strap. With this simplification of the actually somewhat more complex situation, the strap represents a single-span beam loaded upward by the bearing pressure under the exterior footing and supported by downward reactions at the centerlines of the two columns (Fig. 18.19). A width of 30 in. is selected. For a column width of 24 in. this permits beam and column bars to be placed without interference where the two members meet and allows the column forms to be supported on the top surface of the strap. The maximum moment, as determined by equating the shear force to zero, occurs close to the inner edge of the exterior footing. Shear forces are large in the vicinity of the exterior column. The footing is drawn approximately to scale in Fig. 18.18, which also shows the general arrangement of the reinforcement in footings and strap.

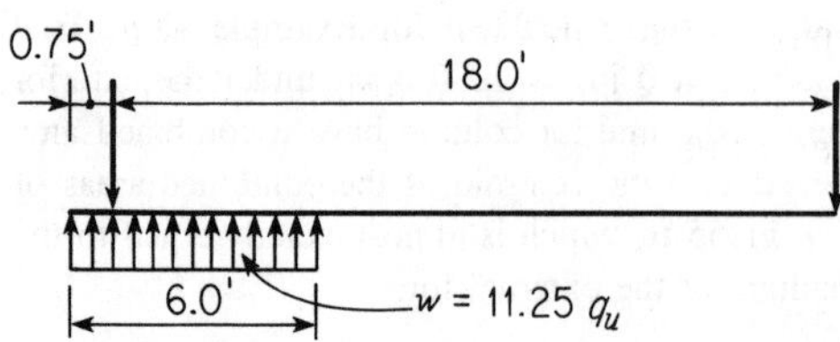

FIGURE 18.19
Forces acting on the strap of Example 18.4.

18.9 STRIP, GRID, AND MAT FOUNDATIONS

As mentioned in Sec. 18.7, in the case of heavily loaded columns, particularly if they are to be supported on relatively weak or uneven soils, continuous foundations are resorted to. They may consist of a continuous strip footing supporting all columns in a given row, or of two sets of such strip footings intersecting at right angles so that they form one continuous *grid foundation* (Fig. 18.12). For even larger loads or weaker soils the strips are made to merge, resulting in a *mat foundation* (Fig. 18.13).

For the design of such continuous foundations it is essential that reasonably realistic assumptions be made regarding the distribution of bearing pressures that act as upward loads on the foundation. For compressible soils it can be assumed in first approximation that the deformation or settlement of the soil at a given location and the bearing pressure at that location are proportional to each other. If columns are spaced at moderate distances and if the strip, grid, or mat foundation is very rigid, the settlements in all portions of the foundation will be substantially the same. This means that the bearing pressure, also known as *subgrade reaction,* will be the same provided that the centroid of the foundation coincides with the resultant of the loads. If they do not coincide, then for such rigid foundations the subgrade reaction can be assumed as linear and determined from statics, in the same manner as discussed for single footings (see Fig. 18.3). In this case all loads, the downward column loads as well as the upward-bearing pressures, are known. Hence, moments and shear forces in the foundation can be found by statics alone. Once these are determined, the design of strip and grid foundations is similar to that of inverted continuous beams and that of mat foundations to that of inverted flat slabs or plates.

On the other hand, if the foundation is relatively flexible and the column spacing large, settlements will no longer be uniform or linear. For one thing, the more heavily loaded columns will cause larger settlements, and thereby larger subgrade reactions, than the lighter ones. Also, since the continuous strip or slab midway between columns will deflect upward relative to the nearby columns, this means that the soil settlement, and thereby the subgrade reaction, will be smaller midway between columns than directly at the columns. This is shown schematically in Fig. 18.20. In this case the subgrade reaction can no longer be assumed as uniform. A reasonably accurate but fairly complex analysis can then be made using the theory of beams on elastic foundations (Ref. 18.7).

A simplified procedure has been developed that covers the most frequent situations of strip and grid foundations (Ref. 18.8). The method first defines the conditions under which a foundation can be regarded as rigid so that uniform or overall linear distribution of subgrade reactions can be assumed. This is the case when the average of two adjacent span lengths in a continuous strip does not exceed $1.75/\lambda$, provided also that the adjacent span and column loads do not differ by more than 20 percent of the larger value. Here

$$\lambda = \sqrt[4]{\frac{k_s b}{3E_c I}} \tag{18.12}$$

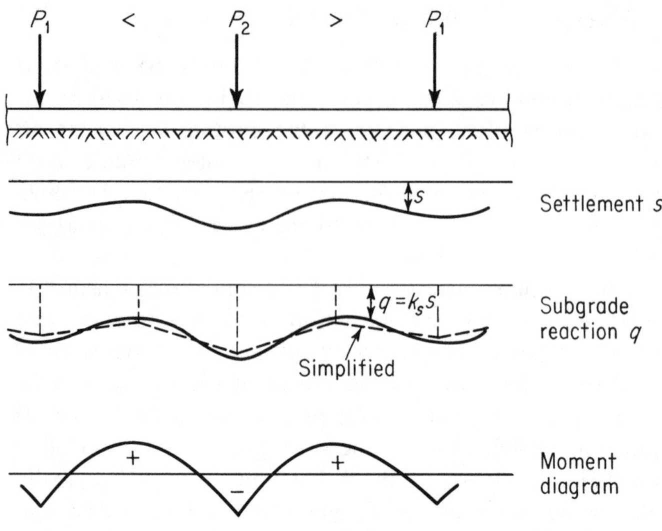

FIGURE 18.20
Strip footing. (*Adapted from Ref. 18.6.*)

where $k_s = Sk'_s$

k'_s = coefficient of subgrade reaction as defined in soils mechanics, basically force per unit area required to produce unit settlement, kips/ft^3

b = width of footing, ft

E_c = modulus of elasticity of concrete, kips/ft^2

I = moment of inertia of footing, ft^4

S = shape factor, being $[(b + 1)/2b]^2$ for granular soils such as sands, and $(n + 0.5)/1.5n$ for cohesive soils such as clays, where n = ratio of longer to shorter side of strip.

If the average of two adjacent spans exceeds $1.75/\lambda$, the foundation is regarded as flexible. Provided that adjacent spans and column loads differ by no more than 20 percent, the complex curvilinear distribution of subgrade reaction can be replaced by a set of equivalent trapezoidally distributed reactions, which are also shown in Fig. 18.20. Reference 18.8 contains fairly simple equations for determining the intensities p of the equivalent pressures under the columns and at the middle of the spans and also gives equations for the positive and negative moments caused by these equivalent subgrade reactions. With this information, the design of continuous strip and grid footings proceeds similarly to that of footings under two columns (Sec. 18.8).

Mat Foundations, likewise, require different approaches, depending on whether they can be classified as rigid or flexible. As in strip footings, if the column spacing is less than $1/\lambda$, the structure may be regarded as rigid, the soil pressure can be assumed as uniformly or linearly distributed, and the design is based on statics. On the other hand, when the foundation is considered flexible as defined above, and if the variation of adjacent column loads and spans is not greater than 20 percent, the same simplified procedure as for strip and grid foundations can be applied to mat foundations. The mat is divided into two sets of mutually perpendicular strip footings of width equal to the distance between midspans, and the distribution of bearing pressures and bending moments is carried out for each strip, as explained before. Once moments are determined, the mat is treated essentially the same as a flat slab or plate, with the reinforcement allocated between column and middle strips as in these slab structures.

This approach is feasible only when columns are located in a regular rectangular grid pattern. When a mat which can be regarded as rigid supports columns at random locations, the subgrade reactions can still be taken as uniform or as linearly distributed and the mat analyzed by statics. If it is a flexible mat that supports such randomly located columns, the design is based on the theory of plates on elastic foundation. An outlines of this procedure is found in Ref. 18.8.

18.10 PILE CAPS

If the bearing capacity of the upper soil layers is insufficient for a spread foundation, but firmer strata are available at greater depth, piles are used to transfer the loads to these deeper strata. Piles are generally arranged in groups or clusters, one under each column. The group is capped by a spread footing or cap that distributes the column load to all piles in the group. These pile caps are in most ways very similar to footings on soil, except for two features. For one, reactions on caps act as concentrated loads at the individual piles, rather than as distributed pressures. For another, if the total of all pile reactions in a cluster is divided by the area of the footing to obtain an equivalent uniform pressure (for purposes of comparison only), it is found that this equivalent pressure is considerably higher in pile caps than for spread footings. This means that moments, and particularly shears, are also correspondingly larger, which requires greater footing depths than for a spread footing of similar horizontal dimensions. In order to spread the load evenly to all piles, it is in any event advisable to provide ample rigidity, i.e., depth for pile caps.

Allowable bearing capacities of piles R_a are obtained from soil exploration, pile-driving energy, and test loadings, and their determination is not within the scope of the present book (see Refs. 18.1 and 18.2). As in spread footings, the effective portion of R_a available to resist the unfactored column loads is the allowable pile reaction less the weight of footing, backfill, and surcharge per pile. That is,

$$R_e = R_a - W_f \tag{18.13}$$

where W_f is the total weight of footing, fill, and surcharge divided by the number of piles.

Once the available or effective pile reaction R_e is determined, the number of piles in a concentrically loaded cluster is the integer next larger than

$$n = \frac{D + L}{R_e}$$

As far as the effects of wind, earthquake moments at the foot of the columns, and safety against overturning are concerned, design considerations are the same as in Sec. 18.4 on spread footings. These effects generally produce an eccentrically loaded pile cluster in which different piles carry different loads. The number and location of piles in such a cluster is determined by successive approximation from the condition that the load on the most heavily loaded pile shall not exceed the allowable pile reaction R_a. With a linear distribution of pile loads due to bending, the maximum pile reaction is

$$R_{\max} = \frac{P}{n} + \frac{M}{I_{pg}/c} \tag{18.14}$$

where P is the maximum load (including weight of cap, backfill, etc.) and M the moment to be resisted by the pile group, both referred to the bottom of the cap; I_{pg} is the moment of inertia of the entire pile group about the centroidal axis about which bending occurs; and c is the distance from that axis to the extreme pile. $I_{pg} = \Sigma_1^n(1 \times y_n^2)$; i.e., it is the moment of inertia of n piles, each counting as one unit and located a distance y_n from the described centroidal axis.

Piles are generally arranged in tight patterns, which minimizes the cost of the caps, but they cannot be placed closer than conditions of driving and of undisturbed carrying capacity will permit. A spacing of about 3 times the butt (top) diameter of the pile but no less than 2 ft 6 in. is customary. Commonly, piles with allowable reactions of 30 to 70 tons are spaced at 3 ft 0 in. (Ref. 18.6).

The *design* of footings on piles is similar to that of single-column footings. One approach is to design the cap for the pile reactions calculated for the factored column loads. For a concentrically loaded cluster this would give $R_u = (1.4D + 1.7L)/n$. However, since the number of piles was taken as the next larger integral according to Eq. (18.14), determining R_u in this manner can lead to a design where the strength of the cap is less than the capacity of the pile group. It is therefore recommended that the pile reaction for strength design be taken as

$$R_u = R_e \times \text{average load factor} \tag{18.15}$$

where the average load factor $= (1.4D + 1.7L)/(D + L)$. In this manner the cap is designed to be capable of developing the full allowable capacity of the pile group. Details of a typical pile cap are shown in Fig. 18.21.

As in single-column spread footings, the depth of the pile cap is usually governed by shear. In this regard both punching or two-way shear and flexural or one-way shear need to be considered. The critical sections are the same as given in Sec. 18.6*a*. The difference is that shears on caps are caused by concentrated pile reactions rather than by distributed bearing pressures. This poses the question of how to calculate shear if the critical section intersects the circumference of one or more piles. For this case ACI Code 15.5.3 accounts for the fact that pile reaction is not really a point load, but is distributed over the pile-bearing area. Correspondingly, for piles with diameters d_p, it stipulates as follows:

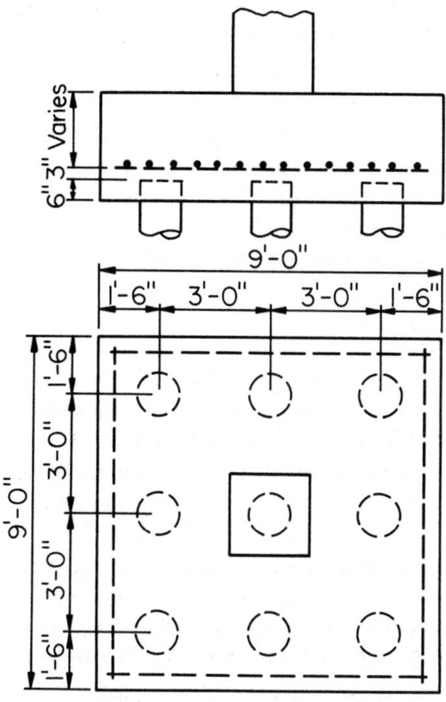

FIGURE 18.21
Typical single-column footing on piles (pile cap).

Computation of shear on any section through a footing on piles shall be in accordance with the following:

(*a*) The entire reaction from any pile whose center is located $d_p/2$ or more outside this section shall be considered as producing shear on that section.

(*b*) The reaction from any pile whose center is located $d_p/2$ or more inside the section shall be considered as producing no shear on that section.

(*c*) For intermediate positions of the pile center, the portion of the pile reaction to be considered as producing shear on the section shall be based on straight-line interpolation between the full value at $d_p/2$ outside the section and zero at $d_p/2$ inside the section.

In addition to checking two-way and one-way shear, as just discussed, punching shear must also be investigated for the individual pile. Particularly in caps on a small number of heavily loaded piles, it is this possibility of a pile punching upward through the cap that may govern the required depth. The critical perimeter for this action, again, is located at a distance $d/2$ outside the upper edge of the pile. However, for relatively deep caps and closely spaced piles, critical perimeters around adjacent piles may overlap. In this case, fracture, if any, would undoubtedly occur along an outward-slanting surface around both adjacent piles. For such situations the critical perimeter is so located that its length is a minimum, as shown for two adjacent piles in Fig. 18.22.

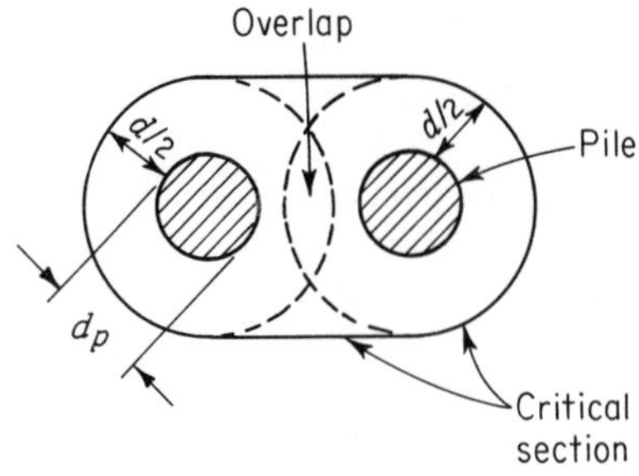

FIGURE 18.22
Critical section for punching shear with closely spaced piles.

REFERENCES

18.1. R. B. Peck, W. E. Hanson, and T. H. Thornburn, *Foundation Engineering,* 2d ed., John Wiley and Sons, Inc., New York, 1974.

18.2. K. Terzaghi and R. B. Peck, *Soil Mechanics in Engineering Practice,* 2d ed., John Wiley and Sons, Inc., New York, 1967.

18.3. A. N. Talbot, "Reinforced Concrete Wall Footings and Column Footings," Univ. Ill. Eng. Exp. Stn. Bull. 67, 1913.

18.4. F. E. Richart, "Reinforced Concrete Wall and Column Footings," *J. ACI,* vol. 45, 1948, pp. 97 and 237.

18.5. E. Hognestad, "Shearing Strength of Reinforced Column Footings," *J. ACI,* vol. 50, 1953, p. 189.

18.6. F. Kramrisch, "Footings," chap. 5 in M. Fintel (ed.), *Handbook of Concrete Engineering,* 2d ed., Van Nostrand Reinhold, New York, 1985.

18.7. M. Heteneyi, *Beams on Elastic Foundations,* Univ. of Michigan Press, Ann Arbor, 1946.

18.8. "Suggested Design Procedures for Combined Footings and Mats," Report of ACI Committee 436, *J. ACI,* vol. 63, no. 10, 1966, pp. 1041–1057.

PROBLEMS

18.1 A continuous strip footing is to be located concentrically under a 12 in. wall that delivers service loads $D = 25{,}000$ lb/ft and $L = 15{,}000$ lb/ft to the top of the footing. The bottom of the footing will be 4 ft below the final ground surface. The soil has a density of 120 pcf and allowable bearing capacity of 8000 psf. Material strengths are $f_c' = 3000$ psi and $f_y = 60{,}000$ psi. Find (*a*) the required width of the footing, (*b*) the required effective and total depths, based on shear, and (*c*) the required flexural steel area.

18.2 An interior column for a tall concrete structure carries total service loads $D = 500$ kips and $L = 514$ kips. The column is 22×22 in. in cross section and is reinforced with twelve No. 11 bars centered 3 in. from the column faces (equal number of bars each face). For the column, $f_c' = 4000$ psi and $f_y = 60{,}000$ psi. The column will be supported on a square footing, with the bottom of the footing 6 ft below grade. Design the footing, determining all concrete dimensions and amount and placement of all reinforcement, including length and placement of dowel steel. No shear reinforcement is permitted. The allowable soil-bearing pressure is 8000 psf. Material strengths for the footing are $f_c' = 3000$ psi and $f_y = 60{,}000$ psi.

18.3 Two interior columns for a high-rise concrete structure are spaced 15 ft apart, and each carries service loads $D = 500$ kips and $L = 514$ kips. The columns are to be 22 in. square in cross section, and will each be reinforced with twelve No. 11 bars centered 3 in. from the column faces, with an equal number of bars at each face. For the column, $f_c' = 4000$ psi and $f_y = 60{,}000$ psi. The columns will be supported on a rectangular combined footing with a long-side dimension twice that of the short side. The allowable soil-bearing pressure is 8000 psf. The bottom of the footing will be 6 ft below grade. Design the footing for these columns, using $f_c' = 3000$ psi and $f_y = 60{,}000$ psi. Specify all reinforcement, including length and placement of footing, bars and dowel steel.

18.4 A pile cap is to be designed to distribute a concentric force from a single column to a nine-pile group, with geometry as shown in Fig. 18.21. The cap will carry calculated dead load and service live load of 280 kips and 570 kips respectively from a 19 in. square concrete column reinforced with six No. 14 bars. The permissible load per pile at service load is 100 kips, and the pile diameter is 16 in. Find the required effective and total depths of the pile cap and the required reinforcement. Check all relevant aspects of the design, including the development length for the reinforcement and transfer of forces at the base of the column. Material strengths for the column are $f_c' = 4000$ psi and $f_y = 60{,}000$ psi, and for the pile cap are $f_c' = 3000$ psi and $f_y = 60{,}000$ psi.

18.5 Complete the design of Example 18.4 and determine all dimensions and reinforcement for the strap footing. Compare the total volume of concrete of the strap footing of Example 18.4 with that of the rectangular combined footing of Example 18.3. It will be found that the strap footing is significantly more economical of material (although forming would be more costly). This economy of material would increase with increasing distance between columns.

CHAPTER 19

RETAINING WALLS

19.1 FUNCTION AND TYPES OF RETAINING WALLS

Retaining walls are used to hold back masses of earth or other loose material where conditions make it impossible to let those masses assume their natural slopes. Such conditions occur when the width of an excavation, cut, or embankment is restricted by conditions of ownership, use of the structure, or economy. For example, in railway or highway construction the width of the right of way is fixed, and the cut or embankment must be contained within that width. Similarly, the basement walls of buildings must be located within the property and must retain the soil surrounding the basement.

Free-standing retaining walls, as distinct from those which form parts of structures, such as basement walls, are of various types, the most common of which are shown in Fig. 19.1. The gravity wall (Fig. 19.1*a*) retains the earth entirely by its own weight. The reinforced concrete cantilever wall (Fig. 19.1*b*) consists of the vertical arm that retains the earth and is held in position by the footing or base slab. In this case, the weight of the fill on top of the heel, in addition to the weight of the wall, contributes to the stability of the structure. Since the arm represents a vertical cantilever, its required thickness increases rapidly with increasing height. To reduce the bending moments in vertical walls of great height, counterforts are used spaced at distances from each other equal to or slightly larger than one-half of the height (Fig. 19.1*c*). Property rights or other restrictions sometimes make it necessary to place the wall at the forward edge of the base slab, i.e., to omit the toe. Whenever it is possible, toe extensions of one-third to one-fourth of the width of the base provide a more economical solution.

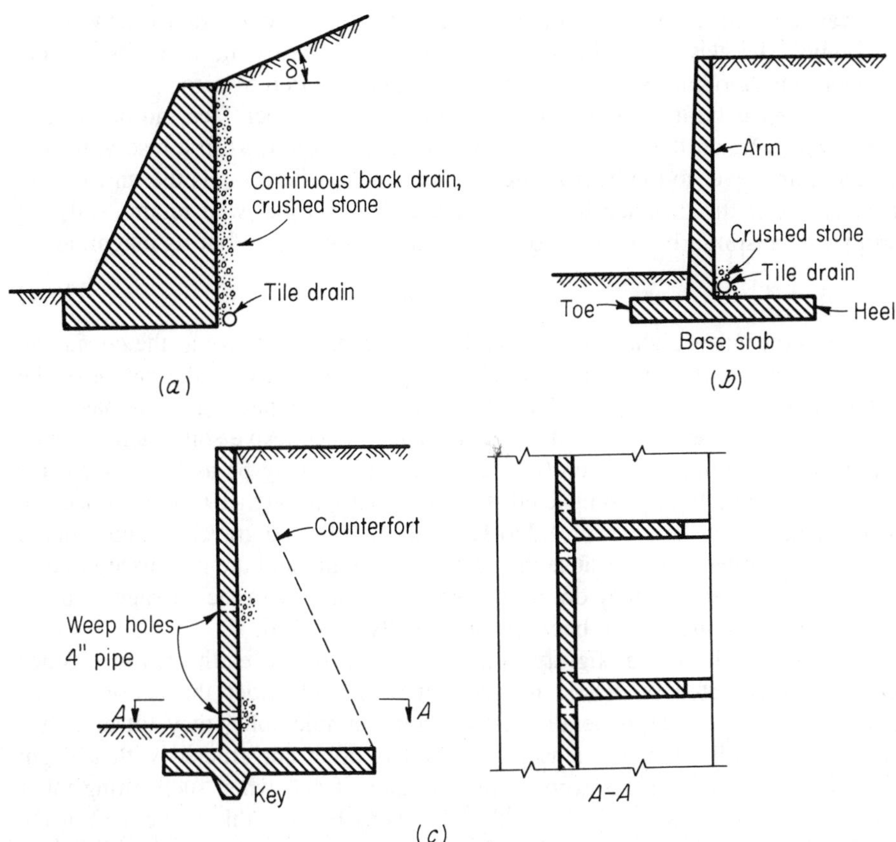

FIGURE 19.1
Types of retaining walls and backdrains: (*a*) gravity wall; (*b*) cantilever wall; (*c*) counterfort wall.

Which of the three types of walls is appropriate in a given case depends on a variety of conditions, such as local availability and price of construction materials and property rights. In general, gravity walls are economical only for relatively low walls, possibly up to about 10 ft. Cantilever walls are economical for heights from 10 to 20 ft, while counterforts are used for greater heights.

19.2 EARTH PRESSURE

In their physical behavior soils and other granular masses occupy a position intermediate between liquids and solids. If sand is poured from a dump truck, it flows, but, unlike a frictionless liquid, it will not assume a horizontal surface. It maintains itself in a stable heap with sides subtending with the horizontal an *angle of repose* whose tangent is roughly equal to the coefficient of intergranular friction. If a pit is dug in clay soil, its sides can usually be made vertical over considerable depths without support; i.e., the clay will behave like a solid and will retain the shape it is given. If, however, the pit is flooded, the sides will give way, and, in many cases, the saturated clay will be converted nearly into a true liquid. The clay is capable of maintaining its shape by means of its internal cohesion, but flooding reduces that cohesion greatly, often to zero.

If a wall is built in contact with a solid, such as a rock face, no pressure is exerted on it. If, on the other hand, a wall retains a liquid, as in a reservoir, it is subject at any level to the hydrostatic pressure $w_w h$, where w_w is the unit weight of water and h the distance from the surface. If a vertical wall retains soil, the earth pressure similarly increases proportionally to the depth, but its magnitude is

$$p_h = C_0 w h \tag{19.1}$$

where w is the unit weight of the soil and C_0 is a constant known as the *coefficient of earth pressure at rest*. The value of C_0 depends not only on the nature of the backfill but also on the method of depositing and compacting it. It has been determined experimentally that, for uncompacted noncohesive soils such as sands and gravels, C_0 ranges between 0.4 and 0.5, while it may be as high as 0.8 for the same soils in a highly compacted state (Refs 19.1 through 19.3). For cohesive soils, C_0 may be of the order of 0.7 to 1.0. Clean sands and gravels are considered superior to all other soils because they are free-draining and are not susceptible to frost action and because they do not become less stable with the passage of time. For this reason, noncohesive backfills are usually specified.

Usually, walls move slightly under the action of the earth pressure. Since walls are constructed of elastic material, they deflect under the action of the pressure, and since they generally rest on compressible soils, they tilt and shift away from the fill. (For this reason, the wall is often constructed with a slight batter toward the fill on the exposed face so that, if and when such tilting takes place, it does not appear evident to the observer.) Even if this movement at the top of the wall is only of the order of a fraction of a percent of the wall height ($\frac{1}{2}$ to $\frac{1}{10}$ percent according to Ref. 19.2), the rest pressure is materially decreased by it.

If the wall moves away from the fill, a sliding plane ab (Fig. 19.2) forms in the soil mass, and the wedge abc, sliding along that plane, exerts pressure against the wall. Here the angle ϕ is known as the *angle of internal friction*; i.e., its tangent is equal to the coefficient of intergranular friction, which can be determined by appropriate laboratory tests. The corresponding pressure is known as the *active earth pressure*. If, on the other hand, the wall is pushed against the fill, a sliding plane ad is formed, and the wedge acd is pushed upward by the

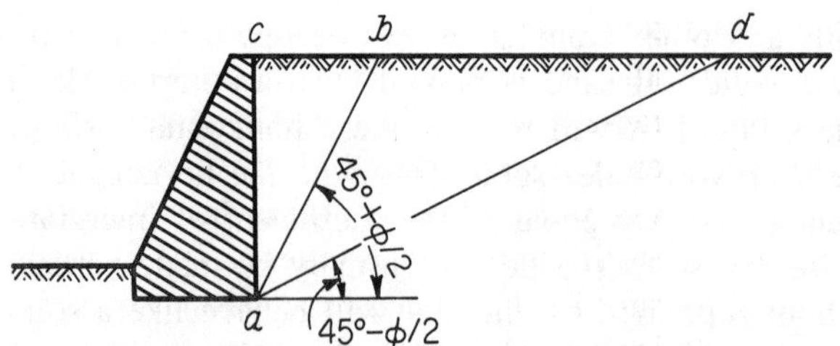

FIGURE 19.2
Basis of active and passive earth pressure determination.

wall along that plane. The pressure that this larger wedge exerts against the wall is known as the *passive earth pressure*. (This latter case will also occur at the left face of the gravity wall of Fig. 19.1 when this wall yields slightly to the left under the pressure of the fill.)

The magnitude of these pressures has been analyzed by Rankine, Coulomb, and others. If the soil surface subtends an angle δ with the horizontal (Fig. 19.1*a*), then, according to Rankine, the *coefficient for active earth pressure* is

$$C_a = \cos\delta \frac{\cos\delta - \sqrt{\cos^2\delta - \cos^2\phi}}{\cos\delta + \sqrt{\cos^2\delta - \cos^2\phi}} \tag{19.2}$$

and the *coefficient for passive pressure* is

$$C_p = \cos\delta \frac{\cos\delta + \sqrt{\cos^2\delta - \cos^2\phi}}{\cos\delta - \sqrt{\cos^2\delta - \cos^2\phi}} \tag{19.3}$$

For the frequent case of horizontal surface, that is, $\delta = 0$ (Fig. 19.2), for active pressure,

$$C_{ah} = \frac{1 - \sin\phi}{1 + \sin\phi} \tag{19.4}$$

and for passive pressure,

$$C_{ph} = \frac{1 + \sin\phi}{1 - \sin\phi} \tag{19.5}$$

Rankine's theory is valid only for noncohesive soils such as sand and gravel but, with corresponding adjustments, can also be used successfully for cohesive clay soils.

From Eqs. (19.1) to (19.5) it is seen that the earth pressure at a given depth h depends on the inclination of the surface δ, the unit weight w, and the angle of friction ϕ. The first two of these are easily determined, while little agreement has yet been reached as to the proper values of ϕ. For the ideal case of a dry, noncohesive fill, ϕ could be determined by laboratory tests and then used in the formulas. This is impossible for clays, only part of whose resistance is furnished by intergranular friction while the rest is due to internal cohesion. For this reason, their actual ϕ values are often increased by an arbitrary amount to account implicitly for the added cohesion. However, this is often unsafe since, as was shown by the example of the flooded pit, cohesion may vanish almost completely due to saturation and inundation.

In addition, fills behind retaining walls are rarely uniform, and, what is more important, they are rarely dry. Proper drainage of the fill is vitally important to reduce pressures (see Sec. 19.6), but even in a well-drained fill the pressure will temporarily increase during heavy storms or sudden thaws. This is due to the fact that even though the drainage may successfully remove the water as fast as it appears, its movement through the fill toward the drains causes additional pressure (seepage pressure). In addition, frost action and other influences may temporarily

Table 19.1 Unit weights, effective angles of internal friction ϕ, and coefficients of friction with concrete f

Soil	Unit weight, pcf	ϕ, degrees	f
1. Sand or gravel without fine particles, highly permeable	110–120	33–40	0.5–0.6
2. Sand or gravel with silt mixture, low permeability	120–130	25–35	0.4–0.5
3. Silty sand, sand and gravel with high clay content	110–120	23–30	0.3–0.4
4. Medium or stiff clay	100–120	25–35[a]	0.2–0.4
5. Soft clay, silt	90–110	20–25[a]	0.2–0.3

[a] For saturated conditions ϕ for clays and silts may be close to zero.

increase its value over that of the theoretical active pressure. Many walls that were designed without regard to these factors have failed, been displaced, or cracked.

It is good practice, therefore, to select conservative values for ϕ, considerably smaller than the actual test values, in all cases except where extraordinary and usually expensive precautions are taken to keep the fill dry under all conditions. An example of recommended earth-pressure values, which are quite conservative, though based on extensive research and practical experience, can be found in Ref. 19.2. Less conservative values are often used in practical designs, but these should be employed with caution in view of the fact that occasional trouble has been encountered with walls so designed.

Table 19.1 gives representative values for w and ϕ often used in engineering practice. (Note that the ϕ values take no account of probable additional pressures due to porewater, seepage, frost, etc.)The table also contains values for the coefficient of friction f between concrete and various soils. The values of ϕ for soils 3 through 5 may be quite unconservative; under saturated conditions, clays and silts may become entirely liquid (that is, $\phi = 0$). Soils of type 1 or 2 should be used for backfills of retaining walls wherever possible.

19.3 EARTH PRESSURE FOR COMMON CONDITIONS OF LOADING

In computing earth pressures on walls, three common conditions of loading are most often met: (1) horizontal surface of fill at the top of the wall, (2) inclined surface of fill sloping up and back from the top of the wall, and (3) horizontal surface of fill carrying a uniformly distributed additional load (surcharge) such as from goods in a storage yard or traffic on a road.

The increase in pressure caused by uniform surcharge s (case 3) is computed by converting its load into an equivalent, imaginary height of earth h' above the top of the wall such that

$$h' = \frac{s}{w} \tag{19.6}$$

and measuring the depth to a given point on the wall from this imaginary surface. This amounts to replacing h with $(h + h')$ in Eq. (19.1).

The distributions of pressure for cases 1 to 3 are shown in Fig. 19.3. The total earth thrust P per linear foot of wall is evidently equal to the area under the pressure distribution figure, and its line of action passes through the centroid of the pressure. Figure 19.3 gives information, computed in this manner, on magnitude, point of action, and direction of P for these three cases.

Occasionally retaining walls must be built for conditions in which the groundwater level is above the base of the wall, either permanently or seasonally. In that case the pressure of the soil *above* groundwater is determined as usual. The part of the wall *below* groundwater is subject to the sum of the water pressure and the earth pressure. The former is equal to the full hydrostatic pressure $p_w = w_w h_w$, where w_w and h_w are, respectively, the unit weight of water and the distance from the groundwater level to the point on the wall. The additional pressure of the soil below the groundwater level is computed from Eq. (19.1), where, however, for the portion of the soil below water, w is replaced with $w - w_w$, while h, as usual, is measured from the soil surface. That is, for submerged soil, buoyancy reduces the effective weight in the indicated manner. Pressures of this magnitude, which are considerably larger than those of drained soil, will also occur temporarily after heavy rainstorms or thaws in walls without provision for drainage, or if drains have become clogged.

The seeming simplicity of the determination of earth pressure as here indicated should not lull the designer into a false sense of security and certainty. No theory is more accurate than the assumptions on which it is based. Actual soil pressures are affected by irregularities of soil properties, porewater and drainage conditions, and climatic and other factors that cannot be expressed in formulas. This situation, on the one hand, indicates that involved refinements of theoretical earth pressure determinations, as sometimes attempted, are of little practical value.

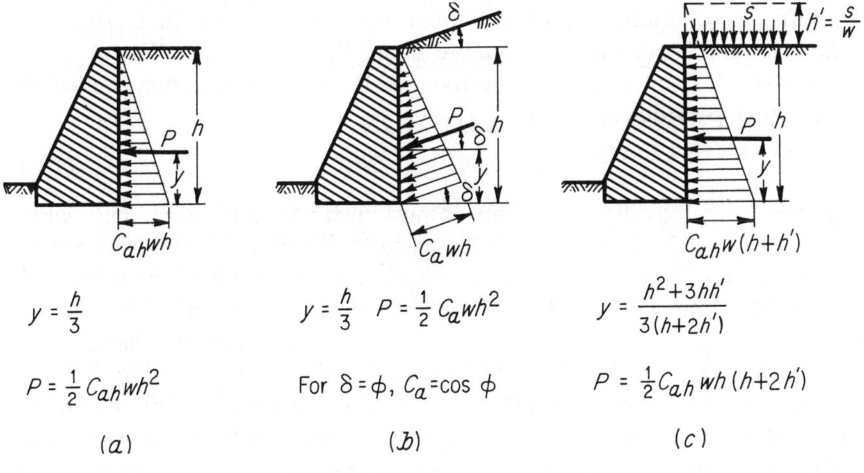

FIGURE 19.3
Earth pressures for (*a*) horizontal surface; (*b*) sloping surface; (*c*) horizontal surface with surcharge *s*.

On the other hand, the design of a retaining wall is seldom a routine procedure, since the local conditions that affect pressures and safety vary from one locality to another.

19.4 EXTERNAL STABILITY

A wall may fail in two different ways: (1) its individual parts may not be strong enough to resist the acting forces, such as when a vertical cantilever wall is cracked by the earth pressure acting on it, and (2) the wall as a whole may be bodily displaced by the earth pressure, without breaking up internally. To design against the first possibility requires the determination of the necessary dimensions, thicknesses, and reinforcement to resist the moments and shears; this procedure, then, is in no way different from that of determining required dimensions and reinforcement of other types of concrete structures. The usual load factors and strength reduction factors of the ACI Code may be applied (see Sec. 19.5).

To safeguard the wall against bodily displacements, i.e., to ensure its external stability, requires special consideration. Consistent with current practice in geotechnical engineering, the stability investigation is based on actual earth pressures (as nearly as they may be determined) and on computed or estimated service dead and live loads, all without load factors. Computed bearing pressures are compared against allowable values, and overall factors of safety evaluated by comparing resisting forces to maximum loads acting under service conditions.

A wall, such as that of Fig. 19.4, together with the soil mass *ijkl* that rests on the base slab, may be bodily displaced by the earth thrust P that acts on the plane *ak* by *sliding* along the plane *ab*. Such sliding is resisted by the friction between the soil and footing along the same plane. To forestall motion, the forces that resist sliding must exceed those that tend to produce sliding; a factor of safety of 1.5 is generally assumed satisfactory in this connection.

In Fig. 19.4, the force that tends to produce sliding is the horizontal component P_h of the total earth thrust P. The resisting friction force is fR_v, where f is the coefficient of friction between the concrete and soil (see Table 19.1) and R_v is the vertical component of the total resultant R; that is, $R_v = W + P_v$ (W = weight of wall plus soil resting on the footing, P_v = vertical component of P). Hence, to provide sufficient safety,

$$f(W + P_v) \geq 1.5P_h \tag{19.7}$$

Actually, for the wall to slide to the left, it must push with it the earth *nmb*, which gives rise to the passive earth pressure indicated by the triangle *rmb*. This passive pressure represents a further resisting force that could be added to the left side of Eq. (19.7). However, this should be done only if the proper functioning of this added resistance is ensured. For that purpose the fill *ghmv* must be placed before the backfill *ijkl* is put in place and must be secure against later removal by scour or other means throughout the lifetime of the wall. If these conditions are not met, it is better not to count on the additional resistance of the passive pressure.

If the required sliding resistance cannot be developed by these means, a key wall *cdef* can be used to increase horizontal resistance. In this case sliding, if it

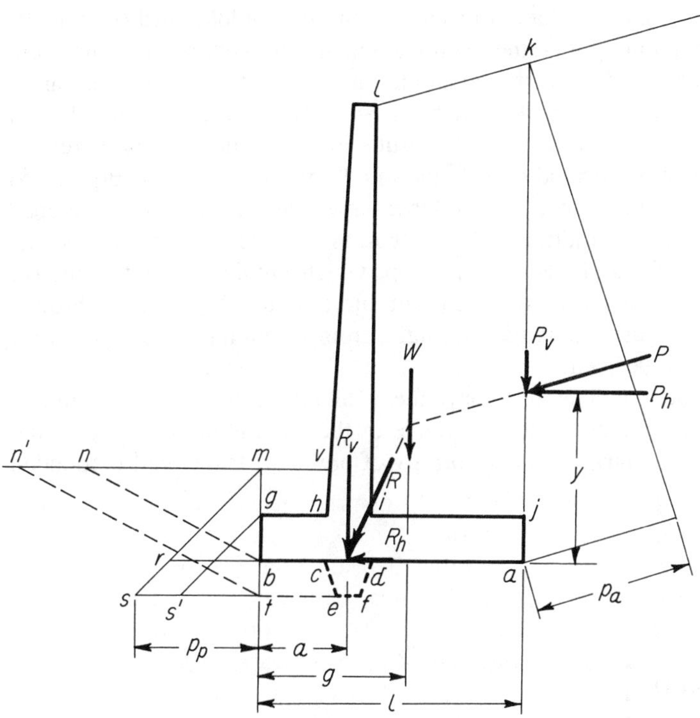

FIGURE 19.4
External stability of a cantilever wall.

occurs, takes place along the planes *ad* and *tf*. While along *ad* and *ef* the friction coefficient *f* applies, sliding along *te* occurs within the soil mass. The coefficient of friction that applies in this portion is consequently tan ϕ, where the value of ϕ may be taken from the next to last column in Table 19.1. In this situation sliding of the front soil occurs upward along tn' so that, if the front fill is secure, the corresponding resistance from passive soil pressure is represented by the pressure triangle *stm*. If doubt exists as to the reliability of the fill above the toe, the free surface should more conservatively be assumed at the top level of the footing, in which case the passive pressure is represented by the triangle $s'tg$.

Next, it is necessary to ensure that the pressure under the footing does not exceed the *permissible bearing pressure* for the particular soil. Let *a* (Fig. 19.4) be the distance from the front edge *b* to the intersection of the resultant with the base plane, and let R_v be the vertical component of *R*. (This intersection need not be located beneath the vertical arm, as shown, even though an economical wall generally results if it is so located.) Then the base plane *ab*, 1 ft wide longitudinally, is subject to a normal force R_v and to a moment about the centroid $(l/2 - a)R_v$. When these values are substituted in the usual formula for bending plus axial force

$$q_{\substack{max \\ min}} = \frac{N}{A} \pm \frac{Mc}{I} \tag{19.8}$$

it will be found that if the resultant is located within the middle third ($a > l/3$), compression will act throughout the section, and the maximum and minimum pressures can be computed from the equations in Fig. 19.5*a*. If the resultant is located just at the edge of the middle third ($a = l/3$), the pressure distribution is as shown in Fig. 19.5*b*, and Eq. (19.8) results in the formula given there.

If the resultant were located outside the middle third ($a < l/3$), Eq. (19.8) would indicate tension at and near point a. Obviously, tension cannot be developed between soil and a concrete footing that merely rests on it. Hence, in this case the pressure distribution of Fig. 19.5*c* will develop, which implies a slight lifting off the soil of the rear part of the footing. Equilibrium requires that R_v pass through the centroid of the pressure distribution triangle, from which the formula for q for this case can easily be derived.

It is good practice, in general, to have the resultant located within the middle third. This will not only reduce the magnitude of the maximum bearing pressure but will also prevent too large a nonuniformity of pressure. If the wall is founded

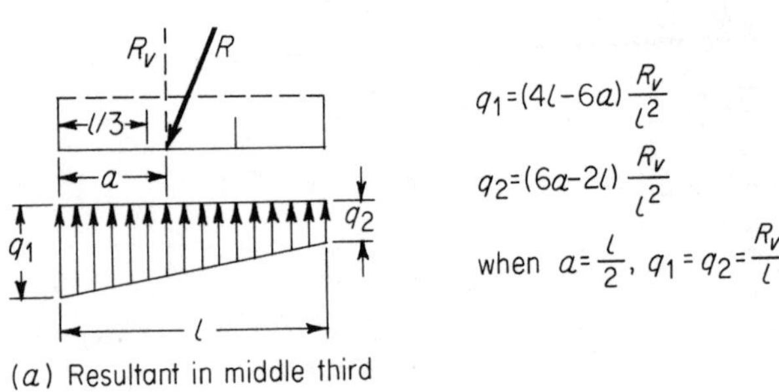

(*a*) Resultant in middle third

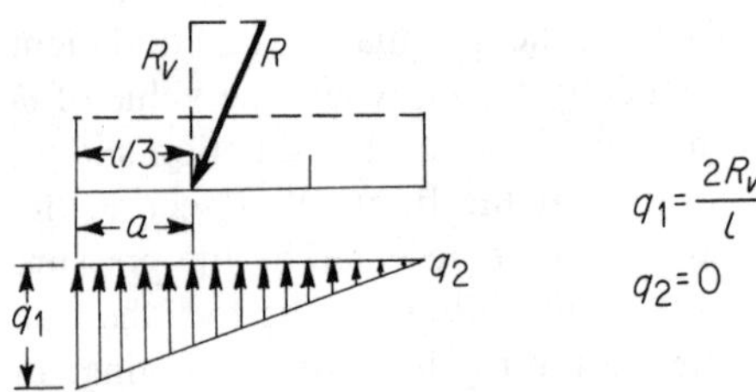

(*b*) Resultant at edge of middle third

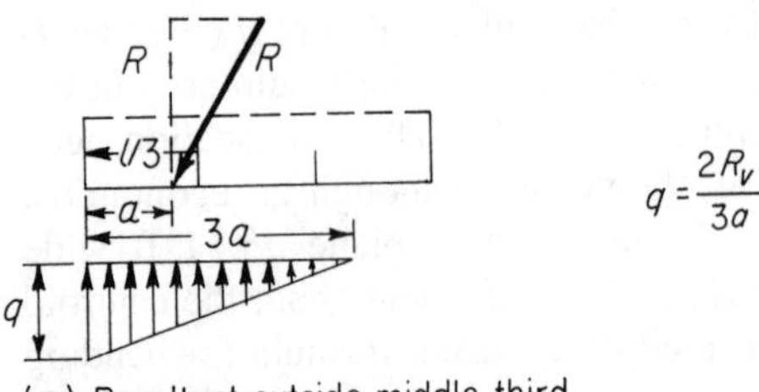

(*c*) Resultant outside middle third

FIGURE 19.5
Bearing pressures for different locations of resultant.

on a highly compressible soil, such as certain clays, a pressure distribution as in Fig. 19.5*b* would result in a much larger settlement of the toe than of the heel, with a corresponding tilting of the wall. In a foundation on such a soil the resultant, therefore, should strike at or very near the center of the footing. If the foundation is on very incompressible soil, such as well-compacted gravel or rock, the resultant can be allowed to fall outside the middle third (Fig. 19.5*c*).

A third mode of failure is the possibility of the wall *overturning* bodily around the front edge *b*. For this to occur, the overturning moment yP_h about point *b* would have to be larger than the restoring moment $(Wg + P_v l)$ in Fig. 19.4, which is the same as saying that the resultant would have to strike outside the edge *b*. If, as is mostly the case, the resultant strikes within the middle third, adequate safety against overturning exists, and no special check need be made. If the resultant is located outside the middle third, a factor of safety of at least 1.5 should be maintained against overturning; i.e., the restoring moment should be at least 1.5 times the overturning moment.

19.5 BASIS OF STRUCTURAL DESIGN

In the investigation of a retaining wall for external stability, described in Sec. 19.4, it is the current practice to base the calculations on actual earth pressures, and on computed or estimated service dead and live loads, all with load factors of 1.0 (i.e., without load increase to account for a hypothetical overload condition). Computed soil bearing pressures, for service load conditions, are compared with allowable values set suitably lower than ultimate bearing values. Factors of safety against overturning and sliding are established, based on service load conditions.

On the other hand, the structural design of a retaining wall should be consistent with methods used for all other types of members, and thus should be based on factored loads in recognition of the possibility of an increase above service loading. ACI Code load factors relating to structural design of retaining walls are summarized as follows:

1. If resistance to lateral earth pressure *H* is included in the design, together with dead loads *D* and live loads *L*, the required strength *U* shall be at least equal to

$$U = 1.4D + 1.7L + 1.7H$$

2. Where *D* or *L* reduce the effect of *H*, the required strength *U* shall be at least equal to

$$U = 0.9D + 1.7H$$

3. For any combination of *D*, *L*, and *H*, the required strength shall not be less than

$$U = 1.4D + 1.7L$$

While the ACI Code approach to load factor design is logical and relatively easy to apply to members in buildings, its application to structures that are to resist earth pressures is not so easy. Many alternative combinations of factored dead and live loads and lateral pressures are possible. Dead loads such as the weight of the concrete should be multiplied by 0.9 where they reduce design moments, such as for the toe slab of a cantilevered retaining wall, but should be multiplied by 1.4 where they increase moments, such as for the heel slab. The dead load of the earth over the heel should be multiplied by 1.4, according to the ACI Code, but to do so would not be consistent with specifying the lateral pressure of $1.7H$, which depends, at least in part, on the lateral thrust from the earth over the heel slab. Obviously, no two factored load states could be obtained concurrently. For each combination of factored loads, different reactive soil pressures will be produced under the structure, requiring a new determination of those pressures for each alternative combination. Furthermore, there is no reason to believe that soil pressure would continue to be linearly distributed at the overload stage, or would increase in direct proportion to the load increase; information on soil pressure distributions at incipient failure is incomplete.

Necessarily, a somewhat simplified view of load factor design must be adopted in designing retaining walls. The following procedure appears to be consistent with the intent of the ACI Code. Where errors or inconsistencies are introduced, results will be on the safe side.

Lateral earth pressures will be considered to be live loads, and a factor of 1.7 applied. In general, the reactive pressure of the soil under the structure at the factored load stage will be taken equal to 1.7 times the soil pressure found for service load conditions in the stability analysis.† For cantilever retaining walls, the calculated dead load of the toe slab, which causes moments acting in the opposite sense to those produced by the upward soil reaction, will be multiplied by a factor of 0.9. For the heel slab, the required moment capacity will be based on the dead load of the heel slab itself, plus the earth directly above it, both multiplied by 1.4. Surcharge, if present, will be treated as live load with a load factor of 1.7. The upward pressure of the soil under the heel slab will be taken equal to zero, recognizing that for the severe overload stage a nonlinear pressure distribution will probably be obtained, with most of the reaction concentrated near the toe. Similar assumptions appear to be reasonable in designing counterfort walls.

19.6 DRAINAGE AND OTHER DETAILS

Such failures or damage to retaining walls as have occasionally occurred were due, in most cases, to one of two causes: overloading of the soil under the wall with consequent forward tipping or insufficient drainage of the backfill. In the latter case, hydrostatic pressure from porewater accumulated during or after rainstorms greatly increases the thrust on the wall; in addition, in subfreezing weather, ice pressure of considerable magnitude can develop in such poorly drained soils. The

† These reactions are caused by the assumed factored load condition and have no direct relationship to ultimate soil bearing values or pressure distributions.

two causes are often interconnected, since large thrusts correspondingly increase the bearing pressure under the footing.

Allowable bearing pressures should be selected with great care. It is necessary, for this purpose, to investigate not only the type of soil immediately underlying the footing, but also the deeper layers. Unless reliable information is available at the site, subsurface borings should be made to a depth at least equal to the height of the wall. The foundation must be laid below *frost depth*, which amounts to 4 to 5 ft and more in the northern states, in order to ensure against heaving by the freezing of soils containing moisture.

Drainage can be provided in various ways. *Weep holes* consisting of 6 or 8 in. pipe embedded in the wall, as shown in Fig. 19.1*c*, are usually spaced horizontally at 5 to 10 ft. In addition to the bottom row, additional rows should be provided in walls of substantial height. To facilitate drainage and prevent clogging, 1 ft^3 or more of crushed stone is placed at the rear end of each weeper. Care must be taken that the outflow from the weep holes is carried off safely so as not to seep into and soften the soil underneath the wall. To prevent this, instead of weepers, *longitudinal drains* embedded in crushed stone or gravel can be provided along the rear face of the wall (Fig. 19.1*b*) at one or more levels; the drains discharge at the ends of the wall or at a few intermediate points. The most efficient drainage is provided by a *continuous back drain* consisting of a layer of gravel or crushed stone covering the entire rear face of the wall (Fig. 19.1*a*), with discharge at the ends. Such drainage is expensive, however, unless appropriate material is cheaply available at the site. Wherever possible the surface of the fill should be covered with a layer of low permeability and, in the case of a horizontal surface, should be laid with a slight slope away from the wall toward a gutter or other drainage.

In long walls provision must be made against damage caused by *expansion* or *contraction* from temperate changes and shrinkage. The AASHTO specification requires that for gravity as well as reinforced concrete walls expansion joints be made at intervals of 90 ft or less, and contraction joints at not more than 30 ft (Ref. 19.4). The same specifications provide that in reinforced concrete walls horizontal temperature reinforcement of not less than $\frac{1}{8}$ in^2 per foot of depth be provided adjacent to the exposed surface. Similar provisions are found in Ref. 19.5.

19.7 EXAMPLE: DESIGN OF A GRAVITY RETAINING WALL

A gravity wall is to retain a bank 11 ft 6 in. high whose horizontal surface is subject to a live load surcharge of 400 psf. The soil is a sand-and-gravel mixture with a rather moderate amount of fine, silty particles. It can, therefore, be assumed to be in class 2 of Table 19.1, with the following characteristics: unit weight $w = 120$ pcf, $\phi = 30°$ (with adequate drainage to be provided), and base friction coefficient $f = 0.5$. With sin 30° $= 0.5$, from Eqs. (19.4) and (19.5), the soil pressure coefficients are $C_{ah} = 0.333$ and $C_{ph} = 3.0$. The allowable bearing pressure is assumed to be 8000 psf. This coarse-grained soil has little compressibility, so that the resultant can be allowed to strike near the outer-third point (see Sec. 19.4). The weight of the concrete is $w_c = 150$ pcf.

The optimum design of any retaining wall is a matter of successive approximation. Reasonable dimensions are assumed based on experience, and the various conditions of stability are checked for these dimensions. On the basis of a first trial, dimensions are readjusted, and one or two additional trials usually result in a favorable design. In the following, only the final design is analyzed in detail. The final dimensions are shown in Fig. 19.6.

The equivalent height of surcharge is $h' = 400/120 = 3.33$ ft. From Fig. 19.3c the total earth thrust is

$$P = 1/2 \times 0.333 \times 120 \times 15 \times 21.67 = 6500 \text{ lb}$$

and its distance from the base is $y = (225 + 150)/(3 \times 21.67) = 5.77$ ft. Hence, the overturning moment $M_o = 6500 \times 5.77 = 37{,}500$ ft-lb. To compute the weight W and its restoring moment M_r about the edge of the toe, individual weights are taken, as shown in Fig. 19.6. With x representing the distance of the line of action of each subweight from the front edge, the following computation results:

Component weights	W, lb	x, ft	$M_r = xW$, ft-lb
$W_1: 10 \times 2 \times 150$	3,000	5.0	15,000
$W_2: 1.5 \times 13 \times 150$	2,930	1.5	4,400
$W_3: 7/2 \times 13 \times 150$	6,830	4.58	31,300
$W_4: 7/2 \times 13 \times 120$	5,460	6.92	37,800
$W_5: 0.75 \times 13 \times 120$	1,170	9.63	11,270
Total	19,390		99,770

The distance of the resultant from the front edge is

$$a = \frac{99{,}770 - 37{,}500}{19{,}390} = 3.21 \text{ ft}$$

which is just outside the middle third. The safety factor against overturning, $99{,}770/37{,}500 = 2.66$, is ample. From Fig. 19.5c the maximum soil pressure is $q = (2 \times 19{,}390)/(3 \times 3.21) = 4030$ psf.

The above computations were made for the case in which the surcharge extends only to the rear edge of the wall, point a of Fig. 19.6. If the surcharge extends forward to point b, the following modifications are obtained:

$$W = 19{,}390 + 400 \times 7.75 = 22{,}490 \text{ lb}$$

$$M_r = 99{,}770 + 400 \times 7.75 \times 6.13 = 118{,}770 \text{ ft-lb}$$

$$a = \frac{118{,}770 - 37{,}500}{22{,}490} = 3.61 \text{ ft}$$

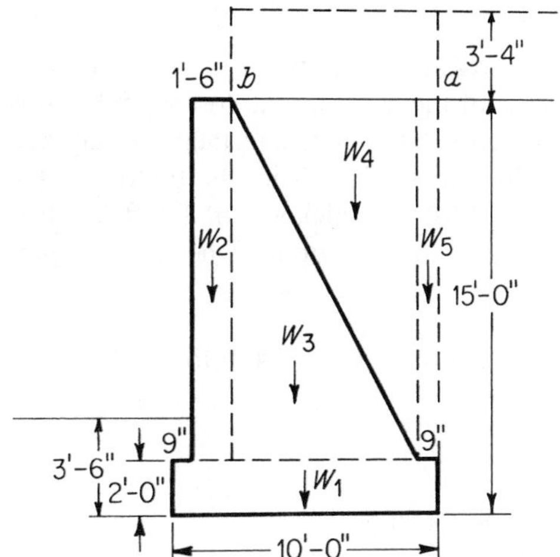

FIGURE 19.6
Gravity retaining wall.

This is inside the middle third, and, from Fig. 19.5*a*, the maximum bearing pressure is

$$q_1 = \frac{(40.0 - 21.7)\ 22{,}490}{100} = 4120 \text{ psf}$$

The situation most conducive to sliding is obtained when the surcharge extends only to point *a*, since additional surcharge between *a* and *b* would increase the total weight and the corresponding resisting friction. The friction force is

$$F = 0.5 \times 19{,}390 = 9695 \text{ lb}$$

Additionally, sliding is resisted by the passive earth pressure on the front of the wall. Although the base plane is 3.5 ft below grade, the top layer of soil cannot be relied upon to furnish passive pressure, since it is frequently loosened by roots and the like, or it could be scoured out by cloudbursts. For this reason the top 1.5 ft will be discounted in computing the passive pressure, which then becomes

$$P_p = 1/2wh^2C_{ph} = 1/2 \times 120 \times 4 \times 3.0 = 720 \text{ lb}$$

The safety factor against sliding, (9695 + 720)/6500 = 1.6, is but slightly larger than the required value 1.5, indicating a favorable design.

19.8 EXAMPLE: DESIGN OF A CANTILEVER RETAINING WALL

A cantilever wall is to be designed for the situation of the gravity wall in Sec. 19.7. A concrete with $f'_c = 3000$ psi and steel with $f_y = 60{,}000$ psi will be used.

a. Preliminary Design

To facilitate computation of weights for checking the stability of the wall, it is advantageous first to ascertain the thickness of the arm and the footing.† For this purpose the thickness of the footing is roughly estimated, and then the required thickness of the arm is determined at its bottom section. With the bottom of the footing at 3.5 ft below grade and an estimated footing thickness of 1.5 ft, the free height of the arm is 13.5 ft. Hence, with respect to the bottom of the arm (see Fig. 19.3),

$$P = 1/2 \times 0.333 \times 120 \times 13.5 \times 20.16 = 5440 \text{ lb}$$

$$y = \frac{183 + 135}{3 \times 20.16} = 5.25 \text{ ft}$$

$$M_u = 1.7 \times 5440 \times 5.25 = 48{,}600 \text{ ft-lb}$$

For the given grades of concrete and steel, the maximum permitted steel ratio $\rho_{max} = 0.0160$. For economy and ease of bar placement, a ratio of about half the maximum, or 0.008, will be used. Then from Graph A.1*b* of App. A,

$$\frac{M_u}{\phi b d^2} = 430$$

For a unit length of the wall ($b = 12$ in.), with $\phi = 0.90$ as usual for flexure, the required effective depth is

$$d = \sqrt{\frac{48{,}600 \times 12}{0.90 \times 12 \times 430}} = 11.2 \text{ in.}$$

A protective cover of 2 in. is required for concrete exposed to earth. Thus, estimating the bar diameter to be 1 in., the minimum required thickness of the arm at the base is 13.7 in. This will be increased to 16 in., because the cost of the excess concrete in such structures is usually more than balanced by the simultaneous saving in steel. The arm is then checked for shear at a distance *d* above the base, or 12.5 ft below the top of the wall:

$$P = 1/2 \times 0.333 \times 120 \times 12.5 \times 19.16 = 4900 \text{ lb}$$

$$V_u = 1.7 \times 4900 = 8330 \text{ lb}$$

$$\phi V_c = 2\phi \sqrt{f_c'} bd$$

$$= 2 \times 0.85 \sqrt{3000} \times 12 \times 13.5$$

$$= 15{,}100 \text{ lb}$$

confirming that the arm is more than adequate to resist the factored shear force.

† Valuable guidance will be provided for the inexperienced designer by tabulated designs such as those found in Ref. 19.6 and by the sample calculations of Ref. 19.7.

The thickness of the base is usually the same or slightly larger than that at the bottom of the arm. Hence, the estimated 1.5 ft need not be revised. Since the moment in the arm decreases with increasing distance from the base and is zero at the top, the arm thickness at the top will be made 8 in. It is now necessary to assume lengths of heel and toe slabs and to check the stability for these assumed dimensions. Intermediate trials are omitted here, and the final dimensions are shown in Fig. 19.7*a*. Trial computations have shown that safety against sliding can be achieved only by an excessively long heel or by a key. The latter, requiring the smaller concrete volume, has been adopted.

b. Stability Investigation

Weights and moments about the front edge are as follows:

Component weights	W, lb	x, ft	M_r, ft-lb
$W_1 : 0.67 \times 13.5 \times 150$	1,360	4.08	5,550
$W_2 : 0.67 \times 0.5 \times 13.5 \times 150$	680	4.67	3,180
$W_3 : 9.75 \times 1.5 \times 150$	2,190	4.88	10,700
$W_4 : 1.33 \times 1.25 \times 150$	250	4.42	1,100
$W_5 : 3.75 \times 2 \times 120$	900	1.88	1,690
$W_6 : 0.67 \times 0.5 \times 13.5 \times 120$	540	4.86	2,620
$W_7 : 4.67 \times 13.5 \times 120$	7,570	7.42	56,200
Total	13,490		81,040

The total soil pressure on the plane *ac* is the same as for the gravity wall of Sec. 19.7, that is, $P = 6500$ lb, and the overturning moment is

$$M_o = 37{,}500 \text{ ft-lb}$$

The distance of the resultant from the front edge is

$$a = \frac{81{,}040 - 37{,}500}{13{,}490} = 3.23 \text{ ft}$$

which locates the resultant barely outside the middle third. The corresponding maximum soil pressure at the toe, from Fig. 19.5, is

$$q = 2 \times \frac{13{,}490}{3 \times 3.23} = 2780 \text{ psf}$$

The factor of safety against overturning, 81,040/37,500 = 2.16, is ample.

To check the safety against sliding, remember (Sec. 19.4) that if sliding occurs it proceeds between concrete and soil along the heel and key (i.e., length *ae* in Fig. 19.4), but takes place within the soil in front of the key (i.e., along length *te* in Fig. 19.4). Consequently, the coefficient of friction that applies for

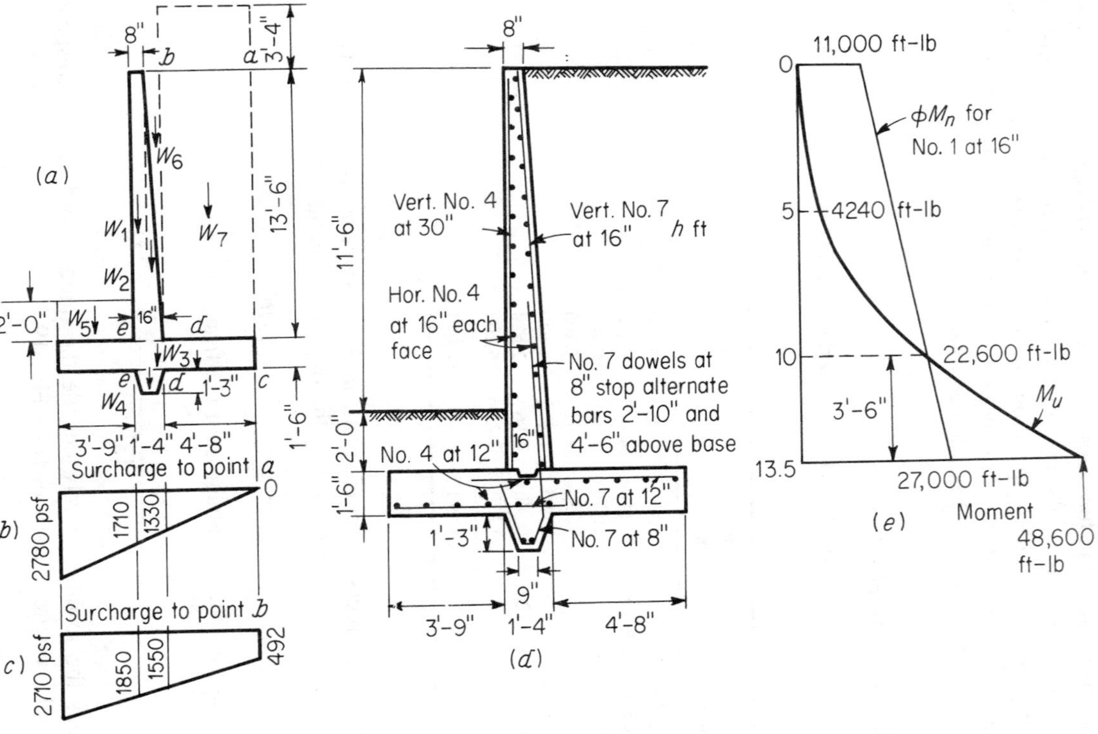

FIGURE 19.7
Cantilever retaining wall: (*a*) cross section; (*b*) bearing pressure with surcharge to *a*; (*c*) bearing pressure with surcharge to *b*; (*d*) reinforcement; (*e*) moment variation with height.

the former length is $f = 0.5$, while for the latter it is equal to the internal soil friction, i.e., tan 30° = 0.577.

The bearing pressure distribution is shown in Fig. 19.7*b*. Since the resultant is at a distance $a = 3.23$ ft from the front, i.e., nearly at the middle third, it is assumed that the bearing pressure becomes zero exactly at the edge of the heel, as shown in Fig. 19.7*b*.

The resisting force is then computed as the sum of the friction forces of the rear and front portion, plus the passive soil pressure in front of the wall. For the latter, as in Sec. 19.7, the top 1.5 ft layer of soil will be discounted as unreliable. Hence,

Friction, toe:	$(2780 + 1710) \times 0.5 \times 3.75 \times 0.577 = 4860$ lb
Friction, heel and key:	$1710 \times 0.5 \times 6 \times 0.5 = 2570$ lb
Passive earth pressure:	$0.5 \times 120 \times 3.25^2 \times 3.0 = 1910$ lb
Total resistance to sliding:	$= 9340$ lb

The factor of safety against sliding, 9340/6500 = 1.44, is only 4 percent below the recommended value of 1.5 and can be regarded as adequate.

The computations hold for the case in which the surcharge extends from the right to point *a* above the edge of the heel. The other case of load distribution, in which the surcharge is placed over the entire surface of the fill up to point *b*, evidently does not change the earth pressure on the plane *ac*. It does, however, add to the sum of the vertical forces and increases both the restoring moment M_r and the friction along the base. Consequently, the danger of sliding or overturning is greater when the surcharge extends only to *a*, for which situation these two cases have been checked and found adequate. In view of the added vertical load, however, the bearing pressure is largest when the surface is loaded to *b*. For this case,

$$W = 13{,}490 + 400 \times 5.33 = 15{,}600 \text{ lb}$$

$$M_r = 81{,}040 + 400 \times 5.33 \times 7.09 = 96{,}200 \text{ ft-lb}$$

$$a = \frac{96{,}200 - 37{,}500}{15{,}600} = 3.76 \text{ ft}$$

which places the resultant inside the middle third. Hence, from Fig. 19.5,

$$q_1 = (39.0 - 22.5)\frac{15{,}600}{9.75^2} = 2710 \text{ psf}$$

$$q_2 = (22.5 - 19.5)\frac{15{,}600}{9.75^2} = 492 \text{ psf}$$

which is far below the allowable pressure of 8000 psf. The corresponding bearing pressure distribution is shown in Fig. 19.7*c*.

The external stability of the wall has now been ascertained, and it remains to determine the required reinforcement and to check internal resistances.

c. Arm and Key

The moment at the bottom section of the arm has previously been determined as $M_u = 48{,}600$ ft-lb, and a wall thickness of 16 in. at the bottom and 8 in. at the top has been selected. With a concrete cover of 2 in. clear, $d = 16.0 - 2.0 - 0.5 = 13.5$ in. Then

$$\frac{M_u}{\phi b d^2} = \frac{48{,}600 \times 12}{0.9 \times 12 \times 183} = 295$$

From Graph A.1*b* of App. A, with $f_y = 60{,}000$ psi and $f_c' = 3000$ psi, the required steel ratio $\rho = 0.0052$ and $A_s = 0.0052 \times 12 \times 13.5 = 0.84$ in^2/ft. The required area is provided by No. 7 bars at 8 in. center to center.

The bending moment in the arm decreases rapidly with increasing distance from the bottom. For this reason only part of the main reinforcement is needed at higher elevations, and alternate bars will be discontinued where no longer needed. To determine the cutoff point, the moment diagram for the arm has been drawn by computing bending moments at two intermediate levels, 10 ft and 5 ft from the top. These two moments, determined in the same manner as that at the base of the arm, were found to be 22,600 and 4240 ft-lb respectively. The resisting moment provided by alternate bars, i.e., by No. 7 bars at 16 in. center to center, at the bottom of the arm is

$$\phi M_n = \frac{0.90 \times 0.46 \times 60{,}000}{12}(13.50 - 0.45) = 27{,}000 \text{ ft-lb}$$

At the top, $d = 8.0 - 2.5 = 5.5$ in., and the resisting moment of the same bars is only $\phi M_n = 27{,}000(5.5/13.5) = 11{,}000$ ft-lb. Hence, the straight line drawn in Fig. 19.7*e* indicates the resisting moment provided at any elevation by half the number of main bars. The intersection of this line with the moment diagram at a distance of 3 ft 6 in. from the bottom represent the point above which alternate bars are no longer needed. ACI Code 12.10.3 specifies that any bar shall be extended beyond the point at which it is no longer needed to carry flexural stress for a distance equal to d or 12 bar diameters, whichever is greater. In the arm, at a distance of 3 ft 6 in. from the bottom, $d = 11.4$ in., while 12 bar diameters for No. 7 bars are equal to 10.5 in. Hence, half the bars can be discontinued 12 in. above the point where no longer needed, or a distance of 4 ft 6 in. above the base. This exceeds the minimum required development length of 26 in. above the base.

In order to facilitate construction, the footing is poured first, and a construction joint is provided at the base of the arm, as shown in Fig. 19.7*d*. The main bars of the arm, therefore, end at the top of the base slab, and dowels are placed in the latter to be spliced with them. It will be realized that the integrity of the arm depends entirely on the integrity of the splicing of these tension bars. Splicing all tension bars in one section by simple contact splices can easily lead to splitting of the concrete owing to the stress concentrations at the ends of the spliced bars. One way to avoid this difficulty is to weld all splices; this will entail considerable extra cost.

In this particular wall, another way of placing the reinforcing offers a more economical solution. Because alternate bars in the arm can be discontinued at a distance of 4 ft 6 in. above the base, alternate dowels will be carried up from the top of the base to that distance. These need not be spliced at all, because above that level only alternate No. 7 bars, 16 in. on centers, are needed. These latter bars are placed full length over the entire height of the arm and are spliced at the bottom with alternate shorter dowels. By this means, only 50 percent of the bars needed at the bottom of the arm are spliced; this is not objectionable.

For splices of deformed bars in tension, at sections where the ratio of steel provided to steel required is less than 2 and where no more than 50 percent of the steel is spliced, the ACI Code requires a Class B splice of length equal to 1.3 times the development length of the bar (see Sec. 5.11*a*). The development length of the No. 7 bars for given material strengths is 26 in., and so the required splice length is $1.3 \times 26 = 33.8$ in. Alternate dowels will be carried 2 ft 10 in. above the base for splicing with the corresponding bars in the arm, while the intervening dowels are carried to 4 ft 6 in. above the base and terminated without splicing as being no longer needed to carry moments.

According to the ACI Code, main flexural reinforcement is not to be terminated in a tension zone unless one of three conditions is satisfied: (1) shear at the cutoff point does not exceed two-thirds that permitted, (2) certain excess shear reinforcement is provided, or (3) the continuing reinforcement provides double the area required for flexure at the cutoff point. It is easily confirmed that the shear 4 ft 6 in. above the base is well below two-thirds the value that can be carried by the concrete; thus main bars can be terminated as planned. The minimum tensile steel ratio $200/f_y$ specified by the ACI Code, strictly speaking, does not apply to retaining walls, which may be considered to be slabs. However, because the integrity of the retaining wall depends absolutely on the vertical bars, it appears prudent to use this limit in such cases. The actual steel ratio provided by the No. 7 bars at 16 in. spacing, with $d = 10.8$ in. just above the cutoff point, is 0.0035, just above the minimum value of $200/60{,}000 = 0.0033$. A final ACI Code requirement is that the maximum spacing of the primary flexural reinforcement exceed neither 3 times the wall thickness nor 18 in.; these restrictions are satisfied as well.

Since the dowels had to be extended at least partly into the key to produce the necessary length of embedment, they were bent as shown to provide reinforcement for the key. The exact force which the key must resist is difficult to determine, since probably the major part of the force acting on the portion of the soil in front of the key is transmitted to it through friction along the base of the footing. The relatively strong reinforcement of the key by means of the extended dowels is considered sufficient to prevent separation from the footing.

The sloping sides of the key were provided in order to facilitate excavation without loosening the adjacent soil. This is necessary in order to ensure proper functioning of the key.

In addition to the main steel in the stem, reinforcement is needed to control shrinkage and temperature cracking. Calculations will be based on the average wall thickness of 12 in. In the vertical direction, not less than 0.0012 times the gross concrete area must be provided, with not less than one-half that amount adjacent to the exposed face. This requirement is met using No. 4 bars 30 in. on centers. In the horizontal direction, the required steel area is 0.0020 times the gross concrete area, again with not less than one-half at the outside face. No. 4 bars 16 in. on centers, each face, will be used as shown in Fig. 19.7*d*.

d. Toe Slab

The toe slab acts as a cantilever projecting outward from the face of the stem. It must resist the upward pressures of Fig. 19.7*b* or *c* and the downward load of the toe slab itself, each multiplied by appropriate load factors. The downward load of the earth fill over the toe will be neglected because it is subject to possible erosion or removal. A load factor of 1.7 will be applied to the service load bearing pressures. Comparison of the pressures of Fig. 19.7*b* and *c* indicates that for the toe slab, the more severe loading case results from surcharge to *b*. Because the self-weight of the toe slab tends to reduce design moments and shears, it will be multiplied by a load factor of 0.9. Thus the factored load moment at the outer face of the stem is

$$M_u = 1.7\left(\frac{2710}{2} \times 3.75^2 \times \frac{2}{3} + \frac{1850}{2} \times 3.75^2 \times \frac{1}{3}\right) - 0.9\left(225 \times 3.75^2 \times \frac{1}{2}\right)$$

$$= 27{,}600 \text{ ft-lb}$$

For concrete cast against and permanently exposed to earth, a minimum protective cover for steel of 3 in. is required; if the bar diameter is about 1 in. the effective depth will be 18.0 − 3.0 − 0.5 = 14.5 in. Thus for a 12 in. strip of toe slab.

$$\frac{M_u}{\phi bd^2} = \frac{27{,}600 \times 12}{0.9 \times 12 \times 210} = 146$$

Graph A.1*b* of App. A shows that, for this value, the required steel ratio would be slightly below the minimum of 200/60,000 = 0.0033. A somewhat thinner base slab appears possible. However, moments in the heel slab are yet to be investigated, as well as shears in both the toe and heel, and the trial depth of 18 in. will be retained tentatively. The required flexural steel:

$$A_s = 0.0033 \times 12 \times 14.5 = 0.57 \text{ in}^2/\text{ft}$$

is provided by No. 7 bars 12 in. on centers. The required length of embedment for these bars past the exterior face of the stem is the full development length of 26 in. subject to a reduction factor of 0.8, reflecting the 12 in. lateral spacing; thus they will be continued 21 in. past the face of the wall, as shown in Fig. 19.7*d*.

Shear will be checked at a distance d = 1.21 ft from the face of the stem (2.54 ft from the end of the toe), according to the usual ACI Code procedures.

The service load bearing pressure at that location (with reference to Fig. 19.7*c*) is 2130 psf, and the factored load shear is

$$\begin{aligned} V_u &= 1.7(2710 \times 1/2 \times 2.54 + 2130 \times 1/2 \times 2.54) - 0.9(225 \times 2.54) \\ &= 9940 \text{ lb} \end{aligned}$$

The design shear strength of the concrete is

$$\phi V_c = 2 \times 0.85 \sqrt{3000} \times 12 \times 14.5 = 16{,}200 \text{ lb}$$

well above the required value V_u.

e. Heel Slab

The heel slab, too, acts as a cantilever, projecting in this case from the back face of the stem and loaded by surcharge, earth fill, and its own weight. The upward reaction of the soil will be neglected here, for reasons given earlier. Applying appropriate load factors, the moment to be resisted is

$$\begin{aligned} M_u &= 1.7(400 \times 4.67^2 \times 1/2) + 1.4(1620 \times 4.67^2 \times 1/2 + 225 \times 4.67^2 \times 1/2) \\ &= 35{,}600 \text{ ft-lb} \end{aligned}$$

Thus

$$\frac{M_u}{\phi b d^2} = \frac{35{,}600 \times 12}{0.9 \times 12 \times 210} = 188$$

From Graph A.1*b* the required steel ratio is almost exactly the minimum value of 0.0033. The required steel area will again be provided using No. 7 bars 12 in. on centers, as shown in Fig. 19.7*d*. These bars are classed as top bars, as they have more than 12 in. of concrete below; thus the required length of embedment to the left of the inside face of the stem is $26 \times 1.3 \times 0.8 = 27$ in.

According to normal ACI Code procedures, the first critical section for shear would be a distance d from the face of support. However, the justification for this provision of the ACI Code is the presence, in the usual case, of vertical compressive stress near a support which tends to decrease the likelihood of shear failure in that region. However, the cantilevered heel slab is essentially hung from the bottom of the stem by the flexural tensile steel in the stem, and the concrete compression normally found near a support is absent here. Consequently, the critical section for shear in the heel slab will be taken at the back face of the stem. At that location,

$$\begin{aligned} V_u &= 1.7(400 \times 4.67) + 1.4(1845 \times 4.67) \\ &= 15{,}240 \text{ lb} \end{aligned}$$

The design shear strength provided by the concrete is the same as for the toe slab:

$$\phi V_c = 16{,}200 \text{ lb}$$

Because this is only 6 percent in excess of the required value V_u, no reduction in thickness of the base slab, considered earlier, will be made.

The base slab is well below grade and will not be subjected to the extremes of temperature that will be imposed on the stem concrete. Consequently, the requirement for crack control steel in the direction perpendicular to the main reinforcement is much less. No. 4 bars 12 in. on centers will be provided, at one face only, placed as shown in Fig. 19.7*d*. These bars serve chiefly as spacers for the main flexural reinforcement.

19.9 COUNTERFORT RETAINING WALLS

The external stability of a counterfort retaining wall is determined in the same manner as in the examples of Secs. 19.7 and 19.8. The toe slab represents a cantilever built in along the front face of the wall, loaded upward by the bearing pressure, exactly as in the cantilever wall of Sec. 19.8. Reinforcement is provided by bars *a* in Fig. 19.8.

A panel of the vertical wall between two counterforts is a slab acted upon by the horizontal earth pressure and supported along three sides, i.e., at the two counterforts and the base slab, while the fourth side, the top edge, is not supported. The earth pressure increases with distance from the free surface. The determination

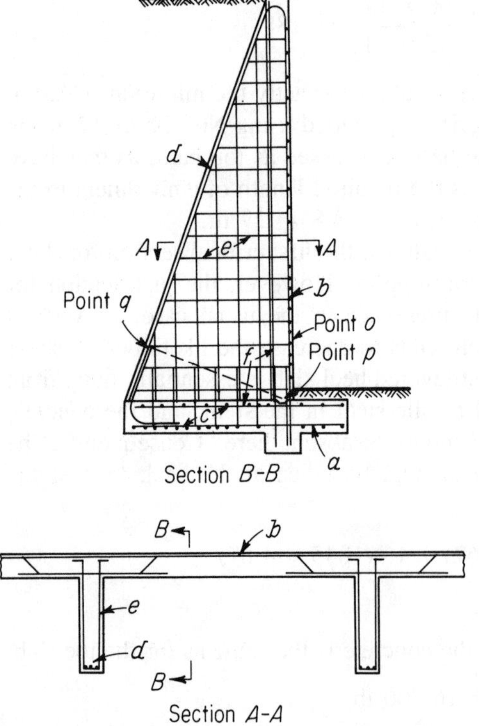

FIGURE 19.8
Details of counterfort retaining wall.

of moments and shears in such a slab supported on three sides and nonuniformly loaded is rather involved. It is customary in the design of such walls to disregard the support of the vertical wall by the base slab and to design it as if it were a continuous slab spanning horizontally between counterforts. This procedure is conservative, because the moments obtained by this approximation are larger than those corresponding to the actual conditions of support, particularly in the lower part of the wall. Hence, for very large installations, significant savings may be achieved by a more accurate analysis. The best computational tool for this is the Hillerborg *strip method*, a plasticity-based theory for design of slabs described in detail in Chap. 15. Alternatively, results of elastic analysis are tabulated for a range of variables in Ref. 19.8.

Slab moments are determined for strips 1 ft wide spanning horizontally, usually for the strip at the bottom of the wall and for three or four equally spaced additional strips at higher elevations. The earth pressure on the different strips decreases with increasing elevation and is determined from Eq. (19.1). Moment values for the bottom strips may be reduced to account for the fact that additional support is provided by the base slab. Horizontal bars b (Fig. 19.8) are provided as required with increased spacing or decreased diameter corresponding to the smaller moments. Alternate bars are bent to provide for the negative moments at the counterforts, or additional straight bars are placed for them.

The heel slab is supported, as is the wall slab, i.e., by the counterforts and at the wall. It is loaded downward by the weight of the fill resting on it, its own weight, and such surcharge as there may be. This load is partially counteracted by the bearing pressure. As in the vertical wall, a simplified analysis consists in neglecting the influence of the support along the third side and in determining moments and shears for strips parallel to the wall, each strip representing a continuous beam supported at the counterforts. With horizontal soil surface, the downward load is constant for the entire heel, whereas the upward load from the bearing pressure is usually smallest at the rear edge and increases frontward. For this reason, the span moments are positive (compression on top) and the support moments negative in the rear portion of the heel. Near the wall the bearing pressure often exceeds the vertical weights, resulting in a net upward load. The signs of the moments are correspondingly reversed, and steel must be placed accordingly. Bars c are provided for these moments.

The counterforts are wedge-shaped cantilevers built in at the bottom in the base slab. They support the wall slab and, therefore, are loaded by the total soil pressure over a length equal to the distance center to center between counterforts. They act as a T beam of which the wall slab is the flange and the counterfort the stem. The maximum bending moment is that of the total earth pressure, taken about the bottom of the wall slab. This moment is held in equilibrium by the force in the bars d and hence the effective depth for bending is the perpendicular distance pq from the center of bars d to the center of the bottom section of the wall slab. Since the moment decreases rapidly in the upper parts of the counterfort, part of the bars d can be discontinued.

In regard to shear, the authors suggest the horizontal section oq a distance a above the base slab as a conservative location for checking adequacy. Modification of the customary shear computation is required for wedge-shaped members (see Sec. 4.7). Usually concrete alone is sufficient to carry the shear, although bars *e* act as stirrups and can be used for resisting excess shear.

The main purpose of bars *e* is to counteract the pull of the wall slab, and they are thus designed for the full reaction of this slab.

The remaining bars of Fig. 19.8 serve as shrinkage reinforcement, except that bars *f* have an important additional function. It will be recalled that the wall and heel slabs are supported on three sides. Even though they were designed as if supported only by the counterforts, they develop moments where they join. The resulting tension in and near the reentrant corner should be provided for by bars *f*.

The question of reinforcing bar details, always important, is particularly so for corners subject to substantial bending moments, such as are present for both cantilever and counterfort retaining walls. Valuable suggestions are found in Ref. 19.9.

REFERENCES

19.1. R. B. Peck, W. E. Hanson, and T. H. Thornburn, *Foundation Engineering*, 2d ed., John Wiley and Sons, Inc., New York, 1974.
19.2. K. Terzaghi and R. B. Peck, *Soil Mechanics in Engineering Practice*, 2d ed., John Wiley and Sons, Inc., New York, 1967.
19.3. W. C. Huntington, *Earth Pressures and Retaining Walls*, John Wiley and Sons, Inc., New York, 1957.
19.4. "Standard Specifications for Highway Bridges," 13th ed., American Association of State Highway and Transportation Officials (AASHTO), Washington, DC, 1983.
19.5. *Manual for Railway Engineering*, American Railway Engineering Association (AREA), Washington, DC, 1973.
19.6. *CRSI Handbook*, 6th ed., Concrete Reinforcing Steel Institute (CRSI), Chicago, 1984.
19.7. M. Fintel, *Handbook of Concrete Engineering*, 2d ed., Van Nostrand Reinhold Co., New York, 1985.
19.8. "Rectangular Concrete Tanks," Publication No. 1S003D, Portland Cement Association, Skokie, IL, 1963.
19.9. I. H. E. Nilsson and A. Losberg, "Reinforced Concrete Corners and Joints Subjected to Bending Moment," *J. Struct. Div. ASCE*, vol. 102, no. ST6, 1976, pp. 1229–1254.

PROBLEMS

19.1 A cantilever retaining wall is to be designed with geometry as indicated in Fig. P19.1. Backfill material will be well-drained gravel having unit weight $w = 120$ pcf, internal friction angle $\phi = 33°$, and friction factor against the concrete base $f = 0.55$. Backfill placed in front of the toe will have the same properties and will be well compacted. The final grade behind the wall will be level with the top of the wall, with no surcharge. At the lower level, it will be 3 ft above the top of the base slab. To improve sliding resistance, a key will be used, tentatively projecting to a depth 4 ft below the top of the base slab. (This dimension may be modified if necessary.)

(*a*) Based on a stability investigation, select wall geometry suitable for the specified conditions. For a first trial, place the outer face of the wall $\frac{1}{3}$ the width of the base slab back from the toe.

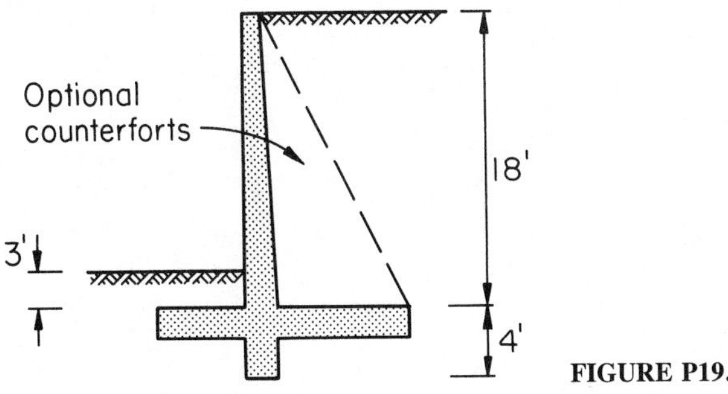

FIGURE P19.1

(*b*) Prepare the complete structural design, specifying size, placement, and cutoff points for all reinforcement. Materials have strengths $f_c' = 4000$ psi and $f_y = 60{,}000$ psi. Allowable soil bearing pressure is 5000 psf.

19.2 Redesign the wall of Problem 19.1 as a counterfort retaining wall. Counterforts are tentatively spaced 12 ft on centers, although this may be modified if desirable. Include all reinforcement details, including rebars in the counterforts.

CHAPTER 20

CONCRETE BUILDING SYSTEMS

20.1 INTRODUCTION

Most of the material in the preceding chapters has pertained to the design of reinforced concrete *structural elements*, e.g., slabs, columns, beams, and footings. These elements are combined in various ways to create *structural systems* for buildings and other construction. An important part of the total responsibility of the structural engineer is to select, from many alternatives, the best structural system for the given conditions. The wise choice of structural system is far more important, in its effect on overall economy and serviceability, than refinements in proportioning the individual members. Close cooperation with the architect in the early stages of a project is essential in developing a structure that not only meets functional and esthetic requirements but exploits to the fullest the special advantages of reinforced concrete, which include the following:

Versatility of form. Usually placed in the structure in the fluid state, the material is readily adaptable to a wide variety of architectural and functional requirements.

Durability. With proper concrete protection of the steel reinforcement, the structure will have long life even under highly adverse climatic or environmental conditions.

Fire resistance. With proper protection for the reinforcement, a reinforced concrete structure provides the maximum in fire protection.

Speed of construction. In terms of the entire period, from the date of approval of the contract drawings to the date of completion, a concrete building can often be completed in less time than a steel structure. Although the field erection of a steel building is more rapid, this phase must necessarily be preceded by prefabrication of all parts in the shop.

Cost. In many cases the first cost of a concrete structure is less than that of a comparable steel structure. In almost every case, maintenance costs are less.

Availability of labor and material. It is always possible to make use of local sources of labor, and in many inaccessible areas a nearby source of good aggregate can be found, so that only the cement and reinforcement need be brought in from a remote source.

Two record-setting examples of good building design in concrete are shown in Figs. 20.1 and 20.2.

FIGURE 20.1
View of 311 South Wacker Drive under construction. When completed, it will be the world's tallest concrete building, with total height 946 ft. (*Courtesy of Portland Cement Association.*)

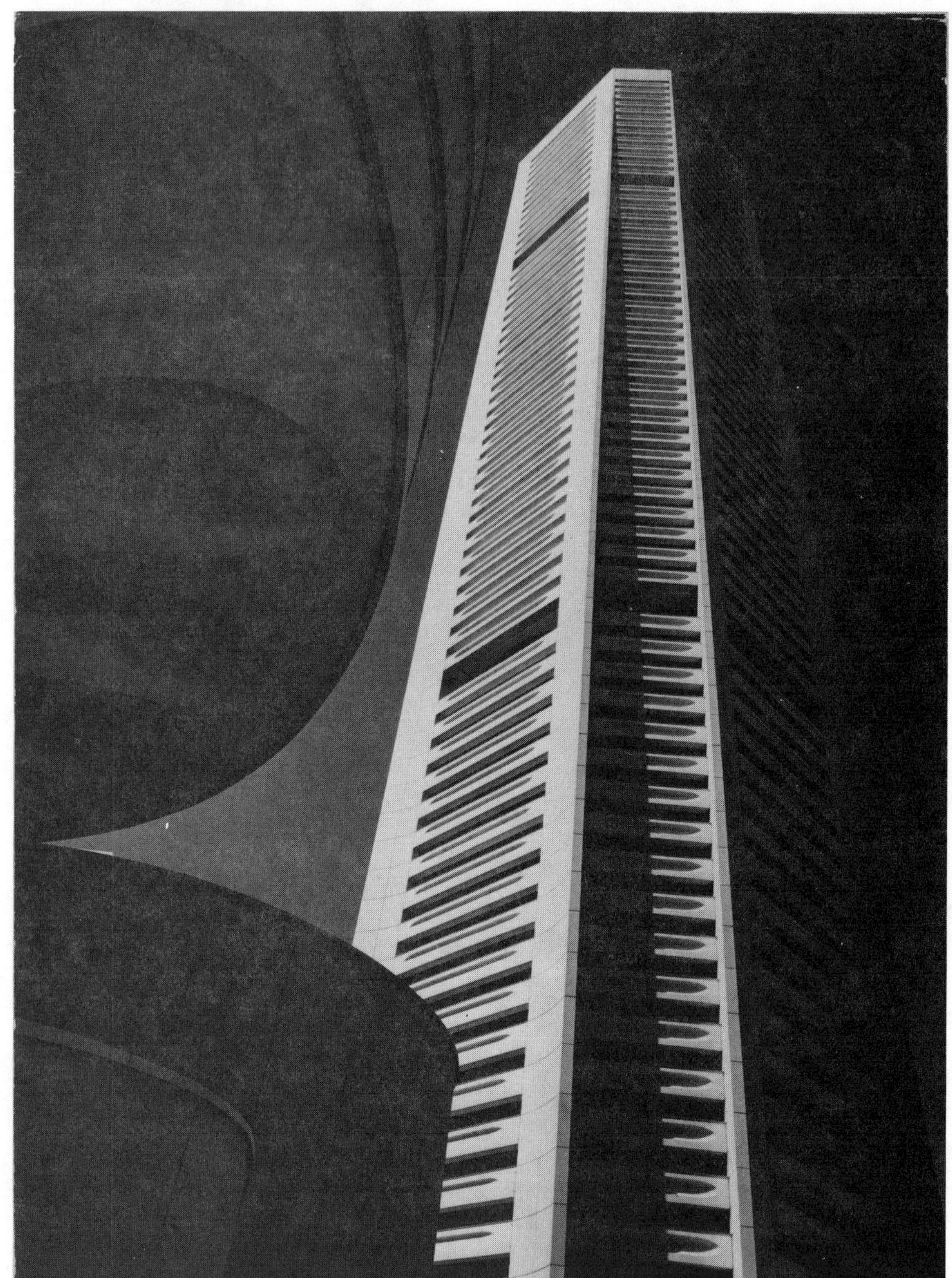

FIGURE 20.2
MLC Centre in Sydney, Australia; at 808 ft, the world's third tallest reinforced concrete building.

20.2 FLOOR AND ROOF SYSTEMS

The types of concrete floor and roof systems are so numerous as to defy concise classification. In steel construction, the designer usually is limited to the use of structural shapes that have been standardized in form and size by the relatively few producers in the field. In reinforced concrete, on the other hand, the engineer has almost complete control over the form of the structural parts of a building. In addition, many small producers of reinforced concrete structural elements and accessories can compete profitably in this field, since plant and equipment requirements are not excessive. This has resulted in the development of a wide variety of concrete systems. Only the more common types can be mentioned in this text.

In general, the commonly used reinforced concrete floor and roof systems can be classified as one-way systems, in which the main reinforcement in each structural element runs in one direction only, and two-way systems, in which the main reinforcement in at least one of the structural elements runs in perpendicular directions. Systems of each type can be identified in the following list:

a. One-way slab supported by monolithic concrete beams
b. One-way slab supported by steel beams (shear connectors may be used for composite action in the direction of the beam span)
c. One-way slab with cold-formed steel decking as form and reinforcement
d. One-way joist floor (also known as ribbed slab)
e. Two-way slab supported by edge beams for each panel
f. Flat slabs, with column capitals or dropped panels or both, but without beams
g. Flat plates, without beams and with no dropped panels or column capitals
h. Two-way joist floors, with or without beams on the column lines

Each of these types will be described briefly in the following sections. Additional information will be found in Refs. 20.1 to 20.3. In addition to the cast-in-place floor and roof systems described in this section, a great variety of precast concrete systems has been devised. Some of these will be described in Sec. 20.6.

a. Monolithic Beam-and-Girder Floors

A beam-and-girder floor consists of a series of parallel beams supported at their extremities by girders, which in turn frame into concrete columns placed at more or less regular intervals over the entire floor area, as shown in Fig. 20.3. This framework is covered by a one-way reinforced concrete slab, the load from which is transmitted first to the beams and then to the girders and columns. The beams are usually spaced so that they come at the midpoints, at the third points, or at the quarter points of the girders. The arrangement of beams and spacing of columns should be determined by economical and practical considerations. These will be affected by the use to which the building is to be put, the size and shape of the ground area, and the load that must be carried. A comparison of a number of trial designs and estimates should be made if the size of the building

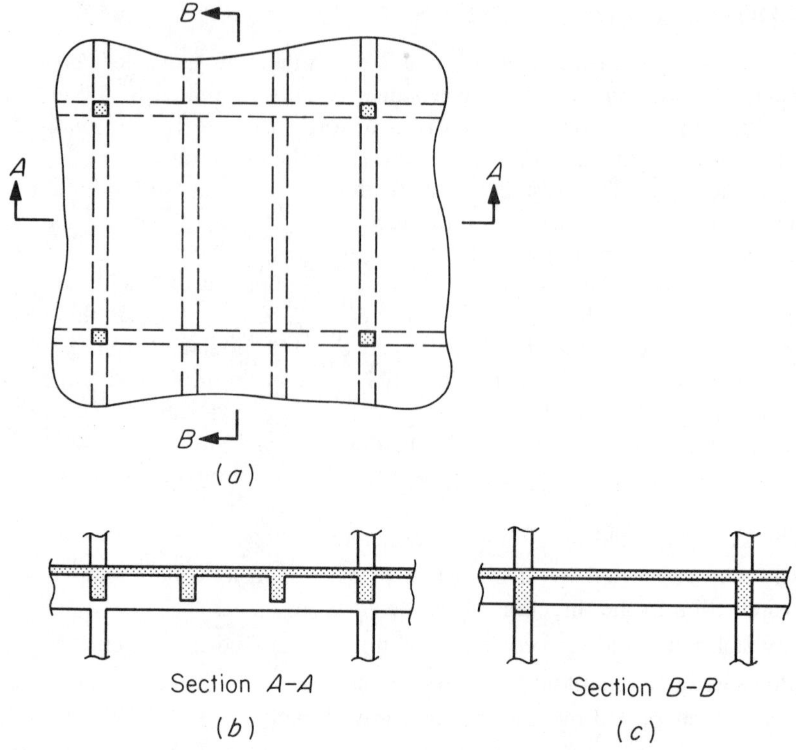

FIGURE 20.3
Framing of beam-and-girder floor: (*a*) plan view; (*b*) section through beams; (*c*) section through girders.

warrants, and the most satisfactory arrangement selected. If the spans in one direction are not long, say 16 ft or less, the beams may be omitted altogether, and the slab, spanning in one direction, can be carried directly by girders spanning in the perpendicular direction on the column lines. Since the slabs, beams, and girders are built monolithically, the beams and girders are designed as T beams and advantage is taken of continuity.

Beam-and-girder floors are adapted to any loads and to any spans that might be encountered in ordinary building construction. The normal maximum spread in live load values is from 40 to 400 psf, and the normal range in column spacings is from 16 to 32 ft.

The design and detailing of the joints where beams or girders frame into building columns should be given careful consideration, particularly for designs in which substantial horizontal loads are to be resisted by rigid frame action of the building. In this case, the column region, within the depth of the beams framing into it, is subjected to significant horizontal shears as well as to axial and flexural loads. Special horizontal column ties must be included to avoid uncontrolled diagonal cracking and disintegration of the concrete, particularly if the joint is subjected to load reversals. Specific recommendations for the design of beam-column joints will be found in Chap. 10 and Ref. 20.4. Joint design for buildings that must resist seismic forces is subject to special ACI Code provisions.

In normal beam-and-girder construction, the depth of the beams or girders is generally two or three times the web width. For light loads, a floor system has been developed in which the beams are omitted in one direction, the one-way slab being carried directly by column-line beams that are very broad and shallow, as shown in Fig. 20.4. These beams, supported directly by the columns, become little more than a thickened portion of slab. This type of construction, in fact, is known as *banded slab* construction, and there are a number of advantages associated with its use. In the direction of the slab span, a haunched member is present, in effect, with the maximum effective depth at the location of greatest negative moment, across the support lines. Negative moments are small at the edge of the haunch, where the depth becomes less, and positive slab moments in the span are reduced as well. Although the flexural steel in the beam (slab-band) will be increased because of the reduced effective depth compared with a beam of normal proportions, cost comparisons have shown that this may be outweighed by savings in slab steel. Other advantages include reduced construction depth, permitting reduction in the overall height of a building, and greater flexibility in locating columns, which may be displaced some distance from the centerline of the slab-bands without significantly changing the structural action of the floor. Formwork is simplified because of the reduction in the number of framing members. For such systems, special attention should be given to design details at the beam-column joint. Transverse top steel may be required to distribute the column reaction over the width of the slab-band. In addition, punching shear failure is possible; this may be investigated using the same methods presented earlier for flat plates (see Sec. 13.6).

b. Composite Construction with Steel Beams

One-way reinforced concrete slabs are also frequently used in buildings for which the columns, beams, and girders are structural steel framing. The slab is normally designed for full continuity over the supporting beams, and the usual methods followed. The spacing of the beams is usually about 6 to 8 ft. Often the slab is poured so that its underside is flush with the underside of the top flange of the supporting beam, as shown in Fig. 20.5. This simplifies construction, since the plywood slab form can be wedged up against the top flange of the beam. Of perhaps greater significance is the fact that positive bracing against lateral buckling is provided for the compression flange of the beam, whereas if the bottom of the slab were flush with the top of the steel, friction alone would provide resistance to lateral movement.

Increasingly, buildings of this type are designed for composite action (see Chap. 17). Shear connectors are welded to the top of the steel beam and are embedded in the concrete slab, as shown in Fig. 20.5*b*. By preventing longitudinal slip between the slab and steel beam in the direction of the beam axis, the combined member is both stronger and stiffer than if composite action were not developed. Restated, this means that for given loads and deflection limits, smaller and lighter steel beams can be used.

FIGURE 20.4
Banded slab floor system: (*a*) interior slab band; (*b*) edge band at exterior columns.

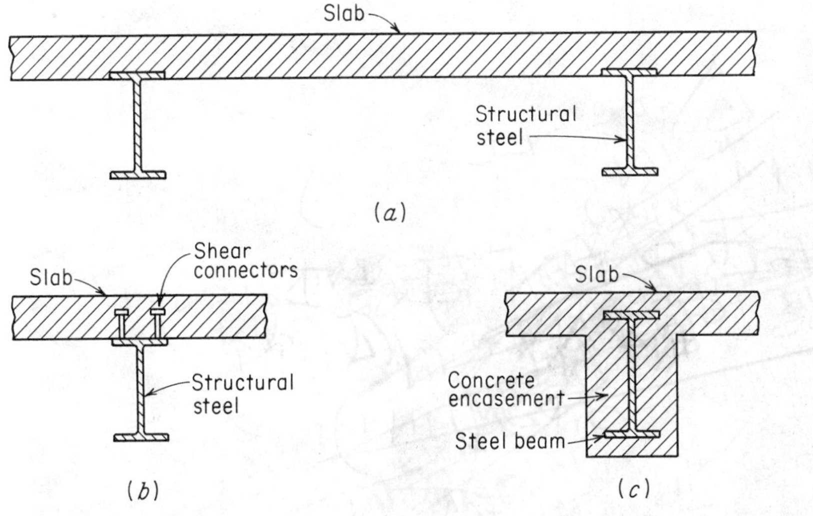

FIGURE 20.5
Composite beam-and-slab floors.

Composite floors may also use encased beams, as in Fig. 20.5*c*, offering the advantage of full fireproofing of the steel, but at the cost of more complicated formwork and possible difficulty in placing the concrete around and under the steel member. Such fully encased beams do not require shear connectors as a rule.

c. Steel Deck Reinforced Composite Slabs

In recent years much use has been made of light-gage cold-formed steel deck panels for floors, over which is cast a concrete floor slab. With suitable attention to details, the slab becomes composite with the steel deck, which then serves not only as form for the slab but as the main tensile flexural steel as well. This type of construction was discussed in detail in Sec. 17.4, and an illustration was shown in Fig. 17.10. Suitable for relatively light floor loading and short spans, composite steel deck reinforced slabs are found in office buildings and apartment buildings, with column-line girders and beams in the perpendicular direction subdividing panels into spans up to about 12 ft. Temporary shoring may be used at the midspan point or third point of the panels to avoid excessive stresses and deflections while the slab is poured, when the steel deck panel alone must carry the load.

d. One-Way Joist Floors

A one-way joist floor consists of a series of small, closely spaced reinforced concrete T beams, framing into monolithically cast concrete girders, which are in turn carried by the building columns. The T beams, called either *joists* or *ribs*, are formed by creating void spaces in what otherwise would be a solid slab. Usually

FIGURE 20.6
Steel forms for one-way joist floor.

these voids are formed using special steel pans, as shown in Fig. 20.6. Concrete is cast between the forms to create the ribs, and poured to a depth over the top of the forms so as to create a thin monolithic slab that becomes the T beam flange.

Since the strength of concrete in tension is small and is commonly neglected in design, elimination of much of the tension concrete in a slab by the use of pan forms results in a saving of weight with little change in the structural characteristics of the slab. Ribbed floors are economical for buildings such as apartment houses, hotels, and hospitals, where the live loads are fairly small and the spans comparatively long. They are not suitable for heavy construction such as in warehouses, printing plants, and heavy manufacturing buildings.

Standard forms for the void spaces between ribs are either 20 or 30 in. wide, and 8, 10, 12, 14, 16, or 20 in. deep. They are tapered in cross section, as shown in Fig. 20.7, generally at a slope of 1 to 12, to facilitate removal. Any joist width can be obtained by varying the width of the soffit (bottom) form. Tapered end pans are used where it is desired to obtain a wider joist near the end supports, such as may be required for high shear or negative bending moment. After the concrete has hardened, the steel pans are removed for reuse.

According to ACI Code 8.11, ribs must not be less than 4 in. wide and may not have a depth greater than 3.5 times the minimum web width. (For easier bar placement and placement of concrete, a minimum web width of 5 in. is desirable.) The clear spacing between ribs (determined by the pan width) must not exceed 30 in. The slab thickness over the top of the pans must not be less than one-twelfth the clear distance between ribs, nor less than 2 in., according to ACI

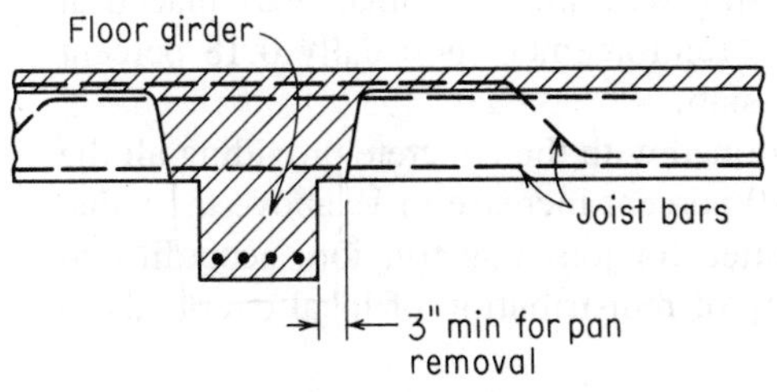

(a) Longitudinal section through joists

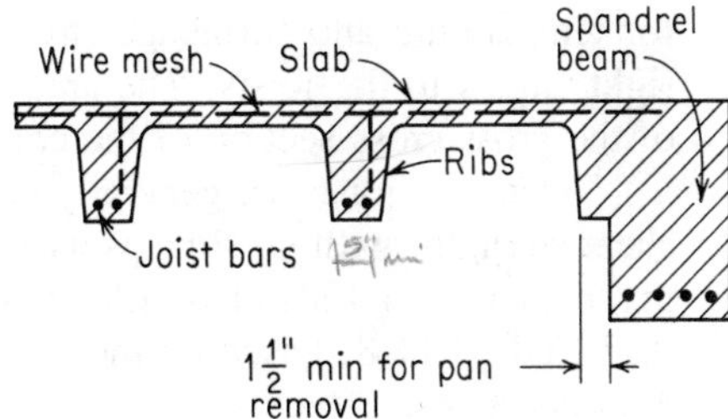

(b) Transverse section through joists

FIGURE 20.7
One-way joist floor cross sections: (*a*) cross section through supporting girder showing end of joists; (*b*) cross section through typical joists.

Code 8.11.6. Table 20.1 gives unit weights, in terms of psf of floor surface, for common combinations of joist width and depth, slab thickness, and form width.

Reinforcement for the joists usually consists of two bars in the positive bending region, with one bar discontinued where no longer needed or bent up to provide a part of the negative steel requirement over the supporting girders. Straight top bars are added over the support to provide for negative bending moment. According to ACI Code 7.13.2, at least one bottom bar must be continuous over the support, or at noncontinuous supports, terminated in a standard hook, as a measure to improve structural integrity in the event of major structural damage.

ACI Code 7.7.1 permits a reduced concrete cover of $\frac{3}{4}$ in. to be used for joist construction, just as for slabs. The thin slab (top flange) is reinforced mainly

Table 20.1 Weight of one-way joist floor systems

3 in. top slab			**4 1/2 in. top slab**		
Depth of pan form, in.	**Width of joist + pan form, in.**	**Weight, psf**	**Depth of pan form, in.**	**Width of joist + pan form, in.**	**Weight, psf**
8	5 + 20	60	8	5 + 20	79
8	5 + 30	54	8	5 + 30	72
10	5 + 20	67	10	5 + 20	85
10	5 + 30	58	10	5 + 30	77
12	5 + 20	74	12	5 + 20	92
12	5 + 30	63	12	5 + 30	82
14	5 + 20	81	14	5 + 20	99
14	5 + 30	68	14	5 + 30	87
16	6 + 20	94	16	6 + 20	113
16	6 + 30	78	16	6 + 30	97
20	6 + 20	111	20	6 + 20	130
20	6 + 30	91	20	6 + 30	109

Source: Adapted from Ref. 20.2.

for temperature and shrinkage stresses, using wire mesh or small bars placed at right angles to the joists. The area of this reinforcement is usually 0.18 percent of the gross cross section of the concrete slab.

One-way joists are generally proportioned with the concrete providing all the shear strength, with no stirrups used. A 10 percent increase in V_c above the value given by Eq. (4.12*a*) or (4.12*b*) is permitted for joist construction, according to ACI Code 8.11.8, based on the possibility of redistribution of local overloads to adjacent joists.

The girders that support the joist are poured monolithically with them and are T beams with cross section depending on the shape of the end pans that form the joists, as is evident from Fig. 20.7*a*. The thin concrete slab directly over the top of the pans is generally neglected in the calculations for the girders, and the T beam flange thickness is taken as the full height of the joists. The flange width can be adjusted to meet requirements by varying the placement of the end pans, but a 3 in. minimum from the T beam web to the end of the pan is required for pan removal.

A type of one-way joist floor system has evolved known as a *joist-band system* in which the joists are supported by broad girders having the same total depth as the joists. Separate beam forms are eliminated, and the same deck forms the soffit of both joists and girders. The simplified formwork, faster construction, level ceiling with no obstructing beams, and reduced overall height of walls, columns, vertical utilities, etc., all combine to achieve overall reduction in cost in most cases.

In one-way joist floors, the thickness of the slab is often controlled by fire resistance requirements. For a rating of 2 hours, for example, the slab must be about $4\frac{1}{2}$ in. thick. If 20 or 30 in. pan forms are used, slab span is small and slab strength is underutilized. This has led to what is known as a *wide module joist system*, or *skip joist system*. Such floors generally have 6 to 8 in. wide ribs at 5 to 6 ft on centers, with $4\frac{1}{2}$ in. top slab. These floors not only provide more efficient use of concrete in the slab, but also require less formwork labor, according to field studies. By ACI Code 8.11.4, wide module joist ribs must be designed as ordinary T beams, because the clear spacing between ribs exceeds the 30 in. maximum for joist construction, and special ACI Code provisions for joists do not apply. Concrete cover for reinforcement is as required for beams, not joists, and the 10 percent increase in V_c does not apply. Often the joists in wide module systems are carried by wide beams on the column lines, the depth of which is the same as that of the joists, to form a joist-band system equivalent to that described earlier.

Useful design information pertaining to one-way joist floors, including extensive load tables, will be found in the *CRSI Handbook* (Ref. 20.2). Suggested bar details and typical design drawings are found in the *ACI Detailing Manual* (Ref. 20.3).

e. Two-Way Edge-Supported Slabs

Two-way solid slabs supported by beams on the column lines on all sides of each slab panel have been discussed in detail in Chaps. 12 and 13. The perimeter beams are usually concrete cast monolithically with the slab, although they may also be structural steel, often encased in concrete for composite action and for improved fire resistance. If the supporting beams are steel, or if they are monolithic concrete and have a depth of about 3 times the slab depth or more, the coefficient method of analysis and design of Chap. 12 can be used. Otherwise, the direct design method or the equivalent frame method described in Chap. 13 should be followed, because it accounts for the interaction between slab and supporting beams in a more rational way.

Two-way solid slab systems are suitable for intermediate to heavy loads on spans up to about 30 ft. This range corresponds closely to that for beamless slabs with dropped panels and column capitals, described in the following section. The latter are often preferred because of the complete elimination of obstructing beams below the slab.

For lighter loads and shorter spans, a two-way solid slab system has evolved in which the column-line beams are wide and shallow, such that a cross section through the floor in either direction resembles the slab-band shown earlier in Fig. 20.4. The result is a two-way slab-band floor that, from below, appears as a paneled ceiling. Advantages are similar to those given earlier for one-way slab-band floors and for joist-band systems.

f. Beamless Flat Slabs with Dropped Panels or Column Capitals

By suitably proportioning and reinforcing the slab, it is possible to eliminate supporting beams altogether. The slab is supported directly on the columns. In a rectangular or square region centered on the columns, the slab may be thickened and the column tops flared, as shown in Fig. 20.8. The thickened slab is termed a *dropped panel* and the column flare is referred to as a *column capital*. Both of these serve a double purpose: they increase the shear strength of the floor system in the critical region around the column and they provide increased effective depth for the flexural steel in the region of high negative bending moment over the support. Beamless systems with dropped panels or column capitals or both are termed *flat slab systems* (although almost all slabs in structural engineering practice are "flat" in the usual sense of the word), and are differentiated from flat plate systems, with absolutely no projections below the slab, which are described in the following section.

In general, flat slab construction is economical for live loads of 100 psf or more and for spans up to about 30 ft. They are widely used for storage warehouses, parking garages, and below-grade structures carrying heavy earth-fill loads, for example. For lighter loads such as in apartment houses, hotels, and office buildings, flat plates (Sec. 20.2) or some form of joist construction (Secs. 20.2d and h) will usually prove less expensive. For spans longer than about 30 ft, beams and girders are used because of the greater stiffness of that form of construction.

Flat slabs may be designed either by the direct design method or the equivalent frame method, both described in detail in Chap. 13.

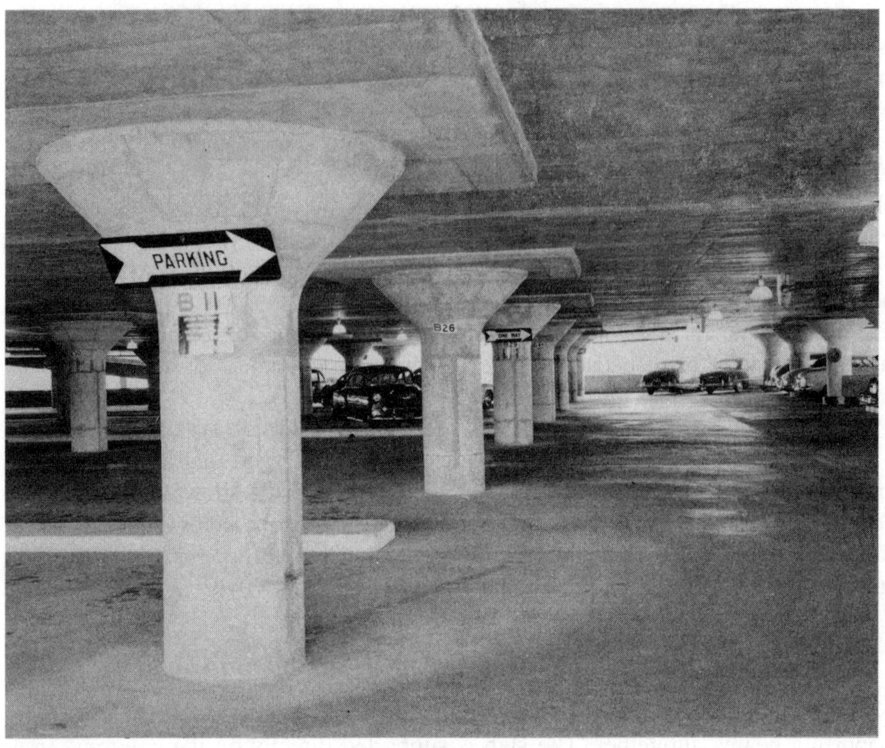

FIGURE 20.8
Flat slab garage floor. (*Courtesy of Portland Cement Association.*)

g. Flat Plate Slabs

A flat plate floor is essentially a flat slab floor with the dropped panels and column capitals omitted, so that a floor of uniform thickness is carried directly by prismatic columns. Flat plate floors have been found to be economical and otherwise advantageous for such uses as apartment buildings, as shown in Fig. 20.9, where the spans are moderate and loads relatively light. The construction depth for each floor is held to the absolute minimum, with resultant savings in the overall height of the building. The smooth underside of the slab can be painted directly and left exposed for ceiling, or plaster can be applied to the concrete. Minimum construction time and low labor costs result from the very simple formwork.

Certain problems associated with flat plate construction require special attention. Shear stresses near the columns may be very high, requiring the use of special forms of slab reinforcement there. The transfer of moments from slab to columns may further increase these shear stresses and requires concentration of negative flexural steel in the region close to the columns. Both these problems are treated in detail in Chap. 13. At the exterior columns, where such shear and

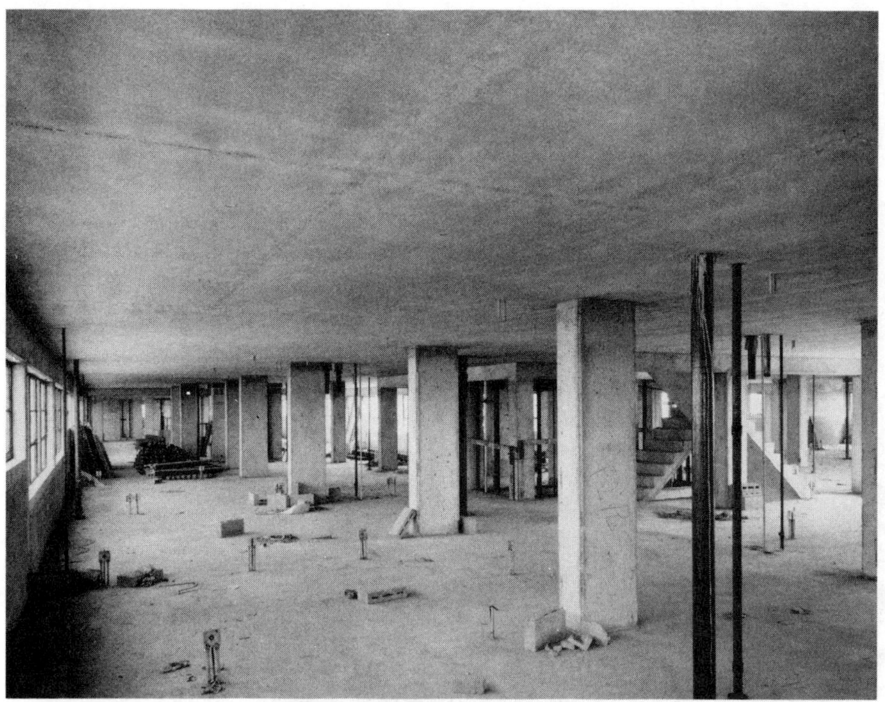

FIGURE 20.9
Flat plate floor construction. (*Courtesy of Portland Cement Association.*)

moment transfer may cause particular difficulty, the design is much improved by extending the slab past the column in a short cantilever.

Some flat plate buildings are constructed by the lift slab method, shown in Fig. 20.10. A casting bed (often doubling as the ground-floor slab) is poured, steel columns are erected and braced, and at ground level successive slabs, which will later become the upper floors, are cast. A membrane or sprayed parting agent is laid down between successive pours so that each slab can be lifted in its turn, starting with the top. Jacks placed atop the columns are connected to threaded rods extending down the faces of the columns and connecting, in turn, to lifting collars embedded in the slabs, as shown in Fig. 13.17*d*. When a slab is in its final position, shear plates are welded to the column below the lifting collar, or other devices are used to transfer the vertical slab reaction. Lifting collars such as those shown in Fig. 13.17*d*, in addition to providing anchorage for the lifting rods, serve to increase the effective size of the support for the slab and consequently improve the shear strength of the slab. The successful erection of structures using the lift slab method requires precise control of the lifting operation at all times, because even slight differences in level of the support collars may drastically change moments and shears in the slab, possibly leading to reversal of loading. Catastrophic accidents have resulted from failure to observe proper care in lifting or to provide adequate lateral bracing for the columns.

FIGURE 20.10
Lift slab construction used with flat plate floors; student dormitory at Clemson University, South Carolina.

h. Two-Way Joist Floors

As in one-way floor systems, the dead weight of two-way slabs can be reduced considerably by creating void spaces in what would otherwise be a solid slab. For the most part, the concrete removed is in tension and ineffective, so the lighter floor has virtually the same structural characteristics as the corresponding solid floor. Voids are usually formed using dome-shaped steel pans that are removed for reuse after the slab has hardened. Forms are placed on a plywood platform as shown in Fig. 20.11. Note in that figure that domes have been omitted near the columns to obtain a solid slab in that region of negative bending moment and high shear. The lower flange of each dome contacts that of the adjacent dome, so that the concrete is poured entirely against a metal surface, resulting in an excellent finished appearance of the slab. A wafflelike appearance (these slabs are sometimes called waffle slabs) is imparted to the underside of the slab, which can be featured to architectural advantage, as shown in Fig. 20.12.

Two-way joist floors are designed following the usual procedures for two-way solid slab systems, as presented in Chap. 13, with the solid regions at the columns considered as dropped panels. Joists in each direction are divided into

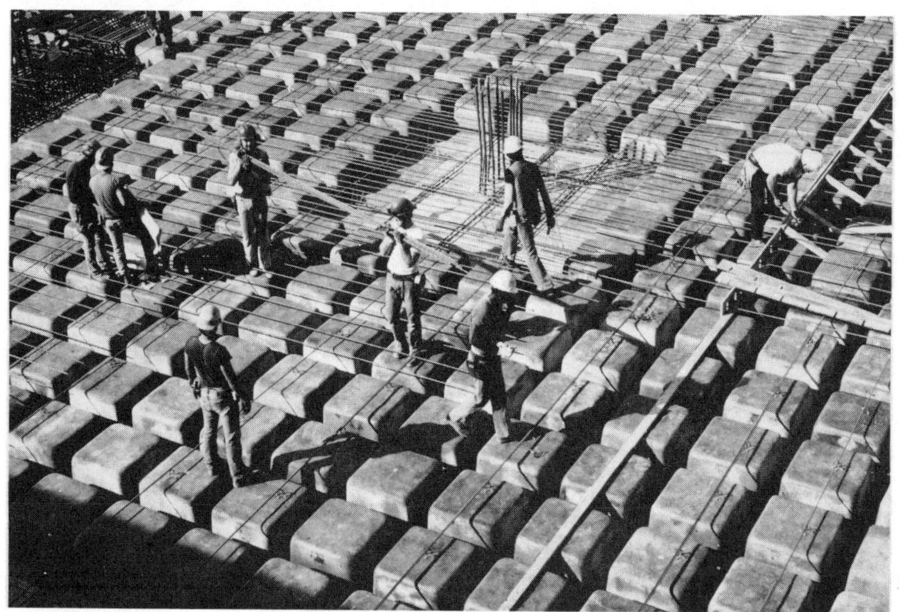

FIGURE 20.11
Two-way joist floor under construction with steel dome forms. (*Courtesy of Ceco Corporation.*)

FIGURE 20.12
Regency House Apartments, San Antonio, with cantilevered two-way joist slab plus integral beams on column lines.

Table 20.2 Equivalent slab thickness and weight of two-way joist floor systems

	3 in. top slab		$4\frac{1}{2}$ in. top slab	
Depth of pan form, in.	Equivalent uniform thickness, in.	Weight, psf	Equivalent uniform thickness, in.	Weight, psf
36 in. module (30 in. pans plus 6 in. ribs)				
8	5.8	73	7.3	92
10	6.7	83	8.2	102
12	7.4	95	9.1	114
14	8.3	106	9.9	120
16	9.1	114	10.6	133
20	10.8	135	12.3	154
24 in. module (19 in. pans plus 5 in. ribs)				
6	5.7	72	7.2	90
8	6.8	85	8.3	103
10	7.3	91	8.8	111
12	8.6	107	10.1	126

Source: Adapted from Ref. 20.2.

column strip joists and middle strip joists, the former including all joists that frame into the solid head. Each joist rib usually includes two bars for positive moment resistance, and one may be discontinued where no longer required. Negative steel is provided by separate straight bars running in each direction over the columns.

In design calculations, the self-weight of two-way joist floors is considered to be uniformly distributed, based on an equivalent slab of uniform thickness having the same volume of concrete as the actual ribbed slab. Equivalent thicknesses and weights are given in Table 20.2 for standard 30 and 19 in. pans of various depths and for either a 3 in. top slab or $4\frac{1}{2}$ in. top slab, based on normal weight concrete of 150 pcf density.

20.3 FLOOR AND ROOF SURFACES

Many types of wearing surfaces are used on concrete floors. Concrete topping $\frac{3}{4}$ to 2 in. thick is frequently used. Such topping may be either monolithic or bonded. Monolithic finishes should be placed within 45 min after the base slab is poured; bonded finishes are placed on a fully hardened base slab. To obtain an effective bond in the latter case, the base slab should be rough. The value of epoxy cement to improve the bond between new and old concrete is well established. Mix proportions for the topping surface vary, and quite often an artificial hardening agent is added.†

† See Portland Cement Association bulletins "Surface Treatments for Concrete Floors" and "Specifications for Heavy-Duty Floor Concrete Floor Finish" for more complete coverage of this important topic.

A wooden floor may be provided for, if desired, by embedding beveled nailing strips or "sleepers," usually two by fours laid flat, in a layer of cinder concrete on top of the main slab. Small vitrified-clay flat tiles embedded in a 2 in. layer of Portland cement mortar also provide a satisfactory floor surface. Asphalt or linoleum tile placed over a smooth cement base is frequently used.

The durability of any wearing surface depends in a large measure upon the method of placing the surface. If not properly constructed, the concrete finish might spall, the wood floor dry-rot, the tiles curl, and the linoleum crack. To ensure the maximum degree of serviceability from a given type of floor surface, a special study of methods that have proved successful for laying that particular type of floor should be made.

The design of roofs is similar to the design of floors. In addition to the structural requirements, the roofs must be impervious to the passage of water, provide for adequate drainage, and furnish protection against condensation.

To provide adequate drainage, the roof slab may be pitched slightly, or a filling of some light material such as cinder concrete, covered with a suitable roofing material, may be used, the thickness of the filling being varied to give the required slope to the roof surface. The amount of slope required for drainage will depend upon the smoothness of the exposed surface. A value of $\frac{1}{8}$ in./ft might be used with a surface of hard tile. Felt and gravel roofs should have a pitch of at least $\frac{1}{4}$ in./ft. Some form of flashing is required along the parapets to prevent the drainage from seeping into the building at the edges of the roof slab.

Condensation can best be guarded against by proper ventilation and insulation. The form of insulation to be used will depend upon the particular class of building under consideration.

Imperviousness can best be provided for by the application of some form of separate roof covering, such as multiple layers of treated felt cemented together and to the slab by means of coal-tar pitch or asphalt, and overlaid with gravel, vitrified tile embedded in asphalt, or any of the commercial types of built-up roofings. Corrugated iron or copper roofings are sometimes placed on reinforced concrete buildings, but they are usually more expensive than other types of coverings.

20.4 WALLS

a. Panel, Curtain, and Bearing Walls

As a general rule, the exterior walls of a reinforced concrete building are supported at each floor by the skeleton framework, their only function being to enclose the building. Such walls are called *panel walls*. They may be made of concrete (often precast), cinder concrete block, brick, tile blocks, or insulated metal panels. The latter may be faced with aluminum, stainless steel, or a porcelain-enamel finish over steel, backed by insulating material and an inner surface sheathing. The thickness of each of these types of panel walls will vary according to the material, type of construction, climatological conditions, and the building requirements governing the particular locality in which the construction takes place.

The pressure of the wind is usually the only load that is considered in determining the structural thickness of a wall panel, although in some cases exterior walls are used as diaphragms to transmit forces caused by horizontal loads down to the building foundations.

Curtain walls are similar to panel walls except that they are not supported at each story by the frame of the building, but are self-supporting. However, they are often anchored to the building frame at each floor to provide lateral support.

A *bearing wall* may be defined as one that carries any vertical load in addition to its own weight. Such walls may be constructed of stone masonry, brick, concrete block, or reinforced concrete. Occasional projections or pilasters add to the strength of the wall and are often used at points of load concentration. In small commercial buildings, bearing walls may be used with economy and expediency. In larger commercial and manufacturing buildings, when the element of time is an important factor, the delay necessary for the erection of the bearing wall and the attending increased cost of construction often dictate the use of some other arrangement.

Bearing walls may be of either single or double thickness, the advantage of the latter type being that the air space between the walls renders the interior of the building less liable to temperature variation and makes the wall itself more nearly moistureproof. On account of the greater gross thickness of the double wall, such construction reduces the available floor space. This feature is often sufficient in itself to warrant the selection of the solid wall unless the factors of condensation and temperature are of great importance. Cavity-wall construction is usually limited to a total height of about 40 ft.

The thickness of bearing walls varies with the height. The New York City Building Code requires that for bearing walls made of solid masonry units (brick, stone, concrete brick, etc.) and not more than 75 ft in height, the thickness of the uppermost 55 ft of height shall be at least 12 in., and the thickness below this shall be at least 16 in., except that for walls in buildings of one, two, and three stories the top-story thickness in each case may be 8 in. In computing wall thicknesses, a maximum story height of 13 ft shall be assumed.

According to ACI Code 14.5, the load capacity of a wall is given by

$$\phi P_{nw} = 0.55\phi f'_c A_g \left[1 - \left(\frac{k\, l_c}{32h}\right)^2\right] \tag{20.1}$$

where ϕP_{nw} = design axial load strength
A_g = gross area of section, in^2
l_c = vertical distance between supports, in.
h = thickness of wall, in.
ϕ = 0.70

and the effective length factor k is taken as 0.8 for walls restrained against rotation at top or bottom or both, 1.0 for walls unrestrained against rotation at both ends, and 2.0 for walls not braced against lateral translation.

In the case of concentrated loads, the length of the wall to be considered as effective for each shall not exceed the center-to-center distance between loads;

nor shall it exceed the width of the bearing plus 4 times the wall thickness. Reinforced concrete bearing walls shall have a thickness of at least $\frac{1}{25}$ times the unsupported height or width, whichever is shorter. Reinforced concrete bearing walls of buildings shall be not less than 4 in. thick.

The area of horizontal reinforcement shall be not less than 0.0025 and that of vertical reinforcement not less than 0.0015 times the area of the concrete. These values may be reduced to 0.0020 and 0.0012 respectively, if the reinforcement is not larger than $\frac{5}{8}$ in. in diameter or consists of welded wire fabric.

b. Basement Walls

In determining the thickness of basement walls, the lateral pressure of the earth, if any, must be considered in addition to other structural features. If it is part of a bearing wall, the lower portion may be designed either as a slab supported by the basement and first floors or as a retaining wall, depending upon the type of construction. If columns and wall beams are available for support, each basement wall panel of reinforced concrete may be designed to resist the earth pressure as a simple slab reinforced in either one or two directions.

A minimum thickness of $7\frac{1}{2}$ in. is specified by ACI Code 14.5.3 for reinforced concrete basement walls. In wet ground a minimum thickness of 12 in. should be used. In any case, the thickness cannot be less than that of the wall above.

Care should be taken to brace a basement wall thoroughly from the inside if the earth is backfilled before the wall has obtained sufficient strength to resist the lateral pressure without such assistance or if it is placed before the first-floor slab is in position.

c. Partition Walls

Interior walls used for the purpose of subdividing the floor area may be made of cinder block, brick, precast concrete, metal lath and plaster, clay tile, or metal. The type of wall selected will depend upon the fire resistance required; flexibility of rearrangement; ease with which electrical conduits, plumbing, etc., can be accommodated; and architectural requirements.

d. Shear Walls

Horizontal forces acting on buildings, e.g., those due to wind or seismic action, can be resisted by different means. Rigid-frame resistance of the structure, augmented by the contribution of ordinary masonry walls and partitions, can provide for wind loads in many cases. However, when heavy horizontal loading is likely, such as would result from an earthquake, reinforced concrete shear walls are used. These may be added solely to resist horizontal forces, or concrete walls enclosing stairways or elevator shafts may also serve as shear walls.

Figure 20.13 shows a building with wind or seismic forces represented by arrows acting on the edge of each floor or roof. The horizontal surfaces act as

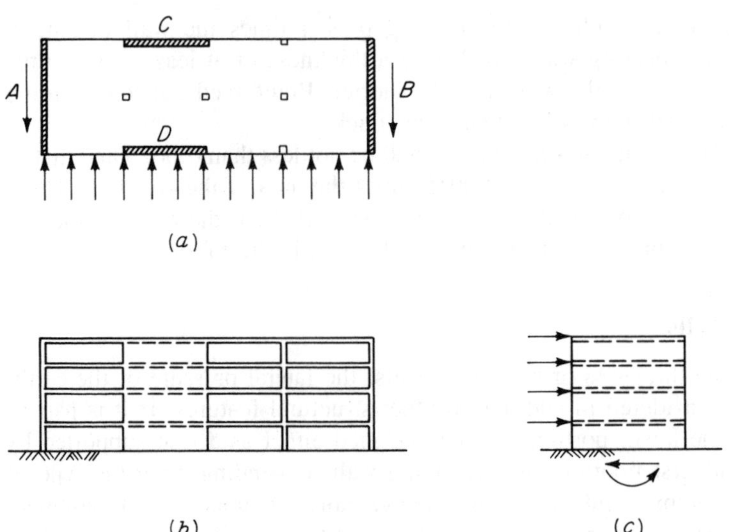

FIGURE 20.13
Building with shear walls subject to horizontal loads: (*a*) typical floor; (*b*) front elevation; (*c*) end elevation.

deep beams to transmit loads to vertical resisting elements *A* and *B*. These shear walls, in turn, act as cantilever beams fixed at their base to carry loads down to the foundation. They are subjected to (1) a variable shear, which reaches a maximum at the base, (2) a bending moment, which tends to cause vertical tension near the loaded edge and compression at the far edge, and (3) a vertical compression due to ordinary gravity loading from the structure. For the building shown, additional shear walls *C* and *D* are provided to resist loads acting in the long direction of the structure.

Shear is apt to be critical for walls of relatively low ratio of height to length. High shear walls are controlled mainly by flexural requirements, particularly if only uniformly distributed reinforcement is used.

Figure 20.14 shows a typical shear wall of height h_w, length l_w, and thickness h. It is assumed to be fixed at its base and loaded horizontally along its left edge. Vertical flexural reinforcement of area A_s is provided at the left edge, with its centroid a distance d from the extreme compression face. To allow for reversal of load, identical reinforcement is provided along the right edge. Horizontal shear reinforcement of area A_v at spacing s_2 is provided, as well as vertical shear reinforcement of area A_h at spacing s_1. Such distributed steel is normally placed in two layers, parallel to the faces of the wall.

The design basis for shear walls, according to ACI Code 11.10, is of the same general form as that used for ordinary beams:

$$V_u \le \phi V_n \tag{20.2}$$

where

$$V_n = V_c + V_s \tag{20.3}$$

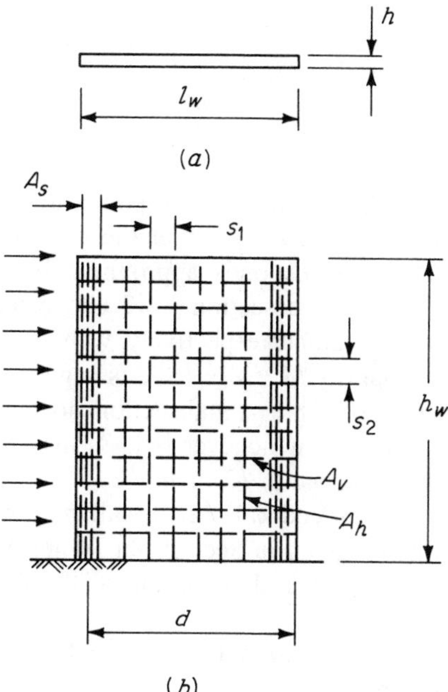

FIGURE 20.14
Geometry and reinforcement of a typical shear wall: (*a*) cross section; (*b*) elevation.

Based on tests (Refs. 20.5 and 20.6) an upper limit has been established on the nominal shear strength for walls:

$$V_n \leq 10\sqrt{f'_c}hd \tag{20.4}$$

In this and all other equations pertaining to the design of shear walls, the distance d may be taken equal to $0.8l_w$. A larger value of d, equal to the distance from the extreme compression face to the center of force of all reinforcement in tension, may be used when determined by a strain compatibility analysis.

The value of V_c, the nominal shear strength provided by the concrete, may be based on the usual equations for beams, according to ACI Code 11.10.5. For walls subject to vertical compression,

$$V_c = 2\sqrt{f'_c}hd \tag{20.5}$$

and for walls subject to vertical tension N_u,

$$V_c = 2\left(1 + \frac{N_u}{500A_g}\right)\sqrt{f'_c}hd \tag{20.6}$$

Here N_u is the factored axial load in pounds, taken negative for tension, and A_g is the gross area of horizontal concrete section in square inches. Alternately, the value of V_c may be based on a more detailed calculation, as the lesser of

$$V_c = 3.3\sqrt{f'_c}hd + \frac{N_u d}{4l_w} \tag{20.7}$$

or

$$V_c = \left[0.6\sqrt{f'_c} + \frac{l_w\left(1.25\sqrt{f'_c} + 0.2N_u/l_w h\right)}{M_u/V_u - l_w/2}\right] hd \tag{20.8}$$

where N_u is negative for tension as before. Equation (20.7) corresponds to the occurrence of a principal tensile stress of approximately $4\sqrt{f'_c}$ at the centroid of the shear-wall cross section. Equation (20.8) corresponds approximately to the occurrence of a flexural tensile stress of $6\sqrt{f'_c}$ at a section $l_w/2$ above the section being investigated. Thus the two equations predict, respectively, web-shear cracking and flexure-shear cracking. When the quantity $M_u/V_u - l_w/2$ is negative, Eq. (20.8) is inapplicable. According to the ACI Code, horizontal sections located closer to the wall base than a distance $l_w/2$ or $h_w/2$, whichever is less, may be designed for the same V_c as that computed at a distance $l_w/2$ or $h_w/2$.

When the factored shear force V_u does not exceed $\phi V_c/2$, a wall may be reinforced according to the minimum requirements given in Sec. 20.4a. When V_u exceeds $\phi V_c/2$, reinforcement for shear is to be provided according to the following requirements.

The nominal shear strength V_s provided by the horizontal wall steel is determined on the same basis as for ordinary beams:

$$V_s = \frac{A_v f_y d}{s_2} \tag{20.9}$$

where A_v = area of horizontal shear reinforcement within vertical distance s_2, in^2

s_2 = vertical distance between horizontal reinforcement, in.

f_y = yield strength of reinforcement, psi

Substituting Eq. (20.9) into Eq. (20.3), then combining with Eq. (20.2), one obtains the equation for the required area of horizontal shear reinforcement within a distance s_2:

$$A_v = \frac{(V_u - \phi V_c)\, s_2}{\phi f_y d} \tag{20.10}$$

The minimum permitted ratio of horizontal shear steel to gross concrete area of vertical section is

$$\rho_h = 0.0025 \tag{20.11}$$

and the maximum spacing s_2 is not to exceed $l_w/5$, $3h$, or 18 in.

Test results indicate that for low shear walls vertical distributed reinforcement is needed as well as horizontal reinforcement. Code provisions require vertical steel of area A_h within a spacing s_1, such that the ratio of vertical steel to gross concrete area of horizontal section will be not less than

$$\rho_n = 0.0025 + 0.5\left(2.5 - \frac{h_w}{l_w}\right)(\rho_h - 0.0025) \tag{20.12}$$

but not less than 0.0025. However, the vertical steel ratio need not be greater than the required horizontal steel ratio. The spacing of the vertical bars is not to exceed $l_w/3$, $3h$, or 18 in.

Walls may be subject to flexural tension due to overturning moment, even when the vertical compression from gravity loads is superimposed. In many but not all cases, vertical steel is provided, concentrated near the wall edges, as in Fig. 20.14. The required steel area can be found by the usual methods for beams.

The dual function of the floors and roofs in buildings with shear walls should be noted. In addition to resisting gravity loads, they must act as deep beams spanning between shear-resisting elements. Because of their proportions, both shearing and flexural stresses are usually quite low. According to the ACI Code, a 25 percent reduction in overall load factor is allowed when wind or earthquake effects are combined with the effects of gravity loads. Consequently, floor and roof reinforcement designed for gravity loads can usually serve as reinforcement for horizontal beam action also, with no increase in bar areas.

20.5 STAIRWAYS

a. Types of Concrete Stairs

The simplest form of reinforced concrete stairway consists of an inclined slab supported at the ends upon beams, with steps formed upon its upper surface. Such a stair slab is usually designed as a simple slab with a span equal to the horizontal distance between supports. This method of design requires steel to be placed only in the direction of the length of the slab. Transverse steel, usually one bar to each tread, is used only to assist in distribution of the load and to provide temperature reinforcement. It sometimes becomes necessary to include a platform slab at one or both ends of the inclined slab. Many successful designs made as outlined above for the simple inclined slab indicate that the effect of the angle that occurs in a slab of this type can safely be disregarded, although reinforcement should be carefully detailed.

It is advisable to keep the unsupported span of a stair slab reasonably short. If no break occurs in the flight between floors, intermediate beams, supported by the structural framework of the building, can be used. If the stair between floors is divided into two or more flights, beams as described above may be used to support the intermediate landing, with these in turn supported as above for the long straight flight, or the intermediate slab may be suspended from a beam at the upper floor level by means of rod hangers. Where conditions permit, the intermediate slab may be supported directly by the exterior walls of the building.

b. Building Code Requirements

The required number of stairways, and many of the details, are governed to a large extent by the provisions of the governing building code. Among other things, these provisions stipulate the maximum distance from the most remote point in the floor area to the stairway, the minimum width of stairway, the maximum height of any straight flight, the maximum height (or rise) of a single step, the minimum distance (the run) between the vertical faces of two consecutive steps, and the required relation between the rise and the run to give safety and convenience in climbing.

In most codes the minimum width of any stair slab and the minimum dimension of any landing are about 44 in. The maximum rise of a stair step is usually specified as about $7\frac{3}{4}$ in., and the minimum run or tread width, exclusive of nosing, $9\frac{1}{2}$ in. A rise of less than $6\frac{1}{2}$ in. is not considered generally satisfactory. In order to give a satisfactory and comfortable ratio of rise to run, various rules have been adopted. One requires that for steps without nosings, the sum of the rise and run shall be $17\frac{1}{2}$ in. but the rise shall not be less than $6\frac{1}{2}$ in. or more than $7\frac{3}{4}$ in.

The maximum height of a straight flight between landings is generally 12 ft, except for stairways serving as exits from places of assembly, where a maximum of 8 ft is normally specified.

The number of stairways is governed by the width of the stair slab, the number of probable occupants on each floor, and the dimensions of the floor area. One typical code specification is that the distance from any point in an open floor area to the nearest stairway or exit shall not exceed 100 ft, that the corresponding distance along corridors in a particular area shall not exceed 125 ft, and that the combined width of all stairways in any story shall be such that the stairs can accommodate at one time the total number of persons occupying the largest floor area served by such stairs above the floor area under consideration on the basis of one person for each 22 in. of stair width and $1\frac{1}{2}$ treads on the stairs and one person for each $3\frac{1}{2}$ ft^2 floor area on the landings and halls within the stairway enclosure.

In all buildings over 40 ft in height, and in all mercantile buildings regardless of height, the required stairways must be completely enclosed by fireproof partitions, and at least one stairway must continue to the roof. An open ornamental stairway can be used from the main entrance floor to the floor next above, provided it is not the only stairway.

c. Construction Details

The usual practice is to construct the stairways after the main structural framework of the building has been completed, in which event recesses should be left in the beams to support the stair slabs, and dowels should be provided to furnish the necessary anchorage. Occasionally, however, the stairs are poured at the same time as the floors, in which event negative moment reinforcement should be furnished over the supports of the stair slab, as in any continuous-beam construction. The steps are usually poured monolithically with the slab, but they may be molded after the main slab is in place. In the latter instance, provision must be made for securing the step to the slab.

20.6 PRECAST CONCRETE FOR BUILDINGS

From the early 1960s to the present, building costs have increased at a considerably faster rate than that of most industrial products. This is true not only for the

United States but for most industrialized nations. One of the main reasons for these high building costs is the large amount of site labor involved in traditional construction processes. Even disregarding cost, the demand for skilled on-site building labor is outrunning supply and will increasingly do so in most industrial and many developing nations. This trend can be slowed or halted only through greater industrialization of construction.

Answering this need, *precast concrete construction* has developed rapidly and continues to grow in importance. Industrialization is achieved by mass production of repetitive and often standardized units: columns, beams, floor and roof elements, wall panels, etc. These are produced in precasting yards under factory conditions. On large jobs precasting yards are sometimes constructed on or adjacent to the site. More frequently, these yards are stationary regional enterprises which supply precast members to sizable areas within reasonable shipping distances, of the order of 200 mi. Advantages of precast construction are less labor per unit because of mechanized series production; use of unskilled local labor, in contrast to skilled mobile construction labor; shorter construction time because site labor mostly involves only foundations and the connecting of the precast units; better quality control and higher concrete strength achievable under factory conditions; and greater independence of construction from weather and season. Disadvantages are the greater cost of transporting precast units as compared with transporting materials and the additional technical problems and costs of site connections of precast elements.

Precast construction is used in all major types of structures: industrial buildings, residential and office buildings, halls of sizable span, bridges, etc. Precast members frequently are prestressed in the casting yard. In the context of the present chapter it is irrelevant whether a precast member is also prestressed. Discussion is focused on types of precast members and precast structures and on methods of connection; these are essentially independent of whether the desired strength of the member was achieved with ordinary reinforcement or by prestressing. A broad discussion of precast construction, which includes planning, design, materials, manufacturing, handling, construction, and inspection, will be found in Ref. 20.7.

a. Types of Precast Members

A number of types of precast units have established themselves. Though not formally standardized at this time, they are widely available, with minor local variations. At the same time, the precasting process is sufficiently adaptable for special shapes developed for particular projects to be produced economically, provided the number of repetitive units is sufficiently large. This is particularly important for exterior wall panels which permit a wide variety of architectural treatments.

Wall panels are made in a considerable variety of shapes, depending on architectural requirements. The most frequent four shapes are shown in Fig. 20.15. These units are produced in one- to four-story-high sections and up to 8 ft in width. They are used either as curtain walls attached to columns and beams or as bearing walls. To improve thermal insulation, sandwich panels are used that consist of

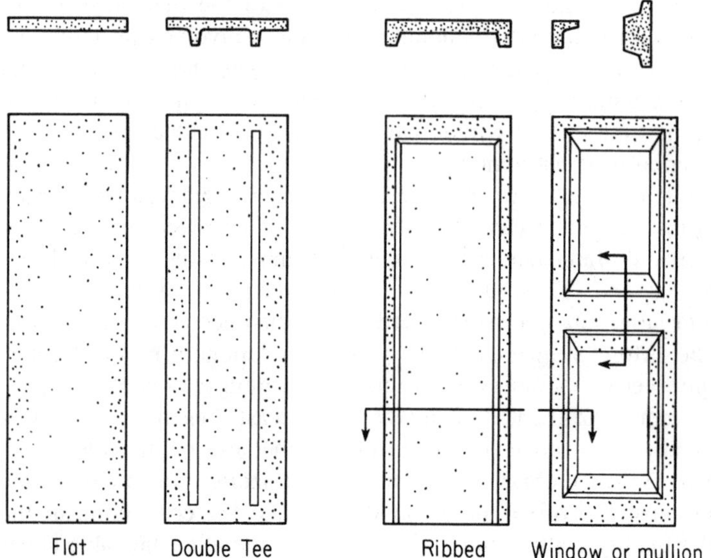

FIGURE 20.15
Precast concrete wall panels.

an insulation core (e.g., foam glass, glass fiber, or expanded plastics) between two layers of normal or lightweight concrete. The two layers must be adequately interconnected through the core to act as one unit. A variety of surface finishes can be produced through the use of special exposed aggregates or of colored cement, sometimes employed in combination. The special design problems which arise in load-bearing wall panels, such as tilt-up construction, are discussed in Ref. 20.8.

Stresses in wall panels are frequently more severe in handling and during erection than in the finished structure, and design must provide for these temporary conditions. Also, control of cracking is of greater importance in wall panels than in other precast units, for appearance more than for safety. To control cracking, the maximum tensile stress in the concrete, calculated by straight-line theory, should not exceed the modulus of rupture of the particular concrete with an adequate margin of safety. A safety factor of 1.65 for normal weight concrete and 2.0 for lightweight concrete is recommended in Ref. 20.8. A wealth of information on precast wall panels is found in Refs. 20.9 to 20.13.

Roof and floor elements are made in a wide variety of shapes adapted to specific conditions, such as span lengths, magnitude of loads, desired fire ratings, appearance, etc. Figure 20.16 shows typical examples of the most common shapes, arranged in approximate order of increasing span length, even though the spans covered by the various configurations overlap widely.

Flat slabs (Fig. 20.16*a*) are usually 4 in. thick, although they are used as thin as $2\frac{1}{2}$ in. when continuous over several spans, and are produced in widths of 4 to 8 ft and in lengths up to 36 ft. Depending on the magnitude of loads and on deflection limitations, they are used over roof and floor spans ranging from 8

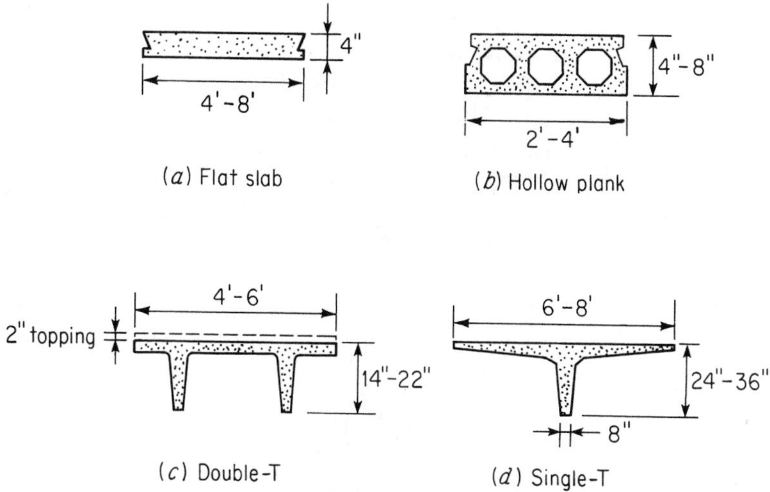

FIGURE 20.16
Precast floor and roof elements.

to about 22 ft. For lower weight and better insulation and to cover longer spans, *hollow planks* (Fig. 20.16*b*) of a variety of shapes are used. Some of these are made by extrusion in special machines. Depths range from 4 to about 8 in., widths from 2 to 4 ft. Again depending on load and deflection requirements, they are used on roof spans from about 16 to 34 ft and on 12 to 26 ft floor spans, which can be augmented to about 30 ft if a 2 in. topping is applied to act monolithically with the hollow plank.

For longer spans *double-T* members (Fig. 20.16*c*) are the most widely used shapes. Usual depths are from 14 to 22 in. They are generally used on roof spans up to about 60 ft. When used as floor members, a concrete topping of at least 2 in. is usually applied to act monolithically with the precast members for spans up to about 50 ft, depending on load and deflection requirements. Finally, *single-T* members are available in dimensions shown in Fig. 20.16*d*, mostly used for roof spans up to 100 ft and more.

In all these units, the member itself or its flange constitutes the roof or floor slab. If the floor or roof proper is made of other material (plywood, gypsum, plank, etc.), it can be supported on *precast joists* in a variety of shapes for spans from about 15 to 60 ft.

For floor and roof units of reinforced rather than prestressed concrete with spans up to 35 ft, Ref. 20.14 contains valuable and authoritative information on design (including a method of designing by test rather than analysis), details, fabrication and erection, materials, and quality control.

The shape of *precast beams* depends chiefly on the manner of framing. If floor and roof members are supported on top of the beams, these are mostly rectangular in shape (Fig. 20.17*a*). To reduce total depth of floor and roof construction, the tops of beams are often made flush with the top surface of the floor elements.

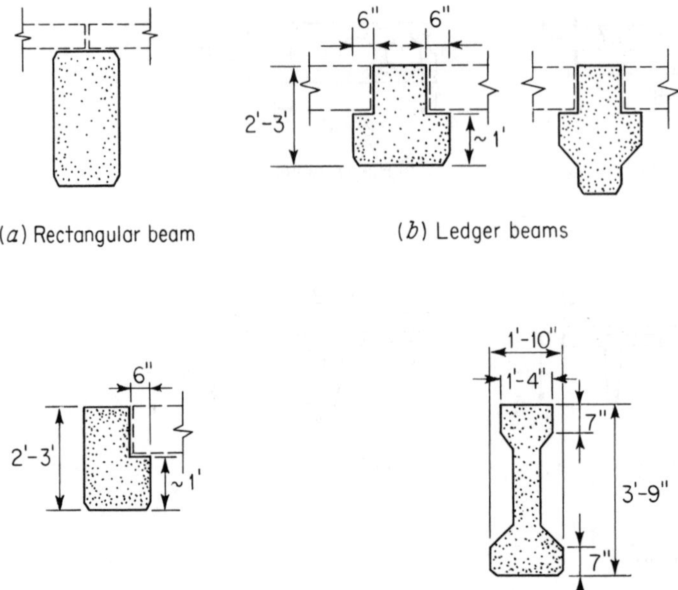

FIGURE 20.17
Precast beams and girders.

To provide bearing, the beams are then constructed as ledger beams (Fig. 20.17*b*) or L beams (Fig. 20.17*c*). Although these shapes pertain to building construction, precast beams or girders are also frequently used in highway bridges. As an example, Fig. 20.17*d* shows one of the various AASHTO bridge girders, so named because they were developed by the American Association of State Highway and Transportation Officials.

If *precast columns* of single-story height are used so that the beams rest on top of the columns, simple prismatic columns are employed which are available in sizes from about 12 × 12 to 24 × 24 in. (Fig. 20.18*a*). In this case, the beams are usually made continuous over the columns. Alternatively, in multistory construction, the columns can be made continuous for up to about six stories. In this case, integral brackets are frequently used to provide a bearing for the beams, as shown in Fig. 20.18*b* (see also Sec. 20.6d). Occasionally T columns are used for direct support of double-T floor members without the use of intermediate beams (Fig. 20.18*c*).

Figures 20.19 to 20.27 illustrate some of the many ways in which precast members have been used. Figure 20.19 shows a long-span single-T girder being lowered into place atop a precast rectangular beam, which in turn rests on a precast rectangular column. The photograph for Fig. 20.20 was taken in a precasting yard producing a variety of L, T, and rectangular shapes. Figure 20.21 shows symmetrical precast I beams, such as are used both for buildings and bridges.

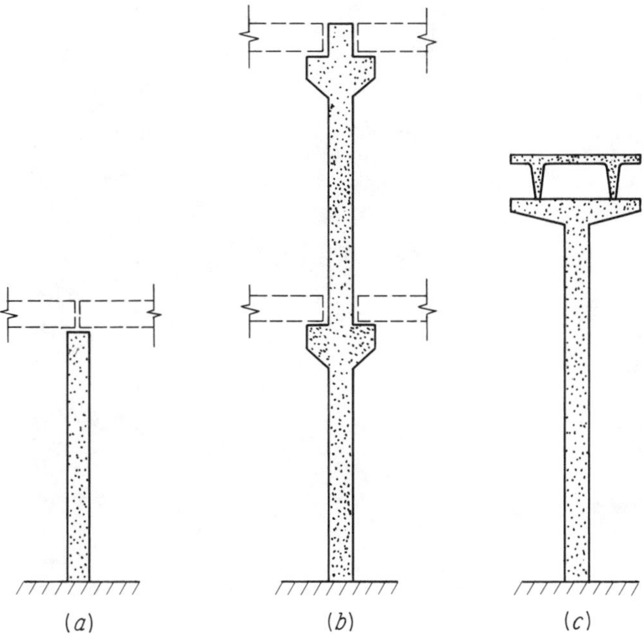

FIGURE 20.18
Precast concrete columns.

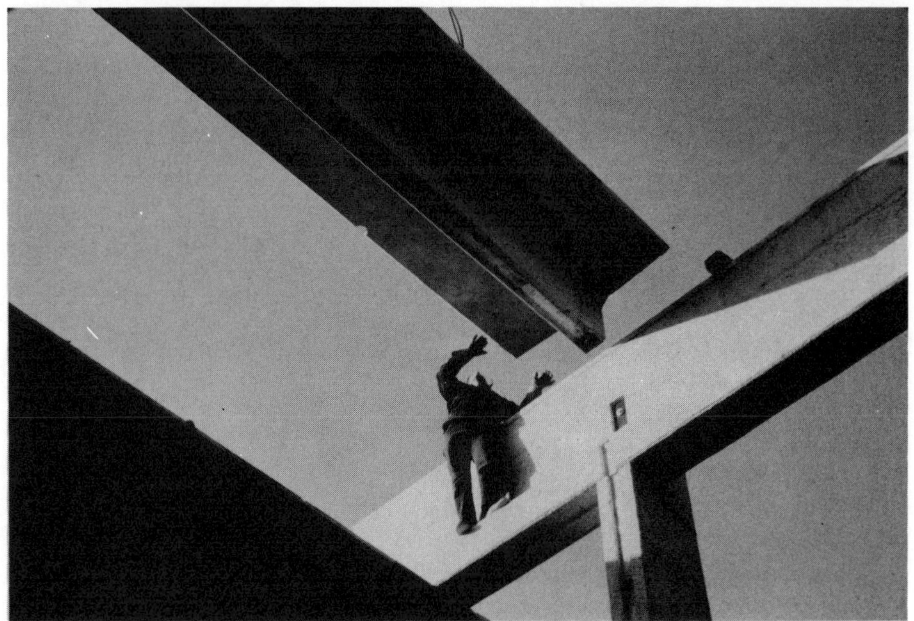

FIGURE 20.19
Long-span precast single-T girder used with precast beams and columns.

FIGURE 20.20
Precast L beam.

FIGURE 20.21
Precast I beams designed for composite action with a deck slab to be cast in place.

The projecting stirrup bars along the top flange will provide secure interlock between the precast beams and a cast-in-place slab added later, ensuring composite action. Figure 20.22 shows a multistory parking garage recently built at Cornell University. The three-story precast columns support L-section and inverted T-section girders. The girders, in turn, carry 60 ft span prestressed single T beams, which provide the deck surface.

Figure 20.23 demonstrates that unusual architectural designs can be realized in precast concrete, as in this all-precast administration building. Wall panels are utilized to produce a curved facade. Wedge-shaped repetitive floor units span freely from the exterior facade to the interior curved beam and column framework. In the insurance building of Fig. 20.24, 99 ft, 4 in. precast girders span between exterior walls supported on four points each and provide six floors of office space entirely free of interior supports. The convention headquarters of Fig. 20.25 combines cast-in-place frames and floor slabs with precast double-T roof beams and precast wall panels of special design. Figure 20.26 shows a 21-story hotel under construction, which, except for the service units, consists entirely of box-shaped, room-sized modules completely prefabricated and stacked on top of each other. Abroad, such precast modules, with plumbing, wiring, and heating preinstalled, are widely used for multistory apartment buildings as an alternative to making similar apartment structures in precast wall, roof, and floor panels, which are more easily shipped but less easily erected than box-shaped modules.

FIGURE 20.22
Precast parking garage at Cornell University.

FIGURE 20.23
All-precast administration building. (*Courtesy of Portland Cement Association.*)

Finally, Fig. 20.27 shows an example of the frequent combined use of structural steel with precast concrete. In this case the framing of an eight-story hotel was done in bolted structural steel, while precast concrete floor and roof planks and precast wall panels were used for all other main structural components. This construction is economical for 6 to 12 story buildings, where it saves both in cost and in construction time. It is one example of the increasingly important combined use of various structural materials and methods.

b. Connections

Cast-in-place reinforced concrete structures, by their very nature, tend to be monolithic and continuous. Connections, in the sense of joining two hitherto separate pieces, rarely occur in that type of construction. Precast structures, on the other hand, resemble steel construction in that the final structure consists of large numbers of prefabricated elements that are connected on the site to form the finished structure. In both types of construction, such connections can be detailed to transmit gravity forces only, or gravity and horizontal forces, or moments in addition to these forces. In the last case a continuous structure is obtained much as in cast-in-place construction, and connections that achieve such continuity by appropriate

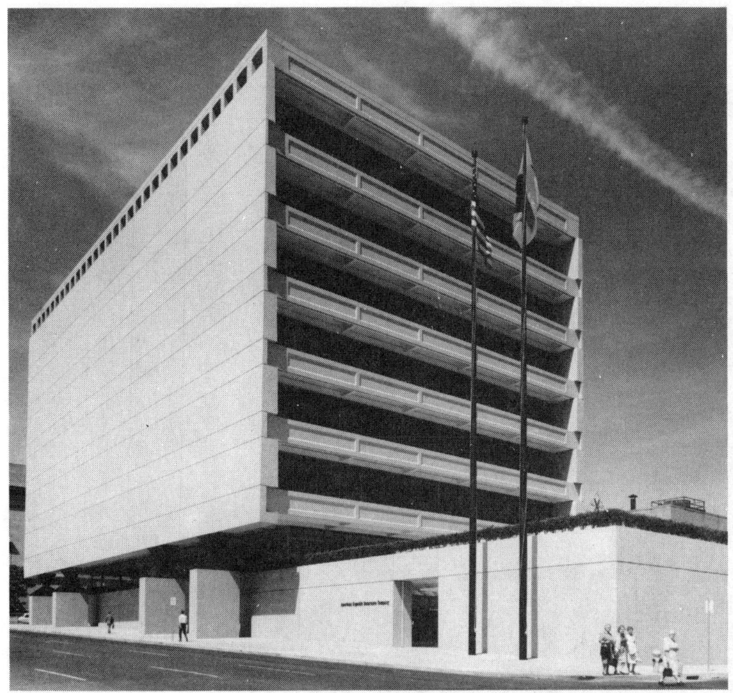

FIGURE 20.24
Precast girders of 99 ft span and 44 in. depth for a column-free interior. (*Courtesy of Portland Cement Association.*)

FIGURE 20.25
Precast roof and wall panels combined with cast-in-place frames and floor slabs. (*Courtesy of Portland Cement Association.*)

FIGURE 20.26
Precast room-sized modules for a 21-story hotel. (*Courtesy of Portland Cement Association.*)

use of special hardware, rebars, and concrete to transmit all tension, compression, and shear stresses are sometimes called *hard* connections. In contrast, connections that transmit reactions in one direction only, analogous to rockers or rollers in steel structures, but permit a limited amount of motion to relieve other forces, such as horizontal reaction components, are sometimes known as *soft* connections (Ref. 20.15). In almost all precast connections, bearing plates are used to ensure distribution and reasonable uniformity of bearing pressures. If these bearing plates are of steel and the plates of the two connected members are suitably joined by welding or otherwise, a hard connection is obtained in the sense that horizontal, as well as vertical, forces are transmitted. On the other hand, bearing plates are in use that permit relief of horizontal forces by limited amounts of motion, such as

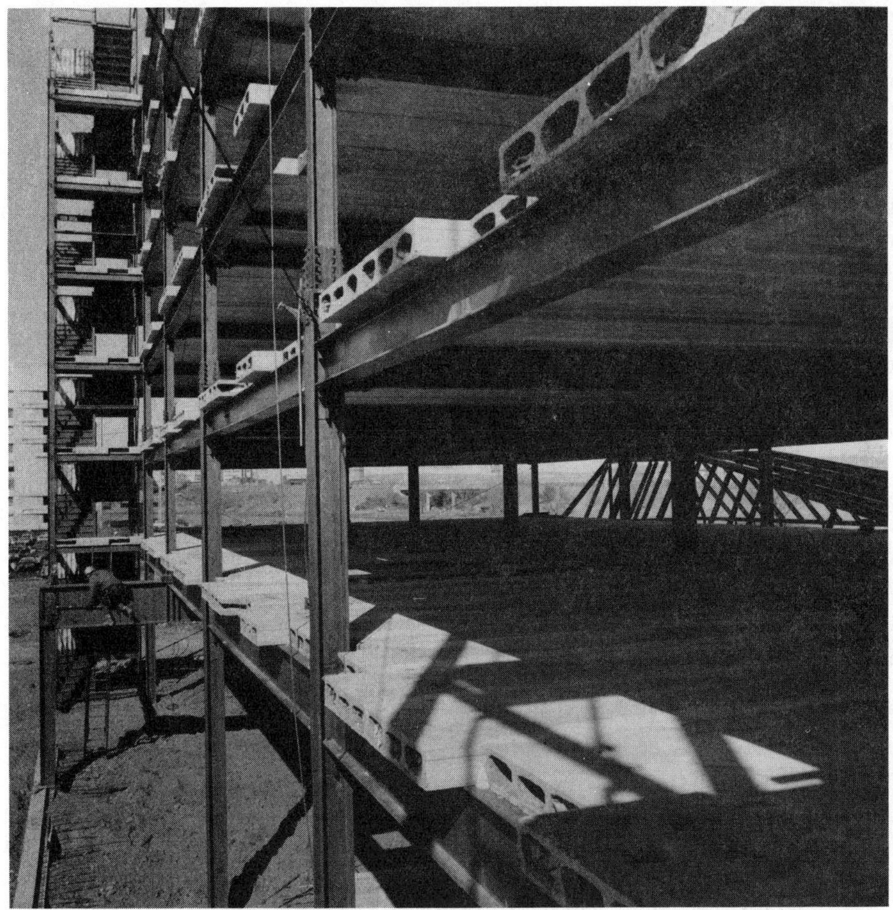

FIGURE 20.27
Steel framing combined with precast concrete floor planks for an eight-story hotel. (*Courtesy of Bethlehem Steel Co.*)

by use of Teflon pads, which effectively eliminate bearing friction, or of pads with low shear rigidity (elastomeric or rubber pads of various types). These transmit gravity loads but permit sizable horizontal deformations.

Precast concrete structures are subject to dimensional changes from creep, shrinkage, and relaxation of prestress in addition to temperature, while in steel structures only temperature changes produce dimensional variations. In the earlier development of precast construction there was a tendency to use soft connections extensively in order to permit these dimensional changes to occur without causing restraint forces in the members, and particularly in the connections. More recent experience with extensive use of soft connections indicates that the resulting structures tend to show insufficient stability against lateral forces such as high wind and particularly earthquake effects. Therefore, those with long experience with precast structures tend to advocate the use of hard connections that result in a high degree of continuity (Ref. 20.15, among others). This trend is also reflected in recent design aids (e.g., Refs. 20.9 and 20.16) and runs parallel to the trend toward continuity in steel structures by means of welding or high tensile bolts. In connections of this sort, provision must be made for resistance to the restraint forces that are caused by the previously described volume changes (see, for example, Ref. 20.9). Considerable information concerning this and other matters relating to connections is found in Ref. 20.16.

Bearing stresses on plain concrete are limited by the ACI Code to $0.85\phi f'_c$, except when the supporting area is wider on all sides than the loaded area A_1. In such a case this value of the permissible bearing stress may be multiplied by $\sqrt{A_2/A_1}$ but not more than 2.0, where A_2 is the maximum portion of the supporting surface that is geometrically similar to and concentric with the loading area (see Sec. 18.6b).

In the design of connections, it is prudent to use load factors that exceed those required for the connected members. This is so because connections are generally subject to high stress concentrations, which preclude the development of much ductility. In contrast, the members connected are likely to possess considerable ductility if designed by usual ACI Code procedures and will give warning of impending collapse if overloading should take place. In addition, imperfections in connection geometry may cause large changes in the magnitude of stresses compared with those assumed in the design.

In designing members according to the ACI Code, load factors of 1.4 and 1.7 are applied to dead and live loads, D and L respectively, to determine the required strength. When volume change effects T are considered, they are normally treated as dead load, and the ultimate load effect U is calculated from the equation $U = 0.75(1.4D + 1.4T + 1.7L)$. The overall reduction factor of 0.75 is introduced recognizing that simultaneous occurrence of worst effects is not likely. However, for design of brackets in precast construction, for example, the ACI Code provides that the tensile force resulting from volume change effects should be included with the live load. Thus a load factor of 1.7 is used. No overall reduction factor is used for brackets.

A wide variety of connection details for precast concrete building components have evolved, only a few of which will be shown here as more or less representative connections. Many additional possibilities are described fully in Ref. 20.16.

Column base connections are generally accomplished using steel base plates that are anchored into the precast column. Figure 20.28*a* shows a column base detail with projecting base plate. Four anchor bolts are used, with double nuts facilitating erection and leveling of the column. Typically about 2 in. of nonshrink grout is used between the top of the pier, footing, or wall and the bottom of the steel base plate. Column reinforcement is welded to the top face of the base plate. Tests have confirmed that such column connections can transmit the full moment for which the column is designed, if properly detailed (Ref. 20.17).

An alternative base detail is shown in Fig. 20.28*b*, with the dimensions of the base plate the same as, or slightly smaller than, the outside column dimensions.

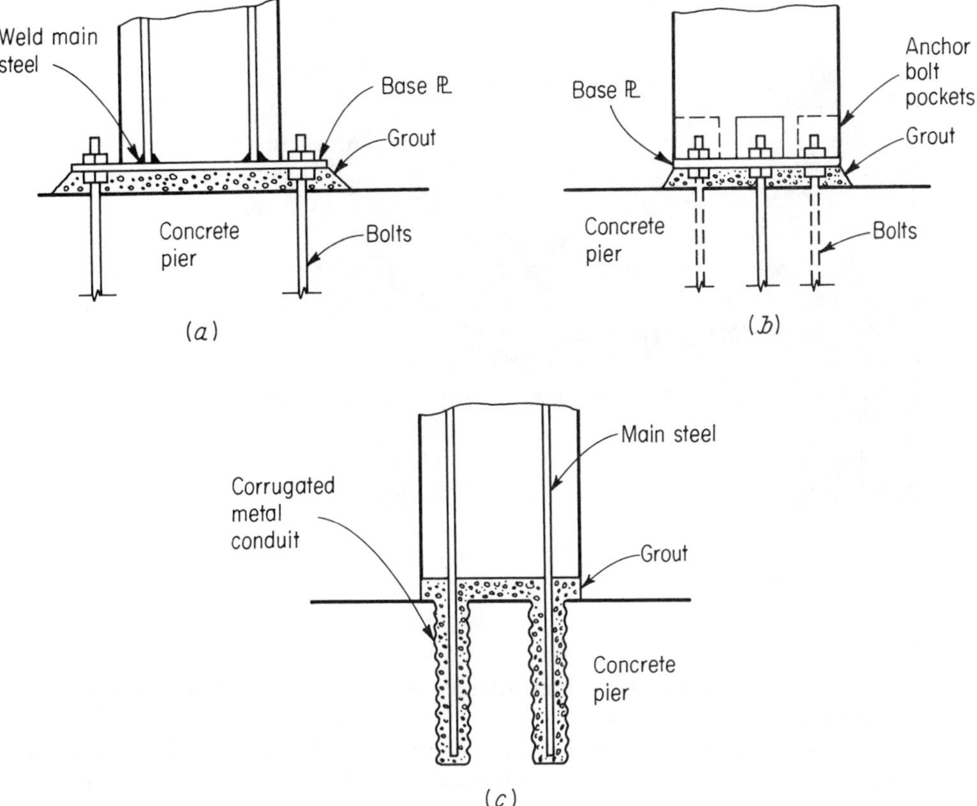

FIGURE 20.28
Column base connections.

Anchor bolt pockets are provided, either centered on the column faces as shown, or located at the corners. Bolt pockets are grouted after the nuts are tightened. Column rebars, not shown here, would be welded to the top face of the base plate as before. Fig. 20.29 shows the baseplate detail, similar to Fig. 20.28*b*, that was used for the precast three-story columns in the parking garage of Fig. 20.22.

In Fig. 20.28*c*, the main column bars project from the ends of the precast member a sufficient distance to develop their strength by bond. The projecting bars are inserted into grout-filled holes cast in the foundation when it is poured.

In all the cases shown, confinement steel should be provided around the anchor bolts in the form of closed ties. A minimum of four No. 3 ties is recommended, placed at 3 in. centers near the top surface of the pier or wall. Tie reinforcement in the columns should be provided as usual.

Figure 20.30 shows several *beam-to-column* connections. In all cases, rectangular beams are shown, but similar details apply to I or T beams. The figure

FIGURE 20.29
Detail at base of precast column.

shows only the basic geometry; and auxiliary reinforcement, anchors, and ties are omitted for the sake of clarity.

Figure 20.30*a* shows a joint detail with a concealed haunch. Well-anchored bearing angles are provided at the column seat and beam end. This type of connection may be used to provide vertical and horizontal reaction components, and with the addition of post-tensioned prestressing, will provide moment resistance as well.

Figure 20.30*b* shows a typical bracket, common for industrial construction where the projecting bracket is not objectionable. The seat angle is welded to rebars anchored in the column. A steel bearing plate is used at the bottom of the beam and anchored into the concrete.

The embedded steel shape of Fig. 20.30*c* is used when it is necessary to avoid projections beyond the face of the column or below the bottom of the beam. A socket is formed in casting the beam, with steel angle or plate at its top, to receive the beam stub.

Finally, Fig. 20.30*d* shows a doweled connection with bars projecting from the column into holes formed in the beam ends. These are grouted after the beams are in position.

Figure 20.31 shows several typical *column-to-column* connections. Figure 20.31*a* shows a detail using anchor bolt pockets and a double-nut system for leveling the upper column. Bolts can also be located at the center of the column faces, as in Fig. 20.28*b*. The detail of Fig. 20.31*b* permits the main steel to be lap-spliced with that from the column below. One of the many possibilities for splicing a column through a continuous beam is shown in Fig. 20.31*c*. Main rebars in both upper and lower columns should be welded to steel cap and base

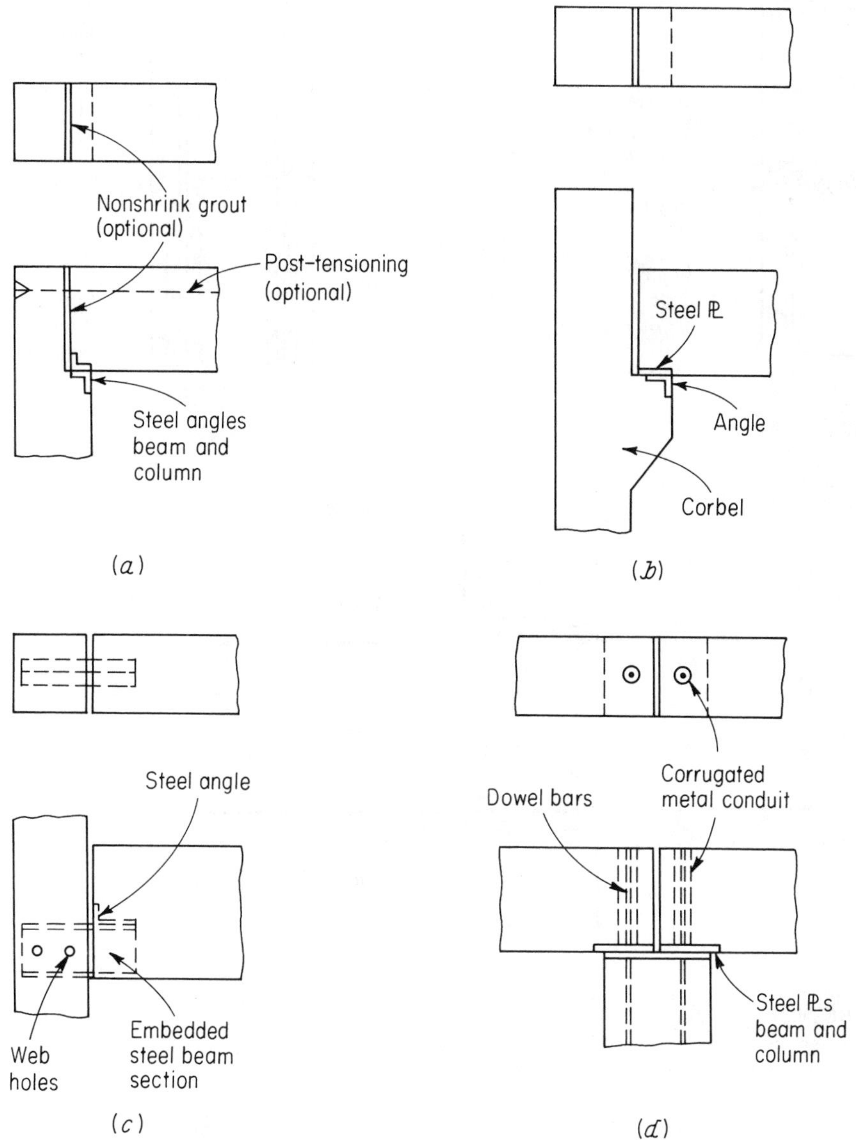

FIGURE 20.30
Beam-to-column connections.

plates to transfer their load, and anchor bolts should be designed with the same consideration. Closely spaced ties must be provided in the columns and in this case in the beam as well, to transfer the load between columns.

Slab-to-beam connections generally use some variation of the detail shown in Fig. 20.32. Support is provided by an L girder (Fig. 20.32*a*) or an inverted

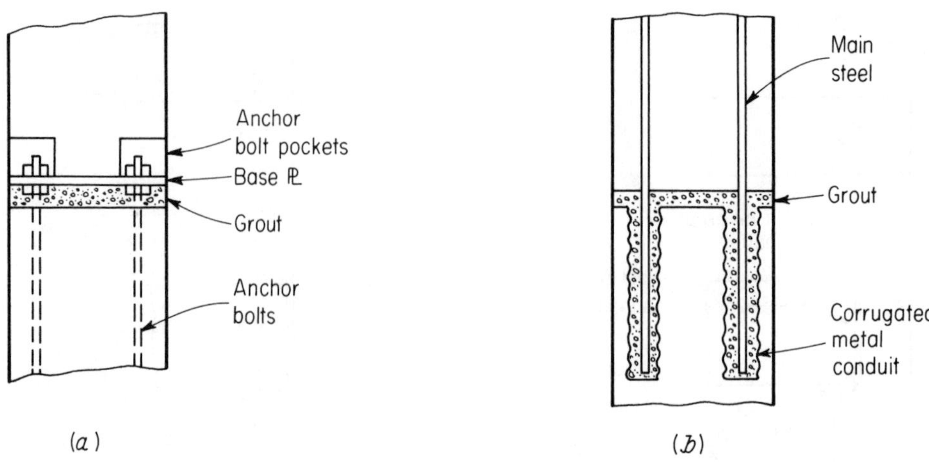

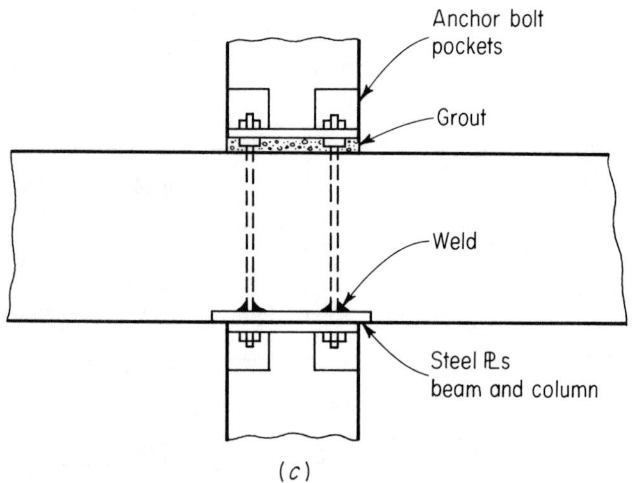

FIGURE 20.31
Column-to-column connections.

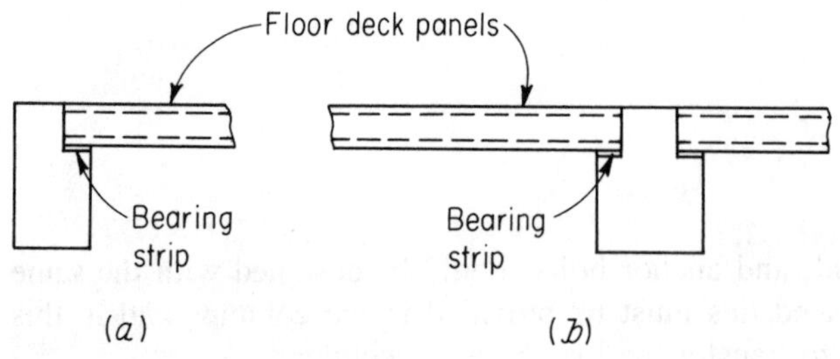

FIGURE 20.32
Slab-to-beam connections.

T girder (Fig. 20.32*b*) that is flush with the top of the precast floor planks. The detail shown is sufficient if no mechanical tie is required between the precast parts. Where positive connection is required, steel plates are set into the top of the members, suitably anchored, and short connecting plates are welded so as to attach the built-in plates.

Basic tools for the design of precast concrete connections are the *shear friction design method* described in detail in Chap. 4 and the *strut-and-tie model* introduced in Chap. 10. Example 4.6 of Sec. 4.10 demonstrated the use of the shear-friction approach to determining the reinforcement for the end-bearing region of a precast concrete girder. The use of both the shear friction method and the strut-and-tie model for joint behavior was shown in Sec. 10.7 of Chap. 10, and Example 10.4 presented the detailed design of a bracket for a precast concrete column. Much additional design information pertaining to precast concrete connection design will be found in Refs. 20.15 through 20.25.

c. Structural Integrity

Precast concrete structures normally lack the joint continuity and high degree of redundancy characteristic of monolithic, cast-in-place reinforced concrete construction. Progressive collapse in the event of abnormal loading, in which the failure of one element leads to the collapse of another, then another, can produce catastrophic results. For this reason, as was mentioned earlier, the use of "soft" connections that rely on friction caused by gravity forces, is generally no longer permitted. Full moment-resisting connections are unusual, but some positive means of connecting members to their supports, with due regard to the need to accommodate dimensional changes associated with creep, shrinkage, and temperature effects, is strongly recommended.

In addition, experience with precast structures has shown that the introduction of special reinforcement in the form of tension ties, though adding little to the cost of construction, can contribute greatly to maintaining structural integrity in the event of extraordinary loading, such as loads caused by extreme winds, earthquake, or explosion. This tension reinforcement is best arranged in a three-dimensional grid, usually on the column lines, tying the floors together vertically and in both horizontal directions. For precast concrete construction, ACI Code 7.13.3 requires that tension ties must be provided in the transverse, longitudinal, and vertical directions of the structure and around its perimeter. Specific details vary widely. Although no specific guidance is offered in either the ACI Code or Commentary regarding steel placement or design forces, valuable suggestions will be found in Refs. 20.16 and 20.26.

20.7 TYPICAL ENGINEERING DRAWINGS FOR BUILDINGS

Design information is conveyed to the builder mainly by engineering drawings. Their preparation is therefore a matter of the utmost importance, and they should be carefully checked by the design engineer to ensure that concrete dimensions and reinforcement agree with the calculations.

Engineering drawings for buildings usually consist of a plan view of each floor showing overall dimensions and locating the main structural elements, cross-sectional views through typical members, and beam and slab schedules that give detailed information on the concrete dimensions and reinforcement in tabular form. Sectional views are usually drawn to a larger scale than the plan and serve to locate the steel and establish cutoff and bend points as well as to define the shape of the member. Usually a separate drawing is included that gives, in the form of schedules and cross sections, the details of columns and footings.

It is wise to include, on each drawing, the material strengths used for the design of the structure, as well as the service live load on which the calculations were based.

Figures 20.33 through 20.35 show typical engineering drawings for a slab-beam-and-girder floor, a one-way joist floor, and a flat plate floor, and Fig. 20.36 shows a typical column schedule.

REFERENCES

20.1 M. Fintel (ed.), *Handbook of Concrete Engineering*, 2d ed., Van Nostrand Reinhold, New York, 1985.

20.2 *CRSI Handbook*, 6th ed., Concrete Reinforcing Steel Institute, Chicago, 1984.

20.3 *ACI Detailing Manual*, American Concrete Institute Publication SP-66, Detroit, 1988.

20.4 "Recommendations for Design of Beam-Column Joints in Monolithic Reinforced Concrete Structures," Reported by ACI Comm. 352, *J. ACI*, vol. 82, no. 3, 1985, pp. 266–283.

20.5 A. E. Cardenas, J. M. Hanson, W. G. Corley, and E. Hognestad, "Design Provisions for Shear Walls," *J. ACI*, vol. 70, no. 3, 1973, pp. 221–230.

20.6 A. E. Cardenas and D. D. Magura, "Strength of High-Rise Shear Walls: Rectangular Cross Section," pp. 119–150 in *Response of Multistory Concrete Structures to Lateral Forces*, ACI Special Publication SP-36, 1973.

20.7 "Precast Structural Concrete in Buildings," Reported by ACI Comm. 512, *J. ACI*, vol. 71, no. 11, 1974, pp. 537–549.

20.8 K. M. Kripanarayanan and M. Fintel, "Analysis and Design of Slender Tilt-Up Reinforced Concrete Wall Panels," *J. ACI*, vol. 71, no. 1, 1974, pp. 20–28.

20.9 *PCI Design Handbook*, 2d ed., Prestressed Concrete Institute, Chicago, 1978.

20.10 "Symposium on Precast Concrete Wall Panels," Special Publication SP-11, American Concrete Institute, Detroit, 1965.

20.11 "Quality Standards and Tests for Precast Concrete Wall Panels," Reported by ACI Comm. 533, *J. ACI*, vol. 66, no. 4, 1969, pp. 270–275.

20.12 "Selection and Use of Materials for Precast Concrete Wall Panels," Reported by ACI Comm. 533, *J. ACI*, vol. 66, no. 10, 1969, pp. 814–820.

20.13 "Design of Precast Concrete Wall Panels," Reported by ACI Comm. 533, *J. ACI*, vol. 68, no. 7, 1971, pp. 504–513.

20.14 "Recommended Practice for Manufactured Reinforced Concrete Floor and Roof Units," Reported by ACI Comm. 512, *J. ACI*, vol. 63, no. 6, 1966, pp. 625–636.

20.15 P. W. Birkland and H. W. Birkland, "Connections in Precast Concrete Construction," *J. ACI*, vol. 63, no. 3, 1966, pp. 345–368.

20.16 "Design and Typical Details of Connections for Precast and Prestressed Concrete," PCI Committee on Connections Details, Prestressed Concrete Institute, Chicago, 1988.

20.17 R. W. LaFraugh and D. D. Magura, "Connections in Precast Concrete Structures—Column Base Plates," *J. ACI*, vol. 11, no. 6, 1966, pp. 18–39.

20.18 R. F. Mast, "Auxiliary Reinforcement in Precast Concrete Connections," *J. Struct. Div. ASCE*, vol. 94, no. ST6, 1968, pp. 1485–1504.

20.19 J. A. Hofbeck, I. O. Ibrahim, and A. H. Mattock, "Shear Transfer in Reinforced Concrete," *J. ACI*, vol. 66, no. 2, 1969, pp. 119–128.

20.20 A. H. Mattock, "Shear Transfer in Concrete Having Reinforcement at an Angle to the Shear Plane," in *Shear in Reinforced Concrete*, ACI Special Publication SP-42, American Concrete Institute, Detroit, 1974.

20.21 A. H. Mattock and N. M. Hawkins, "Shear Transfer in Reinforced Concrete—Recent Research," *J. Prestressed Concr. Inst.*, vol. 17, no. 2, 1972, pp. 55–75.

20.22 A. F. Shaikh, "Proposed Revisions to Shear-Friction Provisions," *J. Prestressed Concr. Inst.*, vol. 23, no. 2, 1978, pp. 12–21.

20.23 L. B. Kriz and C. H. Raths, "Connections in Precast Concrete Structures—Strength of Corbels," *J. Prestressed Concr. Inst.*, vol. 10, no. 1, 1965, pp. 16–47.

20.24 A. H. Mattock, K. C. Chen, and K. Soongswang, "The Behavior of Reinforced Concrete Corbels," *J. Prestressed Concr. Inst.*, vol. 21, no. 2, 1976, pp. 52–77.

20.25 A. H. Mattock, "Design Proposals for Reinforced Concrete Corbels," *J. Prestressed Concr. Inst.*, vol. 21, no. 3, 1976, pp. 18–24.

20.26 "Proposed Design Requirements for Precast Concrete," reported by PCI Committee on Building Code and PCI Technical Activities Committee, *J. Prestressed Concr. Inst.*, vol. 31, no. 6, 1986, pp. 32–47.

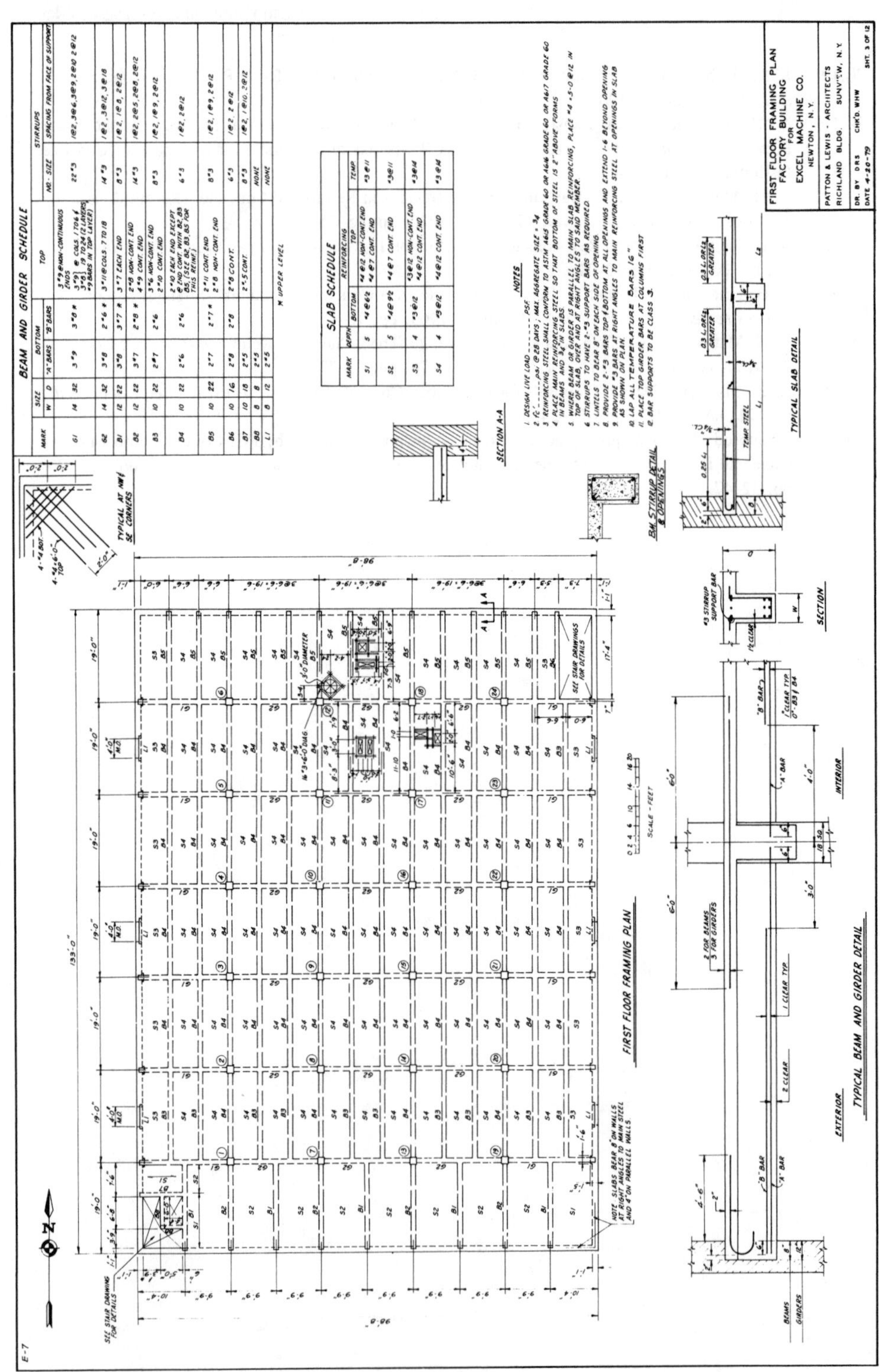

FIGURE 20.33
Engineering drawing of a beam-and-girder floor. (*From Ref. 20.3. Courtesy of American Concrete Institute.*)

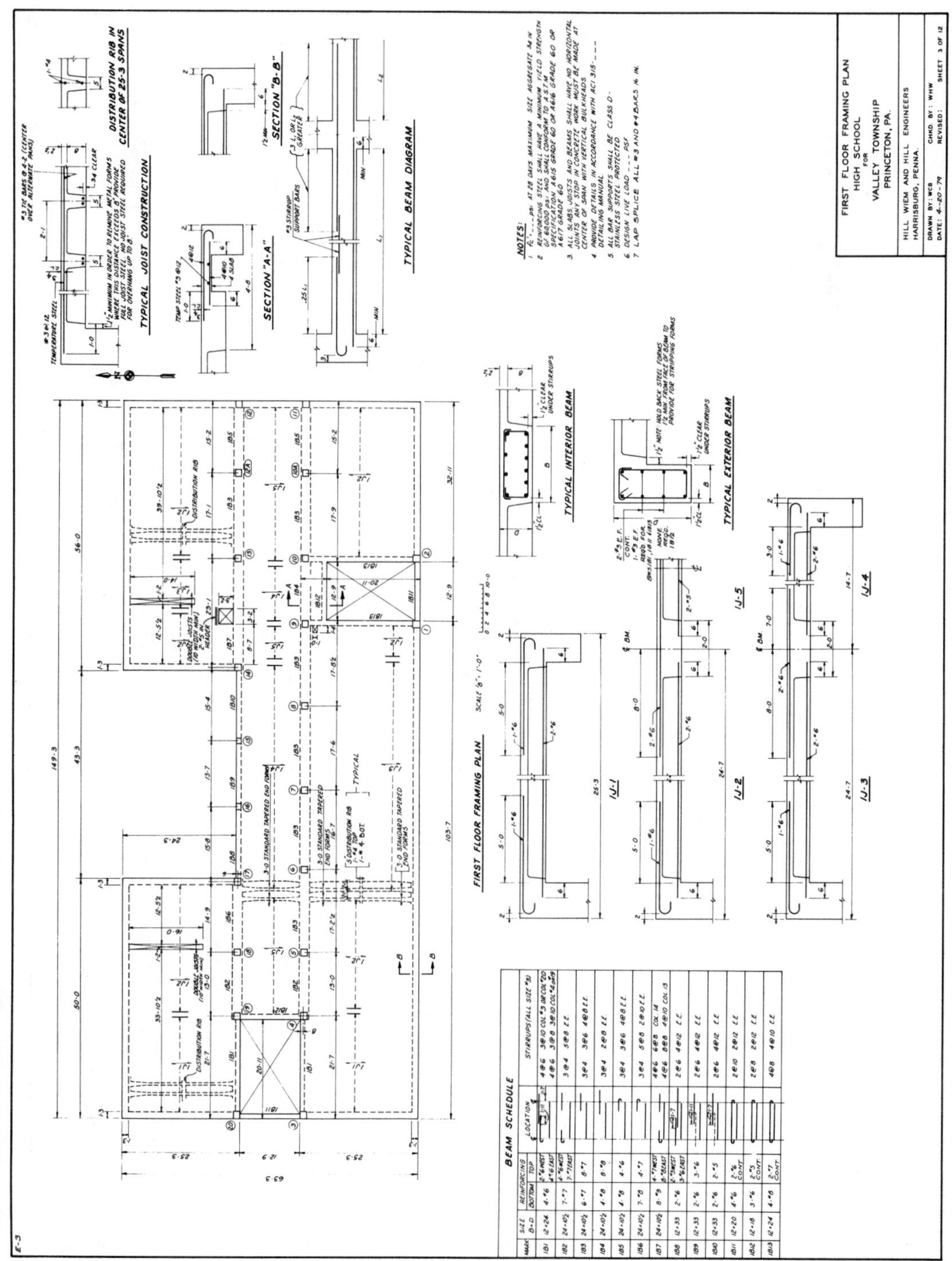

FIGURE 20.34
Engineering drawing of one-way joist floor. (*From Ref. 20.3. Courtesy of American Concrete Institute.*)

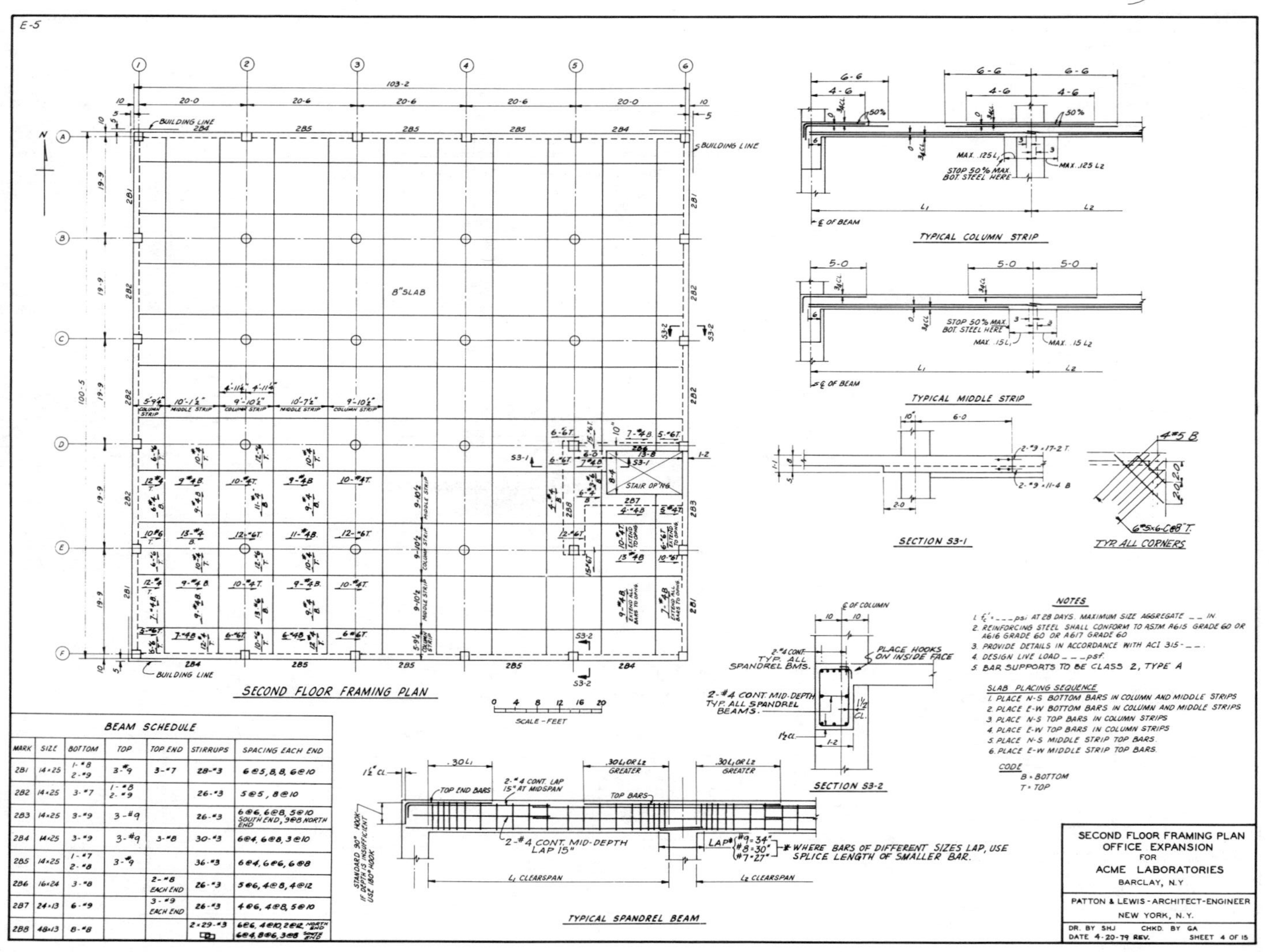

BEAM SCHEDULE

MARK	SIZE	BOTTOM	TOP	TOP END	STIRRUPS	SPACING EACH END
2B1	14×25	1-#8 2-#9	3-#9	3-#7	28-#3	6@5, 8, 8, 6@10
2B2	14×25	3-#7	1-#8 2-#9		26-#3	5@5, 8@10
2B3	14×25	3-#9	3-#9		26-#3	6@6, 6@8, 5@10 SOUTH END, 9@8 NORTH END
2B4	14×25	3-#9	3-#9	3-#8	30-#3	6@4, 6@8, 3@10
2B5	14×25	1-#7 2-#8	3-#9		36-#3	6@4, 6@6, 6@8
2B6	16×24	3-#8		2-#8 EACH END	26-#3	5@6, 4@8, 4@12
2B7	24×13	6-#9		3-#9 EACH END	26-#3	4@6, 4@8, 5@10
2B8	48×13	8-#8			2×29-#3	6@6, 4@10, 2@12, NORTH END 6@4, 8@6, 3@8 SOUTH END

FIGURE 20.35
Engineering drawing of flat plate slab floor. (*From Ref. 20.3. Courtesy of American Concrete Institute.*)

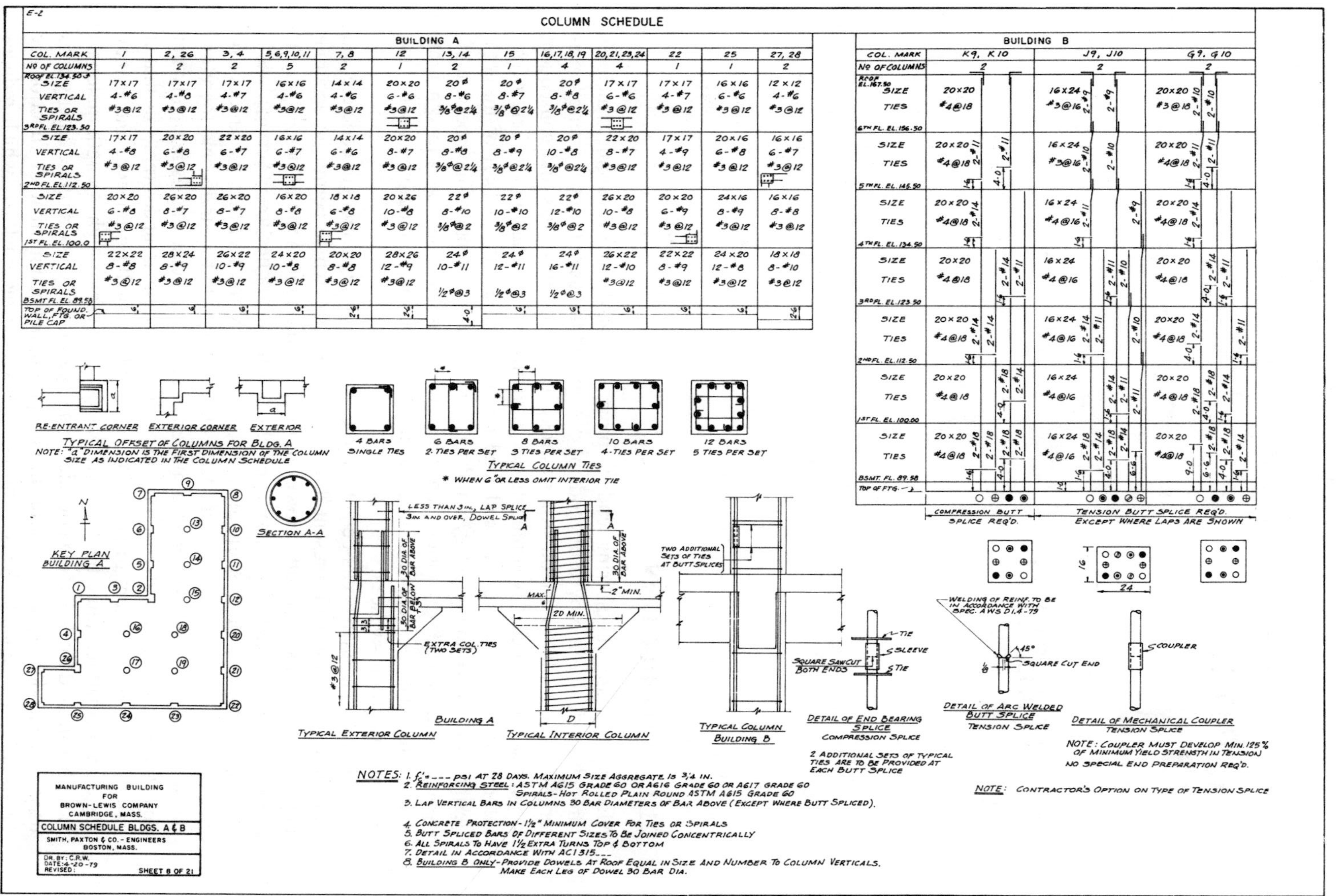

COLUMN SCHEDULE

BUILDING A

COL. MARK	1	2, 26	3, 4	5,6,9,10,11	7, 8	12	13, 14	15	16,17,18,19	20,21,23,24	22	25	27, 28
NO OF COLUMNS	1	2	2	5	2	1	2	1	4	4	1	1	2
ROOF EL. 134.50 — SIZE	17×17	17×17	17×17	16×16	14×14	20×20	20ϕ	20ϕ	20ϕ	17×17	17×17	16×16	12×12
VERTICAL	4-#6	4-#8	4-#7	4-#6	4-#6	6-#6	8-#6	8-#7	8-#8	6-#6	4-#7	6-#6	4-#6
TIES OR SPIRALS	#3@12	#3@12	#3@12	#3@12	#3@12	#3@12	3/8ϕ@2¼	3/8ϕ@2¼	3/8ϕ@2¼	#3@12	#3@12	#3@12	#3@12
3RD FL. EL. 123.50 — SIZE	17×17	20×20	22×20	16×16	14×14	20×20	20ϕ	20ϕ	20ϕ	22×20	17×17	20×16	16×16
VERTICAL	4-#8	6-#8	6-#7	6-#7	6-#6	8-#7	8-#8	8-#9	10-#8	8-#7	4-#9	6-#8	6-#7
TIES OR SPIRALS	#3@12	#3@12	#3@12	#3@12	#3@12	#3@12	3/8ϕ@2¼	3/8ϕ@2¼	3/8ϕ@2¼	#3@12	#3@12	#3@12	#3@12
2ND FL. EL. 112.50 — SIZE	20×20	26×20	26×20	16×20	18×18	20×26	22ϕ	22ϕ	22ϕ	26×20	20×20	24×16	16×16
VERTICAL	6-#8	8-#7	8-#7	8-#8	6-#8	10-#8	8-#10	10-#10	12-#10	10-#8	6-#9	8-#9	8-#8
TIES OR SPIRALS	#3@12	#3@12	#3@12	#3@12	#3@12	#3@12	3/8ϕ@2	3/8ϕ@2	3/8ϕ@2	#3@12	#3@12	#3@12	#3@12
1ST FL. EL. 100.0 — SIZE	22×22	28×24	26×22	24×20	20×20	28×26	24ϕ	24ϕ	24ϕ	26×22	22×22	24×20	18×18
VERTICAL	8-#8	8-#9	10-#9	10-#8	8-#8	12-#9	10-#11	12-#11	16-#11	12-#10	8-#9	12-#8	8-#10
TIES OR SPIRALS	#3@12	#3@12	#3@12	#3@12	#3@12	#3@12	½ϕ@3	½ϕ@3	½ϕ@3	#3@12	#3@12	#3@12	#3@12
BSMT FL. EL. 89.58 — TOP OF FOUND. WALL, FTG. OR PILE CAP	6	6	6	6	2-6	2-6	4-0	6	6	6	6	6	2-6

BUILDING B

COL. MARK	K9, K10	J9, J10	G9, G10
NO OF COLUMNS	2	2	2
ROOF EL. 167.50 — SIZE	20×20	16×24	20×20
TIES	#4@18	#3@16	#3@18
6TH FL. EL. 156.50 — SIZE	20×20	16×24	20×20
TIES	#4@18	#3@16	#4@18
5TH FL. EL. 145.50 — SIZE	20×20	16×24	20×20
TIES	#4@18	#4@16	#4@18
4TH FL. EL. 134.50 — SIZE	20×20	16×24	20×20
TIES	#4@18	#4@16	#4@18
3RD FL. EL. 123.50 — SIZE	20×20	16×24	20×20
TIES	#4@18	#4@16	#4@18
2ND FL. EL. 112.50 — SIZE	20×20	16×24	20×20
TIES	#4@18	#4@16	#4@18
1ST FL. EL. 100.00 — SIZE	20×20	16×24	20×20
TIES	#4@18	#4@16	#4@18
BSMT. FL. 89.58 — TOP OF FTG.			

FIGURE 20.36
Typical building column schedule and details. (*From Ref. 20.3. Courtesy of American Concrete Institute.*)

CHAPTER
21

PRESTRESSED CONCRETE

21.1 INTRODUCTION

Modern structural engineering tends to progress toward more economic structures through gradually improved methods of design and the use of higher strength materials. This results in a reduction of cross-sectional dimensions and consequent weight savings. Such developments are particularly important in the field of reinforced concrete, where the dead load represents a substantial part of the total design load. Also, in multistory buildings, any saving in depth of members, multiplied by the number of stories, can represent a substantial saving in total height, load on foundations, length of heating and electrical ducts, plumbing risers, and wall and partition surfaces.

Significant savings can be achieved by the use of high-strength concrete and steel in conjunction with present-day design methods, which permit an accurate appraisal of member strength. However, there are limitations to this development, due mainly to the interrelated problems of cracking and deflection at service loads. The efficient use of high-strength steel is limited by the fact that the amount of cracking (width and number of cracks) is proportional to the strain, and therefore the stress, in the steel. Although a moderate amount of cracking is normally not objectionable in structural concrete, excessive cracking is undesirable in that it exposes the reinforcement to corrosion, it may be visually offensive, and it may trigger a premature failure by diagonal tension. The use of high-strength materials is further limited by deflection considerations, particularly when refined analysis is used. The slender members that result may permit deflections that are functionally or visually unacceptable. This is further aggravated by cracking, which reduces the flexural stiffness of members.

These limiting features of ordinary reinforced concrete have been largely overcome by the development of prestressed concrete. A prestressed concrete member can be defined as one in which there have been introduced internal stresses of such magnitude and distribution that the stresses resulting from the given external loading are counteracted to a desired degree. Concrete is basically a compressive material, with its strength in tension a low and unreliable value. Prestressing applies a precompression to the member that reduces or eliminates undesirable tensile stresses that would otherwise be present. Cracking under service loads can be minimized or even avoided entirely. Deflections may be limited to an acceptable value; in fact, members can be designed to have zero deflection under the combined effects of service load and prestress force. Deflection and crack control, achieved through prestressing, permit the engineer to make use of efficient and economical high-strength steels in the form of strands, wires, or bars, in conjunction with concretes of much higher strength than normal. Thus prestressing results in overall improvement in performance of structural concrete used for ordinary loads and spans and extends the range of application far beyond old limits, leading not only to much longer spans than previously thought possible, but permitting innovative new structural forms to be employed.

21.2 EFFECTS OF PRESTRESSING

There are at least three alternative ways to look at the prestressing of concrete: (a) as a method of achieving *concrete stress control*, by which the concrete is precompressed so that tension normally resulting from the applied loads is reduced or eliminated, (b) as a means for introducing *equivalent loads* on the concrete member so that the effects of the applied loads are counteracted to the desired degree, and (c) as a *special variation of reinforced concrete* in which prestrained high-strength steel is used, usually in conjunction with high-strength concrete. Each of these viewpoints is useful in the analysis and design of prestressed concrete structures, and they will be illustrated in the following paragraphs.

a. Concrete Stress Control by Prestressing

Many important features of prestressed concrete can be demonstrated by simple examples. Consider first the plain, unreinforced concrete beam shown in Fig. 21.1*a*. It carries a single concentrated load at the center of its span. (The self-weight of the member will be neglected here.) As the load W is gradually applied, longitudinal flexural stresses are induced. If the concrete is stressed only within its elastic range, the flexural stress distribution at midspan will be linear, as shown.

At a relatively low load, the tensile stress in the concrete at the bottom of the beam will reach the tensile strength of the concrete, f_r, and a crack will form. Because no restraint is provided against upward extension of the crack, the beam will collapse without further increase of load.

Now consider an otherwise identical beam, shown in Fig. 21.1*b*, in which a longitudinal axial force P is introduced prior to the vertical loading. The longitudinal prestressing force will produce a uniform axial compression $f_c = P/A_c$,

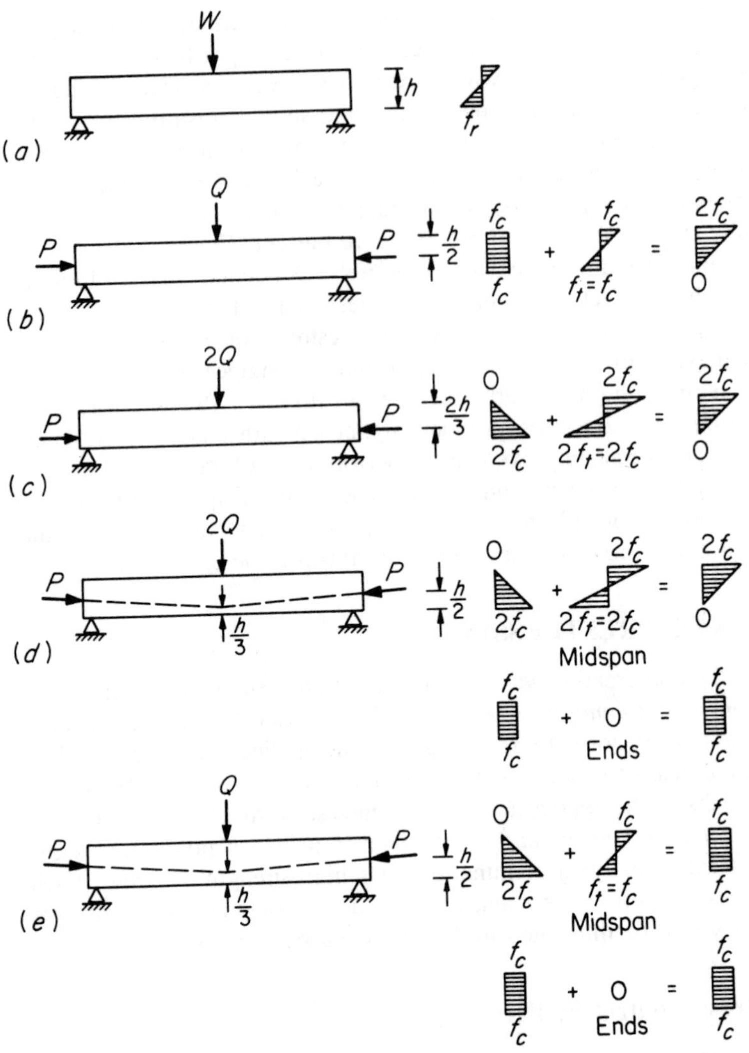

FIGURE 21.1
Alternative schemes for prestressing a rectangular concrete beam: (*a*) plain concrete beam; (*b*) axially prestressed beam; (*c*) eccentrically prestressed beam; (*d*) beam with variable eccentricity; (*e*) balanced load stage for beam with variable eccentricity.

where A_c is the cross-sectional area of the concrete. The force can be adjusted in magnitude so that, when the transverse load Q is applied, the superposition of stresses due to P and Q will result in zero tensile stress at the bottom of the beam as shown. Tensile stress in the concrete may be eliminated in this way or reduced to a specified amount.

But it would be more logical to apply the prestressing force near the bottom of the beam, to compensate more effectively for the load-induced tension. A possible design specification, for example, might be to introduce the maximum compression at the bottom of the beam without causing tension at the top, when only the prestressing force acts. It is easily shown that, for a beam with rectangular cross section, the point of application of the prestressing force should be at the lower third point of the section depth to achieve this. The force P, with the same value as before, but applied with eccentricity $e = h/6$ relative to the concrete centroid, will produce a longitudinal compressive stress distribution varying linearly from zero at the top surface to a maximum of $2f_c = P/A_c + Pec_2/I_c$ at the bottom, where f_c is the concrete stress at the concrete centroid, c_2 is the distance from the concrete centroid to the bottom of the beam, and I_c is the moment of inertia of the cross section. This is shown in Fig. 21.1*c*. The stress at the bottom will be exactly twice the value produced before by axial prestressing.

Consequently, the transverse load can now be twice as great as before, or $2Q$, and still cause no tensile stress. In fact, the final stress distribution resulting from the superposition of load and prestressing force in Fig. 21.1*c* is identical to that of Fig. 21.1*b*, with the same prestressing force, although the load is twice as great. The advantage of eccentric prestressing is obvious.

The methods by which concrete members are prestressed will be discussed in Section 21.3. For present purposes, it is sufficient to know that one practical method of prestressing uses high-strength steel tendons passing through a conduit embedded in the concrete beam. The tendon is anchored, under high tension, at both ends of the beam, thereby causing a longitudinal compressive stress in the concrete. The prestress force of Fig. 21.1*b* and *c* could easily have been applied in this way.

A significant improvement can be made, however, by using a prestressing tendon with variable eccentricity with respect to the concrete centroid, as shown in Fig. 21.1*d*. The load $2Q$ produces a bending moment that varies linearly along the span, from zero at the supports to maximum at midspan. Intuitively, one suspects that the best arrangement of prestressing would produce a *countermoment* that acts in the opposite sense to the load-induced moment and that would vary in the same way. This would be achieved by giving the tendon an eccentricity that varies linearly, from zero at the supports to maximum at midspan. This is shown in Fig. 21.1*d*. The stresses at midspan are the same as those in Fig. 21.1*c*, both when the load $2Q$ acts and when it does not. At the supports, where only the prestress force with zero eccentricity acts, a uniform compression stress f_c is obtained as shown.

For each characteristic load distribution, there is a *best tendon profile* that produces a prestress moment diagram that corresponds to that of the applied load. If the prestress countermoment is made exactly equal and opposite to the load-induced moment, the result is a beam that is subject only to uniform axial compressive stress in the concrete all along the span. Such a beam would be free of flexural cracking, and theoretically it would not be deflected up or down when that particular load is in place, compared to its position as originally cast. Such a result would be obtained for a load of $\frac{1}{2} \times 2Q = Q$, as in Fig. 21.1*e*, for example.

Some important conclusions can be drawn from these simple examples as follows:

1. Prestressing can control or even eliminate concrete tensile stress for specified loading.
2. Eccentric prestress is usually much more efficient than concentric prestress.
3. Variable eccentricity is usually preferable to constant eccentricity, from the viewpoints of both stress control and deflection control.

b. Equivalent Loads

The effect of a change in the vertical alignment of a prestressing tendon is to produce a vertical force on the concrete beam. That force, together with the prestressing force acting at the ends of the beam through the tendon anchorages, can be looked upon as a system of external loads.

In Fig. 21.2*a*, for example, a tendon that applies force P at the centroid of the concrete section at the ends of a beam and that has a uniform slope at angle θ between the ends and midspan introduces a transverse force $2P \sin \theta$ at the point of change of slope at midspan. At the anchorages, the vertical component of the prestressing force is $P \sin \theta$ and the horizontal component is $P \cos \theta$. The horizontal component is very nearly equal to P for the usual flat slope angles. The moment diagram for the beam of Fig. 21.2*a* is seen to have the same form as that for any center-loaded simple span.

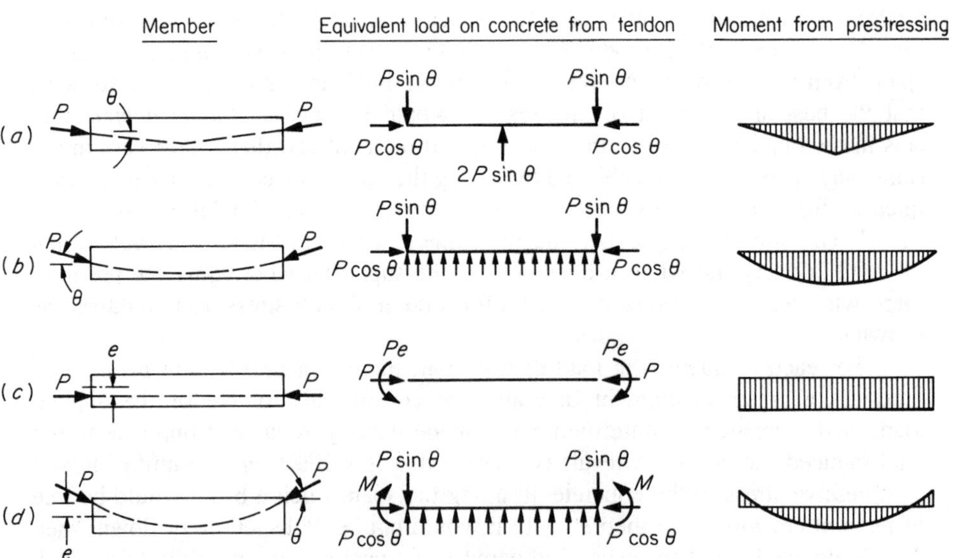

FIGURE 21.2
Equivalent loads and moments produced by prestressing tendons.

The beam of Fig. 21.2*b*, with a curved tendon, is subject to a vertical upward load from the tendon as well as the forces P at each end. The exact distribution of the load depends on the profile of the tendon. A tendon with a parabolic profile, for example, will produce a uniformly distributed load. In this case, the moment diagram will be parabolic, as for a uniformly loaded simple span.

If a straight tendon is used with constant eccentricity, as in Fig. 21.2*c*, there are no vertical forces on the concrete, but the beam is subject to a moment Pe at each end, as well as the axial force P, and a diagram of constant moment results.

The end moment must also be accounted for in the beam of Fig. 21.2*d*, in which a parabolic tendon is used that does not pass through the concrete centroid at the ends of the span. In this case, a uniformly distributed upward load plus end anchorage forces are produced, as in Fig. 21.2*b*, but in addition, the end moments $M = Pe \cos \theta$ must be accounted for.

It may be evident that for any arrangement of applied loads, a tendon profile can be selected so that the equivalent loads acting on the beam from the tendon are just equal and opposite to the applied loads. The result would be a state of pure compressive stress in the concrete, as discussed in somewhat different terms in reference to stress control and Fig. 21.1*e*. An advantage of the equivalent load concept is that it leads the designer to select what is probably the best tendon profile for a particular loading.

c. Prestressed Concrete as a Variation of Reinforced Concrete

In the descriptions of the effects of prestressing in the paragraphs above, it was implied that the prestress force remained constant as the vertical load was introduced, that the concrete responded elastically, and that no concrete cracking occurred. These conditions may prevail up to about the service load level, but if the loads should be increased much beyond that, flexural tensile stresses will eventually exceed the modulus of rupture and cracks will form. Loads can usually be increased much beyond the cracking load in well-designed prestressed beams.

Eventually both the steel and concrete at the cracked section will be stressed into their inelastic range. The condition at incipient failure is shown in Fig. 21.3, which shows a beam carrying a *factored load*, equal to some multiple of the expected service load. The beam undoubtedly would be in a partially cracked state; a possible pattern of flexural cracking is shown in Fig. 21.3*a*.

At the maximum moment section, only the concrete in compression is effective, and all the tension is taken by the steel. The external moment from the applied loads is resisted by the internal force couple $Cz = Tz$. The behavior at this stage is almost identical to that of an ordinary reinforced concrete beam at overload. The main difference is that the very high-strength steel used must be *prestrained* before loads are applied to the beam; otherwise, the high steel stresses would produce excessive concrete cracking and large beam deflections.

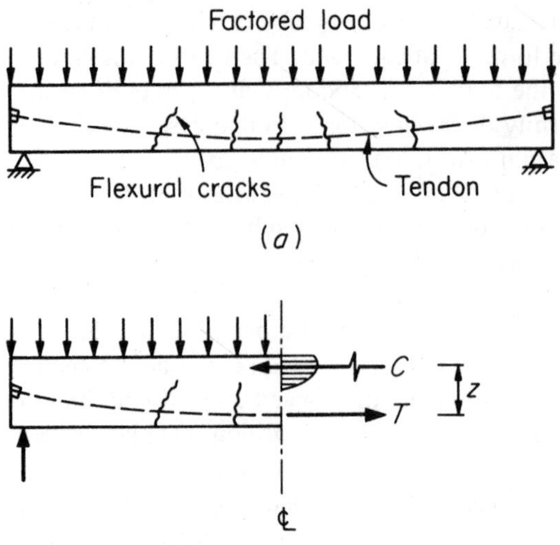

FIGURE 21.3
Prestressed concrete beam at load near flexural failure: (*a*) beam with factored load applied; (*b*) equilibrium of forces on left half of beam.

Each of the three viewpoints described—concrete stress control, equivalent loads, and reinforced concrete using prestrained steel—is useful in the analysis and design of prestressed concrete beams, and none of the three is sufficient in itself. Neither an elastic stress analysis nor an equivalent load analysis provides information about strength or safety margin. However, the stress analysis is helpful in predicting the extent of cracking, and the equivalent load analysis is often the best way to calculate deflections. Strength analysis is essential to evaluate safety against collapse, but it tells nothing about cracking or deflections of the beam under service conditions.

21.3 SOURCES OF PRESTRESS FORCE

Prestress can be applied to a concrete member in many ways. Perhaps the most obvious method of precompressing is the use of jacks reacting against abutments, as shown in Fig. 21.4*a*. Such a scheme has been employed for large projects. Many variations are possible, including replacing the jacks with compression struts after the desired stress in the concrete is obtained or using inexpensive jacks, that remain in place in the structure, in some cases with a cement grout used as the hydraulic fluid. The principal difficulty associated with such a system is that even a slight movement of the abutments will drastically reduce the prestress force.

In most cases the same result is more conveniently obtained by tying the jack bases together with wires or cables, as shown in Fig. 21.4*b*. These wires or cables may be external, located on each side of the beam; more usually they are passed through a hollow conduit embedded in the concrete beam. Usually, one end of the prestressing tendon is anchored, and all the force is applied at the other end. After attainment of the desired amount of prestress force, the tendon

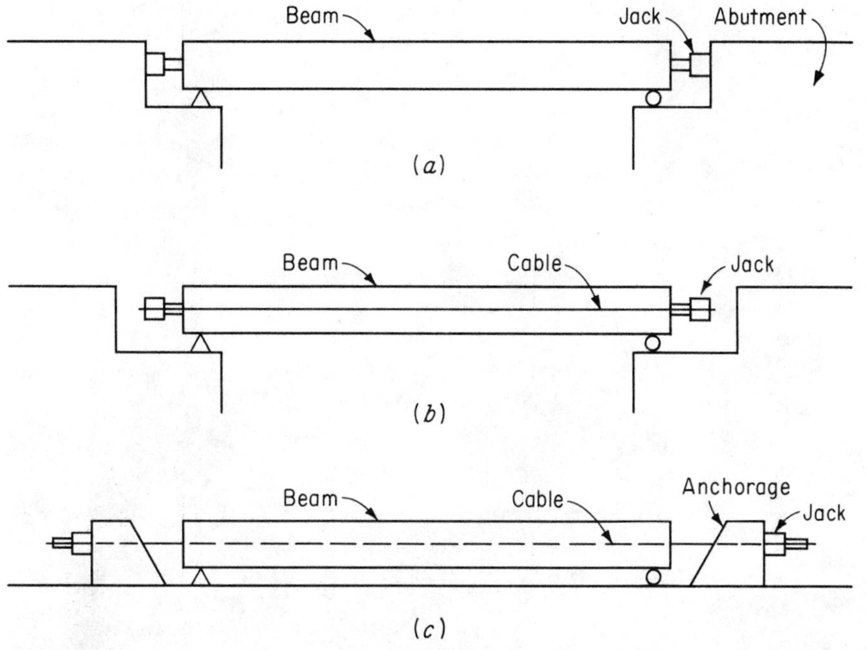

FIGURE 21.4
Prestressing methods: (*a*) post-tensioning by jacking against abutments; (*b*) post-tensioning with jacks reacting against beam; (*c*) pretensioning with tendon stressed between fixed external anchorages.

is wedged against the concrete and the jacking equipment is removed for reuse. Note that in this type of prestressing the entire system is self-contained and is independent of relative displacement of the supports.

Another method of prestressing which is widely used is illustrated by Fig. 21.4*c*. The prestressing strands are tensioned between massive abutments in a casting yard prior to placing the concrete in the beam forms. The beam is poured around the tensioned strands, and after the concrete has attained sufficient strength, the jacking pressure is released. This transfers the prestressing force to the concrete by bond and friction along the strands, chiefly at the outer ends.

Other means for introducing the desired prestressing force have been attempted on an experimental basis. Thermal prestressing can be achieved by preheating the steel by electrical or other means. Anchored against the ends of the concrete beam while in the extended state, the steel cools and tends to contract. The prestress force is developed through the restrained contraction. The use of expanding cement in concrete members has been tried with varying success. The volumetric expansion, restrained by steel strands or by fixed abutments, produces the prestress force.

Most of the patented systems for applying prestress in current use are variations of those shown in Fig. 21.4*b* and *c*. Such systems can generally be classified as *pretensioning* or *post-tensioning* systems. In the case of pretensioning, the tendons are stressed before the concrete is placed, as in Fig. 21.4*c*. This system is

FIGURE 21.5
Massive strand jacking abutment at the end of a long pretensioning bed. (*Courtesy of Concrete Technology Corporation.*)

well suited for mass production, since casting beds can be made several hundred feet long, the entire length cast at once, and individual beams cut to the desired length from the long casting. Fig. 21.5 shows workers using a hydraulic jack to tension strands at the anchorage of a long pretensioning bed. Although each tendon is individually stressed in this case, often large capacity jacks are used to tension all strands simultaneously.

In post-tensioned construction, shown in Fig. 21.4*b*, the tendons are tensioned after the concrete is placed and has acquired its strength. Usually, a hollow conduit or sleeve is provided in the beam, through which the tendon is passed. In some cases, hollow box-section beams are used. The jacking force is usually applied against the ends of the hardened concrete, eliminating the need for massive abutments. In Fig. 21.6, six tendons, each consisting of many individual strands, are being post-tensioned sequentially using a portable hydraulic jack.

FIGURE 21.6
Post-tensioning a bridge girder using a portable jack to stress multistrand tendons. (*Courtesy of Concrete Technology Corporation.*)

A large number of particular systems, steel elements, jacks, and anchorage fittings have been developed in this country and abroad, many of which differ from each other only in minor details (Refs. 21.1 to 21.6). As far as the designer of prestressed concrete structures is concerned, it is unnecessary and perhaps even undesirable to specify in detail the technique that is to be followed and the equipment to be used. It is frequently best to specify only the magnitude and line of action of the prestress force. The contractor is then free, in bidding the work, to receive quotations from several different prestressing subcontractors, with resultant cost savings. It is evident, however, that the designer must have some knowledge of the details of the various systems contemplated for use, so that in selecting cross-sectional dimensions, any one of several systems can be accommodated.

21.4 PRESTRESSING STEELS

Early attempts at prestressing concrete were unsuccessful because steel of ordinary structural strength was used. The low prestress obtainable in such rods was quickly lost due to shrinkage and creep in the concrete.

Such changes in length of concrete have much less effect on prestress force if that force is obtained using highly stressed steel wires or cables. In Fig. 21.7*a*, a concrete member of length L is prestressed using steel bars of ordinary strength stressed to 24,000 psi. With $E_s = 29 \times 10^6$ psi, the unit strain ϵ_s required to

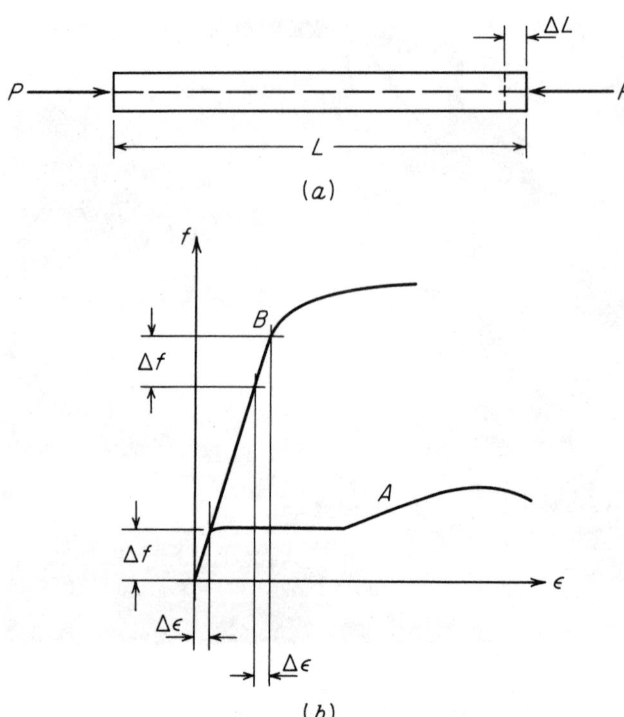

FIGURE 21.7
Loss of prestress due to concrete shrinkage and creep.

produce the desired stress in the steel of 24,000 is

$$\epsilon_s = \frac{\Delta L}{L} = \frac{f_s}{E_s} = \frac{24,000}{29 \times 10^6} = 8.0 \times 10^{-4}$$

However, the long-term strain in the concrete due to shrinkage and creep alone, if the prestress force were maintained over a long period, would be of the order of 8.0×10^{-4} and would be sufficient to completely relieve the steel of all stress.

Alternatively, suppose that the beam is prestressed using high tensile steel stressed to 150,000 psi. The elastic modulus of steel does not vary greatly, and the same value of 29×10^6 psi will be assumed here. Then in this case the unit strain required to produce the desired stress in the steel is

$$\epsilon_s = \frac{150,000}{29 \times 10^6} = 51.7 \times 10^{-4}$$

If shrinkage and creep strain are the same as before, the net strain in the steel after these losses is

$$\epsilon_{\mathrm{s,net}} = (51.7 - 8.0) \times 10^{-4} = 43.7 \times 10^{-4}$$

and the corresponding stress after losses is

$$f_s = \epsilon_{\mathrm{s,net}} E_s = (43.7 \times 10^{-4})(29 \times 10^6) = 127,000 \text{ psi}$$

This represents a stress loss of about 15 percent, compared with 100 percent loss in the beam using ordinary steel. It is apparent that the amount of stress lost because of shrinkage and creep is independent of the original stress in the steel. Therefore, the higher the original stress the lower the percentage loss. This is illustrated graphically by the stress-strain curves of Fig. 21.7*b*. Curve *A* is representative of ordinary reinforcing rods, with a yield stress of 60,000 psi, while curve *B* represents high tensile steel, with an ultimate stress of 250,000 psi. The stress change Δf resulting from a certain change in strain $\Delta\epsilon$ is seen to have much less effect when high steel stress levels are attained. Prestressing of concrete is therefore practical only when steels of very high strength are used.

Prestressing steel is most commonly used in the form of individual wires, stranded cable made up of seven wires, and alloy-steel bars. The physical properties of these have been discussed in Sec. 2.14, and typical stress-strain curves appear in Fig. 2.13.

The tensile stress permitted by ACI Code 18.5 in prestressing wires, strands, or bars is dependent upon the stage of loading. When the jacking force is first applied, a stress of $0.80f_{pu}$ or $0.94f_{py}$ is allowed, whichever is smaller, where f_{pu} is the ultimate strength of the steel and f_{py} is the yield strength. Immediately after transfer of prestress force to the concrete, the permissible stress is $0.74f_{pu}$ or $0.82f_{py}$, whichever is smaller (except at post-tensioning anchorages where the stress is limited to $0.70f_{pu}$). The justification for a higher allowable stress during the stretching operation is that the steel stress is known quite precisely at this stage. Hydraulic jacking pressure and total steel strain are quantities that are easily measured. In addition, if an accidentally deficient tendon should break, it can be replaced; in effect, the tensioning operation is a performance test of the material. The lower values of allowable stress apply after elastic shortening of the concrete, frictional loss, and anchorage slip have taken place, when service loads may be applied. The steel stress is further reduced during the life of the member due to shrinkage and creep in the concrete and relaxation in the steel.

The strength and other characteristics of prestressing wire, strands, and bars vary somewhat between manufacturers, as do methods of grouping tendons and anchoring them. Typical information is given for illustration in Table A.16 of App. A and in Ref. 21.1.

21.5 CONCRETE FOR PRESTRESSED CONSTRUCTION

Ordinarily, concrete of substantially higher compressive strength is used for prestressed structures than for those constructed of ordinary reinforced concrete. Most prestressed construction in the United States at present is designed for a compressive strength between 5000 and 6000 psi. There are several reasons for this:

1. High-strength concrete normally has a higher modulus of elasticity (see Fig. 2.3). This means a reduction in initial elastic strain under application of prestress force and a reduction in creep strain, which is approximately proportional to elastic strain. This results in a reduction in loss of prestress.

2. In post-tensioned construction, high bearing stresses result at the ends of beams where prestressing force is transferred from the tendons to anchorage fittings, which bear directly against the concrete. This problem can be met by increasing the size of the anchorage fitting or by increasing the bearing capacity of the concrete by increasing its compressive strength. The latter is usually more economical.
3. In pretensioned construction, where transfer by bond is customary, the use of high-strength concrete will permit the development of higher bond stresses.
4. A substantial part of the prestressed construction in the United States is precast, with the concrete mixed, placed, and cured under carefully controlled conditions that facilitate obtaining higher strengths.

The strain characteristics of concrete under short-time and sustained loads assume an even greater importance in prestressed structures than in reinforced concrete structures because of the influence of strain on loss of prestress force. Strains due to stress, together with volume changes due to shrinkage and temperature changes, may have considerable influence on prestressed structures. In this connection, it is suggested that the reader review Secs. 2.7 to 2.10, which discuss in some detail the compressive and tensile strengths of concrete under short-time and sustained loads and the changes in concrete volume that occur due to shrinkage and temperature change.

As for prestressing steels, the allowable stresses in the concrete, according to ACI Code 18.4, depend upon the stage of loading. These stresses are given in Table 21.1. Here f'_{ci} is the compressive strength of the concrete at the time of initial prestress, and f'_c the specified compressive strength of the concrete.

Table 21.1 Permissible stresses in concrete in prestressed flexural members

1. Stresses in concrete immediately after prestress transfer, before time-dependent prestress losses, shall not exceed the following:	
a. Extreme fiber stress in compression	$0.60f'_{ci}$
b. Extreme fiber stress in tension except as permitted in (*c*)	$3\sqrt{f'_{ci}}$
c. Extreme fiber stress in tension at ends of simply supported members	$6\sqrt{f'_{ci}}$
Where computed tensile stresses exceed these values, bonded auxiliary reinforcement (nonprestressed or prestressed) shall be provided in the tensile zone to resist the total tensile force in the concrete computed with the assumption of an uncracked section	
2. Stresses in concrete at service loads, after allowance for all prestress losses, shall not exceed the following:	
a. Extreme fiber stress in compression	$0.45f'_c$
b. Extreme fiber stress in tension in precompressed tensile zone	$6\sqrt{f'_c}$
c. Extreme fiber stress in tension in precompressed tensile zone of members, except two-way slab systems, where analysis based on transformed cracked sections and on bilinear moment-deflection relationships shows that immediate and long-time deflections comply with restrictions stated elsewhere in the ACI Code	$12\sqrt{f'_c}$
3. Permissible stresses in concrete given above may be exceeded if it is shown by test or analysis that performance will not be impaired	

21.6 ELASTIC FLEXURAL ANALYSIS

It has been noted earlier in this text that the design of concrete structures may be based either on providing sufficient strength, which would be utilized fully only if the expected loads were increased by an overload factor, or on keeping material stresses within permissible limits when actual service loads act. In the case of ordinary reinforced concrete members, strength design is generally used. Members are proportioned on the basis of strength requirements and then checked for satisfactory service load behavior, notably with respect to deflection and cracking. The design is then modified if necessary.

In the case of prestressed concrete members, present practice is to proportion members so that stresses in the concrete and steel at actual service loads are within permissible limits. These limits are a fractional part of the actual capacities of the materials. There is some logic to this approach, since an important objective of prestressing is to improve the performance of members at service loads. Furthermore, service load requirements often control the amount of prestress force used. Design based on service loads may usually be carried out assuming elastic behavior of both the concrete and the steel, since stresses are relatively low in each.

Regardless of the starting point chosen for the design, a structural member must be satisfactory at all stages of its loading history. Accordingly, prestressed members proportioned on the basis of permissible stresses must also be checked to ensure that sufficient strength is provided should overloads occur, and deflection and cracking under service loads should be investigated. Consistent with most U.S. practice, in this text the design of prestressed concrete beams will initiate with a consideration of stress limits, after which strength and other properties will be checked.

It is convenient to think of prestressing forces as a system of external forces acting on a concrete member, which must be in equilibrium under the action of those forces. Figure 21.8*a* shows a simple-span prestressed beam with curved

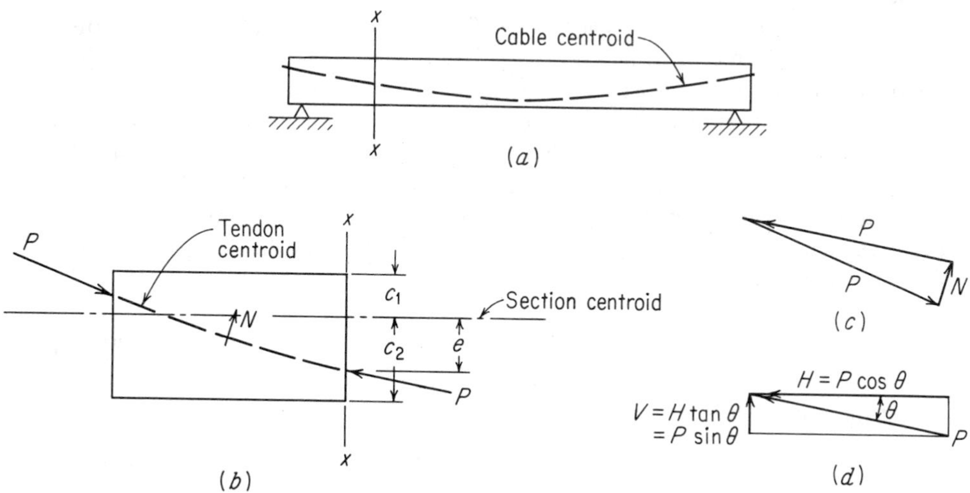

FIGURE 21.8
Prestressing forces acting on concrete.

tendons, typical of many post-tensioned members. The portion of the beam to the left of a vertical cutting plane *x-x* is taken as a free body, with forces acting as shown in Fig. 21.8*b*. The force P at the left end is exerted on the concrete through the tendon anchorage, while the force P at the cutting plane *x-x* results from combined shear and normal stresses acting at the concrete surface at that location. The direction of P is tangent to the curve of the tendon at each location. Note the presence of the force N, acting on the concrete from the tendon, due to tendon curvature. This force will be distributed in some manner along the length of the tendon, the exact distribution depending upon the tendon profile. Its resultant and the direction in which the resultant acts can be found from the force diagram of Fig. 21.8*c*.

It is convenient in working with the prestressing force P to divide it into its components in the horizontal and vertical directions. The horizontal component (Fig. 21.8*d*) is $H = P\cos\theta$, and the vertical component is $V = H\tan\theta = P\sin\theta$, where θ is the angle of inclination of the tendon centroid at the particular section. Since the slope angle is normally quite small, the cosine of θ is very close to unity and it is sufficient for most calculations to take $H = P$.

The magnitude of the prestress force is not constant. The *jacking force* P_j is immediately reduced to what is termed the *initial prestress force* P_i because of elastic shortening of the concrete upon transfer, slip of the tendon as the force is transferred from the jacks to the beam ends, and loss due to friction between the tendon and the concrete (post-tensioning) or between the tendon and cable alignment devices (pretensioning). There is a further reduction of force from P_i to the *effective prestress* P_e, occurring over a long period of time at a gradually decreasing rate, because of concrete creep under the sustained prestress force, concrete shrinkage, and relaxation of stress in the steel. Methods for predicting losses will be discussed in Sec. 21.13. Of primary interest to the designer are the initial prestress P_i immediately after transfer and the final or effective prestress P_e after all losses.

In developing elastic equations for flexural stress, the effects of prestress force, self-weight moment, and dead and live load moments are calculated separately, and the separate stresses are superimposed. When the initial prestress force P_i is applied with an eccentricity e below the centroid of the cross section with area A_c and top and bottom fiber distances c_1 and c_2 respectively, it causes the compressive stress $-P_i/A_c$ and the bending stresses $+P_i e c_1/I_c$ and $-P_i e c_2/I_c$ in the top and bottom fibers respectively (compressive stresses are designated as negative, tensile stresses as positive), as shown in Fig. 21.9*a*. Then, at the top fiber, the stress is

$$f_1 = -\frac{P_i}{A_c} + \frac{P_i e c_1}{I_c} = -\frac{P_i}{A_c}\left(1 - \frac{e c_1}{r^2}\right) \qquad (21.1a)$$

and at the bottom fiber

$$f_2 = -\frac{P_i}{A_c} - \frac{P_i e c_2}{I_c} = -\frac{P_i}{A_c}\left(1 + \frac{e c_2}{r^2}\right) \qquad (21.1b)$$

where r is the radius of gyration of the concrete section. Normally, as the eccentric prestress force is applied, the beam deflects upward. The beam self-weight w_o

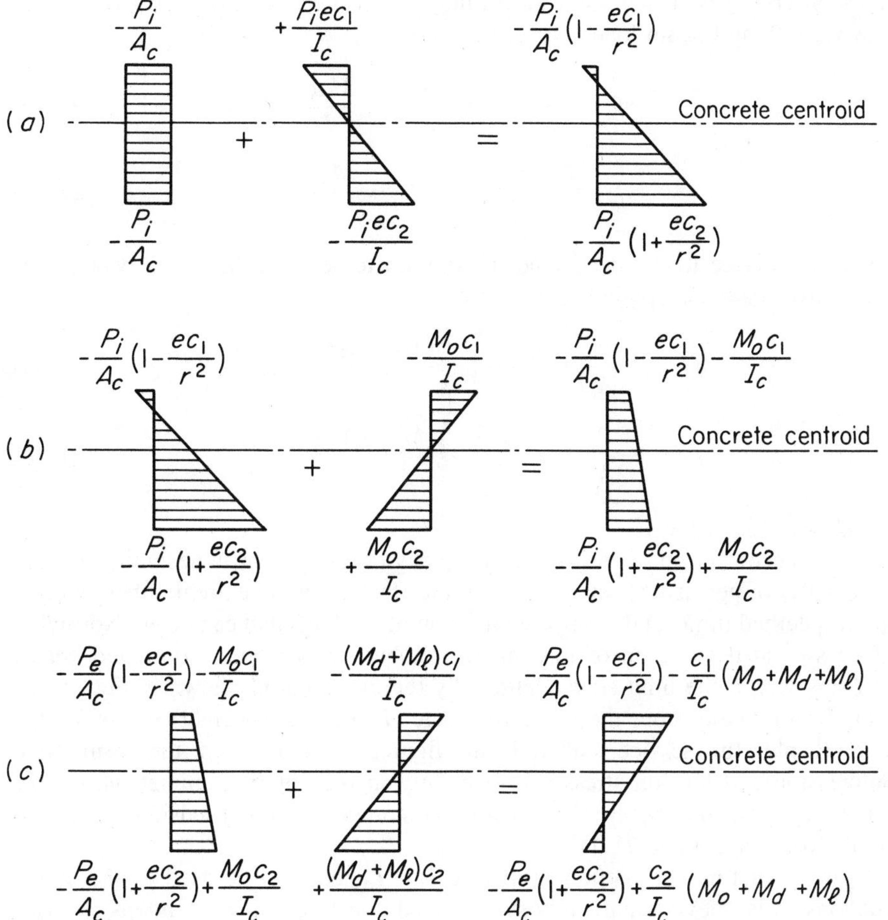

FIGURE 21.9
Concrete stress distributions in beams: (*a*) effect of prestress; (*b*) effect of prestress plus self-weight of beam; (*c*) effect of prestress, self-weight, and external dead and live service loads.

then causes additional moment M_o to act, and the net top and bottom fiber stresses become

$$f_1 = -\frac{P_i}{A_c}\left(1 - \frac{ec_1}{r^2}\right) - \frac{M_o c_1}{I_c} \tag{21.2a}$$

$$f_2 = -\frac{P_i}{A_c}\left(1 + \frac{ec_2}{r^2}\right) + \frac{M_o c_2}{I_c} \tag{21.2b}$$

as shown in Fig. 21.9*b*. At this stage time-dependent losses due to shrinkage, creep, and relaxation commence, and the prestressed force gradually reduces from P_i to P_e. It is usually acceptable to assume that all such losses occur prior to the application of service loads, since the concrete stresses at service loads will be critical after losses, not before. Accordingly, the stresses in the top and bottom fiber, with P_e and beam load acting, become

$$f_1 = -\frac{P_e}{A_c}\left(1 - \frac{e c_1}{r^2}\right) - \frac{M_o c_1}{I_c} \tag{21.3a}$$

$$f_2 = -\frac{P_e}{A_c}\left(1 + \frac{e c_2}{r^2}\right) + \frac{M_o c_2}{I_c} \tag{21.3b}$$

When full service loads (dead load in addition to self-weight of the beam, plus service live load) are applied, the stresses are

$$f_1 = -\frac{P_e}{A_c}\left(1 - \frac{e c_1}{r^2}\right) - \frac{(M_o + M_d + M_l)c_1}{I_c} \tag{21.4a}$$

$$f_2 = -\frac{P_e}{A_c}\left(1 + \frac{e c_2}{r^2}\right) + \frac{(M_o + M_d + M_l)c_2}{I_c} \tag{21.4b}$$

as shown in Fig. 21.9*c*.

It is necessary, in reviewing the adequacy of a beam (or in designing a beam on the basis of permissible stresses), that the stresses in the extreme fibers remain within specified limits under any combination of loadings that can occur. Normally, the stresses at the section of maximum moment, in a properly designed beam, must stay within the limit states defined by the distributions shown in Fig. 21.10 as the beam passes from the unloaded stage (P_i plus self-weight) to the loaded stage (P_e plus full service loads). In the figure, f_{ci} and f_{ti} are the permissible compression and tension stresses respectively in the concrete immediately after transfer, and f_{cs} and f_{ts} are the permissible compression and tension stresses at service loads (see Table 21.1).

In calculating the section properties A_c, I_c, etc., to be used in the above equations, it is relevant that, in post-tensioned construction, the tendons are usually grouted in the conduits after tensioning. Before grouting, stresses should be based on the net section with holes deducted. After grouting, the transformed section should be used with holes considered filled with concrete and with the steel replaced with an equivalent area of concrete. However, it is satisfactory, unless the holes are quite large, to compute section properties on the basis of the gross concrete section. Similarly, while in pretensioned beams the properties of

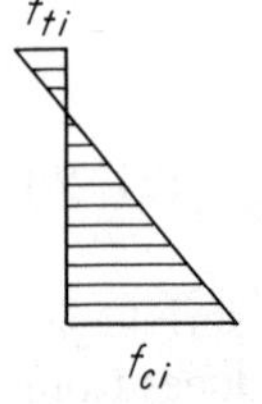

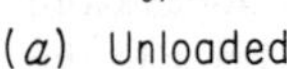

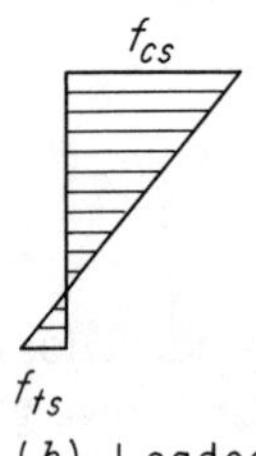

FIGURE 21.10
Stress limits: (*a*) unloaded beam, with initial prestress plus self-weight; (*b*) loaded beam, with effective prestress, self-weight, and full service load.

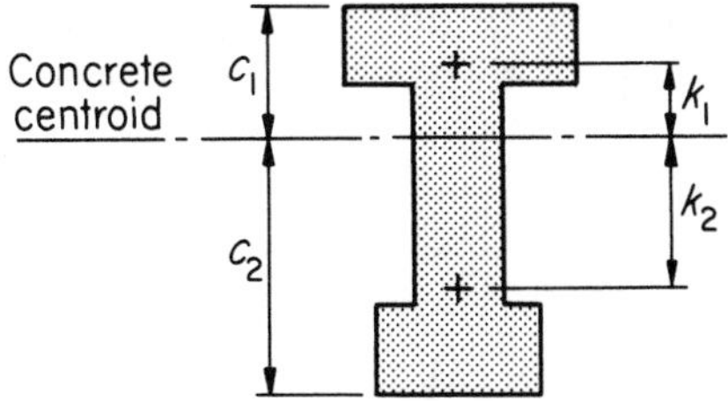

FIGURE 21.11
Location of kern points.

the transformed section should be used, it makes little difference if calculations are based on properties of the gross concrete section.†

It is useful to establish the location of the upper and lower *kern points* of a cross section. These are defined as the limiting points inside which the prestress force resultant may be applied without causing tension anywhere in the cross section. Their locations are obtained by writing the expression for the tensile fiber stress due to application of an eccentric prestress force acting alone and setting this expression equal to zero to solve for the required eccentricity. In Fig. 21.11, to locate the upper kern-point distance k_1 from the neutral axis, let the prestress force resultant P act at that point. Then the bottom fiber stress is

$$f_2 = -\frac{P}{A_c}\left(1 + \frac{ec_2}{r^2}\right) = 0$$

Thus with

$$1 + \frac{ec_2}{r^2} = 0$$

one obtains the corresponding eccentricity

$$e = k_1 = -\frac{r^2}{c_2} \tag{21.5a}$$

Similarly, the lower kern-point distance k_2 is

$$k_2 = \frac{r^2}{c_1} \tag{21.5b}$$

The region between these two limiting points is known as the *kern*, or in some cases the *core*, of the section.

† ACI Code 18.2.6 contains the following provision: "In computing section properties prior to bonding of prestressing tendons, the effect of loss of area due to open ducts shall be considered." It is noted in ACI Commentary 18.2.6 that "If the effect of the open duct area on design is deemed negligible, section properties may be based on total area. In pretensioned members and in post-tensioned members after grouting, section properties may be based on gross sections, net sections, or effective sections using the transformed areas of bonded tendons and nonprestressed reinforcement."

Example 21.1 Pretensioned I beam with constant eccentricity. A simply supported symmetrical I beam shown in cross section in Fig. 21.12*a* will be used on a 40 ft simple span. It has the following section properties:

Moment of inertia: $I_c = 12{,}000 \text{ in}^4$

Concrete area: $A_c = 176 \text{ in}^2$

Radius of gyration: $r^2 = 68.2 \text{ in}^2$

Section modulus: $S = 1000 \text{ in}^3$

Self-weight: $w_o = 0.183 \text{ k/ft}$

and is to carry a superimposed dead plus live load of 0.750 k/ft in addition to its own weight. The beam will be pretensioned with multiple seven-wire strands with centroid at a constant eccentricity of 7.91 in. The prestress force P_i immediately after transfer will be 158 kips; after time-dependent losses the force will reduce to $P_e = 134$ kips. The design strength of the concrete $f_c' = 5000$ psi, and at the time of prestressing the strength will be $f_{ci}' = 3750$ psi. Calculate the concrete flexural stresses at the midspan section of the beam at the time of transfer, and after all losses with full service load in place. Compare with ACI allowable stresses.

Solution. Stresses in the concrete resulting from the initial prestress force of 158 kips may be found by Eqs. (21.1*a*) and (21.1*b*):

$$f_1 = -\frac{158{,}000}{176}\left(1 - \frac{7.91 \times 12}{68.2}\right) = +352 \text{ psi}$$

$$f_2 = -\frac{158{,}000}{176}\left(1 + \frac{7.91 \times 12}{68.2}\right) = 2147 \text{ psi}$$

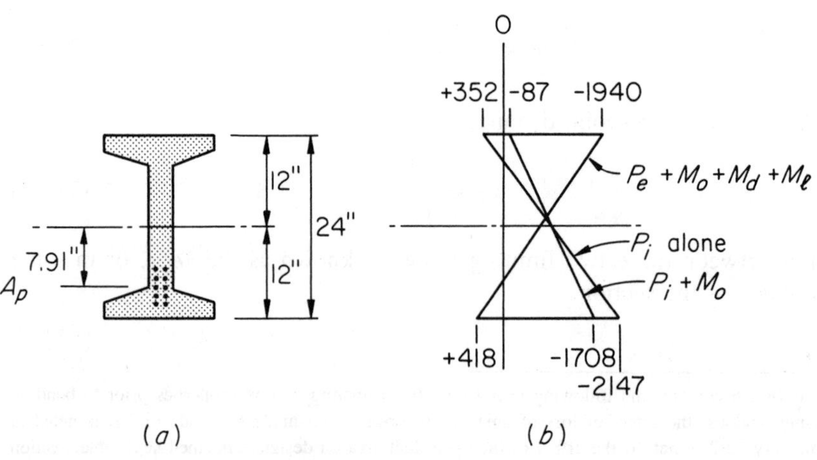

FIGURE 21.12
Design example: pretensioned I beam.

The self-weight of the beam causes the immediate superposition of a moment of

$$M_o = 0.183 \times \frac{40^2}{8} = 36.6 \text{ ft-kips}$$

and corresponding stresses of $36{,}600 \times 12/1000 = 439$ psi so that the net stresses at the top and bottom of the concrete section due to initial prestress and self-weight, from Eqs. (21.2*a*) and 21.2*b*), are

$$f_1 = +352 - 439 = -87 \text{ psi}$$

$$f_2 = -2147 + 439 = -1708 \text{ psi}$$

After losses, the prestress force is reduced to 134 kips, and the concrete stresses due to that force plus self-weight are

$$f_1 = +352 \times \frac{134}{158} - 439 = 140 \text{ psi}$$

$$f_2 = -2147 \times \frac{134}{158} + 439 = -1382 \text{ psi}$$

The superimposed load of 0.750 kips/ft produces a midspan moment of $M_d + M_l = 0.750 \times 40^2/8 = 150$ ft-kips, and the corresponding stresses of $150{,}000 \times 12/1000 = 1800$ psi compression and tension at the top and bottom of the beam, respectively. Thus the service load stresses at the top and bottom faces are

$$f_1 = -140 - 1800 = -1940 \text{ psi}$$

$$f_2 = -1382 + 1800 = +418 \text{ psi}$$

Concrete stresses at midspan are shown in Fig. 21.12*b*. According to the ACI Code (see Table 21.1), the permitted stresses in the concrete are

$$\text{Tension at transfer}: f_{ti} = 3\sqrt{3750} = +184 \text{ psi}$$

$$\text{Compression at transfer}: f_{ci} = 0.60 \times 3750 = -2250 \text{ psi}$$

$$\text{Tension at service load}: f_{ts} = 6\sqrt{5000} = +424 \text{ psi}$$

$$\text{Compression at service load}: f_{cs} = 0.45 \times 5000 = -2250 \text{ psi}$$

At the initial stage, with prestress plus self-weight in place, the actual compression of 1708 psi is well below the limit of 2250 psi, and no tension acts at the top, although 184 psi is allowed. While more prestress force or more eccentricity might be suggested to more fully utilize the section, to attempt to do so in this beam, with constant eccentricity, would violate limits at the support, where self-weight moment is zero. It is apparent that at the supports, the initial prestress force acting alone produces tension of 352 psi at the top of the beam, barely below the permitted value of $6\sqrt{3750} = 367$, so very little improvement can be made. Finally, at full service load, the tension of 418 psi is just under the allowed 424 psi, and compression of 1940 psi is well below the permitted 2250 psi.

21.7 FLEXURAL STRENGTH

In an ordinary reinforced concrete beam, the stress in the tensile steel and the compressive force in the concrete increase in proportion to the applied moment up to and somewhat beyond service load, with the distance between the two internal stress resultants remaining essentially constant. In contrast to this behavior, in a prestressed beam, increased moment is resisted by a proportionate increase in the distance between the compressive and tensile resultant forces, the compressive resultant moving upward as the load is increased. The magnitude of the internal forces remains nearly constant up to, and usually somewhat beyond, service loads.

This situation is changed drastically upon flexural tensile cracking of the prestressed beam. When the concrete cracks, there is a sudden increase in the stress in the steel as the tension that was formerly carried by the concrete is transferred to it. After cracking, the prestressed beam behaves essentially like an ordinary reinforced concrete beam. The compressive resultant cannot continue to move upward indefinitely, and increasing moment must be accompanied by a nearly proportionate increase in steel stress and compressive force. The strength of a prestressed beam can therefore be predicted by the same methods developed for ordinary reinforced concrete beams, with modifications to account for (a) the different shape of the stress-strain curve for prestressing steel, as compared with that for ordinary rebars, and (b) the tensile strain already present in the prestressing steel before the beam is loaded.

Highly accurate predictions of the flexural strength of prestressed beams can be made based on a *strain compatibility analysis* that accounts for these factors in a rational and explicit way (Ref. 21.1). For ordinary design purposes, certain approximate relationships have been derived. ACI Code 18.7 and the accompanying ACI Commentary 18.7 include approximate equations for flexural strength that will be summarized in the following paragraphs.

a. Stress in the Prestressed Steel at Flexural Failure

When a prestressed concrete beam fails in flexure, the prestressing steel is at a stress f_{ps} that is higher than the effective prestress f_{pe} but below the ultimate tensile strength f_{pu}. If the effective prestress $f_{pe} = P_e/A_{ps}$ is not less than $0.50f_{pu}$, ACI Code 18.7.2 permits use of certain approximate equations for f_{ps}. These equations appear quite complex as they are presented in the ACI Code, mainly because they are written in general form to account for differences in type of prestressing steel and to apply to beams in which non-prestressed bar reinforcement may be included in the flexural tension zone or the compression region or both. Separate equations are given for members with bonded tendons and unbonded tendons because, in the latter case, the increase in steel stress at the maximum moment section as the beam is overloaded is much less than if the steel were bonded throughout its length.

For the basic case, in which prestressed steel provides all of the flexural reinforcement, the ACI Code equations can be stated in simplified form as follows:

1. For members with bonded tendons:

$$f_{ps} = f_{pu}\left(1 - \frac{\gamma_p}{\beta_1}\frac{\rho_p f_{pu}}{f'_c}\right) \tag{21.6}$$

in which $\rho_p = A_{ps}/bd_p$, d_p = effective depth to the prestressing steel centroid, β_1 = the familiar relations between stress block depth and depth to the neutral axis, and γ_p is a factor that depends on the type of prestressing steel used, as follows:

$\gamma_p = 0.55$ for f_{py}/f_{pu} not less than 0.80 (high-strength bars)

$\gamma_p = 0.40$ for f_{py}/f_{pu} not less than 0.85 (ordinary strand)

$\gamma_p = 0.28$ for f_{py}/f_{pu} not less than 0.90 (low-relaxation strand)

2. For members with unbonded tendons and with a span-depth ratio of 35 or less (this includes most beams),

$$f_{ps} = f_{pe} + 10{,}000 + \frac{f'_c}{100\rho_p} \tag{21.7}$$

but not greater than f_{py} and not greater than f_{pe} + 60,000 psi.

3. For members with unbonded tendons and with span-depth ratio greater than 35 (applying to many slabs),

$$f_{ps} = f_{pe} + 10{,}000 + \frac{f'_c}{300\rho_p} \tag{21.8}$$

but not greater than f_{py} and not greater than f_{pe} + 30,000 psi.

b. Nominal Flexural Strength and Design Strength

With the stress in the prestressed tensile steel when the member fails in flexure established by Eq. (21.6), (21.7), or (21.8), the nominal flexural strength can be calculated by methods and equations that correspond directly with those used for ordinary reinforced concrete beams. For rectangular cross sections, or flanged sections such as I or T beams in which the stress block depth is equal to or less than the average flange thickness, the nominal flexural strength is

$$M_n = A_{ps} f_{ps}\left(d_p - \frac{a}{2}\right) \tag{21.9}$$

where

$$a = \frac{A_{ps} f_{ps}}{0.85 f'_c b} \tag{21.10}$$

Eqs. (21.9) and (21.10) can be combined as follows:

$$M_n = \rho_p f_{ps} b d_p^2 \left(1 - 0.588 \frac{\rho_p f_{ps}}{f_c'}\right) \tag{21.11}$$

In all cases, the *flexural design strength* is taken equal to ϕM_n, where $\phi = 0.90$ for flexure as usual.

If the stress block depth exceeds the average flange thickness, the method for calculating flexural strength is exactly analogous to that used for ordinary reinforced concrete I and T beams. The total prestressed tensile steel area is divided into two parts for computational purposes. The first part A_{pf}, acting at the stress f_{ps}, provides a tensile force to balance the compression in the overhanging parts of the flange. Thus

$$A_{pf} = 0.85 \frac{f_c'}{f_{ps}} (b - b_w) h_f \tag{21.12}$$

The remaining prestressed steel area

$$A_{pw} = A_{ps} - A_{pf} \tag{21.13}$$

provides tension to balance the compression in the web. The total resisting moment is the sum of the contributions of the two force couples:

$$M_n = A_{pw} f_{ps} \left(d_p - \frac{a}{2}\right) + A_{pf} f_{ps} \left(d_p - \frac{h_f}{2}\right) \tag{21.14a}$$

or

$$M_n = A_{pw} f_{ps} \left(d_p - \frac{a}{2}\right) + 0.85 f_c'(b - b_w) h_f \left(d_p - \frac{h_f}{2}\right) \tag{21.14b}$$

in which

$$a = \frac{A_{pw} f_{ps}}{0.85 f_c' b_w} \tag{21.15}$$

As before, the design strength is taken as ϕM_n, where $\phi = 0.90$.

If, after a prestressed beam is designed by elastic methods at service loads, it has inadequate strength to provide the required safety margin at the factored overload stage, nonprestressed rebars can be added on the tension side and will work in combination with the prestressed steel to provide the needed strength. Such nonprestressed steel, with area A_s, can be assumed to act at its yield stress f_y, to contribute a tension force at ultimate moment of $A_s f_y$. The reader should consult ACI Code 18.7 and ACI Commentary 18.7 for equations for prestressed steel stress at failure and for flexural strength, which are a direct extensions of those given above.

c. Limits for Reinforcement

For ordinary reinforced concrete beams, an *upper limit* is placed on the tensile steel ratio, equal to 0.75 times the balanced steel ratio, in order to ensure that, if flexural failure should occur, it would be a ductile failure with extensive cracking and large deflections before eventual collapse. A corresponding provision is found in ACI Code 18.8.1 for prestressed beams. Stated in simplest terms, the provision is that

$$\frac{0.85a}{d_p} \leq 0.36\beta_1 \tag{21.16}$$

If the prestressed beam does not meet the requirement of Eq. (21.16), it is considered to be *overreinforced*, and alternative equations are given in ACI Commentary 18.8.2 for computing the flexural strength.

It will be recalled that a *minimum tensile steel ratio* equal to $200/f_y$ is imposed for ordinary reinforced concrete beams, in order that a beam be safe against sudden failure upon the formation of flexural cracks. For prestressed beams, because of the same concern, it is required by ACI Code 18.8.3 that the total tensile reinforcement must be adequate to support a factored load at least 1.2 times the cracking load of the beam, calculated on the basis of a modulus of rupture of $7.5\sqrt{f_c'}$.

d. Minimum Bonded Reinforcement

To control cracking in beams and one-way prestressed slabs with *unbonded tendons*, some bonded reinforcement must be added in the form of nonprestressed rebars, uniformly distributed over the tension zone as close as permissible to the tension face. According to ACI Code 18.9.2 the minimum amount of such reinforcement is

$$A_s = 0.004A \tag{21.17}$$

where A is the area of that part of the cross section between the flexural tension face and the centroid of the gross concrete cross section.

Example 21.2 Flexural strength of pretensioned I beam. The prestressed I beam shown in cross section in Fig. 21.13 is pretensioned using seven ordinary stress-relieved Grade 250 $\frac{1}{2}$ in. diameter strands, carrying effective prestress f_{pe} = 143 ksi. Concrete strength is f_c' = 4000 psi. Calculate the design strength of the beam.

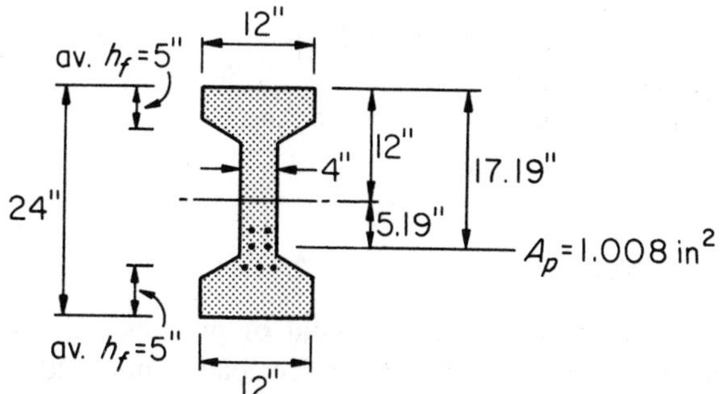

FIGURE 21.13
Post-tensioned beam of Example 18.2.

Solution. The effective prestress in the strands of 143 ksi is well above $0.50 \times 250 = 125$ ksi, confirming that the approximate ACI equations are applicable. The tensile steel ratio is

$$\rho_p = \frac{1.008}{12 \times 17.19} = 0.0049$$

and the steel stress f_{ps} when the beam fails in flexure is found from Eq. (21.6) to be

$$f_{ps} = 250\left(1 - \frac{0.40}{0.85}\frac{0.0049 \times 250}{4}\right) = 214 \text{ ksi}$$

Next it is necessary to check whether the stress block depth is greater or less than the average flange thickness of 5 in. On the assumption that it is not greater than the flange thickness, Eq. (21.10) is used:

$$a = \frac{1.008 \times 214}{0.85 \times 4 \times 12} = 5.29 \text{ in.}$$

It is concluded from this trial calculation that a actually exceeds h, so the trial calculation is not valid and equations for flanged members must be used. The steel that acts with the overhanging flanges is found from Eq. (21.12) to be

$$A_{pf} = \frac{0.85 \times 4\,(12 - 4)\,5}{214} = 0.636 \text{ in}^2$$

and from Eq. (21.13),

$$A_{pw} = 1.008 - 0.636 = 0.372 \text{ in}^2$$

The actual stress block depth is now found from Eq. (21.15):

$$a = \frac{0.372 \times 214}{0.85 \times 4 \times 4} = 5.85 \text{ in.}$$

A check should now be made to determine if the beam can be considered underreinforced. From Eq. (21.16),

$$\frac{0.85 \times 5.85}{17.19} = 0.289$$

This is less than $0.36\beta_1 = 0.36 \times 0.85 = 0.306$, confirming that this can be considered to be an underreinforced prestressed beam. The nominal flexural strength, from Eq. (21.14b), is

$$\begin{aligned} M_n &= 0.372 \times 214\,(17.19 - 2.93) + 0.85 \times 4\,(12 - 4)\,5\,(17.19 - 2.50) \\ &= 3133 \text{ in-kips} = 261 \text{ ft-kips} \end{aligned}$$

and, finally, the design strength is $\phi M_n = 235$ ft-kips.

21.8 PARTIAL PRESTRESSING

Early in the development of prestressed concrete, the goal of prestressing was the complete elimination of concrete tensile stress at service load. This kind of design, in which the service load tensile stress limit $f_{ts} = 0$, is referred to as *full prestressing*.

While full prestressing offers many advantages over nonprestressed construction, some problems can arise. Heavily prestressed beams, particularly those for which full live load is seldom in place, may have excessively large upward deflection, or camber, which will increase with time because of concrete creep under the eccentric prestress force. Fully prestressed beams may also have a tendency for severe longitudinal shortening, causing large restraint forces unless special provision is made to permit free movement at one end of each span. If shortening is permitted to occur freely, prestress losses due to elastic and creep deformation may be large. Furthermore, if heavily prestressed beams are overloaded to failure, they may fail in a sudden and brittle mode, with little warning before collapse.

Today there is general recognition of the advantages of *partial prestressing*, in which flexural tensile stress and some limited cracking is permitted under full service load. That full load may be infrequently applied. Typically many beams carry only dead load much of the time, or dead load plus only part of the service live load. Under these conditions, a partially prestressed beam normally would not be subject to flexural tension, and cracks that form occasionally, when the full live load is in place, would close completely when that live load is removed. Controlled cracks prove no more objectionable in prestressed concrete structures than in reinforced concrete structures. With partial prestressing, excessive camber and troublesome axial shortening is avoided. Should overloading occur, there will be ample warning of distress, with extensive cracking and large deflections (Ref. 21.7 to 21.9).

Although the amount of prestressing steel is reduced in partially prestressed beams compared with fully prestressed beams, a proper safety margin still must be maintained, and, to achieve the necessary flexural strength, partially prestressed beams usually will require additional tensile reinforcement in the form of ordinary nonprestressed reinforcing bars. In fact, partially prestressed beams are often defined as beams in which (a) flexural cracking is permitted at full service load and (b) the main flexural tension reinforcement includes both prestressed and nonprestressed steel. Analysis indicates, and tests confirm, that such nonprestressed steel is fully stressed to f_y at flexural failure.

Partially prestressed beams are permitted by most design specifications. While the ACI Code does not mention partial prestressing explicitly, ACI Code 18.4.2 permits flexural tensile stress of $6\sqrt{f_c'}$ in ordinary design and a *nominal* flexural tension of $12\sqrt{f_c'}$, well above the usual modulus of rupture, if deflections are calculated by methods that account for cracking and if concrete cover requirements are increased. Also, section 18.4.3 of the ACI Code includes the statement that these stresses may be exceeded if it is shown by test or analysis that performance will not be impaired. Furthermore, equations for flexural strength in the ACI Code and Commentary account explicitly for the presence of both nonprestressed and prestressed flexural tension steel.

Apart from the technical advantages, there are usually significant economic advantages associated with partial prestressing. Rebars are less expensive than high tensile strength prestressing steel, and labor costs for bar placement are generally less than for placing and stressing tendons.

The choice of a suitable degree of prestress is governed by a number of factors. These include the nature of the loading (for example, highway or railroad bridges, storage warehouses, etc.), the ratio of live to dead load, the frequency of occurrence of the full load, and the presence of a corrosive environment.

Presently, the most common practice in the United States is to limit service load tension to $12\sqrt{f'_c}$, but in Europe, particularly in Switzerland and Germany, partial prestressing is much more fully utilized. Frequently the criterion is that no tension is permitted under full dead load. In this case, the beam usually will show some cracking when only a small percentage of the live load is in place, and at full live load, stresses are found based on a cracked section analysis, as for reinforced concrete (Refs. 21.10 and 21.11).

21.9 FLEXURAL DESIGN BASED ON CONCRETE STRESS LIMITS

As in reinforced concrete, problems in prestressed concrete can be separated generally as review problems or design problems. For the former, with the applied loads, the concrete cross section, steel area, and the amount and point of application of the prestress force known, Eqs. (21.1) to (21.4) permit the direct calculation of the resulting concrete stresses. The equations of Sec. 21.7 will predict the flexural strength. However, if the dimensions of a concrete section, the steel area and centroid location, and the amount of prestress are to be found—given the loads, limiting stresses, and required strength—the problem is complicated by the many interrelated variables.

There are at least three practical approaches to the flexural design of a prestressed concrete member. Some engineers prefer to assume a concrete section, calculate the required prestress force and eccentricities for what will probably be the controlling load stage, then check the stresses at all stages using the preceding equations, and finally check the flexural strength. The trial section is then revised if necessary. If a beam is to be chosen from a limited number of standard shapes, as is often the case for shorter spans and ordinary loads, this procedure is probably best. For longer spans, a more efficient design may result by designing the cross section so that the specified concrete stress limits of Table 21.1 are closely matched. This cross section, close to "ideal" from the limit stress viewpoint, may then be modified to meet functional requirements (e.g., providing a broad top flange for a bridge deck) or to meet strength requirements, if necessary. Equations facilitating this approach will be developed in this section. A third method of design is based on load balancing, using the concept of equivalent loads (see Sec. 21.2b). A trial section is chosen, after which the prestress force and tendon profile are selected to provide uplift forces as to just balance a specified load. Modifications may then be made, if needed, to satisfy stress limits or strength requirements. This third approach will be developed in Sec. 21.12.

Notation is established pertaining to the allowable concrete stresses at limiting stages as follows:

f_{ci} = allowable compressive stress immediately after transfer

f_{ti} = allowable tensile stress immediately after transfer

f_{cs} = allowable compressive stress at service load, after all losses

f_{ts} = allowable tensile stress at service load, after all losses

The values of these limit stresses are normally set by specification (see Table 21.1).

a. Beams with Variable Eccentricity

For a typical beam in which the tendon eccentricity is permitted to vary along the span, flexural stress distributions in the concrete at the maximum moment section are shown in Fig. 21.14*a*. The eccentric prestress force, having an initial value of P_i, produces the linear distribution (1). Because of the upward camber

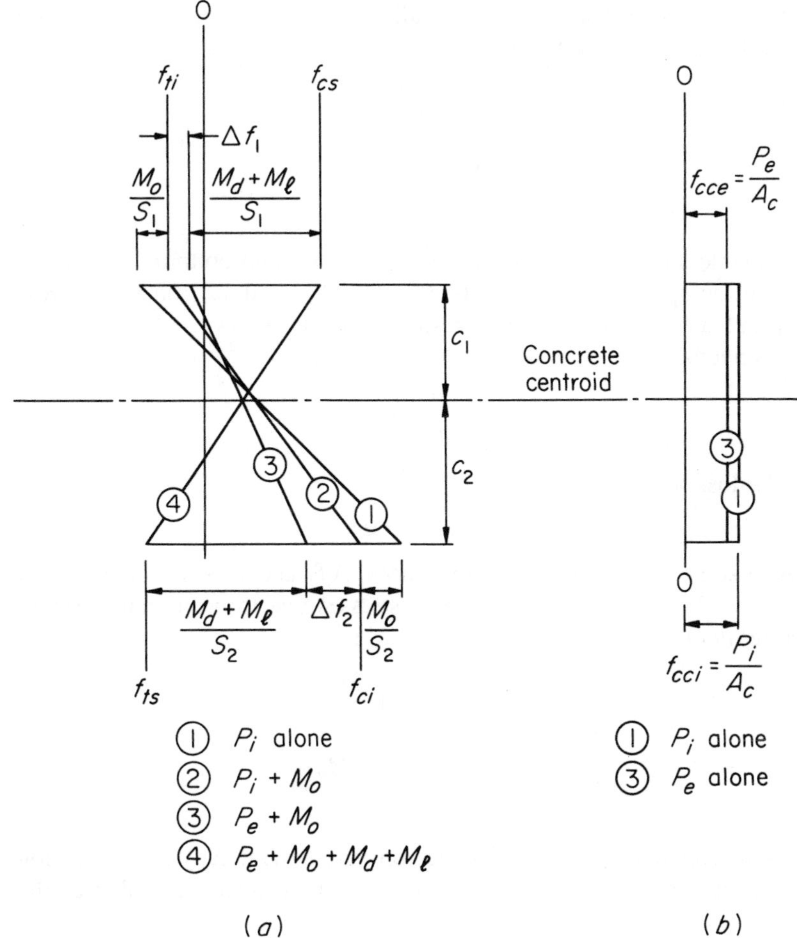

FIGURE 21.14
Flexural stress distributions for beams with variable eccentricity: (*a*) maximum moment section; (*b*) support section.

of the beam as that force is applied, the self-weight of the member is immediately introduced, the flexural stresses resulting from the moment M_o are superimposed, and the distribution (2) is the first that is actually attained. At this stage, the tension at the top surface is not to exceed f_{ti} and the compression at the bottom surface is not to exceed f_{ci}, as suggested by Fig. 21.14*a*.

It will be assumed that all the losses occur at this stage, and that the stress distribution gradually changes to distribution (3). The losses produce a reduction of tension in the amount Δf_1 at the top surface, and a reduction of compression in the amount Δf_2 at the bottom surface.

As the superimposed dead load moment M_d and the service live load moment M_l are introduced, the associated flexural stresses, when superimposed on stresses already present, produce distribution (4). At this stage, the tension at the bottom surface must not be greater than f_{ts}, and the compression at the top of the section must not exceed f_{cs} as shown.

The requirements for the sections moduli S_1 and S_2 with respect to the top and bottom surfaces, respectively, are

$$S_1 \geq \frac{M_d + M_l}{f_{1r}} \tag{a}$$

$$S_2 \geq \frac{M_d + M_l}{f_{2r}} \tag{b}$$

where the available stress ranges f_{1r} and f_{2r} at the top and bottom face can be calculated from the specified stress limits f_{ti}, f_{cs}, f_{ts}, and f_{ci}, once the stress changes Δf_1 and Δf_2, associated with prestress loss, are known.

The effectiveness ratio R is defined as

$$R = \frac{P_e}{P_i} \tag{21.18}$$

Thus, the loss in prestress force is

$$P_i - P_e = (1 - R)P_i \tag{21.19}$$

The changes in stress at the top and bottom faces, Δf_1 and Δf_2, as losses occur, are equal to $(1 - R)$ times the corresponding stresses due to the initial prestress force P_i *acting alone*:

$$\Delta f_1 = (1 - R)\left(f_{ti} + \frac{M_o}{S_1}\right) \tag{c}$$

$$\Delta f_2 = (1 - R)\left(-f_{ci} + \frac{M_o}{S_2}\right) \tag{d}$$

where Δf_1 is a reduction of tension at the top surface and Δf_2 is a reduction of compression at the bottom surface.† Thus, the stress ranges available as the

† Note that the stress limits such as f_{ti} and other specific points along the stress axis are considered signed quantities, whereas stress changes such as M_o/S_1 and Δf_2 are taken as absolute values.

superimposed load moments $M_d + M_l$ are applied are

$$\begin{aligned} f_{1r} &= f_{ti} - \Delta f_1 - f_{cs} \\ &= R f_{ti} - (1 - R)\frac{M_o}{S_1} - f_{cs} \end{aligned} \tag{e}$$

and

$$\begin{aligned} f_{2r} &= f_{ts} - f_{ci} - \Delta f_2 \\ &= f_{ts} - R f_{ci} - (1 - R)\frac{M_o}{S_2} \end{aligned} \tag{f}$$

The minimum acceptable value of S_1 is thus established:

$$S_1 \geq \frac{M_d + M_l}{R f_{ti} - (1 - R)\frac{M_o}{S_1} - f_{cs}}$$

or

$$S_1 \geq \frac{(1 - R)M_o + M_d + M_l}{R f_{ti} - f_{cs}} \tag{21.20}$$

Similarly, the minimum value of S_2 is

$$S_2 \geq \frac{(1 - R)M_o + M_d + M_l}{f_{ts} - R f_{ci}} \tag{21.21}$$

The cross section must be selected to provide at least these values of S_1 and S_2. Furthermore, since $I_c = S_1 c_1 = S_2 c_2$, the centroidal axis must be located such that

$$\frac{c_1}{c_2} = \frac{S_2}{S_1} \tag{g}$$

or in terms of the total section depth $h = c_1 + c_2$

$$\frac{c_1}{h} = \frac{S_2}{S_1 + S_2} \tag{21.22}$$

From Fig. 21.14a, the concrete centroidal stress under initial conditions is given by

$$f_{cci} = f_{ti} - \frac{c_1}{h}(f_{ti} - f_{ci}) \tag{21.23}$$

The initial prestress force is easily obtained by multiplying the value of the concrete centroidal stress by the concrete cross-sectional area A_c.

$$P_i = A_c f_{cci} \tag{21.24}$$

The eccentricity of the prestress force may be found by considering the flexural stresses that must be imparted by the bending moment $P_i e$. With reference to Fig. 21.14, the flexural stress at the top surface of the beam resulting from the eccentric prestress force alone is

$$\frac{P_i e}{S_1} = (f_{ti} - f_{cci}) + \frac{M_o}{S_1} \tag{h}$$

from which the required eccentricity is

$$e = (f_{ti} - f_{cci})\frac{S_1}{P_i} + \frac{M_o}{P_i} \tag{21.25}$$

To summarize the design process in determining the best cross section and the required prestress force and eccentricity based on stress limitations: the required section moduli with respect to the top and bottom surfaces of the member are found from Eqs. (21.20) and (21.21) with the centroidal axis located by Eq. (21.22). Concrete dimensions are chosen to satisfy these requirements as nearly as possible. The concrete centroidal stress for this ideal section is given by Eq. (21.23), the desired initial prestress force is found by Eq. (21.24), and its eccentricity by Eq. (21.25).

In practical situations, very seldom will the concrete section chosen have exactly the required values of S_1 and S_2 as found by this method, nor will the concrete centroid be exactly at the theoretically ideal level. Rounding upward of concrete dimensions, provision of broad flanges for functional reasons, or the use of standardized cross-sectional shapes will normally result in a member whose section properties will exceed the minimum requirements. In such a case, the stresses in the concrete as the member passes from the unloaded stage to the full service load stage will stay within the allowable limits, but the limit stresses will not be obtained exactly. An infinite number of combinations of prestress force and eccentricity will satisfy requirements. Usually the design requiring the lowest value of prestress force, and the largest practical eccentricity, will be the most economical.

The stress distributions of Fig. 21.14*a*, on which the design equations are based, apply at the maximum moment section of the member. Elsewhere, M_o is less and, consequently, the prestress eccentricity or the force must be reduced if the stress limits f_{ti} and f_{ci} are not to be exceeded. In many cases, tendon eccentricity is reduced to zero at the support sections, where all moments due to transverse load are zero. In this case, the stress distributions of Fig. 21.14*b* are obtained. The stress in the concrete is uniformly equal to the centroidal value f_{cci} under conditions of initial prestress and f_{cce} after losses.

Example 21.3 Design of beam with variable eccentricity tendons. A post-tensioned prestressed concrete beam is to carry a live load of 1000 plf and superimposed dead load of 500 plf, in addition to its own weight, on a 40 ft simple span. Normal density concrete will be used with design strength $f'_c = 6000$ psi. It is estimated that, at the time of transfer, the concrete will have attained 70 percent of its ultimate strength, or 4200 psi. Time-dependent losses may be assumed at 15 percent of the initial prestress, giving an effectiveness ratio of 0.85. Determine the required concrete dimensions, magnitude of prestress force, and eccentricity of the steel centroid based on ACI stress limitations as given in Secs. 21.4 and 21.5.

Solution. Referring to Table 21.1, we obtain the following stress limits:

$$f_{ci} = -0.60 \times 4200 = 2520 \text{ psi}$$

$$f_{ti} = 3\sqrt{4200} = +195 \text{ psi}$$

$$f_{cs} = -0.45 \times 6000 = -2700 \text{ psi}$$

$$f_{ts} = 6\sqrt{6000} = +465 \text{ psi}$$

The self-weight of the girder will be estimated at 250 plf. The moments due to transverse loading are

$$M_o = \frac{1}{8} \times 0.250 \times 40^2 = 50 \text{ ft-kips}$$

$$M_d + M_l = \frac{1}{8} \times 1.500 \times 40^2 = 300 \text{ ft-kips}$$

The required section moduli with respect to the top and bottom surfaces of the concrete beam are found from Eqs. (21.20) and (21.21):

$$S_1 \geq \frac{(1-R)M_o + M_d + M_l}{Rf_{ti} - f_{cs}} = \frac{(0.15 \times 50 + 300)12{,}000}{0.85 \times 195 + 2700} = 1288 \text{ in}^3$$

$$S_2 \geq \frac{(1-R)M_o + M_d + M_l}{f_{ts} - Rf_{ci}} = \frac{(0.15 \times 50 + 300)12{,}000}{465 + 0.85 \times 2520} = 1415 \text{ in}^3$$

These values are so nearly the same that a symmetrical beam will be adopted. The 28-in. depth I-section shown in Fig. 21.15*a* will meet the requirements and has the following properties:

$$I_c = 19{,}904 \text{ in}^4$$

$$S = 1422 \text{ in}^3$$

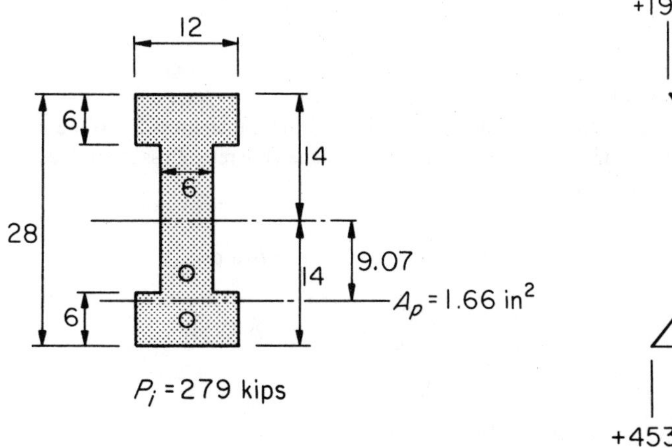

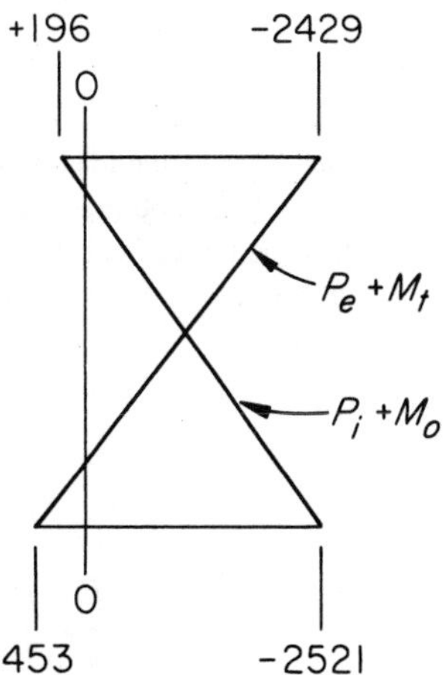

FIGURE 21.15
Design example of beam with variable eccentricity of tendons: (*a*) cross section dimensions; (*b*) concrete stresses at midspan.

$$A_c = 240 \text{ in}^2$$
$$r^2 = 82.9 \text{ in}^2$$
$$w_o = 250 \text{ plf (as assumed)}$$

Next, the concrete centroidal stress is found from Eq. (21.23):

$$f_{cci} = f_{ti} - \frac{c_1}{h}(f_{ti} - f_{ci}) = 195 - \frac{1}{2}(195 + 2520) = -1163 \text{ psi}$$

and from Eq. (21.24) the initial prestress force is

$$P_i = A_c f_{cci} = 240 \times 1.163 = 279 \text{ kips}$$

From Eq. (21.25) the required tendon eccentricity at the maximum moment section of the beam is

$$e = (f_{ti} - f_{cci})\frac{S_1}{P_i} + \frac{M_o}{P_i} = (195 + 1163)\frac{1422}{279{,}000} + \frac{50 \times 12{,}000}{279{,}000}$$
$$= 9.07 \text{ in.}$$

Elsewhere along the span the eccentricity will be reduced in order that the concrete stress limits not be violated.

The required initial prestress force of 279 kips will be provided using tendons consisting of $\frac{1}{4}$ in. diameter stress-relieved wires (see Sec. 2.14). The minimum tensile strength is $f_{pu} = 240$ ksi, and the yield strength may be taken as $f_{py} = 0.85 \times f_{pu} = 204$ ksi. According to the ACI Code (see Sec. 21.5), the permissible stress in the wire immediately after transfer must not exceed $0.82 f_{py} = 168$ ksi or $0.74 f_{pu} = 178$ ksi. The first criterion controls. The required area of prestressed steel is

$$A_p = \frac{279}{168} = 1.66 \text{ in}^2$$

The cross-sectional area of one $\frac{1}{4}$ in. diameter wire is 0.0491 in^2; hence, the number of wires required is

$$\text{Number of wires} = \frac{1.66}{0.0491} = 34$$

Two 17-wire tendons will be used, as shown in Fig. 21.15*a*.

It is good practice to check the calculations by confirming that stress limits are not exceeded at critical load stages. The top and bottom surface concrete stresses produced, in this case, by the separate loadings are:

$$P_i: f_1 = -\frac{279{,}000}{240}\left(1 - \frac{9.07 \times 14}{82.9}\right) = +618 \text{ psi}$$

$$f_2 = -\frac{279{,}000}{240}\left(1 + \frac{9.07 \times 14}{82.9}\right) = -2943 \text{ psi}$$

$$P_e: f_1 = 0.85 \times 618 = 525 \text{ psi}$$

$$f_2 = 0.85(-2943) = -2501 \text{ psi}$$

$$M_o: f_1 = -\frac{50 \times 12{,}000}{1422} = -422 \text{ psi}$$

$$f_2 = +422 \text{ psi}$$

$$M_d + M_l : f_1 = -\frac{300 \times 12{,}000}{1422} = -2532\text{psi}$$

$$f_2 = +2532 \text{ psi}$$

Thus, when the initial prestress force of 279 kips is applied and the beam self-weight acts, the top and bottom stresses in the concrete at midspan are, respectively:

$$f_1 = +618 - 422 = +196 \text{ psi}$$
$$f_2 = -2943 + 422 = -2521 \text{ psi}$$

When the prestress force has reduced to its effective value of 237 kips and the full service load is applied, the concrete stresses are:

$$f_1 = +525 - 422 - 2532 = -2429 \text{ psi}$$

$$f_2 = -2501 + 422 + 2532 = +453 \text{ psi}$$

These limiting stress distributions are shown in Fig. 21.15*b*. Comparison with the specified limit stresses confirms that the design is satisfactory.

b. Beams with Constant Eccentricity

The design method presented in the previous section was based on stress conditions at the maximum moment section of a beam, with the maximum value of moment M_o resulting from self-weight immediately superimposed. If P_i and e were to be held constant along the span, as is often convenient in pretensioned prestressed construction, then the stress limits f_{ti} and f_{ci} would be exceeded elsewhere along the span, where M_o is less than its maximum value. To avoid this condition, the constant eccentricity must be less than that given by Eq. (21.25). Its maximum value is given by conditions at the support of a simple span, where M_o is zero.

Figure 21.16 shows the flexural stress distributions at the support and midspan sections for a beam with constant eccentricity. In this case, the stress limits f_{ti} and f_{ci} are not to be violated when the eccentric prestress moment acts alone, as at the supports. The stress changes Δf_1 and Δf_2 as losses occur are equal to $(1 - R)$ times the top and bottom surface stresses, respectively, due to initial prestress alone:

$$\Delta f_1 = (1 - R)(f_{ti}) \qquad (a)$$
$$\Delta f_2 = (1 - R)(-f_{ci}) \qquad (b)$$

In this case, the available stress ranges between limit stresses must provide for the effect of M_o as well as M_d and M_l, as seen from Fig. 21.16*a*, and are

$$f_{1r} = f_{ti} - \Delta f_1 - f_{cs}$$
$$= R f_{ti} - f_{cs} \qquad (c)$$

$$f_{2r} = f_{ts} - f_{ci} - \Delta f_2$$
$$= f_{ts} - R f_{ci} \qquad (d)$$

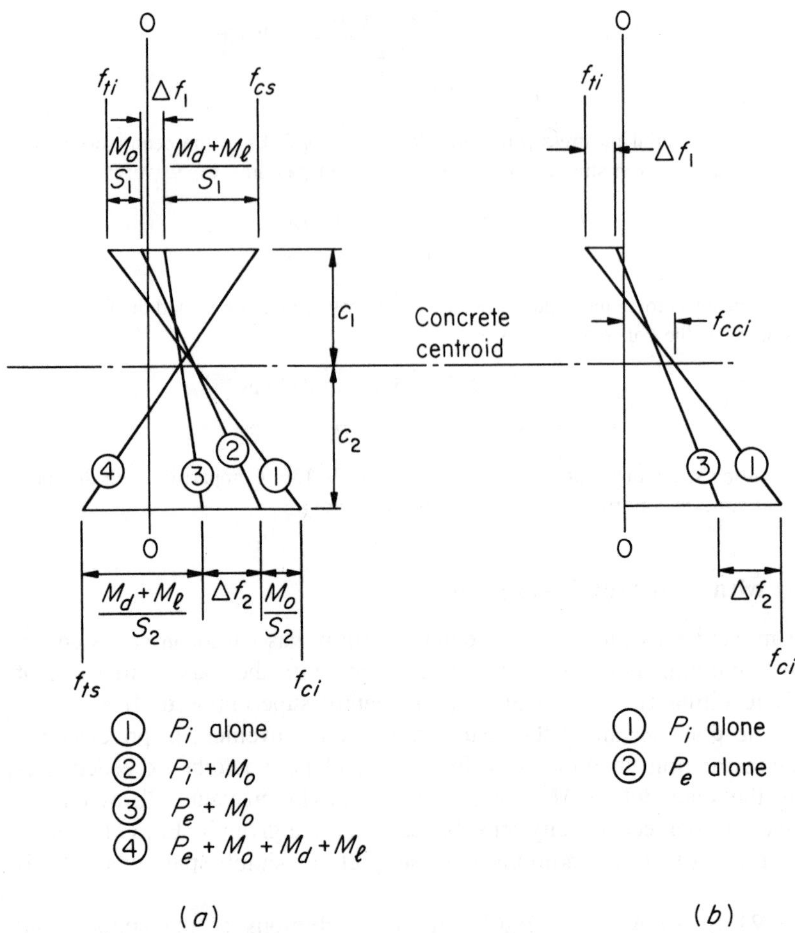

FIGURE 21.16
Flexural stress distributions for beam with constant eccentricity of tendons: (*a*) maximum moment section; (*b*) support section.

and the requirements on the section moduli are that

$$S_1 \geq \frac{M_o + M_d + M_l}{R f_{ti} - f_{cs}} \tag{21.26}$$

$$S_2 \geq \frac{M_o + M_d + M_l}{f_{ts} - R f_{ci}} \tag{21.27}$$

The concrete centroidal stress may be found by Eq. (21.23) and the initial prestress force by Eq. (21.24) as before. However, the expression for required eccentricity differs. In this case, referring to Fig. 21.16*b*,

$$\frac{P_i e}{S_1} = f_{ti} - f_{cci} \tag{e}$$

from which the required eccentricity is

$$e = (f_{ti} - f_{cci})\frac{S_1}{P_i} \tag{21.28}$$

A significant difference between beams with variable eccentricity and those with constant eccentricity will be noted by comparing Eqs. (21.20) and (21.21) with the corresponding Eqs. (21.26) and (21.27). In the first case, the section modulus requirement is governed mainly by the superimposed load moments M_d and M_l. Almost all of the self-weight is carried "free," that is, without increasing section modulus or prestress force, by the simple expedient of increasing the eccentricity along the span by the amount M_o/P_i. In the second case, the eccentricity is controlled by conditions at the supports, where M_o is zero, and the full moment M_o due to self-weight must be included in determining section moduli. Nevertheless, beams with constant eccentricity are often used for practical reasons.

Example 21.4 Design of beam with constant eccentricity tendons. The beam of the preceding example is to be redesigned using straight tendons with constant eccentricity. All other design criteria are the same as before. At the supports, a temporary concrete tensile stress of $6\sqrt{f'_{ci}} = 390$ psi is permitted.

Solution. Anticipating a somewhat less efficient beam, the dead load estimate will be increased to 270 plf in this case. The resulting moment M_o is 54 ft-kips. The moment due to superimposed dead load and live load is 300 ft-kips as before.

Using Eqs. (21.26) and (21.27), the requirements for section moduli are

$$S_1 \geq \frac{M_o + M_d + M_l}{Rf_{ti} - f_{cs}} = \frac{(54 + 300)12{,}000}{0.85 \times 390 + 2700} = 1401 \text{ in}^3$$

$$S_2 \geq \frac{M_o + M_d + M_l}{f_{ts} - Rf_{ci}} = \frac{(54 + 300)12{,}000}{465 + 0.85 \times 2520} = 1629 \text{ in}^3$$

Once again, a symmetrical section will be chosen. Flange dimensions and web width will be kept unchanged compared with the previous example, but in this case a beam depth of 30.5 in. is required. The dimensions of the cross section are shown in Fig. 21.17*a*. The following properties are obtained:

$$I_c = 25{,}207 \text{ in}^4$$

$$S = 1653 \text{ in}^3$$

$$A_c = 255 \text{ in}^2$$

$$r^2 = 98.9 \text{ in}^2$$

$$w_o = 266 \text{ plf (close to the assumed value)}$$

The concrete centroidal stress, from Eq. (21.23), is

$$f_{cci} = f_{ti} - \frac{c_1}{h}(f_{ti} - f_{ci}) = 390 - \frac{1}{2}(390 + 2520) = -1065 \text{ psi}$$

and, from Eq. (21.24), the initial prestress force is

$$P_i = A_c f_{cci} = 255 \times 1.065 = 272 \text{ kips}$$

From Eq. (21.28), the required constant eccentricity is

$$e = (f_{ti} - f_{cci})\frac{S_1}{P_i} = (390 + 1065)\frac{1653}{272{,}000} = 8.84 \text{ in.}$$

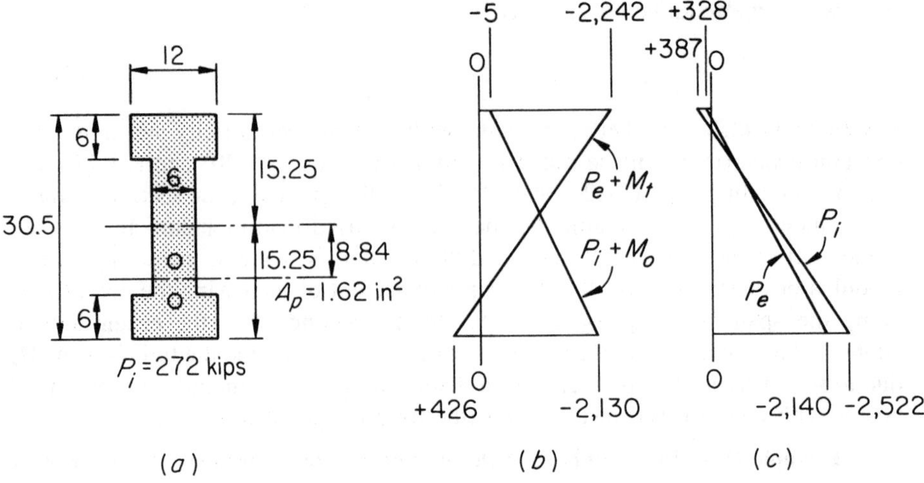

FIGURE 21.17
Design example of beam with constant eccentricity of tendons: (*a*) cross section dimensions; (*b*) stresses at midspan; (*c*) stresses at supports.

Again, two tendons will be used to provide the required force P_i, each composed of multiple $\frac{1}{4}$ in. diameter wires. With the maximum permissible stress in the wires of 168 ksi, the total required steel area is

$$A_p = \frac{272}{168} = 1.62 \text{ in}^2$$

A total of 34 wires is required as before, 17 in each tendon.

The calculations will be checked by verifying concrete stresses at the top and bottom of the beam for the critical load stages. The component stress contributions are

$$P_i : f_1 = -\frac{272{,}000}{255}\left(1 - \frac{8.84 \times 15.25}{98.9}\right) = +387 \text{ psi}$$

$$f_2 = -\frac{272{,}000}{255}\left(1 + \frac{8.84 \times 15.25}{98.9}\right) = -2522 \text{ psi}$$

$$P_e : f_1 = 0.85 \times 387 = +328 \text{ psi}$$

$$f_2 = 0.85(-2522) = -2144 \text{ psi}$$

$$M_o : f_1 = -\frac{54 \times 12{,}000}{1653} = -392 \text{ psi}$$

$$f_2 = +392 \text{ psi}$$

$$M_d + M_l : f_1 = -\frac{300 \times 12{,}000}{1653} = -2178 \text{ psi}$$

$$f_2 = +2178 \text{ psi}$$

Superimposing the appropriate stress contributions, the stress distributions in the concrete at midspan and at the supports are obtained, as shown in Figs. 21.17*b* and 21.17*c*, respectively. When the initial prestress force of 272 kips acts alone, as at the supports, the stresses at the top and bottom surfaces are

$$f_1 = +387 \text{ psi}$$
$$f_2 = -2522 \text{ psi}$$

After losses, the prestress force is reduced to 231 kips and the support stresses are reduced accordingly. At midspan the beam weight is immediately superimposed, and stresses resulting from P_i plus M_o are

$$f_1 = +387 - 392 = -5 \text{ psi}$$
$$f_2 = -2522 + 392 = -2130 \text{ psi}$$

When the full service load acts, together with P_e, the midspan stresses are

$$f_1 = +328 - 392 - 2178 = -2242 \text{ psi}$$
$$f_2 = -2144 + 392 + 2178 = +426 \text{ psi}$$

If we check against the specified limiting stresses, it is evident that the design is satisfactory in this respect at the critical load stages and locations.

21.10 SHAPE SELECTION

One of the special features of prestressed concrete design is the freedom to select cross-section proportions and dimensions to suit the special requirements of the job at hand. The member depth can be changed, the web thickness modified, and the flange widths and thicknesses varied independently to produce a beam with nearly ideal proportions for a given case.

Several common shapes are shown in Fig. 21.18. Some of these are standardized and mass produced, employing reusable steel or fiberglass forms. Others are individually proportioned for large and important works. The double T (Fig. 21.18*a*) is probably the most widely used cross section in U.S. prestressed

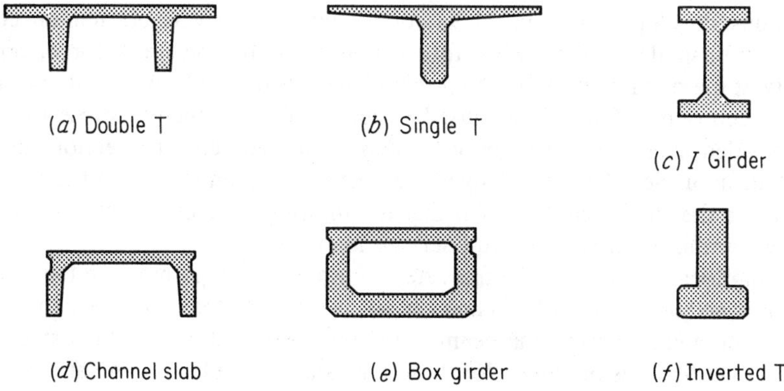

FIGURE 21.18
Typical beam cross sections.

construction. A flat surface is provided, 4 to 8 ft wide. Slab thicknesses and web depth vary, depending upon requirements. Spans to 60 ft are not unusual. The single T (Fig. 21.18*b*) is more appropriate for longer spans, to 120 ft, and heavier loads. The I section (Fig. 21.18*c*) is widely used for bridge spans and roof girders up to about 120 ft, while the channel slab (Fig. 21.18*d*) is suitable for floors in the intermediate-span range. The box girder (Fig. 21.18*e*) is advantageous for bridges of intermediate and major span. The inverted-T section (Fig. 21.18*f*) provides a bearing ledge to carry the ends of precast deck members spanning in the perpendicular direction.

As indicated, the cross section may be symmetrical or unsymmetrical. An unsymmetrical section is a good choice (1) if the available stress ranges f_{1r} and f_{2r} at the top and bottom surfaces are not the same; (2) if the beam must provide a flat, useful surface as well as offering load-carrying capacity; (3) if the beam is to become a part of composite construction, with the cast-in-place slab acting together with precast web; or (4) if the beam must provide supporting surfaces as in Fig. 21.18*f*. In addition, T sections provide increased flexural strength, since the internal arm of the resisting couple at maximum design load is greater than for rectangular sections.

Generally speaking, I, T, and box sections with relatively thin webs and flanges are more efficient than members with thicker parts. However, several factors limit the gain in efficiency that may be obtained in this way. These include the instability of very thin overhanging compression parts, the vulnerability of thin parts to breakage in handling (in the case of precast construction), and the practical difficulty of depositing concrete for very thin elements. The designer must also recognize the need for providing adequate spacing and concrete protection for tendons and anchorages, the importance of construction depth limitations, and the need for lateral stability if the beam is not braced by other members against buckling (Ref. 21.12).

21.11 TENDON PROFILES

The equations developed in Sec. 21.9a for members with variable tendon eccentricity establish the requirements for section modulus, prestress force, and eccentricity at the maximum moment section of the member. Elsewhere along the span, the eccentricity of the steel must be reduced if the concrete stress limits for the unloaded stage are not to be exceeded. (Alternatively, the section must be increased, as in Sec. 21.9b.) Conversely, there is a minimum eccentricity, or upper limit for the steel centroid, such that the limiting concrete stresses are not exceeded when the beam is in the full service load stage.

Limiting locations for the prestress steel centroid at any point along the span can be established using Eqs. (21.2) and (21.4), which give the values of concrete stress at the top and bottom of the beam in the unloaded and service load stages, respectively. The stresses produced for those load stages should be compared with the limiting stresses applicable in a particular case, such as the ACI stress limits of Table 21.1. This permits a solution for tendon eccentricity e as a function of distance x along the span.

To indicate that both eccentricity e and moments M_o or M_t are functions of distance x from the support, they will be written as $e(x)$ and $M_o(x)$ or $M_t(x)$, respectively. In writing statements of inequality, it is convenient to designate tensile stress as larger than zero and compressive stress as smaller than zero. Thus, $+450 > -1350$, and $-600 > -1140$, for example.

Considering first the unloaded stage, we find the tensile stress at the top of the beam must not exceed f_{ti}. From Eq. (21.2*a*)

$$f_{ti} \geq -\frac{P_i}{A_c}\left(1 - \frac{e(x)c_1}{r^2}\right) - \frac{M_o(x)}{S_1} \tag{a}$$

Solving for the maximum eccentricity, we obtain

$$e(x) \leq \frac{f_{ti}S_1}{P_i} + \frac{S_1}{A_c} + \frac{M_o(x)}{P_i} \tag{21.29}$$

At the bottom of the unloaded beam, the stress must not exceed the limiting initial compression. From Eq. (21.2*b*)

$$f_{ci} \leq -\frac{P_i}{A_c}\left(1 + \frac{e(x)c_2}{r^2}\right) + \frac{M_o(x)}{S_2} \tag{b}$$

hence, the second lower limit for the steel centroid is

$$e(x) \leq -\frac{f_{ci}S_2}{P_i} - \frac{S_2}{A_c} + \frac{M_o(x)}{P_i} \tag{21.30}$$

Now considering the member in the fully loaded stage, the upper limit values for the eccentricity may be found. From Eq. (21.4*a*)

$$f_{cs} \leq -\frac{P_e}{A_c}\left(1 - \frac{e(x)c_1}{r^2}\right) - \frac{M_t(x)}{S_1} \tag{c}$$

from which

$$e(x) \geq \frac{f_{cs}S_1}{P_e} + \frac{S_1}{A_c} + \frac{M_t(x)}{P_e} \tag{21.31}$$

and using Eq. (21.4*b*)

$$f_{ts} \geq -\frac{P_e}{A_c}\left(1 + \frac{e(x)c_2}{r^2}\right) + \frac{M_t(x)}{S_2} \tag{d}$$

from which

$$e(x) \geq -\frac{f_{ts}S_2}{P_e} - \frac{S_2}{A_c} + \frac{M_t(x)}{P_e} \tag{21.32}$$

Using Eqs. (21.29) and (21.30), the lower limit of tendon eccentricity is established at successive points along the span. Then, using Eqs. (21.31) and (21.32), the corresponding upper limit is established. This upper limit may well

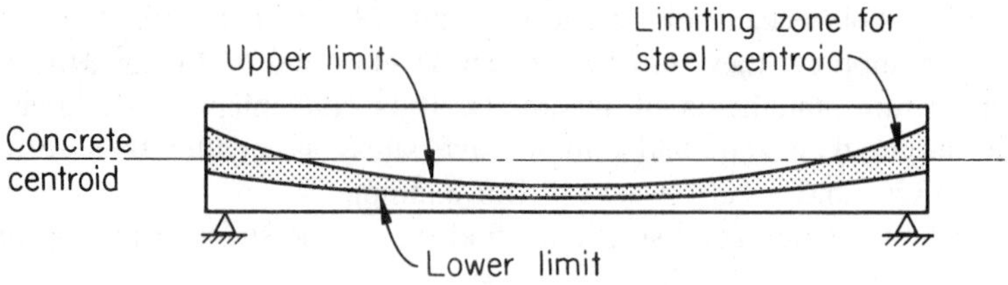

FIGURE 21.19
Typical limiting zone for centroid of prestressing steel.

be negative, indicating that the tendon centroid may be above the concrete centroid at that location.

It is often convenient to plot the envelope of acceptable tendon profiles, as has been done in Fig. 21.19, for a typical case in which both dead and live loads are uniformly distributed. Any tendon centroid falling completely within the shaded zone would be satisfactory from the point of view of concrete stress limits. It should be emphasized that it is only the tendon centroid that must be within the shaded zone; individual cables are often outside of it.

The tendon profile actually used is often a parabolic curve or a catenary in the case of post-tensioned beams. The duct containing the prestressing steel is draped to the desired shape and held in that position by wiring it to the transverse web reinforcement, after which the concrete may be poured. In pretensioned beams, *deflected tendons* are often used. The cables are held down at midspan, at the third points, or at the quarter points of the span and held up at the ends, so that a smooth curve is approximated to a greater or lesser degree.

In practical cases, it is often not necessary to make a centroid zone diagram, such as is shown in Fig. 21.19. By placing the centroid at its known location at midspan, at or close to the concrete centroid at the supports, and with a near-parabolic shape between those control points, satisfaction of the limiting stress requirements is ensured. With nonprismatic beams, beams in which a curved concrete centroidal axis is employed, or with continuous beams, diagrams such as Fig. 21.19 are a great aid.

21.12 FLEXURAL DESIGN BASED ON LOAD BALANCING

It was pointed out in Sec. 21.2b that the effect of a change in the alignment of a prestressing tendon in a beam is to produce a vertical force on the beam at that location. Prestressing a member with curved or deflected tendons thus has the effect of introducing a set of equivalent loads, and these may be treated just as any other loads in finding moments or deflections. Each particular tendon profile produces its own unique set of equivalent forces. Typical tendon profiles, with corresponding equivalent loads and moment diagrams, were illustrated in Fig. 21.2. Both Fig. 21.2 and Sec. 21.2b should be reviewed carefully.

The equivalent load concept offers an alternative approach to the determination of required prestress force and eccentricity. The prestress force and tendon profile can be established so that external loads that will act are exactly counteracted by the vertical forces resulting from prestressing. The net result, for that particular set of external loads, is that the beam is subjected only to axial compression and no bending moment. The selection of the load to be balanced is left to the judgement of the designer. Often the balanced load chosen is the sum of the self-weight and superimposed dead load.

The design approach described in this section was introduced in the United States by T. Y. Lin in 1963 and is known as the *load-balancing method*. The fundamentals will be illustrated in the context of the simply supported, uniformly loaded beam of Fig. 21.20a. The beam is to be designed for a balanced load consisting of its own weight w_o, the superimposed dead load w_d, and some fractional part of the live load denoted by $k_b w_l$. Since the external load is uniformly distributed, it is reasonable to adopt a tendon having a parabolic shape. It is easily shown that a parabolic tendon will produce a uniformly distributed upward load equal to

$$w_p = \frac{8Py}{l^2} \tag{21.33}$$

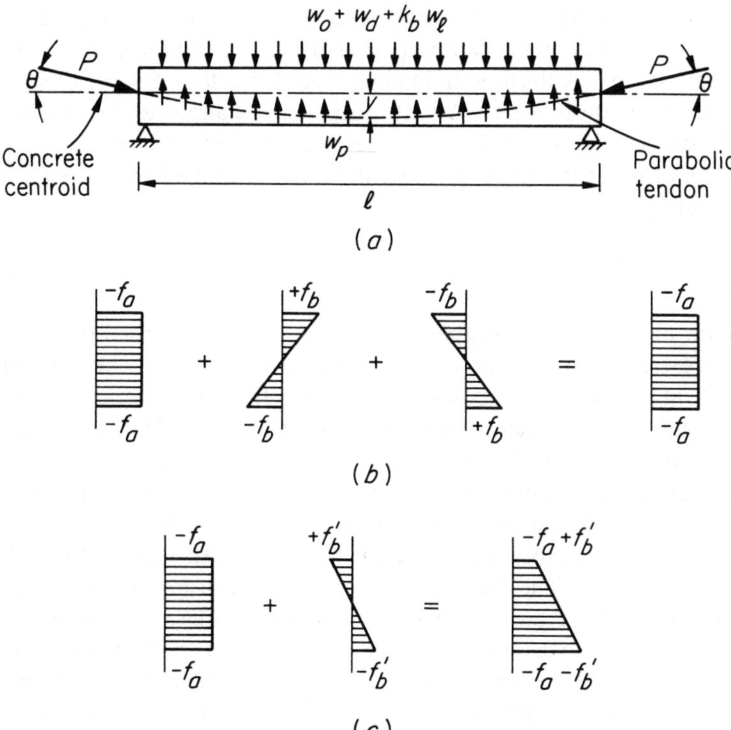

FIGURE 21.20
Load balancing for uniformly loaded beam: (*a*) external and equivalent loads; (*b*) concrete stresses resulting from axial and bending effects of prestress plus bending resulting from balanced external load; (*c*) concrete stresses resulting when load $k_b w_l$ is removed.

where P is the magnitude of the prestress force, y is the maximum sag of the tendon measured with respect to the chord between its end points, and l is the span.

If the downward load exactly equals the upward load from the tendon, these two loads cancel and no bending stress is produced, as shown in Fig. 21.20*b*. The bending stresses due to prestress eccentricity are equal and opposite to the bending stresses resulting from the external load. The net resulting stress is uniform compression f_a equal to that produced by the axial force $P \cos \theta$. Excluding consideration of time-dependent effects, the beam would show no vertical deflection.

If the live load is removed or increased, then bending stresses and deflections will result because of the *unbalanced* portion of the load. Stresses resulting from this differential loading must be calculated and superimposed on the axial compression to obtain the net stresses for the unbalanced state. If we refer to Fig. 21.20*c*, the bending stresses f_b' resulting from removal of the partial live loading are superimposed on the uniform compressive stress f_a, resulting from the combination of eccentric prestress force and full balanced load to produce the final stress distribution shown.

Loads other than uniformly distributed would lead naturally to the selection of other tendon configurations. For example, if the external load consisted of a single concentration at midspan, a deflected tendon such as that of Fig. 21.2*a* would be chosen, with maximum eccentricity at midspan, varying linearly to zero eccentricity at the supports. A third-point loading would lead the designer to select a tendon deflected at the third points. A uniformly loaded cantilever beam would best be stressed using a tendon in which the eccentricity varied parabolically, from zero at the free end to y at the fixed support, in which case the upward reaction of the tendon would be

$$w_p = \frac{2Py}{l^2} \tag{21.34}$$

It should be clear that, for simple spans designed by the load-balancing concept, it is necessary for the tendon to have zero eccentricity at the supports because the moment due to superimposed loads is zero there. Any tendon eccentricity would produce an unbalanced moment (in itself an equivalent load) equal to the horizontal component of the prestress force times its eccentricity. At the simply supported ends, the requirement of zero eccentricity must be retained.

In practice, the load-balancing method of design starts with selection of a trial beam cross section, based on experience and judgment. An appropriate span-depth ratio is often applied. The tendon profile is selected using the maximum available eccentricity, and the prestress force is calculated. The trial design may then be checked to ensure that concrete stresses are within the allowable limits should the live load be totally absent or fully in place, when bending stresses will be superimposed on the axial compressive stresses. There is no assurance that the section will be adequate for these load stages, nor that adequate strength will be provided should the member be overloaded. Revision may be necessary.

It should further be observed that obtaining a uniform compressive concrete stress at the balanced load stage does not ensure that the member will have zero deflection at this stage. The reason for this is that the uniform stress distribution is made up of two parts: that from the eccentric prestress force and that from the external loads. The prestress force varies with time because of shrinkage, creep, and relaxation, changing the vertical deflection associated with the prestress force. Concurrently, the beam will experience creep deflection under the combined effects of the diminishing prestress force and the external loads, a part of which may be sustained and a part of which may be short-term. However, if load balancing is carried out based on the effective prestress force P_e plus self-weight and external dead load only, the result may be near-zero deflection for that combination.

The load-balancing method provides the engineer with a useful tool. For simple spans, it leads the designer to choose a sensible tendon profile and focuses attention very early on the matter of deflection. But the most important advantages become evident in the design of indeterminate prestressed members, including both continuous beams and two-way slabs. For such cases, at least for one unique loading, the member carries only axial compression but no bending. This greatly simplifies the analysis.

Example 21.5 Beam design initiating with load balancing. A post-tensioned beam is to be designed to carry a uniformly distributed load over a 30 ft span, as illustrated by Fig. 21.21. In addition to its own weight, it must carry a dead load of 150 plf and a

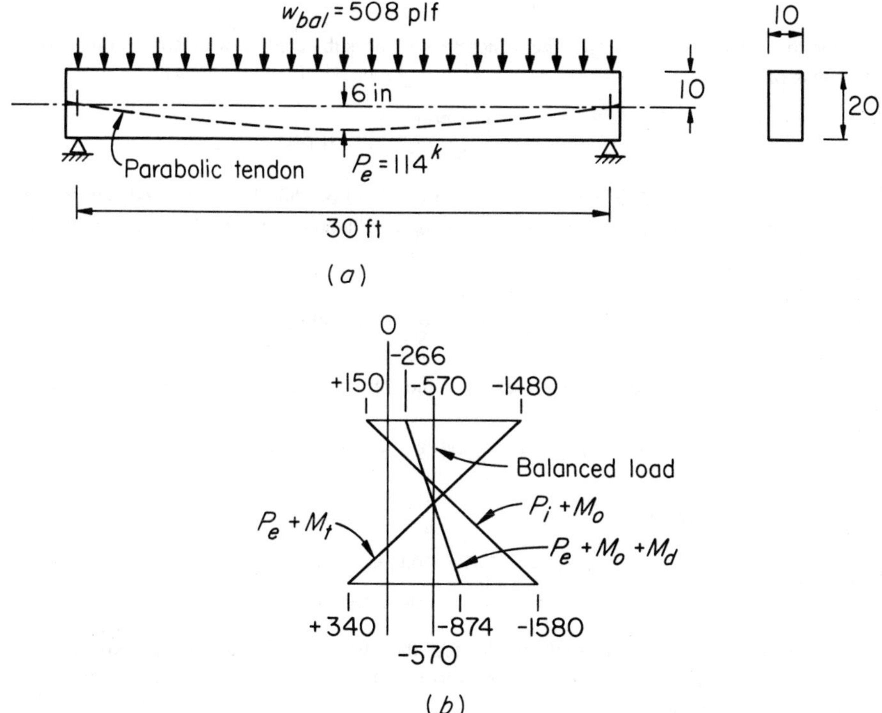

FIGURE 21.21
Example of design by load balancing: (*a*) beam profile and cross section; (*b*) flexural stresses at maximum moment section.

service live load of 600 plf. Concrete strength of 4000 psi will be attained at 28 days; at time of transfer of prestress force, the strength will be 3,000 psi. Prestress loss may be assumed at 20 percent of P_i. On the basis that about one quarter of the live load will be sustained over a substantial time period, k_b of 0.25 will be used in determining the balanced load.

Solution. On the basis of an arbitrarily chosen span-depth ratio of 18, a trial section of 20 in. total depth is selected, having a 10 in. width. The calculated self-weight of the beam is 208 plf and the selected load to be balanced is

$$w_{\text{bal}} = w_o + w_d + k_b w_l = 208 + 150 + 150 = 508 \text{ plf}$$

Based on a minimum concrete cover from the steel centroid to the bottom face of the beam of 4 in., the maximum eccentricity that can be used for the 20 in. trial section is 6 in. A parabolic tendon will be used to produce a uniformly distributed upward tendon reaction. To equilibrate the sustained downward loading, the prestress force P_e after losses, from Eq. (21.33), should be

$$P_e = \frac{w_{\text{bal}} l^2}{8y} = \frac{508 \times 900}{8 \times 0.5} = 114{,}000 \text{ lb}$$

and the corresponding initial prestress force is

$$P_i = \frac{P_e}{R} = \frac{114{,}000}{0.8} = 143{,}000 \text{ lb}$$

For the balanced load stage, the concrete will be subjected to a uniform compressive stress of

$$f_{\text{bal}} = \frac{114{,}000}{200} = -570 \text{ psi}$$

as shown in Fig. 21.21*b*. Should the partial live load of 150 plf be removed, the stresses to be superimposed on f_{bal} result from a net *upward* load of 150 plf. The section modulus for the trial beam is 667 in^3 and

$$M_{\text{unbal}} = 150 \times \frac{900}{8} = 16{,}900 \text{ ft-lb}$$

Hence, the unbalanced bending stresses at the top and bottom faces are

$$f_{\text{unbal}} = 16{,}900 \times \frac{12}{667} = 304 \text{ psi}$$

Thus, the net stresses are

$$f_1 = -570 + 304 = -266 \text{ psi}$$
$$f_2 = -570 - 304 = -874 \text{ psi}$$

Similarly, if the *full* live load should act, the stresses to be superimposed are those resulting from a net *downward* load of 450 plf. The resulting stresses in the concrete at full service load are

$$f_1 = -570 - 910 = -1480 \text{ psi}$$
$$f_2 = -570 + 910 = +340 \text{ psi}$$

Stresses in the concrete with live load absent and live load fully in place are shown in Fig. 21.21*b*.

It is also necessary to investigate the stresses in the initial unloaded stage, when the member is subjected to P_i plus moment due to its own weight.

$$M_o = 208 \times \frac{900}{8} = 23{,}400 \text{ ft-lb}$$

Hence, in the initial stage:

$$f_1 = -\frac{143{,}000}{200}\left(1 - \frac{6 \times 10}{33.35}\right) - \frac{23{,}400 \times 12}{667} = +150 \text{ psi}$$

$$f_2 = -\frac{143{,}000}{200}\left(1 + \frac{6 \times 10}{33.35}\right) + \frac{23{,}400 \times 12}{667} = -1580 \text{ psi}$$

The stresses in the unloaded and full service load stages must be checked against these permitted by Code. With $f'_c = 4000$ psi and $f'_{ci} = 3000$ psi, the permitted stresses are:

$$f_{ti} = +165 \text{ psi} \qquad f_{ts} = +380 \text{ psi}$$
$$f_{ci} = -1800 \text{ psi} \qquad f_{cs} = -1800 \text{ psi}$$

The actual stresses, shown in Fig. 21.21*b*, are within these limits and acceptably close, and no revision will be made in the trial 10 × 20 in. cross section on the basis of stress limits.

The ultimate flexural strength of the members must now be checked, to insure that an adequate factor of safety against collapse has been provided. The required P_i of 143,000 lb will be provided using stranded Grade 250 cable, with $f_{pu} = 250{,}000$ psi and $f_{py} = 212{,}000$ psi. Referring to Sec. 21.4, the initial stress immediately after transfer must not exceed $0.74 \times 250{,}000 = 185{,}000$ psi, or $0.82 \times 212{,}000 = 174{,}000$ psi, which controls in this case. Accordingly, the required area of tendon steel is

$$A_p = 143{,}000/174{,}000 = 0.82 \text{ in}^2$$

This will be provided using eight $\frac{7}{16}$ in. strands, giving an actual area of 0.864 in^2 (Table A.16). The resulting stresses at the initial and final stages are

$$f_{pi} = \frac{143{,}000}{0.864} = 166{,}000 \text{ psi}$$

$$f_{pe} = \frac{114{,}000}{0.864} = 132{,}000 \text{ psi}$$

Using the ACI approximate equation for steel stress at failure [see Eq. (21.6)], with $\rho_p = 0.864/160 = 0.0054$, and $\gamma_p = 0.40$ for the ordinary Grade 250 tendons, the stress f_{ps} is given by

$$f_{ps} = f_{pu}\left(1 - \frac{\gamma_p}{\beta_1}\frac{\rho_p f_{pu}}{f'_c}\right)$$

$$= 250\left(1 - \frac{0.40}{0.85}\frac{0.0054 \times 250}{4}\right)$$

$$= 210 \text{ ksi}$$

Then,

$$a = \frac{A_{ps} f_{ps}}{0.85 f'_c b}$$

$$= \frac{0.864 \times 210}{0.85 \times 4 \times 10} = 5.34 \text{ in.}$$

The nominal flexural strength is

$$M_n = A_p f_{ps}\left(d - \frac{a}{2}\right) = 0.864 \times 210{,}000\left(16 - \frac{5.34}{2}\right)\frac{1}{12}$$

$$= 202{,}000 \text{ ft-lb}$$

This must be reduced by the factor $\phi = 0.90$ as usual to obtain design strength:

$$\phi M_n = 0.90 \times 202{,}000 = 182{,}000 \text{ ft-lb}$$

It will be recalled that the ACI load factors with respect to dead and live loads are, respectively, 1.4 and 1.7. The safety factor obtained in the present case will be evaluated with respect to the service live load moment of 67,500 ft-lb, assuming that dead loads may be 1.4 times the calculated values, in keeping with the ACI requirements. Accordingly,

$$182 = 1.4(23.4 + 16.9) + F_l(67.5)$$

$$F_l = 1.86$$

This exceeds the ACI minimum of 1.7 and the design is judged satisfactory.

21.13 LOSS OF PRESTRESS

As discussed in Sec. 21.6, the initial prestress force P_i immediately after transfer is less than the jacking force P_j because of the elastic shortening of the concrete, slip at the anchorages, and frictional losses along the tendons. The force is reduced further, after a period of many months or even years, due to length changes resulting from shrinkage and creep of the concrete and relaxation of the highly stressed steel; eventually it attains its effective value P_e. In the preceding articles of this chapter, losses were accounted for, making use of an assumed effectiveness ratio $R = P_e/P_i$.

The estimation of losses can be made on several different levels. In most cases, in practical design, detailed calculation of losses is not necessary. A value for R may be assumed, based either on experience or on any of several empirical expressions which are widely used. For cases where greater accuracy is required, it is necessary to estimate the separate losses, taking account of the special conditions of member geometry, material properties, and construction methods that apply. Accuracy of loss estimation can be improved still further by accounting for the interdependence of time-dependent losses, using the summation of losses in a sequence of discrete time steps. These methods will be discussed briefly in the following paragraphs.

Table 21.2 Estimate of prestress losses

Type of prestressing steel	Total loss, psi	
	$f_c' = 4000$ psi	$f_c' = 5000$ psi
Pretensioning strand	—	45,000
Post-tensioning wire or strand[a]	32,000	33,000
Post-tensioning bars[a]	22,000	23,000

[a] Losses due to friction are excluded. Friction losses should be computed according to Sec. 21.13b.

Source: From Ref. 21.13.

a. Lump-Sum Estimates of Losses

It was recognized very early in the development of prestressed concrete that there was a need for approximate expressions to be used to estimate prestress losses in routine design. Many thousands of successful prestressed structures have been built based on such estimates, and where member sizes, spans, materials, construction procedures, amount of prestress force, and environmental conditions are not out of the ordinary, this approach is satisfactory. For such conditions, the American Association of State Highway and Transportation Officials (AASHTO, Ref. 21.13) has recommended the values of Table 21.2. It should be noted that losses due to friction are *not* included in the values for post-tensioned members. These may be calculated separately by the equations of Sec. 21.13b below.

The AASHTO recommended losses of Table 21.2 include losses due to elastic shortening, creep, shrinkage, and relaxation (see below). Thus for comparison with R values for estimating losses, such as were employed for the preceding examples, which included only the time-dependent losses due to shrinkage, creep, and relaxation, elastic shortening losses should be estimated by the methods of Sec. 21.13b and deducted from the total.

b. Estimate of Separate Losses

For cases where lump-sum estimates of loss are inadequate, e.g., for members of unusual proportions, exceptional span, or lightweight concrete, a separate estimate of individual losses should be made. Such an analysis is complicated by the interdependence of time-dependent losses. For example, the relaxation of stress in the tendons is affected by length changes due to creep of concrete. Rate of creep, in turn, is altered by change in tendon stress. In the following six subsections, losses are treated as if they occurred independently, although certain arbitrary adjustments are included to account for the interdependence of time-dependent losses. If greater refinement is necessary, a step-by-step approach, like that mentioned in Sec. 21.13c, may be used (see also Refs. 21.14 and 21.15).

(1) Slip at the anchorages As the load is transferred to the anchorage device in post-tensioned construction, a slight inward movement of the tendon will occur as the wedges seat themselves and as the anchorage itself deforms under stress. The amount of movement will vary greatly, depending on the type of anchorage and on construction techniques. The amount of movement due to seating and stress deformation associated with any particular type of anchorage is best established by test. Once this amount ΔL is determined, the stress loss is easily calculated from

$$\Delta f_{s,\text{slip}} = \frac{\Delta L}{L} E_s \tag{21.35}$$

It is significant to note that the amount of slip is nearly independent of the cable length. For this reason, the stress loss will be large for short tendons and relatively small for long tendons. The practical consequence of this is that it is most difficult to post-tension short tendons with any degree of accuracy.

(2) Elastic shortening of the concrete In pretensioned members, as the tendon force is transferred from the fixed abutments to the concrete beam, elastic instantaneous compressive strain will take place in the concrete, tending to reduce the stress in the bonded prestressing steel. The steel stress loss is

$$\Delta f_{s,\text{elastic}} = E_s \frac{f_c}{E_c} = n f_c \tag{21.36}$$

where f_c is the concrete stress at the level of the steel centroid immediately after prestress is applied:

$$f_c = -\frac{P_i}{A_c}\left(1 + \frac{e^2}{r^2}\right) + \frac{M_o e}{I_c} \tag{21.37}$$

If the tendons are placed with significantly different effective depths, the stress loss in each should be calculated separately.

In computing f_c by Eq. (21.37), the prestress force used should be that after the losses being calculated have occurred. It is usually adequate to estimate this as about 10 percent less than P_j.

In post-tensioned members, if all the strands are tensioned at one time, there will be no loss due to elastic shortening, because this shortening will occur as the jacking force is applied and before the prestressing force is measured. On the other hand, if various strands are tensioned sequentially, the stress loss in each strand will vary, being a maximum in the first strand tensioned and zero in the last strand. In most cases, it is sufficiently accurate to calculate the loss in the first strand and to apply one-half that value to all strands.

(3) Frictional losses Losses due to friction, as the tendon is stressed in post-tensioned members, are usually separated for convenience into two parts: curvature friction and wobble friction. The first is due to intentional bends in the tendon profile as specified and the second to the unintentional variation of the tendon from its intended profile. It is apparent that even a "straight" tendon duct will have some unintentional misalignment so that wobble friction must always be considered in post-tensioned work. Usually, curvature friction must be considered as well. The force at the jacking end of the tendon P_o, required to produce the force P_x at any point x along the tendon, can be found from the expression

$$P_o = P_x \epsilon^{KL+\mu\alpha} \tag{21.38a}$$

where
ϵ = base of natural logarithms
L = tendon length from jacking end to point x
α = angular change of tendon from jacking end to point x, radians
K = wobble friction coefficient, lb/lb per ft
μ = curvature friction coefficient

There has been much research on frictional losses in prestressed construction, particularly with regard to the values of K and μ. These vary appreciably, depending on construction methods and materials used. The values in Table 21.3, from the ACI Code Commentary, may be used as a guide.

If one accepts the approximation that the normal pressure on the duct causing the frictional force results from the undiminished initial tension all the way around the curve, the following simplified expression for loss in tension is obtained:

$$P_o = P_x(1 + KL + \mu\alpha) \tag{21.38b}$$

where α is the angle between the tangents at the ends. The ACI Code permits the use of the simplified form if the value of $KL + \mu\alpha$ is not greater than 0.3.

The loss of prestress for the entire tendon length can be computed by segments, with each segment assumed to consist of either a circular arc or a length of tangent.

(4) Creep of concrete Shortening of concrete under sustained load has been discussed in Sec. 2.7. It can be expressed in terms of the creep coefficient C_c. Creep shortening may be several times the initial elastic shortening, and it is evident that it will result in loss of prestress force. The stress loss can be calculated from

$$\Delta f_{s,\text{creep}} = C_c n f_c \tag{21.39}$$

Table 21.3 Friction coefficients for post-tensioned tendons

Type of tendon	Wobble coefficient K, per ft	Curvature coefficient μ
Grouted tendons in metal sheathing		
Wire tendons	0.0010–0.0015	0.15–0.25
High strength bars	0.0001–0.0006	0.08–0.30
Seven-wire strand	0.0005–0.0020	0.15–0.25
Unbonded tendons		
Mastic-coated wire tendons	0.0010–0.0020	0.05–0.15
Mastic-coated seven-wire strand	0.0010–0.0020	0.05–0.15
Pregreased wire tendons	0.0003–0.0020	0.05–0.15
Pregreased seven-wire strand	0.0003–0.0020	0.05–0.15

Values of C_c for different concrete strengths for average conditions of humidity are given in Table 2.1.

In Eq. (21.39), the concrete stress f_c to be used is that at the level of the steel centroid, when the eccentric prestress force plus all sustained loads are acting. Equation (21.37) can be used, except that the moment M_o should be replaced by the moment due to *all* dead loads plus that due to any portion of the live load that may be considered sustained.

It should be noted that the prestress force causing creep is not constant but diminishes with the passage of time due to relaxation of the steel, shrinkage of the concrete, and length changes associated with creep itself. To account for this, it is recommended that the prestress force causing creep be assumed at 10 percent less than the initial value P_i.

(5) Shrinkage of concrete It is apparent that a decrease in the length of a member due to shrinkage of the concrete will be just as detrimental as length changes due to stress, creep, or other causes. As discussed in Sec. 2.10, the shrinkage strain ϵ_{sh} may vary between about 0.0004 and 0.0008. A typical value of 0.0006 may be used in lieu of specific data. The steel stress loss resulting from shrinkage is

$$\Delta f_{s,\text{shrink}} = \epsilon_{sh} E_s \tag{21.40}$$

Only that part of the shrinkage which occurs after transfer of prestress force to the concrete need be considered. For pretensioned members, transfer commonly takes place only 24 hours after pouring the concrete, and nearly all the shrinkage takes place after that time. However, post-tensioned beams are seldom stressed at an earlier age than 7 days and often much later than that. About 15 percent of ultimate shrinkage may occur by the age of 7 days, under typical conditions, and about 40 percent by the age of 28 days.

(6) Relaxation of steel The phenomenon of relaxation, similar to creep, was discussed in Sec. 2.14. Loss in stress due to relaxation will vary depending upon the stress in the steel, and may be estimated using Eq. (2.11). To allow for the gradual reduction of steel stress resulting from the combined effects of creep, shrinkage, and relaxation, the relaxation calculation can be based on a prestress force 10 percent less than P_i.

It is interesting to observe that the largest part of the relaxation loss occurs shortly after the steel is stretched. For stresses of $0.80 f_{pu}$ and higher, even a very short period of loading will produce substantial relaxation, and this in turn will reduce the relaxation that will occur later at a lower stress level. The relaxation rate can thus be artificially accelerated by temporary overtensioning. This technique has sometimes been employed to reduce losses.

c. Loss Estimation by the Time-Step Method

The loss calculations of the preceding paragraphs recognized the interdependence of creep, shrinkage, and relaxation losses in an approximate way, by an arbitrary reduction of 10 percent of the initial prestress force P_i to obtain the force for which creep and relaxation losses were calculated. For cases requiring greater accuracy, losses can be calculated for discrete time steps over the period of interest. The prestress force causing losses during any time step is taken equal to the value at the end of the preceding time step, accounting for losses due to all causes up to that time. Accuracy can be improved to any desired degree by reducing the length and increasing the number of time steps.

A step-by-step method developed by the Committee on Prestress Losses of the Prestressed Concrete Institute uses only a small number of time steps and consequently is suitable for use in calculations by electronic calculator as well as by computer (Ref. 21.14).

21.14 SHEAR, DIAGONAL TENSION, AND WEB REINFORCEMENT

In prestressed concrete beams at service load, there are two factors that greatly reduce the intensity of diagonal tensile stress, compared with that which would exist if no prestress force were present. The first of these results from the combination of longitudinal compressive stress and shearing stress. An ordinary tensile-reinforced-concrete beam under load is shown in Fig. 21.22*a*. The stresses acting on a small element of the beam taken near the neutral axis and near the support are shown in (*b*). It is found by means of Mohr's circle of stress (*c*) that the principal stresses act at 45° to the axis of the beam (*d*) and are numerically equal to the shear stress intensity; thus

$$t_1 = t_2 = v \qquad (a)$$

Now suppose that the same beam, with the same loads, is subjected to a precompression stress in the amount c, as shown in Fig. 21.23*a* and *b*. From Mohr's

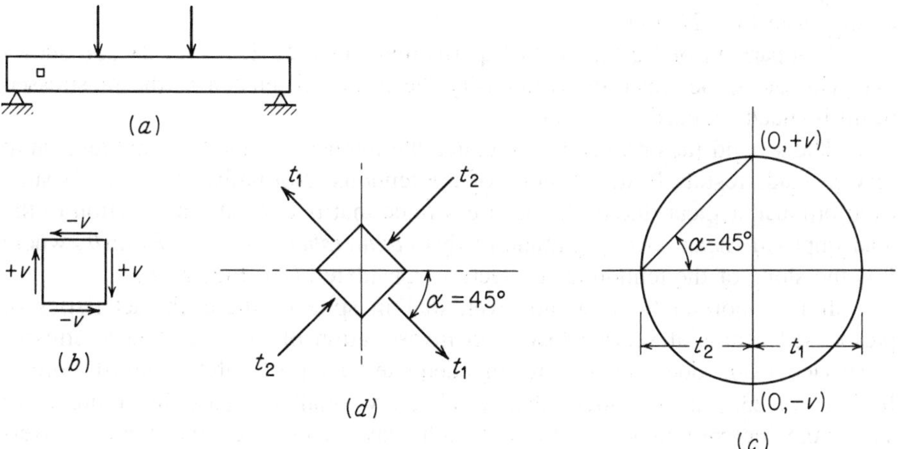

FIGURE 21.22
Principal stress analysis for reinforced concrete beam.

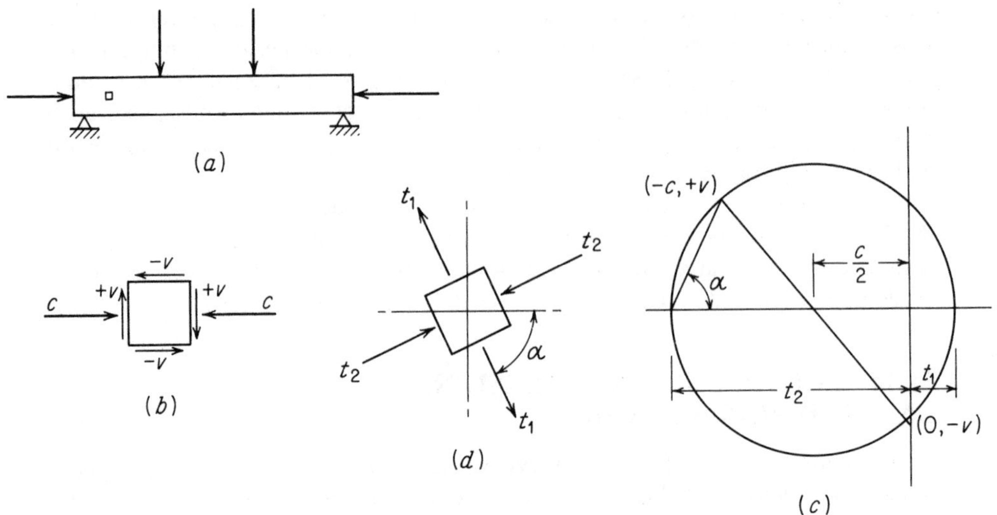

FIGURE 21.23
Principal stress analysis for prestressed concrete beam.

circle (Fig. 21.23*c*), the principal tensile stress is

$$t_1 = -\frac{c}{2} + \sqrt{v^2 + \left(\frac{c}{2}\right)^2} \qquad (b)$$

and the direction of the principal tension with respect to the beam axis is

$$\tan 2\alpha = \frac{2v}{c} \qquad (c)$$

as shown in Fig. 21.23*d*.

Comparison of Eq. (*a*) with Eq. (*b*) and Figs. 21.22*c* with 21.23*c* shows that with the same shear stress intensity the principal tension in the prestressed beam is much reduced.

The second factor working to reduce the intensity of the diagonal tension at service loads results from the slope of the tendons. Normally, this slope is such as to produce a shear due to the prestress force that is opposite in direction to the load-imposed shear. The magnitude of this *countershear* is $V_p = P_e \sin\theta$, where θ is the slope of the tendon at the section considered (see Fig. 21.8).

It is important to note, however, that in spite of these characteristics of prestressed beams at service loads, an investigation of diagonal tensile stresses at service loads does not ensure an adequate factor of safety against failure. In Fig. 21.23*c*, it is evident that a relatively small decrease in compressive stress and increase in shear stress, which may occur when the beam is overloaded, will produce a disproportionate, large increase in the resulting principal tension. In addition to this effect, if the countershear of inclined tendons is used to reduce design shear, its contribution does not increase directly with load but much more slowly (see Sec. 21.7). Consequently, a small increase in total shear may produce a large increase in the net shear for which the beam must be designed. For these two reasons, it is necessary to base design for diagonal tension in prestressed beams on conditions at factored load rather than at service load. The study of principal stresses in the uncracked prestressed beam is significant only in predicting the load at which the first diagonal crack forms.

At loads near failure, a prestressed beam is usually extensively cracked and behaves much like an ordinary reinforced concrete beam. Accordingly, many of the procedures and equations developed in Sec. 4.5 for the design of web reinforcement for nonprestressed beams can be applied to prestressed beams also. Shear design is based on the relation

$$V_u \leq \phi V_n \tag{21.41}$$

where V_u is the total shear force applied to the section at factored loads and V_n is the nominal shear strength, equal to the sum of the contributions of the concrete and web reinforcement:

$$V_n = V_c + V_s \tag{21.42}$$

The strength reduction factor ϕ is taken equal to 0.85 for shear.

In computing the factored load shear V_u, the first critical section is assumed to be at a distance $h/2$ from the face of a support, and sections located a distance less than $h/2$ are designed for the shear computed at $h/2$.

The shear force V_c resisted by the concrete after cracking has occurred is taken equal to the shear that caused the first diagonal crack. Two types of diagonal cracks have been observed in tests of prestressed concrete beams:

1. **Flexure-shear cracks**, occurring at nominal shear V_{ci}, start as nearly vertical flexural cracks at the tension face of the beam, then spread diagonally upward (under the influence of diagonal tension) toward the compression face. These are common in beams with a low value of prestress force.
2. **Web-shear cracks**, occurring at nominal shear V_{cw}, start in the web due to high diagonal tension, then spread diagonally both upward and downward. These are often found in beams with thin webs with high prestress force.

On the basis of extensive tests, it was established that the shear causing flexure-shear cracking can be found from the expression

$$V_{ci} = 0.6\sqrt{f'_c}b_w d + V_{\mathrm{cr},o+d+l} \tag{a}$$

where $V_{\mathrm{cr},o+d+l}$ is the shear force, due to total load, at which the flexural crack forms at the section considered, and $0.6\sqrt{f'_c}b_w d$ represents an additional shear force required to transform the flexural crack into an inclined crack.

While self-weight is generally uniformly distributed, the superimposed dead and live loads may have any distribution. Consequently, it is convenient to separate the total shear into V_o caused by the beam load (without load factor) and V_{cr}, the additional shear force, due to superimposed dead and live loads, corresponding to flexural cracking. Thus

$$V_{ci} = 0.6\sqrt{f_c'}b_w d + V_o + V_{cr} \tag{b}$$

The shear V_{cr} due to superimposed loads can then be found conveniently from

$$V_{cr} = \frac{V_{d+l}}{M_{d+l}}M_{cr} \tag{c}$$

where V_{d+l}/M_{d+l}, the ratio of superimposed dead and live load shear to moment, remains constant as the load increases to cracking load, and

$$M_{cr} = \frac{I_c}{c_2}(6\sqrt{f_c'} + f_{2pe} - f_{2o}) \tag{21.43}$$

where c_2 = distance from concrete centroid to tension face
f_{2pe} = compressive stress at tension face resulting from effective prestress force alone
f_{2o} = bottom-fiber stress due to beam self-weight†

The first term in the parenthesis is a conservative estimate of the modulus of rupture. The bottom-fiber stress due to self-weight is subtracted here because self-weight is considered separately in Eq. (*b*). Thus Eq. (*b*) becomes

$$V_{ci} = 0.6\sqrt{f_c'}b_w d + V_o + \frac{V_{d+l}}{M_{d+l}}M_{cr} \tag{21.44}$$

Tests indicate that V_{ci} need not be taken less than $1.7\sqrt{f_c'}b_w d$. The value of d need not be taken less than $0.80h$ for this and all other equations relating to shear, according to the ACI Code, unless specifically noted otherwise.

The shear force causing web-shear cracking can be found from an exact principal stress calculation, in which the principal tensile stress is set equal to the direct tensile capacity of the concrete (conservatively taken equal to $4\sqrt{f_c'}$ according to the ACI Code). Alternatively, the ACI Code permits use of the approximate expression

$$V_{cw} = (3.5\sqrt{f_c'} + 0.3f_{pc})b_w d + V_p \tag{21.45}$$

in which f_{pc} is the compressive stress in the concrete, after losses, at the centroid of the concrete section (or at the junction of the web and the flange when the centroid lies in the flange) and V_p is the vertical component of the effective prestress force.

After V_{ci} and V_{cw} have been calculated, then V_c, the shear resistance provided by the concrete, is taken equal to the smaller of the two values.

Calculating M_{cr}, V_{ci}, and V_{cw} for a prestressed beam is a tedious matter because many of the parameters vary along the member axis. In practical cases,

† All stresses are used with absolute value here, consistent with ACI convention.

the required quantities may be found at discrete intervals along the span, such as at $l/2$, $l/3$, $l/6$, and at $h/2$ from the support face, and stirrups spaced accordingly.

To shorten the calculation required, the ACI Code includes, as a conservative alternative to the above procedure, an equation for finding the concrete shear resistance V_c directly:

$$V_c = \left(0.6\sqrt{f_c'} + 700\frac{V_u d}{M_u}\right) b_w d \tag{21.46}$$

in which M_u is the bending moment occurring simultaneously with shear force V_u, but $V_u d/M_u$ is not to be taken greater than 1.0. When this equation is used, V_c need not be taken less than $2\sqrt{f_c'}b_w d$ and must not be taken greater than $5\sqrt{f_c'}b_w d$. While Eq. (21.46) is temptingly easy to use and may be adequate for uniformly loaded members of minor importance, its use is apt to result in highly uneconomical designs for I beams of medium and long span and for composite construction (Ref. 21.16).

When shear reinforcement perpendicular to the axis of the beam is used, its contribution to shear strength of a prestressed beam is

$$V_s = \frac{A_v f_y d}{s} \tag{21.47}$$

the same as for a nonprestressed member. According to the ACI Code, the value of V_s must not be taken larger than $8\sqrt{f_c'}b_w d$.

The total nominal shear strength V_n is found by summing the contributions of the steel and concrete, as indicated by Eq. (21.42):

$$V_n = \frac{A_v f_y d}{s} + V_c \tag{21.48}$$

Then, from Eq. (21.41),

$$V_u = \phi V_n = \phi(V_s + V_c)$$

from which

$$V_u = \phi\left(\frac{A_v f_y d}{s} + V_c\right) \tag{21.49}$$

The required cross-section area of one stirrup, A_v, can be calculated by suitable transposition of Eq. (21.49).

$$A_v = \frac{(V_u - \phi V_c)s}{\phi f_y d} \tag{21.50}$$

Normally, in practical design, the engineer will select a trial stirrup size, for which the required spacing is found. Thus, a more convenient form of the last equation is

$$s = \frac{\phi A_v f_y d}{V_u - \phi V_c} \tag{21.51}$$

At least a certain minimum area of shear reinforcement is to be provided in all prestressed concrete members, where the total factored shear force is greater than one half the shear strength ϕV_c provided by the concrete. Exceptions are made for slabs and footings, concrete-joist floor construction, and certain very shallow beams, according to the ACI Code. The minimum area of shear reinforcement to be provided in all other cases is to be taken equal to the smaller of

$$A_v = 50\frac{b_w s}{f_y} \tag{21.52}$$

and

$$A_v = \frac{A_p}{80}\frac{f_{pu}}{f_y}\frac{s}{d}\sqrt{\frac{d}{b_w}} \tag{21.53}$$

in which A_p is the cross-sectional area of the prestressing steel, f_{pu} is the ultimate tensile strength of the prestressing steel, and all other terms are as defined above.

The ACI Code contains certain restrictions on the maximum spacing of web reinforcement to ensure that any potential diagonal crack will be crossed by at least a minimum amount of web steel. For prestressed members, this maximum spacing is not to exceed the smaller of $0.75h$ or 24 in. If the value V_s exceeds $4\sqrt{f_c'}b_w d$, these limits are reduced by half.

Example 21.6. The unsymmetrical I beam shown in Fig. 21.24 carries an effective prestress force of 288 kips and supports a superimposed dead load of 345 plf and service live load of 900 plf, in addition to its own weight of 255 plf, on a 50 ft simple span. At the maximum-moment section, the effective depth to the main steel is 24.5 in. (eccentricity 11.4 in.). The wires are deflected upward 15 ft from the support, and eccentricity is reduced linearly to zero at the support. If concrete having strength $f_c' = 5000$ psi and stirrups with $f_y = 40{,}000$ psi are used, and if the prestressed wires have strength $f_{pu} = 275$ ksi, what is the required stirrup spacing at a point 10 ft from the support?

Solution. For a cross section of the given dimensions, it is easily confirmed that $I_c = 24{,}200$ in^4 and $r^2 = I_c/A_c = 99$ in^2. At a distance 10 ft from the support centerline, the tendon eccentricity is

$$e = 11.4 \times \frac{10}{15} = 7.6 \text{ in.}$$

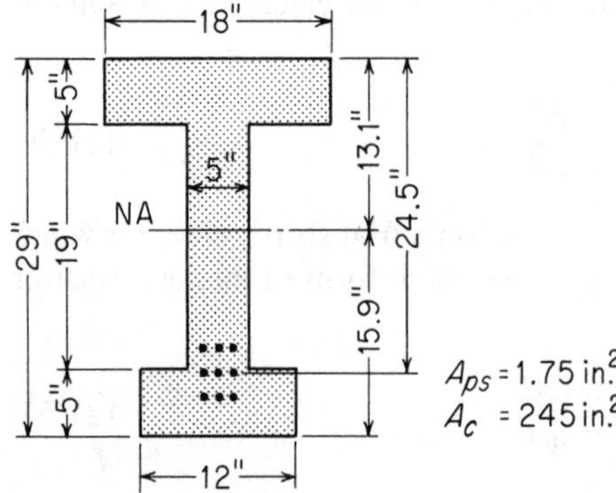

FIGURE 21.24
Post-tensioned beam of Example 21.6.

corresponding to an effective depth d from the compression face of 20.7 in. According to the ACI Code, the larger value of $d = 0.80 \times 29 = 23.2$ in. will be used. Calculation of V_{ci} is based on Eqs. (21.43) and (21.44). The bottom-fiber stress due to effective prestress acting alone is

$$f_{2pe} = -\frac{P_e}{A_c}\left(1 + \frac{ec_2}{r^2}\right) = -\frac{288{,}000}{245}\left(1 + \frac{7.6 \times 15.9}{99}\right) = -2600 \text{ psi}$$

The moment and shear at the section due to beam load alone are, respectively,

$$M_{o,10} = \frac{w_g x}{2}(l - x) = 0.255 \times 5 \times 40 = 51 \text{ ft-kips}$$

$$V_{o,10} = w_g\left(\frac{1}{2} - x\right) = 0.255 \times 15 = 3.8 \text{ kips}$$

and the bottom fiber stress due to this load is

$$f_{2o} = \frac{51 \times 12{,}000 \times 15.9}{24{,}200} = 402 \text{ psi}$$

Then, from Eq. (21.43),

$$M_{\text{cr}} = \frac{24{,}200(425 + 2600 - 402)}{15.9 \times 12} = 333{,}000 \text{ ft-lb}$$

The ratio of superimposed load shear to moment at the section is

$$\frac{V_{d+l}}{M_{d+l}} = \frac{l - 2x}{x(l - x)} = \frac{30}{400} = 0.075 \text{ ft}^{-1}$$

Equation (21.44) is then used to determine the shear force at which flexure-shear cracks can be expected to form:

$$V_{ci} = [0.6\sqrt{5000}(5 \times 23.2) + 3800 + 0.075 \times 330{,}000] \times \frac{1}{1000} = 33.5 \text{ kips}$$

The lower limit of $1.7\sqrt{5000}(5 \times 23.2)/1000 = 13.9$ kips does not control.

Calculation of V_{cw} is based on Eq. (21.45). The slope θ of the tendons at the section under consideration is such that $\sin\theta \approx \tan\theta = 11.4/(15 \times 12) = 0.063$. Consequently, the vertical component of the effective prestress force is $V_p = 0.063 \times 288 = 18.1$ kips. The concrete compressive stress at the section centroid is

$$f_{pc} = \frac{288{,}000}{245} = 1170 \text{ psi}$$

Equation (21.45) can now be used to find the shear at which web-shear cracks should occur:

$$V_{cw} = [(3.5\sqrt{5000} + 0.3 \times 1170)5 \times 23.2 + 18{,}100] \times \frac{1}{1000} = 87.5 \text{ kips}$$

Thus, in the present case,

$$V_c = V_{ci} = 33.5 \text{ kips}$$

At the section considered, the total shear force at factored loads is

$$V_u = 1.4 \times 0.600 \times 15 + 1.7 \times 0.900 \times 15 = 35.6 \text{ kips}$$

When No. 3 U stirrups are used, for which $A_v = 2 \times 0.11 = 0.22$ in^2, the required spacing is found from Eq. (21.51) to be

$$s = \frac{0.85 \times 0.22 \times 40{,}000 \times 23.2}{35{,}600 - 0.85 \times 33{,}500} = 25 \text{ in.}$$

Equation (21.53) is then applied to establish a maximum-spacing criterion

$$0.22 = \frac{1.75}{80} \times \frac{275}{40} \times \frac{s}{23.2}\sqrt{\frac{23.2}{5}} = 0.014s$$

$$s = 15.7 \text{ in.}$$

The other criteria for maximum spacing, three-fourths $\times 29 = 22$ in. and 24 in., do not control here. Open U stirrups will be used, at a spacing of 15 in.

For comparison, the concrete shear will be calculated on the basis of Eq. (21.46). The ratio V_u/M_u is 0.075, and

$$V_c = \left(0.6\sqrt{5000} + 700 \times \frac{0.075}{12} \times 23.2\right)(5 \times 23.2) \times \frac{1}{1000} = 16.7 \text{ kips}$$

The lower and upper limits, $2\sqrt{5000}(5 \times 23.2)/1000 = 16.4$ kips and $5\sqrt{5000}(5 \times 23.2)/1000 = 41.0$ kips, do not control here. Thus, on the basis of V_c obtained from Eq. (21.46), the required spacing of No. 3 U stirrups is

$$s = \frac{0.85 \times 0.22 \times 40{,}000 \times 23.2}{35{,}600 - 0.85 \times 16{,}700} = 8 \text{ in.}$$

For the present case, an I-section beam of intermediate span, more than 3 times the web steel is required at the location investigated if the alternative expression giving V_c directly is used.

21.15 BOND STRESS, TRANSFER LENGTH, AND DEVELOPMENT LENGTH

There are two separate sources of bond stress in prestressed concrete beams: (1) flexural bond, which exists in pretensioned construction between the tendons and the concrete and in grouted post-tensioned construction between the tendons and the grout and between the conduit (if any) and concrete, and (2) prestress transfer bond, generally applicable to pretensioned members only.

a. Flexural Bond

Flexural bond stresses arise due to the change in tension along the tendon resulting from differences in bending moment at adjacent sections. They are proportional to the rate of change of bending moment, hence to the shear force, at a given location along the span. Provided the concrete member is uncracked, flexural bond stress is very low. After cracking, it is higher by an order of magnitude. However, flexural bond stress need not be considered in designing prestressed concrete beams, provided that adequate end anchorage is furnished for the tendon, either in the form of mechanical anchorage (post-tensioning) or strand embedment (pretensioning).

b. Transfer Length and Development Length

For pretensioned beams, when the external jacking force is released, the prestressing force is transferred from the steel to the concrete near the ends of the member by bond, over a distance which is known as the *transfer length*. The transfer length depends upon a number of factors, including the steel stress, the configuration of the steel cross section (e.g., strands vs. wires), the condition of the surface of the steel, and the suddenness with which the jacking force is released. Based on tests of prestressing strand (Ref. 21.17), the effective prestress f_{pe} in the steel may be assumed to act at a transfer length from the end of the member equal to

$$l_t = \frac{f_{pe}}{3} d_b \qquad (a)$$

where l_t = transfer length, in.
d_b = nominal strand diameter, in.
f_{pe} = effective prestress, ksi

The same tests indicate that the additional distance past the original transfer length necessary to develop the failure strength of the steel is closely represented by the expression

$$l_t' = (f_{ps} - f_{pe}) d_b \qquad (b)$$

where the quantity in parentheses is the stress increment above the effective prestress level, in ksi units, to reach the calculated steel stress at failure f_{ps}. Thus the total development length at failure is

$$l_d = l_t + l_t' \qquad (c)$$

or

$$l_d = \left(f_{ps} - \frac{2}{3} f_{pe}\right) d_b \qquad (21.54)$$

The ACI Code does not require that flexural bond stress be checked in either pretensioned or post-tensioned members, but for pretensioned strand it is required that the full development length, given by Eq. (21.54), be provided beyond the critical bending section. Investigation may be limited to those cross sections nearest each end of the member that are required to develop their full flexural strength under the specified ultimate load.

21.16 ANCHORAGE-ZONE DESIGN

In prestressed concrete beams, the prestressing force is introduced as a load concentration acting over a relatively small fraction of the total member depth. For post-tensioned beams with mechanical anchorage, the load is applied at the end face, while for pretensioned beams it is introduced somewhat more gradually

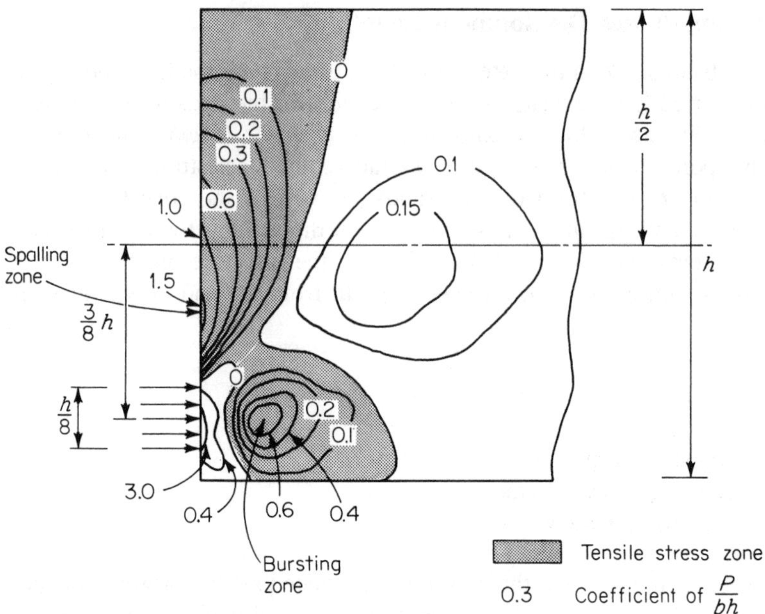

FIGURE 21.25
Contours of equal vertical stress. (*Adapted from Ref. 21.21.*)

over the transfer length. In either case, the compressive stress distribution in the concrete becomes linear, conforming to that dictated by the overall eccentricity of the applied forces, only after a distance from the end roughly equal to the depth of the beam.

This transition of longitudinal compressive stress, from concentrated to linearly distributed, produces transverse (vertical) tensile stresses, which may lead to longitudinal cracking of the member. The pattern and magnitude of the concrete stresses depend on the location and distribution of the concentrated forces applied by the tendons. Numerous studies have been made using the methods of classical elasticity, photoelasticity, and finite element analysis, and typical results are given in Fig. 21.25. Here the beam is loaded uniformly over a height equal to $h/8$ at an eccentricity of $3h/8$. Contour lines are drawn through points of equal vertical tension, with coefficients expressing the ratio of vertical stress to average longitudinal compression. Typically, there are high *bursting stresses* along the axis of the load a short distance inside the end zone and high *spalling stresses* at the loaded face.

In many post-tensioned prestressed I beams, solid end blocks are provided, as shown in Fig. 21.26. While these are often necessary to accommodate end-anchorage hardware, they are of little use in reducing transverse tension or avoiding cracking.

Steel reinforcement for end-zone stresses may be in the form of vertical bars of relatively small diameter and close spacing and should be well anchored at the

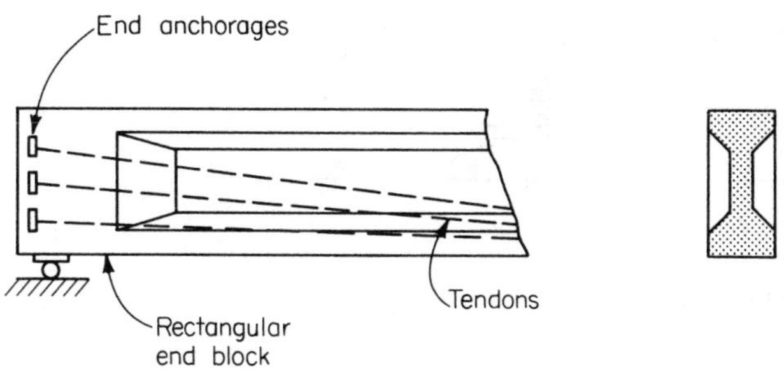

FIGURE 21.26
Post-tensioned I beam with rectangular end block.

top and bottom of the member. Closed stirrups are used commonly, with auxiliary horizontal bars inside the 90° bends.

Rational design of the reinforcement for end zones must recognize that horizontal cracking is likely. If adequate reinforcement is provided, so that the cracks are restricted to a few inches in length and to 0.01 in. or less in width, these cracks will not be detrimental to the performance of the beam either at service load or at the overload stage. It should be noted that end-zone stresses in pretensioned and bonded post-tensioned beams do not increase in proportion to loads. The failure stress f_{ps} in the tendon at beam failure is attained only at the maximum moment section.

For *pretensioned members*, based on tests reported in Ref. 21.18, a very simple equation has been proposed for the design of end-zone reinforcement:

$$A_t = 0.021 \frac{P_i h}{f_s l_t} \tag{21.55}$$

where A_t = total cross-sectional area of stirrups necessary, in^2.
P_i = initial prestress force, lb
h = total member depth, in.
f_s = allowable stress in stirrups, psi
l_t = transfer length, in.

An allowable stress f_s = 20,000 psi has been found in tests to produce acceptably small crack widths. The required reinforcement having total area A_t should be distributed over a length equal to $h/5$ measured from the end face of the beam, and for most efficient crack control the first stirrup should be placed as close to the end face as practical. It is recommended in Ref. 21.18 that vertical reinforcement according to Eq. (21.55) be provided for *all* pretensioned members unless tests or experience indicates that cracking does not occur at service or overload stages.

For *post-tensioned members*, end-zone reinforcement is often designed on the basis of an equilibrium analysis of the cracked anchorage zone (Ref. 21.19).

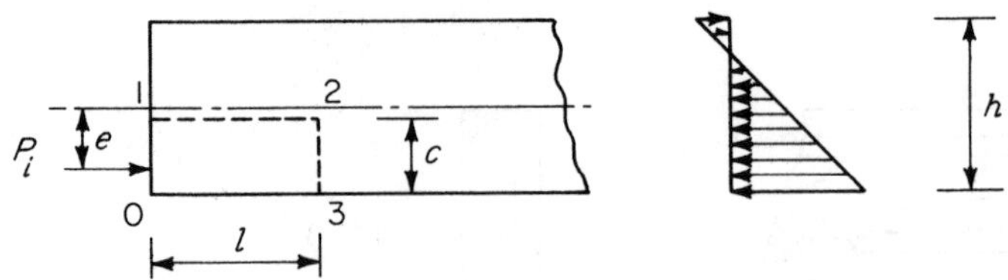

(a) End of beam showing free-body location

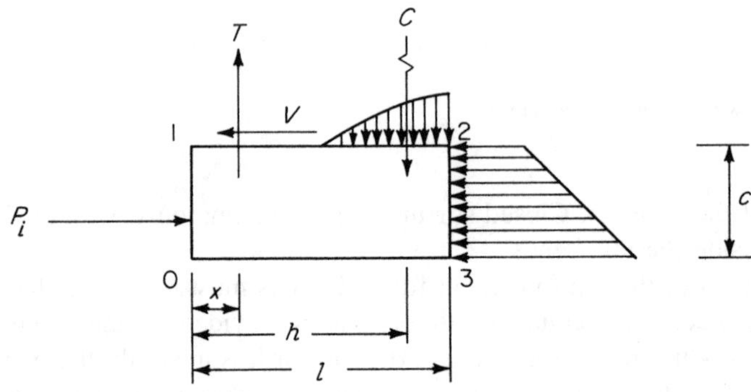

(b) Forces on free body

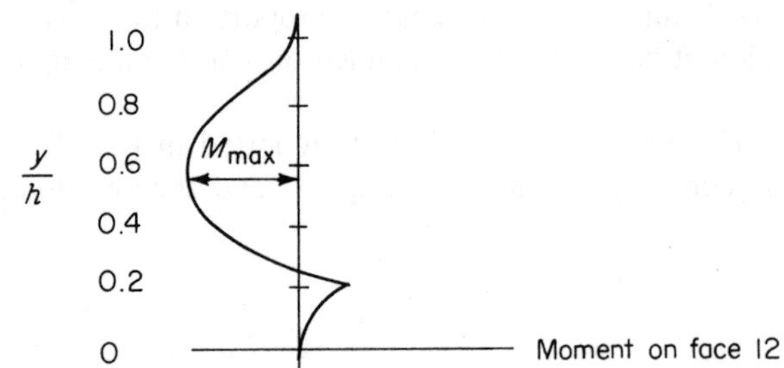

(c) Variation of moment with depth

FIGURE 21.27
End-zone analysis of post-tensioned beam: (*a*) end of beam showing free body; (*b*) forces on free body; (*c*) variation of moment with depth.

Figure 21.27*a* shows the end region of a post-tensioned beam with initial prestress P_i applied at eccentricity e. At some distance l from the end, the compressive stress distribution is linear. Figure 21.27*b* shows the forces and stresses acting on the free body 0-1-2-3 bounded by the edges of the members, by an assumed horizontal crack, and by the inside end face of the anchorage zone.

In general, both a moment and a shear force will be produced on the face 1-2 by the horizontal forces. The shear is resisted by aggregate interlock, and the necessary resisting moment is provided by the tensile force T from the reinforcement and the compressive resultant C from the concrete, assumed to act a distance h from the end face. The height c of the free body, determined by the level of the crack, is established from the condition that the moment due to the horizontal forces will be a maximum at the level at which the crack forms. In practical cases, moments can be calculated at increments of height, starting at the bottom of the beam, and plotted as a function of distance from the bottom. Typical results are shown in Fig. 21.27*c*.

From this analysis, the total required area of steel reinforcement can be found:

$$A_t = \frac{M_{\max}}{f_s(h - x)} \tag{21.56}$$

where f_s is the allowable stress in the stirrups (usually taken as 20,000 psi), x is the distance in inches from the end face to the centroid of the steel, and other terms are as already defined. For post-tensioned beams, stirrups should be provided within a distance $h/2$ from the end face to resist the computed tension.

In addition to vertical tensile stresses causing splitting, end-zone distress may be caused in post-tensioned beams by the high concentration of longitudinal compression under the bearing plates of the anchorages. The bearing stress on the concrete when the initial prestress force P_i acts should not exceed

$$f_b = 0.8f'_{ci}\sqrt{\frac{A_2}{A_1} - 0.2} \qquad \text{and} \qquad \leq 1.25f'_{ci} \tag{21.57a}$$

according to the ACI Code Commentary. After allowance for prestress losses, when P_e acts, the bearing stress should not exceed

$$f_b = 0.6f'_c\sqrt{\frac{A_2}{A_1}} \qquad \text{and} \qquad \leq f'_c \tag{21.57b}$$

In these equations, A_1 is the bearing area of the anchor plate and A_2 is the maximum area of the portion of the anchorage surface that is geometrically similar to, and concentric with, the area of the anchor plate.

Research reported in Ref. 21.20 has confirmed the advisability of adding spiral reinforcement immediately under post-tensioning anchors, a practice that is quite general based on manufacturers' tests. Specific information should be sought for particular systems.

Example 21.7 Design of end-zone reinforcement for post-tensioned beam. End-zone reinforcement is to be designed for the rectangular post-tensioned beam shown in Fig. 21.28*a*. An initial prestress force P_i of 250 kips is applied by two tendons having eccentricity of 10.5 in., producing longitudinal stresses in the concrete that vary linearly from 2153 psi compression at the bottom to 764 psi tension at the top. Closed vertical stirrups will be used as shown in Fig. 21.28*b* at an allowable stress of 20,000 psi.

Solution. For computational purposes the beam will be divided into 3 in. increments of height, and the concrete stress at the center of each increment will be assumed to act uniformly through the depth of that increment. Values are given in Fig. 21.28*c*. The moments from these stresses and from the concentrated tendon forces are computed at 3

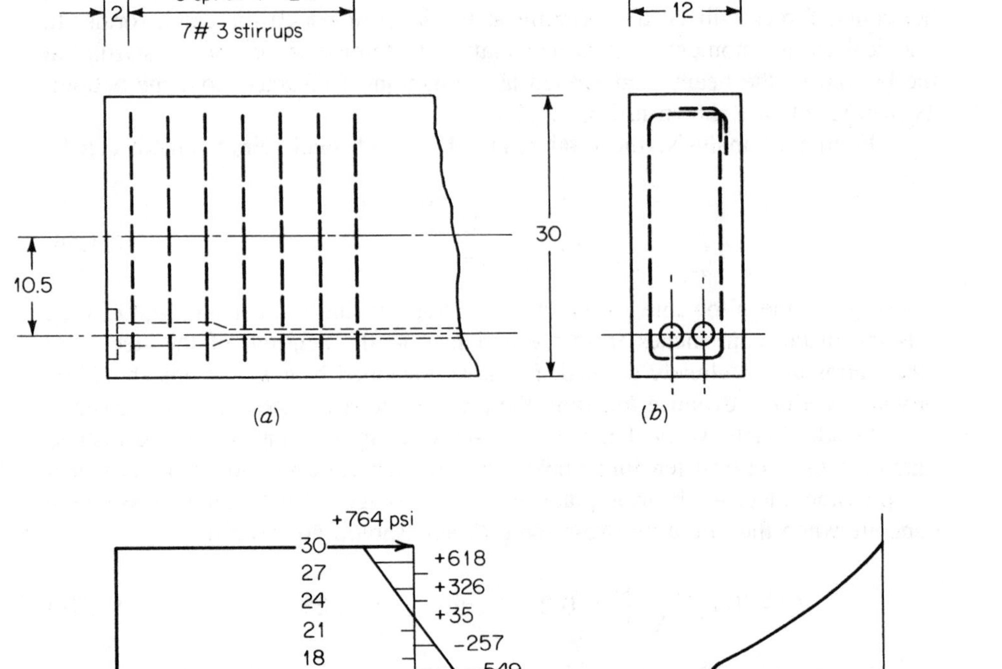

FIGURE 21.28
Design of anchorage zone reinforcement: (*a*) end zone; (*b*) cross section; (*c*) forces and stresses; (*d*) moments.

in. intervals, clockwise moments being taken positive. Results are given in Table 21.4 and are plotted in Fig. 21.28*d*. It is seen that a maximum moment of 388 in-kips is obtained, acting at a distance 15 in. above the bottom of the beam such as to cause tension near the end face (spalling zone).

The centroid of the stirrup forces within a distance $h/2$ from the end face will be assumed to be at $x = 8$ in. Then, from Eq. (21.56), the required cross-sectional area of end-zone reinforcement is

$$A_t = \frac{388{,}000}{20{,}000(30 - 8)} = 0.88 \text{ in}^2$$

Four No. 3 closed-loop stirrups provide an area of $4 \times 0.11 \times 2 = 0.88$ in^2, exactly as required. The first stirrup will be placed 2 in. from the end face, followed by three

Table 21.4 Moments at horizontal section of end-zone design example

Distance from bottom, in.	Moment of concrete stresses, in-kips	Moment of prestress force, in-kips	Net moment, in-kips
0	0	0	0
3	108	0	108
4.5	217	0	217
6	418	−375	43
9	897	−1125	−228
12	1514	−1875	−361
15	2237	−2625	−388
18	3035	−3375	−340
21	3877	−4125	−248
24	4731	−4875	−144
27	5566	−5625	−59
30	6349	−6375	−26

stirrups at 4 in. spacing within the distance $h/2 = 15$ in., as shown in Fig. 21.28*a*. This places the steel centroid 8 in. from the end face as assumed.

A second moment maximum, of opposite sign and of value 217 in-kips, is indicated by Fig. 21.28*d* with associated tension at the level of the prestress force some distance inward from the end face (bursting zone). The tensile force of

$$T = \frac{217{,}000}{22} = 9860 \text{ lb}$$

can be accommodated by three additional stirrups, providing a total of seven No. 3 stirrups at 4 in. spacing. End-zone reinforcement, shown in Fig. 21.28*a*, is thus included through a distance approximately equal to h from the end face.

21.17 DEFLECTION

Deflection of the slender, relatively flexible beams that are made possible by prestressing must be predicted with care. Many members, satisfactory in all other respects, have proved to be unserviceable because of excessive deformation. In some cases, the absolute amount of deflection is excessive. Often, it is the differential deformation between adjacent members (e.g., precast roof-deck units) that has caused problems. More often than not, any difficulties that occur are associated with upward deflection due to the sustained prestress load. Such difficulties are easily avoided by proper consideration in design.

When the prestress force is first applied, a beam will normally camber upward. With the passage of time, concrete shrinkage and creep will cause a gradual reduction of prestress force. In spite of this, the upward deflection usually will increase, due to the differential creep, affecting the highly stressed bottom fibers more than the top. With the application of superimposed dead and live loads, this upward deflection will be partially or completely overcome, and zero or downward deflection obained. Clearly, in computing deformation, careful attention must be paid to the duration of loading.

The prediction of deflection can be approached at any of several levels of accuracy, depending upon the nature and importance of the work. In some cases it is sufficient to place limitations on the span-depth ratio, based on past experience. Generally, deflections must be calculated. (Calculation is required for *all* prestressed members, according to the ACI Code.) The approximate method described here will be found sufficiently accurate for most purposes. In special circumstances, where it is important to obtain the best possible information on deflection at all important load stages, such as for long-span bridges, the only satisfactory approach is to use a summation procedure based on incremental deflection at discrete time steps, as described in Refs. 21.1, 21.21, and 21.22. In this way, the time-dependent changes of prestress force, material properties, and loading can be accounted for to the desired degree of accuracy.

Normally, the deflectons of primary interest are those at the initial stage, when the beam is acted upon by the initial prestress P_i and its own weight, and one or more combinations of load in service, when the prestress force is reduced by losses to the effective value P_e. Deflections are modified by creep under the sustained prestress force and due to all other sustained loads.

The short-term deflection Δ_{pi} due to the initial prestress force P_i can be found based on the variation of prestress moment along the span, making use of the moment-area principles. For statically determinate beams, the ordinates of the moment diagram resulting from the eccentric prestress force are directly proportional to the eccentricity of the steel centroid line with respect to the concrete centroid. For indeterminate beams, eccentricity should be measured to the thrust line rather than to the steel centroid (see Ref. 21.1). In either case, the effect of prestress can also be regarded in terms of equivalent loads and deflections found using familiar deflection equations.

The downward deflection Δ_o due to girder self-weight, which is usually uniformly distributed, is easily found by conventional means. Thus the net deflection obtained immediately upon prestressing is

$$\Delta = -\Delta_{pi} + \Delta_o \tag{21.58}$$

where the negative sign indicates upward displacement.

Long-term deflections due to prestress occur as that force is gradually reducing from P_i to P_e. This can be accounted for in an approximate way by assuming that creep occurs under a constant prestress force equal to the average of the initial and final values. Corresponding to this assumption, the total deflection resulting from prestress alone is

$$\Delta = -\Delta_{pe} - \frac{\Delta_{pi} + \Delta_{pe}}{2} C_c \tag{21.59}$$

where

$$\Delta_{pe} = \Delta_{pi} \frac{P_e}{P_i}$$

and C_c is the creep coefficient for the concrete (see Table 2.1).

The long-term deflection due to self-weight is also increased by creep and can be obtained applying the creep coefficient directly to the instantaneous value.

Thus the total member deflection, after losses and creep deflections, when effective prestress and self-weight act, is

$$\Delta = -\Delta_{pe} - \frac{\Delta_{pi} + \Delta_{pe}}{2} C_c + \Delta_o(1 + C_c) \tag{21.60}$$

The deflection due to superimposed loads can now be added, with the creep coefficient introduced to account for the long-term effect of the sustained loads, to obtain the net deflection at full service loading:

$$\Delta = -\Delta_{pe} - \frac{\Delta_{pi} + \Delta_{pe}}{2} C_c + (\Delta_o + \Delta_d)(1 + C_c) + \Delta_l \tag{21.61}$$

where Δ_d and Δ_l are the immediate deflections due to superimposed dead and live loads respectively.

Since members are mostly or completely uncracked at characteristic load levels of interest, calculations can usually be based on properties of the uncracked concrete section. If the bottom-fiber tension exceeds the modulus of rupture at the load at which deflection is to be found, calculations should be based on a bilinear load-deformation curve or the equivalent, with slopes related to the moment of inertia of the transformed uncracked and cracked sections.

An *effective moment of inertia*, based on a weighted average of the uncracked and cracked section moments of inertia, is generally used (Refs. 21.1, 21.21, and 21.22).

Example 21.8. The 40 ft simply supported T beam shown in Fig. 21.29 is prestressed with a force of 314 kips, using a parabolic tendon which has eccentricity 3 in. above the concrete centroid at the supports and 7.9 in. below the centroid at midspan. After

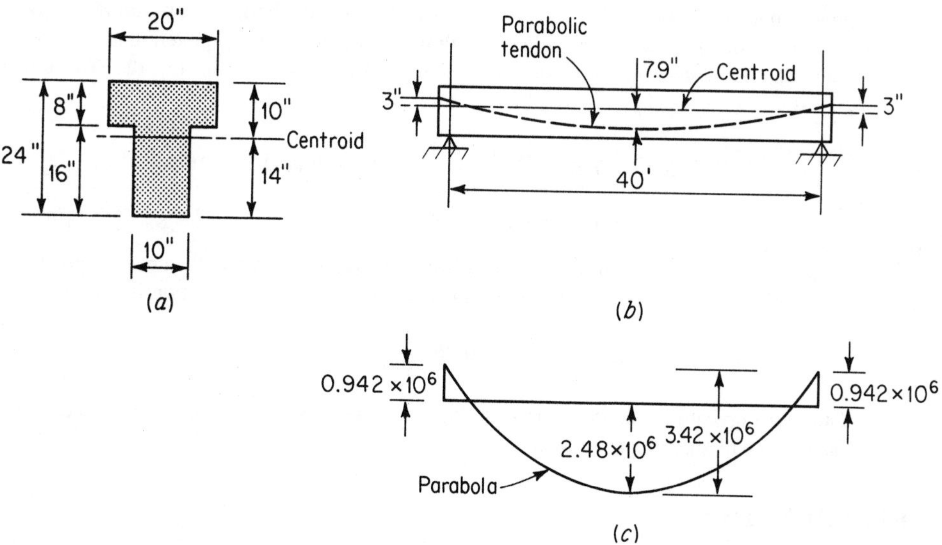

FIGURE 21.29
T beam of Example 21.8: (*a*) cross section; (*b*) tendon profile; (*c*) moment from prestressing force.

time-dependent losses have occurred, this prestress is reduced to 267 kips. In addition to its own weight of 330 plf, the girder must carry a short-term superimposed live load of 900 plf. Estimate the deflection at all critical stages of loading. The creep coefficient $C_c = 2.0$, $E_c = 4 \times 10^6$ psi, and modulus of rupture = 580 psi.

Solution. It is easily confirmed that the stress in the bottom fiber when the beam carries the maximum load to be considered is 80 psi compression. All deflection calculations can therefore be based on the moment of inertia of the gross concrete section, I_c = 15,800 in^4. It is convenient to calculate the deflection due to prestress and that due to girder load separately superimposing the results later. For the eccentricities shown by the tendon profile of Fig. 21.29*b*, the application of P_i = 314 kips causes the moments shown in Fig. 21.29*c* (moments are stated in in-lb units). Applying the second moment-area theorem by taking moments of the *M*/*EI* diagram between midspan and the support, about the support, produces the desired vertical displacement between those two points as follows:

$$\Delta_{pi} = \frac{-(3.42 \times 10^6 \times 240 \times \frac{2}{3} \times 240 \times \frac{5}{8}) + (0.942 \times 10^6 \times 240 \times 120)}{4 \times 10^6 \times 15{,}800} = -0.87 \text{ in.}$$

the minus sign indicating upward deflection due to initial prestress alone. The downward deflection due to the self-weight of the girder is calculated by the well-known equation

$$\Delta_o = \frac{5wl^4}{384EI} = \frac{5 \times 330 \times 40^4 \times 12^4}{384 \times 12 \times 4 \times 10^6 \times 15{,}800} = +0.30 \text{ in.}$$

When these two results are superimposed, the net upward deflection when initial prestress and girder load act together is

$$-\Delta_{pi} + \Delta_o = -0.87 + 0.30 = -0.57 \text{ in.}$$

Shrinkage and creep of the concrete cause gradual reduction of prestress force from P_i = 314 kips to P_e = 267 kips and reduce the bending moment due to prestress proportionately. However, concrete creep acts to increase both the upward-deflection component due to prestress force and the downward-deflection component due to girder load. The net deflection after these changes take place is found from Eq. (21.60), with $\Delta_{pe} = -0.87 \times 267/314 = -0.74$ in.:

$$\Delta = -0.74 - \frac{0.87 + 0.74}{2} \times 2.0 + 0.30(1 + 2.0)$$

$$= -0.74 - 1.61 + 0.90 = -1.45 \text{ in.}$$

In spite of prestress loss, the upward deflection is considerably larger than before. Finally, as the 900 plf short-term superimposed load is applied, the net deflection is

$$\Delta = -1.45 + 0.30\left(\frac{900}{330}\right) = -0.63 \text{ in.}$$

Thus a net upward deflection of about 1/750 times the span is obtained when the member carries its full superimposed load.

REFERENCES

21.1 A. H. Nilson, *Design of Prestressed Concrete*, 2d ed., John Wiley, New York, 1987.
21.2 A. E. Naaman, *Prestressed Concrete Analysis and Design*, McGraw-Hill, New York, 1982.

21.3 T. Y. Lin and N. H. Burns, *Design of Prestressed Concrete Structures*, 3d ed., John Wiley, New York, 1981.

21.4 J. R. Libby, *Modern Prestressed Concrete*, 3d ed., Van Nostrand Reinhold, New York, 1984.

21.5 E. G. Nawy, *Prestressed Concrete*, Prentice Hall, Englewood Cliffs, NJ, 1989.

21.6 M. P. Collins and D. Mitchell, *Prestressed Concrete Structures*, Prentice Hall, Englewood Cliffs, NJ, 1991.

21.7 P. W. Abeles, "Design of Partially Prestressed Concrete Beams," *J. ACI*, vol. 64, no. 10, 1967, pp. 669–677.

21.8 A. H. Nilson, "Discussion of 'Design of Partially Prestressed Concrete Beams' by P. W. Abeles" (Ref. 21.7), *J. ACI*, vol. 65, no. 4, 1968, pp. 345–347.

21.9 P. W. Abeles and B. K. Bardhan-Roy, *Prestressed Concrete Designer's Handbook*, 3d ed., Cement and Concrete Association, London, 1981.

21.10 CEB-FIP Joint Committee, "International Recommendations for the Design and Construction of Prestressed Concrete Structures," Cement and Concrete Association, London, 1970.

21.11 "Code of Practice for the Structural Use of Concrete," CP110, British Standards Institution, London, 1972.

21.12 Y. Guyon, *Limit State Design of Prestressed Concrete*, vols. 1 and 2, John Wiley, New York, 1972.

21.13 "Standard Specifications for Highway Bridges," 14th ed., American Association of State Highway and Transportation Officials, Washington, 1989.

21.14 "Recommendations for Estimating Prestress Losses," reported by PCI Committee on Prestress Losses, *J. Prestressed Concr. Inst.*, vol. 20, no. 4, 1975, pp. 43–75.

21.15 P. Zia, H. K. Preston, N. L. Scott, and E. B. Workman, "Estimating Prestress Losses," *Concr. Int.*, vol. 1, no. 6, 1979, pp. 32–38.

21.16 J. G. MacGregor and J. M. Hanson, "Proposed Changes in Shear Provisions for Reinforced and Prestressed Concrete Beams," *J. ACI*, vol. 66, no. 4, 1969, pp. 276–288.

21.17 N. W. Hanson and P. W. Kaar, "Flexural Bond Tests of Pretensioned Prestressed Beams," *J. ACI*, vol. 30, no. 7, 1959, pp. 783–802.

21.18 W. T. Marshall and A. H. Mattock, "Control of Horizontal Cracking in the Ends of Pretensioned Prestressed Concrete Girders," *J. Prestressed Concr. Inst.*, vol. 5, no. 5, 1962, pp. 56–74.

21.19 P. Gergely and M. A. Sozen, "Design of Anchorage Zone Reinforcement in Prestressed Concrete Beams," *J. Prestressed Concr. Inst.*, vol. 12, no. 2, 1967, pp. 63–75.

21.20 W. C. Stone and J. E. Breen, "Design of Post-Tensioned Girder Anchorage Zones," *J. Prestressed Concr. Inst.*, vol. 29, no. 2, 1984, pp. 28–61.

21.21 "Deflection of Prestressed Concrete Members," Report by ACI Committee 435, *J. ACI*, vol. 60, no. 12, 1963, pp. 1697–1728.

21.22 D. E. Branson, *Deformation of Concrete Structures*, McGraw-Hill, New York, 1977.

PROBLEMS

21.1 A rectangular concrete beam of width $b = 11$ in. and total depth $h = 28$ in. is post-tensioned using a single parabolic tendon having eccentricity 7.8 in. at midspan and 0 in. at the simple support. The initial prestress force $P_i = 334$ kips, and the effectiveness ratio $R = 0.84$. The member is to carry superimposed dead and live loads of 300 and 1000 plf respectively, uniformly distributed over the 40 ft span. Specified concrete strength $f'_c = 5000$ psi and at the time of transfer $f'_{ci} = 4000$ psi. Determine the flexural stress distributions in the concrete at midspan (*a*) for initial conditions before application of superimposed load and (*b*) at full service load. Compare with ACI limit stresses.

21.2 A pretensioned prestressed beam has a rectangular cross section of 6 in. width and 20 in. total depth. It is built using normal density concrete of design strength $f'_c = 4000$ psi and strength at transfer of $f'_{ci} = 3000$ psi. Stress limits are as follows: $f_{ti} = 165$ psi, $f_{ci} = -1800$ psi, $f_{ts} = 380$ psi, and $f_{cs} = -1800$ psi. The effectiveness ratio R may be assumed equal to 0.80. For these conditions, find the initial prestress force P_i and eccentricity e to maximize the superimposed load moment $M_d + M_l$ that can be carried without exceeding stress limits. What uniformly distributed load can be carried on a 30 ft simple span? What tendon profile would you recommend?

21.3 A pretensioned beam is to carry a superimposed dead load of 600 plf and service live load of 1200 plf on a 55 ft simple span. A symmetrical I section with $b = 0.5h$ will be used. Flange thickness $h_f = 0.2h$ and web width $b_w = 0.4b$. The member will be prestressed using Grade 270 strands. Time-dependent losses are estimated at 20 percent of P_i. Normal density concrete will be used, with $f'_c = 5000$ psi and $f'_{ci} = 3000$ psi.

(*a*) Using straight strands, find the required concrete dimensions, prestress force, and eccentricity. Select an appropriate number and size of tendons, and show by sketch their placement in the section.

(*b*) Revise the design of part (*a*) using tendons harped at the third points of the span, with eccentricity reduced to zero at the supports.

(*c*) Comment on your results. In both cases, ACI stress limits are to be applied. You may assume that deflections are not critical and that a tensile stress of $12\sqrt{f'_c}$ is permissible at full service load.

21.4 Establish the required spacing of No. 3 stirrups at a beam cross section subject to factored-load shear V_u of 35.55 kips and moment M_u of 474 ft-kips. Web width $b_w = 5$ in., effective depth $d = 24$ in., and total depth $h = 30$ in. The concrete shear contribution may be based on the approximate relationship of Eq. (21.46). Use $f_y = 40{,}000$ psi for stirrup steel, and take $f'_c = 5000$ psi.

21.5 A symmetrical prestressed I beam having total depth 48 in., flange width 24 in., flange thickness 9.6 in., and web thickness 9.6 in. is to span 70 ft. It is post-tensioned using eighteen one-half in. diameter Grade 270 strands in a single tendon having a parabolic profile, with $e = 18$ in. at midspan and zero at the supports. (The curve can be approximated by a circular arc for loss calculations.) The jacking force $P_j = 618$ kips. Calculate losses due to slip, elastic shortening, friction, creep, shrinkage, and relaxation. Express your results in tabular form as percentages of initial prestress P_i. Creep effects may be assumed to occur under the combination of prestress force plus self-weight. The beam is prestressed when the concrete is aged 7 days. Anchorage slip = 0.25 in., coefficient of strand friction = 0.20, coefficient of wobble friction = 0.0010, creep coefficient = 2.35. Member properties are as follows: $A_c = 737$ in^2, $I_c = 192{,}000$ in^4, $c_1 = c_2 = 24$ in., $f'_c = 5000$ psi, $E_c = 4{,}000{,}000$ psi, $E_s = 27{,}000{,}000$ psi, $w_c = 150$ pcf, and $C_c = 2.65$.

21.6 The concrete T beam shown in Fig. P21.6 is post-tensioned at an initial force $P_i = 229$ kips, which reduces after 1 year to an effective value $P_e = 183$ kips. In addition to its own weight, the beam will carry a superimposed short-term live load of 21.5 kips at midspan. Using the approximate method of Sec. 18.15, find (*a*) the initial deflection of the unloaded girder and (*b*) the deflection at the age of 1 year of the loaded girder. The following data are given: $A_c = 450$ in^2, $c_1 = 8$ in., $I_c = 24{,}600$ in^4, $E_c = 3{,}500{,}000$ psi, $C_c = 2.5$.

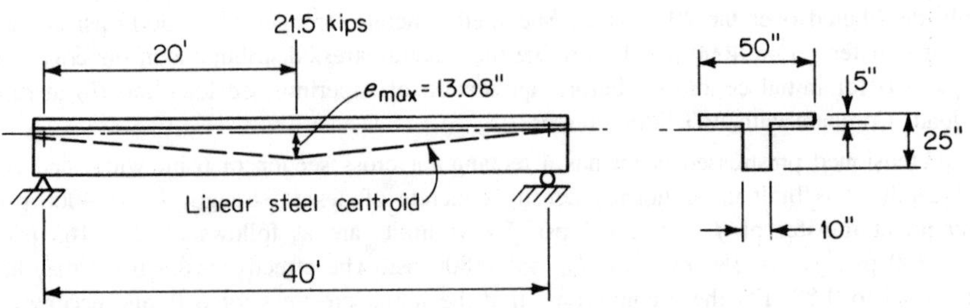

FIGURE P21.6

CHAPTER 22

BRIDGES

22.1 TYPES

Reinforced concrete is particularly well suited for use in bridges of all kinds because of its durability, rigidity, and economy, as well as the comparative ease with which a pleasing appearance can be achieved.

For very short spans, from about 10 to 25 ft, one-way slab bridges are economical. Somewhat longer spans, up to about 50 ft, can be achieved with bridges of this same general type by forming longitudinal voids with fiberboard tubes, thereby reducing dead load. Figure 22.1 shows a voided slab bridge using precast elements. Typical cross sections are shown in Fig. 22.2. Such spans are often prestressed by pretensioned strands, and may be laterally post-tensioned after the units are placed side by side. Such construction offers low initial cost, as a result of standardized plant production, and is characterized by fast, easy erection and low maintenance. For highway spans, an asphalt-wearing surface is normally applied directly to the top face of the concrete units (Refs. 22.1 and 22.2).

Cast-in-place concrete girders such as shown in Fig. 22.3 may be used for spans up to about 100 ft, although their use is less common now than before because of the advantages of precasting for this range of spans. For cast-in-place bridges, the monolithic concrete deck spans as a slab transversely and also provides a broad compression flange for the main girders. Such bridges may be single span and simply supported as shown, or continuous over two or more spans.

Most highway bridges in the United States of medium length, from 60 to about 120 ft, make use either of composite steel-concrete construction or composite prestressed concrete construction. Composite construction with structural steel beams and a concrete deck slab has been discussed in Chap. 17. A continuous composite structure is shown in Fig. 22.4. The concrete deck is made to act integrally with the steel beams through use of *shear connectors* welded to the top flange of the beams and embedded in the slab. Although such a bridge is not strictly a reinforced concrete structure, the design will be presented in some detail in this chapter because of its widespread use. Prestressed concrete bridges

FIGURE 22.1
Precast prestressed voided slab bridge.

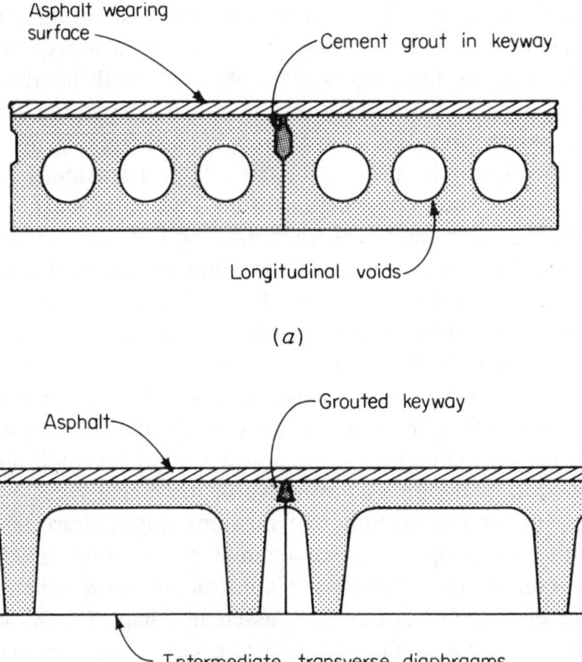

FIGURE 22.2
Precast integral-deck bridge units: (*a*) voided slab; (*b*) channel slab.

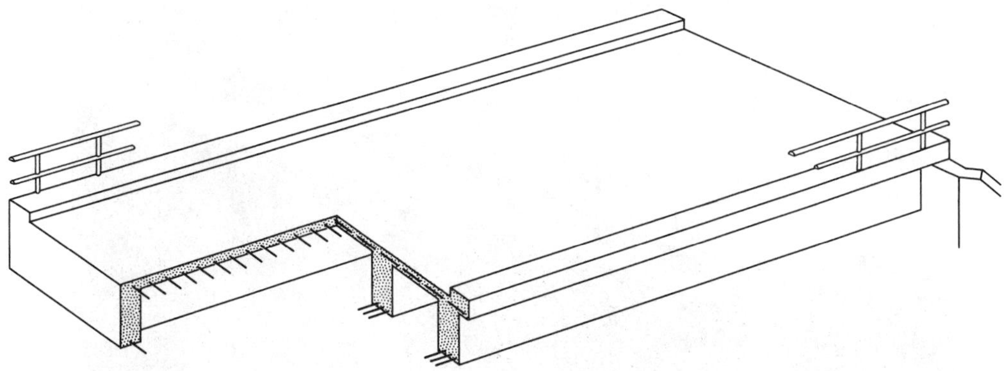

FIGURE 22.3
Cast-in-place deck-girder bridge.

frequently make use of composite action also. Commonly the beams are precast and placed in position by a crane, eliminating the need for obstructing traffic with falsework. The deck slab is then cast in place and locked to the precast units by stirrups that project upward into the slab. Figure 22.5 illustrates construction of this type.

Requirements for long spans have led to the development of segmentally cast post-tensioned prestressed concrete box-girder bridges of the type shown in

FIGURE 22.4
Continuous bridge using cast-in-place concrete slab composite with welded steel girders.

FIGURE 22.5
Standard AASHTO bridge girders, precast and pretensioned, composite with cast-in-place slab.

Fig. 22.6. In typical construction, the work proceeds in two directions from each pier, using the balanced cantilever method shown, and is advanced using either cast-in-place or precast concrete units. Each is post-tensioned to the previously completed construction after placement. Very long spans have been achieved in this way, the longest presently (1990) being close to 1000 ft.

Other possibilities for long-span concrete bridges include the various forms of arches, e.g., the barrel arch of Fig. 22.7 and the parallel arch of Fig. 22.8. Particularly noteworthy, both historically and esthetically, are the bridges designed and built by the Swiss engineer Robert Maillart in the 1920s and 1930s. His Salginatobel Bridge near Zurich is a three-hinged arch with spans of 295 ft (Fig. 22.9). An excellent review of the state of the art of bridge design has been presented by Dr. Fritz Leonhardt (Ref. 22.3), who is himself a bridge designer of international stature.

22.2 DESIGN SPECIFICATIONS

Most highway bridges in the United States are designed according to the requirements of the American Association of State Highway and Transportation Officials (AASHTO specification, Ref. 22.4). Its specifications contain provisions governing loads and load distributions as well as detailed provisions relating to design and construction.

The design of railway bridges is done according to the provisions of the manual of the American Railway Engineering Association (Ref. 22.5). Design requirements for pedestrian crossings and bridges serving special purposes may be established by local, state, or regional codes.

FIGURE 22.6
Houston Ship Channel Bridge, segmentally cast-in-place hollow box girder, prestressed by post-tensioning. The completed structure featured a 750 ft center span flanked by two 375 ft side spans. (*Courtesy of Figg and Muller Engineers, Inc.*)

Design provisions of both the AASHTO specification and the AREA manual are generally patterned after the ACI Code, although there are important differences in preferred methods where alternatives exist. Under the ACI Code, the *strength method of design* using factored loads is preferred, and while *service load design* is permitted, provisions appear only in a brief appendix to the Code. A recent report by the ACI Committee 343 (Ref. 22.6) clearly favors strength design for reinforced concrete bridges. According to the AASHTO specification, *reinforced concrete* may be designed either with reference to service loads and allowable stresses, or alternatively with reference to factored loads and strength.

FIGURE 22.7
Fox River Valley Bridge in Illinois, having five 178 ft barrel arch spans. (*Courtesy of Vogt, Ivers, and Associates, Engineers.*)

FIGURE 22.8
Fixed parallel arch highway bridge. (*Courtesy of Portland Cement Association.*)

FIGURE 22.9
Salginatobel Bridge in the Davos Alps, Switzerland, designed and built in 1930 by Robert Maillart; a three-hinged arch with 295 ft span.

Requirements for strength and serviceability that are a part of the strength design method may be assumed to be satisfied if service load stress limits are met. Service load design provisions and strength design provisions receive equal treatment, with service load design appearing first. *Prestressed concrete*, according to the AASHTO specification, is to be based on strength methods and on behavior at service load stages that may be critical; integrated provisions follow the ACI Code closely. For structural steel, including composite concrete-steel bridges, service load design is the standard method, while load factor design is an acceptable alternative for the design of simple and continuous beam-and-girder structures of moderate length.

Probably most reinforced concrete and composite highway bridges are presently designed by the service load method, in part because of the complexities of load factoring when many combinations of loads are possible (e.g., dead, live, impact, earth pressure, wind, centrifugal force, etc.). There is also a general view that, where repetitive loads act, primary attention should be given to actual stresses existing under normal load conditions.

Reflecting this situation, design examples contained in this chapter will be carried out using service load design, based on stress limits of the AASHTO specification. The reader should experience little difficulty in following the calculations by analogy with the strength design methods used elsewhere in this book.

22.3 LOADS

a. Live Loads

(1) Truck loadings. The live loads of the AASHTO specification consist of standard, idealized trucks or of live loads that are equivalent to a series of trucks. Two systems of loadings are provided: the H loadings, as shown in Fig. 22.10, and the HS loadings, as shown in Fig. 22.11. The H loadings represent a two-axle truck; the HS loadings represent a two-axle tractor plus a single-axle semitrailer.

As shown in the figures, the highway loadings are divided into several classes. The number of the loading indicates the gross weight in tons of the truck or tractor. The gross weight is divided between the front and rear axles, as shown in Figs. 22.10 and 22.11.

Lane loadings, which are to be used when they produce greater stress than the corresponding truck loadings, are shown in Fig. 22.12. In general, the lane loadings, which are designed to approximate the effect of a series of trucks, govern the design of the longer-span bridges.

(2) Other roadway loadings. The possibility that the bridge may be required to carry electric railways, railroad freight cars, military vehicles, or other extraordinary vehicles should be considered in design. The class of such traffic shall determine the loads to be used.

(3) Sidewalk loadings. Sidewalk floors, stringers, and their immediate supports are usually designed for a live load of at least 85 psf of sidewalk area.

(4) Selection of loadings. The AASHTO specifications provide that bridges supporting interstate highways shall be designed for HS20-44 loading or an alternate military loading of two axles, 4 ft apart, each axle weighing 24,000 lb, whichever produces the greatest stress. For other highways that may carry heavy truck traffic the minimum live load shall be HS15-44.

b. Application of Loadings

Some of the more important rules for applying the selected AASHTO loading follow:

1. The lane loading or standard truck loading shall be assumed to occupy a width of 10 ft. These loads shall be placed in 12 ft wide design traffic lanes spaced across the entire bridge roadway width in numbers and positions required to produce the maximum stress. Roadway widths from 20 to 24 ft shall have two design lanes, each equal to one half the roadway width.
2. Each 10 ft lane loading or single standard truck shall be considered as a unit, and fractional load-lane widths or fractional trucks shall not be used.
3. The number and position of lane loadings or truck loadings shall be as specified above and, whether lane loading or truck loading, shall be such as to produce maximum stress subject to the reduction specified below.

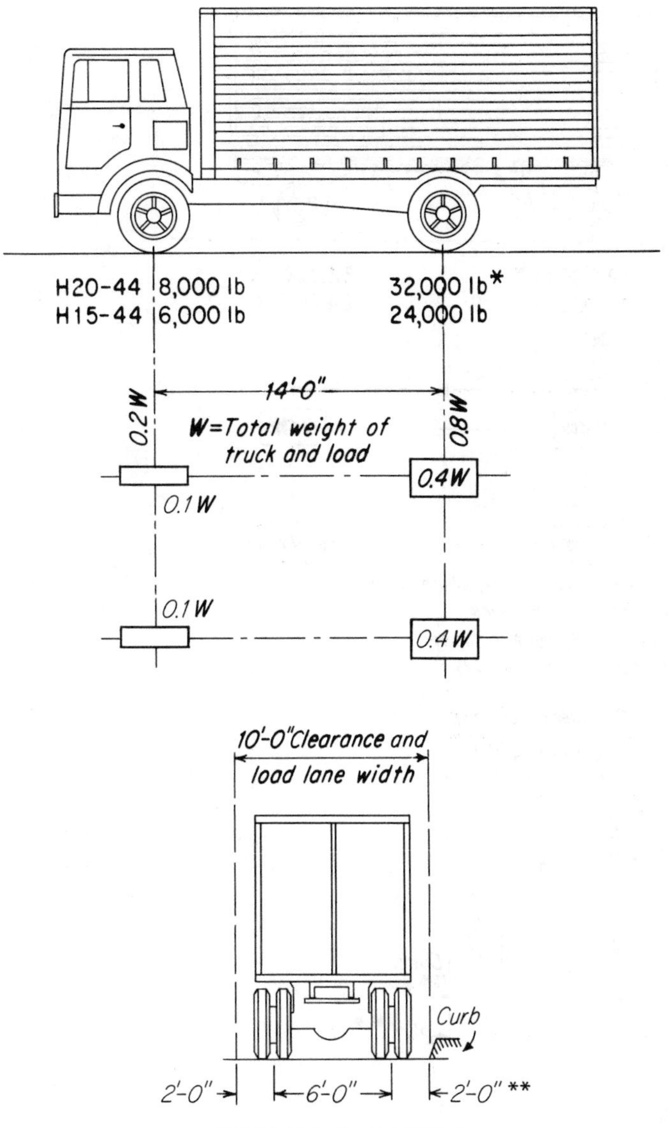

FIGURE 22.10
Standard H truck loading.

* In the design of timber floors for H20 loading, one axle load of 24,000 lb or two axle loads of 16,000 lb each, spaced 4 ft apart, whichever produces the greater stress, may be used instead of the 32,000 lb axle load shown.

** For slab design the centerline of the wheel shall be assumed to be 1 ft from the face of the curb.

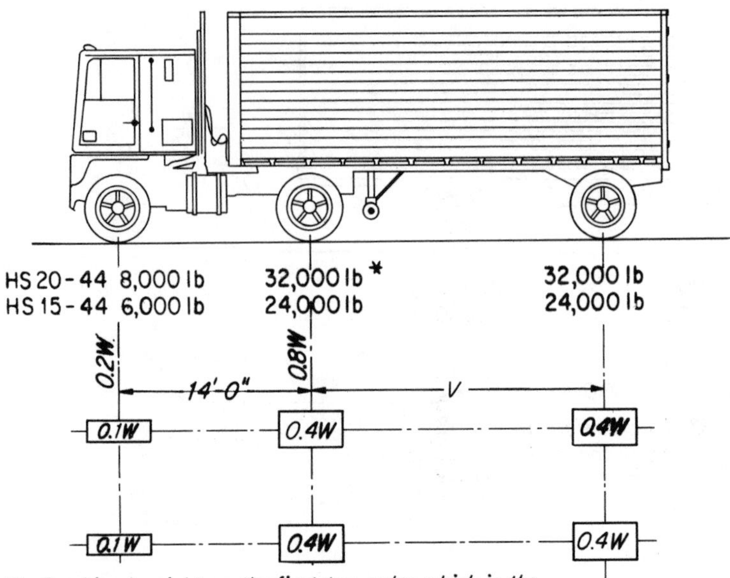

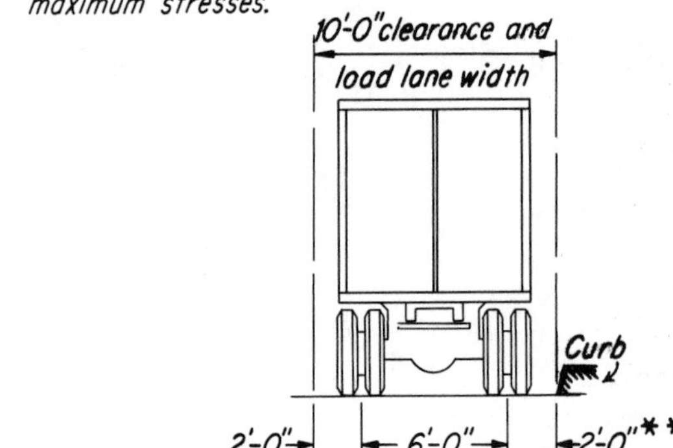

FIGURE 22.11
Standard HS truck loading.

* In the design of timber floors for HS20 loading, one axle load of 24,000 lb or two axle loads of 16,000 lb each, spaced 4 ft apart, whichever produces the greater stress, may be used instead of the 32,000 lb axle load shown.

** For slab design the centerline of the wheel shall be assumed to be 1 ft from the face of the curb.

4. Where maximum stresses are produced in any member by loading any number of traffic lanes simultaneously, the following percentages of the resultant live load stresses shall be used in view of the improbable coincident maximum loading:

One or two lanes	100 percent
Three lanes	90 percent
Four or more lanes	75 percent

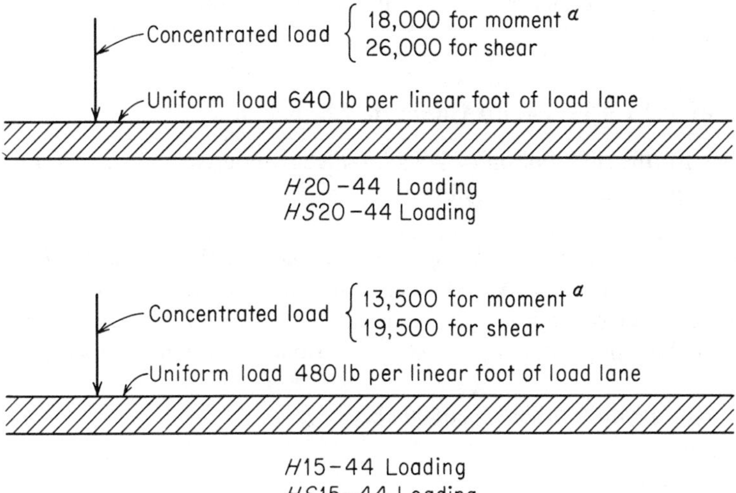

FIGURE 22.12
Equivalent lane loadings. [a] For continuous spans, another concentrated load of equal weight shall be placed in one other span in the series in such a position as to produce maximum negative moment.

c. Impact

Live load stresses due to truck loading (or equivalent lane loading) are increased to allow for vibration and the sudden application of the load. The increase is computed by the formula

$$I = \frac{50}{l + 125} \tag{22.1}$$

where I is the impact fraction of live load stress and l is the loaded length in feet. The maximum impact allowance to be used shall be 30 percent.

d. Distribution of Loads

When a concentrated load is placed on a reinforced concrete slab, the load is distributed over an area larger than the actual contact area. For example, if a concentrated load with a bearing area of 1 ft^2 were placed on the slab of a through-girder bridge, it is reasonable to assume that, on account of the stiffness of the slab, the strips of slab at right angles to the girder and adjacent to the 1 ft strip in direct contact with the load would assist in carrying the load. Similarly, with the beams of a T-beam bridge spaced fairly closely, a concentrated load placed directly over one of the beams would not be carried entirely by that beam, for the concrete slab would be sufficiently rigid to transfer part of the load to adjacent beams.

No distribution is assumed in the direction of the span of the member. The effect of any distribution would be comparatively small.

The following recommendations for the distribution of loads are taken from the AASHTO specification.

e. Distribution of Wheel Loads on Concrete Slabs

The pertinent rules for the distribution of wheel loads on concrete slabs and some additional design requirements are as follows.

(1) Span lengths. For simple spans, the span length shall be the distance center to center of supports but shall not exceed clear span plus thickness of slab.

The following effective span lengths shall be used in calculating the distribution of loads and bending moments for slabs continuous over more than two supports:

Slabs monolithic with beam (without haunches):

S = clear span

Slabs supported on steel stringers:

S = distance between edges of flanges plus one half the stringer flange width

Slabs supported on timber stringers:

S = clear span plus one half the thickness of the stringer

(2) Edge distance of wheel load. In designing slabs, the centerline of the wheel load shall be assumed to be 1 ft from the face of the curb.

(3) Bending moment. Bending moment per foot width of slab shall be calculated according to the methods given under cases 1 and 2 below. In both cases,

$$\begin{aligned} S &= \text{effective span length as defined under above heading, ft} \\ E &= \text{width of slab over which wheel load is distributed, ft} \\ P &= \text{load on one rear wheel of truck} \\ P_{15} &= 12{,}000 \text{ lb for H15 loading} \\ P_{20} &= 16{,}000 \text{ lb for H20 loading} \end{aligned}$$

Case 1: Main reinforcement perpendicular to traffic (spans 2 to 24 ft inclusive). The live load moment for simple spans shall be determined by the following formulas (impact not included):

HS20 loading:

$$\frac{S+2}{32}P_{20} = \text{moment, ft-lb per foot width of slab} \tag{22.2a}$$

HS15 loading:

$$\frac{S+2}{32}P_{15} = \text{moment, ft-lb per foot width of slab} \tag{22.2b}$$

In slabs continuous over three or more supports, a continuity factor of 0.8 shall be applied to the above formulas for both positive and negative moments.

Case 2: Main reinforcement parallel to traffic. Distribution of wheel loads $E = 4 + 0.06S$; maximum 7.0 ft. Lane loads are distributed over a width of $2E$. Longitudinally reinforced slabs shall be designed for the appropriate HS loading.

For simple spans, the maximum live load moment (LLM) per foot width of slab, without impact, is closely approximated by the following formulas:

HS20 loading:

Spans up to and including 50 ft: $LLM = 900S$ ft-lb (22.3*a*)

Spans 50 to 100 ft: $LLM = 1000(1.30S - 20.0)$ ft-lb (22.3*b*)

HS15 loading:

Use three fourths of the values obtained from the formulas for HS20 loading.

Moments in continuous spans shall be determined by suitable analysis using the truck or appropriate lane loading.

(4) Edge beams (longitudinal). Edge beams shall be provided for all slabs having main reinforcement parallel to traffic. The beam may consist of a slab section additionally reinforced, a beam integral with and deeper than the slab, or an integral reinforced section of slab and curb. It shall be designed to resist a live load moment of $0.10PS$, where P is the wheel load in pounds (P_{15} or P_{20}) and S is the span length in feet. The moment as stated is for a freely supported span. It may be reduced 20 percent for continuous spans unless a greater reduction results from a more exact analysis.

(5) Distribution reinforcement. Reinforcement shall be placed in the bottoms of all slabs transverse to the main steel reinforcement to provide for the lateral distribution of the concentrated live loads, except that reinforcement will not be required for culverts or bridge slabs when the depth of the fill over the slab exceeds 2 ft. The amount shall be the percentage of the main-reinforcement steel required for positive moment as given by the following formulas:

For main reinforcement parallel to traffic:

$$\text{Percentage} = \frac{100}{\sqrt{S}} \qquad \text{maximum} = 50\% \tag{22.4a}$$

For main reinforcement perpendicular to traffic:

$$\text{Percentage} = \frac{220}{\sqrt{S}} \qquad \text{maximum} = 67\% \tag{22.4b}$$

where S is the effective span length in feet.

(6) Shear and bond stress in slabs. Slabs designed for bending moment in accordance with the foregoing shall be considered satisfactory in bond and shear.

f. Distribution of Wheel Loads to Stringers, Longitudinal Beams, and Floor Beams

The following rules govern the distribution of wheel loads affecting longitudinal and transverse concrete beams supporting concrete slabs. The AASHTO specification contains additional recommendations for steel and timber flooring, which are not reproduced here.

(1) Position of loads for shear. In calculating end shears and end reactions in transverse floor beams and longitudinal beams and stringers, no longitudinal distribution of the wheel load shall be assumed for the wheel or axle load adjacent to the end at which the stress is being determined. Lateral distribution of the wheel load shall be that produced by assuming the flooring to act as a simple span between stringers or beams. For loads in other positions on the span, the distribution for shear shall be determined by the method prescribed for moment.

(2) Bending moment in stringers and longitudinal beams. In calculating bending moments in longitudinal beams or stringers, no longitudinal distribution of the wheel loads shall be assumed. The lateral distribution shall be determined as follows:

(a) Interior stringers. Supporting concrete floors shall be designed for loads determined in accordance with the following table, in which S is the average spacing of stringers in feet:

Floor system	One traffic lane, fraction of a wheel load to each stringer	Two or more traffic lanes, fraction of a wheel load to each stringer
Concrete slab on steel I-beam stringers and prestressed concrete girders	$S/7.0$ ($S_{max} = 10$ ft)[a]	$S/5.5$ ($S_{max} = 14$ ft)[a]
Concrete slab on concrete stringers	$S/6.0$ ($S_{max} = 6$ ft)[a]	$S/5.0$ ($S_{max} = 10$ ft)[a]
Concrete box girder	$S/8.0$ ($S_{max} = 12$ ft)[a]	$S/7.0$ ($S_{max} = 16$ ft)[a]

[a] If S exceeds the value in parentheses, the load on each stringer shall be the reaction of the wheel loads, under the assumption that the flooring between the stringers acts as a simple beam.

(b) The live load supported by *outside stringers* shall be the reaction of the truck wheels, under the assumption that the flooring acts as a simple beam between stringers.

(c) The *combined capacity* of all the beams in a span shall not be less than the total live and dead load in the panel.

(3) Bending moment in floor beams (transverse). In calculating bending moments in floor beams, no transverse distribution of the wheel loads shall be assumed. If longitudinal stringers are omitted and the concrete floor is supported directly on floor beams, the fraction of wheel load allotted to each floor beam is $S/6$, where S equals the spacing of beams in feet. If S exceeds 6 ft, the load on the beam shall be the reaction of the wheel loads, under the assumption that the flooring between beams acts as a simple beam.

22.4 CONSTRUCTION DETAILS

a. Abutments

Bridge abutments serve to transmit the load from the superstructure to the foundation and act as retaining walls to hold back the earth fill behind them.

A somewhat simplified sketch of a typical abutment is shown in Fig. 22.13. The breast wall provides for the reactions from the bridge girders and also resists the thrust in the direction of the bridge span from the earth backfill. It may be designed to resist this thrust as a retaining wall, cantilevered from the footing, but it is more economical to design for two-way action, recognizing that it will also span horizontally as a slab between wing walls. Its flat top surface provides for fixed or expansion bearings for the main bridge girders.

Wing walls may be cantilevered as shown or carried down to the footing throughout their entire length. They may also be flared outward in plan view, depending upon the need for confinement of the earth fill. The structural action will depend upon the particular geometry, and no general statements can be made. The three-dimensional nature of the breast wall–wing wall–footing system complicates analysis, and conservative approximations are often made in practice. For structures of major importance, model analysis or finite element analysis methods may be used.

Bridge seats are not often built simply as horizontal surfaces, as indicated by the simplified sketch in Fig. 22.13. Instead, they are provided with low back walls to hold back the earth above the bridge seat. A typical detail is shown in cross section in Fig. 22.14. Usually the roadway approach slab is carried on the back wall, in order to minimize problems resulting from the inevitable settlement of the fill.

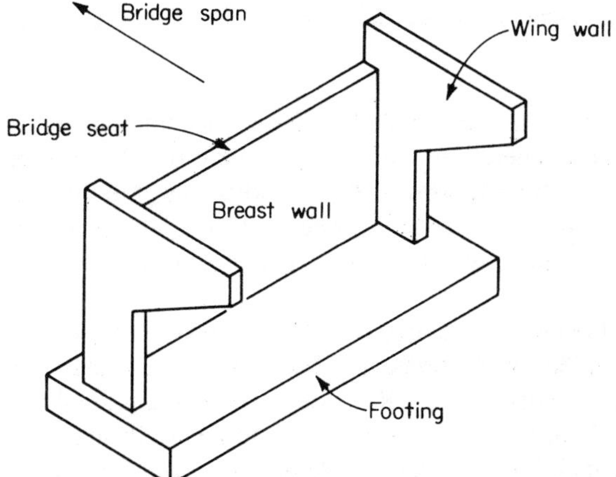

FIGURE 22.13
Typical bridge abutment.

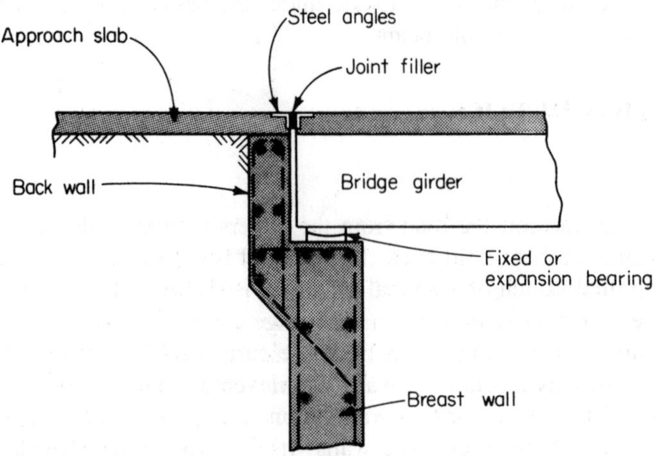

FIGURE 22.14
Detail at bridge seat.

b. Miscellaneous Details

Diaphragms. In deck-girder bridges, thin transverse walls called *diaphragms* are often constructed between the beams at the ends of the bridge. Diaphragms serve two purposes: (1) they furnish lateral support for the beams and (2) when no back wall is built above the bridge seat, they prevent the earth fill from spilling out onto the seat.

Intermediate diaphragms also should be provided between the beams of deck-girder bridges with spans greater than about 40 ft. The intermediate diaphragms are placed at the center or the third points of the span. They serve the purpose of providing lateral stability for the beams, and also provide for lateral distribution of concentrated loads, which may act mainly on one or two girders, thus ensuring that the entire bridge acts as a unit.

Transverse diaphragms are usually from 6 to 8 in. thick, with a nominal amount of reinforcement.

Bearings. Provision must be made in designing bridges to accommodate the length changes that result from temperature variation. In addition, for longer spans, it is necessary to allow for the rotations at the supports that will accompany deflection of the loaded structure. Generally, one end of a bridge girder is fixed against horizontal movement, and provision is made at the far end for displacement.

Spans less than about 50 ft may be arranged to slide on metal plates at one end, and no provision need be made for rotations. For longer spans, expansion is provided for by rollers, rockers, or sliding plates. Rotations may be accommodated at both ends by hinges, curved bearing plates, or pin arrangements. In the past, some very elaborate arrangements of pins, rollers, rockers, and plates have been

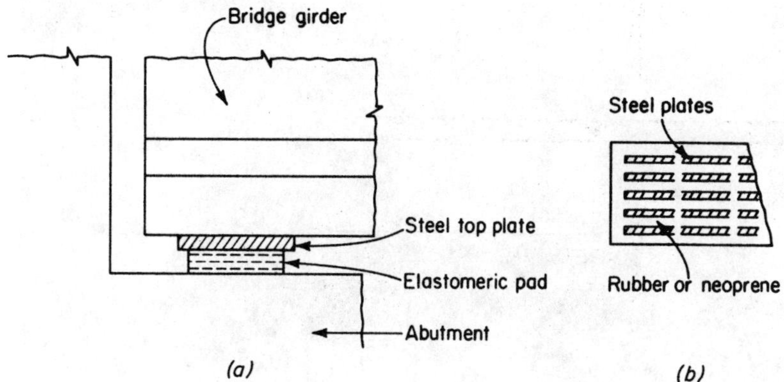

FIGURE 22.15
Elastomeric bearing pad.

used. For these to operate as intended, periodic maintenance is required to keep them clean and free of debris and corrosion. In recent years, for other than minor spans, many highway departments have adopted the use of either elastomeric pads or polytetrafluoroethylene (TFE) bearings, both of which have proved highly effective in accommodating the necessary horizontal and rotational displacements, and are virtually maintenance free.

Elastomeric pads (Fig. 22.15) consist of laminations of elastomer bonded to alternate layers of steel sheets. The elastomer may be either natural rubber or neoprene. All components are molded together into an integral unit, and the ends of the steel plates are covered with elastomer to prevent corrosion. Such units accommodate horizontal movement by shearing deformation of the layers of elastomer.

TFE bearings such as shown in Figs. 22.16 and 22.17 are designed to translate or rotate when a self-lubricating TFE surface slides across a smooth, hard mating surface, preferably of stainless steel. The TFE sliding surface is bonded or mechanically connected to a rigid backup material or is held in position by recessing into the backup material. Provision for rotation may be made with a hinge or radiused sliding surface or by including an elastomeric pad in the assemblage.

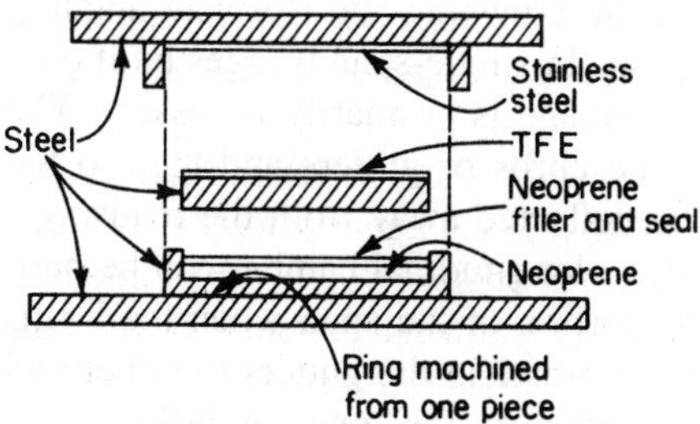

FIGURE 22.16
Polytetrafluoroethylene (TFE) bridge bearing.

FIGURE 22.17
TFE bridge bearing.

Deck joints. Expansion joints in a bridge deck are a principal cause of highway maintenance problems, as a result of corrosion of the joint and the adjacent concrete due to intrusion of water and deicing chemicals, or as a result of structural failure of the joint because of passing loads. Consequently, there is presently a tendency to space such joints at much greater distances than before. For short spans, a simple detail like that in Fig. 22.14 will suffice, in which steel angles are embedded in, and anchored to, the concrete deck and approach slabs, and a bituminous mastic or premolded rubber filler is used, which seals the joint but permits horizontal movement. For widely spaced joints, in which the amount of movement is correspondingly larger, a casting or weldment is generally used, which will maintain a smooth horizontal alignment while sliding longitudinally.

Drainage. Surface water on bridge decks should be disposed of as quickly and as directly as possible. This is accomplished by crowning the roadway about $\frac{1}{8}$ in./ft and pitching the gutters to drain into inlets. In single-span bridges built on a grade, no special provision for longitudinal drainage is ordinarily necessary. The water is carried by the transverse crown to the curbs or gutters and then to the low end of the bridge, where it can easily be deflected away from the roadway.

If the bridge is horizontal longitudinally, a longitudinal camber can be built into the bridge deck by elevating the middle portion of the span when the forms are erected. This camber will serve to carry the water in the gutters to either end of the span. A camber of about $\frac{1}{10}$ in./ft is satisfactory for concrete decks.

22.5 DESIGN OF A SLAB BRIDGE

a. Data and Specifications

The slab bridge of Fig. 22.18 is to be designed according to the following data:

Clear span	15 ft
Clear width	26 ft
Live loading	HS20
Wearing surface	30 psf
Concrete Strength f'_c	3000 psi
Grade 40 reinforcement	

By AASHTO specifications, an allowable concrete stress of

$$f_c = 0.40 f'_c = 1200 \text{ psi}$$

and an allowable steel stress of 20,000 psi will be used.

b. Slab Design

The effective span of the slab is assumed as 15 ft + 1 ft = 16 ft, and a total thickness of 12 in. is selected for trial. The total dead load per square foot is then

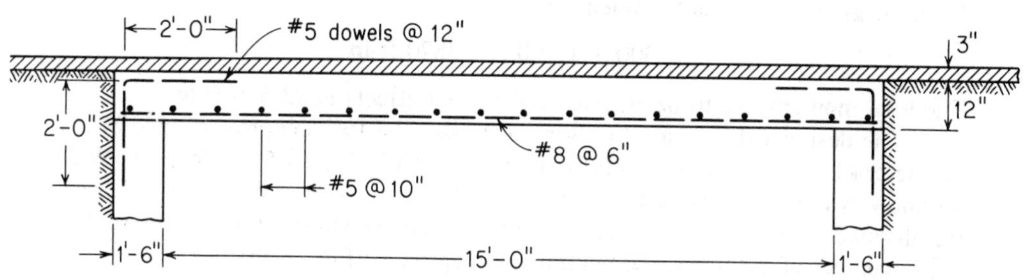

(a) Longitudinal section

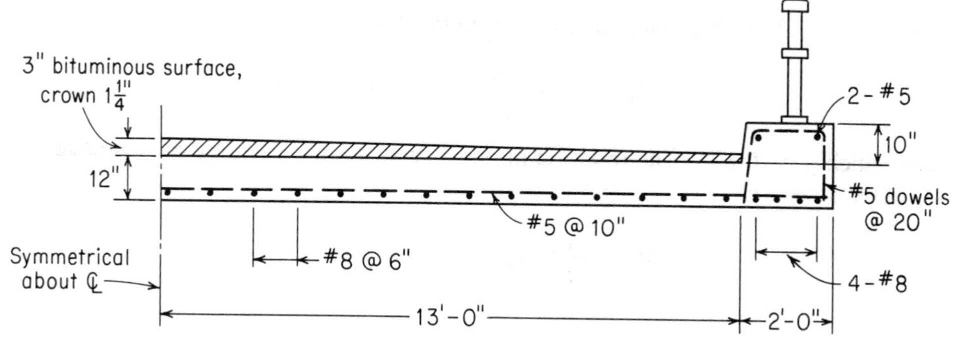

(b) Transverse section

FIGURE 22.18
Details of a slab bridge: (a) longitudinal section; (b) transverse section.

180 psf, and the dead load moment is

$$\frac{1}{8} \times 180 \times 16^2 = 5760 \text{ ft-lb}$$

The load on each rear wheel is 16,000 lb and

$$E = 4 + 0.06S = 4 + 0.06 \times 16 = 4.96 \text{ ft}$$

The load on a unit width of slab is therefore

$$\frac{16{,}000}{4.96} = 3230 \text{ lb}$$

and the live load moment is

$$\frac{PS}{4} = \frac{3230 \times 16}{4} = 12{,}900 \text{ ft-lb}\dagger$$

The impact coefficient is

$$I = \frac{50}{l + 125} = \frac{50}{16 + 125} = 0.355$$

which is greater than the specified maximum of 0.300. The latter will therefore be used, giving an impact moment of

$$0.300 \times 12{,}900 = 3870 \text{ ft-lb}$$

The total moment due to dead, live, and impact effects is 22,530 ft-lb.

The design will be based on service loads, and the slab proportioned based on the cracked elastic cross section, as was shown in Fig. 3.5 of Chap. 3. With an allowable concrete flexural stress of 1200 psi and steel stress of 20,000 psi, transformed with $n = 10$ to an equivalent concrete stress at the level of the reinforcement of 2000 psi, the neutral axis depth (see Fig. 3.5*b*) is

$$kd = \frac{1200}{1200 + 2000} d = 0.375d$$

that is, $k = 0.375$. The internal resisting moment arm is

$$jd = d - \frac{kd}{d} = 0.875d$$

from which $j = 0.875$. The resisting moment capacity, governed by the concrete allowable stress, is

$$M = \frac{f_c}{2} bkdjd = \frac{f_c}{2} kjbd^2$$

†Note that this calculation of live load moments results in a somewhat lower value than the approximate $900S = 14{,}400$ ft-lb.

Thus the minimum permissible effective depth of slab is

$$d = \sqrt{\frac{2M}{f_c kjb}} = \sqrt{\frac{2 \times 22{,}530 \times 12}{1200 \times 0.375 \times 0.875 \times 12}} = 10.7 \text{ in.}$$

With $h = 12$ in., less 1 in. protective covering under the main reinforcement, and less one-half of an assumed bar diameter of 1 in., an effective depth of 10.5 in. is obtained, sufficiently close to the calculated requirement. A 12 in. slab will be used.

The required main reinforcement is

$$A_s = \frac{M}{f_s jd} = \frac{22{,}530 \times 12}{20{,}000 \times 0.875 \times 10.5} = 1.47 \text{ in}^2/\text{ft}$$

which is furnished by No. 8 bars 6 in. on centers ($A_s = 1.57$ in^2/ft). Fifty percent of these bars will be stopped 15 diameters beyond the point at which they are no longer required for moment, or 8 in. short of the face of supports. The remainder will be carried 9 in. into the support. As specified in Sec. 22.3e, the amount of transverse reinforcement required for proper distribution of concentrated loads (also to provide for shrinkage) is

$$\frac{A_s}{\sqrt{S}} = \frac{1.47}{\sqrt{16}} = 0.37 \text{ in}^2$$

No. 5 bars 10 in. on centers are placed directly on top of the longitudinal reinforcement.

To facilitate screeding off the slab, the required curbs will not be made monolithic. Required edge beams will consist of an edge slab section additionally reinforced. The dead load carried by the edge beam is

$$\frac{22 \times 24}{144} 150 = 550 \text{ plf}$$

and the dead load moment is

$$\tfrac{1}{8} \times 550 \times 16^2 = 17{,}600 \text{ ft-lb}$$

The specified live load moment is

$$0.10 P_{20} S = 0.10 \times 16{,}000 \times 16 = 25{,}600 \text{ ft-lb}$$

giving a total moment of 43,200 ft-lb. The resisting moment of the given section is

$$M = \tfrac{1}{2} f_c kjbd^2 = 600 \times 0.375 \times 0.875 \times 24 \times 10.5^2 \times \tfrac{1}{12}$$
$$= 43{,}400 \text{ ft-lb}$$

which is just adequate. The steel required is

$$A_s = \frac{43{,}200 \times 12}{20{,}000 \times 0.875 \times 10.50} = 2.82 \text{ in}^2$$

which will be provided by four No. 8 bars.

According to the AASHTO specification, slabs designed for bending moment may be considered satisfactory in bond and shear. A check of maximum shear stress for the present example will confirm that it is well below the value that can be carried by the concrete.

The crown of the roadway is obtained by varying the thickness of the bituminous wearing surface. For spans less than 40 ft no expansion joints are needed, and a fixed bearing is constructed at each end by placing vertical dowels in the abutment so that they will project up into the concrete deck slab. These are extended horizontally 24 in. to control the cracking that may occur near the edge of the support. The bridge seat is to be coated with a bituminous mixture before the slab is poured. Details are shown in Fig. 22.18.

22.6 DESIGN OF A T BEAM OR DECK-GIRDER BRIDGE

a. Data and Specifications

A T-beam or deck-girder bridge similar to that shown in Fig. 22.3 is to be designed for the following conditions:

Clear span	48 ft
Clear width	29 ft
Live loading	HS20
Concrete strength f_c	3000 psi
Grade 40 reinforcement	

The design is to meet the AASHTO specifications. The bridge will consist of six parallel girders supporting a floor slab, as shown in Fig. 22.19.

b. Slab Design

Since the slab will be poured monolithically with the girders and will be fully continuous, the span will be taken as the clear distance between girders. Under the assumption that the girders will be 14 in. wide, the clear span will be 4 ft 4 in. For a total thickness of slab of 6 in. (including a $\frac{3}{4}$ in. wearing surface), with a 15 psf allowance for possible future protective covering, the total dead load is 90 psf. A coefficient of $\frac{1}{10}$ for positive and negative dead load moments will be assumed in the absence of definite specification values. The dead load positive and negative moments are

$$\tfrac{1}{10} \times 90 \times 4.33^2 = 169 \text{ ft-lb}$$

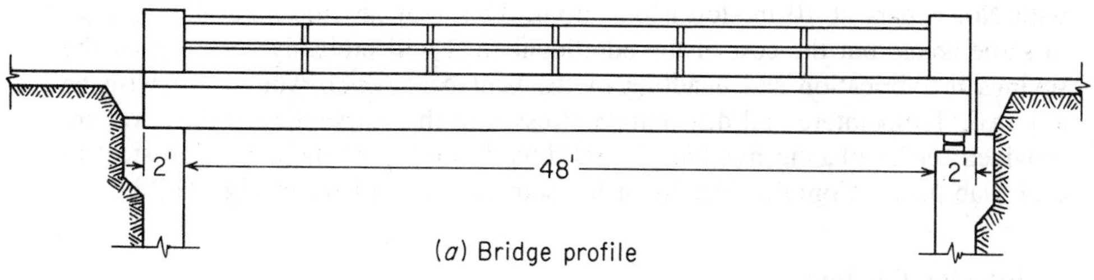

(a) Bridge profile

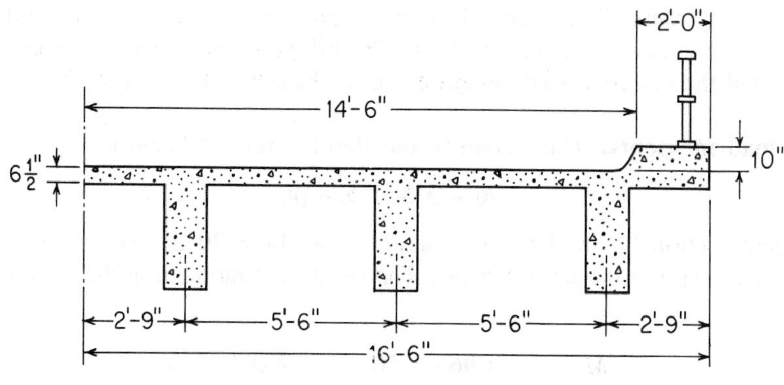

(b) Transverse section

FIGURE 22.19
Deck-girder bridge.

For live load moment computations, case 1 of Sec. 22.3e applies:

$$0.80\,\frac{S+2}{32}\,P_{20} = \frac{0.80 \times 6.33 \times 16{,}000}{32} = 2530 \text{ ft-lb}$$

Since the loaded length is small, the impact coefficient is 0.30, and the impact moment is 760 ft-lb. The total positive and negative moments are 3459 ft-lb. Then

$$d = \sqrt{\frac{2 \times 3459 \times 12}{1200 \times 0.375 \times 0.866 \times 12}} = 4.19 \text{ in}$$

With the addition of 1 in. of protective concrete below the center of an assumed $\frac{3}{4}$ in. bar, and with the $\frac{3}{4}$ in. wearing surface (not considered structurally effective), the total thickness is 6.32 in. A $6\frac{1}{2}$ in. slab will be used, with an effective depth of 4.37 in. The dead load estimate need not be revised and

$$A_s = \frac{3459 \times 12}{20{,}000 \times 0.875 \times 4.37} = 0.54 \text{ in}^2$$

which is furnished by No. 6 bars 10 in. on centers. To avoid excessive bending of bars, and since short spans are involved, straight bars will be used in the slab, with No. 6 bars at 10 in. top and bottom. This will provide a surplus of steel in some areas, but the cost of the additional steel will probably be offset by the savings in fabrication and handling of the bent bars which would otherwise be required. Temperature and distribution stresses in the perpendicular direction are provided for by placing five No. 5 bars directly on top of the main slab steel in each slab panel. Complete details of the slab are shown later in Fig. 22.24.

c. Interior Girders

The interior girders are T beams with a flange width equal to the center-to-center distance of girders. The required stem dimensions are governed by either the maximum moment or maximum shear. The bridge seats will be assumed 2 ft in width, and the effective span length center of bearings taken as 50 ft.

Dead load moments. The weight of the slab per foot of beam is

$$96 \times 5.5 = 528 \text{ plf}$$

The stem section below the slab is assumed as 14 × 30 in, which adds weight of 437 plf, making the total 965 plf. The dead load moment at the center of the span is

$$M_d = \tfrac{1}{8} \times 965 \times 50^2 = 302{,}000 \text{ ft-lb}$$

To determine the points at which some of the horizontal steel may be cut off, it is necessary to compute the moment at some sections between the point of maximum moment and the support. At 10 ft from the support, the dead load moment is 192,800 ft-lb; at 20 ft, 289,000 ft-lb.

Live load moments. The absolute maximum live load moment will occur with an HS20 truck on the bridge in the position† shown in Fig. 22.20. With the distribution of loads as specified in Sec. 22.3f, each interior girder must support 5.50/5.0 = 1.10 wheel loads per wheel. Therefore the load from the rear wheel is 1.1 × 16,000 = 17,600 lb, and that from the front wheel is 1.1 × 4000 = 4400 lb:

$$R_L = \left[17{,}600(36.6 + 22.6) + 4400 \times 8.6\right]\tfrac{1}{50} = 21{,}600 \text{ lb}$$

$$M_{\max} = 21{,}600 \times 27.4 - 17{,}600 \times 14 = 346{,}000 \text{ ft-lb}$$

Ten feet from the left support, the maximum live load moment occurs with the rear trailer wheels at that point and the front wheels 38 ft from the left support. With this position of the loads,

$$R_L = \left[17{,}600(40 + 26) + 4400 \times 12\right]\tfrac{1}{50} = 24{,}400 \text{ lb}$$

$$M_{10} = 24{,}400 \times 10 = 244{,}000 \text{ ft-lb}$$

---

† For a discussion of the position of moving loads for absolute maximum moment, see any elementary text in structural theory.

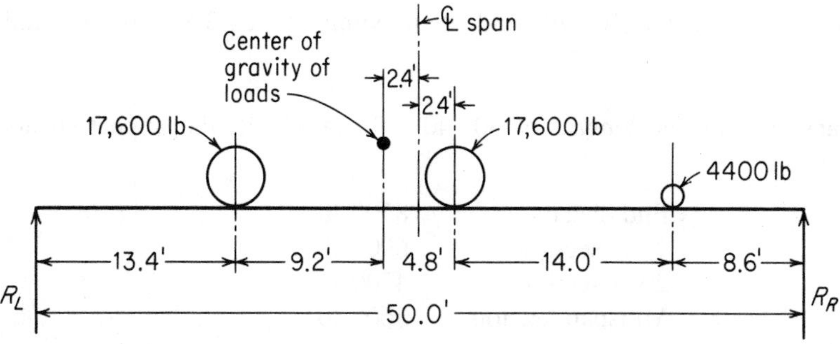

FIGURE 22.20
Position of wheel loads for maximum live load moment.

Similarly,

$$M_{20} = 330{,}000 \text{ ft-lb}$$

Impact moments. In computing impact moments for a girder such as this, it is usual to consider the whole span as the loaded length. The impact coefficient is therefore $50/(50 + 125) = 0.285$, and the impact moments are as follows:

$$M_{max} = 98{,}700 \text{ ft-lb}$$
$$M_{20} = 94{,}100 \text{ ft-lb}$$
$$M_{10} = 69{,}600 \text{ ft-lb}$$

Maximum total moments. The sum of the maximum dead load, live load, and impact moments is 746,000 ft-lb. The total moment at the 20 and 10 ft points are 713,000 and 506,400 ft-lb respectively.

Dead load shears. The maximum dead load shear at the end of the beam is $965 \times 25 = 24{,}100$ lb. Ten feet from the support, the shear is 14,450 lb; 20 ft from the support, it is 4800 lb.

Live load shears. The absolute maximum shear occurs with the truck on the span in the position shown in Fig. 22.21. The maximum live load shear is 30,600 lb.

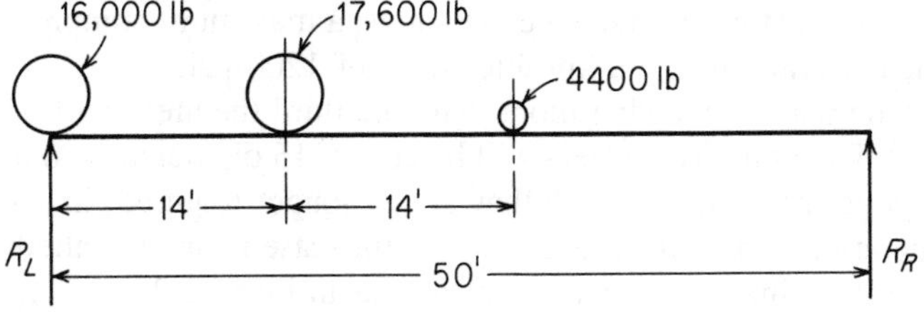

FIGURE 22.21
Position of wheel loads for maximum live load shear.

The maximum shears at the 10, 20, and 25 ft points are 24,300, 16,400, and 12,700 lb respectively.

Impact shears. For loaded lengths of 50, 40, 30, and 25 ft, the impact shears are as follows:

End shear	8700 lb
10 ft section	6900 lb
20 ft section	4700 lb
Midspan section	3600 lb

Total shears. The total shears are as follows:

End shear	63,400 lb
10 ft section	45,650 lb
20 ft section	25,900 lb
Midspan section	16,300 lb

Determination of cross section and steel area. According to the AASHTO specification, the nominal shear stress at service load is to be calculated from $v = V/b_w d$, where b_w = web width. In beams with web reinforcement the nominal shear stress v may not exceed $4.95\sqrt{f_c'}$. In sizing the web area for the present design a maximum shear of $2.95\sqrt{f_c'} = 162$ psi will be used. In this case,

$$b_w d = \frac{V}{v} = \frac{63{,}400}{162} = 391 \text{ in}^2$$

If b_w is taken as 14 in., the required d is 28.0 in. If three rows of No. 11 bars are assumed, with 2 in. clear between rows and 2.5 in. clear below the bottom row to allow for stirrups and concrete protection, a total depth of 34.56 in. is obtained. A 36 in. total depth will be used, resulting in an effective depth of 29.4 in. The depth of the stem below the slab is then $36 - 6\frac{1}{2} = 29\frac{1}{2}$ in. This is so close to the assumed value that the dead load moments need not be revised. The required tensile steel area is then

$$A_s = \frac{M}{f_s(d - t/2)} = \frac{746{,}700 \times 12}{20{,}000(29.4 - 5.75/2)} = 16.90 \text{ in}^2$$

which will be furnished by 11 No. 11 bars.

A check of the maximum stress in the concrete shows a maximum compression of 1190 psi, practically equal to the allowable stress of 1200 psi.

The lower four tensile bars, somewhat more than one third the area, will be extended into the support. The upper two layers will be cut off 15 diameters or one twentieth of the span beyond the point at which they are no longer required, in accordance with the specification. The critical extension in this case is one twentieth of the span, or 30 in. Cutoff points are found graphically as in Fig. 22.22.

Note that extra stirrups must be provided along each terminated bar over a distance from their termination point equal to $0.75d = 22$ in. The excess stirrup area required is $A_v = 60 b_w s/f_y$, and the spacing is not to exceed $d/8\beta_b$, where

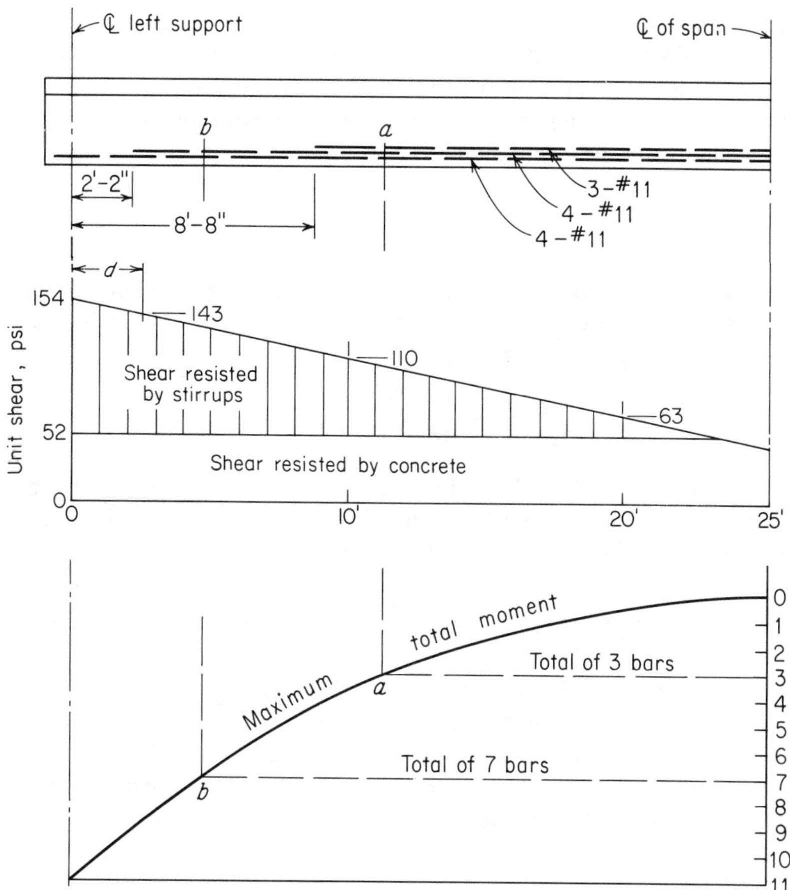

FIGURE 22.22
Shear and moment envelopes and bar cutoff locations for interior girder.

β_b is the ratio of the area of reinforcement cut off to the total area of reinforcement at the section. In the present case, this gives maximum spacings of 13.50 and 7.35 in. at locations *a* and *b* respectively (see Fig. 22.22). The selection of these special stirrups will be deferred pending design of the shear stirrups.

All bars must be extended a full development length, 46 in. in the present case, past locations of maximum stress. Comparison with the selected cutoff points of Fig. 22.22 confirms that this requirement is met for all bars. In addition, the specification requires that at simple supports the development length should not exceed $M/V + l_a$, where M is the computed moment capacity of the remaining tensile bars, V is the maximum shear force at the section, and l_a is the extension of the bars beyond the centerline of the support. For the four No. 11 bars remaining, which have an effective depth of 32.8 in., $M = 312{,}000$ ft-lb. For $l_a = 9$ in., the computed maximum length is 68 in., well above the development length of 46 in. for the No. 11 bars.

Web reinforcement. The portion of the beam through which web reinforcement is required is determined by computing the unit shears at various points on the beam. Results are shown graphically in Fig. 22.22. The concrete resists a unit shear of $0.95\sqrt{f'_c} = 52$ psi, and the remainder is taken by stirrups. The required spacing is calculated from the expression

$$s = \frac{A_v f_s}{(v - v_c)b_w}$$

No. 5 U stirrups will be used. In the region through which web reinforcement is required, the specification calls for a maximum spacing of $d/2 = 15$ in. The excess shear corresponding to that spacing is

$$v - v_c = \frac{A_v f_s}{s b_w} = \frac{0.61 \times 20{,}000}{15 \times 14} = 58 \text{ psi}$$

and the total shear stress is 110 psi. From Fig. 22.22 it is found that this is attained 10 ft 9 in. from the support. At the first critical section for shear, a distance d from the centerline of the support, the required spacing is

$$s = \frac{A_v f_s}{(v - v_c)b_w} = \frac{0.61 \times 20{,}000}{(143 - 52)14} = 9.6 \text{ in.}$$

The first stirrup will be placed 3 in. from the support centerline. This will be followed by 13 stirrups at 9 in. and 12 at 15 in., filling out the distance to the center of the span. The arrangement is shown in Fig. 22.24. In addition to the shear stirrups, one extra No. 5 U stirrup will be included at the cut bar ends at a and b, satisfying both area and spacing requirements at those locations.

d. Exterior Girders

The exterior girders are identical in cross section with the interior girders, because the raised curb section will be poured separately and cannot be counted on to participate in carrying loads.

Moments. In addition to the 965 plf dead load obtained for the interior girders, the safety curb adds 250 plf, producing a total dead load of 1215 plf and a maximum dead load moment at the center of the span of 380,000 ft-lb. A portion of each wheel load that rests on the exterior slab panel is supported by the exterior girder. That portion is obtained by placing the wheels as close to the curb as the clearance diagram will permit and treating the exterior slab panel as a simple beam. The position is shown in Fig. 22.23, and the proportion of the load is $4.25/5.50 = 0.773$. The longitudinal position of the load which will produce the absolute maximum bending moment is the same as for the interior girders. The absolute maximum live load moment can therefore be obtained by direct proportion:

$$M_{max} = \frac{0.773}{1.10} \times 346{,}000 = 243{,}000 \text{ ft-lb}$$

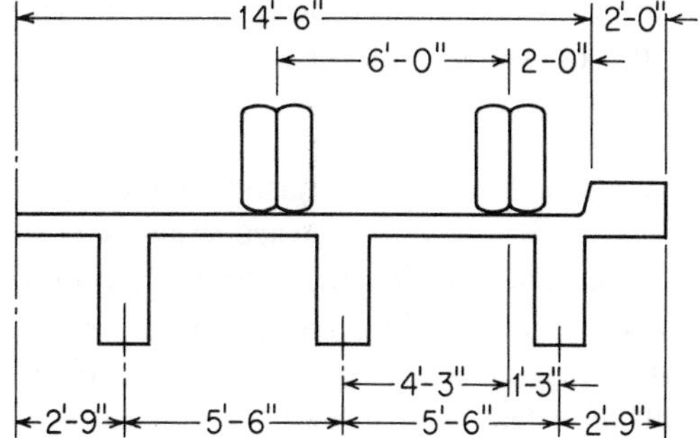

FIGURE 22.23
Lateral position of wheel loads for design of exterior girders.

The impact moment is $0.285 \times 243{,}000 = 69{,}300$ ft-lb, and the total maximum moment is 692,300 ft-lb.

Shears. The maximum dead load shear is

$$V_D = 1215 \times 25 = 30{,}400 \text{ lb}$$

The maximum live load shear is proportional to the maximum live load shear in the interior girders:

$$V_L = \frac{0.773}{1.10} \times 30{,}600 = 21{,}500 \text{ lb}$$

The impact shear at the support is 6100 lb. The total shear at the support is then 58,000 lb. Shears at other points are found similarly.

Determination of cross section and reinforcement. Once the shears and moments due to dead and live loads and impact are obtained, the design of the exterior girders would follow along the lines of that for the interior girders. The specification stipulates that in no case shall an exterior stringer have less carrying capacity than an interior stringer. This controls in the present design, and the exterior girders will duplicate the interior girders.

e. Miscellaneous Details

Diaphragms. A transverse diaphragm will be built between the girders at either end of the bridge. The chief function of these diaphragms is to furnish lateral support to the girders; with some abutment details, the diaphragms also serve to prevent the backfill from spilling out onto the bridge seats. A similar diaphragm will be built between the girders at midspan. Such intermediate diaphragms are required for all spans in excess of 40 ft and serve to ensure that all girders act together in resisting the loads.

Fixed bearing. A fixed bearing is provided at one end. Vertical dowels are placed in the breast wall of the abutment and are bent so as to project into the longitudinal beams or deck slab. Horizontal dowels are embedded in the deck and approach slabs. The diaphragm rests directly on the bridge seat.

Expansion bearing. Elastomeric pads are provided at one end of the span. The bottom of the diaphragm is flush with the bottoms of the beams and is not in contact with the bridge seat.

Waterproofing and drainage. All joints will be filled with a mastic compound to prevent water from seeping through the joints. Removal of surface water will be accomplished by crowning the roadway $1\frac{1}{4}$ in. and pitching the gutters toward the ends by providing a camber of $2\frac{1}{2}$ in. at the center. Besides facilitating drainage, this camber serves to prevent the appearance of sag that would be evident if the girders were at the same level throughout the span.

Full details are shown in Fig. 22.24.

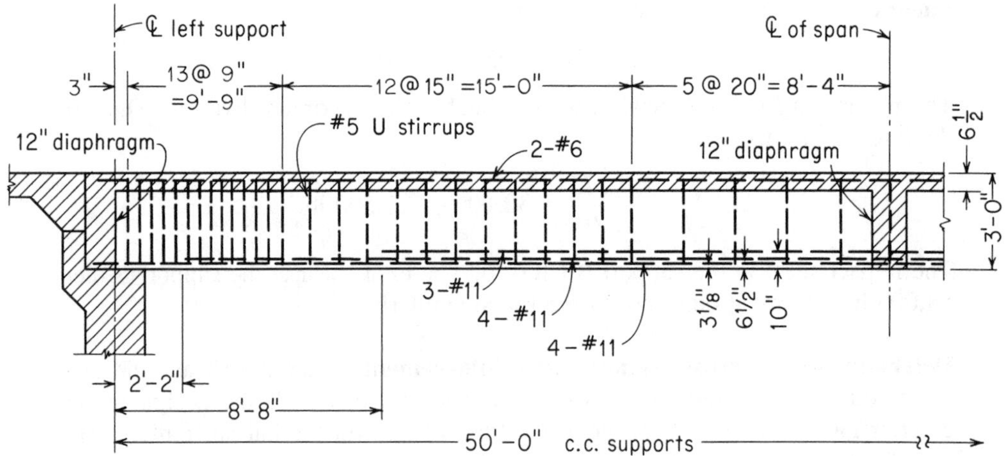

(a) Details of interior girder

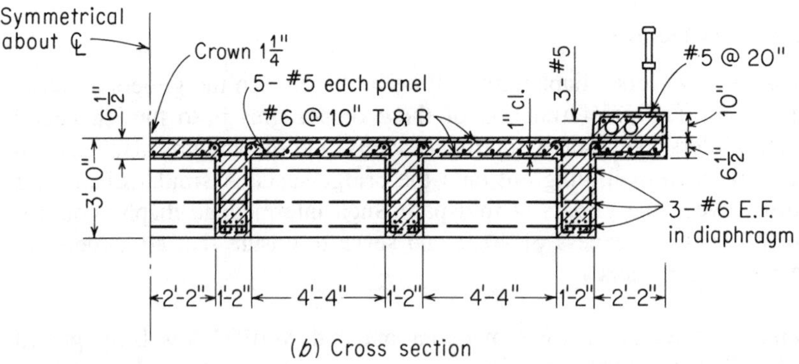

(b) Cross section

FIGURE 22.24
Details of T beam bridge.

22.7 DESIGN OF A COMPOSITE BRIDGE

a. Data and Specifications

A complete bridge similar to that shown in Fig. 22.4 is to be designed for the following conditions:

Span center to center of supports	70 ft
Clear width	28 ft
Live loading	HS20
Wearing surface	30 psf
Concrete strength f_c'	3000 psi
A36 steel beam	
Grade 40 reinforcement	

Welded-stud shear connectors are specified, and a single cover plate will be used on the bottom flange and discontinued where no longer required. The bridge will consist of five parallel girders, as shown in Fig. 22.25. The design is to be based on 2×10^6 cycles of loading.

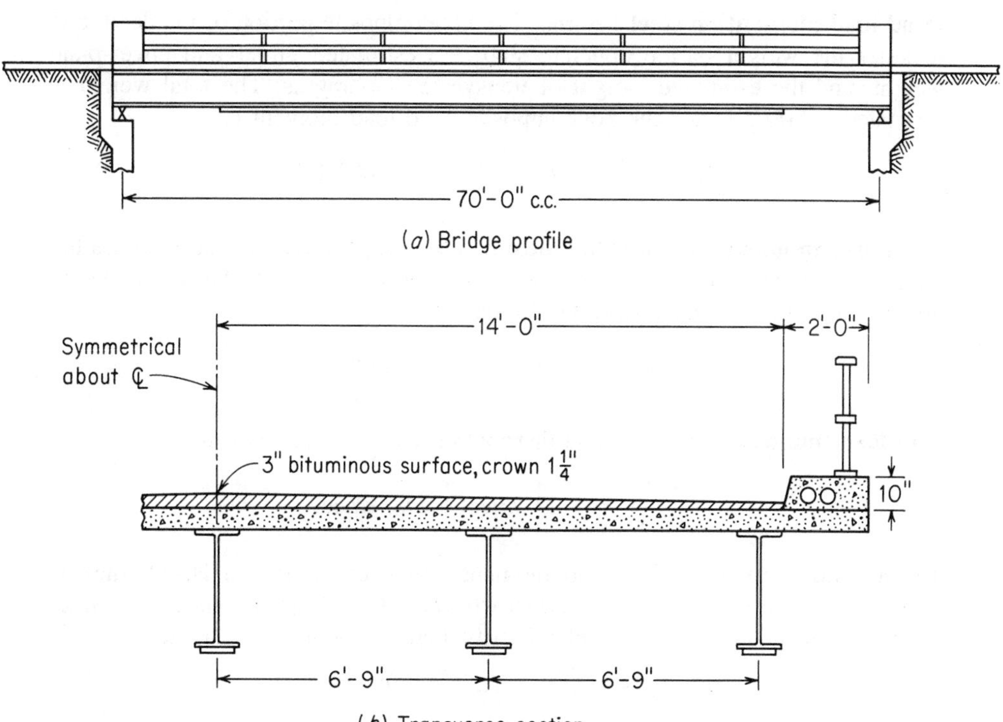

FIGURE 22.25
Composite girder bridge.

b. Slab Design

The design of the slab in the transverse direction follows the procedure presented in Sec. 22.6b. For the given loads and spans, a slab of $6\frac{1}{2}$ in. total thickness is obtained, reinforced with No. 6 bars 8 in. on centers, top and bottom, in the direction transverse to the supporting beams, In the direction parallel to the girders, six No. 6 bars are equally spaced in each slab panel. Details of the slab are shown in Fig. 22.27.

c. Interior Girders

According to the AASHTO specification, the effective width of the concrete compression flange is the smallest of

$$\frac{L}{4} = 70 \times \frac{12}{4} = 210 \text{ in.}$$

$$\text{Center-to-center spacing} = 81 \text{ in.}$$

$$12 \times h_f = 12 \times 6.5 = 78 \text{ in.}$$

The last criterion controls and the effective width will be taken equal to 78 in.

Dead load moment on steel beams. The noncomposite portion of the dead load includes the weight of the concrete slab, the estimated girder and cover-plate weight, and the estimated weight of transverse diaphragms. The total weight is estimated at 800 plf, and the noncomposite dead load moment is

$$M_{ds} = \tfrac{1}{8} \times 800 \times 70^2 = 490{,}000 \text{ ft-lb}$$

Dead load moment on composite beam. The composite dead load includes the weight of bridge rails, safety curbs, and bituminous pavement. These sum up to 300 plf, and the composite dead load moment is

$$M_{dc} = \tfrac{1}{8} \times 300 \times 70^2 = 184{,}000 \text{ ft-lb}$$

Live load moment. The live load distribution factor in this case is

$$\frac{6.75}{5.5} = 1.23 \text{ wheel loads/wheel}$$

The absolute maximum live load moment will occur when an HS20 truck is located with one of its inner 16,000 lb wheels 2.35 ft from the span centerline, and it will occur under that wheel. The live load moment at that point is

$$M_{\max} = 494{,}000 \times 1.23 = 608{,}000 \text{ ft-lb}$$

Impact moment. The impact coefficient in this case is

$$\frac{50}{70 + 125} = 0.256$$

and the maximum impact moment is

$$M_{max} = 608{,}000 \times 0.256 = 156{,}000 \text{ ft-lb}$$

Selection of the cross section. The selection of the cover-plated cross section is necessarily a matter of trial (see Ref. 22.7, however). A section made up of a 36 in. wide flange, 150 lb beam plus a single $10 \times 1\frac{1}{4}$ in. cover plate on the bottom flange, as shown in Fig. 22.26, will be tried. Section properties are calculated in Table 22.1.

The stresses due to various loads are shown in Table 22.2. It is seen that the bottom fiber is stressed to about 90 percent of its allowable stress of 20,000 psi, while the top fiber is understressed by about 25 percent. Since the lightest 36 in. section has been selected, the only alternative would be to use a 33 in. rolled section, which would require a heavier cover plate and which would save little, if any, steel. Consequently, the 36 in. wide flange, 150 lb beam with a cover plate $10 \times 1\frac{1}{4}$ in. will be used. The concrete fiber stress is well within the allowable limit.

The cover plate on the lower flange will be discontinued where it is no longer needed to resist stress. Section properties for the beam without a cover plate are computed as shown in Table 22.3. Once the section moduli of the beam without a cover plate have been found, the point at which the cover plate can be discontinued without exceeding allowable stresses in the remaining section is found by trial to be 12 ft 9 in. from the support centerline. In accordance with specifications, the plate will be continued past that point $1\frac{1}{2}$ times the cover-plate width, or 15 in. in the present case. It will be welded across its end.

The shear connectors are designed for the stress change resulting from application of live and impact loads (the minimum shear when dead loads are excluded being zero for the simple span). At the left support, the maximum live load shear is obtained with one 16,000 lb wheel at the support and 16,000 and 4000 lb wheels at 14 and 28 ft from the support, respectively. After applying the live load distribution factor of 1.23 and the impact coefficient of 0.256, maximum shear $V_{ll+i} = 48{,}200$ lb at this location. Equation (17.1) will be used to calculate the variation of horizontal shear there. Section properties applicable at this location

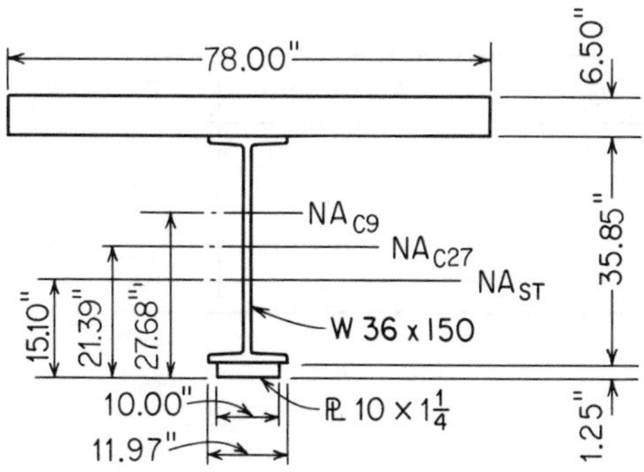

FIGURE 22.26
Girder cross section.

Table 22.1 Properties of section with cover plate

	I_0, in^4	A, in^2	y_b, in.	Ay_b, in^3	d to na, in.	Ad^2, in^4
Steel only						
W36×150	9040	44.2	19.18	848	4.08	736
plate 10×1.25		12.5	0.63	8	14.48	2,620
	9040	56.7		856		3,356
						9,040
						12,396
	$\bar{y} = 856/56.7 = 15.10$			$S_{ts} = 12{,}396/22.01 = 563$ $S_{bs} = 12{,}396/15.10 = 821$		
Composite with $n = 9$						
Steel	12,396	56.7	15.10	856	12.58	8,973
Slab	198	56.4	40.35	2275	12.67	9,054
	12,594	113.1		3131		18,027
						12,594
						30,621
	$\bar{y} = 3{,}131/113.1 = 27.68$			$S_{tc} = 30{,}621/15.92 = 1920$ $S_{ts} = 30{,}621/9.42 = 3250$ $S_{bs} = 30{,}621/27.68 = 1110$		
Composite with $n = 27$						
Steel	12,396	56.7	15.10	856	6.29	2,243
Slab	66	18.8	40.35	759	18.96	6,758
	12,462	75.5		1615		9,001
						12,462
						21,463
	$\bar{y} = 1{,}615/75.5 = 21.39$			$S_{tc} = 21{,}463/22.21 = 970$ $S_{ts} = 21{,}463/15.71 = 1370$ $S_{bs} = 21{,}463/21.39 = 1000$		

Table 22.2 Total stresses

Load	Moment	f_{bs}	f_{ts}	3nf$_c$ or nf$_c$	f_c
M_{ds}	490,000	7,160	10,400		
M_{dc}	184,000	2,210	1,610	2270	80
M_{ll+i}	764,000	8,260	2,820	4780	530
Total	1,438,000	17,630	14,830		610

Table 22.3 Properties of section without cover plate

	I_0, in^4	A, in^2	y_b, in.	Ay_b, in^3	d to na, in.	Ad^2, in^4
Steel only	S_{bs} = 504 in^3			S_{ts} = 504 in^3		
Composite with n = 9						
Steel	9040	44.2	17.93	793	11.87	6,230
Slab	198	56.4	39.10	2205	9.30	4,880
	9238	100.6		2998		11,110
						9,238
						20,348
	$\bar{y}$ = 2998/100.6 = 29.80 in.			S_{tc} = 20,348/12.55 = 1620 in^3 S_{ts} = 20,348/6.05 = 3360 in^3 S_{bs} = 20,348/29.80 = 680 in^3		
Composite with n = 27						
Steel	9040	44.2	17.93	793	6.32	1,760
Slab	66	18.8	39.10	735	14.85	4,150
	9160	63.0		1528		5,910
						9,106
						15,016
	$\bar{y}$ = 1528/63.0 = 24.25 in.			S_{tc} = 15,016/18.10 = 830 in^3 S_{ts} = 15,016/11.60 = 1290 in^3 S_{bs} = 15,016/24.25 = 620 in^3		

are those of the section without a cover plate, with slab transformed on the basis of $n = 9$:

$$S_r = \frac{48{,}000 \times 56.4 \times 9.30}{20{,}348} = 1240 \text{ lb/in.}$$

For the present design, $\frac{3}{4}$ in. studs are selected, and tentatively three studs per transverse row are used, with studs 4 in. high. The allowable shear range for the group is found using Eq. (17.3), with $\alpha = 7850$:

$$\Sigma Z_r = 3 \times 7850 \times 0.75^2 = 13{,}260 \text{ lb}$$

From Eq. (17.4), the required pitch at this location is

$$p = \frac{13{,}260}{1240} = 10.8 \text{ in.}$$

Similar calculations at 7, 14, 21, 28, and 35 ft, using appropriate section properties at each location, indicate required spacings at those points of 12.2, 15.7, 18.5, 22.4, and 28.5 in., respectively. The maximum spacing permitted by AASHTO is 24 in. The stud spacing pattern selected, starting at the left support, is

$$13 \text{ spaces at } 10 \text{ in.} = 10 \text{ ft } 10 \text{ in.}$$

$$12 \text{ spaces at } 15 \text{ in.} = 15 \text{ ft } 0 \text{ in.}$$

$$5 \text{ spaces at } 22 \text{ in.} = 9 \text{ ft } 2 \text{ in.}$$

A strength check must now be made to ensure that a sufficient number of connectors has been provided between the maximum-moment section and the support to develop the horizontal stress resultant for the composite beam. From Eqs. (17.6*a*, *b*),

$$P_1 = 56.7 \times 36{,}000 = 2{,}041{,}000 \text{ lb}$$

$$P_2 = 0.85 \times 3000 \times 78 \times 6.5 = 1{,}290{,}000 \text{ lb}$$

P_2 controls here, being the smaller of the two values. The ultimate strength of one stud connector, from Eq. (17.8), is

$$S_u = 0.4 \times 0.75^2 \sqrt{3000 \times 3{,}120{,}000} = 21{,}800 \text{ lb}$$

and, accordingly, from Eq. (17.5), the required number of connectors to develop the force in the concrete slab is

$$N = \frac{1{,}290{,}000}{0.85 \times 21{,}800} = 70$$

A total of 31 transverse groups of three studs each has been selected as a result of the preceding design, based on fatigue. The resulting total of 93 studs is more than sufficient for strength.

The spacing of the fillet welds joining the cover plate to the lower flange can likewise be increased as the shear-transfer requirement becomes less. The design method follows that for spacing of the welded studs.

d. Exterior Girders

The design of the two exterior girders does not differ materially from that for the interior girders, and details will not be presented here. Dead and live loads will differ somewhat; transverse positioning of live load for maximum effect will follow the procedures outlined for the design of exterior girders for the deck-girder bridge (Sec. 22.6d).

Diaphragms. Transverse diaphragms will be provided at the centerline of supports and at the third points of the span. These will consist of C18 × 42.7 members, field-bolted to vertical ribs welded to the girder webs, as shown in Fig. 22.27. In addition to providing safety against overturning of the girders, such as might be caused by lateral forces on the roadway, these transverse beams provide some distribution of vertical loads between girders, helping to ensure that the bridge acts as an integral unit.

Bearings. A hinged bearing will be provided at one end of each girder, and a rocker bearing at the other, to provide for expansion and contraction of the 70 ft

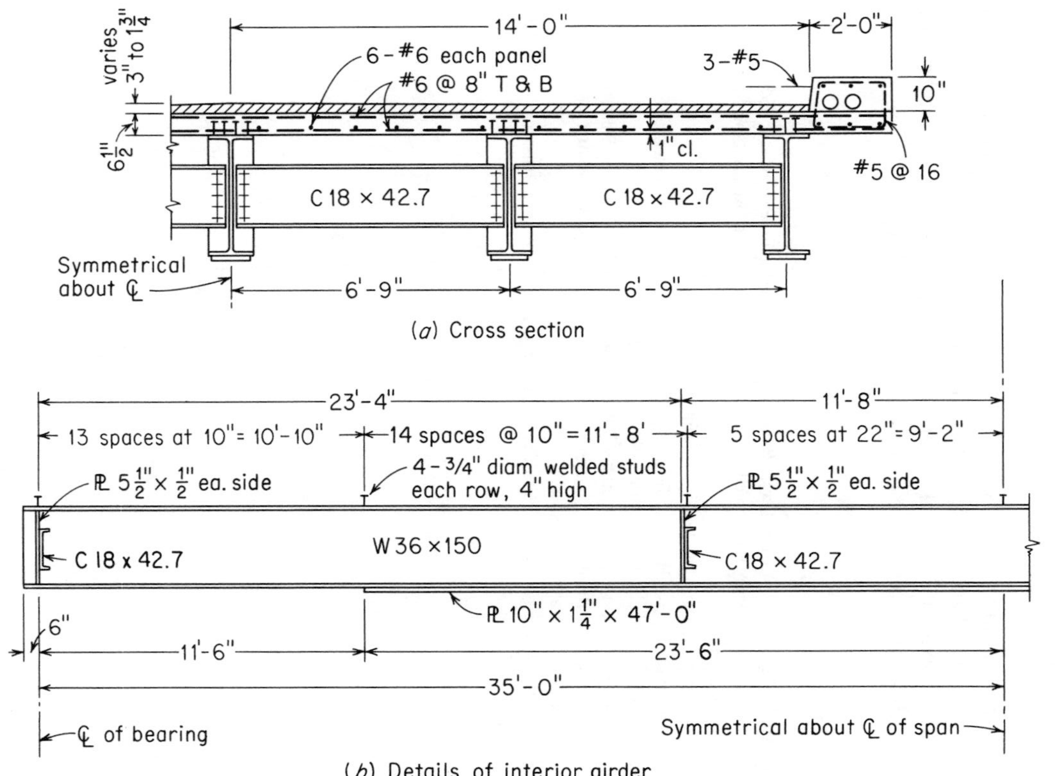

FIGURE 22.27
Details of composite bridge.

span. A joint detail similar to that shown in Fig. 22.14 will be used between the bridge decks and the approach slab.

Drainage. The bituminous surface will be crowned to provide lateral drainage, with a depth of 3 in. at the bridge centerline and a depth of $1\frac{3}{4}$ in. at the curbs. A longitudinal camber of 3 in. will ensure that no water stands on the bridge.

Full details of the composite bridge are shown in Fig. 22.27.

REFERENCES

22.1 "Short Span Bridges: Spans to 100 Feet," Prestressed Concrete Institute, Chicago, 1980.

22.2 "A New Look at Short Span Reinforced Concrete Bridges," Concrete Reinforcing Steel Institute, Schaumburg, IL, 1983.

22.3 F. Leonhardt, *Bridges*, Deutsche Verlags-Anstalt, Stuttgart, 1982.

22.4 "Standard Specifications for Highway Bridges," 14th ed., American Association of State Highway and Transportation Officials, Washington, 1989.

22.5 *Manual for Railway Engineering*, American Railway Engineering Association, Washington, 1973.

22.6 "Analysis and Design of Reinforced Concrete Bridge Structures," Reported by ACI Committee 343, American Concrete Institute, Detroit, 1977.

22.7 "Simple Span Steel Bridges—Composite Beam Design Charts," American Institute of Steel Construction, New York, 1969.

APPENDIX

APPENDIX A DESIGN AIDS

Table A.1 Designations, areas, perimeters, and weights of standard bars

Bar No.[a]	Diameter, in.	Cross-sectional area, in^2	Perimeter, in.	Unit weight per foot, lb
3	$\frac{3}{8} = 0.375$	0.11	1.18	0.376
4	$\frac{1}{2} = 0.500$	0.20	1.57	0.668
5	$\frac{5}{8} = 0.625$	0.31	1.96	1.043
6	$\frac{3}{4} = 0.750$	0.44	2.36	1.502
7	$\frac{7}{8} = 0.875$	0.60	2.75	2.044
8	$1 = 1.000$	0.79	3.14	2.670
9	$1\frac{1}{8} = 1.128$[b]	1.00	3.54	3.400
10	$1\frac{1}{4} = 1.270$[b]	1.27	3.99	4.303
11	$1\frac{3}{8} = 1.410$[b]	1.56	4.43	5.313
14	$1\frac{3}{4} = 1.693$[b]	2.25	5.32	7.650
18	$2\frac{1}{4} = 2.257$[b]	4.00	7.09	13.600

[a] Based on the number of eighths of an inch included in the nominal diameter of the bars. The nominal diameter of a deformed bar is equivalent to the diameter of a plain bar having the same weight per foot as the deformed bar.

[b] Approximate to the nearest $\frac{1}{8}$ in.

Table A.2 Areas of groups of standard bars, in^2

Bar No.	Number of bars 1	2	3	4	5	6	7	8	9	10	11	12
4	0.20	0.39	0.58	0.78	0.98	1.18	1.37	1.57	1.77	1.96	2.16	2.36
5	0.31	0.61	0.91	1.23	1.53	1.84	2.15	2.45	2.76	3.07	3.37	3.68
6	0.44	0.88	1.32	1.77	2.21	2.65	3.09	3.53	3.98	4.42	4.86	5.30
7	0.60	1.20	1.80	2.41	3.01	3.61	4.21	4.81	5.41	6.01	6.61	7.22
8	0.79	1.57	2.35	3.14	3.93	4.71	5.50	6.28	7.07	7.85	8.64	9.43
9	1.00	2.00	3.00	4.00	5.00	6.00	7.00	8.00	9.00	10.00	11.00	12.00
10	1.27	2.53	3.79	5.06	6.33	7.59	8.86	10.12	11.39	12.66	13.92	15.19
11	1.56	3.12	4.68	6.25	7.81	9.37	10.94	12.50	14.06	15.62	17.19	18.75
14	2.25	4.50	6.75	9.00	11.25	13.50	15.75	18.00	20.25	22.50	24.75	27.00
18	4.00	8.00	12.00	16.00	20.00	24.00	28.00	32.00	36.00	40.00	44.00	48.00

Table A.3 Perimeters of groups of standard bars, in.

Bar No.	Number of bars 1	2	3	4	5	6	7	8	9	10	11	12
4	1.6	3.1	4.7	6.2	7.8	9.4	11.0	12.6	14.1	15.7	17.3	18.8
5	2.0	3.9	5.9	7.8	9.8	11.8	13.7	15.7	17.7	19.5	21.6	23.6
6	2.4	4.7	7.1	9.4	11.8	14.1	16.5	18.8	21.2	23.6	25.9	28.3
7	2.8	5.5	8.2	11.0	13.7	16.5	19.2	22.0	24.7	27.5	30.2	33.0
8	3.1	6.3	9.4	12.6	15.7	18.9	22.0	25.1	28.3	31.4	34.6	37.7
9	3.5	7.1	10.6	14.2	17.7	21.3	24.8	28.4	31.9	35.4	39.0	42.5
10	4.0	8.0	12.0	16.0	20.0	23.9	27.9	31.9	35.9	39.9	43.9	47.9
11	4.4	8.9	13.3	17.7	22.2	26.6	31.0	35.4	39.9	44.3	48.7	53.2
14	5.3	10.6	16.0	21.3	26.6	31.9	37.2	42.6	47.9	53.2	58.5	63.8
18	7.1	14.2	21.3	28.4	35.5	42.5	49.6	56.7	63.8	70.9	78.0	85.1

Table A.4 Areas of bars in slabs, in^2/ft

Spacing in.	Bar No. 3	4	5	6	7	8	9	10	11
3	0.44	0.78	1.23	1.77	2.40	3.14	4.00	5.06	6.25
$3\frac{1}{2}$	0.38	0.67	1.05	1.51	2.06	2.69	3.43	4.34	5.36
4	0.33	0.59	0.92	1.32	1.80	2.36	3.00	3.80	4.68
$4\frac{1}{2}$	0.29	0.52	0.82	1.18	1.60	2.09	2.67	3.37	4.17
5	0.26	0.47	0.74	1.06	1.44	1.88	2.40	3.04	3.75
$5\frac{1}{2}$	0.24	0.43	0.67	0.96	1.31	1.71	2.18	2.76	3.41
6	0.22	0.39	0.61	0.88	1.20	1.57	2.00	2.53	3.12
$6\frac{1}{2}$	0.20	0.36	0.57	0.82	1.11	1.45	1.85	2.34	2.89
7	0.19	0.34	0.53	0.76	1.03	1.35	1.71	2.17	2.68
$7\frac{1}{2}$	0.18	0.31	0.49	0.71	0.96	1.26	1.60	2.02	2.50
8	0.17	0.29	0.46	0.66	0.90	1.18	1.50	1.89	2.34
9	0.15	0.26	0.41	0.59	0.80	1.05	1.33	1.69	2.08
10	0.13	0.24	0.37	0.53	0.72	0.94	1.20	1.52	1.87
12	0.11	0.20	0.31	0.44	0.60	0.78	1.00	1.27	1.56

Table A.5 Limiting steel ratios for flexural design

f_y	f'_c	β_1	ρ_b^a	$\rho_{max} = 0.75\rho_b$	$\rho_{min} = \frac{200}{f_y}$
40,000	3000	0.85	0.0371	0.0278	0.0050
	4000	0.85	0.0495	0.0371	0.0050
	5000	0.80	0.0582	0.0437	0.0050
	6000	0.75	0.0655	0.0491	0.0050
	7000	0.70	0.0713	0.0535	0.0050
	8000	0.65	0.0757	0.0568	0.0050
50,000	3000	0.85	0.0275	0.0206	0.0040
	4000	0.85	0.0367	0.0275	0.0040
	5000	0.80	0.0432	0.0324	0.0040
	6000	0.75	0.0486	0.0365	0.0040
	7000	0.70	0.0529	0.0397	0.0040
	8000	0.65	0.0561	0.0421	0.0040
60,000	3000	0.85	0.0214	0.0160	0.0033
	4000	0.85	0.0285	0.0214	0.0033
	5000	0.80	0.0335	0.0251	0.0033
	6000	0.75	0.0377	0.0283	0.0033
	7000	0.70	0.0411	0.0308	0.0033
	8000	0.65	0.0436	0.0327	0.0033
80,000	3000	0.85	0.0141	0.0106	0.0025
	4000	0.85	0.0188	0.0141	0.0025
	5000	0.80	0.0221	0.0166	0.0025
	6000	0.75	0.0249	0.0187	0.0025
	7000	0.70	0.0271	0.0203	0.0025
	8000	0.65	0.0288	0.0216	0.0025

[a] $\rho_b = 0.85\beta_1 \frac{f'_c}{f_y} \frac{87{,}000}{87{,}000 + f_y}$

Table A.6a Flexural resistance factor: $R = \rho f_y \left(1 - 0.588 \frac{\rho f_y}{f'_c}\right)$ psi

ρ	$f_y = 40,000$				$f_y = 60,000$			
	f'_c				f'_c			
	3000	**4000**	**5000**	**6000**	**3000**	**4000**	**5000**	**6000**
.0000	0	0	0	0	0	0	0	0
.0005	20	20	20	20	30	30	30	30
.0010	40	40	40	40	59	59	60	60
.0015	59	59	60	60	88	89	89	89
.0020	79	79	79	79	117	118	118	119
.0025	98	99	99	99	146	147	147	148
.0030	117	118	118	119	174	175	176	177
.0035	136	137	138	138	201	204	205	206
.0040	155	156	157	157	229	232	233	234
.0045	174	175	176	177	256	259	261	263
.0050	192	194	195	196	282	287	289	291
.0055	211	213	214	215	309	314	317	319
.0060	229	232	233	234	335	341	345	347
.0065	247	250	252	253	360	368	372	375
.0070	265	268	271	272	385	394	399	403
.0075	282	287	289	291	410	420	426	430
.0080	300	305	308	310	435	446	453	457
.0085	317	323	326	329	459	472	479	485
.0090	335	341	345	347	483	497	506	511
.0095	352	359	363	366	506	522	532	538
.0100	369	376	381	384	529	547	558	565
.0105	385	394	399	403	552	572	583	591
.0110	402	412	417	421	575	596	609	617
.0115	419	429	435	439	597	620	634	643
.0120	435	446	453	457	618	644	659	669
.0125	451	463	471	476	640	667	684	695
.0130	467	480	488	494	661	691	708	720
.0135	483	497	506	511	681	714	733	746
.0140	499	514	523	529	702	736	757	771
.0145	514	531	540	547	722	759	781	796
.0150	529	547	558	565	741	781	805	821
.0155	545	563	575	582	760	803	828	845
.0160	560	580	592	600	779	825	852	870
.0165	575	596	609	617	798	846	875	894
.0170	589	612	626	635	816	867	898	918
.0175	604	628	642	652	834	888	920	942
.0180	618	644	659	669	851	909	943	966

Table A.6b Flexural resistance factor: $R = \rho f_y \left(1 - 0.588 \dfrac{\rho f_y}{f'_c}\right)$ psi

ρ	$f_y = 40,000$				$f_y = 60,000$			
	f'_c				f'_c			
	3000	4000	5000	6000	3000	4000	5000	6000
.005	192	194	195	196	282	287	289	291
.006	229	232	233	234	335	341	345	347
.007	265	268	271	272	385	394	399	403
.008	300	305	308	310	435	446	453	457
.009	335	341	345	347	483	497	506	511
.010	369	376	381	384	529	547	558	565
.011	402	412	417	421	575	596	609	617
.012	435	446	453	457	618	644	659	669
.013	467	480	488	494	661	691	708	720
.014	499	514	523	529	702	736	757	771
.015	529	547	558	565	741	781	805	821
.016	560	580	592	600	779	825	852	870
.017	589	612	626	635	816	867	898	918
.018	618	644	659	669	851	909	943	966
.019	647	675	692	703	885	949	987	1013
.020	675	706	725	737	918	988	1031	1059
.021	702	736	757	771	949	1027	1073	1104
.022	728	766	789	804		1064	1115	1149
.023	754	796	820	837		1100	1156	1193
.024	779	825	852	870		1135	1196	1237
.025	804	853	882	902		1169	1235	1280
.026	828	881	913	934		1202	1274	1322
.027	851	909	943	966		1234	1311	1363
.028	874	936	972	997		1265	1348	1403
.029	896	962	1002	1028			1384	1443
.030	918	988	1031	1059			1419	1482
.031	939	1014	1059	1089			1453	1521
.032	959	1039	1087	1119			1486	1559
.033	978	1064	1115	1149			1519	1596
.034	997	1088	1142	1179				1632
.035	1016	1112	1170	1208				1668
.036	1034	1135	1196	1237				1703
.037	1051	1158	1222	1265				1737
.038		1180	1248	1294				
.039		1202	1274	1322				
.040	1224	1299	1349					

Table A.7 Parameters k and j for elastic, cracked section beam analysis, where $k = \sqrt{2\rho n + (\rho n)^2} - \rho n$; $j = 1 - \frac{1}{3}k$

	$n = 7$		$n = 8$		$n = 9$		$n = 10$	
ρ	k	j	k	j	k	j	k	j
0.0010	0.112	0.963	0.119	0.960	0.125	0.958	0.132	0.956
0.0020	0.154	0.949	0.164	0.945	0.173	0.942	0.180	0.940
0.0030	0.185	0.938	0.196	0.935	0.207	0.931	0.217	0.928
0.0040	0.210	0.930	0.223	0.926	0.235	0.922	0.246	0.918
0.0050	0.232	0.923	0.246	0.918	0.258	0.914	0.270	0.910
0.0054	0.240	0.920	0.254	0.915	0.267	0.911	0.279	0.907
0.0058	0.247	0.918	0.262	0.913	0.275	0.908	0.287	0.904
0.0062	0.254	0.915	0.269	0.910	0.283	0.906	0.296	0.901
0.0066	0.261	0.913	0.276	0.908	0.290	0.903	0.303	0.899
0.0070	0.268	0.911	0.283	0.906	0.298	0.901	0.311	0.896
0.0072	0.271	0.910	0.287	0.904	0.301	0.900	0.314	0.895
0.0074	0.274	0.909	0.290	0.903	0.304	0.899	0.318	0.894
0.0076	0.277	0.908	0.293	0.902	0.308	0.897	0.321	0.893
0.0078	0.280	0.907	0.296	0.901	0.311	0.896	0.325	0.892
0.0080	0.283	0.906	0.299	0.900	0.314	0.895	0.328	0.891
0.0082	0.286	0.905	0.303	0.899	0.317	0.894	0.331	0.890
0.0084	0.289	0.904	0.306	0.898	0.321	0.893	0.334	0.889
0.0086	0.292	0.903	0.308	0.897	0.324	0.892	0.338	0.887
0.0088	0.295	0.902	0.311	0.896	0.327	0.891	0.341	0.886
0.0090	0.298	0.901	0.314	0.895	0.330	0.890	0.344	0.885
0.0092	0.300	0.900	0.317	0.894	0.332	0.889	0.347	0.884
0.0094	0.303	0.899	0.320	0.893	0.335	0.888	0.350	0.883
0.0096	0.306	0.898	0.323	0.892	0.338	0.887	0.353	0.882
0.0098	0.308	0.897	0.325	0.892	0.341	0.886	0.355	0.882
0.0100	0.311	0.896	0.328	0.891	0.344	0.885	0.358	0.881
0.0104	0.316	0.895	0.333	0.889	0.349	0.884	0.364	0.879
0.0108	0.321	0.893	0.338	0.887	0.354	0.882	0.369	0.877
0.0112	0.325	0.892	0.343	0.886	0.359	0.880	0.374	0.875
0.0116	0.330	0.890	0.348	0.884	0.364	0.879	0.379	0.874
0.0120	0.334	0.889	0.353	0.882	0.369	0.877	0.384	0.872
0.0124	0.339	0.887	0.357	0.881	0.374	0.875	0.389	0.870
0.0128	0.343	0.886	0.362	0.879	0.378	0.874	0.394	0.867
0.0132	0.347	0.884	0.366	0.878	0.383	0.872	0.398	0.867
0.0136	0.351	0.883	0.370	0.877	0.387	0.871	0.403	0.866
0.0140	0.355	0.882	0.374	0.875	0.392	0.869	0.407	0.864
0.0144	0.359	0.880	0.378	0.874	0.396	0.868	0.412	0.863
0.0148	0.363	0.879	0.382	0.873	0.400	0.867	0.416	0.861
0.0152	0.367	0.878	0.386	0.871	0.404	0.865	0.420	0.860
0.0156	0.371	0.876	0.390	0.870	0.408	0.864	0.424	0.859
0.0160	0.374	0.875	0.394	0.869	0.412	0.863	0.428	0.857
0.0170	0.383	0.872	0.403	0.867	0.421	0.860	0.437	0.854
0.0180	0.392	0.869	0.412	0.863	0.430	0.857	0.446	0.851
0.0190	0.400	0.867	0.420	0.860	0.438	0.854	0.455	0.848
0.0200	0.407	0.864	0.428	0.857	0.446	0.851	0.463	0.846

Table A.8 Maximum number of bars as a single layer in beam stems

$\frac{3}{4}$ in. maximum size aggregate, No. 4 stirrups[a]

Bar	Beam width b_w, in.											
No.	**8**	**10**	**12**	**14**	**16**	**18**	**20**	**22**	**24**	**26**	**28**	**30**
5	2	4	5	6	7	8	10	11	12	13	15	16
6	2	3	4	6	7	8	9	10	11	12	14	15
7	2	3	4	5	6	7	8	9	10	11	12	13
8	2	3	4	5	6	7	8	9	10	11	12	13
9	1	2	3	4	5	6	7	8	9	9	10	11
10	1	2	3	4	5	6	6	7	8	9	10	10
11	1	2	3	3	4	5	5	6	7	8	8	9
14	1	2	2	3	3	4	5	5	6	6	7	8
18[b]	1	1	2	2	3	3	4	4	4	5	5	6

1 in. maximum size aggregate, No. 4 stirrups[a]

Bar	Beam width b_w, in.											
No.	**8**	**10**	**12**	**14**	**16**	**18**	**20**	**22**	**24**	**26**	**28**	**30**
5	2	3	4	5	6	7	8	9	10	11	12	13
6	2	3	4	5	6	7	8	9	9	10	11	12
7	1	2	3	4	5	6	7	8	9	10	10	11
8	1	2	3	4	5	6	7	7	8	9	10	11
9	1	2	3	4	5	6	7	7	8	9	9	10
10	1	2	3	4	5	6	6	7	7	8	9	10

[a] Minimum concrete cover assumed to be $1\frac{1}{2}$ in. to the No. 4 stirrup.

[b] For exterior exposures, service load stress must be reduced to satisfy maximum $Z = 145$ (ACI 318-89, Sec. 10.6.4).

Source: Adapted from Ref. 3.7. Used by permission of American Concrete Institute.

Table A.9 Minimum number of bars as a single layer in beam stems governed by crack control requirements of the ACI Code

(a) **Interior exposure**

Minimum number of bars as a single layer in beam stem

Bar	Beam stem width b_w, in.														
No.	8	10	12	14	16	18	20	22	24	26	28	30	32	34	36
3	1	1	1	2	2	2	2	2	2	3	3	3	3	3	3
4	1	1	2	2	2	2	2	2	3	3	3	3	3	3	4
5	1	1	2	2	2	2	2	3	3	3	3	3	3	4	4
6	1	1	2	2	2	2	2	3	3	3	3	3	4	4	4
7	1	2	2	2	2	2	3	3	3	3	3	4	4	4	4
8	1	2	2	2	2	2	3	3	3	3	4	4	4	4	4
9	1	2	2	2	2	3	3	3	3	3	4	4	4	4	5
10	1	2	2	2	2	3	3	3	3	4	4	4	4	5	5
11	(2)	2	2	2	3	3	3	3	4	4	4	4	5	5	5
14	(2)	2	2	2	3	3	3	4	4	4	4	5	5	5	6
18	(2)	(2)	(3)	(3)	3	(4)	4	4	(5)	5	5	6	6	6	7

( ) Maximum permitted is less than minimum number required.

(b) **Exterior exposure**

Minimum number of bars as a single layer in beam stem

Bar	Beam stem width b_w, in.											
No.	8	10	12	14	16	18	20	22	24	26	28	30
5	2	2	2	3	3	3	4	4	4	5	5	5
6	2	2	3	3	3	4	4	4	5	5	5	6
7	2	2	3	3	3	4	4	4	5	5	6	6
8	2	2	3	3	4	4	4	5	5	5	6	6
9	(2)	2	3	3	4	4	5	5	5	6	6	7
10	(2)	(3)	3	3	4	4	5	5	6	6	7	7
11	(2)	(3)	3	(4)	4	5	5	5	6	6	7	7
14	(2)	(3)	(3)	(4)	(4)	(5)	5	(6)	6	(7)	7	8

( ) Maximum permitted is less than minimum number required.

Source: Adapted from Ref. 3.8. Used by permission of Concrete Reinforcing Steel Institute.

Table A.10 Design strength ϕM_n for slab sections 12 in. wide, ft-kips $f_y = 60$ ksi; $\phi M_n = \phi \rho f_y b d^2(1 - 0.59 \rho f_y / f_c')$

		Effective depth d, in.												
f_c'	ρ	**3.0**	**3.5**	**4.0**	**4.5**	**5.0**	**5.5**	**6.0**	**6.5**	**7.0**	**8.0**	**9.0**	**10.0**	**12.0**
3000	0.002	0.9	1.3	1.7	2.1	2.6	3.2	3.8	4.5	5.2	6.7	8.5	10.5	15.2
	0.003	1.4	1.9	2.5	3.2	3.9	4.7	5.6	6.6	7.7	10.0	12.7	15.6	22.5
	0.004	1.9	2.5	3.3	4.2	5.1	6.2	7.4	8.7	10.1	13.2	16.7	20.6	29.6
	0.005	2.3	3.1	4.1	5.1	6.4	7.7	9.1	10.7	12.4	16.3	20.6	25.4	36.6
	0.006	2.7	3.7	4.8	6.1	7.5	9.1	10.8	12.7	14.8	19.3	24.4	30.1	43.4
	0.007	3.1	4.2	5.5	7.0	8.7	10.5	12.5	14.7	17.0	22.2	28.1	34.7	49.9
	0.008	3.5	4.8	6.3	7.9	9.8	11.8	14.1	16.5	19.2	25.0	31.7	39.1	56.3
	0.009	3.9	5.3	7.0	8.8	10.9	13.1	15.6	18.4	21.3	27.8	35.2	43.4	62.6
	0.010	4.3	5.8	7.6	9.6	11.9	14.4	17.1	20.1	23.3	30.5	38.6	47.6	68.6
	0.011	4.7	6.3	8.3	10.5	12.9	15.6	18.6	21.8	25.3	33.1	41.9	51.7	74.4
4000	0.002	1.0	1.3	1.7	2.1	2.7	3.2	3.8	4.5	5.2	6.8	8.6	10.6	15.3
	0.003	1.4	1.9	2.5	3.2	3.9	4.8	5.7	6.7	7.7	10.1	12.8	15.8	22.7
	0.004	1.9	2.6	3.3	4.2	5.2	6.3	7.5	8.8	10.2	13.3	16.9	20.8	30.0
	0.005	2.3	3.2	4.1	5.2	6.5	7.8	9.3	10.9	12.6	16.5	20.9	25.8	37.2
	0.006	2.8	3.8	4.9	6.2	7.7	9.3	11.0	13.0	15.0	19.6	24.9	30.7	44.2
	0.007	3.2	4.3	5.7	7.2	8.9	10.7	12.8	15.0	17.4	22.7	28.7	35.5	51.1
	0.008	3.6	4.9	6.4	8.1	10.0	12.1	14.5	17.0	19.7	25.7	32.5	40.1	57.8
	0.009	4.0	5.5	7.2	9.1	11.2	13.5	16.1	18.9	21.9	28.6	36.2	44.7	64.4
	0.010	4.4	6.0	7.9	10.0	12.3	14.9	17.7	20.8	24.1	31.5	39.9	49.2	70.9
	0.011	4.8	6.6	8.6	10.9	13.4	16.2	19.3	22.7	26.3	34.3	43.4	53.6	77.2
	0.012	5.2	7.1	9.3	11.7	14.5	17.5	20.9	24.5	28.4	37.1	46.9	57.9	83.4
	0.013	5.6	7.6	9.9	12.6	15.5	18.8	22.4	26.2	30.4	39.8	50.3	62.1	89.5
	0.014	6.0	8.1	10.6	13.4	16.6	20.0	23.8	28.0	32.5	42.4	53.6	66.2	95.4
	0.015	6.3	8.6	11.2	14.2	17.6	21.2	25.3	29.7	34.4	45.0	56.9	70.2	101.2
5000	0.002	1.0	1.3	1.7	2.2	2.7	3.2	3.8	4.5	5.2	6.8	8.6	10.6	15.3
	0.003	1.4	1.9	2.5	3.2	4.0	4.8	5.7	6.7	7.8	10.1	12.8	15.9	22.8
	0.004	1.9	2.6	3.4	4.3	5.2	6.3	7.6	8.9	10.3	13.4	17.0	21.0	30.2
	0.005	2.3	3.2	4.2	5.3	6.5	7.9	9.4	11.0	12.8	16.7	21.1	26.0	37.5
	0.006	2.8	3.8	5.0	6.3	7.8	9.4	11.2	13.1	15.2	19.9	25.1	31.0	44.7
	0.007	3.2	4.4	5.7	7.3	9.0	10.9	12.9	15.2	17.6	23.0	29.1	35.9	51.7
	0.008	3.7	5.0	6.5	8.3	10.2	12.3	14.7	17.2	20.0	26.1	33.0	40.8	58.7
	0.009	4.1	5.6	7.3	9.2	11.4	13.8	16.4	19.2	22.3	29.1	36.9	45.5	65.5
	0.010	4.5	6.1	8.0	10.2	12.5	15.2	18.1	21.2	24.6	32.1	40.6	50.2	72.3
	0.011	4.9	6.7	8.8	11.1	13.7	16.6	19.7	23.1	26.8	35.1	44.4	54.8	78.9
	0.012	5.3	7.3	9.5	12.0	14.8	17.9	21.3	25.1	29.1	37.9	48.0	59.3	85.4
	0.013	5.7	7.8	10.2	12.9	15.9	19.3	22.9	26.9	31.2	40.8	51.6	63.7	91.8
	0.014	6.1	8.3	10.9	13.8	17.0	20.6	24.5	28.8	33.4	43.6	55.2	68.1	98.1
	0.015	6.5	8.9	11.6	14.7	18.1	21.9	26.1	30.6	35.5	46.3	58.6	72.4	104.3
	0.016	6.9	9.4	12.3	15.5	19.2	23.2	27.6	32.4	37.5	49.0	62.1	76.6	110.3
	0.017	7.3	9.9	12.9	16.4	20.2	29.1	24.4	34.1	39.6	51.7	65.4	80.8	116.3

Table A.11 Basic development length l_{db} in tension, in.

For No. 11 bars and smaller: $l_{db} = 0.04A_bf_y/\sqrt{f'_c}$

For No. 14 bars: $l_{db} = 0.085f_y/\sqrt{f'_c}$

For No. 18 bars: $l_{db} = 0.125f_y/\sqrt{f'_c}$

Bar no.	f_y ksi	f'_c psi					
		3000	4000	5000	6000	8000	≥10,000
		l_{db} in.					
3	40	3	3	3	2	2	2
	50	4	3	3	3	2	2
	60	5	4	4	3	3	3
4	40	6	5	5	4	4	3
	50	7	6	6	5	4	4
	60	9	8	7	6	5	5
5	40	9	8	7	6	6	5
	50	11	10	9	8	7	6
	60	14	12	11	10	8	7
6	40	13	11	10	9	8	7
	50	16	14	12	11	10	9
	60	19	17	15	14	12	11
7	40	18	15	14	12	11	10
	50	22	19	17	15	13	12
	60	26	23	20	19	16	14
8	40	23	20	18	16	14	13
	50	29	25	22	20	18	16
	60	35	30	27	24	21	19
9	40	29	25	23	21	18	16
	50	37	32	28	26	22	20
	60	44	38	34	31	27	24
10	40	37	32	29	26	23	20
	50	46	40	36	33	28	25
	60	56	48	43	39	34	30
11	40	46	39	35	32	28	25
	50	57	49	44	40	35	31
	60	68	59	53	48	42	37
14	40	62	54	48	44	38	34
	50	77	67	60	55	48	43
	60	93	81	72	66	57	51
18	40	91	79	71	65	56	50
	50	114	99	88	81	70	63
	60	136	119	106	97	84	75

Notes: 1. Development length l_d = basic development length l_{db} times applicable modification factors from Table 5.1.

2. Basic development length, multiplied by the applicable factors of Table 5.1A, must not be less than $0.03d_bf_y/\sqrt{f'_c}$.

3. Basic development length, multiplied by *all* applicable factors of Table 5.1, must not be less than 12 in.

Table A.12 Development length in compression, in.
$l_{db} = 0.02 d_b f_y / \sqrt{f'_c}$ but $\geq 0.0003 d_b f_y$
(Minimum length 8 in. in all cases.)

		f'_c							
		3000		4000		5000		6000	
Bar No.	f_y ksi	Basic l_{db}	Confined	Basic l_{db}	Confined	Basic l_{db}	Confined	Basic l_{db}	Confined
3	40	8	8	8	8	8	8	8	8
	50	8	8	8	8	8	8	8	8
	60	8	8	8	8	8	8	8	8
4	40	8	8	8	8	8	8	8	8
	50	9	8	8	8	8	8	8	8
	60	11	8	9	8	9	8	9	8
5	40	9	8	8	8	8	8	8	8
	50	11	9	10	8	9	8	9	8
	60	14	10	12	9	11	8	11	8
6	40	11	8	9	8	9	8	9	8
	50	14	10	12	9	11	8	11	8
	60	16	12	14	11	14	10	14	10
7	40	13	10	11	8	11	8	11	8
	50	16	12	14	10	13	10	13	10
	60	19	14	17	12	16	12	16	12
8	40	15	11	13	9	12	9	12	9
	50	18	14	16	12	15	11	15	11
	60	22	16	19	14	18	14	18	14
9	40	16	12	14	11	14	10	14	10
	50	21	15	18	13	17	13	17	13
	60	25	19	21	16	20	15	20	15
10	40	19	14	16	12	15	11	15	11
	50	23	17	20	15	19	14	19	14
	60	28	21	24	18	23	17	23	17
11	40	21	15	18	13	17	13	17	13
	50	26	19	22	17	21	16	21	16
	60	31	23	27	20	25	19	25	19
14	40	25	19	21	16	20	15	20	15
	50	31	23	27	20	25	19	25	19
	60	37	28	32	24	30	23	30	23
18	40	33	25	29	21	27	20	27	20
	50	41	31	36	27	34	25	34	25
	60	49	37	43	32	41	30	41	30

Table A.13 Common stock styles of welded wire fabric

Steel designation		Steel area, in^2/ft		Weight (approximate) lb per 100 ft^2
New designation (by *W* number)	Old designation (by steel wire gage)	Longitudinal	Transverse	
Rolls				
6 × 6-W1.4 × W1.4	6 × 6-10 × 10	0.029	0.029	21
6 × 6-W2.0 × W2.0	6 × 6-8 × 8[a]	0.041	0.041	30
6 × 6-W2.9 × W2.9	6 × 6-6 × 6	0.058	0.058	42
6 × 6-W4.0 × W4.0	6 × 6-4 × 4	0.080	0.080	58
4 × 4-W1.4 × W1.4	4 × 4-10 × 10	0.043	0.043	31
4 × 4-W2.0 × W2.0	4 × 4-8 × 8[a]	0.062	0.062	44
4 × 4-W2.9 × W2.9	4 × 4-6 × 6	0.087	0.087	62
4 × 4-W4.0 × W4.0	4 × 4-4 × 4	0.120	0.120	85
Sheets				
6 × 6-W2.9 × W2.9	6 × 6-6 × 6	0.058	0.058	42
6 × 6-W4.0 × W4.0	6 × 6-4 × 4	0.080	0.080	58
6 × 6-W5.5 × W5.5	6 × 6-2 × 2[b]	0.110	0.110	80
4 × 4-W4.0 × W4.0	4 × 4-4 × 4	0.120	0.120	85

[a] Exact W-number size for 8 gage is W2.1.

[b] Exact W-number size for 2 gage is W5.4.

Source: From Ref. 2.38

Table A.14a Coefficients for slabs with variable moment of inertia[a]

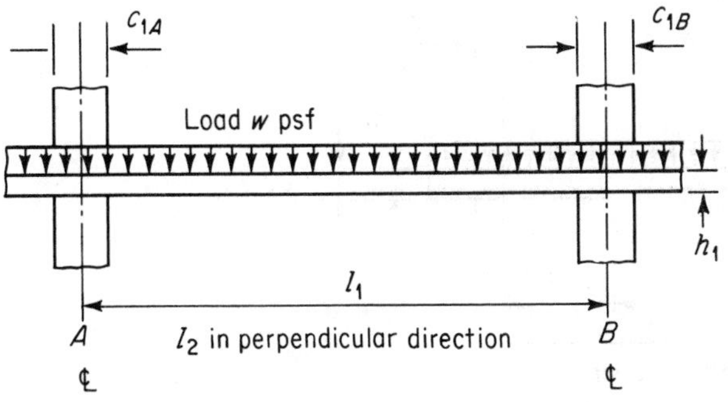

Column dimension		Uniform load FEM = coeff. $(wl_2l_1^2)$		Stiffness factor[b]		Carryover factor	
$\frac{c_{1A}}{l_1}$	$\frac{c_{1B}}{l_1}$	M_{AB}	M_{BA}	k_{AB}	k_{BA}	COF_{AB}	COF_{BA}
0.00	0.00	0.083	0.083	4.00	4.00	0.500	0.500
	0.05	0.083	0.084	4.01	4.04	0.504	0.500
	0.10	0.082	0.086	4.03	4.15	0.513	0.499
	0.15	0.081	0.089	4.07	4.32	0.528	0.498
	0.20	0.079	0.093	4.12	4.56	0.548	0.495
	0.25	0.077	0.097	4.18	4.88	0.573	0.491
0.05	0.05	0.084	0.084	4.05	4.05	0.503	0.503
	0.10	0.083	0.086	4.07	4.15	0.513	0.503
	0.15	0.081	0.089	4.11	4.33	0.528	0.501
	0.20	0.080	0.092	4.16	4.58	0.548	0.499
	0.25	0.078	0.096	4.22	4.89	0.573	0.494
0.10	0.10	0.085	0.085	4.18	4.18	0.513	0.513
	0.15	0.083	0.088	4.22	4.36	0.528	0.511
	0.20	0.082	0.091	4.27	4.61	0.548	0.508
	0.25	0.080	0.095	4.34	4.93	0.573	0.504
0.15	0.15	0.086	0.086	4.40	4.40	0.526	0.526
	0.20	0.084	0.090	4.46	4.65	0.546	0.523
	0.25	0.083	0.094	4.53	4.98	0.571	0.519
0.20	0.20	0.088	0.088	4.72	4.72	0.543	0.543
	0.25	0.086	0.092	4.79	5.05	0.568	0.539
0.25	0.25	0.090	0.090	5.14	5.14	0.563	0.563

[a] Applicable when $c_1/l_1 = c_2/l_2$. For other relationships between these ratios, the constants will be slightly in error.

[b] Stiffness is $K_{AB} = k_{AB}E(l_2h_1^3/12l_1)$ and $K_{BA} = k_{BA}E(l_2h_1^3/12l_1)$.

Table A.14b Coefficients for slabs with variable moment of inertia[a]

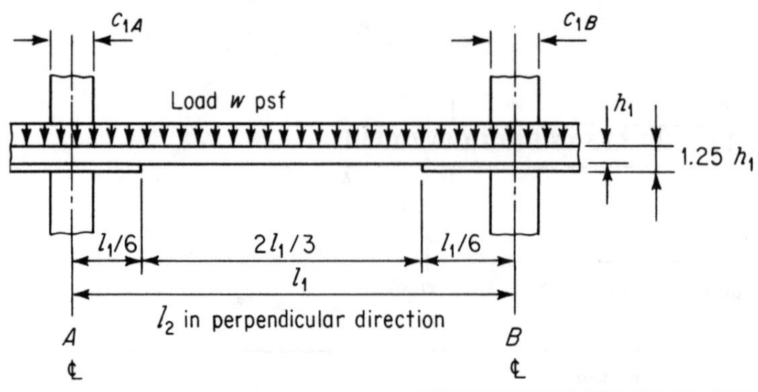

Column dimension		Uniform load FEM = coeff. $(wl_2l_1^2)$		Stiffness factor[b]		Carryover factor	
$\frac{c_{1A}}{l_1}$	$\frac{c_{1B}}{l_1}$	M_{AB}	M_{BA}	k_{AB}	k_{BA}	COF_{AB}	COF_{BA}
0.00	0.00	0.088	0.088	4.78	4.78	0.541	0.541
	0.05	0.087	0.089	4.80	4.82	0.545	0.541
	0.10	0.087	0.090	4.83	4.94	0.553	0.541
	0.15	0.085	0.093	4.87	5.12	0.567	0.540
	0.20	0.084	0.096	4.93	5.36	0.585	0.537
	0.25	0.082	0.100	5.00	5.68	0.606	0.534
0.05	0.05	0.088	0.088	4.84	4.84	0.545	0.545
	0.10	0.087	0.090	4.87	4.95	0.553	0.544
	0.15	0.085	0.093	4.91	5.13	0.567	0.543
	0.20	0.084	0.096	4.97	5.38	0.584	0.541
	0.25	0.082	0.100	5.05	5.70	0.606	0.537
0.10	0.10	0.089	0.089	4.98	4.98	0.553	0.553
	0.15	0.088	0.092	5.03	5.16	0.566	0.551
	0.20	0.086	0.094	5.09	5.42	0.584	0.549
	0.25	0.084	0.099	5.17	5.74	0.606	0.546
0.15	0.15	0.090	0.090	5.22	5.22	0.565	0.565
	0.20	0.089	0.094	5.28	5.47	0.583	0.563
	0.25	0.087	0.097	5.37	5.80	0.604	0.559
0.20	0.20	0.092	0.092	5.55	5.55	0.580	0.580
	0.25	0.090	0.096	5.64	5.88	0.602	0.577
0.25	0.25	0.094	0.094	5.98	5.98	0.598	0.598

[a] Applicable when $c_1/l_1 = c_2/l_2$. For other relationships between these ratios, the constants will be slightly in error.

[b] Stiffness is $K_{AB} = k_{AB}E(l_2h_1^3/12l_1)$ and $K_{BA} = k_{BA}E(l_2h_1^3/12l_1)$.

Table A.14c Coefficients for columns with variable moment of inertia

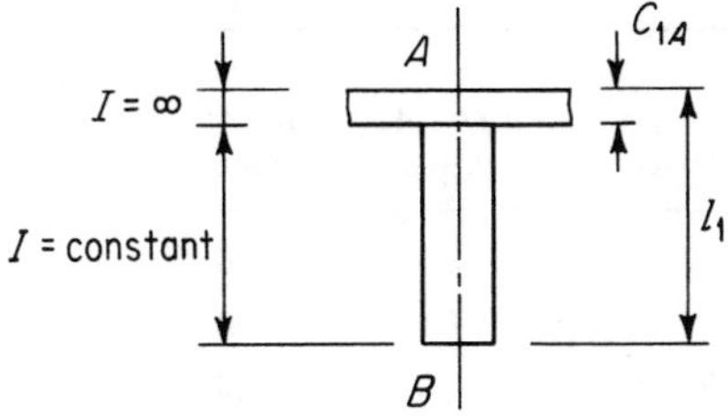

Slab depth c_{1A}/l_1	Uniform load FEM = coeff. $(wl_2l_1^2)$ M_{AB}	M_{BA}	Stiffness factor k_{AB}	k_{BA}	Carryover factor COF_{AB}	COF_{BA}
0.00	0.083	0.083	4.00	4.00	0.500	0.500
0.05	0.100	0.075	4.91	4.21	0.496	0.579
0.10	0.118	0.068	6.09	4.44	0.486	0.667
0.15	0.135	0.060	7.64	4.71	0.471	0.765
0.20	0.153	0.053	9.69	5.00	0.452	0.875
0.25	0.172	0.047	12.44	5.33	0.429	1.000

Table A.15 Size and pitch of spirals, ACI Code

Diameter of column, in.	Out to out of spiral, in.	f'_c 2500	3000	4000	5000
f_y = 40,000					
14, 15	11, 12	$\frac{3}{8}$-2	$\frac{3}{8}$-$1\frac{3}{4}$	$\frac{1}{2}$-$2\frac{1}{2}$	$\frac{1}{2}$-$1\frac{3}{4}$
16	13	$\frac{3}{8}$-2	$\frac{3}{8}$-$1\frac{3}{4}$	$\frac{1}{2}$-$2\frac{1}{2}$	$\frac{1}{2}$-2
17–19	14–16	$\frac{3}{8}$-$2\frac{1}{4}$	$\frac{3}{8}$-$1\frac{3}{4}$	$\frac{1}{2}$-$2\frac{1}{2}$	$\frac{1}{2}$-2
20–23	17–20	$\frac{3}{8}$-$2\frac{1}{4}$	$\frac{3}{8}$-$1\frac{3}{4}$	$\frac{1}{2}$-$2\frac{1}{2}$	$\frac{1}{2}$-2
24–30	21–27	$\frac{3}{8}$-$2\frac{1}{4}$	$\frac{3}{8}$-2	$\frac{1}{2}$-$2\frac{1}{2}$	$\frac{1}{2}$-2
f_y = 60,000					
14, 15	11, 12	$\frac{1}{4}$-$1\frac{3}{4}$	$\frac{3}{8}$-$2\frac{3}{4}$	$\frac{3}{8}$-2	$\frac{1}{2}$-$2\frac{3}{4}$
16–23	13–20	$\frac{1}{4}$-$1\frac{3}{4}$	$\frac{3}{8}$-$2\frac{3}{4}$	$\frac{3}{8}$-2	$\frac{1}{2}$-3
24–29	21–26	$\frac{1}{4}$-$1\frac{3}{4}$	$\frac{3}{8}$-3	$\frac{3}{8}$-$2\frac{1}{4}$	$\frac{1}{2}$-3
30	27	$\frac{1}{4}$-$1\frac{3}{4}$	$\frac{3}{8}$-3	$\frac{3}{8}$-$2\frac{1}{4}$	$\frac{1}{2}$-$3\frac{1}{4}$

Table A.16 Properties of prestressing steels[a]

Seven-wire strand, f_{pu} = 270 ksi

Nominal Diameter (in.)	Area (in^2)	Weight (plf)	$0.7f_{pu}A_p$ (kips)	$0.75f_{pu}A_p$ (kips)	$0.8f_{pu}A_p$ (kips)	$f_{pu}A_p$ (kips)
$\frac{3}{8}$(0.375)	0.085	0.29	16.1	17.2	18.4	23.0
$\frac{7}{16}$(0.438)	0.115	0.40	21.7	23.3	24.8	31.0
$\frac{1}{2}$(0.500)	0.153	0.53	28.9	31.0	33.0	41.3
$\frac{9}{16}$(0.563)	0.192	0.65	36.3	38.9	41.4	51.8
(0.600)	0.215	0.74	40.7	43.5	46.5	58.1

Seven-wire strand, f_{pu} = 250 ksi

Nominal Diameter (in.)	Area (in^2)	Weight (plf)	$0.7f_{pu}A_p$ (kips)	$0.8f_{pu}A_p$ (kips)	$f_{pu}A_p$ (kips)
$\frac{1}{4}$(0.250)	0.036	0.12	6.3	7.2	9.0
$\frac{5}{16}$(0.313)	0.058	0.20	10.2	11.6	14.5
$\frac{3}{8}$(0.375)	0.080	0.27	14.0	16.0	20.0
$\frac{7}{16}$(0.438)	0.108	0.37	18.9	21.6	27.0
$\frac{1}{2}$(0.500)	0.144	0.49	25.2	28.8	36.0
(0.600)	0.215	0.74	37.6	43.0	53.8

Prestressing wire

Diameter	Area (in^2)	Weight (plf)	Ult. strength f_{pu}(ksi)	$0.7f_{pu}A_p$ (kips)	$0.8f_{pu}A_p$ (kips)	$f_{pu}A_p$ (kips)
0.105	0.0087	0.030	279	1.70	1.94	2.43
0.120	0.0114	0.039	273	2.18	2.49	3.11
0.135	0.0143	0.049	268	2.68	3.06	3.83
0.148	0.0173	0.059	263	3.18	3.64	4.55
0.162	0.0206	0.070	259	3.73	4.26	5.33
0.177	0.0246	0.083	255	4.39	5.02	6.27
0.192	0.0289	0.098	250	5.05	5.78	7.22
0.196	0.0302	0.100	250	5.28	6.04	7.55
0.250	0.0491	0.170	240	8.25	9.42	11.78
0.276	0.0598	0.200	235	9.84	11.24	14.05

Smooth prestressing bars, f_{pu} = 145 ksi

Nominal Diameter (in.)	Area (in^2)	Weight (plf)	$0.7f_{pu}A_p$ (kips)	$0.8f_{pu}A_p$ (kips)	$f_{pu}A_p$ (kips)
$\frac{3}{4}$(0.750)	0.442	1.50	44.9	51.3	64.1
$\frac{7}{8}$(0.875)	0.601	2.04	61.0	69.7	87.1
1(1.000)	0.785	2.67	79.7	91.0	113.8
$1\frac{1}{8}$(1.125)	0.994	3.38	100.9	115.3	144.1
$1\frac{1}{4}$(1.250)	1.227	4.17	124.5	142.3	177.9
$1\frac{3}{8}$(1.375)	1.485	5.05	150.7	172.2	215.3

Smooth prestressing bars, f_{pu} = 160 ksi

Nominal Diameter (in.)	Area (in^2)	Weight (plf)	$0.7f_{pu}A_p$ (kips)	$0.8f_{pu}A_p$ (kips)	$f_{pu}A_p$ (kips)
$\frac{3}{4}$(0.750)	0.442	1.50	49.5	56.6	70.7
$\frac{7}{8}$(0.875)	0.601	2.04	67.3	77.0	96.2
1(1.000)	0.785	2.67	87.9	100.5	125.6
$1\frac{1}{8}$(1.125)	0.994	3.38	111.3	127.2	159.0
$1\frac{1}{4}$(1.250)	1.227	4.17	137.4	157.0	196.3
$1\frac{3}{8}$(1.375)	1.485	5.05	166.3	190.1	237.6

[a] Adapted from *PCI Design Handbook*, 3d ed., Prestressed Concrete Institute, Chicago, 1985.

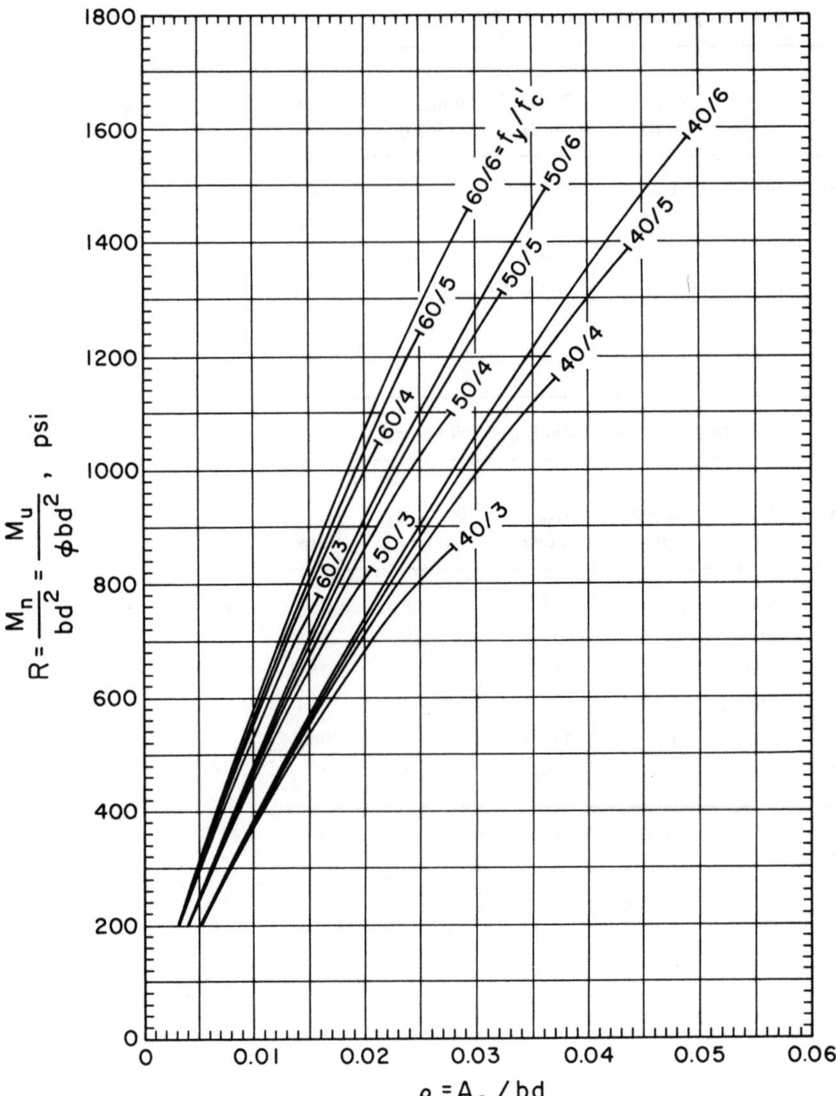

GRAPH A.1*a*
Moment capacity of rectangular sections.

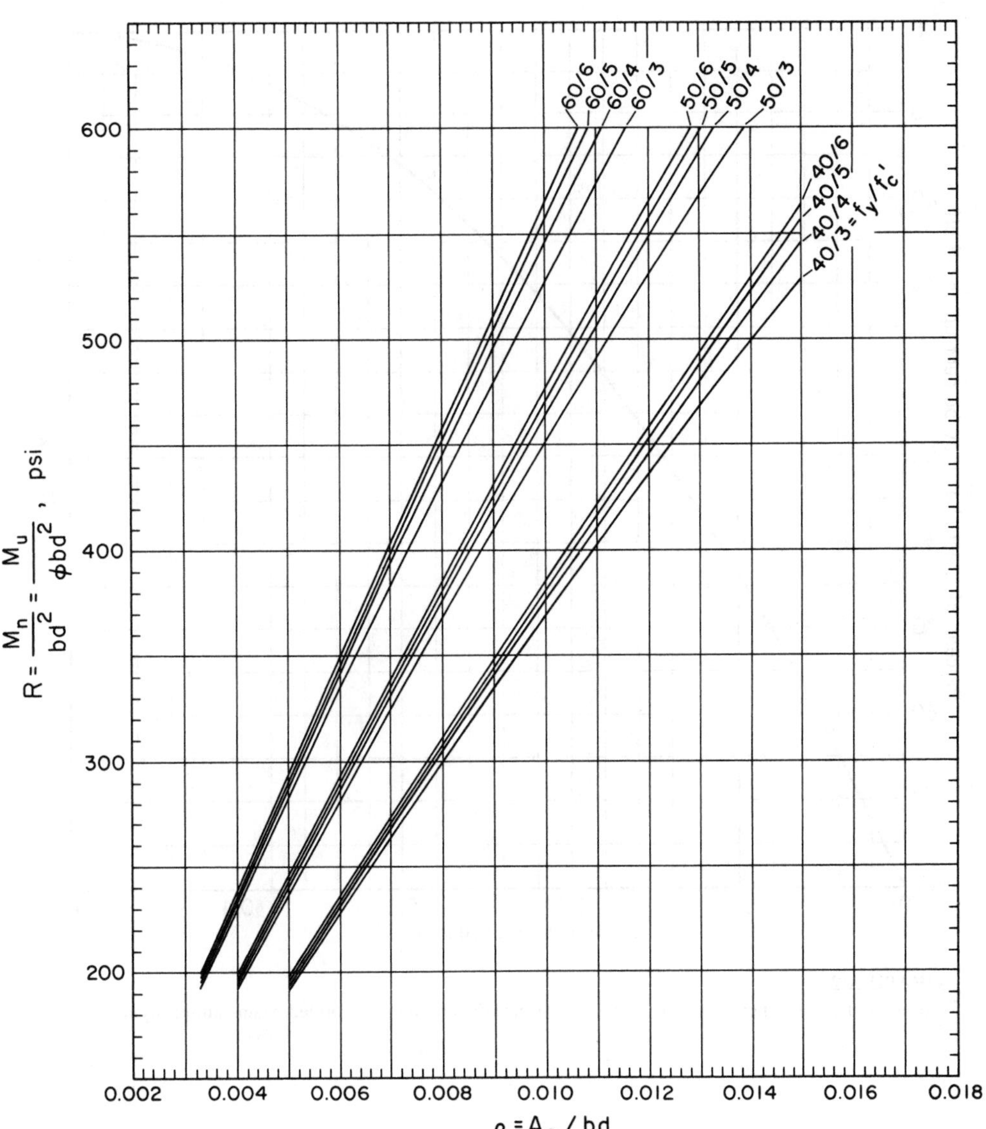

GRAPH A.1*b*
Moment capacity of rectangular sections.

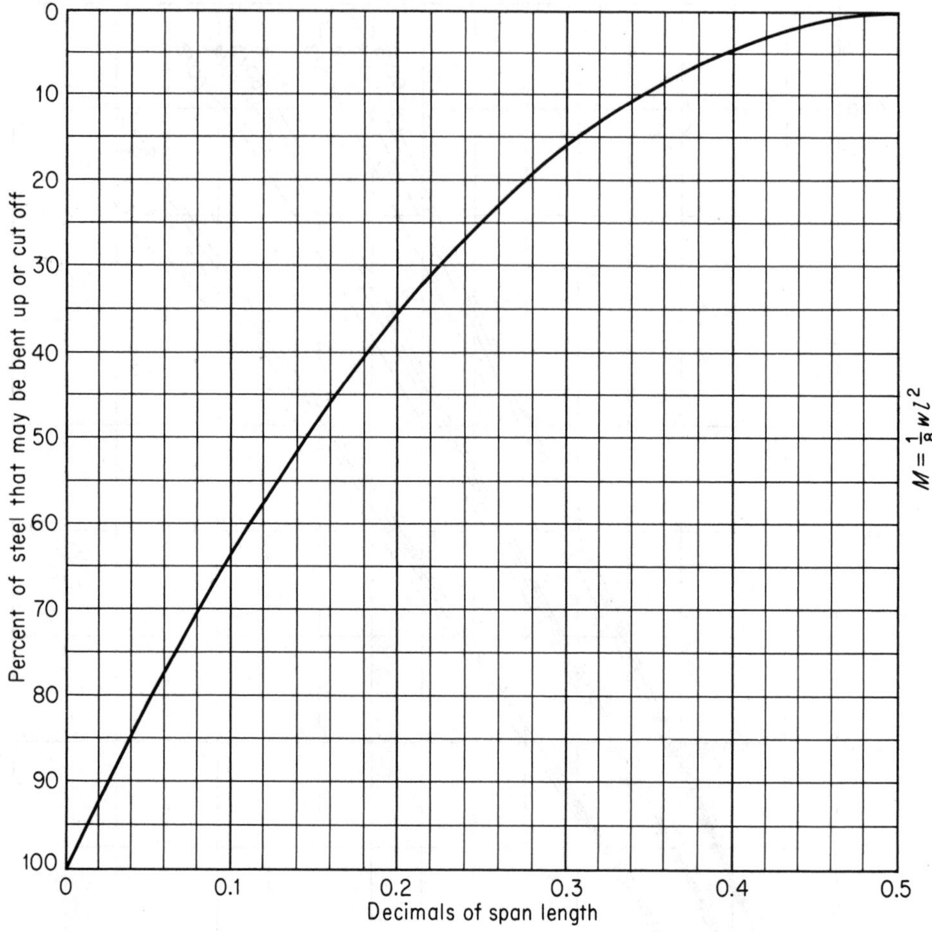

GRAPH A.2

Location of points where bars can be bent up or cut off for simply supported beams uniformly loaded.

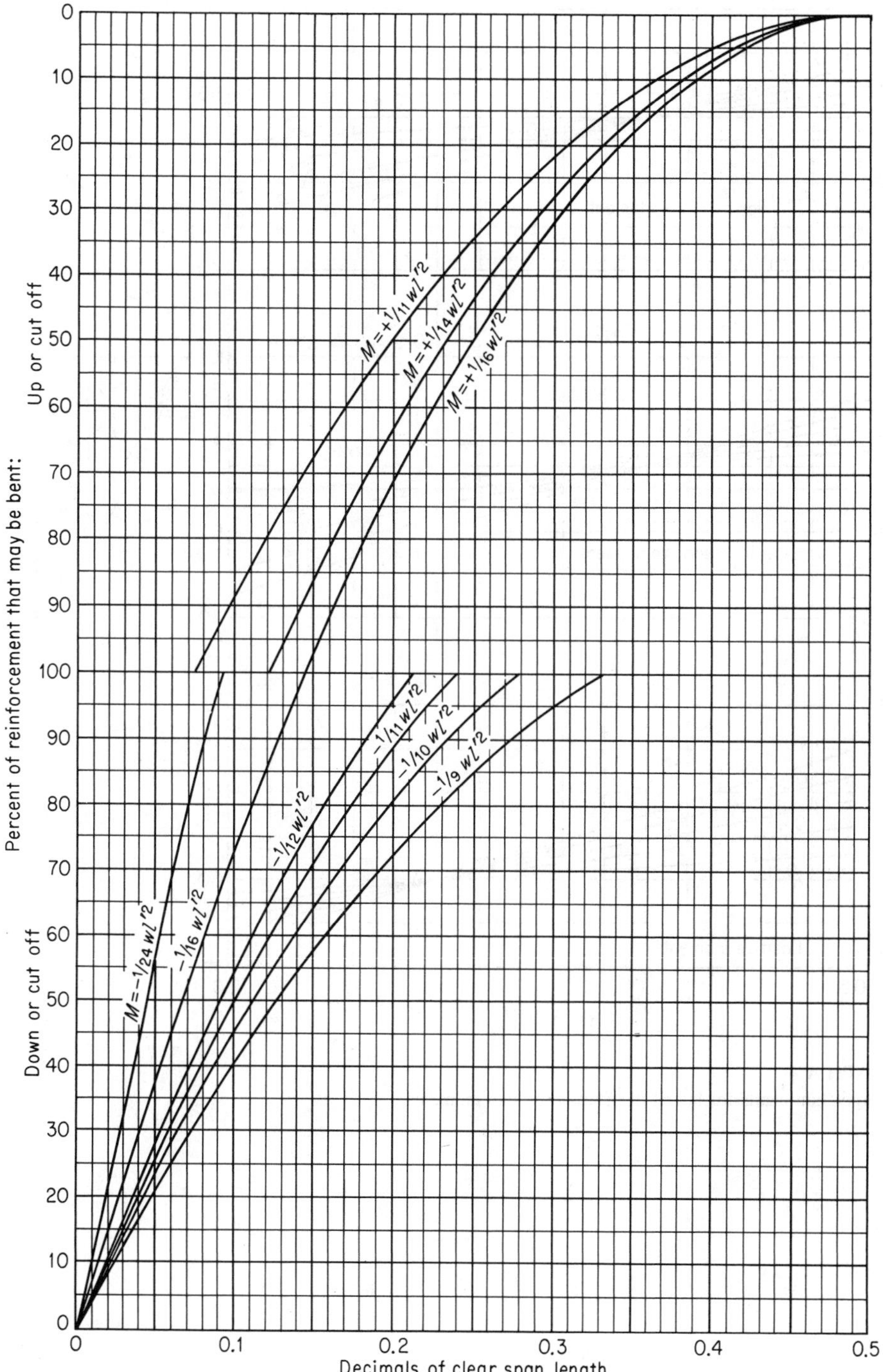

GRAPH A.3
Approximate locations of points where bars can be bent up or down or cut off for continuous beams uniformly loaded and built integrally with their supports according to the coefficients of the ACI Code.

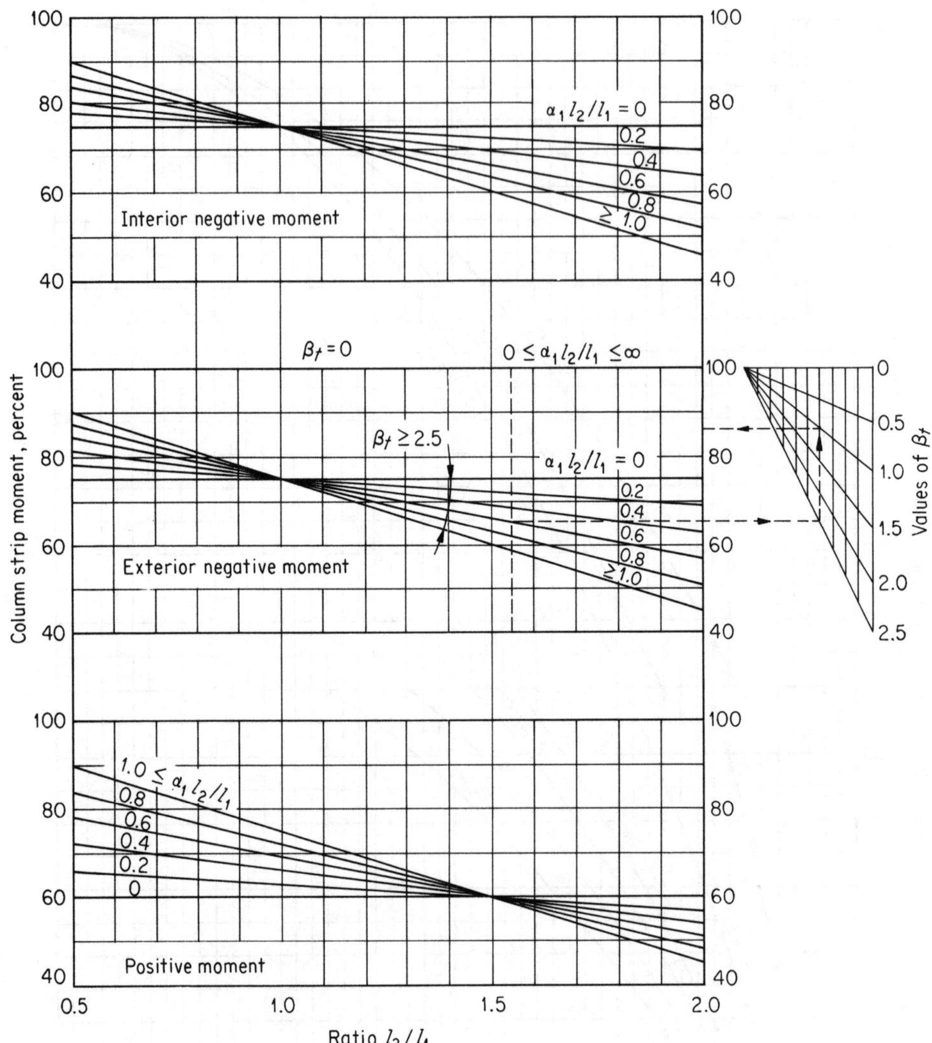

GRAPH A.4
Interpolation charts for lateral distribution of slab moments.

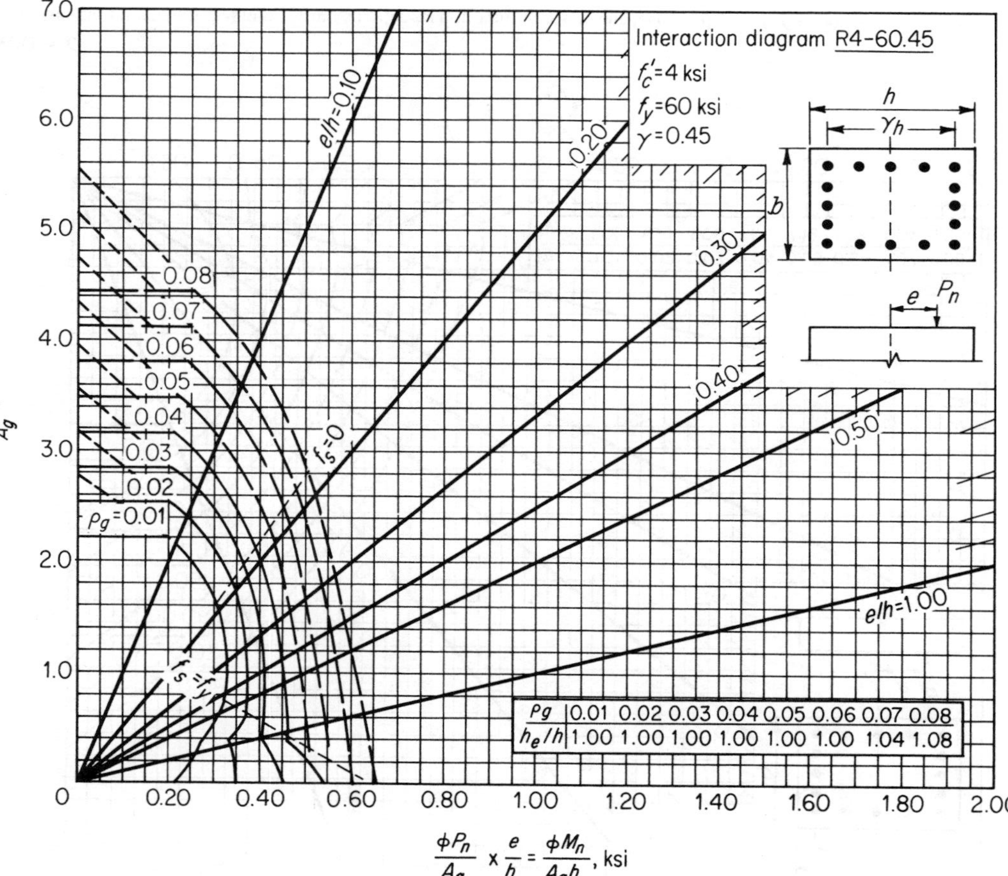

ρ_g	0.01	0.02	0.03	0.04	0.05	0.06	0.07	0.08
h_e/h	1.00	1.00	1.00	1.00	1.00	1.00	1.04	1.08

GRAPH A.5
Column strength interaction diagram for rectangular section with bars on four faces and $\gamma = 0.45$. (*From Ref. 8.7. Courtesy of American Concrete Institute.*)

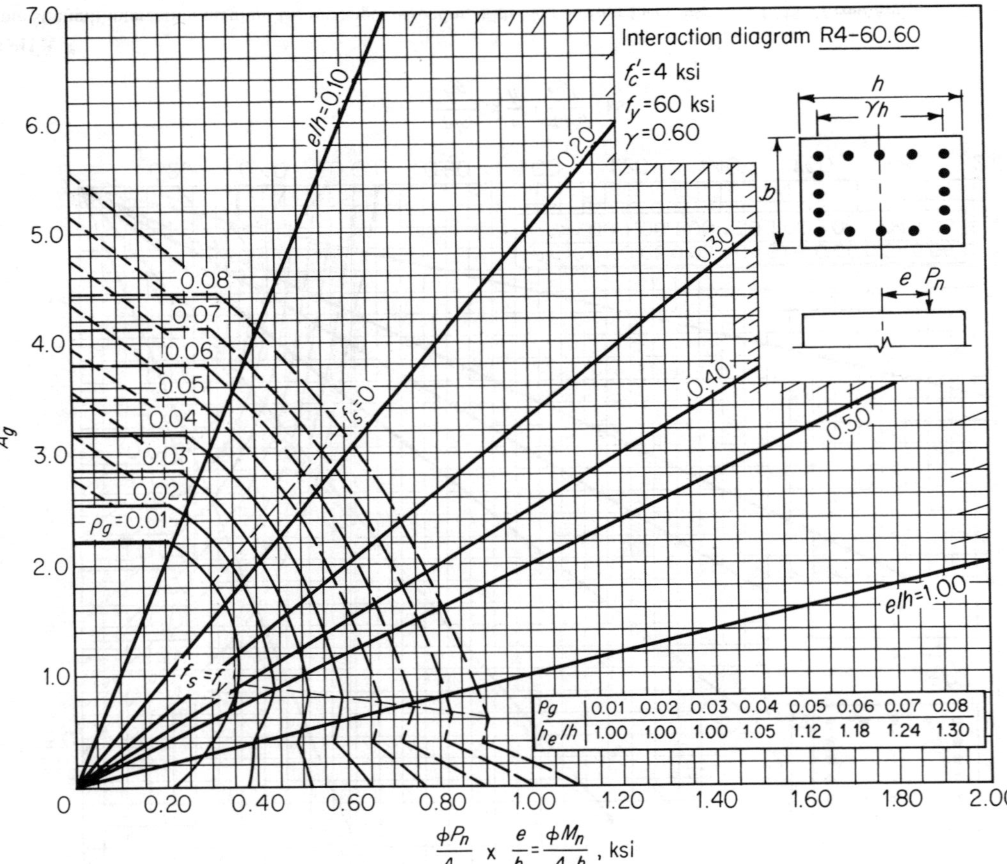

GRAPH A.6
Column strength interaction diagram for rectangular section with bars on four faces and $\gamma = 0.60$. (*From Ref. 8.7. Courtesy of American Concrete Institute.*)

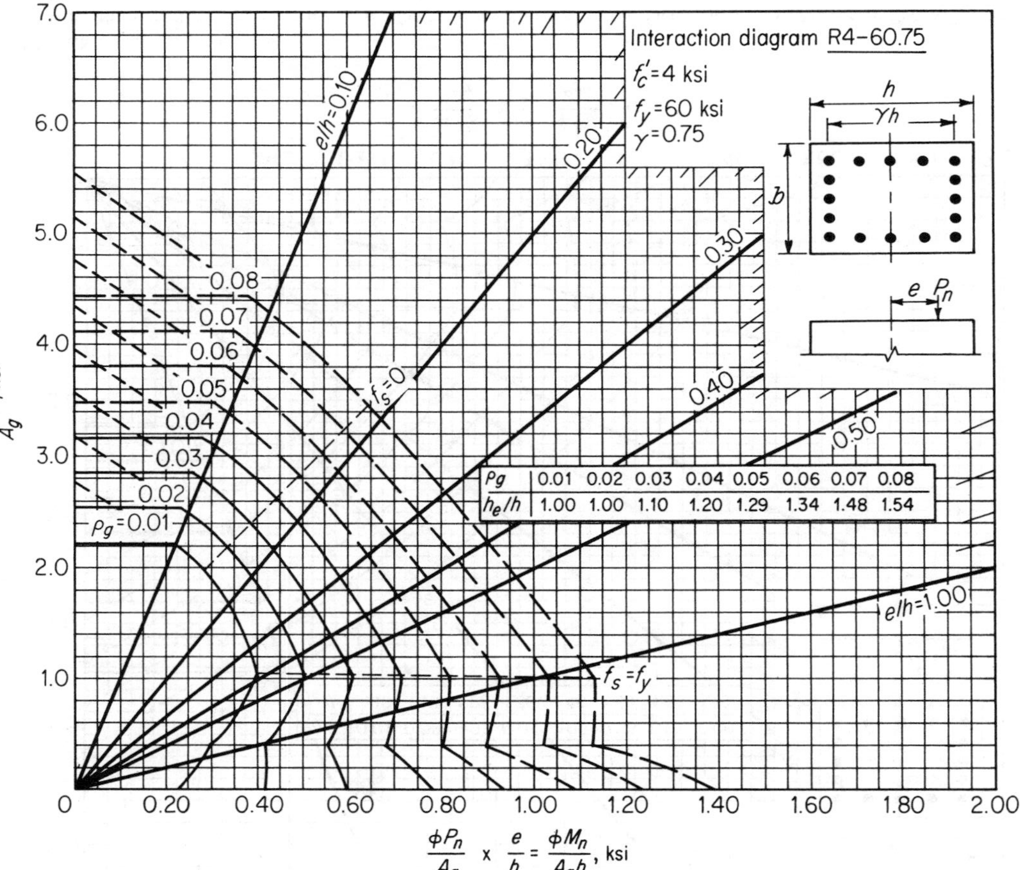

GRAPH A.7
Column strength interaction diagram for rectangular section with bars on four faces and $\gamma = 0.75$. (*From Ref. 8.7. Courtesy of American Concrete Institute.*)

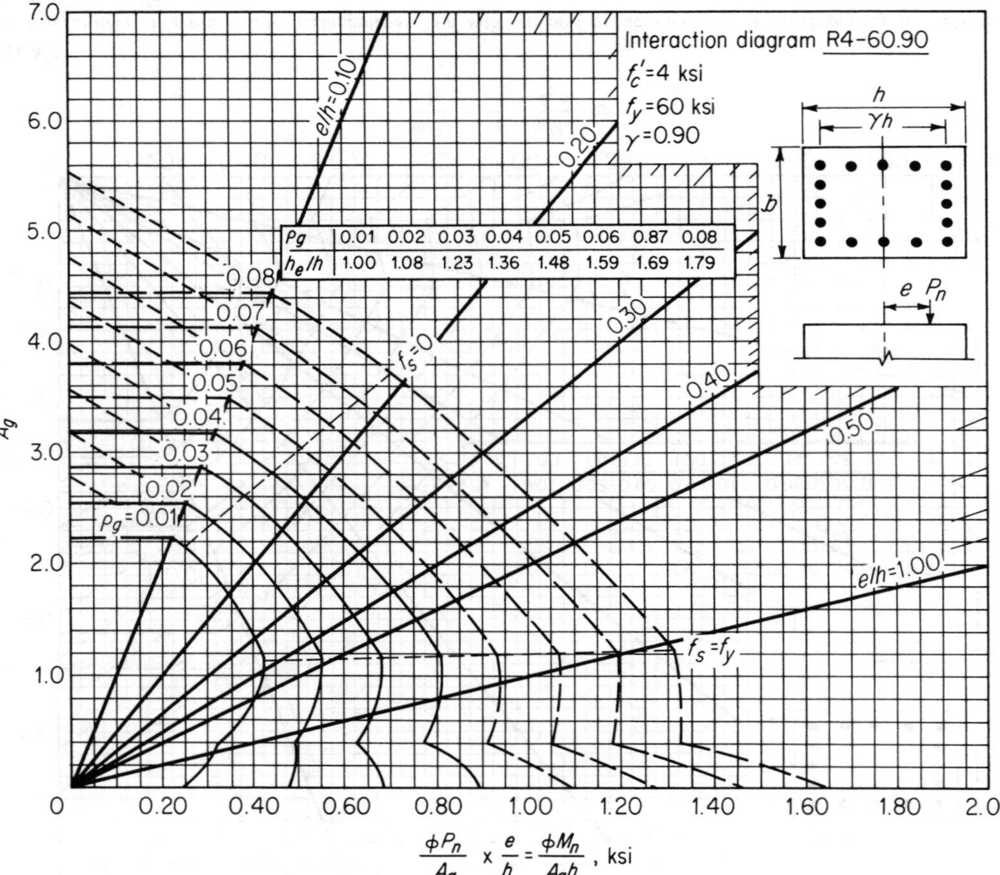

GRAPH A.8

Column strength interaction diagram for rectangular section with bars on four faces and $\gamma = 0.90$. (*From Ref. 8.7. Courtesy of American Concrete Institute.*)

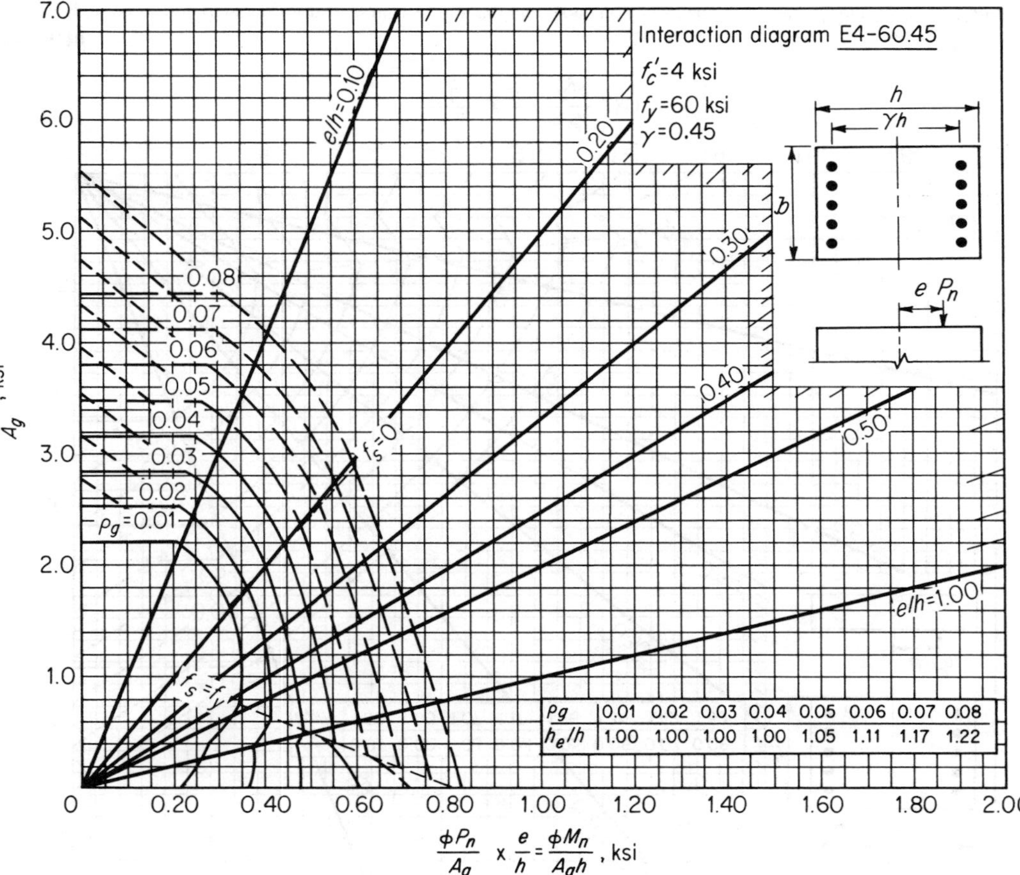

GRAPH A.9
Column strength interaction diagram for rectangular section with bars on end faces and $\gamma = 0.45$. (*From Ref. 8.7. Courtesy of American Concrete Institute.*)

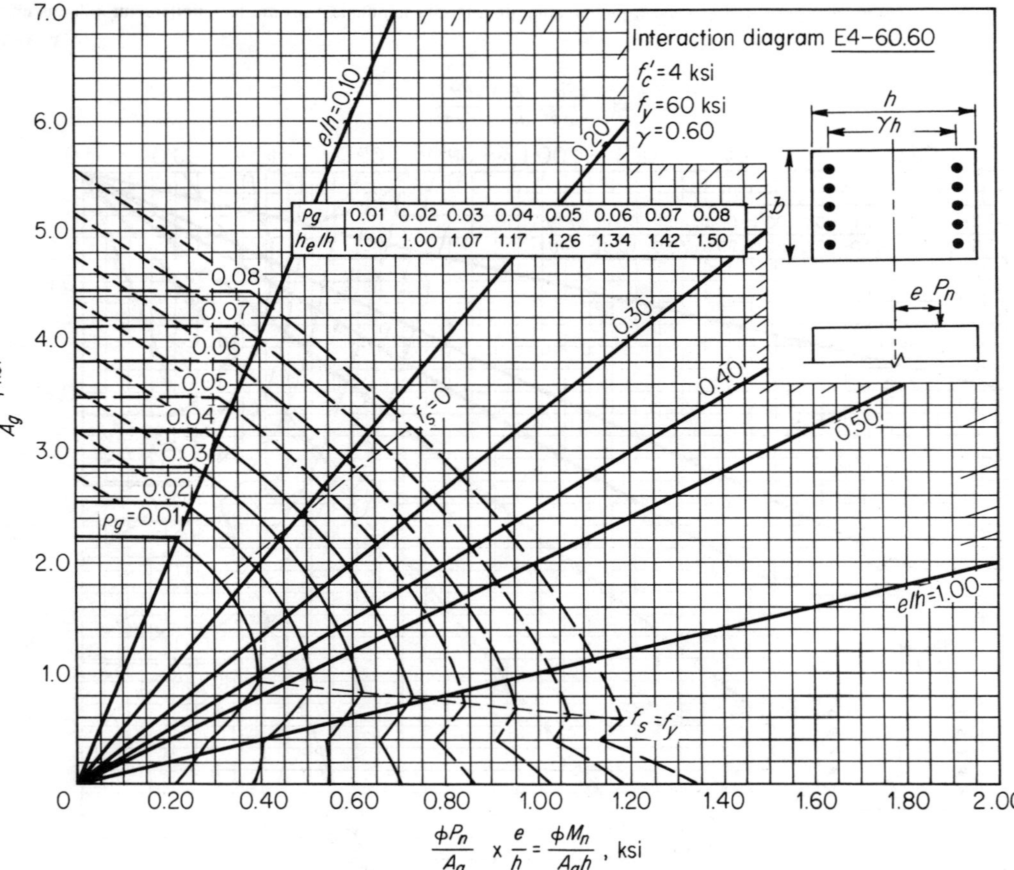

GRAPH A.10
Column strength interaction diagram for rectangular section with bars on end faces and $\gamma = 0.60$. (*From Ref. 8.7. Courtesy of American Concrete Institute.*)

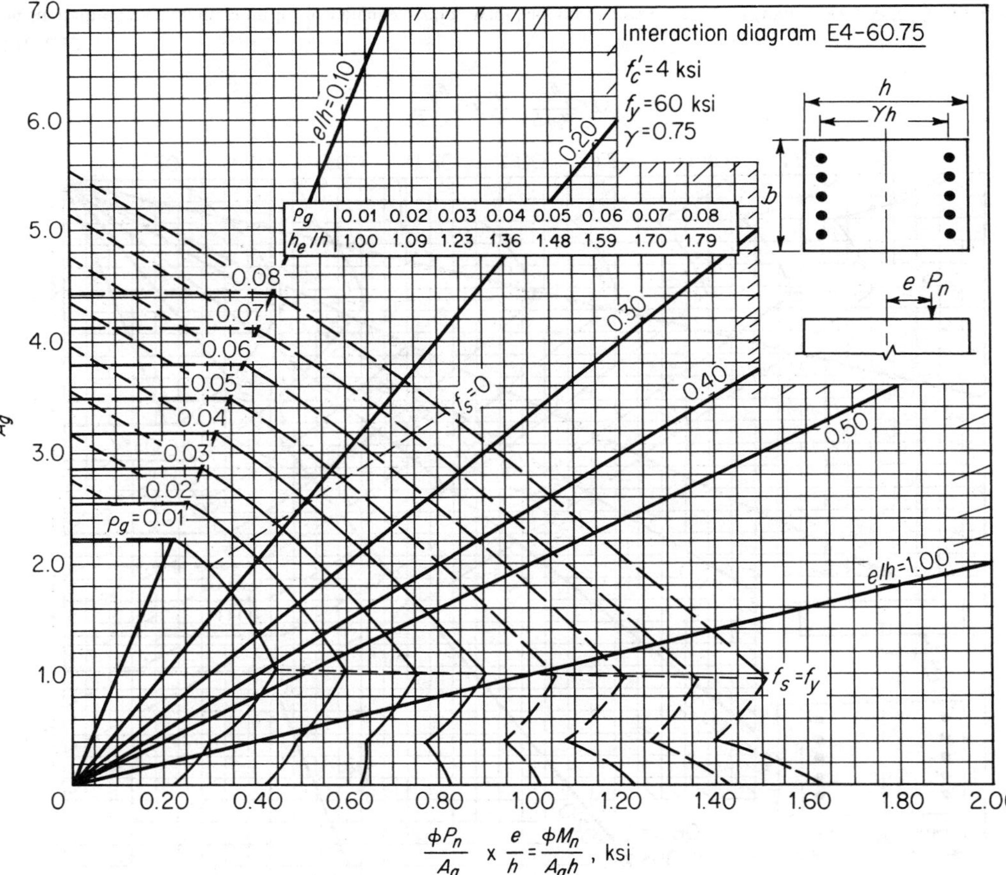

GRAPH A.11

Column strength interaction diagram for rectangular section with bars on end faces and $\gamma = 0.75$. (*From Ref. 8.7. Courtesy of American Concrete Institute.*)

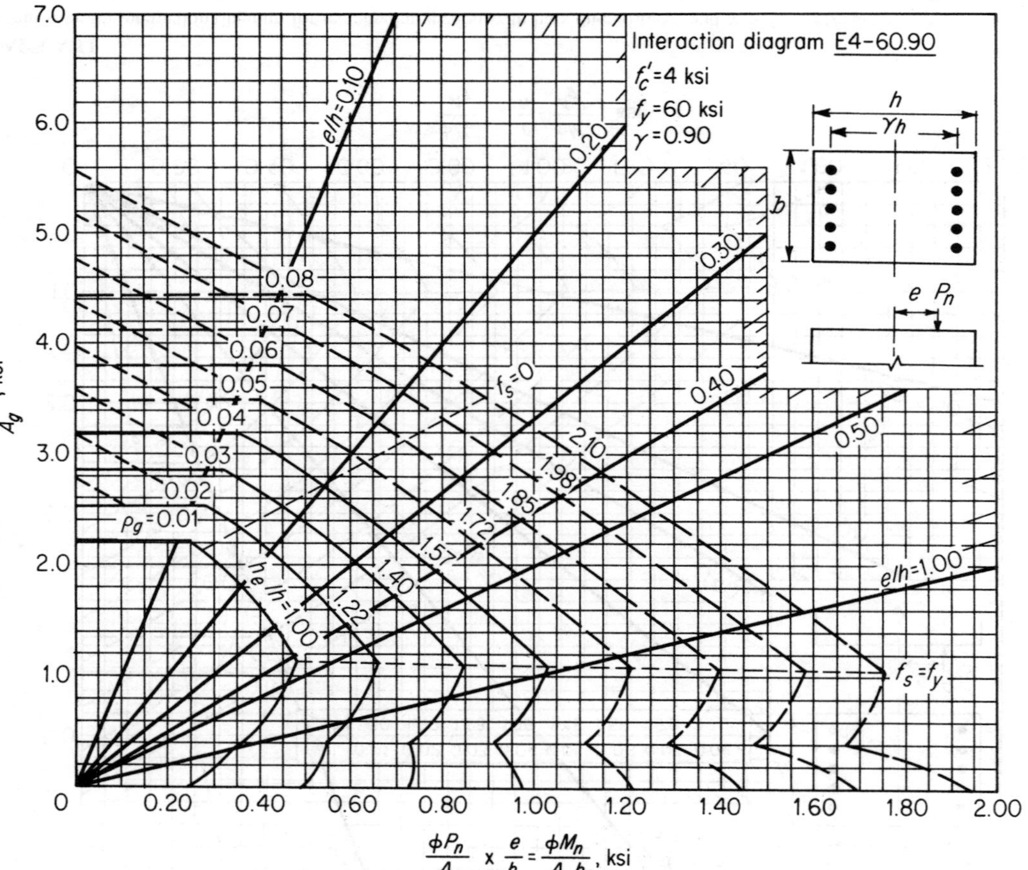

GRAPH A.12

Column strength interaction diagram for rectangular section with bars on end faces and γ = 0.90. (*From Ref. 8.7. Courtesy of American Concrete Institute.*)

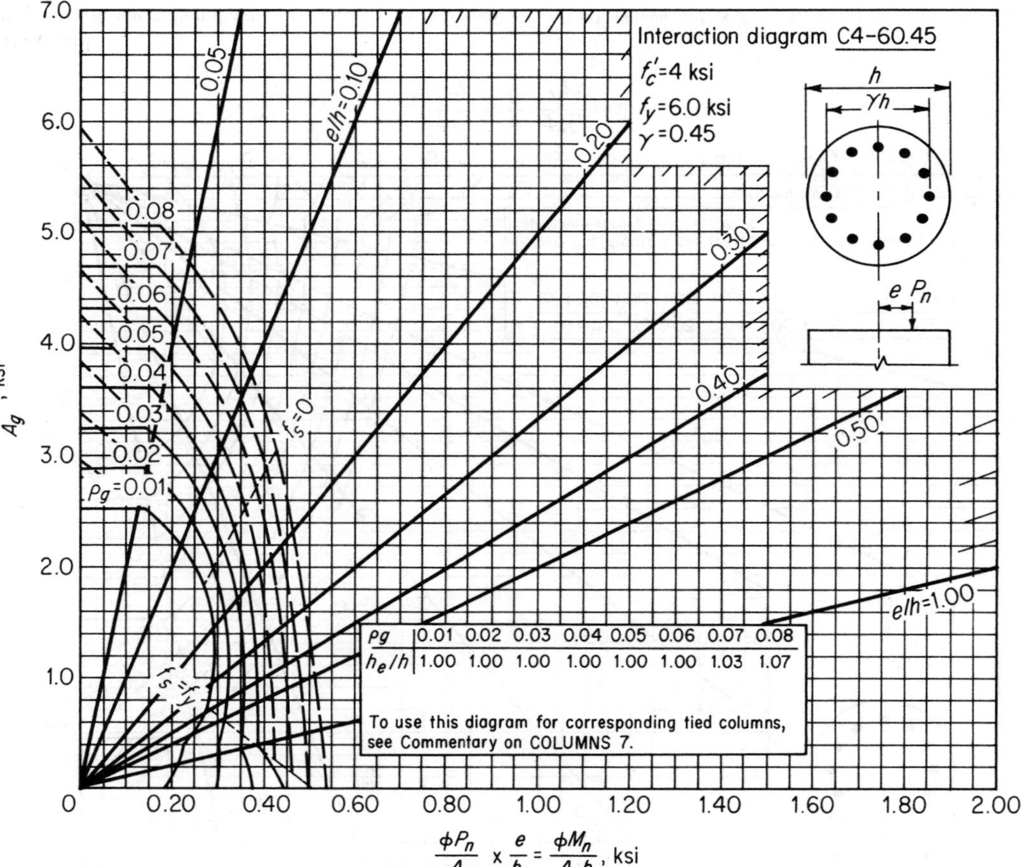

ρ_g	0.01	0.02	0.03	0.04	0.05	0.06	0.07	0.08
h_e/h	1.00	1.00	1.00	1.00	1.00	1.00	1.03	1.07

To use this diagram for corresponding tied columns, see Commentary on COLUMNS 7.

GRAPH A.13
Column strength interaction diagram for circular section with $\gamma = 0.45$. (*From Ref. 8.7. Courtesy of American Concrete Institute.*)

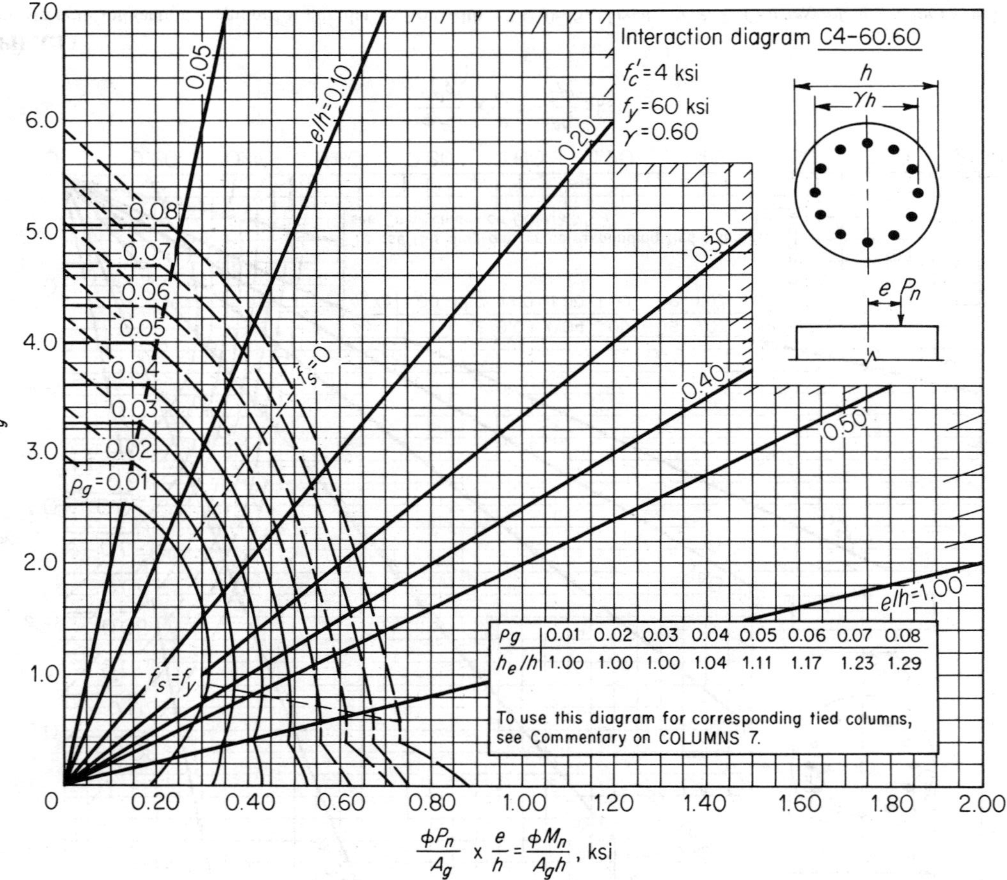

GRAPH A.14
Column strength interaction diagram for circular section with $\gamma = 0.60$. (*From Ref. 8.7. Courtesy of American Concrete Institute*.)

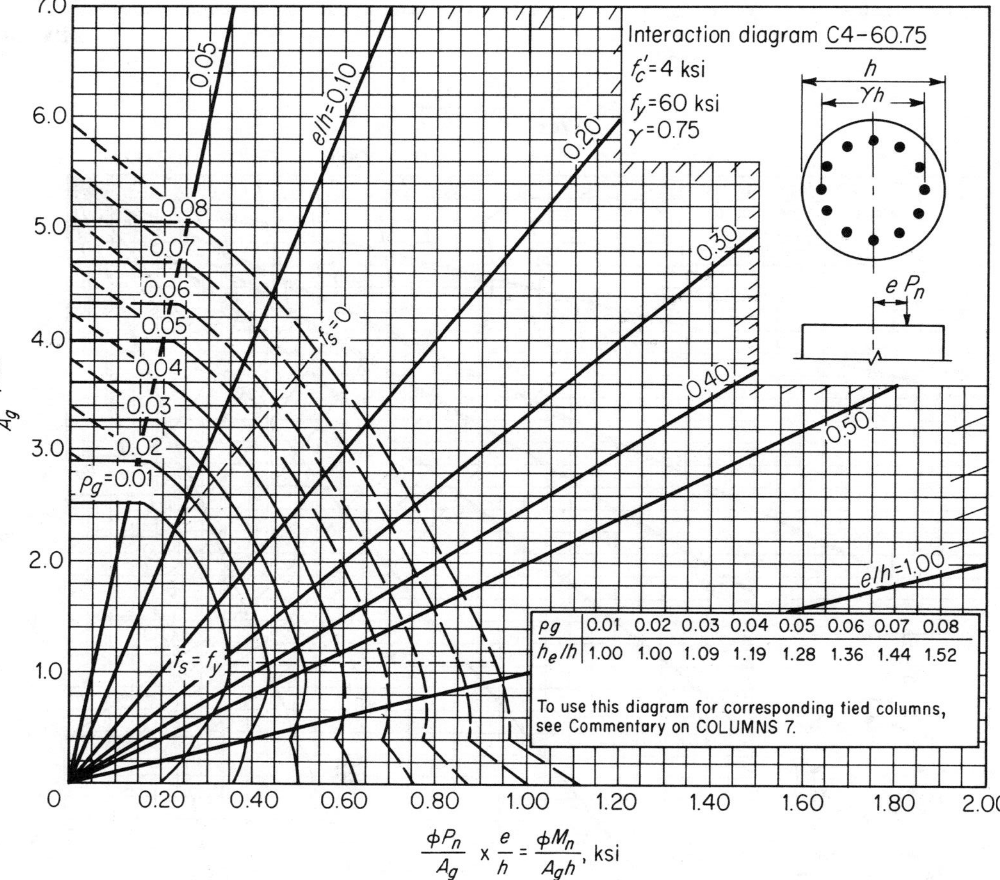

GRAPH A.15
Column strength interaction diagram for circular section with $\gamma = 0.75$. (*From Ref. 8.7. Courtesy of American Concrete Institute.*)

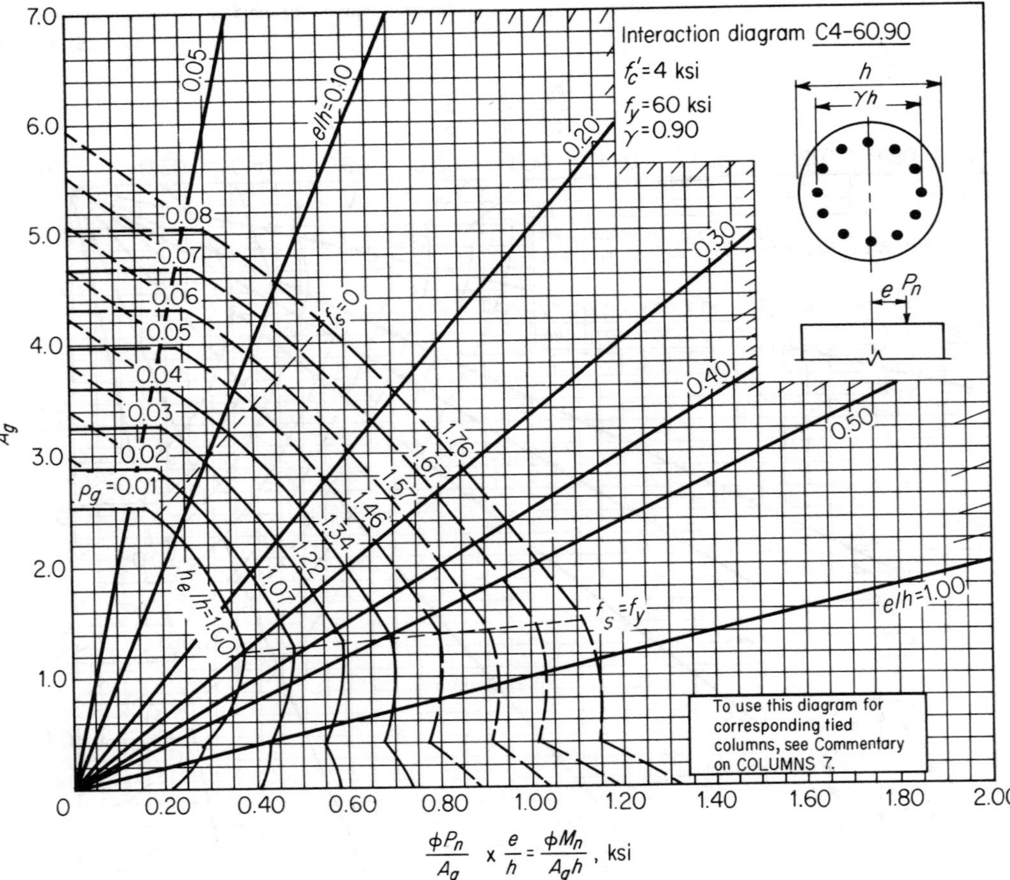

GRAPH A.16
Column strength interaction diagram for circular section with $\gamma = 0.90$. (*From Ref. 8.7. Courtesy of American Concrete Institute.*)

APPENDIX B

SI CONVERSION FACTORS: U.S. CUSTOMARY UNITS TO SI METRIC UNITS

Overall geometry	
Spans	1 ft = 0.3048 m
Displacements	1 in. = 25.4 mm
Surface area	1 ft^2 = 0.0929 m^2
Volume	1 ft^3 = 0.0283 m^3
	1 yd^3 = 0.765 m^3
Structural properties	
Cross-sectional dimensions	1 in. = 25.4 mm
Area	1 in^2 = 645.2 mm^2
Section modulus	1 in^3 = 16.39×10^3 mm^3
Moment of inertia	1 in^4 = 0.4162×10^6 mm^4
Material properties	
Density	1 lb/ft^3 = 16.03 kg/m^3
Modulus and stress	1 lb/in^2 = 0.006895 MPa
	1 kip/in^2 = 6.895 MPa
Loadings	
Concentrated loads	1 lb = 4.448 N
	1 kip = 4.448 kN
Density	1 lb/ft^3 = 0.1571 kN/m^3
Linear loads	1 kip/ft = 14.59 kN/m
Surface loads	1 lb/ft^2 = 0.0479 kN/m^2
	1 kip/ft^2 = 47.9 kN/m^2

Stress and moments

Stress	1 lb/in^2 = 0.006895 MPa
	1 kip/in^2 = 6.895 MPa
Moment or torque	1 ft–lb = 1.356 N–m
	1 ft–kip = 1.356 kN–m

INDEX